NUTRIENTS CATALOG

NUTRIENTS CATALOG

*Vitamins, Minerals, Amino Acids,
Macronutrients — Beneficial Use, Helpers, Inhibitors,
Food Sources, Intake Recommendations,
and Symptoms of Over or Under Use*

by HARVEY NEWSTROM

McFarland & Company, Inc., Publishers
Jefferson, North Carolina, and London

British Library Cataloguing-in-Publication data are available

Library of Congress Cataloguing-in-Publication Data

Newstrom, Harvey.
    The nutrients catalog : vitamins, minerals, amino acids,
macronutrients — beneficial use, helpers, inhibitors, food sources,
intake recommendations, and symptoms of over or under use / by Harvey
Newstrom.
      p. cm.
    Includes bibliographical references.
    ISBN 0-89950-784-0 (lib. bdg. : 50# alk. paper) ∞
    1. Nutrition — Handbooks, manuals, etc. 2. Vitamins in human
nutrition — Handbooks, manuals, etc. 3. Minerals in the body —
Handbooks, manuals, etc. 4. Amino acids in human nutrition —
Handbooks, manuals, etc. 5. Dietary Supplements — Handbooks,
manuals, etc. I. Title.
QP141.N48   1993
613.2'8 — dc20   92-56671
                                                                                 CIP

©1993 Harvey Newstrom. All rights reserved

Manufactured in the United States of America

McFarland & Company, Inc., Publishers
  Box 611, Jefferson, North Carolina 28640

Table of Contents

Preface   xiii
How to Use This Book   xv

PART I: VITAMINS

Vitamin A Complex	1
Provitamin A	1
Vitamin A	6
Vitamin A_1	10
Vitamin A_2	10
Vitamin A Acetate	10
Vitamin A Acid	10
Vitamin A Aldehyde	10
Vitamin A Epoxide	10
Vitamin A Palmitate	10
Monoepoxyvitamin A	10
Neovitamin A	10
Vitamin B Complex	10
Vitamin B	11
Vitamin B_1	11
Vitamin B_1 O, S-Diacetate	23
Vitamin B_1 Disulfide	23

Vitamin B_1 Hydrochloride	23
Vitamin B_1 Mononitrate	23
Vitamin B_1 Phosporic Acid Ester Chloride	23
Vitamin B_1 Phosphoric Acid Ester Salt	23
Vitamin B_1 Propyl Disulfide	23
Vitamin B_1 1,5-Salt	24
Vitamin B_1 Triphosporic Acid Ester	24
Vitamin B_1 Triphosporic Acid Salt	24
Vitamin B_2	24
Vitamin B_2 Complex	24
Vitamin B_2 Phosphate	36
Vitamin B_3 Complex	37
Vitamin B_{3a}	37
Vitamin B_{3b}	43
Vitamin B_4	49
Vitamin B_5	49
Vitamin B_5 Calcium Salt	58
Vitamin B_5 Sodium Salt	58
Vitamin B_6	58
Vitamin B_7	67
Vitamin B_8	67
Vitamin B_9	68
Vitamin B_{10}	68
Vitamin B_{11}	68
Vitamin B_{12}	68
Vitamin B_{12}-60Co	74
Vitamin B_{12}-57Co	74
Vitamin B_{12a}	75
Vitamin B_{12b}	75
Vitamin B_{12c}	75
Vitamin B_{12f}	75
Vitamin B_{12m}	75
Vitamin B_{12p}	75
Vitamin B_{12r}	75
Vitamin B_{12s}	75
Vitamin B_{12}-Zinc Tannate Complex	75
Vitamin B_{12III}	75
psi-Vitamin B_{12}	75
psi-Vitamin B_{12d}	75
Vitamin B_{12} Coenzyme	75

Vitamin B_{13}	76
Vitamin B_{14}	76
Vitamin B_{15}	76
Vitamin $B_{15}H_8$	77
Vitamin B_{16}	77
Vitamin B_{17}	77
Sub-Vitamin B Complex	78
Vitamin B_C	79
Vitamin B_H	85
Vitamin B_H Monophosphate	86
Vitamin B_H Niacinate	86
Vitamin B_P	86
Vitamin B_P Bromide Hexamethylenedicarbamate	87
CDP-Vitamin B_P	87
Vitamin B_P Chloride	87
Vitamin B_P Chloride Carbamate	87
Vitamin B_P Chloride Dihydrogen Phosphate	87
Vitamin B_P Chloride Succinate	87
Vitamin B_P Dehydrocholate	87
Vitamin B_P Dichloride	87
Vitamin B_P Dihydrogen Citrate	87
Vitamin B_P Esterase	88
Vitamin B_P Orotate	88
Vitamin B_P Salicylate	88
Vitamin B_P Theophyllinate	88
Provitamin B_P	88
Vitamin B_T	88
Vitamin B_W	89
Vitamin B_W Sulfoxide	89
Vitamin B_X	90
Vitamin C Complex	90
Antivitamin C	90
Vitamin C	91
Vitamin C Calcium Salt	97
Vitamin C Magnesium Salt	97
Vitamin C Nicotinamide Complex	97
Vitamin C Palmitate	97
Vitamin C Potassium Salt	97
Vitamin C Sodium Salt	97
Vitamin C_2	97

Vitamin C_3	98
Vitamin D Complex	98
Vitamin D_1	99
Vitamin D_2	99
Vitamin D_3	99
Vitamin D_4	99
Vitamin D_5	100
Vitamin D_C	100
Vitamin D_M	100
Vitamin E Complex	100
Alpha-Vitamin E	102
Alpha-Vitamin E Acetate	102
Alpha-Vitamin E Acid Succinate	102
Beta-Vitamin E	102
Gamma-Vitamin E	102
Delta-Vitamin E	102
Epsilon-Vitamin E	102
$Zeta_1$-Vitamin E	102
$Zeta_2$-Vitamin E	102
Eta-Vitamin E	102
Vitamin $E_2(50)$	102
Vitamin F Complex	103
Provitamin F	103
Vitamin F	104
Vitamin F_{99}	105
Vitamin H Complex	105
Vitamin H_1	105
Vitamin H_3	105
Vitamin I	106
Vitamin J	107
Vitamin K Complex	107
Vitamin K_1	108
Vitamin K_1 Oxide	108
Vitamin $K_{1(20)}$	108
Vitamin K_2	108
Vitamin $K_{2(0)}$	108
Vitamin $K_{2(5)}$	108
Vitamin $K_{2(10)}$	108
Vitamin $K_{2(15)}$	108
Vitamin $K_{2(20)}$	108

Vitamin $K_{2(25)}$	108
Vitamin $K_{2(30)}$	108
Vitamin $K_{2(35)}$	108
Vitamin $K_{2(40)}$	109
Vitamin $K_{2(45)}$	109
Vitamin $K_{2(45)}H$	109
Vitamin $K_{2(50)}$	109
Vitamin $K_{2(55)}$	109
Vitamin $K_{2(60)}$	109
Vitamin $K_{2(65)}$	109
Vitamin K_3	109
Vitamin K_3H_2	109
Vitamin K_4	109
Vitamin K_5	109
Vitamin K_6	109
Vitamin K_7	109
Vitamin K_8	109
Vitamin K_9	109
Vitamin $K_9(H)$	109
Vitamin K-S(II)	110
Vitamin L Complex	110
Vitamin L_1	110
Vitamin L_2	110
Vitamin M_i™	110
Vitamin MK Complex	111
Vitamin MK_1	111
Vitamin MK_2	111
Vitamin MK_3	111
Vitamin MK_4	111
Vitamin MK_5	111
Vitamin MK_6	111
Vitamin MK_7	111
Vitamin MK_8	111
Vitamin MK_9	111
Vitamin MK_{10}	111
Vitamin N	112
Vitamin P Complex	112
Vitamin P_1	113
Vitamin P_1 Complex	113
Vitamin P_2	113

Vitamin P_3	113
Vitamin P_4	113
Vitamin Q Complex	113
Vitamin Q	114
Vitamin Q_0	114
Vitamin Q_1	114
Vitamin Q_2	114
Vitamin Q_3	114
Vitamin Q_4	115
Vitamin Q_5	115
Vitamin Q_6	115
Vitamin Q_7	115
Vitamin Q_8	116
Vitamin Q_9	116
Vitamin Q_{10}	116
Vitamin T Complex	118
Vitamin U	118
Vitamin U Bromide	119
Vitamin U Chloride	119
Vitamin U Chick Factor	119
Vitamin V	119
Vitamin W	120
Vitamin X	120
Vitamin Y	121

PART II: MINERALS

Aluminum	123
Antimony	123
Arsenic	124
Barium	124
Beryllium	125
Boron	125
Bromine	126
Cadmium	126
Calcium	126
Cesium	138
Chlorine	139
Trivalent Chromium	139
Hexavalent Chromium	140

Cobalt	140
Copper	141
Fluorine	151
Germanium	151
Gold	152
Iodine	152
Iodine 131	153
Iron	153
Lead	167
Lithium	167
Magnesium	168
Manganese	177
Mercury	186
Molybdenum	186
Nickel	187
Phosphorus	187
Potassium	199
Rubidium	211
Selenium	212
Silicon	212
Silver	213
Sodium	214
Strontium	225
Strontium 90	225
Sulfur	226
Tin	226
Titanium	226
Tungsten	227
Vanadium	227
Zinc	228

PART III: AMINO ACIDS
(Protein Complex)

Alanine	239
Arginine	239
Asparagine	248
Aspartic Acid	249
Cysteine	249
Cystine	257

Glutamate	258
Glutamine	258
Glycine	258
Histidine	259
Hydroxyproline	267
Isoleucine	268
Leucine	276
Lysine	285
Methionine	294
Ornithine	303
Phenylalanine	303
Proline	312
Serine	312
Taurine	313
Threonine	313
Tryptophan	322
Tyrosine	331
Valine	339

PART IV: MACRONUTRIENTS

Carbohydrate	349
Energy	358
Fats	371
Fiber	384
Protein Complex	385
Water	397

Appendices

A: Indications for Use	411
B: Vitamin B Complex Designations	443
C: Nutrient Ratios	445
D: Nutrient Unit Conversions	447

Bibliography	449
Index	461

Preface

This book started in 1983 as my private notes. I had become interested in nutrition and was reading every book I could find on the subject. I started making notes of nutritional data that I could reference. It soon became clear that most of the books were mere pep-talks. While these books convinced the reader of the need for adequate nutrition, they gave little data that could be used. Most books gave only a few symptoms of deficiency for each nutrient, and often these symptoms were so vague that they provided little evidence of actual need. More startling was the fact that most books failed to describe the toxicity symptoms. After convincing the reader that more of certain nutrients was required, the books gave no hint about how to detect an excessive intake of these nutrients. Also lacking were adequate warnings about nutrients that can be harmful under certain conditions, the combinational ratios required by certain sets of nutrients, and alternate names by which nutrients may be known.

I found most books to be incomplete. Some would refer only to nutrients regulated by the FDA. Others would refer only to nutrients that could be purchased in most health food stores. Some would ignore controversial nutrients as if they did not exist, while others imply universal acceptance without mention of the controversy.

As I set out to find a complete reference book that would provide the information that I needed, I found few nutrition books that were actual references. Most were either textbooks, with very introductory material, or case histories with very little data attached. Such books were not set up for reference. It became tedious to search through various volumes to find a reference hidden in the text.

Preface

I finally decided to expand my notes into a full reference handbook of nutrients, a volume in which one might easily look up specific facts. It is not the kind of book that one would read cover to cover, nor is it an introductory text to the subject of nutrition. Instead, it is intended to be a book to be kept on the desk of anyone actively working with nutrition. It is not intended to be used by those without proper nutritional training. It is dangerous to diagnose disease or to prescribe specific nutrients without proper understanding of how nutrition works. This handbook *does not* impart that knowledge.

For the knowledgeable practitioner, this handbook is intended to be a single point of reference. It is to be used to survey information about a particular nutrient, to verify symptoms of deficiency or toxicity, and to find food sources containing a desired nutrient. The practitioner will find quick and easy access to facts and information without having to search through explanatory text.

Harvey Newstrom

How to Use This Book

The parts of this book describe the vitamins, minerals, amino acids, and macronutrients which are essential nutrients. A nutrient is a chemical component of food which must be present in the diet for proper functioning of the body. There are over two dozen vitamins, three dozen minerals, one dozen amino acids, and a half dozen macronutrients that are essential in the diet. If any of these individual nutrients is missing in the diet, its lack will impede a specific set of chemical reactions within the body.

Most of these entries are complete descriptions, but some entries merely direct the reader to another entry. This cross-referencing occurs where different terms may be used for the same nutrient. The most common or widely accepted term will be the one under which a nutrient is listed. Sometimes various terms have been used incorrectly in the literature. A warning to this effect will be placed in parentheses at the beginning of the entry.

Names lists the various names by which a nutrient is known. Some of these are common terms, while others may be chemical identifications. Those which are capitalized are brand names under which the nutrient has been sold.

Classification identifies the common group or groups with which the nutrient is usually categorized. For vitamins, these groups are usually "lipid-soluble vitamin" or "water-soluble vitamin," although a few are classified as an "amino acid" or "hormone precursor" as well. Those in the vitamin B complex are also categorized as "vitamin B complex" or "subvitamin B complex." Also, any vitamin precursor to one of the above will be labeled as a "precursor." For minerals, these groups are usually just "mineral" or "trace mineral" for the required nutrients and "contaminant" for the undesirable ones. The amino acids are classified as an "essential amino acid" or a "non-essential amino acid," with a couple that are considered "semi-essential." Those containing sulfur are also classified as "sulfur-containing amino acids." In addition to the above classifications, any nutrient can be classified as an "antioxidant" if it has such properties.

Forms identifies the various chemical forms of a nutrient. This helps the reader find the various chemical identifications for a nutrient. Some forms are listed under separate entries that give more information about the particular form, which may not apply to all forms of the nutrient. Usually, the separate entry merely references the primary entry for the nutrient.

Deficiency describes the symptoms associated with a diet that is lacking the particular nutrient. This section should be checked to verify a suspected deficiency. If the person does not have many of the symptoms associated with a deficiency, a deficiency of the nutrient in question is unlikely. Remember that there are many causes for each symptom. The ability of a deficiency of a particular nutrient to cause a symptom does not mean that every occurrence of the symptom indicates deficiency.

Side-effects describes the symptoms associated with a diet that contains a large amount of a particular nutrient. Usually the side-effects come about only after an excessive amount of a nutrient is taken, but sometimes they can occur at normal intake levels. Although side-effects are generally considered temporary as opposed to toxicity symptoms, some side-effects can be dangerous and result in permanent injury.

Toxicity describes the symptoms of poisoning associated with an excessive intake of a particular nutrient. These symptoms are usually extremely dangerous and can result in permanent damage. Anyone taking nutritional supplements should consult this section for symptoms to monitor. If such symptoms do occur, the nutrient intake should be lowered. Some nutrients can help detoxify other nutrients. Such detoxifiers are noted in parentheses. Remember that a lack of toxicity symptoms does not indicate that a nutrient is harmless.

A few of the deficiency symptoms, side-effects and toxicity symptoms will be followed by a question mark. This indicates that the symptom has been indicated in some studies but has not been confirmed or generally accepted. Such symptoms are listed as possible, but cannot be considered conclusive.

Some symptoms may appear contradictory, such as weight gain and weight loss, constipation and diarrhea, or increased blood pressure and decreased blood pressure. In these cases the nutrient imbalance is causing regulatory functions in the body to fail such that either symptom is possible as a result, or both symptoms may occur at alternating intervals. Naturally both symptoms cannot occur simultaneously, but either symptom can be invoked as a result of the same nutritional cause.

Inhibitors lists the antagonists for a particular nutrient. These are the substances or conditions that counter the effect of a nutrient, either by reducing its absorption or by neutralizing it within the body. The presence of inhibitors usually increases the need for a nutrient. Inhibitors also can be used to help detoxify a nutrient, although they are not as effective as the nutrients specifically mentioned for detoxifying.

Helpers lists the nutrient co-factors for a particular nutrient. These substances, usually other nutrients, help a nutrient function. They can aid in absorption or work synergistically with the nutrient. The presence of these helpers will usually decrease the need for a nutrient. A lack of these helpers can induce deficiency symptoms of a nutrient, even when adequate levels of the nutrient are present.

Sources indexes the food sources for a nutrient. The food sources are usually expressed as a number of grams (g), milligrams (mg), or micrograms (μg) per every 100 grams (g) of food. Each amount will have a suffix showing the sampling error as specified in the most recent documents available. This is expressed as a " ± " followed by a number. A measurement of "100 ± 0.5" indicates that the actual value may range from 99.5 to 100.5 due to slight imprecision in the sampling methods. No sources that failed to specify an error rate were used. Remember that many foods are eaten in amounts differing from 100 grams; foods eaten in large amounts may be better sources than they appear. Conversely, foods eaten in low amounts may not be as good. Be sure to calculate the total amount of the food eaten.

Restaurant versions are sometimes listed as separate foods. For some foods the restaurant recipes are so radically different that they do not contain the same ingredients. For example, a strawberry milkshake contains milk, ice cream, vanilla and strawberries, whereas a restaurant strawberry milkshake contains soy protein, water, locust bean extract, polysorbate-60, and artificial flavor. Their nutritional values are radically different.

Applications describes the possible uses for a nutrient. These are the ailments for which a nutrient can be applied. Remember that application merely means that the nutrient can be used to counteract deficiency, or may induce a helpful effect. Applications are not cures. Nutrients can only cure the specific symptoms that are directly caused by a deficiency. Nutrients applied to other conditions may effect change, but they cannot fight the root cause. Also remember that the application possibilities are not caused by a lack of the nutrient.

Daily Dosage documents the various recommendations that are made for a particular nutrient. The recommendations come from a variety of sources, and many of them are contradictory. The amount of each nutrient that should be ingested each day is a very controversial topic.

If conversion between different units is common, the conversion ratios are listed in parentheses. Listed next are the United States governmental recommendations. These vary as governmental recognition of nutrients changes over the years. Remember that the United States recommendations do not match the recommendations of other countries, so be sure to check the recommendations of each specific country. Also remember that the RDA recommendations have been revised many times. Be sure to check the date of any references to the RDA to determine which version is being referenced.

"Pre-1958 MDR" refers to the pre-1958 Minimum Daily Requirements, the original governmental recommendations for the minimum dosage required to prevent deficiency.

"1958 RDA" indicates the 1958 Recommended Dietary Allowances, which were the original governmental recommendations for adequate daily dosage. They superseded the MDR.

"1974 PCS" refers to the 1974 Permissive Composition of Supplements, the governmental recommendations for the allowable dosage of nutrient supplements.

"1974 RDA" dosages are the 1974 Recommended Dietary Allowances, an intermediate change between the original governmental recommendations and the current recommendations for adequate daily dosage.

"1974 USRDA" figures are the 1974 United States Recommended Daily Allowances, the specific dosages that the government required to be used to calculate the percent of U.S. RDA on food nutrient labels.

"1980 RDA" refers to the 1980 Recommended Dietary Allowances, the former governmental recommendations for the adequate daily dosage.

"1980 USRDA" indicates the 1980 United States Recommended Daily Allowances, the specific dosages that the government required to be used to calculate the percent of U.S. RDA on food nutrient labels.

"1980 SADDI" dosages are those designated in the 1980 Safe and Adequate Daily Dietary Intakes in an appendix to the 1980 RDA. These were not as officially endorsed as the primary recommendations of the RDA.

"1989 RDA" refers to the 1989 Recommended Dietary Allowances, the governmental recommendations for adequate daily dosage that were in effect at the time of this writing.

"1989 USRDA" indicates the 1989 United States Recommended Daily Allowances, the current specific dosages that the government requires to be used to calculate the percent of U.S. RDA on food nutrient labels, as of this date.

"1989 SADDI" dosages are those listed as Safe and Adequate Daily Dietary Intakes in an appendix to the 1989 RDA. These are not as officially endorsed as the primary recommendations of the RDA.

"Nutritional" dosages are those used for general nutrition, and the most common dosages found in over-the-counter vitamin supplements.

"Therapeutic" dosages are those most often recommended to combat specific deficiencies or used for specific applications.

"Experimental" are the highest dosages known to be used for any purpose, and the dosages suggested by those trying to maximize their possible intake. Experimental dosages are extremely dangerous and should only be attempted by knowledgeable professionals. Constant medical testing is required to monitor internal levels.

"Toxic" is the lowest level commonly expected to induce toxicity. Lower levels can cause toxicity in children or those not at optimum health. Dosages under the toxic level are not guaranteed to be safe. For those nutrients for which no toxic dosage is known up to some level, remember that this merely means the toxic level has not been discovered or recognized. It does not mean that the toxic level is above the referenced amount; nor does it mean that the nutrient is safe at all levels.

Warnings describe specific problems and special considerations with a particular nutrient. Also listed are any special conditions that might make taking a nutrient dangerous. Anyone taking nutrients should read these warnings. Remember that this list of warnings is not complete. There are many complicated dangers that are too involved to list here. Do not assume that the absence of a specific warning means that a nutrient is safe for a specific condition.

Appendix A lists the symptoms that will direct the reader to specific nutrients. These can be deficiency symptoms, toxicity symptoms, side-effects, warnings for particular nutrients, or special conditions that may influence nutrition. When any of these symptoms occur, the reader should look up the referenced nutrient for more information.

Appendix B compares various terminology systems for the vitamin B complex nutrients. This allows the reader to cross-reference this book with other works that use different terminology. "Erroneous References" are also listed showing some mistakes in nomenclature made by early researchers that have been commonly reproduced in later books.

Appendix C delineates the proper dietary ratios that should be maintained among various nutrients for optimum utilization.

Appendix D contains the formulas for converting one form of unit measurements to another.

The bibliography lists the sources that were used in compiling this reference. Not all the sources are equal in accuracy, however. Do not assume that all sections of a book are valid just because it is referenced here. Many books were slanted toward one viewpoint, so the information in them must be considered biased.

Above all, this book is intended to be informative. It will not modify the reader's beliefs or level of expertise. This book cannot replace a proper practitioner, nor does it intend to alleviate the responsibility of any person to obtain proper care.

PART I

Vitamins

VITAMIN A COMPLEX

(Different forms of vitamin A. See provitamin A, vitamin A_1, vitamin A_2, vitamin A acetate, vitamin A acid, vitamin A aldehyde, vitamin A epoxide, vitamin A palmitate, and neovitamin A.)

PROVITAMIN A

(Provitamin A is internally converted to vitamin A.)

Names: provitamin A, vitamin A_0, carotene, beta-carotene
Classification: lipid-soluble vitamin precursor, antioxidant
Forms: alpha-carotene (only half as potent as the beta-carotene form of provitamin A), beta-carotene (most potent form of provitamin A), gamma-carotene (low potency), delta-carotene/epsilon, psi-carotene
Deficiency: decreased cancer resistance, lost source of vitamin A
Side-effects: skin yellowing, carotene deposition in tissues
Toxicity: none known up to 150,000 iu
Inhibitors: ferrous sulfate, polyunsaturated fats, laxatives, mineral oil
Helpers: vitamin B complex, vitamin B_P, vitamin C, vitamin D, vitamin E, calcium, phosphorus, zinc
Sources (iu per 100 grams of food): red palm oil (90000 ± 1000), freeze-dried red sweet peppers (77250 ± 31.3), freeze-dried chives (68250 ± 62.5), paprika (63650 ± 25), freeze-dried parsley (63214 ± 35.7), cayenne pepper (37450 ± 25), dried parsley (30300 ± 50), chili powder (30267 ± 16.7), raw carrots (28129 ± 0.7), carrot juice (25751 ± 0.3), boiled carrots (24554 ± 0.6), canned pumpkin (22056 ± 0.4), baked sweet potato (21822 ± 0.4),

frozen and boiled carrots (17701 ± 0.7), tomato powder (17247 ± 0.5), boiled sweet potato (17054 ± 0.3), frozen and baked sweet potato (16410 ± 0.6), raw dandelion greens (14000 ± 1.8), canned carrots (13767 ± 0.7), basil (13100 ± 50), canned red chili peppers (11893 ± 0.7), boiled dandelion greens (11700 ± 1), raw mustard spinach (9900 ± 0.7), boiled lamb's-quarters (9700 ± 0.6), raw garden cress (9300 ± 2), canned spinach (8777 ± 0.5), boiled pokeberry shoots (8700 ± 0.6), canned pumpkin pie mix (8299 ± 0.4), boiled mustard spinach (8200 ± 0.6), boiled spinach (8190 ± 0.6), raw vine spinach (8000 ± 0.5), canned sweet potato (7983 ± 0.3), frozen and boiled turnip greens (7976 ± 0.6), frozen and boiled spinach (7784 ± 0.5), boiled garden cress (7700 ± 0.7), raw turnip greens (7600 ± 1.8), boiled kale (7400 ± 0.8), dried apricots (7240 ± 1.4), savory (7200 ± 50), boiled butternut squash (7001 ± 0.5), raw spinach (6714 ± 1.8), raw chives (6400 ± 16.7), frozen and boiled kale (6354 ± 0.8), raw swamp cabbage (6300 ± 0.9), freeze-dried green sweet peppers (6250 ± 31.3), baked hubbard squash (6035 ± 0.5), frozen and boiled collards (5981 ± 0.6), raw red sweet peppers (5700 ± 1), boiled turnip greens (5499 ± 0.7), raw laver/nori seaweed (5202 ± 0.5), boiled swamp cabbage (5200 ± 1), raw parsley (5200 ± 1.7), oregano (5200 ± 25), boiled jute potherb (5186 ± 1.2), frozen and boiled turnip greens and turnips (5161 ± 0.5), boiled beet greens (5100 ± 0.7), raw spring onions with tops (5000 ± 1), marjoram (4800 ± 50), raw watercress (4700 ± 2.9), frozen and boiled mustard greens (4469 ± 0.7), wheat flakes (4464 ± 1.8), corn flakes (4464 ± 1.8), bran flakes (4464 ± 1.8), bran (4464 ± 1.8), raw new zealand spinach (4400 ± 1.8), candied sweet potato (4190 ± 0.5), sage (4100 ± 50), boiled hubbard squash (4005 ± 0.4), raw dock (4000 ± 0.7), raw chickory greens (4000 ± 0.6), mango (3894 ± 0.2), boiled red sweet peppers (3760 ± 0.7), bay leaf (3700 ± 50), boiled new zealand spinach (3622 ± 0.6), canned turnip greens (3586 ± 0.4), raw celtuce (3500 ± 0.5), boiled dock (3474 ± 0.5), raisin bran (3378 ± 1.4), tarragon (3350 ± 25), frozen and boiled red sweet peppers (3343 ± 0.5), frozen and boiled butternut squash (3339 ± 0.4), cantaloupe (3224 ± 0.3), soybean oil and cottonseed oil diet margarine tub (3180 ± 10), corn oil diet margarine tub (3180 ± 10), corn oil margarine tub (3160 ± 10), boiled swiss chard (3139 ± 0.6), soybean oil and cottonseed oil margarine tub (3100 ± 10), soybean oil margarine tub (3100 ± 10), safflower oil margarine tub (3100 ± 10), soybean oil and cottonseed oil margarine stick (3100 ± 10), soybean oil margarine stick (3100 ± 10), safflower oil and soybean oil margarine stick (3100 ± 10), corn oil margarine stick (3100 ± 10), soybean oil and cottonseed oil margarine liquid (3100 ± 10), boiled mustard greens (3031 ± 0.7), raw chinese cabbage (3000 ± 1.4), raw coriander (2775 ± 12.5), boiled amaranth (2770 ± 0.8), apricots (2612 ± 0.5), raw romaine lettuce (2600 ± 1.8), boiled chinese cabbage (2568 ± 0.6), tomato paste (2469 ± 0.4), pumpkin pie (2465 ± 0.4), yellow passion fruit juice (2410 ± 0.2), sweet potato pie (2395 ± 0.4), boiled collards (2220 ± 0.3), japanese persimmon (2167 ± 0.3), raw endive (2052 ± 2), papaya (2014 ± 0.2), boiled scotch kale (1994 ± 0.8), dried prunes (1987 ± 0.6), poultry seasoning (1950 ± 25), raw looseleaf lettuce (1900 ± 1.8), dried rosemary (1900 ± 25), frozen and boiled broccoli (1892 ± 0.5), boiled purslane (1852 ± 0.9), cooked tahitian taro (1765 ± 0.7), canned jalapeño peppers (1700 ± 0.7), raw broccoli (1541 ± 1.1), loquats (1528 ± 0.5), pitanga (1500 ± 0.3), boiled broccoli (1409 ± 0.6), boiled blackeye pea pods (1400 ± 1.1), chili sauce (1400 ± 3.3), tomato catsup (1400 ± 3.3), tomato purée (1361 ± 0.2), boiled red tomato (1353 ± 0.4), cumin seed (1350 ± 25), apricot nectar (1316 ± 0.2), raw red tomato (1133 ±

I. Vitamins — PROVITAMIN A

0.4), dried ginko nuts (1107 ± 1.8), boiled pumpkin (1082 ± 0.4), raw red chili peppers (1076 ± 1.1), shortbread cookie (1063 ± 6.3), raw savoy cabbage (1000 ± 1.4), curry powder (1000 ± 25), tomato sauce (980 ± 0.4), maypo (974 ± 0.3), raw butterhead lettuce (973 ± 3.3), toaster pastry (964 ± 1), marinara sauce (961 ± 0.2), tangerine (920 ± 0.6), cooked plantain (909 ± 0.3), boiled savoy cabbage (889 ± 0.7), instant oatmeal (855 ± 0.3), mandarin oranges (852 ± 0.4), barbecue sauce (850 ± 3.1), oheloberries (830 ± 0.4), boiled asparagus (829 ± 0.6), frozen and boiled asparagus (818 ± 0.8), guava (792 ± 0.6), raw green chili peppers (769 ± 1.1), canned green peas (768 ± 0.6), raw acerola (766 ± 0.5), nectarine (736 ± 0.4), ground-cherries (720 ± 0.4), boiled brussels sprouts (719 ± 0.6), purple passion fruit juice (717 ± 0.2), mace (700 ± 25), passion fruit (700 ± 2.8), tempeh (686 ± 0.6), frozen and boiled green peas (668 ± 0.6), boiled green beans (666 ± 0.8), raw green tomato (641 ± 0.4), raw green peas (640 ± 0.6), coleslaw (635 ± 0.8), raw florida avocado (612 ± 0.2), raw california avocado (612 ± 0.3), canned green chili peppers (610 ± 0.7), canned peeled red tomatoes (604 ± 0.4), elderberries (600 ± 0.3), boiled green peas (598 ± 0.6), frozen and boiled brussels sprouts (588 ± 0.6), boiled okra (575 ± 0.6), ginko nuts (564 ± 1.8), dried japanese persimmon (559 ± 1.5), tomato juice (556 ± 0.3), restaurant tomato juice (556 ± 0.4), canned stewed red tomatoes (555 ± 0.4), cloves (550 ± 25), canned zucchini in tomato juice (539 ± 0.4), peach (534 ± 0.6), raw green sweet peppers (530 ± 1), frozen and boiled green beans (528 ± 0.7), canned red sweet peppers (520 ± 0.7), dried and frozen tofu (518 ± 2.9), bran muffins (515 ± 1.3), frozen and boiled okra (514 ± 0.5), acerola juice (509 ± 0.2), yellow whole ground corn meal (508 ± 0.4), allspice (500 ± 25), carambola (493 ± 0.4), yellow bolted corn meal (484 ± 0.4), unenriched yellow degermed corn meal (442 ± 0.4), enriched yellow degermed corn meal (442 ± 0.4), cherry pie (441 ± 0.4), red grapefruit juice (440 ± 0.2), frozen and boiled zucchini (431 ± 0.4), baked acorn squash (428 ± 0.5), bread stuffing (420 ± 0.3), tangerine juice (420 ± 0.2), orange peel (417 ± 8.3), pickled ripe manzanillo/mission olives (413 ± 10.9), sapote (410 ± 0.2), caraway seed (400 ± 25), boiled green sweet peppers (388 ± 0.7), dried pumpkin and squash seeds (386 ± 1.8), watermelon (366 ± 0.3), raw wakame seaweed (360 ± 0.5), cream puff (354 ± 0.4), canned green beans (349 ± 0.7), pickled rip sevillano/ascolano olives (346 ± 3.1), pineapple chiffon pie (346 ± 0.6), bloody mary (343 ± 0.3), yellow corn flour (342 ± 0.4), raw zucchini (340 ± 0.8), rose apple (339 ± 0.5), raw crookneck squash (338 ± 0.8), raw iceberg lettuce (330 ± 2.5), popover roll (325 ± 1.3), plum (323 ± 0.8), canned white corn (319 ± 0.5), chocolate chiffon pie (309 ± 0.6), cooked prunes (306 ± 0.5), cinnamon (300 ± 25), dried black walnuts (300 ± 1.8), whole ground corn meal muffins (300 ± 1.3), corn meal muffins (300 ± 1.3), kumquats (300 ± 2.6), jackfruit (297 ± 0.5), boiled succotash (294 ± 0.5), frozen and boiled green sweet peppers (290 ± 0.5), gooseberries (290 ± 0.3), boiled crookneck squash (288 ± 0.6), roselle (286 ± 0.9), milk chocolate (286 ± 1.8), dry roasted macadamia nuts (282 ± 1.8), hamburger pickle relish (271 ± 1.8), chocolate cream pie (264 ± 0.4), butterscotch pie (263 ± 0.4), pickled green olives (261 ± 1.1), boiled dishcloth gourd (260 ± 0.6), pink grapefruit (259 ± 0.4), boiled acorn squash (258 ± 0.4), peach nectar (258 ± 0.2), banana custard pie (254 ± 0.4), frozen and boiled yellow corn (249 ± 0.6), canned yellow corn (241 ± 0.5), boiled zucchini (240 ± 0.6), dried pistachio nuts (236 ± 1.8), canned yellow corn with red and green peppers (232 ± 0.4), frozen and boiled succotash (231 ± 0.6), valencia orange (230 ± 0.4), mammy apple

(230 ± 0.5), black currants (230 ± 0.9), canned shellie beans (228 ± 0.4), custard pie (228 ± 0.4), coconut custard pie (228 ± 0.4), blueberry muffins (225 ± 1.3), rusk crackers (222 ± 5.6), boiled yellow corn (217 ± 0.6), cherries (215 ± 0.7), toasted soybean nuts (204 ± 1.8), raw chinese chestnuts (204 ± 1.8), pumpkin pie spice (200 ± 25), black pepper (200 ± 25), roasted soybean nuts (200 ± 0.6), orange juice (200 ± 0.2), frozen and boiled crookneck squash (195 ± 0.5), frozen and boiled fordhook lima beans (191 ± 0.6), waffles (187 ± 0.7), pineapple custard pie (184 ± 0.4), navel orange (183 ± 0.4), kiwi fruit (175 ± 0.7), lemon chiffon pie (173 ± 0.6), vegetarian baked beans (171 ± 0.2), french toast (171 ± 0.8), yellow hominy with red and green peppers (169 ± 0.2), hot dog pickle relish (168 ± 1.8), boiled and frozen baby lima beans (167 ± 0.6), firm raw tofu (166 ± 0.4), blackberries (165 ± 0.7), boiled artichoke (164 ± 0.6), boiled green soybeans (156 ± 0.6), canned green sweet peppers (156 ± 0.7), raw sprouted alfalfa seeds (155 ± 1.5), pecan pie (155 ± 0.5), peach pie (155 ± 0.5), ginger (150 ± 25), fennel seed (150 ± 25), canned succotash (146 ± 0.4), boiled artichoke hearts (144 ± 0.6), figs (142 ± 1), canned succotash with creamed corn (141 ± 0.4), lemon meringue pie (139 ± 0.4), bread and butter pickles (133 ± 3.3), cooked wild long grain rice (133 ± 0.4), corn pone (133 ± 0.6), cornbread (133 ± 0.6), dried figs (133 ± 0.3), frozen creamed yellow corn (130 ± 0.4), raspberries (130 ± 0.4), dried pecans (129 ± 1.8), raw celery (128 ± 1.3), raw green cabbage (126 ± 1.4), dried english/persian walnuts (125 ± 1.8), canned crookneck squash (120 ± 0.5), soybean flour (120 ± 0.6), white currants (120 ± 0.9), red currants (120 ± 0.9), boiled cardoon (118 ± 0.5), raw kelp/kombu/tangle seaweed (116 ± 0.5), animal crackers (115 ± 1.9), yellow hominy (114 ± 0.2), sugar cookie (113 ± 3.1), frozen and boiled yellow snap beans (112 ± 0.7), raw scallop squash (111 ± 0.8), pancakes (111 ± 1.9), papaya nectar (111 ± 0.2), boiled/baked spaghetti squash (110 ± 0.6), roasted soybean flour (109 ± 0.6), sour pickles (108 ± 0.8), dill pickles (108 ± 0.8), boiled celery (108 ± 0.7), raw frozen rhubarb (107 ± 0.4), boiled sprouted green peas (107 ± 0.5), danish pastry (107 ± 1.2), canned yellow snap beans (104 ± 0.7), apple brown betty (102 ± 0.2), nutmeg (100 ± 25), american grapes (100 ± 0.5), blueberries (100 ± 0.3), canned creamed yellow corn (97 ± 0.4), tequila sunrise (97 ± 0.3), raw leeks (96 ± 1.9), blackberry pie (93 ± 0.4), miso (87 ± 0.4), boiled scallop squash (86 ± 0.6), dried japanese chestnuts (86 ± 1.8), raw tofu (85 ± 0.4), boiled green cabbage (85 ± 0.7), boiled yellow snap beans (81 ± 0.8), bananas (81 ± 0.4), restaurant orange juice (78 ± 0.4), frozen and boiled black-eyed peas (75 ± 0.6), roasted japanese chestnuts (75 ± 1.8), thompson seedless grapes (73 ± 0.3), european grapes (73 ± 0.3), raw eggplant (71 ± 1.2), frozen and cooked sweetened rhubarb (69 ± 0.4), wax beans (68 ± 0.4), dried filberts (68 ± 1.8), sweet pickles (67 ± 3.3), yellow mustard seed (67 ± 16.7), boysenberries (67 ± 0.4), boiled eggplant (65 ± 1), dried sesame kernels (63 ± 6.3), screwdriver (62 ± 0.2), unsweetened baking chocolate (61 ± 1.8), corn germ (60 ± 0.5), sapodilla (60 ± 0.3), cooked unenriched yellow degermed corn meal (58 ± 0.2), cooked enriched yellow degermed corn meal (58 ± 0.2), raw apple with skin (54 ± 0.4), prickly pear (51 ± 0.5), dried dates (51 ± 0.6), rhubarb pie (51 ± 0.4), dill seed (50 ± 25), celery seed (50 ± 25), dried sunflower seeds (50 ± 1.8), lemon peel (50 ± 8.3), canned butter beans (48 ± 0.4), cheese crackers with peanut butter (48 ± 1.2), boiled leeks (46 ± 1.9), boiled chayote (46 ± 0.6), cranberry sauce (46 ± 0.5), raw cucumber (44 ± 1), golden seedless raisins (44 ± 0.5), cooked apple without skin (44 ± 0.3), raw apple without skin (44 ± 0.4), dried pilinuts (43 ± 1.8),

I. Vitamins

PROVITAMIN A

strawberry pie (43 ± 0.5), gingersnaps cookie (43 ± 7.1), stir-fried sprouted lentils (41 ± 0.5), quince (41 ± 0.5), raw red cabbage (40 ± 1.4), low fat soybean flour (40 ± 0.6), defatted soybean flour (40 ± 0.5), jujube (40 ± 0.5), honeydew melon (40 ± 0.5), crab apples (40 ± 0.5), carissa (40 ± 2.5), breadfruit (40 ± 0.5), defatted soy meal (39 ± 0.4), oatmeal cookie (38 ± 3.8), raw japanese chestnuts (36 ± 1.8), semi-sweet chocolate (36 ± 1.8), bittersweet chocolate (36 ± 1.8), boiled kohlrabi (35 ± 0.6), loganberries (35 ± 0.3), blueberry pie (34 ± 0.4), boiled mungo beans (31 ± 0.3), tamarind (30 ± 0.4), casaba melon (30 ± 0.3), dried pinyon pine nuts (29 ± 1.8), raw european chestnuts (29 ± 1.8), lemon (29 ± 0.9), french fries (29 ± 0.6), semi-sweet bakers chocolate (29 ± 1.8), strawberries (28 ± 0.3), boiled chickpeas (27 ± 0.3), boiled red cabbage (27 ± 0.7), boiled butterbur (27 ± 0.5), frozen and boiled turnips (25 ± 0.5), hummus (25 ± 0.2), roasted european chestnuts (25 ± 1.8), mulberries (25 ± 0.4), red beans (24 ± 0.4), boiled mung beans (24 ± 0.2), canned chickpeas (24 ± 0.2), canned sprouted mung beans (23 ± 0.8), dry roasted soybean nuts (23 ± 0.6), pineapple (23 ± 0.3), frozen and boiled cauliflower (22 ± 0.6), frozen and boiled onions (21 ± 0.5), raw sprouted mung beans (21 ± 1), german sweet bakers chocolate (21 ± 1.8), poi (20 ± 0.4), raw jerusalem artichoke (20 ± 0.7), raw bamboo shoots (20 ± 0.7), pear (20 ± 0.3), lemon juice (20 ± 0.2), chamomile tea (20 ± 0.3), frozen white corn (19 ± 0.7), apple pie (19 ± 0.4), canned sauerkraut (18 ± 0.4), stir-fried sprouted soybeans (17 ± 0.5), pineapple pie (17 ± 0.4), oatmeal (17 ± 0.3), boiled yardlong bean (16 ± 0.3), raw cauliflower (16 ± 1), boiled black-eyed peas (15 ± 0.3), boiled cauliflower (15 ± 0.8), boiled broad beans (15 ± 0.3), carob flour (15 ± 0.5), falafel (14 ± 1), boiled and steamed japanese chestnuts (14 ± 1.8), cooked sprouted mung beans (13 ± 0.8), canned black-eyed peas (13 ± 0.2), boiled beets (13 ± 0.6), steamed sprouted soybeans (11 ± 1.1), whole dried sesame seeds (11 ± 5.6), oil roasted macadamia nuts (11 ± 1.8), applesauce (11 ± 0.4), onion rings (11 ± 0.6), boiled mothbeans (10 ± 0.3), boiled catjang black-eyed peas (10 ± 0.3), raw cassava (10 ± 0.5), canned broad beans (10 ± 0.2), lime (10 ± 0.7), white grapefruit (10 ± 0.4), cherimoya (10 ± 0.1), lime juice (10 ± 0.2), white grapefruit juice (10 ± 0.2), boiled soybeans (9 ± 0.3), boiled lentils (8 ± 0.3), canned bamboo shoots (8 ± 0.4), seedless raisins (8 ± 0.5), grape juice (8 ± 0.2), raisin pie (8 ± 0.4), whiskey sour (8 ± 0.6), raw radish (7 ± 1.1), boiled split peas (7 ± 0.3), boiled black turtle beans (6 ± 0.3), boiled black beans (6 ± 0.3), pickled beets (6 ± 0.4), boiled adzuki beans (6 ± 0.2), sugar apple (6 ± 0.3), pineapple juice (5 ± 0.2), canned chinese water chestnuts (4 ± 0.7), canned black turtle beans (4 ± 0.2), roasted chinese chestnuts (4 ± 1.8), java plum (4 ± 0.4), prune juice (4 ± 0.2), restaurant cranberry juice cocktail (4 ± 0.3), boiled french beans (3 ± 0.3), chocolate syrup (3 ± 1.3), daiquiri (3 ± 0.8), boiled yellow beans (2 ± 0.3), boiled pinto beans (2 ± 0.3), boiled pigeon peas (2 ± 0.3), canned navy beans (2 ± 0.2), boiled navy beans (2 ± 0.3), soursop (2 ± 0.2), piña colada (2 ± 0.4), canned pinto beans (1 ± 0.2), canned great northern beans (1 ± 0.2), boiled great northern beans (1 ± 0.3), apple juice (1 ± 0.2), restaurant hot chocolate (1 ± 0.2), tom collins (1 ± 0.2), gin & tonic (1 ± 0.2)

Applications: cancer, antioxidant therapy, sun sensitivity, free radical oxidation, life span extension, radiation, decreased radiation resistance, aging

Daily Dosage (1 retinol equivalent = 6 μg beta-carotene = 5 iu = 5 usp): pre–1958 MDR – 4000 iu as vitamin A; 1958 RDA – 5000 iu as vitamin A; 1974 PCS – 2500–5000 iu as vitamin A; 1974 RDA – 4000–5000 iu as vitamin A; 1974 USRDA – 5000 iu as vitamin A; 1980 RDA – 4000–5000 iu as vitamin

A; 1980 USRDA—5000 iu as vitamin A; 1989 RDA—4000-5000 iu as vitamin A; 1989 USRDA—5000 iu as vitamin A; nutritional—15,000-25,000 iu; therapeutic—50,000-125,000 iu; experimental—125,000 iu +; toxic—none known up to 150,000 iu

Warnings: Diabetics cannot convert carotene into vitamin A. Increased amounts can cause side-effects. See side-effects symptoms.

VITAMIN A

Names: vitamin A, antixerophthalmic factor/vitamin, anti-infective factor/vitamin, lard factor, axerohpthol, biosterol, oleovitamin A

Classification: lipid-soluble vitamin, antioxidant

Forms: vitamin A_1/retinol, vitamin A acetate, vitamin A acid, vitamin A aldehyde, vitamin A epoxide, vitamin A palmitate, vitamin A_2/dehydroretinol, neovitamin A, provitamin A

Deficiency: night blindness, itchy eyes, dry eyes, deterioration of tear ducts, bloodshot eyes, sties, thickening of cornea, softening of cornea, acne, skin rashes, rough skin, dry skin, skin eczema, enlarged follicles, dry hair, brittle hair, appetite decreased, deterioration of gastrointestinal tract, deterioration of salivary glands, decreased sense of smell, fatigue, weight decreased, deterioration of growth, immunity decreased, infections, sterility, decreased vitamin C, decreased cancer resistance, slow healing, increased menstrual bleeding, birth defects, deterioration of lactation

Side-effects: see *toxicity* symptoms

Toxicity: (Detoxified by vitamin C) dry skin, itching, profuse sweating, peeling skin, scaling skin, thickening skin, rashes, sore lips, sparse hair, coarse hair, hair decreased, brittle hair, brittle nails, eyebrow hair decreased, increase cranium pressure, blurred vision, protruding eyes, irregular menses, headache, emotional agitation, nausea, abdominal pain, appetite decreased, diarrhea, muscle weakness, fatigue, bone thickening, bone demineralization, aching bones, aching joints, enlarged spleen, enlarged liver, enlarged kidneys, retarded growth, decreased immunity, development of some cancers

Inhibitors: mineral oil, strenuous physical activities during digestion, ferrous sulfate, dicumarol, laxatives, steroids, PBC/polychlorinated biphenyl, sodium benzoate, nitrites, aflatoxin, DDT, dielrin, diarrhea, bilary dysfunction, pancreatic dysfunction, celiac disease, vitamin D deficiency

Helpers: vitamin B complex, vitamin B_P, vitamin C, vitamin D, vitamin E, calcium, phosphorus, zinc

Sources (iu per 100 grams of food): polar bear liver (50000000 ± 5000000), shark liver oil (12000000 ± 500000), cod liver oil (200000 ± 50000), broiled lamb liver (74511 ± 1.1), raw duck liver (39907 ± 0.5), fried beef liver (36105 ± 0.5), braised beef liver (35679 ± 0.5), raw goose liver (30998 ± 0.5), braised pork liver (17997 ± 0.5), pork liver-cheese (17489 ± 1.3), simmered chicken liver (16375 ± 0.5), pork liver sausage (14050 ± 2.8), simmered turkey liver (12581 ± 0.5), fried chicken giblets (11929 ± 0.5), simmered chicken giblets (7431 ± 0.5), simmered turkey giblets (6036 ± 0.5), sweet butter (3300 ± 3.3), butter (3060 ± 3.3), whipped butter (2900 ± 4.5), dry instant skim milk (2370 ± 0.5), raw whale meat (1860 ± 0.5), chicken egg yolk (1841 ± 2.9), whipping cream, heavy (1467 ± 3.3), cream cheese (1446 ± 1.8), port du salut cheese (1350 ± 1.8), duck egg (1329 ± 0.7), limburger cheese (1296 ± 1.8), havarti cheese (1289 ± 1.8), pimento cheese spread (1279 ± 1.8), simmered beef kidneys (1241 ± 0.5), gruyere cheese (1236 ± 1.8), american cheese (1225 ± 1.8), fontina cheese (1189 ± 1.8), neufchatel cheese (1146 ± 1.8), muenster cheese (1136 ± 1.8), cooked eel (1136 ± 0.6), whipping cream, light (1127 ± 3.3), brick cheese (1096 ± 1.8), cheddar cheese (1071 ± 1.8), caraway cheese (1068 ± 1.8), roquefort cheese (1061 ± 1.8), tilsit cheese (1057 ± 1.8), colby cheese (1046 ± 1.8), raw eel (1042 ± 0.6), cheshire cheese (996 ± 1.8), monterey cheese (961 ± 1.8), medium cream (940 ± 3.3), camembert cheese (936 ± 1.8), edam cheese (929 ± 1.8), dry whole milk (922 ± 1.6), swiss cheese (857 ± 1.8), provolone cheese (825 ± 1.8), mozzarella cheese (804 ± 1.8), cheese soufflé (800 ± 0.5), sour cream (792 ± 4.2), canned chicken liver pâté (732 ± 1.8), bleu cheese (729 ± 1.8), parmesan cheese (700 ± 10), brie cheese (675 ± 1.8), chili (657 ± 0.2), gouda cheese (654 ± 1.8), chicken egg (626 ± 1), cheese pizza (625 ± 0.4), romano cheese (579 ± 1.8), taco (519 ± 0.6), evaporated lowfat milk (496 ± 0.4), ricotta cheese (490 ± 0.4), o'brien potatoes (481 ± 0.3), prune pudding (462 ± 0.4), pepperoni pizza (443 ± 0.4), half and half cream (433 ± 3.3), evaporated skim milk (391 ± 1.6), pineapple upside-down cake (388 ± 0.7), eggnog (352 ± 0.2), sweetened condensed milk (329 ± 1.3), cheeseburger (304 ± 0.4), bread pudding (302 ± 0.2), quail egg (300 ± 5.6), tapioca cream pudding (291 ± 0.3), smoked cisco (282 ± 0.6), pound cake (267 ± 1.7), au gratin potatoes (264 ± 0.4), pickled herring (260 ± 3.3), braised pork kidneys (260 ± 0.5), cheesecake (254 ± 0.6), yellow corn pudding (246 ± 0.2), bacon cheeseburger (245 ± 0.3), evaporated whole milk (243 ± 0.4), cooked sturgeon (242 ± 0.6), human milk (239 ± 1.6), boston cream pie (214 ± 0.5), raw sturgeon (211 ± 0.6), roasted duck with skin (210 ± 0.5), potato salad (209 ± 0.4), lowfat 1% milk (205 ± 0.2), lowfat 2% milk (205 ± 0.2), skim milk (204 ± 0.2), protein fortified lowfat 1% milk (203 ± 0.2), protein fortified lowfat 2% milk (203 ± 0.2), protein fortified skim milk (203 ± 0.2), roasted dark meat chicken with skin (201 ± 0.5), caramel cake (200 ± 0.5), dry buttermilk (200 ± 7.1), lowfat 1% chocolate milk (200 ± 0.2), lowfat 2% chocolate milk (200 ± 0.2), fried chicken neck with skin (192 ± 1.4), sponge cake (189 ± 0.8), simmered chicken gizzard (188 ± 0.5), stewed dark meat chicken with skin (186 ± 0.5), goat milk (185 ± 0.2), simmered turkey gizzard (185 ± 0.5), indian buffalo milk (178 ± 0.2), cottage cheese (171 ± 1.8), caramel sundae (170 ± 0.3), mashed potatoes (169 ± 0.5), roasted chicken thigh with skin (165 ± 0.8), raw pheasant without skin (165 ± 0.5), vanilla pudding (161 ± 0.2), simmered chicken neck with skin (161 ± 1.3), roast beef sandwich (160 ± 0.3), roasted chicken wing with skin (159 ± 1.5), stewed chicken thigh with skin (151 ± 0.7), chocolate pudding (150 ± 0.2), cottage pudding (148 ± 0.9), sheep milk (147 ± 0.2), malted milk (142 ± 0.2), hot fudge sundae (140 ± 0.3), strawberry sundae (140 ± 0.3), chicken and turkey salad (139 ± 1.8), scalloped potatoes (135 ± 0.4), stewed chicken wing with skin (133 ± 1.3), restaurant vanilla milkshake (130 ± 0.2), hot chocolate (127 ± 0.2), whole milk (126 ± 0.2), fried chicken wing with skin (125 ± 1.6), baked beans with pork and tomato sauce (124 ± 0.2), whole yogurt (123 ± 0.2), whole chocolate malted milk (123 ± 0.2), pancakes with butter and syrup (122 ± 0.3), fried chicken back with skin (122 ± 0.7), simmered chicken neck without skin (122 ± 2.8), whole chocolate milk (121 ± 0.2), burrito (120 ± 0.3), ham and cheese sandwich (120 ± 0.3), restaurant strawberry milkshake (120 ± 0.2), carob milk (120 ± 0.2), raw bluefish (119 ± 0.6), potato pancakes (117 ± 0.7), fruitcake, dark (116 ± 1.2), vanilla milkshake (114 ± 0.2), baked beans with pork and sweet sauce (114 ± 0.2), raw wolffish (113 ± 0.6), roasted light meat chicken with skin (110 ± 0.5), fried dark meat chicken with skin (104 ± 0.5), braised lamb heart (103 ± 0.3), brownies (100 ± 2.5), pork pickle and pimento loaf (100 ± 1.8), roasted chicken drumstick with skin (100 ± 1), light mayonnaise (100 ± 3.6), mayonnaise (100 ± 3.6), yellow cake (99 ±

VITAMIN A

0.7), fried chicken thigh with skin (98 ± 0.8), stewed light meat chicken with skin (96 ± 0.5), light cream (93 ± 1.7), restaurant chocolate milkshake (93 ± 0.2), cooked kingfish (93 ± 0.5), roasted chicken breast with skin (93 ± 0.5), raw clams (91 ± 0.6), breaded and fried clams (91 ± 0.6), stewed chicken drumstick with skin (91 ± 0.9), devil's food cake (88 ± 0.8), breaded and fried fish (88 ± 0.7), cooked black bass (86 ± 0.4), chocolate milkshake (86 ± 0.2), fried chicken drumstick with skin (84 ± 1), stewed chicken breast with skin (82 ± 0.5), hamburger sandwich (79 ± 0.5), fried dark meat chicken without skin (79 ± 0.5), fried chicken (78 ± 0.6), roasted duck without skin (77 ± 0.5), roasted dark meat chicken without skin (72 ± 0.5), raw shark (71 ± 0.6), fruitcake, light (70 ± 1.2), roasted goose with skin (70 ± 0.5), noodles (69 ± 0.3), stewed dark meat chicken without skin (69 ± 0.5), fried light meat chicken with skin (68 ± 0.5), sardines (67 ± 2.1), cooked rockfish (66 ± 0.6), lowfat yogurt (66 ± 0.2), roasted chicken thigh without skin (65 ± 1), fish sandwich (64 ± 0.3), cooked sea bass (64 ± 0.6), cooked sockeye salmon (62 ± 0.6), stewed chicken drumstick without skin (59 ± 1.1), raw sockeye salmon (58 ± 0.6), raw quail without skin (57 ± 0.5), raw rockfish (56 ± 0.6), smoked whitefish (56 ± 0.6), raw sea bass (55 ± 0.6), cooked halibut (54 ± 0.6), cooked mackerel (54 ± 0.6), battered and fried shark (54 ± 0.6), canned sockeye salmon (53 ± 0.6), fried chicken breast with skin (50 ± 0.5), raw mackerel (49 ± 0.6), cooked whelk (48 ± 0.6), raw halibut (47 ± 0.6), dried and salted cod (42 ± 0.6), cooked mullet (42 ± 0.6), tuna salad spread (42 ± 0.9), cooked swordfish (41 ± 0.6), braised beef lungs (39 ± 0.5), kippered herring (38 ± 1.3), white cake (37 ± 0.6), dry skim milk (37 ± 1.7), raw mullet (36 ± 0.6), raw pink salmon (35 ± 0.6), raw swordfish (35 ± 0.6), cooked whiting (34 ± 0.6), buttermilk (33 ± 0.2), soy milk (32 ± 0.2), cooked herring (31 ± 0.6), raw ling (31 ± 0.6), fried light meat chicken without skin (30 ± 0.5), raw chum salmon (29 ± 0.6), raw whiting (29 ± 0.6), roasted light meat chicken without skin (29 ± 0.5), raw herring (28 ± 0.6), raw yellowtail (28 ± 0.6), simmered chicken heart (28 ± 0.5), simmered turkey heart (28 ± 0.5), tuna salad (27 ± 0.2), stewed light meat chicken without skin (27 ± 0.5), gefiltefish with broth (26 ± 1.2), cooked lobster (26 ± 0.6), smoked chinook salmon (26 ± 0.6), raw whelk (26 ± 0.6), cooked pike (25 ± 0.6), fried chicken breast without skin (23 ± 0.6), smoked haddock (22 ± 0.6), cooked rainbow trout (22 ± 0.6), braised pork heart (22 ± 0.4), raw pike (21 ± 0.6), roasted chicken breast without skin (21 ± 0.6), raw rainbow trout (20 ± 0.6), cooked haddock (19 ± 0.6), stewed chicken breast without skin (19 ± 0.5), raw trout (18 ± 0.6), raw haddock (16 ± 0.6), canned pink salmon (16 ± 0.6), whey (16 ± 0.2), cooked clams (14 ± 0.6), canned cod (14 ± 0.6), cooked cod (14 ± 0.6), cooked perch (14 ± 0.6), apple tapioca pudding (12 ± 0.2), raw cod (12 ± 0.6), cooked flatfish (12 ± 0.6), raw perch (12 ± 0.6), raw pollock (11 ± 0.6), raw turbot (11 ± 0.6), braised pork boston blade braised including fat (10 ± 0.5), braised pork center rib including fat (10 ± 0.5), fried pork loin blade including fat (10 ± 0.5), roasted pork rump including fat (10 ± 0.5), braised pork spareribs (10 ± 0.5), braised pork top loin including fat (10 ± 0.5), cooked carp (9 ± 0.6), raw flatfish (9 ± 0.6), braised pork arm braised including fat (9 ± 0.5), broiled pork boston blade including fat (9 ± 0.5), braised pork boston blade excluding fat (9 ± 0.5), braised pork center loin including fat (9 ± 0.5), broiled pork center loin including fat (9 ± 0.5), fried pork center loin including fat (9 ± 0.5), braised pork center loin excluding fat (9 ± 0.5), fried pork center rib including fat (9 ± 0.5), braised pork loin including fat (9 ± 0.5), broiled pork loin including fat (9 ± 0.5), broiled pork loin blade including fat (9 ± 0.5), roasted pork rump excluding fat (9 ± 0.5), broiled pork sirloin including fat (9 ± 0.5), fried pork top loin including fat (9 ± 0.5), braised pork top loin excluding fat (9 ± 0.5), raw pork jowl (9 ± 0.5), raw carp (8 ± 0.6), breaded and fried catfish (8 ± 0.6), cooked alaska king crab (8 ± 0.6), roasted pork arm including fat (8 ± 0.5), braised pork arm excluding fat (8 ± 0.5), roasted pork boston blade including fat (8 ± 0.5), broiled pork boston blade excluding fat (8 ± 0.5), roasted pork center loin including fat (8 ± 0.5), broiled pork center loin excluding fat (8 ± 0.5), fried pork center loin excluding fat (8 ± 0.5), roasted pork center loin excluding fat (8 ± 0.5), roasted pork center rib including fat (8 ± 0.5), fried pork center rib excluding fat (8 ± 0.5), roasted pork center rib excluding fat (8 ± 0.5), roasted pork leg including fat (8 ± 0.5), roasted pork loin including fat (8 ± 0.5), braised pork loin excluding fat (8 ± 0.5), broiled pork loin excluding fat (8 ± 0.5), broiled pork loin blade excluding fat (8 ± 0.5), fried pork loin blade excluding fat (8 ± 0.5), roasted pork loin blade excluding fat (8 ± 0.5), roasted pork shank including fat (8 ± 0.5), roasted pork shoulder including fat (8 ± 0.5), roasted pork sirloin including fat (8 ± 0.5), braised pork sirloin excluding fat (8 ± 0.5), broiled pork sirloin excluding fat (8 ± 0.5), broiled pork top loin including fat (8 ± 0.5), roasted pork top loin including fat (8 ± 0.5), fried pork top loin excluding fat (8 ± 0.5), roasted pork top loin excluding fat (8 ± 0.5), raw alaska king crab (7 ± 0.6), roasted pork arm excluding fat (7 ± 0.5), roasted pork boston blade excluding fat (7 ± 0.5), broiled pork center rib including fat (7 ± 0.5), roasted pork leg excluding fat (7 ± 0.5), roasted pork loin excluding fat (7 ± 0.5), braised pork loin blade including fat (7 ± 0.5), roasted pork shank excluding fat (7 ± 0.5), roasted pork shoulder excluding fat (7 ± 0.5), roasted pork sirloin excluding fat (7 ± 0.5), roasted pork tenderloin (7 ± 0.5), filled milk (7 ± 0.2), skim yogurt (7 ± 0.2), broiled pork center rib excluding fat (6 ± 0.5), broiled pork top loin excluding fat (6 ± 0.5), turkey salad (6 ± 0.9), braised pork loin blade excluding fat (5 ± 0.5), custard (4 ± 0.2), braised pork center rib excluding fat (3 ± 0.5)

Applications: bladder inflammation, angina pectoris, hardening of the arteries, atherosclerosis, diabetes, hemophilia, jaundice, mononucleosis, stroke, fracture, softening of bones, rickets, celiac disease, colitis, diarrhea, alcohol cravings, epilepsy, meningitis, ear infection, deterioration of vision, bitot spots, cataracts, conjunctivitis, eyestrain, glaucoma, night blindness, gallstones, cystic fibrosis, goiter, overactive thyroid, inflammation of prostate, swollen glands, hair deterioration, fever, headache, sinusitis, congestive heart failure, myocardial infarction, constipation, hemorrhoids, worms, arthritis, gout, kidney stones, kidney inflammation, varicose veins, cirrhosis of liver, hepatitis, allergies, asthma, bronchitis, common cold, croup, emphysema, hay fever, influenza, tuberculosis, canker sore, bad breath, muscular dystrophy, nail deterioration, vaginal itching, abscess, acne, athlete's foot, bedsores, boils, burns, carbuncle, dandruff, skin inflammation, dry skin, skin eczema, skin impetigo, skin psoriasis, shingles, skin ulcers, warts, gastritis, peptic ulcer, pyorrhea, tooth and gum deterioration, chicken pox, fatigue, infection, kwashiorkor, measles, pregnancy, rheumatic fever, inflammation of nose, scurvy, stress, cancer, antioxidant therapy, sun sensitivity, free radical oxidation, life span extension, radiation, decreased radiation resistance, aging

Daily Dosage (1 retinol equivalent = 1 μg retinol = 5 iu = 5 usp): pre–1958 MDR—4000 iu; 1958 RDA—5000 iu; 1974 PCS—2500–5000 iu; 1980 RDA—4000–5000 iu; 1980 USRDA—5000 iu; 1989 RDA—4000–5000 iu; 1989

VITAMIN A

USRDA — 5000 iu; nutritional — 15,000–25,000 iu; therapeutic — 50,000–100,000 iu; experimental — 100,000 iu; toxic — 50,000 iu +
Warnings: Increased amounts are dangerous. See toxicity symptoms.

VITAMIN A_0 *see* **provitamin A**

VITAMIN A_1

(Specific form of vitamin A)

Names: vitamin A_1, retinol, Acon, Afaxin, Agiolan, Alphalin, Anatola, Aoral, Apexol, Apostavit, Atav, Avibon, Avita, Avitol, Axerol, Dohyfral A, Epiteliol, Nio-A-Let, Prepalin, Testavol, Vaflol, Vi-Alpha, Vitpex, Vogan, Vogan-Neu

VITAMIN A_2

(Specific form of vitamin A)

Forms: dehydroretinol

VITAMIN A ACETATE

(Specific form of vitamin A)

VITAMIN A ACID

(Specific form of vitamin A)

Names: vitamin A acid, retinoic acid, tretinoin, Aberel, Airol, Aknoten, Cordes Vas, Dermairol, Epi-Aberel, Eudyna, Retin-A
Applications: premature aging of skin (This is the form of vitamin A found in Retin-A.)

VITAMIN A ALDEHYDE

(Specific form of vitamin A)

VITAMIN A EPOXIDE

(Specific form of vitamin A)

Names: vitamin A epoxide, hepaxanthin, monoepoxyvitamin A

VITAMIN A PALMITATE

(Specific form of vitamin A)

MONOEPOXYVITAMIN A *see* **vitamin A epoxide**

NEOVITAMIN A

(Specific form of vitamin A)

Names: neovitamin A, 5–cis-vitamin A

OLEOVITAMIN A *see* **vitamin A**

FACTOR AN *see* **vitamin B_W sulfoxide**

VITAMIN B COMPLEX

(Different water soluble vitamins. See vitamin B_1, vitamin B_2, vitamin B_{3a}, vitamin B_{3b}, vitamin B_4, vitamin B_5, vitamin B_6, vitamin B_7 vitamin B_8, vitamin B_9, vitamin B_{10}, vitamin B_{11}, vitamin B_{12}, vitamin B_{13}, vitamin B_{14}, vitamin B_{15}, vitamin B_{16}, and vitamin B_{17}. Usually includes the sub-vitamin B complex. Sometimes vitamin L, vitamin N, or vitamin Q_{10} are erroneously included.)

VITAMIN B

(Found to be a mixture. Later subdivided into separate vitamins. See vitamin B complex. The term "vitamin B" was originally used for vitamin B_1.)

FACTOR B *see* **vitamin B_{12p}**

VITAMIN B_1

Names: vitamin B_1, thiamine, aneurin, polyneuramin, oryzamin, antineuritic factor/vitamin, antiberiberi factor/vitamin
Classification: water-soluble vitamin, vitamin B complex, antioxidant
Forms: thiamine disulfide, thiamine hydrochloride, thiamine mononitrate, thiamine phosphoric acid ester chloride, thiamine phosphoric acid ester phosphate salt, thiamine 1,5–salt, vitamin B_1 propyl disulfide, thiamine triphosphoric acid ester, thiamine triphosphoric acid salt, thiamine O,S-diacetate
Deficiency: beriberi, brain deterioration, memory decreased, depression, emotional agitation, emotional deterioration, vision decreased, inflammation of optic nerve, mental health deterioration, central nervous system deterioration, deterioration of reflexes, increased pyruvic acid in blood, tingling toes, burning feet, decreased sense of touch, fatigue, digestion deterioration, appetite decreased, difficulty in digesting carbohydrates, constipation, immunity decreased, decreased protein synthesis, abdominal/chest pains, cardiac deterioration, decreased blood pressure, varicose veins, bluish skin color, labored breathing, tender leg muscles, decreased cancer resistance

Side-effects: see toxicity symptoms
Toxicity: muscle tremors, fluid accumulation in tissues, skin inflammation, nervousness, heart palpitation, allergies, altered thyroid and insulin production, vitamin B_6 decreased
Inhibitors: cooking, fever, hyperthyroidism, liver deterioration, digestion deterioration, tannin/tannic acid in tea or betel nuts, chlorogenic acid in coffee, vitamin B_6 decreased, vitamin B_{12} deficiency, baking soda, estrogen
Helpers: vitamin B complex, vitamin B_2, vitamin B_3, vitamin B_C, vitamin C, vitamin E, manganese, sulphur
Sources (mg per 100 grams of food): brewers yeast (15.82 ± 0.018), torula yeast (14.18 ± 0.018), rice polish (3.2 ± 0.009), granola bar (2.92 ± 0.021), dried spirulina seaweed (2.38 ± 0.005), dry bakers yeast (2.36 ± 0.018), dried sunflower seeds (2.32 ± 0.018), rice bran (2.21 ± 0.005), corn germ (1.7 ± 0.005), toasted wheat germ (1.68 ± 0.018), bran flakes (1.43 ± 0.018), freeze-dried parsley (1.43 ± 0.357), bran (1.43 ± 0.018), corn flakes (1.43 ± 0.018), wheat flakes (1.32 ± 0.018), fried pork center loin excluding fat (1.25 ± 0.005), freeze-dried chives (1.25 ± 0.625), dried pinyon pine nuts (1.25 ± 0.018), freeze-dried red sweet peppers (1.25 ± 0.313), freeze-dried green sweet peppers (1.25 ± 0.313), toasted sesame kernels (1.21 ± 0.018), sesame butter (1.2 ± 0.033), broiled pork center loin excluding fat (1.15 ± 0.005), raisin bran (1.08 ± 0.014), roasted canned lean ham (1.04 ± 0.005), broiled pork sirloin excluding fat (1.03 ± 0.005), fried pork center loin including fat (1.02 ± 0.005), broiled pork center loin including fat (1 ± 0.005), savory (1 ± 0.5), dried coriander leaf (1 ± 0.5), sage (1 ± 0.5), dried brazil nuts (1 ± 0.018), broiled pork loin excluding fat (0.97 ± 0.005),

VITAMIN B$_1$

canned ham (0.96 ± 0.005), roasted pork tenderloin (0.94 ± 0.005), unheated lean ham (0.93 ± 0.005), dried pilinuts (0.93 ± 0.018), roasted pork center loin excluding fat (0.91 ± 0.005), tomato powder (0.91 ± 0.005), hard pork salami (0.9 ± 0.05), broiled pork sirloin including fat (0.9 ± 0.005), broiled pork center rib excluding fat (0.89 ± 0.005), broiled pork top loin excluding fat (0.89 ± 0.005), braised pork center loin excluding fat (0.88 ± 0.005), unheated ham (0.86 ± 0.005), dried pecans (0.86 ± 0.018), broiled pork loin including fat (0.84 ± 0.005), canned lean ham (0.84 ± 0.005), roasted pork center loin including fat (0.83 ± 0.005), roasted canned ham (0.82 ± 0.005), dried japanese chestnuts (0.82 ± 0.018), dried pignolia pine nuts (0.82 ± 0.018), dried pistachio nuts (0.82 ± 0.018), grilled canadian bacon (0.81 ± 0.011), roasted pork loin excluding fat (0.8 ± 0.005), roasted pork sirloin excluding fat (0.8 ± 0.005), broiled pork center rib including fat (0.79 ± 0.005), whole dried sesame seeds (0.78 ± 0.056), fried pork center rib excluding fat (0.77 ± 0.005), fried pork top loin excluding fat (0.77 ± 0.005), braised pork center loin including fat (0.77 ± 0.005), broiled pork loin blade excluding fat (0.76 ± 0.005), roasted pork rump excluding fat (0.76 ± 0.005), broiled pork top loin including fat (0.76 ± 0.005), broiled pork boston blade excluding fat (0.75 ± 0.005), unheated canadian bacon (0.75 ± 0.009), roasted lean ham (0.75 ± 0.005), peanut flour (0.75 ± 0.125), dried sesame kernels (0.75 ± 0.063), pork sausage (0.74 ± 0.019), fried pork loin blade excluding fat (0.74 ± 0.005), roasted pork sirloin including fat (0.74 ± 0.005), braised pork sirloin excluding fat (0.74 ± 0.005), roasted cured pork arm excluding fat (0.73 ± 0.005), roasted ham (0.73 ± 0.005), millit (0.73 ± 0.005), roasted pork loin including fat (0.72 ± 0.005), compressed bakers yeast (0.71 ± 0.018), pork link sausage (0.71 ± 0.007), pork luxury loaf (0.71 ± 0.018), minced ham lunch meat (0.71 ± 0.024), roasted pork rump including fat (0.71 ± 0.005), defatted soybean flour (0.7 ± 0.005), cooked bacon (0.69 ± 0.004), braised pork loin excluding fat (0.69 ± 0.005), roasted pork leg excluding fat (0.69 ± 0.005), defatted soy meal (0.69 ± 0.004), dried peanuts (0.68 ± 0.018), broiled pork boston blade including fat (0.67 ± 0.005), yellow mustard seed (0.67 ± 0.167), poppy seed (0.67 ± 0.167), corn tortilla (0.67 ± 0.017), broiled pork loin blade including fat (0.66 ± 0.005), braised pork sirloin including fat (0.66 ± 0.005), fried pork center rib including fat (0.64 ± 0.005), roasted pork center rib excluding fat (0.64 ± 0.005), roasted pork top loin excluding fat (0.64 ± 0.005), italian pork sausage (0.63 ± 0.007), roasted pork shank excluding fat (0.63 ± 0.005), roasted pork leg including fat (0.63 ± 0.005), fried pork top loin including fat (0.63 ± 0.005), chopped ham lunch meat (0.62 ± 0.024), roasted cured pork arm including fat (0.61 ± 0.005), fried pork loin blade including fat (0.61 ± 0.005), ham and cheese roll (0.61 ± 0.018), braised pork loin including fat (0.61 ± 0.005), dark rye flour (0.61 ± 0.004), hard pork and beef salami (0.6 ± 0.05), braised pork arm excluding fat (0.6 ± 0.005), roasted racoon (0.59 ± 0.005), roasted pork boston blade excluding fat (0.59 ± 0.005), roasted pork center rib including fat (0.59 ± 0.005), braised pork top loin excluding fat (0.59 ± 0.005), braised pork center rib excluding fat (0.59 ± 0.005), roasted pork shoulder excluding fat (0.58 ± 0.005), roasted pork arm excluding fat (0.58 ± 0.005), roasted pork shank including fat (0.58 ± 0.005), roasted pork top loin including fat (0.58 ± 0.005), dark buckwheat flour (0.58 ± 0.005), soybean flour (0.58 ± 0.006), pork mothers loaf (0.57 ± 0.024), pork beerwurst salami (0.57 ± 0.022), roasted pork loin blade excluding fat (0.57 ± 0.005), dry roasted red pistachio nuts (0.57 ± 0.018), raw goose

I. Vitamins

VITAMIN B$_1$

liver (0.56 ± 0.005), braised pork heart (0.56 ± 0.004), roasted pork boston blade including fat (0.55 ± 0.005), braised pork boston blade excluding fat (0.55 ± 0.005), braised pork loin blade excluding fat (0.55 ± 0.005), whole wheat flour (0.55 ± 0.004), roasted pork shoulder including fat (0.54 ± 0.005), braised pork arm braised including fat (0.54 ± 0.005), braised pork center rib including fat (0.53 ± 0.005), pork bologna (0.52 ± 0.022), roasted pork loin blade including fat (0.52 ± 0.005), roasted pork arm including fat (0.52 ± 0.005), braised pork top loin including fat (0.52 ± 0.005), pork bratwurst (0.51 ± 0.006), braised pork boston blade braised including fat (0.51 ± 0.005), dinner rolls (0.5 ± 0.018), pork and beef honey loaf (0.5 ± 0.018), polish pork sausage (0.5 ± 0.018), oregano (0.5 ± 0.25), dried rosemary (0.5 ± 0.25), paprika (0.5 ± 0.25), cayenne pepper (0.5 ± 0.25), caraway seed (0.5 ± 0.25), dill seed (0.5 ± 0.25), fennel seed (0.5 ± 0.25), mace (0.5 ± 0.25), cumin seed (0.5 ± 0.25), curry powder (0.5 ± 0.25), nutmeg (0.5 ± 0.25), onion powder (0.5 ± 0.25), dehydrated onion flakes (0.5 ± 0.036), dried filberts (0.5 ± 0.018), hamburger bun (0.5 ± 0.013), hot dog bun (0.5 ± 0.013), broiled lamb liver (0.49 ± 0.011), braised pork loin blade including fat (0.49 ± 0.005), pita bread (0.47 ± 0.013), raw irish moss seaweed (0.47 ± 0.005), dried and frozen tofu (0.47 ± 0.029), unheated ham patties (0.46 ± 0.008), roasted pork blade roll (0.46 ± 0.005), rice flour (0.46 ± 0.004), roasted japanese chestnuts (0.46 ± 0.018), white bread (0.46 ± 0.021), english muffin (0.46 ± 0.009), raw wild rice (0.45 ± 0.003), enriched white degermed corn meal (0.44 ± 0.004), enriched yellow degermed corn meal (0.44 ± 0.004), all purpose enriched wheat flour (0.44 ± 0.004), enriched wheat bread (0.44 ± 0.004), toasted english muffin (0.44 ± 0.01), dry buttermilk (0.43 ± 0.071), ham salad (0.43 ± 0.018), dry roasted pistachio nuts (0.43 ± 0.018), dried ginko nuts (0.43 ± 0.018), tamarind (0.43 ± 0.004), dry roasted soybean nuts (0.43 ± 0.006), dry roasted peanuts (0.43 ± 0.018), oil roasted cashews (0.43 ± 0.018), dry instant skim milk (0.42 ± 0.005), stir-fried sprouted soybeans (0.42 ± 0.005), potato flour (0.42 ± 0.006), braised pork spareribs (0.41 ± 0.005), yellow corn pudding (0.41 ± 0.002), roasted soybean flour (0.41 ± 0.006), cheese crackers (0.4 ± 0.033), braised pork kidneys (0.4 ± 0.005), dry skim milk (0.4 ± 0.017), cracked wheat bread (0.4 ± 0.02), mixed grain bread (0.4 ± 0.02), french rolls (0.4 ± 0.01), rye bread (0.4 ± 0.02), french bread (0.39 ± 0.018), berliner (0.39 ± 0.022), pork and beef peppered loaf (0.39 ± 0.018), pork and beef picnic loaf (0.39 ± 0.018), raw pork jowl (0.39 ± 0.005), dried english/persian walnuts (0.39 ± 0.018), canned pork lunch meat (0.38 ± 0.024), pork and beef sausage (0.38 ± 0.038), sorghum grain (0.38 ± 0.005), white whole ground corn meal (0.38 ± 0.004), yellow whole ground corn meal (0.38 ± 0.004), low fat soybean flour (0.38 ± 0.006), sweet roll (0.38 ± 0.012), bagels (0.38 ± 0.009), raw bacon (0.37 ± 0.007), whole wheat roll (0.36 ± 0.018), italian bread (0.36 ± 0.018), oatmeal bread (0.36 ± 0.018), raw japanese chestnuts (0.36 ± 0.018), dried macadamia nuts (0.36 ± 0.018), whole wheat bread (0.36 ± 0.02), toasted rye bread (0.36 ± 0.023), graham crackers (0.36 ± 0.036), raw trout (0.35 ± 0.006), barbecue loaf (0.35 ± 0.022), grilled ham patties (0.35 ± 0.008), pork and beef knockwurst (0.34 ± 0.007), pumpernickel bread (0.34 ± 0.016), raw reindeer (0.33 ± 0.005), pork and beef pepperoni (0.33 ± 0.083), dried shitake mushrooms (0.33 ± 0.033), chili powder (0.33 ± 0.167), garlic powder (0.33 ± 0.167), toasted cracked wheat bread (0.33 ± 0.024), toasted whole wheat bread (0.33 ± 0.024), saltine crackers (0.33 ± 0.083), braised pork tongue

(0.32 ± 0.005), pork and beef old fashioned loaf (0.32 ± 0.018), pork and beef lunch meat (0.32 ± 0.018), ham and cheese spread (0.32 ± 0.018), almond meal (0.32 ± 0.018), toasted pumpernickel bread (0.32 ± 0.018), soybean protein concentrate (0.32 ± 0.018), dry roasted macadamia nuts (0.32 ± 0.018), oil roasted sunflower seeds (0.32 ± 0.018), oil roasted spanish peanuts (0.32 ± 0.018), toaster pastry (0.32 ± 0.01), cashew butter (0.32 ± 0.018), dry roasted pecans (0.32 ± 0.018), raisin bread (0.32 ± 0.02), pretzels (0.32 ± 0.018), whole grain rye crackers (0.31 ± 0.038), white bolted corn meal (0.3 ± 0.004), yellow bolted corn meal (0.3 ± 0.004), instant oatmeal (0.3 ± 0.003), boiled winged beans (0.3 ± 0.003), pork olive loaf (0.29 ± 0.018), pork pickle and pimento loaf (0.29 ± 0.018), oil roasted virginia peanuts (0.29 ± 0.018), dried european chestnuts (0.29 ± 0.018), oil roasted peanuts (0.29 ± 0.018), frozen and boiled green peas (0.29 ± 0.006), ry-krisp crackers (0.29 ± 0.036), toasted raisin bread (0.29 ± 0.024), raw quail without skin (0.28 ± 0.005), pork liver sausage (0.28 ± 0.028), pork liverwurst (0.28 ± 0.028), dry whole milk (0.28 ± 0.016), maypo (0.28 ± 0.003), cheese pizza (0.28 ± 0.004), dried oriental radish (0.28 ± 0.009), dried and salted cod (0.27 ± 0.006), pepperoni pizza (0.27 ± 0.004), taco shell (0.27 ± 0.045), tostada shell (0.27 ± 0.045), boiled hyacinth beans (0.27 ± 0.003), raw green peas (0.27 ± 0.006), braised pork liver (0.26 ± 0.005), beef and pork salami (0.26 ± 0.022), pork and beef brotwurst (0.26 ± 0.007), pork and beef link sausage with american cheese (0.26 ± 0.012), pork and beef link sausage (0.26 ± 0.007), roast beef sandwich (0.26 ± 0.003), roasted duck without skin (0.26 ± 0.005), boiled green soybeans (0.26 ± 0.006), frozen and boiled black-eyed peas (0.26 ± 0.006), boiled green peas (0.26 ± 0.006), bran muffins (0.25 ± 0.013), boiled pink beans (0.25 ± 0.003), freeze-dried shallots (0.25 ± 0.125), boiled peanuts (0.25 ± 0.016), roasted european chestnuts (0.25 ± 0.018), raw european chestnuts (0.25 ± 0.018), fenugreek seed (0.25 ± 0.125), shredded wheat (0.25 ± 0.018), shortbread cookie (0.25 ± 0.063), raw venison (0.24 ± 0.006), hot dog (0.24 ± 0.006), chicken egg yolk (0.24 ± 0.029), hamburger sandwich (0.24 ± 0.005), waffles (0.24 ± 0.007), danish pastry (0.24 ± 0.012), boiled black beans (0.24 ± 0.003), pork and beef kielbasa/kolbassy (0.23 ± 0.019), cheeseburger (0.23 ± 0.004), oatmeal cookie (0.23 ± 0.038), raw cassava (0.23 ± 0.005), boiled black turtle beans (0.23 ± 0.003), dry breadcrumbs (0.22 ± 0.005), pork and beef luncheon sausage (0.22 ± 0.022), stir-fried sprouted lentils (0.22 ± 0.005), raw spirulina seaweed (0.22 ± 0.005), boiled yellow corn (0.22 ± 0.006), boiled sprouted green peas (0.22 ± 0.005), raw garlic (0.22 ± 0.056), braised lamb heart (0.21 ± 0.003), fried beef liver (0.21 ± 0.005), pork livercheese (0.21 ± 0.013), cooked sockeye salmon (0.21 ± 0.006), burrito (0.21 ± 0.003), dried pumpkin and squash seeds (0.21 ± 0.018), enriched biscuits (0.21 ± 0.018), dried black walnuts (0.21 ± 0.018), ginko nuts (0.21 ± 0.018), oil roasted macadamia nuts (0.21 ± 0.018), pot barley pearled (0.21 ± 0.003), steamed sprouted soybeans (0.21 ± 0.011), boiled cranberry beans (0.21 ± 0.003), boiled yardlong bean (0.21 ± 0.003), dry roasted cashews (0.21 ± 0.018), dried almonds (0.21 ± 0.018), almond paste (0.21 ± 0.018), puffed wheat (0.21 ± 0.036), whiskey sour (0.21 ± 0.018), dried jujube (0.21 ± 0.005), corn meal muffins (0.2 ± 0.013), pie crust (0.2 ± 0.003), braised beef liver (0.2 ± 0.005), beef and pork frankfurter (0.2 ± 0.011), fish sandwich (0.2 ± 0.003), raw sockeye salmon (0.2 ± 0.006), dry cocoa powder (0.2 ± 0.1), raw jerusalem artichoke (0.2 ± 0.007), white corn flour (0.2 ± 0.004), yellow corn flour (0.2 ± 0.004), boiled black-eyed peas (0.2 ± 0.003), boiled navy beans (0.2 ± 0.003), pancakes (0.19 ± 0.019), simmered beef kidneys (0.19 ± 0.005), cooked eel (0.19 ± 0.006), boiled pinto beans (0.19 ± 0.003), boiled yellow beans (0.19 ± 0.003), boiled split peas (0.19 ± 0.003), whole ground corn meal muffins (0.18 ± 0.013), braised beef pancreas (0.18 ± 0.005), raw mackerel (0.18 ± 0.006), raw wolffish (0.18 ± 0.006), pork and beef spread (0.18 ± 0.018), french toast (0.18 ± 0.008), raw dandelion greens (0.18 ± 0.018), raw chinese chestnuts (0.18 ± 0.018), peanut brittle (0.18 ± 0.018), soybean protein isolate (0.18 ± 0.018), dried watermelon seeds (0.18 ± 0.018), french fried potatoes (0.18 ± 0.01), summer sausage (0.17 ± 0.022), beef and pork summer sausage (0.17 ± 0.022), beef and pork bologna (0.17 ± 0.022), roasted duck with skin (0.17 ± 0.005), yellow cake (0.17 ± 0.007), cooked wild long grain rice (0.17 ± 0.004), boiled arrowhead (0.17 ± 0.042), orange peel (0.17 ± 0.083), boiled succotash (0.17 ± 0.005), boiled lentils (0.17 ± 0.003), baked acorn squash (0.17 ± 0.005), roasted leg of lamb excluding fat (0.16 ± 0.006), roasted veal leg (0.16 ± 0.006), roasted muskrat (0.16 ± 0.005), peach pie (0.16 ± 0.005), pecan pie (0.16 ± 0.005), cooked mackerel (0.16 ± 0.006), duck egg (0.16 ± 0.007), noodles (0.16 ± 0.003), pineapple upside-down cake (0.16 ± 0.007), raw lotus root (0.16 ± 0.006), natto (0.16 ± 0.006), boiled mung beans (0.16 ± 0.002), boiled baby lima beans (0.16 ± 0.003), boiled catjang black-eyed peas (0.16 ± 0.003), boiled kidney beans (0.16 ± 0.003), boiled great northern beans (0.16 ± 0.003), boiled lima beans (0.16 ± 0.003), boiled soybeans (0.16 ± 0.003), firm raw tofu (0.16 ± 0.004), seedless raisins (0.16 ± 0.005), soy milk (0.16 ± 0.002), roasted lamb shoulder excluding fat (0.15 ± 0.006), roasted leg of lamb including fat (0.15 ± 0.006), popover roll (0.15 ± 0.013), blueberry muffins (0.15 ± 0.013), cottage pudding (0.15 ± 0.009), brownies (0.15 ± 0.025), simmered chicken liver (0.15 ± 0.005), raw eel (0.15 ± 0.006), white cake (0.15 ± 0.006), raw chinese water chestnuts (0.15 ± 0.008), stir-fried sprouted mung beans (0.15 ± 0.008), light rye flour (0.15 ± 0.005), tomato paste (0.15 ± 0.004), boiled pigeon peas (0.15 ± 0.003), boiled mungo beans (0.15 ± 0.003), canned great northern beans (0.15 ± 0.002), fried tofu (0.15 ± 0.038), vegetarian baked beans (0.15 ± 0.002), lobster paste (0.14 ± 0.071), broiled lamb loin chop excluding fat (0.14 ± 0.01), broiled lamb rib chop excluding fat (0.14 ± 0.012), raw frog legs (0.14 ± 0.005), fruitcake, dark (0.14 ± 0.012), simmered beef heart (0.14 ± 0.005), simmered beef shank excluding fat (0.14 ± 0.005), fried beef sirloin excluding fat (0.14 ± 0.005), braised beef flank excluding fat (0.14 ± 0.005), braised beef flank including fat (0.14 ± 0.005), braised pork spleen (0.14 ± 0.005), raw yellowtail (0.14 ± 0.006), sponge cake (0.14 ± 0.008), cornbread (0.14 ± 0.006), corn pone (0.14 ± 0.006), gingersnaps cookie (0.14 ± 0.071), pancakes with butter and syrup (0.14 ± 0.003), dried acorns (0.14 ± 0.018), roasted chinese chestnuts (0.14 ± 0.018), boiled and steamed japanese chestnuts (0.14 ± 0.018), raw bamboo shoots (0.14 ± 0.007), unenriched corn meal (0.14 ± 0.004), unenriched yellow degermed corn meal (0.14 ± 0.004), falafel (0.14 ± 0.01), oil roasted almonds (0.14 ± 0.018), dry roasted almonds (0.14 ± 0.018), toasted almonds (0.14 ± 0.018), canned navy beans (0.14 ± 0.002), canned black turtle beans (0.14 ± 0.002), potato chips (0.14 ± 0.018), boiled okra (0.14 ± 0.006), toasted white bread (0.14 ± 0.024), puffed rice (0.14 ± 0.036), macaroni (0.14 ± 0.004), spaghetti (0.14 ± 0.004), roasted veal rib roast (0.13 ± 0.006), roasted lamb shoulder including fat (0.13 ± 0.006), broiled lamb loin chop including fat (0.13 ± 0.007), kippered herring (0.13 ± 0.013), fried beef brains (0.13 ± 0.005), beef sum-

VITAMIN B$_1$

mer sausage (0.13 ± 0.022), simmered beef shank including fat (0.13 ± 0.005), broiled beef sirloin excluding fat (0.13 ± 0.005), broiled beef tenderloin excluding fat (0.13 ± 0.005), breaded and fried shrimp (0.13 ± 0.006), pork and beef mortadella (0.13 ± 0.033), tempeh (0.13 ± 0.006), potato pancakes (0.13 ± 0.007), devil's food cake (0.13 ± 0.008), cooked instant white rice (0.13 ± 0.003), boiled dandelion greens (0.13 ± 0.01), sugar cookie (0.13 ± 0.031), boiled lupins (0.13 ± 0.003), poi (0.13 ± 0.004), crunchy peanut butter (0.13 ± 0.016), creamy peanut butter (0.13 ± 0.031), boiled french beans (0.13 ± 0.003), almond butter (0.13 ± 0.031), french fries (0.13 ± 0.006), microwaved potato without skin (0.13 ± 0.003), ham and cheese sandwich (0.12 ± 0.003), roasted opossum (0.12 ± 0.005), broiled lamb rib chop including fat (0.12 ± 0.007), cooked herring (0.12 ± 0.006), fried beef sirloin including fat (0.12 ± 0.005), broiled beef top round excluding fat (0.12 ± 0.005), broiled beef tenderloin including fat (0.12 ± 0.005), angel food cake (0.12 ± 0.008), boiled adzuki beans (0.12 ± 0.002), light pearled barley (0.12 ± 0.003), boiled mothbeans (0.12 ± 0.003), boiled lotus root (0.12 ± 0.006), frozen and fried hash brown potatoes (0.12 ± 0.006), raw watercress (0.12 ± 0.029), boiled chickpeas (0.12 ± 0.003), boiled white beans (0.12 ± 0.003), canned green peas (0.12 ± 0.006), frozen and french fried potatoes without skin (0.12 ± 0.01), microwaved potato with skin (0.12 ± 0.002), instant corn grits (0.12 ± 0.004), turkey egg (0.11 ± 0.006), cooked kingfish (0.11 ± 0.005), quail egg (0.11 ± 0.056), canned smoked goose liver pâté (0.11 ± 0.018), turkey summer sausage (0.11 ± 0.018), beef lunch meat (0.11 ± 0.018), fried beef top round excluding fat (0.11 ± 0.005), broiled beef rib eye excluding fat (0.11 ± 0.005), broiled beef flank excluding fat (0.11 ± 0.005), broiled beef flank including fat (0.11 ± 0.005), roasted beef tenderloin excluding fat (0.11 ± 0.005), broiled beef sirloin including fat (0.11 ± 0.005), broiled beef top round including fat (0.11 ± 0.005), broiled porterhouse steak excluding fat (0.11 ± 0.005), broiled T-bone steak excluding fat (0.11 ± 0.005), beef pastrami (0.11 ± 0.018), raw ling (0.11 ± 0.006), taco (0.11 ± 0.006), onion rings (0.11 ± 0.006), fried chicken back with skin (0.11 ± 0.007), dry roasted sunflower seeds (0.11 ± 0.018), rusk crackers (0.11 ± 0.056), kumquats (0.11 ± 0.026), raw acorns (0.11 ± 0.018), breadfruit (0.11 ± 0.005), cooked taro (0.11 ± 0.008), cooked parboiled white rice (0.11 ± 0.003), cooked white rice (0.11 ± 0.002), sugar apple (0.11 ± 0.003), boiled beet greens (0.11 ± 0.007), toasted soybean nuts (0.11 ± 0.018), raw romaine lettuce (0.11 ± 0.018), oil roasted valencia peanuts (0.11 ± 0.018), raw california avocado (0.11 ± 0.003), raw florida avocado (0.11 ± 0.002), canned kidney beans (0.11 ± 0.002), raw mushrooms (0.11 ± 0.014), tangerine (0.11 ± 0.006), baked potato with skin (0.11 ± 0.002), apple pie (0.11 ± 0.004), oatmeal (0.11 ± 0.003), seeded raisins (0.11 ± 0.005), ground-cherries (0.11 ± 0.004), blackstrap molasses (0.1 ± 0.025), maltex (0.1 ± 0.003), fried chicken giblets (0.1 ± 0.005), fried beef top round including fat (0.1 ± 0.005), broiled beef round excluding fat (0.1 ± 0.005), roasted beef tip excluding fat (0.1 ± 0.005), broiled porterhouse steak including fat (0.1 ± 0.005), bacon cheeseburger (0.1 ± 0.003), anchovy canned in olive oil (0.1 ± 0.025), chocolate cream pie (0.1 ± 0.004), fried chicken thigh with skin (0.1 ± 0.008), fried dark meat chicken with skin (0.1 ± 0.005), boiled lamb's-quarters (0.1 ± 0.006), cherimoya (0.1 ± 0.001), raw shallots (0.1 ± 0.05), raw laver/nori seaweed (0.1 ± 0.005), roasted soybean nuts (0.1 ± 0.006), boiled spinach (0.1 ± 0.006), frozen and boiled okra (0.1 ± 0.005), boiled broad beans (0.1 ± 0.003), frozen and boiled brussels sprouts (0.1 ± 0.006), boiled asparagus (0.1 ± 0.006), canned white beans (0.1 ± 0.002), boiled brussels sprouts (0.1 ± 0.006), canned pinto beans (0.1 ± 0.002), raw eggplant (0.1 ± 0.012), baked/boiled tropical yam (0.1 ± 0.007), raw carrots (0.1 ± 0.007), dried dates (0.1 ± 0.006), boiled acorn squash (0.1 ± 0.004), baked potato without skin (0.1 ± 0.003), boiled potato without skin (0.1 ± 0.004), frozen and boiled potatoes without skin (0.1 ± 0.005), american grapes (0.1 ± 0.005), braised/roasted/stewed veal chuck (0.09 ± 0.006), light and dark meat turkey roll (0.09 ± 0.009), light meat turkey roll (0.09 ± 0.009), fruitcake, light (0.09 ± 0.012), raw whale meat (0.09 ± 0.005), roasted goose without skin (0.09 ± 0.005), braised pork pancreas (0.09 ± 0.005), raw herring (0.09 ± 0.006), simmered chicken giblets (0.09 ± 0.005), broiled beef rib excluding fat (0.09 ± 0.005), broiled beef rib eye including fat (0.09 ± 0.005), beef salami (0.09 ± 0.022), broiled beef round including fat (0.09 ± 0.005), roasted beef round tip including fat (0.09 ± 0.005), roasted beef tenderloin including fat (0.09 ± 0.005), beef honey roll sausage (0.09 ± 0.022), roasted eye of beef round excluding fat (0.09 ± 0.005), broiled T-bone steak including fat (0.09 ± 0.005), broiled beef top loin excluding fat (0.09 ± 0.005), beef beerwurst salami (0.09 ± 0.022), raw flatfish (0.09 ± 0.006), cooked cod (0.09 ± 0.006), raw pork stomach (0.09 ± 0.005), fried dark meat chicken without skin (0.09 ± 0.005), miso (0.09 ± 0.004), mashed potatoes (0.09 ± 0.005), boiled jute potherb (0.09 ± 0.012), boiled blackeye pea pods (0.09 ± 0.011), cooked brown rice (0.09 ± 0.003), raw savoy cabbage (0.09 ± 0.014), orange juice (0.09 ± 0.002), hummus (0.09 ± 0.002), boiled parsnips (0.09 ± 0.006), valencia orange (0.09 ± 0.004), raw sprouted alfalfa seeds (0.09 ± 0.015), navel orange (0.09 ± 0.004), raw red chili peppers (0.09 ± 0.011), raw green chili peppers (0.09 ± 0.011), raw potato without skin (0.09 ± 0.004), pineapple (0.09 ± 0.003), carrot juice (0.09 ± 0.003), european grapes (0.09 ± 0.003), thompson seedless grapes (0.09 ± 0.003), roasted beaver (0.08 ± 0.006), canned goose liver pâté (0.08 ± 0.038), cooked rainbow trout (0.08 ± 0.006), raw chum salmon (0.08 ± 0.006), roasted goose with skin (0.08 ± 0.005), sardines (0.08 ± 0.021), simmered beef brains (0.08 ± 0.005), braised beef chuck arm pot roast excluding fat (0.08 ± 0.005), roasted beef rib roasted excluding fat (0.08 ± 0.005), broiled beef rib including fat (0.08 ± 0.005), cooked flatfish (0.08 ± 0.006), braised beef bottom round excluding fat (0.08 ± 0.005), braised beef chuck blade roast excluding fat (0.08 ± 0.005), roasted eye of beef round including fat (0.08 ± 0.005), braised pork lungs (0.08 ± 0.005), broiled beef top loin including fat (0.08 ± 0.005), braised pork brains (0.08 ± 0.005), chicken egg (0.08 ± 0.01), cooked clams (0.08 ± 0.006), canned cod (0.08 ± 0.006), raw cod (0.08 ± 0.006), enchilada (0.08 ± 0.002), raw pheasant without skin (0.08 ± 0.005), cooked grouper (0.08 ± 0.006), sweetened condensed milk (0.08 ± 0.013), malted milk (0.08 ± 0.002), fried chicken breast without skin (0.08 ± 0.006), fried chicken breast with skin (0.08 ± 0.005), fried chicken drumstick with skin (0.08 ± 0.01), roasted chicken drumstick with skin (0.08 ± 0.01), fried light meat chicken with skin (0.08 ± 0.005), roasted chicken thigh without skin (0.08 ± 0.01), fried chicken neck with skin (0.08 ± 0.014), fried chicken (0.08 ± 0.006), lemon meringue pie (0.08 ± 0.004), potato salad (0.08 ± 0.004), o'brien potatoes (0.08 ± 0.003), raw pepeao (0.08 ± 0.005), light buckwheat flour (0.08 ± 0.005), mandarin oranges (0.08 ± 0.004), steamed hawaii mountain yam (0.08 ± 0.007), raw garden cress (0.08 ± 0.02), raw endive (0.08 ± 0.02), boiled broccoli (0.08 ± 0.006), raw cauliflower (0.08 ± 0.01), raw leeks

(0.08 ± 0.019), raw sprouted mung beans (0.08 ± 0.01), canned black-eyed peas (0.08 ± 0.002), restaurant orange juice (0.08 ± 0.004), boiled green beans (0.08 ± 0.008), boiled yellow snap beans (0.08 ± 0.008), raw scallop squash (0.08 ± 0.008), raw zucchini (0.08 ± 0.008), boiled mushrooms (0.08 ± 0.006), baked hubbard squash (0.08 ± 0.005), raw red sweet peppers (0.08 ± 0.01), raw green sweet peppers (0.08 ± 0.01), boiled eggplant (0.08 ± 0.01), raw tofu (0.08 ± 0.004), raw spring onions with tops (0.08 ± 0.01), boiled red tomato (0.08 ± 0.004), hash brown potatoes (0.08 ± 0.006), dried prunes (0.08 ± 0.006), watermelon (0.08 ± 0.003), canned chicken liver pâté (0.07 ± 0.018), milk chocolate (0.07 ± 0.018), braised/broiled veal loin (0.07 ± 0.006), braised/broiled veal round with rump (0.07 ± 0.006), raw rainbow trout (0.07 ± 0.006), turkey salami (0.07 ± 0.009), light meat chicken roll (0.07 ± 0.005), breaded and fried catfish (0.07 ± 0.006), chicken roll (0.07 ± 0.009), chicken frankfurter (0.07 ± 0.011), cooked flounder (0.07 ± 0.005), tomato catsup (0.07 ± 0.033), russian salad dressing (0.07 ± 0.033), simmered pork tail (0.07 ± 0.005), mince pie (0.07 ± 0.004), cooked pike (0.07 ± 0.006), raw grouper (0.07 ± 0.006), roasted light meat chicken without skin (0.07 ± 0.005), simmered chicken heart (0.07 ± 0.005), simmered turkey heart (0.07 ± 0.005), pickled herring (0.07 ± 0.033), braised beef shortribs excluding fat (0.07 ± 0.005), braised beef chuck arm pot roast including fat (0.07 ± 0.005), cooked whiting (0.07 ± 0.006), well done broiled extra lean ground beef (0.07 ± 0.005), braised beef brisket excluding fat (0.07 ± 0.005), roasted beef rib including fat (0.07 ± 0.005), braised beef bottom round including fat (0.07 ± 0.005), well done fried extra lean ground beef (0.07 ± 0.005), well done baked lean ground beef (0.07 ± 0.005), braised beef chuck blade roast including fat (0.07 ± 0.005), raw turbot (0.07 ± 0.006), tilsit cheese (0.07 ± 0.018), chopped smoked beef (0.07 ± 0.018), brie cheese (0.07 ± 0.018), gruyere cheese (0.07 ± 0.018), cooked halibut (0.07 ± 0.006), turkey frankfurter (0.07 ± 0.011), battered and fried shark (0.07 ± 0.006), limburger cheese (0.07 ± 0.018), breaded and fried fish (0.07 ± 0.007), gefiltefish with broth (0.07 ± 0.012), sheep milk (0.07 ± 0.002), fried light meat chicken without skin (0.07 ± 0.005), stewed chicken drumstick without skin (0.07 ± 0.011), roasted chicken breast without skin (0.07 ± 0.006), roasted chicken breast with skin (0.07 ± 0.005), half and half cream (0.07 ± 0.033), roasted dark meat chicken without skin (0.07 ± 0.005), roasted dark meat chicken with skin (0.07 ± 0.005), au gratin potatoes (0.07 ± 0.004), scalloped potatoes (0.07 ± 0.004), coleslaw (0.07 ± 0.008), boiled pokeberry shoots (0.07 ± 0.006), red beans (0.07 ± 0.004), chili sauce (0.07 ± 0.033), tomato sauce (0.07 ± 0.004), frozen creamed yellow corn (0.07 ± 0.004), elderberries (0.07 ± 0.003), tomato purée (0.07 ± 0.002), soursop (0.07 ± 0.002), raw chickory (0.07 ± 0.011), raw turnip greens (0.07 ± 0.018), raw spinach (0.07 ± 0.018), raw parsley (0.07 ± 0.017), frozen and boiled asparagus (0.07 ± 0.008), boiled and frozen baby lima beans (0.07 ± 0.006), frozen and boiled fordhook lima beans (0.07 ± 0.006), raw butterhead lettuce (0.07 ± 0.033), raw broccoli (0.07 ± 0.011), screwdriver (0.07 ± 0.002), frozen and boiled succotash (0.07 ± 0.006), raw coconut (0.07 ± 0.011), baked sweet potato (0.07 ± 0.004), frozen and baked sweet potato (0.07 ± 0.006), frozen and boiled yellow corn (0.07 ± 0.006), boiled butternut squash (0.07 ± 0.005), boiled rutabaga (0.07 ± 0.006), dried coconut (0.07 ± 0.018), dried figs (0.07 ± 0.003), canned potato without skin (0.07 ± 0.006), bread pudding (0.06 ± 0.002), coconut custard pie (0.06 ± 0.004), cooked black bass

VITAMIN B$_1$

(0.06 ± 0.004), cooked alaska king crab (0.06 ± 0.006), raw flounder (0.06 ± 0.005), raw sole (0.06 ± 0.005), cooked snapper (0.06 ± 0.006), raw pike (0.06 ± 0.006), raw bluefish (0.06 ± 0.006), well done broiled lean ground beef (0.06 ± 0.005), well done fried lean ground beef (0.06 ± 0.005), raw whiting (0.06 ± 0.006), braised beef brisket including fat (0.06 ± 0.005), medium-rare broiled extra lean ground beef (0.06 ± 0.005), medium-rare fried extra lean ground beef (0.06 ± 0.005), breaded fried squid (0.06 ± 0.006), raw halibut (0.06 ± 0.006), vienna sausage (0.06 ± 0.031), raw anchovy (0.06 ± 0.006), chili (0.06 ± 0.002), roasted turkey dark meat without skin (0.06 ± 0.005), roasted turkey light meat without skin (0.06 ± 0.005), roasted turkey dark meat with skin (0.06 ± 0.005), roasted turkey light meat with skin (0.06 ± 0.005), restaurant chocolate milkshake (0.06 ± 0.002), roasted light meat chicken with skin (0.06 ± 0.005), roasted chicken thigh with skin (0.06 ± 0.008), fried chicken wing with skin (0.06 ± 0.016), stewed dark meat chicken without skin (0.06 ± 0.005), stewed chicken thigh with skin (0.06 ± 0.007), simmered chicken neck without skin (0.06 ± 0.028), raw wakame seaweed (0.06 ± 0.005), boiled salsify (0.06 ± 0.007), boiled garden cress (0.06 ± 0.007), raisin bun (0.06 ± 0.008), raw chickory greens (0.06 ± 0.006), raw celtuce (0.06 ± 0.005), canned white corn (0.06 ± 0.005), raw green tomato (0.06 ± 0.004), apple brown betty (0.06 ± 0.002), cooked enriched white degermed corn meal (0.06 ± 0.002), cooked enriched yellow degermed corn meal (0.06 ± 0.002), tangerine juice (0.06 ± 0.002), casaba melon (0.06 ± 0.003), guava (0.06 ± 0.006), mango (0.06 ± 0.002), figs (0.06 ± 0.01), boiled artichoke (0.06 ± 0.006), frozen and boiled spinach (0.06 ± 0.005), raw green cabbage (0.06 ± 0.014), boiled cauliflower (0.06 ± 0.008), boiled artichoke hearts (0.06 ± 0.006), carob flour (0.06 ± 0.005), pineapple juice (0.06 ± 0.002), teriyaki sauce (0.06 ± 0.028), boiled scallop squash (0.06 ± 0.006), raw red cabbage (0.06 ± 0.014), raw onions (0.06 ± 0.006), boiled red sweet peppers (0.06 ± 0.007), boiled green sweet peppers (0.06 ± 0.007), raw red tomato (0.06 ± 0.004), barbecue sauce (0.06 ± 0.031), raw spiny dogfish (0.05 ± 0.005), braised/stewed veal breast plate (0.05 ± 0.006), cheese soufflé (0.05 ± 0.005), barbados molasses (0.05 ± 0.025), raw dungeness crab (0.05 ± 0.006), custard pie (0.05 ± 0.004), sweet potato pie (0.05 ± 0.004), raw catfish (0.05 ± 0.006), bread stuffing (0.05 ± 0.003), cooked rockfish (0.05 ± 0.006), stewed rabbit (0.05 ± 0.004), light molasses (0.05 ± 0.025), raw alaska king crab (0.05 ± 0.006), bread sticks (0.05 ± 0.025), raw snapper (0.05 ± 0.006), cream of wheat (0.05 ± 0.003), cream of rice (0.05 ± 0.003), simmered turkey liver (0.05 ± 0.005), simmered turkey giblets (0.05 ± 0.005), cooked whelk (0.05 ± 0.006), braised beef spleen (0.05 ± 0.005), smoked cisco (0.05 ± 0.006), raw pollock (0.05 ± 0.006), braised beef shortribs including fat (0.05 ± 0.005), medium-rare broiled lean ground beef (0.05 ± 0.005), medium-rare fried lean ground beef (0.05 ± 0.005), cooked swordfish (0.05 ± 0.006), turkey breast meat summer sausage (0.05 ± 0.024), turkey pastrami (0.05 ± 0.009), smoked beef sausage (0.05 ± 0.012), well done baked extra lean ground beef (0.05 ± 0.005), turkey ham (0.05 ± 0.009), medium-rare baked lean ground beef (0.05 ± 0.005), smoked haddock (0.05 ± 0.005), beef frankfurter (0.05 ± 0.009), raw shark (0.05 ± 0.006), skim yogurt (0.05 ± 0.002), indian buffalo milk (0.05 ± 0.002), restaurant vanilla milkshake (0.05 ± 0.002), whole chocolate malted milk (0.05 ± 0.002), chocolate milkshake (0.05 ± 0.002), restaurant strawberry milkshake (0.05 ± 0.002), stewed chicken drumstick with skin (0.05 ± 0.009), stewed chicken breast with skin (0.05 ± 0.005), stewed

VITAMIN B_1

dark meat chicken with skin (0.05 ± 0.005), stewed chicken wing with skin (0.05 ± 0.013), evaporated whole milk (0.05 ± 0.004), simmered chicken neck with skin (0.05 ± 0.013), goat milk (0.05 ± 0.002), baked beans with pork and sweet sauce (0.05 ± 0.002), baked beans with pork and tomato sauce (0.05 ± 0.002), raw vine spinach (0.05 ± 0.005), cooked sprouted mung beans (0.05 ± 0.008), raw welsh onions (0.05 ± 0.005), refried beans (0.05 ± 0.002), carissa (0.05 ± 0.025), canned stewed red tomatoes (0.05 ± 0.004), canned butter beans (0.05 ± 0.004), bulgur (0.05 ± 0.004), boiled savoy cabbage (0.05 ± 0.007), black currants (0.05 ± 0.009), frozen and boiled butternut squash (0.05 ± 0.004), raw kelp/kombu/tangle seaweed (0.05 ± 0.005), frozen and boiled collards (0.05 ± 0.006), boysenberries (0.05 ± 0.004), frozen and boiled broccoli (0.05 ± 0.005), raw iceberg lettuce (0.05 ± 0.025), canned lima beans (0.05 ± 0.002), frozen and boiled turnip greens (0.05 ± 0.006), cooked plantain (0.05 ± 0.003), loganberries (0.05 ± 0.003), raw crookneck squash (0.05 ± 0.008), restaurant tomato juice (0.05 ± 0.004), tomato juice (0.05 ± 0.003), boiled green cabbage (0.05 ± 0.007), tamari sauce (0.05 ± 0.009), shoyu sauce (0.05 ± 0.009), boiled kale (0.05 ± 0.008), boiled scotch kale (0.05 ± 0.008), frozen and boiled kale (0.05 ± 0.008), boiled sweet potato (0.05 ± 0.003), frozen and boiled red sweet peppers (0.05 ± 0.005), frozen and boiled green sweet peppers (0.05 ± 0.005), blueberries (0.05 ± 0.003), plum (0.05 ± 0.008), cheshire cheese (0.04 ± 0.018), fontina cheese (0.04 ± 0.018), tapioca cream pudding (0.04 ± 0.003), banana custard pie (0.04 ± 0.004), raw lingcod (0.04 ± 0.006), raw chinook salmon (0.04 ± 0.006), pineapple chiffon pie (0.04 ± 0.006), pineapple custard pie (0.04 ± 0.004), raw rockfish (0.04 ± 0.006), raw octopus (0.04 ± 0.006), hamburger pickle relish (0.04 ± 0.018), hot dog pickle relish (0.04 ± 0.018), honeydew melon (0.04 ± 0.005), pineapple pie (0.04 ± 0.004), gouda cheese (0.04 ± 0.018), well done broiled ground beef (0.04 ± 0.005), smoked whitefish (0.04 ± 0.006), well done fried ground beef (0.04 ± 0.005), well done baked ground beef (0.04 ± 0.005), braised beef lungs (0.04 ± 0.005), lebanon bologna (0.04 ± 0.022), raw swordfish (0.04 ± 0.006), medium-rare baked extra lean ground beef (0.04 ± 0.005), swiss cheese (0.04 ± 0.018), turkey bologna (0.04 ± 0.018), edam cheese (0.04 ± 0.018), provolone cheese (0.04 ± 0.018), cooked shrimp (0.04 ± 0.006), beef bologna (0.04 ± 0.022), cooked haddock (0.04 ± 0.006), camembert cheese (0.04 ± 0.018), bleu cheese (0.04 ± 0.018), raw haddock (0.04 ± 0.006), pork headcheese (0.04 ± 0.018), cheddar cheese (0.04 ± 0.018), pimento cheese spread (0.04 ± 0.018), american cheese (0.04 ± 0.018), roquefort cheese (0.04 ± 0.018), havarti cheese (0.04 ± 0.018), lowfat yogurt (0.04 ± 0.002), cheesecake (0.04 ± 0.006), eggnog (0.04 ± 0.002), protein fortified lowfat 2% milk (0.04 ± 0.002), protein fortified skim milk (0.04 ± 0.002), protein fortified lowfat 1% milk (0.04 ± 0.002), cream cheese (0.04 ± 0.018), strawberry sundae (0.04 ± 0.003), caramel sundae (0.04 ± 0.003), hot fudge sundae (0.04 ± 0.003), chicken and turkey salad (0.04 ± 0.018), skim milk (0.04 ± 0.002), lowfat 1% milk (0.04 ± 0.002), lowfat 2% milk (0.04 ± 0.002), whole milk (0.04 ± 0.002), hot chocolate (0.04 ± 0.002), carob milk (0.04 ± 0.002), lowfat 1% chocolate milk (0.04 ± 0.002), lowfat 2% chocolate milk (0.04 ± 0.002), whole chocolate milk (0.04 ± 0.002), caraway cheese (0.04 ± 0.018), whey (0.04 ± 0.002), stewed chicken breast without skin (0.04 ± 0.005), stewed light meat chicken without skin (0.04 ± 0.005), stewed light meat chicken with skin (0.04 ± 0.005), turkey salad (0.04 ± 0.009), dried longans (0.04 ± 0.005), cooked tahitian taro (0.04 ± 0.007), custard (0.04 ± 0.002), bittersweet chocolate (0.04 ± 0.018), cooked shitake mushrooms (0.04 ± 0.007), cream puff (0.04 ± 0.004), raw new zealand spinach (0.04 ± 0.018), sweet chocolate (0.04 ± 0.018), animal crackers (0.04 ± 0.019), raw swamp cabbage (0.04 ± 0.009), unenriched biscuits (0.04 ± 0.018), raw dock (0.04 ± 0.007), boiled burdock root (0.04 ± 0.004), boiled mustard greens (0.04 ± 0.007), boiled chinese cabbage (0.04 ± 0.006), marinara sauce (0.04 ± 0.002), frozen white corn (0.04 ± 0.007), boiled dishcloth gourd (0.04 ± 0.006), canned zucchini in tomato juice (0.04 ± 0.004), boiled yambean (0.04 ± 0.005), frozen and boiled turnips (0.04 ± 0.005), boiled kohlrabi (0.04 ± 0.006), white grapefruit juice (0.04 ± 0.002), red grapefruit juice (0.04 ± 0.002), raw ginger root (0.04 ± 0.021), frozen and boiled mustard greens (0.04 ± 0.007), canned peeled red tomatoes (0.04 ± 0.004), gooseberries (0.04 ± 0.003), raw looseleaf lettuce (0.04 ± 0.018), red currants (0.04 ± 0.009), white currants (0.04 ± 0.009), frozen and boiled yellow snap beans (0.04 ± 0.007), frozen and boiled green beans (0.04 ± 0.007), tequila sunrise (0.04 ± 0.003), pummelo (0.04 ± 0.003), boiled turnip greens (0.04 ± 0.007), canned cranberry beans (0.04 ± 0.002), boiled beets (0.04 ± 0.006), canned yellow corn (0.04 ± 0.005), boiled leeks (0.04 ± 0.019), boiled crookneck squash (0.04 ± 0.006), bananas (0.04 ± 0.004), canned sweet potato (0.04 ± 0.003), cantaloupe (0.04 ± 0.003), boiled zucchini (0.04 ± 0.006), boiled carrots (0.04 ± 0.006), frozen and boiled crookneck squash (0.04 ± 0.005), raw cucumber (0.04 ± 0.01), boiled onions (0.04 ± 0.005), boiled red cabbage (0.04 ± 0.007), unsweetened baking chocolate (0.04 ± 0.018), boiled hubbard squash (0.04 ± 0.004), boiled/baked spaghetti squash (0.04 ± 0.006), frozen and boiled zucchini (0.04 ± 0.004), dried sweetened flaked coconut (0.04 ± 0.02), cherries (0.04 ± 0.007), semi-sweet bakers chocolate (0.04 ± 0.018), vanilla pudding (0.03 ± 0.002), boston cream pie (0.03 ± 0.005), pound cake (0.03 ± 0.017), gingerbread (0.03 ± 0.008), butterscotch pie (0.03 ± 0.004), pumpkin pie (0.03 ± 0.004), raisin pie (0.03 ± 0.004), tuna salad (0.03 ± 0.002), simmered beef tongue (0.03 ± 0.005), medium-rare broiled ground beef (0.03 ± 0.005), medium-rare fried ground beef (0.03 ± 0.005), medium-rare baked ground beef (0.03 ± 0.005), simmered chicken gizzard (0.03 ± 0.005), simmered turkey gizzard (0.03 ± 0.005), cooked corned beef brisket (0.03 ± 0.005), breaded and fried scallops (0.03 ± 0.016), vanilla milkshake (0.03 ± 0.002), whole yogurt (0.03 ± 0.002), filled milk (0.03 ± 0.002), roasted chicken wing with skin (0.03 ± 0.015), evaporated skim milk (0.03 ± 0.016), buttermilk (0.03 ± 0.002), evaporated lowfat milk (0.03 ± 0.004), restaurant coleslaw (0.03 ± 0.005), longans (0.03 ± 0.005), boiled new zealand spinach (0.03 ± 0.006), mulberries (0.03 ± 0.004), raw chinese cabbage (0.03 ± 0.014), boiled dock (0.03 ± 0.005), boiled swiss chard (0.03 ± 0.006), boiled purslane (0.03 ± 0.009), canned yellow corn with red and green peppers (0.03 ± 0.004), boiled pumpkin (0.03 ± 0.004), frozen and boiled turnip greens and turnips (0.03 ± 0.005), canned shellie beans (0.03 ± 0.004), pickled beets (0.03 ± 0.004), pitanga (0.03 ± 0.003), canned jalapeño peppers (0.03 ± 0.007), boiled chayote (0.03 ± 0.006), canned red sweet peppers (0.03 ± 0.007), canned green sweet peppers (0.03 ± 0.007), boiled calabash gourd (0.03 ± 0.007), wax beans (0.03 ± 0.004), wheat cake (0.03 ± 0.004), carambola (0.03 ± 0.004), crab apples (0.03 ± 0.005), dried sweetened shredded coconut (0.03 ± 0.005), boiled waxgourd (0.03 ± 0.006), coconut cream (0.03 ± 0.002), coconut milk (0.03 ± 0.002), canned sauerkraut (0.03 ± 0.004), pomegranate (0.03 ± 0.003), raspber-

VITAMIN B_1

ries (0.03 ± 0.004), blackberries (0.03 ± 0.007), coconut water (0.03 ± 0.002), papaya (0.03 ± 0.002), raw celeriac (0.03 ± 0.005), canned chickpeas (0.03 ± 0.002), canned succotash with creamed corn (0.03 ± 0.004), frozen and boiled cauliflower (0.03 ± 0.006), canned succotash (0.03 ± 0.004), soy sauce (0.03 ± 0.009), bloody mary (0.03 ± 0.003), lemon juice (0.03 ± 0.002), pink grapefruit (0.03 ± 0.004), frozen and boiled carrots (0.03 ± 0.007), canned sprouted mung beans (0.03 ± 0.008), piña colada (0.03 ± 0.004), lemon (0.03 ± 0.009), white grapefruit (0.03 ± 0.004), raw celery (0.03 ± 0.013), lime (0.03 ± 0.007), boiled turnips (0.03 ± 0.006), japanese persimmon (0.03 ± 0.003), apricots (0.03 ± 0.005), raw frozen rhubarb (0.03 ± 0.004), boiled celery (0.03 ± 0.007), grape juice (0.03 ± 0.002), cranberry sauce (0.03 ± 0.005), kiwi fruit (0.03 ± 0.007), cooked prunes (0.03 ± 0.005), jackfruit (0.03 ± 0.005), prune pudding (0.02 ± 0.004), chocolate pudding (0.02 ± 0.002), chocolate chiffon pie (0.02 ± 0.006), lemon chiffon pie (0.02 ± 0.006), cheese crackers with peanut butter (0.02 ± 0.012), caramel cake (0.02 ± 0.005), simmered pork ears (0.02 ± 0.005), raw monkfish (0.02 ± 0.006), surimi shrimp (0.02 ± 0.006), strawberry pie (0.02 ± 0.005), rhubarb pie (0.02 ± 0.004), blackberry pie (0.02 ± 0.004), blueberry pie (0.02 ± 0.004), cherry pie (0.02 ± 0.004), canned pink salmon (0.02 ± 0.006), raw whelk (0.02 ± 0.006), smoked chinook salmon (0.02 ± 0.006), canned corned beef (0.02 ± 0.005), raw squid (0.02 ± 0.006), raw shrimp (0.02 ± 0.006), ricotta cheese (0.02 ± 0.004), candied sweet potato (0.02 ± 0.005), boiled amaranth (0.02 ± 0.008), canned shitake mushrooms (0.02 ± 0.004), boiled swamp cabbage (0.02 ± 0.01), boiled bamboo shoots (0.02 ± 0.004), sweetened coconut cream (0.02 ± 0.002), mammy apple (0.02 ± 0.005), roselle (0.02 ± 0.009), rose apple (0.02 ± 0.005), boiled cardoon (0.02 ± 0.005), loquats (0.02 ± 0.005), canned bamboo shoots (0.02 ± 0.004), okara tofu (0.02 ± 0.008), cooked unenriched corn meal (0.02 ± 0.002), cooked unenriched yellow degermed corn meal (0.02 ± 0.002), quince (0.02 ± 0.005), lime juice (0.02 ± 0.002), raw acerola (0.02 ± 0.005), acerola juice (0.02 ± 0.002), canned spinach (0.02 ± 0.005), canned creamed yellow corn (0.02 ± 0.004), canned broad beans (0.02 ± 0.002), strawberries (0.02 ± 0.003), frozen and boiled onions (0.02 ± 0.005), canned pumpkin (0.02 ± 0.004), canned crookneck squash (0.02 ± 0.005), pear (0.02 ± 0.003), boiled collards (0.02 ± 0.003), tuna salad (0.02 ± 0.009), frozen and cooked sweetened rhubarb (0.02 ± 0.004), peach (0.02 ± 0.006), daiquiri (0.02 ± 0.008), raw apple without skin (0.02 ± 0.003), cooked apple without skin (0.02 ± 0.003), applesauce (0.02 ± 0.004), prune juice (0.02 ± 0.002), raw oriental radish (0.02 ± 0.011), jujube (0.02 ± 0.005), apple juice (0.02 ± 0.002), corn grits (0.02 ± 0.002), dry dessert wine (0.02 ± 0.008), sweet dessert wine (0.02 ± 0.008), manhattan (0.02 ± 0.009), raw smelt (0.01 ± 0.005), canned bonito (0.01 ± 0.005), canned tuna in oil (0.01 ± 0.006), simmered pork feet (0.01 ± 0.005), pickled pork feet (0.01 ± 0.005), brown sugar (0.01 ± 0.003), surimi scallops (0.01 ± 0.006), canned sockeye salmon (0.01 ± 0.006), raw milkfish (0.01 ± 0.006), cooked lobster (0.01 ± 0.006), raw cuttlefish (0.01 ± 0.006), raw beef tripe (0.01 ± 0.005), raw scallops (0.01 ± 0.006), beer (0.01 ± 0.001), red table wine (0.01 ± 0.005), dried lychees (0.01 ± 0.005), dried agar seaweed (0.01 ± 0.005), canned pumpkin pie mix (0.01 ± 0.004), lychees (0.01 ± 0.005), prickly pear (0.01 ± 0.005), canned red chili peppers (0.01 ± 0.007), canned green chili peppers (0.01 ± 0.007), oheloberries (0.01 ± 0.004), canned chinese water chestnuts (0.01 ± 0.007), white hominy (0.01 ± 0.002), sapote (0.01 ± 0.002), raw agar

seaweed (0.01 ± 0.005), boiled butterbur (0.01 ± 0.005), tea (0.01 ± 0.002), java plum (0.01 ± 0.004), canned turnip greens (0.01 ± 0.004), canned green beans (0.01 ± 0.007), canned yellow snap beans (0.01 ± 0.007), canned carrots (0.01 ± 0.007), nectarine (0.01 ± 0.004), golden seedless raisins (0.01 ± 0.005), raw apple with skin (0.01 ± 0.004), farina (0.01 ± 0.003), papaya nectar (0.01 ± 0.002), apricot nectar (0.01 ± 0.002), restaurant cranberry juice cocktail (0.01 ± 0.003), chamomile tea (0.01 ± 0.003)
Applications: anemia, diabetes, constipation, diarrhea, alcohol cravings, Bell's palsy, mental illness, multiple sclerosis, neuritis, Ménière's syndrome, deterioration of vision, night blindness, fever, headache, congestive heart failure, worms, leg cramp, sciatica, influenza, shingles, digestion upset, pellagra, stress, cancer, antioxidant therapy, sun sensitivity, free radical oxidation, life span extension, radiation, decreased radiation resistance, aging
Daily Dosage (at least 100 mg vitamin B_1 per every 1000 mg vitamin B_{3a}): pre–1958 MDR — 1.0 mg; 1958 RDA — 1.6 mg; 1974 PCS — 0.75–2.25 mg; 1974 RDA — 1.0–1.5 mg; 1974 USRDA — 1.5 mg; 1980 RDA — 1.0–1.5 mg; 1980 USRDA — 1.5 mg; 1989 RDA — 1.0–1.5 mg; 1989 USRDA — 1.5 mg; nutritional — 10–50 mg; therapeutic — 200–500 mg; experimental — 1000–4000 mg; toxic — 10,000 mg +
Warnings: Increased amounts are dangerous. See toxicity symptoms.

VITAMIN B_1 O,S-DIACETATE

(lipid-soluble form of vitamin B_1)

Names: vitamin B_1 O,S-diacetate, Acetiamine, Thianeuron
Classification: lipid-soluble vitamin, vitamin B complex, antioxidant

VITAMIN B_1 DISULFIDE

(Specific form of vitamin B_1)

Names: vitamin B_1 disulfide, thimaine disulfide, aneurine disulfide, Neolamin

VITAMIN B_1 HYDROCHLORIDE

(Specific form of vitamin B_1)

Names: vitamin B_1 hydrochloride, thiamine hydrochloride, thiamine chloride hydrochloride, aneurine hydrochloride, thiaminium chloride hydrochloride, Bedome, Begiolan, Benerva, Bequin, Berin, Betabion hydrochloride, Betalin S, Betaxin, Bethiazine, Bevitex, Bewon, Biuno, Bivatin, Bivita, Clotiamina, Metabolin, Thiadoxine, Thiavit, Tiamidon, Tiaminal, Vitaneuron

VITAMIN B_1 MONONITRATE

(Specific form of vitamin B_1)

Names: vitamin B_1 mononitrate, thiamine mononitrate, aneurine mononitrate, Betabion mononitrate

VITAMIN B_1 PHOSPHORIC ACID ESTER CHLORIDE

(Specific form of vitamin B_1)

Names: vitamin B_1 phosphoric acid ester chloride, thiamine phosphoric acid ester chloride, thiamine orthophosphate ester chloride

VITAMIN B_1 PHOSPHORIC ACID ESTER SALT

(Specific form of vitamin B_1)

Names: vitamin B_1 phosphoric acid ester salt, thimaine phosphoric acid ester salt, thiamine monophosphate ester phosphoric acid salt, Umbeon

VITAMIN B_1 PROPYL DISULFIDE

(lipid-soluble form of vitamin B_1)

VITAMIN B$_2$

Names: vitamin B$_1$ propyl disulfide, thiamine propyl disulfide, DTPT, TPD, Prosultiamine, Alinamin, Aneurimec, Ausovit B1, Betatron, Binova, Ditiovit, Liponeurina, Marineurina, Orobetina, Proneurin, Sintotiamina, Tipidi

VITAMIN B$_1$ 1,5–SALT

(Specific form of vitamin B1)

Names: vitamin B$_1$ 1,5–salt, thiamine 1,5–salt, aneurin-1,5 salt, thiamine chloride naphthalene-1,5–disulfonic acid salt

VITAMIN B$_1$ TRIPHOSPHORIC ACID ESTER

(Specific form of vitamin B$_1$)

Names: vitamin B$_1$ triphosphoric acid ester, thiamine triphosphoric acid ester, thiamine triphosphate ester

VITAMIN B$_1$ TRIPHOSPHORIC ACID SALT

(Specific form of vitamin B$_1$)

Names: vitamin B$_1$ triphosphoric acid salt, thiamin triphosphoric acid salt, thiamine triphosphate salt

VITAMIN B$_2$

VITAMIN B$_2$ Complex

(After vitamin B was split into vitamin B$_1$ and vitamin B$_2$, it was discovered that there were many more vitamins in the complex. The B complex vitamins discovered after and including vitamin B$_2$ were called the vitamin B$_2$ complex.)

VITAMIN B$_2$

Names: vitamin B$_2$, riboflavin, vitamin G, lyochrome, lactoflavin, hepatoflavin, ovoflavin, uroflavin, Beflavine, Flavaxin, Ribipca
Classification: water-soluble vitamin, vitamin B complex, antioxidant cofactor
Forms: riboflavin, riboflavin phosphate
Deficiency: retarded growth, cracks around mouth, mouth sores, scaling face skin, frequent bloodshot eyes, light sensitivity, burning eyes, itching eyes, grainy feeling on inside of eyelids, eye fatigue, dilated eyes, sties, need of bright light to see, bloodshot eyes, burning hands, burning feet, vaginal itching, genital rashes, inability to urinate, anemia, appetite decreased, weight decreased, digestion deterioration, fatigue, depression, increased emotional agitation, emotional deterioration, dizziness, trembling, decreased antibody production, hair decreased, oily skin, skin inflammation, baldness, purplish tongue, sore tongue, deterioration of protein utilization
Side-effects: turns urine yellow; *see toxicity* symptoms
Toxicity: itching, numbness, skin burning sensations, skin prickling sensations
Inhibitors: light, alkalies, water, sulfa drugs, estrogen, alcohol, antibiotics, fats, phenothiazines (tranquilizers)
Helpers: vitamin B$_3$, vitamin B$_6$, vitamin C, phosphorus, fiber
Sources (mg per 100 grams of food): dry bakers yeast (5.46 ± 0.018), broiled lamb liver (5.11 ± 0.011), torula yeast (5.11 ± 0.018), brewers yeast (4.32 ± 0.018), fried beef liver (4.14 ± 0.005), braised beef liver (4.1 ± 0.005), simmered beef kidneys (4.06 ± 0.005), dried spirulina seaweed (3.67 ± 0.005), pork livercheese (2.24 ± 0.013), braised pork liver (2.2 ± 0.005), freeze-dried parsley (2.14 ± 0.357), dried parsley (2 ± 0.5), paprika (2 ± 0.25), simmered chicken liver (1.75 ± 0.005), dry instant skim milk (1.75 ± 0.005), almond meal

(1.71 ± 0.018), braised pork heart (1.71 ± 0.004), compressed bakers yeast (1.68 ± 0.018), braised pork kidneys (1.59 ± 0.005), dry skim milk (1.57 ± 0.017), pork liver sausage (1.56 ± 0.028), simmered beef heart (1.54 ± 0.005), fried chicken giblets (1.52 ± 0.005), wheat flakes (1.5 ± 0.018), bran flakes (1.43 ± 0.018), bran (1.43 ± 0.018), corn flakes (1.43 ± 0.018), canned chicken liver pâté (1.43 ± 0.018), dry buttermilk (1.43 ± 0.071), simmered turkey liver (1.42 ± 0.005), dried shitake mushrooms (1.27 ± 0.033), freeze-dried red sweet peppers (1.25 ± 0.313), freeze-dried green sweet peppers (1.25 ± 0.313), freeze-dried chives (1.25 ± 0.625), dry whole milk (1.22 ± 0.016), soybean flour (1.16 ± 0.006), canned potato without skin (1.11 ± 0.006), raisin bran (1.08 ± 0.014), pork liverwurst (1.06 ± 0.028), braised lamb heart (1.03 ± 0.003), dried coriander leaf (1 ± 0.5), cayenne pepper (1 ± 0.25), oil roasted almonds (1 ± 0.018), simmered chicken giblets (0.95 ± 0.005), roasted soybean flour (0.94 ± 0.006), raw cuttlefish (0.91 ± 0.006), simmered turkey giblets (0.9 ± 0.005), raw goose liver (0.89 ± 0.005), simmered turkey heart (0.88 ± 0.005), toasted wheat germ (0.82 ± 0.018), dried almonds (0.79 ± 0.018), quail egg (0.78 ± 0.056), tomato powder (0.76 ± 0.005), dry roasted soybean nuts (0.76 ± 0.006), corn germ (0.75 ± 0.005), almond paste (0.75 ± 0.018), simmered chicken heart (0.74 ± 0.005), toasted raisin bread (0.71 ± 0.024), raw reindeer (0.68 ± 0.005), chili powder (0.67 ± 0.167), dried oriental radish (0.67 ± 0.009), braised pork pancreas (0.66 ± 0.005), raisin bread (0.64 ± 0.02), almond butter (0.63 ± 0.031), raw wild rice (0.63 ± 0.003), roquefort cheese (0.61 ± 0.018), toasted pumpernickel bread (0.61 ± 0.018), dry roasted almonds (0.61 ± 0.018), toasted almonds (0.61 ± 0.018), raw irish moss seaweed (0.59 ± 0.005), dried lychees (0.57 ± 0.005), brie cheese (0.54 ± 0.018), pumpernickel bread (0.53 ± 0.016), roasted racoon (0.52 ± 0.005), rice flour (0.52 ± 0.004), braised pork tongue (0.51 ± 0.005), ry-krisp crackers (0.5 ± 0.036), caraway seed (0.5 ± 0.25), coriander seed (0.5 ± 0.25), dill seed (0.5 ± 0.25), fennel seed (0.5 ± 0.25), mace (0.5 ± 0.25), black pepper (0.5 ± 0.25), peanut flour (0.5 ± 0.125), cloves (0.5 ± 0.25), turmeric (0.5 ± 0.25), saltine crackers (0.5 ± 0.083), cumin seed (0.5 ± 0.25), curry powder (0.5 ± 0.25), limburger cheese (0.5 ± 0.018), camembert cheese (0.5 ± 0.018), braised beef pancreas (0.49 ± 0.005), raw venison (0.48 ± 0.006), sesame butter (0.47 ± 0.033), corn tortilla (0.47 ± 0.017), turkey egg (0.47 ± 0.006), roasted duck without skin (0.47 ± 0.005), carob flour (0.47 ± 0.005), toasted sesame kernels (0.46 ± 0.018), breaded fried squid (0.46 ± 0.006), raw mushrooms (0.46 ± 0.014), caraway cheese (0.46 ± 0.018), raw laver/nori seaweed (0.45 ± 0.005), chili (0.45 ± 0.002), broiled pork boston blade excluding fat (0.44 ± 0.005), toasted cracked wheat bread (0.43 ± 0.024), turkey summer sausage (0.43 ± 0.018), sweetened condensed milk (0.42 ± 0.013), broiled pork loin excluding fat (0.42 ± 0.005), raw squid (0.41 ± 0.006), ham and cheese sandwich (0.41 ± 0.003), chicken egg yolk (0.41 ± 0.029), cooked mackerel (0.41 ± 0.006), cheese crackers (0.4 ± 0.033), cracked wheat bread (0.4 ± 0.02), mixed grain bread (0.4 ± 0.02), parmesan cheese (0.4 ± 0.1), dry cocoa powder (0.4 ± 0.1), duck egg (0.4 ± 0.007), broiled pork sirloin excluding fat (0.4 ± 0.005), dried japanese chestnuts (0.39 ± 0.018), beef and pork salami (0.39 ± 0.022), cheddar cheese (0.39 ± 0.018), colby cheese (0.39 ± 0.018), monterey cheese (0.39 ± 0.018), edam cheese (0.39 ± 0.018), bleu cheese (0.39 ± 0.018), romano cheese (0.39 ± 0.018), rice polish (0.39 ± 0.009), roasted goose without skin (0.39 ± 0.005), braised pork boston blade excluding fat (0.39 ± 0.005), roasted pork tenderloin (0.39 ± 0.005), roasted beaver (0.38 ± 0.006), roasted opossum (0.38 ± 0.005), toasted white bread (0.38 ± 0.024), toasted english muffin (0.38 ± 0.01), braised pork spareribs (0.38 ± 0.005), broiled pork boston blade including fat (0.38 ± 0.005), broiled pork loin blade excluding fat (0.38 ± 0.005), millit (0.38 ± 0.005), fried pork loin blade excluding fat (0.37 ± 0.005), roasted pork boston blade excluding fat (0.37 ± 0.005), pimento cheese spread (0.36 ± 0.018), dried european chestnuts (0.36 ± 0.018), dried jujube (0.36 ± 0.005), sheep milk (0.36 ± 0.002), toasted rye bread (0.36 ± 0.023), american cheese (0.36 ± 0.018), brick cheese (0.36 ± 0.018), tilsit cheese (0.36 ± 0.018), swiss cheese (0.36 ± 0.018), gouda cheese (0.36 ± 0.018), milk chocolate (0.36 ± 0.018), french rolls (0.36 ± 0.01), broiled pork loin including fat (0.36 ± 0.005), roasted pork rump excluding fat (0.36 ± 0.005), braised pork arm excluding fat (0.36 ± 0.005), braised pork loin excluding fat (0.36 ± 0.005), roasted pork arm excluding fat (0.36 ± 0.005), roasted pork loin excluding fat (0.36 ± 0.005), roasted pork shoulder excluding fat (0.36 ± 0.005), anchovy canned in olive oil (0.35 ± 0.025), summer sausage (0.35 ± 0.022), simmered beef tongue (0.35 ± 0.005), broiled pork sirloin including fat (0.35 ± 0.005), fried pork center rib excluding fat (0.35 ± 0.005), braised pork sirloin excluding fat (0.35 ± 0.005), fried pork top loin excluding fat (0.35 ± 0.005), roasted pork leg excluding fat (0.35 ± 0.005), toaster pastry (0.34 ± 0.01), raw spirulina seaweed (0.34 ± 0.005), roasted pork loin blade excluding fat (0.34 ± 0.005), roasted pork shank excluding fat (0.34 ± 0.005), yellow mustard seed (0.33 ± 0.167), poppy seed (0.33 ± 0.167), cooked oysters (0.33 ± 0.006), raw chives (0.33 ± 0.167), maypo (0.33 ± 0.003), simmered turkey gizzard (0.33 ± 0.005), dried pepeao (0.33 ± 0.042), kippered herring (0.33 ± 0.013), hamburger bun (0.33 ± 0.013), hot dog bun (0.33 ± 0.013), raw trout (0.33 ± 0.006), fried beef sirloin excluding fat (0.33 ± 0.005), braised pork boston blade braised including fat (0.33 ± 0.005), roasted pork rump including fat (0.33 ± 0.005), roasted pork boston blade including fat (0.33 ± 0.005), fried pork center loin excluding fat (0.33 ± 0.005), roasted pork sirloin excluding fat (0.33 ± 0.005), braised pork loin blade excluding fat (0.33 ± 0.005), roasted ham (0.33 ± 0.005), canned smoked goose liver pâté (0.32 ± 0.018), dried pumpkin and squash seeds (0.32 ± 0.018), pork and beef peppered loaf (0.32 ± 0.018), braised pork lungs (0.32 ± 0.005), evaporated whole milk (0.32 ± 0.004), rye bread (0.32 ± 0.02), havarti cheese (0.32 ± 0.018), muenster cheese (0.32 ± 0.018), provolone cheese (0.32 ± 0.018), dinner rolls (0.32 ± 0.018), dry roasted red pistachio nuts (0.32 ± 0.018), chicken egg (0.32 ± 0.01), english muffin (0.32 ± 0.009), waffles (0.32 ± 0.007), raw mackerel (0.32 ± 0.006), well done broiled extra lean ground beef (0.32 ± 0.005), roasted beef tenderloin excluding fat (0.32 ± 0.005), roasted goose with skin (0.32 ± 0.005), broiled pork loin blade including fat (0.32 ± 0.005), braised pork top loin excluding fat (0.32 ± 0.005), roasted pork loin including fat (0.32 ± 0.005), roasted pork shoulder including fat (0.32 ± 0.005), broiled pork center rib excluding fat (0.32 ± 0.005), broiled pork top loin excluding fat (0.32 ± 0.005), braised pork center rib excluding fat (0.32 ± 0.005), canned goose liver pâté (0.31 ± 0.038), roasted leg of lamb excluding fat (0.31 ± 0.006), roasted veal rib roast (0.31 ± 0.006), evaporated skim milk (0.31 ± 0.016), sweet roll (0.31 ± 0.012), well done baked extra lean ground beef (0.31 ± 0.005), braised pork arm braised including fat (0.31 ± 0.005), broiled pork center loin excluding fat (0.31 ± 0.005), roasted pork center rib excluding fat (0.31 ± 0.005), roasted pork leg including fat (0.31 ± 0.005), roasted pork sirloin including fat (0.31 ± 0.005), roasted pork top loin excluding fat (0.31 ± 0.005), hard pork salami (0.3 ± 0.05), braised beef spleen (0.3 ± 0.005), hard pork and beef salami (0.3 ± 0.05), beef summer sausage (0.3 ± 0.022), beef and pork summer sausage (0.3 ± 0.022), well done fried extra lean ground beef (0.3 ± 0.005), broiled beef tenderloin excluding fat (0.3 ± 0.005), broiled beef sirloin excluding fat (0.3 ± 0.005), braised pork loin including fat (0.3 ± 0.005), roasted pork arm including fat (0.3 ± 0.005), roasted pork shank including fat (0.3 ± 0.005), dry breadcrumbs (0.3 ± 0.005), roasted pork loin blade including fat (0.3 ± 0.005), braised pork sirloin including fat (0.3 ± 0.005), lobster paste (0.29 ± 0.071), shredded wheat (0.29 ± 0.018), broiled lamb loin chop excluding fat (0.29 ± 0.01), braised/roasted/stewed veal chuck (0.29 ± 0.006), boiled beet greens (0.29 ± 0.007), pork luxury loaf (0.29 ± 0.018), pork and beef old fashioned loaf (0.29 ± 0.018), raw quail without skin (0.29 ± 0.005), boiled sprouted green peas (0.29 ± 0.005), boiled mushrooms (0.29 ± 0.006), cooked herring (0.29 ± 0.006), graham crackers (0.29 ± 0.036), dried and frozen tofu (0.29 ± 0.029), white bread (0.29 ± 0.021), gruyere cheese (0.29 ± 0.018), cheshire cheese (0.29 ± 0.018), oil roasted sunflower seeds (0.29 ± 0.018), bagels (0.29 ± 0.009), braised beef chuck arm pot roast excluding fat (0.29 ± 0.005), fried pork loin blade including fat (0.29 ± 0.005), roasted pork blade roll (0.29 ± 0.005), broiled lamb rib chop excluding fat (0.28 ± 0.012), light and dark meat turkey roll (0.28 ± 0.009), roasted lamb shoulder excluding fat (0.28 ± 0.006), raw garden cress (0.28 ± 0.02), bran muffins (0.28 ± 0.013), boiled soybeans (0.28 ± 0.003), simmered chicken neck without skin (0.28 ± 0.028), hot dog (0.28 ± 0.006), low fat soybean flour (0.28 ± 0.006), braised beef chuck blade roast excluding fat (0.28 ± 0.005), fried beef top round excluding fat (0.28 ± 0.005), roasted beef tenderloin including fat (0.28 ± 0.005), fried pork center loin including fat (0.28 ± 0.005), fried pork center rib including fat (0.28 ± 0.005), fried pork top loin including fat (0.28 ± 0.005), roasted pork center rib including fat (0.28 ± 0.005), broiled pork center rib including fat (0.28 ± 0.005), braised pork loin blade including fat (0.28 ± 0.005), cooked bacon (0.28 ± 0.004), roasted leg of lamb including fat (0.27 ± 0.006), evaporated lowfat milk (0.27 ± 0.004), chicken egg white (0.27 ± 0.015), medium-rare broiled extra lean ground beef (0.27 ± 0.005), roasted beef tip excluding fat (0.27 ± 0.005), broiled beef top round excluding fat (0.27 ± 0.005), broiled beef tenderloin including fat (0.27 ± 0.005), fried beef sirloin including fat (0.27 ± 0.005), roasted duck with skin (0.27 ± 0.005), braised pork center rib including fat (0.27 ± 0.005), braised pork top loin including fat (0.27 ± 0.005), broiled pork center loin including fat (0.27 ± 0.005), braised pork center loin excluding fat (0.27 ± 0.005), broiled pork top loin including fat (0.27 ± 0.005), roasted pork top loin including fat (0.27 ± 0.005), raw anchovy (0.26 ± 0.006), boiled pokeberry shoots (0.26 ± 0.006), boiled lamb's-quarters (0.26 ± 0.006), barbecue loaf (0.26 ± 0.022), braised pork spleen (0.26 ± 0.005), fried beef brains (0.26 ± 0.005), pork link sausage (0.26 ± 0.007), hamburger sandwich (0.26 ± 0.005), pork sausage (0.26 ± 0.019), pita bread (0.26 ± 0.013), medium-rare fried extra lean ground beef (0.26 ± 0.005), braised beef bottom round excluding fat (0.26 ± 0.005), broiled beef top round including fat (0.26 ± 0.005), broiled beef sirloin including fat (0.26 ± 0.005), roasted pork center loin excluding fat (0.26 ± 0.005), rice bran (0.26 ± 0.005), roasted canned ham (0.26 ± 0.005), enriched white degermed corn meal (0.26 ± 0.004), enriched yellow degermed corn meal (0.26 ± 0.004), all purpose enriched wheat flour (0.26 ± 0.004), enriched wheat bread (0.26 ± 0.004), french bread (0.25 ± 0.018), dry roasted pistachio nuts (0.25 ± 0.018), dried sunflower seeds (0.25 ± 0.018), dry roasted sunflower seeds (0.25 ± 0.018), turkey ham (0.25 ± 0.009), turkey pastrami (0.25 ± 0.009), braised/broiled veal loin (0.25 ± 0.006), braised/broiled veal round with rump (0.25 ± 0.006), breaded and fried clams (0.25 ± 0.006), raw frog legs (0.25 ± 0.005), raw dandelion greens (0.25 ± 0.018), roasted japanese chestnuts (0.25 ± 0.018), pork and beef honey loaf (0.25 ± 0.018), pork and beef picnic loaf (0.25 ± 0.018), pork pickle and pimento loaf (0.25 ± 0.018), pork olive loaf (0.25 ± 0.018), shortbread cookie (0.25 ± 0.063), fenugreek seed (0.25 ± 0.125), raw coriander (0.25 ± 0.125), port du salut cheese (0.25 ± 0.018), mozzarella cheese (0.25 ± 0.018), pretzels (0.25 ± 0.018), unsweetened baking chocolate (0.25 ± 0.018), fried chicken neck with skin (0.25 ± 0.014), popover roll (0.25 ± 0.013), french toast (0.25 ± 0.008), braised beef bottom round including fat (0.25 ± 0.005), roasted beef round tip including fat (0.25 ± 0.005), fried beef top round including fat (0.25 ± 0.005), broiled porterhouse steak excluding fat (0.25 ± 0.005), broiled T-bone steak excluding fat (0.25 ± 0.005), fried dark meat chicken without skin (0.25 ± 0.005), defatted soybean flour (0.25 ± 0.005), roasted canned lean ham (0.25 ± 0.005), unheated ham (0.25 ± 0.005), roasted turkey dark meat without skin (0.25 ± 0.005), defatted soy meal (0.25 ± 0.004), miso (0.25 ± 0.004), roasted lamb shoulder including fat (0.24 ± 0.006), roasted veal leg (0.24 ± 0.006), braised/stewed veal breast plate (0.24 ± 0.006), raw pork jowl (0.24 ± 0.005), dried and salted cod (0.24 ± 0.006), simmered chicken gizzard (0.24 ± 0.005), cheese pizza (0.24 ± 0.004), pepperoni pizza (0.24 ± 0.004), italian pork sausage (0.24 ± 0.007), raw herring (0.24 ± 0.006), toasted whole wheat bread (0.24 ± 0.024), simmered chicken neck with skin (0.24 ± 0.013), danish pastry (0.24 ± 0.012), fried chicken thigh with skin (0.24 ± 0.008), fried chicken back with skin (0.24 ± 0.007), braised beef chuck arm pot roast including fat (0.24 ± 0.005), medium-rare baked extra lean ground beef (0.24 ± 0.005), well done baked lean ground beef (0.24 ± 0.005), well done broiled lean ground beef (0.24 ± 0.005), well done fried lean ground beef (0.24 ± 0.005), cheese soufflé (0.24 ± 0.005), fried dark meat chicken with skin (0.24 ± 0.005), braised pork center loin including fat (0.24 ± 0.005), roasted pork center loin including fat (0.24 ± 0.005), roasted turkey dark meat with skin (0.24 ± 0.005), restaurant chocolate milkshake (0.24 ± 0.002), light meat turkey roll (0.23 ± 0.009), broiled lamb loin chop including fat (0.23 ± 0.007), roasted cured pork arm excluding fat (0.23 ± 0.005), pork and beef brotwurst (0.23 ± 0.007), pork and beef kielbasa/kolbassy (0.23 ± 0.019), boiled spinach (0.23 ± 0.006), raw wakame seaweed (0.23 ± 0.005), skim yogurt (0.23 ± 0.002), whole grain rye crackers (0.23 ± 0.038), corn meal muffins (0.23 ± 0.013), stewed chicken drumstick without skin (0.23 ± 0.011), roasted chicken thigh without skin (0.23 ± 0.01), angel food cake (0.23 ± 0.008), braised beef chuck blade roast including fat (0.23 ± 0.005), broiled beef round excluding fat (0.23 ± 0.005), roasted dark meat chicken without skin (0.23 ± 0.005), canned lean ham (0.23 ± 0.005), canned ham (0.23 ± 0.005), raw lotus root (0.22 ± 0.006), pork and beef luncheon sausage (0.22 ± 0.022), braised pork brains (0.22 ± 0.005), berliner (0.22 ± 0.022), cooked rainbow trout (0.22 ± 0.006), roast beef sandwich (0.22 ± 0.003), lowfat yogurt (0.22 ± 0.002), rusk crackers (0.22 ± 0.056), whole dried sesame seeds (0.22 ± 0.056), pancakes (0.22 ± 0.019), fried chicken drumstick with skin (0.22 ± 0.01), braised beef brisket excluding fat (0.22 ± 0.005), medium-rare fried lean

VITAMIN B$_2$

ground beef (0.22 ± 0.005), broiled beef rib eye excluding fat (0.22 ± 0.005), broiled porterhouse steak including fat (0.22 ± 0.005), unheated lean ham (0.22 ± 0.005), dried agar seaweed (0.22 ± 0.005), dark rye flour (0.22 ± 0.004), chocolate milkshake (0.22 ± 0.002), puffed wheat (0.21 ± 0.036), sardines (0.21 ± 0.021), italian bread (0.21 ± 0.018), oatmeal bread (0.21 ± 0.018), broiled lamb rib chop including fat (0.21 ± 0.007), raw clams (0.21 ± 0.006), cooked whelk (0.21 ± 0.006), roasted muskrat (0.21 ± 0.005), boiled broccoli (0.21 ± 0.006), cooked tahitian taro (0.21 ± 0.007), beef lunch meat (0.21 ± 0.018), ham and cheese spread (0.21 ± 0.018), dried pinyon pine nuts (0.21 ± 0.018), cheeseburger (0.21 ± 0.004), cream cheese (0.21 ± 0.018), fontina cheese (0.21 ± 0.018), neufchatel cheese (0.21 ± 0.018), enriched biscuits (0.21 ± 0.018), dry roasted cashews (0.21 ± 0.018), roasted chicken drumstick with skin (0.21 ± 0.01), roasted chicken thigh with skin (0.21 ± 0.008), white cake (0.21 ± 0.006), medium-rare broiled lean ground beef (0.21 ± 0.005), well done broiled ground beef (0.21 ± 0.005), well done fried ground beef (0.21 ± 0.005), broiled beef rib excluding fat (0.21 ± 0.005), roasted beef rib roasted excluding fat (0.21 ± 0.005), broiled beef round including fat (0.21 ± 0.005), simmered beef shank excluding fat (0.21 ± 0.005), broiled T-bone steak including fat (0.21 ± 0.005), roasted dark meat chicken with skin (0.21 ± 0.005), barbados molasses (0.2 ± 0.025), blackstrap molasses (0.2 ± 0.025), breaded and fried oysters (0.2 ± 0.006), raw pepeao (0.2 ± 0.005), protein fortified lowfat 2% milk (0.2 ± 0.002), protein fortified skim milk (0.2 ± 0.002), malted milk (0.2 ± 0.002), whole wheat bread (0.2 ± 0.02), blueberry muffins (0.2 ± 0.013), sponge cake (0.2 ± 0.008), yellow cake (0.2 ± 0.007), well done baked ground beef (0.2 ± 0.005), medium-rare fried ground beef (0.2 ± 0.005), broiled beef rib eye including fat (0.2 ± 0.005), braised beef shortribs excluding fat (0.2 ± 0.005), simmered beef shank including fat (0.2 ± 0.005), broiled beef top loin excluding fat (0.2 ± 0.005), stewed dark meat chicken without skin (0.2 ± 0.005), roasted lean ham (0.2 ± 0.005), raw ling (0.19 ± 0.006), raw pollock (0.19 ± 0.006), roasted cured pork arm including fat (0.19 ± 0.005), tomato paste (0.19 ± 0.004), boiled jute potherb (0.19 ± 0.012), minced ham lunch meat (0.19 ± 0.024), chopped ham lunch meat (0.19 ± 0.024), natto (0.19 ± 0.006), stir-fried sprouted soybeans (0.19 ± 0.005), raw chinese water chestnuts (0.19 ± 0.008), raw rainbow trout (0.19 ± 0.006), golden seedless raisins (0.19 ± 0.005), caramel sundae (0.19 ± 0.003), hot fudge sundae (0.19 ± 0.003), pancakes with butter and syrup (0.19 ± 0.003), eggnog (0.19 ± 0.002), pork bratwurst (0.19 ± 0.006), bread pudding (0.19 ± 0.002), protein fortified lowfat 1% milk (0.19 ± 0.002), restaurant strawberry milkshake (0.19 ± 0.002), pork mothers loaf (0.19 ± 0.024), canned pork lunch meat (0.19 ± 0.024), grilled canadian bacon (0.19 ± 0.011), stewed chicken drumstick with skin (0.19 ± 0.009), stewed chicken thigh with skin (0.19 ± 0.007), canned sockeye salmon (0.19 ± 0.006), canned pink salmon (0.19 ± 0.006), broiled beef flank including fat (0.19 ± 0.005), braised beef flank excluding fat (0.19 ± 0.005), broiled beef flank excluding fat (0.19 ± 0.005), medium-rare baked lean ground beef (0.19 ± 0.005), medium-rare broiled ground beef (0.19 ± 0.005), ricotta cheese (0.19 ± 0.004), coconut custard pie (0.19 ± 0.004), vanilla milkshake (0.19 ± 0.002), custard (0.19 ± 0.002), taco shell (0.18 ± 0.045), tostada shell (0.18 ± 0.045), turkey bologna (0.18 ± 0.018), dried pignolia pine nuts (0.18 ± 0.018), dried pistachio nuts (0.18 ± 0.018), turkey frankfurter (0.18 ± 0.011), cooked sockeye salmon (0.18 ± 0.006), raw chum salmon

VITAMIN B$_2$

(0.18 ± 0.006), raw european chestnuts (0.18 ± 0.018), raw chinese chestnuts (0.18 ± 0.018), raw japanese chestnuts (0.18 ± 0.018), dried ginko nuts (0.18 ± 0.018), raw spinach (0.18 ± 0.018), ham and cheese roll (0.18 ± 0.018), roasted european chestnuts (0.18 ± 0.018), chopped smoked beef (0.18 ± 0.018), pork headcheese (0.18 ± 0.018), pork and beef link sausage (0.18 ± 0.007), stir-fried sprouted mung beans (0.18 ± 0.008), seeded raisins (0.18 ± 0.005), beef pastrami (0.18 ± 0.018), boiled salsify (0.18 ± 0.007), falafel (0.18 ± 0.01), strawberry sundae (0.18 ± 0.003), tapioca cream pudding (0.18 ± 0.003), bacon cheeseburger (0.18 ± 0.003), restaurant vanilla milkshake (0.18 ± 0.002), hot chocolate (0.18 ± 0.002), bittersweet chocolate (0.18 ± 0.018), cashew butter (0.18 ± 0.018), oil roasted cashews (0.18 ± 0.018), whole ground corn meal muffins (0.18 ± 0.013), turkey salami (0.18 ± 0.009), unheated canadian bacon (0.18 ± 0.009), devil's food cake (0.18 ± 0.008), grilled ham patties (0.18 ± 0.008), braised beef flank including fat (0.18 ± 0.005), broiled beef top loin including fat (0.18 ± 0.005), stewed dark meat chicken with skin (0.18 ± 0.005), pork and beef pepperoni (0.17 ± 0.083), popcorn (0.17 ± 0.083), raw oysters (0.17 ± 0.006), lemon peel (0.17 ± 0.083), orange peel (0.17 ± 0.083), pork bologna (0.17 ± 0.022), pork beerwurst salami (0.17 ± 0.022), lebanon bologna (0.17 ± 0.022), beef honey roll sausage (0.17 ± 0.022), beef salami (0.17 ± 0.022), boiled dandelion greens (0.17 ± 0.01), frozen and boiled spinach (0.17 ± 0.005), dried prunes (0.17 ± 0.006), raw beef tripe (0.17 ± 0.005), simmered beef brains (0.17 ± 0.005), burrito (0.17 ± 0.003), lowfat 1% milk (0.17 ± 0.002), lowfat 1% chocolate milk (0.17 ± 0.002), sour cream (0.17 ± 0.042), cottage pudding (0.17 ± 0.009), cooked shitake mushrooms (0.17 ± 0.007), braised beef brisket including fat (0.17 ± 0.005), cooked corned beef brisket (0.17 ± 0.005), broiled beef rib including fat (0.17 ± 0.005), roasted beef rib including fat (0.17 ± 0.005), roasted eye of beef round excluding fat (0.17 ± 0.005), cream puff (0.17 ± 0.004), chocolate cream pie (0.17 ± 0.004), raw dungeness crab (0.16 ± 0.006), canned oysters (0.16 ± 0.006), smoked cisco (0.16 ± 0.006), instant oatmeal (0.16 ± 0.003), boiled garden cress (0.16 ± 0.007), pork and beef link sausage with american cheese (0.16 ± 0.012), boiled green soybeans (0.16 ± 0.006), boiled artichoke (0.16 ± 0.006), lowfat 2% milk (0.16 ± 0.002), lowfat 2% chocolate milk (0.16 ± 0.002), vanilla pudding (0.16 ± 0.002), whole milk (0.16 ± 0.002), whole chocolate malted milk (0.16 ± 0.002), whole chocolate milk (0.16 ± 0.002), buttermilk (0.16 ± 0.002), tamari sauce (0.16 ± 0.009), medium-rare baked ground beef (0.16 ± 0.005), roasted eye of beef round including fat (0.16 ± 0.005), custard pie (0.16 ± 0.004), whey (0.16 ± 0.002), pork and beef sausage (0.15 ± 0.038), raw sockeye salmon (0.15 ± 0.006), raw kelp/kombu/tangle seaweed (0.15 ± 0.005), pork and beef knockwurst (0.15 ± 0.007), boiled green peas (0.15 ± 0.006), raw pheasant without skin (0.15 ± 0.005), tamarind (0.15 ± 0.004), roasted soybean nuts (0.15 ± 0.006), canned corned beef (0.15 ± 0.005), carob milk (0.15 ± 0.002), oatmeal cookie (0.15 ± 0.038), unheated ham patties (0.15 ± 0.008), braised beef shortribs including fat (0.15 ± 0.005), dark buckwheat flour (0.15 ± 0.005), sorghum grain (0.15 ± 0.005), breaded and fried shrimp (0.14 ± 0.006), cooked rainbow smelt (0.14 ± 0.006), longans (0.14 ± 0.005), raw spring onions with tops (0.14 ± 0.01), boiled amaranth (0.14 ± 0.008), braised beef lungs (0.14 ± 0.005), raw new zealand spinach (0.14 ± 0.018), potato flour (0.14 ± 0.006), boiled sweet potato (0.14 ± 0.003), pork and beef lunch meat (0.14 ± 0.018), canned spinach (0.14 ± 0.005), dried pecans

(0.14 ± 0.018), toasted soybean nuts (0.14 ± 0.018), dried english/persian walnuts (0.14 ± 0.018), dried apricots (0.14 ± 0.014), prune pudding (0.14 ± 0.004), indian buffalo milk (0.14 ± 0.002), skim milk (0.14 ± 0.002), goat milk (0.14 ± 0.002), gingersnaps cookie (0.14 ± 0.071), pork and beef spread (0.14 ± 0.018), polish pork sausage (0.14 ± 0.018), sweet chocolate (0.14 ± 0.018), whole wheat roll (0.14 ± 0.018), dried acorns (0.14 ± 0.018), dried brazil nuts (0.14 ± 0.018), dried peanuts (0.14 ± 0.018), oil roasted valencia peanuts (0.14 ± 0.018), dried watermelon seeds (0.14 ± 0.018), soybean protein concentrate (0.14 ± 0.018), smoked beef sausage (0.14 ± 0.012), fruitcake, dark (0.14 ± 0.012), shoyu sauce (0.14 ± 0.009), cornbread (0.14 ± 0.006), corn pone (0.14 ± 0.006), cooked black bass (0.14 ± 0.004), fish sandwich (0.14 ± 0.003), chocolate pudding (0.14 ± 0.002), whole yogurt (0.14 ± 0.002), dried sesame kernels (0.13 ± 0.063), pickled herring (0.13 ± 0.033), granola bar (0.13 ± 0.021), raw cusk (0.13 ± 0.006), cooked perch (0.13 ± 0.006), light meat chicken roll (0.13 ± 0.005), raw green peas (0.13 ± 0.006), purple passion fruit juice (0.13 ± 0.002), pork and beef mortadella (0.13 ± 0.033), baked sweet potato (0.13 ± 0.004), beef bologna (0.13 ± 0.022), beef and pork bologna (0.13 ± 0.022), beef beerwurst salami (0.13 ± 0.022), raw chickory (0.13 ± 0.011), cheesecake (0.13 ± 0.006), yellow corn pudding (0.13 ± 0.002), banana custard pie (0.13 ± 0.004), whipping cream, heavy (0.13 ± 0.033), whipping cream, light (0.13 ± 0.033), medium cream (0.13 ± 0.033), half and half cream (0.13 ± 0.033), vienna sausage (0.13 ± 0.031), sugar cookie (0.13 ± 0.031), creamy peanut butter (0.13 ± 0.031), fried chicken wing with skin (0.13 ± 0.016), crunchy peanut butter (0.13 ± 0.016), potato pancakes (0.13 ± 0.007), fried chicken breast without skin (0.13 ± 0.006), breaded and fried catfish (0.13 ± 0.006), cooked kingfish (0.13 ± 0.005), fried light meat chicken with skin (0.13 ± 0.005), fried chicken breast with skin (0.13 ± 0.005), fried light meat chicken without skin (0.13 ± 0.005), roasted turkey light meat with skin (0.13 ± 0.005), roasted turkey light meat without skin (0.13 ± 0.005), boiled winged beans (0.13 ± 0.003), chicken roll (0.12 ± 0.009), raw lingcod (0.12 ± 0.006), raw rainbow smelt (0.12 ± 0.006), cooked flatfish (0.12 ± 0.006), raw smelt (0.12 ± 0.005), raw pork stomach (0.12 ± 0.005), frozen and boiled brussels sprouts (0.12 ± 0.006), raw watercress (0.12 ± 0.029), sugar apple (0.12 ± 0.003), frozen and boiled collards (0.12 ± 0.006), boiled asparagus (0.12 ± 0.006), raw sprouted mung beans (0.12 ± 0.01), frozen and boiled okra (0.12 ± 0.005), raw sprouted alfalfa seeds (0.12 ± 0.015), raw california avocado (0.12 ± 0.003), raw florida avocado (0.12 ± 0.002), raw chinook salmon (0.12 ± 0.006), sweet potato pie (0.12 ± 0.004), canned pumpkin pie mix (0.12 ± 0.004), canned black turtle beans (0.12 ± 0.002), cooked swordfish (0.12 ± 0.006), onion rings (0.12 ± 0.006), filled milk (0.12 ± 0.002), animal crackers (0.12 ± 0.019), roasted chicken wing with skin (0.12 ± 0.015), fruitcake, light (0.12 ± 0.012), roasted chicken breast without skin (0.12 ± 0.006), roasted light meat chicken with skin (0.12 ± 0.005), roasted chicken breast with skin (0.12 ± 0.005), stewed chicken breast with skin (0.12 ± 0.005), roasted light meat chicken without skin (0.12 ± 0.005), stewed light meat chicken without skin (0.12 ± 0.005), stewed chicken breast without skin (0.12 ± 0.005), whole wheat flour (0.12 ± 0.004), noodles (0.12 ± 0.003), chicken salad (0.11 ± 0.018), roasted chinese chestnuts (0.11 ± 0.018), dried macadamia nuts (0.11 ± 0.018), dry roasted pecans (0.11 ± 0.018), dried pilinuts (0.11 ± 0.018), dried black walnuts (0.11 ± 0.018), chicken frankfurter (0.11 ± 0.011), raw catfish (0.11 ± 0.006), smoked whitefish (0.11 ±

VITAMIN B$_2$

0.006), smoked chinook salmon (0.11 ± 0.006), raw whelk (0.11 ± 0.006), raw perch (0.11 ± 0.006), enchilada (0.11 ± 0.002), raw broccoli (0.11 ± 0.011), raw turnip greens (0.11 ± 0.018), raw swamp cabbage (0.11 ± 0.009), kumquats (0.11 ± 0.026), raw garlic (0.11 ± 0.056), passion fruit (0.11 ± 0.028), beef and pork frankfurter (0.11 ± 0.011), raw romaine lettuce (0.11 ± 0.018), beef frankfurter (0.11 ± 0.009), frozen and boiled kale (0.11 ± 0.008), boiled new zealand spinach (0.11 ± 0.006), ginko nuts (0.11 ± 0.018), au gratin potatoes (0.11 ± 0.004), cherimoya (0.11 ± 0.001), jellied corned beef loaf (0.11 ± 0.018), ham salad (0.11 ± 0.018), pineapple upside-down cake (0.11 ± 0.007), semi-sweet bakers chocolate (0.11 ± 0.018), unenriched biscuits (0.11 ± 0.018), raw acorns (0.11 ± 0.018), dried coconut (0.11 ± 0.018), dried filberts (0.11 ± 0.018), dry roasted macadamia nuts (0.11 ± 0.018), oil roasted macadamia nuts (0.11 ± 0.018), dry roasted peanuts (0.11 ± 0.018), oil roasted peanuts (0.11 ± 0.018), oil roasted virginia peanuts (0.11 ± 0.018), soybean protein isolate (0.11 ± 0.018), raisin bun (0.11 ± 0.008), fried chicken (0.11 ± 0.006), tempeh (0.11 ± 0.006), boston cream pie (0.11 ± 0.005), stewed light meat chicken with skin (0.11 ± 0.005), white whole ground corn meal (0.11 ± 0.004), yellow whole ground corn meal (0.11 ± 0.004), medium molasses (0.1 ± 0.025), sorghum (0.1 ± 0.024), breaded and fried scallops (0.1 ± 0.016), raw parsley (0.1 ± 0.017), raw dock (0.1 ± 0.007), raw cassava (0.1 ± 0.005), mulberries (0.1 ± 0.004), frozen and boiled asparagus (0.1 ± 0.008), raw chickory greens (0.1 ± 0.006), yellow passion fruit juice (0.1 ± 0.002), cooked sprouted mung beans (0.1 ± 0.008), boiled green beans (0.1 ± 0.008), boiled yellow snap beans (0.1 ± 0.008), frozen and boiled green peas (0.1 ± 0.006), bananas (0.1 ± 0.004), cooked prunes (0.1 ± 0.005), lemon meringue pie (0.1 ± 0.004), boiled yellow beans (0.1 ± 0.003), brownies (0.1 ± 0.025), turkey breast meat summer sausage (0.1 ± 0.024), pound cake (0.1 ± 0.017), stewed chicken wing with skin (0.1 ± 0.013), soy sauce (0.1 ± 0.009), gingerbread (0.1 ± 0.008), raw bacon (0.1 ± 0.007), chocolate chiffon pie (0.1 ± 0.006), dried dates (0.1 ± 0.006), butterscotch pie (0.1 ± 0.004), pumpkin pie (0.1 ± 0.004), canned shitake mushrooms (0.1 ± 0.004), firm raw tofu (0.1 ± 0.004), cooked halibut (0.09 ± 0.006), battered and fried shark (0.09 ± 0.006), canned bonito (0.09 ± 0.005), raw red chili peppers (0.09 ± 0.011), raw green chili peppers (0.09 ± 0.011), raw chinese cabbage (0.09 ± 0.014), frozen and boiled broccoli (0.09 ± 0.005), raw welsh onions (0.09 ± 0.005), boiled dock (0.09 ± 0.005), raspberries (0.09 ± 0.004), boiled swiss chard (0.09 ± 0.006), boiled blackeye pea pods (0.09 ± 0.011), stir-fried sprouted lentils (0.09 ± 0.005), scalloped potatoes (0.09 ± 0.004), boiled purslane (0.09 ± 0.009), plum (0.09 ± 0.008), boiled succotash (0.09 ± 0.005), seedless raisins (0.09 ± 0.005), boiled pinto beans (0.09 ± 0.003), raw swordfish (0.09 ± 0.006), pineapple chiffon pie (0.09 ± 0.006), pineapple custard pie (0.09 ± 0.004), dried figs (0.09 ± 0.003), boiled broad beans (0.09 ± 0.003), red beans (0.09 ± 0.004), bread stuffing (0.09 ± 0.003), canned blue crab (0.08 ± 0.006), canned tuna in oil (0.08 ± 0.006), raw bluefish (0.08 ± 0.006), raw wolffish (0.08 ± 0.006), cooked rockfish (0.08 ± 0.006), raw turbot (0.08 ± 0.006), raw flatfish (0.08 ± 0.006), boiled brussels sprouts (0.08 ± 0.006), boiled kale (0.08 ± 0.008), boiled swamp cabbage (0.08 ± 0.01), canned green peas (0.08 ± 0.006), canned yellow corn (0.08 ± 0.005), canned yellow corn with red and green peppers (0.08 ± 0.004), raw endive (0.08 ± 0.02), raw whale meat (0.08 ± 0.005), boiled pumpkin (0.08 ± 0.004), cooked crayfish (0.08 ± 0.006), cooked pike (0.08 ± 0.006), canned black-eyed peas

I. Vitamins

VITAMIN B$_2$

(0.08 ± 0.002), cooked flounder (0.08 ± 0.005), boiled lentils (0.08 ± 0.003), cooked clams (0.08 ± 0.006), canned cod (0.08 ± 0.006), cooked cod (0.08 ± 0.006), boiled mungo beans (0.08 ± 0.003), boiled arrowhead (0.08 ± 0.042), fried tofu (0.08 ± 0.038), turkey salad (0.08 ± 0.009), simmered pork chitterlings (0.08 ± 0.005), apple pie (0.08 ± 0.004), white bolted corn meal (0.08 ± 0.004), yellow bolted corn meal (0.08 ± 0.004), macaroni (0.08 ± 0.004), spaghetti (0.08 ± 0.004), puffed rice (0.07 ± 0.036), gefiltefish with broth (0.07 ± 0.012), raw scallops (0.07 ± 0.006), raw rockfish (0.07 ± 0.006), raw halibut (0.07 ± 0.006), cooked lobster (0.07 ± 0.006), stewed rabbit (0.07 ± 0.004), instant corn grits (0.07 ± 0.004), dehydrated onion flakes (0.07 ± 0.036), lychees (0.07 ± 0.005), strawberries (0.07 ± 0.003), coleslaw (0.07 ± 0.008), boiled turnip greens (0.07 ± 0.007), frozen and boiled turnip greens (0.07 ± 0.006), raw celtuce (0.07 ± 0.005), raw looseleaf lettuce (0.07 ± 0.018), tomato catsup (0.07 ± 0.033), chili sauce (0.07 ± 0.033), boiled and steamed japanese chestnuts (0.07 ± 0.018), canned white corn (0.07 ± 0.005), frozen and boiled yellow snap beans (0.07 ± 0.007), frozen and boiled green beans (0.07 ± 0.007), frozen and boiled turnip greens and turnips (0.07 ± 0.005), russian salad dressing (0.07 ± 0.033), raw butterhead lettuce (0.07 ± 0.033), canned succotash with creamed corn (0.07 ± 0.004), boiled yellow corn (0.07 ± 0.006), frozen and boiled succotash (0.07 ± 0.006), raw bamboo shoots (0.07 ± 0.007), raw jerusalem artichoke (0.07 ± 0.007), prune juice (0.07 ± 0.002), lemon chiffon pie (0.07 ± 0.006), frozen and boiled yellow corn (0.07 ± 0.006), taco (0.07 ± 0.006), raw cod (0.07 ± 0.006), canned kidney beans (0.07 ± 0.002), sweet butter (0.07 ± 0.033), butter (0.07 ± 0.033), chicken and turkey salad (0.07 ± 0.018), german sweet bakers chocolate (0.07 ± 0.018), semi-sweet chocolate (0.07 ± 0.018), oil roasted spanish peanuts (0.07 ± 0.018), light cream (0.07 ± 0.017), cheese crackers with peanut butter (0.07 ± 0.012), caramel cake (0.07 ± 0.005), peach pie (0.07 ± 0.005), pecan pie (0.07 ± 0.005), light rye flour (0.07 ± 0.005), simmered pork ears (0.07 ± 0.005), simmered pork tail (0.07 ± 0.005), pot barley pearled (0.07 ± 0.003), boiled cranberry beans (0.07 ± 0.003), boiled pink beans (0.07 ± 0.003), soy milk (0.07 ± 0.002), boiled adzuki beans (0.07 ± 0.002), raw milkfish (0.06 ± 0.006), raw monkfish (0.06 ± 0.006), raw shark (0.06 ± 0.006), cooked whiting (0.06 ± 0.006), cooked alaska king crab (0.06 ± 0.006), boiled burdock root (0.06 ± 0.004), vegetarian baked beans (0.06 ± 0.002), raw acerola (0.06 ± 0.005), acerola juice (0.06 ± 0.002), raw red sweet peppers (0.06 ± 0.01), guava (0.06 ± 0.006), raw green sweet peppers (0.06 ± 0.01), raw cauliflower (0.06 ± 0.01), elderberries (0.06 ± 0.003), tomato purée (0.06 ± 0.002), frozen and boiled cauliflower (0.06 ± 0.006), mango (0.06 ± 0.002), canned sweet potato (0.06 ± 0.003), boiled mustard greens (0.06 ± 0.007), boiled chinese cabbage (0.06 ± 0.006), boiled red tomato (0.06 ± 0.004), o'brien potatoes (0.06 ± 0.003), canned turnip greens (0.06 ± 0.004), prickly pear (0.06 ± 0.005), frozen and boiled fordhook lima beans (0.06 ± 0.006), tomato sauce (0.06 ± 0.004), marinara sauce (0.06 ± 0.002), european grapes (0.06 ± 0.003), thompson seedless grapes (0.06 ± 0.003), raw carrots (0.06 ± 0.007), potato salad (0.06 ± 0.004), steamed sprouted soybeans (0.06 ± 0.011), frozen and baked sweet potato (0.06 ± 0.006), cherries (0.06 ± 0.007), frozen white corn (0.06 ± 0.007), boiled and frozen baby lima beans (0.06 ± 0.006), refried beans (0.06 ± 0.002), canned green beans (0.06 ± 0.007), canned yellow snap beans (0.06 ± 0.007), raw crayfish (0.06 ± 0.006), raw pike (0.06 ± 0.006), canned pumpkin (0.06 ± 0.004), canned shellie beans (0.06 ± 0.004),

VITAMIN B$_2$

baked beans with pork and sweet sauce (0.06 ± 0.002), coconut water (0.06 ± 0.002), figs (0.06 ± 0.01), frozen and boiled black-eyed peas (0.06 ± 0.006), boiled chickpeas (0.06 ± 0.003), boiled french beans (0.06 ± 0.003), boiled great northern beans (0.06 ± 0.003), boiled kidney beans (0.06 ± 0.003), boiled navy beans (0.06 ± 0.003), boiled split peas (0.06 ± 0.003), boiled yardlong bean (0.06 ± 0.003), canned great northern beans (0.06 ± 0.002), boiled mung beans (0.06 ± 0.002), canned pinto beans (0.06 ± 0.002), teriyaki sauce (0.06 ± 0.028), boiled peanuts (0.06 ± 0.016), canned sprouted mung beans (0.06 ± 0.008), simmered pork feet (0.06 ± 0.005), white corn flour (0.06 ± 0.004), yellow corn flour (0.06 ± 0.004), boiled black beans (0.06 ± 0.003), boiled pigeon peas (0.06 ± 0.003), light molasses (0.05 ± 0.025), breaded and fried fish (0.05 ± 0.007), raw lobster (0.05 ± 0.006), canned tuna in water (0.05 ± 0.006), cooked eel (0.05 ± 0.006), raw whiting (0.05 ± 0.006), smoked haddock (0.05 ± 0.006), cooked haddock (0.05 ± 0.006), raw alaska king crab (0.05 ± 0.006), raw flounder (0.05 ± 0.005), raw sole (0.05 ± 0.005), cream of wheat (0.05 ± 0.003), boiled lupins (0.05 ± 0.003), black currants (0.05 ± 0.009), kiwi fruit (0.05 ± 0.007), boiled cauliflower (0.05 ± 0.008), boiled scotch kale (0.05 ± 0.008), red currants (0.05 ± 0.009), white currants (0.05 ± 0.009), carissa (0.05 ± 0.025), boiled green cabbage (0.05 ± 0.007), soursop (0.05 ± 0.002), raw red tomato (0.05 ± 0.004), boiled okra (0.05 ± 0.006), frozen and boiled mustard greens (0.05 ± 0.007), boiled parsnips (0.05 ± 0.006), blueberries (0.05 ± 0.003), cooked plantain (0.05 ± 0.003), baked hubbard squash (0.05 ± 0.005), carrot juice (0.05 ± 0.003), raw crookneck squash (0.05 ± 0.008), hummus (0.05 ± 0.002), peach (0.05 ± 0.006), boiled artichoke hearts (0.05 ± 0.006), frozen and boiled crookneck squash (0.05 ± 0.005), raw iceberg lettuce (0.05 ± 0.025), canned creamed yellow corn (0.05 ± 0.004), canned succotash (0.05 ± 0.004), american grapes (0.05 ± 0.005), boiled carrots (0.05 ± 0.006), pickled beets (0.05 ± 0.004), baked beans with pork and tomato sauce (0.05 ± 0.002), canned broad beans (0.05 ± 0.002), boiled black-eyed peas (0.05 ± 0.003), boiled catjang black-eyed peas (0.05 ± 0.003), canned navy beans (0.05 ± 0.002), bread sticks (0.05 ± 0.025), honey (0.05 ± 0.024), chocolate syrup (0.05 ± 0.013), unenriched corn meal (0.05 ± 0.004), unenriched yellow degermed corn meal (0.05 ± 0.004), boiled bamboo shoots (0.05 ± 0.004), raw tofu (0.05 ± 0.004), light pearled barley (0.05 ± 0.003), boiled black turtle beans (0.05 ± 0.003), boiled lima beans (0.05 ± 0.003), boiled baby lima beans (0.05 ± 0.003), raw shrimp (0.04 ± 0.006), cooked shrimp (0.04 ± 0.006), surimi shrimp (0.04 ± 0.006), raw octopus (0.04 ± 0.006), raw eel (0.04 ± 0.006), raw haddock (0.04 ± 0.006), maltex (0.04 ± 0.003), sweetened coconut cream (0.04 ± 0.002), jujube (0.04 ± 0.005), canned red chili peppers (0.04 ± 0.007), canned green chili peppers (0.04 ± 0.007), navel orange (0.04 ± 0.004), valencia orange (0.04 ± 0.004), potato chips (0.04 ± 0.018), pitanga (0.04 ± 0.003), strawberry pie (0.04 ± 0.005), raw green tomato (0.04 ± 0.004), raw radish (0.04 ± 0.011), boiled rutabaga (0.04 ± 0.006), blackberries (0.04 ± 0.007), raw potato without skin (0.04 ± 0.004), pineapple (0.04 ± 0.003), mammy apple (0.04 ± 0.005), canned jalapeño peppers (0.04 ± 0.007), canned stewed red tomatoes (0.04 ± 0.004), raw leeks (0.04 ± 0.019), frozen and french fried potatoes without skin (0.04 ± 0.01), roselle (0.04 ± 0.009), ground-cherries (0.04 ± 0.004), apricots (0.04 ± 0.005), boiled collards (0.04 ± 0.003), boiled chayote (0.04 ± 0.006), candied sweet potato (0.04 ± 0.005), french fries (0.04 ± 0.006), boiled dishcloth gourd (0.04 ± 0.006), boiled crookneck squash (0.04 ± 0.006), mashed potatoes (0.04 ± 0.005), oheloberries (0.04 ± 0.004), nectarine (0.04 ± 0.004), raw ginger root (0.04 ± 0.021), boiled leeks (0.04 ± 0.019), hamburger pickle relish (0.04 ± 0.018), boiled zucchini (0.04 ± 0.006), raw celeriac (0.04 ± 0.005), frozen creamed yellow corn (0.04 ± 0.004), poi (0.04 ± 0.004), frozen and boiled zucchini (0.04 ± 0.004), pear (0.04 ± 0.003), frozen and boiled carrots (0.04 ± 0.007), rhubarb pie (0.04 ± 0.004), boysenberries (0.04 ± 0.004), canned zucchini in tomato juice (0.04 ± 0.004), frozen and boiled butternut squash (0.04 ± 0.004), raw yellowtail (0.04 ± 0.006), mince pie (0.04 ± 0.004), apple brown betty (0.04 ± 0.002), canned cranberry beans (0.04 ± 0.002), dried sweetened flaked coconut (0.04 ± 0.02), peanut brittle (0.04 ± 0.018), soda crackers (0.04 ± 0.018), hot dog pickle relish (0.04 ± 0.018), tuna salad spread (0.04 ± 0.009), light buckwheat flour (0.04 ± 0.005), pickled pork feet (0.04 ± 0.005), canned butter beans (0.04 ± 0.004), boiled hyacinth beans (0.04 ± 0.003), boiled white beans (0.04 ± 0.003), restaurant hot chocolate (0.04 ± 0.002), grape juice (0.04 ± 0.002), cooked enriched white degermed corn meal (0.04 ± 0.002), cooked enriched yellow degermed corn meal (0.04 ± 0.002), canned white beans (0.04 ± 0.002), yellow hominy with red and green peppers (0.03 ± 0.002), boiled red sweet peppers (0.03 ± 0.007), boiled green sweet peppers (0.03 ± 0.007), papaya (0.03 ± 0.002), pummelo (0.03 ± 0.003), raw red cabbage (0.03 ± 0.014), orange juice (0.03 ± 0.002), raw green cabbage (0.03 ± 0.014), canned red sweet peppers (0.03 ± 0.007), canned green sweet peppers (0.03 ± 0.007), frozen and boiled red sweet peppers (0.03 ± 0.005), frozen and boiled green sweet peppers (0.03 ± 0.005), boiled red cabbage (0.03 ± 0.007), mandarin oranges (0.03 ± 0.004), raw savoy cabbage (0.03 ± 0.014), breadfruit (0.03 ± 0.005), gooseberries (0.03 ± 0.003), honeydew melon (0.03 ± 0.005), rose apple (0.03 ± 0.005), raw scallop squash (0.03 ± 0.008), restaurant tomato juice (0.03 ± 0.004), tomato juice (0.03 ± 0.003), boiled savoy cabbage (0.03 ± 0.007), boiled yambean (0.03 ± 0.005), loganberries (0.03 ± 0.003), boiled oriental radish (0.03 ± 0.007), quince (0.03 ± 0.005), canned peeled red tomatoes (0.03 ± 0.004), microwaved potato without skin (0.03 ± 0.003), microwaved potato with skin (0.03 ± 0.002), canned sauerkraut (0.03 ± 0.004), baked potato with skin (0.03 ± 0.002), baked/boiled tropical yam (0.03 ± 0.007), boiled turnips (0.03 ± 0.006), raw zucchini (0.03 ± 0.008), frozen and boiled potatoes without skin (0.03 ± 0.005), raw celery (0.03 ± 0.013), boiled calabash gourd (0.03 ± 0.007), boiled hubbard squash (0.03 ± 0.004), frozen and fried hash brown potatoes (0.03 ± 0.006), hash brown potatoes (0.03 ± 0.006), pomegranate (0.03 ± 0.003), cooked taro (0.03 ± 0.008), boiled celery (0.03 ± 0.007), raw frozen rhubarb (0.03 ± 0.004), wax beans (0.03 ± 0.004), boiled/baked spaghetti squash (0.03 ± 0.006), frozen and boiled turnips (0.03 ± 0.005), canned chickpeas (0.03 ± 0.002), human milk (0.03 ± 0.016), canned carrots (0.03 ± 0.007), canned crookneck squash (0.03 ± 0.005), frozen and cooked sweetened rhubarb (0.03 ± 0.004), boiled cardoon (0.03 ± 0.005), applesauce (0.03 ± 0.004), canned chinese water chestnuts (0.03 ± 0.007), raisin pie (0.03 ± 0.004), dried japanese persimmon (0.03 ± 0.015), red table wine (0.03 ± 0.005), bulgur (0.03 ± 0.004), wheat cake (0.03 ± 0.004), brown sugar (0.03 ± 0.003), canned lima beans (0.03 ± 0.002), beer (0.03 ± 0.001), oatmeal (0.02 ± 0.003), white hominy (0.02 ± 0.002), yellow hominy (0.02 ± 0.002), boiled kohlrabi (0.02 ± 0.006), lemon (0.02 ± 0.009), cantaloupe (0.02 ± 0.003), cranberry juice cocktail (0.02 ± 0.002), restaurant orange juice

VITAMIN B$_2$

(0.02 ± 0.004), pink grapefruit (0.02 ± 0.004), white grapefruit juice (0.02 ± 0.002), red grapefruit juice (0.02 ± 0.002), white grapefruit (0.02 ± 0.004), tangerine (0.02 ± 0.006), tangerine juice (0.02 ± 0.002), raw oriental radish (0.02 ± 0.011), carambola (0.02 ± 0.004), sapote (0.02 ± 0.002), tequila sunrise (0.02 ± 0.003), casaba melon (0.02 ± 0.003), boiled butternut squash (0.02 ± 0.005), sapodilla (0.02 ± 0.003), cranberry sauce (0.02 ± 0.005), bloody mary (0.02 ± 0.003), baked potato without skin (0.02 ± 0.003), boiled scallop squash (0.02 ± 0.006), pineapple juice (0.02 ± 0.002), french fried potatoes (0.02 ± 0.01), watermelon (0.02 ± 0.003), sour pickles (0.02 ± 0.008), crab apples (0.02 ± 0.005), japanese persimmon (0.02 ± 0.003), boiled potato without skin (0.02 ± 0.004), dill pickles (0.02 ± 0.008), frozen and boiled onions (0.02 ± 0.005), peach nectar (0.02 ± 0.002), raw coconut (0.02 ± 0.011), raw cucumber (0.02 ± 0.01), blackberry pie (0.02 ± 0.004), blueberry pie (0.02 ± 0.004), raw eggplant (0.02 ± 0.012), boiled eggplant (0.02 ± 0.01), restaurant coleslaw (0.02 ± 0.005), loquats (0.02 ± 0.005), dried sweetened shredded coconut (0.02 ± 0.005), pineapple pie (0.02 ± 0.004), canned bamboo shoots (0.02 ± 0.004), cooked wild long grain rice (0.02 ± 0.004), boiled mothbeans (0.02 ± 0.003), apple juice (0.02 ± 0.002), hush puppies (0.02 ± 0.011), 53 proof coffee liqueur (0.02 ± 0.01), 63 proof coffee liqueur (0.02 ± 0.01), dry dessert wine (0.02 ± 0.008), sweet dessert wine (0.02 ± 0.008), okara tofu (0.02 ± 0.008), rose table wine (0.02 ± 0.005), raw agar seaweed (0.02 ± 0.005), cherry pie (0.02 ± 0.004), black tea (0.02 ± 0.003), cooked brown rice (0.02 ± 0.003), apricot nectar (0.02 ± 0.002), cooked grouper (0.01 ± 0.006), surimi scallops (0.01 ± 0.006), farina (0.01 ± 0.003), corn grits (0.01 ± 0.002), lemon juice (0.01 ± 0.002), restaurant cranberry juice cocktail (0.01 ± 0.003), screwdriver (0.01 ± 0.002), lime (0.01 ± 0.007), lime juice (0.01 ± 0.002), boiled lotus root (0.01 ± 0.006), boiled butterbur (0.01 ± 0.005), java plum (0.01 ± 0.004), whiskey sour (0.01 ± 0.006), baked acorn squash (0.01 ± 0.005), raw onions (0.01 ± 0.006), boiled acorn squash (0.01 ± 0.004), boiled beets (0.01 ± 0.006), boiled onions (0.01 ± 0.005), raw apple with skin (0.01 ± 0.004), piña colada (0.01 ± 0.004), raw apple without skin (0.01 ± 0.004), pear nectar (0.01 ± 0.002), steamed hawaii mountain yam (0.01 ± 0.007), white table wine (0.01 ± 0.005), chamomile tea (0.01 ± 0.003), restaurant coffee (0.01 ± 0.003), cooked apple without skin (0.01 ± 0.003), cooked parboiled white rice (0.01 ± 0.003), cooked unenriched corn meal (0.01 ± 0.002), cooked unenriched yellow degermed corn meal (0.01 ± 0.002), cooked white rice (0.01 ± 0.002)

Applications: diabetes, diarrhea, multiple sclerosis, neuritis, Parkinson's disease, dizziness, Ménière's syndrome, cataracts, conjunctivitis, glaucoma, night blindness, adrenal deterioration, baldness, worms, arthritis, kidney inflammation, leg cramp, influenza, vaginal itching, acne, bedsores, skin inflammation, skin ulcers, digestion upset, peptic ulcer, alcohol cravings, cancer, pellagra, retarded growth, stress

Daily Dosage: pre–1958 MDR – 1.2 mg; 1958 RDA – 1.8 mg; 1974 PCS – 0.8–2.6 mg; 1974 RDA – 1.1–1.8 mg; 1974 USRDA – 1.7 mg; 1980 RDA – 1.1–1.8 mg; 1980 USRDA – 1.7 mg; 1989 RDA – 1.1–1.8 mg; 1989 USRDA – 1.7 mg; nutritional – 10–25 mg; therapeutic – 100–200 mg; experimental – 500–1000 mg; toxic – 1000 mg

Warnings: Increased amounts can be toxic. See toxicity symptoms.

VITAMIN B$_2$ PHOSPHATE

(Specific form of vitamin B$_2$)

Names: vitamin B$_2$ phosphate, riboflavin phosphate, flavine mononucleotide, cytoflav, coflavinase, alloxazine mononucleotide

VITAMIN B₃ COMPLEX

(Vitamin B₃ exists in two forms that are not quite interchangeable. See vitamin B₃. Sometimes these forms are denoted the "vitamin B₃ complex.")

(The term "vitamin B₃" is sometimes incorrectly used to denote pantothenic acid/vitamin B₅.)
Names: vitamin B₃, niacin, vitamin PP, PP factor, pellagra preventive factor/vitamin, antipellagra factor/vitamin
Classification: water-soluble vitamin, vitamin B complex, antioxidant
Forms: nicotinic acid, niacinamide/nicotinamide (Niacinamide/nicotinamide is a derivative form of nicotinic acid, and does not have all the effects as that form. See vitamin B₃ₐ and vitamin B₃ᵦ.)

PROVITAMIN B₃ *see* **tryptophan**

PROVITAMIN B₃ₐ *see* **tryptophan**

VITAMIN B₃ₐ

(The term "niacin" is sometimes used to refer to vitamin B₃ᵦ as well.)

Names: vitamin B₃ₐ, vitamin B₃, niacin, vitamin PP, PP factor, pellagra preventive factor/vitamin, antipellagra factor/vitamin, Nicacid, Nicagin, Nicobid, Niconacid, Nico-Span, Nicotene, Nicotinipca, Nicyl, Akotin, Daskil, Tinic, Nicolar, Wampocap
Classification: water-soluble vitamin, vitamin B complex, antioxidant
Forms: nicotinic acid

Deficiency: failing vision, hypersensitivity to light, pellagra, skin inflammation, skin hypersensitivity, backaches, bad breath, tender gums, diarrhea, digestion upset, small ulcers, nausea, appetite decreased, dulled sense of taste, hyperacute sense of smell, fatigue, nervousness, emotional instability, depression, alternating depression/emotional agitation, confusion, emotional agitation, hallucinations, headache, memory decreased, lack of sleep, muscular weakness, fatigue, deterioration of amino acid utilization, acne, increased serum cholesterol
Side-effects: fatigue, nausea, itching, burning sensation (These effects are not harmful.) *See toxicity* symptoms
Toxicity: skin deterioration, gout, liver deterioration, headaches, ulcers, blood sugar decreased
Inhibitors: water, sulfa drugs, alcohol, food-processing, sleeping pills, estrogen, sodium nitrate
Helpers: vitamin B complex, vitamin B₁, vitamin B₂, vitamin B₅, vitamin C, phosphorus
Sources (mg per 100 grams of food): rice bran (31.2 ± 0.05), peanut flour (27.5 ± 1.25), bran flakes (17.9 ± 0.18), bran (17.9 ± 0.18), corn flakes (17.9 ± 0.18), wheat flakes (17.9 ± 0.18), paprika (15 ± 2.5), oil roasted spanish peanuts (15 ± 0.18), oil roasted peanuts (15 ± 0.18), oil roasted virginia peanuts (14.6 ± 0.18), dried peanuts (14.3 ± 0.18), oil roasted valencia peanuts (14.3 ± 0.18), dried shitake mushrooms (14 ± 0.33), crunchy peanut butter (13.8 ± 0.16), creamy peanut butter (13.8 ± 0.31), dry roasted peanuts (13.6 ± 0.18), raisin bran (13.5 ± 0.14), dried spirulina seaweed (12.8 ± 0.05), dried ginko nuts (11.8 ± 0.18), puffed wheat (10.7 ± 0.36), savory (10 ± 5), dried coriander leaf (10 ± 5), yellow mustard seed (10 ± 1.67), cayenne pepper (10 ± 2.5), basil (10 ± 5), dried parsley (10 ± 5), tomato powder (9.1 ± 0.05), cheese crackers (8 ± 0.33), freeze-dried parsley (7.1 ± 3.57), dry roasted sunflower seeds (7.1 ± 0.18), chili pow-

der (6.7 ± 1.67), saltine crackers (6.7 ± 0.83), almond meal (6.4 ± 0.18), freeze-dried red sweet peppers (6.3 ± 3.13), freeze-dried green sweet peppers (6.3 ± 3.13), raw wild rice (6.2 ± 0.03), ginko nuts (6.1 ± 0.18), toasted wheat germ (5.7 ± 0.18), toasted sesame kernels (5.4 ± 0.18), shredded wheat (5.4 ± 0.18), sesame butter (5.3 ± 0.33), boiled peanuts (5.3 ± 0.16), dried sesame kernels (5 ± 0.63), corn tortilla (5 ± 0.17), oregano (5 ± 2.5), caraway seed (5 ± 2.5), dill seed (5 ± 2.5), fennel seed (5 ± 2.5), cumin seed (5 ± 2.5), curry powder (5 ± 2.5), turmeric (5 ± 2.5), ginger (5 ± 2.5), tarragon (5 ± 2.5), toasted raisin bread (4.8 ± 0.24), dried sunflower seeds (4.6 ± 0.18), toasted english muffin (4.6 ± 0.1), tempeh (4.6 ± 0.06), boiled mushrooms (4.5 ± 0.06), whole dried sesame seeds (4.4 ± 0.56), soybean flour (4.4 ± 0.06), dried pinyon pine nuts (4.3 ± 0.18), whole wheat flour (4.3 ± 0.04), french bread (4.3 ± 0.18), italian bread (4.3 ± 0.18), toasted whole wheat bread (4.3 ± 0.24), oil roasted sunflower seeds (4.3 ± 0.18), pretzels (4.3 ± 0.18), potato chips (4.3 ± 0.18), toasted white bread (4.3 ± 0.24), pita bread (4.2 ± 0.13), toaster pastry (4.2 ± 0.1), hamburger bun (4 ± 0.13), hot dog bun (4 ± 0.13), mixed grain bread (4 ± 0.2), french rolls (4 ± 0.1), whole wheat bread (4 ± 0.2), raisin bread (4 ± 0.2), raw mushrooms (4 ± 0.14), tamari sauce (4 ± 0.09), dinner rolls (3.9 ± 0.18), sorghum grain (3.9 ± 0.05), toasted pumpernickel bread (3.9 ± 0.18), maypo (3.9 ± 0.03), white bread (3.8 ± 0.21), toasted cracked wheat bread (3.8 ± 0.24), english muffin (3.7 ± 0.09), pot barley pearled (3.7 ± 0.03), dried japanese chestnuts (3.6 ± 0.18), dried pignolia pine nuts (3.6 ± 0.18), toasted rye bread (3.6 ± 0.23), dried almonds (3.6 ± 0.18), peanut brittle (3.6 ± 0.18), dried watermelon seeds (3.6 ± 0.18), oil roasted almonds (3.6 ± 0.18), cheese crackers with peanut butter (3.6 ± 0.12), enriched white degermed corn meal (3.5 ± 0.04), enriched yellow degermed corn meal (3.5 ± 0.04), all purpose enriched wheat flour (3.5 ± 0.04), enriched wheat bread (3.5 ± 0.04), bagels (3.5 ± 0.09), dry breadcrumbs (3.5 ± 0.05), potato flour (3.4 ± 0.06), pumpernickel bread (3.4 ± 0.16), dried oriental radish (3.4 ± 0.09), dried pepeao (3.3 ± 0.42), roasted soybean flour (3.3 ± 0.06), bran muffins (3.3 ± 0.13), shoyu sauce (3.3 ± 0.09), cracked wheat bread (3.2 ± 0.2), rye bread (3.2 ± 0.2), oatmeal bread (3.2 ± 0.18), french fried potatoes (3.2 ± 0.1), tomato paste (3.2 ± 0.04), instant oatmeal (3.1 ± 0.03), almond butter (3.1 ± 0.31), light pearled barley (3.1 ± 0.03), dried lychees (3.1 ± 0.05), dark buckwheat flour (2.9 ± 0.05), whole wheat roll (2.9 ± 0.18), graham crackers (2.9 ± 0.36), almond paste (2.9 ± 0.18), dry roasted almonds (2.9 ± 0.18), toasted almonds (2.9 ± 0.18), puffed rice (2.9 ± 0.36), dried apricots (2.9 ± 0.14), ground-cherries (2.8 ± 0.04), soy sauce (2.8 ± 0.09), dark rye flour (2.7 ± 0.04), hush puppies (2.7 ± 0.11), defatted soybean flour (2.6 ± 0.05), defatted soy meal (2.6 ± 0.04), rice flour (2.6 ± 0.04), canned shitake mushrooms (2.6 ± 0.04), dry roasted macadamia nuts (2.5 ± 0.18), fenugreek seed (2.5 ± 1.25), shortbread cookie (2.5 ± 0.63), dried acorns (2.5 ± 0.18), frozen and fried hash brown potatoes (2.4 ± 0.06), frozen and french fried potatoes without skin (2.4 ± 0.1), bulgur (2.4 ± 0.04), millit (2.3 ± 0.05), yellow passion fruit juice (2.2 ± 0.02), corn germ (2.2 ± 0.05), low fat soybean flour (2.2 ± 0.06), dried dates (2.2 ± 0.06), sweet roll (2.1 ± 0.12), dried macadamia nuts (2.1 ± 0.18), ry-krisp crackers (2.1 ± 0.36), raw green peas (2.1 ± 0.06), danish pastry (2.1 ± 0.12), oil roasted macadamia nuts (2.1 ± 0.18), cooked wild long grain rice (2.1 ± 0.04), french fries (2.1 ± 0.06), hash brown potatoes (2.1 ± 0.06), white whole ground corn meal (2 ± 0.04), yellow whole ground corn meal (2 ± 0.04), boiled green peas (2 ± 0.06), waffles (2 ± 0.07), dry cocoa powder (2 ± 1), blackstrap molasses (2 ± 0.25), tamarind (1.9 ± 0.04), white bolted corn meal (1.9 ± 0.04), yellow bolted corn meal (1.9 ± 0.04), raw california avocado (1.9 ± 0.03), raw florida avocado (1.9 ± 0.02), dried prunes (1.9 ± 0.06), carob flour (1.9 ± 0.05), dried brazil nuts (1.8 ± 0.18), oil roasted cashews (1.8 ± 0.18), cashew butter (1.8 ± 0.18), taco shell (1.8 ± 0.45), tostada shell (1.8 ± 0.45), dried pumpkin and squash seeds (1.8 ± 0.18), enriched biscuits (1.8 ± 0.18), pie crust (1.8 ± 0.03), raw acorns (1.8 ± 0.18), toasted soybean nuts (1.8 ± 0.18), sapote (1.8 ± 0.02), popcorn (1.7 ± 0.83), passion fruit (1.7 ± 0.28), orange peel (1.7 ± 0.83), microwaved potato with skin (1.7 ± 0.02), tomato purée (1.7 ± 0.02), boiled yellow corn (1.6 ± 0.06), microwaved potato without skin (1.6 ± 0.03), baked potato with skin (1.6 ± 0.02), raw wakame seaweed (1.6 ± 0.05), marinara sauce (1.6 ± 0.02), purple passion fruit juice (1.5 ± 0.02), whole grain rye crackers (1.5 ± 0.38), frozen and boiled green peas (1.5 ± 0.06), oatmeal cookie (1.5 ± 0.38), corn meal muffins (1.5 ± 0.13), pancakes (1.5 ± 0.19), french toast (1.5 ± 0.08), boiled mungo beans (1.5 ± 0.03), raw laver/nori seaweed (1.5 ± 0.05), raw potato without skin (1.5 ± 0.04), canned white corn (1.5 ± 0.05), cooked shitake mushrooms (1.5 ± 0.07), dry roasted pistachio nuts (1.4 ± 0.18), raw japanese chestnuts (1.4 ± 0.18), roasted european chestnuts (1.4 ± 0.18), raw cassava (1.4 ± 0.05), dry roasted cashews (1.4 ± 0.18), white corn flour (1.4 ± 0.04), yellow corn flour (1.4 ± 0.04), soybean protein isolate (1.4 ± 0.18), boiled succotash (1.4 ± 0.05), gingersnaps cookie (1.4 ± 0.71), roasted chinese chestnuts (1.4 ± 0.18), roasted soybean nuts (1.4 ± 0.06), baked potato without skin (1.4 ± 0.03), cooked brown rice (1.4 ± 0.03), unsweetened baking chocolate (1.4 ± 0.18), zwieback crackers (1.4 ± 0.71), raw jerusalem artichoke (1.3 ± 0.07), blueberry muffins (1.3 ± 0.13), sugar cookie (1.3 ± 0.31), cherimoya (1.3 ± 0.01), boiled potato without skin (1.3 ± 0.04), frozen and boiled potatoes without skin (1.3 ± 0.05), tomato catsup (1.3 ± 0.33), chili sauce (1.3 ± 0.33), frozen and boiled succotash (1.3 ± 0.06), frozen and boiled yellow corn (1.3 ± 0.06), oyster crackers (1.3 ± 0.63), cooked extra long grain white rice (1.3 ± 0.06), dried and frozen tofu (1.2 ± 0.29), boiled green soybeans (1.2 ± 0.06), stir-fried sprouted lentils (1.2 ± 0.05), raw spirulina seaweed (1.2 ± 0.05), cooked parboiled white rice (1.2 ± 0.03), raw garden cress (1.2 ± 0.2), guava (1.2 ± 0.06), dried pecans (1.1 ± 0.18), dried pistachio nuts (1.1 ± 0.18), dry roasted red pistachio nuts (1.1 ± 0.18), dried filberts (1.1 ± 0.18), stir-fried sprouted soybeans (1.1 ± 0.05), dried english/persian walnuts (1.1 ± 0.18), raw european chestnuts (1.1 ± 0.18), boiled sprouted green peas (1.1 ± 0.05), raw garlic (1.1 ± 0.56), steamed sprouted soybeans (1.1 ± 0.11), boiled lentils (1.1 ± 0.03), stir-fried sprouted mung beans (1.1 ± 0.08), macaroni (1.1 ± 0.04), spaghetti (1.1 ± 0.04), poi (1.1 ± 0.04), onion rings (1.1 ± 0.06), rusk crackers (1.1 ± 0.56), seeded raisins (1.1 ± 0.05), boiled pokeberry shoots (1.1 ± 0.06), tomato sauce (1.1 ± 0.04), frozen and boiled fordhook lima beans (1.1 ± 0.06), teriyaki sauce (1.1 ± 0.28), frozen and boiled red sweet peppers (1.1 ± 0.05), frozen and boiled green sweet peppers (1.1 ± 0.05), bittersweet chocolate (1.1 ± 0.18), frozen white corn (1.1 ± 0.07), canned yellow corn (1.1 ± 0.05), golden seedless raisins (1.1 ± 0.05), soda crackers (1.1 ± 0.18), cooked parboiled extra long grain white rice (1.1 ± 0.06), medium molasses (1 ± 0.25), dry roasted soybean nuts (1 ± 0.06), whole ground corn meal muffins (1 ± 0.13), popover roll (1 ± 0.13), raw chinese water chestnuts (1 ± 0.08), cornbread (1 ± 0.06), corn pone (1 ± 0.06), unenriched corn meal (1 ± 0.04), unenriched yellow degermed corn meal (1 ± 0.04), falafel (1 ± 0.1), cooked instant white rice (1 ± 0.03), cooked white rice (1 ± 0.02), maltex (1 ± 0.03), boiled asparagus (1 ± 0.06), raw carrots (1 ± 0.07), frozen and boiled asparagus (1 ± 0.08), boiled butternut squash (1 ± 0.05), bread sticks (1 ± 0.25), canned yellow corn with red and green peppers (1 ± 0.04), canned broad beans (1 ± 0.02), peach (1 ± 0.06), nectarine (1 ± 0.04), boiled split peas (0.9 ± 0.03), baked acorn squash (0.9 ± 0.05), boiled okra (0.9 ± 0.06), instant corn grits (0.9 ± 0.04), breadfruit (0.9 ± 0.05), sugar apple (0.9 ± 0.03), apple pie (0.9 ± 0.04), boiled lamb's-quarters (0.9 ± 0.06), miso (0.9 ± 0.04), boiled jute potherb (0.9 ± 0.12), boiled blackeye pea pods (0.9 ± 0.11), raw red chili peppers (0.9 ± 0.11), raw green chili peppers (0.9 ± 0.11), soursop (0.9 ± 0.02), canned potato without skin (0.9 ± 0.06), boiled artichoke (0.9 ± 0.06), raw swamp cabbage (0.9 ± 0.09), coconut cream (0.9 ± 0.02), raspberries (0.9 ± 0.04), canned creamed yellow corn (0.9 ± 0.04), jujube (0.9 ± 0.05), boiled winged beans (0.8 ± 0.03), boiled arrowhead (0.8 ± 0.42), seedless raisins (0.8 ± 0.05), boiled pigeon peas (0.8 ± 0.03), frozen and boiled okra (0.8 ± 0.05), boiled parsnips (0.8 ± 0.06), boiled broccoli (0.8 ± 0.06), raw sprouted mung beans (0.8 ± 0.1), boiled red tomato (0.8 ± 0.04), red beans (0.8 ± 0.04), boiled and frozen baby lima beans (0.8 ± 0.06), boiled garden cress (0.8 ± 0.07), raisin bun (0.8 ± 0.08), bread stuffing (0.8 ± 0.03), cooked sprouted mung beans (0.8 ± 0.08), boysenberries (0.8 ± 0.04), cooked plantain (0.8 ± 0.03), loganberries (0.8 ± 0.03), boiled scotch kale (0.8 ± 0.08), raw ginger root (0.8 ± 0.21), canned peeled red tomatoes (0.8 ± 0.04), canned sweet potato (0.8 ± 0.03), boiled/baked spaghetti squash (0.8 ± 0.06), coconut milk (0.8 ± 0.02), cooked prunes (0.8 ± 0.05), rose apple (0.8 ± 0.05), prune juice (0.8 ± 0.02), dehydrated onion flakes (0.7 ± 0.36), roasted japanese chestnuts (0.7 ± 0.18), soybean protein concentrate (0.7 ± 0.18), dried european chestnuts (0.7 ± 0.18), frozen and boiled black-eyed peas (0.7 ± 0.06), dried black walnuts (0.7 ± 0.18), boiled yellow beans (0.7 ± 0.03), raw chinese chestnuts (0.7 ± 0.18), boiled baby lima beans (0.7 ± 0.03), boiled catjang black-eyed peas (0.7 ± 0.03), boiled great northern beans (0.7 ± 0.03), boiled and steamed japanese chestnuts (0.7 ± 0.18), raw bamboo shoots (0.7 ± 0.07), boiled adzuki beans (0.7 ± 0.02), boiled mothbeans (0.7 ± 0.03), canned green peas (0.7 ± 0.06), chocolate cream pie (0.7 ± 0.04), boiled broad beans (0.7 ± 0.03), raw turnip greens (0.7 ± 0.18), raw spinach (0.7 ± 0.18), raw parsley (0.7 ± 0.17), raw broccoli (0.7 ± 0.11), dried coconut (0.7 ± 0.18), dried figs (0.7 ± 0.03), canned stewed red tomatoes (0.7 ± 0.04), restaurant tomato juice (0.7 ± 0.04), tomato juice (0.7 ± 0.03), boiled sweet potato (0.7 ± 0.03), hamburger pickle relish (0.7 ± 0.18), semi-sweet bakers chocolate (0.7 ± 0.18), wheat cake (0.7 ± 0.04), canned red chili peppers (0.7 ± 0.07), canned green chili peppers (0.7 ± 0.07), boiled pink beans (0.6 ± 0.03), boiled mung beans (0.6 ± 0.02), boiled kidney beans (0.6 ± 0.03), light rye flour (0.6 ± 0.05), canned black turtle beans (0.6 ± 0.02), boiled french beans (0.6 ± 0.03), boiled brussels sprouts (0.6 ± 0.06), baked/boiled tropical yam (0.6 ± 0.07), raw sprouted alfalfa seeds (0.6 ± 0.15), raw cauliflower (0.6 ± 0.1), boiled green beans (0.6 ± 0.08), boiled yellow snap beans (0.6 ± 0.08), raw scallop squash (0.6 ± 0.08), baked hubbard squash (0.6 ± 0.05), raw red sweet peppers (0.6 ± 0.1), raw green sweet peppers (0.6 ± 0.1), boiled eggplant (0.6 ± 0.1), baked sweet potato (0.6 ± 0.04), frozen and baked sweet potato (0.6 ± 0.06), boiled rutabaga (0.6 ± 0.06), raw chickory greens (0.6 ± 0.06), raw celtuce (0.6 ± 0.05), mango (0.6 ± 0.02), boiled artichoke hearts (0.6 ± 0.06), raw red

VITAMIN B₃ COMPLEX

tomato (0.6 ± 0.04), barbecue sauce (0.6 ± 0.31), cream of wheat (0.6 ± 0.03), frozen and boiled collards (0.6 ± 0.06), frozen and boiled kale (0.6 ± 0.08), honeydew melon (0.6 ± 0.05), frozen and boiled turnips (0.6 ± 0.05), boiled crookneck squash (0.6 ± 0.06), cantaloupe (0.6 ± 0.03), mulberries (0.6 ± 0.04), raw chinese cabbage (0.6 ± 0.14), canned red sweet peppers (0.6 ± 0.07), canned green sweet peppers (0.6 ± 0.07), canned succotash with creamed corn (0.6 ± 0.04), canned succotash (0.6 ± 0.04), apricots (0.6 ± 0.05), boiled amaranth (0.6 ± 0.08), lychees (0.6 ± 0.05), boiled black beans (0.5 ± 0.03), boiled black turtle beans (0.5 ± 0.03), boiled cranberry beans (0.5 ± 0.03), boiled yardlong bean (0.5 ± 0.03), dried jujube (0.5 ± 0.05), boiled black-eyed peas (0.5 ± 0.03), boiled navy beans (0.5 ± 0.03), canned great northern beans (0.5 ± 0.02), canned navy beans (0.5 ± 0.02), boiled lupins (0.5 ± 0.03), boiled chickpeas (0.5 ± 0.03), cooked taro (0.5 ± 0.08), boiled beet greens (0.5 ± 0.07), canned kidney beans (0.5 ± 0.02), boiled spinach (0.5 ± 0.06), frozen and boiled brussels sprouts (0.5 ± 0.06), raw eggplant (0.5 ± 0.12), boiled acorn squash (0.5 ± 0.04), pineapple (0.5 ± 0.03), raw zucchini (0.5 ± 0.08), frozen creamed yellow corn (0.5 ± 0.04), elderberries (0.5 ± 0.03), raw green tomato (0.5 ± 0.04), cooked enriched white degermed corn meal (0.5 ± 0.02), cooked enriched yellow degermed corn meal (0.5 ± 0.02), boiled cauliflower (0.5 ± 0.08), raw vine spinach (0.5 ± 0.05), refried beans (0.5 ± 0.02), canned butter beans (0.5 ± 0.04), frozen and boiled butternut squash (0.5 ± 0.04), raw kelp/kombu/tangle seaweed (0.5 ± 0.05), frozen and boiled turnip greens (0.5 ± 0.06), raw crookneck squash (0.5 ± 0.08), boiled kale (0.5 ± 0.08), plum (0.5 ± 0.08), boiled chinese cabbage (0.5 ± 0.06), canned zucchini in tomato juice (0.5 ± 0.04), canned cranberry beans (0.5 ± 0.02), bananas (0.5 ± 0.04), boiled carrots (0.5 ± 0.06), pumpkin pie (0.5 ± 0.04), boiled purslane (0.5 ± 0.09), kiwi fruit (0.5 ± 0.07), cherry pie (0.5 ± 0.04), canned crookneck squash (0.5 ± 0.05), prickly pear (0.5 ± 0.05), canned carrots (0.5 ± 0.07), rice polish (0.4 ± 0.09), dried pilinuts (0.4 ± 0.18), boiled hyacinth beans (0.4 ± 0.03), boiled pinto beans (0.4 ± 0.03), raw lotus root (0.4 ± 0.06), boiled lima beans (0.4 ± 0.03), boiled soybeans (0.4 ± 0.03), firm raw tofu (0.4 ± 0.04), vegetarian baked beans (0.4 ± 0.02), raw romaine lettuce (0.4 ± 0.18), hummus (0.4 ± 0.02), carrot juice (0.4 ± 0.03), light buckwheat flour (0.4 ± 0.05), raw endive (0.4 ± 0.2), raw leeks (0.4 ± 0.19), canned black-eyed peas (0.4 ± 0.02), milk chocolate (0.4 ± 0.18), mince pie (0.4 ± 0.04), raw chickory (0.4 ± 0.11), raw coconut (0.4 ± 0.11), boiled salsify (0.4 ± 0.07), apple brown betty (0.4 ± 0.02), casaba melon (0.4 ± 0.03), figs (0.4 ± 0.1), frozen and boiled spinach (0.4 ± 0.05), boiled scallop squash (0.4 ± 0.06), cream of rice (0.4 ± 0.03), raw welsh onions (0.4 ± 0.05), black currants (0.4 ± 0.09), frozen and boiled broccoli (0.4 ± 0.05), pineapple chiffon pie (0.4 ± 0.06), pineapple custard pie (0.4 ± 0.04), hot dog pickle relish (0.4 ± 0.18), pineapple pie (0.4 ± 0.04), cooked tahitian taro (0.4 ± 0.07), raw new zealand spinach (0.4 ± 0.18), sweet chocolate (0.4 ± 0.18), animal crackers (0.4 ± 0.19), unenriched biscuits (0.4 ± 0.18), raw dock (0.4 ± 0.07), boiled mustard greens (0.4 ± 0.07), boiled kohlrabi (0.4 ± 0.06), raw looseleaf lettuce (0.4 ± 0.18), frozen and boiled yellow snap beans (0.4 ± 0.07), frozen and boiled green beans (0.4 ± 0.07), boiled turnip greens (0.4 ± 0.07), boiled leeks (0.4 ± 0.19), boiled zucchini (0.4 ± 0.06), frozen and boiled crookneck squash (0.4 ± 0.05), raw cucumber (0.4 ± 0.1), frozen and boiled zucchini (0.4 ± 0.04), dried sweetened flaked coconut (0.4 ± 0.2), cherries (0.4 ± 0.07), boiled new zealand spinach (0.4 ± 0.06), boiled dock (0.4 ± 0.05), boiled pumpkin (0.4 ±

0.04), canned jalapeño peppers (0.4 ± 0.07), boiled chayote (0.4 ± 0.06), boiled calabash gourd (0.4 ± 0.07), carambola (0.4 ± 0.04), dried sweetened shredded coconut (0.4 ± 0.05), blackberries (0.4 ± 0.07), raw celeriac (0.4 ± 0.05), bloody mary (0.4 ± 0.03), frozen and boiled carrots (0.4 ± 0.07), jackfruit (0.4 ± 0.05), strawberry pie (0.4 ± 0.05), candied sweet potato (0.4 ± 0.05), boiled swamp cabbage (0.4 ± 0.1), mammy apple (0.4 ± 0.05), roselle (0.4 ± 0.09), raw acerola (0.4 ± 0.05), acerola juice (0.4 ± 0.02), canned spinach (0.4 ± 0.05), beer (0.4 ± 0.01), canned pumpkin pie mix (0.4 ± 0.04), canned chinese water chestnuts (0.4 ± 0.07), semi-sweet chocolate (0.4 ± 0.18), german sweet bakers chocolate (0.4 ± 0.18), dried japanese persimmon (0.3 ± 0.15), peach pie (0.3 ± 0.05), pecan pie (0.3 ± 0.05), boiled lotus root (0.3 ± 0.06), canned pinto beans (0.3 ± 0.02), american grapes (0.3 ± 0.05), raw savoy cabbage (0.3 ± 0.14), navel orange (0.3 ± 0.04), european grapes (0.3 ± 0.03), thompson seedless grapes (0.3 ± 0.03), coleslaw (0.3 ± 0.08), coconut custard pie (0.3 ± 0.04), raw green cabbage (0.3 ± 0.14), raw red cabbage (0.3 ± 0.14), boiled red sweet peppers (0.3 ± 0.07), boiled green sweet peppers (0.3 ± 0.07), custard pie (0.3 ± 0.04), sweet potato pie (0.3 ± 0.04), boiled green cabbage (0.3 ± 0.07), blueberries (0.3 ± 0.03), banana custard pie (0.3 ± 0.04), boiled burdock root (0.3 ± 0.04), boiled yambean (0.3 ± 0.05), frozen and boiled mustard greens (0.3 ± 0.07), gooseberries (0.3 ± 0.03), boiled red cabbage (0.3 ± 0.07), boiled hubbard squash (0.3 ± 0.04), raisin pie (0.3 ± 0.04), longans (0.3 ± 0.05), boiled swiss chard (0.3 ± 0.06), frozen and boiled turnip greens and turnips (0.3 ± 0.05), pickled beets (0.3 ± 0.04), pitanga (0.3 ± 0.03), wax beans (0.3 ± 0.04), boiled waxgourd (0.3 ± 0.06), pomegranate (0.3 ± 0.03), papaya (0.3 ± 0.02), frozen and boiled cauliflower (0.3 ± 0.06), white grapefruit (0.3 ± 0.04), raw celery (0.3 ± 0.13), boiled turnips (0.3 ± 0.06), boiled celery (0.3 ± 0.07), grape juice (0.3 ± 0.02), rhubarb pie (0.3 ± 0.04), blackberry pie (0.3 ± 0.04), blueberry pie (0.3 ± 0.04), boiled bamboo shoots (0.3 ± 0.04), boiled cardoon (0.3 ± 0.05), canned pumpkin (0.3 ± 0.04), oheloberries (0.3 ± 0.04), java plum (0.3 ± 0.04), canned turnip greens (0.3 ± 0.04), apricot nectar (0.3 ± 0.02), peach nectar (0.3 ± 0.02), chocolate syrup (0.3 ± 0.13), sapodilla (0.2 ± 0.03), raw irish moss seaweed (0.2 ± 0.05), boiled white beans (0.2 ± 0.03), valencia orange (0.2 ± 0.04), restaurant orange juice (0.2 ± 0.04), raw tofu (0.2 ± 0.04), raw spring onions with tops (0.2 ± 0.1), watermelon (0.2 ± 0.03), pineapple juice (0.2 ± 0.02), canned lima beans (0.2 ± 0.02), boiled dishcloth gourd (0.2 ± 0.06), white grapefruit juice (0.2 ± 0.02), red grapefruit juice (0.2 ± 0.02), red currants (0.2 ± 0.09), white currants (0.2 ± 0.09), tequila sunrise (0.2 ± 0.03), pummelo (0.2 ± 0.03), boiled beets (0.2 ± 0.06), butterscotch pie (0.2 ± 0.04), canned shellie beans (0.2 ± 0.04), canned sauerkraut (0.2 ± 0.04), pink grapefruit (0.2 ± 0.04), canned sprouted mung beans (0.2 ± 0.08), raw frozen rhubarb (0.2 ± 0.04), chocolate chiffon pie (0.2 ± 0.06), lemon chiffon pie (0.2 ± 0.06), loquats (0.2 ± 0.05), okara tofu (0.2 ± 0.08), quince (0.2 ± 0.05), strawberries (0.2 ± 0.03), boiled collards (0.2 ± 0.03), frozen and cooked sweetened rhubarb (0.2 ± 0.04), applesauce (0.2 ± 0.04), raw oriental radish (0.2 ± 0.11), corn grits (0.2 ± 0.02), dry dessert wine (0.2 ± 0.08), sweet dessert wine (0.2 ± 0.08), manhattan (0.2 ± 0.09), brown sugar (0.2 ± 0.03), dried agar seaweed (0.2 ± 0.05), papaya nectar (0.2 ± 0.02), 53 proof coffee liqueur (0.2 ± 0.1), 63 proof coffee liqueur (0.2 ± 0.1), raw radish (0.2 ± 0.11), restaurant coffee (0.2 ± 0.03), coffee (0.2 ± 0.03), whiskey sour (0.1 ± 0.06), tangerine (0.1 ± 0.06), oatmeal (0.1 ± 0.03), canned white beans (0.1 ± 0.02), lemon meringue pie (0.1 ± 0.04), raw

I. Vitamins

pepeao (0.1 ± 0.05), steamed hawaii mountain yam (0.1 ± 0.07), screwdriver (0.1 ± 0.02), tangerine juice (0.1 ± 0.02), raw onions (0.1 ± 0.06), cream puff (0.1 ± 0.04), boiled onions (0.1 ± 0.05), crab apples (0.1 ± 0.05), coconut water (0.1 ± 0.02), canned chickpeas (0.1 ± 0.02), lemon juice (0.1 ± 0.02), piña colada (0.1 ± 0.04), lemon (0.1 ± 0.09), lime (0.1 ± 0.07), japanese persimmon (0.1 ± 0.03), cranberry sauce (0.1 ± 0.05), canned bamboo shoots (0.1 ± 0.04), cooked unenriched corn meal (0.1 ± 0.02), cooked unenriched yellow degermed corn meal (0.1 ± 0.02), lime juice (0.1 ± 0.02), frozen and boiled onions (0.1 ± 0.05), pear (0.1 ± 0.03), raw apple without skin (0.1 ± 0.04), cooked apple without skin (0.1 ± 0.03), apple juice (0.1 ± 0.02), red table wine (0.1 ± 0.05), raw agar seaweed (0.1 ± 0.05), boiled butterbur (0.1 ± 0.05), canned green beans (0.1 ± 0.07), canned yellow snap beans (0.1 ± 0.07), raw apple with skin (0.1 ± 0.04), farina (0.1 ± 0.03), restaurant cranberry juice cocktail (0.1 ± 0.03), rose table wine (0.1 ± 0.05), yellow hominy with red and green peppers (0.1 ± 0.02), boiled oriental radish (0.1 ± 0.07), pear nectar (0.1 ± 0.02), white table wine (0.1 ± 0.05)

Applications: hardening of the arteries, atheroclerosis, increased cholesterol, diabetes, hemophilia, high blood pressure, vein deterioration, diarrhea, dizziness, epilepsy, headache, lack of sleep, mental illness, multiple sclerosis, neuritis, Parkinson's disease, Ménière's syndrome, conjunctivitis, night blindness, baldness, constipation, arthritis, tuberculosis, canker sore, bad breath, acne, bedsores, skin inflammation, digestion upset, pyorrhea, alcohol cravings, stress, cancer, antioxidant therapy, sun sensitivity, free radical oxidation, life span extension, radiation, decreased radiation resistance, aging

Daily Dosage (1 niacin equivalent = 1 mg niacin = 1 mg niacinamide = 60 mg tryptophan): (at least 1000 mg vitamin C, 100 mg vitamin B_1 and 100 mg vitamin B_6 per every 1000 mg.) 1974 PCS — 10–30 mg; 1974; RDA — 13–19 mg; 1974 USRDA — 20 mg; 1980 RDA — 13–19 mg; 1980 USRDA — 20 mg; 1989 RDA — 13–19 mg; 1989 USRDA — 20 mg; nutritional — 20–30 mg; therapeutic — 1000–3000 mg; experimental — 3000–11000 mg; toxic — 10,000 mg +

Warnings: Caution with severe diabetes, glaucoma, peptic ulcers, decreased liver, increased blood pressure and gout. Causes flushing feeling which is not harmful. Acidity should be countered with antacid. Increased amounts are dangerous. See toxicity symptoms.

VITAMIN B_{3b}

(The term "niacin" sometimes used incorrectly for this form.)

Names: vitamin B_{3b}, vitamin B_3, niacinamide, nicotinamide, nicotinic acid amide, vitamin PP, Benicot, Aminicotin, Vi-Nicotyl, Dipegyl, Nicamindon, Nicotamide, Nicotilamide, Pelonin Amide, Amide PP, Nicofort, Niozymin, Pelmine

Classification: water-soluble vitamin, vitamin B complex, antioxidant

Forms: niacinamide/nicotinamide

Deficiency: failing vision, hypersensitivity to light, pellagra, backaches, bad breath, tender gums, diarrhea, digestion upset, nausea, appetite decreased, dulled sense of taste, hyperacute sense of smell, fatigue, nervousness, emotional instability, depression, alternating depression/emotional agitation, confusion, emotional agitation, hallucinations, memory decreased, lack of sleep, muscular weakness, fatigue, deterioration of amino acid utilization

Side-effects: fatigue, nausea, itching, burning sensation (These effects are not harmful, and do not occur as easily as with vitamin B_{3a}.) *See toxicity* symptoms

Toxicity: skin deterioration, gout, liver deterioration, headaches, ulcers, blood sugar decreased (These effects do not occur as easily as with vitamin B_{3a}.)

VITAMIN B₃ COMPLEX

Inhibitors: water, sulfa drugs, alcohol, food-processing, sleeping pills, estrogen
Helpers: vitamin B complex, vitamin B_1, vitamin B_2, vitamin B_5, vitamin C, phosphorus
Sources (mg per 100 grams of food): torula yeast (45 ± 0.18), brewers yeast (38.2 ± 0.18), dry bakers yeast (37.1 ± 0.18), broiled lamb liver (24.9 ± 0.11), anchovy canned in olive oil (20 ± 0.25), fried chicken breast without skin (14.8 ± 0.06), fried beef liver (14.4 ± 0.05), raw anchovy (14 ± 0.06), fried chicken breast with skin (13.8 ± 0.05), roasted chicken breast without skin (13.7 ± 0.06), fried light meat chicken without skin (13.4 ± 0.05), roasted chicken breast with skin (12.8 ± 0.05), roasted light meat chicken without skin (12.4 ± 0.05), fried light meat chicken with skin (12 ± 0.05), pork livercheese (11.8 ± 0.13), cooked swordfish (11.8 ± 0.06), canned tuna in oil (11.6 ± 0.06), compressed bakers yeast (11.4 ± 0.18), stewed rabbit (11.3 ± 0.04), roasted light meat chicken with skin (11.1 ± 0.05), fried chicken giblets (11 ± 0.05), braised beef liver (10.7 ± 0.05), canned bonito (9.8 ± 0.05), raw swordfish (9.6 ± 0.06), raw mackerel (9.1 ± 0.06), braised pork liver (8.4 ± 0.05), stewed chicken breast without skin (8.4 ± 0.05), pork liver sausage (8.3 ± 0.28), raw quail without skin (8.2 ± 0.05), roasted veal leg (8.1 ± 0.06), turkey breast meat summer sausage (8.1 ± 0.24), raw chinook salmon (7.9 ± 0.06), roasted veal rib roast (7.8 ± 0.06), stewed chicken breast with skin (7.8 ± 0.05), stewed light meat chicken without skin (7.8 ± 0.05), dried and salted cod (7.5 ± 0.06), canned chicken liver pâté (7.5 ± 0.18), fried chicken back with skin (7.4 ± 0.07), cooked bacon (7.3 ± 0.04), cooked halibut (7.2 ± 0.06), fried dark meat chicken without skin (7.1 ± 0.05), light meat turkey roll (7 ± 0.09), braised pork loin excluding fat (6.9 ± 0.05), fried chicken thigh with skin (6.9 ± 0.08), stewed light meat chicken with skin (6.9 ± 0.05), braised pork center loin excluding fat (6.8 ± 0.05), grilled canadian bacon (6.8 ± 0.11), cooked mackerel (6.8 ± 0.06), raw yellowtail (6.8 ± 0.06), fried dark meat chicken with skin (6.8 ± 0.05), raw pheasant without skin (6.8 ± 0.05), roasted turkey light meat without skin (6.8 ± 0.05), roasted chicken wing with skin (6.8 ± 0.15), cooked sockeye salmon (6.7 ± 0.06), fried chicken (6.7 ± 0.06), roasted dark meat chicken without skin (6.6 ± 0.05), fried chicken wing with skin (6.6 ± 0.16), canned pink salmon (6.6 ± 0.06), braised pork top loin excluding fat (6.5 ± 0.05), braised pork center rib excluding fat (6.5 ± 0.05), raw goose liver (6.5 ± 0.05), roasted chicken thigh without skin (6.5 ± 0.1), well done broiled ground beef (6.5 ± 0.05), well done fried ground beef (6.5 ± 0.05), raw milkfish (6.5 ± 0.06), raw venison (6.4 ± 0.06), braised lamb heart (6.4 ± 0.03), braised/roasted/stewed veal chuck (6.4 ± 0.06), roasted dark meat chicken with skin (6.4 ± 0.05), roasted turkey light meat with skin (6.3 ± 0.05), roasted chicken thigh with skin (6.3 ± 0.08), roasted ham (6.2 ± 0.05), roasted leg of lamb excluding fat (6.2 ± 0.06), unheated canadian bacon (6.1 ± 0.09), broiled lamb loin chop excluding fat (6.1 ± 0.1), fried chicken drumstick with skin (6.1 ± 0.1), stewed chicken drumstick without skin (6.1 ± 0.11), fried pork center loin excluding fat (6 ± 0.05), broiled pork loin excluding fat (6 ± 0.05), hard pork salami (6 ± 0.5), roasted pork loin excluding fat (6 ± 0.05), braised pork center loin including fat (6 ± 0.05), braised pork loin including fat (6 ± 0.05), braised pork heart (6 ± 0.04), simmered beef kidneys (6 ± 0.05), broiled beef top round excluding fat (6 ± 0.05), roasted chicken drumstick with skin (6 ± 0.1), raw bluefish (6 ± 0.06), well done broiled lean ground beef (6 ± 0.05), braised pork arm excluding fat (5.9 ± 0.05), simmered beef shank excluding fat (5.9 ± 0.05), braised pork spleen (5.9 ± 0.05), broiled beef top round including fat (5.9 ± 0.05), well done

VITAMIN B₃ COMPLEX

broiled extra lean ground beef (5.9 ± 0.05), raw halibut (5.9 ± 0.06), simmered turkey liver (5.9 ± 0.05), well done baked ground beef (5.9 ± 0.05), braised pork kidneys (5.8 ± 0.05), raw sockeye salmon (5.8 ± 0.06), broiled lamb rib chop excluding fat (5.8 ± 0.12), medium-rare broiled ground beef (5.8 ± 0.05), medium-rare fried ground beef (5.8 ± 0.05), canned tuna in water (5.8 ± 0.06), braised pork sirloin excluding fat (5.7 ± 0.05), roasted pork sirloin excluding fat (5.6 ± 0.05), braised pork loin blade excluding fat (5.6 ± 0.05), braised pork center rib including fat (5.6 ± 0.05), roasted lamb shoulder excluding fat (5.6 ± 0.06), braised beef spleen (5.6 ± 0.05), broiled pork center loin excluding fat (5.5 ± 0.05), roasted pork center loin excluding fat (5.5 ± 0.05), braised pork top loin including fat (5.5 ± 0.05), braised pork spareribs (5.5 ± 0.05), raw reindeer (5.5 ± 0.05), roasted leg of lamb including fat (5.5 ± 0.06), simmered beef shank including fat (5.5 ± 0.05), fried beef top round excluding fat (5.5 ± 0.05), well done baked lean ground beef (5.5 ± 0.05), well done fried lean ground beef (5.5 ± 0.05), canned sockeye salmon (5.5 ± 0.06), roasted pork loin including fat (5.4 ± 0.05), roasted pork center rib excluding fat (5.4 ± 0.05), roasted pork top loin excluding fat (5.4 ± 0.05), sardines (5.4 ± 0.21), braised/broiled veal loin (5.4 ± 0.06), braised/broiled veal round with rump (5.4 ± 0.06), well done fried extra lean ground beef (5.4 ± 0.05), well done baked extra lean ground beef (5.4 ± 0.05), unheated ham (5.3 ± 0.05), broiled pork loin including fat (5.3 ± 0.05), canned lean ham (5.3 ± 0.05), roasted canned ham (5.3 ± 0.05), braised pork tongue (5.3 ± 0.05), broiled beef top loin excluding fat (5.3 ± 0.05), fried chicken neck with skin (5.3 ± 0.14), light meat chicken roll (5.3 ± 0.05), chicken roll (5.3 ± 0.09), tuna salad spread (5.3 ± 0.09), fried pork center loin including fat (5.2 ± 0.05), broiled pork center rib excluding fat (5.2 ± 0.05), broiled pork top loin excluding fat (5.2 ± 0.05), fried pork center rib excluding fat (5.2 ± 0.05), fried pork top loin excluding fat (5.2 ± 0.05), roasted pork sirloin including fat (5.2 ± 0.05), braised pork arm braised including fat (5.2 ± 0.05), medium-rare broiled lean ground beef (5.2 ± 0.05), roasted duck without skin (5.1 ± 0.05), broiled lamb loin chop including fat (5.1 ± 0.07), smoked haddock (5.1 ± 0.06), broiled pork center loin including fat (5 ± 0.05), roasted pork center loin including fat (5 ± 0.05), roasted pork rump excluding fat (5 ± 0.05), braised pork sirloin including fat (5 ± 0.05), hard pork and beef salami (5 ± 0.5), pork and beef pepperoni (5 ± 0.83), turkey summer sausage (5 ± 0.18), beef pastrami (5 ± 0.18), medium-rare broiled extra lean ground beef (5 ± 0.05), roasted canned lean ham (4.9 ± 0.05), roasted pork leg excluding fat (4.9 ± 0.05), roasted pork shank excluding fat (4.9 ± 0.05), roasted pork center rib including fat (4.9 ± 0.05), broiled beef flank excluding fat (4.9 ± 0.05), fried beef top round including fat (4.9 ± 0.05), stewed chicken thigh with skin (4.9 ± 0.07), broiled pork sirloin excluding fat (4.8 ± 0.05), unheated lean ham (4.8 ± 0.05), roasted cured pork arm excluding fat (4.8 ± 0.05), roasted pork top loin including fat (4.8 ± 0.05), braised pork loin blade including fat (4.8 ± 0.05), roasted duck with skin (4.8 ± 0.05), broiled beef rib eye excluding fat (4.8 ± 0.05), broiled beef flank including fat (4.8 ± 0.05), medium-rare fried lean ground beef (4.8 ± 0.05), medium-rare baked ground beef (4.8 ± 0.05), roasted pork tenderloin (4.7 ± 0.05), broiled pork center rib including fat (4.7 ± 0.05), broiled pork loin blade excluding fat (4.7 ± 0.05), roasted pork rump including fat (4.7 ± 0.05), roasted pork loin blade excluding fat (4.7 ± 0.05), roasted lamb shoulder including fat (4.7 ± 0.06), light and dark meat turkey roll (4.7 ± 0.09), broiled beef top loin including fat (4.7 ± 0.05), medium-rare fried extra lean ground beef (4.7 ± 0.05), stewed dark meat chicken without skin (4.7 ± 0.05), smoked chinook salmon (4.7 ± 0.06), broiled pork top loin including fat (4.6 ± 0.05), pork link sausage (4.6 ± 0.07), roasted pork leg including fat (4.6 ± 0.05), braised beef flank excluding fat (4.6 ± 0.05), broiled lamb rib chop including fat (4.6 ± 0.07), broiled porterhouse steak excluding fat (4.6 ± 0.05), broiled T-bone steak excluding fat (4.6 ± 0.05), chopped smoked beef (4.6 ± 0.18), braised/stewed veal breast plate (4.6 ± 0.06), cooked haddock (4.6 ± 0.06), fried pork loin blade excluding fat (4.5 ± 0.05), roasted pork shank including fat (4.5 ± 0.05), raw pork jowl (4.5 ± 0.05), cooked eel (4.5 ± 0.06), simmered chicken liver (4.5 ± 0.05), braised beef flank including fat (4.5 ± 0.05), kippered herring (4.5 ± 0.13), raw pork stomach (4.5 ± 0.05), simmered turkey giblets (4.5 ± 0.05), stewed dark meat chicken with skin (4.5 ± 0.05), stewed chicken wing with skin (4.5 ± 0.13), pork sausage (4.4 ± 0.19), fried pork center rib including fat (4.4 ± 0.05), fried pork top loin including fat (4.4 ± 0.05), broiled pork sirloin including fat (4.3 ± 0.05), broiled pork boston blade excluding fat (4.3 ± 0.05), minced ham lunch meat (4.3 ± 0.24), roasted pork boston blade excluding fat (4.3 ± 0.05), roasted pork shoulder excluding fat (4.3 ± 0.05), roasted pork arm excluding fat (4.3 ± 0.05), braised pork boston blade excluding fat (4.3 ± 0.05), pepperoni pizza (4.3 ± 0.04), summer sausage (4.3 ± 0.22), fried beef sirloin excluding fat (4.3 ± 0.05), beef summer sausage (4.3 ± 0.22), broiled beef sirloin excluding fat (4.3 ± 0.05), broiled beef rib eye including fat (4.3 ± 0.05), beef honey roll sausage (4.3 ± 0.22), medium-rare baked lean ground beef (4.3 ± 0.05), lebanon bologna (4.3 ± 0.22), broiled pork loin blade including fat (4.2 ± 0.05), italian pork sausage (4.2 ± 0.07), roasted pork loin blade including fat (4.2 ± 0.05), broiled beef round excluding fat (4.2 ± 0.05), roasted goose with skin (4.2 ± 0.05), stewed chicken drumstick with skin (4.2 ± 0.09), medium-rare baked extra lean ground beef (4.2 ± 0.05), roasted cured pork arm including fat (4.1 ± 0.05), simmered beef heart (4.1 ± 0.05), cooked herring (4.1 ± 0.06), broiled porterhouse steak including fat (4.1 ± 0.05), roasted goose without skin (4.1 ± 0.05), simmered chicken giblets (4.1 ± 0.05), roasted beef rib roasted excluding fat (4.1 ± 0.05), braised beef bottom round excluding fat (4.1 ± 0.05), roasted lean ham (4 ± 0.05), broiled pork boston blade including fat (4 ± 0.05), roasted pork boston blade including fat (4 ± 0.05), roasted pork shoulder including fat (4 ± 0.05), braised pork boston blade braised including fat (4 ± 0.05), roast beef sandwich (4 ± 0.03), braised beef pancreas (4 ± 0.05), simmered chicken gizzard (4 ± 0.05), fried pork loin blade including fat (3.9 ± 0.05), pork bologna (3.9 ± 0.22), roasted pork arm including fat (3.9 ± 0.05), hamburger sandwich (3.9 ± 0.05), beef and pork summer sausage (3.9 ± 0.22), broiled beef tenderloin excluding fat (3.9 ± 0.05), broiled beef sirloin including fat (3.9 ± 0.05), broiled T-bone steak including fat (3.9 ± 0.05), braised beef bottom round including fat (3.9 ± 0.05), simmered chicken neck without skin (3.9 ± 0.28), cooked rockfish (3.9 ± 0.06), turkey bologna (3.9 ± 0.18), chopped ham lunch meat (3.8 ± 0.24), hot dog (3.8 ± 0.06), fried beef brains (3.8 ± 0.05), broiled beef rib excluding fat (3.8 ± 0.05), roasted eye of beef round excluding fat (3.8 ± 0.05), braised beef brisket excluding fat (3.8 ± 0.05), turkey frankfurter (3.8 ± 0.11), raw haddock (3.8 ± 0.06), fried beef sirloin including fat (3.7 ± 0.05), roasted beef tip excluding fat (3.7 ± 0.05), broiled beef round including fat (3.7 ± 0.05), braised beef chuck arm pot roast excluding fat (3.7 ± 0.05), pork luxury loaf (3.6 ± 0.18), ham and cheese roll (3.6 ± 0.18), polish pork sausage (3.6 ± 0.18), broiled beef tenderloin including fat (3.6 ± 0.05), beef lunch meat (3.6 ± 0.18), roasted turkey dark meat without skin (3.6 ± 0.05), cheese pizza (3.5 ± 0.04), beef and pork salami (3.5 ± 0.22), pork and beef luncheon sausage (3.5 ± 0.22), raw eel (3.5 ± 0.06), roasted beef round tip including fat (3.5 ± 0.05), beef beerwurst salami (3.5 ± 0.22), roasted eye of beef round including fat (3.5 ± 0.05), turkey salami (3.5 ± 0.09), roasted turkey dark meat with skin (3.5 ± 0.05), turkey pastrami (3.5 ± 0.09), turkey ham (3.5 ± 0.09), roasted beef tenderloin excluding fat (3.4 ± 0.05), simmered chicken neck with skin (3.4 ± 0.13), pork mothers loaf (3.3 ± 0.24), canned pork lunch meat (3.3 ± 0.24), pork and beef brotwurst (3.3 ± 0.07), cheeseburger (3.3 ± 0.04), bacon cheeseburger (3.3 ± 0.03), braised pork brains (3.3 ± 0.05), simmered turkey heart (3.3 ± 0.05), roasted beef rib including fat (3.3 ± 0.05), raw pollock (3.3 ± 0.06), smoked beef sausage (3.3 ± 0.12), canned ham (3.2 ± 0.05), pork bratwurst (3.2 ± 0.06), pork and beef honey loaf (3.2 ± 0.18), pork and beef peppered loaf (3.2 ± 0.18), grilled ham patties (3.2 ± 0.08), pork and beef link sausage (3.2 ± 0.07), braised pork pancreas (3.2 ± 0.05), raw herring (3.2 ± 0.06), braised beef shortribs excluding fat (3.2 ± 0.05), raw dungeness crab (3.2 ± 0.06), raw rockfish (3.2 ± 0.06), unheated ham patties (3.1 ± 0.08), pork and beef sausage (3.1 ± 0.38), burrito (3.1 ± 0.03), breaded and fried shrimp (3.1 ± 0.06), roasted beef tenderloin including fat (3.1 ± 0.05), broiled beef rib including fat (3.1 ± 0.05), chicken frankfurter (3.1 ± 0.11), braised beef chuck arm pot roast including fat (3.1 ± 0.05), cooked black bass (3.1 ± 0.04), simmered turkey gizzard (3.1 ± 0.05), pork beerwurst salami (3 ± 0.22), berliner (3 ± 0.22), beef salami (3 ± 0.22), braised beef brisket including fat (3 ± 0.05), cooked corned beef brisket (3 ± 0.05), cooked crayfish (2.9 ± 0.06), pork and beef lunch meat (2.9 ± 0.18), cooked kingfish (2.9 ± 0.05), raw flatfish (2.9 ± 0.06), raw shark (2.9 ± 0.06), chicken salad (2.9 ± 0.18), raw bacon (2.8 ± 0.07), pork and beef knockwurst (2.8 ± 0.07), pork and beef link sausage with american cheese (2.8 ± 0.12), simmered chicken heart (2.8 ± 0.05), battered and fried shark (2.8 ± 0.06), raw cusk (2.7 ± 0.06), pork and beef kielbasa/kolbassy (2.7 ± 0.19), beef and pork frankfurter (2.7 ± 0.11), pork and beef mortadella (2.7 ± 0.33), braised beef chuck blade roast excluding fat (2.7 ± 0.05), beef and pork bologna (2.6 ± 0.22), breaded fried squid (2.6 ± 0.06), cooked shrimp (2.6 ± 0.06), beef bologna (2.6 ± 0.22), raw shrimp (2.6 ± 0.06), cooked oysters (2.5 ± 0.06), cooked perch (2.5 ± 0.06), pork and beef picnic loaf (2.5 ± 0.18), pork and beef old fashioned loaf (2.5 ± 0.18), canned smoked goose liver pâté (2.5 ± 0.18), cooked cod (2.5 ± 0.06), cooked clams (2.5 ± 0.06), canned cod (2.5 ± 0.06), cooked flounder (2.5 ± 0.05), braised beef shortribs including fat (2.5 ± 0.05), beef frankfurter (2.5 ± 0.09), braised beef lungs (2.5 ± 0.05), raw crayfish (2.4 ± 0.06), roasted pork blade roll (2.4 ± 0.05), raw ling (2.4 ± 0.06), braised beef chuck blade roast including fat (2.4 ± 0.05), smoked cisco (2.4 ± 0.06), smoked whitefish (2.4 ± 0.06), canned corned beef (2.4 ± 0.05), canned goose liver pâté (2.3 ± 0.38), barbecue loaf (2.2 ± 0.22), simmered beef brains (2.2 ± 0.05), cooked flatfish (2.2 ± 0.06), breaded and fried catfish (2.2 ± 0.06), raw turbot (2.2 ± 0.06), simmered beef tongue (2.2 ± 0.05), breaded and fried clams (2.1 ± 0.06), ham salad (2.1 ± 0.18), ham and cheese spread (2.1 ± 0.18), pork pickle and pimento loaf (2.1 ± 0.18), fish sandwich (2.1 ± 0.03), raw wolffish (2.1 ± 0.06), potato pancakes (2.1 ± 0.07), raw cod (2.1 ± 0.06), raw catfish (2.1 ± 0.06), raw octopus (2.1 ± 0.06), turkey salad (2.1 ± 0.09), raw squid (2.1 ± 0.06), raw perch (2 ± 0.06), cooked whelk (2 ± 0.06), braised pork lungs (1.9 ± 0.05), vienna sausage (1.9 ± 0.31), raw

VITAMIN B₃ COMPLEX

lingcod (1.9 ± 0.06), raw clams (1.8 ± 0.06), cooked rainbow smelt (1.8 ± 0.06), pork olive loaf (1.8 ± 0.18), pork and beef spread (1.8 ± 0.18), chicken and turkey salad (1.8 ± 0.18), jellied corned beef loaf (1.8 ± 0.18), taco (1.7 ± 0.06), raw flounder (1.7 ± 0.05), raw sole (1.7 ± 0.05), breaded and fried oysters (1.6 ± 0.06), ham and cheese sandwich (1.6 ± 0.03), cooked whiting (1.6 ± 0.06), breaded and fried scallops (1.6 ± 0.16), yellow cake (1.5 ± 0.07), pancakes with butter and syrup (1.5 ± 0.03), breaded and fried fish (1.5 ± 0.07), raw rainbow smelt (1.4 ± 0.06), raw lobster (1.4 ± 0.06), canned blue crab (1.4 ± 0.06), dry buttermilk (1.4 ± 0.71), white cake (1.4 ± 0.06), raw smelt (1.4 ± 0.05), raw oysters (1.3 ± 0.06), canned oysters (1.3 ± 0.06), cooked alaska king crab (1.3 ± 0.06), raw whiting (1.3 ± 0.06), noodles (1.2 ± 0.03), raw frog legs (1.2 ± 0.05), devil's food cake (1.2 ± 0.08), angel food cake (1.2 ± 0.08), raw cuttlefish (1.2 ± 0.06), raw scallops (1.2 ± 0.06), pineapple upside-down cake (1.1 ± 0.07), cottage pudding (1.1 ± 0.09), sponge cake (1.1 ± 0.08), simmered pork tail (1.1 ± 0.05), scalloped potatoes (1.1 ± 0.04), raw alaska king crab (1.1 ± 0.06), bleu cheese (1.1 ± 0.18), pork headcheese (1.1 ± 0.18), raw whelk (1.1 ± 0.06), cooked lobster (1.1 ± 0.06), yellow corn pudding (1 ± 0.02), dry skim milk (1 ± 0.17), mashed potatoes (1 ± 0.05), o'brien potatoes (1 ± 0.03), au gratin potatoes (1 ± 0.04), chili (1 ± 0.02), dry instant skim milk (0.9 ± 0.05), potato salad (0.9 ± 0.04), gingerbread (0.8 ± 0.08), gjetost cheese (0.7 ± 0.18), fruitcake, dark (0.7 ± 0.12), fruitcake, light (0.7 ± 0.12), russian salad dressing (0.7 ± 0.33), camembert cheese (0.7 ± 0.18), roquefort cheese (0.7 ± 0.18), hot fudge sundae (0.7 ± 0.03), dry whole milk (0.6 ± 0.16), strawberry sundae (0.6 ± 0.03), caramel sundae (0.6 ± 0.03), brownies (0.5 ± 0.25), malted milk (0.5 ± 0.02), bread pudding (0.5 ± 0.02), baked beans with pork and tomato sauce (0.5 ± 0.02), cheesecake (0.5 ± 0.06), prune pudding (0.5 ± 0.04), simmered pork ears (0.5 ± 0.05), simmered pork feet (0.5 ± 0.05), honey (0.5 ± 0.24), cooked grouper (0.4 ± 0.06), raw grouper (0.4 ± 0.06), tilsit cheese (0.4 ± 0.18), brie cheese (0.4 ± 0.18), limburger cheese (0.4 ± 0.18), sheep milk (0.4 ± 0.02), cooked snapper (0.4 ± 0.06), baked beans with pork and sweet sauce (0.4 ± 0.02), caraway cheese (0.4 ± 0.18), pickled pork feet (0.4 ± 0.05), surimi scallops (0.4 ± 0.06), sweetened condensed milk (0.3 ± 0.13), whole chocolate malted milk (0.3 ± 0.02), goat milk (0.3 ± 0.02), pound cake (0.3 ± 0.17), evaporated skim milk (0.3 ± 0.16), human milk (0.3 ± 0.16), soy milk (0.2 ± 0.02), restaurant chocolate milkshake (0.2 ± 0.02), cheese soufflé (0.2 ± 0.05), raw snapper (0.2 ± 0.06), restaurant vanilla milkshake (0.2 ± 0.02), restaurant strawberry milkshake (0.2 ± 0.02), evaporated whole milk (0.2 ± 0.04), hot chocolate (0.2 ± 0.02), boston cream pie (0.2 ± 0.05), vanilla milkshake (0.2 ± 0.02), evaporated lowfat milk (0.2 ± 0.04), duck egg (0.1 ± 0.07), skim yogurt (0.1 ± 0.02), indian buffalo milk (0.1 ± 0.02), chocolate milkshake (0.1 ± 0.02), tapioca cream pudding (0.1 ± 0.03), lowfat yogurt (0.1 ± 0.02), eggnog (0.1 ± 0.02), protein fortified lowfat 2% milk (0.1 ± 0.02), protein fortified skim milk (0.1 ± 0.02), protein fortified lowfat 1% milk (0.1 ± 0.02), skim milk (0.1 ± 0.02), lowfat 1% milk (0.1 ± 0.02), lowfat 2% milk (0.1 ± 0.02), whole milk (0.1 ± 0.02), carob milk (0.1 ± 0.02), lowfat 1% chocolate milk (0.1 ± 0.02), lowfat 2% chocolate milk (0.1 ± 0.02), whole chocolate milk (0.1 ± 0.02), whey (0.1 ± 0.02), custard (0.1 ± 0.02), vanilla pudding (0.1 ± 0.02), whole yogurt (0.1 ± 0.02), filled milk (0.1 ± 0.02), chocolate pudding (0.1 ± 0.02), caramel cake (0.1 ± 0.05), surimi shrimp (0.1 ± 0.06), ricotta cheese (0.1 ± 0.04), raw beef tripe (0.1 ± 0.05), simmered pork chitterlings (0.1 ± 0.05), light cream (0.1 ± 0.17)

Applications: hardening of the arteries, atheroclerosis, increased cholesterol, diabetes, hemophilia, high blood pressure, vein deterioration, diarrhea, dizziness, epilepsy, headache, lack of sleep, mental illness, multiple sclerosis, neuritis, Parkinson's disease, Ménière's syndrome, conjunctivitis, night blindness, constipation, arthritis, tuberculosis, canker sore, bad breath, bedsores, digestion upset, pyorrhea, alcohol cravings, stress, cancer, antioxidant therapy, sun sensitivity, free radical oxidation, life span extension, radiation, decreased radiation resistance, aging
Daily Dosage (1 niacin equivalent = 1 mg niacin = 1 mg niacinamide = 60 mg tryptophan): 1974 PCS — 10–30 mg as vitamin B_3; 1974 RDA — 13–19 mg as vitamin B_3; 1974 USRDA — 20 mg as vitamin B_3; 1980 RDA — 13–19 mg as vitamin B_3; 1980 USRDA — 20 mg as vitamin B_3; 1989 RDA — 13–19 mg as vitamin B_3; 1989 USRDA — 20 mg as vitamin B_3; nutritional — 20–30 mg; therapeutic — ? (Vitamin B_{3a} is used for therapy.); experimental — ? (Vitamin B_{3a} is used for experimental effects.); toxic — none known up to 10000 mg
Warnings: Caution with severe diabetes, glaucoma, peptic ulcers, liver deterioration, increased blood pressure and gout. Causes flushing feeling which is not harmful. Acidity should be countered with antacid. Does not help skin deterioration, gout, liver deterioration, headaches, ulcers, and blood sugar decreased as does vitamin B_{3a}. Does not lower blood cholesterol levels as does vitamin B_{3a}. Increased amounts are dangerous. See toxicity symptoms.

VITAMIN B_4

(The term "vitamin B_4" is sometimes used to denote vitamin B_W.)
(The term "vitamin B_4" is sometimes used to denote vitamin B_P.)

Names: vitamin B_4, adenine, Leuco-4
Classification: nucleic acid, formerly water-soluble vitamin, formerly vitamin B complex
Forms: Found to be amino acid adenine. Early isolations were possibly mixed with vitamin B_2, vitamin B_6, arginine, cystine and glycine.
Deficiency: muscular weakness in rats and chicks, possibly similar symptoms in humans
Side-effects: unknown
Toxicity: unknown
Inhibitors: unknown
Helpers: vitamin B_{3a}, vitamin B_{3b}
Sources: all plants and animals, DNA, RNA
Applications: research on heredity, research on virus diseases, research on cancer
Daily Dosage: 1974 PCS — none; 1989 SADDI — none; 1989 RDA — none; 1989 USRDA — none; nutritional — unknown; therapeutic — unknown; experimental — unknown; toxic — unknown
Warnings: none

VITAMIN B_5

(The term "vitamin B_5" is sometimes used incorrectly to denote vitamin B_{3a}.)

Names: vitamin B_5, pantothenic acid, pantothenate, filtrate factor, chick antidermatitis factor
Classification: water-soluble vitamin, vitamin B complex, antioxidant
Forms: pantothenic acid, calcium pantothenate, sodium pantothenate
Deficiency: decreased blood pressure, fatigue, weakness, depression, emotional deterioration, blood sugar decreased, numb/tingling hands, numb/tingling feet, kidney hemorrhage, decreased antibody production, infections, decreased balance, decreased coordination, decreased reflexes, constipation, appetite decreased, nausea, bile salts formation deterioration, digestion

VITAMIN B$_5$

deterioration, duodenal ulcers, abdominal cramps, leg muscle cramps, adrenal decreased, sensitivity to insulin, convulsions, fatigue, spinal cord deterioration, cornea deterioration, fetal abnormalities, testicle deterioration, growth failure, ulcers, headaches, lack of sleep, lung infections, grey hair, coma, fatigue, stress, emotional agitation, skin deterioration, slow healing, decreased radiation resistance
Side-effects: see *toxicity* symptoms
Toxicity: liver impairment
Inhibitors: heat, food-processing, canning, caffeine, sulfa drugs, sleeping pills, estrogen, alcohol, ferrous sulfate, methyl bromide
Helpers: vitamin B complex, vitamin B$_6$, vitamin B$_{12}$, vitamin B$_C$, vitamin B$_W$, vitamin C, sulfur
Sources (mg per 100 grams of food): simmered turkey liver (5.96 ± 0.005), fried beef liver (5.92 ± 0.005), simmered chicken liver (5.41 ± 0.005), braised pork liver (4.77 ± 0.005), braised pork pancreas (4.74 ± 0.005), braised beef liver (4.57 ± 0.005), fried chicken giblets (4.45 ± 0.005), chicken egg yolk (4.41 ± 0.029), braised beef pancreas (4.25 ± 0.005), tomato powder (3.76 ± 0.005), dry skim milk (3.57 ± 0.017), pork livercheese (3.53 ± 0.013), dried spirulina seaweed (3.48 ± 0.005), simmered turkey giblets (3.46 ± 0.005), pork liver sausage (3.39 ± 0.028), dry instant skim milk (3.23 ± 0.005), raw abalone (3 ± 0.006), dry buttermilk (3 ± 0.071), simmered chicken giblets (2.96 ± 0.005), pork liverwurst (2.94 ± 0.028), braised pork kidneys (2.87 ± 0.005), freeze-dried parsley (2.86 ± 0.357), dried peanuts (2.82 ± 0.018), simmered turkey heart (2.72 ± 0.005), simmered chicken heart (2.65 ± 0.005), braised pork heart (2.47 ± 0.004), dry whole milk (2.28 ± 0.016), raw mushrooms (2.2 ± 0.014), boiled mushrooms (2.17 ± 0.006), oil roasted peanuts (2.11 ± 0.018), defatted soybean flour (2 ± 0.005), defatted soy meal (1.98 ± 0.004), raw trout (1.94 ± 0.006), braised pork brains (1.82 ± 0.005), low fat soybean flour (1.82 ± 0.006), bleu cheese (1.75 ± 0.018), roquefort cheese (1.75 ± 0.018), bran (1.75 ± 0.018), dried pecans (1.75 = 0.018), simmered beef kidneys (1.69 ± 0.005), dried and salted cod (1.67 ± 0.006), pork and beef pepperoni (1.67 ± 0.083), chicken egg (1.66 ± 0.01), raw lobster (1.64 ± 0.006), soybean flour (1.59 ± 0.006), barbecue loaf (1.57 ± 0.022), roasted duck without skin (1.5 ± 0.005), camembert cheese (1.39 ± 0.018), oil roasted spanish peanuts (1.39 ± 0.018), oil roasted virginia peanuts (1.39 ± 0.018), oil roasted valencia peanuts (1.39 ± 0.018), dry roasted peanuts (1.39 ± 0.018), toasted wheat germ (1.39 ± 0.018), dried ginko nuts (1.36 ± 0.018), dehydrated onion flakes (1.36 ± 0.036), stewed chicken drumstick without skin (1.3 ± 0.011), roasted turkey dark meat without skin (1.29 ± 0.005), fried dark meat chicken without skin (1.26 ± 0.005), dry roasted cashews (1.25 ± 0.018), freeze-dried shallots (1.25 ± 0.125), fried chicken drumstick with skin (1.22 ± 0.01), roasted chicken drumstick with skin (1.21 ± 0.01), roasted dark meat chicken without skin (1.21 ± 0.005), roasted soybean flour (1.21 ± 0.006), oil roasted cashews (1.21 ± 0.018), cashew butter (1.21 ± 0.018), fried chicken thigh with skin (1.19 ± 0.008), roasted chicken thigh without skin (1.19 ± 0.01), stir-fried sprouted soybeans (1.19 ± 0.005), limburger cheese (1.18 ± 0.018), dried filberts (1.18 ± 0.018), roasted turkey dark meat with skin (1.16 ± 0.005), fried dark meat chicken with skin (1.16 ± 0.005), ry-krisp crackers (1.14 ± 0.036), roasted dark meat chicken with skin (1.11 ± 0.005), roasted chicken thigh with skin (1.11 ± 0.008), hard pork and beef salami (1.1 ± 0.05), roasted duck with skin (1.1 ± 0.005), fried chicken back with skin (1.1 ± 0.007), canned jalapeño peppers (1.07 ± 0.007), cooked bacon (1.06 ± 0.004), fried chicken breast without skin (1.03 ± 0.006), fried light meat

chicken without skin (1.03 ± 0.005), fried pork center loin including fat (1.01 ± 0.005), fried chicken breast with skin (1 ± 0.005), roasted light meat chicken without skin (0.97 ± 0.005), roasted chicken breast without skin (0.97 ± 0.006), fried light meat chicken with skin (0.97 ± 0.005), fried chicken neck with skin (0.97 ± 0.014), crunchy peanut butter (0.97 ± 0.016), raw california avocado (0.97 ± 0.003), raw florida avocado (0.97 ± 0.002), beef salami (0.96 ± 0.022), raw pheasant without skin (0.96 ± 0.005), roasted chicken breast with skin (0.94 ± 0.005), creamy peanut butter (0.94 ± 0.031), roasted light meat chicken with skin (0.93 ± 0.005), potato pancakes (0.93 ± 0.007), dried european chestnuts (0.93 ± 0.018), raw endive (0.92 ± 0.02), broiled pork loin excluding fat (0.91 ± 0.005), broiled pork sirloin excluding fat (0.91 ± 0.005), roasted chicken wing with skin (0.91 ± 0.015), braised pork spleen (0.89 ± 0.005), stewed dark meat chicken without skin (0.89 ± 0.005), fried chicken wing with skin (0.88 ± 0.016), boiled yellow corn (0.88 ± 0.006), simmered beef heart (0.87 ± 0.005), beef and pork salami (0.87 ± 0.022), smoked chinook salmon (0.87 ± 0.006), raw mackerel (0.86 ± 0.006), stewed chicken drumstick with skin (0.86 ± 0.009), shredded wheat (0.86 ± 0.018), toasted whole wheat bread (0.86 ± 0.024), simmered turkey gizzard (0.85 ± 0.005), raw bluefish (0.82 ± 0.006), pork pickle and pimento loaf (0.82 ± 0.018), broiled pork boston blade excluding fat (0.82 ± 0.005), broiled pork loin blade excluding fat (0.82 ± 0.005), french toast (0.82 ± 0.008), dried coconut (0.82 ± 0.018), pork and beef kielbasa/kolbassy (0.81 ± 0.019), boiled peanuts (0.81 ± 0.016), fried pork loin blade excluding fat (0.8 ± 0.005), sponge cake (0.8 ± 0.008), waffles (0.8 ± 0.007), pork olive loaf (0.79 ± 0.018), fried pork center loin excluding fat (0.79 ± 0.005), fried pork center rib excluding fat (0.79 ± 0.005), fried pork top loin excluding fat (0.79 ± 0.005), pork link sausage (0.78 ± 0.007), stewed chicken thigh with skin (0.78 ± 0.007), dried dates (0.78 ± 0.006), pork and beef link sausage with american cheese (0.77 ± 0.012), roasted pork blade roll (0.77 ± 0.005), roasted pork center loin including fat (0.77 ± 0.005), stewed dark meat chicken with skin (0.77 ± 0.005), raw mullet (0.76 ± 0.006), broiled pork sirloin including fat (0.76 ± 0.005), sweetened condensed milk (0.76 ± 0.013), tomato paste (0.76 ± 0.004), wheat flakes (0.75 ± 0.018), braised pork spareribs (0.75 ± 0.005), broiled pork loin including fat (0.75 ± 0.005), roasted pork rump excluding fat (0.75 ± 0.005), broiled pork center rib excluding fat (0.75 ± 0.005), broiled pork top loin excluding fat (0.75 ± 0.005), evaporated skim milk (0.75 ± 0.016), pork sausage (0.74 ± 0.019), pork bologna (0.74 ± 0.022), braised pork center loin excluding fat (0.74 ± 0.005), dried apricots (0.74 ± 0.014), steamed sprouted soybeans (0.74 ± 0.011), roasted canned ham (0.73 ± 0.005), roasted ham (0.72 ± 0.005), whole wheat bread (0.72 ± 0.02), simmered chicken gizzard (0.71 ± 0.005), brie cheese (0.71 ± 0.018), braised pork sirloin excluding fat (0.71 ± 0.005), braised pork loin excluding fat (0.7 ± 0.005), raw shark (0.69 ± 0.006), broiled pork boston blade including fat (0.69 ± 0.005), broiled pork center loin excluding fat (0.69 ± 0.005), roasted pork shank excluding fat (0.69 ± 0.005), roasted pork tenderloin (0.69 ± 0.005), pork and beef picnic loaf (0.68 ± 0.018), pork and beef honey loaf (0.68 ± 0.018), roasted pork rump including fat (0.68 ± 0.005), roasted pork center rib including fat (0.68 ± 0.005), roasted turkey light meat without skin (0.68 ± 0.005), boiled sprouted green peas (0.68 ± 0.005), canned yellow corn (0.68 ± 0.005), roasted pork leg excluding fat (0.67 ± 0.005), braised pork arm excluding fat (0.67 ± 0.005), simmered chicken neck without skin (0.67 ± 0.028), toasted cracked wheat bread (0.67 ± 0.024), braised pork lungs (0.66 ± 0.005), broiled pork loin blade including fat (0.66 ± 0.005), braised pork top loin excluding fat (0.66 ± 0.005), braised pork center rib excluding fat (0.66 ± 0.005), french fried potatoes (0.66 ± 0.01), frozen and french fried potatoes without skin (0.66 ± 0.01), raw herring (0.65 ± 0.006), roasted cured pork arm excluding fat (0.65 ± 0.005), braised pork loin blade excluding fat (0.65 ± 0.005), roasted pork center loin excluding fat (0.65 ± 0.005), baked sweet potato (0.65 ± 0.004), raw rainbow smelt (0.64 ± 0.006), pork and beef lunch meat (0.64 ± 0.018), roasted pork loin excluding fat (0.64 ± 0.005), braised pork boston blade excluding fat (0.64 ± 0.005), skim yogurt (0.64 ± 0.002), mixed grain bread (0.64 ± 0.02), dried english/persian walnuts (0.64 ± 0.018), boiled lentils (0.64 ± 0.003), sardines (0.63 ± 0.021), roasted pork sirloin excluding fat (0.63 ± 0.005), broiled pork center rib including fat (0.63 ± 0.005), roasted turkey light meat with skin (0.63 ± 0.005), evaporated whole milk (0.63 ± 0.004), freeze-dried red sweet peppers (0.63 ± 0.313), freeze-dried green sweet peppers (0.63 ± 0.313), dried sesame kernels (0.63 ± 0.063), braised beef lungs (0.62 ± 0.005), braised pork center loin including fat (0.62 ± 0.005), toasted english muffin (0.62 ± 0.01), chopped smoked beef (0.61 ± 0.018), pork and beef old fashioned loaf (0.61 ± 0.018), ham and cheese spread (0.61 ± 0.018), broiled pork top loin including fat (0.61 ± 0.005), raw sockeye salmon (0.61 ± 0.006), parmesan cheese (0.6 ± 0.1), roasted pork loin blade including fat (0.6 ± 0.005), fried pork center rib including fat (0.6 ± 0.005), bran muffins (0.6 ± 0.013), cracked wheat bread (0.6 ± 0.02), microwaved potato without skin (0.6 ± 0.003), boiled split peas (0.6 ± 0.003), pomegranate (0.6 ± 0.003), raw yellowtail (0.59 ± 0.006), fried pork loin blade including fat (0.59 ± 0.005), roasted pork arm excluding fat (0.59 ± 0.005), broiled pork center loin including fat (0.59 ± 0.005), roasted pork leg including fat (0.59 ± 0.005), roasted pork shank including fat (0.59 ± 0.005), braised pork sirloin including fat (0.59 ± 0.005), fried pork top loin including fat (0.59 ± 0.005), lowfat yogurt (0.59 ± 0.002), boiled parsnips (0.59 ± 0.006), raw turbot (0.58 ± 0.006), raw wolffish (0.58 ± 0.006), roasted pork shoulder excluding fat (0.58 ± 0.005), roasted pork center rib excluding fat (0.58 ± 0.005), roasted pork top loin excluding fat (0.58 ± 0.005), cheesecake (0.58 ± 0.006), raw sprouted alfalfa seeds (0.58 ± 0.015), fried beef brains (0.57 ± 0.005), simmered beef brains (0.57 ± 0.005), beef and pork summer sausage (0.57 ± 0.022), turkey breast meat summer sausage (0.57 ± 0.024), gruyere cheese (0.57 ± 0.018), roasted pork boston blade excluding fat (0.57 ± 0.005), braised pork loin including fat (0.57 ± 0.005), roasted pork sirloin including fat (0.57 ± 0.005), roasted canned lean ham (0.57 ± 0.005), neufchatel cheese (0.57 ± 0.018), stewed chicken breast without skin (0.57 ± 0.005), stewed light meat chicken without skin (0.57 ± 0.005), mashed potatoes (0.57 ± 0.005), roasted european chestnuts (0.57 ± 0.018), stir-fried sprouted lentils (0.57 ± 0.005), roasted cured pork arm including fat (0.56 ± 0.005), braised pork arm braised including fat (0.56 ± 0.005), boiled succotash (0.56 ± 0.005), baked potato without skin (0.56 ± 0.003), frozen and baked sweet potato (0.56 ± 0.006), roasted pork loin including fat (0.55 ± 0.005), braised pork center rib including fat (0.55 ± 0.005), stewed chicken breast with skin (0.55 ± 0.005), baked potato with skin (0.55 ± 0.002), raw broccoli (0.55 ± 0.011), dinner rolls (0.54 ± 0.018), beef lunch meat (0.54 ± 0.018), pork and beef peppered loaf (0.54 ± 0.018), pork luxury loaf (0.54 ± 0.018), ham and cheese roll (0.54 ± 0.018), stewed light meat chicken with skin (0.54 ± 0.005), potato salad (0.54 ±

I. Vitamins

VITAMIN B5

0.004), oatmeal cookie (0.54 ± 0.038), granola bar (0.54 ± 0.021), toasted pumpernickel bread (0.54 ± 0.018), raw european chestnuts (0.54 ± 0.018), roasted pork loin blade excluding fat (0.53 ± 0.005), unheated canadian bacon (0.53 ± 0.009), braised pork top loin including fat (0.53 ± 0.005), simmered chicken neck with skin (0.53 ± 0.013), baked beans with pork and tomato sauce (0.53 ± 0.002), hamburger bun (0.53 ± 0.013), hot dog bun (0.53 ± 0.013), canned sweet potato (0.53 ± 0.003), boiled sweet potato (0.53 ± 0.003), simmered beef tongue (0.52 ± 0.005), lebanon bologna (0.52 ± 0.022), braised pork loin blade including fat (0.52 ± 0.005), scalloped potatoes (0.52 ± 0.004), raw flatfish (0.51 ± 0.006), roasted pork boston blade including fat (0.51 ± 0.005), grilled canadian bacon (0.51 ± 0.011), english muffin (0.51 ± 0.009), boiled potato without skin (0.51 ± 0.004), provolone cheese (0.5 ± 0.018), roast beef sandwich (0.5 ± 0.003), roasted pork shoulder including fat (0.5 ± 0.005), pimento cheese spread (0.5 ± 0.018), american cheese (0.5 ± 0.018), roasted pork top loin including fat (0.5 ± 0.005), stewed chicken wing with skin (0.5 ± 0.013), puffed wheat (0.5 ± 0.036), dried japanese chestnuts (0.5 ± 0.018), toasted rye bread (0.5 ± 0.023), graham crackers (0.5 ± 0.036), hash brown potatoes (0.5 ± 0.006), orange peel (0.5 ± 0.083), baked acorn squash (0.5 ± 0.005), raw green tomato (0.5 ± 0.004), broiled beef top round excluding fat (0.49 ± 0.005), canned lean ham (0.49 ± 0.005), roasted pork arm including fat (0.49 ± 0.005), restaurant strawberry milkshake (0.49 ± 0.002), broiled beef top round including fat (0.48 ± 0.005), beef honey roll sausage (0.48 ± 0.022), pork mothers loaf (0.48 ± 0.024), canned pork lunch meat (0.48 ± 0.024), pork beerwurst salami (0.48 ± 0.022), chocolate cream pie (0.48 ± 0.004), toasted raisin bread (0.48 ± 0.024), toasted white bread (0.48 ± 0.024), french rolls (0.48 ± 0.01), dry roasted soybean nuts (0.48 ± 0.006), roasted beef tip excluding fat (0.47 ± 0.005), unheated lean ham (0.47 ± 0.005), yellow cake (0.47 ± 0.007), pumpernickel bread (0.47 ± 0.016), boiled baby lima beans (0.47 ± 0.003), fried beef top round excluding fat (0.46 ± 0.005), roasted eye of beef round excluding fat (0.46 ± 0.005), polish pork sausage (0.46 ± 0.018), pork and beef sausage (0.46 ± 0.038), dried almonds (0.46 ± 0.018), dried prunes (0.46 ± 0.006), toasted soybean nuts (0.46 ± 0.018), breadfruit (0.46 ± 0.005), roasted beef tenderloin excluding fat (0.45 ± 0.005), italian pork sausage (0.45 ± 0.007), unheated ham (0.45 ± 0.005), frozen and fried hash brown potatoes (0.45 ± 0.006), roasted soybean nuts (0.45 ± 0.006), baked hubbard squash (0.45 ± 0.005), broiled beef flank excluding fat (0.44 ± 0.005), broiled beef flank including fat (0.44 ± 0.005), roasted beef rib roasted excluding fat (0.44 ± 0.005), well done broiled lean ground beef (0.44 ± 0.005), pork and beef link sausage (0.44 ± 0.007), o'brien potatoes (0.44 ± 0.003), raisin bread (0.44 ± 0.02), rye bread (0.44 ± 0.02), french fries (0.44 ± 0.006), tomato purée (0.44 ± 0.002), french bread (0.43 ± 0.018), fried beef sirloin excluding fat (0.43 ± 0.005), roasted beef round tip including fat (0.43 ± 0.005), swiss cheese (0.43 ± 0.018), pork and beef spread (0.43 ± 0.018), cheddar cheese (0.43 ± 0.018), almond paste (0.43 ± 0.018), boiled mungo beans (0.43 ± 0.003), dried figs (0.43 ± 0.003), pita bread (0.42 ± 0.013), fried beef top round including fat (0.42 ± 0.005), well done broiled extra lean ground beef (0.42 ± 0.005), braised beef bottom round excluding fat (0.42 ± 0.005), roasted eye of beef round including fat (0.42 ± 0.005), cooked corned beef brisket (0.42 ± 0.005), eggnog (0.42 ± 0.002), restaurant vanilla milkshake (0.42 ± 0.002), white bread (0.42 ± 0.021), boiled lima beans (0.42 ± 0.003), simmered beef shank excluding fat (0.41 ± 0.005), broiled beef round excluding fat (0.41 ± 0.005), raw crayfish (0.41 ± 0.006), raw swordfish (0.41 ± 0.006), sheep milk (0.41 ± 0.002), dried and frozen tofu (0.41 ± 0.029), boiled mung beans (0.41 ± 0.002), boiled black-eyed peas (0.41 ± 0.003), boiled chayote (0.41 ± 0.006), roasted beef tenderloin including fat (0.4 ± 0.005), braised beef bottom round including fat (0.4 ± 0.005), roasted lean ham (0.4 ± 0.005), devil's food cake (0.4 ± 0.008), turkey salad (0.4 ± 0.009), boiled yardlong bean (0.4 ± 0.003), dried sweetened flaked coconut (0.4 ± 0.02), canned pumpkin (0.4 ± 0.004), italian bread (0.39 ± 0.018), broiled beef sirloin excluding fat (0.39 ± 0.005), pork and beef luncheon sausage (0.39 ± 0.022), canned ham (0.39 ± 0.005), whole yogurt (0.39 ± 0.002), restaurant chocolate milkshake (0.39 ± 0.002), au gratin potatoes (0.39 ± 0.004), potato chips (0.39 ± 0.018), boiled catjang black-eyed peas (0.39 ± 0.003), raw turnip greens (0.39 ± 0.018), black currants (0.39 ± 0.009), simmered beef shank including fat (0.38 ± 0.005), braised beef flank excluding fat (0.38 ± 0.005), braised beef chuck arm pot roast excluding fat (0.38 ± 0.005), braised beef flank including fat (0.38 ± 0.005), broiled beef rib excluding fat (0.38 ± 0.005), well done fried ground beef (0.38 ± 0.005), broiled beef tenderloin excluding fat (0.38 ± 0.005), medium-rare broiled lean ground beef (0.38 ± 0.005), protein fortified lowfat 2% milk (0.38 ± 0.002), protein fortified skim milk (0.38 ± 0.002), whey (0.38 ± 0.002), tamari sauce (0.38 ± 0.009), sweet roll (0.38 ± 0.012), raw potato without skin (0.38 ± 0.004), raw sprouted mung beans (0.38 ± 0.01), well done broiled ground beef (0.37 ± 0.005), fried beef sirloin including fat (0.37 ± 0.005), broiled beef top loin excluding fat (0.37 ± 0.005), protein fortified lowfat 1% milk (0.37 ± 0.002), oatmeal bread (0.36 ± 0.018), gouda cheese (0.36 ± 0.018), raw clams (0.36 ± 0.006), broiled beef round including fat (0.36 ± 0.005), well done fried lean ground beef (0.36 ± 0.005), braised beef brisket excluding fat (0.36 ± 0.005), tilsit cheese (0.36 ± 0.018), beef and pork frankfurter (0.36 ± 0.011), raw perch (0.36 ± 0.006), tempeh (0.36 ± 0.006), chocolate milkshake (0.36 ± 0.002), cornbread (0.36 ± 0.006), corn pone (0.36 ± 0.006), bagels (0.36 ± 0.009), puffed rice (0.36 ± 0.036), boiled butternut squash (0.36 ± 0.005), canned potato without skin (0.36 ± 0.006), boiled/baked spaghetti squash (0.36 ± 0.006), raw pollock (0.35 ± 0.006), broiled beef sirloin including fat (0.35 ± 0.005), roasted beef rib including fat (0.35 ± 0.005), braised beef chuck blade roast excluding fat (0.35 ± 0.005), broiled beef tenderloin including fat (0.35 ± 0.005), medium-rare broiled extra lean ground beef (0.35 ± 0.005), beef beerwurst salami (0.35 ± 0.022), well done baked extra lean ground beef (0.35 ± 0.005), raw bacon (0.35 ± 0.007), cheese pizza (0.35 ± 0.004), pepperoni pizza (0.35 ± 0.004), frozen and boiled brussels sprouts (0.35 ± 0.006), braised beef shortribs excluding fat (0.34 ± 0.005), broiled beef rib eye excluding fat (0.34 ± 0.005), medium-rare fried ground beef (0.34 ± 0.005), broiled porterhouse steak excluding fat (0.34 ± 0.005), well done baked lean ground beef (0.34 ± 0.005), strawberries (0.34 ± 0.003), medium-rare broiled ground beef (0.33 ± 0.005), broiled T-bone steak excluding fat (0.33 ± 0.005), broiled beef top loin including fat (0.33 ± 0.005), raw halibut (0.33 ± 0.006), skim milk (0.33 ± 0.002), sour cream (0.33 ± 0.042), pineapple upside-down cake (0.33 ± 0.007), saltine crackers (0.33 ± 0.083), shoyu sauce (0.33 ± 0.009), raw spirulina seaweed (0.33 ± 0.005), boiled beet greens (0.33 ± 0.007), raw chives (0.33 ± 0.167), lemon peel (0.33 ± 0.083), braised beef chuck arm pot roast including fat (0.32 ± 0.005), medium-rare fried lean ground beef (0.32 ± 0.005),

VITAMIN B$_5$

pork and beef knockwurst (0.32 ± 0.007), pork bratwurst (0.32 ± 0.006), ham salad (0.32 ± 0.018), raw ling (0.32 ± 0.006), lowfat 1% milk (0.32 ± 0.002), lowfat 2% milk (0.32 ± 0.002), whole milk (0.32 ± 0.002), hot chocolate (0.32 ± 0.002), boiled pigeon peas (0.32 ± 0.003), raw new zealand spinach (0.32 ± 0.018), smoked cisco (0.31 ± 0.006), broiled beef rib eye including fat (0.31 ± 0.005), broiled beef rib including fat (0.31 ± 0.005), broiled porterhouse steak including fat (0.31 ± 0.005), unheated ham patties (0.31 ± 0.008), hamburger sandwich (0.31 ± 0.005), goat milk (0.31 ± 0.002), tomato sauce (0.31 ± 0.004), instant corn grits (0.31 ± 0.004), baked/boiled tropical yam (0.31 ± 0.007), canned succotash (0.31 ± 0.004), boiled carrots (0.31 ± 0.006), raw coconut (0.31 ± 0.011), raw acerola (0.31 ± 0.005), raw red cabbage (0.31 ± 0.014), braised beef chuck blade roast including fat (0.3 ± 0.005), beef frankfurter (0.3 ± 0.009), carob milk (0.3 ± 0.002), lowfat 1% chocolate milk (0.3 ± 0.002), lowfat 2% chocolate milk (0.3 ± 0.002), whole chocolate milk (0.3 ± 0.002), filled milk (0.3 ± 0.002), raw parsley (0.3 ± 0.017), boiled pink beans (0.3 ± 0.003), boiled acorn squash (0.3 ± 0.004), figs (0.3 ± 0.01), boiled hubbard squash (0.3 ± 0.004), canned chickpeas (0.3 ± 0.002), braised beef brisket including fat (0.29 ± 0.005), broiled T-bone steak including fat (0.29 ± 0.005), edam cheese (0.29 ± 0.018), brick cheese (0.29 ± 0.018), chopped ham lunch meat (0.29 ± 0.024), havarti cheese (0.29 ± 0.018), cream cheese (0.29 ± 0.018), chicken and turkey salad (0.29 ± 0.018), malted milk (0.29 ± 0.002), whole chocolate malted milk (0.29 ± 0.002), hush puppies (0.29 ± 0.011), lemon meringue pie (0.29 ± 0.004), gingersnaps cookie (0.29 ± 0.071), pretzels (0.29 ± 0.018), dry roasted macadamia nuts (0.29 ± 0.018), falafel (0.29 ± 0.01), boiled broccoli (0.29 ± 0.006), boiled red tomato (0.29 ± 0.004), boiled chickpeas (0.29 ± 0.003), boiled pinto beans (0.29 ± 0.003), hummus (0.29 ± 0.002), gooseberries (0.29 ± 0.003), raw watercress (0.29 ± 0.029), well done fried extra lean ground beef (0.29 ± 0.005), cooked lobster (0.28 ± 0.006), raw shrimp (0.28 ± 0.006), raw cusk (0.28 ± 0.006), white cake (0.28 ± 0.006), soy sauce (0.28 ± 0.009), frozen and boiled potatoes without skin (0.28 ± 0.005), canned great northern beans (0.28 ± 0.002), boiled turnip greens (0.28 ± 0.007), white grapefruit (0.28 ± 0.004), pink grapefruit (0.28 ± 0.004), well done baked ground beef (0.27 ± 0.005), medium-rare baked lean ground beef (0.27 ± 0.005), medium-rare baked extra lean ground beef (0.27 ± 0.005), grilled ham patties (0.27 ± 0.008), half and half cream (0.27 ± 0.033), buttermilk (0.27 ± 0.002), whipping cream, heavy (0.27 ± 0.033), whipping cream, light (0.27 ± 0.033), medium cream (0.27 ± 0.033), boiled great northern beans (0.27 ± 0.003), frozen and boiled broccoli (0.27 ± 0.005), frozen and boiled zucchini (0.27 ± 0.004), beef bologna (0.26 ± 0.022), beef and pork bologna (0.26 ± 0.022), enchilada (0.26 ± 0.002), miso (0.26 ± 0.004), boiled brussels sprouts (0.26 ± 0.006), boiled black turtle beans (0.26 ± 0.003), bananas (0.26 ± 0.004), boiled new zealand spinach (0.26 ± 0.006), canned lima beans (0.26 ± 0.002), braised beef shortribs including fat (0.25 ± 0.005), medium-rare fried extra lean ground beef (0.25 ± 0.005), raw pork jowl (0.25 ± 0.005), strawberry sundae (0.25 ± 0.003), yellow corn pudding (0.25 ± 0.002), oil roasted almonds (0.25 ± 0.018), almond butter (0.25 ± 0.031), dry roasted almonds (0.25 ± 0.018), toasted almonds (0.25 ± 0.018), shortbread cookie (0.25 ± 0.063), dried brazil nuts (0.25 ± 0.018), soursop (0.25 ± 0.002), boysenberries (0.25 ± 0.004), restaurant tomato juice (0.25 ± 0.004), tomato juice (0.25 ± 0.003), boiled navy beans (0.25 ± 0.003), raw cucumber (0.25 ± 0.01), navel orange

(0.25 ± 0.004), sapodilla (0.25 ± 0.003), valencia orange (0.25 ± 0.004), raw eel (0.24 ± 0.006), cheeseburger (0.24 ± 0.004), caramel sundae (0.24 ± 0.003), onion rings (0.24 ± 0.006), chicken egg white (0.24 ± 0.015), frozen and boiled succotash (0.24 ± 0.006), raspberries (0.24 ± 0.004), frozen and boiled okra (0.24 ± 0.005), cooked sprouted mung beans (0.24 ± 0.008), loganberries (0.24 ± 0.003), raw red tomato (0.24 ± 0.004), apricots (0.24 ± 0.005), boiled black beans (0.24 ± 0.003), boiled cranberry beans (0.24 ± 0.003), blackberries (0.24 ± 0.007), angel food cake (0.23 ± 0.008), human milk (0.23 ± 0.016), sugar apple (0.23 ± 0.003), cooked plantain (0.23 ± 0.003), boiled yellow beans (0.23 ± 0.003), carrot juice (0.23 ± 0.003), boiled red cabbage (0.23 ± 0.007), boiled white beans (0.23 ± 0.003), medium-rare baked ground beef (0.22 ± 0.005), toaster pastry (0.22 ± 0.01), frozen and boiled yellow corn (0.22 ± 0.006), teriyaki sauce (0.22 ± 0.028), boiled kidney beans (0.22 ± 0.003), boiled french beans (0.22 ± 0.003), canned succotash with creamed corn (0.22 ± 0.004), papaya (0.22 ± 0.002), lime (0.22 ± 0.007), cranberry sauce (0.22 ± 0.005), raw whelk (0.21 ± 0.006), raw whiting (0.21 ± 0.006), port du salut cheese (0.21 ± 0.018), pork headcheese (0.21 ± 0.018), colby cheese (0.21 ± 0.018), cottage cheese (0.21 ± 0.018), hot fudge sundae (0.21 ± 0.003), raw japanese chestnuts (0.21 ± 0.018), boiled okra (0.21 ± 0.006), raw ginger root (0.21 ± 0.021), frozen and boiled black-eyed peas (0.21 ± 0.006), raw looseleaf lettuce (0.21 ± 0.018), acerola juice (0.21 ± 0.002), watermelon (0.21 ± 0.003), soybean oil and cottonseed oil margarine liquid (0.2 ± 0.1), corn tortilla (0.2 ± 0.017), instant oatmeal (0.2 ± 0.003), boiled artichoke (0.2 ± 0.006), boiled artichoke hearts (0.2 ± 0.006), tangerine (0.2 ± 0.006), oatmeal (0.2 ± 0.003), raw oysters (0.19 ± 0.006), breaded and fried fish (0.19 ± 0.007), minced ham lunch meat (0.19 ± 0.024), indian buffalo milk (0.19 ± 0.002), raw carrots (0.19 ± 0.007), canned black-eyed peas (0.19 ± 0.002), canned white beans (0.19 ± 0.002), lemon (0.19 ± 0.009), orange juice (0.19 ± 0.002), muenster cheese (0.18 ± 0.018), jellied corned beef loaf (0.18 ± 0.018), bacon cheeseburger (0.18 ± 0.003), burrito (0.18 ± 0.003), caraway cheese (0.18 ± 0.018), corn flakes (0.18 ± 0.018), ginko nuts (0.18 ± 0.018), unsweetened baking chocolate (0.18 ± 0.018), canned creamed yellow corn (0.18 ± 0.004), boiled and frozen baby lima beans (0.18 ± 0.006), canned black turtle beans (0.18 ± 0.002), plum (0.18 ± 0.008), boiled soybeans (0.18 ± 0.003), raw celery (0.18 ± 0.013), boiled asparagus (0.17 ± 0.006), frozen and boiled asparagus (0.17 ± 0.008), peach (0.17 ± 0.006), canned peeled red tomatoes (0.17 ± 0.004), canned navy beans (0.17 ± 0.002), noodles (0.16 ± 0.003), guava (0.16 ± 0.006), frozen and boiled fordhook lima beans (0.16 ± 0.006), nectarine (0.16 ± 0.004), boiled broad beans (0.16 ± 0.003), mango (0.16 ± 0.002), pineapple (0.16 ± 0.003), frozen and boiled butternut squash (0.16 ± 0.004), bloody mary (0.16 ± 0.003), frozen and boiled carrots (0.16 ± 0.007), boiled swiss chard (0.16 ± 0.006), orange juice (0.16 ± 0.004), raw cod (0.15 ± 0.006), restaurant coleslaw (0.15 ± 0.005), boiled green peas (0.15 ± 0.006), boiled celery (0.15 ± 0.007), fried tofu (0.15 ± 0.038), raw scallops (0.14 ± 0.006), maypo (0.14 ± 0.003), taco (0.14 ± 0.006), mayonnaise (0.14 ± 0.036), tamarind (0.14 ± 0.004), frozen and boiled green peas (0.14 ± 0.006), golden seedless raisins (0.14 ± 0.005), apple pie (0.14 ± 0.004), raw cauliflower (0.14 ± 0.01), boiled rutabaga (0.14 ± 0.006), canned kidney beans (0.14 ± 0.002), boiled spinach (0.14 ± 0.006), elderberries (0.14 ± 0.003), canned cranberry beans (0.14 ± 0.002), canned carrots (0.14 ± 0.007), canned pinto beans (0.14 ± 0.002), raw green cabbage (0.14 ± 0.014), boiled turnips (0.14 ± 0.006), raw spring onions with tops (0.14 ± 0.01), raw onions (0.14 ± 0.006), lime juice (0.14 ± 0.002), raw haddock (0.13 ± 0.006), braised pork boston blade braised including fat (0.13 ± 0.005), coleslaw (0.13 ± 0.008), canned green peas (0.13 ± 0.006), boiled crookneck squash (0.13 ± 0.006), cantaloupe (0.13 ± 0.003), boiled cauliflower (0.13 ± 0.008), firm raw tofu (0.13 ± 0.004), cherries (0.13 ± 0.007), screwdriver (0.13 ± 0.002), restaurant hot chocolate (0.12 ± 0.002), canned broad beans (0.12 ± 0.002), frozen and boiled collards (0.12 ± 0.006), boiled onions (0.12 ± 0.005), smoked whitefish (0.11 ± 0.006), pancakes with butter and syrup (0.11 ± 0.003), cooked wild long grain rice (0.11 ± 0.004), whole dried sesame seeds (0.11 ± 0.056), macaroni (0.11 ± 0.004), spaghetti (0.11 ± 0.004), semi-sweet bakers chocolate (0.11 ± 0.018), raw scallop squash (0.11 ± 0.008), raw crookneck squash (0.11 ± 0.008), boiled mustard greens (0.11 ± 0.007), boiled zucchini (0.11 ± 0.006), baked beans with pork and sweet sauce (0.1 ± 0.002), raw green peas (0.1 ± 0.006), cooked prunes (0.1 ± 0.005), frozen and boiled crookneck squash (0.1 ± 0.005), mammy apple (0.1 ± 0.005), blueberries (0.1 ± 0.003), frozen and boiled cauliflower (0.1 ± 0.006), pineapple juice (0.1 ± 0.002), lemon juice (0.1 ± 0.002), maltex (0.09 ± 0.003), vegetarian baked beans (0.09 ± 0.002), boiled beets (0.09 ± 0.006), canned sauerkraut (0.09 ± 0.004), quince (0.09 ± 0.005), raw radish (0.09 ± 0.011), boiled green beans (0.08 ± 0.008), boiled yellow snap beans (0.08 ± 0.008), boiled eggplant (0.08 ± 0.01), raw zucchini (0.08 ± 0.008), frozen and boiled spinach (0.08 ± 0.005), boiled scallop squash (0.08 ± 0.006), frozen and boiled onions (0.08 ± 0.005), cream of wheat (0.07 ± 0.003), pickled herring (0.07 ± 0.033), mozzarella cheese (0.07 ± 0.018), light mayonnaise (0.07 ± 0.036), soybean protein isolate (0.07 ± 0.018), raw red chili peppers (0.07 ± 0.011), raw green chili peppers (0.07 ± 0.011), soybean protein concentrate (0.07 ± 0.018), boiled and steamed japanese chestnuts (0.07 ± 0.018), raw spinach (0.07 ± 0.018), raw eggplant (0.07 ± 0.012), frozen and boiled turnip greens (0.07 ± 0.006), german sweet bakers chocolate (0.07 ± 0.018), boiled green cabbage (0.07 ± 0.007), red currants (0.07 ± 0.009), white currants (0.07 ± 0.009), raw frozen rhubarb (0.07 ± 0.004), pear (0.07 ± 0.003), cranberry juice cocktail (0.07 ± 0.002), pork and beef brotwurst (0.06 ± 0.007), beer (0.06 ± 0.001), frozen and boiled yellow snap beans (0.06 ± 0.007), frozen and boiled green beans (0.06 ± 0.007), raw tofu (0.06 ± 0.004), papaya nectar (0.06 ± 0.002), raw apple with skin (0.06 ± 0.004), farina (0.06 ± 0.003), restaurant cranberry juice cocktail (0.06 ± 0.003), soy milk (0.05 ± 0.002), carob flour (0.05 ± 0.005), seedless raisins (0.05 ± 0.005), boiled scotch kale (0.05 ± 0.008), frozen and boiled kale (0.05 ± 0.008), boiled kale (0.05 ± 0.008), canned crookneck squash (0.05 ± 0.005), canned spinach (0.05 ± 0.005), frozen and cooked sweetened rhubarb (0.05 ± 0.004), applesauce (0.05 ± 0.004), raw apple without skin (0.05 ± 0.004), cooked apple without skin (0.05 ± 0.003), raw iceberg lettuce (0.05 ± 0.025), red table wine (0.04 ± 0.005), raw red sweet peppers (0.04 ± 0.01), raw green sweet peppers (0.04 ± 0.01), grape juice (0.04 ± 0.002), canned turnip greens (0.04 ± 0.004), boiled collards (0.04 ± 0.003), whiskey sour (0.04 ± 0.006), coconut water (0.04 ± 0.002), rose table wine (0.03 ± 0.005), european grapes (0.03 ± 0.003), thompson seedless grapes (0.03 ± 0.003), boiled red sweet peppers (0.03 ± 0.007), boiled green sweet peppers (0.03 ± 0.007), chocolate syrup (0.03 ± 0.013), dry dessert wine (0.03 ± 0.008), sweet dessert wine (0.03 ±

0.008), tuna salad spread (0.02 ± 0.009), frozen and boiled red sweet peppers (0.02 ± 0.005), frozen and boiled green sweet peppers (0.02 ± 0.005), american grapes (0.02 ± 0.005), manhattan (0.02 ± 0.009), white table wine (0.02 ± 0.005), daiquiri (0.02 ± 0.008), frozen and boiled mustard greens (0.01 ± 0.007), chamomile tea (0.01 ± 0.003), instant tea (0.01 ± 0.002)

Applications: anemia, blood sugar decreased, bladder inflammation, fracture, diarrhea, epilepsy, dizziness, lack of sleep, mental illness, multiple sclerosis, neuritis, cataracts, burning and tingling sensations, adrenal deterioration, baldness, headache, worms, arthritis, gout, leg cramp, veins deterioration, allergies, asthma, tuberculosis, muscular dystrophy, acne, skin psoriasis, gastritis, digestion upset, nausea, alcohol cravings, depression, fatigue, infection, retarded growth, stress, cancer, antioxidant therapy, sun sensitivity, free radical oxidation, life span extension, radiation, decreased radiation resistance, aging

Daily Dosage (at least 1 mg vitamin B_C and 100 mg vitamin B_W per every 300 mg vitamin B_5): 1974 PCS — 5–15 mg; 1980 RDA — none; 1980 SADDI — 4–7 mg; 1980 USRDA — none; 1989 RDA — none; 1989 SADDI — 4–7 mg; 1989 USRDA — none; nutritional — 20–50 mg; therapeutic — 500–1000 mg; experimental — 3000 mg; toxic — 10,000 mg +

Warnings: Acidity should be countered with antacid. Increased amounts are dangerous. See toxicity symptoms.

VITAMIN B$_5$ CALCIUM SALT

(Specific form of vitamin B5)

Names: vitamin B_5 calcium salt, calcium pantothenate, pantothenic acid calcium salt, Calpanate, Galamila, Pantholin

VITAMIN B$_5$ SODIUM SALT

(Specific form of vitamin B_5)

Names: vitamin B_5 sodium salt, sodium pantothenate, Panthoject

---

VITAMIN B$_6$

Names: vitamin B_6, adermin, yeast eluate factor, factor Y, vitamin Y
Classification: water-soluble vitamin, vitamin B complex, antioxidant, coenzyme factor
Forms: pyridoxine, pyridoxamine, pyridoxal
Deficiency (Rebound Deficiency can occur if large doses are halted abruptly.): convulsions, abnormal brain waves, confusion, mental retardation in infants, mental health deterioration, nervousness, lack of sleep, deterioration of dream recall, nervous system deterioration, nerve pain, pain sensitivity, muscle pain, numbness, temporary paralysis of limb, cramps, tendon/ligament difficulties, increased connective tissue, arthritis, unnecessary swellings, water retention, increased urination, increased xanthurenic acid in urine, increased vitamin C in urine, nausea, motion sickness, anemia, blood sugar decreased, decreased glucose tolerance, fatigue, depression, stress, emotional agitation, deterioration of protein synthesis, decreased protein supply to brain, cracks around mouth, cracks around eyes, glossy tongue, scaling lips, inflamed gums, mouth sores, immunity decreased, greasy/cheesy oozing of skin, oily hair, skin inflammation, dandruff, hair decreased, visual deterioration, decreased hemoglobin production, heart disease, increased cholesterol, atherosclerosis, coenzyme pyridoxal phosphate levels decreased, decreased lymph cell count
Side-effects: increased dream recall
Toxicity: decreased feeling in extremities, suppressed lactation, emotional agitation

VITAMIN B$_6$

Inhibitors: long storage, canning, roasting or stewing of meat, water, food processing, alcohol, estrogen, alkalis, ultraviolet light, penicillamine, cortisone, prednisone, Isoniazid/INH

Helpers: vitamin B$_1$, vitamin B$_2$, vitamin B$_5$, vitamin F, vitamin C, magnesium, potassium, sodium

Sources (mg per 100 grams of food): freeze-dried red sweet peppers (2.5 ± 0.313), freeze-dried green sweet peppers (2.5 ± 0.313), dried dill weed (2 ± 0.5), wheat flakes (1.79 ± 0.018), bran flakes (1.79 ± 0.018), bran (1.79 ± 0.018), corn flakes (1.79 ± 0.018), dehydrated onion flakes (1.57 ± 0.036), freeze-dried shallots (1.5 ± 0.125), fried beef liver (1.43 ± 0.005), freeze-dried parsley (1.43 ± 0.357), corn germ (1.41 ± 0.005), raisin bran (1.35 ± 0.014), dried chervil (1 ± 0.5), dried parsley (1 ± 0.5), toasted wheat germ (1 ± 0.018), restaurant hot chocolate (1 ± 0.002), braised beef liver (0.91 ± 0.005), dried and salted cod (0.86 ± 0.006), whole dried sesame seeds (0.78 ± 0.056), raw goose liver (0.77 ± 0.005), raw pheasant without skin (0.74 ± 0.005), dried european chestnuts (0.68 ± 0.018), cooked whelk (0.65 ± 0.006), fried chicken breast without skin (0.64 ± 0.006), fried light meat chicken without skin (0.63 ± 0.005), fried chicken giblets (0.61 ± 0.005), fried beef top round excluding fat (0.61 ± 0.005), dried filberts (0.61 ± 0.018), hard pork salami (0.6 ± 0.05), roasted light meat chicken without skin (0.6 ± 0.005), roasted chicken breast without skin (0.59 ± 0.006), simmered chicken liver (0.58 ± 0.005), fried chicken breast with skin (0.58 ± 0.005), bananas (0.58 ± 0.004), braised pork liver (0.57 ± 0.005), defatted soybean flour (0.57 ± 0.005), defatted soy meal (0.57 ± 0.004), dried english/persian walnuts (0.57 ± 0.018), broiled beef top round excluding fat (0.56 ± 0.005), fried beef top round including fat (0.55 ± 0.005), roasted chicken breast with skin (0.55 ± 0.005), broiled pork sirloin excluding fat (0.54 ± 0.005), broiled beef top round including fat (0.54 ± 0.005), fried light meat chicken with skin (0.54 ± 0.005), roasted turkey light meat without skin (0.54 ± 0.005), braised pork sirloin excluding fat (0.53 ± 0.005), simmered beef kidneys (0.52 ± 0.005), simmered turkey liver (0.52 ± 0.005), low fat soybean flour (0.52 ± 0.006), roasted light meat chicken with skin (0.52 ± 0.005), fried beef sirloin excluding fat (0.5 ± 0.005), fried pork center loin excluding fat (0.5 ± 0.005), hard pork and beef salami (0.5 ± 0.05), broiled beef round excluding fat (0.5 ± 0.005), roasted european chestnuts (0.5 ± 0.018), potato chips (0.5 ± 0.018), cooked mullet (0.49 ± 0.006), canned ham (0.48 ± 0.005), canned chickpeas (0.48 ± 0.002), pork livercheese (0.47 ± 0.013), roasted goose without skin (0.47 ± 0.005), broiled pork center loin excluding fat (0.47 ± 0.005), roasted turkey light meat with skin (0.47 ± 0.005), braised pork kidneys (0.46 ± 0.005), soybean flour (0.46 ± 0.006), tomato powder (0.46 ± 0.005), broiled pork loin excluding fat (0.46 ± 0.005), cooked mackerel (0.46 ± 0.006), roasted pork shank excluding fat (0.46 ± 0.005), unheated lean ham (0.46 ± 0.005), braised pork loin excluding fat (0.45 ± 0.005), roasted pork loin excluding fat (0.45 ± 0.005), roasted pork leg excluding fat (0.45 ± 0.005), broiled beef sirloin excluding fat (0.45 ± 0.005), braised pork center loin excluding fat (0.45 ± 0.005), roasted pork center loin excluding fat (0.45 ± 0.005), roasted canned lean ham (0.45 ± 0.005), canned lean ham (0.45 ± 0.005), grilled canadian bacon (0.45 ± 0.011), broiled pork sirloin including fat (0.44 ± 0.005), fried pork center rib excluding fat (0.44 ± 0.005), fried pork top loin excluding fat (0.44 ± 0.005), roasted pork loin blade excluding fat (0.44 ± 0.005), broiled beef tenderloin excluding fat (0.44 ± 0.005), broiled beef round including fat (0.44 ± 0.005), crunchy peanut butter (0.44 ± 0.016), ry-krisp crackers

(0.43 ± 0.036), broiled pork loin blade excluding fat (0.43 ± 0.005), kippered herring (0.43 ± 0.013), raw mullet (0.42 ± 0.006), roasted pork tenderloin (0.42 ± 0.005), roasted pork sirloin excluding fat (0.42 ± 0.005), fried beef sirloin including fat (0.42 ± 0.005), broiled beef top loin excluding fat (0.42 ± 0.005), broiled beef flank including fat (0.42 ± 0.005), broiled beef flank excluding fat (0.42 ± 0.005), instant oatmeal (0.42 ± 0.003), raw milkfish (0.42 ± 0.006), braised pork arm excluding fat (0.41 ± 0.005), roasted pork arm excluding fat (0.41 ± 0.005), braised pork loin blade excluding fat (0.41 ± 0.005), fried chicken wing with skin (0.41 ± 0.016), roasted chicken wing with skin (0.41 ± 0.015), fried pork loin blade excluding fat (0.4 ± 0.005), roasted pork shoulder excluding fat (0.4 ± 0.005), raw mackerel (0.4 ± 0.006), broiled pork center rib excluding fat (0.4 ± 0.005), broiled pork top loin excluding fat (0.4 ± 0.005), roasted pork center rib excluding fat (0.4 ± 0.005), roasted pork top loin excluding fat (0.4 ± 0.005), roasted beef tip excluding fat (0.4 ± 0.005), broiled beef tenderloin including fat (0.4 ± 0.005), broiled pork center loin including fat (0.4 ± 0.005), broiled beef sirloin including fat (0.4 ± 0.005), broiled porterhouse steak excluding fat (0.4 ± 0.005), roasted pork center loin including fat (0.4 ± 0.005), broiled beef rib eye excluding fat (0.4 ± 0.005), roasted lean ham (0.4 ± 0.005), cooked halibut (0.4 ± 0.006), raw bluefish (0.4 ± 0.006), smoked haddock (0.4 ± 0.006), hummus (0.4 ± 0.002), braised pork heart (0.39 ± 0.004), maypo (0.39 ± 0.003), braised pork top loin excluding fat (0.39 ± 0.005), braised pork center rib excluding fat (0.39 ± 0.005), roasted pork leg including fat (0.39 ± 0.005), roasted pork shank including fat (0.39 ± 0.005), fried pork center loin including fat (0.39 ± 0.005), fried beef brains (0.39 ± 0.005), broiled T-bone steak excluding fat (0.39 ± 0.005), stewed chicken drumstick without skin (0.39 ± 0.011), raw european chestnuts (0.39 ± 0.018), unheated canadian bacon (0.39 ± 0.009), raw cusk (0.39 ± 0.006), smoked whitefish (0.39 ± 0.006), oil roasted peanuts (0.39 ± 0.018), broiled pork loin including fat (0.38 ± 0.005), toaster pastry (0.38 ± 0.01), roasted beef tenderloin excluding fat (0.38 ± 0.005), roasted pork loin including fat (0.38 ± 0.005), roasted pork sirloin including fat (0.38 ± 0.005), braised pork center loin including fat (0.38 ± 0.005), tomato paste (0.38 ± 0.004), roasted eye of beef round excluding fat (0.38 ± 0.005), creamy peanut butter (0.38 ± 0.031), potato pancakes (0.38 ± 0.007), cooked swordfish (0.38 ± 0.006), dry skim milk (0.37 ± 0.017), carob flour (0.37 ± 0.005), roasted goose with skin (0.37 ± 0.005), roasted beef round tip including fat (0.37 ± 0.005), fried dark meat chicken without skin (0.37 ± 0.005), roasted cured pork arm excluding fat (0.37 ± 0.005), simmered beef shank excluding fat (0.37 ± 0.005), broiled beef top loin including fat (0.37 ± 0.005), dried spirulina seaweed (0.36 ± 0.005), braised pork loin including fat (0.36 ± 0.005), roasted pork loin blade including fat (0.36 ± 0.005), braised pork sirloin including fat (0.36 ± 0.005), braised beef bottom round excluding fat (0.36 ± 0.005), roasted turkey dark meat without skin (0.36 ± 0.005), roasted dark meat chicken without skin (0.36 ± 0.005), broiled beef rib eye including fat (0.36 ± 0.005), braised beef flank excluding fat (0.36 ± 0.005), chopped smoked beef (0.36 ± 0.018), roasted soybean flour (0.35 ± 0.006), braised pork spareribs (0.35 ± 0.005), cooked herring (0.35 ± 0.006), roasted pork center rib including fat (0.35 ± 0.005), pork link sausage (0.35 ± 0.007), roasted chicken thigh without skin (0.35 ± 0.01), fried chicken drumstick with skin (0.35 ± 0.01), broiled porterhouse steak including fat (0.35 ± 0.005), roasted chicken drumstick with skin (0.35 ± 0.01), broiled beef rib excluding fat (0.35 ±

0.005), braised beef flank including fat (0.35 ± 0.005), pork beerwurst salami (0.35 ± 0.022), roasted eye of beef round including fat (0.35 ± 0.005), microwaved potato with skin (0.35 ± 0.002), baked potato with skin (0.35 ± 0.002), dry instant skim milk (0.34 ± 0.005), simmered chicken giblets (0.34 ± 0.005), broiled pork loin blade including fat (0.34 ± 0.005), roasted beef tenderloin including fat (0.34 ± 0.005), broiled pork center rib including fat (0.34 ± 0.005), broiled pork top loin including fat (0.34 ± 0.005), roasted pork top loin including fat (0.34 ± 0.005), braised beef bottom round including fat (0.34 ± 0.005), unheated ham (0.34 ± 0.005), fried chicken thigh with skin (0.34 ± 0.008), simmered beef shank including fat (0.34 ± 0.005), stewed chicken breast without skin (0.34 ± 0.005), raw whelk (0.34 ± 0.006), raw halibut (0.34 ± 0.006), cooked haddock (0.34 ± 0.006), pork liver sausage (0.33 ± 0.028), simmered turkey giblets (0.33 ± 0.005), poppy seed (0.33 ± 0.167), raw chives (0.33 ± 0.167), roasted pork shoulder including fat (0.33 ± 0.005), braised beef chuck arm pot roast excluding fat (0.33 ± 0.005), fried pork center rib including fat (0.33 ± 0.005), fried pork top loin including fat (0.33 ± 0.005), pork sausage (0.33 ± 0.019), italian pork sausage (0.33 ± 0.007), broiled T-bone steak including fat (0.33 ± 0.005), chopped ham lunch meat (0.33 ± 0.024), stewed light meat chicken without skin (0.33 ± 0.005), turkey breast meat summer sausage (0.33 ± 0.024), raw swordfish (0.33 ± 0.006), simmered turkey heart (0.32 ± 0.005), simmered chicken heart (0.32 ± 0.005), well done broiled extra lean ground beef (0.32 ± 0.005), pork luxury loaf (0.32 ± 0.018), braised pork loin blade including fat (0.32 ± 0.005), braised pork center rib including fat (0.32 ± 0.005), pork and beef honey loaf (0.32 ± 0.018), pork and beef picnic loaf (0.32 ± 0.018), well done fried lean ground beef (0.32 ± 0.005), fried dark meat chicken with skin (0.32 ± 0.005), roasted turkey dark meat with skin (0.32 ± 0.005), golden seedless raisins (0.32 ± 0.005), toasted soybean nuts (0.32 ± 0.018), dried coconut (0.32 ± 0.018), microwaved potato without skin (0.32 ± 0.003), dry whole milk (0.31 ± 0.016), broiled pork boston blade excluding fat (0.31 ± 0.005), roasted ham (0.31 ± 0.005), well done fried extra lean ground beef (0.31 ± 0.005), braised pork top loin including fat (0.31 ± 0.005), raw herring (0.31 ± 0.006), fried chicken back with skin (0.31 ± 0.007), roasted chicken thigh with skin (0.31 ± 0.008), roasted dark meat chicken with skin (0.31 ± 0.005), raw ling (0.31 ± 0.006), fried chicken (0.31 ± 0.006), raw haddock (0.31 ± 0.006), corn tortilla (0.3 ± 0.017), roasted pork boston blade excluding fat (0.3 ± 0.005), roasted pork rump excluding fat (0.3 ± 0.005), beef and pork summer sausage (0.3 ± 0.022), roasted canned ham (0.3 ± 0.005), well done broiled lean ground beef (0.3 ± 0.005), braised beef brisket excluding fat (0.3 ± 0.005), well done broiled ground beef (0.3 ± 0.005), roasted beef rib roasted excluding fat (0.3 ± 0.005), tempeh (0.3 ± 0.006), baked potato without skin (0.3 ± 0.003), dry buttermilk (0.29 ± 0.071), chicken egg yolk (0.29 ± 0.029), pork and beef peppered loaf (0.29 ± 0.018), well done baked extra lean ground beef (0.29 ± 0.005), dried and frozen tofu (0.29 ± 0.029), fried pork loin blade including fat (0.29 ± 0.005), braised beef chuck blade roast excluding fat (0.29 ± 0.005), well done baked ground beef (0.29 ± 0.005), minced ham lunch meat (0.29 ± 0.024), broiled beef rib including fat (0.29 ± 0.005), dried peanuts (0.29 ± 0.018), frozen and boiled brussels sprouts (0.29 ± 0.006), stewed chicken breast with skin (0.29 ± 0.005), raw red chili peppers (0.29 ± 0.011), raw green chili peppers (0.29 ± 0.011), roasted pork arm including fat (0.28 ± 0.005), bran muffins (0.28 ± 0.013), cooked bacon (0.28 ± 0.004), turkey pastrami (0.28 ± 0.009), braised beef chuck arm pot roast including fat (0.28 ± 0.005), medium-rare fried lean ground beef (0.28 ± 0.005), braised beef short-ribs excluding fat (0.28 ± 0.005), raw pollock (0.28 ± 0.006), roasted cured pork arm including fat (0.28 ± 0.005), raw california avocado (0.28 ± 0.003), raw florida avocado (0.28 ± 0.002), smoked chinook salmon (0.28 ± 0.006), cooked clams (0.28 ± 0.006), canned cod (0.28 ± 0.006), cooked cod (0.28 ± 0.006), hash brown potatoes (0.28 ± 0.006), braised pork boston blade excluding fat (0.27 ± 0.005), broiled pork boston blade including fat (0.27 ± 0.005), roasted pork rump including fat (0.27 ± 0.005), roasted pork boston blade including fat (0.27 ± 0.005), braised pork arm braised including fat (0.27 ± 0.005), medium-rare broiled extra lean ground beef (0.27 ± 0.005), medium-rare fried extra lean ground beef (0.27 ± 0.005), well done fried ground beef (0.27 ± 0.005), medium-rare broiled ground beef (0.27 ± 0.005), smoked cisco (0.27 ± 0.006), stewed light meat chicken with skin (0.27 ± 0.005), boiled potato without skin (0.27 ± 0.004), duck egg (0.26 ± 0.007), summer sausage (0.26 ± 0.022), beef summer sausage (0.26 ± 0.022), barbecue loaf (0.26 ± 0.022), well done baked lean ground beef (0.26 ± 0.005), medium-rare broiled lean ground beef (0.26 ± 0.005), pork bologna (0.26 ± 0.022), lebanon bologna (0.26 ± 0.022), beef honey roll sausage (0.26 ± 0.022), dried prunes (0.26 ± 0.006), raw potato without skin (0.26 ± 0.004), french fries (0.26 ± 0.006), brie cheese (0.25 ± 0.018), roasted duck without skin (0.25 ± 0.005), turkey summer sausage (0.25 ± 0.018), shredded wheat (0.25 ± 0.018), turkey ham (0.25 ± 0.009), pork olive loaf (0.25 ± 0.018), fried chicken neck with skin (0.25 ± 0.014), braised beef chuck blade roast including fat (0.25 ± 0.005), dry roasted cashews (0.25 ± 0.018), ham and cheese roll (0.25 ± 0.018), cashew butter (0.25 ± 0.018), oil roasted cashews (0.25 ± 0.018), braised beef brisket including fat (0.25 ± 0.005), roasted beef rib including fat (0.25 ± 0.005), dried brazil nuts (0.25 ± 0.018), oil roasted valencia peanuts (0.25 ± 0.018), baked sweet potato (0.25 ± 0.004), raw turnip greens (0.25 ± 0.018), dry roasted peanuts (0.25 ± 0.018), oil roasted virginia peanuts (0.25 ± 0.018), seedless raisins (0.25 ± 0.005), raw cod (0.25 ± 0.006), oil roasted spanish peanuts (0.25 ± 0.018), raw garden cress (0.24 ± 0.02), boiled spinach (0.24 ± 0.006), medium-rare fried ground beef (0.24 ± 0.005), simmered beef brains (0.24 ± 0.005), boiled sweet potato (0.24 ± 0.003), cooked flatfish (0.24 ± 0.006), raw cauliflower (0.24 ± 0.01), cooked plantain (0.24 ± 0.003), boiled carrots (0.24 ± 0.006), frozen and french fried potatoes without skin (0.24 ± 0.01), baked/boiled tropical yam (0.24 ± 0.007), french fried potatoes (0.24 ± 0.01), braised pork tongue (0.23 ± 0.005), braised pork boston blade braised including fat (0.23 ± 0.005), boiled soybeans (0.23 ± 0.003), cooked corned beef brisket (0.23 ± 0.005), medium-rare baked ground beef (0.23 ± 0.005), stewed chicken wing with skin (0.23 ± 0.013), elderberries (0.23 ± 0.003), mashed potatoes (0.23 ± 0.005), cooked carp (0.22 ± 0.006), dry roasted soybean nuts (0.22 ± 0.006), beef and pork salami (0.22 ± 0.022), miso (0.22 ± 0.004), medium-rare baked extra lean ground beef (0.22 ± 0.005), pork and beef luncheon sausage (0.22 ± 0.022), berliner (0.22 ± 0.022), cooked sockeye salmon (0.22 ± 0.006), braised beef shortribs including fat (0.22 ± 0.005), cooked prunes (0.22 ± 0.005), dried figs (0.22 ± 0.003), carrot juice (0.22 ± 0.003), simmered beef heart (0.21 ± 0.005), camembert cheese (0.21 ± 0.018), pork and beef old fashioned loaf (0.21 ± 0.018), roasted pork blade roll (0.21 ± 0.005), stewed dark meat chicken without skin (0.21 ± 0.005), pork bratwurst (0.21 ± 0.006), raw spinach (0.21 ± 0.018), tamari sauce (0.21 ± 0.009), boiled green peas (0.21 ± 0.006), roasted soybean nuts (0.21 ± 0.006), pork and beef lunch meat (0.21 ± 0.018), raw flatfish (0.21 ± 0.006), o'brien potatoes (0.21 ± 0.003), boiled cauliflower (0.21 ± 0.008), parmesan cheese (0.2 ± 0.1), anchovy canned in olive oil (0.2 ± 0.025), whole wheat bread (0.2 ± 0.02), medium-rare baked lean ground beef (0.2 ± 0.005), turkey frankfurter (0.2 ± 0.011), sugar apple (0.2 ± 0.003), raw red cabbage (0.2 ± 0.014), raw savoy cabbage (0.2 ± 0.014), frozen and boiled potatoes without skin (0.2 ± 0.005), baked acorn squash (0.2 ± 0.005), raw carp (0.19 ± 0.006), canned potato without skin (0.19 ± 0.006), toasted whole wheat bread (0.19 ± 0.024), pork and beef kielbasa/kolbassy (0.19 ± 0.019), roast beef sandwich (0.19 ± 0.003), boiled broccoli (0.19 ± 0.006), pork mothers loaf (0.19 ± 0.024), canned pork lunch meat (0.19 ± 0.024), stewed chicken drumstick with skin (0.19 ± 0.009), seeded raisins (0.19 ± 0.005), raw sockeye salmon (0.19 ± 0.006), dried dates (0.19 ± 0.006), turkey salad (0.19 ± 0.009), gefiltefish with broth (0.19 ± 0.012), canned sweet potato (0.19 ± 0.003), boiled okra (0.19 ± 0.006), toasted pumpernickel bread (0.18 ± 0.018), braised beef pancreas (0.18 ± 0.005), bleu cheese (0.18 ± 0.018), roasted duck with skin (0.18 ± 0.005), pork pickle and pimento loaf (0.18 ± 0.018), beef lunch meat (0.18 ± 0.018), stewed chicken thigh with skin (0.18 ± 0.007), turkey bologna (0.18 ± 0.018), pork headcheese (0.18 ± 0.018), pork and beef link sausage (0.18 ± 0.007), beef pastrami (0.18 ± 0.018), dried pecans (0.18 ± 0.018), polish pork sausage (0.18 ± 0.018), dry roasted macadamia nuts (0.18 ± 0.018), scalloped potatoes (0.18 ± 0.004), boiled brussels sprouts (0.18 ± 0.006), boiled lentils (0.18 ± 0.003), boiled turnip greens (0.18 ± 0.007), boiled pink beans (0.18 ± 0.003), cooked whiting (0.18 ± 0.006), frozen and baked sweet potato (0.18 ± 0.006), baked hubbard squash (0.18 ± 0.005), boiled onions (0.18 ± 0.005), pork liverwurst (0.17 ± 0.028), simmered chicken neck without skin (0.17 ± 0.028), sardines (0.17 ± 0.021), grilled ham patties (0.17 ± 0.008), stewed dark meat chicken with skin (0.17 ± 0.005), pork and beef pepperoni (0.17 ± 0.083), lemon peel (0.17 ± 0.083), orange peel (0.17 ± 0.083), beef salami (0.17 ± 0.022), dried apricots (0.17 ± 0.014), shoyu sauce (0.17 ± 0.009), raw green peas (0.17 ± 0.006), beef and pork bologna (0.17 ± 0.022), beef beerwurst salami (0.17 ± 0.022), au gratin potatoes (0.17 ± 0.004), raw parsley (0.17 ± 0.017), raw ginger root (0.17 ± 0.021), pumpernickel bread (0.16 ± 0.016), raw laver/nori seaweed (0.16 ± 0.005), simmered beef tongue (0.16 ± 0.005), bacon cheeseburger (0.16 ± 0.003), pork and beef knockwurst (0.16 ± 0.007), raw broccoli (0.16 ± 0.011), boiled pinto beans (0.16 ± 0.003), raw red sweet peppers (0.16 ± 0.01), raw green sweet peppers (0.16 ± 0.01), boiled navy beans (0.16 ± 0.003), boiled peanuts (0.16 ± 0.016), boiled lima beans (0.16 ± 0.003), raw yellowtail (0.16 ± 0.006), dried sweetened flaked coconut (0.16 ± 0.02), raw onions (0.16 ± 0.006), frozen and boiled spinach (0.15 ± 0.005), unheated ham patties (0.15 ± 0.008), raw bacon (0.15 ± 0.007), taco (0.15 ± 0.006), tomato purée (0.15 ± 0.002), raw carrots (0.15 ± 0.007), raw whiting (0.15 ± 0.006), boiled red cabbage (0.15 ± 0.007), boiled savoy cabbage (0.15 ± 0.007), roquefort cheese (0.14 ± 0.018), boiled beet greens (0.14 ± 0.007), raw anchovy (0.14 ± 0.006), braised pork brains (0.14 ± 0.005), puffed wheat (0.14 ± 0.036), ham and cheese spread (0.14 ± 0.018), soybean protein concentrate (0.14 ± 0.018), boiled asparagus (0.14 ± 0.006), ham salad (0.14 ± 0.018), soy sauce (0.14 ± 0.009), boiled kale (0.14 ± 0.008), cooked pike (0.14 ± 0.006), guava (0.14 ± 0.006), mango

(0.14 ± 0.002), potato salad (0.14 ± 0.004), boiled chickpeas (0.14 ± 0.003), breaded and fried fish (0.14 ± 0.007), boiled scotch kale (0.14 ± 0.008), watermelon (0.14 ± 0.003), chili (0.13 ± 0.002), boiled sprouted green peas (0.13 ± 0.005), pork and beef brotwurst (0.13 ± 0.007), canned corned beef (0.13 ± 0.005), pork and beef mortadella (0.13 ± 0.033), beef bologna (0.13 ± 0.022), vienna sausage (0.13 ± 0.031), beef and pork frankfurter (0.13 ± 0.011), boiled yellow beans (0.13 ± 0.003), frozen and boiled broccoli (0.13 ± 0.005), coleslaw (0.13 ± 0.008), canned succotash with creamed corn (0.13 ± 0.004), vegetarian baked beans (0.13 ± 0.002), cooked shrimp (0.13 ± 0.006), canned sauerkraut (0.13 ± 0.004), frozen and fried hash brown potatoes (0.13 ± 0.006), boiled butternut squash (0.13 ± 0.005), hush puppies (0.13 ± 0.011), mixed grain bread (0.12 ± 0.02), simmered turkey gizzard (0.12 ± 0.005), simmered chicken gizzard (0.12 ± 0.005), falafel (0.12 ± 0.01), pork and beef link sausage with american cheese (0.12 ± 0.012), smoked beef sausage (0.12 ± 0.012), yellow corn pudding (0.12 ± 0.002), raw watercress (0.12 ± 0.029), frozen and boiled collards (0.12 ± 0.006), beef frankfurter (0.12 ± 0.009), frozen and boiled fordhook lima beans (0.12 ± 0.006), raw pike (0.12 ± 0.006), figs (0.12 ± 0.01), boiled great northern beans (0.12 ± 0.003), boiled kidney beans (0.12 ± 0.003), frozen and boiled carrots (0.12 ± 0.007), roasted pumpkin and squash seeds (0.11 ± 0.018), jackfruit (0.11 ± 0.005), dried almonds (0.11 ± 0.018), quail egg (0.11 ± 0.056), almond paste (0.11 ± 0.018), simmered chicken neck with skin (0.11 ± 0.013), pork and beef spread (0.11 ± 0.018), enchilada (0.11 ± 0.002), jellied corned beef loaf (0.11 ± 0.018), frozen and boiled green peas (0.11 ± 0.006), boiled succotash (0.11 ± 0.005), chicken and turkey salad (0.11 ± 0.018), european grapes (0.11 ± 0.003), thompson seedless grapes (0.11 ± 0.003), boiled and frozen baby lima beans (0.11 ± 0.006), boiled french beans (0.11 ± 0.003), canned great northern beans (0.11 ± 0.002), teriyaki sauce (0.11 ± 0.028), frozen and boiled mustard greens (0.11 ± 0.007), raw crookneck squash (0.11 ± 0.008), american grapes (0.11 ± 0.005), raw shrimp (0.11 ± 0.006), frozen and boiled red sweet peppers (0.11 ± 0.005), frozen and boiled green sweet peppers (0.11 ± 0.005), raw scallop squash (0.11 ± 0.008), restaurant tomato juice (0.11 ± 0.004), tomato juice (0.11 ± 0.003), canned carrots (0.11 ± 0.007), cantaloupe (0.11 ± 0.003), restaurant coleslaw (0.11 ± 0.005), boiled acorn squash (0.11 ± 0.004), toasted cracked wheat bread (0.1 ± 0.024), chicken egg (0.1 ± 0.01), sweet roll (0.1 ± 0.012), hamburger sandwich (0.1 ± 0.005), cheese pizza (0.1 ± 0.004), cheeseburger (0.1 ± 0.004), burrito (0.1 ± 0.003), canned spinach (0.1 ± 0.005), cornbread (0.1 ± 0.006), corn pone (0.1 ± 0.006), raw sprouted mung beans (0.1 ± 0.01), firm raw tofu (0.1 ± 0.004), frozen and boiled yellow corn (0.1 ± 0.006), frozen and boiled crookneck squash (0.1 ± 0.005), boiled black-eyed peas (0.1 ± 0.003), canned navy beans (0.1 ± 0.002), boiled crookneck squash (0.1 ± 0.006), boiled red sweet peppers (0.1 ± 0.007), boiled green sweet peppers (0.1 ± 0.007), boiled hubbard squash (0.1 ± 0.004), pomegranate (0.1 ± 0.003), boiled/baked spaghetti squash (0.1 ± 0.006), pineapple juice (0.1 ± 0.002), raw eggplant (0.1 ± 0.012), raw mushrooms (0.09 ± 0.014), toasted rye bread (0.09 ± 0.023), cooked oysters (0.09 ± 0.006), boiled mushrooms (0.09 ± 0.006), raw pork jowl (0.09 ± 0.005), canned oysters (0.09 ± 0.006), boiled artichoke (0.09 ± 0.006), breaded and fried shrimp (0.09 ± 0.006), frozen and boiled kale (0.09 ± 0.008), frozen and boiled succotash (0.09 ± 0.006), frozen and boiled cauliflower (0.09 ± 0.006), baked beans with pork and sweet sauce

(0.09 ± 0.002), frozen and boiled black-eyed peas (0.09 ± 0.006), boiled yardlong bean (0.09 ± 0.003), boiled parsnips (0.09 ± 0.006), boiled catjang black-eyed peas (0.09 ± 0.003), boiled rutabaga (0.09 ± 0.006), pineapple (0.09 ± 0.003), boiled white beans (0.09 ± 0.003), raw green cabbage (0.09 ± 0.014), canned peeled red tomatoes (0.09 ± 0.004), raw zucchini (0.09 ± 0.008), canned lima beans (0.09 ± 0.002), lemon (0.09 ± 0.009), boiled scallop squash (0.09 ± 0.006), tuna salad (0.08 ± 0.002), cracked wheat bread (0.08 ± 0.02), braised pork lungs (0.08 ± 0.005), rye bread (0.08 ± 0.02), pepperoni pizza (0.08 ± 0.004), hot fudge sundae (0.08 ± 0.003), pork and beef sausage (0.08 ± 0.038), oatmeal cookie (0.08 ± 0.038), onion rings (0.08 ± 0.006), plum (0.08 ± 0.008), fried tofu (0.08 ± 0.038), cooked lobster (0.08 ± 0.006), boiled cranberry beans (0.08 ± 0.003), canned pinto beans (0.08 ± 0.002), cooked eel (0.08 ± 0.006), boiled artichoke hearts (0.08 ± 0.006), boiled black turtle beans (0.08 ± 0.003), boiled baby lima beans (0.08 ± 0.003), jujube (0.08 ± 0.005), boiled zucchini (0.08 ± 0.006), canned white beans (0.08 ± 0.002), gooseberries (0.08 ± 0.003), boiled eggplant (0.08 ± 0.01), oil roasted almonds (0.07 ± 0.018), dry roasted almonds (0.07 ± 0.018), toasted almonds (0.07 ± 0.018), limburger cheese (0.07 ± 0.018), cheddar cheese (0.07 ± 0.018), colby cheese (0.07 ± 0.018), edam cheese (0.07 ± 0.018), pimento cheese spread (0.07 ± 0.018), american cheese (0.07 ± 0.018), brick cheese (0.07 ± 0.018), swiss cheese (0.07 ± 0.018), gouda cheese (0.07 ± 0.018), muenster cheese (0.07 ± 0.018), provolone cheese (0.07 ± 0.018), dinner rolls (0.07 ± 0.018), waffles (0.07 ± 0.007), graham crackers (0.07 ± 0.036), gruyere cheese (0.07 ± 0.018), port du salut cheese (0.07 ± 0.018), mozzarella cheese (0.07 ± 0.018), malted milk (0.07 ± 0.002), tamarind (0.07 ± 0.004), fish sandwich (0.07 ± 0.003), medium cream (0.07 ± 0.033), half and half cream (0.07 ± 0.033), pineapple upside-down cake (0.07 ± 0.007), boiled broad beans (0.07 ± 0.003), puffed rice (0.07 ± 0.036), frozen and boiled turnip greens (0.07 ± 0.006), raw looseleaf lettuce (0.07 ± 0.018), canned kidney beans (0.07 ± 0.002), boiled mung beans (0.07 ± 0.002), boiled black beans (0.07 ± 0.003), black currants (0.07 ± 0.009), red currants (0.07 ± 0.009), white currants (0.07 ± 0.009), boiled green cabbage (0.07 ± 0.007), baked beans with pork and tomato sauce (0.07 ± 0.002), raw eel (0.07 ± 0.006), navel orange (0.07 ± 0.004), valencia orange (0.07 ± 0.004), raw radish (0.07 ± 0.011), frozen and boiled butternut squash (0.07 ± 0.004), loganberries (0.07 ± 0.003), tangerine (0.07 ± 0.006), bloody mary (0.07 ± 0.003), frozen and boiled onions (0.07 ± 0.005), cottage cheese (0.07 ± 0.018), almond butter (0.06 ± 0.031), breaded fried squid (0.06 ± 0.006), raw squid (0.06 ± 0.006), french rolls (0.06 ± 0.01), evaporated skim milk (0.06 ± 0.016), braised pork spleen (0.06 ± 0.005), french toast (0.06 ± 0.008), breaded and fried oysters (0.06 ± 0.006), sponge cake (0.06 ± 0.008), cheesecake (0.06 ± 0.006), boiled green beans (0.06 ± 0.008), boiled yellow snap beans (0.06 ± 0.008), raspberries (0.06 ± 0.004), canned green peas (0.06 ± 0.006), canned yellow corn (0.06 ± 0.005), boiled mungo beans (0.06 ± 0.003), strawberries (0.06 ± 0.003), frozen and boiled yellow snap beans (0.06 ± 0.007), frozen and boiled green beans (0.06 ± 0.007), boiled yellow corn (0.06 ± 0.006), canned pumpkin (0.06 ± 0.004), soursop (0.06 ± 0.002), canned creamed yellow corn (0.06 ± 0.004), blackberries (0.06 ± 0.007), apricots (0.06 ± 0.005), grape juice (0.06 ± 0.002), boiled turnips (0.06 ± 0.006), cranberry sauce (0.06 ± 0.005), raw cucumber (0.06 ± 0.01), barbecue sauce (0.06 ± 0.031), toasted raisin bread (0.05 ± 0.024), sweetened condensed milk (0.05 ± 0.013), toasted white bread (0.05 ± 0.024), evaporated whole milk (0.05 ± 0.004), evaporated lowfat milk (0.05 ± 0.004), restaurant chocolate milkshake (0.05 ± 0.002), skim yogurt (0.05 ± 0.002), lowfat yogurt (0.05 ± 0.002), protein fortified lowfat 2% milk (0.05 ± 0.002), protein fortified skim milk (0.05 ± 0.002), eggnog (0.05 ± 0.002), protein fortified lowfat 1% milk (0.05 ± 0.002), restaurant strawberry milkshake (0.05 ± 0.002), restaurant vanilla milkshake (0.05 ± 0.002), raw oysters (0.05 ± 0.006), lowfat 1% milk (0.05 ± 0.002), lowfat 2% milk (0.05 ± 0.002), whole chocolate malted milk (0.05 ± 0.002), carob milk (0.05 ± 0.002), goat milk (0.05 ± 0.002), boiled winged beans (0.05 ± 0.003), canned black turtle beans (0.05 ± 0.002), canned black-eyed peas (0.05 ± 0.002), boiled split peas (0.05 ± 0.003), boiled pigeon peas (0.05 ± 0.003), raw red tomato (0.05 ± 0.004), raw iceberg lettuce (0.05 ± 0.025), canned succotash (0.05 ± 0.004), canned broad beans (0.05 ± 0.002), raw tofu (0.05 ± 0.004), boysenberries (0.05 ± 0.004), canned cranberry beans (0.05 ± 0.002), canned crookneck squash (0.05 ± 0.005), beer (0.05 ± 0.001), tequila sunrise (0.05 ± 0.003), lemon juice (0.05 ± 0.002), raw apple with skin (0.05 ± 0.004), raw apple without skin (0.05 ± 0.004), cooked apple without skin (0.05 ± 0.003), raisin bread (0.04 ± 0.02), toasted english muffin (0.04 ± 0.01), havarti cheese (0.04 ± 0.018), english muffin (0.04 ± 0.009), braised beef spleen (0.04 ± 0.005), white bread (0.04 ± 0.021), bagels (0.04 ± 0.009), french bread (0.04 ± 0.018), pretzels (0.04 ± 0.018), unsweetened baking chocolate (0.04 ± 0.018), italian bread (0.04 ± 0.018), oatmeal bread (0.04 ± 0.018), cream cheese (0.04 ± 0.018), neufchatel cheese (0.04 ± 0.018), yellow cake (0.04 ± 0.007), pancakes with butter and syrup (0.04 ± 0.003), ricotta cheese (0.04 ± 0.004), vanilla milkshake (0.04 ± 0.002), hot chocolate (0.04 ± 0.002), lowfat 1% chocolate milk (0.04 ± 0.002), chocolate cream pie (0.04 ± 0.004), lowfat 2% chocolate milk (0.04 ± 0.002), whole milk (0.04 ± 0.002), whole chocolate milk (0.04 ± 0.002), skim milk (0.04 ± 0.002), raw chickory (0.04 ± 0.011), raw pork stomach (0.04 ± 0.005), frozen and boiled okra (0.04 ± 0.005), filled milk (0.04 ± 0.002), raw endive (0.04 ± 0.02), instant corn grits (0.04 ± 0.004), soy milk (0.04 ± 0.002), boiled collards (0.04 ± 0.003), candied sweet potato (0.04 ± 0.005), frozen and boiled zucchini (0.04 ± 0.004), pummelo (0.04 ± 0.003), quince (0.04 ± 0.005), red table wine (0.04 ± 0.005), restaurant orange juice (0.04 ± 0.004), pink grapefruit (0.04 ± 0.004), white grapefruit (0.04 ± 0.004), sapodilla (0.04 ± 0.003), raw coconut (0.04 ± 0.011), screwdriver (0.04 ± 0.002), lime juice (0.04 ± 0.002), java plum (0.04 ± 0.004), boiled beets (0.04 ± 0.006), raw spirulina seaweed (0.03 ± 0.005), hamburger bun (0.03 ± 0.013), hot dog bun (0.03 ± 0.013), pita bread (0.03 ± 0.013), chocolate milkshake (0.03 ± 0.002), white cake (0.03 ± 0.006), caramel sundae (0.03 ± 0.003), strawberry sundae (0.03 ± 0.003), devil's food cake (0.03 ± 0.008), buttermilk (0.03 ± 0.002), whey (0.03 ± 0.002), whole yogurt (0.03 ± 0.002), raw sprouted alfalfa seeds (0.03 ± 0.015), apple pie (0.03 ± 0.004), light cream (0.03 ± 0.017), boiled red tomato (0.03 ± 0.004), canned turnip greens (0.03 ± 0.004), cherries (0.03 ± 0.007), coconut water (0.03 ± 0.002), blueberries (0.03 ± 0.003), maltex (0.03 ± 0.003), raw celery (0.03 ± 0.013), boiled celery (0.03 ± 0.007), apple juice (0.03 ± 0.002), rose table wine (0.03 ± 0.005), cream of rice (0.03 ± 0.003), angel food cake (0.02 ± 0.008), braised beef lungs (0.02 ± 0.005), indian buffalo milk (0.02 ± 0.002), frozen and boiled asparagus (0.02 ± 0.008), lemon meringue pie (0.02 ± 0.004), peach (0.02 ± 0.006), nectarine (0.02 ± 0.004), pear (0.02 ± 0.003), papaya (0.02 ± 0.002),

raw frozen rhubarb (0.02 ± 0.004), frozen and cooked sweetened rhubarb (0.02 ± 0.004), applesauce (0.02 ± 0.004), oatmeal (0.02 ± 0.003), cooked wild long grain rice (0.02 ± 0.004), corn grits (0.02 ± 0.002), restaurant cranberry juice cocktail (0.02 ± 0.003), whiskey sour (0.02 ± 0.006), noodles (0.01 ± 0.003), macaroni (0.01 ± 0.004), spaghetti (0.01 ± 0.004), raw acerola (0.01 ± 0.005), raw celeriac (0.01 ± 0.005), farina (0.01 ± 0.003), white table wine (0.01 ± 0.005), papaya nectar (0.01 ± 0.002)
Applications: anemia, increased cholesterol, diabetes, blood sugar decreased, jaundice, pernicious anemia, bladder inflammation, colitis, diarrhea, Bell's palsy, epilepsy, lack of sleep, mental illness, multiple sclerosis, neuritis, Parkinson's disease, dizziness, conjunctivitis, inflammation of prostate, baldness, dandruff, headache, celiac disease, hemorrhoids, worms, arthritis, kidney stones, asthma, common cold, influenza, tuberculosis, bad breath, muscular dystrophy, rheumatism, inflammation of prostate, vaginal itching, acne, skin inflammation, skin eczema, skin psoriasis, shingles, gastritis, digestion upset, nausea of pregnancy, pyorrhea, alcohol cravings, skin edema, weight increased, stress, cancer, antioxidant therapy, sun sensitivity, free radical oxidation, life span extension, radiation, decreased radiation resistance, aging
Daily Dosage (at least 100 mg per every 1000 mg vitamin B_{3a}): 1974 PCS—1.0–3.0 mg; 1974 RDA—1.8–2.2 mg; 1974 USRDA—2.0 mg; 1980 RDA—1.8–2.2 mg; 1980 USRDA—2.0 mg; 1989 RDA—1.6–2.0 mg; 1989 USRDA—2.0 mg; nutritional—10–50 mg; therapeutic—200–500 mg; experimental—1000–1750 mg; toxic—2000 mg +
Warnings: Pregnant women should check with doctors before taking supplements over 50 mg. Should not be taken by anyone under L-dopa treatment for Parkinson's disease. Not for people taking tryptophan for depression, as it will convert to vitamin B_3. Increased amounts are dangerous. See toxicity symptoms

PROVITAMIN B_6 *see* **vitamin B_{13}**

VITAMIN B_7

(The term "vitamin B_7" is sometimes used to denote vitamin B_{3a}.)
(The term "vitamin B_7" is sometimes used to denote vitamin B_W.)

Names: vitamin B_7, vitamin I, rice polish factor
Classification: possible water-soluble vitamin, possible vitamin B complex
Forms: (not isolated to a specific chemical)
Deficiency: digestion deterioration in pigeons, possibly similar symptoms in humans
Side-effects: unknown
Toxicity: unknown
Inhibitors: unknown
Helpers: unknown
Sources: rice polishings
Applications: unknown
Daily Dosage: 1974 PCS—none; 1989 SADDI—none; 1989 RDA—none; 1989 USRDA—none; nutritional—unknown; therapeutic—unknown; experimental—unknown; toxic—unknown
Warnings: none

VITAMIN B_8

(Found to be the nucleic acid adenylic acid.)
(The term "vitamin B_8" sometimes used to denote vitamin B_H.)

Names: vitamin B_8, adenylic acid, muscle adenylic acid, ergadenylic acid,

t-adenylic acid, AMP, NSC-20264, Adenyl, Cardiomone, Cardiomone sodium salt, Lycedan, My-B-Den, Myoston, Phosaden
Classification: nucleic acid, formerly water-soluble vitamin, formerly vitamin B complex
Forms: adenylic acid, adenosine monophosphate
Deficiency: decreased RNA synthesis, decreased ADP and ATP synthesis, deterioration of breakdown of carbohydrates, deterioration of breakdown of proteins, deterioration of breakdown of fats, deterioration of hormone function
Side-effects: unknown
Toxicity: unknown
Inhibitors: unknown
Helpers: unknown
Sources: plants, animals, DNA, RNA
Applications: unknown
Daily Dosage: 1974 PCS — none; 1989 SADDI — none; 1989 RDA — none; 1989 USRDA — none; nutritional — unknown; therapeutic — unknown; experimental — unknown; toxic — unknown
Warnings: none

VITAMIN B_9

(The original vitamin B_9 was found to be a mixture. Those B complex vitamins discovered after vitamin B_9 are sometimes denoted "vitamin B_9 complex.")
(The term "vitamin B_9" sometimes used to denote vitamin B_C.)
Classification: formerly water-soluble vitamin, formerly vitamin B complex. now defunct
Forms: later found to be a mixture of vitamin B complex vitamins

VITAMIN B_{10}

(The term "vitamin B_{10}" is sometimes used to denote vitamin B_X.)

Names: vitamin B_{10}, factor R, vitamin R
Classification: formerly water-soluble vitamin, formerly vitamin B complex, now defunct
Forms: found to be vitamin B_C compounds possibly mixed with pteroylmonoglutamic acid and vitamin B_{12}
Deficiency: deterioration of growth, deterioration of feather growth and deterioration of blood in chicks, possibly similar symptoms in humans

VITAMIN B_{11}

Names: vitamin B_{11}, factor S, vitamin S
Classification: formerly water-soluble vitamin, formerly vitamin B complex
Forms: found to be vitamin B_C compounds possibly mixed with pteroylmonoglutamic acid and vitamin B_{12}
Deficiency: deterioration of growth and deterioration of blood and skin, possibly relating to peptide streptogenin activity? (in chicks, possibly similar symptoms in humans)

VITAMIN B_{12}

Names: vitamin B_{12}, cobalamin, erythrotin, antipernicious factor, extrinsic factor, animal protein factor, factor X, Antipernicin, Bedoce, Bedodeka, Bedoz, Behepan, Berubi, Berubigen, Betalin-12, Bevatine-12, Bevidox, Bexii, Bexil, Biocobalamine, Biocres, Bitevan, B-Telve, B-Twelv, Byladoce, Claretin-12, Cobalin, Cobamin, Cobamine, Cobione, Covit, Crystamin, Cycobemin, Cycolamin, Cykobeminet, Cytacon, Cytamen, Cytobion, Distivit, Dobetin, Docemine, Docigram, Docivit, Dodecabee, Dodecavite, Dodex, Ducobee, Duodecibin, Embiol, Emociclina, Eritrone, Erycytol, Erythrotin, Euhaemon, Fresmin, Hemo-B-Doze, Hemomin, Hepagon, Hepavis, Hepcovite, Hydoxamin, Hydroxobase, Macra-

VITAMIN B_{12}

bin, Megabion, Megalovel, Milbedoce, Millevit, Nagravon, Normocytin, Peraemon, Pernaevit, Pernipur, Plecyamin, Poyamin, Redamina, Redisol, Rhodacryst, Rubesol, Rubivitan, Rubramin, Rubripca, Rubrocitol, Sytobex, Vibalt, Vibisone, Virubra, Vitarubin, Vita-Rubra, Vitral
Classification: water-soluble vitamin, vitamin B complex
Forms: cyanocobalamin, vitamin B_{12a}/hydroxycobalamin/vitamin OHB_{12}, vitamin B_{12b}/aquocobalamin, vitamin B_{12c}/nitritocobalamin, vitamin B_{12f}, vitamin B_{12m}, vitamin B_{12p}/etiocobalamin/factor B, vitamin B_{12r}/cob (II)alamin, vitamin B_{12s}/cob (I)alamin/hydridocobalamin, vitamin B_{12III}, psi-vitamin B_{12}, psi-vitamin B_{12d}
Deficiency: (Rebound Deficiency can occur if large doses are halted abruptly.) pernicious anemia, large/abnormal blood cells, decreased blood clotting, menstrual deterioration, heart palpitation, deterioration of growth, deterioration of carbohydrate metabolism, weakness, weak pulse, numb/tingling hands, numb/tingling feet, stiffness, fatigue, depression, emotional agitation, memory decreased, emotional deterioration, deterioration of nervous system, brain deterioration, difficult walking, decreased reflexes, nervousness, paleness, sore tongue, red tongue, difficult swallowing, diarrhea, body odor, dandruff, deterioration of folate utilization, hangover, blood iron level deceptively increased, sterility, immunity decreased, decreased color perception
Side-effects: color dreaming
Toxicity: none known up to 1000 μg long term or 100,000 μg short term
Inhibitors: acids, alkalies, water, sunlight, alcohol, estrogen, sleeping pills, neomycin, methotrexate, cholestyramine, colchicine, sodium aminosalicylate, slow-release potassium chloride, metaformin, phenformin, sodium, nitroprusside, chloramphenicol, codeine, oral anti-diabetic agents, aspirin, aspirin substitutes, hydralazine
Helpers: vitamin B_6, vitamin B_{13}, vitamin B_{17}, vitamin B_C, vitamin B_H, vitamin B_P, vitamin B_W, vitamin C, potassium, sodium, sorbitol
Sources (μg per 100 grams of food): fried beef liver (111.8 ± 0.005), braised beef liver (71 ± 0.005), raw grouper (60 ± 0.006), raw duck liver (54 ± 0.005), simmered beef kidneys (51.3 ± 0.005), raw clams (49.45 ± 0.006), simmered turkey liver (47.5 ± 0.005), breaded and fried clams (40.27 ± 0.006), cooked oysters (38.27 ± 0.006), roasted light meat chicken without skin (34 ± 0.005), pork livercheese (24.55 ± 0.013), simmered turkey giblets (24.03 ± 0.005), pork liver sausage (20.11 ± 0.028), simmered chicken liver (19.39 ± 0.005), raw oysters (19.13 ± 0.006), canned oysters (19.13 ± 0.006), cooked mackerel (19 ± 0.006), kippered herring (18.7 ± 0.013), braised pork liver (18.67 ± 0.005), cooked whelk (18.14 ± 0.006), braised pork pancreas (17.07 ± 0.005), braised beef pancreas (16.6 ± 0.005), breaded and fried oysters (15.64 ± 0.006), fried beef brains (15.2 ± 0.005), simmered beef heart (14.3 ± 0.005), raw herring (13.67 ± 0.006), pork liverwurst (13.44 ± 0.028), fried chicken giblets (13.31 ± 0.005), cooked herring (13.14 ± 0.006), simmered chicken giblets (10.14 ± 0.005), dried and salted cod (10 ± 0.006), canned smoked goose liver pâté (9.5 ± 0.018), raw whelk (9.07 ± 0.006), sardines (8.96 ± 0.021), raw mackerel (8.71 ± 0.006), simmered beef brains (8.6 ± 0.005), raw trout (7.79 ± 0.006), braised pork kidneys (7.79 ± 0.005), cooked blue crab (7.31 ± 0.006), simmered chicken heart (7.29 ± 0.005), simmered turkey heart (7.15 ± 0.005), crab cakes (5.93 ± 0.008), simmered beef tongue (5.9 ± 0.005), cooked sockeye salmon (5.8 ± 0.006), beef summer sausage (5.61 ± 0.022), summer sausage (5.52 ± 0.022), duck egg (5.4 ± 0.007), raw bluefish (5.39 ± 0.006), bran flakes (5.36 ± 0.018), wheat flakes (5.36 ± 0.018), braised beef spleen (5.02 ± 0.005), beef and pork summer sausage (4.61 ±

0.022), pickled herring (4.27 ± 0.033), smoked cisco (4.26 ± 0.006), turkey summer sausage (4.11 ± 0.018), raisin bran (4.05 ± 0.014), dry skim milk (4.03 ± 0.017), dry instant skim milk (3.99 ± 0.005), cooked rainbow smelt (3.96 ± 0.006), beef lunch meat (3.93 ± 0.018), raw striped bass (3.82 ± 0.006), chicken egg yolk (3.82 ± 0.029), braised pork heart (3.79 ± 0.004), simmered beef shank excluding fat (3.79 ± 0.005), fried beef sirloin excluding fat (3.69 ± 0.005), simmered beef shank including fat (3.59 ± 0.005), dry buttermilk (3.57 ± 0.071), beef and pork salami (3.48 ± 0.022), braised beef shortribs excluding fat (3.46 ± 0.005), cooked crayfish (3.46 ± 0.006), raw rainbow smelt (3.44 ± 0.006), fried beef top round excluding fat (3.43 ± 0.005), braised beef flank excluding fat (3.41 ± 0.005), raw milkfish (3.4 ± 0.006), braised beef chuck arm pot roast excluding fat (3.4 ± 0.005), braised beef flank including fat (3.36 ± 0.005), broiled beef rib eye excluding fat (3.32 ± 0.005), well done broiled ground beef (3.28 ± 0.005), broiled beef rib excluding fat (3.28 ± 0.005), smoked chinook salmon (3.26 ± 0.006), smoked whitefish (3.26 ± 0.006), dry whole milk (3.25 ± 0.016), fried beef sirloin including fat (3.21 ± 0.005), raw pollock (3.19 ± 0.006), fried beef top round including fat (3.18 ± 0.005), cooked lobster (3.11 ± 0.006), broiled beef rib eye including fat (3.06 ± 0.005), broiled beef flank excluding fat (3.05 ± 0.005), beef salami (3.04 ± 0.022), broiled beef flank including fat (3.02 ± 0.005), well done fried ground beef (3 ± 0.005), raw eel (3 ± 0.006), raw cuttlefish (3 ± 0.006), broiled beef round excluding fat (2.98 ± 0.005), medium-rare broiled ground beef (2.93 ± 0.005), braised beef chuck arm pot roast including fat (2.93 ± 0.005), roasted beef rib roasted excluding fat (2.92 ± 0.005), well done baked ground beef (2.9 ± 0.005), roasted beef tip excluding fat (2.89 ± 0.005), cooked eel (2.88 ± 0.006), broiled beef sirloin excluding fat (2.85 ± 0.005), hard pork salami (2.8 ± 0.05), broiled beef rib including fat (2.79 ± 0.005), roasted beef tenderloin excluding fat (2.77 ± 0.005), braised pork spleen (2.76 ± 0.005), broiled beef round including fat (2.75 ± 0.005), roasted beef round tip including fat (2.74 ± 0.005), well done broiled lean ground beef (2.72 ± 0.005), medium-rare fried ground beef (2.71 ± 0.005), raw crayfish (2.71 ± 0.006), broiled beef sirloin including fat (2.66 ± 0.005), berliner (2.65 ± 0.022), braised beef shortribs including fat (2.62 ± 0.005), cooked whiting (2.6 ± 0.006), braised beef lungs (2.59 ± 0.005), well done fried lean ground beef (2.58 ± 0.005), lebanon bologna (2.57 ± 0.022), broiled beef tenderloin excluding fat (2.57 ± 0.005), well done broiled extra lean ground beef (2.56 ± 0.005), roasted beef tenderloin including fat (2.56 ± 0.005), braised beef brisket excluding fat (2.55 ± 0.005), roasted beef rib including fat (2.51 ± 0.005), cooked flatfish (2.51 ± 0.006), broiled beef top round excluding fat (2.48 ± 0.005), braised beef bottom round excluding fat (2.47 ± 0.005), braised beef chuck blade roast excluding fat (2.47 ± 0.005), broiled beef tenderloin including fat (2.45 ± 0.005), broiled beef top round including fat (2.44 ± 0.005), braised beef bottom round including fat (2.4 ± 0.005), braised pork tongue (2.39 ± 0.005), medium-rare broiled lean ground beef (2.35 ± 0.005), beef honey roll sausage (2.35 ± 0.022), medium-rare baked ground beef (2.34 ± 0.005), pork and beef pepperoni (2.33 ± 0.083), well done fried extra lean ground beef (2.32 ± 0.005), raw whiting (2.31 ± 0.006), medium-rare fried lean ground beef (2.27 ± 0.005), broiled porterhouse steak excluding fat (2.27 ± 0.005), broiled T-bone steak excluding fat (2.27 ± 0.005), well done baked lean ground beef (2.26 ± 0.005), braised beef brisket including fat (2.23 ± 0.005), braised beef chuck blade roast including fat (2.23 ±

0.005), raw turbot (2.2 ± 0.006), medium-rare broiled extra lean ground beef (2.17 ± 0.005), roasted eye of beef round excluding fat (2.17 ± 0.005), broiled porterhouse steak including fat (2.16 ± 0.005), tilsit cheese (2.14 ± 0.018), broiled T-bone steak including fat (2.12 ± 0.005), roasted eye of beef round including fat (2.1 ± 0.005), pork and beef brotwurst (2.06 ± 0.007), raw wolffish (2.04 ± 0.006), braised pork lungs (2.03 ± 0.005), cooked swordfish (2.02 ± 0.006), turkey breast meat summer sausage (2 ± 0.024), broiled beef top loin excluding fat (2 ± 0.005), medium-rare fried extra lean ground beef (2 ± 0.005), turkey pastrami (2 ± 0.009), pork and beef peppered loaf (2 ± 0.018), pork and beef luncheon sausage (1.96 ± 0.022), beef beerwurst salami (1.96 ± 0.022), broiled beef top loin including fat (1.94 ± 0.005), simmered chicken gizzard (1.94 ± 0.005), hard pork and beef salami (1.9 ± 0.05), simmered turkey gizzard (1.9 ± 0.005), breaded and fried shrimp (1.87 ± 0.006), well done baked extra lean ground beef (1.86 ± 0.005), smoked beef sausage (1.86 ± 0.012), beef pastrami (1.79 ± 0.018), turkey ham (1.79 ± 0.009), medium-rare baked lean ground beef (1.77 ± 0.005), cooked bacon (1.76 ± 0.004), raw swordfish (1.75 ± 0.006), chopped smoked beef (1.75 ± 0.018), pork sausage (1.74 ± 0.019), medium-rare baked extra lean ground beef (1.73 ± 0.005), pork and beef link sausage with american cheese (1.72 ± 0.012), swiss cheese (1.71 ± 0.018), barbecue loaf (1.7 ± 0.022), brie cheese (1.68 ± 0.018), pork link sausage (1.63 ± 0.007), cooked corned beef brisket (1.63 ± 0.005), pork and beef kielbasa/kolbassy (1.62 ± 0.019), canned corned beef (1.62 ± 0.005), gruyere cheese (1.61 ± 0.018), smoked haddock (1.6 ± 0.006), turkey bologna (1.57 ± 0.018), edam cheese (1.57 ± 0.018), raw beef tripe (1.54 ± 0.005), port du salut cheese (1.54 ± 0.018), raw carp (1.53 ± 0.006), beef frankfurter (1.53 ± 0.009), raw scallops (1.53 ± 0.006), raw flatfish (1.52 ± 0.006), braised beef thymus (1.51 ± 0.005), pork and beef link sausage (1.51 ± 0.007), pork and beef picnic loaf (1.5 ± 0.018), provolone cheese (1.5 ± 0.018), muenster cheese (1.5 ± 0.018), raw shark (1.49 ± 0.006), cooked shrimp (1.49 ± 0.006), pork and beef mortadella (1.47 ± 0.033), beef bologna (1.43 ± 0.022), braised pork brains (1.42 ± 0.005), cooked haddock (1.39 ± 0.006), pork luxury loaf (1.39 ± 0.018), cooked halibut (1.36 ± 0.006), beef and pork bologna (1.35 ± 0.022), chicken egg (1.34 ± 0.01), pork and beef old fashioned loaf (1.32 ± 0.018), breaded and fried scallops (1.32 ± 0.016), camembert cheese (1.32 ± 0.018), raw yellowtail (1.31 ± 0.006), italian pork sausage (1.3 ± 0.007), turkey frankfurter (1.29 ± 0.011), pork and beef lunch meat (1.29 ± 0.018), beef and pork frankfurter (1.29 ± 0.011), raw squid (1.29 ± 0.006), pork olive loaf (1.29 ± 0.018), jellied corned beef loaf (1.29 ± 0.018), brick cheese (1.29 ± 0.018), bleu cheese (1.25 ± 0.018), breaded fried squid (1.22 ± 0.006), battered and fried shark (1.21 ± 0.006), raw haddock (1.2 ± 0.006), bacon cheeseburger (1.2 ± 0.003), raw halibut (1.19 ± 0.006), pork and beef knockwurst (1.18 ± 0.007), pork pickle and pimento loaf (1.18 ± 0.018), maypo (1.17 ± 0.003), raw shrimp (1.16 ± 0.006), cooked perch (1.15 ± 0.006), pork and beef spread (1.14 ± 0.018), broiled pork boston blade excluding fat (1.13 ± 0.005), roasted cured pork arm excluding fat (1.11 ± 0.005), pork and beef honey loaf (1.11 ± 0.018), unheated ham patties (1.09 ± 0.008), broiled pork loin excluding fat (1.08 ± 0.005), braised pork spareribs (1.08 ± 0.005), pork headcheese (1.07 ± 0.018), limburger cheese (1.07 ± 0.018), roasted canned ham (1.06 ± 0.005), pork mothers loaf (1.05 ± 0.024), raw cusk (1.05 ± 0.006), cooked cod (1.05 ± 0.006), cooked clams (1.05 ± 0.006), canned cod (1.05 ± 0.006), roasted pork blade roll

VITAMIN B_{12}

(1.05 ± 0.005), broiled pork loin blade excluding fat (1.04 ± 0.005), broiled pork boston blade including fat (1.04 ± 0.005), fried pork loin blade excluding fat (1 ± 0.005), polish pork sausage (1 ± 0.018), raw perch (1 ± 0.006), vienna sausage (1 ± 0.031), breaded and fried fish (1 ± 0.007), raw pork stomach (0.99 ± 0.005), broiled pork loin including fat (0.98 ± 0.005), roasted pork boston blade excluding fat (0.96 ± 0.005), minced ham lunch meat (0.95 ± 0.024), pork bratwurst (0.95 ± 0.006), broiled pork loin blade including fat (0.94 ± 0.005), roasted pork loin excluding fat (0.93 ± 0.005), roasted cured pork arm including fat (0.93 ± 0.005), raw bacon (0.93 ± 0.007), raw lobster (0.93 ± 0.006), roasted pork boston blade including fat (0.91 ± 0.005), pork bologna (0.91 ± 0.022), raw cod (0.91 ± 0.006), enchilada (0.9 ± 0.002), anchovy canned in olive oil (0.9 ± 0.025), braised pork boston blade excluding fat (0.9 ± 0.005), roast beef sandwich (0.9 ± 0.003), chopped ham lunch meat (0.9 ± 0.024), canned pork lunch meat (0.9 ± 0.024), fish sandwich (0.9 ± 0.003), roasted pork shoulder excluding fat (0.88 ± 0.005), roasted pork loin including fat (0.87 ± 0.005), fried pork loin blade including fat (0.87 ± 0.005), pork beerwurst salami (0.87 ± 0.022), braised pork loin excluding fat (0.84 ± 0.005), raw pheasant without skin (0.84 ± 0.005), broiled pork sirloin excluding fat (0.84 ± 0.005), braised pork boston blade braised including fat (0.84 ± 0.005), tempeh (0.84 ± 0.006), gefiltefish with broth (0.83 ± 0.012), unheated ham (0.83 ± 0.005), roasted pork shoulder including fat (0.83 ± 0.005), braised pork loin blade excluding fat (0.82 ± 0.005), canned lean ham (0.82 ± 0.005), raw pork jowl (0.82 ± 0.005), ham and cheese roll (0.82 ± 0.018), cheddar cheese (0.82 ± 0.018), colby cheese (0.82 ± 0.018), roasted pork loin blade excluding fat (0.8 ± 0.005), broiled pork sirloin including fat (0.8 ± 0.005), hamburger sandwich (0.8 ± 0.005), cheeseburger (0.8 ± 0.004), braised pork loin including fat (0.79 ± 0.005), ham salad (0.79 ± 0.018), roasted pork sirloin excluding fat (0.78 ± 0.005), roasted pork arm excluding fat (0.78 ± 0.005), canned ham (0.78 ± 0.005), grilled canadian bacon (0.77 ± 0.011), fried pork center loin including fat (0.77 ± 0.005), roasted pork sirloin including fat (0.76 ± 0.005), braised pork loin blade including fat (0.76 ± 0.005), roasted pork loin blade including fat (0.76 ± 0.005), unheated lean ham (0.75 ± 0.005), ham and cheese spread (0.75 ± 0.018), broiled pork center loin excluding fat (0.74 ± 0.005), roasted pork arm including fat (0.74 ± 0.005), fried pork center loin excluding fat (0.73 ± 0.005), roasted pork rump excluding fat (0.73 ± 0.005), roasted pork leg excluding fat (0.72 ± 0.005), braised pork arm excluding fat (0.71 ± 0.005), broiled pork center loin including fat (0.71 ± 0.005), roasted canned lean ham (0.71 ± 0.005), roasted pork shank excluding fat (0.71 ± 0.005), roasted pork rump including fat (0.71 ± 0.005), sheep milk (0.71 ± 0.002), pimento cheese spread (0.71 ± 0.018), american cheese (0.71 ± 0.018), roasted ham (0.7 ± 0.005), broiled pork center rib excluding fat (0.7 ± 0.005), broiled pork top loin excluding fat (0.7 ± 0.005), roasted pork leg including fat (0.7 ± 0.005), grilled ham patties (0.7 ± 0.008), braised pork arm braised including fat (0.69 ± 0.005), roasted pork shank including fat (0.69 ± 0.005), cooked grouper (0.69 ± 0.006), broiled pork center rib including fat (0.68 ± 0.005), mozzarella cheese (0.68 ± 0.018), unheated canadian bacon (0.67 ± 0.009), braised pork sirloin excluding fat (0.67 ± 0.005), broiled pork top loin including fat (0.67 ± 0.005), braised pork sirloin including fat (0.66 ± 0.005), roasted lean ham (0.65 ± 0.005), roquefort cheese (0.64 ± 0.018), cottage cheese (0.64 ± 0.018), raw anchovy (0.62 ± 0.006), fried pork center rib excluding fat (0.62 ± 0.005), fried pork top loin excluding fat (0.62 ± 0.005), fried pork

VITAMIN B$_{12}$

I. Vitamins

center rib including fat (0.62 ± 0.005), braised pork center loin excluding fat (0.61 ± 0.005), braised pork center loin including fat (0.61 ± 0.005), fried pork top loin including fat (0.61 ± 0.005), skim yogurt (0.61 ± 0.002), roasted pork center loin excluding fat (0.6 ± 0.005), roasted pork center loin including fat (0.6 ± 0.005), burrito (0.6 ± 0.003), havarti cheese (0.57 ± 0.018), roasted pork center rib including fat (0.56 ± 0.005), roasted pork top loin including fat (0.56 ± 0.005), raw ling (0.56 ± 0.006), lowfat yogurt (0.56 ± 0.002), roasted pork center rib excluding fat (0.55 ± 0.005), roasted pork top loin excluding fat (0.55 ± 0.005), roasted pork tenderloin (0.55 ± 0.005), braised pork center rib including fat (0.54 ± 0.005), braised pork top loin including fat (0.54 ± 0.005), braised pork top loin excluding fat (0.52 ± 0.005), braised pork center rib excluding fat (0.52 ± 0.005), vanilla milkshake (0.52 ± 0.002), taco (0.51 ± 0.006), sponge cake (0.5 ± 0.008), chili (0.5 ± 0.002), cheesecake (0.49 ± 0.006), waffles (0.48 ± 0.007), pork and beef sausage (0.46 ± 0.038), canned blue crab (0.46 ± 0.006), sweetened condensed milk (0.45 ± 0.013), eggnog (0.45 ± 0.002), french toast (0.45 ± 0.008), protein fortified lowfat 2% milk (0.43 ± 0.002), protein fortified skim milk (0.43 ± 0.002), protein fortified lowfat 1% milk (0.43 ± 0.002), cream cheese (0.43 ± 0.018), roasted duck without skin (0.4 ± 0.005), cheese pizza (0.4 ± 0.004), hot fudge sundae (0.4 ± 0.003), strawberry sundae (0.4 ± 0.003), caramel sundae (0.4 ± 0.003), chicken and turkey salad (0.39 ± 0.018), malted milk (0.39 ± 0.002), skim milk (0.38 ± 0.002), roasted turkey light meat without skin (0.37 ± 0.005), roasted turkey dark meat without skin (0.37 ± 0.005), lowfat 1% milk (0.37 ± 0.002), whole yogurt (0.37 ± 0.002), chocolate cream pie (0.37 ± 0.004), fried chicken breast without skin (0.36 ± 0.006), fried light meat chicken without skin (0.36 ± 0.005), roasted turkey dark meat with skin (0.36 ± 0.005), restaurant vanilla milkshake (0.36 ± 0.002), indian buffalo milk (0.36 ± 0.002), lowfat 2% milk (0.36 ± 0.002), whole milk (0.36 ± 0.002), fried chicken breast with skin (0.35 ± 0.005), roasted turkey light meat with skin (0.35 ± 0.005), whole chocolate malted milk (0.35 ± 0.002), hot chocolate (0.35 ± 0.002), roasted chicken breast without skin (0.34 ± 0.006), stewed chicken drumstick without skin (0.34 ± 0.011), restaurant chocolate milkshake (0.34 ± 0.002), carob milk (0.34 ± 0.002), lowfat 1% chocolate milk (0.34 ± 0.002), lowfat 2% chocolate milk (0.34 ± 0.002), whole chocolate milk (0.34 ± 0.002), filled milk (0.34 ± 0.002), ricotta cheese (0.34 ± 0.004), roasted chicken breast with skin (0.33 ± 0.005), fried light meat chicken with skin (0.33 ± 0.005), fried dark meat chicken without skin (0.33 ± 0.005), fried chicken drumstick with skin (0.33 ± 0.01), roasted chicken drumstick with skin (0.33 ± 0.01), half and half cream (0.33 ± 0.033), sour cream (0.33 ± 0.042), roasted light meat chicken with skin (0.32 ± 0.005), roasted dark meat chicken without skin (0.32 ± 0.005), chocolate milkshake (0.32 ± 0.002), fried chicken thigh with skin (0.31 ± 0.008), roasted chicken thigh without skin (0.31 ± 0.01), noodles (0.31 ± 0.003), restaurant strawberry milkshake (0.31 ± 0.002), onion rings (0.31 ± 0.006), fried dark meat chicken with skin (0.3 ± 0.005), roasted duck with skin (0.3 ± 0.005), pepperoni pizza (0.3 ± 0.004), roasted chicken wing with skin (0.29 ± 0.015), roasted dark meat chicken with skin (0.29 ± 0.005), roasted chicken thigh with skin (0.29 ± 0.008), potato pancakes (0.29 ± 0.007), caraway cheese (0.29 ± 0.018), light mayonnaise (0.29 ± 0.036), neufchatel cheese (0.29 ± 0.018), fried chicken back with skin (0.28 ± 0.007), fried chicken wing with skin (0.28 ± 0.016), raw sockeye salmon (0.28 ± 0.006), whey (0.28 ± 0.002), fried chicken neck with skin (0.25 ±

VITAMIN B$_{12}$

0.014), yellow cake (0.25 ± 0.007), evaporated skim milk (0.25 ± 0.016), stewed chicken breast without skin (0.23 ± 0.005), stewed light meat chicken without skin (0.23 ± 0.005), bran muffins (0.23 ± 0.013), stewed dark meat chicken without skin (0.22 ± 0.005), devil's food cake (0.22 ± 0.008), buttermilk (0.22 ± 0.002), stewed chicken breast with skin (0.21 ± 0.005), stewed chicken drumstick with skin (0.21 ± 0.009), evaporated lowfat milk (0.21 ± 0.004), miso (0.21 ± 0.004), stewed light meat chicken with skin (0.2 ± 0.005), fried chicken (0.2 ± 0.006), stewed dark meat chicken with skin (0.2 ± 0.005), au gratin potatoes (0.2 ± 0.004), hush puppies (0.2 ± 0.011), restaurant coleslaw (0.2 ± 0.005), whipping cream, heavy (0.2 ± 0.033), whipping cream, light (0.2 ± 0.033), restaurant hot chocolate (0.2 ± 0.002), medium cream (0.2 ± 0.033), soybean oil and cottonseed oil margarine liquid (0.2 ± 0.1), stewed chicken thigh with skin (0.19 ± 0.007), stewed chicken wing with skin (0.18 ± 0.013), simmered chicken neck without skin (0.17 ± 0.028), evaporated whole milk (0.17 ± 0.004), cornbread (0.17 ± 0.006), corn pone (0.17 ± 0.006), lemon meringue pie (0.16 ± 0.004), turkey salad (0.15 ± 0.009), potato salad (0.15 ± 0.004), scalloped potatoes (0.14 ± 0.004), gingersnaps cookie (0.14 ± 0.071), mayonnaise (0.14 ± 0.036), simmered chicken neck with skin (0.13 ± 0.013), pancakes with butter and syrup (0.1 ± 0.003), white cake (0.1 ± 0.006), yellow corn pudding (0.09 ± 0.002), pineapple upside-down cake (0.08 ± 0.007), o'brien potatoes (0.08 ± 0.003), oatmeal cookie (0.08 ± 0.038), goat milk (0.07 ± 0.002), chicken egg white (0.06 ± 0.015), mashed potatoes (0.05 ± 0.005), angel food cake (0.03 ± 0.008), human milk (0.03 ± 0.016), light cream (0.03 ± 0.017), candied sweet potato (0.03 ± 0.005), coleslaw (0.03 ± 0.008), baked beans with pork and sweet sauce (0.02 ± 0.002), beer (0.02 ± 0.001), baked beans with pork and tomato sauce (0.01 ± 0.002), cooked wild long grain rice (0.01 ± 0.004), red table wine (0.01 ± 0.005), rose table wine (0.01 ± 0.005), internal synthesis from intestinal bacteria

Applications: anemia, angina pectoris, hardening of the arteries, atherosclerosis, diabetes, blood sugar decreased, pernicious anemia, thinning of bones, epilepsy, lack of sleep, multiple sclerosis, neuritis, dizziness, adrenal deterioration, celiac disease, worms, arthritis, bursitis, cirrhosis of liver, allergies, asthma, tuberculosis, muscular dystrophy, pellagra, skin psoriasis, shingles, skin ulcers, gastritis, peptic ulcer, alcohol cravings, weight increased

Daily Dosage: 1974 PCS — 3–9 µg; 1974 RDA — 3 µg; 1974 USRDA — 6 µg; 1980 RDA — 3 µg; 1980 USRDA — 6 µg; 1989 RDA — 2 µg; 1989 USRDA — 6 µg; nutritional — 25–100 µg; therapeutic — 250–500 µg; experimental — 500–1000 µg; toxic — none known up to 1,000 µg long term or 100,000 µg short term

Warnings: none

RADIOACTIVE VITAMIN B$_{12}$
see **vitamin B$_{12}$–60Co and vitamin B$_{12}$–57Co**

VITAMIN B$_{12}$–60Co

(radioactive form of vitamin B$_{12}$)

Names: vitamin B$_{12}$–60Co, radioactive vitamin B$_{12}$, radioactive cyanocobalamin, radiocyanocobalamin, Racobalamin, Rubratope

Applications: diagnostic aid with tracking of radioacitve isotope

Warnings: radioactive, half-life 5.2 years

VITAMIN B$_{12}$–57Co

(radioactive form of vitamin B$_{12}$)

Names: vitamin B$_{12}$–57Co, radioactive vitamin B$_{12}$, radioactive cyanocobala-

min, radiocyanocobalamin, Racobalamin, Rubratope
Applications: diagnostic aid with tracking of radioacitve isotope
Warnings: radioactive, half-life 270 days, preferred over vitamin B_{12}–60Co

VITAMIN B_{12a}

(Specific form of vitamin B_{12})

Names: vitamin B_{12a}, hydroxycobalamin, vitamin OHB_{12}, Alpha Cobione, Alpha-Ruvite, Axion, Axlon, Ciplamin H, Cobalex, Codroxomin, Depogamma, Docelan, Docevita, Droxomin, Ducobee-Hy, Duradoce, Duralta-12, Hydrogrisevit, Hydrovit, Hyxobamine, Idrogriseovit, Lyovit-H, Neo-Betalin 12, Neo-Cytamen, Neo-Macrabin, Neo-Rojamin, OH-Duphar, Oxobemin, Primabalt RP, Oxolamine, alpha-Redisol, Redisol H, Sytobex-H, Vitadurin

VITAMIN B_{12b}

(Specific form of vitamin B_{12})

Names: vitamin B_{12b}, aquocobalamin, aquocobamide, hydroxocobalamin
Sources: Streptomyces aureofaciens cultures

VITAMIN B_{12c}

(Specific form of vitamin B_{12})

Names: vitamin B_{12c}, nitrosocobalamin, nitrocobalamin, nitritocobalamin
Sources: Streptomyces griseus cultures

VITAMIN B_{12f}

(Specific form of vitamin B_{12})

VITAMIN B_{12m}

(Specific form of vitamin B_{12})

VITAMIN B_{12p}

(Specific form of vitamin B_{12})
(The term "etiocobalamin" is sometimes used to denote any vitamin B_{12} lacking the nucleotide group.)

Names: vitamin B_{12p}, etiocobalamin, factor B, cobinamide dicyanide

VITAMIN B_{12r}

(Specific form of vitamin B_{12})

Names: vitamin B_{12r}, cob (II)alamin

VITAMIN B_{12s}

(Specific form of vitamin B_{12})

Names: vitamin B_{12s}, cob (I)alamin, hydridocobalamin

VITAMIN B_{12}–ZINC TANNATE COMPLEX

(Specific form of vitamin B_{12})

Names: vitamin B_{12}–zinc tannate complex, cyanocobalamin zinc tannate complex, zinc-vitamin B_{12}–tannate
Applications: long acting injectable vitamin B_{12} preparations

VITAMIN B_{12III}

(Specific form of vitamin B_{12})

psi-VITAMIN B_{12}

(Specific form of vitamin B_{12})

psi-VITAMIN B_{12d}

(Specific form of vitamin B_{12})

VITAMIN B_{12} COENZYME

(Specific form of vitamin B_{12})

VITAMIN B_{13}

Classification: coenzyme
Names: vitamin B_{12} coenzyme, vitamin OHB_{12}, cobamamide, cobamamidum, coenzyme B_{12}, DBC, Actimide, Ademide, Anabasi, Betarin, Calomide, Cobalion, Cobaltamin S, Cobanzyme, Cobazymase, Dolonevran, Enzicoba, Heraclene, Hi-Fresmin, Hycobal, Indusil, Ripresil, Sabalamin, Xobaline
Applications: This is the form actually used in the human body.

VITAMIN B_{13}

Names: vitamin B_{13}, whey factor, orotic acid, animal galactose factor, Oropur, Orotyl, provitamin B_6
Classification: water-soluble vitamin, water-soluble vitamin precursor, vitamin B complex
Forms: orotic acid
Deficiency: anemia, large abnormal red blood cells, heart disease, heart arrhythmia, skin deterioration, skin eczema, skin psoriasis, weight increased, crystals in urine, cell deterioration, growth deterioration, mental retardation, liver deterioration, premature aging, decreased vitamin B_{12} usage, immunity decreased, decreased vitamin B_C usage
Side-effects: none known
Toxicity: none known
Inhibitors: water, sunlight
Helpers: calcium, magnesium, potassium, zinc
Sources: root vegetables, whey
Applications: multiple sclerosis, anemia, liver deterioration
Daily Dosage: 1974 PCS — none; 1989 SADDI — none; 1989 RDA — none; 1989 USRDA — none; nutritional — 10 mg; therapeutic — 100 mg; experimental — 1000 mg; toxic — unknown
Warnings: none

VITAMIN B_{14}

Classification: water-soluble vitamin, vitamin B complex
Forms: not isolated to specific chemical
Deficiency: deterioration of red blood cell production by bone marrow, pernicious anemia, accelerated reproduction of cancer cells
Side-effects: unknown
Toxicity: unknown
Inhibitors: unknown
Helpers: unknown
Sources: yeast, organ meats, whole grains, legumes, eggs
Applications: anemia, cancer
Daily Dosage: 1974 PCS — none; 1989 SADDI — none; 1989 RDA — none; 1989 USRDA — none; nutritional — none; therapeutic — unknown; experimental — unknown; toxic — unknown
Warnings: unknown

VITAMIN B_{15}

(controversial, considered fraudulent by FDA, most research done in the U.S.S.R.)
(The terms "vitamin B_{15}" and "pangamic acid" are sometimes used to denote diisopropylamine dichloroacetate.)

Names: vitamin B_{15}, pangamic acid, calcium pangamate, sodium pangamate, dimethylglycine, DMG, trimethylglycine, TMG
Classification: water-soluble vitamin, vitamin B complex
Forms: dimethylglycine, trimethylglycine, vitamin $B_{15}H_8$
Deficiency: headaches, heart deterioration, heart disease, chest pains, shortness of breath, stress, lack of sleep, shortened cell life, premature aging, angina, decreased blood oxygenation, increased blood cholesterol, hardening of arteries, circulatory deterioration, insufficient oxygen to cells, skin deterioration, slow wound healing, immunity

decreased, glandular deterioration, nervous system deterioration, mental health deterioration, fatigue, decreased protein synthesis
Side-effects: flushing of skin?
Toxicity: none known up to 100,000 mg
Inhibitors: water, sunlight
Helpers: vitamin B complex, vitamin C, vitamin E
Sources (mg per 100 grams of food): rice bran (200 ± 5), corn (150 ± 5), oats (110 ± 5), wheat germ (70 ± 5), barley (12 ± 0.5), apricot kernels, yeast, liver, corn grits, sunflower seeds, pumpkin seeds, oat grits, wheat bran, whole grain cereals
Applications: angina pectoris, atherosclerosis, increased cholesterol, high blood pressure, multiple sclerosis, autism, headache, diabetes mellitus, cirrhosis of liver, asthma, emphysema, alcohol cravings, cancer, hepatitis, decreased oxygen, rheumatic fever, rheumatism
Daily Dosage: 1989 SADDI – none; 1989 RDA – none; 1989 USRDA – none; nutritional – 2.5–15 mg; therapeutic – 75–250 mg; experimental – 500–1000 mg; toxic – none known up to 100,000 mg
Warnings: acidity of pangamic acid form should be countered with antacid

VITAMIN $B_{15}H_8$

(Specific form of vitamin B_{15})

VITAMIN B_{16}

Names: vitamin B_{16}
Classification: possible water-soluble vitamin, possible vitamin B complex
Forms: unknown
Deficiency: (being researched in Russia)
Side-effects: unknown
Toxicity: unknown
Inhibitors: unknown
Helpers: unknown
Sources: unknown
Applications: unknown
Daily Dosage: 1974 PCS – none; 1989 SADDI – none; 1989 RDA – none; 1989 USRDA – none; nutritional – unknown; therapeutic – unknown; experimental – unknown; toxic – unknown
Warnings: unknown

VITAMIN B_{17}

(controversial, considered fraudulent by FDA, illegal in most states)
(The term "laetrile" is sometimes used to designate mandelonitrile ß-glucuronide.)

Names: vitamin B_{17}, amygdalin, laetrile, NSC-15780, aprikern, nitrilosides
Classification: water-soluble vitamin, vitamin B complex
Forms: amygdalin
Deficiency: cancer, deterioration of thyroid action, sickle cell anemia, decreased red blood cell count, decreased levels of hemoglobin, increased blood pressure, unregulated blood pressure, fetor, rheumatic diseases, joint pain, limited movement of joints, appetite decreased, digestion deterioration, flatulence, bad breath, tooth decay
Side-effects: *see toxicity* symptoms
Toxicity: (Usually due to cyanide poisoning.) acrid taste, choking feeling, nausea, numbness, chest tightness, anxiety, dizziness, confusion, headache, convulsions, frothing of the mouth, incontinence, rapid/irregular/feeble pulse
Inhibitors: none known
Helpers: vitamin A, vitamin B complex, vitamin B_{15}, vitamin C, vitamin E, cysteine, methionine
Sources (mg per 100 grams of food): black limas (5100 ± 50), bitter almonds (4200 ± 50), Burma white limas (3600 ± 50), bean sprouts (2000 ± 50), green lima beans (1700 ± 50), wild cherry bark (1300 ± 50), bitter cassava (940 ± 5), flax seeds (900 ± 5), quince

seeds (680 ± 5), wild blackberry (500 +), elderberry (500 +), elderberry (500 +), apple seeds (500 +), apricot seeds (500 +), cherry seeds (500 +), nectarines seeds (500 +), peach seed (500 +), pear seed (500 +), plum seed (500 +), prune seed (500 +), fava seed (500 +), fava beans (500 +), bitter almond (500 +), macadamia nuts (500 +), bamboo sprouts (500 +), alfalfa leaves (500 +), buckwheat (340 ± 5), vetch (310 ± 5), guava seeds (190 ± 5), United States white limas (170 ± 5), boysenberry (100 +), currant (100 +), gooseberry (100 +), huckleberry (100 +), loganberry (100 +), mulberry (100 +), quince (100 +), raspberry (100 +), alfalfa sprout (100 +), buckwheat (100 +), flax seed (100 +), miller (100 +), squash seed (100 +), mung bean sprouts (100 +), garbanzo beans (100 +), lentils (100 +), kidney beans (34 ± 0.5), navy beans (34 ± 0.5), black-eyed peas (34 ± 0.5), young almonds (30 ± 0.5), mature almonds (15 ± 0.5), commercial blackberry, cranberry, black beans, green peas, sweet potatoes/yams, cashews, beet tops, spinach, watercress
Applications: cancer, sickle cell anemia, high blood pressure, decreased blood pressure, rheumatic diseases, thyroid deterioration, digestion deterioration, tooth decay, fatigue
Daily Dosage: 1989 SADDI — none; 1989 RDA — none; 1989 USRDA — none; nutritional — 250–1000 mg; therapeutic — 3000–5000 mg; experimental — 10,000 mg; toxic — 1000 mg +
Warnings: Natural sources contain enzyme beta-glucosidase which can break down vitamin B_{17} into toxins if the natural state is corrupted for a prolonged period of time (such as a slurry setting overnight in a blender). Dry heat (roasting) destroys this enzyme. Increased amounts are dangerous. See toxicity symptoms.

SUB-VITAMIN B COMPLEX

(A secondary set of different B complex vitamins. See vitamin B_C, vitamin B_H, vitamin B_P, vitamin B_T, vitamin B_W, and vitamin B_X.)

VITAMIN B_C

Names: vitamin M, vitamin B_C, factor U, folic acid, folacin, falsaure, Cytofol, Folacin, PGA, Foldine, Folaemin, Foliamin, Folicet, Folipac, Folettes, Folsan, Folvite, Incafolic, Millafol, citrovorum factor, factor CF, CF, liver lactobacillus casei factor, factor LC
Classification: water-soluble vitamin, sub-vitamin B complex
Forms: folic acid
Deficiency: retarded growth, anemia, large abnormal blood cells, numb/tingling hands, numb/tingling feet, slow/weak pulse, decrease in number of white blood cells, immunity decreased, hair decreased, weakness, skin pallor, inflamed tongue, sore tongue, depression, gastrointestinal upsets, diarrhea
Side-effects: vivid dreaming, inhibited zinc absorption
Toxicity: gastrointestinal upsets, emotional agitation, nervousness, altered sleep pattern, allergic skin reactions?
Inhibitors: water, sulfa drugs, sunlight, estrogen, food-processing, heat, aspirin, methotrexate
Helpers: vitamin B complex, vitamin B_5, vitamin B_{13}, vitamin B_{12}, vitamin B_W, vitamin C
Sources (μg per 100 grams of food): freeze-dried parsley (1571 ± 35.7), simmered chicken liver (770 ± 0.5), simmered turkey liver (666 ± 0.5), pie crust (611 ± 0.3), low fat soybean flour (410 ± 0.6), fried chicken giblets (379 ± 0.5), simmered chicken giblets (376 ± 0.5), bran flakes (357 ± 1.8), bran (357 ± 1.8), corn flakes (357 ± 1.8), toasted wheat germ (357 ± 1.8),

I. Vitamins

VITAMIN B_C

soybean flour (345 ± 0.6), simmered turkey giblets (345 ± 0.5), soybean protein concentrate (339 ± 1.8), defatted soybean flour (305 ± 0.5), defatted soy meal (302 ± 0.4), raisin bran (270 ± 1.4), freeze-dried red sweet peppers (250 ± 31.3), freeze-dried green sweet peppers (250 ± 31.3), dry roasted macadamia nuts (250 ± 1.8), oil roasted sunflower seeds (239 ± 1.8), toasted soybean nuts (229 ± 1.8), roasted soybean flour (227 ± 0.6), fried beef liver (220 ± 0.5), braised beef liver (217 ± 0.5), roasted soybean nuts (212 ± 0.6), boiled black-eyed peas (208 ± 0.3), boiled cranberry beans (207 ± 0.3), dry roasted soybean nuts (205 ± 0.6), raw turnip greens (193 ± 1.8), raw spinach (193 ± 1.8), raw parsley (183 ± 1.7), boiled lentils (181 ± 0.3), raw kelp/kombu/tangle seaweed (180 ± 0.5), soybean protein isolate (175 ± 1.8), boiled pinto beans (172 ± 0.3), boiled chickpeas (172 ± 0.3), boiled pink beans (168 ± 0.3), dehydrated onion flakes (164 ± 3.6), braised pork liver (163 ± 0.5), boiled mung beans (159 ± 0.2), chicken egg yolk (153 ± 2.9), boiled baby lima beans (150 ± 0.3), boiled black beans (149 ± 0.3), dry roasted peanuts (146 ± 1.8), boiled spinach (146 ± 0.6), boiled yardlong bean (146 ± 0.3), frozen and boiled okra (146 ± 0.5), raw endive (144 ± 2), boiled catjang black-eyed peas (142 ± 0.3), frozen and boiled black-eyed peas (141 ± 0.6), boiled navy beans (140 ± 0.3), raw romaine lettuce (136 ± 1.8), frozen and boiled asparagus (135 ± 0.8), boiled kidney beans (129 ± 0.3), oil roasted valencia peanuts (125 ± 1.8), oil roasted virginia peanuts (125 ± 1.8), oil roasted spanish peanuts (125 ± 1.8), tomato powder (120 ± 0.5), boiled artichoke (119 ± 0.6), boiled turnip greens (118 ± 0.7), dried european chestnuts (111 ± 1.8), boiled pigeon peas (111 ± 0.3), oil roasted peanuts (107 ± 1.8), frozen and boiled spinach (107 ± 0.5), dried peanuts (104 ± 1.8), boiled broad beans (104 ± 0.3), boiled great northern beans (102 ± 0.3), frozen and boiled brussels sprouts (101 ± 0.6), freeze-dried shallots (100 ± 12.5), whole dried sesame seeds (100 ± 5.6), simmered beef kidneys (98 ± 0.5), boiled asparagus (98 ± 0.6), canned spinach (98 ± 0.5), dried and frozen tofu (94 ± 2.9), boiled mungo beans (94 ± 0.3), crunchy peanut butter (91 ± 1.6), corn germ (90 ± 0.5), boiled and frozen baby lima beans (89 ± 0.6), frozen and boiled fordhook lima beans (87 ± 0.6), instant oatmeal (85 ± 0.3), boiled black turtle beans (85 ± 0.3), boiled lima beans (83 ± 0.3), creamy peanut butter (81 ± 3.1), boiled yellow beans (81 ± 0.3), canned great northern beans (81 ± 0.2), boiled white beans (81 ± 0.3), toaster pastry (80 ± 1), simmered chicken heart (80 ± 0.5), duck egg (80 ± 0.7), simmered turkey heart (79 ± 0.5), falafel (78 ± 1), canned cranberry beans (77 ± 0.2), frozen and boiled collards (76 ± 0.6), boiled peanuts (75 ± 1.6), boiled french beans (75 ± 0.3), raw butterhead lettuce (73 ± 3.3), dried filberts (71 ± 1.8), roasted european chestnuts (71 ± 1.8), dry roasted cashews (71 ± 1.8), raw broccoli (70 ± 1.1), boiled broccoli (69 ± 0.6), dried english/persian walnuts (68 ± 1.8), cashew butter (68 ± 1.8), oil roasted cashews (68 ± 1.8), canned chickpeas (67 ± 0.2), raw cauliflower (66 ± 1), raw leeks (65 ± 1.9), raw california avocado (65 ± 0.3), raw green peas (65 ± 0.6), pepperoni pizza (65 ± 0.4), canned white beans (65 ± 0.2), boiled split peas (65 ± 0.3), raw european chestnuts (64 ± 1.8), brie cheese (64 ± 1.8), camembert cheese (64 ± 1.8), boiled green peas (64 ± 0.6), mixed grain bread (64 ± 2), oil roasted almonds (64 ± 1.8), dry roasted almonds (64 ± 1.8), toasted almonds (64 ± 1.8), boysenberries (64 ± 0.4), almond butter (63 ± 3.1), toasted whole wheat bread (62 ± 2.4), chicken egg (62 ± 1), raw sprouted mung beans (62 ± 1), canned navy beans (62 ± 0.2), dried pistachio nuts (61 ± 1.8), dried almonds (61 ± 1.8), canned black turtle beans (61 ± 0.2), boiled brussels sprouts

(60 ± 0.6), canned pinto beans (60 ± 0.2), hummus (59 ± 0.2), frozen and boiled green peas (59 ± 0.6), boiled parsnips (58 ± 0.6), french rolls (58 ± 1), dried watermelon seeds (57 ± 1.8), frozen and boiled broccoli (57 ± 0.5), almond paste (57 ± 1.8), raw green cabbage (57 ± 1.4), limburger cheese (57 ± 1.8), whole wheat bread (56 ± 2), raw iceberg lettuce (55 ± 2.5), boiled soybeans (54 ± 0.3), raw florida avocado (53 ± 0.2), simmered chicken gizzard (53 ± 0.5), boiled beets (53 ± 0.6), tempeh (52 ± 0.6), boiled cauliflower (52 ± 0.8), simmered turkey gizzard (52 ± 0.5), canned black-eyed peas (51 ± 0.2), fenugreek seed (50 ± 12.5), rykrisp crackers (50 ± 3.6), dry skim milk (50 ± 1.7), shredded wheat (50 ± 1.8), roquefort cheese (50 ± 1.8), canned lima beans (50 ± 0.2), canned yellow corn (50 ± 0.5), dry instant skim milk (49 ± 0.5), canned kidney beans (49 ± 0.2), potato chips (46 ± 1.8), boiled okra (46 ± 0.6), cheese pizza (46 ± 0.4), boiled yellow corn (46 ± 0.6), toasted rye bread (45 ± 2.3), canned green peas (45 ± 0.6), canned creamed yellow corn (45 ± 0.4), canned succotash with creamed corn (44 ± 0.4), boiled artichoke hearts (44 ± 0.6), restaurant orange juice (44 ± 0.4), dry roasted pecans (43 ± 1.8), dry buttermilk (43 ± 7.1), bran muffins (43 ± 1.3), braised pork kidneys (41 ± 0.5), frozen and boiled cauliflower (41 ± 0.6), canned turnip greens (41 ± 0.4), rye bread (40 ± 2), dried pecans (39 ± 1.8), restaurant coleslaw (39 ± 0.5), dinner rolls (39 ± 1.8), frozen and boiled turnip greens (39 ± 0.6), valencia orange (39 ± 0.4), havarti cheese (39 ± 1.8), dry whole milk (38 ± 1.6), baked beans with pork and sweet sauce (38 ± 0.2), toasted raisin bread (38 ± 2.4), toasted white bread (38 ± 2.4), toasted english muffin (38 ± 1), hamburger bun (38 ± 1.3), hot dog bun (38 ± 1.3), bleu cheese (36 ± 1.8), puffed wheat (36 ± 3.6), boiled sprouted green peas (36 ± 0.5), raisin bread (36 ± 2), raw sprouted alfalfa seeds (36 ± 1.5), light mayonnaise (36 ± 3.6), mayonnaise (36 ± 3.6), screwdriver (35 ± 0.2), navel orange (34 ± 0.4), boiled green beans (34 ± 0.8), boiled yellow snap beans (34 ± 0.8), miso (33 ± 0.4), frozen and boiled succotash (33 ± 0.6), canned broad beans (33 ± 0.2), white bread (33 ± 2.1), canned green beans (32 ± 0.7), canned yellow snap beans (32 ± 0.7), wheat flakes (32 ± 1.8), canned succotash (32 ± 0.4), english muffin (32 ± 0.9), french bread (32 ± 1.8), italian bread (32 ± 1.8), oatmeal bread (32 ± 1.8), raw scallop squash (31 ± 0.8), sweet roll (31 ± 1.2), fried tofu (31 ± 3.8), french fried potatoes (30 ± 1), potato pancakes (29 ± 0.7), carob flour (29 ± 0.5), firm raw tofu (29 ± 0.4), roast beef sandwich (28 ± 0.3), pork liverwurst (28 ± 2.8), french toast (28 ± 0.8), french fries (27 ± 0.6), coleslaw (27 ± 0.8), hush puppies (27 ± 1.1), raw radish (27 ± 1.1), raw coconut (27 ± 1.1), cooked plantain (26 ± 0.3), loganberries (26 ± 0.3), yellow corn pudding (25 ± 0.2), fish sandwich (25 ± 0.3), raw red chili peppers (24 ± 1.1), raw green chili peppers (24 ± 1.1), vegetarian baked beans (24 ± 0.2), bagels (24 ± 0.9), boiled leeks (23 ± 1.9), baked sweet potato (23 ± 0.4), frozen and baked sweet potato (23 ± 0.6), raw crookneck squash (23 ± 0.8), frozen and boiled yellow corn (23 ± 0.6), pineapple juice (23 ± 0.2), baked beans with pork and tomato sauce (23 ± 0.2), sponge cake (23 ± 0.8), teriyaki sauce (22 ± 2.8), raw zucchini (22 ± 0.8), boiled scallop squash (21 ± 0.6), brick cheese (21 ± 1.8), gouda cheese (21 ± 1.8), puffed rice (21 ± 3.6), pita bread (21 ± 1.3), corn tortilla (20 ± 1.7), raw red cabbage (20 ± 1.4), raw onions (20 ± 0.6), boiled butternut squash (20 ± 0.5), restaurant tomato juice (20 ± 0.4), tomato juice (20 ± 0.3), boiled crookneck squash (20 ± 0.6), raw mushrooms (20 ± 1.4), boiled green cabbage (20 ± 0.7), tangerine (20 ± 0.6), bananas (19 ± 0.4), tamari sauce (19 ± 0.9), baked acorn squash (19 ± 0.5), waffles (19 ± 0.7), cooked oysters

(18 ± 0.6), boiled mushrooms (18 ± 0.6), cheddar cheese (18 ± 1.8), edam cheese (18 ± 1.8), port du salut cheese (18 ± 1.8), cheesecake (18 ± 0.6), canned sweet potato (17 ± 0.3), baked hubbard squash (17 ± 0.5), bacon cheeseburger (17 ± 0.3), cantaloupe (17 ± 0.3), hamburger sandwich (17 ± 0.5), burrito (17 ± 0.3), raw eggplant (17 ± 1.2), boiled zucchini (17 ± 0.6), strawberries (17 ± 0.3), saltine crackers (17 ± 8.3), frozen and french fried potatoes without skin (16 ± 1), baked/boiled tropical yam (16 ± 0.7), shoyu sauce (16 ± 0.9), raw red sweet peppers (16 ± 1), raw green sweet peppers (16 ± 1), cheeseburger (16 ± 0.4), canned pink salmon (15 ± 0.6), cooked whiting (15 ± 0.6), boiled rutabaga (15 ± 0.6), oatmeal cookie (15 ± 3.8), boiled eggplant (15 ± 1), raw tofu (15 ± 0.4), chicken egg white (15 ± 1.5), raw spring onions with tops (14 ± 1), boiled carrots (14 ± 0.6), taco (14 ± 0.6), raw carrots (14 ± 0.7), soy sauce (14 ± 0.9), boiled kale (14 ± 0.8), boiled scotch kale (14 ± 0.8), frozen and boiled kale (14 ± 0.8), graham crackers (14 ± 3.6), bloody mary (14 ± 0.3), cottage cheese (14 ± 1.8), breaded and fried oysters (14 ± 0.6), pretzels (14 ± 1.8), gingersnaps cookie (14 ± 7.1), raw trout (13 ± 0.6), fried beef top round excluding fat (13 ± 0.5), fried chicken (13 ± 0.6), raw potato without skin (13 ± 0.4), sardines (13 ± 2.1), raw whiting (13 ± 0.6), frozen and boiled crookneck squash (13 ± 0.5), onion rings (13 ± 0.6), frozen and boiled onions (13 ± 0.5), raw cucumber (13 ± 1), lemon juice (13 ± 0.2), shortbread cookie (13 ± 6.3), cooked whelk (12 ± 0.6), broiled beef top round excluding fat (12 ± 0.5), broiled beef top round including fat (12 ± 0.5), roasted pork leg excluding fat (12 ± 0.5), microwaved potato with skin (12 ± 0.2), microwaved potato without skin (12 ± 0.3), well done baked lean ground beef (12 ± 0.5), dried dates (12 ± 0.6), boiled onions (12 ± 0.5), boiled red cabbage (12 ± 0.7), breaded and fried fish (12 ± 0.7), chili (12 ± 0.2), pineapple upside-down cake (12 ± 0.7), canned pumpkin (12 ± 0.4), skim yogurt (12 ± 0.2), yellow cake (12 ± 0.7), pink grapefruit (12 ± 0.4), strawberry sundae (12 ± 0.3), fried beef top round including fat (11 ± 0.5), braised beef bottom round excluding fat (11 ± 0.5), baked potato with skin (11 ± 0.2), braised beef bottom round including fat (11 ± 0.5), braised beef chuck arm pot roast excluding fat (11 ± 0.5), well done broiled extra lean ground beef (11 ± 0.5), dried coconut (11 ± 1.8), well done broiled lean ground beef (11 ± 0.5), well done baked extra lean ground beef (11 ± 0.5), well done baked ground beef (11 ± 0.5), boiled sweet potato (11 ± 0.3), dried apricots (11 ± 1.4), frozen and boiled carrots (11 ± 0.7), boiled acorn squash (11 ± 0.4), cooked lobster (11 ± 0.6), muenster cheese (11 ± 1.8), provolone cheese (11 ± 1.8), gruyere cheese (11 ± 1.8), sweetened condensed milk (11 ± 1.3), lowfat yogurt (11 ± 0.2), unsweetened baking chocolate (11 ± 1.8), neufchatel cheese (11 ± 1.8), candied sweet potato (11 ± 0.5), canned sprouted mung beans (10 ± 0.8), piña colada (10 ± 0.4), fried beef sirloin excluding fat (10 ± 0.5), broiled beef round excluding fat (10 ± 0.5), broiled beef sirloin excluding fat (10 ± 0.5), roasted pork leg including fat (10 ± 0.5), simmered beef shank excluding fat (10 ± 0.5), well done fried lean ground beef (10 ± 0.5), well done fried extra lean ground beef (10 ± 0.5), well done broiled ground beef (10 ± 0.5), well done fried ground beef (10 ± 0.5), roasted duck without skin (10 ± 0.5), frozen and boiled red sweet peppers (10 ± 0.5), frozen and boiled green sweet peppers (10 ± 0.5), canned carrots (10 ± 0.7), boiled red sweet peppers (10 ± 0.7), boiled green sweet peppers (10 ± 0.7), boiled hubbard squash (10 ± 0.4), pineapple (10 ± 0.3), lemon (10 ± 0.9), raw oysters (10 ± 0.6), boiled winged beans (10 ± 0.3), raw red tomato (10 ± 0.4), canned crookneck squash (10 ± 0.5), white grapefruit (10 ± 0.4), devil's food cake (10 ± 0.8),

raw celery (10 ± 1.3), canned sockeye salmon (9 ± 0.6), lime (9 ± 0.7), cooked mullet (9 ± 0.6), broiled beef round including fat (9 ± 0.5), fried beef sirloin including fat (9 ± 0.5), broiled pork center rib excluding fat (9 ± 0.5), broiled pork top loin excluding fat (9 ± 0.5), roasted pork center rib excluding fat (9 ± 0.5), roasted pork top loin excluding fat (9 ± 0.5), broiled beef sirloin including fat (9 ± 0.5), stewed chicken drumstick without skin (9 ± 1.1), fried dark meat chicken without skin (9 ± 0.5), roasted turkey dark meat without skin (9 ± 0.5), braised beef flank excluding fat (9 ± 0.5), braised beef flank including fat (9 ± 0.5), simmered beef shank including fat (9 ± 0.5), roasted turkey dark meat with skin (9 ± 0.5), baked potato without skin (9 ± 0.3), braised beef chuck arm pot roast including fat (9 ± 0.5), medium-rare fried lean ground beef (9 ± 0.5), medium-rare broiled extra lean ground beef (9 ± 0.5), medium-rare fried extra lean ground beef (9 ± 0.5), medium-rare broiled ground beef (9 ± 0.5), boiled potato without skin (9 ± 0.4), medium-rare broiled lean ground beef (9 ± 0.5), medium-rare fried ground beef (9 ± 0.5), medium-rare baked ground beef (9 ± 0.5), medium-rare baked extra lean ground beef (9 ± 0.5), medium-rare baked lean ground beef (9 ± 0.5), scalloped potatoes (9 ± 0.4), canned oysters (9 ± 0.6), evaporated skim milk (9 ± 1.6), boiled turnips (9 ± 0.6), chocolate cream pie (9 ± 0.4), boiled red tomato (9 ± 0.4), maltex (9 ± 0.3), lemon meringue pie (9 ± 0.4), japanese persimmon (8 ± 0.3), fried pork center rib excluding fat (8 ± 0.5), fried pork top loin excluding fat (8 ± 0.5), raw mullet (8 ± 0.6), broiled beef top loin excluding fat (8 ± 0.5), broiled beef flank including fat (8 ± 0.5), broiled beef flank excluding fat (8 ± 0.5), roasted beef tip excluding fat (8 ± 0.5), broiled porterhouse steak excluding fat (8 ± 0.5), broiled beef rib eye excluding fat (8 ± 0.5), braised pork top loin excluding fat (8 ± 0.5), braised pork center rib excluding fat (8 ± 0.5), broiled T-bone steak excluding fat (8 ± 0.5), roasted beef tenderloin excluding fat (8 ± 0.5), roasted dark meat chicken without skin (8 ± 0.5), roasted pork center rib including fat (8 ± 0.5), roasted chicken thigh without skin (8 ± 1), fried chicken drumstick with skin (8 ± 1), roasted chicken drumstick with skin (8 ± 1), broiled beef rib excluding fat (8 ± 0.5), fried chicken thigh with skin (8 ± 0.8), fried dark meat chicken with skin (8 ± 0.5), fried chicken back with skin (8 ± 0.7), braised beef brisket excluding fat (8 ± 0.5), roasted beef rib roasted excluding fat (8 ± 0.5), hash brown potatoes (8 ± 0.6), mashed potatoes (8 ± 0.5), o'brien potatoes (8 ± 0.3), frozen and boiled potatoes without skin (8 ± 0.5), turkey salad (8 ± 0.9), au gratin potatoes (8 ± 0.4), cornbread (8 ± 0.6), corn pone (8 ± 0.6), boiled/baked spaghetti squash (8 ± 0.6), breaded and fried shrimp (8 ± 0.6), malted milk (8 ± 0.2), apricots (8 ± 0.5), evaporated whole milk (8 ± 0.4), evaporated lowfat milk (8 ± 0.4), frozen and boiled zucchini (8 ± 0.4), caramel sundae (8 ± 0.3), angel food cake (8 ± 0.8), braised beef lungs (8 ± 0.5), raw frozen rhubarb (8 ± 0.4), sour cream (8 ± 4.2), romano cheese (7 ± 1.8), broiled beef tenderloin excluding fat (7 ± 0.5), broiled beef tenderloin including fat (7 ± 0.5), smoked whitefish (7 ± 0.6), roasted eye of beef round excluding fat (7 ± 0.5), roasted beef round tip including fat (7 ± 0.5), broiled beef top loin including fat (7 ± 0.5), broiled beef rib eye including fat (7 ± 0.5), broiled porterhouse steak including fat (7 ± 0.5), roasted eye of beef round including fat (7 ± 0.5), roasted beef tenderloin including fat (7 ± 0.5), broiled pork center rib including fat (7 ± 0.5), broiled pork top loin including fat (7 ± 0.5), roasted pork top loin including fat (7 ± 0.5), broiled T-bone steak including fat (7 ± 0.5), roasted dark meat chicken with skin (7 ± 0.5), braised beef shortribs excluding fat (7 ± 0.5), roasted beef rib including fat

(7 ± 0.5), simmered beef brains (7 ± 0.5), dried figs (7 ± 0.3), stewed dark meat chicken without skin (7 ± 0.5), pork and beef lunch meat (7 ± 1.8), canned potato without skin (7 ± 0.6), stewed chicken drumstick with skin (7 ± 0.9), tuna salad (7 ± 0.2), pimento cheese spread (7 ± 1.8), american cheese (7 ± 1.8), swiss cheese (7 ± 1.8), mozzarella cheese (7 ± 1.8), vanilla milkshake (7 ± 0.2), whole yogurt (7 ± 0.2), boiled celery (7 ± 0.7), pear (7 ± 0.3), whipping cream, heavy (7 ± 3.3), whipping cream, light (7 ± 3.3), fried abalone (6 ± 0.6), broiled pork sirloin excluding fat (6 ± 0.5), roasted turkey light meat without skin (6 ± 0.5), fried pork center loin excluding fat (6 ± 0.5), broiled pork center loin excluding fat (6 ± 0.5), roasted turkey light meat with skin (6 ± 0.5), roasted pork shank excluding fat (6 ± 0.5), roasted pork loin excluding fat (6 ± 0.5), canned lean ham (6 ± 0.5), roasted pork tenderloin (6 ± 0.5), roasted pork sirloin excluding fat (6 ± 0.5), fried beef brains (6 ± 0.5), raw whelk (6 ± 0.6), fried pork center rib including fat (6 ± 0.5), fried pork top loin including fat (6 ± 0.5), braised pork center rib including fat (6 ± 0.5), braised pork top loin including fat (6 ± 0.5), roasted chicken thigh with skin (6 ± 0.8), roasted pork rump excluding fat (6 ± 0.5), braised beef chuck blade roast excluding fat (6 ± 0.5), broiled beef rib including fat (6 ± 0.5), braised pork rump including fat (6 ± 0.5), fried chicken neck with skin (6 ± 1.4), braised beef brisket including fat (6 ± 0.5), roasted duck with skin (6 ± 0.5), stewed chicken thigh with skin (6 ± 0.7), simmered chicken neck without skin (6 ± 2.8), stewed dark meat chicken with skin (6 ± 0.5), potato salad (6 ± 0.4), hot fudge sundae (6 ± 0.3), protein fortified lowfat 2% milk (6 ± 0.2), protein fortified skim milk (6 ± 0.2), protein fortified lowfat 1% milk (6 ± 0.2), whole chocolate malted milk (6 ± 0.2), boiled collards (6 ± 0.3), white cake (6 ± 0.6), blueberries (6 ± 0.3), indian buffalo milk (6 ± 0.2), whiskey sour (6 ± 0.6), tuna salad spread (6 ± 0.9), human milk (6 ± 1.6), raw abalone (5 ± 0.6), canned tuna in water (5 ± 0.6), fried chicken breast without skin (5 ± 0.6), braised pork sirloin excluding fat (5 ± 0.5), canned ham (5 ± 0.5), broiled pork loin excluding fat (5 ± 0.5), braised pork loin excluding fat (5 ± 0.5), braised pork center loin excluding fat (5 ± 0.5), roasted canned lean ham (5 ± 0.5), broiled pork sirloin including fat (5 ± 0.5), roasted pork loin blade excluding fat (5 ± 0.5), broiled pork loin blade excluding fat (5 ± 0.5), braised pork arm excluding fat (5 ± 0.5), roasted pork arm excluding fat (5 ± 0.5), braised pork loin blade excluding fat (5 ± 0.5), roasted pork shoulder excluding fat (5 ± 0.5), broiled pork center loin including fat (5 ± 0.5), roasted pork shank including fat (5 ± 0.5), fried pork center loin including fat (5 ± 0.5), broiled pork loin including fat (5 ± 0.5), roasted pork loin including fat (5 ± 0.5), roasted pork sirloin including fat (5 ± 0.5), broiled pork boston blade excluding fat (5 ± 0.5), roasted pork boston blade excluding fat (5 ± 0.5), roasted canned ham (5 ± 0.5), cooked bacon (5 ± 0.4), braised pork boston blade excluding fat (5 ± 0.5), braised beef chuck blade roast including fat (5 ± 0.5), braised beef shortribs including fat (5 ± 0.5), canned pork lunch meat (5 ± 2.4), simmered beef tongue (5 ± 0.5), lowfat 1% milk (5 ± 0.2), lowfat 2% milk (5 ± 0.2), carob milk (5 ± 0.2), hot chocolate (5 ± 0.2), lowfat 1% chocolate milk (5 ± 0.2), lowfat 2% chocolate milk (5 ± 0.2), whole milk (5 ± 0.2), whole chocolate milk (5 ± 0.2), skim milk (5 ± 0.2), filled milk (5 ± 0.2), chocolate milkshake (5 ± 0.2), apple pie (5 ± 0.4), frozen and cooked sweetened rhubarb (5 ± 0.4), chocolate syrup (5 ± 1.3), black tea (5 ± 0.3), cream of wheat (4 ± 0.3), fried light meat chicken without skin (4 ± 0.5), roasted light meat chicken without skin (4 ± 0.5), fried chicken breast with skin (4 ± 0.5), fried light

meat chicken with skin (4 ± 0.5), unheated lean ham (4 ± 0.5), grilled canadian bacon (4 ± 1.1), braised pork heart (4 ± 0.4), maypo (4 ± 0.3), unheated canadian bacon (4 ± 0.9), braised pork center loin including fat (4 ± 0.5), roasted cured pork arm excluding fat (4 ± 0.5), braised pork loin including fat (4 ± 0.5), roasted pork loin blade including fat (4 ± 0.5), braised pork sirloin including fat (4 ± 0.5), braised pork spareribs (4 ± 0.5), pork beerwurst salami (4 ± 2.2), broiled pork loin blade including fat (4 ± 0.5), roasted pork shoulder including fat (4 ± 0.5), pork luxury loaf (4 ± 1.8), braised pork loin blade including fat (4 ± 0.5), pork and beef picnic loaf (4 ± 1.8), beef and pork summer sausage (4 ± 2.2), pork and beef peppered loaf (4 ± 1.8), roasted pork arm including fat (4 ± 0.5), broiled pork boston blade including fat (4 ± 0.5), roasted pork boston blade including fat (4 ± 0.5), braised pork arm braised including fat (4 ± 0.5), beef summer sausage (4 ± 2.2), pork bologna (4 ± 2.2), lebanon bologna (4 ± 2.2), dried prunes (4 ± 0.6), pork olive loaf (4 ± 1.8), dried brazil nuts (4 ± 1.8), carrot juice (4 ± 0.3), pork and beef old fashioned loaf (4 ± 1.8), pork pickle and pimento loaf (4 ± 1.8), pork headcheese (4 ± 1.8), beef and pork bologna (4 ± 2.2), beef beerwurst salami (4 ± 2.2), raw yellowtail (4 ± 0.6), dried sweetened flaked coconut (4 ± 2), beef bologna (4 ± 2.2), beef and pork frankfurter (4 ± 1.1), cooked shrimp (4 ± 0.6), beef frankfurter (4 ± 0.9), jellied corned beef loaf (4 ± 1.8), chicken and turkey salad (4 ± 1.8), european grapes (4 ± 0.3), thompson seedless grapes (4 ± 0.3), american grapes (4 ± 0.5), raw shrimp (4 ± 0.6), restaurant chocolate milkshake (4 ± 0.2), pancakes with butter and syrup (4 ± 0.3), cherries (4 ± 0.7), nectarine (4 ± 0.4), oatmeal (4 ± 0.3), noodles (4 ± 0.3), semi-sweet bakers chocolate (4 ± 1.8), german sweet bakers chocolate (4 ± 1.8), roasted chicken breast without skin (3 ± 0.6), roasted chicken breast with skin (3 ± 0.5), roasted light meat chicken with skin (3 ± 0.5), fried chicken wing with skin (3 ± 1.6), roasted chicken wing with skin (3 ± 1.5), roasted lean ham (3 ± 0.5), unheated ham (3 ± 0.5), stewed chicken breast without skin (3 ± 0.5), stewed light meat chicken without skin (3 ± 0.5), golden seedless raisins (3 ± 0.5), stewed chicken breast with skin (3 ± 0.5), roasted cured pork arm including fat (3 ± 0.5), stewed light meat chicken with skin (3 ± 0.5), seedless raisins (3 ± 0.5), stewed chicken wing with skin (3 ± 1.3), roasted pork blade roll (3 ± 0.5), seeded raisins (3 ± 0.5), simmered chicken neck with skin (3 ± 1.3), grape juice (3 ± 0.2), restaurant strawberry milkshake (3 ± 0.2), restaurant vanilla milkshake (3 ± 0.2), raw apple with skin (3 ± 0.4), cream of rice (3 ± 0.3), peach (3 ± 0.6), cooked wild long grain rice (3 ± 0.4), raw beef tripe (2 ± 0.5), roasted goose with skin (2 ± 0.5), smoked chinook salmon (2 ± 0.6), smoked cisco (2 ± 0.6), simmered beef heart (2 ± 0.5), gefiltefish with broth (2 ± 1.2), watermelon (2 ± 0.3), plum (2 ± 0.8), cranberry sauce (2 ± 0.5), soy milk (2 ± 0.2), red table wine (2 ± 0.5), macaroni (2 ± 0.4), spaghetti (2 ± 0.4), farina (2 ± 0.3), papaya nectar (2 ± 0.2), daiquiri (2 ± 0.8), apricot nectar (1 ± 0.2), restaurant hot chocolate (1 ± 0.2), roasted pork center loin excluding fat (1 ± 0.5), roasted pork center loin including fat (1 ± 0.5), raw bluefish (1 ± 0.6), fried pork loin blade including fat (1 ± 0.5), braised pork boston blade braised including fat (1 ± 0.5), cooked carp (1 ± 0.6), raw bacon (1 ± 0.7), raw pork jowl (1 ± 0.5), eggnog (1 ± 0.2), raw apple without skin (1 ± 0.4), cooked apple without skin (1 ± 0.3), instant corn grits (1 ± 0.4), whey (1 ± 0.2), rose table wine (1 ± 0.5), applesauce (1 ± 0.4), restaurant cranberry juice cocktail (1 ± 0.3), chamomile tea (1 ± 0.3), tom collins (1 ± 0.2)

Applications: anemia, leukemia, pernicious anemia, diarrhea, alcohol cravings, mental illness, adrenal deteri-

oration, baldness, hardening of the arteries, atherosclerosis, celiac disease, diverticulitis, arthritis, emphysema, nail deterioration, skin psoriasis, skin ulcers, gastritis, digestion upset, alcohol cravings, bruises, fatigue, kwashiorkor, pellagra, scurvy, stress, tonsillitis
Daily Dosage: (at least 1 mg vitamin B_C per every 300 mg vitamin B_5) 1974 PCS — 200–400 µg; 1974 RDA — 400 µg; 1974 USRDA — 400 µg; 1980 RDA — 400 µg; 1980 USRDA — 400 µg; 1989 RDA — 200 µg; 1989 USRDA — 200 µg; nutritional — 400–800 µg; therapeutic — 5000–10,000 µg; experimental — 10,000–15,000 µg; toxic — 15,000 µg +
Warnings: Not recommended for people with history of convulsive deterioration or hormone related cancer. May inhibit sulfonamides. Can mask symptoms of vitamin B_{12} deficiency while deterioration continues. Acidity should be countered with antacid. Increased amounts are dangerous. See toxicity symptoms.

VITAMIN B_H

Names: vitamin B_H, inositol, inosite, myoinositol, mesoinositol, i-inositol, mesoinosite, cyclohexanehexol, cyclohexitol, hexahydroxycyclohexane, muscle sugar, meat sugar, phaseomannite, dambose, nucite, bios I, rat antispectacled eye factor, mouse antialopecia factor
Classification: water-soluble vitamin, sub-vitamin B complex, antioxidant
Forms: inositol, inositol monophosphate, inositol niacinate
Deficiency: hair decreased, skin eczema, increased cholesterol, lack of sleep
Side-effects: none known
Toxicity: none known up to 15,000 mg
Inhibitors: caffeine, water, sulfa drugs, estrogen, food-processing, alcohol
Helpers: vitamin B complex, vitamin B_{12}, vitamin B_P, vitamin F

Sources (mg per 100 grams of food): lecithin (2200 ± 50), tea leaves (1000 ± 50), wheat germ (770 ± 5), rice (700 ± 5), navy beans (500 ± 5), rice bran (460 ± 5), rice polishings (454 ± 0.5), pork (410 ± 5), cooked barley (390 ± 5), rice germ (370 ± 5), whole wheat (370 ± 5), veal (340 ± 5), brewer's yeast (270 ± 5), torula yeast (270 ± 5), oatmeal (270 ± 5), beef (260 ± 5), blackeyed peas (240 ± 5), garbanzo beans (240 ± 5), alfalfa (210 ± 5), orange (210 ± 5), soy flour (205 ± 0.5), soybeans (200 ± 5), roasted peanuts (180 ± 5), peanut butter (180 ± 5), lima beans (170 ± 5), blackstrap molasses (170 ± 5), green peas (162 ± 0.5), split peas (150 ± 5), grapefruit (150 ± 5), sunflower seeds (150 ± 5), lentils (130 ± 5), raisins (120 ± 5), cantaloupe (120 ± 5), brown rice (119 ± 0.5), orange juice (117 ± 0.5), whole wheat flour (110 ± 0.5), peaches (96 ± 0.5), cabbage (95 ± 0.5), cauliflower (95 ± 0.5), human milk (92 ± 0.5), onion (88 ± 0.5), fruits (80 ± 0.5), sprouts (70 ± 0.5), whole wheat bread (67 ± 0.5), sweet potatoes (66 ± 0.5), watermelon (64 ± 0.5), strawberries (60 ± 0.5), lamb (57 ± 0.5), lettuce (55 ± 0.5), beef liver (51 ± 0.5), corn (50 ± 0.5), carrots (48 ± 0.5), chicken (48 ± 0.5), tomatoes (46 ± 0.5), turnip greens (46 ± 0.5), oysters (44 ± 0.5), egg (33 ± 0.5), apples (31 ± 0.5), vegetables (30 ± 0.5), potatoes (29 ± 0.5), cheeses (25 ± 0.5), eggs (22 ± 0.5), salmon (18 ± 0.5), halibut (17 ± 0.5), milk (13 ± 0.5), intestinal flora, internal conversion of lecithin
Applications: hardening of the arteries, atherosclerosis, increased cholesterol, stroke, constipation, dizziness, glaucoma, baldness, cirrhosis of liver, asthma, gastritis, weight increased, cancer, antioxidant therapy, sun sensitivity, free radical oxidation, life span extension, radiation, decreased radiation resistance, aging
Daily Dosage: 1989 SADDI — none; 1989 RDA — none; 1989 USRDA — none; nutritional — 500–1000 mg; therapeutic — 1000–3000 mg; experimental

—5000–15000 mg; toxic—none known up to 15,000 mg
Warnings: none

VITAMIN B$_H$ MONOPHOSPHATE

(Specific form of vitamin B vitamin B$_H$)

VITAMIN B$_H$ NIACINATE

(Specific form of vitamin B$_H$ combined with vitamin B$_3$)

Names: vitamin B$_H$ niacinate, inositol niacinate, hexanicotinoyl inositol, inositol hexanicotinate, mesoinositol hexanicotinate, Dilcit, Dilexpal, Mesotal, Esantene, Hamovannid, Hexanicit, Hexanicotol, Hexopal, Linodil, Mesonex, Palohex
Applications: peripheral vasodilator (See vitamin B$_{3a}$.)

VITAMIN B$_P$

Names: vitamin J, vitamin B$_p$, choline, amanitine, bilineurine, bursine, fagine, gossypine, luridine, sincaline, vidine
Classification: water-soluble vitamin, sub-vitamin B complex, antioxidant
Forms: choline, choline chloride, choline dehydrocholate, choline dihydrogen citrate, choline esterase, choline salicylate, choline theophyllinate
Deficiency: memory decreased, fatty liver, ulcers, increased blood pressure, kidney hemorrhage, immunity decreased, cirrhosis of liver, fatty deterioration of liver, hardening of arteries, Alzheimer's disease?, vitamin B$_6$ decreased
Side-effects: see toxicity symptoms
Toxicity: nausea, diarrhea, abdominal cramps, vitamin B$_6$ decreased
Inhibitors: water, sulfa drugs, estrogen, food-processing, alcohol, sleeping pills
Helpers: vitamin A, vitamin B$_{12}$, vitamin B$_C$, vitamin B$_H$, vitamin F, vitamin H$_3$, lithium, manganese
Sources (mg per 100 grams of food): lecithin (2200 ± 50), egg yolk (1490 ± 5), liver (550 ± 5), caviar (540 ± 5), whole egg (504 ± 0.5), wheat germ (406 ± 0.5), soybeans (340 ± 5), rice germ (300 ± 5), black-eyed peas (257 ± 0.5), torula yeast (250 ± 5), garbanzo beans (245 ± 0.5), brewer's yeast (240 ± 5), lentils (223 ± 0.5), sunflower seeds (220 ± 5), split peas (201 ± 0.5), rice bran (170 ± 5), roasted peanuts (162 ± 0.5), oatmeal (156 ± 0.5), blackstrap molasses (150 ± 5), peanut butter (145 ± 0.5), bran (143 ± 0.5), alfalfa (140 ± 5), barley (139 ± 0.5), asparagus (130 ± 5), ham (122 ± 0.5), brown rice (112 ± 0.5), lamb (110 ± 0.5), flax seeds (110 ± 0.5), veal (104 ± 0.5), rice polishings (102 ± 0.5), carrots (95 ± 0.5), whole wheat cereal (94 ± 0.5), corn (92 ± 0.5), molasses (86 ± 0.5), pork (77 ± 0.5), trout (77 ± 0.5), beef (75 ± 0.5), green peas (75 ± 0.5), sweet potatoes (66 ± 0.5), pecans (50 ± 0.5), cheddar cheese (48 ± 0.5), green beans (42 ± 0.5), potatoes (29 ± 0.5), cabbage (23 ± 0.5), spinach (22 ± 0.5), textured vegetable protein (20.5 ± 0.25), milk (15 ± 0.5), orange juice (12 ± 0.5), butter (5 ± 0.5), vegetable oils (5.0 ± 0.05), egg white (2.0 ± 0.05), apple (1.0 ± 0.05), internal methionine conversion, internal synthesis from vitamin H$_3$, internal synthesis from lecithin
Applications: angina pectoris, increased cholesterol, hepatitis, blood sugar decreased, stroke, dizziness, multiple sclerosis, glaucoma, hyperthyroidism, hair deterioration, hardening of the arteries, atherosclerosis, high blood pressure, constipation, cirrhosis of liver, asthma, muscular dystrophy, skin eczema, alcohol cravings, cancer, antioxidant therapy, sun sensitivity, free radical oxidation, life span extension, radiation, decreased radiation resistance, aging
Daily Dosage: 1989 SADDI—none; 1989 RDA—none; 1989 USRDA—

none; nutritional — 500–1000 mg; therapeutic — 1000–3000 mg; experimental — 5000–15,000 mg; toxic — 20,000 mg +

Warnings: Don't take during depressive period of manic-depressive psychosis. Increased amounts are dangerous. See toxicity symptoms.

VITAMIN Bp BROMIDE HEXAMETHYLENEDICARBAMATE

(Specific form of vitamin Bp)

Names: vitamin Bp bromide hexamethylenedicarbamate, choline bromide hexamethylenedicarbamate, hexacarbacholine bromide, hexamethylenedicarbamic acid choline bromide diester, BC 16, Imbretil

CDP-VITAMIN Bp

(Specific form of vitamin Bp)

Names: CDP-vitamin Bp, CDP-choline, citicoline, cytidine diphosphate choline ester, Audes, Cereb, Colite, Corenalin, Cyscholin, Difosfocin, Emicholin, Ensign, Haocolin, Hornbest, Neucolis, Nicholin, Nicolin, Niticolin, Recognan, Rexort, Sinkron, Suncholin
Applications: cerebral circulation deterioration

VITAMIN Bp CHLORIDE

(Specific form of vitamin Bp)

Names: vitamin Bp chloride, choline chloride, Biocolina, Hepacholine, Lipotril

VITAMIN Bp CHLORIDE CARBAMATE

(Specific form of vitamin Bp)

Names: vitamin Bp chloride carbamate, choline chloride carbamate, Carbachol, carbamylcholine chloride, carbocholine, Carcholin, Moryl, Doryl, Coletyl, Lentin

VITAMIN Bp CHLORIDE DIHYDROGEN PHOSPHATE

(Specific form of vitamin Bp)

Names: vitamin Bp chloride dihydrogen phosphate, choline chloride dihydrogen phosphate, phosphorylcholine chloride, choline phosphate chloride, choline chloride phosphate, choline phosphoric acid ester, choline phosphoric acid ester chloride

VITAMIN Bp CHLORIDE SUCCINATE

(Specific form of vitamin Bp)

Names: vitamin Bp chloride succinate, choline chloride succinate, suxamethonium chloride, choline succinate dichloride, Anectine chloride, Scoline chloride, Lysthenon, Midarine, Quelicin chloride, Sucostrin chloride, Ultrapal chloride, Succicuran

VITAMIN Bp DEHYDROCHOLATE

(Specific form of vitamin Bp)

Names: vitamin Bp dehydrocholate, choline dehydrocholate, dehydrocholic acid salt of choline, Biscolan

VITAMIN Bp DICHLORIDE

(Specific form of vitamin Bp)

Names: vitamin Bp dichloride, choline dichloride, clorocholine chloride, AC 38555, CCC, Cycocel, Cycogan

VITAMIN Bp DIHYDROGEN CITRATE

(Specific form of vitamin Bp)

Names: vitamin B$_P$ dihydrogen citrate, choline dihydrogen citrate, Chothyn, Cirrocolina, Citracholine, Neurotropan

VITAMIN B$_P$ ESTERASE

(Specific form of vitamin B$_P$)

Forms: vitamin B$_P$ esterase, acetylcholinesterase, butyrylcholinesterase

VITAMIN B$_P$ OROTATE

(Specific form of vitamin B$_P$, combined with vitamin B$_{13}$)

Names: vitamin B$_P$ orotate, choline orotate, Cholergol

VITAMIN B$_P$ SALICYLATE

(Specific form of vitamin B$_P$)

Names: vitamin B$_P$ salicylate, choline salicylate acid salt, salicylic acid choline salt, Actasal, Arret, Arthropan, Artrobione, Mundisal

VITAMIN B$_P$ THEOPHYLLINATE

(Specific form of vitamin B$_P$)

Names: vitamin B$_P$ theophyllinate, choline theophyllinate, theophylline cholinate, theophylline salt of choline, oxtriphylline, oxytrimethylline, Theokolin, Teofilcolina, Filoral, Cholinophylline, Choledyl, Theoxylline, Soliphylline

PROVITAMIN B$_P$

(See methionine)
(The term "provitamin B$_P$" is also used for DMAE/dimethylaminoethanol and DEAE/diethylaminoethanol, which can be converted to vitamin B$_P$ internally.)

VITAMIN B$_T$

(The term "B$_T$" has been used as an abbreviation for biotin.)

Names: vitamin B$_T$, carnitine
Classification: water-soluble vitamin, sub-vitamin B complex, nonessential amino acid
Forms: L-carnitine
Deficiency: decreased metabolizing of fatty acids, angina, heart disease, enlarged heart, heart malfunctions, fibrillation, increased keytones in blood, fatigue, muscle weakness, muscle cramps, muscle pain, fat accumulation in muscles, increased blood cholesterol, increased blood lipids, increased blood triglycerides, decreased hemoglobin levels in blood, decreased number of red blood cells, male sterility
Side-effects: see toxicity symptoms
Toxicity: pupil dilation, mouth watering, diarrhea, nausea
Inhibitors: heat, D-carnitine, dialysis
Helpers: vitamin B$_3$, vitamin B$_5$, vitamin B$_6$, vitamin C, iron, methionine, lysine
Sources (mg per 100 grams of food): sheep (210 ± 0.5), lamb (78 ± 0.5), fishes (70.0 ± 0.05), beef (64 ± 0.5), chicken (7.5 ± 0.05), lamb liver (2.6 ± 0.05), yeasts (2.5 ± 0.05), yeast (2.4 ± 0.05), cow milk (2.0 ± 0.05), alfalfa (2.0 ± 0.05), wheat (1.0 ± 0.05), wheat germ (1.0 ± 0.05), chicken liver (0.6 ± 0.05), bread (0.2 ± 0.05), cauliflower (0.1 ± 0.05), peanuts (0.1 ± 0.05), internal conversion of lysine with the help of methionine
Applications: angina pectoris, heart disease, heart deterioration, cardiac ischemia, chronic decreased oxygen in blood, kidney disease, cirrhosis, muscle disease, nervous system deterioration, underactive thyroid, weight increased, male sterility, Duchenne-type muscular dystrophy, myotonic dystrophy, limb-girdlemuscular dystrophy, dialysis-induced anemia
Daily Dosage: 1989 SADDI — none;

I. Vitamins

1989 RDA — none; 1989 USRDA — none; nutritional — 50–100 mg; therapeutic — 250–500 mg; experimental — 1000–2000 mg; toxic — none known up to 5000 mg
Warnings: Increased amounts are toxic. See toxicity symptoms.

PROVITAMIN B$_T$ *see* **lysine**

VITAMIN B$_W$

Names: vitamin B$_w$, vitamin H, vitamin H$_1$, bacterial vitamin H$_1$, biotin, coenzyme R, anti-egg white injury factor, factor X, nucite, bios II, Bioepiderm
Classification: water-soluble vitamin, sub-vitamin B complex, coenzyme
Forms: biotin, biotin sulfoxide
Deficiency: (Rebound Deficiency can occur if large doses are halted abruptly.) anemia, fatigue, depression, appetite decreased, nausea, sore tongue, cracked tongue, inflamed tongue, skin pallor, grayish mucous membranes, skin rashes, skin changes, skin inflammation, skin eczema, increase skin sensitivity, skin burning sensations, skin prickling sensations, hair decreased, lack of sleep, retarded wound healing, immunity decreased, muscle pain, increased cholesterol, heart conditions, paralysis
Side-effects: none known
Toxicity: none known up to 10,000 µg
Inhibitors: raw egg whites, water, sulfa drugs, estrogen, food-processing, alcohol, antibiotics, nitrous acid, oxygen, formaldehyde, chloramine T, strong acid or alkali
Helpers: vitamin B$_5$, vitamin B$_{12}$, vitamin B$_C$, vitamin C, sulphur
Sources (µg per 100 grams of food): rye (330 ± 5), royal jelly (290 ± 5), brewer's yeast (200 ± 5), lamb liver (127 ± 0.5), torula yeast (100 ± 5), pork liver (100 ± 5), butter (100 ± 5), beef liver (96 ± 0.5), soy flour (70 ± 0.5), sunflower seeds (70 ± 5), rice (70 ± 0.5), soybeans (61 ± 0.5), rice bran (60 ± 0.5), rice germ (58 ± 0.5), rice polishings (57 ± 0.5), egg yolk (52 ± 0.5), green peas (42 ± 0.5), peanut butter (39 ± 0.5), walnuts (37 ± 0.5), roasted peanuts (34 ± 0.5), garbanzo beans (32 ± 0.5), barley (31 ± 0.5), alfalfa (30 ± 0.5), tuna (30 ± 0.5), cashews (30 ± 0.5), pecans (27 ± 0.5), oatmeal (24 ± 0.5), canned sardines (24 ± 0.5), whole egg (22 ± 0.5), black-eyed peas (21 ± 0.5), corn (21 ± 0.5), mackerel (20 ± 0.5), clams (20 ± 0.5), wheat germ (20 ± 0.5), blackstrap molasses (20 ± 0.5), bee pollen (20 ± 0.5), split peas (18 ± 0.5), almonds (18 ± 0.5), beans (17 ± 0.5), cauliflower (17 ± 0.5), mushrooms (16 ± 0.5), whole wheat cereal (16 ± 0.5), canned salmon (15 ± 0.5), textured vegetable protein (15 ± 0.5), bran (14 ± 0.5), lentils (13 ± 0.5), brown rice (12 ± 0.5), chicken (10 ± 0.5), coconut (10 ± 0.5), lamb (10 ± 0.5), medium molasses (10 ± 0.5), turkey (10 ± 0.5), shrimp (10 ± 0.5), oysters (8.7 ± 0.05), halibut (8.0 ± 0.05), sprouts (8.0 ± 0.05), egg white (7.0 ± 0.05), goat milk (6.3 ± 0.05), pork (6.2 ± 0.05), avocados (6.0 ± 0.05), vegetables (5.0 ± 0.05), cow milk (4.7 ± 0.05), bananas (4.4 ± 0.05), yams (4.3 ± 0.05), beef (4.0 ± 0.05), cheeses (3.6 ± 0.05), fruits (2.5 ± 0.05), human milk (1.0 ± 0.05), apples (1.0 ± 0.05), potatoes (0.1 ± 0.05)
Applications: baldness, muscle pains, skin inflammation, skin eczema, infant skin inflammation, depression
Daily Dosage: (at least 100 mg per every 300 mg vitamin B$_5$) 1974 PCS — 150–450 µg; 1980 RDA — none; 1980 SADDI — 100–200 µg; 1980 USRDA — none; 1989 RDA — none; 1989 SADDI — 100–200 µg; 1989 USRDA — none; nutritional — 300–500 µg; therapeutic — 1000 µg; experimental — 5000–10,000 µg; toxic — none known up to 10,000 µg
Warnings: none

VITAMIN B$_W$ SULFOXIDE

(ineffective form of vitamin B$_W$)

Names: vitamin B_W sulfoxide, biotin sulfoxide, AN factor
Sources: *Aspergillus niger* cultures, milk residue concentrates, internal synthesis

VITAMIN B_X

(The term "vitamin B_X" has been used to denote vitamin B_5)

Names: vitamin H_2, vitamin B_X, vitamin H', PABA, para-aminobenzoic acid, chromotrichia factor, anti-chromotrichia factor, trichochromogenic factor, anticanitic vitamin, Amben, Paraminol, Sunbrella
Classification: water-soluble vitamin, sub-vitamin B complex, antioxidant
Forms: PABA/para-aminobenzoic acid
Deficiency: nervousness, depression, fatigue, hallucinations, digestion upsets, dry hair, graying hair, balding?, decreased sun tolerance, decreased arsenic and antimony resistance, decreased ozone resistance
Side-effects: see toxicity symptoms
Toxicity: fatty changes in liver, fatty changes in kidneys, fatty changes in heart, nausea
Inhibitors: water, sulfa drugs, food-processing, alcohol, estrogen
Helpers: vitamin B complex, vitamin B_C, vitamin C, vitamin H_3
Sources (mg per 100 grams of food): sunflower seeds (62.0 ± 0.05), liver (0.62 ± 0.005), brewer's yeast (0.49 ± 0.005), wheat germ (0.037 ± 0.0005), mushrooms, bran, cabbage, oats, spinach, whole milk, eggs, internal conversion of vitamin B_C, internal conversion of vitamin H_3
Applications: anemia, constipation, baldness, headache, burns, sunburn, skin vitiligo, cancer, antioxidant therapy, sun sensitivity, free radical oxidation, life span extension, radiation, decreased radiation resistance, aging
Daily Dosage: 1989 SADDI—none; 1989 RDA—none; 1989 USRDA—none; nutritional—50–100 mg; therapeutic—500–1000 mg; experimental—2000–3000 mg; toxic—5,000 mg + ?
Warnings: May inhibit sulfonamides. Acidity should be countered with antacid. Increased amounts are dangerous. See toxicity symptoms.

VITAMIN C COMPLEX

(Vitamin C and the bioflavinoids. See vitamin C_1, vitamin C_2, vitamin C_3, vitamin P_1, vitamin P_1 complex, vitamin P_2, vitamin P_3, vitamin P_4.)

ANTIVITAMIN C

(Neutralizes vitamin C on contact.)

Names: antivitamin C, dehydroascorbic acid
Classification: water-soluble antivitamin, oxidant
Forms: dehydroascorbic acid
Deficiency: (not required)
Side-effects: see toxicity symptoms
Toxicity: vitamin C decreased
Inhibitors: antioxidants, provitamin A, vitamin A, vitamin C, vitamin E, vitamin P
Sources: vitamin C mixed with water that has been left standing for many hours

Applications: none known
Daily Dosage: 1974 PCS — none; 1989 SADDI — none; 1989 RDA — none; 1989 USRDA — none; nutritional — none; therapeutic — none; experimental — none; toxic — any amount
Warnings: Vitamin C will eventually convert to antivitamin C if mixed with water. Increased amounts are dangerous. See toxicity symptoms.

VITAMIN C

Names: vitamin C, vitamin C_1, ascorbic acid, hexuronic acid, antiscrobutic vitamin, cevitamic acid, Cebid, Cebion, Cantaxin, Cevalin, Cevatine, Cevimin, Cevitex, Cewin, Cipca, Cebicure, C-Vimin, Cevitamin, Testascorbic, Allercorb, Cecon, Ce-Vi-Sol, Ascorin, Ascorteal, Cegiolan, Adenex, Ascorvit, Cevex, Lemascorb, Ciamin, Hybrin, Vitacee, Cantan, Catavin C, Celin, Cenetone, Cescorbat, Cereon, Cergona, Cetemican, Cetamid, Planavit C, Colascor, Concemin, Duoscorb, Scorbacid, Davitamon C, Proscorbin, Redoxon, Scorbu-C, Ribena, Vicelat, Vitacin, Vitacimin, Vitascorbol, Xitrix, Cevitan, Laroscorbine
Classification: water-soluble vitamin, antioxidant
Forms: ascorbic acid, ascorbic acid calcium salt, ascorbic acid magnesium salt, ascorbic acid nicotinamide complex, ascorbic acid potassium salt, ascorbic acid sodium salt, ascorbyl palmitate
Deficiency: (Rebound Deficiency can occur if large doses are halted abruptly.) scurvy, general swelling, dry skin, small skin hemorrhages, easy bruising, nosebleeds, bleeding gums, hemorrhaging, increased cholesterol levels, weak teeth in children, loose teeth, weak bones in children, separation of ends of long bones, swollen joints, aching joints, aching muscles, aching extremities, dry hair, swelling hair follicles, hardening hair follicles, hair decreased, drying tear glands, appetite decreased, weakness, fatigue, depression, stressed, emotional agitation, shortness of breath, muscle weakness, anemia, weight decreased, digestion deterioration, slow healing, immunity decreased, decreased cancer resistance, gallstones, decreased heat tolerance, decreased cold tolerance, back trouble
Side-effects: transient colic, diarrhea, gas, increased urination, vitamin B_{12} decreased, vitamin B_C decreased
Toxicity: kidney stones?, skin rashes, increased urinary oxalate excretion, deterioration to growing bone
Inhibitors: ginseng, smoking, water, cooking, heat, light, oxygen, aspirin, aspirin substitutes, ferrous sulfate, barbiturates, adrenaline, estrogen, stilbestrol, ammonium chloride, antihistamines, thyroid, atripine, sulfonamides, antacids, anticoagulants, cortisone, prednisone, diuretics, antidepressants, indomethacin, sodium salicylate, sodium nitrate, theobromine sodium salicylate, methenamine
Helpers: vitamin B_2, vitamin C_2, vitamin C_3, vitamin P, vitamin T, calcium, magnesium
Sources (mg per 100 grams of food): rose hips (3000 ± 500), freeze-dried red sweet peppers (1875 ± 31.3), freeze-dried green sweet peppers (1875 ± 31.3), raw acerola (1678 ± 0.5), acerola juice (1600 ± 0.2), freeze-dried chives (625 ± 62.5), dried coriander leaf (300 ± 50), raw red chili peppers (242 ± 1.1), raw green chili peppers (242 ± 1.1), dried parsley (200 ± 50), raw red sweet peppers (190 ± 1), guava (183 ± 0.6), dried lychees (183 ± 0.5), black currants (180 ± 0.9), freeze-dried parsley (143 ± 35.7), lemon peel (133 ± 8.3), orange peel (133 ± 8.3), raw mustard spinach (131 ± 0.7), raw green sweet peppers (128 ± 1), tomato powder (117 ± 0.5), boiled red sweet peppers (112 ± 0.7), boiled green sweet peppers (112 ± 0.7), raw vine spinach (102 ± 0.5), basil (100 ± 50), cloves (100 ± 25), cooked blue crab (100 ± 0.6), kiwi fruit (99 ± 0.7), raw broccoli (93 ± 1.1), raw parsley (90 ± 1.7),

longans (84 ± 0.5), boiled pokeberry shoots (82 ± 0.6), dehydrated onion flakes (79 ± 3.6), raw cauliflower (72 ± 1), lychees (72 ± 0.5), jujube (69 ± 0.5), persimmon (68 ± 2), raw garden cress (68 ± 2), canned red chili peppers (68 ± 0.7), canned green chili peppers (68 ± 0.7), chili powder (67 ± 16.7), raw chives (67 ± 16.7), boiled mustard spinach (66 ± 0.6), boiled broccoli (63 ± 0.6), boiled brussels sprouts (62 ± 0.6), papaya (62 ± 0.2), dried japanese chestnuts (61 ± 1.8), raw turnip greens (61 ± 1.8), pummelo (61 ± 0.3), raw red cabbage (57 ± 1.4), navel orange (57 ± 0.4), strawberries (57 ± 0.3), raw swamp cabbage (55 ± 0.9), boiled cauliflower (55 ± 0.8), bran (54 ± 1.8), corn flakes (54 ± 1.8), wheat flakes (54 ± 1.8), boiled kohlrabi (54 ± 0.6), lemon (53 ± 0.9), boiled scotch kale (52 ± 0.8), allspice (50 ± 25), cinnamon (50 ± 25), paprika (50 ± 25), cayenne pepper (50 ± 25), dried rosemary (50 ± 25), turmeric (50 ± 25), braised beef spleen (50 ± 0.5), orange juice (50 ± 0.2), raw green cabbage (49 ± 1.4), valencia orange (49 ± 0.4), raw dock (48 ± 0.7), raw cassava (48 ± 0.5), canned red sweet peppers (47 ± 0.7), canned green sweet peppers (47 ± 0.7), raw chinese cabbage (46 ± 1.4), raw spring onions with tops (46 ± 1), frozen and boiled brussels sprouts (46 ± 0.6), lemon juice (46 ± 0.2), raw lotus root (44 ± 0.6), potato chips (43 ± 1.8), raw european chestnuts (43 ± 1.8), cantaloupe (43 ± 0.3), cranberry juice cocktail (43 ± 0.2), boiled kale (42 ± 0.8), tomato paste (42 ± 0.4), raw watercress (41 ± 2.9), red currants (41 ± 0.9), white currants (41 ± 0.9), boiled amaranth (41 ± 0.8), frozen and boiled red sweet peppers (41 ± 0.5), frozen and boiled green sweet peppers (41 ± 0.5), carissa (40 ± 2.5), raw green peas (40 ± 0.6), frozen and boiled broccoli (40 ± 0.5), raw laver/nori seaweed (39 ± 0.5), restaurant orange juice (39 ± 0.4), cooked tahitian taro (38 ± 0.7), pink grapefruit (38 ± 0.4), white grapefruit juice (38 ± 0.2), red grapefruit juice (38 ± 0.2), kumquats (37 ± 2.6), boiled lamb's-quarters (37 ± 0.6), raw chinese chestnuts (36 ± 1.8), raw dandelion greens (36 ± 1.8), broiled lamb liver (36 ± 1.1), mulberries (36 ± 0.4), elderberries (36 ± 0.3), sugar apple (36 ± 0.3), pork bologna (35 ± 2.2), boiled red cabbage (35 ± 0.7), mandarin oranges (35 ± 0.4), restaurant cranberry juice cocktail (35 ± 0.3), tomato purée (35 ± 0.2), raw garlic (33 ± 5.6), boiled jute potherb (33 ± 1.2), coleslaw (33 ± 0.8), braised beef lungs (33 ± 0.5), white grapefruit (33 ± 0.4), raw savoy cabbage (31 ± 1.4), tangerine (31 ± 0.6), frozen and boiled cauliflower (31 ± 0.6), screwdriver (31 ± 0.2), tangerine juice (31 ± 0.2), hard pork and beef salami (30 ± 5), pork beerwurst salami (30 ± 2.2), lime (30 ± 0.7), braised beef thymus (30 ± 0.5), purple passion fruit juice (30 ± 0.2), minced ham lunch meat (29 ± 2.4), raw japanese chestnuts (29 ± 1.8), roasted japanese chestnuts (29 ± 1.8), dried ginko nuts (29 ± 1.8), raw new zealand spinach (29 ± 1.8), raw spinach (29 ± 1.8), pork and beef brotwurst (29 ± 0.7), breadfruit (29 ± 0.5), lime juice (29 ± 0.2), passion fruit (28 ± 2.8), boiled turnip greens (28 ± 0.7), dried longans (28 ± 0.5), gooseberries (28 ± 0.3), mango (28 ± 0.2), pork and beef mortadella (27 ± 3.3), beef and pork frankfurter (27 ± 1.1), frozen and boiled collards (27 ± 0.6), boiled lotus root (27 ± 0.6), raw welsh onions (27 ± 0.5), pitanga (27 ± 0.3), canned sweet potato (27 ± 0.3), pork and beef knockwurst (26 ± 0.7), boiled mustard greens (26 ± 0.7), boiled chinese cabbage (26 ± 0.6), boiled dock (26 ± 0.5), freeze-dried shallots (25 ± 12.5), ham and cheese roll (25 ± 1.8), pork and beef peppered loaf (25 ± 1.8), roasted european chestnuts (25 ± 1.8), raw romaine lettuce (25 ± 1.8), beef frankfurter (25 ± 0.9), frozen and boiled asparagus (25 ± 0.8), frozen and boiled kale (25 ± 0.8), boiled beet greens (25 ± 0.7), strawberry pie (25 ± 0.5), raspberries (25 ± 0.4), baked sweet potato (25 ± 0.4), boiled green cabbage (24 ± 0.7), boiled garden cress (24 ± 0.7), raw chickory greens (24 ± 0.6), braised pork liver (24 ± 0.5), raw green tomato (24 ± 0.4), pork and beef kielbasa/kolbassy (23 ± 1.9), raw oriental radish (23 ± 1.1), fried beef liver (23 ± 0.5), braised beef liver (23 ± 0.5), honeydew melon (23 ± 0.5), summer sausage (22 ± 2.2), beef bologna (22 ± 2.2), beef and pork bologna (22 ± 2.2), lebanon bologna (22 ± 2.2), beef summer sausage (22 ± 2.2), beef and pork summer sausage (22 ± 2.2), raw radish (22 ± 1.1), boiled rutabaga (22 ± 0.6), frozen and boiled turnip greens (22 ± 0.6), rose apple (22 ± 0.5), chopped smoked beef (21 ± 1.8), pork headcheese (21 ± 1.8), pork and beef honey loaf (21 ± 1.8), pork luxury loaf (21 ± 1.8), blackberries (21 ± 0.7), carambola (21 ± 0.4), boiled red tomato (21 ± 0.4), boiled asparagus (20 ± 0.6), braised beef pancreas (20 ± 0.5), raw celtuce (20 ± 0.5), raw potato without skin (20 ± 0.4), sapote (20 ± 0.2), soursop (20 ± 0.2), chopped ham lunch meat (19 ± 2.4), pork and beef link sausage with american cheese (19 ± 1.2), pork and beef link sausage (19 ± 0.7), potato flour (19 ± 0.6), boiled butterbur (19 ± 0.5), tequila sunrise (19 ± 0.3), pork and beef old fashioned loaf (18 ± 1.8), pork and beef picnic loaf (18 ± 1.8), raw looseleaf lettuce (18 ± 1.8), raw scallop squash (18 ± 0.8), boiled swiss chard (18 ± 0.6), tomato juice (18 ± 0.4), raw red tomato (18 ± 0.4), restaurant tomato juice (18 ± 0.3), yellow passion fruit juice (18 ± 0.2), barbecue loaf (17 ± 2.2), beef honey roll sausage (17 ± 2.2), pork and beef luncheon sausage (17 ± 2.2), beef beerwurst salami (17 ± 2.2), beef salami (17 ± 2.2), boiled black-eyed pea pods (17 ± 1.1), boiled dandelion greens (17 ± 1), boiled green soybeans (17 ± 0.6), boiled sweet potato (17 ± 0.3), boiled swamp cabbage (16 ± 1), stir-fried sprouted mung beans (16 ± 0.8), boiled savoy cabbage (16 ± 0.7), boiled new zealand spinach (16 ± 0.6), boiled okra (16 ± 0.6), simmered chicken liver (16 ± 0.5), boiled yambean (16 ± 0.5), o'brien potatoes (16 ± 0.3), casaba melon (16 ± 0.3), loganberries (16 ± 0.3), boiled oriental radish (15 ± 0.7), quince (15 ± 0.5), boiled butternut squash (15 ± 0.5), canned peeled red tomatoes (15 ± 0.4), canned turnip greens (15 ± 0.4), pineapple (15 ± 0.3), sapodilla (15 ± 0.3), microwaved potato without skin (15 ± 0.3), microwaved potato with skin (15 ± 0.2), beef lunch meat (14 ± 1.8), pork and beef lunch meat (14 ± 1.8), pork pickle and pimento loaf (14 ± 1.8), dried european chestnuts (14 ± 1.8), ginko nuts (14 ± 1.8), boiled green peas (14 ± 0.6), cranberry sauce (14 ± 0.5), mammy apple (14 ± 0.5), prickly pear (14 ± 0.5), braised pork brains (14 ± 0.5), canned spinach (14 ± 0.5), java plum (14 ± 0.4), canned sauerkraut (14 ± 0.4), bloody mary (14 ± 0.3), tomato catsup (13 ± 3.3), chili sauce (13 ± 3.3), beef and pork salami (13 ± 2.2), raw sprouted mung beans (13 ± 1), frozen and boiled mustard greens (13 ± 0.7), canned jalapeño peppers (13 ± 0.7), frozen and boiled fordhook lima beans (13 ± 0.6), boiled parsnips (13 ± 0.6), natto (13 ± 0.6), dried jujube (13 ± 0.5), stir-fried sprouted lentils (13 ± 0.5), frozen and boiled spinach (13 ± 0.5), tomato sauce (13 ± 0.4), canned stewed red tomatoes (13 ± 0.4), blueberries (13 ± 0.3), baked potato without skin (13 ± 0.3), marinara sauce (13 ± 0.2), baked potato with skin (13 ± 0.2), raw leeks (12 ± 1.9), frozen and french fried potatoes without skin (12 ± 1), roselle (12 ± 0.9), baked/ boiled tropical yam (12 ± 0.7), whiskey sour (12 ± 0.6), boiled turnips (12 ± 0.6), braised pork spleen (12 ± 0.5), frozen and boiled okra (12 ± 0.5), stir-fried sprouted soybeans (12 ± 0.5), maypo (12 ± 0.3), pork liver sausage (11 ± 2.8), pork olive loaf (11 ± 1.8), canned chicken liver pâté (11 ± 1.8), boiled and steamed japanese chestnuts (11 ± 1.8), raw chickory (11 ± 1.1), cooked sprouted mung beans (11 ± 0.8), boiled scallop squash (11 ± 0.6), braised pork kidneys (11 ± 0.5), baked acorn squash (11 ± 0.5), scalloped potatoes (11 ± 0.4), ground-cherries (11 ± 0.4), european grapes (11 ± 0.3), thompson seedless grapes (11 ± 0.3), cooked plantain (11 ± 0.3), pineapple juice (11 ± 0.2), raw shallots (10 ± 5), french fried potatoes (10 ± 1), boiled purslane (10 ± 0.9), boiled green beans (10 ± 0.8), boiled yellow snap beans (10 ± 0.8), raw carrots (10 ± 0.7), frozen and boiled green peas (10 ± 0.6), boiled spinach (10 ± 0.6), boiled waxgourd (10 ± 0.6), apricots (10 ± 0.5), canned white corn (10 ± 0.5), dried spirulina seaweed (10 ± 0.5), baked hubbard squash (10 ± 0.5), au gratin potatoes (10 ± 0.4), potato salad (10 ± 0.4), boiled collards (10 ± 0.3), berliner (9 ± 2.2), dry whole milk (9 ± 1.6), raw sprouted alfalfa seeds (9 ± 1.5), steamed sprouted soybeans (9 ± 1.1), plum (9 ± 0.8), raw zucchini (9 ± 0.8), frozen and boiled yellow snap beans (9 ± 0.7), frozen and boiled green beans (9 ± 0.7), raw onions (9 ± 0.6), canned green peas (9 ± 0.6), frozen and baked sweet potato (9 ± 0.6), fried chicken giblets (9 ± 0.5), canned yellow corn (9 ± 0.5), frozen and boiled potatoes without skin (9 ± 0.5), frozen and boiled turnip greens and turnips (9 ± 0.5), bananas (9 ± 0.4), canned yellow corn with red and green peppers (9 ± 0.4), carrot juice (9 ± 0.3), watermelon (9 ± 0.3), cherimoya (9 ± 0.1), raw endive (8 ± 2), raw celery (8 ± 1.3), raw crookneck squash (8 ± 0.8), sour pickles (8 ± 0.8), boiled calabash gourd (8 ± 0.7), boiled chayote (8 ± 0.6), simmered chicken giblets (8 ± 0.5), crab apples (8 ± 0.5), braised pork lungs (8 ± 0.5), boiled succotash (8 ± 0.5), japanese persimmon (8 ± 0.3), raw california avocado (8 ± 0.3), raw florida avocado (8 ± 0.2), hummus (8 ± 0.2), soybean oil mayonnaise (7 ± 3.6), russian salad dressing (7 ± 3.3), raw butterhead lettuce (7 ± 3.3), dried shitake mushrooms (7 ± 3.3), bread and butter pickles (7 ± 3.3), sweet pickles (7 ± 3.3), jellied corned beef loaf (7 ± 1.8), ham and cheese spread (7 ± 1.8), ham salad (7 ± 1.8), toasted wheat germ (7 ± 1.8), dry skim milk (7 ± 1.7), cherries (7 ± 0.7), frozen white corn (7 ± 0.7), peach (7 ± 0.6), boiled artichoke hearts (7 ± 0.6), raw quail without skin (7 ± 0.5), jackfruit (7 ± 0.5), boiled sprouted green peas (7 ± 0.5), frozen and boiled crookneck squash (7 ± 0.5), candied sweet potato (7 ± 0.5), boiled potato without skin (7 ± 0.4), boiled acorn squash (7 ± 0.4), boiled hubbard squash (7 ± 0.4), canned succotash with creamed corn (7 ± 0.4), barbecue sauce (6 ± 3.1), dill pickles (6 ± 0.8), raw cuttlefish (6 ± 0.6), french fries (6 ± 0.6), boiled beets (6 ± 0.6), boiled yellow corn (6 ± 0.6), boiled dishcloth gourd (6 ± 0.6), boiled and frozen baby lima beans (6 ± 0.6), canned potato without skin (6 ± 0.6), frozen and fried hash brown potatoes (6 ± 0.6), hash brown potatoes (6 ± 0.6), boiled crookneck squash (6 ± 0.6), frozen and boiled succotash (6 ± 0.6), raw whale meat (6 ± 0.5), mashed potatoes (6 ± 0.5), raw pheasant without skin (6 ± 0.5), braised pork pancreas (6 ± 0.5), boiled onions (6 ± 0.5), raw apple with skin (6 ± 0.4), oheloberries (6 ± 0.4), pomegranate (6 ± 0.3), refried beans (6 ± 0.2), citrus marmalade (5 ± 2.5), raw iceberg lettuce (5 ± 2.5), bran muffins (5 ± 1.3), cooked taro (5 ± 0.8), raw chinese water chestnuts (5 ± 0.8), pineapple upside-down cake (5 ± 0.7), boiled celery (5 ± 0.7), raw squid (5 ± 0.6), breaded fried squid (5 ± 0.6), cheesecake (5 ± 0.6), dry roasted soybean nuts (5 ± 0.6), boiled artichoke (5 ± 0.6), dry instant skim milk (5 ± 0.5), seeded raisins (5 ± 0.5), frozen and boiled onions (5 ± 0.5), piña colada (5 ± 0.4), nectarine (5 ± 0.4), canned creamed yellow corn (5 ± 0.4), boiled pumpkin (5 ± 0.4), raw frozen rhubarb (5 ± 0.4), canned succotash (5 ± 0.4), wax beans (5 ± 0.4), peach nectar (5 ± 0.2), raw ginger root (4 ± 2.1), boiled leeks (4 ± 1.9), beef pastrami (4 ± 1.8), dried pecans (4 ± 1.8), dried pinyon pine nuts (4 ± 1.8), toasted soybean nuts (4 ± 1.8), dried english/persian walnuts (4 ± 1.8), hamburger pickle relish (4 ± 1.8), raw coconut (4 ± 1.1), raw cucumber (4 ± 1), raw bamboo shoots (4 ± 0.7), canned green beans (4 ± 0.7), canned yellow snap beans (4 ± 0.7), raw jerusalem artichoke (4 ± 0.7), boiled salsify (4 ± 0.7), raw crayfish (4 ± 0.6), cooked crayfish (4 ± 0.6), raw chinook salmon (4 ± 0.6), dried and salted cod (4 ± 0.6), cooked pike (4 ± 0.6), cooked rainbow trout (4 ± 0.6), raw pike (4 ± 0.6), raw rainbow trout (4 ± 0.6), dried prunes (4 ± 0.6), boiled mushrooms (4 ± 0.6), boiled zucchini (4 ± 0.6), boiled/baked spaghetti squash (4 ± 0.6), american grapes (4 ± 0.5), corn germ (4 ± 0.5), raw celeriac (4 ± 0.5), frozen and boiled turnips (4 ± 0.5), blackberry pie (4 ± 0.4), sweet potato pie (4 ± 0.4), raw apple without skin (4 ± 0.4), frozen creamed yellow corn (4 ± 0.4), poi (4 ± 0.4), canned pumpkin (4 ± 0.4), canned pumpkin pie mix (4 ± 0.4), frozen and boiled zucchini (4 ± 0.4), pear (4 ± 0.3), sheep milk (4 ± 0.2), prune juice (4 ± 0.2), canned chickpeas (4 ± 0.2), human milk (3 ± 1.6), dried apricots (3 ± 1.4), raw mushrooms (3 ± 1.4), pork livercheese (3 ± 1.3), sweetened condensed milk (3 ± 1.3), canned carrots (3 ± 0.7), frozen and boiled carrots (3 ± 0.7), boiled carrots (3 ± 0.6), cooked prunes (3 ± 0.5), golden seedless raisins (3 ± 0.5), seedless raisins (3 ± 0.5), fried beef brains (3 ± 0.5), raw beef tripe (3 ± 0.5), raw wakame seaweed (3 ± 0.5), canned crookneck squash (3 ± 0.5), blueberry pie (3 ± 0.4), lemon meringue pie (3 ± 0.4), rhubarb pie (3 ± 0.4), boysenberries (3 ± 0.4), tamarind (3 ± 0.4), pickled beets (3 ± 0.4), frozen and cooked sweetened rhubarb (3 ± 0.4), canned shellie beans (3 ± 0.4), canned zucchini in tomato juice (3 ± 0.4), frozen and boiled butternut squash (3 ± 0.4), yellow corn pudding (3 ± 0.2), baked beans with pork and tomato sauce (3 ± 0.2), baked beans with pork and sweet sauce (3 ± 0.2), papaya nectar (3 ± 0.2), coconut cream (3 ± 0.2), coconut milk (3 ± 0.2), coconut water (3 ± 0.2), canned black turtle beans (3 ± 0.2), canned black-eyed peas (3 ± 0.2), raw eggplant (2 ± 1.2), figs (2 ± 1), falafel (2 ± 1), boiled eggplant (2 ± 1), daiquiri (2 ± 0.8), raw yellowtail (2 ± 0.6), lemon chiffon pie (2 ± 0.6), roasted soybean nuts (2 ± 0.6), frozen and boiled yellow corn (2 ± 0.6), frozen and boiled black-eyed peas (2 ± 0.6), cooked flounder (2 ± 0.5), simmered turkey liver (2 ± 0.5), simmered turkey giblets (2 ± 0.5), simmered chicken gizzard (2 ± 0.5), simmered turkey gizzard (2 ± 0.5), simmered chicken heart (2 ± 0.5), simmered turkey heart (2 ± 0.5), canned corned beef (2 ± 0.5), simmered beef heart (2 ± 0.5), braised pork tongue (2 ± 0.5), boiled cardoon (2 ± 0.5), cheese pizza (2 ± 0.4), prune pudding (2 ± 0.4), pepperoni pizza (2 ± 0.4), evaporated whole milk (2 ± 0.4), braised pork heart (2 ± 0.4), applesauce (2 ± 0.4), caramel sundae (2 ± 0.3), roast beef sandwich (2 ± 0.3), hot fudge sundae (2 ± 0.3), strawberry sundae (2 ± 0.3), pancakes with butter and syrup (2 ± 0.3), boiled lentils (2 ± 0.3), boiled pinto beans (2 ± 0.3), boiled soybeans (2 ± 0.3), boiled yellow beans (2 ± 0.3), eggnog (2 ± 0.2), indian buffalo milk (2 ± 0.2), tuna salad (2 ± 0.2), tom collins (2 ± 0.2), canned broad beans (2 ± 0.2), italian pork sausage (1 ± 0.7), pork link sausage (1 ± 0.7), canned chinese water chestnuts (1 ± 0.7), raw coho salmon (1 ± 0.6), cooked coho salmon (1 ± 0.6), pork bratwurst (1 ± 0.6), taco (1 ± 0.6), cooked mullet (1 ± 0.6), cooked swordfish (1 ± 0.6), raw mullet (1 ± 0.6), raw swordfish (1 ± 0.6), cooked herring (1 ± 0.6), raw herring (1 ± 0.6), cooked clams (1 ± 0.6), canned cod (1 ± 0.6), cooked cod (1 ± 0.6), raw cod (1 ± 0.6), cooked carp (1 ± 0.6), raw carp (1 ± 0.6), pineapple chiffon pie (1 ± 0.6), onion rings (1 ± 0.6), simmered beef tongue (1 ± 0.5), simmered beef kidneys (1 ± 0.5), hamburger sandwich (1 ± 0.5), restaurant coleslaw (1 ± 0.5), loquats (1 ± 0.5), simmered beef brains (1 ± 0.5), dried sweetened shredded coconut (1 ± 0.5), raw pepeao (1 ± 0.5), raw spirulina seaweed (1 ± 0.5), evaporated lowfat milk (1 ± 0.4), cheeseburger (1 ± 0.4), banana custard pie (1 ± 0.4), mince pie (1 ± 0.4), pineapple pie (1 ± 0.4), pineapple custard pie (1 ± 0.4), raisin pie (1 ± 0.4), canned bamboo shoots (1 ± 0.4), cooked wild long grain rice (1 ± 0.4), tapioca cream pudding (1 ± 0.3), bacon cheeseburger (1 ± 0.3), burrito (1 ± 0.3), ham and cheese sandwich (1 ± 0.3), dried figs (1 ± 0.3), boiled broad beans (1 ± 0.3), boiled chickpeas (1 ± 0.3), boiled black-eyed peas (1 ± 0.3), boiled catjang black-eyed peas (1 ± 0.3), boiled french beans (1 ± 0.3), boiled great northern beans (1 ± 0.3), boiled kidney beans (1 ± 0.3), boiled mothbeans (1 ± 0.3), boiled mungo beans (1 ± 0.3), boiled navy beans (1 ± 0.3), boiled split peas (1 ± 0.3), boiled yardlong bean (1 ± 0.3), chili (1 ± 0.2), bread pudding (1 ± 0.2), lowfat 1% milk (1 ± 0.2), lowfat 2% milk (1 ± 0.2), skim milk (1 ± 0.2), protein fortified lowfat 1% milk (1 ± 0.2), protein fortified lowfat 2% milk (1 ± 0.2), protein fortified skim milk (1 ± 0.2), lowfat 1% chocolate milk (1 ± 0.2), lowfat 2% chocolate milk (1 ± 0.2), goat milk (1 ± 0.2), vanilla pudding (1 ± 0.2), malted milk (1 ± 0.2), restaurant vanilla milkshake (1 ± 0.2), hot chocolate (1 ± 0.2), whole milk (1 ± 0.2), whole chocolate malted milk (1 ± 0.2), whole chocolate milk (1 ± 0.2), restaurant strawberry milkshake (1 ± 0.2), carob milk (1 ± 0.2), lowfat yogurt (1 ± 0.2), buttermilk (1 ± 0.2), filled milk (1 ± 0.2), skim yogurt (1 ± 0.2), apple brown betty (1 ± 0.2), apple juice (1 ± 0.2), pear nectar (1 ± 0.2), canned cranberry beans (1 ± 0.2), canned great northern beans (1 ± 0.2), canned kidney beans (1 ± 0.2), boiled mung beans (1 ± 0.2), canned navy beans (1 ± 0.2), canned pinto beans (1 ± 0.2)

Applications: bladder inflammation, alcohol cravings, anemia, angina pectoris, hardening of the arteries, bruising, increased cholesterol, diabetes, hemophilia, high blood pressure, blood sugar decreased, jaundice, leukemia, mononucleosis, pernicious anemia, vein deterioration, stroke, varicose veins, fracture, softening of bones, thinning of bones, rickets, celiac disease, colitis, cystic fibrosis, diarrhea, worms, dizziness, epilepsy, fatigue, decreased oxygen, lack of sleep, meningitis, mental illness, multiple sclerosis, Parkinson's disease, shingles, ear infection, deterioration of vision, cataracts, conjunctivitis, eyestrain glaucoma, vision and focus deterioration, gallstones, adrenal deterioration, goiter, inflammation of prostate, swollen glands, baldness, hair deterioration, headache, constipation, hemorrhoids, arthritis, bursitis, gout, kidney stones, kidney inflammation, leg cramp, cirrhosis of liver, hepatitis, allergies, bronchitis, common cold, croup, emphysema, hay fever, influenza, pneumonia, tuberculosis, canker sore, bad breath, muscular dystrophy, rheumatism, abscess, acne, athlete's foot, bedsores, boils, bruises, burns, carbuncle, skin eczema, skin impetigo, skin psoriasis, scurvy, gastritis, peptic ulcer, pyorrhea, tooth and gum deterioration, backache, beriberi, chicken pox, fever, infection, influenza, kwashiorkor, weight increased, pregnancy, rheumatic fever, male sterility, AIDS, stress, cancer, antioxidant therapy, sun sensitivity, free radical oxidation, life span extension, radiation, decreased radiation resistance, aging

Daily Dosage: (at least 1000 mg vitamin C per every 1000 mg vitamin B_{3a} and at least 3000 mg vitamin C per every 1000 mg cysteine) pre–1958 MDR – 30 mg; 1958 RDA – 75 mg; 1974 PCS – 30–90 mg; 1974 RDA – 45–55 mg; 1974 USRDA – 55 mg; 1980 RDA – 50–60 mg; 1980 USRDA – 60 mg; 1989 RDA – 50–60 mg; 1989 USRDA – 60 mg; nutritional – 500–1000 mg; therapeutic – 3000–7000 mg; experimental – 10,000–25,000 mg; toxic – 25,000 mg

Warnings: May reverse anticoagulant activity of warfarin. Diabetics and heart patients may require less medications. Can change results of lab tests for sugar in blood and urine. Can give false negative in test for blood in stool. Acidity should be countered by antacid. Increased amounts are dangerous. See toxicity symptoms.

VITAMIN C_1 see vitamin C

VITAMIN C CALCIUM SALT

(Specific form of vitamin C)

Names: vitamin C calcium salt, ascorbic acid calcium salt, calcium ascorbate

VITAMIN C MAGNESIUM SALT

(Specific form of vitamin C)

Names: vitamin C magnesium salt, ascorbic acid magnesium salt, magnesium ascorbate

VITAMIN C NICOTINAMIDE COMPLEX

(Specific form of vitamin C, combined with vitamin B_{3b})

Names: vitamin C nicotinamide complex, ascorbic acid nicotinamide complex, nicotinamide ascorbate, niacinamide ascorbate, vitamin B_3 ascorbate, vitamin B_{3b} ascorbate, merpress, Nicoscorbine, Nicastubine
Applications: vitamin B_{3b} therapy

VITAMIN C PALMITATE

(fat-soluble form of vitamin C)

Names: ascorbyl palmitate, vitamin C palmitate, vitamin C palmitate, fat-soluble vitamin C, lipid-soluble vitamin C
Classification: lipid-soluble vitamin, antioxidant
Forms: ascorbyl palmitate
Inhibitors: laxatives, mineral oil
Applications: fat-soluble anti-oxidant therapy, cancer, antioxidant therapy, sun sensitivity, free radical oxidation, life span extension, radiation, decreased radiation resistance, aging
Daily Dosage: (1 mg vitamin C palmitate = 0.42 mg vitamin C activity) pre-1958 MDR — 71 mg as vitamin C; 1958 RDA — 179 mg as vitamin C; 1974 PCS — 71-214 mg as vitamin C; 1974 RDA — 107-131 mg as vitamin C; 1974 USRDA — 131 mg as vitamin C; 1980 RDA — 119-143 mg as vitamin C; 1980 USRDA — 143 mg as vitamin C; 1989 RDA — 119-143 mg as vitamin C; 1989 USRDA — 143 mg as vitamin C; nutritional — 200-500 mg; therapeutic — 1000-2000 mg; experimental — 3000-5000 mg; toxic — none known up to 5000 mg

VITAMIN C POTASSIUM SALT

(Specific form of vitamin C)

Names: vitamin C potassium salt, ascorbic acid potassium salt, potassium ascorbate

VITAMIN C SODIUM SALT

(Specific form of vitamin C)

Names: vitamin C sodium salt, ascorbic acid sodium salt, sodium ascorbate

VITAMIN C_2

(Specific form of vitamin P)

Names: vitamin C_2, vitamin J, catechol, catechol (flavin)
Classification: water-soluble vitamin, bioflavinoid, antioxidant
Forms: catechol (flavin)
Sources: orange juice, lemon juice, cabbage
Applications: vitamin C therapy, converts dehydro-ascorbic acid into ascorbic acid, thyroid hyperplasia, pneumonia, vitamin P therapy

VITAMIN C₃

(Specific form of vitamin P)

Warnings: Do not confuse with catechol (pyrocatechol), which is also called catechol.

Classification: water-soluble vitamin, bioflavinoid, antioxidant
Deficiency: fragile blood vessels, easy bruising, nose bleeds, aggregation of blood cells, predisposition for blood clots, decreased vitamin C availability, decreased vitamin A absorption, decreased levels of epinephrine, varicose veins, hemorrhoids, decreased antiviral activity, decreased frostbite resistance
Side-effects: none known
Toxicity: none known up to 3 grams
Applications: protection from cigarette smoke, vitamin C therapy, converts dehydroascorbic acid into ascorbic acid, thyroid hyperplasia, pneumonia, vitamin P therapy

FACTOR CF *see* **vitamin B_C**

VITAMIN D COMPLEX

(Different forms of vitamin D. See vitamin D_1, vitamin D_2, vitamin D_3, vitamin D_4, vitamin D_5, vitamin D_C, vitamin D_M.)

Names: vitamin D, rachitamin, rachitasterol, antirachitic factor
Classification: lipid-soluble vitamin, hormone precursor
Forms: vitamin D_1/lumisterol, vitamin D_2/ergocalciferol/viosterol/activated ergosterol, vitamin D_3/cholecalciferol/ergosterol/activated 7-dehydrocholesterol, vitamin D_4/dihydrotachysterol/22:23–dihydrovitamin D2, vitamin D_5/irradiated 7-dehydrositosterol, vitamin D_C, vitamin D_M
Deficiency: rickets, bone demineralization, softening bones, softening teeth, decreased calcium absorption, decreased phosphorus absorption, increased lead in blood, increased calcium in blood, muscular/nervous weakness, decreased muscle tone, fatigue, cramps, constipation, emotional agitation, increased serum cholesterol, decreased kidney function
Side-effects: none known
Toxicity: (Detoxified by vitamin C)
D_1, D_2, D_4, D_5, D_C, D_M: thirst, increased urination, urgency of urination, kidney failure, diarrhea, nausea, appetite decreased, weight decreased, headache, depression, sunstroke?, calcium salts deposits in blood vessels, calcium salts deposits in liver, calcium salts deposits in lung, calcium salts deposits in kidney, calcium salts deposits in stomach, calcium salts deposits in skin, depleted magnesium levels, weakness, constipation, aches, stiffness, abnormal bone growth in children, numb/tingling bones and fingertips, increased blood pressure
D_3: none known
Inhibitors: mineral oil, smog, laxatives, phenturide, promidone, pheytoin, phenobarbitone, the hypnotic glutethimide, corticosteroids, anticonvulsants
Helpers: vitamin A, vitamin B_P, vita-

min C, vitamin F, vitamin T, calcium, phosphorus
Sources (iu per 100 grams of food): synthetically produced *D_1, D_2:* butter (90 ± 0.5), cow milk (40 ± 0.5), cheeses (30 ± 0.5), cream (15 ± 0.5), cottage cheese (4.0 ± 0.05)
D_3: tuna liver oil (10,000,000), cod liver oil (20,000), canned sardines (500 ± 50), salmon (350 ± 5), tuna (250 ± 5), egg yolk (160 ± 5), shrimp (150 ± 5), mushrooms (150 ± 5), shrimp (150 ± 5), sunflower seeds (92 ± 0.5), liver (50 ± 0.5), eggs (48 ± 0.5), mushrooms (40 ± 0.5), oysters (10.0 ± 0.05), corn oil (9.0 ± 0.05), human milk (6.0 ± 0.05), cow milk (4.0 ± 0.05), bee pollen (1.6 ± 0.05), bass (1.0 ± 0.05), sunshine (10000 ± 500/day in summer)
D_4: irradiating 22:23–dihydroergosterol
D_5: irradiating 7–dehydrositosterol
Applications: bladder inflammation, increased cholesterol, diabetes, fracture, softening of bones, thinning of bones, rickets, epilepsy, meningitis, Bitiot's spots, cataracts, eyestrain, glaucoma, vision and focus deterioration, cystic fibrosis, gallstones, fever, celiac disease, constipation, worms, arthritis, leg cramp, sciatica, cirrhosis of liver, jaundice, allergies, bronchitis, common cold, emphysema, tuberculosis, canker sores, muscle spasm, vaginal itching, acne, bedsores, burns, carbuncles, skin eczema, skin psoriasis, shingles, pyorrhea, aging, alcohol cravings, backache, cancer, fatigue, lack of sleep, kwashiorkor, pregnancy, rheumatic fever, stress
Daily Dosage: (10 µg cholecalciferol = 400 iu = 400 usp) pre–1958 MDR – 400 iu; 1958 RDA – 400 iu; 1974 PCS – 200–400 iu; 1974 RDA – 200–400 iu; 1974 USRDA – 400 iu; 1980 RDA – 200–400 iu; 1980 USRDA – 400 iu; 1989 RDA – 200–400 iu; 1989 USRDA – 400 iu; nutritional – 400–1000 iu; therapeutic – 1500–5000 iu; experimental – 10,000–50,000 iu; toxic – 25,000 iu +

Warnings: Increased amounts are dangerous. See toxicity symptoms.

VITAMIN D_1

(Specific form of vitamin D)

Names: vitamin D_1, lumisterol
Forms: calciferol

VITAMIN D_2

(synthetic form of vitamin D)

Names: vitamin D_2, oleovitamin D_2, ergocalciferol, viosterol, activated ergosterol, Drisdol, D-Tracetten, Divit Urto, Ostelin, Condol, Ergorone, Davitin, Metadee, Mina D_2, Mulsiferol, Mykostin, Radsterin, Shock-Ferol, Dee-Ron, Decaps, Deltalin, De-Rat Concentrate, Deratol, Hi-Deratol, Detalup, Diactol, Doral, Vio-D, Ertron, Infron, Radiostol, Sterogyl, Fortodyl
Forms: ergocalciferol / viosterol / activated ergosterol

VITAMIN D_3

(internal form of vitamin D)

Names: vitamin D_3, ergosterol, activated 7-dehydrocholesterol, oleovitamin D_3, cholecalciferol, CC, Duphafral D_3 1000, Delsterol, Deparal, Ebivit, Micro-Dee, Neo Dohyfral D_3, Provitina, Ricketon, Trivitan, D_3–Vicotrat, Vi-De-3–hydrosol, Vigantol, Vigorsan
Forms: cholecalciferol / ergosterol / activated 7-dehydrocholesterol

VITAMIN D_4

(ineffective form of vitamin D)

VITAMIN D₅

Names: vitamin D₄, oleovitamin D₄, 22:23–dihydrovitamin D₂, 22,23–dihydroergocalciferol
Forms: dihydrotachysterol

VITAMIN D₅

(Specific form of vitamin D)
Names: vitamin D₅, irradiated 7–dehydrositosterol
Forms: irradiated 7–dehydrositosterol

VITAMIN D_C

(Specific form of vitamin D)

VITAMIN D_M

(Specific form of vitamin D)

VITAMIN E COMPLEX

(Different forms of vitamin E. See alpha-vitamin E, alpha-vitamin E acetate, alpha-vitamin E acid succinate, beta-vitamin E, gamma-vitamin E, deltavitamin E, epsilon-vitamin E, zeta₁–vitamin E, zeta₂-vitamin E, eta-vitamin E, vitamin E₂(50))

Names: vitamin E, vitamin E₁, vitamin E complex, antisterility factor/vitamin, factor X, Eprolin-S, Epsilan, Ephynal, Syntopherol, E-Vimin, Eviphero, Etavit, Phytogermine, Profecundin, Tokopharm, Viteolin, Escorb, Vacuals, Covitol, Evion
Classification: lipid-soluble vitamin, antioxidant
Forms: alpha-tocopherol, alpha-tocopherol acetate, alpha-tocopherol acid succinate, beta-tocopherol, gamma-tocopherol, delta-tocopherol, epsilon-tocopherol, zeta1–tocopherol, zeta2–tocopherol, eta-tocopherol, vitamin E₂(50)/tocoquinone-10/tocopherylquinone
Deficiency: (Rebound Deficiency can occur if large doses are halted abruptly.) anemia, degenerating muscles, muscular weakness, sore muscles, charley horse, ceroid deposits in muscle, degenerating endocrine glands, sterility, degenerating peripheral vascular system, nervous system deterioration, angina, decreased survival time of red blood cells, red blood cell rupture, increased creatine in urine, acne, increased need for oxygen, slow healing, remaining scars, remaining stretch marks, wrinkled skin, decreased ozone resistance, decreased radiation resistance, decreased cancer resistance
Side-effects: see toxicity symptoms
Toxicity: headache, nausea, dizziness, decreased basal metabolism rate, fatigue, muscle weakness, decreased thyroid hormone levels, increased levels of blood triglycerides in women, slightly increased blood fats, thinner blood, increased bleeding, decreased blood sugar, blurred vision, chapped lips, inflamed mouth, digestion deterioration, decreased sexual organ function, decreased ability to convert provitamin A to vitamin A

Inhibitors: ferrous sulfate, vitamin B_P, heat, oxygen, freezing temperatures, food-processing, chlorine, mineral oil, laxatives, estrogen, thyroid hormones
Helpers: vitamin A, vitamin B complex, vitamin B_1, vitamin B_H, vitamin C, vitamin F, manganese, selenium, phosphorus
Sources (iu per 100 grams of food): wheat germ oil (216 ± 0.5), wheat germ (160 ± 0.5), sunflower seeds (90 ± 0.5), sunflower seed oil (88 ± 0.5), safflower oil (72 ± 0.5), almonds (48 ± 0.5), sesame oil (45 ± 0.5), cottonseed oil (44 ± 0.5), safflower nuts (35 ± 0.5), peanut oil (34 ± 0.5), sunflower seeds (31 ± 0.5), wheat (30 ± 0.5), corn oil (29 ± 0.5), walnuts (22 ± 0.5), hazelnuts (21 ± 0.5), apricot oil (21 ± 0.5), peanuts (18 ± 0.5), olive oil (18 ± 0.5), soybean oil (14 ± 0.5), roasted peanuts (13 ± 0.5), margarine (13 ± 0.5), mayonnaise (12.0 ± 0.05), peanut butter (11 ± 0.5), cabbage (7.8 ± 0.05), almond oil (7.5 ± 0.05), Brazil nuts (6.5 ± 0.05), cod liver oil (5.4 ± 0.05), cashews (5.1 ± 0.05), soy lecithin (4.8 ± 0.05), butter (3.6 ± 0.05), spinach (3.2 ± 0.05), oatmeal (3.0 ± 0.05), bran (3.0 ± 0.05), asparagus (2.9 ± 0.05), salmon (2.5 ± 0.05), brown rice (2.5 ± 0.05), whole rye (2.3 ± 0.05), vegetable oils (2.3 ± 0.05), dark rye bread (2.2 ± 0.05), broccoli (2.0 ± 0.05), pecans (1.9 ± 0.05), rye & wheat crackers (1.9 ± 0.05), parsley (1.8 ± 0.05), barley (1.7 ± 0.05), corn (1.7 ± 0.05), avocados (1.5 ± 0.05), whole wheat bread (1.4 ± 0.05), bee pollen (1.3 ± 0.05), lard (1.0 ± 0.05), carrots (1.0 ± 0.05), cheeses (1.0 ± 0.05), leeks (1.0 ± 0.05), coconut (1.00 ± 0.05), peas (0.99 ± 0.005), walnuts (0.92 ± 0.005), bananas (0.88 ± 0.005), eggs (0.83 ± 0.005), tomatoes (0.72 ± 0.005), cream (0.70 ± 0.05), coconut oil (0.60 ± 0.05), liver (0.60 ± 0.05), apples (0.60 ± 0.05), halibut (0.60 ± 0.05), shrimp (0.50 ± 0.05), carrots (0.50 ± 0.05), yeasts (0.40 ± 0.05), turkey (0.30 ± 0.05), vegetables (0.30 ± 0.05), fruits (0.30 ± 0.05), lamb (0.29 ± 0.05), human milk (0.23 ± 0.05), chicken (0.20 ± 0.05), medium molasses (0.20 ± 0.05), beef (0.20 ± 0.05), beans (0.20 ± 0.05), pork (0.20 ± 0.05), lamb (0.20 ± 0.05), cottage cheese (0.100 ± 0.005), cow milk (0.060 ± 0.005)
Applications: bladder inflammation, anemia, angina pectoris, hardening of the arteries, atherosclerosis, bruising, coronary thrombosis, diabetes, high blood pressure, pernicious anemia, veins deterioration, stroke, phlebitis deterioration, varicose veins, thinning of bones, colitis, epilepsy, mental illness, multiple sclerosis, Parkinson's disease, Ménière's syndrome, deterioration of vision, cataracts, eyestrain, gallstones, cystic fibrosis, hyperthyroidism, inflammation of prostate, baldness, dandruff, headache, sinusitis, congestive heart failure, myocardial infarction, celiac disease, constipation, hemorrhoids, arthritis, bursitis, gout, kidney stones, kidney inflammation, leg cramp, sciatica, allergies, bronchitis, common cold, emphysema, hay fever, muscular dystrophy, rheumatism, pregnancy miscarriage, vaginal itching, abscess, acne, athlete's foot, bedsores, boils, burns, carbuncle, skin impetigo, skin ulcers, warts, gastritis, peptic ulcer, backache, measles, weight increased, pregnancy, sunburn, cancer, antioxidant therapy, sun sensitivity, free radical oxidation, life span extension, radiation, decreased radiation resistance, aging
Daily Dosage: (1 alpha-tocopherol equivalent = 1 mg alpha-tocopherol = 3 iu = 3 usp) 1974 PCS — 15–45 iu; 1974 RDA — 12–15 iu; 1974 USRDA — 15 iu; 1980 RDA — 24–30 iu; 1980 RDA — 24–30 iu; 1980 USRDA — 30 iu; 1989 RDA — 24–30 iu; 1989 USRDA — 30 iu; nutritional — 100–400 iu; therapeutic — 500–1000 iu; experimental — 2000–3000 iu; toxic — 3000 iu
Warnings: Overactive thyroids, diabetes, increased blood pressure, rheumatic fever requires cautious vitamin E intake. Consult physician with rheumatic heart fever before taking. Can ele-

vate blood pressure in hypertensives. Slow increasing dosage can eventually lower blood pressure. Diabetics may be able to reduce insulin levels. (See physician.) Decreases should be gradual. Increased amounts are toxic. See toxicity symptoms.

ALPHA-VITAMIN E

(Specific form of vitamin E)

Names: alpha-vitamin E, alpha-tocopherol
Forms: alpha-tocopherol, alpha-tocopherol acetate, alpha-tocopherol acid succinate

ALPHA-VITAMIN E ACETATE

(Specific form of alpha-vitamin E)

Names: alpha-vitamin E acetate, vitamin E acetate, alpha-tocopherol acetate
Forms: alpha-tocopherol acetate

ALPHA-VITAMIN E ACID SUCCINATE

(Specific form of alpha-vitamin E)

Names: alpha-vitamin E acid succinate, vitamin E acid succinate, alpha-tocopherol acid succinate
Forms: alpha-tocopherol acid succinate

BETA-VITAMIN E

(Specific form of vitamin E)

Names: beta-vitamin E, beta-tocopherol, cumotocopherol, neotocopherol, p-xylotocopherol
Forms: beta-tocopherol

GAMMA-VITAMIN E

(Specific form of vitamin E)

Names: gamma-vitamin E, gamma-tocopherol, o-xylotocopherol

Forms: gamma-tocopherol

DELTA-VITAMIN E

(Specific form of vitamin E)

Names: delta-vitamin E, delta-tocopherol
Forms: delta-tocopherol

EPSILON-VITAMIN E

(Specific form of vitamin E)

Names: epsilon-vitamin E, epsilon-tocopherol
Forms: epsilon-tocopherol

ZETA$_1$–VITAMIN E

(Specific form of vitamin E)

Names: zeta$_1$–vitamin E, zeta$_1$–tocopherol
Forms: zeta$_1$–tocopherol

ZETA$_2$–VITAMIN E

(Specific form of vitamin E)

Names: zeta$_2$–vitamin E, zeta$_2$–tocopherol
Forms: zeta$_2$–tocopherol

ETA-VITAMIN E

(Specific form of vitamin E)

Names: eta-vitamin E, eta-tocopherol
Forms: eta-tocopherol

VITAMIN E$_2$(50)

(Specific form of vitamin E)

Names: vitamin E$_2$(50), vitamin E$_2$, tocoquinone, tocopherylquinone
Forms: tocoquinone-10/tocopherylquinone

VITAMIN F COMPLEX

(Different essential fatty acids. See provitamin F and vitamin F)
(The FDA disallows the term "vitamin F" for advertising purposes since some fast food franchises started advertising "vitamin enriched" for all foods fried in oil, which is a source of added vitamin F.)

PROVITAMIN F

(Specific form of vitamin F)

Names: vitamin F_0, provitamin F, essential fatty acids, EFA
Classification: essential fatty acid, lipid-soluble vitamin
Forms: linoleic acid, cis-linoleic acid, gamma-linolenic acid, linolenic acid, oleic acid, methyl linoleate, ethyl linoleate, cyclohexylamide/linolexamide, Clinolamide
Deficiency: lost source of vitamin F
Side-effects: see toxicity symptoms
Toxicity: metabolism decreased, abnormal weight gain
Inhibitors: saturated fats, saturated fatty acids, nonessential fatty acids, trans fatty acids, laxatives, mineral oil, alcohol, diabetes, aging, cholesterol, heat, deodorizing oils, hydrogenating oils, oxygen, ferrous sulfate, viral infections, radiation, cancer, vitamin C decreased, zinc decreased
Helpers: vitamin A, vitamin B_{3a}, vitamin B_6, vitamin B_{12}, vitamin B_{15}, vitamin B_T, vitamin C, vitamin D, vitamin E, magnesium, phosphorus, selenium, zinc, polyunsaturated fats
Sources (mg per 100 grams of food):
cis-linoleic acid: evening primrose oil (72000 ± 500)
linoleic acid: safflower oil (77000 ± 500), sunflower oil (60000 ± 500), corn oil (54000 ± 500), soy oil (52000 ± 500), walnut oil (48000 ± 500), wheat germ oil (44000 ± 500), sesame oil (42000 ± 500), cottonseed oil (35000 ± 500), sunflower seeds (30000 ± 500), walnuts (29000 ± 500), peanut oil (25000 ± 500), Brazil nuts (23000 ± 500), margarine (22000 ± 500), sesame seeds (20000 ± 500), pecans (14000 ± 500), peanuts (12000 ± 500), almonds (11000 ± 500), lard (10000 ± 500), olive oil (10000 ± 500), hazelnuts (9300 ± 50), linseed oil (8000 ± 50), wheat germ (4400 ± 50), cashews (3200 ± 50), butter (2700 ± 50), oats (2600 ± 50), spices (2400 ± 50), olives (2200 ± 50), cocoa butter (2100 ± 50), egg yolk (2100 ± 50), coconut oil (2000 ± 50), pork (2000 ± 50), avocados (1900 ± 50), liver (1500 ± 50), soybeans (1400 ± 50), turkey (1300 ± 50), wheat (1200 ± 50), chicken (1200 ± 50), tuna (1200 ± 50), rye (1200 ± 50), lamb (1000 ± 50), cheeses (850 ± 5), eggs (780 ± 5), rice (660 ± 5), barley (620 ± 5), corn (520 ± 5), beans (450 ± 5), herring (390 ± 5), veal (290 ± 5), human milk (270 ± 5), beef (240 ± 5), salmon (230 ± 5), lima beans (220 ± 5), spinach (200 ± 5), goat milk (200 ± 5), cow milk (140 ± 5), cream (130 ± 5), venison (120 ± 5), yogurt (49 ± 0.5), fruits (20 ± 0.5), sole (7.9 ± 0.05), haddock (2.2 ± 0.05), internal conversion of oleic acid
gamma-linolenic acid: evening primrose oil (9000)
linolenic acid: sesame oil (67000 ± 500), linseed oil (52000 ± 500), sesame seeds (32000 ± 500), soy oil (7200 ± 50), spices (7100 ± 50), walnut oil

VITAMIN F

(6000 ± 50), cottonseed oil (4100 ± 50), walnuts (3600 ± 50), soybeans (1300 ± 50), safflower oil (1000 ± 50), trout (1000 ± 50), egg yolk (930 ± 5), pecans (920 ± 5), beans (750 ± 5), smelt (370 ± 5), herring (310 ± 5), sablefish (200 ± 5), lamb (190 ± 5), sole (160 ± 5), millet (160 ± 5), sardines (140 ± 5), lima beans (100 ± 5), beef (95 ± 0.5), spinach (80 ± 0.5), corn (68 ± 0.5), rye (60 ± 0.5), wheat (51 ± 0.5), milk (40 ± 0.5), oysters (30 ± 0.5), tuna (21 ± 0.5), rice (17 ± 0.5), clams (14 ± 0.5), barley (5.5 ± 0.05), tomatoes (2.8 ± 0.05), cod (2.8 ± 0.05)
oleic acid: olive oil (75000 ± 500), almond oil (69000 ± 500), peanut oil (54000 ± 500), hazelnuts (50000 ± 500), lard (46000 ± 500), pecans (44000 ± 500), cocoa butter (38000 ± 500), almonds (37000 ± 500), corn oil (35000 ± 500), cashews (34000 ± 500), cottonseed oil (30000 ± 500), soy oil (27000 ± 500), butter (27000 ± 500), peanuts (26000 ± 500), Brazil nuts (21000 ± 500), sesame oil (21000 ± 500), walnuts (21000 ± 500), safflower oil (20000 ± 500), beef (18000 ± 500), wheat germ oil (15000 ± 500), sunflower oil (12000 ± 500), sesame seeds (10000 ± 500), cheeses (9500 ± 50), pork (9000 ± 50), coconut oil (8600 ± 50), olives (7600 ± 50), avocados (7400 ± 50), lamb (7100 ± 50), sunflower seeds (6000 ± 50), eggs (6000 ± 50), flax seeds (5100 ± 50), soybeans (4900 ± 50), veal (4700 ± 50), salmon (4200 ± 50), haddock (3500 ± 50), coconut (3100 ± 50), turkey (2600 ± 50), oats (2200 ± 50), spices (1800 ± 50), wheat germ (1500 ± 50), cream (1300 ± 50), egg yolk (1200 ± 50), chicken (1200 ± 50), tuna (1200 ± 50), cow milk (1200 ± 50), human milk (1100 ± 50), millet (670 ± 5), rice (660 ± 5), corn (340 ± 5), rye (280 ± 5), wheat (210 ± 5), barley (200 ± 5), beans (120 ± 5), spinach (120 ± 5), fruits (100 ± 5), lima beans (45 ± 0.5)
Applications: wounds, vitamin F therapy
Daily Dosage: (1 g fats = 9 calories) 1989 SADDI—none; 1989 RDA—none; 1989 USRDA—none; nutritional—1–2g; therapeutic—10g; experimental—10–20g; toxic—25g
Warnings: Keep levels low with gallstones, digestion upset and skin psoriasis. Increased amounts are toxic. See toxicity symptoms.

VITAMIN F

Names: vitamin F, vitamin F_1, arachidonic acid, essential fatty acids, EFA
Classification: essential fatty acid, lipid-soluble vitamin
Forms: arachidonic acid
Deficiency: growth cessation, skin eczema, acne, dry skin, dandruff, dry hair, dull hair, brittle hair, hair decreased, nail deterioration, soft nails, brittle nails, flaking nails, dry eyes, dry mouth, diarrhea, allergies, varicose veins, weight decreased, weight increased, gallstones, decreased radiation resistance, heart disease, cancer, deterioration of skin, sterility, swollen joints, liver deterioration, fatigue, emotional agitation, decreased immunity, decreased T-cell blood count
Side-effects: see toxicity symptoms
Toxicity: metabolism decreased, abnormal weight gain
Inhibitors: saturated fats, saturated fatty acids, nonessential fatty acids, trans fatty acids, alcohol, diabetes, aging, cholesterol, heat, deodorizing oils, hydrogenating oils, oxygen, ferrous sulfate, viral infections, radiation, cancer, vitamin C decreased, zinc decreased
Helpers: vitamin A, vitamin B_{3a}, vitamin B_6, vitamin B_{12}, vitamin B_{15}, vitamin B_T, vitamin C, vitamin D, vitamin E, magnesium, phosphorus, selenium, zinc, polyunsaturated fats
Sources (mg per 100 grams of food):
arachidonic acid: walnut oil (1600), walnuts (960), whitefish (480), mackerel (470), sole (320), herring (310), sar-

dines (280), swordfish (180), lamb (95), tuna (94), mussels (90), scallops (81), beef (38), human milk (30), algae (13), cod (9.6), haddock (1.6), internal conversion of linoleic acid
Applications: colitis, diarrhea, mental illness, multiple sclerosis, Ménière's syndrome, premenstrual syndrome, emotional agitation, hyperkinesis, minimal brain dysfunction, multiple sclerosis, inflammation of prostate, coronary thrombosis, increased blood pressure, angina, insulin therapy, constipation, arthritis, inflammation, leg cramp, asthma, bronchitis, acne, skin inflammation, skin eczema, skin psoriasis, tooth and gum deterioration, allergies, common cold, weight increased, weight decreased, hangover, depression, breast cancer
Daily Dosage: (1 g fats = 9 calories) 1989 SADDI — none; 1989 RDA — none; 1989 USRDA — none; nutritional — 1–2 g; therapeutic — 10 g; experimental — 10–20 g; toxic — 25 g
Warnings: Keep levels low with gallstones, digestion upset and skin psoriasis. Increased amounts are toxic. See toxicity symptoms.

VITAMIN F_1 see **vitamin F**

VITAMIN F_{99}

Classification: essential fatty acid, lipid-soluble vitamin
Forms: concentrated mixture of linoleic acid, linolenic acid and arachidonic acid

VITAMIN G see **vitamin B_2**

FACTOR GT see **trivalent chromium under MINERALS**

VITAMIN H COMPLEX

(Different vitamins that were later classified into the sub-vitamin B complex. See vitamin B_C, vitamin B_H, vitamin B_P, vitamin B_W, and vitamin B_X.)

VITAMIN H see **vitamin B_W**

VITAMIN H_1

(Term "vitamin H_1" sometimes used incorrectly to denote vitamin B_6 or vitamin B_X.)
(See vitamin B_W.)

VITAMIN H' see **vitamin B_X**

VITAMIN H_2 see **vitamin B_X**

VITAMIN H_3

(Very controversial. Essential need in human nutrition doubted by most researchers. Breaks down into the nutrients vitamin B_X and diethylaminoethanol, a provitamin form of vitamin B_P.)

Names: vitamin H_3, gerovital, gerovital H_3, GH_3, KH_3, para-aminobenzoyl-diethylaminoethanol hydrochloride, para-aminobenzoate hydrochloride, Novocain, Ethocaine, Neocaine, Synca-

ine, Scurocaine, Allocaine, Alocaine, Anesthesol, Anestil, Cetain, Isocaine-Asid, Isocaine-Heisler, Atoxicocaine, Naucaine, Bernacaine, Irocaine, Juvocaine, Jenacain, Kerocaine, Paracain, Planocaine, Aminocaine, Eugerase, Sevicaine, Topokain, Westocaine
Classification: nutritional product, anesthetic, drug, vitamin percursor
Forms: procaine HCL/para-aminobenzoic diethylaminoethanol hydrochloride
Deficiency: arthritis, inflammation of arteries, hardening of the brain arteries, trophic ulcers, digestion upset, baldness, senile Parkinsonism, decreased hearing, ringing in ears, noises in ears, blurred vision, decreased vision, increased blood pressure, defective heart conditions, irregular heart rhythms, cardiospasm, arterial fibrillation, angina pectoris, varicose veins, asthma, decreased cell regeneration, skin inflammation, skin hives, skin edema, wrinkled skin, schizophrenia, skin ichthyosis, senile keratosism, dermatosclerosis, skin psoriasis, rashes, skin leukoderma, decreased hair color, decreased muscle tone, failing memory, oxygen decreased, increased cholesterol

Side-effects: see **toxicity** symptoms.
Toxicity: skin hives, nausea, decreased pulse rate, convulsions
Inhibitors: enzyme cholinesterase, caffeine
Helpers: vitamin B_{12}, vitamin B_C, vitamin B_X, vitamin F, vitamin P, chlorine, potassium, potassium metabisulfate, sodium, benzoic acid (preservative), dibasic phosphate
Sources: commercial procaine preparations
Applications: arthritis, rheumatism, digestion deterioration, senile Parkinsonism, decreased hearing, hearing deterioration, circulatory deterioration, heart deterioration, asthma, skin deterioration, hair deterioration, mental health deterioration, emotional deterioration, schizophrenia, hyperkinetic children
Daily Dosage: 1989 SADDI – none; 1989 RDA – none; 1989 USRDA – none; nutritional – none; therapeutic – 200 mg; experimental – 500 mg; toxic – ?
Warnings: Increased amounts are dangerous. See toxicity symptoms.

VITAMIN I

(Term "vitamin I" sometimes used for vitamin B_W.)
(Term "vitamin I" was proposed to denote a possible internal combination of vitamin A and vitamin E.)
(Term "vitamin I" was also once used for the now defunct vitamin B_7.)

FACTOR I *see* **vitamin B_7**

VITAMIN J

(Term "vitamin J" sometimes used for vitamin C_2.)
(See vitamin B_P.)

VITAMIN K COMPLEX

(Different forms of vitamin K. See vitamin K_1, vitamin K_2, vitamin K_3, vitamin K_4, vitamin K_5, vitamin K_6, vitamin K_7, vitamin K_8, vitamin K_9, vitamin $K_9(H)$, vitamin K-S (II), vitamin MK_1, vitamin MK_2, vitamin MK_3, vitamin MK_4, vitamin MK_5, vitamin MK_6, vitamin MK_7, vitamin MK_8, vitamin MK_9 and vitamin MK_{10}.)

Names: vitamin K, antihemorrhagic factor/vitamin
Classification: lipid-soluble vitamin
Forms: vitamin K_1/vitamin $K_{1(20)}$/phytonadione, vitamin K_2, vitamin $K_{2(0)}$, vitamin $K_{2(5)}$, vitamin $K_{2(10)}$, vitamin $K_{2(15)}$, vitamin $K_{2(20)}$, vitamin $K_{2(25)}$, vitamin $K_{2(30)}$, vitamin $K_{2(35)}$, vitamin $K_{2(40)}$, vitamin $K_{2(45)}$, vitamin $K_{2(50)}$, menaquinone, vitamin K_3/menodoine/menaphthone, vitamin K_4/menadiol, vitamin K_5, vitamin K_6, vitamin K_7, vitamin K_8, vitamin K_9, vitamin $K_9(H)$, vitamin K-S (II), vitamin MK_1, vitamin MK_2, vitamin MK_3, vitamin MK_4, vitamin MK_5, vitamin MK_6, vitamin MK_7, vitamin MK_8, vitamin MK_9, vitamin MK_{10}
Deficiency: decreased blood clotting, nose bleeds, increased blood pressure, hemorrhage, diarrhea
Side-effects: vitamin K_1, vitamin K_2: none known
Toxicity: vitamin K_3, vitamin K_4, vitamin K_5, vitamin K_6, vitamin K_7, vitamin K_8, vitamin K_9, vitamin $K_9(H)$, vitamin K-S (II): flushing, sweating, chest constriction, severe neural symptoms, increased bilirubin in blood, red cell breakdown, anemia, sweats, flushing, yellowing of skin
Inhibitors: blood thinner Dicumarol, anticoagulants, X-rays, radiation, frozen foods, aspirin, aspirin substitutes, air pollution, mineral oil, antibiotics, neomycin, sweating, food processing, laxatives
Helpers: vitamin H_3, intestinal bacteria
Sources (µg per 100 grams of food): *K_1, K_2:* cheddar (22000 ± 500), Camembert (16000 ± 500), brussels sprouts (1500 ± 50), soy lecithin (1200 ± 50), pork (950 ± 5), turnip greens (650 ± 5), alfalfa (520 ± 5), oats (490 ± 5), spinach (330 ± 5), soybeans (300 ± 5), cabbage (250 ± 5), broccoli (200 ± 5), lettuce (129 ± 0.5), beef liver (92 ± 0.5), potatoes (80 ± 0.5), bran (69 ± 0.5), watercress (57 ± 0.5), asparagus (57 ± 0.5), peas (44 ± 0.5), coffee (39 ± 0.5), cheese (35 ± 0.5), beef (35 ± 0.5), wheat germ (33 ± 0.5), butter (30 ± 0.5), tomatoes (27 ± 0.5), pork liver (25 ± 0.5), honey (25 ± 0.5), whole

VITAMIN K$_1$

wheat (17 ± 0.5), strawberries (13 ± 0.5), eggs (12 ± 0.5), carrots (10 ± 0.5), corn (10 ± 0.5), mushrooms (8.3 ± 0.05), peaches (7.9 ± 0.05), corn oil (7.1 ± 0.05), beef (7 ± 0.5), chicken liver (7 ± 0.5), raisins (6 ± 0.5), green beans (4.4 ± 0.05), milk (3 ± 0.5), potato (3 ± 0.5), cow milk (2.9 ± 0.05), bananas (2.0 ± 0.05), cauliflower (2.0 ± 0.05), intestinal flora (except after newborn period)
vitamin K$_3$, vitamin K$_4$, vitamin K$_6$, vitamin K$_7$, vitamin K$_8$, vitamin K$_9$, vitamin K$_9$(H), vitamin K-S (II): synthetic
vitamin K$_5$: food preservative
Applications: bruising, hemorrhage, gallstones, cystic fibrosis, celiac disease, worms, cirrhosis of liver, jaundice, skin ulcers, aging, alcohol cravings, cancer, hepatitis, kwashiorkor
Daily Dosage: 1980 RDA — none; 1980 SADDIs — 70–140 µg; 1989 RDA — 80 µg; 1989 USRDA — 80 µg; nutritional — 140–200 µg; therapeutic — 500–1000 µg; experimental — 1000–5000 µg; toxic — variable with form
Warnings: *vitamin K$_3$, vitamin K$_4$, vitamin K$_5$, vitamin K$_6$, vitamin K$_7$, vitamin K$_8$, vitamin K$_9$, vitamin K$_9$(H), and vitamin K-S (II)*: counteract blood thinner Dicumarol. Increased amounts are dangerous. See toxicity symptoms.

VITAMIN K$_1$

(Specific form of vitamin K)

Names: vitamin K$_1$, K-Ject, Konakion, Mephyton, Mono-Kay
Classification: lipid-soluble vitamin
Forms: vitamin K$_1$/Phytonadione, vitamin K$_1$ oxide, vitamin K$_{1(20)}$/Naphthalendione

VITAMIN K$_1$ OXIDE

(Specific form of vitamin K)

Names: vitamin K$_1$ oxide
Formsvitamin K$_1$/phytonadione oxide

VITAMIN K$_{1(20)}$

(Specific form of vitamin K$_1$)
Forms: naphthalenedione

VITAMIN K$_2$

(Specific form of vitamin K)

Classification: lipid-soluble vitamin
Forms: *menaquinone:* vitamin K$_{2(0)}$, vitamin K$_{2(5)}$, vitamin K$_{2(10)}$, vitamin K$_{2(15)}$, vitamin K$_{2(20)}$, vitamin K$_{2(25)}$, vitamin K$_{2(30)}$, vitamin K$_{2(35)}$, vitamin K$_{2(40)}$, vitamin K$_{2(45)}$, vitamin K$_{2(50)}$

VITAMIN K$_{2(0)}$

(Specific form of vitamin K$_2$)

VITAMIN K$_{2(5)}$

(Specific form of vitamin K$_2$)

VITAMIN K$_{2(10)}$

(Specific form of vitamin K$_2$)

VITAMIN K$_{2(15)}$

(Specific form of vitamin K$_2$)

VITAMIN K$_{2(20)}$

(Specific form of vitamin K$_2$)

VITAMIN K$_{2(25)}$

(Specific form of vitamin K$_2$)

VITAMIN K$_{2(30)}$

(Specific form of vitamin K$_2$)

VITAMIN K$_{2(35)}$

(Specific form of vitamin K$_2$)

VITAMIN K$_{2(40)}$

(Specific form of vitamin K$_2$)

VITAMIN K$_{2(45)}$

(Specific form of vitamin K$_2$)

VITAMIN K$_{2(45)}$H

(Specific form of vitamin K$_2$)

VITAMIN K$_{2(50)}$

(Specific form of vitamin K$_2$)

VITAMIN K$_{2(55)}$

(Specific form of vitamin K$_2$)

VITAMIN K$_{2(60)}$

(Specific form of vitamin K$_2$)

VITAMIN K$_{2(65)}$

(Specific form of vitamin K$_2$)

VITAMIN K$_3$

(Specific form of vitamin K)

Classification: lipid-soluble vitamin
Forms: vitamin K$_3$/menodoine/menaphthone

VITAMIN K$_3$H$_2$

(Specific form of vitamin K$_3$)

Classification: formerly lipid-soluble vitamin, now known not to have all of the properties of vitamin K
Forms: naphthalenediol

VITAMIN K$_4$

(Specific form of vitamin K)

Classification: lipid-soluble vitamin for some forms, water-soluble vitamin for some forms
Forms: menadiol diacetate, menadiol dibutyrate, menadiol diphosphate (tetrasodium salt), menadiol disulfate

VITAMIN K$_5$

(Specific form of vitamin K)

Classification: water-soluble vitamin

VITAMIN K$_6$

(Specific form of vitamin K)

Classification: water-soluble vitamin

VITAMIN K$_7$

(Specific form of vitamin K)

Classification: water-soluble vitamin

VITAMIN K$_8$

(Specific form of vitamin K)

VITAMIN K$_9$

(Specific form of vitamin K)

VITAMIN K$_9$(H)

(Specific form of vitamin K$_9$)

VITAMIN K-S (II)

(Specific form of vitamin K)

Classification: lipid-soluble vitamin

VITAMIN L COMPLEX

(Different forms of vitamin L. See vitamin L_1 and vitamin L_2.)

Names: vitamin L, vitamin L complex
Classification: lipid-soluble vitamin, sometimes erroneously included in the vitamin B complex
Forms: vitamin L_1, vitamin L_2
Deficiency: lactation failure
Side-effects: none known
Toxicity: none known
Inhibitors: none known
Helpers: unknown
Sources:
L_1: beef liver extract, yeast
L_2: yeast, liver
Applications: deterioration of lactation
Daily Dosage: 1974 PCS—none; 1989 SADDI—none; 1989 RDA—none; 1989 USRDA—none; nutritional—unknown; therapeutic—unknown; experimental—unknown; toxic—unknown
Warnings: acidity should be countered by antacid

VITAMIN L_1

(Specific form of vitamin L)

Names: vitamin L, vitamin L_1, anthranilic acid, ortho-aminobenzoic acid
Classification: lipid-soluble vitamin, sometimes erroneously included in the vitamin B complex
Forms: ortho-aminobenzoic acid/anthranilic acid
Inhibitors: laxatives, mineral oil

VITAMIN L_2

(Specific form of vitamin L)

Names: vitamin L, vitamin L_2, adenyl thiomethylpentose
Classification: lipid-soluble vitamin, sometimes erroneously included in the vitamin B complex
Forms: adenyl thiomethylpentose
Inhibitors: laxatives, mineral oil

FACTOR LC *see* **vitamin B_C**

VITAMIN M *see* **vitamin B_C**

VITAMIN M_i™

(Trademark for a product developed by a Dr. Hans Nieper)

I. Vitamins   VITAMIN MK$_{10}$

Names: vitamin M$_i$™, Membrane Integrity Factor™, 2-AEP salt
Classification: nutritional product, water-soluble vitamin?, neurotransmitter?
Forms: 2-AEP salt, calcium 2-AEP
Deficiency: deterioration of cell membrane integrity?, decreased cell lifespan?, premature aging?
Side-effects: unknown
Toxicity: unknown
Inhibitors: unknown
Helpers: unknown
Sources: Lögic™ brand Nutritional Supplements distributed by International Marketing Company
Applications: Aging
Daily Dosage: 1974 PCS — none; 1989 SADDI — none; 1989 RDA — none; 1989 USRDA — none; nutritional — none; therapeutic — unknown; experimental — unknown; toxic — unknown
Warnings: unknown

VITAMIN MK COMPLEX

(Specific forms of vitamin K. See vitamin MK$_1$, vitamin MK$_2$, vitamin MK$_3$, vitamin MK$_4$, vitamin MK$_5$, vitamin MK$_6$, vitamin MK$_7$, vitamin MK$_8$, vitamin MK$_9$ and vitamin MK$_{10}$.)

VITAMIN MK$_1$

(Specific form of vitamin K)

VITAMIN MK$_2$

(Specific form of vitamin K)

VITAMIN MK$_3$

(Specific form of vitamin K)

VITAMIN MK$_4$

(Specific form of vitamin K)

VITAMIN MK$_5$

(Specific form of vitamin K)

VITAMIN MK$_6$

(Specific form of vitamin K)

VITAMIN MK$_7$

(Specific form of vitamin K)

VITAMIN MK$_8$

(Specific form of vitamin K)

VITAMIN MK$_9$

(Specific form of vitamin K)

VITAMIN MK$_{10}$

(Specific form of vitamin K)

VITAMIN N

Names: vitamin N, thioctic acid, alpha-lipoic acid, valeric acid, pentanoic acid, protogen A, acetate replacing factor, pyruvate oxidation factor, POF, Biletan, Lipoicin, Thioctacid, Thioctan, Tioctan, Tioctidasi
Classification: lipid-soluble vitamin, sometimes erroneously included in the vitamin B complex
Forms: dithiolan-3 pentanamide / thioctic acid / alpha-lipoic acid
Deficiency: fatigue, decreased ATP production, decreased muscle strength, increased cholesterol, decreased cancer resistance
Side-effects: none known
Toxicity: none known

Inhibitors: laxatives, mineral oil
Helpers: unknown
Sources: stomach extracts, brain, internal synthesis
Applications: liver deterioration, Amanita poisoning
Daily Dosage: 1989 SADDI — none; 1989 RDA — none; 1989 USRDA — none; nutritional — 100–500 mg; therapeutic — 1000–3000 mg; experimental — 5000–10,000 mg; toxic — none known up to 10,000 mg
Warnings: none

VITAMIN OHB$_{12}$ *see* **vitamin B$_{12}$ coenzyme**

VITAMIN P COMPLEX

(Different forms of vitamin P. See vitamin P$_1$, vitamin P$_1$ complex, vitamin P$_2$, vitamin P$_3$, vitamin P$_4$, vitamin C$_2$, vitamin C$_3$.)

Names: vitamin P, vitamin P complex, vitamin C complex, bioflavinoids, flavones, favonals, Arliflav, C.V.P., Pecitrol Veinogene
Classification: water-soluble vitamin, bioflavinoid, antioxidant
Forms: citrin, hesperidin, rutin, troxerutin, esculetin, phenylchroman
Deficiency: fragile blood vessels, easy bruising, nose bleeds, aggregation of blood cells, predisposition for blood clots, decreased vitamin C availability, decreased vitamin A absorption, decreased levels of epinephrine, varicose veins, hemorrhoids, decreased antiviral activity, decreased frostbite resistance

Side-effects: none known
Toxicity: none known up to 3000 mg
Inhibitors: none known
Helpers: vitamin C
Sources: grapes, rose hips, prunes, orange, lemon juice, cherries, black currant, plums, parsley, spinach, grapefruit, cabbage, apricots, peppers, papaya, cantaloupe, tomato, broccoli, blackberry, apples, lettuce, walnuts, potatoes, parsnips, peas, watercress, carrots
Applications: hardening of the arteries, atherosclerosis, bruising, increased cholesterol, hemophilia, high blood pressure, leukemia, stroke, varicose

veins, decreased oxygen in blood, hemorrhoids, arthritis, rheumatic fever, rheumatism, pneumonia, skin ulcers, pyorrhea, scurvy, common cold, cancer, antioxidant therapy, sun sensitivity, free radical oxidation, life span extension, radiation, decreased radiation resistance, aging
Daily Dosage: 1989 SADDI — none; 1989 RDA — none; 1989 USRDA — none; nutritional — 100–500 mg; therapeutic — 500–1000 mg; experimental — 2000–3000 mg; toxic — none known up to 3000 mg
Warnings: none

VITAMIN P_1

(Specific form of vitamin P)
Names: vitamin P_1, rutin, rutoside, melin, phytomelin, eldrin, ilxathin, sophorin, globularicitrin, paliuroside, osyritrin, osyritin, myrticolorin, violaquercitrin, Birutan, Rutabion, Rutozyd, Tanrutin
Classification: water-soluble vitamin, bioflavinoid, antioxidant
Forms: rutin, esculin, esculetin, methylesculin

VITAMIN P_1 COMPLEX

(Specific forms of vitamin P_1)
Names: vitamin P_1, esculin, esculetin, esculoside, biocolorin, enallachrome, polychrome, Escosyl, phenylchroman, cichorigenin
Classification: water-soluble vitamin, bioflavinoid, antioxidant
Forms: esculin, esculetin, methylesculin

VITAMIN P_2

(Specific form of vitamin P)
Names: vitamin P_2, hesperidin, cirantin, hesperetin-7-rutinoside
Classification: water-soluble vitamin, bioflavinoid, antioxidant
Forms: hesperidin

VITAMIN P_3

(Specific form of vitamin P)
Classification: water-soluble vitamin, bioflavinoid, antioxidant

VITAMIN P_4

(Specific form of vitamin P)
Classification: water-soluble vitamin, bioflavinoid, antioxidant
Names: vitamin P_4, factor P-Zyma, venoruton P_4, troxerutin, trioxyethylrutin, tri (hydroxyethyl)rutoside, Posorutin, Ruven, Veinamitol, HR, Paroven, Relvene, Varemoid, Venoruton
Forms: troxerutin
Applications: veins deterioration

VITAMIN PP *see* **vitamin B_3**

VITAMIN Q COMPLEX

(Different forms of vitamin Q. See vitamin Q_1, vitamin Q_2, vitamin Q_3, vitamin Q_4, vitamin Q_5, vitamin Q_6, vitamin Q_7, vitamin Q_8, vitamin Q_9 and vitamin Q_{10})

PROVITAMIN Q *see* **vitamin Q_7, vitamin Q_8 and vitamin Q_9**

VITAMIN Q

(The term "vitamin Q" was originally used by a Dr. Quick for this substance. It was later used for the vitamin Q complex.)
Names: vitamin Q, Quick's vitamin
Classification: formerly vitamin
Forms: Not isolated to a specific chemical, possibly not related to vitamins Q_0 - Q_{10}.
Deficiency: hereditary bleeding disorder telangiectasia, decreased blood clotting, blood clots
Side-effects: none known
Toxicity: none known
Sources: soy beans, clover, alfalfa
Applications: circulatory deterioration, telangiectasia
Warnings: none

VITAMIN Q_0

Names: vitamin Q_0, coenzyme Q_0, CoQ_0, ubiquinone (0), ubichromenol (0)
Classification: possible lipid-soluble vitamin precursor, possible coenzyme precursor, possible antioxidant
Forms: ubiquinone (0), ubichromenol (0)
Deficiency: possible lost source of vitamin/coenzyme Q_{10} in humans, required nutrient for some microorganisms
Side-effects: none known
Toxicity: none known
Inhibitors: laxatives, mineral oil
Helpers: vitamin A, vitamin E, vitamin K, selenium

VITAMIN Q_1

Names: vitamin Q_1, coenzyme Q_1, CoQ_1, ubiquinone (5), ubichromenol (5)
Classification: possible lipid-soluble vitamin precursor, possible coenzyme precursor, possible antioxidant cofactor
Forms: ubiquinone (5), ubichromenol (5)
Deficiency: possible lost source of vitamin/coenzyme Q_{10} in humans, required nutrient for some microorganisms
Side-effects: none known
Toxicity: none known
Inhibitors: laxatives, mineral oil
Helpers: vitamin A, vitamin E, vitamin K, selenium

VITAMIN Q_2

Names: vitamin Q_2, coenzyme Q_2, CoQ_2, ubiquinone (10), ubichromenol (10)
Classification: possible lipid-soluble vitamin precursor, possible coenzyme precursor, possible antioxidant cofactor
Forms: ubiquinone (10), ubichromenol (10)
Deficiency: possible lost source of vitamin/coenzyme Q_{10} in humans, required nutrient for some microorganisms
Side-effects: none known
Toxicity: none known
Inhibitors: laxatives, mineral oil
Helpers: vitamin A, vitamin E, vitamin K, selenium

VITAMIN Q_3

Names: vitamin Q_3, coenzyme Q_3, CoQ_3, ubiquinone (15), ubichromenol (15)

I. Vitamins

Classification: possible lipid-soluble vitamin precursor, possible coenzyme precursor, possible antioxidant cofactor
Forms: ubiquinone (15), ubichromenol (15)
Deficiency: possible lost source of vitamin/coenzyme Q_{10} in humans, required nutrient for some microorganisms
Side-effects: none known
Toxicity: none known
Inhibitors: laxatives, mineral oil
Helpers: vitamin A, vitamin E, vitamin K, selenium

VITAMIN Q_4

Names: vitamin Q_4, coenzyme Q_4, CoQ_4, ubiquinone (20), ubichromenol (20)
Classification: possible lipid-soluble vitamin precursor, possible coenzyme precursor, possible antioxidant cofactor
Forms: ubiquinone (20), ubichromenol (20)
Deficiency: possible lost source of vitamin/coenzyme Q_{10} in humans, required nutrient for some microorganisms
Side-effects: none known
Toxicity: none known
Inhibitors: laxatives, mineral oil
Helpers: vitamin A, vitamin E, vitamin K, selenium

VITAMIN Q_5

Names: vitamin Q_5, coenzyme Q_5, CoQ_5, ubiquinone (25), ubichromenol (25)
Classification: possible lipid-soluble vitamin precursor, possible coenzyme precursor, possible antioxidant cofactor
Forms: ubiquinone (25), ubichromenol (25)
Deficiency: possible lost source of vitamin/coenzyme Q_{10} in humans, required nutrient for some microorganisms

Side-effects: none known
Toxicity: none known
Inhibitors: laxatives, mineral oil
Helpers: vitamin A, vitamin E, vitamin K, selenium

VITAMIN Q_6

Names: vitamin Q_6, coenzyme Q_6, CoQ_6, ubiquinone (30), ubichromenol (30)
Classification: possible lipid-soluble vitamin precursor, possible coenzyme precursor, possible antioxidant cofactor
Forms: ubiquinone (30), ubichromenol (30)
Deficiency: possible lost source of vitamin/coenzyme Q_{10} in humans, required nutrient for some microorganisms and yeast
Side-effects: none known
Toxicity: none known
Inhibitors: laxatives, mineral oil
Helpers: vitamin A, vitamin E, vitamin K, selenium

VITAMIN Q_7

Names: vitamin Q_7, coenzyme Q_7, CoQ_7, ubiquinone (35), ubichromenol (35)
Classification: probable lipid-soluble vitamin precursor, probable coenzyme precursor, probable antioxidant cofactor
Forms: ubiquinone (35), ubichromenol (35)
Deficiency: lost source of vitamin/coenzyme Q_{10} in humans, required nutrient for some microorganisms, yeast and fungi
Side-effects: none known
Toxicity: none known up to 1000 mg
Inhibitors: laxatives, mineral oil
Helpers: vitamin A, vitamin E, vitamin K, selenium

VITAMIN Q_8

Names: vitamin Q_8, coenzyme Q_8, CoQ_8, ubiquinone (40), ubichromenol (40)
Classification: probable lipid-soluble vitamin precursor, probable coenzyme precursor, probable antioxidant cofactor
Forms: ubiquinone (40), ubichromenol (40)
Deficiency: lost source of vitamin/conezyme Q_{10} in humans, required nutrient for fungi and bacteria
Side-effects: none known
Toxicity: none known up to 1000 mg
Inhibitors: laxatives, mineral oil
Helpers: vitamin A, vitamin E, vitamin K, selenium

VITAMIN Q_9

Names: vitamin Q_9, coenzyme Q_9, CoQ_9, ubiquinone (45), ubichromenol (45)
Classification: lipid-soluble vitamin precursor, coenzyme precursor, antioxidant cofactor
Forms: ubiquinone (45), ubichromenol (45)
Deficiency: lost source of vitamin/coenzyme Q_{10} in humans; required nutrient for invertebrates, rats, mice and walleyed pike, fungi, plants and bacteria
Side-effects: none known
Toxicity: none known up to 1000 mg
Inhibitors: laxatives, mineral oil
Helpers: vitamin A, vitamin E, vitamin K, selenium
Sources (mg per 100 grams of food): corn oil (18.6 ± 0.05), wheat germ (10.3 ± 0.05), cottonseed oil (5.9 ± 0.05), raw sweet corn (3.9 ± 0.05), rice bran (3.1 ± 0.05), whole grain corn (2.5 ± 0.05), safflower oil (2.5 ± 0.05), sunflower oil (2.1 ± 0.05), buckwheat (2.1 ± 0.05), millet (1.5 ± 0.05), Job's tears (1.3 ± 0.05), oats (1.3 ± 0.05), barley (1.1 ± 0.05), soybean oil (0.8 ± 0.05), wheat grain (0.67 ± 0.005), olive oil (0.65 ± 0.005), almonds (0.63 ± 0.005), brown rice (0.48 ± 0.005), basella (0.45 ± 0.005), rice bran oil (0.40 ± 0.005), edible burdock (0.36 ± 0.005), kinako (0.3 ± 0.005), well-milled rice (0.26 ± 0.005), pumpkin (0.22 ± 0.005), rapeseed oil (0.21 ± 0.005), wheat germ (0.15 ± 0.005), lettuce (0.14 ± 0.005), cucumber (0.13 ± 0.005), azuki beans (0.1 ± 0.005), chestnuts (0.06 ± 0.005), cattlefish (0.06 ± 0.005), yellowtail (0.03 ± 0.005), broccoli (0.03 ± 0.005), sweet pepper (0.02 ± 0.005), cow milk (0.02 ± 0.005), garlic (0.01 ± 0.005)

PROVITAMIN Q_{10} *see* **vitamin Q_9.**

(The term "provitamin Q_{10}" sometimes is applied to vitamin Q_8 and vitamin Q_7 as well.)

VITAMIN Q_{10} COMPLEX *see* **vitamin Q_1, vitamin Q_2, vitamin Q_3, vitamin Q_4, vitamin Q_5, vitamin Q_6, vitamin Q_7, vitamin Q_8, vitamin Q_9 and vitamin Q_{10}**

VITAMIN Q_{10}

Names: vitamin Q_{10}, coenzyme Q_{10}, CoQ_{10}, CoQ, ubiquinone (50), ubichromenol (50), mitoquinone, SA, Q-275, 272-substance, coenzyme Q_{199}, ubidecarenone, NSC I40865, Adelir, Heartcin, Inokiton, Neuquinone, Taidecanone, Udekinon
Classification: lipid-soluble vitamin, coenzyme, antioxidant
Forms: ubiquinone (50), ubichromenol (50)
Deficiency: (No nutritional requirement in rats, mice, or walleyed pike, yeasts or microorganisms.) deterioration of heart function, cardiac deterioration, heart disease, angina pectoris,

decreased cell energy from mitochondria, fatigue, decreased serum levels of coenzyme Q_{10}, high blood pressure, gum disease, gingivitus, bleeding gums, decreased immunity, decreased levels of immunoglobulin/antibody G, nervousness, digestion deterioration, stomach ulcers, bruises
Side-effects: none known
Toxicity: none known up to 1000 mg
Inhibitors: laxatives, mineral oil
Helpers: vitamin A, vitamin E, vitamin K, selenium
Sources (mg per 100 grams of food): soybean oil (9.2 ± 0.05), rapeseed oil (7.3 ± 0.05), butter (7.1 ± 0.05), sardine (6.4 ± 0.05), mackerel (4.3 ± 0.05), egg (3.7 ± 0.05), sesame oil (3.2 ± 0.05), beef (3.1 ± 0.05), peanuts (2.7 ± 0.05), pork (2.4 ± 0.05), sesame seeds (2.3 ± 0.05), cattlefish (2.3 ± 0.05), chicken (2.1 ± 0.05), cheese (2.1 ± 0.05), horse mackerel (2.0 ± 0.05), yellowtail (2.0 ± 0.05), pistachios (2.0 ± 0.05), walnuts (1.9 ± 0.05), hazelnuts (1.7 ± 0.05), cottonseed oil (1.7 ± 0.05), chestnuts (1.4 ± 0.05), almonds (1.4 ± 0.05), corn oil (1.3 ± 0.05), eel (1.1 ± 0.05), spinach (1.0 ± 0.05), perilla leaf (1.0 ± 0.05), broccoli (0.86 ± 0.005), rapeflower (0.74 ± 0.005), green raw soybeans (0.58 ± 0.005), flatfish (0.55 ± 0.005), rice bran (0.54 ± 0.005), sunflower oil (0.42 ± 0.005), olive oil (0.41 ± 0.005), safflower oil (0.40 ± 0.005), sweet potato (0.36 ± 0.005), wheat germ (0.35 ± 0.005), sweet pepper (0.33 ± 0.005), kinako (0.31 ± 0.005), boiled soybeans (0.29 ± 0.005), azuki beans (0.22 ± 0.005), carrot (0.22 ± 0.005), dry soybeans (0.21 ± 0.005), natto (0.21 ± 0.005), eggplant (0.21 ± 0.005), cabbage (0.16 ± 0.005), millet (0.15 ± 0.005), cauliflower (0.14 ± 0.005), buckwheat (0.13 ± 0.005), chinese cabbage (0.10 ± 0.005), potatoe (0.10 ± 0.005), lard (0.10 ± 0.005), Job's tears (0.07 ± 0.005), milk (0.04 ± 0.005), lard, internal conversion of vitamins/coenzymes Q_7, Q_8, Q_9, and possibly others), internal synthesis from vitamin E
Applications: heart disease, heart attack, angina pectoris, increased blood pressure, decreased oxygen, blood clotting, stroke, antioxidant therapy, free radical oxidation, life span extension, radiation, decreased radiation resistance, aging, leukemia, some forms of cancer, chemotherapy, Adriamycin treatments for cancer, gum diseases, bleeding gums, gingivitus, weight increased, immune system deficiencies, AIDS, muscular dystrophy, multiple sclerosis, Alzheimer's disease, schizophrenia, skin lesions, stomach ulcers, allergies, asthma
Daily Dosage: 1989 SADDI — none; 1989 RDA — none; 1989 USRDA — none; nutritional — 10–50 mg; therapeutic — 100–200 mg; experimental — 500–1000 mg; toxic — none known up to 1000 mg

VITAMIN Q_{199} *see* vitamin Q_{10}

COENZYME R *see* vitamin B_W

VITAMIN R *see* vitamin B_{10}

FACTOR R *see* vitamin B_{10}

VITAMIN S *see* vitamin B_{11}

FACTOR S *see* vitamin B_{11}

VITAMIN T COMPLEX

(Different forms of vitamin T. See vitamin T.)

Names: vitamin T, vitamin T complex, vitamin T Goetsch, Goetsch's vitamin, factor T, termitin, torutilin, Tegotin, Temina
Classification: lipid-soluble vitamin
Forms: mycoine, penicin
Deficiency: some anemias, deterioration of healing, deterioration of vascular tone of veins, hemophilia, decreased blood clotting, fading memory, decreased nerve regeneration, decreased production of blood platelets, mental health deterioration, appetite decreased, deterioration of growth, deterioration of protein utilization, immunity decreased, fatigue, decreased resistance to cold, shortened life span
Side-effects: none known
Toxicity: none known
Inhibitors: laxatives, mineral oil
Helpers: unknown
Sources: sesame seeds, egg yolks, fungi fats, torula yeast, termites, roaches
Applications: circulatory deterioration, hemophilia, nervous system deterioration, scurvy, infantile anorexia, pregnancy
Daily Dosage: 1974 PCS—none; 1989 SADDI—none; 1989 RDA—none; 1989 USRDA—none; nutritional—unknown; therapeutic—unknown; experimental—unknown; toxic—unknown
Warnings: none

VITAMIN T GOETSCH *see* **vitamin T**

VITAMIN U

(The term "vitamin U" was originally used for a substance later found to be vitamin B_C. The term later was used for this substance. See vitamin U chick factor.)

Names: vitamin U, ulcer-preventive factor, anti-gizzard erosion factor, methylmethioninesulfonium chloride, ardesyl, MMSC, Cabagin, Cabagin-U, Epadyn-U, Vitas-U
Classification: water-soluble vitamin
Forms: l-methionine methylsulfonium salt, methylmethioninesulfonium chloride, methylmethioninesulfonium bromide
Deficiency: ulceration of internal organs, stomach ulcers, stomach ailments, gas pains, deterioration of appetite, slow wound healing, increased cholesterol levels
Side-effects: iodine decreased?
Toxicity: none known
Inhibitors: heat, aging, storage
Helpers: vitamin C
Sources: alfalfa, raw cabbage, cabbage juice, celery, fresh greens, raw milk, raw egg yolk, raw greens, cereal grasses, fresh animal fats

I. Vitamins

Applications: digestion deterioration, heart disease, skin deterioration, liver disease, cirrhosis of liver, increased cholesterol
Daily Dosage: 1974 PCS—none; 1989 SADDI—none; 1989 RDA—none; 1989 USRDA—none; nutritional—10; therapeutic—100; experimental—1000; toxic—unknown
Warnings: Increased amounts may cause side-effects. See side-effects symptoms.

VITAMIN U BROMIDE

(Specific form of vitamin U)

Names: vitamin U bromide, methylmethioninesulfonium bromide, Ardesyl
Classification: water-soluble vitamin
Forms: methylmethioninesulfonium bromide

VITAMIN U CHLORIDE

(Specific form of vitamin U)

Names: vitamin U chloride, methylmethioninesulfonium chloride
Classification: water-soluble vitamin

VITAMIN U CHICK FACTOR

(The term "vitamin U" was originally used for vitamin U chick factor, a substance later found to be vitamin B_C. The term later was used for another substance. See vitamin U.)

Names: vitamin U, chick factor
Classification: formerly water-soluble vitamin, formerly vitamin B complex vitamin, now defunct

Forms: discovered to be vitamin B_C

FACTOR U *see* **vitamin B_C**

VITAMIN V

(The term "vitamin V" is sometimes used to denote vitamin B_X.)

Classification: possible water-soluble vitamin, possible nucleic acid
Forms: possibly derivatives of vitamins B_3, nicotinamide adenine dinucleotide/NAD

Deficiency: deterioration of development in chicks, possibly similar symptoms in humans
Side-effects: unknown

Toxicity: unknown
Inhibitors: unknown
Helpers: unknown
Sources: unknown
Applications: unknown
Daily Dosage: 1974 PCS—none; 1989 SADDI—none; 1989 RDA—none; 1989 USRDA—none; nutritional—unknown; therapeutic—unknown; experimental—unknown; toxic—unknown
Warnings: unknown

VITAMIN W

Classification: possible water-soluble vitamin
Forms: possibly vitamin B_W
Deficiency: similar to vitamin B_W deficiency
Side-effects: unknown
Toxicity: unknown
Inhibitors: unknown
Helpers: unknown
Sources: unknown
Applications: unknown
Daily Dosage: unknown
Warnings: unknown

VITAMIN X

(being researched)
(The term "vitamin X" is sometimes used to denote vitamin P.)
(The term "vitamin X" was sometimes used interchangeably with the term "factor X." See factor X.)

Classification: possible water-soluble vitamin
Forms: probably vitamin B_W
Deficiency: similar to vitamin B_w decreased
Side-effects: unknown
Toxicity: unknown
Inhibitors: unknown
Helpers: unknown
Sources: unknown
Applications: unknown
Daily Dosage: 1974 PCS—none; 1989 SADDI—none; 1989 RDA—none; 1989 USRDA—none; nutritional—unknown; therapeutic—unknown; experimental—unknown; toxic—unknown
Warnings: unknown

FACTOR X

(The term "factor X" has been used for virtually every vitamin during the isolation phase of research. Specifically, see vitamin B_{12}, vitamin B_W, vitamin E.)

VITAMIN Y

Classification: possible water-soluble vitamin
Forms: (probably vitamin B_6)
Deficiency: similar to vitamin B_6 deficiency

FACTOR Y *see* **vitamin B_6**

PART II
Minerals

ALUMINUM

Names: aluminum, originally alumium, later aluminium, Al, Al^{+3}, Al^{+++}, element 13
Classification: possible trace mineral, contaminant
Deficiency: (being researched)
Side-effects: see toxicity symptoms
Toxicity: constipation, colic, appetite decreased, nausea, skin ailments, twitching leg muscles, increased perspiration, fatigue, motor paralysis, local numbness, fatty deterioration of kidneys, liver; decreased levels of calcium, phosphorus
Inhibitors: unknown
Helpers: unknown
Sources: aluminum cookware, water pollution, some baking powders, some white flours, some children's aspirin, antacids
Applications: none known
Daily Dosage: 1974 PCS—none; 1989 SADDI—none; 1989 RDA—none; 1989 USRDA—none; nutritional—none; therapeutic—none; experimental—none; toxic—small amounts
Warnings: Small amounts are dangerous. See toxicity symptoms.

ANTIMONY

Names: antimony, Sb, element 51
Classification: possible trace mineral, contaminant
Deficiency: (being researched)
Side-effects: see toxicity symptoms
Toxicity: metallic taste in mouth,

nausea, pain in mouth, pain in throat, pain in stomach, spasms in fingers, spasms in arms, spasms in legs, collapse
Inhibitors: unknown
Helpers: unknown
Sources: enamels
Applications: none known

Daily Dosage: 1974 PCS — none; 1989 SADDI — none; 1989 RDA — none; 1989 USRDA — none; nutritional — unknown; therapeutic — none; experimental — none; toxic — small amounts
Warnings: Small amounts are dangerous. See toxicity symptoms.

ARSENIC

Names: arsenic, As, element 33
Classification: trace mineral, contaminant
Deficiency: deterioration of growth, iron increased in spleen, coarse hair, decreased red blood cell life
Side-effects: see toxicity symptoms
Toxicity: weakness, appetite decreased, thickening of skin on palms/soles, nausea, diarrhea, constipation, congestion of eyes, inflammation of nose, sore throat, laryngitis, sneezing, hoarseness, coughing, increased salivation, inflamed mouth, inflamed tongue, skin rashes, peeling of hands, peeling of feet, yellow jaundice, enlarged liver, kidney deterioration, urination changes, decreased hair, decreased fingernails, tingling hands/feet, burning hands/feet, numb hands/feet, paralysis, anemia, garlic body odor, garlicky breath, increased sweating, itching eyes, watering eyes; increased color on neck, eyelids, nipples, armpits; swelling of eyelids, face, ankles
Inhibitors: vitamin B_X, vitamin N
Helpers: unknown
Sources: air pollution, additives, weed killers
Applications: none known
Daily Dosage: 1974 PCS — none; 1989 SADDI — none; 1989 RDA — none; 1989 USRDA — none; nutritional — unknown; therapeutic — none; experimental — none; toxic — small amounts
Warnings: Small amounts are dangerous. See toxicity symptoms.

BARIUM

Names: barium, Ba, Ba^{+2}, Ba^{++}, element 56
Classification: trace mineral, contaminant
Deficiency: (Trace amounts are required.)
Side-effects: see toxicity symptoms
Toxicity: nausea, cramps, paralysis of arms, paralysis of legs
Inhibitors: unknown

Helpers: unknown
Sources: industry
Applications: medical testing
Daily Dosage: 1974 PCS — none; 1989 SADDI — none; 1989 RDA — none; 1989 USRDA — none; nutritional — unknown; therapeutic — none; experimental — none; toxic — small amounts
Warnings: Small amounts are dangerous. See toxicity symptoms.

BERYLLIUM

Names: beryllium, formerly glucinum, Be, Be^{+2}, Be^{++}, element 4
Classification: possible trace mineral, contaminant
Deficiency: (being researched)
Side-effects: see toxicity symptoms
Toxicity: decreased magnesium levels, difficult breathing, beryllium in organs, organ failure, enzyme system interference
Inhibitors: unknown
Helpers: unknown
Sources: neon signs, electronic devices, steel, many common household products, fumes, smokes
Applications: none known
Daily Dosage: 1974 PCS—none; 1989 SADDI—none; 1989 RDA—none; 1989 USRDA—none; nutritional—unknown; therapeutic—none; experimental—none; toxic—small amounts
Warnings: Small amounts are dangerous. See toxicity symptoms.

BORON

Names: boron, B, element 5
Classification: trace mineral
Deficiency: calcium decreased, magnesium decreased, phosphorus decreased, bone deterioration, decreased estrogen synthesis, decreased vitamin D synthesis, decreased steroid synthesis, breakdown of hormone function, bone demineralization
Side-effects: see toxicity symptoms
Toxicity: nausea, diarrhea, abdominal pains, skin inflammation, muscle spasms, shock, enzyme inhibition
Inhibitors: unknown
Helpers: unknown
Sources (mg per 100 grams of food): soy meal (2.8 ± 0.05), prunes (2.7 ± 0.05), raisins (2.5 ± 0.05), almonds (2.3 ± 0.05), rosehips (1.9 ± 0.05), peanuts (1.8 ± 0.05), hazel nuts (1.6 ± 0.05), vegetables (1.3 ± 0.05), dates (0.92 ± 0.005), honey (0.72 ± 0.005), wine (0.85 ± 0.005), apple sauce (0.279 ± 0.0005), grape juice (0.202 ± 0.0005), apple juice (0.188 ± 0.0005), peaches (0.187 ± 0.0005), broccoli flowers (0.185 ± 0.0005), cherries (0.147 ± 0.0005), pears (0.122 ± 0.0005), dairy (0.11 ± 0.0005), cereals (0.092 ± 0.0005), broccoli stalks (0.089 ± 0.0005), carrots (0.075 ± 0.0005), green beans (0.046 ± 0.0005), orange juice (0.041 ± 0.0005), iceberg lettuce (0.039 ± 0.0005), egg noodles (0.037 ± 0.0005), fish (0.036 ± 0.0005), cornflakes (0.031 ± 0.0005), white bread (0.020 ± 0.0005), vanilla ice cream (0.019 ± 0.0005), canned potatoes (0.017 ± 0.0005), meat (0.016 ± 0.0005), spaghetti noodles (0.006 ± 0.0005), chicken breast (0.005 ± 0.0005), freeze dried coffee (0.005 ± 0.0005), minute rice (0.003 ± 0.0005), 2% fat milk (0.002 ± 0.0005)
Applications: none known
Daily Dosage: 1989 SADDI—none; 1989 RDA—none; 1989 USRDA—none; nutritional—2-3 mg; therapeutic—?; experimental—?; toxic—?
Warnings: Small amounts are dangerous. See toxicity symptoms.

BROMINE

Names: bromine, bromin, Br, Br⁻, element 35
Classification: trace mineral
Deficiency: liver deterioration, deterioration of brain performance, mental abnormalities, adrenal deterioration, thyroid deterioration, pituitary gland deterioration
Side-effects: see toxicity symptoms
Toxicity: acne, coldness of arms and legs, fetid breath, lack of sleep, male sterility, headache, emotional agitation, emotional instability, fatigue, hallucinations, amnesia, confusion
Inhibitors: chlorides, salt, mercurial diuretics
Helpers: unknown
Sources: fuels, animal glands, mussels, kelp, sea plants, seawater
Applications: none known
Daily Dosage: 1974 PCS—none; 1989 SADDI—none; 1989 RDA—none; 1989 USRDA—none; nutritional—unknown; therapeutic—none; experimental—none; toxic—unknown
Warnings: increased amounts are dangerous. See toxicity symptoms.

CADMIUM

Names: cadmium, Cd, element 48
Classification: trace mineral, contaminant
Deficiency: (Trace amounts are required.)
Side-effects: see toxicity symptoms
Toxicity: (Detoxified by zinc) increased blood pressure, headache, kidney deterioration, proteins in urine, muscular weakness, dry throat, nausea, violent gastrointestinal symptoms, tightness in chest, emphysema
Inhibitors: vitamin C, calcium, iron, selenium, zinc
Helpers: unknown
Sources: air pollution, water pollution, cadmium-plated food containers, fungicides, cigarette smoking, coffee, tea, refined grains, water pipes, photographic agents, galvanoplating, pyrotechnics
Applications: none known
Daily Dosage: 1974 PCS—none; 1989 SADDI—none; 1989 RDA—none; 1989 USRDA—none; nutritional—unknown; therapeutic—none; experimental—none; toxic—small amounts
Warnings: increased amounts are dangerous. See toxicity symptoms.

CALCIUM

Names: calcium, Ca, Ca^{+2}, Ca^{++}, element 20
Classification: mineral
Deficiency: cramps, emotional agitation, nervousness, lack of sleep, headaches, cadmium retention, lead

CALCIUM

retention, teeth grinding, muscle spasm/twitching, muscle tendon strain, convulsions, thinning bones, softening bones, softening teeth, late tooth dentition, diminished calcium in blood, cold hands, cold feet, numbness, hydrocele, night sweats, varicose veins, carbuncles, skin eruptions, deep abscesses, chronic oozing ulcers, cracked skin, sore breasts
Side-effects: see toxicity symptoms
Toxicity: decreased gastrointestinal tone, kidney failure, emotional deterioration, increased calcium in blood
Inhibitors: fats, oxalic acid (in chocolate, rhubarb), phytic acid (in grains), aspirin, barbiturates, strong sedatives, aluminum, neomycin, decreased hydrochloric acid, decreased vitamin D, decreased magnesium
Helpers: vitamin A, vitamin C, vitamin D, vitamin F, boron, iron, magnesium, manganese, phosphorus, strontium, hydrochloric acid
Sources (mg per 100 grams of food): rennin (3509 ± 4.5), basil (3000 ± 50), savory (3000 ± 50), dried parsley (1900 ± 50), dried dill weed (1800 ± 50), celery seed (1750 ± 25), dill seed (1600 ± 25), cinnamon (1400 ± 25), parmesan cheese (1380 ± 10), poppy seed (1367 ± 16.7), dry skim milk (1257 ± 1.7), dry instant skim milk (1231 ± 0.5), fennel seed (1200 ± 25), marjoram (1200 ± 50), oregano (1200 ± 25), sage (1200 ± 50), dry buttermilk (1100 ± 7.1), romano cheese (1079 ± 1.8), gruyere cheese (1025 ± 1.8), whole toasted sesame seeds (1004 ± 1.8), cumin seed (1000 ± 25), whole dried sesame seeds (978 ± 5.6), swiss cheese (971 ± 1.8), dry whole milk (913 ± 1.6), tarragon (900 ± 25), freeze-dried chives (875 ± 62.5), dried chervil (800 ± 50), provolone cheese (764 ± 1.8), monterey cheese (757 ± 1.8), dried rosemary (750 ± 25), poultry seasoning (750 ± 25), edam cheese (739 ± 1.8), cheddar cheese (729 ± 1.8), muenster cheese (725 ± 1.8), gouda cheese (707 ± 1.8), tilsit cheese (707 ± 1.8), cloves (700 ± 25), dried coriander leaf (700 ± 50), anise seed (700 ± 25), caraway seed (700 ± 25), colby cheese (693 ± 1.8), blackstrap molasses (685 ± 2.5), caraway cheese (682 ± 1.8), brick cheese (682 ± 1.8), roquefort cheese (671 ± 1.8), port du salut cheese (657 ± 1.8), cheshire cheese (650 ± 1.8), allspice (650 ± 25), coriander seed (650 ± 25), havarti cheese (629 ± 1.8), dried oriental radish (629 ± 0.9), dried agar seaweed (625 ± 0.5), pimento cheese spread (621 ± 1.8), pumpkin pie spice (600 ± 25), yellow mustard seed (567 ± 16.7), fontina cheese (557 ± 1.8), bleu cheese (536 ± 1.8), mozzarella cheese (525 ± 1.8), limburger cheese (504 ± 1.8), feta cheese (500 ± 1.8), curry powder (500 ± 25), bay leaf (500 ± 50), black pepper (450 ± 25), american cheese (443 ± 1.8), torula yeast (429 ± 1.8), almond meal (429 ± 1.8), sesame butter (427 ± 3.3), gjetost cheese (404 ± 1.8), onion powder (400 ± 25), cardamom (400 ± 25), camembert cheese (393 ± 1.8), fried tofu (369 ± 3.8), dried and frozen tofu (365 ± 2.9), soybean protein concentrate (364 ± 1.8), canned smelt (358 ± 0.5), rice bran (355 ± 0.5), carob flour (349 ± 0.5), white pepper (300 ± 25), medium molasses (290 ± 2.5), evaporated skim milk (288 ± 1.6), dry roasted almonds (286 ± 1.8), toasted almonds (286 ± 1.8), sweetened condensed milk (284 ± 1.3), dry roasted soybean nuts (270 ± 0.6), almond butter (269 ± 3.1), dried almonds (268 ± 1.8), evaporated whole milk (261 ± 0.4), boiled lamb's quarters (258 ± 0.6), dehydrated onion flakes (257 ± 3.6), evaporated lowfat milk (250 ± 0.4), barbados molasses (245 ± 2.5), defatted soy meal (243 ± 0.4), defatted soybean flour (241 ± 0.5), oil roasted almonds (236 ± 1.8), chili powder (233 ± 16.7), milk chocolate (232 ± 1.8), almond paste (232 ± 1.8), anchovy canned in olive oil (230 ± 2.5), ham and cheese spread (221 ± 1.8), natto (217 ± 0.6), roselle (216 ± 0.9), brewers yeast (214 ± 1.8), boiled jute potherb (212 ± 1.2), raw mustard spinach (211 ± 0.7), frozen and boiled

collards (211 ± 0.6), boiled amaranth (209 ± 0.8), ricotta cheese (207 ± 0.4), soybean flour (206 ± 0.6), waffles (205 ± 0.7), firm raw tofu (205 ± 0.4), cheese soufflé (201 ± 0.5), mace (200 ± 25), paprika (200 ± 25), turmeric (200 ± 25), oil roasted filberts (200 ± 1.8), nutmeg (200 ± 25), skim yogurt (199 ± 0.2), toasted english muffin (198 ± 1), dry roasted filberts (196 ± 1.8), raw frozen rhubarb (194 ± 0.4), sheep milk (193 ± 0.2), toaster pastry (192 ± 1), raw turnip greens (189 ± 1.8), dried filberts (189 ± 1.8), low fat soybean flour (188 ± 0.6), roasted soybean flour (188 ± 0.6), brie cheese (186 ± 1.8), raw dandelion greens (186 ± 1.8), cheese pizza (183 ± 0.4), lowfat yogurt (183 ± 0.2), soybean protein isolate (179 ± 1.8), dried brazil nuts (179 ± 1.8), raw garlic (178 ± 5.6), freeze-dried shallots (175 ± 12.5), indian buffalo milk (169 ± 0.2), pickled anchovy (168 ± 1.8), raw kelp/kombu/ tangle seaweed (168 ± 0.5), sorghum (167 ± 2.4), orange peel (167 ± 8.3), tomato powder (166 ± 0.5), light molasses (165 ± 2.5), pepperoni pizza (163 ± 0.4), english muffin (161 ± 0.9), dried and salted cod (160 ± 0.6), boiled mustard spinach (158 ± 0.6), wheat flakes (154 ± 1.8), chicken egg yolk (153 ± 2.9), frozen and boiled turnip greens (152 ± 0.6), raw wakame seaweed (150 ± 0.5), fenugreek seed (150 ± 12.5), cayenne pepper (150 ± 25), peanut flour (150 ± 12.5), cooked tahitian taro (149 ± 0.7), raw anchovy (147 ± 0.6), boiled green soybeans (146 ± 0.6), vanilla milkshake (146 ± 0.2), frozen and boiled spinach (146 ± 0.5), maple sugar (146 ± 1.8), dried pilinuts (146 ± 1.8), taco shell (145 ± 4.5), tostada shell (145 ± 4.5), frozen and cooked sweetened rhubarb (145 ± 0.4), dried figs (144 ± 0.3), toasted white bread (143 ± 2.4), freeze-dried parsley (143 ± 35.7), protein fortified lowfat 2% milk (143 ± 0.2), protein fortified skim milk (143 ± 0.2), protein fortified lowfat 1% milk (142 ± 0.2), boiled winged beans (142 ± 0.3), ham and cheese sandwich (140 ± 0.3), corn tortilla (140 ± 1.7), boiled dandelion greens (140 ± 1), dry cocoa powder (140 ± 10), toasted soybean nuts (139 ± 1.8), roasted soybean nuts (138 ± 0.6), cooked perch (138 ± 0.6), boiled turnip greens (138 ± 0.7), frozen and boiled kale (138 ± 0.8), boiled spinach (136 ± 0.6), dried pistachio nuts (136 ± 1.8), hamburger bun (135 ± 1.3), hot dog bun (135 ± 1.3), taco (135 ± 0.6), bran muffins (135 ± 1.3), goat milk (134 ± 0.2), lemon peel (133 ± 8.3), chocolate milkshake (132 ± 0.2), boiled scotch kale (132 ± 0.8), toasted sesame kernels (132 ± 1.8), malted milk (131 ± 0.2), eggnog (130 ± 0.2), raw parsley (130 ± 1.7), turkey frankfurter (129 ± 1.1), filled milk (128 ± 0.2), canned spinach (126 ± 0.5), chicken salad (125 ± 1.8), white bread (125 ± 2.1), freeze-dried red sweet peppers (125 ± 31.3), freeze-dried green sweet peppers (125 ± 31.3), dried sesame kernels (125 ± 6.3), skim milk (123 ± 0.2), lowfat 1% milk (123 ± 0.2), dry breadcrumbs (122 ± 0.5), caramel sundae (122 ± 0.3), restaurant vanilla milkshake (122 ± 0.2), lowfat 2% milk (122 ± 0.2), enriched biscuits (121 ± 1.8), unenriched biscuits (121 ± 1.8), cheeseburger (121 ± 0.4), whole yogurt (121 ± 0.2), brown mustard (120 ± 10), au gratin potatoes (120 ± 0.4), yellow cake (119 ± 0.7), whole milk (119 ± 0.2), hot chocolate (119 ± 0.2), dinner rolls (118 ± 1.8), canned turnip greens (118 ± 0.4), raw watercress (118 ± 2.9), dried sunflower seeds (118 ± 1.8), dried pepeao (117 ± 4.2), vanilla pudding (117 ± 0.2), sour cream (117 ± 4.2), buttermilk (116 ± 0.2), whole chocolate malted milk (115 ± 0.2), lowfat 1% chocolate milk (115 ± 0.2), toasted raisin bread (114 ± 2.4), boiled beet greens (114 ± 0.7), lowfat 2% chocolate milk (114 ± 0.2), carob milk (114 ± 0.2), boiled broccoli (114 ± 0.6), whole ground corn meal muffins (113 ± 1.3), cooked whelk (113 ± 0.6), devil's food cake (113 ± 0.8), restaurant chocolate milkshake (113 ± 0.2), restaurant strawberry milkshake (113 ± 0.2), white cake (112 ± 0.6), custard (112 ± 0.2), whole chocolate milk (112 ± 0.2), pork olive loaf (111 ± 1.8), french toast (111 ± 0.8), burrito (110 ± 0.3), raw vine spinach (109 ± 0.5), bread pudding (109 ± 0.2), cheese crackers (107 ± 3.3), whole wheat roll (107 ± 1.8), danish pastry (107 ± 1.2), raw perch (107 ± 0.6), half and half cream (107 ± 3.3), cornbread (106 ± 0.6), corn pone (106 ± 0.6), raw chinese cabbage (106 ± 1.4), strawberry sundae (106 ± 0.3), corn meal muffins (105 ± 1.3), crab cakes (105 ± 0.8), tapioca cream pudding (105 ± 0.3), raw tofu (105 ± 0.4), dry roasted red pistachio nuts (104 ± 1.8), mixed grain bread (104 ± 2), raisin bread (104 ± 2), cooked blue crab (104 ± 0.6), boiled soybeans (102 ± 0.3), canned blue crab (101 ± 0.6), pancakes (100 ± 1.9), saffron (100 ± 50), raw spinach (100 ± 1.8), ginger (100 ± 25), raw chickory greens (100 ± 0.6), raw coriander (100 ± 12.5), frozen and boiled mustard greens (100 ± 0.7), maple syrup (100 ± 2.5), oil roasted spanish peanuts (100 ± 1.8), turkey egg (99 ± 0.6), pork pickle and pimento loaf (96 ± 1.8), chicken frankfurter (96 ± 1.1), custard pie (96 ± 0.4), chocolate pudding (96 ± 0.2), sweet chocolate (96 ± 1.8), dried english/persian walnuts (96 ± 1.8), frozen and boiled okra (96 ± 0.5), popover roll (95 ± 1.3), coconut custard pie (94 ± 0.4), pickled rip sevillano/ascolano olives (93 ± 3.1), turkey bologna (93 ± 1.8), medium cream (93 ± 3.3), boiled chinese cabbage (93 ± 0.6), tempeh (93 ± 0.6), instant oatmeal (92 ± 0.3), toasted rye bread (91 ± 2.3), raw cuttlefish (91 ± 0.6), cottage pudding (91 ± 0.9), frozen and boiled turnip greens and turnips (91 ± 0.5), raw cassava (91 ± 0.5), gingerbread (90 ± 0.8), boiled white beans (90 ± 0.3), raw blue crab (89 ± 0.6), cooked oysters (89 ± 0.6), pickled rip manzanillo/mission olives (87 ± 10.9), pork and beef old fashioned loaf (86 ± 1.8), ry-krisp crackers (86 ± 3.6), cooked rainbow trout (86 ± 0.6), oil roasted peanuts (86 ± 1.8), oil roasted virginia peanuts (86 ± 1.8), blueberry muffins (85 ± 1.3), oatmeal cookie (85 ± 3.8), cooked black bass (85 ± 0.4), brown sugar (85 ± 0.3), chocolate cream pie (84 ± 0.4), caramel cake (84 ± 0.5), kippered herring (83 ± 1.3), stir-fried sprouted soybeans (82 ± 0.5), bran (82 ± 1.8), toasted pumpernickel bread (82 ± 1.8), pita bread (82 ± 1.3), cream cheese (82 ± 1.8), unsweetened baking chocolate (82 ± 1.8), toasted whole wheat bread (81 ± 2.4), cream puff (81 ± 0.4), raw garden cress (80 ± 2), yellow mustard (80 ± 10), pickled herring (80 ± 3.3), rye bread (80 ± 2), cooked kingfish (80 ± 0.5), raw sunfish (80 ± 0.6), raw freshwater bass (80 ± 0.6), okara tofu (80 ± 0.8), french bread (79 ± 1.8), dried jujube (79 ± 0.5), boiled purslane (78 ± 0.9), boiled collards (78 ± 0.3), bacon cheeseburger (77 ± 0.3), raw swamp cabbage (77 ± 0.9), toasted cracked wheat bread (76 ± 2.4), cooked rainbow smelt (76 ± 0.6), dry roasted macadamia nuts (75 ± 1.8), neufchatel cheese (75 ± 1.8), raisin bun (75 ± 0.8), sugar cookie (75 ± 3.1), butterscotch pie (75 ± 0.4), cooked herring (74 ± 0.6), tamarind (74 ± 0.4), boiled mustard greens (74 ± 0.7), wax beans (73 ± 0.4), cooked pike (73 ± 0.6), canned white beans (73 ± 0.2), whole wheat bread (72 ± 2), pumpernickel bread (72 ± 1.6), pineapple upside-down cake (72 ± 0.7), boiled cardoon (72 ± 0.5), fruitcake, dark (72 ± 1.2), raw irish moss seaweed (72 ± 0.5), boiled kale (72 ± 0.8), lobster paste (71 ± 7.1), raw white sucker (71 ± 0.6), dried japanese chestnuts (71 ± 1.8), dry roasted pistachio nuts (71 ± 1.8), dried macadamia nuts (71 ± 1.8), dry roasted sunflower seeds (71 ± 1.8), raw laver/nori seaweed (70 ± 0.5), boiled navy beans (70 ± 0.3), sweet potato pie (69 ± 0.4), dried european chestnuts (68 ± 1.8), raw looseleaf lettuce (68 ± 1.8), boiled great northern beans (68 ± 0.3), quail egg (67 ± 5.6), miso (67 ± 0.4), saltine crackers (67 ± 8.3), breaded and fried shrimp (67 ± 0.6), fruitcake, light

(67 ± 1.2), boston cream pie (67 ± 0.5), whipping cream, heavy (67 ± 3.3), whipping cream, light (67 ± 3.3), garlic powder (67 ± 16.7), raw rainbow trout (67 ± 0.6), raw chives (67 ± 16.7), banana custard pie (66 ± 0.4), cracked wheat bread (64 ± 2), breaded and fried clams (64 ± 0.6), rhubarb pie (64 ± 0.4), duck egg (64 ± 0.7), boiled french beans (63 ± 0.3), boiled okra (63 ± 0.6), breaded and fried oysters (62 ± 0.6), cooked whiting (62 ± 0.6), boiled yellow beans (62 ± 0.3), pancakes with butter and syrup (61 ± 0.3), oatmeal bread (61 ± 1.8), cottage cheese (61 ± 1.8), cooked lobster (61 ± 0.6), baked beans with pork and sweet sauce (61 ± 0.2), raw flounder (61 ± 0.5), raw sole (61 ± 0.5), rice polish (61 ± 0.9), dried peanuts (61 ± 1.8), dried hickory nuts (61 ± 1.8), soybean oil and cottonseed oil margarine liquid (60 ± 10), chicken egg (60 ± 1), horseradish sauce (60 ± 3.3), raw pollock (60 ± 0.6), raw freshwater drum (60 ± 0.6), cooked halibut (60 ± 0.6), raw rainbow smelt (60 ± 0.6), steamed sprouted soybeans (60 ± 1.1), boiled garden cress (60 ± 0.7), raw spring onions with tops (60 ± 1), cooked alaska king crab (59 ± 0.6), boiled butterbur (59 ± 0.5), pork and beef link sausage with american cheese (58 ± 1.2), granola bar (58 ± 2.1), boiled swiss chard (58 ± 0.6), raw herring (58 ± 0.6), raw leeks (58 ± 1.9), ham and cheese roll (57 ± 1.8), barbecue loaf (57 ± 2.2), cheese crackers with peanut butter (57 ± 1.2), hamburger sandwich (57 ± 0.5), scalloped potatoes (57 ± 0.4), raw new zealand spinach (57 ± 1.8), bittersweet chocolate (57 ± 1.8), oil roasted sunflower seeds (57 ± 1.8), dried black walnuts (57 ± 1.8), boiled peanuts (56 ± 1.6), baked beans with pork and tomato sauce (56 ± 0.2), cheesecake (56 ± 0.6), raw whelk (56 ± 0.6), raw pike (56 ± 0.6), prickly pear (56 ± 0.5), boiled black turtle beans (56 ± 0.3), black currants (55 ± 0.9), boiled blackeye pea pods (55 ± 1.1), pork and beef peppered loaf (54 ± 1.8), whole grain rye crackers (54 ± 3.8), animal crackers (54 ± 1.9), dried watermelon seeds (54 ± 1.8), raw agar seaweed (54 ± 0.5), dry roasted peanuts (54 ± 1.8), oil roasted valencia peanuts (54 ± 1.8), dark rye flour (54 ± 0.4), dried acorns (54 ± 1.8), dried butternuts (54 ± 1.8), raw octopus (53 ± 0.6), falafel (53 ± 1), boiled swamp cabbage (53 ± 1), golden seedless raisins (53 ± 0.5), boiled mungo beans (53 ± 0.3), canned great northern beans (53 ± 0.2), boiled pokeberry shoots (52 ± 0.6), pickled green olives (52 ± 1.1), raw shrimp (52 ± 0.6), hot fudge sundae (52 ± 0.3), cooked carp (52 ± 0.6), raw endive (52 ± 2), maypo (52 ± 0.3), boiled pink beans (52 ± 0.3), raw milkfish (51 ± 0.6), pumpkin pie (51 ± 0.4), raw burbot (51 ± 0.6), frozen and boiled broccoli (51 ± 0.5), raw red cabbage (51 ± 1.4), dried prunes (51 ± 0.6), boiled lupins (51 ± 0.3), bran flakes (50 ± 1.8), italian bread (50 ± 1.8), vegetarian baked beans (50 ± 0.2), hummus (50 ± 0.2), pineapple custard pie (50 ± 0.4), boiled burdock root (50 ± 0.4), boiled cranberry beans (50 ± 0.3), pork and beef brotwurst (49 ± 0.7), battered and fried shark (49 ± 0.6), seedless raisins (49 ± 0.5), boiled chickpeas (49 ± 0.3), smoked haddock (48 ± 0.6), boiled new zealand spinach (48 ± 0.6), raw whiting (48 ± 0.6), raw broccoli (48 ± 1.1), light cream (48 ± 1.7), boiled pinto beans (48 ± 0.3), canned navy beans (47 ± 0.2), refried beans (47 ± 0.2), peach pie (47 ± 0.5), pecan pie (47 ± 0.5), braised pork spareribs (47 ± 0.5), raw halibut (47 ± 0.6), whey (47 ± 0.2), raw shad (47 ± 0.6), boiled salsify (47 ± 0.7), boiled green beans (47 ± 0.8), boiled yellow snap beans (47 ± 0.8), pork and beef picnic loaf (46 ± 1.8), raw alaska king crab (46 ± 0.6), raw dungeness crab (46 ± 0.6), raw clams (46 ± 0.6), boiled and steamed european chestnuts (46 ± 1.8), raw green cabbage (46 ± 1.4), dry roasted cashews (46 ± 1.8), frozen and boiled yellow snap beans (46 ± 0.7), frozen and boiled green beans (46 ± 0.7), dried apricots (46 ± 1.4), oil

II. Minerals

CALCIUM

roasted macadamia nuts (46 ± 1.8), toasted wheat germ (46 ± 1.8), simmered pork feet (45 ± 0.5), pork bratwurst (45 ± 0.6), brownies (45 ± 2.5), raw oysters (45 ± 0.6), canned oysters (45 ± 0.6), dried longans (45 ± 0.5), coleslaw (45 ± 0.8), breaded and fried catfish (44 ± 0.6), restaurant coleslaw (44 ± 0.5), simmered chicken neck without skin (44 ± 2.8), raw lotus root (44 ± 0.6), baked acorn squash (44 ± 0.5), acorn flour (43 ± 1.8), pork mothers loaf (43 ± 2.4), light meat chicken roll (43 ± 0.5), gingersnaps cookie (43 ± 7.1), raw goose liver (43 ± 0.5), corn syrup (43 ± 2.4), dry bakers yeast (43 ± 1.8), oil roasted cashews (43 ± 1.8), dried pumpkin and squash seeds (43 ± 1.8), roasted pumpkin and squash seeds (43 ± 1.8), cashew butter (43 ± 1.8), boiled pigeon peas (43 ± 0.3), raw dock (43 ± 0.7), raw acorns (43 ± 1.8), pork and beef kielbasa/kolbassy (42 ± 1.9), chicken roll (42 ± 0.9), enchilada (42 ± 0.2), breaded and fried scallops (42 ± 1.6), bagels (42 ± 0.9), cooked haddock (42 ± 0.6), cooked pompano (42 ± 0.6), raw trout (42 ± 0.6), boiled rutabaga (42 ± 0.6), kumquats (42 ± 2.6), boiled yardlong bean (42 ± 0.3), raw carp (41 ± 0.6), crunchy peanut butter (41 ± 1.6), boiled butternut squash (41 ± 0.5), whole wheat flour (41 ± 0.4), roast beef sandwich (40 ± 0.3), bread stuffing (40 ± 0.3), light meat turkey roll (40 ± 0.9), raw mullet (40 ± 0.6), raw catfish (40 ± 0.6), cooked snapper (40 ± 0.6), yellow corn pudding (40 ± 0.2), raw scup (40 ± 0.6), raw shallots (40 ± 5), boiled hyacinth beans (40 ± 0.3), wheat gluten (40 ± 0.4), navel orange (40 ± 0.4), valencia orange (40 ± 0.4), breaded fried squid (39 ± 0.6), cooked shrimp (39 ± 0.6), boiled artichoke hearts (39 ± 0.6), shredded wheat (39 ± 1.8), mulberries (39 ± 0.4), sapote (39 ± 0.2), semisweet bakers chocolate (39 ± 1.8), raw celtuce (39 ± 0.5), elderberries (38 ± 0.3), fried abalone (38 ± 0.6), sponge cake (38 ± 0.8), cooked sheepshead (38 ± 0.6), boiled dock (38 ± 0.5), canned pinto beans (37 ± 0.2), boiled parsnips (37 ± 0.6), boiled red cabbage (37 ± 0.7), raw eggplant (37 ± 1.2), roasted japanese chestnuts (36 ± 1.8), pork luxury loaf (36 ± 1.8), graham crackers (36 ± 3.6), fish sandwich (36 ± 0.3), o'brien potatoes (36 ± 0.3), canned pumpkin pie mix (36 ± 0.4), boiled celery (36 ± 0.7), peanut brittle (36 ± 1.8), boiled brussels sprouts (36 ± 0.6), raw romaine lettuce (36 ± 1.8), boiled broad beans (36 ± 0.3), figs (36 ± 1), dried pecans (36 ± 1.8), dry roasted pecans (36 ± 1.8), oil roasted pecans (36 ± 1.8), raisin bran (35 ± 1.4), canned black turtle beans (35 ± 0.2), raw celery (35 ± 1.3), tomato paste (35 ± 0.4), citrus marmalade (35 ± 2.5), frozen and baked sweet potato (35 ± 0.6), pot barley pearled (34 ± 0.3), raw ling (34 ± 0.6), raw shark (34 ± 0.6), raw savoy cabbage (34 ± 1.4), jackfruit (34 ± 0.5), dark buckwheat flour (33 ± 0.5), pork sausage (33 ± 1.9), hot dog (33 ± 0.6), bread and butter pickles (33 ± 3.3), red beans (33 ± 0.4), cooked mussels (33 ± 0.6), canned butter beans (33 ± 0.4), canned cranberry beans (33 ± 0.2), canned chickpeas (33 ± 0.2), canned stewed red tomatoes (33 ± 0.4), roasted turkey dark meat with skin (33 ± 0.5), raw haddock (33 ± 0.6), potato flour (33 ± 0.6), boiled green cabbage (33 ± 0.7), dried lychees (33 ± 0.5), dried dates (33 ± 0.6), lime (33 ± 0.7), pickled pork feet (32 ± 0.5), light and dark meat turkey roll (32 ± 0.9), breaded and fried croaker (32 ± 0.6), raw abalone (32 ± 0.6), roasted turkey dark meat without skin (32 ± 0.5), raw snapper (32 ± 0.6), simmered beef shank excluding fat (32 ± 0.5), raw squid (32 ± 0.6), frozen and boiled turnips (32 ± 0.5), human milk (32 ± 1.6), raw japanese chestnuts (32 ± 1.8), semisweet chocolate (32 ± 1.8), red currants (32 ± 0.9), white currants (32 ± 0.9), blackberries (32 ± 0.7), canned sauerkraut (31 ± 0.4), fried chicken neck with skin (31 ± 1.4), cooked mullet (31 ± 0.6), cooked crayfish (31 ± 0.6), boiled carrots (31 ± 0.6), creamy peanut butter (31 ± 3.1), boiled leeks (31 ± 1.9), bread sticks (30 ± 2.5), canned shellie beans (30 ± 0.4), simmered beef shank including fat (30 ± 0.5), boiled savoy cabbage (30 ± 0.7), raw sprouted alfalfa seeds (30 ± 1.5), pork link sausage (29 ± 0.7), frozen and boiled carrots (29 ± 0.7), dried toasted coconut (29 ± 1.8), puffed wheat (29 ± 3.6), dried chinese chestnuts (29 ± 1.8), raw european chestnuts (29 ± 1.8), roasted european chestnuts (29 ± 1.8), boiled baby lima beans (29 ± 0.3), rose apple (29 ± 0.5), sorghum grain (28 ± 0.5), pork liverwurst (28 ± 2.8), potato pancakes (28 ± 0.7), mince pie (28 ± 0.4), boiled and frozen baby lima beans (28 ± 0.6), seeded raisins (28 ± 0.5), raw cauliflower (28 ± 1), baked sweet potato (28 ± 0.4), boiled onions (28 ± 0.5), boiled green peas (28 ± 0.6), boiled kidney beans (28 ± 0.3), persimmon (28 ± 2), cooked extra long grain white rice (28 ± 0.6), whipped butter (27 ± 4.5), onion rings (27 ± 0.6), chili (27 ± 0.2), canned kidney beans (27 ± 0.2), raw grouper (27 ± 0.6), simmered pork chitterlings (27 ± 0.5), raw oriental radish (27 ± 1.1), canned peeled red tomatoes (27 ± 0.4), boiled adzuki beans (27 ± 0.2), frozen and boiled onions (27 ± 0.5), boiled cauliflower (27 ± 0.8), boysenberries (27 ± 0.4), boiled mung beans (27 ± 0.2), boiled black beans (27 ± 0.3), boiled crookneck squash (27 ± 0.6), canned jalapeño peppers (26 ± 0.7), dill pickles (26 ± 0.8), raw queen crab (26 ± 0.6), smoked cisco (26 ± 0.6), canned broad beans (26 ± 0.2), sweet roll (26 ± 1.2), mashed potatoes (26 ± 0.5), raw mussels (26 ± 0.6), canned green beans (26 ± 0.7), canned yellow snap beans (26 ± 0.7), canned carrots (26 ± 0.7), candied sweet potato (26 ± 0.5), cooked eel (26 ± 0.6), raw celeriac (26 ± 0.5), cooked tilefish (26 ± 0.6), raw tilefish (26 ± 0.6), simmered chicken neck with skin (26 ± 1.3), boiled lotus root (26 ± 0.6), raw carrots (26 ± 0.7), boiled catjang black-eyed peas (26 ± 0.3), kiwi fruit (26 ± 0.7), canned pumpkin (26 ± 0.4), boiled sprouted green peas (26 ± 0.5), lemon (26 ± 0.9), boiled acorn squash (26 ± 0.4), loganberries (26 ± 0.3), pretzels (25 ± 1.8), oyster crackers (25 ± 6.3), potato chips (25 ± 1.8), shortbread cookie (25 ± 6.3), raw scallops (25 ± 0.6), dried coconut (25 ± 1.8), german sweet bakers chocolate (25 ± 1.8), dried pignolia pine nuts (25 ± 1.8), boiled black-eyed peas (25 ± 0.3), raw onions (25 ± 0.6), gooseberries (25 ± 0.3), boiled calabash gourd (25 ± 0.7), italian pork sausage (24 ± 0.7), gefiltefish with broth (24 ± 1.2), coconut water (24 ± 0.2), fried chicken back with skin (24 ± 0.7), frozen and boiled green peas (24 ± 0.6), raw crayfish (24 ± 0.6), carrot juice (24 ± 0.3), frozen and boiled brussels sprouts (24 ± 0.6), boiled kohlrabi (24 ± 0.6), sugar apple (24 ± 0.3), frozen and boiled black-eyed peas (24 ± 0.6), raw green peas (24 ± 0.6), boiled asparagus (24 ± 0.6), papaya (24 ± 0.2), dried japanese persimmon (24 ± 1.5), cherimoya (23 ± 0.1), lemon chiffon pie (23 ± 0.6), pineapple chiffon pie (23 ± 0.6), chocolate chiffon pie (23 ± 0.6), cooked flounder (23 ± 0.5), boiled turnips (23 ± 0.6), frozen and boiled asparagus (23 ± 0.8), cooked prunes (23 ± 0.5), teriyaki sauce (22 ± 2.8), rusk crackers (22 ± 5.6), prune pudding (22 ± 0.4), raw pompano (22 ± 0.6), canned sweet potato (22 ± 0.3), frozen and boiled fordhook lima beans (22 ± 0.6), raw chinook salmon (22 ± 0.6), boiled/baked spaghetti squash (22 ± 0.6), raw crookneck squash (22 ± 0.8), light rye flour (22 ± 0.5), raspberries (22 ± 0.4), tamari sauce (21 ± 0.9), soda crackers (21 ± 1.8), canned lima beans (21 ± 0.2), cooked clams (21 ± 0.6), canned cod (21 ± 0.6), raw sheepshead (21 ± 0.6), roasted turkey light meat with skin (21 ± 0.5), cooked grouper (21 ± 0.6), boiled artichoke (21 ± 0.6), stewed rabbit (21 ± 0.4), restaurant hot chocolate (21 ± 0.2), dried ginko nuts (21 ± 1.8), boiled sweet potato (21 ± 0.3), sapodilla (21 ± 0.3), jujube (21 ± 0.5), millit (20 ± 0.5), chili sauce (20 ±

II. Minerals

CALCIUM

3.3), pork and beef mortadella (20 ± 3.3), tomato catsup (20 ± 3.3), corn oil diet margarine tub (20 ± 10), soybean oil and cottonseed oil diet margarine tub (20 ± 10), russian salad dressing (20 ± 3.3), butter (20 ± 3.3), bulgur (20 ± 0.4), canned black-eyed peas (20 ± 0.2), canned green peas (20 ± 0.6), french fried potatoes (20 ± 1), pound cake (20 ± 1.7), raw eel (20 ± 0.6), corn oil margarine tub (20 ± 10), safflower oil margarine tub (20 ± 10), soybean oil margarine tub (20 ± 10), soybean oil and cottonseed oil margarine tub (20 ± 10), raw radish (20 ± 1.1), corn oil margarine stick (20 ± 10), safflower oil and soybean oil margarine stick (20 ± 10), soybean oil margarine stick (20 ± 10), soybean oil and cottonseed oil margarine stick (20 ± 10), sweet butter (20 ± 3.3), raw iceberg lettuce (20 ± 2.5), frozen and boiled crookneck squash (20 ± 0.5), guava (20 ± 0.6), cream of wheat (20 ± 0.3), white whole ground corn meal (20 ± 0.4), yellow whole ground corn meal (20 ± 0.4), beef frankfurter (19 ± 0.9), turkey salami (19 ± 0.9), low calorie russian salad dressing (19 ± 3.1), barbecue sauce (19 ± 3.1), surimi shrimp (19 ± 0.6), potato salad (19 ± 0.4), blackberry pie (19 ± 0.4), braised pork tongue (19 ± 0.5), cooked flatfish (19 ± 0.6), roasted turkey light meat without skin (19 ± 0.5), simmered chicken heart (19 ± 0.5), java plum (19 ± 0.4), raw wild rice (19 ± 0.3), boiled lentils (19 ± 0.3), frozen and boiled butternut squash (19 ± 0.4), cooked parboiled white rice (19 ± 0.3), cooked parboiled extra long grain white rice (19 ± 0.6), raw frog legs (18 ± 0.5), raw welsh onions (18 ± 0.5), pork and beef honey loaf (18 ± 1.8), hot dog pickle relish (18 ± 1.8), smoked whitefish (18 ± 0.6), marinara sauce (18 ± 0.2), raisin pie (18 ± 0.4), simmered pork ears (18 ± 0.5), apple brown betty (18 ± 0.2), raw turbot (18 ± 0.6), fried chicken giblets (18 ± 0.5), boiled waxgourd (18 ± 0.6), fried dark meat chicken without skin (18 ± 0.5), raw flatfish (18 ± 0.6), raw sea-trout (18 ± 0.6), cooked taro (18 ± 0.8), raw red chili peppers (18 ± 1.1), raw green chili peppers (18 ± 1.1), raw chinese chestnuts (18 ± 1.8), roasted chinese chestnuts (18 ± 1.8), breadfruit (18 ± 0.5), raw scallop squash (18 ± 0.8), tangerine juice (18 ± 0.2), crab apples (18 ± 0.5), shoyu sauce (17 ± 0.9), pork and beef pepperoni (17 ± 8.3), sour pickles (17 ± 0.8), tuna salad (17 ± 0.2), canned zucchini in tomato juice (17 ± 0.4), simmered beef kidneys (17 ± 0.5), fried dark meat chicken with skin (17 ± 0.5), braised pork loin blade excluding fat (17 ± 0.5), marshmallow (17 ± 8.3), frozen and boiled cauliflower (17 ± 0.6), boiled succotash (17 ± 0.5), raw ginger root (17 ± 2.1), baked hubbard squash (17 ± 0.5), wheat cake (17 ± 0.4), boiled lima beans (17 ± 0.3), frozen and boiled zucchini (17 ± 0.4), pineapple juice (17 ± 0.2), white bolted corn meal (17 ± 0.4), yellow bolted corn meal (17 ± 0.4), popcorn (17 ± 8.3), french rolls (16 ± 1), strawberry pie (16 ± 0.5), broiled lamb liver (16 ± 1.1), fried light meat chicken without skin (16 ± 0.5), fried chicken breast without skin (16 ± 0.6), fried chicken wing with skin (16 ± 1.6), fried light meat chicken with skin (16 ± 0.5), fried chicken breast with skin (16 ± 0.5), braised beef pancreas (16 ± 0.5), braised pork pancreas (16 ± 0.5), french fries (16 ± 0.6), coconut milk (16 ± 0.2), boiled oriental radish (16 ± 0.7), poi (16 ± 0.4), raw pepeao (16 ± 0.5), boiled yambean (16 ± 0.5), light pearled barley (16 ± 0.3), all purpose enriched wheat flour (16 ± 0.4), enriched wheat bread (16 ± 0.4), loquats (16 ± 0.5), boiled scallop squash (16 ± 0.6), canned sprouted mung beans (15 ± 0.8), dried sweetened shredded coconut (15 ± 0.5), roasted dark meat chicken without skin (15 ± 0.5), roasted dark meat chicken with skin (15 ± 0.5), cooked mackerel (15 ± 0.6), roasted chicken wing with skin (15 ± 1.5), roasted light meat chicken without skin (15 ± 0.5), roasted light meat chicken with skin (15 ± 0.5), roasted chicken breast without skin (15 ± 0.6), broiled pork center rib excluding fat (15 ± 0.5), broiled pork top loin excluding fat (15 ± 0.5), raw croaker (15 ± 0.6), raw cod (15 ± 0.6), simmered turkey gizzard (15 ± 0.5), frozen and boiled succotash (15 ± 0.6), frozen and fried hash brown potatoes (15 ± 0.6), tomato purée (15 ± 0.2), raw zucchini (15 ± 0.8), boiled pumpkin (15 ± 0.4), cherries (15 ± 0.7), cane syrup (15 ± 2.5), braised lamb heart (14 ± 0.3), simmered pork tail (14 ± 0.5), stir-fried sprouted lentils (14 ± 0.5), pork headcheese (14 ± 1.8), turkey summer sausage (14 ± 1.8), tomato sauce (14 ± 0.4), safflower oil and soybean oil mayonnaise (14 ± 3.6), soybean oil mayonnaise (14 ± 3.6), cherry pie (14 ± 0.4), zwieback crackers (14 ± 7.1), cooked cod (14 ± 0.6), roasted goose without skin (14 ± 0.5), stewed dark meat chicken without skin (14 ± 0.5), stewed dark meat chicken with skin (14 ± 0.5), roasted chicken breast with skin (14 ± 0.5), braised pork loin blade including fat (14 ± 0.5), roasted pork loin blade excluding fat (14 ± 0.5), raw lingcod (14 ± 0.6), simmered chicken liver (14 ± 0.5), raw spot (14 ± 0.6), compressed bakers yeast (14 ± 1.8), soursop (14 ± 0.2), honeydew melon (14 ± 0.5), Perrier water (14 ± 0.3), american grapes (14 ± 0.5), apricots (14 ± 0.5), strawberries (14 ± 0.3), tangerine (14 ± 0.6), braised pork spleen (13 ± 0.5), raw jerusalem artichoke (13 ± 0.7), stir-fried sprouted mung beans (13 ± 0.8), sweet pickles (13 ± 3.3), lebanon bologna (13 ± 2.2), berliner (13 ± 2.2), summer sausage (13 ± 2.2), pork bologna (13 ± 2.2), pork and beef luncheon sausage (13 ± 2.2), beef and pork salami (13 ± 2.2), low calorie thousand island salad dressing (13 ± 3.3), beef and pork bologna (13 ± 2.2), beef bologna (13 ± 2.2), vienna sausage (13 ± 3.1), low calorie french salad dressing (13 ± 3.1), thousand island salad dressing (13 ± 3.1), turkey salad (13 ± 0.9), pineapple pie (13 ± 0.4), clam liquid (13 ± 0.2), lemon meringue pie (13 ± 0.4), chocolate syrup (13 ± 1.3), well done fried ground beef (13 ± 0.5), fried chicken thigh with skin (13 ± 0.8), cooked sea bass (13 ± 0.6), roasted chicken thigh with skin (13 ± 0.8), braised pork kidneys (13 ± 0.5), broiled pork loin blade excluding fat (13 ± 0.5), fried pork loin blade excluding fat (13 ± 0.5), braised beef chuck blade roast excluding fat (13 ± 0.5), roasted leg of lamb excluding fat (13 ± 0.6), roasted goose with skin (13 ± 0.5), broiled beef rib eye excluding fat (13 ± 0.5), stewed chicken wing with skin (13 ± 1.3), stewed light meat chicken without skin (13 ± 0.5), broiled beef rib eye including fat (13 ± 0.5), braised beef chuck blade roast including fat (13 ± 0.5), stewed light meat chicken with skin (13 ± 0.5), stewed chicken breast with skin (13 ± 0.5), stewed chicken breast without skin (13 ± 0.5), broiled pork center rib including fat (13 ± 0.5), simmered turkey giblets (13 ± 0.5), simmered turkey heart (13 ± 0.5), raw quail without skin (13 ± 0.5), raw pheasant without skin (13 ± 0.5), raw coconut (13 ± 1.1), dried shitake mushrooms (13 ± 3.3), raw green tomato (13 ± 0.4), baked/ boiled tropical yam (13 ± 0.7), raw sprouted mung beans (13 ± 1), raw bamboo shoots (13 ± 0.7), raw cucumber (13 ± 1), boiled split peas (13 ± 0.3), boiled zucchini (13 ± 0.6), boiled chayote (13 ± 0.6), cooked bacon (12 ± 0.4), breaded and fried fish (12 ± 0.7), fried chicken (12 ± 0.6), angel food cake (12 ± 0.8), chicken egg white (12 ± 1.5), well done broiled ground beef (12 ± 0.5), fried chicken drumstick with skin (12 ± 1), roasted chicken drumstick with skin (12 ± 1), raw mackerel (12 ± 0.6), well done broiled lean ground beef (12 ± 0.5), roasted chicken thigh without skin (12 ± 1), raw whale meat (12 ± 0.5), cooked rockfish (12 ± 0.6), well done baked ground beef (12 ± 0.5), stewed chicken drumstick with skin (12 ± 0.9), stewed chicken thigh with skin (12 ± 0.7), well done baked lean ground beef (12 ± 0.5),

II. Minerals

CALCIUM

broiled lamb loin chop excluding fat (12 ± 1), broiled lamb rib chop excluding fat (12 ± 1.2), roasted veal rib roast (12 ± 0.6), roasted lamb shoulder excluding fat (12 ± 0.6), roasted duck without skin (12 ± 0.5), roasted pork loin blade including fat (12 ± 0.5), broiled pork top loin including fat (12 ± 0.5), simmered chicken giblets (12 ± 0.5), braised beef spleen (12 ± 0.5), braised pork center rib exlucding fat (12 ± 0.5), braised pork top loin excluding fat (12 ± 0.5), braised beef shortribs including fat (12 ± 0.5), braised/roasted/stewed veal chuck (12 ± 0.6), braised/stewed veal breast plate (12 ± 0.6), rice flour (12 ± 0.4), dried sweetened flaked coconut (12 ± 2), raw acerola (12 ± 0.5), canned crookneck squash (12 ± 0.5), prune juice (12 ± 0.2), boiled bamboo shoots (12 ± 0.4), white grapefruit (12 ± 0.4), cooked brown rice (12 ± 0.3), cooked wild long grain rice (12 ± 0.4), light buckwheat flour (11 ± 0.5), raw venison (11 ± 0.6), canned chicken liver pâté (11 ± 1.8), raw duck liver (11 ± 0.5), grilled canadian bacon (11 ± 1.1), beef lunch meat (11 ± 1.8), pork and beef lunch meat (11 ± 1.8), roasted cured pork arm excluding fat (11 ± 0.5), pork liver sausage (11 ± 2.8), beef and pork frankfurter (11 ± 1.1), pork and beef spread (11 ± 1.8), jellied corned beef loaf (11 ± 1.8), polish pork sausage (11 ± 1.8), smoked chinook salmon (11 ± 0.6), tuna salad spread (11 ± 0.9), chicken and turkey salad (11 ± 1.8), blueberry pie (11 ± 0.4), pickled beets (11 ± 0.4), canned succotash with creamed corn (11 ± 0.4), canned succotash (11 ± 0.4), fried beef liver (11 ± 0.5), braised beef lungs (11 ± 0.5), stewed chicken drumstick without skin (11 ± 1.1), well done fried lean ground beef (11 ± 0.5), restaurant hash brown potatoes (11 ± 0.6), medium-rare fried ground beef (11 ± 0.5), medium-rare broiled ground beef (11 ± 0.5), medium-rare broiled lean ground beef (11 ± 0.5), fried beef sirloin excluding fat (11 ± 0.5), raw sea bass (11 ± 0.6), fried beef sirloin including fat (11 ± 0.5), broiled pork loin blade including fat (11 ± 0.5), broiled beef sirloin excluding fat (11 ± 0.5), braised/broiled veal round with rump (11 ± 0.6), braised/broiled veal loin (11 ± 0.6), simmered turkey liver (11 ± 0.5), roasted beef rib including fat (11 ± 0.5), broiled beef sirloin including fat (11 ± 0.5), broiled beef rib including fat (11 ± 0.5), roasted leg of lamb including fat (11 ± 0.6), roasted duck with skin (11 ± 0.5), braised beef shortribs excluding fat (11 ± 0.5), roasted lamb shoulder including fat (11 ± 0.6), raw chum salmon (11 ± 0.6), boiled beets (11 ± 0.6), roasted pork center rib excluding fat (11 ± 0.5), roasted pork top loin excluding fat (11 ± 0.5), raw cusk (11 ± 0.6), passion fruit (11 ± 2.8), mammy apple (11 ± 0.5), raw chinese water chestnuts (11 ± 0.8), raw california avocado (11 ± 0.3), jelly beans (11 ± 1.8), cooked sprouted mung beans (11 ± 0.8), cantaloupe (11 ± 0.3), microwaved potato with skin (11 ± 0.2), mandarin oranges (11 ± 0.4), raw florida avocado (11 ± 0.2), quince (11 ± 0.5), boiled and steamed chinese chestnuts (11 ± 1.8), boiled and steamed japanese chestnuts (11 ± 1.8), coconut cream (11 ± 0.2), european grapes (11 ± 0.3), thompson seedless grapes (11 ± 0.3), orange juice (11 ± 0.2), pear (11 ± 0.3), pink grapefruit (11 ± 0.4), hard pork salami (10 ± 5), hard pork and beef salami (10 ± 5), minced ham lunch meat (10 ± 2.4), roasted cured pork arm including fat (10 ± 0.5), pork and beef knockwurst (10 ± 0.7), pork and beef link sausage (10 ± 0.7), medium-rare fried lean ground beef (10 ± 0.5), braised pork loin excluding fat (10 ± 0.5), roasted beef rib roasted excluding fat (10 ± 0.5), broiled beef rib excluding fat (10 ± 0.5), simmered chicken gizzard (10 ± 0.5), roasted pork sirloin excluding fat (10 ± 0.5), fried pork loin blade including fat (10 ± 0.5), braised beef chuck arm pot roast including fat (10 ± 0.5), medium-rare baked ground beef

(10 ± 0.5), raw pork stomach (10 ± 0.5), braised pork liver (10 ± 0.5), braised pork center rib including fat (10 ± 0.5), braised pork top loin including fat (10 ± 0.5), roasted pork center rib including fat (10 ± 0.5), roasted pork top loin including fat (10 ± 0.5), baked potato with skin (10 ± 0.2), papaya nectar (10 ± 0.2), carissa (10 ± 2.5), boiled hubbard squash (10 ± 0.4), acerola juice (10 ± 0.2), mango (10 ± 0.2), noodles (10 ± 0.3), cooked white rice (10 ± 0.2), ground-cherries (9 ± 0.4), beef and pork summer sausage (9 ± 2.2), unheated canadian bacon (9 ± 0.9), beef summer sausage (9 ± 2.2), beef honey roll sausage (9 ± 2.2), pork beerwurst salami (9 ± 2.2), beef salami (9 ± 2.2), smoked beef sausage (9 ± 1.2), turkey pastrami (9 ± 0.9), beef beerwurst salami (9 ± 2.2), turkey ham (9 ± 0.9), restaurant tomato juice (9 ± 0.4), fried beef brains (9 ± 0.5), apple pie (9 ± 0.4), simmered beef brains (9 ± 0.5), braised pork brains (9 ± 0.5), well done broiled extra lean ground beef (9 ± 0.5), roasted pork arm excluding fat (9 ± 0.5), roasted pork loin excluding fat (9 ± 0.5), roasted pork tenderloin (9 ± 0.5), braised beef chuck arm pot roast excluding fat (9 ± 0.5), canned shitake mushrooms (9 ± 0.4), well done baked extra lean ground beef (9 ± 0.5), broiled beef top loin including fat (9 ± 0.5), braised beef brisket including fat (9 ± 0.5), raw rockfish (9 ± 0.6), broiled T-bone steak including fat (9 ± 0.5), roasted pork sirloin including fat (9 ± 0.5), medium-rare baked lean ground beef (9 ± 0.5), fried pork center rib excluding fat (9 ± 0.5), fried pork top loin excluding fat (9 ± 0.5), broiled lamb rib chop including fat (9 ± 0.7), boiled dishcloth gourd (9 ± 0.6), tomato juice (9 ± 0.3), rose table wine (9 ± 0.5), white table wine (9 ± 0.5), grape juice (9 ± 0.2), pitanga (9 ± 0.3), frozen white corn (9 ± 0.7), oatmeal (9 ± 0.3), restaurant orange juice (9 ± 0.4), white grapefruit juice (9 ± 0.2), red grapefruit juice (9 ± 0.2), lime juice (9 ± 0.2), roasted ham (8 ± 0.5), pork livercheese (8 ± 1.3), roasted lean ham (8 ± 0.5), cooked corned beef brisket (8 ± 0.5), unheated ham patties (8 ± 0.8), grilled ham patties (8 ± 0.8), roasted canned ham (8 ± 0.5), surimi scallops (8 ± 0.6), canned bonito (8 ± 0.5), braised pork arm excluding fat (8 ± 0.5), well done fried extra lean ground beef (8 ± 0.5), braised pork lungs (8 ± 0.5), roasted pork shoulder excluding fat (8 ± 0.5), braised pork boston blade excluding fat (8 ± 0.5), roasted pork arm including fat (8 ± 0.5), broiled beef top loin excluding fat (8 ± 0.5), braised pork loin including fat (8 ± 0.5), roasted pork loin including fat (8 ± 0.5), broiled porterhouse steak including fat (8 ± 0.5), broiled beef tenderloin including fat (8 ± 0.5), roasted beef tenderloin including fat (8 ± 0.5), broiled lamb loin chop including fat (8 ± 0.7), frozen and french fried potatoes without skin (8 ± 1), hash brown potatoes (8 ± 0.6), raw monkfish (8 ± 0.6), boiled arrowhead (8 ± 4.2), boiled red tomato (8 ± 0.4), dry dessert wine (8 ± 0.8), sweet dessert wine (8 ± 0.8), canned bamboo shoots (8 ± 0.4), piña colada (8 ± 0.4), red table wine (8 ± 0.5), frozen and boiled red sweet peppers (8 ± 0.5), frozen and boiled green sweet peppers (8 ± 0.5), japanese persimmon (8 ± 0.3), watermelon (8 ± 0.3), screwdriver (8 ± 0.2), macaroni (8 ± 0.4), spaghetti (8 ± 0.4), canned red chili peppers (7 ± 0.7), canned green chili peppers (7 ± 0.7), chipped beef (7 ± 1.8), unheated lean ham (7 ± 0.5), unheated ham (7 ± 0.5), beef pastrami (7 ± 1.8), hamburger pickle relish (7 ± 1.8), roasted pork blade roll (7 ± 0.5), ham salad (7 ± 1.8), italian salad dressing (7 ± 3.3), light mayonnaise (7 ± 3.6), raw bacon (7 ± 0.7), french salad dressing (7 ± 3.6), mayonnaise (7 ± 3.6), frozen creamed yellow corn (7 ± 0.4), bloody mary (7 ± 0.3), braised pork arm braised including fat (7 ± 0.5), broiled pork loin excluding fat (7 ± 0.5), roasted pork boston blade excluding fat (7 ± 0.5), braised beef flank including fat (7 ± 0.5), dried

pinyon pine nuts (7 ± 1.8), medium-rare broiled extra lean ground beef (7 ± 0.5), medium-rare fried extra lean ground beef (7 ± 0.5), braised beef liver (7 ± 0.5), roasted pork shoulder including fat (7 ± 0.5), braised pork boston blade braised including fat (7 ± 0.5), cooked sockeye salmon (7 ± 0.6), broiled porterhouse steak excluding fat (7 ± 0.5), broiled T-bone steak excluding fat (7 ± 0.5), broiled pork loin including fat (7 ± 0.5), roasted pork rump excluding fat (7 ± 0.5), roasted pork leg excluding fat (7 ± 0.5), roasted pork shank excluding fat (7 ± 0.5), broiled beef tenderloin excluding fat (7 ± 0.5), roasted pork rump including fat (7 ± 0.5), raw bluefish (7 ± 0.6), broiled beef round including fat (7 ± 0.5), simmered beef tongue (7 ± 0.5), roasted beef tenderloin excluding fat (7 ± 0.5), braised pork sirloin excluding fat (7 ± 0.5), medium-rare baked extra lean ground beef (7 ± 0.5), roasted veal leg (7 ± 0.6), fried pork center rib including fat (7 ± 0.5), fried pork top loin including fat (7 ± 0.5), gum drops (7 ± 1.8), braised pork heart (7 ± 0.4), frozen and boiled potatoes without skin (7 ± 0.5), steamed hawaii mountain yam (7 ± 0.7), raw red tomato (7 ± 0.4), raw potato without skin (7 ± 0.4), boiled potato without skin (7 ± 0.4), maltex (7 ± 0.3), apricot nectar (7 ± 0.2), lemon juice (7 ± 0.2), raw apple with skin (7 ± 0.4), cranberry sauce (7 ± 0.5), oheloberries (7 ± 0.4), pineapple (7 ± 0.3), puffed rice (7 ± 3.6), cider vinegar (7 ± 3.3), canned potato without skin (6 ± 0.6), canned lean ham (6 ± 0.5), canned ham (6 ± 0.5), roasted canned lean ham (6 ± 0.5), cooked swordfish (6 ± 0.6), broiled pork boston blade excluding fat (6 ± 0.5), broiled beef flank excluding fat (6 ± 0.5), broiled beef flank including fat (6 ± 0.5), braised beef brisket excluding fat (6 ± 0.5), braised beef flank excluding fat (6 ± 0.5), roasted pork center loin excluding fat (6 ± 0.5), fried beef top round including fat (6 ± 0.5), roasted pork boston blade including fat (6 ± 0.5), sim- mered beef heart (6 ± 0.5), roasted beef round tip including fat (6 ± 0.5), broiled beef top round excluding fat (6 ± 0.5), broiled beef top round including fat (6 ± 0.5), roasted eye of beef round including fat (6 ± 0.5), roasted pork leg including fat (6 ± 0.5), roasted pork shank including fat (6 ± 0.5), braised pork center loin excluding fat (6 ± 0.5), braised pork sirloin including fat (6 ± 0.5), braised beef bottom round including fat (6 ± 0.5), braised pork center loin including fat (6 ± 0.5), raw sockeye salmon (6 ± 0.6), whiskey sour (6 ± 0.6), blueberries (6 ± 0.3), tequila sunrise (6 ± 0.3), boiled eggplant (6 ± 1), raw red sweet peppers (6 ± 1), raw green sweet peppers (6 ± 1), apple juice (6 ± 0.2), raw mushrooms (6 ± 1.4), bananas (6 ± 0.4), white corn flour (6 ± 0.4), yellow corn flour (6 ± 0.4), enriched white degermed corn meal (6 ± 0.4), unenriched corn meal (6 ± 0.4), enriched yellow degermed corn meal (6 ± 0.4), unenriched yellow degermed corn meal (6 ± 0.4), peach (6 ± 0.6), soy sauce (5 ± 0.9), turkey breast meat summer sausage (5 ± 2.4), chopped ham lunch meat (5 ± 2.4), canned pork lunch meat (5 ± 2.4), canned white corn (5 ± 0.5), canned yellow corn (5 ± 0.5), instant corn grits (5 ± 0.4), raw swordfish (5 ± 0.6), fried pork center loin excluding fat (5 ± 0.5), broiled pork center loin excluding fat (5 ± 0.5), broiled pork boston blade including fat (5 ± 0.5), fried pork center loin including fat (5 ± 0.5), fried beef top round excluding fat (5 ± 0.5), roasted beef tip excluding fat (5 ± 0.5), broiled beef round excluding fat (5 ± 0.5), roasted pork center loin including fat (5 ± 0.5), roasted eye of beef round excluding fat (5 ± 0.5), broiled pork sirloin excluding fat (5 ± 0.5), broiled pork sirloin including fat (5 ± 0.5), canned tuna in oil (5 ± 0.6), braised beef bottom round excluding fat (5 ± 0.5), tom collins (5 ± 0.2), casaba melon (5 ± 0.3), peach nectar (5 ± 0.2), microwaved potato without skin (5 ± 0.3), beer (5 ± 0.1), honey

(5 ± 2.4), baked potato without skin (5 ± 0.3), ale (5 ± 0.1), malt liquor (5 ± 0.1), boiled mushrooms (5 ± 0.6), carambola (5 ± 0.4), cooked apple without skin (5 ± 0.3), lychees (5 ± 0.5), purple passion fruit juice (4 ± 0.2), canned red sweet peppers (4 ± 0.7), canned green sweet peppers (4 ± 0.7), corn flakes (4 ± 1.8), canned yellow corn with red and green peppers (4 ± 0.4), yellow hominy (4 ± 0.2), white hominy (4 ± 0.2), yellow hominy with red and green peppers (4 ± 0.2), broiled pork center loin including fat (4 ± 0.5), raw pork jowl (4 ± 0.5), orange soda (4 ± 0.1), soy milk (4 ± 0.2), canned chinese water chestnuts (4 ± 0.7), ginko nuts (4 ± 1.8), yellow passion fruit juice (4 ± 0.2), pear nectar (4 ± 0.2), applesauce (4 ± 0.4), restaurant cranberry juice cocktail (4 ± 0.3), pummelo (4 ± 0.3), boiled red sweet peppers (4 ± 0.7), boiled green sweet peppers (4 ± 0.7), raw apple without skin (4 ± 0.4), nectarine (4 ± 0.4), canned creamed yellow corn (3 ± 0.4), apple tapioca pudding (3 ± 0.2), bourbon & soda (3 ± 0.4), boiled mothbeans (3 ± 0.3), daiquiri (3 ± 0.8), cola (3 ± 0.1), cranberry juice cocktail (3 ± 0.2), cooked shitake mushrooms (3 ± 0.7), pomegranate (3 ± 0.3), cream of rice (3 ± 0.3), plum (3 ± 0.8), cooked instant white rice (3 ± 0.3), boiled yellow corn (2 ± 0.6), lemon-lime soda (2 ± 0.1), 53 proof coffee liqueur (2 ± 1), 63 proof coffee liqueur (2 ± 1), cooked plantain (2 ± 0.3), frozen and boiled yellow corn (2 ± 0.6), gin & tonic (2 ± 0.2), manhattan (2 ± 0.9), instant tea (2 ± 0.2), restaurant coffee (2 ± 0.3), chamomile tea (2 ± 0.3), farina (2 ± 0.3), coffee (2 ± 0.3), cooked enriched white degermed corn meal (1 ± 0.2), cooked unenriched corn meal (1 ± 0.2), cooked enriched yellow degermed corn meal (1 ± 0.2), cooked unenriched yellow degermed corn meal (1 ± 0.2), sweetened coconut cream (1 ± 0.2), martini (1 ± 0.7), longans (1 ± 0.5)

Applications: anemia, diabetes, hemophilia, pernicious anemia, fracture, softening of bones, thinning of bones, rickets, colitis, diarrhea, dizziness, epilepsy, lack of sleep, mental illness, Parkinson's disease, Ménière's syndrome, cataracts, fever, hardening of the arteries, atherosclerosis, high blood pressure, celiac disease, constipation, hemorrhoids, worms, arthritis, kidney inflammation, leg cramp, allergies, common cold, tuberculosis, general muscle cramps, muscle spasm, nail deterioration, acne, peptic ulcer, pyorrhea, tooth and gum deterioration, aging, fever, weight increased, sunburn

Daily Dosage: (at least 50 mg magnesium per every 100 mg and at least 1300 mg but not more than 2000 mg phosphorus per every 1000 mg) 1974 PCS — 125–1500 mg; 1980 RDA — 800–1200 mg; 1980 USRDA — 1000 mg; 1989 RDA — 800–1200 mg; 1989 USRDA — 1000 mg; nutritional — 1000–1500 mg; therapeutic — 2000–2500 mg; experimental — 3000–4000 mg; toxic — 5000 mg +

Warnings: Keep levels low with kidney stones. Increased amounts are dangerous. See toxicity symptoms.

CESIUM

Names: cesium, caesium, Cs, Cs^+, element 55
Classification: possible trace mineral, contaminant
Deficiency: (being researched)
Side-effects: see toxicity symptoms
Toxicity: cesium poisoning
Inhibitors: unknown

II. Minerals

Helpers: rubidium
Sources: industry
Applications: none known
Daily Dosage: 1974 PCS — none; 1989 SADDI — none; 1989 RDA — none; 1989 USRDA — none; nutritional — unknown; therapeutic — none; experimental — none; toxic — small amounts
Warnings: Small amounts are dangerous.

CHLORINE

Names: chlorine, chlorin, Cl, Cl⁻, element 17
Classification: mineral, contaminant
Deficiency: salt cravings, digestion deterioration, hair decreased, decreased teeth, muscle cramps, granulation of eyelids, hay fever, watery eyes, runny nose, blistering skin eczema, warts
Side-effects: *see toxicity* symptoms
Toxicity: decreased cancer resistance, increased blood pressure, anemia, weak pulse, difficult breathing
Inhibitors: unknown
Helpers: unknown
Sources: salt, kelp, dulse, sea greens, leafy greens, rye flour, ripe olives, seafood, meats
Applications: diarrhea, nausea
Daily Dosage: 1980 RDA — none; 1980 SADDI — 1700–5100 mg; 1989 RDA — none; 1989 SADDI — 1700–5100 mg; nutritional — 3000–9000 mg (as chloride); therapeutic — 10,000 mg; experimental — ?; toxic — ?
Warnings: increased amounts are dangerous. See toxicity symptoms.

TRIVALENT CHROMIUM

Names: chromium, trivalent chromium, Cr, element 24, GTF, glucose tolerance factor
Classification: trace mineral
Deficiency: decreased glucose tolerance, corneal opacities, raised cholesterol levels in blood, increased incidence of plaques on aorta, hardening of arteries, increased blood pressure, increased cholesterol formation in liver, deterioration of growth
Side-effects: none known
Toxicity: none known up to 5000 µg
Inhibitors: unknown
Helpers: vitamin B₃, vitamin C
Sources (µg per 100 grams of food): brewer's yeast (112 ± 0.5), beef round (57 ± 0.5), calf liver (55 ± 0.5), whole wheat bread (42 ± 0.5), wheat bran (38 ± 0.5), rye bread (38 ± 0.5), fresh chili (30 ± 0.5), oysters (26 ± 0.5), potatoes (24 ± 0.5), wheat germ (23 ± 0.5), green pepper (19 ± 0.5), hen's eggs (16 ± 0.5), chicken (15 ± 0.5), apple (14 ± 0.5), butter (13 ± 0.5), parsnips (13 ± 0.5), cornmeal (12 ± 0.5), lamb chop (12 ± 0.5), scallops (11 ± 0.5), Swiss cheese (11 ± 0.5), banana (10 ± 0.5), spinach (10 ± 0.5), pork chop (10 ± 0.5), carrots (9 ± 0.5), dry navy beans (8 ± 0.5), shrimp (7 ± 0.5), lettuce (7 ± 0.5), orange (5 ± 0.5), lobster tail (5 ± 0.5), blueberries (5 ± 0.5), green beans (4 ± 0.5), cabbage

(4 ± 0.5), mushrooms (4 ± 0.5), beer (3 ± 0.5), strawberries (3 ± 0.5), milk (1 ± 0.5)
Applications: diabetes, kwashiorkor
Daily Dosage: 1980 RDA — none; 1980 SADDI — 50–200 μg; 1989 RDA — none; 1989 SADDI — 50–200 μg; nutritional — 200 μg; therapeutic — 300 μg; experimental — ?; toxic: 5000 μg +
Warnings: none

HEXAVALENT CHROMIUM

Names: chromium, hexavalent chromium, Cr, element 24
Classification: contaminant
Deficiency: (no nutritional requirement)
Side-effects: see toxicity symptoms
Toxicity: gastrointestinal hemorrhage, lung cancer, esophagus cancer, skin ulcers
Inhibitors: vitamin C
Helpers: unknown
Sources: water pollution, air pollution, cigarette smoke
Applications: none known
Daily Dosage: 1974 PCS — none; 1989 SADDI — none; 1989 RDA — none; 1989 USRDA — none; nutritional — none; therapeutic — none; experimental — none; toxic — small amounts
Warnings: Small amounts are dangerous. See toxicity symptoms.

COBALT

Names: cobalt, Co, Co^{+2}, Co^{++}, Co^{+3}, Co^{+++}, element 27
Classification: trace mineral
Deficiency: anemia, vitamin B_{12} decreased, retarded growth
Side-effects: see toxicity symptoms
Toxicity: decreased cardiac output, cardiac enlargement, heart disease, enlarged thyroid
Inhibitors: acids, alkalies, water, sunlight, alcohol, estrogen, sleeping pills, neomycin, methotrexate, cholestyramine, colchicine, sodium aminosalicylate, slow-release potassium chloride, metaformin, phenoformin, sodium, nitroprusside, chloramphenicol, codeine, oral anti-diabetic agents, aspirin, aspirin substitutes, hydralazine
Helpers: copper, iron, zinc
Sources: green leafy vegetables, fruits, organ meats, oysters, clams, poultry, milk
Applications: pernicious anemia
Daily Dosage: 1974 PCS — none; 1989 SADDI — none; 1989 RDA — none; 1989 USRDA — none; nutritional — unknown; therapeutic — none; experimental — none; toxic — unknown
Warnings: increased amounts are dangerous. See toxicity symptoms.

COPPER

Names: copper, Cu, Cu^+, Cu^{+2}, Cu^{++}, element 29
Classification: trace mineral
Deficiency: anemia, red blood cell rupture, faulty bone development, bone disease, weakness, difficulty in breathing, faulty nerve development, decreased sense of taste, skin eczema, skin sores, abnormal skin pigmentation, abnormal hair pigmentation
Side-effects: see toxicity symptoms
Toxicity: (Detoxified by manganese, molybdenum, selenium and zinc) abdominal pain, enlargement of liver, liver cirrhosis, enlargement of spleen, increased blood pressure, increased fat in feces, lack of sleep, jaundice, emotional agitation, decreased zinc to brain
Inhibitors: unknown
Helpers: cobalt, iron, zinc
Sources (mg per 100 grams of food): cooked oysters (8.922 ± 0.0006), raw goose liver (8.193 ± 0.0005), raw duck liver (5.962 ± 0.0005), low fat soybean flour (5.08 ± 0.0006), canned oysters (4.461 ± 0.0006), raw oysters (4.461 ± 0.0006), breaded and fried oysters (4.294 ± 0.0006), whole dried sesame seeds (4.078 ± 0.0056), defatted soybean flour (4.065 ± 0.0005), soybean flour (2.92 ± 0.0006), fried beef liver (2.822 ± 0.0005), braised beef liver (2.789 ± 0.0005), unsweetened baking chocolate (2.782 ± 0.0018), whole toasted sesame seeds (2.504 ± 0.0018), dry roasted cashews (2.25 ± 0.0018), roasted soybean flour (2.221 ± 0.0006), cashew butter (2.221 ± 0.0018), oil roasted cashews (2.2 ± 0.0018), breaded fried squid (2.114 ± 0.0006), cooked whelk (2.06 ± 0.0006), defatted soy meal (2 ± 0.0004), cooked lobster (1.94 ± 0.0006), raw squid (1.891 ± 0.0006), dry roasted sunflower seeds (1.857 ± 0.0018), oil roasted sunflower seeds (1.829 ± 0.0018), peanut flour (1.8 ± 0.0125), dried brazil nuts (1.796 ± 0.0018), dried sunflower seeds (1.779 ± 0.0018), raw lobster (1.664 ± 0.0006), sesame butter (1.613 ± 0.0033), soybean protein isolate (1.6 ± 0.0018), oil roasted filberts (1.6 ± 0.0018), dry roasted filberts (1.586 ± 0.0018), dried filberts (1.532 ± 0.0018), cooked mussels (1.494 ± 0.0006), dry roasted macadamia nuts (1.429 ± 0.0018), dried pumpkin and squash seeds (1.407 ± 0.0018), dried english/persian walnuts (1.407 ± 0.0018), roasted pumpkin and squash seeds (1.404 ± 0.0018), freeze-dried red sweet peppers (1.375 ± 0.0313), freeze-dried green sweet peppers (1.375 ± 0.0313), dried japanese chestnuts (1.332 ± 0.0018), oil roasted peanuts (1.293 ± 0.0018), oil roasted virginia peanuts (1.271 ± 0.0018), dry roasted pecans (1.254 ± 0.0018), toasted almonds (1.246 ± 0.0018), oil roasted almonds (1.243 ± 0.0018), dry roasted almonds (1.243 ± 0.0018), tomato powder (1.241 ± 0.0005), dry roasted pistachio nuts (1.229 ± 0.0018), oil roasted pecans (1.214 ± 0.0018), dried pistachio nuts (1.207 ± 0.0018), dried pecans (1.2 ± 0.0018), cooked alaska king crab (1.188 ± 0.0006), dried and frozen tofu (1.176 ± 0.0029), bran (1.157 ± 0.0018), toasted soybean nuts (1.079 ± 0.0018), dry roasted soybean nuts (1.079 ± 0.0006), semi-sweet bakers chocolate (1.05 ± 0.0018), dried pinyon pine nuts (1.05 ± 0.0018), dried pignolia pine nuts (1.039 ± 0.0018), dried black walnuts (1.036 ± 0.0018), raw whelk (1.031 ± 0.0006), dried peanuts (1.018 ± 0.0018), soybean protein concentrate (0.975 ± 0.0018), ry-krisp crackers (0.964 ± 0.0036), dried almonds (0.957 ± 0.0018), braised beef spleen (0.924 ± 0.0005), raw alaska king crab (0.922 ± 0.0006), almond butter (0.9 ± 0.0031), almond paste (0.846 ± 0.0018), oil roasted valencia

peanuts (0.839 ± 0.0018), dried acorns (0.829 ± 0.0018), roasted soybean nuts (0.828 ± 0.0006), dried toasted coconut (0.825 ± 0.0018), dried longans (0.807 ± 0.0005), dried coconut (0.807 ± 0.0018), boiled winged beans (0.773 ± 0.0003), canned blue crab (0.76 ± 0.0006), bran flakes (0.75 ± 0.0018), rice bran (0.75 ± 0.0005), dried hickory nuts (0.75 ± 0.0018), simmered beef heart (0.74 ± 0.0005), roasted japanese chestnuts (0.739 ± 0.0018), german sweet bakers chocolate (0.718 ± 0.0018), braised pork kidneys (0.683 ± 0.0005), simmered beef kidneys (0.68 ± 0.0005), raw dungeness crab (0.674 ± 0.0006), shredded wheat (0.671 ± 0.0018), dry roasted peanuts (0.671 ± 0.0018), tempeh (0.67 ± 0.0006), raw blue crab (0.669 ± 0.0006), natto (0.667 ± 0.0006), dried european chestnuts (0.661 ± 0.0018), oil roasted spanish peanuts (0.661 ± 0.0018), cooked blue crab (0.645 ± 0.0006), braised pork liver (0.634 ± 0.0005), dried lychees (0.631 ± 0.0005), toasted wheat germ (0.629 ± 0.0018), raw acorns (0.629 ± 0.0018), simmered turkey heart (0.627 ± 0.0005), acorn flour (0.621 ± 0.0018), crab cakes (0.61 ± 0.0008), dried chinese chestnuts (0.6 ± 0.0018), raw quail without skin (0.594 ± 0.0005), tomato paste (0.592 ± 0.0004), creamy peanut butter (0.588 ± 0.0031), raw cuttlefish (0.587 ± 0.0006), raisin bran (0.578 ± 0.0014), carob flour (0.571 ± 0.0005), raw japanese chestnuts (0.571 ± 0.0018), simmered turkey liver (0.56 ± 0.0005), cooked crayfish (0.559 ± 0.0006), dried ginko nuts (0.543 ± 0.0018), stir-fried sprouted soybeans (0.527 ± 0.0005), crunchy peanut butter (0.516 ± 0.0016), roasted european chestnuts (0.514 ± 0.0018), braised pork heart (0.508 ± 0.0004), chocolate syrup (0.505 ± 0.0013), boiled mushrooms (0.504 ± 0.0006), simmered chicken heart (0.502 ± 0.0005), corn germ (0.5 ± 0.0005), boiled peanuts (0.5 ± 0.0016), boiled and steamed european chestnuts (0.479 ± 0.0018), wheat flakes (0.468 ± 0.0018), cheese crackers (0.467 ± 0.0033), dried butternuts (0.457 ± 0.0018), raw european chestnuts (0.454 ± 0.0018), dried japanese persimmon (0.441 ± 0.0015), miso (0.437 ± 0.0004), raw crayfish (0.436 ± 0.0006), raw coconut (0.436 ± 0.0011), raw octopus (0.435 ± 0.0006), dried prunes (0.43 ± 0.0006), freeze-dried parsley (0.429 ± 0.0357), dried apricots (0.429 ± 0.0014), fried chicken giblets (0.422 ± 0.0005), puffed wheat (0.414 ± 0.0036), dehydrated onion flakes (0.414 ± 0.0036), refried beans (0.411 ± 0.0002), boiled soybeans (0.407 ± 0.0003), fried tofu (0.4 ± 0.0038), dried sweetened flaked coconut (0.396 ± 0.002), roasted chinese chestnuts (0.393 ± 0.0018), simmered turkey giblets (0.391 ± 0.0005), toasted whole wheat bread (0.39 ± 0.0024), clam liquid (0.389 ± 0.0002), pork livercheese (0.384 ± 0.0013), toasted english muffin (0.38 ± 0.001), firm raw tofu (0.378 ± 0.0004), coconut cream (0.378 ± 0.0002), freeze-dried shallots (0.375 ± 0.0125), simmered chicken liver (0.37 ± 0.0005), raw chinese chestnuts (0.368 ± 0.0018), raw pork stomach (0.365 ± 0.0005), golden seedless raisins (0.363 ± 0.0005), potato pancakes (0.359 ± 0.0007), breaded and fried clams (0.356 ± 0.0006), boiled chickpeas (0.352 ± 0.0003), raw turnip greens (0.35 ± 0.0018), raw clams (0.344 ± 0.0006), whole wheat bread (0.344 ± 0.002), boiled hyacinth beans (0.341 ± 0.0003), anchovy canned in olive oil (0.34 ± 0.0025), stir-fried sprouted lentils (0.337 ± 0.0005), microwaved potato with skin (0.334 ± 0.0002), steamed sprouted soybeans (0.33 ± 0.0011), toasted cracked wheat bread (0.319 ± 0.0024), taco shell (0.318 ± 0.0045), tostada shell (0.318 ± 0.0045), smoked whitefish (0.315 ± 0.0006), dried figs (0.313 ± 0.0003), dried sweetened shredded coconut (0.313 ± 0.0005), english muffin (0.311 ± 0.0009), seedless raisins (0.309 ± 0.0005), baked potato with skin (0.305 ± 0.0002), oil roasted macadamia nuts (0.304 ± 0.0018), seeded raisins (0.302 ± 0.0005), corn tortilla (0.3 ± 0.0017), dried macadamia nuts (0.3 ± 0.0018), raw sunfish (0.3 ± 0.0006), boiled adzuki beans (0.298 ± 0.0002), boiled navy beans (0.295 ± 0.0003), dried dates (0.288 ± 0.0006), boiled white beans (0.287 ± 0.0003), mixed grain bread (0.284 ± 0.002), raw wakame seaweed (0.284 ± 0.0005), ginko nuts (0.279 ± 0.0018), roasted goose without skin (0.276 ± 0.0005), breaded and fried shrimp (0.274 ± 0.0006), boiled catjang black-eyed peas (0.271 ± 0.0003), boiled pink beans (0.271 ± 0.0003), boiled black turtle beans (0.269 ± 0.0003), boiled pigeon peas (0.269 ± 0.0003), boiled black-eyed peas (0.268 ± 0.0003), coconut milk (0.266 ± 0.0002), raw california avocado (0.266 ± 0.0003), dried jujube (0.265 ± 0.0005), roasted goose with skin (0.264 ± 0.0005), raw shrimp (0.264 ± 0.0006), raw laver/nori seaweed (0.264 ± 0.0005), braised pork brains (0.263 ± 0.0005), pickled rip manzanillo/mission olives (0.261 ± 0.0109), falafel (0.259 ± 0.001), boiled broad beans (0.259 ± 0.0003), raw potato without skin (0.259 ± 0.0004), boiled pinto beans (0.257 ± 0.0003), simmered chicken giblets (0.255 ± 0.0005), baked beans with pork and tomato sauce (0.254 ± 0.0002), boiled and steamed chinese chestnuts (0.254 ± 0.0018), boiled turnip greens (0.253 ± 0.0007), boiled lentils (0.251 ± 0.0003), boiled beet greens (0.251 ± 0.0007), raw florida avocado (0.251 ± 0.0002), boiled great northern beans (0.247 ± 0.0003), boiled kidney beans (0.242 ± 0.0003), simmered beef brains (0.24 ± 0.0005), pork liver sausage (0.239 ± 0.0028), devil's food cake (0.237 ± 0.0008), microwaved potato without skin (0.237 ± 0.0003), sweetened coconut cream (0.236 ± 0.0002), boiled lima beans (0.235 ± 0.0003), simmered pork chitterlings (0.233 ± 0.0005), canned white beans (0.232 ± 0.0002), raw freshwater drum (0.232 ± 0.0006), roasted duck without skin (0.231 ± 0.0005), boiled cranberry beans (0.231 ± 0.0003), boiled lupins (0.231 ± 0.0003), smoked chinook salmon (0.231 ± 0.0006), fried abalone (0.228 ± 0.0006), pickled rip sevillano/ ascolano olives (0.228 ± 0.0031), hummus (0.228 ± 0.0002), roasted duck with skin (0.227 ± 0.0005), boiled yardlong bean (0.225 ± 0.0003), braised beef lungs (0.221 ± 0.0005), oatmeal bread (0.221 ± 0.0018), simmered beef tongue (0.22 ± 0.0005), fried beef brains (0.22 ± 0.0005), beef and pork salami (0.217 ± 0.0022), boiled baby lima beans (0.215 ± 0.0003), smoked cisco (0.215 ± 0.0006), baked potato without skin (0.215 ± 0.0003), bran muffins (0.213 ± 0.0013), raw anchovy (0.211 ± 0.0006), carissa (0.21 ± 0.0025), boiled black beans (0.209 ± 0.0003), canned navy beans (0.208 ± 0.0002), baked sweet potato (0.208 ± 0.0004), boiled and steamed japanese chestnuts (0.207 ± 0.0018), vegetarian baked beans (0.206 ± 0.0002), potato chips (0.204 ± 0.0018), french bread (0.2 ± 0.0018), italian bread (0.2 ± 0.0018), raw burbot (0.2 ± 0.0006), boiled and frozen baby lima beans (0.197 ± 0.0006), raw abalone (0.196 ± 0.0006), tomato sauce (0.196 ± 0.0004), gefiltefish with broth (0.195 ± 0.0012), raw white sucker (0.195 ± 0.0006), raw tofu (0.193 ± 0.0004), cooked shrimp (0.193 ± 0.0006), cooked prunes (0.193 ± 0.0005), canned black turtle beans (0.192 ± 0.0002), canned tuna in water (0.192 ± 0.0006), sardines (0.188 ± 0.0021), raw trout (0.188 ± 0.0006), jackfruit (0.187 ± 0.0005), boiled yellow beans (0.186 ± 0.0003), cream puff (0.185 ± 0.0004), frozen and boiled black-eyed peas (0.184 ± 0.0006), frozen and baked sweet potato (0.183 ± 0.0006), boiled split peas (0.181 ± 0.0003), canned spinach (0.18 ± 0.0005), canned lima beans (0.18 ± 0.0002), burrito (0.18 ± 0.0003), broiled beef tenderloin excluding fat (0.179 ± 0.0005), pita bread (0.179 ± 0.0013), boiled succotash (0.179 ± 0.0005), hash brown potatoes (0.179 ± 0.0006), cooked rainbow smelt (0.178 ± 0.0006), canned succotash with creamed corn (0.178 ± 0.0004), dried and salted cod (0.176 ± 0.0006), raw green peas (0.176 ± 0.0006), hot dog bun (0.175 ± 0.0013), boiled spinach (0.174 ± 0.0006), canned chickpeas (0.174 ± 0.0002), simmered turkey gizzard (0.173 ± 0.0005), boiled green peas (0.173 ± 0.0006), raw red chili peppers (0.173 ± 0.0011), raw green chili peppers (0.173 ± 0.0011), simmered beef shank excluding fat (0.172 ± 0.0005), frozen and boiled asparagus (0.172 ± 0.0008), puffed rice (0.171 ± 0.0036), cooked bacon (0.17 ± 0.0004), longans (0.169 ± 0.0005), roasted beef tenderloin excluding fat (0.167 ± 0.0005), broiled beef tenderloin including fat (0.167 ± 0.0005), saltine crackers (0.167 ± 0.0083), boiled potato without skin (0.167 ± 0.0004), hamburger bun (0.165 ± 0.0013), braised pork boston blade excluding fat (0.165 ± 0.0005), braised beef chuck arm pot roast excluding fat (0.164 ± 0.0005), boiled mothbeans (0.164 ± 0.0003), frozen and french fried potatoes without skin (0.164 ± 0.001), tomato purée (0.163 ± 0.0002), scalloped potatoes (0.163 ± 0.0004), cooked swordfish (0.162 ± 0.0006), chipped beef (0.161 ± 0.0018), simmered beef shank including fat (0.161 ± 0.0005), braised pork arm excluding fat (0.161 ± 0.0005), boiled sweet potato (0.161 ± 0.0003), roasted turkey dark meat without skin (0.16 ± 0.0005), chili (0.16 ± 0.0002), canned great northern beans (0.16 ± 0.0002), roasted pork tenderloin (0.16 ± 0.0005), hard pork salami (0.16 ± 0.005), raw sprouted mung beans (0.16 ± 0.001), au gratin potatoes (0.16 ± 0.0004), raw sprouted alfalfa seeds (0.158 ± 0.0015), toasted raisin bread (0.157 ± 0.0024), toasted white bread (0.157 ± 0.0024), boiled mung beans (0.156 ± 0.0002), canned sprouted mung beans (0.156 ± 0.0008), boiled scotch kale (0.155 ± 0.0008), boiled kale (0.155 ± 0.0008), cooked corned beef brisket (0.154 ± 0.0005), frozen and fried hash brown potatoes (0.153 ± 0.0006), roasted beef tenderloin including fat (0.152 ± 0.0005), summer sausage (0.152 ± 0.0022), baked/boiled tropical yam (0.151 ± 0.0007), toaster pastry (0.15 ± 0.001), roasted turkey dark meat with skin (0.15 ± 0.0005), frozen and boiled turnip greens (0.15 ± 0.0006), canned kidney beans (0.15 ± 0.0002), boiled collards (0.15 ± 0.0003), raw irish moss seaweed (0.149 ± 0.0005), fried beef sirloin excluding fat (0.149 ± 0.0005), braised beef chuck blade roast excluding fat (0.148 ± 0.0005), lychees (0.148 ± 0.0005), broiled beef sirloin excluding fat (0.146 ± 0.0005), pretzels (0.146 ± 0.0018), roasted ham (0.145 ± 0.0005), tuna salad (0.145 ± 0.0002), broiled porterhouse steak excluding fat (0.143 ± 0.0005), broiled T-bone steak excluding fat (0.143 ± 0.0005), turkey summer sausage (0.143 ± 0.0018), braised pork boston blade braised including fat (0.143 ± 0.0005), pork and beef spread (0.143 ± 0.0018), braised pork spareribs (0.142 ± 0.0005), canned cranberry beans (0.142 ± 0.0002), chicken egg yolk (0.141 ± 0.0029), cooked rainbow trout (0.141 ± 0.0006), frozen and boiled spinach (0.141 ± 0.0005), cooked mullet (0.141 ± 0.0006), pork sausage (0.141 ± 0.0019), marinara sauce (0.141 ± 0.0002), raisin bread (0.14 ± 0.002), canned jalapeño peppers (0.14 ± 0.0007), canned pinto beans (0.14 ± 0.0002), blackberries (0.14 ± 0.0007), boiled mungo beans (0.139 ± 0.0003), frozen and boiled green peas (0.139 ± 0.0006), raw rainbow smelt (0.139 ± 0.0006), canned sweet potato (0.139 ± 0.0003), light cream (0.138 ± 0.0017), white bread (0.138 ± 0.0021), braised pork arm braised including fat (0.138 ± 0.0005), french fried potatoes (0.138 ± 0.001), boiled parsnips (0.138 ± 0.0006), broiled beef sirloin including fat (0.137 ± 0.0005), mashed potatoes (0.137 ± 0.0005), fried beef sirloin including fat (0.135 ± 0.0005), kippered herring (0.135 ± 0.0013), maltex (0.135 ± 0.0003), boiled carrots (0.135 ± 0.0006), braised beef bottom round excluding fat (0.134 ± 0.0005), tamari sauce (0.134 ± 0.0009), braised pork spleen (0.133 ± 0.0005), broiled porterhouse steak including fat (0.133 ± 0.0005), braised pork sirloin excluding fat (0.133 ± 0.0005), braised beef chuck arm pot roast including fat (0.132 ± 0.0005), fried beef top round excluding fat (0.131 ± 0.0005), fried chicken neck with skin (0.131 ± 0.0014), broiled pork boston blade excluding fat (0.131 ± 0.0005), canned tuna in oil (0.131 ± 0.0006), braised beef bottom round including fat (0.13 ± 0.0005), raw kelp/kombu/tangle seaweed (0.13 ± 0.0005), roast beef sandwich (0.13 ± 0.0003), broiled T-bone steak including fat (0.13 ± 0.0005), hot dog (0.13 ± 0.0006), enchilada (0.13 ± 0.0002), taco (0.13 ± 0.0006), roasted canned ham (0.13 ± 0.0005), canned red sweet peppers (0.13 ± 0.0007), canned green sweet peppers (0.13 ± 0.0007), quince (0.13 ± 0.0005), raw spinach (0.129 ± 0.0018), gingersnaps cookie (0.129 ± 0.0071), roasted pork boston blade excluding fat (0.129 ± 0.0005), roasted pork shoulder excluding fat (0.129 ± 0.0005), o'brien potatoes (0.129 ± 0.0003), simmered chicken neck without skin (0.128 ± 0.0028), braised pork loin blade excluding fat (0.128 ± 0.0005), roasted pork arm excluding fat (0.128 ± 0.0005), roasted cured pork arm excluding fat (0.128 ± 0.0005), raw swordfish (0.126 ± 0.0006), roasted beef tip excluding fat (0.125 ± 0.0005), fried beef top round including fat (0.125 ± 0.0005), braised pork loin excluding fat (0.125 ± 0.0005), pork headcheese (0.125 ± 0.0018), pork pickle and pimento loaf (0.125 ± 0.0018), braised beef flank excluding fat (0.124 ± 0.0005), braised beef flank including fat (0.123 ± 0.0005), broiled beef top round excluding fat (0.123 ± 0.0005), cooked sprouted mung beans (0.123 ± 0.0008), beef salami (0.122 ± 0.0022), cooked sheepshead (0.122 ± 0.0006), broiled beef top round including fat (0.121 ± 0.0005), roasted beef round tip including fat (0.121 ± 0.0005), graham crackers (0.121 ± 0.0036), braised beef brisket excluding fat (0.12 ± 0.0005), pineapple upside-down cake (0.12 ± 0.0007), cheese pizza (0.12 ± 0.0004), soy milk (0.12 ± 0.0002), carambola (0.12 ± 0.0004), braised beef chuck blade roast including fat (0.119 ± 0.0005), chocolate cream pie (0.119 ± 0.0004), toasted rye bread (0.118 ± 0.0023), roasted pork boston blade including fat (0.118 ± 0.0005), cooked herring (0.118 ± 0.0006), potato salad (0.118 ± 0.0004), broiled pork boston blade including fat (0.117 ± 0.0005), canned black-eyed peas (0.117 ± 0.0002), loganberries (0.117 ± 0.0003), shoyu sauce (0.116 ± 0.0009), braised pork sirloin including fat (0.116 ± 0.0005), pickled beets (0.116 ± 0.0004), pork and beef kielbasa/kolbassy (0.115 ± 0.0019), roasted pork shoulder including fat (0.115 ± 0.0005), boiled french beans (0.115 ± 0.0003), roasted cured pork arm including fat (0.113 ± 0.0005), pear (0.113 ± 0.0003), japanese persimmon (0.113 ± 0.0003), roasted pork arm including fat (0.112 ± 0.0005), canned stewed red tomatoes (0.112 ± 0.0004), raw eggplant (0.112 ± 0.0012), raw rainbow trout (0.111 ± 0.0006), raw mushrooms (0.111 ± 0.0014), simmered chicken gizzard (0.11 ± 0.0005), braised pork pancreas (0.11 ± 0.0005), canned peeled red tomatoes (0.11 ± 0.0004), pineapple (0.11 ± 0.0003), mango (0.11 ± 0.0002), braised pork loin blade including fat (0.109 ± 0.0005), braised pork loin including fat (0.109 ± 0.0005), roasted pork rump excluding fat (0.109 ± 0.0005), canned broad beans (0.109 ± 0.0002), restaurant hash brown potatoes (0.109 ± 0.0006), canned succotash (0.109 ± 0.0004), oatmeal cookie (0.108 ± 0.0038), roasted pork leg excluding fat (0.108 ± 0.0005), roasted pork shank excluding fat (0.108 ± 0.0005), boiled eggplant

(0.108 ± 0.001), braised beef shortribs excluding fat (0.107 ± 0.0005), broiled beef round excluding fat (0.107 ± 0.0005), broiled beef top loin excluding fat (0.107 ± 0.0005), beef lunch meat (0.107 ± 0.0018), canned pumpkin (0.107 ± 0.0004), pickled herring (0.107 ± 0.0033), roasted pork sirloin excluding fat (0.107 ± 0.0005), pork and beef peppered loaf (0.107 ± 0.0018), red currants (0.107 ± 0.0009), white currants (0.107 ± 0.0009), turkey ham (0.105 ± 0.0009), broiled pork sirloin excluding fat (0.105 ± 0.0005), kumquats (0.105 ± 0.0026), rye bread (0.104 ± 0.002), broiled beef top loin including fat (0.104 ± 0.0005), raw red sweet peppers (0.104 ± 0.001), raw green sweet peppers (0.104 ± 0.001), braised pork center loin excluding fat (0.104 ± 0.0005), canned carrots (0.104 ± 0.0007), bananas (0.104 ± 0.0004), boiled green beans (0.103 ± 0.0008), boiled yellow snap beans (0.103 ± 0.0008), roasted pork rump including fat (0.103 ± 0.0005), boiled crookneck squash (0.103 ± 0.0006), guava (0.103 ± 0.0006), candied sweet potato (0.102 ± 0.0005), canned pink salmon (0.102 ± 0.0006), raw crookneck squash (0.102 ± 0.0008), raw scallop squash (0.102 ± 0.0008), well done fried extra lean ground beef (0.101 ± 0.0005), breaded and fried catfish (0.101 ± 0.0006), roasted pork loin excluding fat (0.101 ± 0.0005), broiled pork loin blade excluding fat (0.101 ± 0.0005), tomato juice (0.101 ± 0.0003), french rolls (0.1 ± 0.001), broiled beef rib eye excluding fat (0.1 ± 0.0005), hamburger sandwich (0.1 ± 0.0005), cheeseburger (0.1 ± 0.0004), roasted eye of beef round excluding fat (0.1 ± 0.0005), teriyaki sauce (0.1 ± 0.0028), baked beans with pork and sweet sauce (0.1 ± 0.0002), grilled ham patties (0.1 ± 0.0008), hard pork and beef salami (0.1 ± 0.005), braised pork center rib exlucding fat (0.1 ± 0.0005), braised pork top loin excluding fat (0.1 ± 0.0005), pork luxury loaf (0.1 ± 0.0018), roasted pork leg including fat (0.1 ± 0.0005), canned chinese water chestnuts (0.1 ± 0.0007), french fries (0.1 ± 0.0006), raw endive (0.1 ± 0.002), boiled asparagus (0.1 ± 0.0006), restaurant tomato juice (0.1 ± 0.0004), braised beef shortribs including fat (0.099 ± 0.0005), roasted eye of beef round including fat (0.099 ± 0.0005), roasted pork loin blade excluding fat (0.099 ± 0.0005), roasted pork sirloin including fat (0.099 ± 0.0005), unheated ham (0.099 ± 0.0005), roasted beef rib roasted excluding fat (0.098 ± 0.0005), broiled beef flank excluding fat (0.098 ± 0.0005), broiled beef flank including fat (0.098 ± 0.0005), broiled beef rib eye including fat (0.098 ± 0.0005), fried pork loin blade excluding fat (0.098 ± 0.0005), roasted pork shank including fat (0.098 ± 0.0005), broiled pork loin excluding fat (0.098 ± 0.0005), canned zucchini in tomato juice (0.098 ± 0.0004), well done baked extra lean ground beef (0.097 ± 0.0005), broiled beef round including fat (0.097 ± 0.0005), simmered chicken neck with skin (0.097 ± 0.0013), soy sauce (0.097 ± 0.0009), frozen and boiled okra (0.097 ± 0.0005), boiled red tomato (0.097 ± 0.0004), raw red cabbage (0.097 ± 0.0014), canned sauerkraut (0.096 ± 0.0004), cherries (0.096 ± 0.0007), roasted beef rib including fat (0.095 ± 0.0005), pork mothers loaf (0.095 ± 0.0024), broiled pork sirloin including fat (0.095 ± 0.0005), raw mussels (0.094 ± 0.0006), broiled beef rib excluding fat (0.094 ± 0.0005), braised beef brisket including fat (0.094 ± 0.0005), cooked mackerel (0.094 ± 0.0006), raw catfish (0.094 ± 0.0006), braised pork center loin including fat (0.094 ± 0.0005), raw freshwater bass (0.093 ± 0.0006), polish pork sausage (0.093 ± 0.0018), pork bratwurst (0.093 ± 0.0006), roasted pork loin including fat (0.093 ± 0.0005), ham and cheese spread (0.093 ± 0.0018), well done broiled ground beef (0.092 ± 0.0005), well done baked lean ground beef (0.092 ±

0.0005), broiled beef rib including fat (0.092 ± 0.0005), fried chicken back with skin (0.092 ± 0.0007), raw herring (0.092 ± 0.0006), french toast (0.091 ± 0.0008), onion rings (0.091 ± 0.0006), well done fried ground beef (0.09 ± 0.0005), raw beef tripe (0.09 ± 0.0005), braised pork center rib including fat (0.09 ± 0.0005), roasted pork loin blade including fat (0.09 ± 0.0005), broiled pork loin blade including fat (0.09 ± 0.0005), raw green tomato (0.09 ± 0.0004), boiled cauliflower (0.09 ± 0.0008), european grapes (0.09 ± 0.0003), thompson seedless grapes (0.09 ± 0.0003), pineapple juice (0.09 ± 0.0002), braised beef pancreas (0.089 ± 0.0005), fried chicken thigh with skin (0.089 ± 0.0008), fried dark meat chicken without skin (0.089 ± 0.0005), beef and pork frankfurter (0.089 ± 0.0011), fried pork center loin excluding fat (0.089 ± 0.0005), broiled pork loin including fat (0.089 ± 0.0005), apricots (0.089 ± 0.0005), well done fried lean ground beef (0.088 ± 0.0005), fried dark meat chicken with skin (0.088 ± 0.0005), braised pork top loin including fat (0.088 ± 0.0005), lebanon bologna (0.087 ± 0.0022), medium-rare fried extra lean ground beef (0.087 ± 0.0005), beef summer sausage (0.087 ± 0.0022), beef honey roll sausage (0.087 ± 0.0022), beef and pork summer sausage (0.087 ± 0.0022), beef and pork bologna (0.087 ± 0.0022), pork and beef luncheon sausage (0.087 ± 0.0022), barbecue loaf (0.087 ± 0.0022), berliner (0.087 ± 0.0022), well done baked ground beef (0.086 ± 0.0005), black currants (0.086 ± 0.0009), baked acorn squash (0.086 ± 0.0005), boiled okra (0.086 ± 0.0006), boiled zucchini (0.086 ± 0.0006), noodles (0.085 ± 0.0003), bagels (0.084 ± 0.0009), canned sockeye salmon (0.084 ± 0.0006), canned lean ham (0.084 ± 0.0005), breadfruit (0.084 ± 0.0005), well done broiled extra lean ground beef (0.083 ± 0.0005), canned turnip greens (0.083 ± 0.0004), boiled brussels sprouts (0.083 ± 0.0006), fried pork loin blade including fat (0.083 ± 0.0005), frozen and boiled brussels sprouts (0.083 ± 0.0006), boiled scallop squash (0.083 ± 0.0006), dinner rolls (0.082 ± 0.0018), medium-rare broiled ground beef (0.082 ± 0.0005), hot dog pickle relish (0.082 ± 0.0018), hamburger pickle relish (0.082 ± 0.0018), canned green peas (0.082 ± 0.0006), broiled pork center loin excluding fat (0.082 ± 0.0005), piña colada (0.082 ± 0.0004), medium-rare fried ground beef (0.081 ± 0.0005), italian pork sausage (0.081 ± 0.0007), roasted chicken thigh without skin (0.081 ± 0.001), roasted pork center loin excluding fat (0.081 ± 0.0005), bacon cheeseburger (0.08 ± 0.0003), fried chicken drumstick with skin (0.08 ± 0.001), roasted dark meat chicken without skin (0.08 ± 0.0005), fish sandwich (0.08 ± 0.0003), stewed chicken drumstick without skin (0.08 ± 0.0011), boysenberries (0.08 ± 0.0004), fried pork center loin including fat (0.08 ± 0.0005), broiled pork center rib excluding fat (0.08 ± 0.0005), broiled pork top loin excluding fat (0.08 ± 0.0005), canned crookneck squash (0.08 ± 0.0005), hot fudge sundae (0.08 ± 0.0003), creme de menthe (0.08 ± 0.001), roasted lean ham (0.079 ± 0.0005), roasted pork center rib excluding fat (0.078 ± 0.0005), roasted pork top loin excluding fat (0.078 ± 0.0005), frozen and boiled potatoes without skin (0.078 ± 0.0005), cooked pompano (0.078 ± 0.0006), well done broiled lean ground beef (0.077 ± 0.0005), medium-rare fried lean ground beef (0.077 ± 0.0005), roasted dark meat chicken with skin (0.077 ± 0.0005), roasted chicken thigh with skin (0.077 ± 0.0008), roasted chicken drumstick with skin (0.077 ± 0.001), roasted pork center loin including fat (0.077 ± 0.0005), breaded and fried scallops (0.077 ± 0.0016), raw red tomato (0.077 ± 0.0004), roasted pork blade roll (0.076 ± 0.0005), broiled pork center loin including fat (0.076 ± 0.0005), raw pink salmon (0.076 ±

0.0006), minced ham lunch meat (0.076 ± 0.0024), medium-rare baked extra lean ground beef (0.075 ± 0.0005), stewed dark meat chicken without skin (0.075 ± 0.0005), turkey salad (0.075 ± 0.0009), ham and cheese roll (0.075 ± 0.0018), fried pork center rib excluding fat (0.075 ± 0.0005), fried pork top loin excluding fat (0.075 ± 0.0005), broiled pork center rib including fat (0.075 ± 0.0005), ham salad (0.075 ± 0.0018), raw mackerel (0.074 ± 0.0006), roasted pork center rib including fat (0.074 ± 0.0005), pork bologna (0.074 ± 0.0022), unheated lean ham (0.074 ± 0.0005), broiled pork top loin including fat (0.074 ± 0.0005), raspberries (0.074 ± 0.0004), cooked carp (0.073 ± 0.0006), roasted pork top loin including fat (0.073 ± 0.0005), frozen and boiled crookneck squash (0.073 ± 0.0005), frozen and boiled carrots (0.073 ± 0.0007), jujube (0.073 ± 0.0005), apricot nectar (0.073 ± 0.0002), nectarine (0.073 ± 0.0004), medium-rare baked lean ground beef (0.072 ± 0.0005), pork link sausage (0.072 ± 0.0007), raw whitefish (0.072 ± 0.0006), stewed chicken thigh with skin (0.071 ± 0.0007), pork and beef honey loaf (0.071 ± 0.0018), pork and beef old fashioned loaf (0.071 ± 0.0018), unheated ham patties (0.071 ± 0.0008), pork and beef picnic loaf (0.071 ± 0.0018), pork and beef brotwurst (0.071 ± 0.0007), boiled red sweet peppers (0.071 ± 0.0007), boiled green sweet peppers (0.071 ± 0.0007), medium-rare baked ground beef (0.07 ± 0.0005), medium-rare broiled extra lean ground beef (0.07 ± 0.0005), light and dark meat turkey roll (0.07 ± 0.0009), stewed chicken drumstick with skin (0.07 ± 0.0009), pork and beef link sausage with american cheese (0.07 ± 0.0012), figs (0.07 ± 0.001), gooseberries (0.07 ± 0.0003), raw pheasant without skin (0.069 ± 0.0005), boiled broccoli (0.069 ± 0.0006), bloody mary (0.069 ± 0.0003), boiled red cabbage (0.069 ± 0.0007), peach nectar (0.069 ± 0.0002), corn flakes (0.068 ± 0.0018), stewed dark meat chicken with skin (0.068 ± 0.0005), prune juice (0.068 ± 0.0002), canned pumpkin pie mix (0.068 ± 0.0004), frozen and boiled yellow snap beans (0.068 ± 0.0007), frozen and boiled green beans (0.068 ± 0.0007), fried pork center rib including fat (0.068 ± 0.0005), fried pork top loin including fat (0.068 ± 0.0005), peach (0.068 ± 0.0006), waffles (0.067 ± 0.0007), pork and beef mortadella (0.067 ± 0.0033), canned ham (0.067 ± 0.0005), cooked sockeye salmon (0.067 ± 0.0006), crab apples (0.067 ± 0.0005), pear nectar (0.067 ± 0.0002), maypo (0.066 ± 0.0003), medium-rare broiled lean ground beef (0.066 ± 0.0005), chicken egg (0.066 ± 0.001), lime (0.066 ± 0.0007), cooked plantain (0.066 ± 0.0003), cooked coho salmon (0.065 ± 0.0006), breaded and fried croaker (0.065 ± 0.0006), cooked pike (0.065 ± 0.0006), raw bacon (0.065 ± 0.0007), boiled butternut squash (0.065 ± 0.0005), restaurant chocolate milkshake (0.065 ± 0.0002), canned corned beef (0.064 ± 0.0005), raw shad (0.064 ± 0.0006), shortbread cookie (0.063 ± 0.0063), fried chicken wing with skin (0.063 ± 0.0016), chopped ham lunch meat (0.062 ± 0.0024), jellied corned beef loaf (0.061 ± 0.0018), boiled artichoke hearts (0.061 ± 0.0006), boiled artichoke (0.061 ± 0.0006), blueberries (0.061 ± 0.0003), raw spring onions with tops (0.06 ± 0.001), cornbread (0.06 ± 0.0006), corn pone (0.06 ± 0.0006), frozen and boiled succotash (0.06 ± 0.0006), canned yellow corn with red and green peppers (0.06 ± 0.0004), fried chicken (0.06 ± 0.0006), cheesecake (0.06 ± 0.0006), strawberry sundae (0.06 ± 0.0003), pork and beef link sausage (0.059 ± 0.0007), frozen and boiled mustard greens (0.059 ± 0.0007), pork and beef knockwurst (0.059 ± 0.0007), fried light meat chicken with skin (0.058 ± 0.0005), whole chocolate milk (0.058 ± 0.0002), cranberry sauce (0.058 ± 0.0005), raw parsley (0.057 ± 0.0017), canned potato without skin (0.057 ± 0.0006), fried chicken breast with skin (0.057 ± 0.0005), raw zucchini (0.057 ± 0.0008), roasted chicken wing with skin (0.056 ± 0.0015), raw carp (0.056 ± 0.0006), boiled beets (0.056 ± 0.0006), navel orange (0.056 ± 0.0004), frozen and boiled fordhook lima beans (0.055 ± 0.0006), frozen and boiled collards (0.055 ± 0.0006), oatmeal (0.055 ± 0.0003), raw chum salmon (0.055 ± 0.0006), turkey salami (0.054 ± 0.0009), fried light meat chicken without skin (0.054 ± 0.0005), raw butterfish (0.054 ± 0.0006), havarti cheese (0.054 ± 0.0018), cooked sturgeon (0.053 ± 0.0006), turkey pastrami (0.053 ± 0.0009), beef frankfurter (0.053 ± 0.0009), fried chicken breast without skin (0.053 ± 0.0006), roasted light meat chicken with skin (0.053 ± 0.0005), apple pie (0.053 ± 0.0004), grilled canadian bacon (0.053 ± 0.0011), raw bluefish (0.053 ± 0.0006), raw scallops (0.053 ± 0.0006), pork beerwurst salami (0.052 ± 0.0022), boiled yellow corn (0.052 ± 0.0006), boiled acorn squash (0.052 ± 0.0004), raw sockeye salmon (0.052 ± 0.0006), canned creamed yellow corn (0.052 ± 0.0004), rose table wine (0.052 ± 0.0005), cooked tilefish (0.052 ± 0.0006), raw mullet (0.051 ± 0.0006), raw coho salmon (0.051 ± 0.0006), raw pike (0.051 ± 0.0006), raw scup (0.051 ± 0.0006), raw pollock (0.051 ± 0.0006), restaurant vanilla milkshake (0.051 ± 0.0002), sponge cake (0.05 ± 0.0008), pancakes with butter and syrup (0.05 ± 0.0003), roasted light meat chicken without skin (0.05 ± 0.0005), roasted chicken breast with skin (0.05 ± 0.0005), roasted canned lean ham (0.05 ± 0.0005), pork olive loaf (0.05 ± 0.0018), caramel sundae (0.05 ± 0.0003), 80 proof rum (0.05 ± 0.0012), restaurant orange juice (0.05 ± 0.0004), restaurant hot chocolate (0.05 ± 0.0002), white grapefruit (0.05 ± 0.0004), roasted chicken breast without skin (0.049 ± 0.0006), strawberries (0.049 ± 0.0003), sweet roll (0.048 ± 0.0012), roasted turkey light meat with skin (0.048 ± 0.0005), frozen and boiled kale (0.048 ± 0.0008), canned yellow corn (0.048 ± 0.0005), turkey breast meat summer sausage (0.048 ± 0.0024), pummelo (0.048 ± 0.0003), raw carrots (0.047 ± 0.0007), frozen and boiled zucchini (0.047 ± 0.0004), boiled hubbard squash (0.047 ± 0.0004), unheated canadian bacon (0.046 ± 0.0009), carrot juice (0.046 ± 0.0003), raw spot (0.046 ± 0.0006), cooked snapper (0.046 ± 0.0006), dry dessert wine (0.046 ± 0.0008), sweet dessert wine (0.046 ± 0.0008), cooked grouper (0.045 ± 0.0006), stewed chicken wing with skin (0.045 ± 0.0013), raw broccoli (0.045 ± 0.0011), raw yellowtail (0.045 ± 0.0006), baked hubbard squash (0.045 ± 0.0005), turkey frankfurter (0.044 ± 0.0011), yellow cake (0.044 ± 0.0007), stewed light meat chicken with skin (0.044 ± 0.0005), stewed meat chicken without skin (0.044 ± 0.0005), stewed chicken breast with skin (0.044 ± 0.0005), frozen and boiled red sweet peppers (0.044 ± 0.0005), frozen and boiled green sweet peppers (0.044 ± 0.0005), orange juice (0.044 ± 0.0002), pink grapefruit (0.044 ± 0.0004), beef bologna (0.043 ± 0.0022), stewed chicken breast without skin (0.043 ± 0.0005), frozen and boiled broccoli (0.043 ± 0.0005), yellow corn pudding (0.043 ± 0.0002), applesauce (0.043 ± 0.0004), daiquiri (0.043 ± 0.0008), smoked haddock (0.042 ± 0.0006), roasted turkey light meat without skin (0.042 ± 0.0005), battered and fried shark (0.042 ± 0.0006), raw croaker (0.042 ± 0.0006), tequila sunrise (0.042 ± 0.0003), cantaloupe (0.042 ± 0.0003), plum (0.042 ± 0.0008), raw sturgeon (0.041 ± 0.0006), raw dolphinfish (0.041 ± 0.0006), light meat chicken roll (0.041 ± 0.0005), raw chinook salmon (0.041 ± 0.0006), raw tilefish (0.041 ± 0.0006), raw apple with skin (0.041 ± 0.0004), cooked wild long grain rice

(0.04 ± 0.0004), white cake (0.04 ± 0.0006), cooked whiting (0.04 ± 0.0006), raw pork jowl (0.04 ± 0.0005), raw onions (0.04 ± 0.0006), coconut water (0.04 ± 0.0002), raw radish (0.04 ± 0.0011), american grapes (0.04 ± 0.0005), loquats (0.04 ± 0.0005), raw cucumber (0.04 ± 0.001), boiled onions (0.04 ± 0.0005), 53 proof coffee liqueur (0.04 ± 0.001), 63 proof coffee liqueur (0.04 ± 0.001), beef beerwurst salami (0.039 ± 0.0022), canned green beans (0.038 ± 0.0007), canned yellow snap beans (0.038 ± 0.0007), barbecue sauce (0.038 ± 0.0031), angel food cake (0.038 ± 0.0008), canned pork lunch meat (0.038 ± 0.0024), raw pompano (0.038 ± 0.0006), valencia orange (0.037 ± 0.0004), screwdriver (0.037 ± 0.0002), raw turbot (0.036 ± 0.0006), turkey bologna (0.036 ± 0.0018), pork and beef lunch meat (0.036 ± 0.0018), lemon (0.036 ± 0.0009), frozen and boiled butternut squash (0.036 ± 0.0004), cooked rockfish (0.036 ± 0.0006), cooked clams (0.036 ± 0.0006), canned cod (0.036 ± 0.0006), cooked cod (0.036 ± 0.0006), boiled rutabaga (0.036 ± 0.0006), light mayonnaise (0.036 ± 0.0036), mayonnaise (0.036 ± 0.0036), light meat turkey roll (0.035 ± 0.0009), cooked halibut (0.035 ± 0.0006), chicken roll (0.035 ± 0.0009), raw celery (0.035 ± 0.0013), boiled/baked spaghetti squash (0.035 ± 0.0006), cooked apple without skin (0.035 ± 0.0003), raw milkfish (0.034 ± 0.0006), cream of rice (0.034 ± 0.0003), cooked haddock (0.033 ± 0.0006), cooked perch (0.033 ± 0.0006), raw shark (0.033 ± 0.0006), frozen and boiled yellow corn (0.033 ± 0.0006), mandarin oranges (0.033 ± 0.0004), white grapefruit juice (0.033 ± 0.0002), red grapefruit juice (0.033 ± 0.0002), raw cauliflower (0.032 ± 0.001), raw flatfish (0.032 ± 0.0006), watermelon (0.032 ± 0.0003), raw striped bass (0.031 ± 0.0006), raw sheepshead (0.031 ± 0.0006), raw whiting (0.031 ± 0.0006), raw seatrout (0.031 ± 0.0006), boiled celery (0.031 ± 0.0007), raw apple without skin (0.031 ± 0.0004), cream of wheat (0.03 ± 0.0003), restaurant coleslaw (0.03 ± 0.0005), raw iceberg lettuce (0.03 ± 0.0025), whiskey sour (0.03 ± 0.0006), cola (0.03 ± 0.0001), lime juice (0.03 ± 0.0002), cooked eel (0.029 ± 0.0006), raw rockfish (0.029 ± 0.0006), tangerine (0.029 ± 0.0006), raw wolffish (0.029 ± 0.0006), lemon juice (0.029 ± 0.0002), boiled green cabbage (0.028 ± 0.0007), raw cod (0.028 ± 0.0006), raw monkfish (0.028 ± 0.0006), grape juice (0.028 ± 0.0002), raw snapper (0.028 ± 0.0006), manhattan (0.028 ± 0.0009), raw halibut (0.027 ± 0.0006), raw lingcod (0.027 ± 0.0006), frozen and cooked sweetened rhubarb (0.027 ± 0.0004), chicken egg white (0.027 ± 0.0015), raw haddock (0.026 ± 0.0006), raw perch (0.026 ± 0.0006), cooked flatfish (0.026 ± 0.0006), tangerine juice (0.025 ± 0.0002), whole chocolate malted milk (0.025 ± 0.0002), malted milk (0.025 ± 0.0002), raw eel (0.024 ± 0.0006), frozen and boiled cauliflower (0.024 ± 0.0006), cooked sea bass (0.024 ± 0.0006), frozen and boiled onions (0.024 ± 0.0005), lemon meringue pie (0.023 ± 0.0004), coleslaw (0.023 ± 0.0008), raw green cabbage (0.023 ± 0.0014), raw frozen rhubarb (0.023 ± 0.0004), apple juice (0.022 ± 0.0002), restaurant strawberry milkshake (0.022 ± 0.0002), white table wine (0.021 ± 0.0005), 94 proof gin (0.021 ± 0.0012), 94 proof rum (0.021 ± 0.0012), 94 proof vodka (0.021 ± 0.0012), 94 proof whiskey (0.021 ± 0.0012), 100 proof gin (0.021 ± 0.0012), 100 proof rum (0.021 ± 0.0012), 100 proof vodka (0.021 ± 0.0012), 100 proof whiskey (0.021 ± 0.0012), 86 proof whiskey (0.021 ± 0.0012), boiled sprouted green peas (0.02 ± 0.0005), spaghetti (0.02 ± 0.0004), macaroni (0.02 ± 0.0004), raw grouper (0.02 ± 0.0006), red table wine (0.02 ± 0.0005), raw butterhead lettuce (0.02 ± 0.0033), orange soda (0.02 ±

0.0001), tuna salad spread (0.019 ± 0.0009), raw sea bass (0.019 ± 0.0006), raw cusk (0.018 ± 0.0006), restaurant cranberry juice cocktail (0.018 ± 0.0003), papaya (0.016 ± 0.0002), rose apple (0.016 ± 0.0005), chamomile tea (0.015 ± 0.0003), papaya nectar (0.013 ± 0.0002), cranberry juice cocktail (0.013 ± 0.0002), corn grits (0.012 ± 0.0002), farina (0.011 ± 0.0003), instant corn grits (0.01 ± 0.0004), lemon-lime soda (0.01 ± 0.0001), black tea (0.01 ± 0.0003), restaurant coffee (0.01 ± 0.0003), 80 proof vodka (0.01 ± 0.0012), carob milk (0.009 ± 0.0002), beer (0.009 ± 0.0001), instant tea (0.008 ± 0.0002), coffee (0.007 ± 0.0003), martini (0.006 ± 0.0007), 90 proof gin (0.005 ± 0.0012)

Applications: anemia, leukemia, thinning of bones, baldness, bedsores, skin edema

Daily Dosage: (at least 14 mg zinc per every 1 mg copper) 1974 PCS – 1–3 mg; 1980 RDA – none; 1980 SADDI – 2–3 mg; 1989 RDA – none; 1989 SADDI – 2–3 mg; nutritional – 2–5 mg; therapeutic – 5–10 mg; experimental – 10–20 mg; toxic – 25 mg +

Warnings: Wilson's disease makes one easily susceptible to copper poisoning. Increased amounts are dangerous. See toxicity symptoms.

FLUORINE

Names: fluorine, fluorin, F, F⁻, Fl, element 9
Classification: trace mineral
Deficiency: predisposition to cavities, thinning of bones
Side-effects: see toxicity symptoms
Toxicity: (Detoxified by vitamin C) mottling of teeth, pitting of permanent teeth, late dentition, projecting bone growth of spine, arthritis, white blood cell abnormalities, muscular pain, muscle tendon strain, deterioration of heart muscle, weak pulse, varicose veins, lesions in gastrointestinal tract, nausea, cramps in abdomen, convulsions, lesions in nervous system, lesions in eyes, carbuncles, cracked skin
Inhibitors: aluminum salts of fluoride
Helpers: unknown
Sources: tea, coffee, fluoridated water, bone meal, air pollution, insecticides, rodent poisons
Applications: thinning of bones, tooth decay, tooth and gum deterioration
Daily Dosage: 1980 RDA – none; 1980 SADDI – 1.5–4.0 mg as fluoride; 1989 RDA – none; 1989 SADDI – 1.5–4.0 mg as fluoride; nutritional – ?; therapeutic – 1.1–2.2 mg; experimental – ?; toxic – ?
Warnings: increased amounts are dangerous. See toxicity symptoms.

GERMANIUM

Names: germanium, Ge, element 32, predicted as "eka-silicon"
Classification: trace mineral
Deficiency: decreased immunity, de-

creased cell oxygenation, decreased interferon production, bone deterioration, heart deterioration, increased blood pressure
Side-effects: none known
Toxicity: none known
Applications: increased cholesterol in blood, increased blood pressure, heart disease, rheumatoid arthritis, cancer, leukemia, sarcoma, thinning of bones, emotional deterioration, Reynaud's disease, failing vision, apoplexy, glandular myoma
Daily Dosage: 1989 SADDI — none; 1989 RDA — none; 1989 USRDA — none; nutritional — 2–6 mg; therapeutic — 25 mg; experimental — 25 + mg; toxic — none known
Warnings: none

GOLD

Classification: possible trace mineral, contaminant
Names: gold, Au, element 79
Deficiency: (being researched)
Side-effects: see toxicity symptoms
Toxicity: heavy metal poisoning
Inhibitors: unknown
Helpers: unknown
Sources: gold eating utensils
Applications: none known
Daily Dosage: 1974 PCS — none; 1989 SADDI — none; 1989 RDA — none; 1989 USRDA — none; nutritional — unknown; therapeutic — none; experimental — none; toxic — small amounts
Warnings: Small amounts are dangerous.

IODINE

Names: iodine, iodin, I, I⁻, element 53
Classification: trace mineral
Deficiency: cold feet, goiter, fatigue, skin eczema, dry hair, brittle hair, arrested physical/mental development, deafness and muteness
Side-effects: see toxicity symptoms
Toxicity: fatigue, headaches, weight gain, dry skin, acne, sensitive to cold, thin nails, brittle nails, rapid pulse, irregular menstrual bleeding, increased salivation
Inhibitors: food-processing, nutrient-deterioration of soil
Helpers: unknown
Sources (μg per 100 grams of food)
clams (90 ± 0.5), shrimp (65 ± 0.5), haddock (62 ± 0.5), halibut (56 ± 0.5), oysters (50 ± 0.5), salmon (50 ± 0.5), canned sardines (37 ± 0.5), beef liver (19 ± 0.5), pineapple (16 ± 0.5), canned tuna (16 ± 0.5), eggs (14 ± 0.5), peanuts (11 ± 0.5), whole wheat bread (11 ± 0.5), cheddar cheese (11 ± 0.5), pork (10 ± 0.5), lettuce (10 ± 0.5), spinach (9 ± 0.5), green peppers (9 ± 0.5), butter (9 ± 0.5), milk (7 ± 0.5), cream (6 ± 0.5), cottage cheese (6 ± 0.5), beef (6 ± 0.5), lamb (3 ± 0.5), raisins (3 ± 0.5)
Applications: angina pectoris, hair deterioration, hardening of the arteries,

atherosclerosis, arthritis, goiter, hyperthyroidism, underactive thyroid, cretinism, decreased physical and mental vigor
Daily Dosage: 1974 PCS — 75–225 µg; 1980 RDA — 150 µg; 1980 USRDA — 150 µg; 1989 RDA — 150 µg; 1989 USRDA — 150 µg; nutritional — 150–200 µg; therapeutic — 150–200 µg; experimental — ?; toxic — ?
Warnings: increased amounts are dangerous. See toxicity symptoms.

IODINE 131

Names: iodine 131, radioactive iodine, I-131
Classification: contaminant
Deficiency: (no nutritional requirement)
Side-effects: see toxicity symptoms
Toxicity: Thyroid cancer
Inhibitors: intake of normal iodine
Helpers: decreased levels of normal iodine
Sources: contaminated fruits, contaminated vegetables, nuclear power air pollution
Applications: none known
Daily Dosage: 1974 PCS — none; 1989 SADDI — none; 1989 RDA — none; 1989 USRDA — none; nutritional — none; therapeutic — none; experimental — none; toxic — small amounts
Warnings: Small amounts are dangerous. See toxicity symptoms.

IRON

Names: iron, Fe, Fe^{+2}, Fe^{++}, Fe^{+3}, Fe^{+++}, element 26
Classification: trace mineral
Deficiency: anemia, increased menses, nosebleeds, headache, hair decreased, dry scaling lips, immunity decreased, muscle weakness, fatigue, depression, dizziness, bodily weakness, difficulty in swallowing, thin nails, edges of nails turned up, intestinal diseases, continuous diarrhea, constipation, overall itching
Side-effects: see toxicity symptoms
Toxicity: bronze discolored skin, dizziness, decreased weight, headache, shortness of breath, fatigue, increased urination, internal inflammation
Inhibitors: caffeine, phosphoproteins in eggs, phytates in unleavened whole wheat, neomycin, blood decreased
Helpers: vitamin B_6, vitamin B_{12}, vitamin B_C, vitamin C, vitamin E, vitamin T, calcium, cobalt, copper, phosphorus, hydrochloric acid
Sources (mg per 100 grams of food): dried parsley (127 ± 0.5), cumin seed (69.5 ± 0.25), basil (59 ± 0.5), freeze-dried parsley (53.57 ± 0.357), savory (53 ± 0.5), marjoram (50 ± 0.5), dried dill weed (49 ± 0.5), turmeric (45.5 ± 0.25), celery seed (45 ± 0.25), cinnamon (44 ± 0.25), braised beef spleen (39.36 ± 0.005), anise seed (39 ± 0.25), oregano (33 ± 0.25), fenugreek seed (31 ± 0.125), raw duck liver (30.53 ± 0.005), black pepper (30.5 ± 0.25),

curry powder (29.5 ± 0.25), bran flakes (28.93 ± 0.018), dried spirulina seaweed (28.5 ± 0.005), poultry seasoning (26.5 ± 0.25), tarragon (26 ± 0.25), bay leaf (26 ± 0.5), paprika (25 ± 0.25), dried coriander leaf (25 ± 0.5), boiled calabash gourd (24.66 ± 0.007), braised pork spleen (22.2 ± 0.005), dried agar seaweed (21.4 ± 0.005), freeze-dried chives (20 ± 0.625), sage (20 ± 0.5), torula yeast (19.64 ± 0.018), dried chervil (19 ± 0.5), fennel seed (18.5 ± 0.25), broiled lamb liver (18 ± 0.011), braised pork liver (17.92 ± 0.005), brewers yeast (17.5 ± 0.018), dried rosemary (17.5 ± 0.25), potato flour (17.2 ± 0.006), caraway seed (17 ± 0.25), dill seed (17 ± 0.25), white pepper (17 ± 0.25), pumpkin pie spice (17 ± 0.25), dry bakers yeast (16.43 ± 0.018), braised pork lungs (16.41 ± 0.005), bran (16.07 ± 0.018), blackstrap molasses (16 ± 0.025), wheat flakes (15.89 ± 0.018), dried pumpkin and squash seeds (15.18 ± 0.018), roasted pumpkin and squash seeds (15.14 ± 0.018), whole toasted sesame seeds (14.96 ± 0.018), whole dried sesame seeds (14.56 ± 0.056), soybean protein isolate (14.5 ± 0.018), coriander seed (14.5 ± 0.25), cardamom (14 ± 0.25), raw clams (13.98 ± 0.006), breaded and fried clams (13.92 ± 0.006), defatted soy meal (13.7 ± 0.004), cooked oysters (13.4 ± 0.006), sorghum (12.38 ± 0.024), chili powder (12.33 ± 0.167), raisin bran (12.16 ± 0.014), dry cocoa powder (12 ± 0.1), mace (12 ± 0.25), yellow mustard seed (11 ± 0.167), pork livercheese (10.82 ± 0.013), soybean protein concentrate (10.79 ± 0.018), freeze-dried red sweet peppers (10.63 ± 0.313), freeze-dried green sweet peppers (10.63 ± 0.313), ginger (10.5 ± 0.25), firm raw tofu (10.47 ± 0.004), fried chicken giblets (10.32 ± 0.005), cooked whelk (10.06 ± 0.006), dried and frozen tofu (9.71 ± 0.029), pork liver sausage (9.33 ± 0.028), dried pignolia pine nuts (9.32 ± 0.018), canned chicken liver pâté (9.29 ± 0.018), defatted soybean flour (9.24 ± 0.005), toasted wheat germ (9.21 ± 0.018), simmered chicken heart (9.03 ± 0.005), cloves (9 ± 0.25), sesame butter (8.93 ± 0.033), raw irish moss seaweed (8.9 ± 0.005), poppy seed (8.67 ± 0.167), almond meal (8.61 ± 0.018), natto (8.6 ± 0.006), simmered chicken liver (8.47 ± 0.005), saffron (8 ± 0.5), toasted sesame kernels (7.89 ± 0.018), rice bran (7.88 ± 0.005), corn germ (7.8 ± 0.005), simmered turkey liver (7.8 ± 0.005), dried sesame kernels (7.75 ± 0.063), simmered beef heart (7.51 ± 0.005), dried watermelon seeds (7.5 ± 0.018), simmered beef kidneys (7.31 ± 0.005), unsweetened baking chocolate (7 ± 0.018), cayenne pepper (7 ± 0.25), breaded and fried oysters (6.95 ± 0.006), simmered turkey heart (6.89 ± 0.005), dried sunflower seeds (6.86 ± 0.018), dried pistachio nuts (6.86 ± 0.018), millit (6.8 ± 0.005), oil roasted sunflower seeds (6.79 ± 0.018), braised beef liver (6.77 ± 0.005), cooked mussels (6.72 ± 0.006), dried oriental radish (6.72 ± 0.009), simmered turkey giblets (6.71 ± 0.005), canned oysters (6.71 ± 0.006), raw oysters (6.7 ± 0.006), allspice (6.5 ± 0.25), simmered chicken giblets (6.44 ± 0.005), corn flakes (6.43 ± 0.018), pork liverwurst (6.39 ± 0.028), soybean flour (6.38 ± 0.006), fried beef liver (6.28 ± 0.005), raw parsley (6.2 ± 0.017), dried pepeao (6.17 ± 0.042), cream puff (6.15 ± 0.004), dry roasted cashews (6.07 ± 0.018), raw cuttlefish (6.01 ± 0.006), medium molasses (6 ± 0.025), low fat soybean flour (5.99 ± 0.006), braised pork heart (5.83 ± 0.004), roasted soybean flour (5.81 ± 0.006), chicken egg yolk (5.59 ± 0.029), freeze-dried shallots (5.5 ± 0.125), simmered turkey gizzard (5.44 ± 0.005), dried longans (5.4 ± 0.005), braised beef lungs (5.4 ± 0.005), raw tofu (5.36 ± 0.004), raw octopus (5.31 ± 0.006), raw reindeer (5.3 ± 0.005), braised pork kidneys (5.29 ± 0.005), boiled soybeans (5.14 ± 0.003), cashew butter (5.11 ± 0.018), raw whelk (5.04 ± 0.006), toasted almonds (5 ± 0.018), compressed bakers yeast (5 ± 0.018), bittersweet chocolate (5 ± 0.018), braised pork tongue (4.99 ± 0.005), fried tofu (4.85 ± 0.038), puffed wheat (4.79 ± 0.036), corn tortilla (4.73 ± 0.017), dried apricots (4.71 ± 0.014), anchovy canned in olive oil (4.65 ± 0.025), boiled hyacinth beans (4.58 ± 0.003), chipped beef (4.57 ± 0.018), tomato powder (4.56 ± 0.005), dark rye flour (4.53 ± 0.004), toasted soybean nuts (4.5 ± 0.018), raw quail without skin (4.5 ± 0.005), light molasses (4.5 ± 0.025), sorghum grain (4.4 ± 0.005), boiled winged beans (4.33 ± 0.003), dry roasted red pistachio nuts (4.29 ± 0.018), shredded wheat (4.29 ± 0.018), raw wild rice (4.19 ± 0.003), simmered chicken gizzard (4.15 ± 0.005), oil roasted cashews (4.14 ± 0.018), turkey egg (4.1 ± 0.006), cream of wheat (4.1 ± 0.003), dried butternuts (4.07 ± 0.018), toaster pastry (4 ± 0.01), dry roasted soybean nuts (3.95 ± 0.006), raw mussels (3.95 ± 0.006), dry roasted sunflower seeds (3.93 ± 0.018), roasted soybean nuts (3.9 ± 0.006), toasted whole wheat bread (3.9 ± 0.024), fried beef sirloin excluding fat (3.9 ± 0.005), oil roasted almonds (3.89 ± 0.018), dry roasted almonds (3.86 ± 0.018), simmered beef shank excluding fat (3.86 ± 0.005), duck egg (3.86 ± 0.007), whole grain rye crackers (3.85 ± 0.038), corn syrup (3.81 ± 0.024), fried abalone (3.8 ± 0.006), braised beef chuck arm pot roast excluding fat (3.79 ± 0.005), dried almonds (3.71 ± 0.018), simmered pork chitterlings (3.7 ± 0.005), almond butter (3.69 ± 0.031), boiled white beans (3.69 ± 0.003), braised beef chuck blade roast excluding fat (3.68 ± 0.005), quail egg (3.67 ± 0 .056), roasted beef tenderloin excluding fat (3.66 ± 0.005), simmered beef shank including fat (3.61 ± 0.005), dry breadcrumbs (3.6 ± 0.005), raw cassava (3.6 ± 0.005), broiled beef tenderloin excluding fat (3.58 ± 0.005), dried brazil nuts (3.57 ± 0.018), dried pilinuts (3.57 ± 0.018), instant oatmeal (3.57 ± 0.003), boiled spinach (3.57 ± 0.006), cheese crackers (3.53 ± 0.033), braised/roasted/stewed veal chuck (3.53 ± 0.006), toasted raisin bread (3.52 ± 0.024), nutmeg (3.5 ± 0.25), maypo (3.5 ± 0.003), braised beef flank excluding fat (3.47 ± 0.005), oil roasted filberts (3.46 ± 0.018), braised beef bottom round excluding fat (3.46 ± 0.005), toasted english muffin (3.46 ± 0.01), light cream (3.45 ± 0.017), whole wheat bread (3.44 ± 0.02), dry roasted filberts (3.43 ± 0.018), dried toasted coconut (3.43 ± 0.018), dried japanese chestnuts (3.43 ± 0.018), roasted veal rib roast (3.41 ± 0.006), falafel (3.41 ± 0.01), braised beef flank including fat (3.4 ± 0.005), raw jerusalem artichoke (3.4 ± 0.007), simmered beef tongue (3.39 ± 0.005), brown sugar (3.38 ± 0.003), broiled beef sirloin excluding fat (3.36 ± 0.005), braised beef shortribs excluding fat (3.36 ± 0.005), dried coconut (3.36 ± 0.018), semi-sweet bakers chocolate (3.36 ± 0.018), pickled rip sevillano/ascolano olives (3.33 ± 0.031), whole wheat flour (3.33 ± 0.004), popcorn (3.33 ± 0.083), boiled lentils (3.33 ± 0.003), dried filberts (3.32 ± 0.018), braised/stewed veal breast plate (3.31 ± 0.006), dried peanuts (3.29 ± 0.018), french bread (3.29 ± 0.018), mixed grain bread (3.28 ± 0.02), baked beans with pork and tomato sauce (3.28 ± 0.002), pickled rip manzanillo/mission olives (3.26 ± 0.109), braised beef bottom round including fat (3.25 ± 0.005), fried beef sirloin including fat (3.25 ± 0.005), broiled beef tenderloin including fat (3.25 ± 0.005), raw anchovy (3.25 ± 0.006), toasted white bread (3.24 ± 0.024), sour pickles (3.23 ± 0.008), dry roasted pistachio nuts (3.21 ± 0.018), almond paste (3.21 ± 0.018), raw abalone (3.19 ± 0.006), roasted beef tenderloin including fat (3.19 ± 0.005), braised/broiled veal round with rump (3.18 ± 0.006), braised/broiled veal loin (3.18 ± 0.006), granola bar (3.17 ± 0.021), fried beef top round excluding fat

(3.15 ± 0.005), bran muffins (3.15 ± 0.013), oatmeal cookie (3.15 ± 0.038), ry-krisp crackers (3.14 ± 0.036), cooked crayfish (3.14 ± 0.006), toasted pumpernickel bread (3.14 ± 0.018), boiled mothbeans (3.14 ± 0.003), boiled jute potherb (3.14 ± 0.012), raisin bread (3.12 ± 0.02), dried black walnuts (3.11 ± 0.018), raw dandelion greens (3.11 ± 0.018), dried pinyon pine nuts (3.11 ± 0.018), stir-fried sprouted lentils (3.1 ± 0.005), toasted rye bread (3.09 ± 0.023), broiled porterhouse steak excluding fat (3.08 ± 0.005), cooked shrimp (3.08 ± 0.006), braised beef chuck arm pot roast including fat (3.07 ± 0.005), toasted cracked wheat bread (3.05 ± 0.024), boiled catjang black-eyed peas (3.05 ± 0.003), broiled beef sirloin including fat (3.01 ± 0.005), broiled T-bone steak excluding fat (3 ± 0.005), well done baked ground beef (2.99 ± 0.005), canned white beans (2.99 ± 0.002), hamburger bun (2.98 ± 0.013), hot dog bun (2.98 ± 0.013), tomato paste (2.98 ± 0.004), braised beef chuck blade roast including fat (2.96 ± 0.005), well done baked extra lean ground beef (2.96 ± 0.005), dinner rolls (2.96 ± 0.018), french rolls (2.96 ± 0.01), roasted beef tip excluding fat (2.94 ± 0.005), boiled kidney beans (2.94 ± 0.003), carob flour (2.94 ± 0.005), sardines (2.92 ± 0.021), enriched wheat bread (2.92 ± 0.004), all purpose enriched wheat flour (2.92 ± 0.004), rice flour (2.9 ± 0.004), enriched white degermed corn meal (2.9 ± 0.004), enriched yellow degermed corn meal (2.9 ± 0.004), chopped smoked beef (2.89 ± 0.018), boiled chickpeas (2.89 ± 0.003), broiled beef top round excluding fat (2.88 ± 0.005), roasted goose without skin (2.87 ± 0.005), fried beef top round including fat (2.86 ± 0.005), boiled black turtle beans (2.85 ± 0.003), raw kelp/kombu/tangle seaweed (2.85 ± 0.005), roasted goose with skin (2.83 ± 0.005), white bread (2.83 ± 0.021), roasted veal leg (2.82 ± 0.006), english muffin (2.82 ± 0.009), peach pie (2.82 ± 0.005), pecan pie (2.82 ± 0.005), broiled beef top round including fat (2.81 ± 0.005), tamarind (2.8 ± 0.004), canned jalapeño peppers (2.79 ± 0.007), braised beef brisket excluding fat (2.77 ± 0.005), well done broiled extra lean ground beef (2.77 ± 0.005), dark buckwheat flour (2.76 ± 0.005), pumpernickel bread (2.75 ± 0.016), turkey ham (2.75 ± 0.009), well done broiled ground beef (2.74 ± 0.005), miso (2.74 ± 0.004), well done fried extra lean ground beef (2.73 ± 0.005), rye bread (2.72 ± 0.02), roasted beef round tip including fat (2.71 ± 0.005), well done fried ground beef (2.71 ± 0.005), raw spinach (2.71 ± 0.018), pot barley pearled (2.7 ± 0.003), roasted duck without skin (2.7 ± 0.005), roasted duck with skin (2.7 ± 0.005), braised pork pancreas (2.69 ± 0.005), broiled beef round excluding fat (2.69 ± 0.005), roast beef sandwich (2.69 ± 0.003), cracked wheat bread (2.68 ± 0.02), garlic powder (2.67 ± 0.167), saltine crackers (2.67 ± 0.083), broiled porterhouse steak including fat (2.66 ± 0.005), well done baked lean ground beef (2.66 ± 0.005), beef and pork salami (2.65 ± 0.022), bagels (2.65 ± 0.009), taco shell (2.64 ± 0.045), tostada shell (2.64 ± 0.045), boiled yardlong bean (2.64 ± 0.003), oatmeal bread (2.64 ± 0.018), graham crackers (2.64 ± 0.036), italian bread (2.64 ± 0.018), braised beef pancreas (2.61 ± 0.005), roasted beef rib roasted excluding fat (2.61 ± 0.005), boiled pinto beans (2.61 ± 0.003), simmered chicken neck without skin (2.61 ± 0.028), wax beans (2.6 ± 0.004), seeded raisins (2.59 ± 0.005), ham and cheese sandwich (2.57 ± 0.003), broiled beef rib eye excluding fat (2.57 ± 0.005), fruitcake, dark (2.56 ± 0.012), broiled T-bone steak including fat (2.54 ± 0.005), broiled beef flank excluding fat (2.53 ± 0.005), broiled beef rib excluding fat (2.52 ± 0.005), summer sausage (2.52 ± 0.022), persimmon (2.52 ± 0.02), dried and salted cod (2.51 ± 0.006), boiled black-eyed peas (2.51 ±

0.003), dry roasted macadamia nuts (2.5 ± 0.018), onion powder (2.5 ± 0.25), broiled beef flank including fat (2.5 ± 0.005), boiled green soybeans (2.5 ± 0.006), semi-sweet chocolate (2.5 ± 0.018), peanut brittle (2.5 ± 0.018), bacon cheeseburger (2.49 ± 0.003), boiled yellow beans (2.48 ± 0.003), well done fried lean ground beef (2.48 ± 0.005), boiled navy beans (2.48 ± 0.003), lebanon bologna (2.48 ± 0.022), dried prunes (2.48 ± 0.006), gefiltefish with broth (2.48 ± 0.012), broiled beef top loin excluding fat (2.47 ± 0.005), dried english/persian walnuts (2.46 ± 0.018), raw crayfish (2.45 ± 0.006), well done broiled lean ground beef (2.45 ± 0.005), medium-rare fried ground beef (2.45 ± 0.005), cooked rainbow trout (2.44 ± 0.006), medium-rare broiled ground beef (2.44 ± 0.005), raw coconut (2.44 ± 0.011), dried european chestnuts (2.43 ± 0.018), dried macadamia nuts (2.43 ± 0.018), broiled beef round including fat (2.42 ± 0.005), fried chicken neck with skin (2.42 ± 0.014), pita bread (2.42 ± 0.013), raw shrimp (2.41 ± 0.006), medium-rare baked ground beef (2.41 ± 0.005), boiled baby lima beans (2.4 ± 0.003), raw dock (2.4 ± 0.007), whole wheat roll (2.39 ± 0.018), boiled lima beans (2.39 ± 0.003), white whole ground corn meal (2.38 ± 0.004), yellow whole ground corn meal (2.38 ± 0.004), tamari sauce (2.38 ± 0.009), chicken salad (2.36 ± 0.018), medium-rare fried extra lean ground beef (2.36 ± 0.005), beef lunch meat (2.36 ± 0.018), medium-rare broiled extra lean ground beef (2.35 ± 0.005), broiled beef rib eye including fat (2.34 ± 0.005), roasted turkey dark meat without skin (2.33 ± 0.005), dried chinese chestnuts (2.32 ± 0.018), braised beef shortribs including fat (2.31 ± 0.005), boiled pink beans (2.3 ± 0.003), canned spinach (2.3 ± 0.005), oil roasted spanish peanuts (2.29 ± 0.018), german sweet bakers chocolate (2.29 ± 0.018), coconut cream (2.29 ± 0.002), simmered chicken neck with skin (2.29 ± 0.013), gingersnaps cookie (2.29 ± 0.071), medium-rare baked extra lean ground beef (2.28 ± 0.005), tempeh (2.27 ± 0.006), roasted turkey dark meat with skin (2.27 ± 0.005), beef summer sausage (2.26 ± 0.022), boiled amaranth (2.26 ± 0.008), boiled swiss chard (2.26 ± 0.006), dry roasted peanuts (2.25 ± 0.018), roasted leg of lamb excluding fat (2.24 ± 0.006), broiled beef top loin including fat (2.24 ± 0.005), dried figs (2.24 ± 0.003), fried beef brains (2.22 ± 0.005), beef honey roll sausage (2.22 ± 0.022), simmered beef brains (2.21 ± 0.005), dry roasted pecans (2.21 ± 0.018), hamburger sandwich (2.2 ± 0.005), braised beef brisket including fat (2.19 ± 0.005), medium-rare fried lean ground beef (2.18 ± 0.005), raw pork stomach (2.18 ± 0.005), raw wakame seaweed (2.18 ± 0.005), beef salami (2.17 ± 0.022), dried hickory nuts (2.14 ± 0.018), oil roasted pecans (2.14 ± 0.018), dried pecans (2.14 ± 0.018), roasted japanese chestnuts (2.14 ± 0.018), boiled great northern beans (2.13 ± 0.003), frozen and boiled black-eyed peas (2.12 ± 0.006), raw leeks (2.12 ± 0.019), roasted beef rib including fat (2.11 ± 0.005), medium-rare broiled lean ground beef (2.11 ± 0.005), pepperoni pizza (2.1 ± 0.004), boiled black beans (2.09 ± 0.003), boiled cranberry beans (2.09 ± 0.003), medium-rare baked lean ground beef (2.09 ± 0.005), broiled beef rib including fat (2.08 ± 0.005), cheeseburger (2.08 ± 0.004), chocolate syrup (2.08 ± 0.013), canned corned beef (2.08 ± 0.005), seedless raisins (2.08 ± 0.005), boiled dock (2.08 ± 0.005), jellied corned beef loaf (2.07 ± 0.018), french toast (2.06 ± 0.008), braised pork boston blade excluding fat (2.05 ± 0.005), broiled lamb loin chop excluding fat (2.04 ± 0.01), beef and pork summer sausage (2.04 ± 0.022), shoyu sauce (2.02 ± 0.009), chicken frankfurter (2 ± 0.011), peanut flour (2 ± 0.125), turkey summer sausage (2 ± 0.018), light pearled barley (2 ± 0.003), boiled adzuki beans (2 ± 0.002), brown mustard (2 ± 0.1), brownies (2 ± 0.025), yellow mustard (2 ± 0.1), bread and butter pickles (2 ± 0.033), raw coriander (2 ± 0.125), waffles (1.97 ± 0.007), boiled and frozen baby lima beans (1.96 ± 0.006), pretzels (1.96 ± 0.018), braised pork arm excluding fat (1.95 ± 0.005), roasted eye of beef round excluding fat (1.95 ± 0.005), raw beef tripe (1.95 ± 0.005), frozen and boiled turnip greens (1.94 ± 0.006), oil roasted peanuts (1.93 ± 0.018), beef pastrami (1.93 ± 0.018), chicken egg (1.92 ± 0.01), boiled scotch kale (1.92 ± 0.008), crunchy peanut butter (1.91 ± 0.016), raw rainbow trout (1.91 ± 0.006), dried sweetened shredded coconut (1.91 ± 0.005), stir-fried sprouted mung beans (1.9 ± 0.008), cooked kingfish (1.9 ± 0.005), canned black turtle beans (1.9 ± 0.002), boiled beet greens (1.9 ± 0.007), roasted lamb shoulder excluding fat (1.88 ± 0.006), shortbread cookie (1.88 ± 0.063), raw spring onions with tops (1.88 ± 0.01), hot dog (1.87 ± 0.006), broiled lamb rib chop excluding fat (1.86 ± 0.012), cooked corned beef brisket (1.86 ± 0.005), danish pastry (1.86 ± 0.012), raw agar seaweed (1.86 ± 0.005), braised pork spareribs (1.85 ± 0.005), canned navy beans (1.85 ± 0.002), mulberries (1.85 ± 0.004), roasted eye of beef round including fat (1.84 ± 0.005), braised pork brains (1.82 ± 0.005), oil roasted macadamia nuts (1.82 ± 0.018), creamy peanut butter (1.81 ± 0.031), sweet roll (1.81 ± 0.012), boiled dandelion greens (1.81 ± 0.01), white bolted corn meal (1.8 ± 0.004), yellow bolted corn meal (1.8 ± 0.004), dried jujube (1.8 ± 0.005), canned lima beans (1.8 ± 0.002), raw laver/nori seaweed (1.8 ± 0.005), golden seedless raisins (1.79 ± 0.005), white corn flour (1.79 ± 0.004), yellow corn flour (1.79 ± 0.004), smoked beef sausage (1.77 ± 0.012), refried beans (1.77 ± 0.002), corn meal muffins (1.75 ± 0.013), boiled and steamed european chestnuts (1.75 ± 0.018), boiled mungo beans (1.74 ± 0.003), boiled mushrooms (1.74 ± 0.006), dried shitake mushrooms (1.73 ± 0.033), braised pork boston blade braised including fat (1.72 ± 0.005), teriyaki sauce (1.72 ± 0.028), devil's food cake (1.72 ± 0.008), pie crust (1.72 ± 0.003), canned smelt (1.7 ± 0.005), dried lychees (1.7 ± 0.005), oil roasted virginia peanuts (1.68 ± 0.018), dried sweetened flaked coconut (1.68 ± 0.02), sponge cake (1.68 ± 0.008), raw swamp cabbage (1.68 ± 0.009), raw garlic (1.67 ± 0.056), raw chives (1.67 ± 0.167), boiled sprouted green peas (1.67 ± 0.005), marshmallow (1.67 ± 0.083), baked beans with pork and sweet sauce (1.66 ± 0.002), roasted leg of lamb including fat (1.65 ± 0.006), turkey pastrami (1.65 ± 0.009), beef bologna (1.65 ± 0.022), oil roasted valencia peanuts (1.64 ± 0.018), coconut milk (1.64 ± 0.002), fried chicken back with skin (1.63 ± 0.007), fruitcake, light (1.63 ± 0.012), raw mackerel (1.62 ± 0.006), braised pork loin blade excluding fat (1.62 ± 0.005), cooked bacon (1.61 ± 0.004), dried ginko nuts (1.61 ± 0.018), braised pork arm braised including fat (1.61 ± 0.005), passion fruit (1.61 ± 0.028), spaghetti (1.61 ± 0.004), macaroni (1.61 ± 0.004), roasted pork boston blade excluding fat (1.6 ± 0.005), turkey salami (1.6 ± 0.009), grilled ham patties (1.6 ± 0.008), canned pinto beans (1.6 ± 0.002), chili (1.6 ± 0.002), elderberries (1.6 ± 0.003), cooked carp (1.59 ± 0.006), potato pancakes (1.59 ± 0.007), gingerbread (1.59 ± 0.008), pineapple upside-down cake (1.59 ± 0.007), turkey frankfurter (1.58 ± 0.011), frozen and boiled green peas (1.58 ± 0.006), dehydrated onion flakes (1.57 ± 0.036), canned great northern beans (1.57 ± 0.002), hummus (1.57 ± 0.002), cooked mackerel (1.56 ± 0.006), cooked tahitian taro (1.56 ± 0.007), burrito (1.55 ± 0.003), boiled green peas (1.55 ± 0.006), canned cranberry beans (1.55 ± 0.002),

II. Minerals

IRON

roasted pork tenderloin (1.54 ± 0.005), roasted chinese chestnuts (1.54 ± 0.018), black currants (1.54 ± 0.009), roasted pork shoulder excluding fat (1.52 ± 0.005), boiled succotash (1.52 ± 0.005), beef beerwurst salami (1.52 ± 0.022), caramel cake (1.52 ± 0.005), beef and pork bologna (1.52 ± 0.022), frozen and boiled spinach (1.52 ± 0.005), italian pork sausage (1.51 ± 0.007), raw mustard spinach (1.51 ± 0.007), canned turnip greens (1.51 ± 0.004), kippered herring (1.5 ± 0.013), stewed rabbit (1.5 ± 0.004), whole ground corn meal muffins (1.5 ± 0.013), fried chicken thigh with skin (1.5 ± 0.008), fried dark meat chicken with skin (1.5 ± 0.005), raw frog legs (1.5 ± 0.005), hard pork and beef salami (1.5 ± 0.05), popover roll (1.5 ± 0.013), blueberry muffins (1.5 ± 0.013), red beans (1.5 ± 0.004), canned butter beans (1.5 ± 0.004), frozen and fried hash brown potatoes (1.5 ± 0.006), noodles (1.5 ± 0.003), braised beef thymus (1.49 ± 0.005), raw trout (1.49 ± 0.006), raw freshwater bass (1.49 ± 0.006), fried dark meat chicken without skin (1.49 ± 0.005), boiled broad beans (1.49 ± 0.003), simmered pork ears (1.49 ± 0.005), roasted lean ham (1.48 ± 0.005), pancakes (1.48 ± 0.019), cottage pudding (1.48 ± 0.009), soy sauce (1.48 ± 0.009), raw green peas (1.47 ± 0.006), roselle (1.47 ± 0.009), canned sauerkraut (1.47 ± 0.004), pork and beef kielbasa/kolbassy (1.46 ± 0.019), polish pork sausage (1.46 ± 0.018), pork and beef link sausage (1.46 ± 0.007), raw japanese chestnuts (1.46 ± 0.018), breaded and fried catfish (1.44 ± 0.006), lobster paste (1.43 ± 0.071), enriched biscuits (1.43 ± 0.018), sweet chocolate (1.43 ± 0.018), pork and beef luncheon sausage (1.43 ± 0.022), raw chinese chestnuts (1.43 ± 0.018), soda crackers (1.43 ± 0.018), enchilada (1.43 ± 0.002), hot dog pickle relish (1.43 ± 0.018), maple sugar (1.43 ± 0.018), roasted pork arm excluding fat (1.42 ± 0.005), roasted pork boston blade including fat (1.42 ± 0.005), taco (1.42 ± 0.006), beef frankfurter (1.42 ± 0.009), cooked herring (1.41 ± 0.006), cooked mullet (1.41 ± 0.006), roasted turkey light meat with skin (1.41 ± 0.005), cooked flounder (1.4 ± 0.005), smoked haddock (1.4 ± 0.006), braised pork loin excluding fat (1.4 ± 0.005), boiled mung beans (1.4 ± 0.002), pork and beef mortadella (1.4 ± 0.033), raisin bun (1.4 ± 0.008), canned pumpkin (1.39 ± 0.004), raw looseleaf lettuce (1.39 ± 0.018), sugar cookie (1.38 ± 0.031), roasted canned ham (1.37 ± 0.005), stewed chicken thigh with skin (1.37 ± 0.007), roasted dark meat chicken with skin (1.36 ± 0.005), pork and beef honey loaf (1.36 ± 0.018), stewed dark meat chicken without skin (1.36 ± 0.005), yellow cake (1.36 ± 0.007), frozen and boiled fordhook lima beans (1.36 ± 0.006), baked potato with skin (1.36 ± 0.002), broiled pork boston blade excluding fat (1.35 ± 0.005), roasted turkey light meat without skin (1.35 ± 0.005), fried chicken drumstick with skin (1.35 ± 0.01), canned chickpeas (1.35 ± 0.002), boiled artichoke hearts (1.35 ± 0.006), roasted ham (1.34 ± 0.005), cooked haddock (1.34 ± 0.006), cheese pizza (1.34 ± 0.004), roasted chicken thigh with skin (1.34 ± 0.008), frozen and french fried potatoes without skin (1.34 ± 0.01), braised pork sirloin excluding fat (1.33 ± 0.005), bulgur (1.33 ± 0.004), roasted dark meat chicken without skin (1.33 ± 0.005), roasted chicken drumstick with skin (1.33 ± 0.01), light and dark meat turkey roll (1.33 ± 0.009), stewed chicken drumstick with skin (1.33 ± 0.009), pork mothers loaf (1.33 ± 0.024), pork and beef pepperoni (1.33 ± 0.083), boiled swamp cabbage (1.33 ± 0.01), frozen and boiled turnip greens and turnips (1.33 ± 0.005), sweet pickles (1.33 ± 0.033), raw garden cress (1.32 ± 0.02), steamed sprouted soybeans (1.32 ± 0.011), raw white sucker (1.31 ± 0.006), roasted pork shoulder including fat (1.31 ± 0.005), roasted chicken thigh without skin (1.31 ± 0.01), fish sandwich (1.31 ± 0.003), stewed dark meat chicken with skin (1.31 ± 0.005), prune pudding (1.31 ± 0.004), hard pork salami (1.3 ± 0.05), stewed chicken drumstick without skin (1.3 ± 0.011), braised pork loin blade including fat (1.3 ± 0.005), okara tofu (1.3 ± 0.008), pickled green olives (1.3 ± 0.011), carissa (1.3 ± 0.025), turkey bologna (1.29 ± 0.018), boiled split peas (1.29 ± 0.003), pork bratwurst (1.28 ± 0.006), broiled lamb loin chop including fat (1.27 ± 0.007), boiled green beans (1.27 ± 0.008), boiled yellow snap beans (1.27 ± 0.008), braised pork center rib exlucding fat (1.26 ± 0.005), braised pork top loin excluding fat (1.26 ± 0.005), breaded and fried shrimp (1.26 ± 0.006), pork sausage (1.26 ± 0.019), light meat turkey roll (1.26 ± 0.009), roasted chicken wing with skin (1.26 ± 0.015), canned potato without skin (1.26 ± 0.006), raw red sweet peppers (1.26 ± 0.01), raw green sweet peppers (1.26 ± 0.01), cooked wild long grain rice (1.26 ± 0.004), roasted pork loin blade excluding fat (1.25 ± 0.005), boiled arrowhead (1.25 ± 0.042), pork and beef old fashioned loaf (1.25 ± 0.018), fried chicken wing with skin (1.25 ± 0.016), oyster crackers (1.25 ± 0.063), raw carp (1.24 ± 0.006), microwaved potato with skin (1.24 ± 0.002), canned kidney beans (1.23 ± 0.002), raw mushrooms (1.23 ± 0.014), chocolate chiffon pie (1.23 ± 0.006), fried light meat chicken with skin (1.21 ± 0.005), potato chips (1.21 ± 0.018), cream cheese (1.21 ± 0.018), acorn flour (1.21 ± 0.018), boiled brussels sprouts (1.21 ± 0.006), hamburger pickle relish (1.21 ± 0.018), raw sunfish (1.2 ± 0.006), boiled lupins (1.2 ± 0.003), pickled herring (1.2 ± 0.033), raw shallots (1.2 ± 0.05), raw vine spinach (1.2 ± 0.005), raw red chili peppers (1.2 ± 0.011), raw green chili peppers (1.2 ± 0.011), boiled pokeberry shoots (1.2 ± 0.006), fried chicken breast with skin (1.19 ± 0.005), roasted pork arm including fat (1.19 ± 0.005), cooked perch (1.18 ± 0.006), roasted lamb shoulder including fat (1.18 ± 0.006), pork headcheese (1.18 ± 0.018), raw california avocado (1.18 ± 0.003), prune juice (1.18 ± 0.002), broiled pork boston blade including fat (1.17 ± 0.005), barbecue loaf (1.17 ± 0.022), berliner (1.17 ± 0.022), braised pork loin including fat (1.16 ± 0.005), pork link sausage (1.16 ± 0.007), raw lotus root (1.16 ± 0.006), beef and pork frankfurter (1.16 ± 0.011), dried dates (1.16 ± 0.006), cooked rainbow smelt (1.15 ± 0.006), roasted pork loin excluding fat (1.15 ± 0.005), raw pheasant without skin (1.15 ± 0.005), pork and beef sausage (1.15 ± 0.038), roasted pork rump excluding fat (1.14 ± 0.005), fried chicken breast without skin (1.14 ± 0.006), fried light meat chicken without skin (1.14 ± 0.005), roasted light meat chicken with skin (1.14 ± 0.005), boiled broccoli (1.14 ± 0.006), raw dolphinfish (1.13 ± 0.006), braised pork sirloin including fat (1.13 ± 0.005), cooked grouper (1.13 ± 0.006), stewed chicken wing with skin (1.13 ± 0.013), candied sweet potato (1.13 ± 0.005), roasted pork leg excluding fat (1.12 ± 0.005), frozen and boiled collards (1.12 ± 0.006), frozen and boiled mustard greens (1.12 ± 0.007), boiled leeks (1.12 ± 0.019), roasted pork shank excluding fat (1.11 ± 0.005), raw herring (1.11 ± 0.006), battered and fried shark (1.11 ± 0.006), rusk crackers (1.11 ± 0.056), boiled pigeon peas (1.11 ± 0.003), raw romaine lettuce (1.11 ± 0.018), raw turnip greens (1.11 ± 0.018), cooked prunes (1.11 ± 0.005), pancakes with butter and syrup (1.1 ± 0.003), roasted pork sirloin excluding fat (1.09 ± 0.005), roasted pork center loin excluding fat (1.09 ± 0.005), bread pudding (1.09 ± 0.002), unenriched corn meal (1.09 ± 0.004), unenriched yellow degermed corn meal (1.09 ± 0.004), broiled pork loin blade excluding fat (1.08 ± 0.005), crab cakes (1.08 ± 0.008), light rye flour (1.08 ± 0.005), chocolate cream pie (1.08 ± 0.004), boiled french beans (1.08 ±

0.003), white cake (1.08 ± 0.006), dill pickles (1.08 ± 0.008), cooked halibut (1.07 ± 0.006), milk chocolate (1.07 ± 0.018), braised pork center rib including fat (1.07 ± 0.005), pork luxury loaf (1.07 ± 0.018), pork and beef link sausage with american cheese (1.07 ± 0.012), roasted pork loin blade including fat (1.07 ± 0.005), pork and beef peppered loaf (1.07 ± 0.018), dried acorns (1.07 ± 0.018), puffed rice (1.07 ± 0.036), jelly beans (1.07 ± 0.018), canned shitake mushrooms (1.06 ± 0.004), canned sockeye salmon (1.06 ± 0.006), cooked black bass (1.06 ± 0.004), roasted light meat chicken without skin (1.06 ± 0.005), roasted chicken breast with skin (1.06 ± 0.005), unheated ham patties (1.06 ± 0.008), cornbread (1.06 ± 0.006), corn pone (1.06 ± 0.006), canned pumpkin pie mix (1.06 ± 0.004), roasted pork rump including fat (1.05 ± 0.005), cheese soufflé (1.05 ± 0.005), raw haddock (1.05 ± 0.006), cooked swordfish (1.04 ± 0.006), fried pork loin blade excluding fat (1.04 ± 0.005), broiled lamb rib chop including fat (1.04 ± 0.007), pork pickle and pimento loaf (1.04 ± 0.018), pork and beef picnic loaf (1.04 ± 0.018), raw european chestnuts (1.04 ± 0.018), boiled chinese cabbage (1.04 ± 0.006), roasted chicken breast without skin (1.03 ± 0.006), braised pork top loin including fat (1.03 ± 0.005), pork and beef brotwurst (1.03 ± 0.007), raw mullet (1.02 ± 0.006), light buckwheat flour (1.02 ± 0.005), mince pie (1.02 ± 0.004), breaded fried squid (1.01 ± 0.006), roasted pork loin including fat (1.01 ± 0.005), parmesan cheese (1 ± 0.1), fried pork center loin excluding fat (1 ± 0.005), roasted pork center rib excluding fat (1 ± 0.005), roasted pork top loin excluding fat (1 ± 0.005), roasted pork leg including fat (1 ± 0.005), roasted cured pork arm excluding fat (1 ± 0.005), roasted pork sirloin including fat (1 ± 0.005), boiled peanuts (1 ± 0.016), canned bonito (1 ± 0.005), tuna salad (1 ± 0.002), ginko nuts (1 ± 0.018), canned broad beans (1 ± 0.002), bread stuffing (1 ± 0.003), red currants (1 ± 0.009), white currants (1 ± 0.009), ground-cherries (1 ± 0.004), sapote (1 ± 0.002), maple syrup (1 ± 0.025), canned shellie beans (0.99 ± 0.004), unheated ham (0.99 ± 0.005), roasted pork center loin including fat (0.99 ± 0.005), raw catfish (0.98 ± 0.006), stewed light meat chicken with skin (0.98 ± 0.005), canned black-eyed peas (0.98 ± 0.002), frozen and boiled turnips (0.98 ± 0.005), roasted pork shank including fat (0.97 ± 0.005), light meat chicken roll (0.97 ± 0.005), raw sprouted alfalfa seeds (0.97 ± 0.015), raw shad (0.96 ± 0.006), chicken roll (0.96 ± 0.009), boiled and steamed chinese chestnuts (0.96 ± 0.018), braised pork center loin excluding fat (0.95 ± 0.005), roasted cured pork arm including fat (0.95 ± 0.005), canned green peas (0.95 ± 0.006), yellow hominy (0.94 ± 0.002), canned lean ham (0.94 ± 0.005), turkey salad (0.94 ± 0.009), frozen and boiled kale (0.94 ± 0.008), broiled pork loin excluding fat (0.93 ± 0.005), ham and cheese roll (0.93 ± 0.018), stewed light meat chicken without skin (0.93 ± 0.005), roasted european chestnuts (0.93 ± 0.018), baked acorn squash (0.93 ± 0.005), tomato purée (0.93 ± 0.002), raisin pie (0.93 ± 0.004), broiled pork center loin excluding fat (0.92 ± 0.005), raw perch (0.92 ± 0.006), roasted canned lean ham (0.92 ± 0.005), stewed chicken breast with skin (0.92 ± 0.005), onion rings (0.92 ± 0.006), raw rainbow smelt (0.91 ± 0.006), cooked blue crab (0.91 ± 0.006), broiled pork loin blade including fat (0.91 ± 0.005), raw burbot (0.91 ± 0.006), raw freshwater drum (0.91 ± 0.006), pork and beef knockwurst (0.91 ± 0.007), boiled kale (0.91 ± 0.008), bread sticks (0.9 ± 0.025), boiled lotus root (0.9 ± 0.006), raw sprouted mung beans (0.9 ± 0.01), raw chickory greens (0.9 ± 0.006), apple pie (0.9 ± 0.004), canned green beans (0.9 ± 0.007), canned yellow snap beans (0.9 ± 0.007), cooked coho salmon (0.89 ± 0.006), broiled pork sirloin excluding fat (0.89 ± 0.005), roasted pork center rib including fat (0.89 ± 0.005), roasted pork blade roll (0.89 ± 0.005), frozen and boiled succotash (0.89 ± 0.006), raw broccoli (0.89 ± 0.011), canned sweet potato (0.89 ± 0.003), roasted pork top loin including fat (0.88 ± 0.005), stewed chicken breast without skin (0.88 ± 0.005), raw grouper (0.88 ± 0.006), butterscotch pie (0.88 ± 0.004), vienna sausage (0.88 ± 0.031), poi (0.88 ± 0.004), cooked white rice (0.88 ± 0.002), barbecue sauce (0.88 ± 0.031), boiled red sweet peppers (0.88 ± 0.007), boiled green sweet peppers (0.88 ± 0.007), cooked extra long grain white rice (0.88 ± 0.006), canned chinese water chestnuts (0.87 ± 0.007), breaded and fried croaker (0.86 ± 0.006), pork and beef lunch meat (0.86 ± 0.018), lemon chiffon pie (0.86 ± 0.006), pineapple chiffon pie (0.86 ± 0.006), canned pink salmon (0.85 ± 0.006), smoked chinook salmon (0.85 ± 0.006), angel food cake (0.85 ± 0.008), boysenberries (0.85 ± 0.004), raw striped bass (0.84 ± 0.006), canned blue crab (0.84 ± 0.006), raw halibut (0.84 ± 0.006), fried pork center loin including fat (0.84 ± 0.005), raw shark (0.84 ± 0.006), raw cusk (0.84 ± 0.006), french fries (0.84 ± 0.006), raw endive (0.84 ± 0.02), frozen and boiled potatoes without skin (0.84 ± 0.005), braised pork center loin including fat (0.83 ± 0.005), canned ham (0.83 ± 0.005), orange peel (0.83 ± 0.083), lemon peel (0.83 ± 0.083), fried pork loin blade including fat (0.82 ± 0.005), frozen and boiled yellow snap beans (0.82 ± 0.007), frozen and boiled green beans (0.82 ± 0.007), grilled canadian bacon (0.81 ± 0.011), broiled pork center rib excluding fat (0.81 ± 0.005), broiled pork top loin excluding fat (0.81 ± 0.005), raw swordfish (0.81 ± 0.006), broiled pork loin including fat (0.81 ± 0.005), breaded and fried scallops (0.81 ± 0.016), broiled pork center loin including fat (0.81 ± 0.005), chopped ham lunch meat (0.81 ± 0.024), tuna salad spread (0.81 ± 0.009), hash brown potatoes (0.81 ± 0.006), fried pork center rib excluding fat (0.8 ± 0.005), fried pork top loin excluding fat (0.8 ± 0.005), raw flounder (0.8 ± 0.005), raw sole (0.8 ± 0.005), cooked parboiled white rice (0.8 ± 0.003), raw chinese cabbage (0.8 ± 0.014), marinara sauce (0.8 ± 0.002), canned red sweet peppers (0.8 ± 0.007), canned green sweet peppers (0.8 ± 0.007), boiled mustard spinach (0.8 ± 0.006), sapodilla (0.8 ± 0.003), ham and cheese spread (0.79 ± 0.018), colby cheese (0.79 ± 0.018), canned yellow corn with red and green peppers (0.79 ± 0.004), pork and beef spread (0.79 ± 0.018), boiled garden cress (0.79 ± 0.007), raw acorns (0.79 ± 0.018), boiled turnip greens (0.79 ± 0.007), raw new zealand spinach (0.79 ± 0.018), cooked instant white rice (0.79 ± 0.003), broiled pork sirloin including fat (0.78 ± 0.005), pork bologna (0.78 ± 0.022), boiled purslane (0.78 ± 0.009), boiled burdock root (0.77 ± 0.004), tomato sauce (0.77 ± 0.004), raw pink salmon (0.76 ± 0.006), unheated lean ham (0.76 ± 0.005), minced ham lunch meat (0.76 ± 0.024), french fried potatoes (0.76 ± 0.01), raw potato without skin (0.76 ± 0.004), cooked alaska king crab (0.75 ± 0.006), cooked parboiled extra long grain white rice (0.75 ± 0.006), lemon meringue pie (0.75 ± 0.004), strawberry pie (0.75 ± 0.005), raw blue crab (0.74 ± 0.006), pork beerwurst salami (0.74 ± 0.022), dried japanese persimmon (0.74 ± 0.015), frozen and boiled brussels sprouts (0.74 ± 0.006), instant corn grits (0.74 ± 0.004), cooked taro (0.73 ± 0.008), boiled cardoon (0.73 ± 0.005), canned stewed red tomatoes (0.73 ± 0.004), broiled pork center rib including fat (0.72 ± 0.005), maltex (0.72 ± 0.003), raw coho salmon (0.71 ± 0.006), raw chinook salmon (0.71 ± 0.006), monterey cheese (0.71 ± 0.018), cooked pike (0.71 ± 0.006), cheese crackers with peanut butter (0.71 ± 0.012), canned pork lunch meat (0.71 ± 0.024), safflower oil and soybean oil mayonnaise (0.71 ± 0.036), soybean oil mayonnaise (0.71 ± 0.036), canned crookneck squash (0.71 ± 0.005), broiled pork top loin including fat (0.7 ± 0.005), coconut custard pie (0.7 ± 0.004), boiled blackeye pea pods (0.7 ± 0.011), boiled lamb's quarters (0.7 ± 0.006), boiled mustard greens (0.7 ± 0.007), quince (0.7 ± 0.005), mammy apple (0.7 ± 0.005), cheddar cheese (0.68 ± 0.018), raw squid (0.68 ± 0.006), oatmeal (0.68 ± 0.003), canned zucchini in tomato juice (0.68 ± 0.004), rhubarb pie (0.68 ± 0.004), cooked sheepshead (0.67 ± 0.006), cooked pompano (0.67 ± 0.006), unheated canadian bacon (0.67 ± 0.009), pound cake (0.67 ± 0.017), chili sauce (0.67 ± 0.033), tomato catsup (0.67 ± 0.033), frozen and boiled okra (0.67 ± 0.005), russian salad dressing (0.67 ± 0.033), horseradish sauce (0.67 ± 0.033), low calorie thousand island salad dressing (0.67 ± 0.033), cider vinegar (0.67 ± 0.033), canned tuna in oil (0.66 ± 0.006), fried pork center rib including fat (0.66 ± 0.005), fried pork top loin including fat (0.66 ± 0.005), boiled asparagus (0.66 ± 0.006), boiled new zealand spinach (0.66 ± 0.006), raw ling (0.65 ± 0.006), potato salad (0.65 ± 0.004), cooked sprouted mung beans (0.65 ± 0.008), feta cheese (0.64 ± 0.018), cooked eel (0.64 ± 0.006), au gratin potatoes (0.64 ± 0.004), restaurant hash brown potatoes (0.64 ± 0.006), loganberries (0.64 ± 0.003), canned carrots (0.64 ± 0.007), frozen and boiled asparagus (0.63 ± 0.008), low calorie russian salad dressing (0.63 ± 0.031), thousand island salad dressing (0.63 ± 0.031), low calorie french salad dressing (0.63 ± 0.031), canned succotash with creamed corn (0.62 ± 0.004), boiled beets (0.62 ± 0.006), boiled carrots (0.62 ± 0.006), fried chicken (0.61 ± 0.006), ham salad (0.61 ± 0.018), custard pie (0.61 ± 0.004), boiled yellow corn (0.61 ± 0.006), frozen and boiled broccoli (0.61 ± 0.005), chicken and turkey salad (0.61 ± 0.018), canned peeled red tomatoes (0.61 ± 0.004), canned tuna in water (0.6 ± 0.006), surimi shrimp (0.6 ± 0.006), yellow hominy with red and green peppers (0.6 ± 0.002), raw pompano (0.6 ± 0.006), raw bacon (0.6 ± 0.007), raw chinese water chestnuts (0.6 ± 0.008), jackfruit (0.6 ± 0.005), sugar apple (0.6 ± 0.003), boiled red tomato (0.6 ± 0.004), soursop (0.6 ± 0.002), boiled butternut squash (0.6 ± 0.005), apple brown betty (0.6 ± 0.002), lime (0.6 ± 0.007), lemon (0.6 ± 0.009), white hominy (0.59 ± 0.002), raw alaska king crab (0.59 ± 0.006), blueberry pie (0.59 ± 0.004), boiled parsnips (0.58 ± 0.006), soy milk (0.58 ± 0.002), raw cauliflower (0.58 ± 0.01), coleslaw (0.58 ± 0.008), cooked plantain (0.58 ± 0.003), restaurant tomato juice (0.58 ± 0.004), tomato juice (0.58 ± 0.003), frozen and boiled butternut squash (0.58 ± 0.004), roquefort cheese (0.57 ± 0.018), scalloped potatoes (0.57 ± 0.004), boiled pumpkin (0.57 ± 0.004), raw green cabbage (0.57 ± 0.014), blackberries (0.57 ± 0.007), raspberries (0.57 ± 0.004), boiled artichoke (0.56 ± 0.006), yellow corn pudding (0.56 ± 0.002), boiled acorn squash (0.56 ± 0.004), boiled sweet potato (0.56 ± 0.003), raw pepeao (0.56 ± 0.005), raw chum salmon (0.55 ± 0.006), cooked sockeye salmon (0.55 ± 0.006), raw pike (0.55 ± 0.006), raw celtuce (0.55 ± 0.005), apricots (0.55 ± 0.005), restaurant coleslaw (0.54 ± 0.005), provolone cheese (0.54 ± 0.018), pork olive loaf (0.54 ± 0.018), boiled salsify (0.54 ± 0.007), raw eggplant (0.54 ± 0.012), breadfruit (0.54 ± 0.005), boiled and steamed japanese chestnuts (0.54 ± 0.018), raw scup (0.53 ± 0.006), cooked rockfish (0.53 ± 0.006), sweet potato pie (0.53 ± 0.004), banana custard pie (0.53 ± 0.004), pumpkin pie (0.53 ± 0.004), canned succotash (0.53 ± 0.004), frozen and baked sweet potato (0.53 ± 0.006), raw florida avocado

(0.53 ± 0.002), frozen and boiled red sweet peppers (0.52 ± 0.005), frozen and boiled green sweet peppers (0.52 ± 0.005), raw eel (0.51 ± 0.006), smoked whitefish (0.51 ± 0.006), wheat cake (0.51 ± 0.004), cooked brown rice (0.51 ± 0.003), baked/boiled tropical yam (0.51 ± 0.007), frozen and boiled crookneck squash (0.51 ± 0.005), raw green tomato (0.51 ± 0.004), blackberry pie (0.51 ± 0.004), sweetened coconut cream (0.51 ± 0.002), pineapple pie (0.51 ± 0.004), raw chickory (0.51 ± 0.011), brie cheese (0.5 ± 0.018), chocolate pudding (0.5 ± 0.002), citrus marmalade (0.5 ± 0.025), raw bamboo shoots (0.5 ± 0.007), raw carrots (0.5 ± 0.007), cherimoya (0.5 ± 0.001), raw ginger root (0.5 ± 0.021), raw iceberg lettuce (0.5 ± 0.025), canned red chili peppers (0.5 ± 0.007), canned green chili peppers (0.5 ± 0.007), acerola juice (0.5 ± 0.002), raw butterfish (0.49 ± 0.006), boston cream pie (0.49 ± 0.005), raw red cabbage (0.49 ± 0.014), cooked clams (0.48 ± 0.006), canned cod (0.48 ± 0.006), raw bluefish (0.48 ± 0.006), raw yellowtail (0.48 ± 0.006), smoked cisco (0.48 ± 0.006), cooked cod (0.48 ± 0.006), cheesecake (0.48 ± 0.006), raw crookneck squash (0.48 ± 0.008), frozen and boiled carrots (0.48 ± 0.007), raw celery (0.48 ± 0.013), frozen and boiled zucchini (0.48 ± 0.004), raw red tomato (0.48 ± 0.004), jujube (0.48 ± 0.005), honey (0.48 ± 0.024), dry whole milk (0.47 ± 0.016), raw sockeye salmon (0.47 ± 0.006), o'brien potatoes (0.47 ± 0.003), boiled rutabaga (0.47 ± 0.006), baked hubbard squash (0.47 ± 0.005), raw sheepshead (0.46 ± 0.006), raw pollock (0.46 ± 0.006), baked sweet potato (0.46 ± 0.004), carrot juice (0.46 ± 0.003), boiled okra (0.45 ± 0.006), pineapple custard pie (0.44 ± 0.004), canned sprouted mung beans (0.44 ± 0.008), cooked shitake mushrooms (0.44 ± 0.007), pimento cheese spread (0.43 ± 0.018), edam cheese (0.43 ± 0.018), muenster cheese (0.43 ± 0.018), raw celeriac (0.43 ± 0.005), steamed hawaii mountain yam (0.43 ± 0.007), raw zucchini (0.43 ± 0.008), red table wine (0.43 ± 0.005), cooked whiting (0.42 ± 0.006), custard (0.42 ± 0.002), tapioca cream pudding (0.42 ± 0.003), raw pork jowl (0.42 ± 0.005), canned yellow corn (0.42 ± 0.005), boiled cauliflower (0.42 ± 0.008), cooked enriched white degermed corn meal (0.42 ± 0.002), cooked enriched yellow degermed corn meal (0.42 ± 0.002), raw rockfish (0.41 ± 0.006), microwaved potato without skin (0.41 ± 0.003), kiwi fruit (0.41 ± 0.007), frozen and boiled cauliflower (0.41 ± 0.006), raw oriental radish (0.41 ± 0.011), pickled beets (0.41 ± 0.004), boiled collards (0.41 ± 0.003), coffee (0.41 ± 0.003), raw smelt (0.4 ± 0.005), stir-fried sprouted soybeans (0.4 ± 0.005), boiled kohlrabi (0.4 ± 0.006), raw savoy cabbage (0.4 ± 0.014), raw scallop squash (0.4 ± 0.008), honeydew melon (0.4 ± 0.005), casaba melon (0.4 ± 0.003), american cheese (0.39 ± 0.018), cooked lobster (0.39 ± 0.006), boiled green cabbage (0.39 ± 0.007), cooked sea bass (0.38 ± 0.006), turkey breast meat summer sausage (0.38 ± 0.024), raw cod (0.38 ± 0.006), ricotta cheese (0.38 ± 0.004), animal crackers (0.38 ± 0.019), canned creamed yellow corn (0.38 ± 0.004), boiled savoy cabbage (0.38 ± 0.007), cherries (0.38 ± 0.007), strawberries (0.38 ± 0.003), boiled waxgourd (0.38 ± 0.006), rose table wine (0.38 ± 0.005), apricot nectar (0.38 ± 0.002), kumquats (0.37 ± 0.026), bloody mary (0.37 ± 0.003), apple juice (0.37 ± 0.002), pineapple (0.37 = 0.003), raw cisco (0.36 ± 0.006), raw whitefish (0.36 ± 0.006), raw croaker (0.36 ± 0.006), raw dungeness crab (0.36 ± 0.006), unenriched biscuits (0.36 ± 0.018), frozen white corn (0.36 ± 0.007), boiled zucchini (0.36 ± 0.006), boiled crookneck squash (0.36 ± 0.006), boiled dishcloth gourd (0.36 ± 0.006), raw onions (0.36 ± 0.006), boiled red cabbage (0.36 ± 0.007), yellow passion fruit juice (0.36 ± 0.002),

II. Minerals

light mayonnaise (0.36 ± 0.036), figs (0.36 ± 0.01), mayonnaise (0.36 ± 0.036), olive oil (0.36 ± 0.036), gum drops (0.36 ± 0.018), raw flatfish (0.35 ± 0.006), baked potato without skin (0.35 ± 0.003), boiled eggplant (0.35 ± 0.01), crab apples (0.35 ± 0.005), applesauce (0.35 ± 0.004), raw whiting (0.34 ± 0.006), canned white corn (0.34 ± 0.005), cherry pie (0.34 ± 0.004), frozen and boiled onions (0.34 ± 0.005), papaya nectar (0.34 ± 0.002), dry skim milk (0.33 ± 0.017), cooked flatfish (0.33 ± 0.006), boiled/baked spaghetti squash (0.33 ± 0.006), raw monkfish (0.32 ± 0.006), bleu cheese (0.32 ± 0.018), camembert cheese (0.32 ± 0.018), raw lingcod (0.32 ± 0.006), raw spot (0.32 ± 0.006), raw milkfish (0.32 ± 0.006), boiled scallop squash (0.32 ± 0.006), canned bamboo shoots (0.32 ± 0.004), white table wine (0.32 ± 0.005), surimi scallops (0.31 ± 0.006), dry instant skim milk (0.31 ± 0.005), cooked tilefish (0.31 ± 0.006), restaurant chocolate milkshake (0.31 ± 0.002), hot chocolate (0.31 ± 0.002), chocolate milkshake (0.31 ± 0.002), boiled potato without skin (0.31 ± 0.004), lychees (0.31 ± 0.005), gooseberries (0.31 ± 0.003), guava (0.31 ± 0.006), bananas (0.31 ± 0.004), breaded and fried fish (0.3 ± 0.007), frozen and boiled yellow corn (0.3 ± 0.006), prickly pear (0.3 ± 0.005), pomegranate (0.3 ± 0.003), dry buttermilk (0.29 ± 0.071), raw scallops (0.29 ± 0.006), raw sea bass (0.29 ± 0.006), neufchatel cheese (0.29 ± 0.018), vegetarian baked beans (0.29 ± 0.002), boiled yambean (0.29 ± 0.005), coconut water (0.29 ± 0.002), raw radish (0.29 ± 0.011), american grapes (0.29 ± 0.005), evaporated skim milk (0.28 ± 0.016), loquats (0.28 ± 0.005), boiled hubbard squash (0.28 ± 0.004), raw frozen rhubarb (0.28 ± 0.004), raw butterhead lettuce (0.27 ± 0.033), raw seatrout (0.27 ± 0.006), mashed potatoes (0.27 ± 0.005), whipped butter (0.27 ± 0.045), raw cucumber (0.27 ± 0.01), mandarin oranges (0.27 ± 0.004), tequila sunrise (0.27 ± 0.003), carob milk (0.26 ± 0.002), carambola (0.26 ± 0.004), european grapes (0.26 ± 0.003), thompson seedless grapes (0.26 ± 0.003), pineapple juice (0.26 ± 0.002), pear nectar (0.26 ± 0.002), gouda cheese (0.25 ± 0.018), brick cheese (0.25 ± 0.018), raw tilefish (0.25 ± 0.006), cottage cheese (0.25 ± 0.018), pear (0.25 ± 0.003), cooked snapper (0.24 ± 0.006), lowfat 1% chocolate milk (0.24 ± 0.002), lowfat 2% chocolate milk (0.24 ± 0.002), whole chocolate milk (0.24 ± 0.002), boiled bamboo shoots (0.24 ± 0.004), purple passion fruit juice (0.24 ± 0.002), grape juice (0.24 ± 0.002), dry dessert wine (0.24 ± 0.008), sweet dessert wine (0.24 ± 0.008), evaporated lowfat milk (0.23 ± 0.004), strawberry sundae (0.23 ± 0.003), boiled chayote (0.23 ± 0.006), human milk (0.23 ± 0.016), boiled turnips (0.22 ± 0.006), piña colada (0.22 ± 0.004), fontina cheese (0.21 ± 0.018), tilsit cheese (0.21 ± 0.018), cheshire cheese (0.21 ± 0.018), cantaloupe (0.21 ± 0.003), cooked unenriched corn meal (0.21 ± 0.002), cooked unenriched yellow degermed corn meal (0.21 ± 0.002), frozen and cooked sweetened rhubarb (0.21 ± 0.004), eggnog (0.2 ± 0.002), frozen creamed yellow corn (0.2 ± 0.004), boiled onions (0.2 ± 0.005), butter (0.2 ± 0.033), sweet butter (0.2 ± 0.033), orange juice (0.2 ± 0.002), white grapefruit juice (0.2 ± 0.002), red grapefruit juice (0.2 ± 0.002), tangerine juice (0.2 ± 0.002), corn grits (0.2 ± 0.002), raw acerola (0.2 ± 0.005), pitanga (0.2 ± 0.003), cranberry sauce (0.2 ± 0.005), apple tapioca pudding (0.2 ± 0.002), evaporated whole milk (0.19 ± 0.004), whole chocolate malted milk (0.19 ± 0.002), java plum (0.19 ± 0.004), cooked apple without skin (0.19 ± 0.003), peach nectar (0.19 ± 0.002), swiss cheese (0.18 ± 0.018), mozzarella cheese (0.18 ± 0.018), sweetened condensed milk (0.18 ± 0.013), raw snapper (0.18 ± 0.006), raw watercress (0.18 ± 0.029), watermelon (0.18 ± 0.003), raw apple with skin (0.18 ± 0.004), blueberries (0.17 ± 0.003), cream of rice (0.16 ± 0.003), restaurant cranberry juice cocktail (0.16 ± 0.002), hot fudge sundae (0.15 ± 0.003), boiled oriental radish (0.15 ± 0.007), japanese persimmon (0.15 ± 0.003), nectarine (0.15 ± 0.004), daiquiri (0.15 ± 0.008), cranberry juice cocktail (0.15 ± 0.003), rice polish (0.14 ± 0.009), havarti cheese (0.14 ± 0.018), limburger cheese (0.14 ± 0.018), caramel sundae (0.14 ± 0.003), boiled celery (0.13 ± 0.007), longans (0.13 ± 0.005), mango (0.13 ± 0.002), indian buffalo milk (0.12 ± 0.002), navel orange (0.12 ± 0.004), pummelo (0.12 ± 0.003), pink grapefruit (0.12 ± 0.004), 80 proof rum (0.12 ± 0.012), lemon-lime soda (0.12 ± 0.001), malted milk (0.11 ± 0.002), restaurant strawberry milkshake (0.11 ± 0.002), peach (0.11 ± 0.006), plum (0.11 ± 0.008), tangerine (0.11 ± 0.006), sheep milk (0.1 ± 0.002), vanilla milkshake (0.1 ± 0.002), restaurant orange juice (0.1 ± 0.004), boiled butterbur (0.1 ± 0.005), papaya (0.1 ± 0.002), white granulated sugar (0.1 ± 0.003), raw wolffish (0.09 ± 0.006), skim yogurt (0.09 ± 0.002), restaurant vanilla milkshake (0.09 ± 0.002), valencia orange (0.09 ± 0.004), oheloberries (0.09 ± 0.004), manhattan (0.09 ± 0.009), martini (0.09 ± 0.007), lowfat yogurt (0.08 ± 0.002), sour cream (0.08 ± 0.042), screwdriver (0.08 ± 0.002), whiskey sour (0.08 ± 0.006), chamomile tea (0.08 ± 0.003), powdered sugar (0.08 ± 0.004), creme de menthe (0.08 ± 0.01), half and half cream (0.07 ± 0.033), medium cream (0.07 ± 0.033), restaurant hot chocolate (0.07 ± 0.002), rose apple (0.07 ± 0.005), raw apple without skin (0.07 ± 0.004), orange soda (0.07 ± 0.001), protein fortified lowfat 2% milk (0.06 ± 0.002), protein fortified skim milk (0.06 ± 0.002), protein fortified lowfat 1% milk (0.06 ± 0.002), whey (0.06 ± 0.002), white grapefruit (0.06 ± 0.004), 53 proof coffee liqueur (0.06 ± 0.01), 63 proof coffee liqueur (0.06 ± 0.01), goat milk (0.05 ± 0.002), filled milk (0.05 ± 0.002), lowfat 1% milk (0.05 ± 0.002), whole yogurt (0.05 ± 0.002), lowfat 2% milk (0.05 ± 0.002), whole milk (0.05 ± 0.002), buttermilk (0.05 ± 0.002), cola (0.05 ± 0.001), 94 proof gin (0.05 ± 0.012), 94 proof rum (0.05 ± 0.012), 94 proof vodka (0.05 ± 0.012), 94 proof whiskey (0.05 ± 0.012), 100 proof gin (0.05 ± 0.012), 100 proof rum (0.05 ± 0.012), 100 proof vodka (0.05 ± 0.012), 100 proof whiskey (0.05 ± 0.012), skim milk (0.04 ± 0.002), chicken egg white (0.03 ± 0.015), beer (0.03 ± 0.001), lime juice (0.03 ± 0.002), lemon juice (0.03 ± 0.002), tea (0.03 ± 0.002), farina (0.02 ± 0.003), 86 proof whiskey (0.02 ± 0.012), black tea (0.02 ± 0.003), instant tea (0.02 ± 0.002), malt liquor (0.01 ± 0.001), ale (0.01 ± 0.001), restaurant coffee (0.01 ± 0.003)

Applications: anemia, diabetes, leukemia, menstruation, pernicious anemia, colitis, diarrhea, alcohol cravings, celiac disease, colitis, worms, gout, kidney inflammation, tuberculosis, nail deterioration, pregnancy, scurvy, skin ulcers, gastritis, peptic ulcer, tooth and gum deterioration, aging, bruises, cancer

Daily Dosage: 1974 PCS — 9–27 mg; 1980 RDA — 10–18 mg; 1980 USRDA — 18 mg; 1989 RDA — 10–15 mg; 1989 USRDA — 15 mg; nutritional — 25–100 mg; therapeutic — 100–200 mg; experimental — 300 mg; toxic — ?

Warnings: Do not take with sickle-cell anemia, iron increased in skin or anemia. Increased amounts are dangerous. See toxicity symptoms.

LEAD

Names: lead, Pb, element 82
Classification: contaminant, trace mineral
Deficiency: (Trace amounts are required.)
Side-effects: see toxicity symptoms
Toxicity: (Detoxified by vitamin C, calcium) appetite decreased, constipation, metallic taste in mouth, nausea, slight albumin in urine, protein in urine, heartburn, upset stomach, stomach deterioration, slight weight decreased, hearing changes, balance deterioration, mental sluggishness, decreased I.Q., destruction of nerve tissue, tingling, numbness, paralysis, hallucinations, convulsions, coma, swallowing difficulties, vision deterioration, breathing deterioration, voice changes, very fine twitching of face muscles, slight trembling in hands, slight trembling in fingers, headaches, dizziness, fatigue, joint pain, vague aching, neck/shoulder pains, occasional cramps, occasional stiffness, anemia, small spontaneous bruising, anxiety, depression, emotional agitation, lack of sleep
Inhibitors: vitamin C, vitamin E, vitamin N, calcium, iron, selenium, zinc
Helpers: unknown
Sources: air pollution, motor vehicle exhaust, water pollution, insecticides, food contamination
Applications: none known
Daily Dosage: 1974 PCS — none; 1989 SADDI — none; 1989 RDA — none; 1989 USRDA — none; nutritional — unknown; therapeutic — none; experimental — none; toxic — small amounts
Warnings: Small amounts are dangerous. See toxicity symptoms. There is enough lead in Earth's atmosphere to cause toxicity by breathing. No amounts are required in food.

LITHIUM

Names: lithium, Li, Li^+, element 3
Classification: trace mineral, psychotropic drug as lithium carbonate
Deficiency: depression, alcohol cravings, unregulated conversion of essential fatty acids to prostaglandins, unstable serotonin neurotransmitter levels, decreased acetylcholine receptors, decreased lymphocyte levels, increased suppressor cell activity
Side-effects: see toxicity symptoms
Toxicity: birth defects, body fluid imbalances, sodium replacement in tissues
Inhibitors: unknown
Helpers: vitamin B_P, sodium
Sources: metallurgy, nuclear technology, ocean, rock formations
Applications: depression, alcohol cravings, mental imbalances, emotional agitation
Daily Dosage: 1974 PCS — none; 1980 RDA — none; 1980 USRDA — none; 1989 RDA — none; 1989 USRDA — none; nutritional — 50 µg-3 mg; therapeutic — 900–2100 mg as lithium carbonate; experimental — 2500 + mg as lithium carbonate; toxic — varies
Warnings: Do not take with severe kidney deterioration or cardiovascular disease, or while using diuretics or during pregnancy. Small amounts are dangerous. See toxicity symptoms.

MAGNESIUM

Names: magnesium, Mg, Mg^{+2}, Mg^{++}, element 12
Classification: mineral
Deficiency: emotional agitation, nervousness, stress, bone deformities, deterioration of teeth, calcium deposits in tissues, sodium deposits in tissues, fluid retention in tissues, deterioration of kidneys, deterioration of skin, fatigue, decreased blood levels of calcium, potassium and magnesium, deterioration of heart endocardium, bone muscle, cardiac muscle, cramps, knotting of muscle fibers, mental changes, dizziness, nerve pain, shooting pains, colic, increased blood pressure, stillbirths, female sterility
Side-effects: see toxicity symptoms
Toxicity: slow heartbeat, fatigue, weakness, muscle tremors, decreased reflexes, anesthesia, coma, diarrhea
Inhibitors: diuretics, alcohol, streptomycin
Helpers: vitamin B_6, vitamin C, vitamin D, boron, calcium, phosphorus, protein
Sources (mg per 100 grams of food): dried agar seaweed (770 ± 0.5), corn germ (672 ± 0.5), basil (600 ± 50), dried pumpkin and squash seeds (543 ± 1.8), roasted pumpkin and squash seeds (543 ± 1.8), dried watermelon seeds (521 ± 1.8), savory (500 ± 50), dried dill weed (500 ± 50), celery seed (450 ± 25), soybean flour (428 ± 0.6), dry cocoa powder (420 ± 10), fennel seed (400 ± 25), cumin seed (400 ± 25), dried coriander leaf (400 ± 50), bran (379 ± 1.8), peanut flour (375 ± 12.5), roasted soybean flour (369 ± 0.6), whole toasted sesame seeds (361 ± 1.8), freeze-dried parsley (357 ± 35.7), dried sunflower seeds (357 ± 1.8), whole dried sesame seeds (356 ± 5.6), toasted sesame kernels (350 ± 1.8), dried sesame kernels (350 ± 6.3), yellow mustard seed (333 ± 16.7), toasted wheat germ (325 ± 1.8), soybean protein concentrate (314 ± 1.8), toasted almonds (311 ± 1.8), dry roasted almonds (307 ± 1.8), oil roasted almonds (307 ± 1.8), defatted soy meal (306 ± 0.4), oil roasted filberts (304 ± 1.8), unsweetened baking chocolate (304 ± 1.8), dried parsley (300 ± 50), poppy seed (300 ± 16.7), sage (300 ± 50), tarragon (300 ± 25), cloves (300 ± 25), coriander seed (300 ± 25), almond butter (300 ± 3.1), dried almonds (300 ± 1.8), dry roasted filberts (300 ± 1.8), defatted soybean flour (290 ± 0.5), dried filberts (289 ± 1.8), dry roasted cashews (264 ± 1.8), almond paste (261 ± 1.8), cashew butter (261 ± 1.8), oil roasted cashews (257 ± 1.8), dill seed (250 ± 25), caraway seed (250 ± 25), curry powder (250 ± 25), cardamom (250 ± 25), rykrisp crackers (243 ± 3.6), dried butternuts (239 ± 1.8), dried pinyon pine nuts (239 ± 1.8), low fat soybean flour (230 ± 0.6), dried brazil nuts (229 ± 1.8), dry roasted soybean nuts (228 ± 0.6), dried black walnuts (204 ± 1.8), marjoram (200 ± 50), oregano (200 ± 25), anise seed (200 ± 25), black pepper (200 ± 25), paprika (200 ± 25), turmeric (200 ± 25), nutmeg (200 ± 25), dried spirulina seaweed (195 ± 0.5), oil roasted peanuts (189 ± 1.8), oil roasted virginia peanuts (189 ± 1.8), freeze-dried red sweet peppers (188 ± 31.3), freeze-dried green sweet peppers (188 ± 31.3), bran flakes (186 ± 1.8), dried peanuts (182 ± 1.8), tomato powder (178 ± 0.5), fenugreek seed (175 ± 12.5), toasted soybean nuts (175 ± 1.8), dried hickory nuts (175 ± 1.8), dry roasted peanuts (175 ± 1.8), creamy peanut butter (175 ± 3.1), cooked whelk (173 ± 0.6), dried oriental radish (171 ± 0.9), dried english/persian walnuts (171 ± 1.8), oil roasted spanish peanuts (168 ± 1.8), millit (162 ± 0.5), dried pistachio nuts (161 ± 1.8), oil

roasted valencia peanuts (161 ± 1.8), crunchy peanut butter (159 ± 1.6), dried rosemary (150 ± 25), poultry seasoning (150 ± 25), allspice (150 ± 25), onion powder (150 ± 25), mace (150 ± 25), cayenne pepper (150 ± 25), ginger (150 ± 25), puffed wheat (150 ± 3.6), roasted soybean nuts (145 ± 0.6), semisweet bakers chocolate (143 ± 1.8), dried pepeao (142 ± 4.2), dried chinese chestnuts (139 ± 1.8), dry roasted pecans (136 ± 1.8), chili powder (133 ± 16.7), dried and salted cod (133 ± 0.6), dried shitake mushrooms (133 ± 3.3), dry roasted pistachio nuts (132 ± 1.8), dry roasted sunflower seeds (132 ± 1.8), shredded wheat (132 ± 1.8), oil roasted pecans (132 ± 1.8), raisin bran (130 ± 1.4), oil roasted sunflower seeds (129 ± 1.8), dried pecans (129 ± 1.8), raw kelp/kombu/tangle seaweed (121 ± 0.5), dry instant skim milk (118 ± 0.5), dried japanese chestnuts (118 ± 1.8), dried macadamia nuts (118 ± 1.8), oil roasted macadamia nuts (118 ± 1.8), natto (115 ± 0.6), wheat flakes (111 ± 1.8), acorn flour (111 ± 1.8), dry skim milk (110 ± 1.7), toasted whole wheat bread (110 ± 2.4), cooked oysters (108 ± 0.6), raw wakame seaweed (107 ± 0.5), cooked halibut (107 ± 0.6), boiled mothbeans (104 ± 0.3), boiled peanuts (103 ± 1.6), raw dock (103 ± 0.7), raw tofu (102 ± 0.4), dry buttermilk (100 ± 7.1), dried chervil (100 ± 50), pumpkin pie spice (100 ± 25), bay leaf (100 ± 50), white pepper (100 ± 25), freeze-dried shallots (100 ± 12.5), taco shell (100 ± 4.5), tostada shell (100 ± 4.5), dry roasted macadamia nuts (100 ± 1.8), boiled yardlong bean (98 ± 0.3), cooked mackerel (98 ± 0.6), stir-fried sprouted soybeans (96 ± 0.5), boiled catjang black-eyed peas (96 ± 0.3), german sweet bakers chocolate (96 ± 1.8), firm raw tofu (94 ± 0.4), sesame butter (93 ± 3.3), dehydrated onion flakes (93 ± 3.6), dried toasted coconut (93 ± 1.8), dried coconut (93 ± 1.8), roasted chinese chestnuts (93 ± 1.8), tamarind (92 ± 0.4), whole wheat bread (92 ± 2), boiled dock (89 ± 0.5), boiled spinach (88 ± 0.6), bran muffins (88 ± 1.3), boiled soybeans (86 ± 0.3), boiled swiss chard (86 ± 0.6), raw whelk (86 ± 0.6), raw chinese chestnuts (86 ± 1.8), prickly pear (85 ± 0.5), dry whole milk (84 ± 1.6), raw halibut (84 ± 0.6), dried acorns (82 ± 1.8), falafel (82 ± 1), boiled hyacinth beans (82 ± 0.3), raw spinach (79 ± 1.8), toasted pumpernickel bread (79 ± 1.8), canned spinach (76 ± 0.5), dried european chestnuts (75 ± 1.8), raw mackerel (75 ± 0.6), boiled yellow beans (74 ± 0.3), raw swamp cabbage (71 ± 0.9), anchovy canned in olive oil (70 ± 2.5), tempeh (70 ± 0.6), boiled black beans (70 ± 0.3), pumpernickel bread (69 ± 1.6), frozen and boiled spinach (68 ± 0.5), boiled beet greens (68 ± 0.7), corn tortilla (67 ± 1.7), boiled purslane (67 ± 0.9), garlic powder (67 ± 16.7), raw chives (67 ± 16.7), raw pollock (67 ± 0.6), raw agar seaweed (67 ± 0.5), raw cassava (66 ± 0.5), boiled pink beans (65 ± 0.3), raw acorns (64 ± 1.8), roasted japanese chestnuts (64 ± 1.8), boiled jute potherb (63 ± 1.2), boiled white beans (63 ± 0.3), boiled mungo beans (63 ± 0.3), chocolate syrup (63 ± 1.3), fried tofu (62 ± 3.8), raw ling (62 ± 0.6), potato chips (61 ± 1.8), teriyaki sauce (61 ± 2.8), parmesan cheese (60 ± 10), boiled broccoli (60 ± 0.6), steamed sprouted soybeans (60 ± 1.1), dried and frozen tofu (59 ± 2.9), dried figs (59 ± 0.3), boiled navy beans (59 ± 0.3), cooked flatfish (59 ± 0.6), boiled okra (58 ± 0.6), breaded and fried oysters (58 ± 0.6), breaded and fried scallops (58 ± 1.6), boiled scotch kale (57 ± 0.8), boiled and steamed chinese chestnuts (57 ± 1.8), cooked kingfish (56 ± 0.5), boiled french beans (56 ± 0.3), boiled pinto beans (56 ± 0.3), boiled and frozen baby lima beans (56 ± 0.6), raw scallops (56 ± 0.6), boiled amaranth (55 ± 0.8), boiled winged beans (55 ± 0.3), raw oysters (55 ± 0.6), fried abalone (55 ± 0.6), carob flour (54 ± 0.5), boiled lupins (54 ± 0.3), smoked haddock (54 ± 0.6), boiled and steamed european chestnuts (54 ± 1.8), canned oysters (54 ± 0.6), dried ginko nuts (54 ± 1.8), boiled baby lima beans (53 ± 0.3), boiled black-eyed peas (53 ± 0.3), boiled succotash (53 ± 0.5), cooked sea bass (53 ± 0.6), rice flour (53 ± 0.4), boiled adzuki beans (52 ± 0.2), raw turbot (52 ± 0.6), roselle (51 ± 0.9), frozen and boiled okra (51 ± 0.5), canned white beans (51 ± 0.2), canned great northern beans (51 ± 0.2), cooked haddock (51 ± 0.6), tomato paste (51 ± 0.4), dried sweetened shredded coconut (51 ± 0.5), cinnamon (50 ± 25), cooked tahitian taro (50 ± 0.7), boiled great northern beans (50 ± 0.3), boiled cranberry beans (50 ± 0.3), raw japanese chestnuts (50 ± 1.8), boiled arrowhead (50 ± 4.2), boiled black turtle beans (49 ± 0.3), raw shark (49 ± 0.6), frozen and boiled black-eyed peas (49 ± 0.6), mixed grain bread (48 ± 2), boiled chickpeas (48 ± 0.3), raw abalone (48 ± 0.6), boiled mung beans (48 ± 0.2), canned navy beans (47 ± 0.2), raw dungeness crab (46 ± 0.6), dried apricots (46 ± 1.4), dried longans (46 ± 0.5), boiled pigeon peas (46 ± 0.3), kippered herring (45 ± 1.3), dried prunes (45 ± 0.6), boiled kidney beans (45 ± 0.3), taco (44 ± 0.6), battered and fried shark (44 ± 0.6), boiled lima beans (44 ± 0.3), dried sweetened flaked coconut (44 ± 2), raw parsley (43 ± 1.7), boiled cardoon (43 ± 0.5), boiled broad beans (43 ± 0.3), miso (42 ± 0.4), baked acorn squash (42 ± 0.5), dried lychees (42 ± 0.5), raw ginger root (42 ± 2.1), cooked cod (42 ± 0.6), raw spot (42 ± 0.6), raw anchovy (41 ± 0.6), cooked herring (41 ± 0.6), breaded and fried croaker (41 ± 0.6), boiled sprouted green peas (41 ± 0.5), cooked clams (41 ± 0.6), canned cod (41 ± 0.6), raw sea bass (41 ± 0.6), burrito (40 ± 0.3), breaded and fried shrimp (40 ± 0.6), tamari sauce (40 ± 0.9), raw croaker (40 ± 0.6), raw california avocado (40 ± 0.3), soybean protein isolate (39 ± 1.8), cooked perch (39 ± 0.6), canned blue crab (39 ± 0.6), cooked rainbow trout (39 ± 0.6), cooked rainbow smelt (39 ± 0.6), raw new zealand spinach (39 ± 1.8), boiled burdock root (39 ± 0.4), refried beans (39 ± 0.2), breaded fried squid (39 ± 0.6), boiled artichoke hearts (39 ± 0.6), raw haddock (39 ± 0.6), boiled green peas (39 ± 0.6), canned lima beans (39 ± 0.2), sardines (38 ± 2.1), devil's food cake (38 ± 0.8), oatmeal cookie (38 ± 3.8), toasted cracked wheat bread (38 ± 2.4), cooked carp (38 ± 0.6), cooked mussels (38 ± 0.6), cooked grouper (38 ± 0.6), dried jujube (37 ± 0.5), jackfruit (37 = 0.5), coconut milk (37 ± 0.2), swiss cheese (36 ± 1.8), raw dandelion greens (36 ± 1.8), cracked wheat bread (36 ± 2), oatmeal bread (36 ± 1.8), raw shrimp (36 ± 0.6), cooked snapper (36 ± 0.6), graham crackers (36 ± 3.6), boiled lentils (36 ± 0.3), boiled split peas (36 ± 0.3), boiled beets (36 ± 0.6), cooked lobster (35 ± 0.6), baked beans with pork and tomato sauce (35 ± 0.2), golden seedless raisins (35 ± 0.5), cooked sheepshead (35 ± 0.6), canned black turtle beans (35 ± 0.2), dried dates (35 ± 0.6), stir-fried sprouted lentils (35 ± 0.5), fried beef top round excluding fat (35 ± 0.5), canned pink salmon (34 ± 0.6), raw blue crab (34 ± 0.6), baked beans with pork and sweet sauce (34 ± 0.2), cooked shrimp (34 ± 0.6), raw mussels (34 ± 0.6), frozen and boiled fordhook lima beans (34 ± 0.6), french fried potatoes (34 ± 1), shoyu sauce (34 = 0.9), french fries (34 ± 0.6), cooked rockfish (34 ± 0.6), raw florida avocado (34 ± 0.2), cooked swordfish (34 ± 0.6), broiled pork sirloin excluding fat (34 ± 0.5), canned tuna in oil (34 ± 0.6), raw whitefish (33 ± 0.6), crab cakes (33 ± 0.8), cooked blue crab (33 ± 0.6), saltine crackers (33 ± 8.3), seedless raisins (33 ± 0.5), enchilada (33 ± 0.2), raw squid (33 ± 0.6), cooked mullet (33 ± 0.6), cooked tilefish (33 ± 0.6), raw green peas (33 ± 0.6), fried beef sirloin excluding fat (33 ± 0.5), raw bluefish (33 ± 0.6), raw turnip greens (32 ± 1.8), raw herring (32 ± 0.6), raw burbot (32 ± 0.6), vegetarian baked beans (32 ± 0.2), boiled new zealand spinach

(32 ± 0.6), cooked pompano (32 ± 0.6), canned cranberry beans (32 ± 0.2), raw snapper (32 ± 0.6), cooked crayfish (32 ± 0.6), raw european chestnuts (32 ± 1.8), roasted european chestnuts (32 ± 1.8), potato pancakes (32 ± 0.7), canned broad beans (32 ± 0.2), dried japanese persimmon (32 ± 1.5), raw sheepshead (32 ± 0.6), raw flatfish (32 ± 0.6), raw seatrout (32 ± 0.6), raw cod (32 ± 0.6), broiled beef sirloin excluding fat (32 ± 0.5), raw cusk (32 ± 0.6), chipped beef (32 ± 1.8), fried pork center loin excluding fat (32 ± 0.5), boiled yellow corn (32 ± 0.6), cooked plantain (32 ± 0.3), frozen and boiled collards (31 ± 0.6), indian buffalo milk (31 ± 0.2), raw perch (31 ± 0.6), raw sunfish (31 ± 0.6), raw freshwater bass (31 ± 0.6), raw white sucker (31 ± 0.6), raw rainbow trout (31 ± 0.6), raw freshwater drum (31 ± 0.6), raw rainbow smelt (31 ± 0.6), boiled swamp cabbage (31 ± 1), raw shad (31 ± 0.6), canned kidney beans (31 ± 0.2), raw grouper (31 ± 0.6), boiled artichoke (31 ± 0.6), fried chicken breast without skin (31 ± 0.6), raw coconut (31 ± 1.1), cooked sockeye salmon (31 ± 0.6), fried beef top round including fat (31 ± 0.5), broiled beef top round excluding fat (31 ± 0.5), raw chickory greens (30 ± 0.6), raw flounder (30 ± 0.5), raw sole (30 ± 0.5), sapote (30 ± 0.2), simmered beef shank excluding fat (30 ± 0.5), seeded raisins (30 ± 0.5), kiwi fruit (30 ± 0.7), cooked flounder (30 ± 0.5), cooked taro (30 ± 0.8), fried chicken breast with skin (30 ± 0.5), broiled pork center rib excluding fat (30 ± 0.5), broiled pork top loin excluding fat (30 ± 0.5), broiled beef tenderloin excluding fat (30 ± 0.5), broiled beef top round including fat (30 ± 0.5), broiled pork center loin excluding fat (30 ± 0.5), canned sockeye salmon (29 ± 0.6), provolone cheese (29 ± 1.8), monterey cheese (29 ± 1.8), edam cheese (29 ± 1.8), cheddar cheese (29 ± 1.8), muenster cheese (29 ± 1.8), gouda cheese (29 ± 1.8), roquefort cheese (29 ± 1.8), toasted raisin bread (29 ± 2.4), hummus (29 ± 0.2), raw carp (29 ± 0.6), boiled butternut squash (29 ± 0.5), boiled parsnips (29 ± 0.6), raw savoy cabbage (29 ± 1.4), canned chickpeas (29 ± 0.2), frozen and boiled green peas (29 ± 0.6), fried light meat chicken without skin (29 ± 0.5), roasted chicken breast without skin (29 ± 0.6), broiled pork loin excluding fat (29 ± 0.5), broiled porterhouse steak excluding fat (29 ± 0.5), broiled T-bone steak excluding fat (29 ± 0.5), roasted pork rump excluding fat (29 ± 0.5), bananas (29 ± 0.4), broiled pork sirloin including fat (29 ± 0.5), ginko nuts (29 ± 1.8), evaporated skim milk (28 ± 1.6), raw mullet (28 ± 0.6), raw celtuce (28 ± 0.5), simmered beef shank including fat (28 ± 0.5), raw tilefish (28 ± 0.6), canned black-eyed peas (28 ± 0.2), roasted turkey light meat without skin (28 ± 0.5), roasted chicken breast with skin (28 ± 0.5), broiled beef sirloin including fat (28 ± 0.5), passion fruit (28 ± 2.8), fried pork center rib excluding fat (28 ± 0.5), fried pork top loin excluding fat (28 ± 0.5), canned bonito (28 ± 0.5), braised pork sirloin excluding fat (28 ± 0.5), broiled pork boston blade excluding fat (28 ± 0.5), broiled beef round excluding fat (28 ± 0.5), toasted rye bread (27 ± 2.3), cooked whiting (27 ± 0.6), raw leeks (27 ± 1.9), breaded and fried catfish (27 ± 0.6), canned pinto beans (27 ± 0.2), raw sprouted alfalfa seeds (27 ± 1.5), raw pompano (27 ± 0.6), frozen and boiled crookneck squash (27 ± 0.5), fried light meat chicken with skin (27 ± 0.5), roasted light meat chicken without skin (27 ± 0.5), broiled beef rib eye excluding fat (27 ± 0.5), fried beef sirloin including fat (27 ± 0.5), microwaved potato with skin (27 ± 0.2), baked potato with skin (27 ± 0.2), broiled beef top loin excluding fat (27 ± 0.5), broiled beef tenderloin including fat (27 ± 0.5), roasted beef tenderloin excluding fat (27 ± 0.5), raw swordfish (27 ± 0.6), roasted beef tip excluding fat (27 ± 0.5), roasted eye of beef round excluding fat (27 ± 0.5),

sweetened condensed milk (26 ± 1.3), cheese pizza (26 ± 0.4), frozen and boiled turnip greens (26 ± 0.6), cornbread (26 ± 0.6), corn pone (26 ± 0.6), pita bread (26 ± 1.3), okara tofu (26 ± 0.8), boiled green beans (26 ± 0.8), boiled yellow snap beans (26 ± 0.8), roasted turkey light meat with skin (26 ± 0.5), raw lingcod (26 ± 0.6), raw rockfish (26 ± 0.6), roasted pork rump including fat (26 ± 0.5), fried pork center loin including fat (26 ± 0.5), colby cheese (25 ± 1.8), brick cheese (25 ± 1.8), bleu cheese (25 ± 1.8), evaporated lowfat milk (25 ± 0.4), raw coriander (25 ± 12.5), chocolate cream pie (25 ± 0.4), black currants (25 ± 0.9), italian bread (25 ± 1.8), raw broccoli (25 ± 1.1), roast beef sandwich (25 ± 0.3), raw catfish (25 ± 0.6), boiled acorn squash (25 ± 0.4), pretzels (25 ± 1.8), coconut water (25 ± 0.2), raw crayfish (25 ± 0.6), fried chicken giblets (25 ± 0.5), fried dark meat chicken without skin (25 ± 0.5), breadfruit (25 ± 0.5), raw pepeao (25 ± 0.5), roasted light meat chicken with skin (25 ± 0.5), roasted goose without skin (25 ± 0.5), broiled pork loin blade excluding fat (25 ± 0.5), broiled pork center rib including fat (25 ± 0.5), fried chicken (25 ± 0.6), stewed chicken drumstick without skin (25 ± 1.1), roasted beef rib roasted excluding fat (25 ± 0.5), broiled beef rib excluding fat (25 ± 0.5), well done broiled extra lean ground beef (25 ± 0.5), roasted pork tenderloin (25 ± 0.5), broiled porterhouse steak including fat (25 ± 0.5), broiled pork loin including fat (25 ± 0.5), roasted pork leg excluding fat (25 ± 0.5), roasted pork shank excluding fat (25 ± 0.5), braised pork heart (25 ± 0.4), broiled beef flank excluding fat (25 ± 0.5), simmered beef heart (25 ± 0.5), roasted beef round tip including fat (25 ± 0.5), roasted eye of beef round including fat (25 ± 0.5), braised beef bottom round excluding fat (25 ± 0.5), microwaved potato without skin (25 ± 0.3), baked potato without skin (25 ± 0.3), canned yellow corn with red and green peppers (25 ± 0.4), broiled pork center loin including fat (25 ± 0.5), evaporated whole milk (24 ± 0.4), toasted english muffin (24 ± 1), toasted white bread (24 ± 2.4), raw watercress (24 ± 2.9), rye bread (24 ± 2), braised pork spareribs (24 ± 0.5), raw goose liver (24 ± 0.5), roasted turkey dark meat without skin (24 ± 0.5), boiled crookneck squash (24 ± 0.6), fried chicken back with skin (24 ± 0.7), frozen and boiled brussels sprouts (24 ± 0.6), marinara sauce (24 ± 0.2), raw red chili peppers (24 ± 1.1), raw green chili peppers (24 ± 1.1), fried dark meat chicken with skin (24 ± 0.5), french rolls (24 ± 1), poi (24 ± 0.4), tomato purée (24 ± 0.2), fried pork loin blade excluding fat (24 ± 0.5), broiled beef rib eye including fat (24 ± 0.5), cooked bacon (24 ± 0.4), well done broiled lean ground beef (24 ± 0.5), broiled pork top loin including fat (24 ± 0.5), braised pork loin excluding fat (24 ± 0.5), roasted pork sirloin excluding fat (24 ± 0.5), braised beef chuck arm pot roast excluding fat (24 ± 0.5), broiled beef top loin including fat (24 ± 0.5), broiled T-bone steak including fat (24 ± 0.5), oatmeal (24 ± 0.3), well done fried extra lean ground beef (24 ± 0.5), broiled beef round including fat (24 ± 0.5), broiled beef flank including fat (24 ± 0.5), braised beef flank excluding fat (24 ± 0.5), raw sockeye salmon (24 ± 0.6), broiled pork boston blade including fat (24 ± 0.5), waffles (23 ± 0.7), bacon cheeseburger (23 ± 0.3), roasted turkey dark meat with skin (23 ± 0.5), boiled savoy cabbage (23 ± 0.7), canned pumpkin (23 ± 0.4), canned sweet potato (23 ± 0.3), raw scallop squash (23 ± 0.8), braised pork pancreas (23 ± 0.5), roasted dark meat chicken without skin (23 ± 0.5), roasted chicken thigh with skin (23 ± 0.8), braised beef chuck blade roast excluding fat (23 ± 0.5), stewed chicken breast without skin (23 ± 0.5), roasted chicken drumstick with skin (23 ± 1), roasted chicken thigh without skin (23 ± 1), braised pork center rib excluding fat (23 ± 0.5), braised pork top loin excluding fat (23 ± 0.5), fried beef liver (23 ± 0.5), well done fried lean ground beef (23 ± 0.5), raw chinese water chestnuts (23 ± 0.8), roasted beef tenderloin including fat (23 ± 0.5), braised beef flank including fat (23 ± 0.5), maltex (23 ± 0.3), braised beef brisket excluding fat (23 ± 0.5), braised pork sirloin including fat (23 ± 0.5), braised beef bottom round including fat (23 ± 0.5), canned yellow corn (23 ± 0.5), raw garlic (22 ± 5.6), boiled turnip greens (22 ± 0.7), hot chocolate (22 ± 0.2), frozen and boiled yellow snap beans (22 ± 0.7), frozen and boiled green beans (22 ± 0.7), raw lotus root (22 ± 0.6), raw trout (22 ± 0.6), raw scup (22 ± 0.6), boiled lotus root (22 ± 0.6), loganberries (22 ± 0.3), raw crookneck squash (22 ± 0.8), smoked whitefish (22 ± 0.6), baked hubbard squash (22 ± 0.5), roasted dark meat chicken with skin (22 ± 0.5), frozen and boiled succotash (22 ± 0.6), raw zucchini (22 ± 0.8), well done fried ground beef (22 ± 0.5), roasted goose with skin (22 ± 0.5), stewed light meat chicken without skin (22 ± 0.5), stewed chicken breast with skin (22 ± 0.5), simmered turkey heart (22 ± 0.5), breaded and fried fish (22 ± 0.7), well done broiled ground beef (22 ± 0.5), fried chicken drumstick with skin (22 ± 1), restaurant hash brown potatoes (22 ± 0.6), braised beef shortribs excluding fat (22 ± 0.5), roasted pork loin excluding fat (22 ± 0.5), well done baked extra lean ground beef (22 ± 0.5), roasted pork sirloin including fat (22 ± 0.5), roasted ham (22 ± 0.5), braised pork arm excluding fat (22 ± 0.5), braised pork boston blade excluding fat (22 ± 0.5), frozen and french fried potatoes without skin (22 ± 1), roasted pork leg including fat (22 ± 0.5), braised pork center loin excluding fat (22 ± 0.5), chopped smoked beef (21 ± 1.8), caraway cheese (21 ± 1.8), cheshire cheese (21 ± 1.8), pimento cheese spread (21 ± 1.8), limburger cheese (21 ± 1.8), american cheese (21 ± 1.8), camembert cheese (21 ± 1.8), white bread (21 ± 2.1), dinner rolls (21 ± 1.8), canned turnip greens (21 ± 0.4), pork and beef old fashioned loaf (21 ± 1.8), french bread (21 ± 1.8), pork and beef peppered loaf (21 ± 1.8), hot fudge sundae (21 ± 0.3), maypo (21 ± 0.3), frozen and boiled broccoli (21 ± 0.5), raw whiting (21 ± 0.6), boiled rutabaga (21 ± 0.6), pork luxury loaf (21 ± 1.8), boiled brussels sprouts (21 ± 0.6), sugar apple (21 ± 0.3), braised beef pancreas (21 ± 0.5), roasted chicken wing with skin (21 ± 1.5), simmered chicken liver (21 ± 0.5), raw sprouted mung beans (21 ± 1), boiled zucchini (21 ± 0.6), well done baked lean ground beef (21 ± 0.5), grilled canadian bacon (21 ± 1.1), medium-rare broiled lean ground beef (21 ± 0.5), broiled pork loin blade including fat (21 ± 0.5), roasted pork center rib excluding fat (21 ± 0.5), roasted pork top loin excluding fat (21 ± 0.5), hash brown potatoes (21 ± 0.6), raw monkfish (21 ± 0.6), medium-rare broiled extra lean ground beef (21 ± 0.5), medium-rare fried extra lean ground beef (21 ± 0.5), fried pork center rib including fat (21 ± 0.5), fried pork top loin including fat (21 ± 0.5), raw potato without skin (21 ± 0.4), puffed rice (21 ± 3.6), roasted canned lean ham (21 ± 0.5), roasted pork center loin excluding fat (21 ± 0.5), roasted pork shank including fat (21 ± 0.5), hamburger bun (20 ± 1.3), hot dog bun (20 ± 1.3), malted milk (20 ± 0.2), cheeseburger (20 ± 0.4), au gratin potatoes (20 ± 0.4), cheese crackers (20 ± 3.3), raw chinese cabbage (20 ± 1.4), instant oatmeal (20 ± 0.3), raw spring onions with tops (20 ± 1), hamburger sandwich (20 ± 0.5), bagels (20 ± 0.9), fish sandwich (20 ± 0.3), frozen and baked sweet potato (20 ± 0.6), baked sweet potato (20 ± 0.4), chili (20 ± 0.2), boiled kohlrabi (20 ± 0.6), cooked prunes (20 ± 0.5), braised pork tongue (20 ± 0.5), simmered chicken heart (20 ± 0.5), tuna salad (20 ± 0.2), braised pork loin blade excluding fat (20 ± 0.5), stewed dark meat chicken without skin (20 ± 0.5), soursop (20 ± 0.2), stewed light meat chicken with skin (20 ± 0.5), raw pheasant without skin (20 ± 0.5), roasted duck without skin (20 ± 0.5), simmered chicken giblets (20 ± 0.5), medium-rare fried ground beef (20 ± 0.5), medium-rare broiled ground beef (20 ± 0.5), broiled beef rib including fat (20 ± 0.5), hard pork salami (20 ± 5), hard pork and beef salami (20 ± 5), medium-rare fried lean ground beef (20 ± 0.5), simmered chicken gizzard (20 ± 0.5), braised pork center rib including fat (20 ± 0.5), roasted pork arm excluding fat (20 ± 0.5), boiled dishcloth gourd (20 ± 0.6), roasted pork shoulder excluding fat (20 ± 0.5), braised pork loin including fat (20 ± 0.5), roasted pork boston blade excluding fat (20 ± 0.5), braised beef liver (20 ± 0.5), skim yogurt (19 ± 0.2), english muffin (19 ± 0.9), eggnog (19 ± 0.2), scalloped potatoes (19 ± 0.4), light meat chicken roll (19 ± 0.5), pork sausage (19 ± 1.9), blackberries (19 ± 0.7), fried chicken neck with skin (19 ± 1.4), pork link sausage (19 ± 0.7), onion rings (19 ± 0.6), sweet roll (19 ± 1.2), boiled asparagus (19 ± 0.6), fried chicken wing with skin (19 ± 1.6), boiled scallop squash (19 ± 0.6), simmered turkey gizzard (19 ± 0.5), tomato sauce (19 ± 0.4), well done baked ground beef (19 ± 0.5), stewed chicken drumstick with skin (19 ± 0.9), stewed chicken thigh with skin (19 ± 0.7), braised beef spleen (19 ± 0.5), canned succotash (19 ± 0.4), roasted beef rib including fat (19 ± 0.5), fried pork loin blade including fat (19 ± 0.5), braised beef chuck arm pot roast including fat (19 ± 0.5), braised pork top loin including fat (19 ± 0.5), roasted pork center rib including fat (19 ± 0.5), roasted pork loin including fat (19 ± 0.5), unheated ham (19 ± 0.5), braised pork boston blade braised including fat (19 ± 0.5), boiled potato without skin (19 ± 0.4), braised pork center loin including fat (19 ± 0.5), turkey breast meat summer sausage (19 ± 2.4), roasted pork center loin including fat (19 ± 0.5), soy milk (19 ± 0.2), havarti cheese (18 ± 1.8), mozzarella cheese (18 ± 1.8), feta cheese (18 ± 1.8), ham and cheese spread (18 ± 1.8), raw frozen rhubarb (18 ± 0.4), sheep milk (18 ± 0.2), toaster pastry (18 ± 1), low-fat yogurt (18 ± 0.2), chicken egg yolk (18 ± 2.9), frozen and boiled kale (18 ± 0.8), turkey frankfurter (18 ± 1.1), caramel sundae (18 ± 0.3), whole chocolate malted milk (18 ± 0.2), pork olive loaf (18 ± 1.8), french toast (18 ± 0.8), pork pickle and pimento loaf (18 ± 1.8), boiled kale (18 ± 0.8), ham and cheese roll (18 ± 1.8), boiled salsify (18 ± 0.7), chicken roll (18 ± 0.9), mulberries (18 ± 0.4), o'brien potatoes (18 ± 0.3), light and dark meat turkey roll (18 ± 0.9), mashed potatoes (18 ± 0.5), italian pork sausage (18 ± 0.7), raspberries (18 ± 0.4), canned green peas (18 ± 0.6), pork and beef honey loaf (18 ± 1.8), hot dog pickle relish (18 ± 1.8), simmered beef kidneys (18 ± 0.5), turkey summer sausage (18 ± 1.8), stewed dark meat chicken with skin (18 ± 0.5), braised pork kidneys (18 ± 0.5), braised beef chuck blade roast including fat (18 ± 0.5), baked/boiled tropical yam (18 ± 0.7), raw acerola (18 ± 0.5), smoked chinook salmon (18 ± 0.6), boiled and steamed japanese chestnuts (18 ± 1.8), roasted pork top loin including fat (18 ± 0.5), noodles (18 ± 0.3), unheated canadian bacon (18 ± 0.9), turkey ham (18 ± 0.9), braised beef brisket including fat (18 ± 0.5), macaroni (18 ± 0.4), spaghetti (18 ± 0.4), beef pastrami (18 ± 1.8), braised pork arm braised including fat (18 ± 0.5), roasted pork shoulder including fat (18 ± 0.5), roasted pork boston blade including fat (18 ± 0.5), frozen and boiled yellow corn (18 ± 0.6), orange peel (17 ± 8.3), lemon peel (17 ± 8.3), restaurant chocolate milkshake (17 ± 0.2), strawberry sundae (17 ± 0.3), pineapple upside-down cake (17 ± 0.7), duck egg (17 ± 0.7), barbecue loaf (17 ± 2.2), simmered chicken neck without skin (17 ± 2.8), pork and beef pepperoni (17 ± 8.3), boiled yambean (17 ± 0.5), frozen and fried hash brown potatoes (17 ± 0.6), braised pork loin blade including fat (17 ± 0.5), roasted pork loin blade excluding fat (17 ± 0.5), raw jerusalem artichoke (17 ± 0.7), lebanon bologna (17 ± 2.2), simmered turkey giblets (17 ± 0.5), beef honey roll sausage (17 ± 2.2), medium-rare baked lean ground beef (17 ± 0.5), roasted canned ham (17 ± 0.5), roasted pork arm including fat (17 ± 0.5), unheated lean ham (17 ± 0.5), simmered beef tongue (17 ± 0.5), medium-rare baked extra lean ground beef (17 ± 0.5), canned lean ham (17 ± 0.5), yellow passion fruit juice (17 ± 0.2), canned creamed yellow corn (17 ± 0.4), protein fortified lowfat 2% milk (16 ± 0.2), protein fortified skim milk (16 ± 0.2), protein fortified lowfat 1% milk (16 ± 0.2), choco- late milkshake (16 ± 0.2), raisin bread (16 ± 2), raw endive (16 ± 2), pork and beef brotwurst (16 ± 0.7), light meat turkey roll (16 ± 0.9), canned pumpkin pie mix (16 ± 0.4), figs (16 ± 1), hot dog (16 ± 0.6), raw oriental radish (16 ± 1.1), boysenberries (16 ± 0.4), smoked cisco (16 ± 0.6), java plum (16 ± 0.4), roasted cured pork arm excluding fat (16 ± 0.5), pickled beets (16 ± 0.4), roasted duck with skin (16 ± 0.5), yellow cake (15 ± 0.7), pork and beef kielbasa/kolbassy (15 ± 1.9), yellow corn pudding (15 ± 0.2), boiled green cabbage (15 ± 0.7), boiled leeks (15 ± 1.9), raw carrots (15 ± 0.7), potato salad (15 ± 0.4), turkey salad (15 ± 0.9), stewed chicken wing with skin (15 ± 1.3), roasted pork loin blade including fat (15 ± 0.5), braised beef shortribs including fat (15 ± 0.5), simmered turkey liver (15 ± 0.5), cooked sprouted mung beans (15 ± 0.8), medium-rare baked ground beef (15 ± 0.5), carissa (15 ± 2.5), fried beef brains (15 ± 0.5), canned corned beef (14 ± 0.5), tilsit cheese (14 ± 1.8), fontina cheese (14 ± 1.8), goat milk (14 ± 0.2), lowfat 1% milk (14 ± 0.2), lowfat 2% milk (14 ± 0.2), whole milk (14 ± 0.2), white cake (14 ± 0.6), danish pastry (14 ± 1.2), turkey bologna (14 ± 1.8), boiled mustard greens (14 ± 0.7), breaded and fried clams (14 ± 0.6), pancakes with butter and syrup (14 ± 0.3), raw red cabbage (14 ± 1.4), pork and beef picnic loaf (14 ± 1.8), raw green cabbage (14 ± 1.4), pork bratwurst (14 ± 0.6), restaurant coleslaw (14 ± 0.5), pork mothers loaf (14 ± 2.4), gingersnaps cookie (14 ± 7.1), frozen and boiled turnips (14 ± 0.5), raw cauliflower (14 ± 1), carrot juice (14 ± 0.3), turkey salami (14 ± 0.9), canned zucchini in tomato juice (14 ± 0.4), pineapple juice (14 ± 0.2), prune juice (14 ± 0.2), beef lunch meat (14 ± 1.8), pork and beef lunch meat (14 ± 1.8), polish pork sausage (14 ± 1.8), minced ham lunch meat (14 ± 2.4), roasted cured pork arm including fat (14 ± 0.5), braised pork liver (14 ± 0.5), boiled hubbard squash (14 ± 0.4), smoked beef sausage (14 ± 1.2), turkey pastrami (14 ± 0.9), simmered beef brains (14 ± 0.5), roasted lean ham (14 ± 0.5), boiled red tomato (14 ± 0.4), pineapple (14 ± 0.3), canned ham (14 ± 0.5), raw red sweet peppers (14 ± 1), raw green sweet peppers (14 ± 1), chopped ham lunch meat (14 ± 2.4), cooked shitake mushrooms (14 ± 0.7), raw chickory (13 ± 1.1), frozen and cooked sweetened rhubarb (13 ± 0.4), filled milk (13 ± 0.2), lowfat 1% chocolate milk (13 ± 0.2), lowfat 2% chocolate milk (13 ± 0.2), carob milk (13 ± 0.2), restaurant strawberry milkshake (13 ± 0.2), whole chocolate milk (13 ± 0.2), half and half cream (13 ± 3.3), frozen and boiled mustard greens (13 ± 0.7), raw celery (13 ± 1.3), red currants (13 ± 0.9), white currants (13 ± 0.9), canned sauerkraut (13 ± 0.4), boiled carrots (13 ± 0.6), canned green beans (13 ± 0.7), canned yellow snap beans (13 ± 0.7), simmered chicken neck with skin (13 ± 1.3), shortbread cookie (13 ± 6.3), frozen and boiled asparagus (13 ± 0.8), pork and beef mortadella (13 ± 3.3), frozen and boiled zucchini (13 ± 0.4), loquats (13 ± 0.5), berliner (13 ± 2.2), summer sausage (13 ± 2.2), pork bologna (13 ± 2.2), pork and beef luncheon sausage (13 ± 2.2), beef and pork salami (13 ± 2.2), beef and pork bologna (13 ± 2.2), beef bologna (13 ± 2.2), canned crookneck squash (13 ± 0.5), beef and pork summer sausage (13 ± 2.2), beef summer sausage (13 ± 2.2), pork beerwurst salami (13 ± 2.2), beef salami (13 ± 2.2), beef beerwurst salami (13 ± 2.2), red table wine (13 ± 0.5), roasted pork blade roll (13 ± 0.5), canned potato without skin (13 ± 0.6), boiled eggplant (13 ± 1), boiled mushrooms (13 ± 0.6), restaurant vanilla milkshake (12 ± 0.2), vanilla milkshake (12 ± 0.2), frozen and boiled turnip greens and turnips (12 ± 0.5), chicken egg (12 ± 1), pork and beef link sausage with american cheese (12 ± 1.2), raw eggplant (12 ± 1.2), boiled celery (12 ± 0.7), canned stewed red tomatoes (12 ± 0.4), canned peeled red tomatoes (12 ± 0.4), canned jalapeño peppers (12 ± 0.7), raw celeriac (12 ± 0.5), restaurant hot chocolate (12 ± 0.2), cherries (12 ± 0.7), tangerine (12 ± 0.6), raw cucumber (12 ± 1), angel food cake (12 ± 0.8), pork and beef knockwurst (12 ± 0.7), pork and beef link sausage (12 ± 0.7), acerola juice (12 ± 0.2), braised pork brains (12 ± 0.5), pitanga (12 ± 0.3), white grapefruit juice (12 ± 0.2), red grapefruit juice (12 ± 0.2), cooked corned beef brisket (12 ± 0.5), braised pork lungs (12 ± 0.5), ricotta cheese (11 ± 0.4), skim milk (11 ± 0.2), whole yogurt (11 ± 0.2), buttermilk (11 ± 0.2), boiled chinese cabbage (11 ± 0.6), boiled collards (11 ± 0.3), raw looseleaf lettuce (11 ± 1.8), cheesecake (11 ± 0.6), kumquats (11 ± 2.6), navel orange (11 ± 0.4), sponge cake (11 ± 0.8), boiled red cabbage (11 ± 0.7), boiled cauliflower (11 ± 0.8), candied sweet potato (11 ± 0.5), boiled calabash gourd (11 ± 0.7), pork headcheese (11 ± 1.8), strawberries (11 ± 0.3), clam liquid (11 ± 0.2), raw green tomato (11 ± 0.4), boiled chayote (11 ± 0.6), pork liver sausage (11 ± 2.8), beef and pork frankfurter (11 ± 1.1), jellied corned beef loaf (11 ± 1.8), cantaloupe (11 ± 0.3), mandarin oranges (11 ± 0.4), orange juice (11 ± 0.2), restaurant tomato juice (11 ± 0.4), tomato juice (11 ± 0.3), white table wine (11 ± 0.5), pork livercheese (11 ± 1.3), watermelon (11 ± 0.3), ham salad (11 ± 1.8), frozen and boiled potatoes without skin (11 ± 0.5), raw red tomato (11 ± 0.4), raw mushrooms (11 ± 1.4), canned red sweet peppers (11 ± 0.7), canned green sweet peppers (11 ± 0.7), corn flakes (11 ± 1.8), coleslaw (10 ± 0.8), valencia orange (10 ± 0.4), frozen and boiled carrots (10 ± 0.7), boiled onions (10 ± 0.5), simmered pork chitterlings (10 ± 0.5), raw onions (10 ± 0.6), gooseberries (10 ± 0.3), gefiltefish with broth (10 ± 1.2), papaya (10 ± 0.2), boiled/baked spaghetti squash (10 ± 0.6), boiled sweet potato (10 ± 0.3), jujube (10 ± 0.5), raw iceberg lettuce (10 ± 2.5), guava (10 ± 0.6), braised beef lungs (10 ± 0.5), rose table wine (10 ± 0.5), restaurant orange juice (10 ± 0.4), grilled ham patties (10 ± 0.8), steamed hawaii mountain yam (10 ± 0.7), canned pork lunch meat (10 ± 2.4), lychees (10 ± 0.5), boiled red sweet peppers (10 ± 0.7), boiled green sweet peppers (10 ± 0.7), longans (10 ± 0.5), raw clams (9 ± 0.6), raw radish (9 ± 1.1), frozen and boiled butternut squash (9 ± 0.4), frozen and boiled cauliflower (9 ± 0.6), boiled oriental radish (9 ± 0.7), boiled pumpkin (9 ± 0.4), chicken egg white (9 ± 1.5), white grapefruit (9 ± 0.4), cooked wild long grain rice (9 ± 0.4), mango (9 ± 0.2), grape juice (9 ± 0.2), unheated ham patties (9 ± 0.8), japanese persimmon (9 ± 0.3), raw bacon (9 ± 0.7), carambola (9 ± 0.4), pork and beef sausage (8 ± 3.8), raw beef tripe (8 ± 0.5), sour cream (8 ± 4.2), boiled butterbur (8 ± 0.5), whey (8 ± 0.2), frozen and boiled onions (8 ± 0.5), canned carrots (8 ± 0.7), boiled turnips (8 ± 0.6), tangerine juice (8 ± 0.2), canned sprouted mung beans (8 ± 0.8), apricots (8 ± 0.5), fried chicken thigh with skin (8 ± 0.8), quince (8 ± 0.5), pink grapefruit (8 ± 0.4), apple pie (8 ± 0.4), dry dessert wine (8 ± 0.8), sweet dessert wine (8 ± 0.8), screwdriver (8 ± 0.2), casaba melon (8 ± 0.3), nectarine (8 ± 0.4), medium cream (7 ± 3.3), cream cheese (7 ± 1.8), pickled herring (7 ± 3.3), neufchatel cheese (7 ± 1.8), whipping cream, heavy (7 ± 3.3), whipping cream, light (7 ± 3.3), cottage cheese (7 ± 1.8), raw romaine lettuce (7 ± 1.8), simmered pork ears (7 ± 0.5), simmered pork tail (7 ± 0.5), pork and beef spread (7 ± 1.8), frozen and boiled red sweet peppers (7 ± 0.5), frozen and boiled green sweet peppers (7 ± 0.5), hamburger pickle relish (7 ± 1.8), bloody mary (7 ± 0.3), lemon juice (7 ± 0.2), tequila sunrise (7 ± 0.3), peach (7 ± 0.6), restaurant coffee (7 ± 0.3), barbecue sauce (6 ± 3.1), crab apples (6 ± 0.5), vienna sausage (6 ± 3.1), lemon meringue pie (6 ± 0.4), european grapes (6 ± 0.3), thompson seedless grapes (6 ± 0.3), lime juice (6 ± 0.2), oheloberries (6 ± 0.4), beer (6 ± 0.1), pummelo (6 ± 0.3), plum (6 ± 0.8), coffee (6 ± 0.3), simmered pork feet (5 ± 0.5), rose apple (5 ± 0.5), american grapes (5 ± 0.5), pear (5 ± 0.3), canned bamboo shoots (5 ± 0.4), apricot nectar (5 ± 0.2), cranberry sauce (5 ± 0.5), blueberries (5 ± 0.3), soy sauce (5 ± 0.9), corn grits (5 ± 0.2), pickled pork feet (4 ± 0.5), cream of wheat (4 ± 0.3), beef frankfurter (4 ± 0.9), tuna salad spread (4 ± 0.9), raw apple with skin (4 ± 0.4), whiskey sour (4 ± 0.6), instant corn grits (4 ± 0.4), peach nectar (4 ± 0.2), canned chinese water chestnuts (4 ± 0.7), light cream (3 ± 1.7), human milk (3 ± 1.6), raw bamboo shoots (3 ± 0.7), boiled bamboo shoots (3 ± 0.4), papaya nectar (3 ± 0.2), apple juice (3 ± 0.2), cooked apple without skin (3 ± 0.3), raw pork jowl (3 ± 0.5), applesauce (3 ± 0.4), raw apple without skin (3 ± 0.4), cranberry juice cocktail (3 ± 0.2), cream of rice (3 ± 0.3), black tea (3 ± 0.3), raw laver/ nori seaweed (2 ± 0.5), pear nectar (2 ± 0.2), restaurant cranberry juice cocktail (2 ± 0.3), daiquiri (2 ± 0.8), 53 proof coffee liqueur (2 ± 1), 63 proof coffee liqueur (2 ± 1), manhattan (2 ± 0.9), instant tea (2 ± 0.2), farina (2 ± 0.3), Perrier water (1 ± 0.3), canned succotash with creamed corn (1 ± 0.4), tom collins (1 ± 0.2), orange soda (1 ± 0.1), bourbon & soda (1 ± 0.4), cola (1 ± 0.1), lemon-lime soda (1 ± 0.1), gin & tonic (1 ± 0.2), chamomile tea (1 ± 0.3), martini (1 ± 0.7), tea (1 ± 0.2)

Applications: hardening of the arteries, atherosclerosis, increased cholesterol, diabetes, high blood pressure, fracture, thinning of bones, rickets, colitis, diarrhea, alcohol cravings, epilepsy, mental illness, multiple sclerosis, nervousness, neuritis, Parkinson's disease, celiac disease, arthritis, kidney stones, kidney inflammation, leg cramp, exaggerated muscular response, skin psoriasis, nausea, backache, kwashiorkor, weight increased

Daily Dosage: (at least 50 mg per 100 mg calcium) 1974 PCS — 100–600 mg; 1980 RDA — 300–400 mg; 1980 USRDA — 400 mg; 1989 RDA — 280–400 mg; 1989 USRDA — 400 mg; nutritional — 300–1000 mg; therapeutic — 1000–1500 mg; experimental — 2000–3000 mg; toxic — 3000 mg +

Warnings: Do not take over 3000 mg daily with kidney malfunction. Increased amounts are dangerous. See toxicity symptoms.

MANGANESE

Names: manganese, Mn, element 25
Classification: trace mineral

Deficiency: retarded growth, decreased reproduction ability, deterioration of

MANGANESE

glucose tolerance, decreased blood clotting, bone deformities, weight decreased, skin inflammation, nausea, slow hair growth, changed hair color, decreased levels of blood cholesterol, decreased muscular coordination, dizziness, ear noises, hearing decreased
Side-effects: see *toxicity* symptoms
Toxicity: muscle tremors, mobility decreased, muscle stiffness, decreased facial expression, weakness, decreased iron utilization
Inhibitors: unknown
Helpers: vitamin B complex, vitamin B_1, vitamin E, calcium, phosphorus
Sources (mg per 100 grams of food): toasted wheat germ (20.207 ± 0.0018), dried butternuts (6.654 ± 0.0018), peanut flour (4.9 ± 0.0125), dry roasted pecans (4.775 ± 0.0018), oil roasted pecans (4.621 ± 0.0018), dried pecans (4.571 ± 0.0018), dried black walnuts (4.332 ± 0.0018), dried agar seaweed (4.3 ± 0.0005), soybean protein concentrate (4.189 ± 0.0018), defatted soy meal (3.8 ± 0.0004), dried japanese chestnuts (3.764 ± 0.0018), dried and frozen tofu (3.688 ± 0.0029), low fat soybean flour (3.08 ± 0.0006), defatted soybean flour (3.018 ± 0.0005), dried english/persian walnuts (2.939 ± 0.0018), dried toasted coconut (2.839 ± 0.0018), dried coconut (2.786 ± 0.0018), dried chinese chestnuts (2.639 ± 0.0018), whole toasted sesame seeds (2.532 ± 0.0018), dried sweetened shredded coconut (2.475 ± 0.0005), whole dried sesame seeds (2.456 ± 0.0056), whole wheat bread (2.376 ± 0.002), almond butter (2.356 ± 0.0031), oil roasted spanish peanuts (2.354 ± 0.0018), dried almonds (2.307 ± 0.0018), soybean flour (2.275 ± 0.0006), dry roasted soybean nuts (2.184 ± 0.0006), roasted soybean nuts (2.158 ± 0.0006), dry roasted sunflower seeds (2.139 ± 0.0018), oil roasted filberts (2.136 ± 0.0018), dry roasted filberts (2.121 ± 0.0018), oil roasted sunflower seeds (2.111 ± 0.0018), roasted japanese chestnuts (2.093 ± 0.0018), dry roasted peanuts (2.082 ± 0.0018), roasted soybean flour (2.076 ± 0.0006), dried sunflower seeds (2.05 ± 0.0018), dried filberts (2.046 ± 0.0018), almond paste (2.039 ± 0.0018), toasted almonds (2.032 ± 0.0018), oil roasted virginia peanuts (2.007 ± 0.0018), dry roasted almonds (2.004 ± 0.0018), oil roasted almonds (2.004 ± 0.0018), tomato powder (1.951 ± 0.0005), toasted cracked wheat bread (1.919 ± 0.0024), freeze-dried red sweet peppers (1.875 ± 0.0313), freeze-dried green sweet peppers (1.875 ± 0.0313), crunchy peanut butter (1.866 ± 0.0016), oil roasted valencia peanuts (1.721 ± 0.0018), pineapple (1.649 ± 0.0003), raw chinese chestnuts (1.625 ± 0.0018), raw japanese chestnuts (1.614 ± 0.0018), creamy peanut butter (1.606 ± 0.0031), natto (1.528 ± 0.0006), puffed rice (1.521 ± 0.0036), wheat flakes (1.518 ± 0.0018), raw blue crab (1.506 ± 0.0006), raw coconut (1.5 ± 0.0011), soybean protein isolate (1.493 ± 0.0018), fried tofu (1.492 ± 0.0038), tempeh (1.43 ± 0.0006), mixed grain bread (1.4 ± 0.002), raw wakame seaweed (1.4 ± 0.0005), dried japanese persimmon (1.391 ± 0.0015), dehydrated onion flakes (1.386 ± 0.0036), freeze-dried parsley (1.357 ± 0.0357), dried european chestnuts (1.318 ± 0.0018), coconut cream (1.304 ± 0.0002), blackberries (1.292 ± 0.0007), freeze-dried shallots (1.275 ± 0.0125), oil roasted peanuts (1.254 ± 0.0018), loganberries (1.247 ± 0.0003), boiled winged beans (1.199 ± 0.0003), roasted european chestnuts (1.196 ± 0.0018), firm raw tofu (1.181 ± 0.0004), stir-fried sprouted soybeans (1.133 ± 0.0005), dried peanuts (1.129 ± 0.0018), boiled and steamed chinese chestnuts (1.114 ± 0.0018), boiled chickpeas (1.03 ± 0.0003), boiled peanuts (1.022 ± 0.0016), frozen and boiled okra (1.021 ± 0.0005), raspberries (1.013 ± 0.0004), pineapple juice (0.99 ± 0.0002), raw laver/nori seaweed (0.988 ± 0.0005), raw european chestnuts (0.964 ± 0.0018), frozen and

boiled spinach (0.942 ± 0.0005), boiled spinach (0.936 ± 0.0006), coconut milk (0.916 ± 0.0002), boiled okra (0.911 ± 0.0006), raw spinach (0.896 ± 0.0018), raw freshwater bass (0.889 ± 0.0006), boiled and steamed european chestnuts (0.868 ± 0.0018), oatmeal bread (0.861 ± 0.0018), miso (0.859 ± 0.0004), raw trout (0.851 ± 0.0006), boiled soybeans (0.824 ± 0.0003), oil roasted cashews (0.818 ± 0.0018), sweetened coconut cream (0.815 ± 0.0002), boiled and frozen baby lima beans (0.813 ± 0.0006), frozen and boiled black-eyed peas (0.791 ± 0.0006), dried brazil nuts (0.786 ± 0.0018), boiled carrots (0.753 ± 0.0006), american grapes (0.718 ± 0.0005), steamed sprouted soybeans (0.711 ± 0.0011), raw rainbow smelt (0.7 ± 0.0006), raw sunfish (0.7 ± 0.0006), raw rainbow trout (0.7 ± 0.0006), raw burbot (0.7 ± 0.0006), raw freshwater drum (0.7 ± 0.0006), falafel (0.69 ± 0.001), frozen and baked sweet potato (0.665 ± 0.0006), frozen and boiled collards (0.664 ± 0.0006), canned succotash with creamed corn (0.645 ± 0.0004), boiled white beans (0.636 ± 0.0003), raw tofu (0.605 ± 0.0004), canned chickpeas (0.604 ± 0.0002), canned spinach (0.597 ± 0.0005), red table wine (0.597 ± 0.0005), raw white sucker (0.588 ± 0.0006), boiled and steamed japanese chestnuts (0.586 ± 0.0018), boiled baby lima beans (0.585 ± 0.0003), oatmeal (0.585 ± 0.0003), boiled adzuki beans (0.573 ± 0.0002), hummus (0.569 ± 0.0002), baked sweet potato (0.56 ± 0.0004), boiled pinto beans (0.556 ± 0.0003), boiled navy beans (0.556 ± 0.0003), french bread (0.55 ± 0.0018), boiled pink beans (0.548 ± 0.0003), boysenberries (0.547 ± 0.0004), boiled mothbeans (0.527 ± 0.0003), boiled green peas (0.525 ± 0.0006), boiled great northern beans (0.518 ± 0.0003), boiled lima beans (0.516 ± 0.0003), canned white beans (0.515 ± 0.0002), carob flour (0.508 ± 0.0005), raisin bread (0.504 ± 0.002), stir-fried sprouted lentils (0.502 ± 0.0005), boiled pigeon peas (0.501 ± 0.0003), raw clams (0.5 ± 0.0006), boiled lentils (0.494 ± 0.0003), baked beans with pork and tomato sauce (0.49 ± 0.0002), pita bread (0.489 ± 0.0013), boiled yardlong bean (0.487 ± 0.0003), italian bread (0.479 ± 0.0018), boiled kidney beans (0.477 ± 0.0003), frozen and boiled turnip greens (0.476 ± 0.0006), boiled black-eyed peas (0.475 ± 0.0003), boiled catjang black-eyed peas (0.473 ± 0.0003), raw turnip greens (0.464 ± 0.0018), white table wine (0.459 ± 0.0005), boiled yellow beans (0.455 ± 0.0003), canned sweet potato (0.455 ± 0.0003), canned carrots (0.451 ± 0.0007), raw oysters (0.45 ± 0.0006), frozen and boiled kale (0.449 ± 0.0008), raw whelk (0.447 ± 0.0006), potato chips (0.446 ± 0.0018), boiled black beans (0.444 ± 0.0003), fried beef liver (0.423 ± 0.0005), boiled broad beans (0.421 ± 0.0003), raw endive (0.42 ± 0.002), boiled scotch kale (0.417 ± 0.0008), boiled kale (0.415 ± 0.0008), frozen and boiled green peas (0.414 ± 0.0006), braised beef liver (0.413 ± 0.0005), boiled mungo beans (0.412 ± 0.0003), corn tortilla (0.41 ± 0.0017), raw green peas (0.41 ± 0.0006), canned great northern beans (0.408 ± 0.0002), frozen and boiled carrots (0.405 ± 0.0007), boiled split peas (0.396 ± 0.0003), potato pancakes (0.389 ± 0.0007), dried figs (0.388 ± 0.0003), boiled french beans (0.382 ± 0.0003), chocolate syrup (0.376 ± 0.0013), canned navy beans (0.375 ± 0.0002), raw agar seaweed (0.373 ± 0.0005), baked beans with pork and sweet sauce (0.371 ± 0.0002), frozen and boiled yellow snap beans (0.371 ± 0.0007), frozen and boiled green beans (0.371 ± 0.0007), raw irish moss seaweed (0.37 ± 0.0005), boiled cranberry beans (0.37 ± 0.0003), canned succotash (0.366 ± 0.0004), canned lima beans (0.363 ± 0.0002), grape juice (0.36 ± 0.0002), japanese persimmon (0.355 ± 0.0003), vegetarian baked beans (0.346 ± 0.0002), dry roasted pistachio nuts (0.339 ± 0.0018), boiled turnip greens (0.338 ± 0.0007), boiled sweet potato (0.337 ± 0.0003), dried pistachio nuts (0.332 ± 0.0018), boiled black turtle beans (0.327 ± 0.0003), boiled sprouted green peas (0.325 ± 0.0005), frozen and boiled broccoli (0.325 ± 0.0005), frozen and boiled brussels sprouts (0.321 ± 0.0006), frozen and boiled fordhook lima beans (0.311 ± 0.0006), golden seedless raisins (0.308 ± 0.0005), seedless raisins (0.308 ± 0.0005), dried jujube (0.305 ± 0.0005), canned green peas (0.304 ± 0.0006), braised pork liver (0.3 ± 0.0005), boiled mung beans (0.298 ± 0.0002), dried dates (0.298 ± 0.0006), simmered chicken liver (0.297 ± 0.0005), frozen and french fried potatoes without skin (0.296 ± 0.001), frozen and boiled mustard greens (0.295 ± 0.0007), boiled green beans (0.294 ± 0.0008), boiled yellow snap beans (0.294 ± 0.0008), boiled parsnips (0.294 ± 0.0006), microwaved potato with skin (0.292 ± 0.0002), strawberries (0.29 ± 0.0003), canned broad beans (0.288 ± 0.0002), canned black-eyed peas (0.283 ± 0.0002), blueberries (0.282 ± 0.0003), frozen and boiled succotash (0.28 ± 0.0006), canned turnip greens (0.277 ± 0.0004), dried apricots (0.274 ± 0.0014), boiled artichoke hearts (0.273 ± 0.0006), boiled artichoke (0.273 ± 0.0006), seeded raisins (0.267 ± 0.0005), raw potato without skin (0.263 ± 0.0004), frozen and boiled crookneck squash (0.263 ± 0.0005), black currants (0.255 ± 0.0009), simmered turkey liver (0.25 ± 0.0005), dried longans (0.248 ± 0.0005), boiled collards (0.246 ± 0.0003), boiled broccoli (0.245 ± 0.0006), raw california avocado (0.244 ± 0.0003), boiled beets (0.24 ± 0.0006), raw red chili peppers (0.238 ± 0.0011), raw green chili peppers (0.238 ± 0.0011), dried lychees (0.234 ± 0.0005), canned black turtle beans (0.233 ± 0.0002), spaghetti (0.23 ± 0.0004), raw broccoli (0.23 ± 0.0011), frozen and boiled zucchini (0.23 ± 0.0004), canned pinto beans (0.229 ± 0.0002), baked potato with skin (0.229 ± 0.0002), boiled brussels sprouts (0.227 ± 0.0006), frozen and fried hash brown potatoes (0.223 ± 0.0006), fried chicken giblets (0.222 ± 0.0005), dried ginko nuts (0.221 ± 0.0018), dried prunes (0.22 ± 0.0006), instant tea (0.219 ± 0.0002), canned kidney beans (0.217 ± 0.0002), boiled crookneck squash (0.213 ± 0.0006), braised beef pancreas (0.208 ± 0.0005), boiled asparagus (0.208 ± 0.0006), raw cauliflower (0.204 ± 0.001), pork liver-cheese (0.2 ± 0.0013), raw kelp/kombu/tangle seaweed (0.2 ± 0.0005), canned cranberry beans (0.2 ± 0.0002), macaroni (0.2 ± 0.0004), canned green beans (0.2 ± 0.0007), canned yellow snap beans (0.2 ± 0.0007), noodles (0.2 ± 0.0003), braised pork pancreas (0.198 ± 0.0005), jackfruit (0.197 ± 0.0005), boiled yellow corn (0.194 ± 0.0006), restaurant cranberry juice cocktail (0.193 ± 0.0003), french fried potatoes (0.192 ± 0.001), raw sprouted alfalfa seeds (0.188 ± 0.0015), raw sprouted mung beans (0.188 ± 0.001), red currants (0.186 ± 0.0009), white currants (0.186 ± 0.0009), simmered beef kidneys (0.185 ± 0.0005), frozen and boiled asparagus (0.185 ± 0.0008), frozen and boiled potatoes without skin (0.185 ± 0.0005), frozen and boiled yellow corn (0.18 ± 0.0006), raw red cabbage (0.18 ± 0.0014), boiled zucchini (0.178 ± 0.0006), boiled cauliflower (0.177 ± 0.0008), simmered turkey giblets (0.175 ± 0.0005), simmered chicken giblets (0.17 ± 0.0005), raw florida avocado (0.17 ± 0.0002), microwaved potato without skin (0.17 ± 0.0003), soy milk (0.17 ± 0.0002), scalloped potatoes (0.166 ± 0.0004), au gratin potatoes (0.161 ± 0.0004), baked potato without skin (0.161 ± 0.0003), raw parsley (0.16 ± 0.0017), raw green cabbage (0.16 ± 0.0014), raw scallop squash (0.157 ± 0.0008), raw crookneck squash (0.157 ± 0.0008), cranberry sauce (0.157 ± 0.0005), cranberry juice cock-

MANGANESE

tail (0.157 ± 0.0002), pork liver sausage (0.156 ± 0.0028), boiled rutabaga (0.153 ± 0.0006), hash brown potatoes (0.153 ± 0.0006), boiled red tomato (0.153 ± 0.0004), bananas (0.152 ± 0.0004), prune juice (0.151 ± 0.0002), raw iceberg lettuce (0.15 ± 0.0025), frozen and boiled cauliflower (0.15 ± 0.0006), braised pork kidneys (0.149 ± 0.0005), loquats (0.148 ± 0.0005), guava (0.144 ± 0.0006), cream of rice (0.144 ± 0.0003), gooseberries (0.144 ± 0.0003), raw carrots (0.142 ± 0.0007), cooked sprouted mung beans (0.14 ± 0.0008), boiled potato without skin (0.14 ± 0.0004), raw red sweet peppers (0.14 ± 0.001), raw green sweet peppers (0.14 ± 0.001), raw eggplant (0.139 ± 0.0012), raw celery (0.135 ± 0.0013), boiled eggplant (0.135 ± 0.001), raw butterhead lettuce (0.133 ± 0.0033), raw onions (0.133 ± 0.0006), carrot juice (0.13 ± 0.0003), boiled green cabbage (0.129 ± 0.0007), boiled red cabbage (0.129 ± 0.0007), boiled scallop squash (0.128 ± 0.0006), raw zucchini (0.128 ± 0.0008), figs (0.128 ± 0.001), turkey summer sausage (0.125 ± 0.0018), boiled celery (0.123 ± 0.0007), raw red tomato (0.122 ± 0.0004), o'brien potatoes (0.121 ± 0.0003), dry dessert wine (0.119 ± 0.0008), sweet dessert wine (0.119 ± 0.0008), cooked apple without skin (0.118 ± 0.0003), crab apples (0.115 ± 0.0005), boiled mushrooms (0.115 ± 0.0006), ginko nuts (0.114 ± 0.0018), mashed potatoes (0.114 ± 0.0005), apple juice (0.113 ± 0.0002), chicken egg yolk (0.112 ± 0.0029), boiled onions (0.112 ± 0.0005), sardines (0.108 ± 0.0021), simmered chicken heart (0.107 ± 0.0005), rose table wine (0.105 ± 0.0005), raw whiting (0.104 ± 0.0006), potato salad (0.101 ± 0.0004), raw green tomato (0.1 ± 0.0004), simmered turkey gizzard (0.098 ± 0.0005), cooked prunes (0.098 ± 0.0005), canned crookneck squash (0.097 ± 0.0005), canned potato without skin (0.097 ± 0.0006), coleslaw (0.097 ± 0.0008), boiled red sweet peppers (0.097 ± 0.0007), boiled green sweet peppers (0.097 ± 0.0007), raw frozen rhubarb (0.097 ± 0.0004), frozen and boiled red sweet peppers (0.097 ± 0.0005), frozen and boiled green sweet peppers (0.097 ± 0.0005), cherries (0.093 ± 0.0007), simmered turkey heart (0.092 ± 0.0005), raw scallops (0.091 ± 0.0006), braised pork brains (0.085 ± 0.0005), kumquats (0.084 ± 0.0026), jujube (0.084 ± 0.0005), italian pork sausage (0.082 ± 0.0007), corn flakes (0.082 ± 0.0018), carambola (0.082 ± 0.0004), raw dungeness crab (0.08 ± 0.0006), apricots (0.079 ± 0.0005), tomato juice (0.077 ± 0.0003), pear (0.076 ± 0.0003), braised beef spleen (0.075 ± 0.0005), applesauce (0.075 ± 0.0004), gefiltefish with broth (0.074 ± 0.0012), clam liquid (0.074 ± 0.0002), canned sprouted mung beans (0.073 ± 0.0008), braised pork heart (0.073 ± 0.0004), frozen and cooked sweetened rhubarb (0.073 ± 0.0004), raw radish (0.071 ± 0.0011), hard pork salami (0.07 ± 0.005), pork sausage (0.07 ± 0.0019), pork mothers loaf (0.067 ± 0.0024), canned yellow corn (0.067 ± 0.0005), pork and beef peppered loaf (0.064 ± 0.0018), simmered chicken gizzard (0.062 ± 0.0005), raw cucumber (0.062 ± 0.001), cooked lobster (0.061 ± 0.0006), raw jerusalem artichoke (0.06 ± 0.0007), breadfruit (0.06 ± 0.0005), simmered beef heart (0.059 ± 0.0005), european grapes (0.058 ± 0.0003), thompson seedless grapes (0.058 ± 0.0003), lebanon bologna (0.057 ± 0.0022), beef and pork salami (0.057 ± 0.0022), raw lobster (0.055 ± 0.0006), lychees (0.055 ± 0.0005), roasted lean ham (0.054 ± 0.0005), fried chicken neck with skin (0.053 ± 0.0014), longans (0.052 ± 0.0005), raw shrimp (0.051 ± 0.0006), simmered chicken neck without skin (0.05 ± 0.0028), fried chicken back with skin (0.05 ± 0.0007), polish pork sausage (0.05 ± 0.0018), bloody mary (0.049 ± 0.0003), beef salami (0.048 ± 0.0022), whole chocolate milk (0.048 ± 0.0002), plum (0.048 ± 0.0008), cantaloupe (0.047 ± 0.0003), peach (0.047 ± 0.0006), beef lunch meat (0.046 ± 0.0018), pork bratwurst (0.046 ± 0.0006), braised pork spleen (0.045 ± 0.0005), simmered chicken neck with skin (0.045 ± 0.0013), raw apple with skin (0.045 ± 0.0004), pork and beef link sausage (0.044 ± 0.0007), nectarine (0.044 ± 0.0004), chamomile tea (0.044 ± 0.0003), pork luxury loaf (0.043 ± 0.0018), pork and beef luncheon sausage (0.043 ± 0.0022), beef and pork bologna (0.043 ± 0.0022), chopped ham lunch meat (0.043 ± 0.0024), chicken egg (0.042 ± 0.001), raw shad (0.042 ± 0.0006), manhattan (0.042 ± 0.0009), roasted pork arm excluding fat (0.041 ± 0.0005), cooked bacon (0.041 ± 0.0004), roasted ham (0.041 ± 0.0005), creme de menthe (0.04 ± 0.001), raw abalone (0.04 ± 0.0006), pickled herring (0.04 ± 0.0033), frozen and boiled onions (0.04 ± 0.0005), roasted pork tenderloin (0.039 ± 0.0005), barbecue loaf (0.039 ± 0.0022), fried dark meat chicken with skin (0.039 ± 0.0005), berliner (0.039 ± 0.0022), pork and beef brotwurst (0.039 ± 0.0007), canned creamed yellow corn (0.039 ± 0.0004), restaurant chocolate milkshake (0.039 ± 0.0002), pork and beef kielbasa/kolbassy (0.038 ± 0.0019), roasted pork leg excluding fat (0.037 ± 0.0005), watermelon (0.037 ± 0.0003), tangerine juice (0.037 ± 0.0002), ham and cheese spread (0.036 ± 0.0018), pork olive loaf (0.036 ± 0.0018), pork and beef spread (0.036 ± 0.0018), malted milk (0.036 ± 0.0002), whole chocolate malted milk (0.036 ± 0.0002), raw alaska king crab (0.035 ± 0.0006), beef honey roll sausage (0.035 ± 0.0022), fried chicken thigh with skin (0.035 ± 0.0008), pork bologna (0.035 ± 0.0022), raw eel (0.035 ± 0.0006), raw squid (0.035 ± 0.0006), simmered beef brains (0.035 ± 0.0005), raw herring (0.035 ± 0.0006), raw spot (0.035 ± 0.0006), raw scup (0.035 ± 0.0006), roasted pork shank excluding fat (0.034 ± 0.0005), cooked shrimp (0.034 ± 0.0006), smoked whitefish (0.034 ± 0.0006), roasted pork arm including fat (0.033 ± 0.0005), fried dark meat chicken without skin (0.033 ± 0.0005), beef frankfurter (0.033 ± 0.0009), unheated lean ham (0.033 ± 0.0005), minced ham lunch meat (0.033 ± 0.0024), tangerine (0.032 ± 0.0006), jellied corned beef loaf (0.032 ± 0.0018), braised pork center loin excluding fat (0.032 ± 0.0005), roasted pork leg including fat (0.032 ± 0.0005), pork and beef honey loaf (0.032 ± 0.0018), pork and beef old fashioned loaf (0.032 ± 0.0018), fried beef brains (0.032 ± 0.0005), unheated ham (0.031 ± 0.0005), vienna sausage (0.031 ± 0.0031), roasted cured pork arm excluding fat (0.03 ± 0.0005), beef and pork summer sausage (0.03 ± 0.0022), pork beerwurst salami (0.03 ± 0.0022), pear nectar (0.03 ± 0.0002), fried chicken drumstick with skin (0.029 ± 0.001), roasted canned ham (0.029 ± 0.0005), pork and beef picnic loaf (0.029 ± 0.0018), ham and cheese roll (0.029 ± 0.0018), pork and beef lunch meat (0.029 ± 0.0018), pork pickle and pimento loaf (0.029 ± 0.0018), rose apple (0.029 ± 0.0005), roasted pork shank including fat (0.028 ± 0.0005), roasted pork sirloin excluding fat (0.028 ± 0.0005), fried chicken wing with skin (0.028 ± 0.0016), grilled canadian bacon (0.028 ± 0.0011), braised pork center loin including fat (0.027 ± 0.0005), pork and beef mortadella (0.027 ± 0.0033), navel orange (0.027 ± 0.0004), mango (0.027 ± 0.0002), coffee (0.027 ± 0.0003), simmered beef tongue (0.026 ± 0.0005), roasted pork shoulder excluding fat (0.026 ± 0.0005), roasted pork rump excluding fat (0.026 ± 0.0005), beef bologna (0.026 ± 0.0022), fried light meat chicken with skin (0.026 ± 0.0005), roasted pork sirloin including fat (0.025 ± 0.0005), canned lean ham (0.025 ± 0.0005), raw octopus (0.025 ± 0.0006), cooked pompano (0.025 ± 0.0006), raw sturgeon (0.025 ±

0.0006), raw croaker (0.025 ± 0.0006), raw haddock (0.025 ± 0.0006), roasted pork rump including fat (0.024 ± 0.0005), roasted cured pork arm including fat (0.024 ± 0.0005), roasted canned lean ham (0.024 ± 0.0005), canned pork lunch meat (0.024 ± 0.0024), raw monkfish (0.024 ± 0.0006), roasted turkey dark meat without skin (0.023 ± 0.0005), roasted turkey dark meat with skin (0.023 ± 0.0005), roasted pork blade roll (0.023 ± 0.0005), pork and beef link sausage with american cheese (0.023 ± 0.0012), canned ham (0.023 ± 0.0005), unheated canadian bacon (0.023 ± 0.0009), valencia orange (0.023 ± 0.0004), raw apple without skin (0.023 ± 0.0004), cooked corned beef brisket (0.022 ± 0.0005), roasted pork shoulder including fat (0.022 ± 0.0005), beef and pork frankfurter (0.022 ± 0.0011), cooked mullet (0.022 ± 0.0006), well done baked ground beef (0.021 ± 0.0005), roasted chicken drumstick with skin (0.021 ± 0.001), stewed dark meat chicken without skin (0.021 ± 0.0005), roasted dark meat chicken with skin (0.021 ± 0.0005), roasted chicken thigh with skin (0.021 ± 0.0008), roasted light meat chicken with skin (0.021 ± 0.0005), fried chicken breast without skin (0.021 ± 0.0006), raw bluefish (0.021 ± 0.0006), cooked sheepshead (0.021 ± 0.0006), smoked cisco (0.021 ± 0.0006), simmered beef shank excluding fat (0.02 ± 0.0005), well done baked extra lean ground beef (0.02 ± 0.0005), well done broiled ground beef (0.02 ± 0.0005), broiled beef flank excluding fat (0.02 ± 0.0005), broiled beef flank including fat (0.02 ± 0.0005), fried beef top round excluding fat (0.02 ± 0.0005), stewed chicken drumstick without skin (0.02 ± 0.0011), broiled pork center rib excluding fat (0.02 ± 0.0005), broiled pork top loin excluding fat (0.02 ± 0.0005), roasted turkey light meat without skin (0.02 ± 0.0005), roasted turkey light meat with skin (0.02 ± 0.0005), fried light meat chicken without skin (0.02 ± 0.0005), raw lingcod (0.02 ± 0.0006), white grapefruit juice (0.02 ± 0.0002), red grapefruit juice (0.02 ± 0.0002), braised beef chuck arm pot roast excluding fat (0.019 ± 0.0005), well done broiled extra lean ground beef (0.019 ± 0.0005), fried beef sirloin excluding fat (0.019 ± 0.0005), braised beef flank excluding fat (0.019 ± 0.0005), braised beef flank including fat (0.019 ± 0.0005), well done fried ground beef (0.019 ± 0.0005), stewed chicken drumstick with skin (0.019 ± 0.0009), roasted chicken thigh without skin (0.019 ± 0.001), stewed dark meat chicken with skin (0.019 ± 0.0005), stewed chicken thigh with skin (0.019 ± 0.0007), raw swordfish (0.019 ± 0.0006), peach nectar (0.019 ± 0.0002), 94 proof gin (0.019 ± 0.0012), 94 proof rum (0.019 ± 0.0012), 94 proof vodka (0.019 ± 0.0012), 94 proof whiskey (0.019 ± 0.0012), 100 proof gin (0.019 ± 0.0012), 100 proof rum (0.019 ± 0.0012), 100 proof vodka (0.019 ± 0.0012), 100 proof whiskey (0.019 ± 0.0012), braised beef chuck blade roast excluding fat (0.018 ± 0.0005), simmered beef shank including fat (0.018 ± 0.0005), braised beef shortribs excluding fat (0.018 ± 0.0005), well done baked lean ground beef (0.018 ± 0.0005), well done fried extra lean ground beef (0.018 ± 0.0005), braised beef bottom round excluding fat (0.018 ± 0.0005), fried beef top round including fat (0.018 ± 0.0005), braised pork loin excluding fat (0.018 ± 0.0005), havarti cheese (0.018 ± 0.0018), roasted chicken wing with skin (0.018 ± 0.0015), stewed chicken wing with skin (0.018 ± 0.0013), pork headcheese (0.018 ± 0.0018), stewed light meat chicken without skin (0.018 ± 0.0005), stewed light meat chicken with skin (0.018 ± 0.0005), roasted chicken breast with skin (0.018 ± 0.0005), stewed chicken breast without skin (0.018 ± 0.0005), stewed chicken breast with skin (0.018 ± 0.0005), roasted beef tip excluding fat (0.017 ± 0.0005), braised beef brisket excluding fat (0.017 ± 0.0005), broiled beef sirloin excluding fat (0.017 ± 0.0005), well done broiled lean ground beef (0.017 ± 0.0005), broiled beef tenderloin excluding fat (0.017 ± 0.0005), broiled beef top round excluding fat (0.017 ± 0.0005), medium-rare broiled ground beef (0.017 ± 0.0005), braised beef bottom round including fat (0.017 ± 0.0005), medium-rare fried ground beef (0.017 ± 0.0005), braised pork arm excluding fat (0.017 ± 0.0005), medium-rare baked ground beef (0.017 ± 0.0005), braised pork sirloin excluding fat (0.017 ± 0.0005), roasted dark meat chicken without skin (0.017 ± 0.0005), broiled pork center rib including fat (0.017 ± 0.0005), roasted light meat chicken without skin (0.017 ± 0.0005), roasted chicken breast without skin (0.017 ± 0.0006), raw pheasant without skin (0.017 ± 0 .0005), pummelo (0.017 ± 0.0003), corn grits (0.017 ± 0.0002), whiskey sour (0.017 ± 0.0006), broiled beef rib eye excluding fat (0.016 ± 0.0005), roasted beef rib roasted excluding fat (0.016 ± 0.0005), braised beef chuck arm pot roast including fat (0.016 ± 0.0005), broiled beef rib excluding fat (0.016 ± 0.0005), well done fried lean ground beef (0.016 ± 0.0005), braised pork boston blade excluding fat (0.016 ± 0.0005), medium-rare broiled extra lean ground beef (0.016 ± 0.0005), medium-rare fried extra lean ground beef (0.016 ± 0.0005), broiled beef top round including fat (0.016 ± 0.0005), fried beef sirloin including fat (0.016 ± 0.0005), broiled beef top loin excluding fat (0.016 ± 0.0005), roasted beef tenderloin excluding fat (0.016 ± 0.0005), braised pork loin blade excluding fat (0.016 ± 0.0005), roasted eye of beef round excluding fat (0.016 ± 0.0005), broiled beef round excluding fat (0.016 ± 0.0005), roasted pork loin excluding fat (0.016 ± 0.0005), roasted pork center loin excluding fat (0.016 ± 0.0005), broiled pork top loin including fat (0.016 ± 0.0005), raw mullet (0.016 ± 0.0006), raw flatfish (0.016 ± 0.0006), cooked snapper (0.016 ± 0.0006), raw rockfish (0.016 ± 0.0006), smoked chinook salmon (0.016 ± 0.0006), braised beef chuck blade roast including fat (0.015 ± 0.0005), roasted beef round tip including fat (0.015 ± 0.0005), broiled beef sirloin including fat (0.015 ± 0.0005), broiled porterhouse steak excluding fat (0.015 ± 0.0005), broiled T-bone steak excluding fat (0.015 ± 0.0005), medium-rare baked extra lean ground beef (0.015 ± 0.0005), broiled beef tenderloin including fat (0.015 ± 0.0005), roasted eye of beef round including fat (0.015 ± 0.0005), broiled beef round including fat (0.015 ± 0.0005), braised pork loin including fat (0.015 ± 0.0005), braised beef lungs (0.015 ± 0.0005), fried chicken breast with skin (0.015 ± 0.0005), raw wolffish (0.015 ± 0.0006), raw butterfish (0.015 ± 0.0006), raw catfish (0.015 ± 0.0006), raw mackerel (0.015 ± 0.0006), raw pink salmon (0.015 ± 0.0006), cooked tilefish (0.015 ± 0.0006), raw yellowtail (0.015 ± 0.0006), raw perch (0.015 ± 0.0006), raw chum salmon (0.015 ± 0.0006), raw pollock (0.015 ± 0.0006), raw dolphinfish (0.015 ± 0.0006), raw chinook salmon (0.015 ± 0.0006), raw seatrout (0.015 ± 0.0006), raw cod (0.015 ± 0.0006), raw shark (0.015 ± 0.0006), raw coho salmon (0.015 ± 0.0006), raw halibut (0.015 ± 0.0006), raw striped bass (0.015 ± 0.0006), raw sea bass (0.015 ± 0.0006), raw cusk (0.015 ± 0.0006), restaurant strawberry milkshake (0.015 ± 0.0002), broiled beef rib eye including fat (0.014 ± 0.0005), medium-rare broiled lean ground beef (0.014 ± 0.0005), medium-rare fried lean ground beef (0.014 ± 0.0005), medium-rare baked lean ground beef (0.014 ± 0.0005), braised beef brisket including fat (0.014 ± 0.0005), broiled porterhouse steak including fat (0.014 ± 0.0005), braised pork spareribs (0.014 ± 0.0005), broiled beef top loin including fat (0.014 ±

MANGANESE

0.0005), roasted beef tenderloin including fat (0.014 ± 0.0005), roasted pork boston blade excluding fat (0.014 ± 0.0005), braised pork arm braised including fat (0.014 ± 0.0005), canned corned beef (0.014 ± 0.0005), roasted pork loin including fat (0.014 ± 0.0005), braised pork sirloin including fat (0.014 ± 0.0005), roasted pork center loin including fat (0.014 ± 0.0005), raw sockeye salmon (0.014 ± 0.0006), raw grouper (0.014 ± 0.0006), restaurant vanilla milkshake (0.014 ± 0.0002), orange juice (0.014 ± 0.0002), 86 proof whiskey (0.014 ± 0.0012), roasted beef rib including fat (0.013 ± 0.0005), broiled beef rib including fat (0.013 ± 0.0005), braised beef shortribs including fat (0.013 ± 0.0005), braised pork boston blade braised including fat (0.013 ± 0.0005), broiled T-bone steak including fat (0.013 ± 0.0005), braised pork loin blade including fat (0.013 ± 0.0005), roasted pork loin blade excluding fat (0.013 ± 0.0005), raw pompano (0.013 ± 0.0006), raw sheepshead (0.013 ± 0.0006), raw snapper (0.013 ± 0.0006), papaya nectar (0.013 ± 0.0002), white grapefruit (0.013 ± 0.0004), roasted pork boston blade including fat (0.012 ± 0.0005), braised pork center rib exlucding fat (0.012 ± 0.0005), braised pork top loin excluding fat (0.012 ± 0.0005), fried pork center loin excluding fat (0.012 ± 0.0005), cooked grouper (0.012 ± 0.0006), beer (0.012 ± 0.0001), roasted pork loin blade including fat (0.011 ± 0.0005), broiled pork center loin excluding fat (0.011 ± 0.0005), roasted pork center rib excluding fat (0.011 ± 0.0005), roasted pork top loin excluding fat (0.011 ± 0.0005), raw tilefish (0.011 ± 0.0006), papaya (0.011 ± 0.0002), screwdriver (0.011 ± 0.0002), broiled pork loin excluding fat (0.01 ± 0.0005), broiled pork sirloin excluding fat (0.01 ± 0.0005), braised pork center rib including fat (0.01 ± 0.0005), braised pork top loin including fat (0.01 ± 0.0005), pink grapefruit (0.01 ± 0.0004), broiled pork boston blade excluding fat (0.009 ± 0.0005), broiled pork loin blade excluding fat (0.009 ± 0.0005), fried pork loin blade excluding fat (0.009 ± 0.0005), broiled pork loin including fat (0.009 ± 0.0005), broiled pork sirloin including fat (0.009 ± 0.0005), fried pork center loin including fat (0.009 ± 0.0005), roasted pork center rib including fat (0.009 ± 0.0005), broiled pork center loin including fat (0.009 ± 0.0005), roasted pork top loin including fat (0.009 ± 0.0005), broiled pork boston blade including fat (0.008 ± 0.0005), broiled pork loin blade including fat (0.008 ± 0.0005), lime juice (0.008 ± 0.0002), lemon juice (0.008 ± 0.0002), fried pork loin blade including fat (0.007 ± 0.0005), fried pork center rib excluding fat (0.007 ± 0.0005), fried pork top loin excluding fat (0.007 ± 0.0005), raw bacon (0.007 ± 0.0007), fried pork center rib including fat (0.006 ± 0.0005), fried pork top loin including fat (0.006 ± 0.0005), chicken egg white (0.006 ± 0.0015), raw pork jowl (0.005 ± 0.0005), carob milk (0.002 ± 0.0002).

Applications: diabetes, multiple sclerosis, allergies, asthma, fatigue
Daily Dosage: 1980 RDA — none; 1980 SADDI — 2.5–5.0 mg; 1989 RDA — none; 1989 SADDI — 2.5–5.0 mg; nutritional — 2.5–9.0 mg; therapeutic — 10 mg; experimental — ?; toxic — ?
Warnings: increased amounts are dangerous. See toxicity symptoms.

MERCURY

Names: mercury, quicksilver, Hg, element 80
Classification: contaminant
Deficiency: (no nutritional requirement)
Side-effects: see toxicity symptoms
Toxicity: (Detoxified by selenium) loose teeth, bad breath, metallic taste in mouth, colitis, urinary deterioration, appetite decreased, anemia, increased blood pressure, decreased blood pressure, fatigue, nervousness, anxiety, depression, emotional agitation, lack of sleep, trembling, fatigue, hallucinations, joint pains, changes in walking, vision changes, mental health deterioration, paralysis, numbness, fever, chills, coughing, nausea, stomach pains; sore mouth, gums and throat
Inhibitors: vitamin E, vitamin N, selenium
Helpers: unknown
Sources: air pollution, water pollution, fungicides
Applications: none known
Daily Dosage: 1974 PCS — none; 1989 SADDI — none; 1989 RDA — none; 1989 USRDA — none; nutritional — none; therapeutic — none; experimental — none; toxic — small amounts
Warnings: Small amounts are dangerous. See toxicity symptoms.

MOLYBDENUM

Names: molybdenum, Mb, element 42
Classification: trace mineral
Deficiency: anemia, fatigue, less urine formation, increased fatty acid oxidation, decreased cancer resistance
Side-effects: see toxicity symptoms
Toxicity: copper decreased, gout, bone disease, diarrhea, anemia, decreased growth rate
Inhibitors: tungsten
Helpers: unknown
Sources (μg per 100 grams of food): lentils (155 ± 0.5), beef liver (135 ± 0.5), split peas (130 ± 0.5), cauliflower (120 ± 0.5), green peas (110 ± 0.5), brewer's yeast (109 ± 0.5), wheat germ (100 ± 0.5), spinach (100 ± 0.5), beef kidney (77 ± 0.5), brown rice (75 ± 0.5), garlic (70 ± 0.5), oats (60 ± 0.5), eggs (53 ± 0.5), rye bread (50 ± 0.5), corn (45 ± 0.5), barley (42 ± 0.5), fish (40 ± 0.5), whole wheat (36 ± 0.5), whole wheat bread (32 ± 0.5), chicken (32 ± 0.5), cottage cheese (31 ± 0.5), beef (30 ± 0.5), potatoes (30 ± 0.5), onions (25 ± 0.5), peanuts (25 ± 0.5), coconut (25 ± 0.5), pork (25 ± 0.5), lamb (24 ± 0.5), green beans (21 ± 0.5), crab (19 ± 0.5), molasses (19 ± 0.5), cantaloupe (16 ± 0.5), apricots (14 ± 0.5), raisins (10 ± 0.5), butter (10 ± 0.5), strawberries (7 ± 0.5), carrots (5 ± 0.5), cabbage (5 ± 0.5), whole milk (3 ± 0.5), goat's milk (1 ± 0.5)
Applications: anemia
Daily Dosage: 1980 RDA — none; 1980 SADDI — 150–500 μg; 1989 RDA — none; 1989 SADDI — 150–500 μg; nutritional — 150–500 μg; therapeutic — 500–1000 μg; experimental — 5000–15,000 μg; toxic — 15,000 μg
Warnings: increased amounts are toxic. See toxicity symptoms.

NICKEL

Names: nickel, formerly coppernickel, Ni, element 28
Classification: trace mineral, contaminant
Deficiency: hormone imbalance, gland deterioration, thyroid deterioration, adrenal deterioration, deterioration of prolactin regulation, deterioration of growth, deterioration of pigmentation, blood abnormalities, decreased hematocrit, increased blood cholesterol levels, fatigue, coarse hair, deterioration of RNA/DNA production, deterioration of cell membrane integrity
Side-effects: see toxicity symptoms
Toxicity: sore gums, sore tongue, small red lumps on skin, dizziness, nausea, cough, shortness of breath, low grade fever, chronic asthma
Inhibitors: unknown
Helpers: unknown
Sources (μg per 100 grams of food): dry soybeans (700 ± 50), dry beans (500 ± 50), soy flour (410 ± 5), lentils (310 ± 5), split peas (250 ± 5), green peas (175 ± 5), green beans (153 ± 0.5), oats (150 ± 5), walnuts (132 ± 0.5), hazelnuts (122 ± 0.5), buckwheat (100 ± 5), barley (90 ± 5), corn (90 ± 5), parsley (90 ± 5), whole wheat (38 ± 0.5), spinach (35 ± 0.5), fish (30 ± 0.5), cucumber (27 ± 0.5), liver (26 ± 0.5), rye bread (25 ± 0.5), pork (25 ± 0.5), carrots (25 ± 0.5), eggs (24 ± 0.5), cabbage (22 ± 0.5), tomatoes (20 ± 0.5), onions (20 ± 0.5), potatoes (18 ± 0.5), beef (16 ± 0.5), apricots (16 ± 0.5), oranges (16 ± 0.5), cheese (15 ± 0.5), watermelon (15 ± 0.5), lettuce (14 ± 0.5), apples (13 ± 0.5), whole wheat bread (12 ± 0.5), beets (12 ± 0.5), pears (12 ± 0.5), grapes (8 ± 0.5), radishes (8 ± 0.5), pine nuts (6 ± 0.5), lamb (6 ± 0.5), milk (3 ± 0.5), air pollution, car exhaust
Applications: none known
Daily Dosage: 1980 RDA — none; 1980 SADDI — none; 1989 RDA — none; 1989 SADDI — none; nutritional — 50–75 μg; therapeutic — 100 μg; experimental — unknown; toxic — unknown
Warnings: increased amounts are dangerous. See toxicity symptoms.

PHOSPHORUS

Names: phosphorus, P, P^{-2}, P^{-}, element 15
Classification: mineral
Deficiency: appetite decreased, fatigue, nervous system deterioration, demineralization of bones, demineralization of teeth, cold hands and feet, numbness, hydrocele, sore breasts, night sweats, continuous diarrhea, constipation, faint/rapid pulse, nosebleeds, increased menses, improper fat digestion, deterioration of memory, anxiety, lack of sleep, cramps, nerve pain, shooting pains, colic, low fevers, skin edema, depression, gall bladder deterioration
Side-effects: see toxicity symptoms
Toxicity: calcium decreased, calcium decreased in urine, kidney failure, increased phosphorus in blood
Inhibitors: aluminum
Helpers: vitamin A, vitamin D, vitamin F, boron, calcium, iron, manganese, protein

Sources (mg per 100 grams of food): brewers yeast (1775 ± 1.8), torula yeast (1736 ± 1.8), corn germ (1587 ± 0.5), dry bakers yeast (1307 ± 1.8), dried pumpkin and squash seeds (1189 ± 1.8), roasted pumpkin and squash seeds (1189 ± 1.8), dry roasted sunflower seeds (1171 ± 1.8), toasted wheat germ (1161 ± 1.8), oil roasted sunflower seeds (1154 ± 1.8), dry instant skim milk (985 ± 0.5), dry skim milk (967 ± 1.7), dried and salted cod (951 ± 0.6), bran (943 ± 1.8), yellow mustard seed (933 ± 16.7), almond meal (929 ± 1.8), dry buttermilk (871 ± 7.1), soybean protein concentrate (839 ± 1.8), poppy seed (800 ± 16.7), parmesan cheese (800 ± 10), toasted sesame kernels (786 ± 1.8), dry whole milk (775 ± 1.6), dried sesame kernels (775 ± 6.3), soybean protein isolate (775 ± 1.8), romano cheese (768 ± 1.8), dried watermelon seeds (768 ± 1.8), american cheese (754 ± 1.8), pimento cheese spread (754 ± 1.8), peanut flour (750 ± 12.5), sesame butter (733 ± 3.3), dried sunflower seeds (714 ± 1.8), defatted soy meal (701 ± 0.4), basil (700 ± 50), dry cocoa powder (700 ± 10), defatted soybean flour (674 ± 0.5), ry-krisp crackers (664 ± 3.6), dry roasted soybean nuts (649 ± 0.6), whole toasted sesame seeds (646 ± 1.8), whole dried sesame seeds (633 ± 5.6), gruyere cheese (614 ± 1.8), swiss cheese (611 ± 1.8), dried brazil nuts (607 ± 1.8), caraway seed (600 ± 25), low fat soybean flour (593 ± 0.6), dried pilinuts (582 ± 1.8), freeze-dried parsley (571 ± 35.7), broiled lamb liver (571 ± 1.1), toasted almonds (557 ± 1.8), dry roasted almonds (557 ± 1.8), oil roasted almonds (554 ± 1.8), gouda cheese (554 ± 1.8), celery seed (550 ± 25), edam cheese (543 ± 1.8), dark rye flour (536 ± 0.4), cooked carp (532 ± 0.6), dried almonds (529 ± 1.8), almond butter (525 ± 3.1), cheddar cheese (518 ± 1.8), oil roasted peanuts (514 ± 1.8), dried pignolia pine nuts (514 ± 1.8), dried pistachio nuts (511 ± 1.8), oil roasted virginia peanuts (507 ± 1.8), tilsit cheese (507 ± 1.8), chicken egg yolk (506 ± 2.9), provolone cheese (504 ± 1.8), dried parsley (500 ± 50), dried dill weed (500 ± 50), freeze-dried chives (500 ± 62.5), cumin seed (500 ± 25), fennel seed (500 ± 25), ham and cheese spread (500 ± 1.8), caraway cheese (496 ± 1.8), bran flakes (496 ± 1.8), dry roasted cashews (496 ± 1.8), soybean flour (494 ± 0.6), sardines (492 ± 2.1), dry roasted pistachio nuts (482 ± 1.8), dried and frozen tofu (482 ± 2.9), roasted soybean flour (476 ± 0.6), muenster cheese (475 ± 1.8), dried black walnuts (471 ± 1.8), cheshire cheese (468 ± 1.8), cashew butter (464 ± 1.8), fried beef liver (461 ± 0.5), colby cheese (461 ± 1.8), brick cheese (457 ± 1.8), almond paste (454 ± 1.8), dried butternuts (454 ± 1.8), braised beef pancreas (453 ± 0.5), gjetost cheese (450 ± 1.8), anise seed (450 ± 25), monterey cheese (450 ± 1.8), dry roasted macadamia nuts (436 ± 1.8), oil roasted cashews (432 ± 1.8), raw carp (414 ± 0.6), havarti cheese (407 ± 1.8), braised beef liver (404 ± 0.5), garlic powder (400 ± 16.7), unsweetened baking chocolate (400 ± 1.8), compressed bakers yeast (400 ± 1.8), limburger cheese (396 ± 1.8), roquefort cheese (396 ± 1.8), bleu cheese (393 ± 1.8), fried beef top round excluding fat (392 ± 0.5), dried peanuts (389 ± 1.8), raw cuttlefish (387 ± 0.6), oil roasted spanish peanuts (386 ± 1.8), fried beef brains (386 ± 0.5), whole grain rye crackers (385 ± 3.8), creamy peanut butter (375 ± 3.1), mozzarella cheese (375 ± 1.8), whole wheat flour (372 ± 0.4), canned smelt (370 ± 0.5), raisin bran (370 ± 1.4), toasted soybean nuts (368 ± 1.8), port du salut cheese (364 ± 1.8), braised beef thymus (364 ± 0.5), roasted soybean nuts (363 ± 0.6), dry roasted peanuts (357 ± 1.8), shredded wheat (357 ± 1.8), puffed wheat (357 ± 3.6), simmered beef brains (352 ± 0.5), paprika (350 ± 25), curry powder (350 ± 25), coriander seed (350 ± 25), onion powder (350 ± 25), wheat flakes (350 ± 1.8), camembert cheese (350 ± 1.8), cooked sheepshead (349 ± 0.6), dark buckwheat flour (347 ± 0.5), cooked flounder (344 ± 0.5), feta cheese (343 ± 1.8), cooked pompano (341 ± 0.6), dried hickory nuts (339 ± 1.8), raw wild rice (339 ± 0.3), cooked swordfish (338 ± 0.6), cooked bacon (335 ± 0.4), oil roasted filberts (329 ± 1.8), dry roasted filberts (329 ± 1.8), cooked crayfish (329 ± 0.6), canned pink salmon (328 ± 0.6), canned sockeye salmon (326 ± 0.6), kippered herring (325 ± 1.3), dried english/persian walnuts (321 ± 1.8), cooked rainbow trout (320 ± 0.6), oil roasted valencia peanuts (318 ± 1.8), dried filberts (318 ± 1.8), crunchy peanut butter (316 ± 1.6), freeze-dried red sweet peppers (313 ± 31.3), freeze-dried green sweet peppers (313 ± 31.3), raw sheepshead (313 ± 0.6), simmered chicken liver (312 ± 0.5), millit (311 ± 0.5), roasted goose without skin (309 ± 0.5), dry roasted pecans (307 ± 1.8), raw quail without skin (307 ± 0.5), simmered beef kidneys (306 ± 0.5), braised beef spleen (305 ± 0.5), cooked herring (304 ± 0.6), turmeric (300 ± 25), dried chervil (300 ± 50), dried coriander leaf (300 ± 50), dehydrated onion flakes (300 ± 3.6), dill seed (300 ± 25), oil roasted pecans (300 ± 1.8), toasted whole wheat bread (300 ± 2.4), dried pecans (296 ± 1.8), tomato powder (295 ± 0.5), cooked rainbow smelt (295 ± 0.6), grilled canadian bacon (294 ± 1.1), dried shitake mushrooms (293 ± 3.3), braised pork pancreas (291 ± 0.5), pot barley pearled (290 ± 0.3), bittersweet chocolate (289 ± 1.8), cooked flatfish (289 ± 0.6), roasted pork tenderloin (288 ± 0.5), sorghum grain (287 ± 0.5), cooked kingfish (287 ± 0.5), fried chicken giblets (286 ± 0.5), cooked halibut (285 ± 0.6), cooked whiting (285 ± 0.6), roasted pork rump excluding fat (285 ± 0.5), cooked mussels (285 ± 0.6), fried tofu (285 ± 3.8), raw chum salmon (284 ± 0.6), popcorn (283 ± 8.3), braised pork spleen (283 ± 0.5), cooked whelk (282 ± 0.6), whole wheat roll (282 ± 1.8), roasted ham (281 ± 0.5), roasted pork leg excluding fat (281 ± 0.5), cooked pike (281 ± 0.6), cooked alaska king crab (280 ± 0.6), broiled pork loin excluding fat (279 ± 0.5), cooked oysters (278 ± 0.6), cooked mackerel (278 ± 0.6), roasted pork shank excluding fat (278 ± 0.5), bran muffins (278 ± 1.3), cooked perch (276 ± 0.6), cooked eel (276 ± 0.6), freeze-dried shallots (275 ± 12.5), fenugreek seed (275 ± 12.5), cooked sockeye salmon (275 ± 0.6), granola bar (275 ± 2.1), raw smelt (272 ± 0.5), raw shad (272 ± 0.6), braised beef bottom round excluding fat (272 ± 0.5), simmered turkey liver (272 ± 0.5), dried ginko nuts (271 ± 1.8), fried pork center rib excluding fat (271 ± 0.5), fried pork top loin excluding fat (271 ± 0.5), roasted goose with skin (270 ± 0.5), raw duck liver (269 ± 0.5), braised beef chuck arm pot roast excluding fat (268 ± 0.5), chili powder (267 ± 16.7), fried beef sirloin excluding fat (267 ± 0.5), braised beef flank excluding fat (267 ± 0.5), canned tuna in oil (267 ± 0.6), broiled pork center rib excluding fat (266 ± 0.5), broiled pork top loin excluding fat (266 ± 0.5), fried pork center loin excluding fat (265 ± 0.5), broiled pork sirloin excluding fat (265 ± 0.5), raw swordfish (264 ± 0.6), simmered beef shank excluding fat (263 ± 0.5), fried beef top round including fat (262 ± 0.5), braised beef flank including fat (261 ± 0.5), braised pork spareribs (261 ± 0.5), raw goose liver (261 ± 0.5), cooked clams (260 ± 0.6), canned cod (260 ± 0.6), canned blue crab (260 ± 0.6), roasted pork rump including fat (260 ± 0.5), whole wheat bread (260 ± 2), stewed rabbit (259 ± 0.4), raw crayfish (258 ± 0.6), ham and cheese roll (257 ± 1.8), roasted pork center rib excluding fat (256 ± 0.5), roasted pork top loin excluding fat (256 ± 0.5), white whole ground corn meal (256 ± 0.4), yellow whole ground corn meal (256 ± 0.4), sweetened condensed milk (255 ± 1.3), braised beef bottom round including fat (255 ± 0.5), roasted pork loin excluding fat (254 ± 0.5), turkey summer sausage (254 ± 1.8), smoked haddock (252 ± 0.6), roasted pork sirloin excluding fat (252 ± 0.5), raw rainbow trout (251 ± 0.6), braised pork center loin excluding fat (251 ± 0.5), breaded fried squid (251 ± 0.6), tarragon (250 ± 25), cayenne pepper (250 ± 25), anchovy canned in olive oil (250 ± 2.5), braised pork center rib exlucding fat (250 ± 0.5), braised pork top loin excluding fat (250 ± 0.5), toasted pumpernickel bread (250 ± 1.8), nutmeg (250 ± 25), simmered beef heart (250 ± 0.5), raw venison (249 ± 0.6), raw seatrout (249 ± 0.6), cooked sea bass (248 ± 0.6), roasted veal rib roast (248 ± 0.6), roasted pork arm excluding fat (247 ± 0.5), unheated ham (247 ± 0.5), roasted pork leg including fat (247 ± 0.5), fried chicken breast without skin (247 ± 0.6), broiled beef top round excluding fat (246 ± 0.5), simmered beef shank including fat (246 ± 0.5), boiled soybeans (245 ± 0.3), broiled pork loin blade excluding fat (245 ± 0.5), raw trout (245 ± 0.6), cooked mullet (244 ± 0.6), broiled pork center loin excluding fat (244 ± 0.5), broiled pork boston blade excluding fat (244 ± 0.5), broiled beef sirloin excluding fat (244 ± 0.5), roasted canned ham (243 ± 0.5), roasted cured pork arm excluding fat (243 ± 0.5), roasted beef tip excluding fat (242 ± 0.5), unheated canadian bacon (242 ± 0.9), cooked haddock (241 ± 0.6), braised pork liver (241 ± 0.5), braised pork kidneys (240 ± 0.5), broiled beef top round including fat (239 ± 0.5), braised pork loin excluding fat (239 ± 0.5), roasted pork shank including fat (239 ± 0.5), braised beef brisket excluding fat (239 ± 0.5), broiled beef tenderloin excluding fat (238 ± 0.5), roasted leg of lamb excluding fat (238 ± 0.6), cooked black bass (238 ± 0.4), broiled beef round excluding fat (237 ± 0.5), fried pork loin blade excluding fat (237 ± 0.5), cooked tilefish (236 ± 0.6), roasted beef tenderloin excluding fat (236 ± 0.5), raw herring (236 ± 0.6), broiled pork loin including fat (235 ± 0.5), breaded and fried scallops (235 ± 1.6), braised beef shortribs excluding fat (235 ± 0.5), braised beef chuck blade roast excluding fat (235 ± 0.5), fried chicken breast with skin (233 ± 0.5), braised pork sirloin excluding fat (232 ± 0.5), milk chocolate (232 ± 1.8), braised lamb heart (231 ± 0.3), roasted pork shoulder excluding fat (231 ± 0.5), braised/broiled veal round with rump (231 ± 0.6), raw rainbow smelt (231 ± 0.6), fried light meat chicken without skin (231 ± 0.5), hard pork salami (230 ± 5), raw pheasant without skin (230 ± 0.5), roasted pork sirloin including fat (229 ± 0.5), raw blue crab (229 ± 0.6), turkey breast meat summer sausage (229 ± 2.4), simmered chicken giblets (229 ± 0.5), pork liverwurst (228 ± 2.8), cooked rockfish (228 ± 0.6), roasted chicken breast without skin (228 ± 0.6), raw bluefish (227 ± 0.6), broiled pork center rib including fat (227 ± 0.5), taco shell (227 ± 4.5), tostada shell (227 ± 4.5), braised pork arm excluding fat (226 ± 0.5), roasted eye of beef round excluding fat (226 ± 0.5), broiled pork sirloin including fat (226 ± 0.5), braised/broiled veal loin (225 ± 0.6), roasted pork center rib including fat (224 ± 0.5), canned lean ham (224 ± 0.5), fried beef sirloin including fat (223 ± 0.5), white bolted corn meal (223 ± 0.4), yellow bolted corn meal (223 ± 0.4), quail egg (222 ± 5.6), raw halibut (222 ± 0.6), raw whiting (222 ± 0.6), raw mullet (221 ± 0.6), raw pollock (221 ± 0.6), roasted beef round tip including fat (221 ± 0.5), roasted pork loin including fat (221 ± 0.5), roasted cured pork arm including fat (221 ± 0.5), raw squid (221 ± 0.6), raw pike (220 ± 0.6), duck egg (220 ± 0.7), braised pork brains (220 ± 0.5), rice flour (220 ± 0.4), pumpernickel bread (219 ± 1.6), roasted pork center loin excluding fat (219 ± 0.5), roasted pork top loin including fat (219 ± 0.5), broiled pork top loin including fat (219 ± 0.5), roasted pork boston blade excluding fat (219 ± 0.5), raw scallops (219 ± 0.6), roasted turkey light meat without skin (219 ± 0.5), roasted lamb shoulder excluding fat (219 ± 0.6), raw alaska king crab (219 ± 0.6), raw alewife (218 ± 0.5), broiled beef flank excluding fat (218 ± 0.5), broiled beef top loin excluding fat (218 ± 0.5), broiled beef sirloin including fat (218 ± 0.5), unheated lean ham (218 ± 0.5), broiled lamb loin chop excluding fat (218 ± 1), breaded and fried shrimp (218 ± 0.6), broiled beef tenderloin including fat (217 ± 0.5), braised beef chuck arm pot roast including fat (217 ± 0.5), stir-fried sprouted soybeans (216 ± 0.5), breaded and fried croaker (216 ± 0.6), raw mackerel (216 ± 0.6), raw perch (216 ± 0.6), roasted light meat chicken without skin (216 ± 0.5), broiled beef flank including fat (215 ± 0.5), raw sockeye salmon (215 ± 0.6), breaded and fried catfish (215 ± 0.6), braised pork center loin including fat (215 ± 0.5), raw eel (215 ± 0.6), whole ground corn meal muffins (215 ± 1.3), pancakes with butter and syrup (215 ± 0.3), dried toasted coconut (214 ± 1.8), braised pork loin blade excluding fat (214 ± 0.5), braised pork boston blade excluding fat (214 ± 0.5), fried pork center loin including fat (214 ± 0.5), roasted chicken breast with skin (214 ± 0.5), broiled porterhouse steak excluding fat (213 ± 0.5), roasted beef rib roasted excluding fat (213 ± 0.5), raw catfish (213 ± 0.6), crab cakes (213 ± 0.8), fried light meat chicken with skin (213 ± 0.5), cheese crackers (213 ± 3.3), broiled lamb rib chop excluding fat (212 ± 1.2), mixed grain bread (212 ± 2), pickled anchovy (211 ± 1.8), dried coconut (211 ± 1.8), raw white sucker (211 ± 0.6), broiled beef round including fat (211 ± 0.5), broiled pork center loin including fat (211 ± 0.5), broiled pork boston blade including fat (211 ± 0.5), raw shark (211 ± 0.6), braised pork center rib including fat (210 ± 0.5), evaporated lowfat milk (210 ± 0.4), roasted canned lean ham (209 ± 0.5), raw croaker (209 ± 0.6), broiled T-bone steak excluding fat (208 ± 0.5), broiled beef rib eye excluding fat (208 ± 0.5), roasted eye of beef round including fat (208 ± 0.5), fried pork center rib including fat (208 ± 0.5), roasted turkey light meat with skin (208 ± 0.5), roasted leg of lamb including fat (208 ± 0.6), pork livercheese (208 ± 1.3), fried pork top loin including fat (207 ± 0.5), roasted beef tenderloin including fat (207 ± 0.5), tempeh (206 ± 0.6), cooked blue crab (206 ± 0.6), roasted pork arm including fat (206 ± 0.5), raw shrimp (206 ± 0.6), simmered turkey heart (205 ± 0.5), raw cod (204 ± 0.6), braised pork top loin including fat (204 ± 0.5), raw cusk (204 ± 0.6), oil roasted macadamia nuts (204 ± 1.8), roasted turkey dark meat without skin (204 ± 0.5), simmered turkey giblets (204 ± 0.5), dried oriental radish (203 ± 0.9), roasted duck without skin (203 ± 0.5), broiled beef rib excluding fat (202 ± 0.5), broiled pork loin blade including fat (202 ± 0.5), evaporated whole milk (202 ± 0.4), cooked snapper (201 ± 0.6), raw lingcod (201 ± 0.6), roasted pork loin blade excluding fat (201 ± 0.5), bacon cheeseburger (201 ± 0.3), rennin (200 ± 4.5), savory (200 ± 50), black pepper (200 ± 25), saffron (200 ± 50), cardamom (200 ± 25), marjoram (200 ± 50), boiled arrowhead (200 ± 4.2), semi-sweet bakers chocolate (200 ± 1.8), raw burbot (200 ± 0.6), raw freshwater bass (200 ± 0.6), roasted pork shoulder including fat (200 ± 0.5), roasted light meat chicken with skin (200 ± 0.5), white pepper (200 ± 25), bulgur (200 ± 0.4), raw snapper (199 ± 0.6), raw ling (199 ± 0.6), braised pork loin including fat (199 ± 0.5), simmered chicken heart (199 ± 0.5), raw mussels (198 ± 0.6), turkey pastrami (198 ± 0.9), boiled peanuts (197 ± 1.6), dried longans (196 ± 0.5), braised pork sirloin including fat (196 ± 0.5), roasted pork center loin including fat (196 ± 0.5), roasted lean ham (196 ± 0.5), roasted turkey dark meat with skin (196 ± 0.5), raw pompano (195 ± 0.6), raw flounder (195 ±

0.5), raw sole (195 ± 0.5), broiled beef top loin including fat (195 ± 0.5), roasted pork boston blade including fat (195 ± 0.5), cheese soufflé (195 ± 0.5), evaporated skim milk (194 ± 1.6), raw sea bass (194 ± 0.6), battered and fried shark (194 ± 0.6), canned bonito (193 ± 0.5), falafel (192 ± 1), toaster pastry (192 ± 1), braised pork arm braised including fat (191 ± 0.5), well done broiled ground beef (191 ± 0.5), braised beef chuck blade roast including fat (191 ± 0.5), well done broiled extra lean ground beef (190 ± 0.5), firm raw tofu (190 ± 0.4), broiled porterhouse steak including fat (189 ± 0.5), well done fried ground beef (189 ± 0.5), turkey ham (189 ± 0.9), light pearled barley (189 ± 0.3), brie cheese (189 ± 1.8), broiled beef rib eye including fat (188 ± 0.5), breaded and fried clams (188 ± 0.6), raw haddock (188 ± 0.6), raw tilefish (187 ± 0.6), fried dark meat chicken without skin (187 ± 0.5), fried chicken thigh with skin (187 ± 0.8), lobster paste (186 ± 7.1), raw octopus (186 ± 0.6), raw spot (186 ± 0.6), pork luxury loaf (186 ± 1.8), braised pork lungs (186 ± 0.5), well done fried extra lean ground beef (185 ± 0.5), pork sausage (185 ± 1.9), cooked lobster (185 ± 0.6), light rye flour (185 ± 0.5), raw flatfish (184 ± 0.6), braised pork boston blade braised including fat (184 ± 0.5), stewed chicken drumstick without skin (184 ± 1.1), braised beef brisket including fat (184 ± 0.5), turkey frankfurter (184 ± 1.1), boiled yellow beans (183 ± 0.3), roasted chicken thigh without skin (183 ± 1), corn tortilla (183 ± 1.7), chopped smoked beef (182 ± 1.8), well done broiled lean ground beef (182 ± 0.5), fried pork loin blade including fat (182 ± 0.5), light meat turkey roll (182 ± 0.9), dried lychees (181 ± 0.5), raw dungeness crab (181 ± 0.6), well done fried lean ground beef (181 ± 0.5), boiled yardlong bean (181 ± 0.3), cheese crackers with peanut butter (181 ± 1.2), boiled lentils (180 ± 0.3), raw sunfish (180 ± 0.6), raw freshwater drum (180 ± 0.6), cheese pizza (180 ± 0.4), waffles (180 ± 0.7), chicken egg (180 = 1), dried european chestnuts (179 ± 1.8), roasted dark meat chicken without skin (179 ± 0.5), potato flour (178 ± 0.6), raw rockfish (178 ± 0.6), braised pork heart (178 ± 0.4), tuna salad (178 ± 0.2), braised beef lungs (178 ± 0.5), broiled T-bone steak including fat (177 ± 0.5), pork and beef link sausage with american cheese (177 ± 1.2), braised pork loin blade including fat (176 ± 0.5), fried dark meat chicken with skin (176 ± 0.5), fried chicken drumstick with skin (176 ± 1), chipped beef (175 ± 1.8), canned ham (175 ± 0.5), fried chicken (175 ± 0.6), roasted chicken drumstick with skin (175 ± 1), unenriched biscuits (175 ± 1.8), enriched biscuits (175 ± 1.8), natto (174 ± 0.6), raw anchovy (174 ± 0.6), braised pork tongue (174 ± 0.5), roasted chicken thigh with skin (174 ± 0.8), roasted pork loin blade including fat (173 ± 0.5), broiled lamb loin chop including fat (172 ± 0.7), roasted lamb shoulder including fat (172 ± 0.6), dried japanese chestnuts (171 ± 1.8), pork and beef peppered loaf (171 ± 1.8), medium-rare fried ground beef (171 ± 0.5), turkey egg (170 ± 0.6), breaded and fried fish (170 ± 0.7), italian pork sausage (170 ± 0.7), medium-rare broiled ground beef (170 ± 0.5), well done baked ground beef (170 ± 0.5), corn meal muffins (170 ± 1.3), raw clams (169 ± 0.6), roasted beef rib including fat (169 ± 0.5), boiled chickpeas (168 ± 0.3), toasted rye bread (168 ± 2.3), roasted dark meat chicken with skin (168 ± 0.5), boiled adzuki beans (167 ± 0.2), light and dark meat turkey roll (167 ± 0.9), pork liver sausage (167 ± 2.8), boiled great northern beans (166 ± 0.3), boiled pink beans (165 ± 0.3), taco (165 ± 0.6), broiled beef rib including fat (165 ± 0.5), fried chicken back with skin (165 ± 0.7), stewed chicken breast without skin (165 ± 0.5), pork and beef old fashioned loaf (164 ± 1.8), well done baked lean ground beef (164 ± 0.5), smoked chinook salmon (164 ± 0.6), raw milkfish (162 ± 0.6), raw grouper (162 ± 0.6), pork link sausage (162 ± 0.7), well done baked extra lean ground beef (162 ± 0.5), braised beef shortribs including fat (162 ± 0.5), medium-rare broiled extra lean ground beef (161 ± 0.5), boiled pinto beans (160 ± 0.3), medium-rare fried extra lean ground beef (160 ± 0.5), medium-rare fried lean ground beef (159 ± 0.5), breaded and fried oysters (159 ± 0.6), stewed light meat chicken without skin (159 ± 0.5), boiled green soybeans (158 ± 0.6), medium-rare broiled lean ground beef (158 ± 0.5), sheep milk (158 ± 0.2), ricotta cheese (158 ± 0.4), dried chinese chestnuts (157 ± 1.8), boiled navy beans (157 ± 0.3), chopped ham lunch meat (157 ± 2.4), minced ham lunch meat (157 ± 2.4), light meat chicken roll (157 ± 0.5), broiled lamb rib chop including fat (157 ± 0.7), turkey bologna (157 ± 1.8), dinner rolls (157 ± 1.8), raw irish moss seaweed (157 ± 0.5), raw yellowtail (156 ± 0.6), raw garlic (156 ± 5.6), boiled black-eyed peas (156 ± 0.3), skim yogurt (156 ± 0.2), boiled mungo beans (156 ± 0.3), teriyaki sauce (156 ± 2.8), chicken roll (156 ± 0.9), roasted duck with skin (156 ± 0.5), roasted pork blade roll (156 ± 0.5), stewed chicken breast with skin (156 ± 0.5), raw pork stomach (155 ± 0.5), cheeseburger (155 ± 0.4), simmered chicken gizzard (155 ± 0.5), potato chips (154 ± 1.8), semi-sweet chocolate (154 ± 1.8), beef pastrami (154 ± 1.8), stir-fried sprouted lentils (153 ± 0.5), boiled winged beans (153 ± 0.3), miso (153 ± 0.4), boiled black turtle beans (152 ± 0.3), lebanon bologna (152 ± 2.2), smoked cisco (151 ± 0.6), braised/roasted/stewed veal chuck (151 ± 0.6), oregano (150 ± 25), ginger (150 ± 25), poultry seasoning (150 ± 25), boiled mothbeans (150 ± 0.3), roasted chicken wing with skin (150 ± 1.5), fried chicken wing with skin (150 ± 1.6), unheated ham patties (149 ± 0.8), roast beef sandwich (148 ± 0.3), pork bratwurst (148 ± 0.6), raw frog legs (147 ± 0.5), pork and beef honey loaf (146 ± 1.8), pork and beef kielbasa/kolbassy (146 ± 1.9), stewed light meat chicken with skin (146 ± 0.5), fish sandwich (145 ± 0.3), hot fudge sundae (144 ± 0.3), lowfat yogurt (144 ± 0.2), rye bread (144 ± 2), raw whale meat (144 ± 0.5), pork pickle and pimento loaf (143 ± 1.8), sweet chocolate (143 ± 1.8), stewed dark meat chicken without skin (143 ± 0.5), toasted cracked wheat bread (143 ± 2.4), raw bacon (143 ± 0.7), cooked grouper (142 ± 0.6), boiled kidney beans (142 ± 0.3), boiled catjang black-eyed peas (142 ± 0.3), simmered beef tongue (142 ± 0.5), raw whelk (141 ± 0.6), dry breadcrumbs (141 ± 0.5), pancakes (141 ± 1.9), hard pork and beef salami (140 ± 5), boiled black beans (140 ± 0.3), caramel sundae (140 ± 0.3), stewed chicken drumstick with skin (140 ± 0.9), popover roll (140 ± 1.3), brown mustard (140 ± 10), wheat gluten (140 ± 0.4), dried macadamia nuts (139 ± 1.8), pork bologna (139 ± 2.2), polish pork sausage (139 ± 1.8), canned oysters (139 ± 0.6), raw oysters (139 ± 0.6), neufchatel cheese (139 ± 1.8), cooked cod (138 ± 0.6), oatmeal cookie (138 ± 3.8), braised/stewed veal breast plate (138 ± 0.6), stewed chicken thigh with skin (138 ± 0.7), medium-rare baked ground beef (137 ± 0.5), toasted english muffin (136 ± 1), boiled cranberry beans (136 ± 0.3), steamed sprouted soybeans (136 ± 1.1), cooked shrimp (136 ± 0.6), canned great northern beans (135 ± 0.2), beef honey roll sausage (135 ± 2.2), burrito (135 ± 0.3), brownies (135 ± 2.5), canned navy beans (134 ± 0.2), pork and beef brotwurst (134 ± 0.7), french toast (134 ± 0.8), fried chicken neck with skin (133 ± 1.4), raw queen crab (133 ± 0.6), stewed dark meat chicken with skin (133 ± 0.5), blueberry muffins (133 ± 1.3), smoked whitefish (132 ± 0.6), cottage cheese (132 ± 1.8), barbecue loaf (130 ± 2.2), berliner (130 ± 2.2), pork olive loaf (129 ± 1.8), german sweet bakers chocolate

(129 ± 1.8), pork and beef picnic loaf (129 ± 1.8), pork mothers loaf (129 ± 2.4), tamari sauce (129 ± 0.9), boiled lupins (128 ± 0.3), medium-rare baked lean ground beef (128 ± 0.5), simmered turkey gizzard (128 ± 0.5), simmered chicken neck without skin (128 ± 2.8), cracked wheat bread (128 ± 2), boiled baby lima beans (127 ± 0.3), devil's food cake (127 ± 0.8), chocolate milkshake (126 ± 0.2), chocolate syrup (126 ± 1.3), ginko nuts (125 ± 1.8), boiled broad beans (125 ± 0.3), oatmeal bread (125 ± 1.8), cooked corned beef brisket (125 ± 0.5), medium-rare baked extra lean ground beef (124 ± 0.5), toasted white bread (124 ± 2.4), frozen and boiled black-eyed peas (122 ± 0.6), coconut cream (122 ± 0.2), pork and beef luncheon sausage (122 ± 2.2), rusk crackers (122 ± 5.6), beef lunch meat (121 ± 1.8), graham crackers (121 ± 3.6), ham salad (121 ± 1.8), simmered chicken neck with skin (121 ± 1.3), boiled pigeon peas (120 ± 0.3), boiled hyacinth beans (120 ± 0.3), stewed chicken wing with skin (120 ± 1.3), dried spirulina seaweed (118 ± 0.5), french fries (118 ± 0.6), boiled green peas (118 ± 0.6), dried apricots (117 ± 1.4), dried pepeao (117 ± 4.2), boiled succotash (117 ± 0.5), pork and beef pepperoni (117 ± 8.3), baked beans with pork and tomato sauce (117 ± 0.2), beef and pork salami (117 ± 2.2), indian buffalo milk (117 ± 0.2), custard (117 ± 0.2), malted milk (116 ± 0.2), coconut custard pie (116 ± 0.4), golden seedless raisins (115 ± 0.5), vanilla milkshake (115 ± 0.2), cornbread (115 ± 0.6), corn pone (115 ± 0.6), animal crackers (115 ± 1.9), cottage pudding (115 ± 0.9), fruitcake, dark (114 ± 1.2), fruitcake, light (114 ± 1.2), bread pudding (114 ± 0.2), tamarind (113 ± 0.4), boiled white beans (113 ± 0.3), au gratin potatoes (113 ± 0.4), raw coconut (113 ± 1.1), summer sausage (113 ± 2.2), beef salami (113 ± 2.2), custard pie (113 ± 0.4), shortbread cookie (113 ± 6.3), english muffin (112 ± 0.9), boiled and frozen baby lima beans (112 ± 0.6), protein fortified lowfat 2% milk (112 ± 0.2), protein fortified skim milk (112 ± 0.2), hummus (112 ± 0.2), boiled lima beans (111 ± 0.3), goat milk (111 ± 0.2), protein fortified lowfat 1% milk (111 ± 0.2), canned corned beef (111 ± 0.5), shoyu sauce (110 ± 0.9), strawberry sundae (110 ± 0.3), microwaved potato without skin (109 ± 0.3), hamburger sandwich (109 ± 0.5), eggnog (109 ± 0.2), chocolate cream pie (109 ± 0.4), tapioca cream pudding (109 ± 0.3), pork and beef sausage (108 ± 3.8), canned black turtle beans (108 ± 0.2), raw green peas (108 ± 0.6), hot chocolate (108 ± 0.2), white bread (108 ± 2.1), roasted european chestnuts (107 ± 1.8), pork and beef link sausage (107 ± 0.7), cream cheese (107 ± 1.8), french bread (107 ± 1.8), yellow cake (107 ± 0.7), dried sweetened shredded coconut (106 ± 0.5), microwaved potato with skin (105 ± 0.2), baked beans with pork and sweet sauce (105 ± 0.2), canned kidney beans (105 ± 0.2), turkey salami (105 ± 0.9), smoked beef sausage (105 ± 1.2), acorn flour (104 ± 1.8), dried acorns (104 ± 1.8), roasted chinese chestnuts (104 ± 1.8), vegetarian baked beans (104 ± 0.2), pork beerwurst salami (104 ± 2.2), beef summer sausage (104 ± 2.2), potato pancakes (103 ± 0.7), raw mushrooms (103 ± 1.4), peach pie (103 ± 0.5), pecan pie (103 ± 0.5), maypo (103 ± 0.3), boiled french beans (102 ± 0.3), boiled yellow corn (102 ± 0.6), restaurant chocolate milkshake (102 ± 0.2), restaurant vanilla milkshake (102 ± 0.2), hot dog (102 ± 0.6), lowfat 1% chocolate milk (102 ± 0.2), lowfat 2% chocolate milk (102 ± 0.2), danish pastry (102 ± 1.2), skim milk (101 ± 0.2), boston cream pie (101 ± 0.5), cloves (100 ± 25), allspice (100 ± 25), boiled and steamed european chestnuts (100 ± 1.8), sage (100 ± 50), raw lotus root (100 ± 0.6), pumpkin pie spice (100 ± 25), dried jujube (100 ± 0.5), mace (100 ± 25), bay leaf (100 ± 50), gingerbread (100 ± 0.8), toasted raisin bread (100 ± 2.4), dried

II. Minerals

PHOSPHORUS

sweetened flaked coconut (100 ± 2), boiled mung beans (100 ± 0.2), coconut milk (100 ± 0.2), grilled ham patties (100 ± 0.8), beef and pork summer sausage (100 ± 2.2), whole chocolate malted milk (100 ± 0.2), restaurant strawberry milkshake (100 ± 0.2), whole chocolate milk (100 ± 0.2), pork and beef mortadella (100 ± 3.3), saltine crackers (100 ± 8.3), pita bread (100 ± 1.3), puffed rice (100 ± 3.6), bread sticks (100 ± 2.5), sugar cookie (100 ± 3.1), boiled split peas (99 ± 0.3), pork and beef knockwurst (99 ± 0.7), enriched white degermed corn meal (99 ± 0.4), unenriched corn meal (99 ± 0.4), enriched yellow degermed corn meal (99 ± 0.4), unenriched yellow degermed corn meal (99 ± 0.4), chocolate pudding (98 ± 0.2), chocolate chiffon pie (98 ± 0.6), sponge cake (98 ± 0.8), seedless raisins (97 ± 0.5), filled milk (97 ± 0.2), raw tofu (97 ± 0.4), raw chinese chestnuts (96 ± 1.8), beef beerwurst salami (96 ± 2.2), lowfat 1% milk (96 ± 0.2), peanut brittle (96 ± 1.8), onion rings (95 ± 0.6), whole yogurt (95 ± 0.2), lowfat 2% milk (95 ± 0.2), enriched wheat bread (95 ± 0.4), caramel cake (95 ± 0.5), french fried potatoes (94 ± 1), roasted japanese chestnuts (93 ± 1.8), raw european chestnuts (93 ± 1.8), boiled burdock root (93 ± 0.4), whole milk (93 ± 0.2), soy sauce (93 ± 0.9), half and half cream (93 ± 3.3), italian bread (93 ± 1.8), canned pinto beans (92 ± 0.2), canned white beans (91 ± 0.2), raisin bun (91 ± 0.8), beef and pork bologna (91 ± 2.2), turkey salad (91 ± 0.9), vanilla pudding (91 ± 0.2), canned chickpeas (90 ± 0.2), frozen and boiled green peas (90 ± 0.6), buttermilk (89 ± 0.2), carob milk (89 ± 0.2), soda crackers (89 ± 1.8), pretzels (89 ± 1.8), light buckwheat flour (88 ± 0.5), raisin bread (88 ± 2), beef frankfurter (88 ± 0.9), oyster crackers (88 ± 6.3), cheesecake (88 ± 0.6), boiled mushrooms (87 ± 0.6), beef bologna (87 ± 2.2), all purpose enriched wheat flour (87 ± 0.4), pickled herring (87 ± 3.3), frozen and french fried potatoes without skin (86 ± 1), enchilada (86 ± 0.2), canned cranberry beans (86 ± 0.2), pork and beef lunch meat (86 ± 1.8), raw pork jowl (86 ± 0.5), blackstrap molasses (85 ± 2.5), refried beans (85 ± 0.2), tuna salad spread (85 ± 0.9), white cake (85 ± 0.6), beef and pork frankfurter (84 ± 1.1), sweet potato pie (84 ± 0.4), sour cream (83 ± 4.2), hamburger bun (83 ± 1.3), hot dog bun (83 ± 1.3), lemon chiffon pie (83 ± 0.6), restaurant hash brown potatoes (82 ± 0.6), banana custard pie (82 ± 0.4), french rolls (82 ± 1), canned pork lunch meat (81 ± 2.4), butterscotch pie (81 ± 0.4), yellow mustard (80 ± 10), red beans (80 ± 0.4), pound cake (80 ± 1.7), raw wakame seaweed (80 ± 0.5), white corn flour (79 ± 0.4), yellow corn flour (79 ± 0.4), tomato paste (79 ± 0.4), carob flour (79 ± 0.5), dried japanese persimmon (79 ± 1.5), dried prunes (79 ± 0.6), raw beef tripe (79 ± 0.5), canned broad beans (79 ± 0.2), boiled lotus root (78 ± 0.6), raw jerusalem artichoke (77 ± 0.7), pineapple chiffon pie (77 ± 0.6), cooked taro (76 ± 0.8), canned butter beans (76 ± 0.4), oatmeal (76 ± 0.3), seeded raisins (75 ± 0.5), raw japanese chestnuts (75 ± 1.8), jellied corned beef loaf (75 ± 1.8), instant oatmeal (75 ± 0.3), canned lima beans (74 ± 0.2), gefiltefish with broth (74 ± 1.2), frozen white corn (73 ± 0.7), medium cream (73 ± 3.3), wheat cake (73 ± 0.4), cooked brown rice (73 ± 0.3), boiled jute potherb (72 ± 1.2), frozen and fried hash brown potatoes (72 ± 0.6), chili (72 ± 0.2), maltex (72 ± 0.3), boiled amaranth (71 ± 0.8), zwieback crackers (71 ± 7.1), sweet roll (71 ± 1.2), medium molasses (70 ± 2.5), raw cassava (70 ± 0.5), canned black-eyed peas (70 ± 0.2), raw sprouted alfalfa seeds (70 ± 1.5), boiled parsnips (69 ± 0.6), pumpkin pie (69 ± 0.4), dried figs (68 ± 0.3), boiled and steamed chinese chestnuts (68 ± 1.8), passion fruit (67 ± 2.8), raw chives (67 ± 16.7), frozen and boiled succotash (67 ± 0.6), canned green peas (67 ± 0.6), bagels (67 ± 0.9), cooked tahitian taro (66 ± 0.7), raw broccoli (66 ± 1.1), raw celeriac (66 ± 0.5), bread stuffing (66 ± 0.3), pineapple custard pie (65 ± 0.4), frozen and boiled fordhook lima beans (64 ± 0.6), raw dandelion greens (64 ± 1.8), canned white corn (64 ± 0.5), canned yellow corn (64 ± 0.5), corn flakes (64 ± 1.8), raw chinese water chestnuts (63 ± 0.8), raw dock (63 ± 0.7), scalloped potatoes (63 ± 0.4), pineapple upside-down cake (63 ± 0.7), canned yellow corn with red and green peppers (62 ± 0.4), boiled artichoke (61 ± 0.6), okara tofu (61 ± 0.8), pork and beef spread (61 ± 1.8), pork headcheese (61 ± 1.8), raw shallots (60 ± 5), boiled asparagus (60 ± 0.6), boiled artichoke hearts (60 ± 0.6), whipping cream, light (60 ± 3.3), whipping cream, heavy (60 ± 3.3), citrus marmalade (60 ± 2.5), raw bamboo shoots (59 ± 0.7), raw watercress (59 ± 2.9), black currants (59 ± 0.9), canned succotash with creamed corn (59 ± 0.4), noodles (59 ± 0.3), raw laver/nori seaweed (58 ± 0.5), baked potato with skin (57 ± 0.2), gingersnaps cookie (57 ± 7.1), yellow corn pudding (57 ± 0.2), cooked parboiled white rice (57 ± 0.3), boiled spinach (56 ± 0.6), boiled okra (56 ± 0.6), boiled brussels sprouts (56 ± 0.6), boiled salsify (56 ± 0.7), cooked parboiled extra long grain white rice (56 ± 0.6), frozen and boiled asparagus (55 ± 0.8), frozen and boiled broccoli (55 ± 0.5), canned succotash (55 ± 0.4), baked sweet potato (54 ± 0.4), frozen and boiled brussels sprouts (54 ± 0.6), raw sprouted mung beans (54 ± 1), chili sauce (53 ± 3.3), tomato catsup (53 ± 3.3), raw vine spinach (52 ± 0.5), dried agar seaweed (52 ± 0.5), boiled dock (52 ± 0.5), potato salad (52 ± 0.4), frozen creamed yellow corn (51 ± 0.4), canned creamed yellow corn (51 ± 0.4), barbados molasses (50 ± 2.5), raw spinach (50 ± 1.8), cinnamon (50 ± 25), dried rosemary (50 ± 25), baked potato without skin (50 ± 0.3), o'brien potatoes (50 ± 0.3), vienna sausage (50 ± 3.1), spaghetti (50 ± 0.4), macaroni (50 ± 0.4), pie crust (50 ± 0.3), raw welsh onions (49 ± 0.5), baked/boiled tropical yam (49 ± 0.7), boiled garden cress (49 ± 0.7), canned sweet potato (49 ± 0.3), boiled rutabaga (49 ± 0.6), boiled blackeye pea pods (49 ± 1.1), soy milk (49 ± 0.2), simmered pork feet (48 ± 0.5), frozen and boiled spinach (48 ± 0.5), frozen and boiled yellow corn (48 ± 0.6), light cream (48 ± 1.7), simmered pork tail (47 ± 0.5), raw chickory greens (47 ± 0.6), mashed potatoes (47 ± 0.5), boiled broccoli (47 ± 0.6), simmered pork chitterlings (47 ± 0.5), boiled lamb's quarters (46 ± 0.6), raw potato without skin (46 ± 0.4), raw cauliflower (46 ± 1), raw romaine lettuce (46 ± 1.8), frozen and boiled okra (46 ± 0.5), whey (46 ± 0.2), light molasses (45 ± 2.5), baked acorn squash (45 ± 0.5), boiled kohlrabi (45 ± 0.6), frozen and baked sweet potato (44 ± 0.6), canned spinach (44 ± 0.5), raw red chili peppers (44 ± 1.1), raw green chili peppers (44 ± 1.1), raw carrots (44 ± 0.7), canned pumpkin pie mix (44 ± 0.4), raw acorns (43 ± 1.8), raw turnip greens (43 ± 1.8), boiled swamp cabbage (43 ± 1), red currants (43 ± 0.9), white currants (43 ± 0.9), raw savoy cabbage (43 ± 1.4), raw red cabbage (43 ± 1.4), raw california avocado (42 ± 0.3), carrot juice (42 ± 0.3), frozen and boiled crookneck squash (42 ± 0.5), boiled dandelion greens (42 ± 1), raw kelp/kombu/tangle seaweed (42 ± 0.5), kiwi fruit (41 ± 0.7), hash brown potatoes (41 ± 0.6), boiled mustard greens (41 ± 0.7), cherimoya (40 ± 0.1), groundcherries (40 ± 0.4), boiled beet greens (40 ± 0.7), dried dates (40 ± 0.6), raw parsley (40 ± 1.7), steamed hawaii mountain yam (40 ± 0.7), tomato purée (40 ± 0.2), boiled potato without skin (40 ± 0.4), boiled zucchini (40 ± 0.6), raisin pie (40 ± 0.4), russian salad dressing (40 ± 3.3), soybean oil and cottonseed oil margarine liquid (40 ± 10), lemon meringue pie (40 ± 0.4), raw florida avocado (39 ± 0.2), raw celtuce (39 ± 0.5), raw swamp cabbage (39 ± 0.9), boiled green beans (39 ± 0.8), boiled yellow snap beans (39 ± 0.8), elderberries (39 ± 0.3), boiled crookneck squash (39 ± 0.6), poi (39 ± 0.4), hot dog pickle relish (39 ± 1.8), boiled purslane (38 ± 0.9), boiled scotch kale (38 ± 0.8), mulberries (38 ± 0.4), mince pie (38 ± 0.4), low calorie russian salad dressing (38 ± 3.1), raw chinese cabbage (37 ± 1.4), roselle (37 ± 0.9), dried pinyon pine nuts (36 ± 1.8), jackfruit (36 ± 0.5), marinara sauce (35 ± 0.2), cooked prunes (35 ± 0.5), boiled cauliflower (35 ± 0.8), raw scallop squash (35 ± 0.8), raw leeks (35 ± 1.9), pickled pork feet (34 ± 0.5), canned pumpkin (34 ± 0.4), boiled pokeberry shoots (33 ± 0.6), boiled swiss chard (33 ± 0.6), horseradish sauce (33 ± 3.3), prune pudding (33 ± 0.4), frozen and boiled turnip greens (33 ± 0.6), boiled savoy cabbage (33 ± 0.7), tomato sauce (32 ± 0.4), raw spring onions with tops (32 ± 1), sugar apple (32 ± 0.3), raw zucchini (32 ± 0.8), raw eggplant (32 ± 1.2), raw crookneck squash (32 ± 0.8), chicken and turkey salad (32 ± 1.8), coleslaw (32 ± 0.8), canned sprouted mung beans (32 ± 0.8), boiled dishcloth gourd (31 ± 0.6), boiled beets (31 ± 0.6), boiled carrots (31 ± 0.6), lychees (31 ± 0.5), breadfruit (30 ± 0.5), boiled pumpkin (30 ± 0.4), raw ginger root (29 ± 2.1), boiled chinese cabbage (29 ± 0.6), canned zucchini in tomato juice (29 ± 0.4), boiled red tomato (29 ± 0.4), boiled turnip greens (29 ± 0.7), boiled chayote (29 ± 0.6), raw onions (29 ± 0.6), raw new zealand spinach (29 ± 1.8), cooked shitake mushrooms (29 ± 0.7), safflower oil and soybean oil mayonnaise (29 ± 3.6), soybean oil mayonnaise (29 ± 3.6), raw mustard spinach (28 ± 0.7), cooked plantain (28 ± 0.3), sapote (28 ± 0.2), frozen and boiled kale (28 ± 0.8), raw endive (28 ± 2), persimmon (28 ± 2), canned potato without skin (28 ± 0.6), boiled kale (28 ± 0.8), raw green tomato (28 ± 0.4), boiled red cabbage (28 ± 0.7), boiled scallop squash (28 ± 0.6), angel food cake (28 ± 0.8), cooked white rice (28 ± 0.2), bread and butter pickles (27 ± 3.3), soursop (27 ± 0.2), loquats (27 ± 0.5), boiled acorn squash (27 ± 0.4), frozen and boiled collards (27 ± 0.6), gooseberries (27 ± 0.3), boiled sweet potato (27 ± 0.3), boysenberries (27 ± 0.4), cooked sprouted mung beans (27 ± 0.8), whipped butter (27 ± 4.5), frozen and boiled potatoes without skin (26 ± 0.5), boiled butternut squash (26 ± 0.5), guava (26 ± 0.6), candied sweet potato (26 ± 0.5), frozen and boiled turnips (26 ± 0.5), rhubarb pie (26 ± 0.4), frozen and boiled carrots (26 ± 0.7), loganberries (26 ± 0.3), blackberry pie (26 ± 0.4), raw coriander (25 ± 12.5), raw celery (25 ± 1.3), yellow passion fruit juice (25 ± 0.2), prune juice (25 ± 0.2), raw looseleaf lettuce (25 ± 1.8), frozen and boiled zucchini (25 ± 0.4), boiled and steamed japanese chestnuts (25 ± 1.8), strawberry pie (25 ± 0.5), cherry pie (25 ± 0.4), canned bamboo shoots (25 ± 0.4), sorghum (24 ± 2.4), boiled celery (24 ± 0.7), boiled oriental radish (24 ± 0.7), boiled sprouted green peas (24 ± 0.5), prickly pear (24 ± 0.5), raw red tomato (24 ± 0.4), boiled green cabbage (24 ± 0.7), frozen and boiled mustard greens (24 ± 0.7), frozen and boiled cauliflower (24 ± 0.6), frozen and boiled yellow snap beans (24 ± 0.7), frozen and boiled green beans (24 ± 0.7), simmered pork ears (24 ± 0.5), boiled cardoon (23 ± 0.5), baked hubbard squash (23 ± 0.5), jujube (23 ± 0.5), boiled eggplant (23 ± 1), raw green cabbage (23 ± 1.4), raw oriental radish (23 ± 1.1), boiled yambean (23 ± 0.5), canned carrots (23 ± 0.7), boiled onions (23 ± 0.5), apple pie (23 ± 0.4), blueberry pie (23 ± 0.4), dill pickles (22 ± 0.8), raw red sweet peppers (22 ± 1), raw green sweet peppers (22 ± 1), boiled new zealand spinach (22 ± 0.6), sweetened coconut cream (22 ± 0.2), apple brown betty (22 ± 0.2), cooked wild long grain rice (22 ± 0.4), longans (21 ± 0.5), blackberries (21 ± 0.7), kumquats (21 ± 2.6), canned turnip greens (21 ± 0.4), pineapple pie (21 ± 0.4), restaurant hot chocolate (21 ± 0.2), light mayonnaise (21 ± 3.6), boiled bamboo shoots (20 ± 0.4), apricots (20 ± 0.5), coconut water (20 ± 0.2), canned stewed red tomatoes (20 ± 0.4), raw chickory (20 ± 1.1), raw iceberg lettuce (20 ± 2.5), canned red sweet peppers (20 ± 0.7), canned green sweet peppers (20 ± 0.7), canned chinese water chestnuts (20 ± 0.7), low calorie thousand island salad dressing (20 ± 3.3), canned crookneck squash (20 ± 0.5), corn oil margarine tub (20 ± 10), safflower oil margarine tub (20 ± 10), soybean oil margarine tub (20 ± 10), soybean oil and cottonseed oil margarine tub (20 ± 10), corn oil margarine stick (20 ± 10), safflower oil and soybean oil margarine stick (20 ± 10), soybean oil margarine stick (20 ± 10), soybean oil and cottonseed oil margarine stick (20 ± 10), corn oil diet margarine tub (20 ± 10), soybean oil and cottonseed oil diet margarine tub (20 ± 10), butter (20 ± 3.3), sweet butter (20 ± 3.3), cooked instant white rice (19 ± 0.3), bananas (19 ± 0.4), brown sugar (19 ± 0.3), cherries (19 ± 0.7), canned peeled red tomatoes (19 ± 0.4), restaurant tomato juice (19 ± 0.4), tomato juice (19 ± 0.3), navel orange (19 ± 0.4), canned sauerkraut (19 ± 0.4), barbecue sauce (19 ± 3.1), strawberries (19 ± 0.3), boiled turnips (19 ± 0.6), thousand island salad dressing (19 ± 3.1), canned green beans (19 ± 0.7), canned yellow snap beans (19 ± 0.7), boiled mustard spinach (18 ± 0.6), canned red chili peppers (18 ± 0.7), canned green chili peppers (18 ± 0.7), raw radish (18 ± 1.1), pickled beets (18 ± 0.4), canned jalapeño peppers (18 ± 0.7), lime (18 ± 0.7), hamburger pickle relish (18 ± 1.8), canta- loupe (17 ± 0.3), orange peel (17 ± 8.3), pummelo (17 ± 0.3), quince (17 ± 0.5), valencia orange (17 ± 0.4), lemon peel (17 ± 8.3), japanese persimmon (17 ± 0.3), raw cucumber (17 ± 1), java plum (17 ± 0.4), frozen and boiled turnip greens and turnips (17 ± 0.5), cream of rice (17 ± 0.3), boiled waxgourd (17 ± 0.6), cola (17 ± 0.1), orange juice (17 ± 0.2), honeydew melon (16 ± 0.5), nectarine (16 ± 0.4), restaurant orange juice (16 ± 0.4), carambola (16 ± 0.4), lemon (16 ± 0.9), cream of wheat (16 ± 0.3), sour pickles (15 ± 0.8), crab apples (15 ± 0.5), white grapefruit juice (15 ± 0.2), red grapefruit juice (15 ± 0.2), boiled red sweet peppers (15 ± 0.7), boiled green sweet peppers (15 ± 0.7), rose table wine (15 ± 0.5), boiled leeks (15 ± 1.9), pickled green olives (15 ± 1.1), figs (14 ± 1), wax beans (14 ± 0.4), boiled hubbard squash (14 ± 0.4), tangerine juice (14 ± 0.2), screwdriver (14 ± 0.2), bloody mary (14 ± 0.3), frozen and boiled butternut squash (14 ± 0.4), boiled/baked spaghetti squash (14 ± 0.6), red table wine (14 ± 0.5), white table wine (14 ± 0.5), raw pepeao (14 ± 0.5), cooked enriched white degermed corn meal (14 ± 0.2), cooked unenriched corn meal (14 ± 0.2), cooked enriched yellow degermed corn meal (14 ± 0.2), cooked unenriched yellow degermed corn meal (14 ± 0.2), mayonnaise (14 ± 3.6), corn syrup (14 ± 2.4), sweet pickles (13 ± 3.3), purple passion fruit juice (13 ± 0.2), peach (13 ± 0.6), european grapes (13 ± 0.3), thompson seedless grapes (13 ± 0.3), low calorie french salad dressing (13 ± 3.1), frozen and boiled red sweet peppers (13 ± 0.5), frozen and boiled green sweet peppers (13 ± 0.5), human milk (13 ± 1.6), sapodilla (12 ± 0.3), boiled calabash gourd (12 ± 0.7), raspberries (12 ± 0.4), chicken egg white (12 ± 1.5), raw frozen rhubarb (12 ± 0.4), beer (12 ± 0.1), corn grits (12 ± 0.2), instant corn grits (12 ± 0.4), farina (12 ± 0.3), maple sugar (11 ± 1.8), plum (11 ± 0.8), mango (11 ± 0.2), raw acerola (11 ± 0.5), grape juice (11 ± 0.2), raw spirulina seaweed (11 ± 0.5), pear (11 ± 0.3), pitanga (11 ± 0.3), mammy apple (11 ± 0.5), american grapes (10 ± 0.5), maple syrup (10 ± 2.5), tangerine (10 ± 0.6), mandarin oranges (10 ± 0.4), tequila sunrise (10 ± 0.3), boiled collards (10 ± 0.3), dry dessert wine (10 ± 0.8), sweet dessert wine (10 ± 0.8), blueberries (10 ± 0.3), oheloberries (10 ± 0.4), cooked extra long grain white rice (10 ± 0.6), pink grapefruit (9 ± 0.4), watermelon (9 ± 0.3), apricot nectar (9 ± 0.2), acerola juice (9 ± 0.2), pomegranate (8 ± 0.3), white grapefruit (8 ± 0.4), pineapple juice (8 ± 0.2), rose apple (8 ± 0.5), frozen and cooked sweetened rhubarb (8 ± 0.4), cooked apple without skin (8 ± 0.3), cranberry sauce (8 ± 0.5), boiled butterbur (7 ± 0.5), casaba melon (7 ± 0.3), apple juice (7 ± 0.2), raw apple with skin (7 ± 0.4), pineapple (7 ± 0.3), raw apple without skin (7 ± 0.4), lime juice (7 ± 0.2), cider vinegar (7 ± 3.3), piña colada (7 ± 0.4), applesauce (7 ± 0.4), whiskey sour (7 ± 0.6), cream puff (7 ± 0.4), manhattan (7 ± 0.9), daiquiri (7 ± 0.8), low calorie italian salad dressing (7 ± 3.3), italian salad dressing (7 ± 3.3), lemon juice (6 ± 0.2), peach nectar (6 ± 0.2), 53 proof coffee liqueur (6 ± 1), 63 proof coffee liqueur (6 ± 1), carissa (5 ± 2.5), papaya (5 ± 0.2), raw agar seaweed (5 ± 0.5), honey (5 ± 2.4), 80 proof rum (5 ± 1.2), 94 proof gin (5 ± 1.2), 94 proof rum (5 ± 1.2), 94 proof vodka (5 ± 1.2), 94 proof whiskey (5 ± 1.2), 100 proof gin (5 ± 1.2), 100 proof rum (5 ± 1.2), 100 proof vodka (5 ± 1.2), 100 proof whiskey (5 ± 1.2), 86 proof whiskey (5 ± 1.2), 80 proof vodka (5 ± 1.2), apple tapioca pudding (4 ± 0.2), jelly beans (4 ± 1.8), martini (3 ± 0.7), pear nectar (3 ± 0.2), orange soda (3 ± 0.1), raw turbot (2 ± 0.6), frozen and boiled onions (2 ± 0.5), restaurant cranberry juice cocktail (2 ± 0.3), bourbon & soda (2 ± 0.4), coffee (1 ± 0.3), restaurant coffee (1 ± 0.3), black tea (1 ± 0.3), cranberry juice cocktail (1 ± 0.2), instant tea (1 ± 0.2), tea (1 ± 0.2), gin & tonic (1 ± 0.2)

Applications: fracture, softening of bones, thinning of bones, rickets, stunted growth, colitis, mental illness, hardening of the arteries, atherosclerosis, arthritis, leg cramp, tooth and gum deterioration, backache, cancer, pregnancy, stress

Daily Dosage: (at least 1300 mg phosphorus per every 1000 mg calcium, but not more than 2000 mg phosphorus per every 1000 mg calcium) 1974 PCS— 1250–1500 mg; 1980 RDA—800–1200 mg; 1980 USRDA—1000 mg; 1989 RDA—800–1200 mg; 1989 USRDA—1000 mg; nutritional—800–1600 mg; therapeutic—2000 mg; experimental—3000–6000 mg; toxic—6000 mg +

Warnings: increased amounts are dangerous. See toxicity symptoms.

POTASSIUM

Names: potassium, K, K^+, element 19
Classification: mineral
Deficiency: decreased blood sugar, confusion, anxiety, nervous system deterioration, depression, deterioration of memory, ear noises, acne, dry skin, granulation of eyelids, blistering skin eczema, skin eruptions, warts, lack of sleep, digestion upset, gas, constipation, nausea, improper fat digestion, yellow coating on back of tongue, heart deterioration, muscular weakness, pains in extremities, fatigue, muscle cramps, numbness, tingling, paralysis, fatigue, faint/rapid pulse
Side-effects: see toxicity symptoms
Toxicity: increased T-waves, decreased P-waves, cardiac arrest, slow/irregular pulse, decreased blood pressure, diarrhea, anxiety, muscular weakness, numb/tingling hands, numb/tingling feet, numb/tingling tongue
Inhibitors: cortisone and aldosterone drugs, alcohol, coffee, sugar, diuretics, aspirin, rubidium

POTASSIUM

Helpers: vitamin B_6, magnesium, sodium

Sources (mg per 100 grams of food): freeze-dried parsley (6286 ± 35.7), dried parsley (4900 ± 50), basil (4800 ± 50), dried oriental radish (3495 ± 0.9), dried dill weed (3300 ± 50), freeze-dried red sweet peppers (3188 ± 31.3), freeze-dried green sweet peppers (3188 ± 31.3), freeze-dried chives (3000 ± 62.5), blackstrap molasses (2925 ± 2.5), turmeric (2800 ± 25), dried chervil (2800 ± 50), dried coriander leaf (2700 ± 50), low fat soybean flour (2570 ± 0.6), soybean flour (2515 ± 0.6), defatted soy meal (2490 ± 0.4), paprika (2450 ± 25), tarragon (2400 ± 25), defatted soybean flour (2384 ± 0.5), soybean protein concentrate (2204 ± 1.8), torula yeast (2071 ± 1.8), roasted soybean flour (2040 ± 0.6), dry bakers yeast (2021 ± 1.8), tomato powder (1927 ± 0.5), brewers yeast (1918 ± 1.8), cumin seed (1900 ± 25), cayenne pepper (1800 ± 25), dry skim milk (1793 ± 1.7), dry instant skim milk (1705 ± 0.5), fennel seed (1700 ± 25), chili powder (1667 ± 16.7), dry cocoa powder (1640 ± 10), dehydrated onion flakes (1621 ± 3.6), potato flour (1588 ± 0.6), curry powder (1550 ± 25), dried shitake mushrooms (1533 ± 3.3), savory (1500 ± 50), anise seed (1500 ± 25), toasted soybean nuts (1489 ± 1.8), rice polish (1480 ± 0.9), freeze-dried shallots (1475 ± 12.5), dry buttermilk (1471 ± 7.1), roasted soybean nuts (1470 ± 0.6), dried and salted cod (1458 ± 0.6), almond meal (1421 ± 1.8), corn germ (1420 ± 0.5), celery seed (1400 ± 25), caraway seed (1400 ± 25), dried apricots (1377 ± 1.4), rice bran (1370 ± 0.5), dry roasted soybean nuts (1364 ± 0.6), dried spirulina seaweed (1363 ± 0.5), dry whole milk (1331 ± 1.6), potato chips (1318 ± 1.8), black pepper (1300 ± 25), peanut flour (1300 ± 12.5), bran (1250 ± 1.8), dill seed (1250 ± 25), oregano (1250 ± 25), saffron (1200 ± 50), ginger (1200 ± 25), coriander seed (1150 ± 25), cloves (1150 ± 25), dried agar seaweed (1125 ± 0.5), dried lychees (1110 ± 0.5), dried pistachio nuts (1107 ± 1.8), cardamom (1100 ± 25), medium molasses (1065 ± 2.5), garlic powder (1033 ± 16.7), dried ginko nuts (1011 ± 1.8), onion powder (1000 ± 25), allspice (1000 ± 25), dried european chestnuts (1000 ± 1.8), dry roasted pistachio nuts (982 ± 1.8), toasted wheat germ (957 ± 1.8), tomato paste (932 ± 0.4), light molasses (915 ± 2.5), boiled beet greens (908 ± 0.7), dry roasted red pistachio nuts (907 ± 1.8), marjoram (900 ± 50), boiled arrowhead (883 ± 4.2), unsweetened baking chocolate (864 ± 1.8), dry roasted sunflower seeds (861 ± 1.8), dark rye flour (860 ± 0.4), carob flour (827 ± 0.5), seeded raisins (825 ± 0.5), dried pumpkin and squash seeds (818 ± 1.8), roasted pumpkin and squash seeds (818 ± 1.8), dried japanese persimmon (803 ± 1.5), toasted almonds (786 ± 1.8), dry roasted almonds (782 ± 1.8), dried japanese chestnuts (779 ± 1.8), oil roasted spanish peanuts (775 ± 1.8), yellow mustard seed (767 ± 16.7), raw cassava (764 ± 0.5), almond butter (756 ± 3.1), seedless raisins (751 ± 0.5), crunchy peanut butter (747 ± 1.6), golden seedless raisins (746 ± 0.5), dried prunes (745 ± 0.6), dried almonds (743 ± 1.8), dried chinese chestnuts (736 ± 1.8), french fried potatoes (732 ± 1), natto (730 ± 0.6), dried peanuts (729 ± 1.8), boiled and steamed european chestnuts (725 ± 1.8), acorn flour (721 ± 1.8), dried acorns (718 ± 1.8), oil roasted peanuts (714 ± 1.8), dried figs (712 ± 0.3), dried pepeao (708 ± 4.2), potato pancakes (708 ± 0.7), french fries (704 ± 0.6), sage (700 ± 50), fenugreek seed (700 ± 12.5), dried sunflower seeds (700 ± 1.8), cooked whelk (694 ± 0.6), oil roasted almonds (693 ± 1.8), creamy peanut butter (688 ± 3.1), toasted english muffin (686 ± 1), baked/boiled tropical yam (669 ± 0.7), poppy seed (667 ± 16.7), dried longans (658 ± 0.5), dried watermelon seeds (657 ± 1.8), dry roasted peanuts (657 ± 1.8), almond paste (657 ± 1.8), oil roasted virginia peanuts (654 ± 1.8), dried dates (652 ± 0.6), bran flakes (643 ± 1.8), boiled amaranth (641 ± 0.8), dried pinyon pine nuts (636 ± 1.8), cooked pompano (636 ± 0.6), cooked rainbow trout (634 ± 0.6), raw california avocado (634 ± 0.3), tamarind (628 ± 0.4), cooked tahitian taro (622 ± 0.7), bittersweet chocolate (621 ± 1.8), compressed bakers yeast (618 ± 1.8), oil roasted valencia peanuts (611 ± 1.8), dried pignolia pine nuts (607 ± 1.8), dried brazil nuts (607 ± 1.8), whole grain rye crackers (600 ± 3.8), roasted european chestnuts (600 ± 1.8), cooked flounder (587 ± 0.5), raw chinese water chestnuts (584 ± 0.8), falafel (584 ± 1), cooked halibut (576 ± 0.6), dry roasted cashews (571 ± 1.8), stir-fried sprouted soybeans (567 ± 0.5), dried toasted coconut (561 ± 1.8), english muffin (560 ± 0.9), boiled white beans (560 ± 0.3), raw spinach (557 ± 1.8), raw lotus root (556 ± 0.6), cashew butter (554 ± 1.8), boiled jute potherb (551 ± 1.2), raw coriander (550 ± 12.5), cinnamon (550 ± 25), dried rosemary (550 ± 25), pumpkin pie spice (550 ± 25), dried coconut (550 ± 1.8), boiled swiss chard (549 ± 0.6), raw acorns (546 ± 1.8), anchovy canned in olive oil (545 ± 2.5), raw potato without skin (543 ± 0.4), oil roasted cashews (539 ± 1.8), roasted pork tenderloin (538 ± 0.5), raw parsley (537 ± 1.7), cooked coho salmon (534 ± 0.6), raw bamboo shoots (533 ± 0.7), boiled bamboo shoots (533 ± 0.4), dried black walnuts (532 ± 1.8), boiled adzuki beans (532 ± 0.2), dried jujube (531 ± 0.5), cooked clams (528 ± 0.6), canned cod (528 ± 0.6), raw european chestnuts (525 ± 1.8), cooked snapper (522 ± 0.6), cooked rockfish (520 ± 0.6), raisin bran (519 ± 1.4), ginko nuts (518 ± 1.8), boiled soybeans (515 ± 0.3), dried pilinuts (514 ± 1.8), fried beef top round excluding fat (513 ± 0.5), cooked sheepshead (512 ± 0.6), cooked tilefish (512 ± 0.6), boiled pink beans (508 ± 0.3), boiled lima beans (508 ± 0.3), dried english/persian walnuts (507 ± 1.8), braised pork center rib exlucding fat (501 ± 0.5), braised pork top loin excluding fat (501 ± 0.5), poultry seasoning (500 ± 25), toasted pumpernickel bread (496 ± 1.8), raw spot (496 ± 0.6), fruitcake, dark (495 ± 1.2), raw rainbow trout (495 ± 0.6), steamed hawaii mountain yam (494 ± 0.7), breadfruit (490 ± 0.5), oil roasted sunflower seeds (489 ± 1.8), boiled purslane (488 ± 0.9), raw florida avocado (488 ± 0.2), cooked bacon (486 ± 0.4), cooked taro (483 ± 0.8), roasted chinese chestnuts (482 ± 1.8), whole toasted sesame seeds (482 ± 1.8), raw grouper (482 ± 0.6), cooked grouper (474 ± 0.6), oil roasted filberts (471 ± 1.8), dry roasted filberts (468 ± 1.8), boiled pinto beans (468 ± 0.3), whole dried sesame seeds (467 ± 5.6), boiled spinach (466 ± 0.6), cooked plantain (465 ± 0.3), fried beef sirloin excluding fat (465 ± 0.5), fried pork center rib excluding fat (465 ± 0.5), fried pork top loin excluding fat (465 ± 0.5), fried beef top round including fat (459 ± 0.5), cooked oysters (458 ± 0.6), frozen and french fried potatoes without skin (458 ± 1), cooked mullet (458 ± 0.6), fried pork center loin excluding fat (455 ± 0.5), raw chinese chestnuts (454 ± 1.8), canned white beans (454 ± 0.2), boiled dishcloth gourd (453 ± 0.6), ry-krisp crackers (450 ± 3.6), chipped beef (450 ± 1.8), dried filberts (450 ± 1.8), raw halibut (449 ± 0.6), kippered herring (448 ± 1.3), microwaved potato with skin (447 ± 0.2), simmered beef shank excluding fat (447 ± 0.5), dried hickory nuts (443 ± 1.8), braised pork sirloin excluding fat (443 ± 0.5), broiled beef top round excluding fat (442 ± 0.5), broiled pork center rib excluding fat (439 ± 0.5), broiled pork top loin excluding fat (439 ± 0.5), baked acorn squash (437 ± 0.5), frozen and fried hash brown potatoes (436 ± 0.6), raw lingcod (436 ± 0.6), pumpernickel bread (434 ± 1.6), cooked whiting (434 ± 0.6), broiled pork sirloin excluding fat (434 ± 0.5), boiled black turtle beans (433 ± 0.3), raw tilefish (433 ± 0.6), millit (430 ± 0.5), broiled beef top round including fat (429 ± 0.5), raw chum salmon (429 ± 0.6), cooked carp (427 ± 0.6), dried butternuts (425 ± 1.8), marinara sauce (424 ± 0.2), smoked whitefish (424 ± 0.6), braised beef thymus (423 ± 0.5), roasted pork center rib excluding fat (423 ± 0.5), roasted pork top loin excluding fat (423 ± 0.5), raw coho salmon (422 ± 0.6), raw chickory greens (420 ± 0.6), tomato purée (420 ± 0.2), broiled pork center loin excluding fat (420 ± 0.5), cooked herring (419 ± 0.6), braised pork loin excluding fat (419 ± 0.5), broiled beef tenderloin excluding fat (419 ± 0.5), baked potato with skin (418 ± 0.2), raw snapper (418 ± 0.6), broiled pork loin excluding fat (418 ± 0.5), raw ginger root (417 ± 2.1), simmered beef shank including fat (417 ± 0.5), braised pork loin blade excluding fat (417 ± 0.5), raw dolphinfish (416 ± 0.6), smoked haddock (415 ± 0.6), broiled beef round excluding fat (415 ± 0.5), dried sesame kernels (413 ± 6.3), sesame butter (413 ± 3.3), raw cod (413 ± 0.6), braised pork center rib including fat (413 ± 0.5), braised pork boston blade excluding fat (412 ± 0.5), toasted sesame kernels (411 ± 1.8), semi-sweet bakers chocolate (411 ± 1.8), boiled and frozen baby lima beans (411 ± 0.6), microwaved potato without skin (411 ± 0.3), boiled succotash (409 ± 0.5), roasted ham (409 ± 0.5), broiled pork boston blade excluding fat (409 ± 0.5), frozen and boiled fordhook lima beans (408 ± 0.6), broiled pork loin blade excluding fat (408 ± 0.5), broiled porterhouse steak excluding fat (407 ± 0.5), broiled T-bone steak excluding fat (407 ± 0.5), broiled beef flank excluding fat (405 ± 0.5), braised pork arm excluding fat (405 ± 0.5), raw rockfish (405 ± 0.6), raw sheepshead (405 ± 0.6), raw burbot (404 ± 0.6), boiled kidney beans (403 ± 0.3), broiled beef sirloin excluding fat (403 ± 0.5), cooked mackerel (401 ± 0.6), boiled baby lima beans (401 ± 0.3), raw garlic (400 ± 5.6), nutmeg (400 ± 25), mace (400 ± 25), pork and beef peppered loaf (400 ± 1.8), cooked haddock (399 ± 0.6), braised pork top loin including fat (399 ± 0.5), broiled beef flank including fat (398 ± 0.5), raw dandelion greens (396 ± 1.8), dried pecans (396 ± 1.8), au gratin potatoes (396 ± 0.4), bananas (396 ± 0.4), sardines (396 ± 2.1), broiled beef top loin excluding fat (396 ± 0.5), roasted eye of beef round excluding fat (395 ± 0.5), broiled beef rib eye excluding fat (394 ± 0.5), raw chinook salmon (394 ± 0.6), fried pork loin blade excluding fat (394 ± 0.5), refried beans (393 ± 0.2), wheat flakes (393 ± 1.8), roasted beef tenderloin excluding fat (393 ± 0.5), boiled cardoon (392 ± 0.5), raw cusk (392 ± 0.6), boiled great northern beans (391 ± 0.3), baked potato without skin (391 ± 0.3), roasted pork rump excluding fat (391 ± 0.5), raw sockeye salmon (391 ± 0.6), raw dock (390 ± 0.7), milk chocolate (389 ± 1.8), roasted goose without skin (388 ± 0.5), boiled cranberry beans (387 ± 0.3), roasted beef tip excluding fat (386 ± 0.5), grilled canadian bacon (385 ± 1.1), fried beef sirloin including fat (385 ± 0.5), raw shad (384 ± 0.6), boiled pigeon peas (383 ± 0.3), chopped smoked beef (382 ± 1.8), raw anchovy (382 ± 0.6), pork luxury loaf (382 ± 1.8), pork and beef old fashioned loaf (382 ± 1.8), raw pompano (381 ± 0.6), raw white sucker (380 ± 0.6), broiled beef tenderloin including fat (380 ± 0.5), hard pork and beef salami (380 ± 5), raw ling (379 ± 0.6), broiled beef rib excluding fat (379 ± 0.5), canned sockeye salmon (378 ± 0.6), scalloped potatoes (378 ± 0.4), frozen and baked sweet potato (377 ± 0.6), roasted beef rib roasted excluding fat (376 ± 0.5), cooked sockeye salmon (375 ± 0.6), dry roasted pecans (375 ± 1.8), frozen and boiled black-eyed peas (375 ± 0.6), boiled catjang black-eyed peas (375 ± 0.3), canned blue crab (374 ± 0.6), sweetened condensed milk (374 ± 1.3), raw butterfish (374 ± 0.6), chili sauce (373 ± 3.3), roasted pork leg excluding fat (373 ± 0.5), cooked rainbow smelt (372 ± 0.6), braised pork center loin excluding fat (372 ± 0.5), raw bluefish (372 ± 0.6), boiled chinese cabbage (371 ± 0.6), dried macadamia nuts (371 ± 1.8), broiled pork center rib including fat (371 ± 0.5), raw mushrooms (371 ± 1.4), whole wheat flour (370 ± 0.4), tomato sauce (370 ± 0.4), boiled french beans (370 ± 0.3), roasted pork sirloin excluding fat (370 ± 0.5), cooked swordfish (369 ± 0.6), boiled lentils (369 ± 0.3), well done broiled extra lean ground beef (369 ± 0.5), braised pork sirloin including fat (369 ± 0.5), stewed rabbit (368 ± 0.4), boiled navy beans (368 ± 0.3), boiled parsnips (368 ± 0.6), roasted pork center rib including fat (368 ± 0.5), tempeh (367 ± 0.6), roasted pork loin excluding fat (367 ± 0.5), broiled pork sirloin including fat (367 ± 0.5), raw flounder (366 ± 0.5), raw sole (366 ± 0.5), broiled beef round including fat (366 ± 0.5), shredded wheat (364 ± 1.8), cooked sturgeon (364 ± 0.6), oil roasted pecans (364 ± 1.8), fried beef liver (364 ± 0.5), canned lean ham (364 ± 0.5), boiled lotus root (363 ± 0.6), fried pork center loin including fat (363 ± 0.5), boiled split peas (362 ± 0.3), roasted pork center loin excluding fat (362 ± 0.5), roasted eye of beef round including fat (362 ± 0.5), raw trout (361 ± 0.6), raw flatfish (361 ± 0.6), tomato catsup (360 ± 3.3), boiled burdock root (360 ± 0.4), roasted pork shank excluding fat (360 ± 0.5), well done fried extra lean ground beef (360 ± 0.5), broiled beef sirloin including fat (360 ± 0.5), pork sausage (359 ± 1.9), broiled pork center loin including fat (359 ± 0.5), roasted pork top loin including fat (359 ± 0.5), baked hubbard squash (358 ± 0.5), raw mullet (358 ± 0.6), roasted canned ham (357 ± 0.5), broiled pork top loin including fat (357 ± 0.5), raw coconut (356 ± 1.1), raw laver/nori seaweed (356 ± 0.5), raw freshwater bass (356 ± 0.6), raw cauliflower (356 ± 1), roasted pork rump including fat (356 ± 0.5), broiled porterhouse steak including fat (356 ± 0.5), steamed sprouted soybeans (355 ± 1.1), boiled black beans (355 ± 0.3), boiled celery (355 ± 0.7), boiled mushrooms (355 ± 0.6), raw pollock (355 ± 0.6), raw cisco (354 ± 0.6), boiled butterbur (354 ± 0.5), raw cuttlefish (354 ± 0.6), raw dungeness crab (354 ± 0.6), fried beef brains (354 ± 0.5), boiled garden cress (353 ± 0.7), roasted beef round tip including fat (353 ± 0.5), roasted pork boston blade excluding fat (353 ± 0.5), cooked lobster (352 ± 0.6), roasted pork shoulder excluding fat (352 ± 0.5), broiled beef rib eye including fat (352 ± 0.5), fried pork center rib including fat (352 ± 0.5), cooked crayfish (351 ± 0.6), cooked perch (351 ± 0.6), raw sunfish (351 ± 0.6), canned great northern beans (351 ± 0.2), roasted pork arm excluding fat (351 ± 0.5), braised beef flank excluding fat (351 ± 0.5), broiled beef top loin including fat (351 ± 0.5), broiled pork loin including fat (351 ± 0.5), broiled pork boston blade including fat (351 ± 0.5), sorghum grain (350 ± 0.5), passion fruit (350 ± 2.8), puffed wheat (350 ± 3.6), unheated lean ham (350 ± 0.5), fried pork top loin including fat (350 ± 0.5), cooked eel (349 ± 0.6), well done broiled lean ground beef (349 ± 0.5), baked sweet potato (348 ± 0.4), roasted canned lean ham (348 ± 0.5), raw catfish (348 ± 0.6), braised pork boston blade braised including fat (348 ± 0.5), raw whelk (347 ± 0.6), canned spinach (346 ± 0.5), pork and beef honey loaf (346 ± 1.8), roasted pork loin blade excluding fat (346 ± 0.5), raw croaker (345 ± 0.6), braised pork loin including fat (345 ± 0.5), brown sugar (344 ± 0.3), sapote (344 ± 0.2), cooked flatfish (344 ± 0.6), braised beef flank including fat (344 ± 0.5), pork pickle and pimento loaf (343 ± 1.8), roasted beef tenderloin including fat (343 ± 0.5), unheated canadian bacon (342 ± 0.9), raw seatrout (341 ± 0.6), boiled kohlrabi (340 ± 0.6), breaded and fried catfish (340 ± 0.6), breaded and fried croaker (340 ± 0.6), raw red chili peppers (340 ± 1.1), raw green chili peppers (340 ± 1.1), well done fried lean ground beef (340 ± 0.5), broiled T-bone steak including fat (339 ± 0.5), boiled hyacinth beans (337 ± 0.3), dried sweetened shredded coconut (337 ± 0.5), roasted pork sirloin including fat (337 ± 0.5), oil roasted macadamia nuts (336 ± 1.8), raw japanese chestnuts (336 ± 1.8), braised pork arm braised including fat (336 ± 0.5), pork link sausage (335 ± 0.7), cooked prunes (334 ± 0.5), braised pork loin blade including fat (334 ± 0.5), raw carp (333 ± 0.6), canned tuna in oil (333 ± 0.6), kiwi fruit (332 ± 0.7), breaded and fried scallops (332 ± 1.6), cooked pike (332 ± 0.6), restaurant hash brown potatoes (332 ± 0.6), unheated ham (332 ± 0.5), well done fried ground beef (332 ± 0.5), broiled pork loin blade including fat (332 ± 0.5), broiled lamb liver (331 ± 1.1), evaporated skim milk (331 ± 1.6), raw shallots (330 ± 5), raw celtuce (330 ± 0.5), fried chicken giblets (330 ± 0.5), barbecue loaf (330 ± 2.2), semi-sweet chocolate (329 ± 1.8), raw watercress (329 ± 2.9), roasted goose with skin (329 ± 0.5), raw blue crab (329 ± 0.6), roasted pork leg including fat (329 ± 0.5), evaporated lowfat milk (328 ± 0.4), cooked sea bass (328 ± 0.6), boiled potato without skin (328 ± 0.4), raw herring (327 ± 0.6), well done broiled ground beef (327 ± 0.5), breaded and fried clams (326 ± 0.6), canned pink salmon (326 ± 0.6), frozen and boiled brussels sprouts (326 ± 0.6), granola bar (325 ± 2.1), crab cakes (325 ± 0.8), taco (325 ± 0.6), coconut cream (325 ± 0.2), boiled yellow beans (325 ± 0.3), raw broccoli (325 ± 1.1), cooked blue crab (324 ± 0.6), raw carrots (324 ± 0.7), turkey ham (323 ± 0.9), boiled cauliflower (323 ± 0.8), frozen and boiled kale (322 ± 0.8), hash brown potatoes (322 ± 0.6), raw scallops (322 ± 0.6), raw pink salmon (322 ± 0.6), roasted pork center loin including fat (322 ± 0.5), roasted leg of lamb excluding fat (321 ± 0.6), boiled dock (321 ± 0.5), boiled okra (321 ± 0.6), black currants (321 ± 0.9), light buckwheat flour (320 ± 0.5), raw mussels (320 ± 0.6), braised pork spareribs (320 ± 0.5), chopped ham lunch meat (319 ± 2.4), roasted pork loin including fat (318 ± 0.5), pork and beef pepperoni (317 ± 8.3), boiled brussels sprouts (317 ± 0.6), braised pork center loin including fat (317 ± 0.5), broiled lamb loin chop excluding fat (316 ± 1), raw whitefish (316 ± 0.6), raw endive (316 ± 2), canned ham (316 ± 0.5), breaded and fried fish (315 ± 0.7), boiled yardlong bean (315 ± 0.3), raw clams (314 ± 0.6), raw mackerel (314 ± 0.6), canned sweet potato (313 ± 0.3), boiled beets (313 ± 0.6), braised beef shortribs excluding fat (313 ± 0.5), medium-rare broiled extra lean ground beef (313 ± 0.5), roasted pork boston blade including fat (313 ± 0.5), persimmon (312 ± 2), medium-rare fried extra lean ground beef (312 ± 0.5), raw swamp cabbage (311 ± 0.9), boiled and steamed chinese chestnuts (311 ± 1.8), raw haddock (311 ± 0.6), boiled asparagus (310 ± 0.6), minced ham lunch meat (310 ± 2.4), roasted pork shank including fat (310 ± 0.5), cantaloupe (309 ± 0.3), canned black turtle beans (308 ± 0.2), braised beef bottom round excluding fat (308 ± 0.5), roasted veal leg (307 ± 0.6), broiled lamb rib chop excluding fat (305 ± 1.2), roasted veal rib roast (305 ± 0.6), roasted turkey light meat without skin (305 ± 0.5), braised/broiled veal round with rump (304 ± 0.6), boiled mothbeans (304 ± 0.3), italian pork sausage (304 ± 0.7), roasted pork shoulder including fat (304 ± 0.5), evaporated whole milk (303 ± 0.4), jackfruit (303 ± 0.5), canned bonito (302 ± 0.5), broiled beef rib including fat (302 ± 0.5), canned pinto beans (301 ± 0.2), medium-rare broiled lean ground beef (301 ± 0.5), roasted lamb shoulder excluding fat (300 ± 0.6), bay leaf (300 ± 50), baked beans with pork and tomato sauce (300 ± 0.2), lebanon bologna (300 ± 2.2), pork olive loaf (300 ± 1.8), medium-rare fried ground beef (300 ± 0.5), medium-rare fried lean ground beef (299 ± 0.5), frozen and boiled spinach (298 ± 0.5), boiled green beans (298 ± 0.8), boiled yellow snap beans (298 ± 0.8), fried pork loin blade including fat (297 ± 0.5), pot barley pearled (296 ± 0.3), raw turnip greens (296 ± 1.8), vegetarian baked beans (296 ± 0.2), ham and cheese roll (296 ± 1.8), braised/broiled veal loin (295 ± 0.6), apricots (295 ± 0.5), roasted beef rib including fat (294 ± 0.5), whole wheat roll (293 ± 1.8), horseradish sauce (293 ± 3.3), cooked kingfish (293 ± 0.5), roasted pork arm including fat (293 ± 0.5), smoked cisco (293 ± 0.6), roasted pork loin blade including fat (293 ± 0.5), carrot juice (292 ± 0.3), roasted cured pork arm excluding fat (292 ± 0.5), medium-rare broiled ground beef (292 ± 0.5), braised beef bottom round including fat (292 ± 0.5), boiled chickpeas (291 ± 0.3), raw rainbow smelt (291 ± 0.6), beef honey roll sausage (291 ± 2.2), well done baked extra lean ground beef (291 ± 0.5), prune pudding (290 ± 0.4), roasted turkey dark meat without skin (290 ± 0.5), raw romaine lettuce (289 ± 1.8), mashed potatoes (289 ± 0.5), braised beef chuck arm pot roast excluding fat (289 ± 0.5), burrito (288 ± 0.3), canned navy beans (288 ± 0.2), raw swordfish (288 ± 0.6), frozen and boiled potatoes without skin (287 ± 0.5), boiled rutabaga (287 ± 0.6), roasted lean ham (287 ± 0.5), raw scup (287 ± 0.6), braised beef brisket excluding fat (287 ± 0.5), well done baked lean ground beef (286 ± 0.5), boiled oriental radish (285 ± 0.7), raw celery (285 ± 1.3), roasted turkey light meat with skin (285 ± 0.5), canned tuna in water (284 ± 0.6), white whole ground corn meal (284 ± 0.4), yellow whole ground corn meal (284 ± 0.4), roasted leg of lamb including fat (284 ± 0.6), boiled swamp cabbage (284 ± 1), enchilada (284 ± 0.2), boiled butternut squash (284 ± 0.5), stir-fried sprouted lentils (284 ± 0.5), guava (284 ± 0.6), braised beef spleen (284 ± 0.5), raw sturgeon (284 ± 0.6), berliner (283 ± 2.2), pork bologna (283 ± 2.2), boiled salsify (282 ± 0.7), chili (281 ± 0.2), pork and beef brotwurst (281 ± 0.7), elderberries (280 ± 0.3), boiled winged beans (280 ± 0.3), breaded fried squid (279 ± 0.6), german sweet bakers chocolate (279 ± 1.8), soursop (278 ± 0.2), yellow passion fruit juice (278 ± 0.2), boiled black-eyed peas (278 ± 0.3), turkey breast meat summer sausage (276 ± 2.4), prune juice (276 ± 0.2), fried chicken breast without skin (276 ± 0.6), gingerbread (275 ± 0.8), raw freshwater drum (275 ± 0.6), red currants (275 ± 0.9), white currants (275 ± 0.9), raw crayfish (274 ± 0.6), canned zucchini in tomato juice (274 ± 0.4), boiled scotch kale (274 ± 0.8), roasted turkey dark meat with skin (274 ± 0.5), well done baked ground beef (274 ± 0.5), raw eel (273 ± 0.6), raw perch (273 ± 0.6), sweet chocolate (271 ± 1.8), dry roasted macadamia nuts (271 ± 1.8), toasted raisin bread (271 ± 2.4), boiled green peas (271 ± 0.6), pork and beef picnic loaf (271 ± 1.8), summer sausage (270 ± 2.2), raw beef tripe (270 ± 0.5), pork and beef kielbasa/kolbassy (269 ± 1.9), cooked mussels (268 ± 0.6), light and dark meat turkey roll (268 ± 0.9), dried sweetened flaked coconut (268 ± 2), boiled broad beans (268 ± 0.3), boiled sprouted green peas (268 ± 0.5), raw chives (267 ± 16.7), baked beans with pork and sweet sauce (266 ± 0.2), loquats (266 ± 0.5), o'brien potatoes (266 ± 0.3), longans (266 ± 0.5), boiled mung beans (265 ± 0.2), frozen and boiled succotash (265 ± 0.6), raw loosflleaf lettuce (264 ± 1.8), boiled artichoke (264 ± 0.6), boiled acorn squash (263 ± 0.4), coconut milk (263 ± 0.2), boiled artichoke hearts (263 ± 0.6), fried light meat chicken without skin (263 ± 0.5), braised beef chuck blade roast excluding fat (263 ± 0.5), raw pheasant without skin (262 ± 0.5), cooked alaska king crab (261 ± 0.6), bleu cheese (261 ± 1.8), carissa (260 ± 2.5), canned cranberry beans (260 ± 0.2), boiled red tomato (260 ± 0.4), raw butterhead lettuce (260 ± 3.3), pomegranate (259 ± 0.3), raw pike (259 ± 0.6), turkey pastrami (258 ± 0.9), roasted cured pork arm including fat (258 ± 0.5), fried chicken breast with skin (258 ± 0.5), canned kidney beans (257 ± 0.2), papaya (257 ± 0.2), raw black bass (256 ± 0.5), fried chicken (256 ± 0.6), raw spring onions with tops (256 ± 1), roasted chicken breast without skin (256 ± 0.6), raw sea bass (256 ± 0.6), skim yogurt (255 ± 0.2), potato salad (254 ± 0.4), frozen and boiled crookneck squash (253 ± 0.5), boiled zucchini (253 ± 0.6), fried dark meat chicken without skin (253 ± 0.5), roasted duck without skin (252 ± 0.5), frozen and boiled collards (252 ± 0.6), pork beerwurst salami (252 ± 2.2), honeydew melon (251 ± 0.5), hot fudge sundae (250 ± 0.3), coconut water (250 ± 0.2), jujube (250 ± 0.5), light meat turkey roll (249 ± 0.9), boiled yellow corn (249 ± 0.6), raw whiting (249 ± 0.6), white bolted corn meal (248 ± 0.4), yellow bolted corn meal (248 ± 0.4), sugar apple (248 ± 0.3), bran muffins (248 ± 1.3), boiled eggplant (248 ± 1), raw zucchini (248 ± 0.8), roasted light meat chicken without skin (247 ± 0.5), maple sugar (246 ± 1.8), broiled lamb loin chop including fat (246 ± 0.7), braised beef pancreas (246 ± 0.5), raw green cabbage (246 ± 1.4), turkey summer sausage (246 ± 1.8), raw squid (246 ± 0.6), raisin bun (245 ± 0.8), breaded and fried oysters (245 ± 0.6), boiled lupins (245 ± 0.3), cooked cod (245 ± 0.6), stewed chicken drumstick without skin (245 ± 1.1), roasted chicken breast with skin (245 ± 0.5), raw green peas (244 ± 0.6), braised beef chuck arm pot roast including fat (244 ± 0.5), pork and beef luncheon sausage (243 ± 2.2), roasted lamb shoulder including fat (242 ± 0.6), grilled ham patties (242 ± 0.8), turkey salami (242 ± 0.9), canned broad beans (242 ± 0.2), unheated ham patties (240 ± 0.8), canned stewed red tomatoes (240 ± 0.4), raisin bread (240 ± 2), simmered beef brains (240 ± 0.5), roasted dark meat chicken without skin (240 ± 0.5), polish pork sausage (239 ± 1.8), fried light meat chicken with skin (239 ± 0.5), oatmeal cookie (238 ± 3.8), raw turbot (238 ± 0.6), roasted chicken thigh without skin (238 ± 1), raw quail without skin (237 ± 0.5), raw chinese cabbage (237 ± 1.4), braised pork tongue (237 ± 0.5), firm raw tofu (237 ± 0.4), fried chicken thigh with skin (237 ± 0.8), fruit pectin (236 ± 3.6), braised beef liver (235 ± 0.5), lowfat yogurt (234 ± 0.2), frozen and boiled okra (234 ± 0.5), boiled dandelion greens (233 ± 1), fruitcake, light (233 ± 1.2), simmered beef heart (233 ± 0.5), beef pastrami (232 ± 1.8), toasted rye bread (232 ± 2.3), figs (232 ± 1), raw savoy cabbage (231 ± 1.4), boiled mungo beans (231 ± 0.3), raw radish (231 ± 1.1), raw goose liver (230 ± 0.5), boiled pumpkin (230 ± 0.4), beef summer sausage (230 ± 2.2), fried dark meat chicken with skin (230 ± 0.5), beef and pork summer sausage (230 ± 2.2), cheese crackers with peanut butter (229 ± 1.2), canned oysters (229 ± 0.6), raw oysters (229 ± 0.6), canned potato without skin (229 ± 0.6), fried chicken drumstick with skin (229 ± 1), roasted chicken drumstick with skin (229 ± 1), braised beef brisket including fat (229 ± 0.5), light meat chicken roll (228 ± 0.5), boiled kale (228 ± 0.8), teriyaki sauce (228 ± 2.8), cooked black bass (227 ± 0.4), frozen white corn (227 ± 0.7), raw oriental radish (227 ± 1.1), taco shell (227 ± 4.5), tostada shell (227 ± 4.5), boiled carrots (227 ± 0.6), braised pork spleen (227 ± 0.5), roasted light meat chicken with skin (227 ± 0.5), bacon cheeseburger (226 ± 0.3), chicken roll (226 ± 0.9), raw agar seaweed (226 ± 0.5), pork livercheese (226 ± 1.3), fried chicken back with skin (226 ± 0.7), broiled lamb rib chop including fat (225 ± 0.7), roast beef sandwich (225 ± 0.3), breaded and fried shrimp (225 ± 0.6), braised/ roasted/stewed veal chuck (224 ± 0.6), chocolate milkshake (224 ± 0.2), frozen and boiled turnip greens (224 ± 0.6), cherries (224 ± 0.7), pork mothers loaf (224 ± 2.4), medium-rare baked extra lean ground beef (224 ± 0.5), medium-rare baked lean ground beef (224 ± 0.5), braised beef shortribs including fat (224 ± 0.5), duck egg (223 ± 0.7), braised beef chuck blade roast including fat (223 ± 0.5), beef salami (222 ± 2.2), canned peeled red tomatoes (221 ± 0.4), chocolate syrup (221 ± 1.3), roasted chicken thigh with skin (221 ± 0.8), medium-rare baked ground beef (221 ± 0.5), raw wild rice (220 ± 0.3), fish sandwich (220 ± 0.3), restaurant tomato juice (220 ± 0.4), mixed grain bread (220 ± 2), canned lima beans (220 ± 0.2), raw eggplant (220 ± 1.2), tomato juice (220 ± 0.3), roasted dark meat chicken with skin (220 ± 0.5), wax beans (219 ± 0.4), prickly pear (219 ± 0.5), frozen and boiled asparagus (218 ± 0.8), orange peel (217 ± 8.3), pummelo (216 ± 0.3), bread pudding (215 ± 0.2), ham and cheese sandwich (214 ± 0.3), boiled hubbard squash (214 ± 0.4), canned pork lunch meat (214 ± 2.4), okara tofu (213 ± 0.8), tamari sauce (212 ± 0.9), raw crookneck squash (212 ± 0.8), pork bratwurst (212 ± 0.6), nectarine (212 ± 0.4), canned white corn (211 ± 0.5), simmered turkey gizzard (211 ± 0.5), beef lunch meat (211 ± 1.8), casaba melon (210 ± 0.3), braised/stewed veal breast plate (209 ± 0.6), canned butter beans (208 ± 0.4), roselle (207 ± 0.9), raw red tomato (207 ± 0.4), pork and beef link sausage with american cheese (207 ± 1.2), hamburger sandwich (206 ± 0.5), canned pumpkin (206 ± 0.4), raw red cabbage (206 ± 1.4), braised pork heart (206 ± 0.4), caramel sundae (205 ± 0.3), goat milk (205 ± 0.2), boiled green cabbage (205 ± 0.7), roasted duck with skin (204 ± 0.5), rye bread (204 ± 2), raw green tomato (204 ± 0.4), raw alaska king crab (204 ± 0.6), pork and beef lunch meat (204 ± 1.8), banana custard pie (203 ± 0.4), boiled turnip greens (203 ± 0.7), boiled mustard greens (201 ± 0.7), raw pork stomach (201 ± 0.5), dill pickles (200 ± 0.8), toasted whole wheat bread (200 ± 2.4), pork and beef knockwurst (200 ± 0.7), gingersnaps cookie (200 ± 7.1), simmered turkey giblets (200 ± 0.5), pork liver sausage (200 ± 2.8), beef and pork salami (200 ± 2.2), restaurant chocolate milkshake (200 ± 0.2), malted milk (200 ± 0.2), gooseberries (198 ± 0.3), quince (197 ± 0.5), peach (197 ± 0.6), boiled blackeye pea pods (196 ± 1.1), cheeseburger (196 ± 0.4), restaurant coleslaw (196 ± 0.5), blackberries (196 ± 0.7), raw red sweet peppers (196 ± 1), raw green sweet peppers (196 ± 1), frozen and boiled zucchini (195 ± 0.4), braised pork brains (195 ± 0.5), kumquats (195 ± 2.6), mulberries (194 ± 0.4), simmered turkey liver (194 ± 0.5), crab apples (194 ± 0.5), roasted pork blade roll (194 ± 0.5), sapodilla (193 ± 0.3), raisin pie (192 ± 0.4), cheese pizza (192 ± 0.4), hot chocolate (192 ± 0.2), boiled crookneck squash (192 ± 0.6), american grapes (191 ± 0.5), restaurant orange juice (190 ± 0.4), pork and beef link sausage (190 ± 0.7), edam cheese (189 ± 1.8), camembert cheese (189 ± 1.8), turkey bologna (189 ± 1.8), candied sweet potato (189 ± 0.5), whole chocolate malted milk (189 ± 0.2), cheese crackers (187 ± 3.3), stewed chicken breast without skin (187 ± 0.5), chicken and turkey salad (186 ± 1.8), canned yellow corn (186 ± 0.5), european grapes (185 ± 0.3), thompson seedless grapes (185 ± 0.3), raw shrimp (185 ± 0.6), boiled savoy cabbage (184 ± 0.7), boiled sweet potato (184 ± 0.3), stewed chicken drumstick with skin (184 ± 0.9), poi (183 ± 0.4), vanilla milkshake (183 ± 0.2), canned succotash with creamed corn (183 ± 0.4), simmered turkey heart (183 ± 0.5), frozen and boiled turnips (182 ± 0.5), raw chickory (182 ± 1.1), protein fortified lowfat 2% milk (182 ± 0.2), raw scallop squash (182 ± 0.8), coleslaw (182 ± 0.8), roasted chicken wing with skin (182 ± 1.5), restaurant strawberry milkshake (182 ± 0.2), raw leeks (181 ± 1.9), boiled yambean (181 ± 0.5), protein fortified skim milk (181 ± 0.2), boiled peanuts (181 ± 1.6), fried chicken neck with skin (181 ± 1.4), cooked shrimp (181 ± 0.6), stewed dark meat chicken without skin (181 ± 0.5), pepperoni pizza (180 ± 0.4), protein fortified lowfat 1% milk (180 ± 0.2), frozen and boiled broccoli (180 ± 0.5), simmered beef tongue (180 ± 0.5), stewed light meat chicken without skin (180 ± 0.5), shoyu sauce (179 ± 0.9), canned carrots (179 ± 0.7), simmered beef kidneys (179 ± 0.5), simmered chicken gizzard (179 ± 0.5), navel orange (179 ± 0.4), valencia orange (179 ± 0.4), mince pie (178 ± 0.4), tuna salad (178 ± 0.2), indian buffalo milk (178 ± 0.2), beef and pork bologna (178 ± 2.2), tangerine juice (178 ± 0.2), fried chicken wing with skin (178 ± 1.6), smoked beef sausage (177 ± 1.2), strawberry sundae (177 ± 0.3), stewed chicken breast with skin (177 ± 0.5), whole wheat bread (176 ± 2), maple syrup (175 ± 2.5), smoked chinook salmon (175 ± 0.6), beef beerwurst salami (174 ± 2.2), hummus (174 ± 0.2), restaurant vanilla milkshake (174 ± 0.2), raw queen crab (173 ± 0.6), raw celeriac (173 ± 0.5), boiled chayote (173 ± 0.6), corn tortilla (173 ± 1.7), canned green peas (173 ± 0.6), braised beef lungs (173 ± 0.5), waffles (172 ± 0.7), canned chickpeas (172 ± 0.2), canned black-eyed peas (172 ± 0.2), chocolate pudding (171 ± 0.2), hot dog (171 ± 0.6), pineapple upside-down cake (171 ± 0.7), lychees (171 ± 0.5), plum (171 ± 0.8), brownies (170 ± 2.5), boiled calabash gourd (170 ± 0.7), lowfat 1% chocolate milk (170 ± 0.2), canned sauerkraut (170 ± 0.4), lowfat 2% chocolate milk (169 ± 0.2), barbecue sauce (169 ± 3.1), stewed chicken thigh with skin (169 ± 0.7), toaster pastry (168 ± 1), frozen and boiled green peas (168 ± 0.6), braised pork pancreas (168 ± 0.5), lemon peel (167 ± 8.3), whole chocolate milk (167 ± 0.2), beef and pork frankfurter (167 ± 1.1), stewed light meat chicken with skin (167 ± 0.5), onion rings (166 ± 0.6), skim milk (166 ± 0.2), strawberries (166 ± 0.3), stewed dark meat chicken with skin (166 ± 0.5), eggnog (165 ± 0.2), beef frankfurter (165 ± 0.9), american cheese (164 ± 1.8), pimento cheese spread (164 ± 1.8), graham crackers (164 ± 3.6), miso (164 ± 0.4), ham and cheese spread (164 ± 1.8), coconut custard pie (163 ± 0.4), sweet potato pie (163 ± 0.4), low calorie russian salad dressing (163 ± 3.1), canned succotash (163 ± 0.4), boiled broccoli (163 ± 0.6), carambola (163 ± 0.4), white grapefruit juice (162 ± 0.2), red grapefruit juice (162 ± 0.2), yellow corn pudding (161 ± 0.2), whey (161 ± 0.2), japanese persimmon (161 ± 0.3), pumpkin pie (160 ± 0.4), light pearled barley (160 ± 0.3), turkey frankfurter (160 ± 1.1), russian salad dressing (160 ± 3.3), raw iceberg lettuce (160 ± 2.5), pork and beef mortadella (160 ± 3.3), raw shark (160 ± 0.6), rhubarb pie (159 ± 0.4), devil's food cake (158 ± 0.8), frozen and boiled carrots (158 ± 0.7), simmered chicken giblets (158 ± 0.5), zwieback crackers (157 ± 7.1), oatmeal bread (157 ± 1.8), tangerine (157 ± 0.6), beef bologna (157 ± 2.2), rusk crackers (156 ± 5.6), light rye flour (156 ± 0.5), lowfat 1% milk (156 ± 0.2), mango (156 ± 0.2), whole yogurt (155 ± 0.2), battered and fried shark (155 ± 0.6), lowfat 2% milk (155 ± 0.2), tuna salad spread (155 ± 0.9), raw onions (155 ± 0.6), brie cheese (154 ± 1.8), peanut brittle (154 ± 1.8), canned yellow corn with red and green peppers (153 ± 0.4), screwdriver (153 ± 0.2), dry breadcrumbs (152 ± 0.5), whole milk (152 ± 0.2), soy sauce (152 ± 0.9), toasted cracked wheat bread (152 ± 2.4), raspberries (152 ± 0.4), braised pork lungs (151 ± 0.5), buttermilk (151 ± 0.2), boiled onions (151 ± 0.5), popover roll (150 ± 1.3), ham salad (150 ± 1.8), braised pork liver (150 ± 0.5), raw cucumber (150 ± 1), rice flour (149 ± 0.4), pickled beets (148 ± 0.4), raw sprouted mung beans (148 ± 1), white grapefruit (148 ± 0.4), raw pork jowl (148 ± 0.5), custard (146 ± 0.2), raw acerola (146 ± 0.5), canned red sweet peppers (146 ± 0.7), canned green sweet peppers (146 ± 0.7), fried tofu (146 ± 3.8), bloody mary (146 ± 0.3), loganberries (145 ± 0.3), cooked corned beef brisket (145 ± 0.5), carob milk (145 ± 0.2), frozen creamed yellow corn (143 ± 0.4), turkey salad (143 ± 0.9), braised pork kidneys (143 ± 0.5), chocolate cream pie (142 ± 0.4), sour cream (142 ± 4.2), canned turnip greens (141 ± 0.4), soy milk (141 ± 0.2), brown mustard (140 ± 10), yellow mustard (140 ± 10), simmered chicken liver (140 ± 0.5), boiled red cabbage (140 ± 0.7), boiled scallop squash (140 ± 0.6), stewed chicken wing with skin (140 ± 1.3), raw bacon (140 ± 0.7), provolone cheese (139 ± 1.8), filled milk (139 ± 0.2), boysenberries (139 ± 0.4), frozen and boiled mustard greens (139 ± 0.7), frozen and boiled yellow corn (139 ± 0.6), frozen and boiled cauliflower (139 ± 0.6), simmered chicken neck without skin (139 ± 2.8), vanilla pudding (138 ± 0.2), canned pumpkin pie mix (138 ± 0.4), lemon (138 ± 0.9), custard pie (137 ± 0.4), sheep milk (136 ± 0.2), boiled turnips (136 ± 0.6), muenster cheese (136 ± 1.8), brick cheese (136 ± 1.8), canned corned beef (136 ± 0.5), chicken egg white (136 ± 1.5), corn meal muffins (135 ± 1.3), tapioca cream pudding (135 ± 0.3), canned jalapeño peppers (135 ± 0.7), pineapple juice (134 ± 0.2), canned creamed yellow corn (134 ± 0.4), whole ground corn meal muffins (133 ± 1.3), saltine crackers (133 ± 8.3), mandarin oranges (133 ± 0.4), frozen and boiled butternut squash (133 ± 0.4), cracked wheat bread (132 ± 2), french toast (132 ± 0.8), grape juice (132 ± 0.2), simmered chicken heart (132 ± 0.5), red beans (129 ± 0.4), canned shitake mushrooms (129 ± 0.4), raw new zealand spinach (129 ± 1.8), colby cheese (129 ± 1.8), limburger cheese (129 ± 1.8), toasted white bread (129 ± 2.4), dinner rolls (129 ± 1.8), boiled red sweet peppers (129 ± 0.7), boiled green sweet peppers (129 ± 0.7), pink grapefruit (128 ± 0.4), raw spirulina seaweed (127 ± 0.5), half and half cream (127 ± 3.3), pear (125 ± 0.3), lemon juice (124 ± 0.2), peach pie (123 ± 0.5), pecan pie (123 ± 0.5), rose apple (123 ± 0.5), pancakes (122 ± 1.9), cheese soufflé (121 ± 0.5), soda crackers (121 ± 1.8), gouda cheese (121 ± 1.8), cream cheese (121 ± 1.8), raw tofu (121 ± 0.4), boiled and steamed japanese chestnuts (121 ± 1.8), strawberry pie (120 ± 0.5), enriched white degermed corn meal (120 ± 0.4), unenriched corn meal (120 ± 0.4), enriched yellow degermed corn meal (120 ± 0.4), unenriched yellow degermed corn meal (120 ± 0.4), chicken egg (120 ± 1), apple juice (119 ± 0.2), unenriched biscuits (118 ± 1.8), cooked shitake mushrooms (118 ± 0.7), pita bread (118 ± 1.3), canned chinese water chestnuts (117 ± 0.7), boiled/baked spaghetti squash (117 ± 0.6), watermelon (116 ± 0.3), blueberry muffins (115 ± 1.3), raw apple with skin (115 ± 0.4), neufchatel cheese (114 ± 1.8), apricot nectar (114 ± 0.2), puffed rice (114 ± 3.6), french bread (114 ± 1.8), oyster crackers (113 ± 6.3), low calorie thousand island salad dressing (113 ± 3.3), white bread (113 ± 2.1), cornbread (113 ± 0.6), corn pone (113 ± 0.6), medium cream (113 ± 3.3), thousand island salad dressing (113 ± 3.1), pineapple (113 ± 0.3), raw apple without skin (113 ± 0.4), sweet roll (112 ± 1.2), red table wine (112 ± 0.5), frozen and boiled yellow snap beans (112 ± 0.7), frozen and boiled green beans (112 ± 0.7), swiss cheese (111 ± 1.8), italian bread (111 ± 1.8), pork and beef spread (111 ± 1.8), chocolate chiffon pie (110 ± 0.6), canned shellie beans (109 ± 0.4), canned green beans (109 ± 0.7), canned yellow snap beans (109 ± 0.7), lime juice (109 ± 0.2), pancakes with butter and syrup (108 ± 0.3), raw frozen rhubarb (108 ± 0.4), simmered chicken neck with skin (108 ± 1.3), maltex (107 ± 0.3), cherry pie (105 ± 0.4), ricotta cheese (105 ±

0.4), surimi scallops (104 ± 0.6), jellied corned beef loaf (104 ± 1.8), pitanga (103 ± 0.3), white cake (103 ± 0.6), tequila sunrise (103 ± 0.3), boiled new zealand spinach (102 ± 0.6), cooked sprouted mung beans (102 ± 0.8), lime (101 ± 0.7), sweetened coconut cream (101 ± 0.2), frozen and boiled onions (101 ± 0.5), blackberry pie (100 ± 0.4), apple brown betty (100 ± 0.2), cider vinegar (100 ± 3.3), parmesan cheese (100 ± 10), cheddar cheese (100 ± 1.8), white pepper (100 ± 25), pretzels (100 ± 1.8), french rolls (100 ± 1), whipping cream, light (100 ± 3.3), vienna sausage (100 ± 3.1), rose table wine (99 ± 0.5), pineapple chiffon pie (98 ± 0.6), cheesecake (98 ± 0.6), angel food cake (98 ± 0.8), pineapple custard pie (97 ± 0.4), acerola juice (97 ± 0.2), animal crackers (96 ± 1.9), cheshire cheese (96 ± 1.8), yellow cake (96 ± 0.7), canned crookneck squash (96 ± 0.5), frozen and cooked sweetened rhubarb (96 ± 0.4), butterscotch pie (95 ± 0.4), wheat cake (95 ± 0.4), all purpose enriched wheat flour (95 ± 0.4), enriched wheat bread (95 ± 0.4), roquefort cheese (93 ± 1.8), danish pastry (93 ± 1.2), hamburger bun (93 ± 1.3), hot dog bun (93 ± 1.3), boiled collards (93 ± 0.3), corn flakes (93 ± 1.8), dry dessert wine (92 ± 0.8), sweet dessert wine (92 ± 0.8), bread sticks (90 ± 2.5), gefiltefish with broth (90 ± 1.2), cottage pudding (89 ± 0.9), boston cream pie (89 ± 0.5), surimi shrimp (89 ± 0.6), sponge cake (89 ± 0.8), blueberries (89 ± 0.3), raw kelp/ kombu/ tangle seaweed (89 ± 0.5), boiled leeks (88 ± 1.9), maypo (88 ± 0.3), cooked apple without skin (88 ± 0.3), chicken egg yolk (88 ± 2.9), bulgur (87 ± 0.4), cottage cheese (86 ± 1.8), apple pie (85 ± 0.4), gruyere cheese (82 ± 1.8), enriched biscuits (82 ± 1.8), monterey cheese (82 ± 1.8), soybean protein isolate (82 ± 1.8), lemon chiffon pie (81 ± 0.6), low calorie french salad dressing (81 ± 3.1), soybean oil and cottonseed oil margarine liquid (80 ± 10), white table wine (80 ± 0.5), java plum (79 ± 0.4), canned bamboo shoots (79 ± 0.4), hot dog pickle relish (79 ± 1.8), hamburger pickle relish (79 ± 1.8), raw sprouted alfalfa seeds (79 ± 1.5), sugar cookie (75 ± 3.1), bagels (75 ± 0.9), whipping cream, heavy (73 ± 3.3), pineapple pie (72 ± 0.4), frozen and boiled red sweet peppers (72 ± 0.5), frozen and boiled green sweet peppers (72 ± 0.5), piña colada (71 ± 0.4), cranberry sauce (71 ± 0.5), cooked brown rice (70 ± 0.3), restaurant coffee (69 ± 0.3), mozzarella cheese (68 ± 1.8), pickled herring (67 ± 3.3), blueberry pie (65 ± 0.4), caramel cake (64 ± 0.5), tilsit cheese (64 ± 1.8), feta cheese (64 ± 1.8), shortbread cookie (63 ± 6.3), raw irish moss seaweed (63 ± 0.5), frozen and boiled turnip greens and turnips (62 ± 0.5), spaghetti (61 ± 0.4), macaroni (61 ± 0.4), applesauce (61 ± 0.4), wheat gluten (60 ± 0.4), pound cake (60 ± 1.7), bread stuffing (58 ± 0.3), oatmeal (57 ± 0.3), instant oatmeal (56 ± 0.3), coffee (54 ± 0.3), whiskey sour (53 ± 0.6), honey (52 ± 2.4), restaurant hot chocolate (52 ± 0.2), human milk (52 ± 1.6), cream puff (51 ± 0.4), raw wakame seaweed (50 ± 0.5), pie crust (49 ± 0.3), mammy apple (47 ± 0.5), pickled green olives (46 ± 1.1), havarti cheese (46 ± 1.8), lemon meringue pie (44 ± 0.4), noodles (44 ± 0.3), cooked parboiled white rice (43 ± 0.3), cooked parboiled extra long grain white rice (43 ± 0.6), raw pepeao (42 ± 0.5), peach nectar (41 ± 0.2), simmered pork ears (40 ± 0.5), corn oil margarine tub (40 ± 10), safflower oil margarine tub (40 ± 10), soybean oil margarine tub (40 ± 10), soybean oil and cottonseed oil margarine tub (40 ± 10), corn oil margarine stick (40 ± 10), safflower oil and soybean oil margarine stick (40 ± 10), soybean oil margarine stick (40 ± 10), soybean oil and cottonseed oil margarine stick (40 ± 10), oheloberries (39 ± 0.4), black tea (37 ± 0.3), safflower oil and soybean oil mayonnaise (36 ± 3.6), soybean oil mayonnaise (36 ± 3.6), citrus marmalade (35 ± 2.5), cooked wild long grain rice

(35 ± 0.4), pork headcheese (32 ± 1.8), papaya nectar (31 ± 0.2), 53 proof coffee liqueur (29 ± 1), 63 proof coffee liqueur (29 ± 1), cooked extra long grain white rice (28 ± 0.6), cooked white rice (28 ± 0.2), whipped butter (27 ± 4.5), canned sprouted mung beans (27 ± 0.8), apple tapioca pudding (26 ± 0.2), manhattan (26 ± 0.9), cane syrup (25 ± 2.5), beer (25 ± 0.1), cranberry juice cocktail (24 ± 0.2), raw whale meat (22 ± 0.5), malt liquor (22 ± 0.1), daiquiri (22 ± 0.8), corn grits (22 ± 0.2), french salad dressing (21 ± 3.6), ale (21 ± 0.1), light cream (21 ± 1.7), instant corn grits (21 ± 0.4), corn oil diet margarine tub (20 ± 10), soybean oil and cottonseed oil diet margarine tub (20 ± 10), butter (20 ± 3.3), sweet butter (20 ± 3.3), instant tea (20 ± 0.2), cream of rice (20 ± 0.3), yellow hominy with red and green peppers (19 ± 0.2), martini (19 ± 0.7), beef suet (18 ± 1.8), cream of wheat (18 ± 0.3), dried and frozen tofu (18 ± 2.9), restaurant cranberry juice cocktail (18 ± 0.3), cooked enriched white degermed corn meal (16 ± 0.2), cooked unenriched corn meal (16 ± 0.2), cooked enriched yellow degermed corn meal (16 ± 0.2), cooked unenriched yellow degermed corn meal (16 ± 0.2), white hominy (15 ± 0.2), tea (15 ± 0.2), yellow hominy (14 ± 0.2), mayonnaise (14 ± 3.6), distilled vinegar (13 ± 3.3), low calorie italian salad dressing (13 ± 3.3), italian salad dressing (13 ± 3.3), pear nectar (13 ± 0.2), farina (13 ± 0.3), orange soda (10 ± 0.1), simmered pork chitterlings (8 ± 0.5), tom collins (8 ± 0.2), chamomile tea (8 ± 0.3), light mayonnaise (7 ± 3.6), pickled rip sevillano/ascolano olives (6 ± 3.1), boiled waxgourd (6 ± 0.6), corn syrup (5 ± 2.4), gin & tonic (5 ± 0.2), gum drops (4 ± 1.8), powdered sugar (3 ± 0.4), white granulated sugar (3 ± 0.3), bourbon & soda (2 ± 0.4), 80 proof rum (2 ± 1.2), cola (2 ± 0.1), 94 proof gin (2 ± 1.2), 94 proof rum (2 ± 1.2), 94 proof vodka (2 ± 1.2), 94 proof whiskey (2 ± 1.2), 100 proof gin (2 ± 1.2), 100 proof rum (2 ± 1.2), 100 proof vodka (2 ± 1.2), 100 proof whiskey (2 ± 1.2), 86 proof whiskey (2 ± 1.2), lemon-lime soda (1 ± 0.1)

Applications: angina pectoris, diabetes, high blood pressure, mononucleosis, stroke, fracture, colitis, diarrhea, alcohol cravings, lack of sleep, polio, fever, headache, constipation, worms, congestive heart failure, myocardial infarction, arthritis, gout, allergies, decreased muscle activity, muscular dystrophy, rheumatism, acne, burns, skin inflammation, gastritis, tooth and gum deterioration, cancer, fever, stress

Daily Dosage: (1700 mg potassium per every 1000 mg sodium) 1980 RDA — none; 1980 SADDI — 1875–5625 mg; 1989 RDA — none; 1989 SADDI — 1875–5625 mg; nutritional — 1875–5625 mg; therapeutic — 6000–10,000 mg; experimental — ?; toxic — ?

Warnings: increased amounts are dangerous. See toxicity symptoms.

RUBIDIUM

Names: rubidium, Rb, Rb$^+$, element 37
Classification: trace mineral, contaminant
Deficiency: depression, decreased cancer resistance, decreased cell pH, cell dehydration, increased enzyme toxicity, decreased glucose tolerance, decreased peroxidase elimination
Side-effects: see toxicity symptoms

Toxicity: decreased potassium utilization
Inhibitors: unknown
Helpers: cesium
Applications: cancer
Daily Dosage: 1974 PCS — none; 1989 SADDI — none; 1989 RDA — none; 1989 USRDA — none; nutritional — unknown; therapeutic — none; experimental — none; toxic — small amounts
Warnings: Small amounts are dangerous. See toxicity symptoms.

SELENIUM

Names: selenium, Se, Se^{-2}, $Se^{..}$, element 34
Classification: trace mineral
Deficiency: dandruff, decreased tissue elasticity, deterioration of muscle, calcification of muscle, sterility in males, fetal death/reabsorption, decreased ozone resistance
Side-effects: see *toxicity* symptoms
Toxicity: garlicky skin smell, bad breath, hair deterioration, nails deterioration, tooth deterioration, skin inflammation, fatigue, progressive paralysis
Inhibitors: food-processing
Helpers: vitamin E, chromium
Sources (μg per 100 grams of food): butter (144 ± 0.5), smoked herring (141 ± 0.5), smelts (123 ± 0.5), wheat germ (111 ± 0.5), brazil nuts (103 ± 0.5), apple cider vinegar (89 ± 0.5), scallops (77 ± 0.5), barley (66 ± 0.5), whole wheat bread (66 ± 0.5), lobster (65 ± 0.5), bran (63 ± 0.5), shrimps (59 ± 0.5), red swiss chard (57 ± 0.5), oats (56 ± 0.5), clams (55 ± 0.5), king crab (51 ± 0.5), oysters (49 ± 0.5), milk (48 ± 0.5), cod (43 ± 0.5), brown rice (39 ± 0.5), top round steak (34 ± 0.5), lamb (30 ± 0.5), turnips (27 ± 0.5), molasses (26 ± 0.5), garlic (25 ± 0.5), barley (24 ± 0.5), orange juice (19 ± 0.5), gelatin (19 ± 0.5), beer (19 ± 0.5), beef liver (18 ± 0.5), lamb chop (18 ± 0.5), egg yolk (18 ± 0.5), mushrooms (12 ± 0.5), chicken (12 ± 0.5), swiss cheese (10 ± 0.5), cottage cheese (5 ± 0.5), wine (5 ± 0.5), radishes (4 ± 0.5), grape juice (4 ± 0.5), pecans (3 ± 0.5), hazelnuts (2 ± 0.5), almonds (2 ± 0.5), green beans (2 ± 0.5), kidney beans (2 ± 0.5), onion (2 ± 0.5), carrots (2 ± 0.5), cabbage (2 ± 0.5), orange (1 ± 0.5)
Applications: kwashiorkor, cancer, antioxidant therapy, sun sensitivity, free radical oxidation, life span extension, radiation, decreased radiation resistance, aging
Daily Dosage: 1980 RDA — none; 1980 SADDI — 50–200 μg; 1989 RDA — 55–70 μg; 1989 USRDA — 70 μg; nutritional — 50–200 μg; therapeutic — 250–500 μg; experimental — 1000–2000 μg; toxic — 3000 μg +
Warnings: increased amounts are dangerous. See toxicity symptoms.

SILICON

Names: silicon, Si, element 14
Classification: trace mineral
Deficiency: slow healing, angina, fatigue, dull/glazed eyes, decreased

growth, skin pallor, deterioration of memory, deterioration of tooth mineralization, abnormal tooth enamel, deformed bones, bone demineralization, deterioration of bone growth, distorted eye socket development, deterioration of collagen formation, skin flabbiness, decreased skin elasticity, carbuncles, hair falling out, ribbed nails, ingrown nails, deterioration of embryonic development, immunity decreased
Side-effects: none known
Toxicity: none known
Inhibitors: unknown
Helpers: vitamin C, calcium
Sources (mg per 100 grams of food): rice straw (2730 ± 0.5), sugar beet pulp (2311 ± 0.5), rice hulls (2250 ± 0.5), oat hulls (1691 ± 0.5), alfalfa (1274 ± 0.5), wheat straw (1224 ± 0.5), sugar cane pulp (1127 ± 0.5), oat straw (714 ± 0.5), currey powder (180 ± 0.5), wheat bran (172 ± 0.5), soybean meal (168 ± 0.5), guar gum (142 ± 0.5), chondroitin-4-sulphate (118 ± 0.5), citrus pectin (113 ± 0.5), lemon pectin (110 ± 0.5), sodium pectate (110 ± 0.5), sodium polypectate (104 ± 0.5), cotton (11.6 ± 0.5), nutrisoy flour (9.3 ± 0.5), soya fluff (8 ± 0.5), filter paper (5 ± 0.5), wheat flour (2.1 ± 0.5), cellulose powder (0.6 ± 0.5), connective tissue, bone, skin, beer, whole grains, mushrooms, carrots, tomatoes, liver, buckwheat products, seedy figs, wild strawberry, savoy cabbage, hard water, lettuce, onions, dark greens, horsetail, kelp, comfrey, nettles, milk
Applications: atherosclerosis, cardiovascular disease, broken bones, tooth deterioration, premature aging of skin, increased cholesterol, stomach ulcers, arthritis, Paget's disease, intractable sciatica, hair deterioration
Daily Dosage: 1980 RDA – none; 1980 SADDI – none; 1989 RDA – none; 1989 SADDI – none; nutritional – 1000–1500 mg; therapeutic – unknown; experimental – unknown; toxic – none known
Warnings: none

SILVER

Names: silver, Ag, element 47
Classification: possible trace mineral, contaminant
Deficiency: (being researched)
Side-effects: see toxicity symptoms
Toxicity: anemia, ashen-gray coloring of skin, organs, eyelid membranes, pain in throat, pain in stomach, nausea, collapse
Inhibitors: unknown
Helpers: unknown
Sources: disinfectant in community water supplies, silverware
Applications: none known
Daily Dosage: 1974 PCS – none; 1989 SADDI – none; 1989 RDA – none; 1989 USRDA – none; nutritional – unknown; therapeutic – none; experimental – none; toxic – small amounts
Warnings: Small amounts are dangerous. See toxicity symptoms.

SODIUM

Names: sodium, natrium, Na, Na$^+$, element 11
Classification: mineral
Deficiency: appetite decreased, weight decreased, nausea, muscular weakness, muscular cramps, muscle shrinkage, muscular twitching, collapse of blood vessels, headaches, anxiety, confusion, dizziness, salt cravings, hay fever, watery eyes, runny nose, low fevers, skin edema, depression, gall bladder deterioration
Side-effects: see toxicity symptoms
Toxicity: raised blood pressure, coma, potassium decreased
Inhibitors: perspiration, decreased chromium, decreased potassium
Helpers: vitamin D, chromium, potassium
Sources (mg per 100 grams of food): rennin (22300 ± 4.5), dried and salted cod (7027 ± 0.6), shoyu sauce (5714 ± 0.9), soy sauce (5690 ± 0.9), tamari sauce (5586 ± 0.9), teriyaki sauce (3833 ± 2.8), anchovy canned in olive oil (3670 ± 2.5), miso (3646 ± 0.4), chipped beef (3514 ± 1.8), hard pork salami (2260 ± 5), pickled green olives (2013 ± 1.1), pork and beef pepperoni (1867 ± 8.3), parmesan cheese (1860 ± 10), hard pork and beef salami (1860 ± 5), roquefort cheese (1832 ± 1.8), pretzels (1611 ± 1.8), cooked bacon (1595 ± 0.4), pork and beef peppered loaf (1543 ± 1.8), grilled canadian bacon (1530 ± 1.1), pork olive loaf (1504 ± 1.8), caviar (1500 ± 3.1), roasted ham (1500 ± 0.5), pork link sausage (1500 ± 0.7), canned jalapeño peppers (1463 ± 0.7), beef and pork summer sausage (1452 ± 2.2), american cheese (1450 ± 1.8), pimento cheese spread (1446 ± 1.8), turkey breast meat summer sausage (1433 ± 2.4), unheated lean ham (1429 ± 0.5), dill pickles (1428 ± 0.8), bleu cheese (1414 ± 1.8), pork pickle and pimento loaf (1407 ± 1.8), unheated canadian bacon (1402 ± 0.9), beef summer sausage (1378 ± 2.2), chicken frankfurter (1371 ± 1.1), chopped ham lunch meat (1371 ± 2.4), canned red sweet peppers (1369 ± 0.7), canned green sweet peppers (1369 ± 0.7), ham and cheese roll (1361 ± 1.8), sour pickles (1352 ± 0.8), beef lunch meat (1346 ± 1.8), chili sauce (1340 ± 3.3), lebanon bologna (1339 ± 2.2), pork and beef honey loaf (1336 ± 1.8), barbecue loaf (1335 ± 2.2), saltine crackers (1333 ± 8.3), beef honey roll sausage (1322 ± 2.2), unheated ham (1317 ± 0.5), pork and beef lunch meat (1311 ± 1.8), brown mustard (1300 ± 10), berliner (1296 ± 2.2), pork sausage (1293 ± 1.9), canned pork lunch meat (1290 ± 2.4), chopped smoked beef (1275 ± 1.8), pork headcheese (1271 ± 1.8), pork and beef old fashioned loaf (1264 ± 1.8), yellow mustard (1260 ± 10), canned lean ham (1255 ± 0.5), corn flakes (1254 ± 1.8), pork and beef mortadella (1247 ± 3.3), summer sausage (1243 ± 2.2), minced ham lunch meat (1243 ± 2.4), beef pastrami (1243 ± 1.8), canned ham (1240 ± 0.5), pork luxury loaf (1239 ± 1.8), pork beerwurst salami (1239 ± 2.2), roasted cured pork arm excluding fat (1231 ± 0.5), pork livercheese (1224 ± 1.3), romano cheese (1214 ± 1.8), ham and cheese spread (1211 ± 1.8), roasted lean ham (1203 ± 0.5), cheese crackers (1200 ± 3.3), pork bologna (1183 ± 2.2), pork and beef luncheon sausage (1183 ± 2.2), pork and beef picnic loaf (1179 ± 1.8), beef salami (1178 ± 2.2), hot dog pickle relish (1171 ± 1.8), pork liver sausage (1144 ± 2.8), bran (1143 ± 1.8), roasted canned lean ham (1135 ± 0.5), cooked corned beef brisket (1134 ± 0.5), smoked beef sausage (1130 ± 1.2), feta cheese (1129 ± 1.8), pork mothers loaf (1129 ± 2.4), beef and pork frankfurter

(1120 ± 1.1), soda crackers (1114 ± 1.8), turkey summer sausage (1114 ± 1.8), pork and beef brotwurst (1111 ± 0.7), hamburger pickle relish (1100 ± 1.8), unheated ham patties (1091 ± 0.8), pork and beef link sausage with american cheese (1081 ± 1.2), pork and beef kielbasa/kolbassy (1077 ± 1.9), cooked alaska king crab (1072 ± 0.6), roasted cured pork arm including fat (1072 ± 0.5), beef and pork salami (1065 ± 2.2), grilled ham patties (1053 ± 0.8), turkey frankfurter (1049 ± 1.1), dried spirulina seaweed (1048 ± 0.5), tomato catsup (1040 ± 3.3), turkey pastrami (1040 ± 0.9), oyster crackers (1038 ± 6.3), beef frankfurter (1026 ± 0.9), beef beerwurst salami (1026 ± 2.2), pork and beef spread (1025 ± 1.8), low calorie thousand island salad dressing (1020 ± 3.3), smoked whitefish (1019 ± 0.6), beef and pork bologna (1017 ± 2.2), pork and beef knockwurst (1010 ± 0.7), canned corned beef (1006 ± 0.5), cheese crackers with peanut butter (1005 ± 1.2), soybean protein isolate (1004 ± 1.8), turkey salami (998 ± 0.9), turkey ham (991 ± 0.9), beef bologna (983 ± 2.2), edam cheese (979 ± 1.8), roasted pork blade roll (973 ± 0.5), wheat flakes (964 ± 1.8), jellied corned beef loaf (964 ± 1.8), vienna sausage (950 ± 3.1), pork and beef link sausage (944 ± 0.7), bran flakes (943 ± 1.8), roasted canned ham (941 ± 0.5), ham salad (925 ± 1.8), italian pork sausage (922 ± 0.7), corn oil diet margarine tub (920 ± 10), soybean oil and cottonseed oil diet margarine tub (920 ± 10), kippered herring (918 ± 1.3), dry roasted red pistachio nuts (907 ± 1.8), pickled rip sevillano/ascolano olives (895 ± 3.1), russian salad dressing (887 ± 3.3), provolone cheese (886 ± 1.8), polish pork sausage (886 ± 1.8), whole grain rye crackers (885 ± 3.8), low calorie russian salad dressing (881 ± 3.1), pickled herring (873 ± 3.3), raw wakame seaweed (872 ± 0.5), pickled rip manzanillo/mission olives (870 ± 10.9), chili powder (867 ± 16.7), camembert cheese (854 ± 1.8), whipped butter (845 ± 4.5), raw alaska king crab (836 ± 0.6), gouda cheese (829 ± 1.8), butter (820 ± 3.3), limburger cheese (811 ± 1.8), pork and beef sausage (808 ± 3.8), ry-krisp crackers (800 ± 3.6), low calorie french salad dressing (800 ± 3.1), surimi scallops (795 ± 0.6), toasted rye bread (795 ± 2.3), barbecue sauce (794 ± 3.1), turkey bologna (793 ± 1.8), low calorie italian salad dressing (787 ± 3.3), smoked chinook salmon (784 ± 0.6), toasted english muffin (782 ± 1), hot dog (776 ± 0.6), italian salad dressing (773 ± 3.3), smoked haddock (764 ± 0.6), tilsit cheese (761 ± 1.8), boiled peanuts (750 ± 1.6), soybean oil and cottonseed oil margarine liquid (740 ± 10), dry breadcrumbs (736 ± 0.5), toasted whole wheat bread (729 ± 2.4), raisin bran (727 ± 1.4), ham and cheese sandwich (721 ± 0.3), cheshire cheese (707 ± 1.8), surimi shrimp (705 ± 0.6), caraway cheese (700 ± 1.8), bread sticks (700 ± 2.5), tuna salad spread (698 ± 0.9), rye bread (696 ± 2), light mayonnaise (686 ± 3.6), raw bacon (685 ± 0.7), pepperoni pizza (681 ± 0.4), thousand island salad dressing (681 ± 3.1), bread and butter pickles (673 ± 3.3), canned sauerkraut (661 ± 0.4), french salad dressing (657 ± 3.6), taco shell (655 ± 4.5), tostada shell (655 ± 4.5), english muffin (639 ± 0.9), brie cheese (636 ± 1.8), muenster cheese (636 ± 1.8), whole wheat bread (636 ± 2), blueberry muffins (633 ± 1.3), cheddar cheese (629 ± 1.8), marinara sauce (629 ± 0.2), enriched biscuits (625 ± 1.8), unenriched biscuits (625 ± 1.8), toasted pumpernickel bread (621 ± 1.8), colby cheese (611 ± 1.8), gjetost cheese (607 ± 1.8), tomato sauce (605 ± 0.4), hamburger bun (603 ± 1.3), hot dog bun (603 ± 1.3), cheeseburger (600 ± 0.4), dried parsley (600 ± 50), mayonnaise (600 ± 3.6), bulgur (599 ± 0.4), waffles (593 ± 0.7), burrito (592 ± 0.3), fried abalone (591 ± 0.6), toasted white bread (586 ± 2.4), light meat chicken roll (584 ± 0.5), cheese pizza (582 ± 0.4), french bread (582 ± 1.8), light and dark meat turkey roll (582 ± 0.9), chicken roll (581 ± 0.9), enchilada (579 ± 0.2), french rolls (574 ± 1), onion rings (571 ± 0.6), brick cheese (568 ± 1.8), pita bread (566 ± 1.3), whole wheat roll (564 ± 1.8), taco (563 ± 0.6), safflower oil and soybean oil mayonnaise (557 ± 3.6), soybean oil mayonnaise (557 ± 3.6), pork bratwurst (556 ± 0.6), dinner rolls (554 ± 1.8), dry instant skim milk (548 ± 0.5), monterey cheese (543 ± 1.8), pumpernickel bread (541 ± 1.6), port du salut cheese (539 ± 1.8), raw queen crab (539 ± 0.6), italian bread (539 ± 1.8), dry skim milk (537 ± 1.7), potato salad (529 ± 0.4), gefiltefish with broth (524 ± 1.2), havarti cheese (514 ± 1.8), canned bonito (514 ± 0.5), white bread (513 ± 2.1), potato pancakes (511 ± 0.7), pancakes with butter and syrup (510 ± 0.3), roast beef sandwich (504 ± 0.3), sardines (504 ± 2.1), bread stuffing (504 ± 0.3), toasted cracked wheat bread (495 ± 2.4), whole ground corn meal muffins (495 ± 1.3), oatmeal bread (493 ± 1.8), light meat turkey roll (486 ± 0.9), dry buttermilk (486 ± 7.1), smoked cisco (481 ± 0.6), corn meal muffins (480 ± 1.3), potato chips (475 ± 1.8), hamburger sandwich (472 ± 0.5), graham crackers (471 ± 3.6), breaded and fried scallops (465 ± 1.6), toaster pastry (458 ± 1), breaded and fried fish (458 ± 0.7), canned broad beans (454 ± 0.2), shortbread cookie (450 ± 6.3), mince pie (448 ± 0.4), canned navy beans (448 ± 0.2), white cake (444 ± 0.6), bacon cheeseburger (440 ± 0.3), baked beans with pork and tomato sauce (440 ± 0.2), yellow cake (439 ± 0.7), au gratin potatoes (433 ± 0.4), cracked wheat bread (432 ± 2), fried chicken (431 ± 0.6), toasted raisin bread (429 ± 2.4), pancakes (426 ± 1.9), refried beans (423 ± 0.2), bran muffins (420 ± 1.3), breaded and fried oysters (418 ± 0.6), dry roasted macadamia nuts (418 ± 1.8), canned pinto beans (416 ± 0.2), cooked whelk (412 ± 0.6), mixed grain bread (412 ± 2), chili (408 ± 0.2), cottage cheese (407 ± 1.8), sweet roll (405 ± 1.2), neufchatel cheese (404 ± 1.8), tuna salad (402 ± 0.2), vegetarian baked beans (397 ± 0.2), french toast (395 ± 0.8), raisin bun (385 ± 0.8), canned black turtle beans (384 ± 0.2), danish pastry (383 ± 1.2), chicken and turkey salad (382 ± 1.8), cooked lobster (380 ± 0.6), mozzarella cheese (379 ± 1.8), red beans (377 ± 0.4), raisin bread (376 ± 2), canned zucchini in tomato juice (375 ± 0.4), raw cuttlefish (372 ± 0.6), dry whole milk (372 ± 1.6), cooked mussels (368 ± 0.6), fish sandwich (365 ± 0.3), cheese soufflé (364 ± 0.5), breaded and fried clams (364 ± 0.6), restaurant tomato juice (361 ± 0.4), bagels (360 ± 0.9), turkey salad (358 ± 0.9), freeze-dried parsley (357 ± 35.7), canned butter beans (356 ± 0.4), breaded and fried croaker (348 ± 0.6), canned yellow corn with red and green peppers (347 ± 0.4), canned kidney beans (347 ± 0.2), oatmeal cookie (346 ± 3.8), breaded and fried shrimp (344 ± 0.6), gruyere cheese (339 ± 1.8), baked beans with pork and sweet sauce (336 ± 0.2), canned lima beans (336 ± 0.2), scalloped potatoes (335 ± 0.4), canned shellie beans (334 ± 0.4), canned blue crab (333 ± 0.6), canned cranberry beans (332 ± 0.2), crab cakes (330 ± 0.8), sugar cookie (319 ± 3.1), breaded fried squid (306 ± 0.6), gingerbread (305 ± 0.8), animal crackers (304 ± 1.9), cherry pie (304 ± 0.4), cream cheese (300 ± 1.8), raw abalone (300 ± 0.6), canned chickpeas (299 ± 0.2), canned black-eyed peas (299 ± 0.2), cottage pudding (298 ± 0.9), raw dungeness crab (295 ± 0.6), falafel (294 ± 1), mashed potatoes (294 ± 0.5), raw blue crab (293 ± 0.6), yellow hominy (290 ± 0.2), white hominy (288 ± 0.2), custard pie (287 ± 0.4), gingersnaps cookie (286 ± 7.1), raw mussels (286 ± 0.6), raisin pie (285 ± 0.4), canned creamed yellow corn (285 ± 0.4), yellow hominy with red and green peppers (283 ± 0.2), breaded and fried catfish (280 ± 0.6), granola bar (279 ± 2.1), cooked blue crab (279 ± 0.6), dried oriental radish

(278 ± 0.9), canned turnip greens (278 ± 0.4), canned white corn (276 ± 0.5), chocolate cream pie (273 ± 0.4), canned yellow corn (272 ± 0.5), pineapple pie (271 ± 0.4), rhubarb pie (270 ± 0.4), wax beans (270 ± 0.4), angel food cake (268 ± 0.8), blackberry pie (268 ± 0.4), blueberry pie (268 ± 0.4), cornbread (268 ± 0.6), corn pone (268 ± 0.6), devil's food cake (267 ± 0.8), swiss cheese (264 ± 1.8), pickled beets (264 ± 0.4), dried sweetened shredded coconut (262 ± 0.5), lemon chiffon pie (260 ± 0.6), zwieback crackers (257 ± 7.1), pineapple chiffon pie (256 ± 0.6), canned stewed red tomatoes (254 ± 0.4), caramel cake (252 ± 0.5), chocolate chiffon pie (252 ± 0.6), instant corn grits (251 ± 0.4), cloves (250 ± 25), frozen creamed yellow corn (250 ± 0.4), canned green beans (250 ± 0.7), canned yellow snap beans (250 ± 0.7), sponge cake (248 ± 0.8), coconut custard pie (247 ± 0.4), rusk crackers (244 ± 5.6), canned succotash with creamed corn (244 ± 0.4), hummus (243 ± 0.2), canned carrots (241 ± 0.7), boiled beet greens (240 ± 0.7), pineapple upside-down cake (239 ± 0.7), cooked flounder (237 ± 0.5), raw kelp/kombu/tangle seaweed (233 ± 0.5), restaurant coleslaw (232 ± 0.5), bloody mary (224 ± 0.3), cooked oysters (224 ± 0.6), cooked shrimp (224 ± 0.6), cheesecake (222 ± 0.6), peach pie (221 ± 0.5), pecan pie (221 ± 0.5), canned succotash (221 ± 0.4), popover roll (220 ± 1.3), canned green peas (219 ± 0.6), sweet potato pie (218 ± 0.4), cooked clams (218 ± 0.6), canned cod (218 ± 0.6), o'brien potatoes (217 ± 0.3), french fried potatoes (216 ± 1), clam liquid (215 ± 0.2), butterscotch pie (214 ± 0.4), pumpkin pie (214 ± 0.4), canned pumpkin pie mix (207 ± 0.4), raw whelk (206 ± 0.6), bread pudding (201 ± 0.2), cumin seed (200 ± 25), dried dill weed (200 ± 50), banana custard pie (194 ± 0.4), strawberry pie (194 ± 0.5), fruitcake, light (193 ± 1.2), freeze-dried red sweet peppers (188 ± 31.3), freeze-dried green sweet peppers (188 ± 31.3), boston cream pie (186 ± 0.5), lemon meringue pie (186 ± 0.4), pineapple custard pie (186 ± 0.4), boiled swiss chard (180 ± 0.6), corn tortilla (177 ± 1.7), boiled cardoon (176 ± 0.5), brownies (165 ± 2.5), simmered pork ears (165 ± 0.5), prune pudding (164 ± 0.4), roasted soybean nuts (163 ± 0.6), instant oatmeal (162 ± 0.3), raw scallops (161 ± 0.6), fruitcake, dark (158 ± 1.2), fried beef brains (158 ± 0.5), tapioca cream pudding (156 ± 0.3), apple brown betty (153 ± 0.2), apple pie (153 ± 0.4), chicken egg white (152 ± 1.5), celery seed (150 ± 25), raw turbot (149 ± 0.6), raw shrimp (148 ± 0.6), duck egg (146 ± 0.7), raw goose liver (140 ± 0.5), raw ling (135 ± 0.6), simmered beef kidneys (134 ± 0.5), tomato powder (134 ± 0.5), cooked whiting (133 ± 0.6), sweetened condensed milk (129 ± 1.3), raw new zealand spinach (129 ± 1.8), boiled swamp cabbage (122 ± 1), battered and fried shark (121 ± 0.6), brewers yeast (121 ± 1.8), simmered beef brains (120 ± 0.5), caramel sundae (119 ± 0.3), chicken egg (116 ± 1), braised beef thymus (116 ± 0.5), evaporated skim milk (116 ± 1.6), cooked herring (115 ± 0.6), cooked swordfish (115 ± 0.6), sesame butter (113 ± 3.3), fried chicken giblets (113 ± 0.5), raw swamp cabbage (113 ± 0.9), raw oysters (112 ± 0.6), canned oysters (112 ± 0.6), chocolate milkshake (111 ± 0.2), pound cake (110 ± 1.7), cooked enriched white degermed corn meal (110 ± 0.2), cooked unenriched corn meal (110 ± 0.2), cooked enriched yellow degermed corn meal (110 ± 0.2), cooked unenriched yellow degermed corn meal (110 ± 0.2), braised pork tongue (109 ± 0.5), evaporated lowfat milk (109 ± 0.4), boiled new zealand spinach (108 ± 0.6), hot fudge sundae (107 ± 0.3), boiled waxgourd (107 ± 0.6), fried beef liver (106 ± 0.5), evaporated whole milk (106 ± 0.4), cooked flatfish (105 ± 0.6), buttermilk (105 ± 0.2), coconut water (105 ± 0.2), raw anchovy (104 ± 0.6), braised pork arm excluding fat (102 ± 0.5), dried agar seaweed

(102 ± 0.5), cooked kingfish (101 ± 0.5), braised beef lungs (101 ± 0.5), dried watermelon seeds (100 ± 1.8), dried coriander leaf (100 ± 50), fennel seed (100 ± 25), saffron (100 ± 50), raw spirulina seaweed (98 ± 0.5), fried dark meat chicken without skin (97 ± 0.5), milk chocolate (96 ± 1.8), restaurant chocolate milkshake (96 ± 0.2), raw burbot (96 ± 0.6), cooked perch (96 ± 0.6), vanilla milkshake (96 ± 0.2), chocolate syrup (95 ± 1.3), stewed chicken drumstick without skin (95 ± 1.1), blackstrap molasses (95 ± 2.5), horseradish sauce (93 ± 3.3), well done broiled ground beef (93 ± 0.5), well done fried ground beef (93 ± 0.5), braised pork spareribs (93 ± 0.5), roasted dark meat chicken without skin (93 ± 0.5), braised pork brains (91 ± 0.5), fried chicken back with skin (90 ± 0.7), fried chicken drumstick with skin (90 ± 1), roasted chicken drumstick with skin (90 ± 1), raw herring (89 ± 0.6), raw mackerel (89 ± 0.6), raw swordfish (89 ± 0.6), well done broiled lean ground beef (89 ± 0.5), fried dark meat chicken with skin (89 ± 0.5), fried chicken thigh with skin (89 ± 0.8), raw butterfish (88 ± 0.6), braised pork arm braised including fat (88 ± 0.5), roasted chicken thigh without skin (88 ± 1), raw celery (88 ± 1.3), frozen and boiled green peas (88 ± 0.6), raw dolphinfish (87 ± 0.6), cooked haddock (87 ± 0.6), cooked sea bass (87 ± 0.6), well done fried lean ground beef (87 ± 0.5), roasted dark meat chicken with skin (87 ± 0.5), restaurant hash brown potatoes (86 ± 0.6), raw pollock (86 ± 0.6), frozen and boiled spinach (86 ± 0.5), raw wolffish (85 ± 0.6), fried pork center loin excluding fat (85 ± 0.5), ricotta cheese (84 ± 0.4), cooked mackerel (84 ± 0.6), medium-rare fried ground beef (84 ± 0.5), broiled pork boston blade excluding fat (84 ± 0.5), broiled lamb liver (84 ± 1.1), roasted chicken thigh with skin (84 ± 0.8), cream puff (83 ± 0.4), restaurant strawberry milkshake (83 ± 0.2), broiled beef flank excluding fat (83 ± 0.5), medium-rare broiled ground beef (83 ± 0.5), restaurant vanilla milkshake (82 ± 0.2), broiled beef flank including fat (82 ± 0.5), well done broiled extra lean ground beef (82 ± 0.5), roasted chicken wing with skin (82 ± 1.5), raw flatfish (81 ± 0.6), well done fried extra lean ground beef (81 ± 0.5), braised pork loin blade excluding fat (81 ± 0.5), braised pork lungs (81 ± 0.5), malted milk (81 ± 0.2), fried light meat chicken without skin (81 ± 0.5), fried chicken neck with skin (81 ± 1.4), raw sunfish (80 ± 0.6), roasted pork arm excluding fat (80 ± 0.5), braised pork kidneys (80 ± 0.5), custard (79 ± 0.2), raw shark (79 ± 0.6), fried chicken breast without skin (79 ± 0.6), roasted turkey dark meat without skin (79 ± 0.5), raw spinach (79 ± 1.8), cooked cod (78 ± 0.6), broiled pork center loin excluding fat (78 ± 0.5), raw whale meat (78 ± 0.5), fried chicken wing with skin (78 ± 1.6), medium-rare broiled lean ground beef (77 ± 0.5), medium-rare fried lean ground beef (77 ± 0.5), fried beef sirloin excluding fat (77 ± 0.5), broiled pork loin blade excluding fat (77 ± 0.5), skim yogurt (77 ± 0.2), fried light meat chicken with skin (77 ± 0.5), roasted light meat chicken without skin (77 ± 0.5), fried chicken breast with skin (77 ± 0.5), cooked pompano (76 ± 0.6), cooked rockfish (76 ± 0.6), cooked rainbow smelt (76 ± 0.6), roasted pork shoulder excluding fat (76 ± 0.5), roasted turkey dark meat with skin (76 ± 0.5), roasted goose without skin (76 ± 0.5), raw freshwater drum (75 ± 0.6), raw perch (75 ± 0.6), canned pink salmon (75 ± 0.6), canned sockeye salmon (75 ± 0.6), well done baked ground beef (75 ± 0.5), broiled pork boston blade including fat (75 ± 0.5), braised pork boston blade excluding fat (75 ± 0.5), braised pork loin excluding fat (75 ± 0.5), broiled pork loin excluding fat (75 ± 0.5), roasted light meat chicken with skin (75 ± 0.5), stewed chicken drumstick with skin (75 ± 0.9), raw dandelion greens (75 ± 1.8), roasted beef rib roasted excluding fat (74 ± 0.5), fried

pork loin blade excluding fat (74 ± 0.5), stewed dark meat chicken without skin (74 ± 0.5), cooked sheepshead (73 ± 0.6), roasted pork boston blade excluding fat (73 ± 0.5), roasted chicken breast without skin (73 ± 0.6), cooked mullet (72 ± 0.6), raw sheepshead (72 ± 0.6), raw whiting (72 ± 0.6), braised beef brisket excluding fat (72 ± 0.5), braised beef flank excluding fat (72 ± 0.5), fried pork center loin including fat (72 ± 0.5), stewed chicken thigh with skin (72 ± 0.7), braised beef chuck blade roast excluding fat (71 ± 0.5), braised beef flank including fat (71 ± 0.5), well done baked lean ground beef (71 ± 0.5), fried beef top round excluding fat (71 ± 0.5), roasted leg of lamb excluding fat (71 ± 0.6), dried pinyon pine nuts (71 ± 1.8), medium-rare broiled extra lean ground beef (70 ± 0.5), medium-rare fried extra lean ground beef (70 ± 0.5), roasted pork arm including fat (70 ± 0.5), broiled pork center loin including fat (70 ± 0.5), braised beef liver (70 ± 0.5), low-fat yogurt (70 ± 0.2), stewed dark meat chicken with skin (70 ± 0.5), roasted chicken breast with skin (70 ± 0.5), roasted goose with skin (70 ± 0.5), boiled spinach (70 ± 0.6), candied sweet potato (70 ± 0.5), raw freshwater bass (69 ± 0.6), raw striped bass (69 ± 0.6), cooked halibut (69 ± 0.6), broiled beef rib eye excluding fat (69 ± 0.5), broiled beef rib excluding fat (69 ± 0.5), broiled lamb loin chop excluding fat (69 ± 1), roasted pork center loin excluding fat (69 ± 0.5), roasted pork loin excluding fat (69 ± 0.5), braised pork loin blade including fat (69 ± 0.5), raw black bass (68 ± 0.5), cooked crayfish (68 ± 0.6), raw haddock (68 ± 0.6), raw sea bass (68 ± 0.6), broiled beef top loin excluding fat (68 ± 0.5), fried beef sirloin including fat (68 ± 0.5), roasted pork loin blade excluding fat (68 ± 0.5), roasted pork shoulder including fat (68 ± 0.5), stewed chicken wing with skin (68 ± 1.3), raw pink salmon (67 ± 0.6), fried beef top round including fat (67 ± 0.5), broiled lamb rib chop excluding fat (67 ± 1.2), braised pork boston blade braised including fat (67 ± 0.5), roasted pork boston blade including fat (67 ± 0.5), broiled pork center rib excluding fat (67 ± 0.5), broiled pork loin blade including fat (67 ± 0.5), roasted pork tenderloin (67 ± 0.5), broiled pork top loin excluding fat (67 ± 0.5), roasted veal rib roast (67 ± 0.6), simmered chicken neck without skin (67 ± 2.8), simmered chicken gizzard (67 ± 0.5), corn syrup (67 ± 2.4), boiled carrots (67 ± 0.6), dried pepeao (67 ± 4.2), raw irish moss seaweed (67 ± 0.5), cooked sockeye salmon (66 ± 0.6), braised beef chuck arm pot roast excluding fat (66 ± 0.5), broiled porterhouse steak excluding fat (66 ± 0.5), broiled T-bone steak excluding fat (66 ± 0.5), broiled beef sirloin excluding fat (66 ± 0.5), roasted lamb shoulder excluding fat (66 ± 0.6), broiled pork loin including fat (66 ± 0.5), braised/broiled veal round with rump (66 ± 0.6), raw chinese cabbage (66 ± 1.4), tomato paste (66 ± 0.4), vanilla pudding (65 ± 0.2), raw mullet (65 ± 0.6), raw pompano (65 ± 0.6), roasted beef tip excluding fat (65 ± 0.5), braised pork loin including fat (65 ± 0.5), roasted pork rump excluding fat (65 ± 0.5), braised/broiled veal loin (65 ± 0.6), cooked eel (65 ± 0.6), stewed light meat chicken without skin (65 ± 0.5), roasted duck without skin (65 ± 0.5), boiled artichoke hearts (65 ± 0.6), canned shitake mushrooms (65 ± 0.4), cooked carp (64 ± 0.6), raw catfish (64 ± 0.6), raw snapper (64 ± 0.6), well done baked extra lean ground beef (64 ± 0.5), broiled beef rib eye including fat (64 ± 0.5), broiled beef round excluding fat (64 ± 0.5), simmered beef shank excluding fat (64 ± 0.5), roasted pork center loin including fat (64 ± 0.5), roasted pork leg excluding fat (64 ± 0.5), roasted pork shank excluding fat (64 ± 0.5), roasted turkey light meat without skin (64 ± 0.5), simmered turkey liver (64 ± 0.5), boiled celery (64 ± 0.7), braised beef chuck blade roast including fat (63 ± 0.5), roasted beef rib including fat (63 ± 0.5), broiled beef tenderloin excluding fat (63 ± 0.5), broiled beef top loin including fat (63 ± 0.5), broiled beef sirloin including fat (63 ± 0.5), roasted pork loin including fat (63 ± 0.5), simmered beef heart (63 ± 0.5), whole chocolate malted milk (63 ± 0.2), stewed light meat chicken with skin (63 ± 0.5), roasted turkey light meat with skin (63 ± 0.5), roasted eye of beef round excluding fat (62 ± 0.5), roasted beef round tip including fat (62 ± 0.5), simmered beef shank including fat (62 ± 0.5), roasted pork sirloin excluding fat (62 ± 0.5), stewed chicken breast with skin (62 ± 0.5), stewed chicken breast without skin (62 ± 0.5), braised beef brisket including fat (61 ± 0.5), broiled beef rib including fat (61 ± 0.5), broiled beef top round excluding fat (61 ± 0.5), broiled porterhouse steak including fat (61 ± 0.5), broiled beef tenderloin including fat (61 ± 0.5), roasted leg of lamb including fat (61 ± 0.6), broiled pork center rib including fat (61 ± 0.5), fried pork loin blade including fat (61 ± 0.5), roasted pork loin blade including fat (61 ± 0.5), roasted pork rump including fat (61 ± 0.5), lowfat 1% chocolate milk (61 ± 0.2), raw celeriac (61 ± 0.5), cooked black bass (60 ± 0.4), raw bluefish (60 ± 0.6), raw rockfish (60 ± 0.6), raw rainbow smelt (60 ± 0.6), braised beef chuck arm pot roast including fat (60 ± 0.5), medium-rare baked ground beef (60 ± 0.5), broiled beef round including fat (60 ± 0.5), broiled beef top round including fat (60 ± 0.5), broiled T-bone steak including fat (60 ± 0.5), broiled pork sirloin excluding fat (60 ± 0.5), braised beef pancreas (60 ± 0.5), simmered beef tongue (60 ± 0.5), lowfat 2% chocolate milk (60 ± 0.2), whole chocolate milk (60 ± 0.2), strawberry sundae (59 ± 0.3), raw lingcod (59 ± 0.6), cooked coho salmon (59 ± 0.6), cooked tilefish (59 ± 0.6), roasted eye of beef round including fat (59 ± 0.5), roasted beef tenderloin excluding fat (59 ± 0.5), roasted pork leg including fat (59 ± 0.5), roasted pork sirloin including fat (59 ± 0.5), braised pork sirloin excluding fat (59 ± 0.5), broiled pork top loin including fat (59 ± 0.5), protein fortified lowfat 2% milk (59 ± 0.2), protein fortified skim milk (59 ± 0.2), roasted duck with skin (59 ± 0.5), simmered turkey giblets (59 ± 0.5), frozen and boiled carrots (59 ± 0.7), raw seatrout (58 ± 0.6), braised beef shortribs excluding fat (58 ± 0.5), roasted pork shank including fat (58 ± 0.5), protein fortified lowfat 1% milk (58 ± 0.2), simmered chicken giblets (58 ± 0.5), braised beef spleen (57 ± 0.5), filled milk (57 ± 0.2), chocolate pudding (56 ± 0.2), raw flounder (56 ± 0.5), cooked snapper (56 ± 0.6), raw sole (56 ± 0.5), medium-rare baked lean ground beef (56 ± 0.5), roasted beef tenderloin including fat (56 ± 0.5), raw cisco (55 ± 0.6), raw clams (55 ± 0.6), raw croaker (55 ± 0.6), braised pork center loin excluding fat (55 ± 0.5), broiled pork sirloin including fat (55 ± 0.5), simmered turkey heart (55 ± 0.5), yellow corn pudding (55 ± 0.2), raw cod (54 ± 0.6), raw halibut (54 ± 0.6), broiled lamb loin chop including fat (54 ± 0.7), braised pork sirloin including fat (54 ± 0.5), eggnog (54 ± 0.2), simmered turkey gizzard (54 ± 0.5), canned sweet potato (54 ± 0.3), cooked tahitian taro (54 ± 0.7), whey (54 ± 0.2), dry bakers yeast (54 ± 1.8), raw crayfish (53 ± 0.6), raw grouper (53 ± 0.6), cooked grouper (53 ± 0.6), raw tilefish (53 ± 0.6), roasted lamb shoulder including fat (53 ± 0.6), simmered chicken neck with skin (53 ± 1.3), boiled artichoke (53 ± 0.6), frozen and boiled fordhook lima beans (53 ± 0.6), raw shad (52 ± 0.6), raw trout (52 ± 0.6), braised pork center rib exlucding fat (52 ± 0.5), braised pork top loin excluding fat (52 ± 0.5), raw pork stomach (52 ± 0.5), indian buffalo milk (52 ± 0.2), carob milk (52 ± 0.2), apple tapioca pudding (51 ± 0.2), canned tuna in oil (51 ± 0.6), canned tuna in water (51 ± 0.6), raw whitefish (51 ± 0.6), braised beef bottom round including fat (51 ± 0.5), braised beef bottom round excluding fat (51 ± 0.5), braised pork center loin including fat (51 ± 0.5), raw eel (51 ± 0.6), skim milk (51 ± 0.2), raw quail without skin (51 ± 0.5), simmered chicken liver (51 ± 0.5), sour cream (50 ± 4.2), braised beef shortribs including fat (50 ± 0.5), fried pork center rib excluding fat (50 ± 0.5), fried pork top loin excluding fat (50 ± 0.5), lowfat 1% milk (50 ± 0.2), lowfat 2% milk (50 ± 0.2), goat milk (50 ± 0.2), sweetened coconut cream (50 ± 0.2), allspice (50 ± 25), cinnamon (50 ± 25), coriander seed (50 ± 25), curry powder (50 ± 25), fenugreek seed (50 ± 12.5), ginger (50 ± 25), mace (50 ± 25), onion powder (50 ± 25), paprika (50 ± 25), black pepper (50 ± 25), cayenne pepper (50 ± 25), pumpkin pie spice (50 ± 25), dried rosemary (50 ± 25), tarragon (50 ± 25), turmeric (50 ± 25), freeze-dried shallots (50 ± 12.5), boiled turnips (50 ± 0.6), raw carp (49 ± 0.6), cooked pike (49 ± 0.6), raw chum salmon (49 ± 0.6), medium-rare baked extra lean ground beef (49 ± 0.5), broiled lamb rib chop including fat (49 ± 0.7), roasted veal leg (49 ± 0.6), braised pork liver (49 ± 0.5), whole milk (49 ± 0.2), hot chocolate (49 ± 0.2), boiled beets (49 ± 0.6), frozen and boiled collards (49 ± 0.6), dried longans (48 ± 0.5), braised pork center rib including fat (48 ± 0.5), braised/roasted/stewed veal chuck (48 ± 0.6), simmered chicken heart (48 ± 0.5), raw laver/nori seaweed (48 ± 0.5), chicken egg yolk (47 ± 2.9), raw chinook salmon (47 ± 0.6), raw sockeye salmon (47 ± 0.6), braised pork top loin including fat (47 ± 0.5), raw coho salmon (46 ± 0.6), roasted pork center rib excluding fat (46 ± 0.5), roasted pork top loin excluding fat (46 ± 0.5), braised/stewed veal breast plate (46 ± 0.6), raw beef tripe (46 ± 0.5), whole yogurt (46 ± 0.2), raw chickory greens (46 ± 0.6), fried pork center rib including fat (45 ± 0.5), fried pork top loin including fat (45 ± 0.5), boiled scotch kale (45 ± 0.8), boiled lotus root (45 ± 0.6), boiled purslane (45 ± 0.9), frozen and boiled succotash (45 ± 0.6), roasted pork center rib including fat (44 ± 0.5), roasted pork top loin including fat (44 ± 0.5), raw squid (44 ± 0.6), sheep milk (44 ± 0.2), boiled dandelion greens (44 ± 1), raw scup (42 ± 0.6), braised pork pancreas (42 ± 0.5), stewed rabbit (41 ± 0.4), raw lotus root (41 ± 0.6), raw watercress (41 ± 2.9), half and half cream (40 ± 3.3), medium cream (40 ± 3.3), whipping cream, heavy (40 ± 3.3), raw white sucker (40 ± 0.6), rice flour (40 ± 0.4), corn oil margarine tub (40 ± 10), safflower oil margarine tub (40 ± 10), soybean oil margarine tub (40 ± 10), soybean oil and cottonseed oil margarine tub (40 ± 10), raw parsley (40 ± 1.7), raw pike (39 ± 0.6), raw yellowtail (39 ± 0.6), simmered pork chitterlings (39 ± 0.5), dried european chestnuts (39 ± 1.8), dried coconut (39 ± 1.8), dried toasted coconut (39 ± 1.8), toasted sesame kernels (39 ± 1.8), raw turnip greens (39 ± 1.8), rice polish (38 ± 0.9), dried sesame kernels (38 ± 6.3), raw pheasant without skin (37 ± 0.5), gum drops (36 ± 1.8), braised pork heart (36 ± 0.4), dried japanese chestnuts (36 ± 1.8), frozen and boiled turnips (36 ± 0.5), french fries (35 ± 0.6), carob flour (35 ± 0.5), medium molasses (35 ± 2.5), raw carrots (35 ± 0.7), frozen and fried hash brown potatoes (35 ± 0.6), cooked rainbow trout (34 ± 0.6), potato flour (34 ± 0.6), boiled chinese cabbage (34 ± 0.6), marshmallow (33 ± 8.3), whipping cream, light (33 ± 3.3), restaurant hot chocolate (33 ± 0.2), garlic powder (33 ± 16.7), poppy seed (33 ± 16.7), sweet chocolate (32 ± 1.8), peanut brittle (32 ± 1.8), raw cusk (32 ± 0.6), corn germ (31 ± 0.5), brown sugar (30 ± 0.3), frozen and french fried potatoes without skin (30 ± 1), carrot juice (29 ± 0.3), boiled and steamed european chestnuts (29 ± 1.8), raw savoy cabbage (29 ± 1.4), boiled and frozen baby lima beans (29 ± 0.6), boiled turnip greens (29 ± 0.7), raw spot (28 ± 0.6), passion fruit (28 ± 2.8), seeded raisins (28 ± 0.5), tamarind (28 ± 0.4), raw rainbow trout (27 ± 0.6), raw broccoli (27 ± 1.1), canned spinach (27 ± 0.5), raw pork jowl (25 ± 0.5), peanut flour (25 ± 12.5), raw coriander (25 ± 12.5), frozen and boiled mustard greens (25 ± 0.7), dried sweetened flaked coconut (24 ± 2), frozen and boiled broccoli (24 ± 0.5), raw endive (24 ± 2), hash brown potatoes (24 ± 0.6), raw radish (24 ± 1.1), frozen and boiled brussels sprouts (23 ± 0.6), boiled savoy cabbage (23 ± 0.7), coleslaw (23 ± 0.8), boiled kale (23 ± 0.8), boiled brussels sprouts (22 ± 0.6), raw garlic (22 ± 5.6), light cream (21 ± 1.7), boiled amaranth (21 ± 0.8), boiled kohlrabi (21 ± 0.6), dehydrated onion flakes (21 ± 3.6), defatted soybean flour (20 ± 0.5), raw coconut (20 ± 1.1), corn oil margarine stick (20 ± 10), safflower oil and soybean oil margarine stick (20 ± 10), soybean oil margarine stick (20 ± 10), soybean oil and cottonseed oil margarine stick (20 ± 10), boiled dishcloth gourd (20 ± 0.6), frozen and boiled potatoes without skin (20 ± 0.5), raw oriental radish (20 ± 1.1), tomato purée (20 ± 0.2), raw monkfish (19 ± 0.6), creamy peanut butter (19 ± 3.1), boiled green cabbage (19 ± 0.7), boiled collards (19 ± 0.3), boiled catjang black-eyed peas (19 ± 0.3), raw leeks (19 ± 1.9), tom collins (18 ± 0.2), low fat soybean flour (18 ± 0.6), oil roasted cashews (18 ± 1.8), dried peanuts (18 ± 1.8), dried pumpkin and squash seeds (18 ± 1.8), roasted pumpkin and squash seeds (18 ± 1.8), frozen and boiled cauliflower (18 ± 0.6), boiled rutabaga (18 ± 0.6), boiled/baked spaghetti squash (18 ± 0.6), compressed bakers yeast (18 ± 1.8), boiled arrowhead (17 ± 4.2), raw green cabbage (17 ± 1.4), boiled yellow corn (17 ± 0.6), boiled succotash (17 ± 0.5), human milk (16 ± 1.6), crunchy peanut butter (16 ± 1.6), boiled mustard greens (16 ± 0.7), boiled salsify (16 ± 0.7), mammy apple (15 ± 0.5), coconut milk (15 ± 0.2), citrus marmalade (15 ± 2.5), light molasses (15 ± 2.5), frozen and boiled kale (15 ± 0.8), fried tofu (15 ± 3.8), cooked taro (15 ± 0.8), frozen and boiled turnip greens (15 ± 0.6), frozen and boiled turnip greens and turnips (15 ± 0.5), raw chinese water chestnuts (15 ± 0.8), bourbon & soda (14 ± 0.4), maple sugar (14 ± 1.8), pie crust (14 ± 0.3), orange soda (14 ± 0.1), soursop (14 ± 0.2), cashew butter (14 ± 1.8), dry roasted cashews (14 ± 1.8), raw japanese chestnuts (14 ± 1.8), dried ginko nuts (14 ± 1.8), oil roasted peanuts (14 ± 1.8), raw cauliflower (14 ± 1), boiled oriental radish (14 ± 0.7), torula yeast (14 ± 1.8), java plum (13 ± 0.4), soybean flour (13 ± 0.6), roasted soybean flour (13 ± 0.6), soy milk (13 ± 0.2), almond butter (13 ± 3.1), sweet butter (13 ± 3.3), raw ginger root (13 ± 2.1), frozen and boiled yellow snap beans (13 ± 0.7), frozen and boiled green beans (13 ± 0.7), steamed hawaii mountain yam (13 ± 0.7), dried shitake mushrooms (13 ± 3.3), firm raw tofu (13 ± 0.4), boiled sweet potato (13 ± 0.3), raw green tomato (13 ± 0.4), canned peeled red tomatoes (13 ± 0.4), boiled winged beans (13 ± 0.3), casaba melon (12 ± 0.3), honeydew melon (12 ± 0.5), golden seedless raisins (12 ± 0.5), seedless raisins (12 ± 0.5), sapodilla (12 ± 0.3), raw california avocado (12 ± 0.3), boiled jute potherb (12 ± 1.2), boiled leeks (12 ± 1.9), poi (12 ± 0.4), whiskey sour (11 ± 0.6), jelly beans (11 ± 1.8), shredded wheat (11 ± 1.8), dried figs (11 ± 0.3), almond paste (11 ± 1.8), dried almonds (11 ± 1.8), dry roasted almonds (11 ± 1.8), oil roasted almonds (11 ± 1.8), toasted almonds (11 ± 1.8), whole dried sesame seeds (11 ± 5.6), whole toasted sesame seeds (11 ± 1.8), dried english/persian walnuts (11 ± 1.8), raw red cabbage (11 ± 1.4), raw looseleaf lettuce (11 ± 1.8), steamed sprouted soybeans (11 ± 1.1), baked sweet potato (11 ± 0.4), boiled red tomato (11 ± 0.4), tomato juice (10 ± 0.3), mulberries (10 ± 0.4), sugar apple (10 ± 0.3), maple syrup (10 ± 2.5), boiled broccoli (10 ± 0.6), raw iceberg lettuce (10 ± 2.5), boiled mothbeans (10 ± 0.3), cooked sprouted mung beans (10 ± 0.8), boiled parsnips (10 ± 0.6), raw shallots (10 ± 5), okara tofu (10 ± 0.8), lemon-lime soda (9 ± 0.1), dried apricots (9 ± 1.4), cantaloupe (9 ± 0.3), dried jujube (9 ± 0.5), sapote (9 ± 0.2), raw pepeao (9 ± 0.5), raw agar seaweed (9 ± 0.5), canned chinese water chestnuts (9 ± 0.7), baked/boiled tropical yam (9 ± 0.7), 53 proof coffee liqueur (8 ± 1), 63 proof coffee liqueur (8 ± 1), dry dessert wine (8 ± 0.8), sweet dessert wine (8 ± 0.8), boiled adzuki beans (8 ± 0.2), boiled red cabbage (8 ± 0.7), raw cassava (8 ± 0.5), boiled onions (8 ± 0.5), frozen and boiled onions (8 ± 0.5), baked potato with skin (8 ± 0.2), microwaved potato with skin (8 ± 0.2), baked hubbard squash (8 ± 0.5), frozen and baked sweet potato (8 ± 0.6), raw red tomato (8 ± 0.4), puffed wheat (7 ± 3.6), peach nectar (7 ± 0.2), raw acerola (7 ± 0.5), almond meal (7 ± 1.8), dried chinese chestnuts (7 ± 1.8), ginko nuts (7 ± 1.8), oil roasted macadamia nuts (7 ± 1.8), dry roasted peanuts (7 ± 1.8), oil roasted spanish peanuts (7 ± 1.8), oil roasted valencia peanuts (7 ± 1.8), oil roasted virginia peanuts (7 ± 1.8), dried pistachio nuts (7 ± 1.8), dry roasted pistachio nuts (7 ± 1.8), canned bamboo shoots (7 ± 0.4), boiled chickpeas (7 ± 0.3), raw chickory (7 ± 1.1), boiled garden cress (7 ± 0.7), boiled hyacinth beans (7 ± 0.3), raw butterhead lettuce (7 ± 3.3), raw romaine lettuce (7 ± 1.8), boiled mungo beans (7 ± 0.3), raw red chili peppers (7 ± 1.1), raw green chili peppers (7 ± 1.1), microwaved potato without skin (7 ± 0.3), natto (7 ± 0.6), raw tofu (7 ± 0.4), raw wild rice (7 ± 0.3), fruit pectin (7 ± 3.6), piña colada (6 ± 0.4), creme de menthe (6 ± 1), red table wine (6 ± 0.5), papaya nectar (6 ± 0.2), yellow passion fruit juice (6 ± 0.2), blueberries (6 ± 0.3), mandarin oranges (6 ± 0.4), prickly pear (6 ± 0.5), rice bran (6 ± 0.5), raw sprouted alfalfa seeds (6 ± 1.5), boiled cauliflower (6 ± 0.8), frozen and boiled black-eyed peas (6 ± 0.6), boiled french beans (6 ± 0.3), raw sprouted mung beans (6 ± 1), raw potato without skin (6 ± 0.4), tempeh (6 ± 0.6), dried and frozen tofu (6 ± 2.9), frozen and boiled crookneck squash (6 ± 0.5), boiled white beans (6 ± 0.3), boiled yambean (6 ± 0.5), beer (5 ± 0.1), daiquiri (5 ± 0.8), rose table wine (5 ± 0.5), white table wine (5 ± 0.5), cola (5 ± 0.1), carissa (5 ± 2.5), kiwi fruit (5 ± 0.7), kumquats (5 ± 2.6), cooked plantain (5 ± 0.3), roselle (5 ± 0.9), honey (5 ± 2.4), raw florida avocado (5 ± 0.2), boiled broad beans (5 ± 0.3), frozen and boiled yellow corn (5 ± 0.6), boiled okra (5 ± 0.6), raw green peas (5 ± 0.6), boiled pigeon peas (5 ± 0.3), baked potato without skin (5 ± 0.3), boiled potato without skin (5 ± 0.4), canned pumpkin (5 ± 0.4), canned crookneck squash (5 ± 0.5), boiled hubbard squash (5 ± 0.4), canned white beans (5 ± 0.2), boiled yardlong bean (5 ± 0.3), boiled yellow beans (5 ± 0.3), ale (4 ± 0.1), malt liquor (4 ± 0.1), gin & tonic (4 ± 0.2), manhattan (4 ± 0.9), tequila sunrise (4 ± 0.3), bittersweet chocolate (4 ± 1.8), german sweet bakers chocolate (4 ± 1.8), semi-sweet chocolate (4 ± 1.8), semi-sweet bakers chocolate (4 ± 1.8), maltex (4 ± 0.3), apricot nectar (4 ± 0.2), cranberry juice cocktail (4 ± 0.2), pear nectar (4 ± 0.2), prune juice (4 ± 0.2), dried prunes (4 ± 0.6), quince (4 ± 0.5), toasted wheat germ (4 ± 1.8), raw chinese chestnuts (4 ± 1.8), boiled and steamed chinese chestnuts (4 ± 1.8), roasted chinese chestnuts (4 ± 1.8), raw european chestnuts (4 ± 1.8), roasted european chestnuts (4 ± 1.8), boiled and steamed japanese chestnuts (4 ± 1.8), coconut cream (4 ± 0.2), dried filberts (4 ± 1.8), dry roasted filberts (4 ± 1.8), oil roasted filberts (4 ± 1.8), dried macadamia nuts (4 ± 1.8), dried pilinuts (4 ± 1.8), dried pignolia pine nuts (4 ± 1.8), toasted soybean nuts (4 ± 1.8), dried sunflower seeds (4 ± 1.8), dry roasted sunflower seeds (4 ± 1.8), oil roasted sunflower seeds (4 ± 1.8), boiled asparagus (4 ± 0.6), raw bamboo shoots (4 ± 0.7), boiled bamboo shoots (4 ± 0.4), boiled burdock root (4 ± 0.4), boiled butterbur (4 ± 0.5), boiled black-eyed peas (4 ± 0.3), raw dock (4 ± 0.7), boiled eggplant (4 ± 1), canned great northern beans (4 ± 0.2), boiled lupins (4 ± 0.3), cooked shitake mushrooms (4 ± 0.7), raw spring onions with tops (4 ± 1), raw red sweet peppers (4 ± 1), raw green sweet peppers (4 ± 1), frozen and boiled red sweet peppers (4 ± 0.5), frozen and boiled green sweet peppers (4 ± 0.5), baked acorn squash (4 ± 0.5), boiled butternut squash (4 ± 0.5), unsweetened baking chocolate (4 ± 1.8), soybean protein concentrate (4 ± 1.8), martini (3 ± 0.7), black tea (3 ± 0.3), instant tea (3 ± 0.2), maypo (3 ± 0.3), acerola juice (3 ± 0.2), apple juice (3 ± 0.2), grape juice (3 ± 0.2), applesauce (3 ± 0.4), jackfruit (3 ± 0.5), jujube (3 ± 0.5), dried lychees (3 ± 0.5), papaya (3 ± 0.2), dried japanese persimmon (3 ± 1.5), pitanga (3 ± 0.3), pomegranate (3 ± 0.3), light pearled barley (3 ± 0.3), whole wheat flour (3 ± 0.4), frozen and boiled asparagus (3 ± 0.8), boiled black turtle beans (3 ± 0.3), frozen white corn (3 ± 0.7), boiled dock (3 ± 0.5), boiled green beans (3 ± 0.8), boiled yellow snap beans (3 ± 0.8), boiled baby lima beans (3 ± 0.3), raw mushrooms (3 ± 1.4), boiled mushrooms (3 ± 0.6), frozen and boiled okra (3 ± 0.5), raw onions (3 ± 0.6), boiled green peas (3 ± 0.6), boiled sprouted green peas (3 ± 0.5), raw zucchini (3 ± 0.8), 90 proof gin (2 ± 1.2), coffee (2 ± 0.3), restaurant cranberry juice cocktail (2 ± 0.3), Perrier water (2 ± 0.3), boysenberries (2 ± 0.4), breadfruit (2 ± 0.5), carambola (2 ± 0.4), black currants (2 ± 0.9), red currants (2 ± 0.9), white currants (2 ± 0.9), dried dates (2 ± 0.6), figs (2 ± 1), american grapes (2 ± 0.5), european grapes (2 ± 0.3), thompson seedless grapes (2 ± 0.3), guava (2 ± 0.6), lemon (2 ± 0.9), mango (2 ± 0.2), japanese persimmon (2 ± 0.3), cooked prunes (2 ± 0.5), watermelon (2 ± 0.3), defatted soy meal (2 ± 0.4), all purpose enriched wheat flour (2 ± 0.4), enriched wheat bread (2 ± 0.4), wheat cake (2 ± 0.4), wheat gluten (2 ± 0.4), noodles (2 ± 0.3), dry roasted soybean nuts (2 ± 0.6), boiled blackeye pea pods (2 ± 1.1), raw cucumber (2 ± 1), raw eggplant (2 ± 1.2), boiled great northern beans (2 ± 0.3), boiled kidney beans (2 ± 0.3), boiled lentils (2 ± 0.3), boiled lima beans (2 ± 0.3), boiled mung beans (2 ± 0.2), boiled split peas (2 ± 0.3), boiled pink beans (2 ± 0.3), boiled pinto beans (2 ± 0.3), boiled pumpkin (2 ± 0.4), frozen and cooked sweetened rhubarb (2 ± 0.4), raw crookneck squash (2 ± 0.8), raw scallop squash (2 ± 0.8), boiled zucchini (2 ± 0.6), frozen and boiled zucchini (2 ± 0.4), boiled acorn squash (2 ± 0.4), frozen and boiled butternut squash (2 ± 0.4), screwdriver (1 ± 0.2), chamomile tea (1 ± 0.3), cream of rice (1 ± 0.3), cream of wheat (1 ± 0.3), farina (1 ± 0.3), oatmeal (1 ± 0.3), restaurant coffee (1 ± 0.3), restaurant orange juice (1 ± 0.4), tea (1 ± 0.2), white grapefruit juice (1 ± 0.2), red grapefruit juice (1 ± 0.2), lemon juice (1 ± 0.2), lime juice (1 ± 0.2), orange juice (1 ± 0.2), pineapple juice (1 ± 0.2), tangerine juice (1 ± 0.2), raw apple with skin (1 ± 0.4), cooked apple without skin (1 ± 0.3), apricots (1 ± 0.5), bananas (1 ± 0.4), crab apples (1 ± 0.5), cranberry sauce (1 ± 0.5), gooseberries (1 ± 0.3), lime (1 ± 0.7), loganberries (1 ± 0.3), loquats (1 ± 0.5), lychees (1 ± 0.5), oheloberries (1 ± 0.4), navel orange (1 ± 0.4), pear (1 ± 0.3), pineapple (1 ± 0.3), pummelo (1 ± 0.3), strawberries (1 ± 0.3), tangerine (1 ± 0.6), white corn flour (1 ± 0.4), yellow corn flour (1 ± 0.4), white bolted corn meal (1 ± 0.4), enriched white degermed corn meal (1 ± 0.4), unenriched corn meal (1 ± 0.4), white whole ground corn meal (1 ± 0.4), yellow bolted corn meal (1 ± 0.4), enriched yellow degermed corn meal (1 ± 0.4), unenriched yellow degermed corn meal (1 ± 0.4), yellow whole ground corn meal (1 ± 0.4), dark rye flour (1 ± 0.4),

light rye flour (1 ± 0.5), macaroni (1 ± 0.4), spaghetti (1 ± 0.4), powdered sugar (1 ± 0.4), white granulated sugar (1 ± 0.3), boiled black beans (1 ± 0.3), raw celtuce (1 ± 0.5), boiled chayote (1 ± 0.6), boiled cranberry beans (1 ± 0.3), boiled calabash gourd (1 ± 0.7), boiled navy beans (1 ± 0.3), boiled red sweet peppers (1 ± 0.7), boiled green sweet peppers (1 ± 0.7), raw frozen rhubarb (1 ± 0.4), boiled soybeans (1 ± 0.3), boiled crookneck squash (1 ± 0.6), boiled scallop squash (1 ± 0.6)
Applications: diarrhea, adrenal deterioration, cystic fibrosis, leg cramp, tooth and gum deterioration, dehydration, fever, polio
Daily Dosage: (1700 mg potassium per every 1000 mg sodium) 1980 RDA—none; 1980 SADDI—1100–3300 mg; 1989 RDA—none; 1989 SADDI—1100–3300 mg; nutritional—1000–3000 mg; therapeutic—none; experimental—none; toxic—varies
Warnings: Keep levels low with stroke, congestive heart failure, high blood pressure, and skin edema. Increased amounts are dangerous. See toxicity symptoms.

STRONTIUM

Names: strontium, Sr, Sr^{+2}, Sr^{++}, element 38
Classification: trace mineral
Deficiency: increased tooth decay, depletion of bones, decreased toxin resistance, mitochondria deterioration
Side-effects: unknown
Toxicity: unknown
Inhibitors: unknown
Helpers: unknown
Sources: vegetables, grains
Applications: thinning of bones
Daily Dosage: (1700 mg potassium per every 1000 mg sodium) 1980 RDA—none; 1980 SADDI—none; 1989 RDA—none; 1989 SADDI—none; nutritional—1 mg; therapeutic—none; experimental—none
Warnings: none

STRONTIUM 90

Names: strontium 90, radioactive strontium
Classification: contaminant
Deficiency: (no nutritional requirement)
Side-effects: see *toxicity* symptoms
Toxicity: bone cancers, leukemia
Inhibitors: calcium
Helpers: unknown
Sources: milk, meat, eggs, vegetables, pollution
Applications: none known
Daily Dosage: 1974 PCS—none; 1989 SADDI—none; 1989 RDA—none; 1989 USRDA—none; nutritional—none; therapeutic—none; experimental—none; toxic—small amounts
Warnings: Small amounts are dangerous. See toxicity symptoms.

SULFUR

Names: sulfur, sulphur, S, S^{-2}, $S^{..}$, element 16
Classification: trace mineral
Deficiency: dull hair, skin complexion deterioration, skin eruptions, abscesses, chronic oozing ulcers, deterioration of fingernails, decreased bacterial resistance, yellow coating on back of tongue, fatigue, pains in extremities
Side-effects: see toxicity symptoms
Toxicity: (usually only from non-organic sources)
Inhibitors: insufficient protein
Helpers: vitamin B complex, vitamin B_1, vitamin B_5, vitamin B_W, protein
Sources: fish, eggs, cabbage, lean beef, dried beans, brussels sprouts
Applications: worms, arthritis, skin inflammation, skin eczema, skin psoriasis
Daily Dosage: 1974 PCS – none; 1989 SADDI – none; 1989 RDA – none; 1989 USRDA – none; nutritional – unknown; therapeutic – none; experimental – none; toxic – unknown
Warnings: Exessive amounts are dangerous. See toxicity symptoms.

TIN

Names: tin, Sn, element 50
Classification: trace mineral, contaminant
Deficiency: deterioration of growth, deterioration of tooth development, decreased cancer resistance
Side-effects: see toxicity symptoms
Toxicity: shortened life span, tin accumulation in heart and intestine
Inhibitors: unknown
Helpers: unknown
Sources: air pollution, tin cans, "tin can" flavoring in jarred foods
Applications: none known
Daily Dosage: 1980 RDA – none; 1980 SADDI – none; 1989 RDA – none; 1989 SADDI – none; nutritional – 2 mg; therapeutic – 20 mg; experimental – unknown
Warnings: Small amounts are dangerous. See toxicity symptoms.

TITANIUM

Names: titanium, Ti, element 22
Classification: possible trace mineral, contaminant
Deficiency: (being researched)
Side-effects: unknown
Toxicity: titanium poisoning
Inhibitors: unknown
Helpers: unknown
Sources: industry, white paints
Application: none known

Daily Dosage: 1974 PCS — none; 1989 SADDI — none; 1989 RDA — none; 1989 USRDA — none; nutritional — unknown; therapeutic — none; experimental — none; toxic — unknown
Warnings: increased amounts are dangerous. See toxicity symptoms.

TUNGSTEN

Names: tungsten, formerly wolfram, W, element 74
Classification: possible trace mineral, contaminant
Deficiency: (being researched)
Side-effects: see toxicity symptoms
Toxicity: tungsten will replace molybdenum in xanthine, oxidase, and other enzymes, disrupting their functioning
Inhibitors: unknown
Helpers: unknown
Sources: industry
Applications: none known
Daily Dosage: 1974 PCS — none; 1989 SADDI — none; 1989 RDA — none; 1989 USRDA — none; nutritional — unknown; therapeutic — none; experimental — none; toxic — small amounts
Warnings: Small amounts are dangerous. See toxicity symptoms.

VANADIUM

Names: vanadium, V, element 23
Classification: trace mineral, antioxidant
Deficiency: deterioration of growth, deterioration of bone and teeth mineralization, heart disease, raised cholesterol levels in blood, increased triglyceride levels, deterioration of fat metabolism, increased squalene synthetase levels, decreased acetoacetylcoenzyme A levels, tooth deterioration, bone demineralization, abnormal bone growth, liver deterioration, decreased cancer resistence, deterioration of growth rate, deterioration of reproduction systems
Side-effects: see toxicity symptoms
Toxicity: (Detoxified by vitamin C) anemia, green tongue, confusion, inflammation of lung, inflammation of eyes, cataract development
Inhibitors: unknown
Helpers: unknown
Sources (μg per 100 grams of food): buckwheat (100 ± 5), parsley (80 ± 5), soybeans (70 ± 5), safflower oil (64 ± 0.5), eggs (42 ± 0.5), sunflower seed oil (41 ± 0.5), oats (35 ± 0.5), olive oil (30 ± 0.5), sunflower seeds (15 ± 0.5), corn (15 ± 0.5), green beans (14 ± 0.5), corn oil (12 ± 0.5), oysters (11 ± 0.5), peanut oil (11 ± 0.5), carrots (10 ± 0.5), cabbage (10 ± 0.5), garlic (10 ± 0.5), tomatoes (6 ± 0.5), radishes (5 ± 0.5), onions (5 ± 0.5), whole wheat (5 ± 0.5), lobster (4 ± 0.5), beets (4 ± 0.5), apples (3 ± 0.5), plums (2 ± 0.5), lettuce (2 ± 0.5), millet (2 ± 0.5), liver (<1), fish (<1), meat (<1), peas (<0.01), pears (<0.01), milk (<0.01)
Applications: increased cholesterol in

blood, broken bones, neurasthenia, diabetes, cancer, atherosclerosis, free radical oxidation, life span extension, radiation, decreased radiation resistance, antioxidant therapy, manic depressive states
Daily Dosage: 1989 SADDI — none;
1989 RDA — none; 1989 USRDA — none; nutritional — 0.5–2 mg; therapeutic — 5 mg; experimental — ?; toxic — ?
Warnings: increased amounts are dangerous. See toxicity symptoms.

ZINC

Names: zinc, Zn, element 30
Classification: trace mineral
Deficiency: growth retardation, slowed sexual development, testicular deterioration/dysfunction, sterility, decreased gonad function, deterioration of thymus gland, diminished sense of taste, deterioration of protein formation, diminished sense of smell, appetite decreased, fatigue, deterioration of nutrient absorption, liver deterioration, internal inflammation, decreased healing, anemia, decreased development of bone, muscle, nervous system, hair decreased, acne, skin lesions, white spots on nails, deformed nails, immunity decreased, decreased salivation, bad breath
Side-effects: see toxicity symptoms
Toxicity: nausea, abdominal pain, fatigue, dizziness, dehydration, decreased levels of copper, iron decreased, muscular coordination decreased
Inhibitors: corticosteroids, birth control pills, decreased phosphorus
Helpers: vitamin A, vitamin C, vitamin E, calcium, copper, manganese, phosphorus, selenium, cysteine
Sources (mg per 100 grams of food): cooked oysters (181.91 ± 0.006), raw oysters (90.95 ± 0.006), canned oysters (90.95 ± 0.006), breaded and fried oysters (87.13 ± 0.006), toasted wheat germ (16.89 ± 0.018), bran (13.21 ± 0.018), bran flakes (13.21 ± 0.018), corn germ (10.6 ± 0.005), simmered beef shank excluding fat (10.49 ± 0.005), toasted sesame kernels (10.36 ± 0.018), raisin bran (10.27 ± 0.014), braised beef chuck blade roast excluding fat (10.27 ± 0.005), dried sesame kernels (10.25 ± 0.063), poppy seed (9.67 ± 0.167), simmered beef shank including fat (9.67 ± 0.005), braised beef chuck arm pot roast excluding fat (8.66 ± 0.005), basil (8 ± 0.5), braised beef chuck blade roast including fat (7.83 ± 0.005), braised beef shortribs excluding fat (7.8 ± 0.005), whole dried sesame seeds (7.78 ± 0.056), cooked alaska king crab (7.62 ± 0.006), dried pumpkin and squash seeds (7.57 ± 0.018), roasted pumpkin and squash seeds (7.54 ± 0.018), cardamom (7.5 ± 0.25), simmered chicken heart (7.3 ± 0.005), whole toasted sesame seeds (7.25 ± 0.018), roasted beef tip excluding fat (7.07 ± 0.005), celery seed (7 ± 0.25), broiled beef rib eye excluding fat (6.99 ± 0.005), roasted beef rib roasted excluding fat (6.94 ± 0.005), well done baked extra lean ground beef (6.94 ± 0.005), braised beef brisket excluding fat (6.88 ± 0.005), braised beef chuck arm pot roast including fat (6.74 ± 0.005), braised pork liver (6.72 ± 0.005), oil roasted peanuts (6.71 ± 0.018), oil roasted virginia peanuts (6.61 ± 0.018), broiled beef rib excluding fat (6.55 ± 0.005), broiled beef sirloin excluding fat (6.52 ± 0.005), well done baked lean ground beef (6.51 ± 0.005), freeze-dried parsley (6.43 ± 0.357), well done broiled extra

lean ground beef (6.43 ± 0.005), fried beef sirloin excluding fat (6.4 ± 0.005), roasted beef round tip including fat (6.37 ± 0.005), yellow mustard seed (6.33 ± 0.167), fried chicken giblets (6.27 ± 0.005), well done fried extra lean ground beef (6.27 ± 0.005), well done broiled lean ground beef (6.2 ± 0.005), broiled beef rib eye including fat (6.13 ± 0.005), braised beef liver (6.07 ± 0.005), well done baked ground beef (6.07 ± 0.005), braised beef flank excluding fat (6.05 ± 0.005), savory (6 ± 0.5), dried parsley (6 ± 0.5), caraway seed (6 ± 0.25), rice bran (5.98 ± 0.005), raw alaska king crab (5.94 ± 0.006), braised beef flank including fat (5.92 ± 0.005), well done fried lean ground beef (5.91 ± 0.005), baked beans with pork and tomato sauce (5.86 ± 0.002), well done broiled ground beef (5.81 ± 0.005), dry roasted pecans (5.75 ± 0.018), broiled beef sirloin including fat (5.75 ± 0.005), dry roasted cashews (5.68 ± 0.018), ry-krisp crackers (5.64 ± 0.036), well done fried ground beef (5.62 ± 0.005), broiled beef tenderloin excluding fat (5.59 ± 0.005), oil roasted pecans (5.57 ± 0.018), broiled beef top round excluding fat (5.57 ± 0.005), braised pork boston blade excluding fat (5.57 ± 0.005), dried pecans (5.54 ± 0.018), dill seed (5.5 ± 0.25), anise seed (5.5 ± 0.25), braised beef bottom round excluding fat (5.48 ± 0.005), fried beef liver (5.45 ± 0.005), medium-rare broiled extra lean ground beef (5.45 ± 0.005), medium-rare fried extra lean ground beef (5.42 ± 0.005), broiled beef top round including fat (5.4 ± 0.005), broiled porterhouse steak excluding fat (5.4 ± 0.005), broiled T-bone steak excluding fat (5.4 ± 0.005), dry roasted sunflower seeds (5.36 ± 0.018), medium-rare broiled lean ground beef (5.36 ± 0.005), medium-rare baked extra lean ground beef (5.34 ± 0.005), chipped beef (5.32 ± 0.018), oil roasted sunflower seeds (5.29 ± 0.018), simmered turkey heart (5.27 ± 0.005), cashew butter (5.25 ± 0.018), fried beef sirloin including fat (5.25 ± 0.005), broiled beef top loin excluding fat (5.22 ± 0.005), medium-rare fried lean ground beef (5.2 ± 0.005), medium-rare broiled ground beef (5.18 ± 0.005), roasted beef tenderloin excluding fat (5.17 ± 0.005), roasted beef rib including fat (5.17 ± 0.005), dried sunflower seeds (5.14 ± 0.018), braised beef bottom round including fat (5.13 ± 0.005), medium-rare baked lean ground beef (5.1 ± 0.005), medium-rare fried ground beef (5.07 ± 0.005), defatted soy meal (5.06 ± 0.004), simmered pork chitterlings (5.06 ± 0.005), broiled beef tenderloin including fat (5.04 ± 0.005), broiled beef rib including fat (5.03 ± 0.005), braised beef brisket including fat (5.01 ± 0.005), cumin seed (5 ± 0.25), peanut flour (5 ± 0.125), toasted almonds (5 ± 0.018), turmeric (5 ± 0.25), dried chervil (5 ± 0.5), braised pork arm excluding fat (4.97 ± 0.005), dry roasted almonds (4.96 ± 0.018), oil roasted almonds (4.96 ± 0.018), braised pork loin blade excluding fat (4.93 ± 0.005), medium-rare baked ground beef (4.89 ± 0.005), dried and frozen tofu (4.88 ± 0.029), braised beef shortribs including fat (4.88 ± 0.005), oil roasted cashews (4.82 ± 0.018), simmered beef tongue (4.8 ± 0.005), broiled beef flank excluding fat (4.79 ± 0.005), dry roasted soybean nuts (4.77 ± 0.006), roasted eye of beef round excluding fat (4.74 ± 0.005), broiled beef flank including fat (4.71 ± 0.005), broiled porterhouse steak including fat (4.7 ± 0.005), broiled beef round excluding fat (4.68 ± 0.005), dried brazil nuts (4.64 ± 0.018), fried beef top round excluding fat (4.62 ± 0.005), braised pork boston blade braised including fat (4.61 ± 0.005), sesame butter (4.6 ± 0.033), braised pork spareribs (4.6 ± 0.005), broiled beef top loin including fat (4.6 ± 0.005), braised beef pancreas (4.6 ± 0.005), cooked corned beef brisket (4.58 ± 0.005), simmered chicken giblets (4.57 ± 0.005), braised pork tongue

(4.53 ± 0.005), roasted beef tenderloin including fat (4.48 ± 0.005), roasted turkey dark meat without skin (4.46 ± 0.005), broiled T-bone steak including fat (4.46 ± 0.005), dry instant skim milk (4.41 ± 0.005), soybean protein concentrate (4.39 ± 0.018), simmered chicken gizzard (4.38 ± 0.005), roasted pork boston blade excluding fat (4.37 ± 0.005), dried pinyon pine nuts (4.36 ± 0.018), dried hickory nuts (4.36 ± 0.018), simmered chicken liver (4.34 ± 0.005), roasted eye of beef round including fat (4.33 ± 0.005), dried pignolia pine nuts (4.32 ± 0.018), beef pastrami (4.32 ± 0.018), braised pork pancreas (4.29 ± 0.005), raw dungeness crab (4.27 ± 0.006), broiled pork boston blade excluding fat (4.27 ± 0.005), roasted pork shoulder excluding fat (4.24 ± 0.005), simmered beef kidneys (4.22 ± 0.005), cooked blue crab (4.21 ± 0.006), hard pork salami (4.2 ± 0.05), fried beef top round including fat (4.18 ± 0.005), roasted turkey dark meat with skin (4.16 ± 0.005), simmered turkey gizzard (4.16 ± 0.005), braised pork kidneys (4.15 ± 0.005), jellied corned beef loaf (4.14 ± 0.018), broiled beef round including fat (4.13 ± 0.005), crab cakes (4.1 ± 0.008), dry skim milk (4.07 ± 0.017), roasted pork arm excluding fat (4.07 ± 0.005), soybean protein isolate (4.04 ± 0.018), braised pork arm braised including fat (4.04 ± 0.005), canned blue crab (4.01 ± 0.006), coriander seed (4 ± 0.25), curry powder (4 ± 0.25), paprika (4 ± 0.25), ginger (4 ± 0.25), lebanon bologna (4 ± 0.022), swiss cheese (3.96 ± 0.018), gouda cheese (3.96 ± 0.018), chopped smoked beef (3.96 ± 0.018), soybean flour (3.92 ± 0.006), braised pork loin blade including fat (3.85 ± 0.005), roasted pork boston blade including fat (3.82 ± 0.005), edam cheese (3.79 ± 0.018), broiled pork loin blade excluding fat (3.79 ± 0.005), simmered chicken neck without skin (3.78 ± 0.028), braised pork loin excluding fat (3.72 ± 0.005), dry buttermilk (3.71 ± 0.071), pork livercheese (3.71 ± 0.013), toasted soybean nuts (3.68 ± 0.018), simmered turkey giblets (3.68 ± 0.005), fried pork loin blade excluding fat (3.67 ± 0.005), broiled pork boston blade including fat (3.62 ± 0.005), roasted pork shoulder including fat (3.59 ± 0.005), roasted pork loin blade excluding fat (3.59 ± 0.005), roasted soybean flour (3.58 ± 0.006), wheat germ oil (3.57 ± 0.036), canned corned beef (3.57 ± 0.005), braised pork spleen (3.54 ± 0.005), tilsit cheese (3.54 ± 0.018), fontina cheese (3.54 ± 0.018), raw blue crab (3.53 ± 0.006), fennel seed (3.5 ± 0.25), oregano (3.5 ± 0.25), bacon cheeseburger (3.5 ± 0.003), dried black walnuts (3.46 ± 0.018), roasted pork shank excluding fat (3.45 ± 0.005), chicken egg yolk (3.41 ± 0.029), dry whole milk (3.34 ± 0.016), dried peanuts (3.32 ± 0.018), dry roasted peanuts (3.32 ± 0.018), shredded wheat (3.32 ± 0.018), miso (3.32 ± 0.004), roasted pork arm including fat (3.31 ± 0.005), provolone cheese (3.29 ± 0.018), pork and beef peppered loaf (3.29 ± 0.018), cooked whelk (3.26 ± 0.006), roasted pork leg excluding fat (3.26 ± 0.005), cooked bacon (3.26 ± 0.004), beef honey roll sausage (3.26 ± 0.022), parmesan cheese (3.2 ± 0.1), hard pork and beef salami (3.2 ± 0.05), dried butternuts (3.18 ± 0.018), stewed chicken drumstick without skin (3.18 ± 0.011), roasted soybean nuts (3.14 ± 0.006), cheddar cheese (3.14 ± 0.018), simmered beef heart (3.13 ± 0.005), colby cheese (3.11 ± 0.018), braised pork heart (3.09 ± 0.004), simmered turkey liver (3.09 ± 0.005), fried chicken neck with skin (3.08 ± 0.014), oil roasted valencia peanuts (3.07 ± 0.018), braised pork sirloin excluding fat (3.07 ± 0.005), pork luxury loaf (3.07 ± 0.018), almond butter (3.06 ± 0.031), broiled pork loin blade including fat (3.05 ± 0.005), monterey cheese (3.04 ± 0.018), roasted pork loin excluding fat (3.04 ± 0.005), american cheese (3.04 ± 0.018), havarti cheese (3.04 ± 0.018), natto (3.03 ± 0.006), braised pork loin including fat (3.03 ± 0.005), raw lobster (3.02 ± 0.006), roasted pork rump excluding fat (3.01 ± 0.005), dried dill weed (3 ± 0.5), sage (3 ± 0.5), tarragon (3 ± 0.25), roasted pork tenderloin (3 ± 0.005), pimento cheese spread (3 ± 0.018), roasted pork loin blade including fat (2.99 ± 0.005), dried almonds (2.96 ± 0.018), creamy peanut butter (2.94 ± 0.031), roasted cured pork arm excluding fat (2.94 ± 0.005), roasted pork shank including fat (2.93 ± 0.005), feta cheese (2.93 ± 0.018), cooked lobster (2.92 ± 0.006), broiled pork loin excluding fat (2.92 ± 0.005), fried dark meat chicken without skin (2.91 ± 0.005), braised pork center loin excluding fat (2.91 ± 0.005), fried chicken drumstick with skin (2.9 ± 0.01), roasted lean ham (2.88 ± 0.005), roasted chicken drumstick with skin (2.87 ± 0.01), muenster cheese (2.86 ± 0.018), roasted pork leg including fat (2.86 ± 0.005), boiled hyacinth beans (2.85 ± 0.003), pork liver sausage (2.83 ± 0.028), braised pork center rib exlucding fat (2.82 ± 0.005), braised pork top loin excluding fat (2.82 ± 0.005), pork link sausage (2.82 ± 0.007), roasted dark meat chicken without skin (2.8 ± 0.005), dried english/persian walnuts (2.79 ± 0.018), braised beef spleen (2.79 ± 0.005), smoked beef sausage (2.79 ± 0.012), crunchy peanut butter (2.78 ± 0.016), roasted pork rump including fat (2.74 ± 0.005), fried pork loin blade including fat (2.71 ± 0.005), simmered chicken neck with skin (2.71 ± 0.013), bran muffins (2.7 ± 0.013), bleu cheese (2.68 ± 0.018), cooked mussels (2.67 ± 0.006), stewed dark meat chicken without skin (2.66 ± 0.005), stewed chicken drumstick with skin (2.65 ± 0.009), brick cheese (2.64 ± 0.018), turkey ham (2.63 ± 0.009), roasted pork loin including fat (2.62 ± 0.005), almond paste (2.61 ± 0.018), dried japanese chestnuts (2.61 ± 0.018), barbecue loaf (2.61 ± 0.022), fried dark meat chicken with skin (2.6 ± 0.005), roasted duck without skin (2.6 ± 0.005), roasted chicken thigh without skin (2.58 ± 0.01), beef lunch meat (2.57 ± 0.018), summer sausage (2.57 ± 0.022), braised pork sirloin including fat (2.56 ± 0.005), oil roasted filberts (2.54 ± 0.018), dry roasted filberts (2.54 ± 0.018), pork sausage (2.52 ± 0.019), fried chicken thigh with skin (2.52 ± 0.008), roasted cured pork arm including fat (2.51 ± 0.005), nutmeg (2.5 ± 0.25), freeze-dried red sweet peppers (2.5 ± 0.313), freeze-dried green sweet peppers (2.5 ± 0.313), poultry seasoning (2.5 ± 0.25), onion powder (2.5 ± 0.25), cayenne pepper (2.5 ± 0.25), cinnamon (2.5 ± 0.25), roasted canned ham (2.5 ± 0.005), hot dog (2.5 ± 0.006), roasted pork sirloin excluding fat (2.49 ± 0.005), roasted dark meat chicken with skin (2.49 ± 0.005), berliner (2.48 ± 0.022), fried chicken back with skin (2.47 ± 0.007), roasted ham (2.47 ± 0.005), braised pork center loin including fat (2.47 ± 0.005), raw beef tripe (2.47 ± 0.005), defatted soybean flour (2.46 ± 0.005), pork and beef honey loaf (2.46 ± 0.018), anchovy canned in olive oil (2.45 ± 0.025), broiled pork loin including fat (2.45 ± 0.005), roasted pork blade roll (2.45 ± 0.005), braised pork lungs (2.45 ± 0.005), roast beef sandwich (2.44 ± 0.003), dried filberts (2.43 ± 0.018), camembert cheese (2.43 ± 0.018), turkey summer sausage (2.43 ± 0.018), pork and beef luncheon sausage (2.43 ± 0.022), beef beerwurst salami (2.43 ± 0.022), fried pork center loin excluding fat (2.41 ± 0.005), broiled pork center rib excluding fat (2.38 ± 0.005), broiled pork top loin excluding fat (2.38 ± 0.005), italian pork sausage (2.37 ± 0.007), puffed wheat (2.36 ± 0.036), broiled pork sirloin excluding fat (2.36 ± 0.005), roasted chicken thigh with skin (2.35 ± 0.008), braised pork center rib including fat (2.35 ± 0.005), chili powder (2.33 ± 0.167), garlic powder (2.33 ± 0.167), pork and beef pepperoni (2.33 ± 0.083), pork bratwurst (2.31 ± 0.006), ham and cheese spread (2.29 ± 0.018), turkey bologna (2.29 ± 0.018), roasted pork center loin excluding fat (2.28 ± 0.005), braised pork top loin including fat (2.28 ± 0.005), roasted pork sirloin including fat (2.27 ± 0.005), stewed dark meat chicken with skin (2.26 ± 0.005), pork and beef link sausage with american cheese (2.26 ± 0.012), fenugreek seed (2.25 ± 0.125), wheat flakes (2.25 ± 0.018), stewed chicken thigh with skin (2.25 ± 0.007), mozzarella cheese (2.25 ± 0.018), broiled pork center loin excluding fat (2.23 ± 0.005), roasted canned lean ham (2.23 ± 0.005), roasted pork center rib excluding fat (2.22 ± 0.005), roasted pork top loin excluding fat (2.22 ± 0.005), turkey frankfurter (2.22 ± 0.011), pork and beef picnic loaf (2.21 ± 0.018), beef frankfurter (2.18 ± 0.009), beef bologna (2.17 ± 0.022), beef salami (2.17 ± 0.022), dry roasted macadamia nuts (2.14 ± 0.018), limburger cheese (2.14 ± 0.018), unheated ham (2.14 ± 0.005), turkey pastrami (2.14 ± 0.009), pork and beef mortadella (2.13 ± 0.033), beef and pork salami (2.13 ± 0.022), cooked rainbow smelt (2.12 ± 0.006), pork and beef link sausage (2.12 ± 0.007), roquefort cheese (2.11 ± 0.018), stir-fried sprouted soybeans (2.1 ± 0.005), pork and beef brotwurst (2.1 ± 0.007), beef summer sausage (2.09 ± 0.022), cooked eel (2.07 ± 0.006), dried toasted coconut (2.07 ± 0.018), fried pork center rib excluding fat (2.06 ± 0.005), fried pork top loin excluding fat (2.06 ± 0.005), dried coconut (2.04 ± 0.018), roasted turkey light meat without skin (2.04 ± 0.005), roasted turkey light meat with skin (2.04 ± 0.005), roasted pork center loin including fat (2.04 ± 0.005), ham and cheese roll (2.04 ± 0.018), pork and beef kielbassa/kolbassy (2.04 ± 0.019), pork bologna (2.04 ± 0.022), beef and pork summer sausage (2.04 ± 0.022), broiled pork sirloin including fat (2.03 ± 0.005), broiled pork center rib including fat (2.03 ± 0.005), raw pork stomach (2.01 ± 0.005), marjoram (2 ± 0.5), oil roasted spanish peanuts (2 ± 0.018), dried rosemary (2 ± 0.25), mace (2 ± 0.25), pumpkin pie spice (2 ± 0.25), bay leaf (2 ± 0.5), fried tofu (2 ± 0.038), cheeseburger (2 ± 0.004), light and dark meat turkey roll (1.98 ± 0.009), broiled pork top loin including fat (1.97 ± 0.005), fried pork center loin including fat (1.96 ± 0.005), roasted pork center rib including fat (1.96 ± 0.005), polish pork sausage (1.96 ± 0.018), beef and pork bologna (1.96 ± 0.022), raw irish moss seaweed (1.95 ± 0.005), chopped ham lunch meat (1.95 ± 0.024), taco (1.93 ± 0.006), broiled pork center loin including fat (1.93 ± 0.005), unheated lean ham (1.93 ± 0.005), canned lean ham (1.93 ± 0.005), roasted pork top loin including fat (1.92 ± 0.005), cooked carp (1.91 ± 0.006), toasted whole wheat bread (1.9 ± 0.024), hamburger sandwich (1.9 ± 0.005), minced ham lunch meat (1.9 ± 0.024), semi-sweet bakers chocolate (1.89 ± 0.018), grilled ham patties (1.88 ± 0.008), boiled catjang black-eyed peas (1.87 ± 0.003), dehydrated onion flakes (1.86 ± 0.036), toasted cracked wheat bread (1.86 ± 0.024), roasted duck with skin (1.86 ± 0.005), pork and beef sausage (1.85 ± 0.038), boiled peanuts (1.84 ± 0.016), beef and pork frankfurter (1.84 ± 0.011), dried sweetened shredded coconut (1.82 ± 0.005), roasted chicken wing with skin (1.82 ± 0.015), tempeh (1.81 ± 0.006), turkey salami (1.81 ± 0.009), boiled adzuki beans (1.77 ± 0.002), dried macadamia nuts (1.75 ± 0.018), freeze-dried shallots (1.75 ± 0.125), pork and beef old fashioned loaf (1.75 ± 0.018), fried chicken wing with skin (1.75 ± 0.016), breaded fried squid (1.74 ± 0.006), pork beerwurst salami (1.74 ± 0.022), raw cuttlefish (1.73 ± 0.006), raw anchovy (1.72 ± 0.006), tomato powder (1.71 ± 0.005), raw octopus (1.68 ± 0.006), whole wheat bread (1.68 ± 0.02), grilled canadian bacon (1.68 ± 0.011), pork and beef lunch meat (1.68 ± 0.018), cooked crayfish (1.67 ± 0.006), canned ham

(1.66 ± 0.005), pork and beef knockwurst (1.66 ± 0.007), raw rainbow smelt (1.65 ± 0.006), raw whelk (1.64 ± 0.006), braised beef lungs (1.64 ± 0.005), stewed chicken wing with skin (1.63 ± 0.013), vienna sausage (1.63 ± 0.031), raw eel (1.62 ± 0.006), fried pork center rib including fat (1.62 ± 0.005), fried pork top loin including fat (1.62 ± 0.005), stir-fried sprouted lentils (1.6 ± 0.005), raw mussels (1.6 ± 0.006), dried and salted cod (1.59 ± 0.006), firm raw tofu (1.57 ± 0.004), unheated ham patties (1.57 ± 0.008), cooked shrimp (1.56 ± 0.006), raw sunfish (1.55 ± 0.006), light meat turkey roll (1.54 ± 0.009), boiled chickpeas (1.53 ± 0.003), raw squid (1.53 ± 0.006), falafel (1.51 ± 0.01), black pepper (1.5 ± 0.25), white pepper (1.5 ± 0.25), baked beans with pork and sweet sauce (1.5 ± 0.002), chili (1.5 ± 0.002), raw carp (1.48 ± 0.006), braised pork brains (1.48 ± 0.005), canned pork lunch meat (1.48 ± 0.024), cooked swordfish (1.47 ± 0.006), roasted japanese chestnuts (1.46 ± 0.018), breaded and fried clams (1.46 ± 0.006), toasted rye bread (1.45 ± 0.023), boiled winged beans (1.44 ± 0.003), dried chinese chestnuts (1.43 ± 0.018), corn tortilla (1.43 ± 0.017), pork pickle and pimento loaf (1.43 ± 0.018), pork mothers loaf (1.43 ± 0.024), frozen and boiled black-eyed peas (1.42 ± 0.006), duck egg (1.41 ± 0.007), vegetarian baked beans (1.4 ± 0.002), dry roasted pistachio nuts (1.39 ± 0.018), cooked rainbow trout (1.39 ± 0.006), cheese pizza (1.39 ± 0.004), pork olive loaf (1.39 ± 0.018), unheated canadian bacon (1.39 ± 0.009), boiled lupins (1.38 ± 0.003), breaded and fried shrimp (1.38 ± 0.006), boiled white beans (1.37 ± 0.003), dried pistachio nuts (1.36 ± 0.018), burrito (1.36 ± 0.003), refried beans (1.36 ± 0.002), raw clams (1.36 ± 0.006), kippered herring (1.35 ± 0.013), fried beef brains (1.35 ± 0.005), pork headcheese (1.32 ± 0.018), raw crayfish (1.31 ± 0.006), toasted pumpernickel bread (1.29 ± 0.018), boiled black-eyed peas (1.29 ± 0.003), sardines (1.29 ± 0.021), sweetened condensed milk (1.29 ± 0.013), rye bread (1.28 ± 0.02), taco shell (1.27 ± 0.045), tostada shell (1.27 ± 0.045), cooked herring (1.27 ± 0.006), fried light meat chicken without skin (1.27 ± 0.005), boiled lentils (1.26 ± 0.003), fried light meat chicken with skin (1.26 ± 0.005), turkey salad (1.26 ± 0.009), german sweet bakers chocolate (1.25 ± 0.018), simmered beef brains (1.25 ± 0.005), raw green peas (1.24 ± 0.006), raw kelp/kombu/tangle seaweed (1.23 ± 0.005), roasted light meat chicken without skin (1.23 ± 0.005), roasted light meat chicken with skin (1.23 ± 0.005), sponge cake (1.21 ± 0.008), mixed grain bread (1.2 ± 0.02), boiled green peas (1.19 ± 0.006), stewed light meat chicken without skin (1.19 ± 0.005), low fat soybean flour (1.18 ± 0.006), chicken egg (1.18 ± 0.01), ricotta cheese (1.16 ± 0.004), boiled soybeans (1.15 ± 0.003), raw bacon (1.15 ± 0.007), boiled cranberry beans (1.14 ± 0.003), raw swordfish (1.14 ± 0.006), stewed light meat chicken with skin (1.14 ± 0.005), turkey breast meat summer sausage (1.14 ± 0.024), pumpernickel bread (1.13 ± 0.016), boiled black beans (1.12 ± 0.003), oil roasted macadamia nuts (1.11 ± 0.018), canned white beans (1.11 ± 0.002), raw japanese chestnuts (1.11 ± 0.018), raw shrimp (1.11 ± 0.006), raw coconut (1.11 ± 0.011), ham salad (1.11 ± 0.018), hummus (1.1 ± 0.002), fried chicken breast with skin (1.09 ± 0.005), boiled yardlong bean (1.08 ± 0.003), boiled pinto beans (1.08 ± 0.003), raw rainbow trout (1.08 ± 0.006), fried chicken breast without skin (1.08 ± 0.006), potato chips (1.07 ± 0.018), boiled kidney beans (1.07 ± 0.003), pretzels (1.07 ± 0.018), puffed rice (1.07 ± 0.036), boiled yellow beans (1.06 ± 0.003), boiled navy beans (1.06 ± 0.003), breaded and fried scallops (1.06 ± 0.016), canned chickpeas (1.05 ± 0.002), raw laver/nori seaweed

(1.05 ± 0.005), steamed sprouted soybeans (1.04 ± 0.011), pork and beef spread (1.04 ± 0.018), boiled baby lima beans (1.03 ± 0.003), roasted chicken breast with skin (1.02 ± 0.005), boiled broad beans (1.01 ± 0.003), canned sockeye salmon (1.01 ± 0.006), cloves (1 ± 0.25), allspice (1 ± 0.25), dried sweetened flaked coconut (1 ± 0.02), oatmeal cookie (1 ± 0.038), boiled split peas (1 ± 0.003), roasted chicken breast without skin (1 ± 0.006), cheese crackers (1 ± 0.033), raw whitefish (0.99 ± 0.006), raw herring (0.99 ± 0.006), stewed chicken breast without skin (0.97 ± 0.005), raw pheasant without skin (0.97 ± 0.005), skim yogurt (0.97 ± 0.002), coconut cream (0.96 ± 0.002), boiled pink beans (0.96 ± 0.003), oatmeal bread (0.96 ± 0.018), stewed chicken breast with skin (0.96 ± 0.005), raw scallops (0.95 ± 0.006), boiled lima beans (0.95 ± 0.003), cooked mackerel (0.94 ± 0.006), fried abalone (0.94 ± 0.006), frozen and boiled green peas (0.94 ± 0.006), roasted chinese chestnuts (0.93 ± 0.018), canned pink salmon (0.92 ± 0.006), carob flour (0.91 ± 0.005), evaporated skim milk (0.91 ± 0.016), raw sprouted alfalfa seeds (0.91 ± 0.015), stir-fried sprouted mung beans (0.9 ± 0.008), boiled pigeon peas (0.9 ± 0.003), raw chinese chestnuts (0.89 ± 0.018), potato pancakes (0.89 ± 0.007), lowfat yogurt (0.89 ± 0.002), boiled great northern beans (0.88 ± 0.003), cooked mullet (0.88 ± 0.006), toasted english muffin (0.88 ± 0.01), raw whiting (0.88 ± 0.006), waffles (0.87 ± 0.007), boiled mushrooms (0.87 ± 0.006), cooked pike (0.86 ± 0.006), breaded and fried catfish (0.86 ± 0.006), italian bread (0.86 ± 0.018), french bread (0.86 ± 0.018), french toast (0.85 ± 0.008), boiled mung beans (0.84 ± 0.002), canned cranberry beans (0.84 ± 0.002), boiled mungo beans (0.83 ± 0.003), danish pastry (0.83 ± 0.012), gefiltefish with broth (0.83 ± 0.012), raw milkfish (0.82 ± 0.006), raw tofu (0.81 ± 0.004), raw abalone (0.81 ± 0.006), raw bluefish (0.81 ± 0.006), tomato paste (0.8 ± 0.004), devil's food cake (0.8 ± 0.008), raw endive (0.8 ± 0.02), graham crackers (0.79 ± 0.036), pita bread (0.79 ± 0.013), raw wolffish (0.78 ± 0.006), boiled sprouted green peas (0.78 ± 0.005), boiled spinach (0.77 ± 0.006), canned navy beans (0.77 ± 0.002), evaporated whole milk (0.77 ± 0.004), raw butterfish (0.76 ± 0.006), boiled black turtle beans (0.76 ± 0.003), raw burbot (0.75 ± 0.006), raw white sucker (0.75 ± 0.006), maltex (0.75 ± 0.003), dried apricots (0.74 ± 0.014), raw parsley (0.73 ± 0.017), raw pompano (0.72 ± 0.006), raw catfish (0.72 ± 0.006), fried chicken (0.72 ± 0.006), english muffin (0.72 ± 0.009), light meat chicken roll (0.72 ± 0.005), chicken roll (0.72 ± 0.009), chocolate syrup (0.71 ± 0.013), toasted raisin bread (0.71 ± 0.024), toasted white bread (0.71 ± 0.024), dinner rolls (0.71 ± 0.018), canned green peas (0.71 ± 0.006), cottage cheese (0.71 ± 0.018), canned black-eyed peas (0.7 ± 0.002), frozen and boiled spinach (0.69 ± 0.005), cooked pompano (0.69 ± 0.006), canned pinto beans (0.69 ± 0.002), au gratin potatoes (0.69 ± 0.004), dried acorns (0.68 ± 0.018), dried ginko nuts (0.68 ± 0.018), raw pike (0.67 ± 0.006), coconut milk (0.67 ± 0.002), saltine crackers (0.67 ± 0.083), raw freshwater bass (0.66 ± 0.006), raw freshwater drum (0.66 ± 0.006), chocolate cream pie (0.66 ± 0.004), raw trout (0.66 ± 0.006), canned great northern beans (0.65 ± 0.002), canned lima beans (0.65 ± 0.002), acorn flour (0.64 ± 0.018), boiled french beans (0.64 ± 0.003), boiled succotash (0.64 ± 0.005), cooked sheepshead (0.64 ± 0.006), raisin bread (0.64 ± 0.02), boiled collards (0.64 ± 0.003), canned broad beans (0.63 ± 0.002), white bread (0.63 ± 0.021), hamburger bun (0.63 ± 0.013), hot dog bun (0.63 ± 0.013), raw mackerel (0.62 ± 0.006), cooked flatfish (0.62 ± 0.006), frozen and boiled okra

(0.62 ± 0.005), frozen and boiled yellow snap beans (0.62 ± 0.007), frozen and boiled green beans (0.62 ± 0.007), maypo (0.62 ± 0.003), boiled and steamed chinese chestnuts (0.61 ± 0.018), cooked perch (0.61 ± 0.006), sweetened coconut cream (0.6 ± 0.002), french rolls (0.6 ± 0.01), hot fudge sundae (0.6 ± 0.003), boiled mothbeans (0.59 ± 0.003), cornbread (0.59 ± 0.006), corn pone (0.59 ± 0.006), whole yogurt (0.59 ± 0.002), cooked cod (0.58 ± 0.006), cooked clams (0.58 ± 0.006), canned cod (0.58 ± 0.006), toaster pastry (0.58 ± 0.01), roasted european chestnuts (0.57 ± 0.018), boiled and frozen baby lima beans (0.56 ± 0.006), enchilada (0.56 ± 0.002), tuna salad (0.56 ± 0.002), boiled okra (0.55 ± 0.006), canned kidney beans (0.55 ± 0.002), pineapple upside-down cake (0.55 ± 0.007), yellow cake (0.55 ± 0.007), frozen and boiled asparagus (0.55 ± 0.008), cooked sturgeon (0.54 ± 0.006), raw pink salmon (0.54 ± 0.006), raw spinach (0.54 ± 0.018), dried prunes (0.54 ± 0.006), canned black turtle beans (0.54 ± 0.002), raw european chestnuts (0.54 ± 0.018), cream cheese (0.54 ± 0.018), neufchatel cheese (0.54 ± 0.018), cooked halibut (0.53 ± 0.006), cooked rockfish (0.53 ± 0.006), cooked tilefish (0.53 ± 0.006), cooked whiting (0.53 ± 0.006), raw sockeye salmon (0.53 ± 0.006), bagels (0.53 ± 0.009), caramel sundae (0.53 ± 0.003), canned creamed yellow corn (0.53 ± 0.004), half and half cream (0.53 ± 0.033), pickled herring (0.53 ± 0.033), cooked coho salmon (0.52 ± 0.006), raw yellowtail (0.52 ± 0.006), cooked sea bass (0.52 ± 0.006), breaded and fried croaker (0.52 ± 0.006), raw mullet (0.52 ± 0.006), fish sandwich (0.52 ± 0.003), raw spot (0.51 ± 0.006), cooked grouper (0.51 ± 0.006), cooked sockeye salmon (0.51 ± 0.006), mandarin oranges (0.51 ± 0.004), boiled beet greens (0.5 ± 0.007), raw acorns (0.5 ± 0.018), dried figs (0.5 ± 0.003), canned succotash (0.5 ± 0.004), macaroni (0.5 ± 0.004), spaghetti (0.5 ± 0.004), yellow corn pudding (0.5 ± 0.002), shortbread cookie (0.5 ± 0.063), smoked haddock (0.49 ± 0.006), oatmeal (0.49 ± 0.003), hot chocolate (0.49 ± 0.002), raw mushrooms (0.49 ± 0.014), cooked haddock (0.48 ± 0.006), battered and fried shark (0.48 ± 0.006), boiled yellow corn (0.48 ± 0.006), raw perch (0.48 ± 0.006), raw grouper (0.48 ± 0.006), raw scup (0.48 ± 0.006), smoked whitefish (0.48 ± 0.006), sweet roll (0.48 ± 0.012), boiled asparagus (0.48 ± 0.006), chocolate milkshake (0.48 ± 0.002), russian salad dressing (0.47 ± 0.033), raw chum salmon (0.47 ± 0.006), raw pollock (0.47 ± 0.006), canned tuna in oil (0.47 ± 0.006), cooked sprouted mung beans (0.47 ± 0.008), raw dolphinfish (0.46 ± 0.006), canned spinach (0.46 ± 0.005), raw flatfish (0.46 ± 0.006), raw lingcod (0.46 ± 0.006), canned yellow corn (0.46 ± 0.005), eggnog (0.46 ± 0.002), raspberries (0.46 ± 0.004), raw chinook salmon (0.45 ± 0.006), raw seatrout (0.45 ± 0.006), raw cod (0.45 ± 0.006), frozen and boiled succotash (0.45 ± 0.006), protein fortified lowfat 2% milk (0.45 ± 0.002), protein fortified skim milk (0.45 ± 0.002), protein fortified lowfat 1% milk (0.45 ± 0.002), cooked snapper (0.44 ± 0.006), frozen and boiled fordhook lima beans (0.44 ± 0.006), breaded and fried fish (0.44 ± 0.007), raw spring onions with tops (0.44 ± 0.01), onion rings (0.44 ± 0.006), tamari sauce (0.43 ± 0.009), malted milk (0.43 ± 0.002), gingersnaps cookie (0.43 ± 0.071), raw sturgeon (0.42 ± 0.006), raw shark (0.42 ± 0.006), raw california avocado (0.42 ± 0.003), jackfruit (0.42 ± 0.005), raw florida avocado (0.42 ± 0.002), frozen and french fried potatoes without skin (0.42 ± 0.01), whole chocolate malted milk (0.42 ± 0.002), buttermilk (0.42 ± 0.002), cheesecake (0.42 ± 0.006), raw coho salmon (0.41 ± 0.006), raw halibut (0.41 ± 0.006), raw croaker (0.41 ± 0.006), dried japanese persimmon (0.41 ± 0.015), frozen and boiled turnip greens (0.41 ± 0.006), raw rockfish (0.41 ± 0.006), raw broccoli (0.41 ± 0.011), raw monkfish (0.41 ± 0.006), restaurant chocolate milkshake (0.41 ± 0.002), lowfat 1% chocolate milk (0.41 ± 0.002), lowfat 2% chocolate milk (0.41 ± 0.002), whole chocolate milk (0.41 ± 0.002), raw striped bass (0.4 ± 0.006), raw sea bass (0.4 ± 0.006), raw sprouted mung beans (0.4 ± 0.01), scalloped potatoes (0.4 ± 0.004), skim milk (0.4 ± 0.002), french fries (0.39 ± 0.006), raw sheepshead (0.39 ± 0.006), boiled crookneck squash (0.39 ± 0.006), raw potato without skin (0.39 ± 0.004), boiled and steamed japanese chestnuts (0.39 ± 0.018), lowfat 1% milk (0.39 ± 0.002), lowfat 2% milk (0.39 ± 0.002), vanilla milkshake (0.39 ± 0.002), canned chinese water chestnuts (0.39 ± 0.007), raw wakame seaweed (0.38 ± 0.005), raw haddock (0.38 ± 0.006), french fried potatoes (0.38 ± 0.01), raw cusk (0.38 ± 0.006), whole milk (0.38 ± 0.002), boiled green beans (0.37 ± 0.008), boiled yellow snap beans (0.37 ± 0.008), canned yellow corn with red and green peppers (0.37 ± 0.004), dried european chestnuts (0.36 ± 0.018), boiled artichoke hearts (0.36 ± 0.006), shoyu sauce (0.36 ± 0.009), raw shad (0.36 ± 0.006), boiled artichoke (0.36 ± 0.006), ginko nuts (0.36 ± 0.018), raw tilefish (0.36 ± 0.006), microwaved potato with skin (0.36 ± 0.002), frozen and boiled brussels sprouts (0.36 ± 0.006), filled milk (0.36 ± 0.002), carob milk (0.36 ± 0.002), restaurant vanilla milkshake (0.36 ± 0.002), light mayonnaise (0.36 ± 0.036), mayonnaise (0.36 ± 0.036), raw snapper (0.35 ± 0.006), white cake (0.35 ± 0.006), restaurant strawberry milkshake (0.35 ± 0.002), loganberries (0.34 ± 0.003), frozen and boiled yellow corn (0.34 ± 0.006), frozen and boiled crookneck squash (0.33 ± 0.005), microwaved potato without skin (0.33 ± 0.003), golden seedless raisins (0.32 ± 0.005), baked potato with skin (0.32 ± 0.002), boiled brussels sprouts (0.32 ± 0.006), frozen and fried hash brown potatoes (0.32 ± 0.006), raw red chili peppers (0.31 ± 0.011), raw green chili peppers (0.31 ± 0.011), restaurant hash brown potatoes (0.31 ± 0.006), boiled rutabaga (0.31 ± 0.006), smoked chinook salmon (0.31 ± 0.006), potato salad (0.31 ± 0.004), pancakes with butter and syrup (0.31 ± 0.003), soy sauce (0.31 ± 0.009), frozen and boiled broccoli (0.3 ± 0.005), frozen and baked sweet potato (0.3 ± 0.006), o'brien potatoes (0.3 ± 0.003), goat milk (0.3 ± 0.002), canned crookneck squash (0.3 ± 0.005), dried dates (0.29 ± 0.006), baked potato without skin (0.29 ± 0.003), raw scallop squash (0.29 ± 0.008), raw crookneck squash (0.29 ± 0.008), hash brown potatoes (0.29 ± 0.006), baked sweet potato (0.29 ± 0.004), smoked cisco (0.29 ± 0.006), boiled carrots (0.29 ± 0.006), canned green beans (0.29 ± 0.007), canned yellow snap beans (0.29 ± 0.007), corn flakes (0.29 ± 0.018), raw radish (0.29 ± 0.011), dried lychees (0.28 ± 0.005), blackberries (0.28 ± 0.007), mashed potatoes (0.28 ± 0.005), canned potato without skin (0.28 ± 0.006), lemon meringue pie (0.28 ± 0.004), seedless raisins (0.27 ± 0.005), frozen and boiled collards (0.27 ± 0.006), black currants (0.27 ± 0.009), marinara sauce (0.27 ± 0.002), boiled potato without skin (0.27 ± 0.004), canned pumpkin pie mix (0.27 ± 0.004), medium cream (0.27 ± 0.033), whipping cream, light (0.27 ± 0.033), boiled parsnips (0.26 ± 0.006), pickled beets (0.26 ± 0.004), boiled sweet potato (0.26 ± 0.003), canned carrots (0.26 ± 0.007), apricots (0.26 ± 0.005), boiled and steamed european chestnuts (0.25 ± 0.018), boiled beets (0.25 ± 0.006), tomato sauce (0.25 ± 0.004), canned zucchini in tomato juice (0.25 ± 0.004), frozen and boiled potatoes without skin (0.25 ± 0.005), frozen and boiled carrots (0.25 ± 0.007), cooked wild long grain rice (0.25 ± 0.004), sour cream (0.25 ± 0.042), cooked prunes (0.24 ± 0.005),

II. Minerals — ZINC

boiled scallop squash (0.24 ± 0.006), boiled cauliflower (0.24 ± 0.008), boiled scotch kale (0.23 ± 0.008), canned turnip greens (0.23 ± 0.004), soy milk (0.23 ± 0.002), boiled kale (0.23 ± 0.008), red currants (0.23 ± 0.009), white currants (0.23 ± 0.009), raw cucumber (0.23 ± 0.01), guava (0.23 ± 0.006), raw turbot (0.22 ± 0.006), dried longans (0.22 ± 0.005), indian buffalo milk (0.22 ± 0.002), tomato purée (0.22 ± 0.002), boysenberries (0.22 ± 0.004), hot dog pickle relish (0.21 ± 0.018), restaurant coleslaw (0.21 ± 0.005), boiled/ baked spaghetti squash (0.21 ± 0.006), raw butterhead lettuce (0.2 ± 0.033), raw zucchini (0.2 ± 0.008), raw red cabbage (0.2 ± 0.014), prune juice (0.2 ± 0.002), frozen and boiled mustard greens (0.2 ± 0.007), frozen and boiled zucchini (0.2 ± 0.004), angel food cake (0.2 ± 0.008), coleslaw (0.2 ± 0.008), raw iceberg lettuce (0.2 ± 0.025), whipping cream, heavy (0.2 ± 0.033), low calorie french salad dressing (0.19 ± 0.031), dried jujube (0.19 ± 0.005), baked/ boiled tropical yam (0.19 ± 0.007), strawberry sundae (0.19 ± 0.003), raw carrots (0.19 ± 0.007), canned sauerkraut (0.19 ± 0.004), canned jalapeño peppers (0.19 ± 0.007), restaurant hot chocolate (0.19 ± 0.002), baked acorn squash (0.18 ± 0.005), raw turnip greens (0.18 ± 0.018), seeded raisins (0.18 ± 0.005), canned sweet potato (0.18 ± 0.003), boiled zucchini (0.18 ± 0.006), frozen and boiled kale (0.18 ± 0.008), raw cauliflower (0.18 ± 0.01), carrot juice (0.18 ± 0.003), raw red sweet peppers (0.18 ± 0.01), raw green sweet peppers (0.18 ± 0.01), raw celery (0.18 ± 0.013), boiled onions (0.18 ± 0.005), raw onions (0.18 ± 0.006), bananas (0.17 ± 0.004), canned pumpkin (0.17 ± 0.004), raw green cabbage (0.17 ± 0.014), canned red sweet peppers (0.17 ± 0.007), canned green sweet peppers (0.17 ± 0.007), apple pie (0.17 ± 0.004), boiled green cabbage (0.16 ± 0.007), boiled celery (0.16 ± 0.007), canned stewed red tomatoes (0.16 ± 0.004), canned peeled red tomatoes (0.16 ± 0.004), cantaloupe (0.16 ± 0.003), human milk (0.16 ± 0.016), cream of rice (0.16 ± 0.003), boiled broccoli (0.15 ± 0.006), baked hubbard squash (0.15 ± 0.005), boiled eggplant (0.15 ± 0.01), raw eggplant (0.15 ± 0.012), boiled red cabbage (0.15 ± 0.007), candied sweet potato (0.15 ± 0.005), papaya nectar (0.15 ± 0.002), safflower oil and soybean oil mayonnaise (0.14 ± 0.036), soybean oil mayonnaise (0.14 ± 0.036), boiled turnip greens (0.14 ± 0.007), figs (0.14 ± 0.01), restaurant tomato juice (0.14 ± 0.004), tomato juice (0.14 ± 0.003), peach (0.14 ± 0.006), light cream (0.14 ± 0.017), thousand island salad dressing (0.13 ± 0.031), piña colada (0.13 ± 0.004), italian salad dressing (0.13 ± 0.033), cooked plantain (0.13 ± 0.003), boiled butternut squash (0.13 ± 0.005), noodles (0.13 ± 0.003), boiled red tomato (0.13 ± 0.004), strawberries (0.13 ± 0.003), frozen and boiled cauliflower (0.13 ± 0.006), whey (0.13 ± 0.002), cranberry sauce (0.13 ± 0.005), cream of wheat (0.13 ± 0.003), pineapple juice (0.12 ± 0.002), gooseberries (0.12 ± 0.003), boiled red sweet peppers (0.12 ± 0.007), boiled green sweet peppers (0.12 ± 0.007), frozen and boiled butternut squash (0.12 ± 0.004), pear (0.12 ± 0.003), teriyaki sauce (0.11 ± 0.028), boiled acorn squash (0.11 ± 0.004), breadfruit (0.11 ± 0.005), kumquats (0.11 ± 0.026), raw red tomato (0.11 ± 0.004), japanese persimmon (0.11 ± 0.003), carambola (0.11 ± 0.004), hamburger pickle relish (0.11 ± 0.018), blueberries (0.11 ± 0.003), lime (0.1 ± 0.007), coconut water (0.1 ± 0.002), raw frozen rhubarb (0.1 ± 0.004), red table wine (0.1 ± 0.005), clam liquid (0.1 ± 0.002), boiled hubbard squash (0.09 ± 0.004), frozen and boiled onions (0.09 ± 0.005), nectarine (0.09 ± 0.004), bloody mary (0.09 ± 0.003), plum (0.09 ± 0.008), apricot nectar (0.09 ± 0.002), tuna salad spread (0.09 ± 0.009), pineapple (0.08 ±

ZINC

0.003), frozen and cooked sweetened rhubarb (0.08 ± 0.004), barbecue sauce (0.08 ± 0.031), pummelo (0.08 ± 0.003), peach nectar (0.08 ± 0.002), tom collins (0.08 ± 0.002), bourbon & soda (0.08 ± 0.004), lard (0.08 ± 0.038), lemon (0.07 ± 0.009), raw green tomato (0.07 ± 0.004), white table wine (0.07 ± 0.005), watermelon (0.07 ± 0.003), papaya (0.07 ± 0.002), lychees (0.07 ± 0.005), white grapefruit (0.07 ± 0.004), pink grapefruit (0.07 ± 0.004), dry dessert wine (0.07 ± 0.008), sweet dessert wine (0.07 ± 0.008), corn grits (0.07 ± 0.002), instant corn grits (0.07 ± 0.004), restaurant cranberry juice cocktail (0.07 ± 0.003), daiquiri (0.07 ± 0.008), farina (0.07 ± 0.003), orange soda (0.07 ± 0.001), 80 proof rum (0.07 ± 0.012), olive oil (0.07 ± 0.036), cherries (0.06 ± 0.007), navel orange (0.06 ± 0.004), valencia orange (0.06 ± 0.004), rose table wine (0.06 ± 0.005), tequila sunrise (0.06 ± 0.003), european grapes (0.06 ± 0.003), thompson seedless grapes (0.06 ± 0.003), lime juice (0.06 ± 0.002), rose apple (0.06 ± 0.005), whiskey sour (0.06 ± 0.006), pear nectar (0.06 ± 0.002), loquats (0.05 ± 0.005), white grapefruit juice (0.05 ± 0.002), red grapefruit juice (0.05 ± 0.002), orange juice (0.05 ± 0.002), jujube (0.05 ± 0.005), longans (0.05 ± 0.005), grape juice (0.05 ± 0.002), frozen and boiled red sweet peppers (0.05 ± 0.005), frozen and boiled green sweet peppers (0.05 ± 0.005), lemon juice (0.05 ± 0.002), manhattan (0.05 ± 0.009), 94 proof gin (0.05 ± 0.012), 94 proof rum (0.05 ± 0.012), 94 proof vodka (0.05 ± 0.012), 94 proof whiskey (0.05 ± 0.012), 100 proof gin (0.05 ± 0.012), 100 proof rum (0.05 ± 0.012), 100 proof vodka (0.05 ± 0.012), 100 proof whiskey (0.05 ± 0.012), 86 proof whiskey (0.05 ± 0.012), orange juice (0.04 ± 0.004), screwdriver (0.04 ± 0.002), american grapes (0.04 ± 0.005), raw apple with skin (0.04 ± 0.004), cooked apple without skin (0.04 ± 0.003), applesauce (0.04 ± 0.004), raw apple without skin (0.04 ± 0.004), cola (0.04 ± 0.001), lemon-lime soda (0.04 ± 0.001), chicken egg white (0.03 ± 0.015), mango (0.03 ± 0.002), restaurant coffee (0.03 ± 0.003), apple juice (0.03 ± 0.002), instant tea (0.03 ± 0.002), chamomile tea (0.03 ± 0.003), chicken fat (0.02 ± 0.038), tangerine juice (0.02 ± 0.002), beer (0.02 ± 0.001), coffee (0.02 ± 0.003), cranberry juice cocktail (0.02 ± 0.002), black tea (0.02 ± 0.003), 53 proof coffee liqueur (0.02 ± 0.01), 63 proof coffee liqueur (0.02 ± 0.01), tea (0.02 ± 0.002), martini (0.01 ± 0.007)

Applications: hardening of the arteries, atherosclerosis, increased cholesterol, diabetes, Hodgkin's disease, alcohol cravings, inflammation of prostate, decreased sexual activity, burns, wounds, retarded growth, cancer, antioxidant therapy, sun sensitivity, free radical oxidation, life span extension, radiation, decreased radiation resistance, aging

Daily Dosage: (at least 14 mg zinc per every 1 mg copper) 1974 PCS — 7.5–22.5 mg; 1980 RDA — 15 mg; 1980 USRDA — 15 mg; 1989 RDA — 12–15 mg; 1989 USRDA — 15 mg; nutritional — 15–50 mg; therapeutic — 50–150 mg; experimental — 150–200 mg; toxic — 150 mg +

Warnings: increased amounts are dangerous. See toxicity symptoms.

PART III
Amino Acids

ALANINE

Names: alanine, ALA, amino acid A, 2-amino propanoic acid
Forms: 2-amino propanoic acid
Classification: nonessential amino acid
Deficiency: protein decreased
Side-effects: none known
Toxicity: none known
Inhibitors: increased amounts of other amino acids
Helpers: moderate amounts of other amino acids
Sources: internal synthesis
Applications: none known
Daily Dosage: Nutritional — 1–3 g; therapeutic — ?; experimental — ?; toxic — none known up to 3 g
Warnings: none

ARGININE

Names: arginine, ARG, amino acid R, 2-amino-5-guanido-pentanoic acid
Forms: 2-amino-5-guanido-pentanoic acid
Classification: semi-essential amino acid
Deficiency: impotency, sterility, decreased sperm mobility and formation, immunity decreased, protein decreased, disordered carbohydrate metabolism, delayed sexual maturation
Side-effects: see toxicity symptoms
Toxicity: thickening of skin, enlarged joints, larynx growth
Inhibitors: increased amounts of other amino acids

ARGININE

Helpers: manganese, aspartic acid, citrulline, glutamic acid

Sources (mg per every 100 grams of food): torula yeast (10011 ± 1.8), peanut flour (7025 ± 12.5), soybean protein isolate (6671 ± 1.8), roasted pumpkin and squash seeds (5496 ± 1.8), almond meal (5011 ± 1.8), dried watermelon seeds (4968 ± 1.8), cooked whelk (4936 ± 0.6), dried butternuts (4932 ± 1.8), dried pignolia pine nuts (4736 ± 1.8), soybean protein concentrate (4643 ± 1.8), fenugreek seed (4550 ± 25), dried spirulina seaweed (4147 ± 0.5), dried pumpkin and squash seeds (4089 ± 1.8), creamy peanut butter (3831 ± 3.1), dried and salted cod (3759 ± 0.6), dried black walnuts (3714 ± 1.8), oil roasted peanuts (3657 ± 1.8), defatted soybean flour (3647 ± 0.5), low fat soybean flour (3610 ± 0.6), dried peanuts (3507 ± 1.8), defatted soy meal (3487 ± 0.4), oil roasted spanish peanuts (3350 ± 1.8), dried sesame kernels (3325 ± 6.3), oil roasted valencia peanuts (3236 ± 1.8), dried and frozen tofu (3188 ± 2.9), oil roasted virginia peanuts (3093 ± 1.8), dry roasted soybean nuts (3071 ± 0.6), toasted soybean nuts (3054 ± 1.8), crunchy peanut butter (2875 ± 1.6), dry roasted peanuts (2832 ± 1.8), roasted soybean nuts (2733 ± 0.6), roasted soybean flour (2700 ± 0.6), soybean flour (2679 ± 0.6), whole dried sesame seeds (2633 ± 5.6), oil roasted almonds (2586 ± 1.8), toasted almonds (2586 ± 1.8), toasted sesame kernels (2550 ± 1.8), whole toasted sesame seeds (2550 ± 1.8), dried almonds (2532 ± 1.8), sesame butter (2520 ± 3.3), raw whelk (2468 ± 0.6), dried sunflower seeds (2436 ± 1.8), dried brazil nuts (2425 ± 1.8), braised pork center loin excluding fat (2414 ± 0.5), braised pork top loin excluding fat (2391 ± 0.5), oil roasted filberts (2389 ± 1.8), braised pork sirloin excluding fat (2326 ± 0.5), braised pork loin excluding fat (2290 ± 0.5), oil roasted sunflower seeds (2286 ± 1.8), dried pinyon pine nuts (2282 ± 1.8), broiled pork center loin excluding fat (2222 ± 0.5), dried pistachio nuts (2218 ± 1.8), fried beef top round excluding fat (2216 ± 0.5), dried filberts (2186 ± 1.8), braised pork boston blade excluding fat (2163 ± 0.5), fried chicken giblets (2159 ± 0.5), dried english/persian walnuts (2132 ± 1.8), simmered beef shank excluding fat (2129 ± 0.5), simmered turkey gizzard (2114 ± 0.5), dried hickory nuts (2114 ± 1.8), cooked crayfish (2091 ± 0.6), braised beef chuck arm pot roast excluding fat (2087 ± 0.5), roasted turkey light meat without skin (2086 ± 0.5), braised pork center loin including fat (2083 ± 0.5), dry roasted almonds (2071 ± 1.8), dry roasted sunflower seeds (2068 ± 1.8), braised pork loin blade excluding fat (2062 ± 0.5), fried beef sirloin excluding fat (2053 ± 0.5), well done baked extra lean ground beef (2046 ± 0.5), stewed chicken drumstick without skin (2039 ± 1.1), braised pork center rib including fat (2030 ± 0.5), cashew butter (2025 ± 1.8), roasted pork rump excluding fat (2023 ± 0.5), braised pork spareribs (2018 ± 0.5), fried chicken breast without skin (2015 ± 0.6), roasted turkey light meat with skin (2013 ± 0.5), broiled beef top round excluding fat (2003 ± 0.5), fried beef top round including fat (2003 ± 0.5), broiled pork center rib excluding fat (2001 ± 0.5), broiled pork top loin excluding fat (2001 ± 0.5), roasted pork tenderloin (1999 ± 0.5), well done baked lean ground beef (1999 ± 0.5), simmered beef shank including fat (1998 ± 0.5), fried pork center loin excluding fat (1997 ± 0.5), braised beef bottom round excluding fat (1996 ± 0.5), roasted turkey dark meat without skin (1993 ± 0.5), chipped beef (1989 ± 1.8), braised pork sirloin including fat (1987 ± 0.5), roasted pork center loin excluding fat (1978 ± 0.5), fried light meat chicken without skin (1978 ± 0.5), braised pork top loin including fat (1974 ± 0.5), roasted pork leg excluding fat (1966 ± 0.5), braised beef chuck blade roast excluding fat (1963 ± 0.5), broiled pork sirloin excluding fat (1962 ± 0.5), roasted pork center rib excluding fat (1959 ± 0.5), roasted pork top loin excluding fat (1959 ± 0.5), roasted pork shank excluding fat (1959 ± 0.5), fried chicken breast with skin (1954 ± 0.5), simmered chicken gizzard (1950 ± 0.5), broiled beef top round including fat (1948 ± 0.5), well done baked ground beef (1945 ± 0.5), braised beef shortribs excluding fat (1944 ± 0.5), broiled pork center loin including fat (1943 ± 0.5), fried pork center rib excluding fat (1942 ± 0.5), fried pork top loin excluding fat (1942 ± 0.5), roasted turkey dark meat with skin (1936 ± 0.5), braised pork loin including fat (1934 ± 0.5), broiled pork loin excluding fat (1933 ± 0.5), yellow mustard seed (1933 ± 16.7), well done broiled extra lean ground beef (1930 ± 0.5), simmered beef heart (1925 ± 0.5), broiled beef sirloin excluding fat (1919 ± 0.5), roasted pork sirloin excluding fat (1909 ± 0.5), braised pork arm braised including fat (1908 ± 0.5), well done broiled lean ground beef (1904 ± 0.5), well done fried extra lean ground beef (1890 ± 0.5), fried light meat chicken with skin (1888 ± 0.5), almond butter (1888 ± 3.1), braised beef bottom round including fat (1884 ± 0.5), braised pork boston blade braised including fat (1875 ± 0.5), roasted pork rump including fat (1873 ± 0.5), roasted chicken breast without skin (1871 ± 0.6), roasted pork loin excluding fat (1868 ± 0.5), poppy seed (1867 ± 16.7), roasted light meat chicken without skin (1864 ± 0.5), well done fried lean ground beef (1864 ± 0.5), cooked clams (1864 ± 0.6), oil roasted cashews (1864 ± 1.8), cooked bacon (1861 ± 0.4), braised beef brisket excluding fat (1857 ± 0.5), roasted chicken breast with skin (1837 ± 0.5), well done broiled ground beef (1837 ± 0.5), roasted eye of beef round excluding fat (1832 ± 0.5), cooked shrimp (1826 ± 0.6), well done fried ground beef (1823 ± 0.5), roasted beef tip excluding fat (1815 ± 0.5), roasted light meat chicken with skin (1811 ± 0.5), broiled beef top loin excluding fat (1809 ± 0.5), broiled beef round excluding fat (1798 ± 0.5), roasted pork center loin including fat (1795 ± 0.5), canned blue crab (1793 ± 0.6), cooked lobster (1791 ± 0.6), broiled beef tenderloin excluding fat (1785 ± 0.5), breaded and fried shrimp (1784 ± 0.6), broiled porterhouse steak excluding fat (1780 ± 0.5), broiled T-bone steak excluding fat (1778 ± 0.5), raw shrimp (1775 ± 0.6), broiled beef rib eye excluding fat (1772 ± 0.5), roasted pork leg including fat (1771 ± 0.5), braised beef flank excluding fat (1771 ± 0.5), canned tuna in water (1771 ± 0.6), simmered turkey giblets (1768 ± 0.5), cooked blue crab (1765 ± 0.6), roasted pork sirloin including fat (1764 ± 0.5), dry roasted cashews (1764 ± 1.8), roasted pork shoulder excluding fat (1762 ± 0.5), roasted pork center rib including fat (1753 ± 0.5), broiled pork center rib including fat (1749 ± 0.5), broiled pork boston blade excluding fat (1748 ± 0.5), stewed chicken breast without skin (1748 ± 0.5), fried dark meat chicken without skin (1742 ± 0.5), stewed light meat chicken without skin (1742 ± 0.5), roasted chicken wing with skin (1741 ± 1.5), roasted beef tenderloin excluding fat (1740 ± 0.5), roasted beef rib roasted excluding fat (1740 ± 0.5), braised beef flank including fat (1739 ± 0.5), cooked mussels (1736 ± 0.6), fried beef sirloin including fat (1735 ± 0.5), broiled beef sirloin including fat (1731 ± 0.5), broiled pork loin blade excluding fat (1730 ± 0.5), anchovy canned in olive oil (1730 ± 2.5), crab cakes (1730 ± 0.8), roasted pork shank including fat (1727 ± 0.5), simmered chicken giblets (1727 ± 0.5), roasted pork top loin including fat (1721 ± 0.5), braised pork loin blade including fat (1718 ± 0.5), broiled pork sirloin including fat (1717 ± 0.5), simmered turkey heart (1717 ± 0.5), fried chicken back with skin (1715 ± 0.7), medium-rare broiled extra lean ground beef (1715 ± 0.5), roasted pork loin blade excluding fat (1713 ± 0.5), braised beef chuck arm pot roast including fat (1712 ± 0.5), broiled pork top loin including fat (1698 ± 0.5), simmered chicken heart (1694 ± 0.5), roasted pork boston blade excluding fat (1692 ± 0.5), roasted eye of beef round including fat (1691 ± 0.5), cooked alaska king crab (1691 ± 0.6), fried pork loin blade excluding fat (1689 ± 0.5), stewed chicken breast with skin (1689 ± 0.5), roasted chicken drumstick with skin (1685 ± 1), medium-rare fried extra lean ground beef (1685 ± 0.5), broiled pork loin including fat (1681 ± 0.5), fried dark meat chicken with skin (1681 ± 0.5), fried chicken wing with skin (1681 ± 1.6), fried beef liver (1680 ± 0.5), dry roasted filberts (1679 ± 1.8), roasted beef round tip including fat (1673 ± 0.5), canned corned beef (1673 ± 0.5), fried pork center loin including fat (1670 ± 0.5), medium-rare broiled lean ground beef (1669 ± 0.5), boiled lupins (1669 ± 0.3), roasted pork loin including fat (1665 ± 0.5), fried chicken drumstick with skin (1663 ± 1), fried chicken thigh with skin (1656 ± 0.8), medium-rare baked extra lean ground beef (1652 ± 0.5), roasted dark meat chicken without skin (1651 ± 0.5), braised pork pancreas (1642 ± 0.5), raw lobster (1642 ± 0.6), broiled beef tenderloin including fat (1641 ± 0.5), cooked coho salmon (1636 ± 0.6), medium-rare fried lean ground beef (1636 ± 0.5), cooked sockeye salmon (1634 ± 0.6), roasted dark meat chicken with skin (1634 ± 0.5), raw crayfish (1631 ± 0.6), stewed light meat chicken with skin (1629 ± 0.5), broiled beef top loin including fat (1626 ± 0.5), medium-rare broiled ground beef (1625 ± 0.5), roasted cured pork arm excluding fat (1620 ± 0.5), medium-rare baked lean ground beef (1616 ± 0.5), raw queen crab (1616 ± 0.6), boiled peanuts (1616 ± 1.6), medium-rare fried ground beef (1615 ± 0.5), broiled beef round including fat (1614 ± 0.5), braised beef chuck blade roast including fat (1608 ± 0.5), jellied corned beef loaf (1607 ± 1.8), dry roasted pistachio nuts (1607 ± 1.8), broiled beef flank excluding fat (1605 ± 0.5), broiled beef rib eye including fat (1604 ± 0.5), braised pork liver (1603 ± 0.5), roasted pork arm including fat (1598 ± 0.5), raw alaska king crab (1598 ± 0.6), cooked halibut (1596 ± 0.6), canned tuna in oil (1588 ± 0.6), caviar (1588 ± 3.1), broiled porterhouse steak including fat (1586 ± 0.5), braised pork heart (1586 ± 0.4), broiled beef flank including fat (1583 ± 0.5), cooked rainbow trout (1576 ± 0.6), raw blue crab (1576 ± 0.6), stewed chicken drumstick with skin (1574 ± 0.9), cooked snapper (1573 ± 0.6), turkey breast meat summer sausage (1571 ± 2.4), roasted pork shoulder including fat (1568 ± 0.5), garlic powder (1567 ± 16.7), fried pork center rib including fat (1566 ± 0.5), stewed dark meat chicken without skin (1566 ± 0.5), roasted chicken thigh with skin (1566 ± 0.8), roasted goose with skin (1566 ± 0.5), roasted chicken thigh without skin (1565 ± 1), corn germ (1565 ± 0.5), fried pork top loin including fat (1561 ± 0.5), braised pork kidneys (1561 ± 0.5), broiled pork boston blade including fat (1558 ± 0.5), broiled beef rib excluding fat (1557 ± 0.5), cooked sheepshead (1556 ± 0.6), medium-rare baked ground beef (1554 ± 0.5), fried chicken neck with skin (1553 ± 1.4), braised beef pancreas (1548 ± 0.5), roasted beef tenderloin including fat (1546 ± 0.5), roasted pork boston blade including fat (1544 ± 0.5), parmesan cheese (1540 ± 10), braised pork spleen (1539 ± 0.5), dried pilinuts (1539 ± 1.8), braised beef liver (1533 ± 0.5), raw dungeness crab (1521 ± 0.6), hard pork and beef salami (1520 ± 5), cooked swordfish (1519 ± 0.6), broiled T-bone steak including fat (1516 ± 0.5), smoked haddock (1509 ± 0.6), roasted pork loin blade including fat (1505 ± 0.5), almond paste (1504 ± 1.8), freeze-dried chives (1500 ± 62.5), roasted duck without skin (1499 ± 0.5), simmered beef kidneys (1496 ± 0.5), simmered chicken liver (1493 ± 0.5), braised pork tongue (1488 ± 0.5), cooked grouper (1487 ± 0.6), broiled pork loin blade including fat (1486 ± 0.5), cooked mullet (1484 ± 0.6), simmered chicken neck without skin (1483 ± 2.8), stewed dark meat chicken with skin (1477 ± 0.5), cooked pike (1476 ± 0.6), sardines (1475 ± 2.1), kippered herring (1470 ± 1.3), stewed chicken wing with skin (1470 ± 1.3), simmered turkey liver (1469 ± 0.5), cooked tilefish (1465 ± 0.6), stewed chicken thigh with skin (1456 ± 0.7), braised beef spleen (1454 ± 0.5), pork link sausage (1454 ± 0.7), braised beef brisket including fat (1453 ± 0.5), cooked haddock (1451 ± 0.6), cooked flatfish (1446 ± 0.6), braised beef thymus (1440 ± 0.5), simmered pork feet (1440 ± 0.5), cooked rockfish (1438 ± 0.6), roasted beef rib including fat (1435 ± 0.5), raw pheasant without skin (1433 ± 0.5), cooked perch (1429 ± 0.6), cooked mackerel (1427 ± 0.6), cooked pompano (1416 ± 0.6), cooked eel (1415 ± 0.6), cooked sea bass (1414 ± 0.6), fried abalone (1411 ± 0.6), simmered beef tongue (1408 ± 0.5), cooked whiting (1405 ± 0.6), roasted cured pork arm including fat (1403 ± 0.5), smoked whitefish (1400 ± 0.6), onion powder (1400 ± 25), raw yellowtail (1385 ± 0.6), raw quail without skin (1379 ± 0.5), chopped smoked beef (1379 ± 1.8), cooked herring (1378 ± 0.6), fried pork loin blade including fat (1374 ± 0.5), hard pork salami (1370 ± 5), cooked carp (1368 ± 0.6), cooked cod (1366 ± 0.6), braised beef shortribs including fat (1363 ± 0.5), canned cod (1362 ± 0.6), roasted lean ham (1360 ± 0.5), cooked rainbow smelt (1352 ± 0.6), dill seed (1350 ± 25), beef lunch meat (1332 ± 1.8), dry roasted macadamia nuts (1325 ± 1.8), tempeh (1317 ± 0.6), simmered chicken neck with skin (1316 ± 1.3), turkey ham (1314 ± 0.9), dry skim milk (1310 ± 1.7), roasted canned lean ham (1310 ± 0.5), grilled canadian bacon (1309 ± 1.1), lebanon bologna (1304 ± 2.2), boiled soybeans (1291 ± 0.3), breaded fried squid (1291 ± 0.6), roasted ham (1289 ± 0.5), roasted duck with skin (1284 ± 0.5), raw coho salmon (1282 ± 0.6), turkey pastrami (1282 ± 0.9), falafel (1280 ± 1), raw sockeye salmon (1275 ± 0.6), broiled beef rib including fat (1272 ± 0.5), roasted canned ham (1271 ± 0.5), dry instant skim milk (1270 ± 0.5), breaded and fried scallops (1265 ± 1.6), simmered pork ears (1265 ± 0.5), unheated lean ham (1257 ± 0.5), raw shark (1255 ± 0.6), raw abalone (1248 ± 0.6), raw halibut (1245 ± 0.6), raw trout (1242 ± 0.6), cooked sturgeon (1238 ± 0.6), braised beef lungs (1234 ± 0.5), pork and beef pepperoni (1233 ± 8.3), raw rainbow trout (1229 ± 0.6), raw milkfish (1229 ± 0.6), raw snapper (1227 ± 0.6), canned sockeye salmon (1225 ± 0.6), raw scallops (1224 ± 0.6), raw anchovy (1216 ± 0.6), raw sheepshead (1211 ± 0.6), raw chum salmon (1205 ± 0.6), chopped ham lunch meat (1200 ± 2.4), raw chinook salmon (1200 ± 0.6), dehydrated onion flakes (1200 ± 3.6), raw bluefish (1199 ± 0.6), raw pink salmon (1193 ± 0.6), canned pink salmon (1193 ± 0.6), pork luxury loaf (1186 ± 1.8), pork and beef peppered loaf (1186 ± 1.8), raw swordfish (1185 ± 0.6), raw cuttlefish (1185 ± 0.6), italian pork sausage (1184 ± 0.7), simmered pork tail (1173 ± 0.5), raw pollock (1164 ± 0.6), raw sunfish (1161 ± 0.6), pork headcheese (1161 ± 1.8), raw grouper (1159 ± 0.6), pork sausage (1159 ± 1.9), raw mullet (1158 ± 0.6), dry buttermilk (1157 ± 7.1), raw burbot (1156 ± 0.6), dry roasted pecans (1154 ± 1.8), raw pike (1152 ± 0.6), raw duck liver (1148 ± 0.5), fried tofu (1146 ± 3.8), dried coconut (1146 ± 1.8), canned lean ham (1145 ± 0.5), raw whitefish (1142 ± 0.6), unheated ham (1141 ± 0.5), raw cusk (1136 ± 0.6), raw cisco (1136 ± 0.6), raw ling (1136 ± 0.6), raw squid (1136 ± 0.6), chicken egg yolk (1135 ± 2.9), ham and cheese roll (1132 ± 1.8), raw haddock (1131 ± 0.6), pork and beef luncheon sausage (1130 ± 2.2), raw scup (1129 ± 0.6), raw freshwater bass (1128 ± 0.6), raw flatfish (1128 ± 0.6), roasted pork blade roll (1122 ± 0.5), cooked corned beef brisket (1122 ± 0.5), raw rockfish (1122 ± 0.6), dried pecans (1121 ± 1.8), unheated canadian bacon (1119 ± 0.9), raw perch (1115 ± 0.6), raw mackerel (1113 ± 0.6), raw spot (1108 ± 0.6), raw dolphinfish (1107 ± 0.6), raw pompano (1106 ± 0.6), raw eel (1104 ± 0.6), raw sea bass (1104 ± 0.6), raw whiting (1096 ± 0.6), battered and fried shark (1094 ± 0.6), smoked chinook salmon (1094 ± 0.6), pork and beef honey loaf (1089 ± 1.8), raw catfish (1088 ± 0.6), raw octopus (1088 ± 0.6), beef pastrami (1079 ± 1.8), breaded and fried catfish (1076 ± 0.6), raw herring (1075 ± 0.6), raw carp (1067 ± 0.6), raw cod (1066 ± 0.6), raw croaker (1064 ± 0.6), breaded and fried croaker (1061 ± 0.6), raw striped bass (1061 ± 0.6), raw lingcod (1056 ± 0.6), raw rainbow smelt (1055 ± 0.6), canned ham (1051 ± 0.5), firm raw tofu (1050 ± 0.4), raw freshwater drum (1049 ± 0.6), raw tilefish (1047 ± 0.6), raw wolffish (1047 ± 0.6), berliner (1039 ± 2.2), provolone cheese (1036 ± 1.8), raw butterfish (1034 ± 0.6), cooked oysters (1031 ± 0.6), pork and beef mortadella (1027 ± 3.3), minced ham lunch meat (1024 ± 2.4), dried ginko nuts (1018 ± 1.8), pickled pork feet (1014 ± 0.5), raw shad (1013 ± 0.6), oil roasted pecans (1007 ± 1.8), raw white sucker (1004 ± 0.6), pork bologna (1004 ± 2.2), raw goose liver (1003 ± 0.5), raw seatrout (1001 ± 0.6), raw beef tripe (995 ± 0.5), boiled green soybeans (994 ± 0.6), gruyere cheese (986 ± 1.8), breaded and fried clams (985 ± 0.6), smoked cisco (979 ± 0.6), edam cheese (975 ± 1.8), gouda cheese (975 ± 1.8), summer sausage (974 ± 2.2), tuna salad (967 ± 0.2), raw sturgeon (966 ± 0.6), milk chocolate (964 ± 1.8), raw turbot (960 ± 0.6), dry whole milk (953 ± 1.6), pork and beef brotwurst (946 ± 0.7), pork and beef kielbasa/kolbassy (942 ± 1.9), cheddar cheese (941 ± 0.5), swiss cheese (939 ± 1.8), american cheese (939 ± 1.8), pimento cheese spread (939 ± 1.8), monterey cheese (936 ± 1.8), polish pork sausage (936 ± 1.8), pork and beef lunch meat (936 ± 1.8), raw clams (932 ± 0.6), beef salami (930 ± 2.2), raw pork stomach (924 ± 0.5), dried macadamia nuts (911 ± 1.8), natto (909 ± 0.6), colby cheese (907 ± 1.8), pork and beef picnic loaf (900 ± 1.8), basil (900 ± 50), muenster cheese (893 ± 1.8), cheshire cheese (893 ± 1.8), brick cheese (886 ± 1.8), bran (886 ± 1.8), dried toasted coconut (882 ± 1.8), canned pork lunch meat (876 ± 2.4), freeze-dried red sweet peppers (875 ± 31.3), freeze-dried green sweet peppers (875 ± 31.3), smoked beef sausage (872 ± 1.2), raw mussels (868 ± 0.6), raw monkfish (867 ± 0.6), braised pork lungs (863 ± 0.5), tilsit cheese (861 ± 1.8), beef and pork salami (857 ± 2.2), pork and beef sausage (854 ± 3.8), grilled ham patties (852 ± 0.8), beef and pork frankfurter (849 ± 1.1), surimi scallops (848 ± 0.6), pickled herring (847 ± 3.3), mozzarella cheese (843 ± 1.8), simmered pork chitterlings (840 ± 0.5), port du salut cheese (839 ± 1.8), pork and beef link sausage with american cheese (837 ± 1.2), pork livercheese (837 ± 1.3), boiled chickpeas (835 ± 0.3), pork bratwurst (831 ± 0.6), unheated ham patties (826 ± 0.8), surimi shrimp (824 ± 0.6), pork liverwurst (817 ± 2.8), pork and beef old fashioned loaf (811 ± 1.8), pork beerwurst salami (809 ± 2.2), freeze-dried shallots (800 ± 12.5), oil roasted macadamia nuts (796 ± 1.8), chicken egg (776 ± 1), pork liver sausage (767 ± 2.8), beef beerwurst salami (765 ± 2.2), pork and beef link sausage (765 ± 0.7), duck egg (764 ± 0.7), beef bologna (757 ± 2.2), miso (747 ± 0.4), beef frankfurter (744 ± 0.9), boiled split peas (744 ± 0.3), brie cheese (743 ± 1.8), bleu cheese (721 ± 1.8), puffed wheat (721 ± 3.6), camembert cheese (711 ± 1.8), dried chinese chestnuts (711 ± 1.8), pork and beef knockwurst (709 ± 0.7), limburger cheese

(707 ± 1.8), vienna sausage (706 ± 3.1), boiled broad beans (702 ± 0.3), fennel seed (700 ± 25), beef and pork bologna (700 ± 2.2), boiled lentils (697 ± 0.3), fried beef brains (686 ± 0.5), lard (682 ± 1.8), bran flakes (679 ± 1.8), pork pickle and pimento loaf (675 ± 1.8), raw pork jowl (659 ± 0.5), dried shitake mushrooms (647 ± 3.3), braised pork brains (635 ± 0.5), raw garlic (633 ± 5.6), dried acorns (632 ± 1.8), ricotta cheese (631 ± 0.4), stir-fried sprouted soybeans (629 ± 0.5), boiled sprouted green peas (627 ± 0.5), simmered beef brains (604 ± 0.5), boiled white beans (602 ± 0.3), ham salad (600 ± 1.8), stir-fried sprouted lentils (600 ± 0.5), boiled hyacinth beans (598 ± 0.3), frozen and boiled black-eyed peas (595 ± 0.6), gefiltefish with broth (595 ± 1.2), coconut cream (595 ± 0.2), chicken egg white (591 ± 1.5), pork olive loaf (589 ± 1.8), acorn flour (586 ± 1.8), breaded and fried oysters (585 ± 0.6), shredded wheat (579 ± 1.8), boiled cranberry beans (578 ± 0.3), boiled winged beans (576 ± 0.3), boiled yardlong bean (574 ± 0.3), cottage cheese (570 ± 0.2), boiled yellow beans (567 ± 0.3), boiled catjang black-eyed peas (563 ± 0.3), boiled pink beans (561 ± 0.3), boiled black beans (549 ± 0.3), raw coconut (547 ± 1.1), boiled navy beans (539 ± 0.3), raw tofu (538 ± 0.4), dried sweetened flaked coconut (538 ± 0.7), boiled kidney beans (537 ± 0.3), boiled black-eyed peas (535 ± 0.3), raw bacon (529 ± 0.7), puffed rice (529 ± 3.6), boiled great northern beans (516 ± 0.3), raw oysters (515 ± 0.6), canned oysters (515 ± 0.6), boiled pinto beans (509 ± 0.3), boiled black turtle beans (507 ± 0.3), canned broad beans (505 ± 0.2), wheat flakes (500 ± 1.8), boiled baby lima beans (493 ± 0.3), boiled mung beans (492 ± 0.2), boiled mungo beans (491 ± 0.3), boiled adzuki beans (486 ± 0.2), raw acorns (479 ± 1.8), boiled lima beans (478 ± 0.3), raisin bran (475 ± 1.8), dried sweetened shredded coconut (473 ± 0.5), canned navy beans (466 ± 0.2), canned chickpeas (466 ± 0.2), dried european chestnuts (464 ± 1.8), roasted chinese chestnuts (464 ± 1.8), canned great northern beans (456 ± 0.2), dried oriental radish (455 ± 0.9), hummus (451 ± 0.2), canned white beans (449 ± 0.2), boiled and frozen baby lima beans (444 ± 0.6), sweetened coconut cream (442 ± 0.2), boiled french beans (437 ± 0.3), raw chinese chestnuts (436 ± 1.8), potato flour (433 ± 0.6), raw green peas (428 ± 0.6), raw spirulina seaweed (427 ± 0.5), ginko nuts (425 ± 1.8), boiled green peas (423 ± 0.6), frozen and boiled green peas (408 ± 0.6), steamed sprouted soybeans (406 ± 1.1), frozen and boiled fordhook lima beans (406 ± 0.6), boiled pigeon peas (405 ± 0.3), refried beans (386 ± 0.2), neufchatel cheese (382 ± 1.8), coconut milk (376 ± 0.2), canned black turtle beans (373 ± 0.2), dried japanese chestnuts (350 ± 1.8), canned green peas (349 ± 0.6), canned cranberry beans (343 ± 0.2), potato pancakes (339 ± 0.7), gjetost cheese (332 ± 1.8), baked beans with pork and sweet sauce (329 ± 0.2), canned black-eyed peas (328 ± 0.2), canned kidney beans (322 ± 0.2), baked beans with pork and tomato sauce (319 ± 0.2), raw cassava (314 ± 0.5), stir-fried sprouted mung beans (310 ± 0.8), canned lima beans (302 ± 0.2), boiled and steamed chinese chestnuts (300 ± 1.8), vegetarian baked beans (297 ± 0.2), boiled succotash (296 ± 0.5), cream cheese (289 ± 1.8), sweetened condensed milk (287 ± 1.3), raw laver/nori seaweed (285 ± 0.5), canned pinto beans (283 ± 0.2), evaporated skim milk (272 ± 1.6), corn flakes (268 ± 1.8), boiled mothbeans (263 ± 0.3), tomato powder (258 ± 0.5), frozen and boiled succotash (251 ± 0.6), evaporated whole milk (247 ± 0.4), roasted european chestnuts (229 ± 1.8), frozen and boiled brussels sprouts (218 ± 0.6), yellow corn pudding (217 ± 0.2), okara tofu (215 ± 0.8), soy milk (214 ± 0.2), frozen and boiled turnip greens (210 ± 0.6), au gratin pota-

toes (203 ± 0.4), ginger (200 ± 25), raw chives (200 ± 16.7), sheep milk (198 ± 0.2), roasted japanese chestnuts (196 ± 1.8), raw sprouted mung beans (196 ± 1), boiled jute potherb (195 ± 1.2), macaroni (194 ± 0.4), spaghetti (194 ± 0.4), boiled lamb's-quarters (193 ± 0.6), oatmeal (193 ± 0.3), french fried potatoes (192 ± 1), raw shallots (180 ± 5), frozen and boiled spinach (178 ± 0.5), raw european chestnuts (175 ± 1.8), skim yogurt (172 ± 0.2), boiled spinach (168 ± 0.6), frozen and boiled mustard greens (167 ± 0.7), boiled mustard greens (166 ± 0.7), frozen and fried french fried potatoes (164 ± 1), raw spinach (161 ± 1.8), frozen and boiled broccoli (161 ± 0.5), canned spinach (159 ± 0.5), lowfat yogurt (158 ± 0.2), frozen and boiled kale (158 ± 0.8), raw onions (158 ± 0.6), canned succotash with creamed corn (154 ± 0.4), boiled broccoli (154 ± 0.6), boiled brussels sprouts (153 ± 0.6), raw watercress (153 ± 2.9), canned succotash (152 ± 0.4), potato salad (152 ± 0.4), frozen and boiled collards (152 ± 0.6), raw japanese chestnuts (150 ± 1.8), eggnog (149 ± 0.2), frozen and fried hash brown potatoes (149 ± 0.6), raw swamp cabbage (148 ± 0.9), cooked sprouted mung beans (147 ± 0.8), boiled and steamed european chestnuts (146 ± 1.8), raw broccoli (145 ± 1.1), boiled burdock root (144 ± 0.4), steamed hawaii mountain yam (144 ± 0.7), protein fortified skim milk (143 ± 0.2), protein fortified lowfat 2% milk (143 ± 0.2), protein fortified lowfat 1% milk (142 ± 0.2), restaurant vanilla milkshake (140 ± 0.2), dried apricots (140 ± 1.4), frozen and boiled asparagus (138 ± 0.8), raw welsh onions (137 ± 0.5), boiled rutabaga (136 ± 0.6), boiled yellow corn (135 ± 0.6), hot chocolate (132 ± 0.2), dried longans (131 ± 0.5), buttermilk (126 ± 0.2), raw spring onions with tops (126 ± 1), vanilla milkshake (125 ± 0.2), baked/boiled tropical yam (124 ± 0.7), raw chickory greens (124 ± 0.6), skim milk (123 ± 0.2), restaurant chocolate milkshake (123 ± 0.2), frozen and boiled yellow corn (123 ± 0.6), boiled swiss chard (122 ± 0.6), restaurant strawberry milkshake (121 ± 0.2), boiled asparagus (121 ± 0.6), boiled onions (121 ± 0.5), lowfat 2% milk (120 ± 0.2), filled milk (120 ± 0.2), goat milk (119 ± 0.2), lowfat 1% milk (119 ± 0.2), whole milk (119 ± 0.2), scalloped potatoes (118 ± 0.4), boiled swamp cabbage (118 ± 1), coconut water (118 ± 0.2), lowfat 1% chocolate milk (117 ± 0.2), lowfat 2% chocolate milk (116 ± 0.2), whole chocolate milk (115 ± 0.2), hash brown potatoes (115 ± 0.6), indian buffalo milk (114 ± 0.2), raw savoy cabbage (114 ± 1.4), carob milk (113 ± 0.2), microwaved potato with skin (113 ± 0.2), boiled mushrooms (112 ± 0.6), boiled kohlrabi (111 ± 0.6), chocolate milkshake (110 ± 0.2), o'brien potatoes (109 ± 0.3), frozen and boiled turnip greens and turnips (108 ± 0.5), half and half cream (107 ± 3.3), baked potato with skin (106 ± 0.2), boiled kale (106 ± 0.8), boiled amaranth (105 ± 0.8), boiled scotch kale (105 ± 0.8), whole yogurt (104 ± 0.2), raw mushrooms (103 ± 1.4), canned sprouted mung beans (102 ± 0.8), boiled savoy cabbage (101 ± 0.7), light cream (100 ± 3.3), boiled dock (98 ± 0.5), microwaved potato without skin (97 ± 0.3), raw bamboo shoots (97 ± 0.7), raw cauliflower (96 ± 1), raw red chili peppers (96 ± 1.1), raw green chili peppers (96 ± 1.1), beef suet (96 ± 1.8), canned yellow corn with red and green peppers (95 ± 0.4), raw potato without skin (95 ± 0.4), frozen and boiled onions (95 ± 0.5), raw turnip greens (93 ± 1.8), raw wakame seaweed (92 ± 0.5), frozen and boiled potato without skin (91 ± 0.5), baked potato without skin (90 ± 0.3), boiled cauliflower (90 ± 0.8), chocolate syrup (89 ± 1.3), raw romaine lettuce (89 ± 1.8), cooked shitake mushrooms (89 ± 0.7), raw lotus root (88 ± 0.6), medium cream (87 ± 3.3), frozen and boiled okra (87 ± 0.5), boiled chinese cabbage (87 ± 0.6), canned turnip greens (85 ± 0.4), canned yellow corn

(84 ± 0.5), raw chinese cabbage (83 ± 1.4), mashed potatoes (82 ± 0.5), baked hubbard squash (82 ± 0.5), boiled hubbard squash (82 ± 0.4), whipping cream, light (80 ± 3.3), frozen and baked sweet potato (80 ± 0.6), baked sweet potato (80 ± 0.4), raw red cabbage (80 ± 1.4), boiled potato without skin (79 ± 0.4), frozen and boiled cauliflower (78 ± 0.6), boiled okra (78 ± 0.6), canned sweet potato (77 ± 0.3), boiled sweet potato (77 ± 0.3), raw leeks (77 ± 1.9), tomato paste (76 ± 0.4), boiled green beans (76 ± 0.8), boiled yellow snap beans (76 ± 0.8), boiled beet greens (74 ± 0.7), whipping cream, heavy (73 ± 3.3), raw butterhead lettuce (73 ± 3.3), valencia orange (73 ± 0.4), raw chickory (73 ± 1.1), boiled turnip greens (72 ± 0.7), cream of rice (72 ± 0.3), navel orange (72 ± 0.4), canned creamed yellow corn (71 ± 0.4), safflower oil and soybean oil mayonnaise (71 ± 3.6), soybean oil mayonnaise (71 ± 3.6), raw looseleaf lettuce (71 ± 1.8), dried figs (70 ± 0.3), raw vine spinach (70 ± 0.5), coleslaw (70 ± 0.8), raw green cabbage (69 ± 1.4), frozen and boiled butternut squash (68 ± 0.4), rice polish (68 ± 0.9), cooked plantain (66 ± 0.3), cream of wheat (66 ± 0.3), canned potato without skin (66 ± 0.6), dried dates (66 ± 0.6), raw kelp/kombu/tangle seaweed (65 ± 0.5), raw iceberg lettuce (65 ± 2.5), canned bamboo shoots (64 ± 0.4), raw endive (64 ± 2), raw california avocado (63 ± 0.3), baked acorn squash (63 ± 0.5), boiled yambean (61 ± 0.5), raw eggplant (61 ± 1.2), soybean oil and cottonseed oil margarine liquid (60 ± 10), boiled red cabbage (60 ± 0.7), dried japanese persimmon (59 ± 1.5), canned pumpkin (59 ± 0.4), watermelon (59 ± 0.3), farina (58 ± 0.3), canned pumpkin pie mix (58 ± 0.4), boiled bamboo shoots (57 ± 0.4), boiled purslane (57 ± 0.9), boiled and steamed japanese chestnuts (57 ± 1.8), boiled collards (56 ± 0.3), sapote (55 ± 0.2), boiled green cabbage (55 ± 0.7), raw dock (54 ± 0.7), frozen and boiled crookneck squash (54 ± 0.5), frozen and boiled yellow snap beans (54 ± 0.7), frozen and boiled green beans (54 ± 0.7), boiled lotus root (53 ± 0.6), raw scallop squash (51 ± 0.8), boiled butternut squash (50 ± 0.5), raw zucchini (49 ± 0.8), european grapes (49 ± 0.3), frozen and boiled carrots (49 ± 0.7), frozen and boiled zucchini (48 ± 0.4), bananas (47 ± 0.4), corn grits (47 ± 0.2), raw florida avocado (47 ± 0.2), elderberries (47 ± 0.3), grape juice (47 ± 0.2), orange juice (47 ± 0.2), american grapes (46 ± 0.5), canned green beans (46 ± 0.7), canned yellow snap beans (46 ± 0.7), boiled eggplant (46 ± 1), raw celtuce (46 ± 0.5), apricots (45 ± 0.5), frozen and boiled red sweet peppers (45 ± 0.5), frozen and boiled green sweet peppers (45 ± 0.5), boiled carrots (45 ± 0.6), boiled scallop squash (44 ± 0.6), tangerine (44 ± 0.6), canned zucchini in tomato juice (43 ± 0.4), canned red chili peppers (43 ± 0.7), canned green chili peppers (43 ± 0.7), raw carrots (43 ± 0.7), mandarin oranges (43 ± 0.4), raw ginger root (42 ± 2.1), human milk (42 ± 1.6), raw red sweet peppers (42 ± 1), raw green sweet peppers (42 ± 1), boiled leeks (42 ± 1.9), frozen and boiled turnips (41 ± 0.5), butter (40 ± 3.3), raw crookneck squash (40 ± 0.8), candied sweet potato (40 ± 0.5), raw radish (40 ± 1.1), canned red sweet peppers (39 ± 0.7), canned green sweet peppers (39 ± 0.7), boiled oriental radish (39 ± 0.7), boiled pumpkin (39 ± 0.4), boiled crookneck squash (38 ± 0.6), canned jalapeño peppers (38 ± 0.7), boiled acorn squash (37 ± 0.4), screwdriver (37 ± 0.2), persimmon (36 ± 2), cooked taro (36 ± 0.8), longans (35 ± 0.5), raw cucumber (35 ± 1), raw oriental radish (34 ± 1.1), blueberries (34 ± 0.3), tangerine juice (34 ± 0.2), tomato purée (33 ± 0.2), boiled/baked spaghetti squash (33 ± 0.6), raw green tomato (29 ± 0.4), boiled red sweet peppers (29 ± 0.7), boiled green sweet peppers (29 ± 0.7), boiled beets (28 ± 0.6), thompson seedless grapes (28 ± 0.3), whipped butter (27 ± 4.5),

boiled red tomato (27 ± 0.4), boiled zucchini (27 ± 0.6), canned crookneck squash (26 ± 0.5), strawberries (26 ± 0.3), boiled chayote (26 ± 0.6), canned carrots (26 ± 0.7), whey (25 ± 0.2), japanese persimmon (25 ± 0.3), canned peeled red tomatoes (23 ± 0.4), canned stewed red tomatoes (22 ± 0.4), raw red tomato (22 ± 0.4), pickled beets (21 ± 0.4), guava (21 ± 0.6), corn oil margarine stick (20 ± 10), safflower oil and soybean oil margarine stick (20 ± 10), soybean oil margarine stick (20 ± 10), soybean oil and cottonseed oil margarine stick (20 ± 10), corn oil margarine tub (20 ± 10), safflower oil margarine tub (20 ± 10), soybean oil margarine tub (20 ± 10), soybean oil and cottonseed oil margarine tub (20 ± 10), corn oil diet margarine tub (20 ± 10), soybean oil and cottonseed oil diet margarine tub (20 ± 10), raw celery (20 ± 1.3), mango (19 ± 0.2), boiled turnips (19 ± 0.6), peach (18 ± 0.6), figs (18 ± 1), pineapple (18 ± 0.3), sapodilla (17 ± 0.3), white grapefruit juice (16 ± 0.2), tomato juice (15 ± 0.3), boiled celery (15 ± 0.7), plum (14 ± 0.8), loquats (14 ± 0.5), boiled calabash gourd (14 ± 0.7), crab apples (13 ± 0.5), carambola (11 ± 0.4), papaya (10 ± 0.2), ale (9 ± 0.1), cooked apple without skin (8 ± 0.3), raw apple with skin (6 ± 0.4), applesauce (6 ± 0.4), raw apple without skin (5 ± 0.4), pear (4 ± 0.3), coffee (1 ± 0.3), internal conversion of citrulline using aspartic acid and glutamic acid during growing years

Applications: idiopathic hypospermia, wound healing, diminished thymus gland, hypercholesterolaemia, atherosclerosis, weight reduction, body building, growth hormone release

Daily Dosage: Nutritional—0.5–3 g; therapeutic—5 g; experimental—8 g; toxic—15 g

Warnings: Stimulates growth hormone release, which is undesirable for those still growing. Increases schizophrenia symptoms. May increase severity of herpes simplex infections. Increased amounts are toxic. See toxicity symptoms.

ASPARAGINE

Names: asparagine, ASN, amino acid N, amino acid B, 2-amino-butane-1,4 dioic acid-4-amide
Forms: 2-amino-butane-1,4 dioic acid-4-amide
Classification: nonessential amino acid
Deficiency: protein decreased
Side-effects: none known
Toxicity: none known

Inhibitors: increased amounts of other amino acids
Helpers: moderate amounts of other amino acids
Sources: internal synthesis
Applications: none known
Daily Dosage: unknown
Warnings: none

ASPARTIC ACID

Names: aspartic acid, ASP, amino acid P, amino acid D, amino acid B, 2-amino-butane-1,4 dioic acid
Forms: 2-amino-butane-1,4 dioic acid
Classification: nonessential amino acid
Deficiency: protein decreased
Side-effects: none known
Toxicity: none known
Inhibitors: increased amounts of other amino acids
Helpers: moderate amounts of other amino acids
Sources: internal synthesis
Applications: none known
Daily Dosage: Nutritional — 2.5–5 g; therapeutic — ?; experimental — ?; toxic — none known up to 5 g
Warnings: none

CARNITINE

(specific amino acid used in the body as a coenzyme like a vitamin rather than like an animo acid protein)
(see vitamin B_T)

CYSTEINE

Names: cysteine, CYS, amino acid C, 2-amino-3-mercatopropanoic acid
Forms: 2-amino-3-mercatopropanoic acid
Classification: nonessential amino acid, sulfur-containing amino acid
Deficiency: immunity decreased, deterioration of hair growth, deterioration of fingernail growth, premature aging, protein decreased
Side-effects: see toxicity symptoms
Toxicity: bladder/kidney stones
Inhibitors: increased amounts of other amino acids
Helpers: vitamin B_1, vitamin C
Sources (mg per every 100 grams of food): soybean protein isolate (1046 ± 1.8), soybean protein concentrate (886 ± 1.8), defatted soybean flour (757 ± 0.5), low fat soybean flour (750 ± 0.6), braised beef spleen (727 ± 0.5), defatted soy meal (724 ± 0.4), almond meal (718 ± 1.8), fenugreek seed (700 ± 25), torula yeast (679 ± 1.8), peanut flour (675 ± 12.5), dried and salted cod (673 ± 0.6), dried and frozen tofu (665 ± 2.9), dried spirulina seaweed (662 ± 0.5), dry roasted soybean nuts (638 ± 0.6), toasted soybean nuts (636 ± 1.8), yellow mustard seed (633 ± 16.7), roasted soybean nuts (567 ± 0.6), roasted soybean flour (561 ± 0.6), braised pork kidneys (557 ± 0.5), soybean flour (556 ± 0.6),

dried sesame kernels (525 ± 6.3), dried pistachio nuts (521 ± 1.8), braised pork liver (491 ± 0.5), dried butternuts (489 ± 1.8), dried black walnuts (475 ± 1.8), dried sunflower seeds (457 ± 1.8), braised pork center loin excluding fat (451 ± 0.5), caviar (450 ± 3.1), braised pork top loin excluding fat (446 ± 0.5), dried watermelon seeds (443 ± 1.8), dried pignolia pine nuts (443 ± 1.8), fried chicken giblets (437 ± 0.5), stewed chicken drumstick without skin (434 ± 1.1), braised pork sirloin excluding fat (434 ± 0.5), poppy seed (433 ± 16.7), oil roasted sunflower seeds (429 ± 1.8), fried chicken breast without skin (428 ± 0.6), braised pork loin excluding fat (427 ± 0.5), braised pork heart (423 ± 0.4), fried light meat chicken without skin (420 ± 0.5), fried chicken breast with skin (418 ± 0.5), broiled pork center loin excluding fat (415 ± 0.5), roasted pumpkin and squash seeds (411 ± 1.8), fried beef liver (410 ± 0.5), fried light meat chicken with skin (405 ± 0.5), braised pork boston blade excluding fat (404 ± 0.5), roasted chicken breast without skin (397 ± 0.6), roasted light meat chicken without skin (396 ± 0.5), fried beef top round excluding fat (393 ± 0.5), roasted chicken breast with skin (390 ± 0.5), dry roasted sunflower seeds (389 ± 1.8), simmered turkey gizzard (386 ± 0.5), roasted light meat chicken with skin (385 ± 0.5), braised pork loin blade excluding fat (385 ± 0.5), raw quail without skin (380 ± 0.5), dry roasted pistachio nuts (379 ± 1.8), simmered beef heart (378 ± 0.5), fried chicken back with skin (378 ± 0.7), roasted pork rump excluding fat (378 ± 0.5), simmered beef shank excluding fat (377 ± 0.5), braised pork spareribs (377 ± 0.5), braised pork center loin including fat (376 ± 0.5), fried dark meat chicken without skin (375 ± 0.5), roasted cured pork arm excluding fat (375 ± 0.5), freeze-dried red sweet peppers (375 ± 31.3), freeze-dried green sweet peppers (375 ± 31.3), cooked whelk (374 ± 0.6), braised beef liver (374 ± 0.5), broiled pork center rib excluding fat (373 ± 0.5), broiled pork top loin excluding fat (373 ± 0.5), roasted pork tenderloin (373 ± 0.5), fried pork center loin excluding fat (373 ± 0.5), stewed chicken breast without skin (371 ± 0.5), roasted chicken wing with skin (371 ± 1.5), oil roasted almonds (371 ± 1.8), toasted almonds (371 ± 1.8), braised beef chuck arm pot roast excluding fat (370 ± 0.5), stewed light meat chicken without skin (370 ± 0.5), roasted pork center loin excluding fat (369 ± 0.5), roasted pork leg excluding fat (367 ± 0.5), broiled pork sirloin excluding fat (366 ± 0.5), roasted pork center rib excluding fat (366 ± 0.5), roasted pork top loin excluding fat (366 ± 0.5), roasted pork shank excluding fat (366 ± 0.5), fried dark meat chicken with skin (365 ± 0.5), fried beef sirloin excluding fat (364 ± 0.5), braised pork center rib including fat (364 ± 0.5), simmered turkey heart (364 ± 0.5), dried almonds (364 ± 1.8), fried chicken wing with skin (363 ± 1.6), creamy peanut butter (363 ± 3.1), fried pork center rib excluding fat (362 ± 0.5), fried pork top loin excluding fat (362 ± 0.5), broiled pork loin excluding fat (361 ± 0.5), roasted duck without skin (361 ± 0.5), oil roasted spanish peanuts (361 ± 1.8), stewed chicken breast with skin (359 ± 0.5), simmered chicken heart (359 ± 0.5), roasted chicken drumstick with skin (358 ± 1), fried chicken thigh with skin (358 ± 0.8), simmered chicken gizzard (356 ± 0.5), roasted pork sirloin excluding fat (356 ± 0.5), braised pork sirloin including fat (356 ± 0.5), whole dried sesame seeds (356 ± 5.6), broiled beef top round excluding fat (355 ± 0.5), fried beef top round including fat (355 ± 0.5), fried chicken drumstick with skin (355 ± 1), dried brazil nuts (354 ± 1.8), simmered beef shank including fat (354 ± 0.5), braised beef bottom round excluding fat (354 ± 0.5), simmered turkey giblets (354 ± 0.5), braised pork top loin including fat (351 ± 0.5), roasted dark meat chicken without skin (350 ± 0.5), chipped beef (350 ± 1.8), broiled pork center loin including fat (350 ± 0.5), dried english/persian walnuts (350 ± 1.8), boiled mothbeans (350 ± 0.3), roasted pork loin excluding fat (349 ± 0.5), braised beef chuck blade roast excluding fat (348 ± 0.5), roasted dark meat chicken with skin (348 ± 0.5), stewed light meat chicken with skin (347 ± 0.5), canned corned beef (347 ± 0.5), toasted sesame kernels (346 ± 1.8), whole toasted sesame seeds (346 ± 1.8), oil roasted valencia peanuts (346 ± 1.8), oil roasted peanuts (346 ± 1.8), broiled beef top round including fat (345 ± 0.5), braised pork loin including fat (345 ± 0.5), braised beef shortribs excluding fat (344 ± 0.5), simmered chicken giblets (343 ± 0.5), roasted pork rump including fat (342 ± 0.5), braised pork arm braised including fat (341 ± 0.5), broiled beef sirloin excluding fat (340 ± 0.5), sesame butter (340 ± 3.3), fried chicken neck with skin (339 ± 1.4), braised pork boston blade braised including fat (337 ± 0.5), stewed chicken drumstick with skin (335 ± 0.9), cooked clams (335 ± 0.6), braised beef bottom round including fat (334 ± 0.5), dry skim milk (333 ± 1.7), roasted chicken thigh without skin (333 ± 1), roasted chicken thigh with skin (332 ± 0.8), oil roasted virginia peanuts (332 ± 1.8), dried peanuts (332 ± 1.8), braised beef brisket excluding fat (329 ± 0.5), roasted pork shoulder excluding fat (329 ± 0.5), pork livercheese (329 ± 1.3), cashew butter (329 ± 1.8), stir-fried sprouted lentils (328 ± 0.5), simmered chicken liver (327 ± 0.5), broiled pork boston blade excluding fat (326 ± 0.5), roasted eye of beef round excluding fat (325 ± 0.5), roasted pork center loin including fat (325 ± 0.5), dry instant skim milk (324 ± 0.5), broiled pork loin blade excluding fat (323 ± 0.5), roasted beef tip excluding fat (322 ± 0.5), stewed dark meat chicken without skin (322 ± 0.5), simmered turkey liver (322 ± 0.5), broiled beef top loin excluding fat (321 ± 0.5), roasted pork sirloin including fat (321 ± 0.5), roasted pork loin blade excluding fat (320 ± 0.5), roasted pork leg including fat (320 ± 0.5), broiled beef round excluding fat (319 ± 0.5), tempeh (319 ± 0.6), simmered chicken neck without skin (317 ± 2.8), canned tuna in water (316 ± 0.6), broiled beef tenderloin excluding fat (316 ± 0.5), roasted pork boston blade excluding fat (316 ± 0.5), roasted pork center rib including fat (316 ± 0.5), broiled porterhouse steak excluding fat (315 ± 0.5), broiled T-bone steak excluding fat (315 ± 0.5), fried pork loin blade excluding fat (315 ± 0.5), roasted lean ham (315 ± 0.5), roasted turkey light meat with skin (314 ± 0.5), broiled beef rib eye excluding fat (314 ± 0.5), braised beef flank excluding fat (314 ± 0.5), stewed dark meat chicken with skin (314 ± 0.5), stewed chicken wing with skin (313 ± 1.3), braised beef lungs (313 ± 0.5), cooked bacon (312 ± 0.4), broiled pork center rib including fat (312 ± 0.5), cooked mussels (312 ± 0.6), roasted turkey light meat without skin (311 ± 0.5), anchovy canned in olive oil (310 ± 2.5), stewed chicken thigh with skin (310 ± 0.7), roasted pork shank including fat (310 ± 0.5), raw pheasant without skin (309 ± 0.5), crunchy peanut butter (309 ± 1.6), roasted beef tenderloin excluding fat (308 ± 0.5), roasted beef rib roasted excluding fat (308 ± 0.5), braised beef flank including fat (308 ± 0.5), fried beef sirloin including fat (308 ± 0.5), roasted pork top loin including fat (308 ± 0.5), gruyere cheese (307 ± 1.8), broiled beef sirloin including fat (307 ± 0.5), broiled pork sirloin including fat (307 ± 0.5), dried pumpkin and squash seeds (304 ± 1.8), oil roasted cashews (304 ± 1.8), dry roasted peanuts (304 ± 1.8), braised beef chuck arm pot roast including fat (303 ± 0.5), braised pork loin blade including fat (303 ± 0.5), roasted turkey dark meat with skin (302 ± 0.5), dry buttermilk (300 ± 7.1), roasted eye of beef round including fat (300 ± 0.5), broiled pork top loin including fat (300 ± 0.5), broiled pork loin including fat (300 ± 0.5), roasted pork loin including fat (299 ± 0.5), roasted duck with skin (299 ± 0.5), grilled canadian bacon (298 ± 1.1), roasted ham (298 ± 0.5), roasted turkey dark meat without skin (297 ± 0.5), roasted beef round tip including fat (296 ± 0.5), dry roasted almonds (296 ± 1.8), fried pork center loin including fat (294 ± 0.5), roasted cured pork arm including fat (294 ± 0.5), chicken egg yolk (294 ± 2.9), cooked coho salmon (293 ± 0.6), cooked sockeye salmon (293 ± 0.6), swiss cheese (293 ± 1.8), puffed wheat (293 ± 3.6), well done baked extra lean ground beef (291 ± 0.5), broiled beef tenderloin including fat (291 ± 0.5), unheated lean ham (291 ± 0.5), hard pork salami (290 ± 5), simmered beef tongue (290 ± 0.5), chicken egg (290 ± 1), broiled beef top loin including fat (288 ± 0.5), chicken egg white (288 ± 1.5), cooked halibut (286 ± 0.6), broiled beef round including fat (286 ± 0.5), milk chocolate (286 ± 1.8), dry roasted cashews (286 ± 1.8), braised beef chuck blade roast including fat (285 ± 0.5), canned tuna in oil (284 ± 0.6), well done baked lean ground beef (284 ± 0.5), broiled beef flank excluding fat (284 ± 0.5), broiled beef rib eye including fat (284 ± 0.5), duck egg (284 ± 0.7), roasted pork arm including fat (283 ± 0.5), pork and beef luncheon sausage (283 ± 2.2), cooked rainbow trout (282 ± 0.6), cooked snapper (282 ± 0.6), simmered chicken neck with skin (282 ± 1.3), broiled porterhouse steak including fat (281 ± 0.5), parmesan cheese (280 ± 10), broiled beef flank including fat (280 ± 0.5), roasted pork shoulder including fat (280 ± 0.5), cooked sheepshead (279 ± 0.6), broiled pork boston blade including fat (278 ± 0.5), roasted pork boston blade including fat (278 ± 0.5), well done baked ground beef (276 ± 0.5), broiled beef rib excluding fat (276 ± 0.5), dried hickory nuts (275 ± 1.8), well done broiled extra lean ground beef (274 ± 0.5), roasted beef tenderloin including fat (274 ± 0.5), cooked swordfish (272 ± 0.6), smoked haddock (271 ± 0.6), well done broiled lean ground beef (271 ± 0.5), fried pork center rib including fat (271 ± 0.5), fried pork top loin including fat (270 ± 0.5), well done fried extra lean ground beef (269 ± 0.5), broiled T-bone steak including fat (269 ± 0.5), almond butter (269 ± 3.1), cooked crayfish (268 ± 0.6), boiled soybeans (268 ± 0.3), roasted pork loin blade including fat (267 ± 0.5), cooked grouper (266 ± 0.6), cooked mullet (266 ± 0.6), cooked pike (265 ± 0.6), kippered herring (265 ± 1.3), well done fried lean ground beef (265 ± 0.5), cooked tilefish (264 ± 0.6), unheated ham (264 ± 0.5), sardines (263 ± 2.1), well done broiled ground beef (261 ± 0.5), broiled pork loin blade including fat (261 ± 0.5), fried abalone (261 ± 0.6), cooked haddock (260 ± 0.6), hard pork and beef salami (260 ± 5), roasted pork blade roll (260 ± 0.5), cooked flatfish (259 ± 0.6), well done fried ground beef (259 ± 0.5), cooked rockfish (258 ± 0.6), breaded and fried shrimp (258 ± 0.6), braised beef brisket including fat (258 ± 0.5), chopped ham lunch meat (257 ± 2.4), cooked perch (256 ± 0.6), cooked mackerel (256 ± 0.6), unheated canadian bacon (256 ± 0.9), cooked pompano (254 ± 0.6), roasted beef rib including fat (254 ± 0.5), oil roasted filberts (254 ± 1.8), cooked eel (253 ± 0.6), cooked sea bass (253 ± 0.6), cooked whiting (252 ± 0.6), raw duck liver (252 ± 0.5), breaded and fried scallops (252 ± 1.6), smoked whitefish (251 ± 0.6), roasted canned lean ham (250 ± 0.5), pork liver sausage (250 ± 2.8), pork link sausage (249 ± 0.7), raw yellowtail (248 ± 0.6), cooked herring (247 ± 0.6), pork and beef peppered loaf (246 ± 1.8), bran (246 ± 1.8), cooked carp (245 ± 0.6), cooked cod (245 ± 0.6), canned cod (244 ± 0.6), dry whole milk (244 ± 1.6), medium-rare broiled extra lean ground beef (244 ± 0.5), roasted canned ham (243 ± 0.5), chopped smoked beef (243 ± 1.8), cooked rainbow smelt (242 ± 0.6), braised beef shortribs including fat (242 ± 0.5), breaded fried squid (242 ± 0.6), medium-rare fried extra lean ground beef (240 ± 0.5), jellied corned beef loaf (239 ± 1.8), ham and cheese roll (239 ± 1.8), crab cakes (238 ± 0.8), fried tofu (238 ± 3.8), medium-rare broiled lean ground beef (237 ± 0.5), medium-rare baked extra lean ground beef (235 ± 0.5), fried pork loin blade including fat (235 ± 0.5), cooked shrimp (234 ± 0.6), turkey breast meat summer sausage (233 ± 2.4), medium-rare fried lean ground beef (233 ± 0.5), pork and beef pepperoni (233 ± 8.3), raw coho salmon (232 ± 0.6), cooked corned beef brisket (232 ± 0.5), pork and beef lunch meat (232 ± 1.8), bran flakes (232 ± 1.8), dried filberts (232 ± 1.8), canned blue crab (231 ± 0.6), cooked lobster (231 ± 0.6), medium-rare broiled ground beef (231 ± 0.5), medium-rare baked lean ground beef (230 ± 0.5), medium-rare fried ground beef (230 ± 0.5), lebanon bologna (230 ± 2.2), raw sockeye salmon (228 ± 0.6), raw shrimp (228 ± 0.6), pork and beef kielbasa/kolbassy (227 ± 1.9), cooked blue crab (226 ± 0.6), raw shark (225 ± 0.6), broiled beef rib including fat (225 ± 0.5), beef pastrami (225 ± 1.8), pork headcheese (225 ± 1.8), raw halibut (224 ± 0.6), raw trout (224 ± 0.6), raw abalone (224 ± 0.6), fried beef brains (223 ± 0.5), cooked sturgeon (222 ± 0.6), berliner (222 ± 2.2), medium-rare baked ground beef (221 ± 0.5), dry roasted macadamia nuts (221 ± 1.8), raw rainbow trout (220 ± 0.6), raw milkfish (220 ± 0.6), raw snapper (220 ± 0.6), raw goose liver (220 ± 0.5), raw scallops (220 ± 0.6), natto (220 ± 0.6), canned sockeye salmon (219 ± 0.6), canned lean ham (219 ± 0.5), raw anchovy (218 ± 0.6), firm raw tofu (218 ± 0.4), dry roasted pecans (218 ± 1.8), raw sheepshead (216 ± 0.6), raw chum salmon (216 ± 0.6), cooked alaska king crab (216 ± 0.6), breaded and fried croaker (216 ± 0.6), raw chinook salmon (215 ± 0.6), raw bluefish (215 ± 0.6), raw pink salmon (214 ± 0.6), canned pink salmon (214 ± 0.6), canned pork lunch meat (214 ± 2.4), dried pinyon pine nuts (214 ± 1.8), shredded wheat (214 ± 1.8), almond paste (214 ± 1.8), raw cuttlefish (213 ± 0.6), raw swordfish (212 ± 0.6), battered and fried shark (212 ± 0.6), raw lobster (211 ± 0.6), dried pecans (211 ± 1.8), raw crayfish (209 ± 0.6), raw pollock (208 ± 0.6), raw sunfish (208 ± 0.6), raw grouper (208 ± 0.6), raw mullet (207 ± 0.6), raw burbot (207 ± 0.6), raw queen crab (207 ± 0.6), pork and beef mortadella (207 ± 3.3), raw pike (206 ± 0.6), raw whitefish (205 ± 0.6), breaded and fried catfish (205 ± 0.6), raw alaska king crab (205 ± 0.6), turkey pastrami (205 ± 0.9), raw cusk (204 ± 0.6), raw cisco (204 ± 0.6), raw ling (204 ± 0.6), raw haddock (204 ± 0.6), raw squid (204 ± 0.6), breaded and fried clams (204 ± 0.6), dried european chestnuts (204 ± 1.8), raw scup (202 ± 0.6), raw freshwater bass (202 ± 0.6), raw flatfish (202 ± 0.6), raw blue crab (202 ± 0.6), raw rockfish (201 ± 0.6), italian pork sausage (201 ± 0.7), canned ham (201 ± 0.5), raw perch (200 ± 0.6), simmered beef kidneys (200 ± 0.5), beef lunch meat (200 ± 1.8), summer sausage (200 ± 2.2), basil (200 ± 50), fennel seed (200 ± 25), onion powder (200 ± 25), boiled sprouted green peas (200 ± 0.5), raw mackerel (199 ± 0.6), raw spot (198 ± 0.6), raw dolphinfish (198 ± 0.6), raw pompano (198 ± 0.6), raw eel (198 ± 0.6), raw sea bass (198 ± 0.6), simmered beef brains (197 ± 0.5), turkey ham (196 ± 0.9), raw whiting (196 ± 0.6), smoked chinook salmon (196 ± 0.6), pork sausage (196 ± 1.9), raw octopus (196 ± 0.6), beef and pork salami (196 ± 2.2), raw catfish (195 ± 0.6), raw dungeness crab (195 ± 0.6), raw herring (193 ± 0.6), dried shitake mushrooms (193 ± 3.3), boiled lupins (192 ± 0.3), raw carp (191 ± 0.6), raw cod (191 ± 0.6), raw croaker (191 ± 0.6), raw striped bass (191 ± 0.6), beef salami (191 ± 2.2), minced ham lunch meat (190 ± 2.4), raw lingcod (189 ± 0.6), raw rainbow smelt (189 ± 0.6), oil roasted pecans (189 ± 1.8), raw freshwater drum (188 ± 0.6), raw tilefish (188 ± 0.6), raw wolffish (188 ± 0.6), raw whelk (187 ± 0.6), raw butterfish (185 ± 0.6), cooked oysters (185 ± 0.6), raw shad (182 ± 0.6), corn flakes (182 ± 1.8), falafel (182 ± 1), dried chinese chestnuts (182 ± 1.8), wheat flakes (182 ± 1.8), smoked beef sausage (181 ± 1.2), raw white sucker (180 ± 0.6), raw seatrout (179 ± 0.6), dry roasted filberts (179 ± 1.8), grilled ham patties (178 ± 0.8), smoked cisco (175 ± 0.6), vienna sausage (175 ± 3.1), raw sturgeon (173 ± 0.6), raw turbot (172 ± 0.6), tuna salad (172 ± 0.2), unheated ham patties (172 ± 0.8), boiled peanuts (172 ± 1.6), pork bologna (170 ± 2.2), raw beef tripe (168 ± 0.5), raw clams (168 ± 0.6), garlic powder (167 ± 16.7), boiled winged beans (166 ± 0.3), pork and beef picnic loaf (164 ± 1.8), raisin bran (164 ± 1.8), pork and beef brotwurst (163 ± 0.7), polish pork sausage (161 ± 1.8), beef beerwurst salami (161 ± 2.2), pork and beef honey loaf (157 ± 1.8), beef bologna (157 ± 2.2), dehydrated onion flakes (157 ± 3.6), raw mussels (156 ± 0.6), raw monkfish (155 ± 0.6), pork luxury loaf (154 ± 1.8), beef frankfurter (154 ± 0.9), dried japanese chestnuts (154 ± 1.8), pickled herring (153 ± 3.3), pork and beef link sausage with american cheese (151 ± 1.2), pork liverwurst (150 ± 2.8), pork and beef knockwurst (147 ± 0.7), pork olive loaf (146 ± 1.8), dried acorns (146 ± 1.8), american cheese (143 ± 1.8), pimento cheese spread (143 ± 1.8), pork bratwurst (142 ± 0.6), dried coconut (139 ± 1.8), pork and beef sausage (138 ± 3.8), surimi scallops (136 ± 0.6), acorn flour (136 ± 1.8), beef and pork bologna (135 ± 2.2), surimi shrimp (133 ± 0.6), muenster cheese (132 ± 1.8), brick cheese (132 ± 1.8), breaded and fried oysters (131 ± 0.6), beef and pork frankfurter (129 ± 1.1), boiled split peas (127 ± 0.3), frozen and boiled black-eyed peas (126 ± 0.6), cheddar cheese (125 ± 0.5), monterey cheese (125 ± 1.8), colby cheese (121 ± 1.8), boiled chickpeas (119 ± 0.3), provolone cheese (118 ± 1.8), cheshire cheese (118 ± 1.8), mozzarella cheese (118 ± 1.8), pork pickle and pimento loaf (118 ± 1.8), roasted chinese chestnuts (118 ± 1.8), boiled lentils (118 ± 0.3), cottage cheese (116 ± 0.2), brie cheese (114 ± 1.8), boiled green soybeans (113 ± 0.6), potato pancakes (113 ± 0.7), raw dock (113 ± 0.7), gefiltefish with broth (112 ± 1.2), raw tofu (112 ± 0.4), camembert cheese (111 ± 1.8), raw acorns (111 ± 1.8), raw chinese chestnuts (111 ± 1.8), pork beerwurst salami (109 ± 2.2), bleu cheese (107 ± 1.8), pork and beef old fashioned loaf (107 ± 1.8), puffed rice (107 ± 3.6), dried toasted coconut (107 ± 1.8), boiled white beans (106 ± 0.3), pork and beef link sausage (104 ± 0.7), roasted european chestnuts (104 ± 1.8), boiled cranberry beans (102 ± 0.3), raw laver/nori seaweed (100 = 0.5), boiled yellow beans (100 ± 0.3), ricotta cheese (99 ± 0.4), boiled pink beans (99 ± 0.3), raw kelp/kombu/tangle seaweed (98 ± 0.5), boiled broad beans (97 ± 0.3), boiled black beans (96 ± 0.3), dried macadamia nuts (96 ± 1.8), miso (95 ± 0.4), boiled navy beans (95 ± 0.3), boiled hyacinth beans (95 ± 0.3), boiled kidney beans (94 ± 0.3), raw oysters (93 ± 0.6), canned oysters (93 ± 0.6), boiled great northern beans (91 ± 0.3), boiled yardlong bean (91 ± 0.3), raw bacon (90 ± 0.7), boiled catjang black-eyed peas (90 ± 0.3), neufchatel cheese (89 ± 1.8), boiled pinto beans (89 ± 0.3), boiled black turtle beans (89 ± 0.3), boiled baby lima beans (89 ± 0.3), boiled lima beans (86 ± 0.3), oil roasted macadamia nuts (86 ± 1.8), roasted japanese chestnuts (86 ± 1.8), boiled black-eyed peas (85 ± 0.3), canned navy beans (82 ± 0.2), potato flour (81 ± 0.6), boiled and frozen baby lima beans (81 ± 0.6), canned great northern beans (80 ± 0.2), noodles (80 ± 0.3), canned white beans (79 ± 0.2), raw european chestnuts (79 ± 1.8), boiled pigeon peas (78 ± 0.3), boiled french beans (77 ± 0.3), tomato powder (76 ± 0.5), boiled and steamed chinese chestnuts (75 ± 1.8), sweetened condensed milk (74 ± 1.3), frozen and boiled fordhook lima beans (74 ± 0.6), coconut cream (72 ± 0.2), boiled mungo beans (70 ± 0.3), boiled adzuki beans (70 ± 0.2), canned broad beans (70 ± 0.2), evaporated skim milk (69 ± 1.6), stir-fried sprouted soybeans (69 ± 0.5), cream cheese (68 ± 1.8), raw spirulina seaweed (68 ± 0.5), refried beans (68 ± 0.2), boiled lamb's-quarters (68 ± 0.6), raw garlic (67 ± 5.6), canned chickpeas (67 ± 0.2), raw coconut (67 ± 1.1), canned black turtle beans (66 ± 0.2), yellow corn pudding (65 ± 0.2), dried sweetened flaked coconut (65 ± 0.7), raw cassava (65 ± 0.5), raw japanese chestnuts (64 ± 1.8), macaroni (64 ± 0.4), spaghetti (64 ± 0.4), boiled and steamed european chestnuts (64 ± 1.8), oatmeal (64 ± 0.3), evaporated whole milk (63 ± 0.4), hummus (63 ± 0.2), boiled mung beans (62 ± 0.2), dried oriental radish (62 ± 0.9), canned cranberry beans (60 ± 0.2), evaporated lowfat milk (58 ± 0.4), baked beans with pork and sweet sauce (58 ± 0.2), gjetost cheese (57 ± 1.8), dried ginko nuts (57 ± 1.8), lard (57 ± 1.8), canned kidney beans (57 ± 0.2), dried sweetened shredded coconut (57 ± 0.5), raw pork jowl (56 ± 0.5), baked beans with pork and tomato sauce (56 ± 0.2), boiled succotash (55 ± 0.5), canned lima beans (54 ± 0.2), sweetened coconut cream (53 ± 0.2), vegetarian baked beans (52 ± 0.2), canned black-eyed peas (52 ± 0.2), potato salad (51 ± 0.4), ham salad (50 ± 1.8), canned pinto beans (50 ± 0.2), ginger (50 ± 25), dried figs (50 ± 0.3), malted milk (49 ± 0.2), indian buffalo milk (48 ± 0.2), frozen and boiled succotash (47 ± 0.6), soy milk (47 ± 0.2), goat milk (46 ± 0.2), steamed sprouted soybeans (45 ± 1.1), coconut milk (45 ± 0.2), dried dates (45 ± 0.6), au gratin potatoes (44 ± 0.4), okara tofu (44 ± 0.8), protein fortified skim milk (38 ± 0.2), eggnog (38 ± 0.2), frozen and boiled turnip greens (38 ± 0.6), frozen and boiled spinach (38 ± 0.5), frozen and boiled kale (38 ± 0.8), raw red chili peppers (38 ± 1.1), raw green chili peppers (38 ± 1.1), protein fortified lowfat 2% milk (37 ± 0.2), protein fortified lowfat 1% milk (36 ± 0.2), restaurant vanilla milkshake (36 ± 0.2), boiled spinach (36 ± 0.6), raw spinach (36 ± 1.8), sheep milk (35 ± 0.2), frozen and boiled asparagus (35 ± 0.8), frozen and boiled mustard greens (35 ± 0.7), hot chocolate (34 ± 0.2), scalloped potatoes (34 ± 0.4), canned spinach (34 ± 0.5), cream of wheat (34 ± 0.3), boiled mustard greens (34 ± 0.7), boiled green peas (33 ± 0.6), boiled jute potherb (33 ± 1.2), vanilla milkshake (32 ± 0.2), raw green peas (32 ± 0.6), corn grits (32 ± 0.2), farina (32 ± 0.3), skim milk (31 ± 0.2), restaurant chocolate milkshake (31 ± 0.2), restaurant strawberry milkshake (31 ± 0.2), lowfat 2% milk (31 ± 0.2), filled milk (31 ± 0.2), buttermilk (31 ± 0.2), frozen and boiled collards (31 ± 0.6), microwaved potato with skin (31 ± 0.2), boiled asparagus (31 ± 0.6), lowfat 1% milk (30 ± 0.2), whole milk (30 ± 0.2), lowfat 1% chocolate milk (30 ± 0.2), lowfat 2% chocolate milk (30 ± 0.2), frozen and boiled green peas (30 ± 0.6), o'brien potatoes (30 ± 0.3), whole chocolate milk (29 ± 0.2), carob milk (29 ± 0.2), raw swamp cabbage (29 ± 0.9), baked potato with skin (29 ± 0.2), canned succotash with creamed corn (29 ± 0.4), dried japanese persimmon (29 ± 1.5), chocolate milkshake (28 ± 0.2), raw wakame seaweed (28 ± 0.5), canned succotash (28 ± 0.4), half and half cream (27 ± 3.3), boiled yellow corn (27 ± 0.6), light cream (27 ± 3.3), microwaved potato without skin (27 ± 0.3), raw vine spinach (27 ± 0.5), raw leeks (27 ± 1.9), canned green peas (26 ± 0.6), frozen and boiled yellow corn (26 ± 0.6), stir-fried sprouted mung beans (26 ± 0.8), french fried potatoes (26 ± 1), raw potato without skin (26 ± 0.4), boiled amaranth (26 ± 0.8), cooked shitake mushrooms (26 ± 0.7), ginko nuts (25 ± 1.8), frozen and boiled potato without skin (25 ± 0.5), baked potato without skin (25 ± 0.3), boiled and steamed japanese chestnuts (25 ± 1.8), boiled kale (25 ± 0.8), boiled scotch kale (25 ± 0.8), raw cauliflower (24 ± 1), boiled beet greens (24 ± 0.7), frozen and boiled brussels sprouts (23 ± 0.6), mashed potatoes (23 ± 0.5), boiled cauliflower (23 ± 0.8), frozen and fried french fried potatoes (22 ± 1), raw california avocado (22 ± 0.3), frozen and boiled broccoli (22 ± 0.5), boiled swamp cabbage (22 ± 1), raw bamboo shoots (22 ± 0.7), boiled potato without skin (22 ± 0.4), raw lotus root (22 ± 0.6), tomato paste (22 ± 0.4), boiled broccoli (21 ± 0.6), safflower oil and soybean oil mayonnaise (21 ± 3.6), soybean oil mayonnaise (21 ± 3.6), frozen and fried hash brown potatoes (21 ± 0.6), steamed hawaii mountain yam (21 ± 0.7), chocolate syrup (21 ± 1.3), raw onions (21 ± 0.6), medium cream (20 ± 3.3), whipping cream, light (20 ± 3.3), whipping cream, heavy (20 ± 3.3), canned yellow corn with red and green peppers (20 ± 0.4), frozen and boiled turnip greens and turnips (20 ± 0.5), soybean oil and cottonseed oil margarine liquid (20 ± 10), raw broccoli (20 ± 1.1), frozen and boiled okra (20 ± 0.5), persimmon (20 ± 2), frozen and boiled cauliflower (19 ± 0.6), human milk (19 ± 1.6), raw turnip greens (18 ± 1.8), boiled green beans (18 ± 0.8), boiled yellow snap beans (18 ± 0.8), canned potato without skin (18 ± 0.6), raw romaine lettuce (18 ± 1.8), baked/boiled tropical yam (18 ± 0.7), boiled okra (18 ± 0.6), frozen and boiled red sweet peppers (18 ± 0.5), frozen and boiled green sweet peppers (18 ± 0.5), canned red chili peppers (18 ± 0.7), canned green chili peppers (18 ± 0.7), raw sprouted mung beans (17 ± 1), raw florida avocado (17 ± 0.2), raw savoy cabbage (17 ± 1.4), whey (17 ± 0.2), bananas (17 ± 0.4), raw chinese cabbage (17 ± 1.4), canned yellow corn (16 ± 0.5), raw green tomato (16 ± 0.4), raw red sweet peppers (16 ± 1), raw green sweet peppers (16 ± 1), canned red sweet peppers (16 ± 0.7), canned green sweet peppers (16 ± 0.7), boiled chinese cabbage (16 ± 0.6), boiled onions (16 ± 0.5), canned turnip greens (15 ± 0.4), hash brown potatoes (15 ± 0.6), cream of rice (15 ± 0.3), boiled brussels sprouts (15 ± 0.6), boiled savoy cabbage (15 ± 0.7), coleslaw (15 ± 0.8), raw iceberg lettuce (15 ± 2.5), elderberries (15 ± 0.3), boiled leeks (15 ± 1.9), canned jalapeño peppers (15 ± 0.7), baked sweet potato (14 ± 0.4), frozen and baked sweet potato (14 ± 0.6), canned creamed yellow corn (14 ± 0.4), canned bamboo shoots (14 ± 0.4), frozen and boiled crookneck squash (14 ± 0.5), raw looseleaf lettuce (14 ± 1.8), coconut water (14 ± 0.2), boiled red tomato (14 ± 0.4), canned sweet potato (13 ± 0.3), boiled sweet potato (13 ± 0.3), boiled turnip greens (13 ± 0.7), frozen and boiled yellow snap beans (13 ± 0.7), frozen and boiled green beans (13 ± 0.7), baked hubbard squash (13 ± 0.5), boiled hubbard squash (13 ± 0.4), boiled bamboo shoots (13 ± 0.4), raw butterhead lettuce (13 ± 3.3), boiled lotus root (13 ± 0.6), boiled beets (13 ± 0.6), frozen and boiled onions (13 ± 0.5), japanese persimmon (13 ± 0.3), raw scallop squash (12 ± 0.8), raw zucchini (12 ± 0.8), raw endive (12 ± 2), frozen and boiled zucchini (12 ± 0.4), cooked plantain (12 ± 0.3), canned peeled red tomatoes (12 ± 0.4), canned stewed red tomatoes (12 ± 0.4), raw red tomato (12 ± 0.4), boiled red sweet peppers (12 ± 0.7), boiled green sweet peppers (12 ± 0.7), figs (12 ± 1), cooked sprouted mung beans (11 ± 0.8), whole chocolate malted milk (11 ± 0.2), valencia orange (11 ± 0.4), navel orange (11 ± 0.4), european grapes (11 ± 0.3), dried apricots (11 ± 1.4), boiled scallop squash (11 ± 0.6), canned zucchini in tomato juice (11 ± 0.4), boiled collards (11 ± 0.3), frozen

III. Amino Acids

and boiled butternut squash (11 ± 0.4), raw red cabbage (11 ± 1.4), raw crookneck squash (11 ± 0.8), raw green cabbage (11 ± 1.4), boiled rutabaga (11 ± 0.6), cooked taro (11 ± 0.8), american grapes (10 ± 0.5), canned sprouted mung beans (10 ± 0.8), canned green beans (10 ± 0.7), canned yellow snap beans (10 ± 0.7), boiled purslane (10 ± 0.9), baked acorn squash (10 ± 0.5), boiled crookneck squash (10 ± 0.6), boiled yambean (10 ± 0.5), raw celtuce (10 ± 0.5), tomato purée (10 ± 0.2), pickled beets (10 ± 0.4), frozen and boiled carrots (10 ± 0.7), pear (10 ± 0.3), frozen and boiled turnips (9 ± 0.5), boiled red cabbage (9 ± 0.7), breadfruit (9 ± 0.5), boiled carrots (9 ± 0.6), raw ginger root (8 ± 2.1), boiled burdock root (8 ± 0.4), boiled green cabbage (8 ± 0.7), boiled butternut squash (8 ± 0.5), raw carrots (8 ± 0.7), candied sweet potato (7 ± 0.5), boiled kohlrabi (7 ± 0.6), tangerine (7 ± 0.6), blueberries (7 ± 0.3), boiled zucchini (7 ± 0.6), boiled mushrooms (6 ± 0.6), raw mushrooms (6 ± 1.4), raw watercress (6 ± 2.9), peach (6 ± 0.6), mandarin oranges (6 ± 0.4), thompson seedless grapes (6 ± 0.3), canned crookneck squash (6 ± 0.5), boiled acorn squash (6 ± 0.4), loquats (6 ± 0.5), raw eggplant (5 ± 1.2), raw oriental radish (5 ± 1.1), boiled/baked spaghetti squash (5 ± 0.6), plum (5 ± 0.8), boiled oriental radish (5 ± 0.7), raw celery (5 ± 1.3), canned carrots (5 ± 0.7), crab apples (5 ± 0.5), orange juice (5 ± 0.2), strawberries (5 ± 0.3), boiled turnips (4 ± 0.6), boiled eggplant (4 ± 1), raw radish (4 ± 1.1), tomato juice (4 ± 0.3), raw cucumber (4 ± 1), cooked apple without skin (4 ± 0.3), screwdriver (4 ± 0.2), tangerine juice (4 ± 0.2), canned pumpkin (3 ± 0.4), canned pumpkin pie mix (3 ± 0.4), apricots (3 ± 0.5), boiled celery (3 ± 0.7), raw apple with skin (3 ± 0.4), ale (3 ± 0.1), pineapple (2 ± 0.3), boiled pumpkin (2 ± 0.4), watermelon (2 ± 0.3), applesauce (2 ± 0.4), raw apple without skin (2 ± 0.4), coffee (2 ± 0.3), internal conversion of methionine

Applications: cancer, antioxidant therapy, sun sensitivity, free radical oxidation, life span extension, radiation, decreased radiation resistance, aging

Daily Dosage: (at least 3000 mg vitamin C per every 1000 mg cysteine) Nutritional—0.5–1 g; therapeutic—1; experimental—2; toxic—none known up to 2 g

Warnings: May block insulin in diabetics. Do not confuse with cystine. Increased amounts are dangerous. See toxicity symptoms.

CYSTINE

Deficiency: Protein decreased
Classification: nonessential amino acid
Side-effects: none known
Toxicity: none known
Inhibitors: increased amounts of other amino acids
Helpers: moderate amounts of other amino acids
Sources: internal conversion of cysteine
Applications: none known
Warnings: Do not confuse with cysteine.

GLUTAMATE

Names: glutamate, GLN, amino acid Q
Classification: nonessential amino acid
Deficiency: protein decreased
Side-effects: none known
Toxicity: none known
Inhibitors: increased amounts of other amino acids
Helpers: moderate amounts of other amino acids
Sources: internal synthesis using vitamin B_6
Applications: none known
Warnings: none

GLUTAMINE

Names: glutamine, GLN, amino acid Q, amino acid Z, 2-amino-pentane-1,5-dioic acid-5-amide
Forms: 2-amino-pentane-1,5-dioic acid-5-amide
Classification: nonessential amino acid
Deficiency: decreased I.Q., convulsions, retardation, decreased coordination, deterioration of alertness, alcohol cravings, protein decreased
Side-effects: none known
Toxicity: none known
Inhibitors: taking with food
Helpers: moderate amounts of other amino acids
Sources: whole grains, milk, internal conversion of glutamate
Applications: unknown
Daily Dosage: Nutritional — 1 g; therapeutic — 3?; experimental — ?; toxic — none known up to 3 g
Warnings: none

GLYCINE

Names: glycine, GLY, amino acid G, amino acetic acid
Forms: amino acetic acid
Classification: nonessential amino acid
Deficiency: deterioration of growth, liver deterioration, protein decreased
Side-effects: none known
Toxicity: none known
Inhibitors: increased amounts of other amino acids
Helpers: moderate amounts of other amino acids
Sources: internal conversion of serine using vitamin B_C
Applications: unknown
Daily Dosage: Nutritional — 0.5–2 g; therapeutic — ?; experimental — ?; toxic — none known up to 2 g
Warnings: none

HISTIDINE

Names: histidine, HIS, amino acid H, 2-amino-3(imidazole)-propanoic acid
Forms: 2-amino-3(imidazole)-propanoic acid
Classification: semi-essential amino acid
Deficiency: protein decreased, overstimulation, deterioration of myelin sheaths around nerves, hearing deterioration, diminished sexual arousal
Side-effects: see toxicity symptoms
Toxicity: (Detoxified by methionine) suicidal depression
Inhibitors: increased amounts of other amino acids
Helpers: moderate amounts of other amino acids
Sources (mg per every 100 grams of food): torula yeast (3904 ± 1.8), soybean protein isolate (2304 ± 1.8), dried and salted cod (1849 ± 0.6), braised pork center loin excluding fat (1763 ± 0.5), braised pork top loin excluding fat (1747 ± 0.5), braised pork sirloin excluding fat (1699 ± 0.5), braised pork loin excluding fat (1673 ± 0.5), broiled pork center loin excluding fat (1623 ± 0.5), parmesan cheese (1600 ± 10), braised pork boston blade excluding fat (1580 ± 0.5), soybean protein concentrate (1579 ± 1.8), peanut flour (1525 ± 12.5), braised pork loin blade excluding fat (1506 ± 0.5), roasted pork rump excluding fat (1478 ± 0.5), braised pork spareribs (1474 ± 0.5), broiled pork center rib excluding fat (1462 ± 0.5), broiled pork top loin excluding fat (1462 ± 0.5), roasted pork tenderloin (1460 ± 0.5), fried pork center loin excluding fat (1459 ± 0.5), roasted pork center loin excluding fat (1445 ± 0.5), braised pork center loin including fat (1442 ± 0.5), roasted pork leg excluding fat (1436 ± 0.5), broiled pork sirloin excluding fat (1433 ± 0.5), roasted pork center rib excluding fat (1431 ± 0.5), roasted pork top loin excluding fat (1431 ± 0.5), roasted pork shank excluding fat (1431 ± 0.5), fried pork center rib excluding fat (1418 ± 0.5), fried pork top loin excluding fat (1418 ± 0.5), broiled pork loin excluding fat (1412 ± 0.5), braised pork center rib including fat (1395 ± 0.5), dried and frozen tofu (1394 ± 2.9), roasted pork sirloin excluding fat (1394 ± 0.5), braised pork sirloin including fat (1365 ± 0.5), roasted pork loin excluding fat (1364 ± 0.5), broiled pork center loin including fat (1344 ± 0.5), braised pork top loin including fat (1340 ± 0.5), roasted pork rump including fat (1322 ± 0.5), braised pork loin including fat (1321 ± 0.5), braised pork arm braised including fat (1304 ± 0.5), braised pork boston blade braised including fat (1289 ± 0.5), roasted pork shoulder excluding fat (1287 ± 0.5), broiled pork boston blade excluding fat (1277 ± 0.5), defatted soybean flour (1268 ± 0.5), broiled pork loin blade excluding fat (1264 ± 0.5), low fat soybean flour (1255 ± 0.6), roasted pork center loin including fat (1253 ± 0.5), roasted pork loin blade excluding fat (1252 ± 0.5), fenugreek seed (1250 ± 25), roasted pork sirloin including fat (1239 ± 0.5), roasted pork boston blade excluding fat (1236 ± 0.5), fried pork loin blade excluding fat (1233 ± 0.5), roasted pork leg including fat (1230 ± 0.5), roasted pork center rib including fat (1213 ± 0.5), defatted soy meal (1212 ± 0.4), fried beef top round excluding fat (1201 ± 0.5), broiled pork center rib including fat (1198 ± 0.5), roasted pork shank including fat (1186 ± 0.5), roasted pork top loin including fat (1179 ± 0.5), broiled pork sirloin including fat (1174 ± 0.5), simmered beef shank excluding fat (1153 ± 0.5), braised pork loin blade including fat (1150 ± 0.5), broiled pork top loin including fat (1145 ± 0.5),

broiled pork loin including fat (1144 ± 0.5), roasted pork loin including fat (1144 ± 0.5), gruyere cheese (1132 ± 1.8), braised beef chuck arm pot roast excluding fat (1131 ± 0.5), provolone cheese (1129 ± 1.8), almond meal (1121 ± 1.8), fried pork center loin including fat (1116 ± 0.5), fried beef sirloin excluding fat (1112 ± 0.5), dried spirulina seaweed (1085 ± 0.5), broiled beef top round excluding fat (1085 ± 0.5), fried beef top round including fat (1085 ± 0.5), simmered beef shank including fat (1082 ± 0.5), braised beef bottom round excluding fat (1082 ± 0.5), swiss cheese (1079 ± 1.8), roasted pork arm including fat (1076 ± 0.5), roasted pork shoulder including fat (1071 ± 0.5), dry roasted soybean nuts (1067 ± 0.6), roasted pork boston blade including fat (1067 ± 0.5), broiled pork boston blade including fat (1064 ± 0.5), braised beef chuck blade roast excluding fat (1063 ± 0.5), broiled beef top round including fat (1055 ± 0.5), braised beef shortribs excluding fat (1053 ± 0.5), edam cheese (1046 ± 1.8), gouda cheese (1046 ± 1.8), broiled beef sirloin excluding fat (1040 ± 0.5), fried chicken breast without skin (1037 ± 0.6), braised beef bottom round including fat (1021 ± 0.5), fried pork center rib including fat (1020 ± 0.5), fried light meat chicken without skin (1018 ± 0.5), roasted pork loin blade including fat (1016 ± 0.5), fried pork top loin including fat (1014 ± 0.5), toasted soybean nuts (1011 ± 1.8), braised beef brisket excluding fat (1006 ± 0.5), roasted eye of beef round excluding fat (993 ± 0.5), broiled pork loin blade including fat (989 ± 0.5), roasted beef tip excluding fat (983 ± 0.5), dry skim milk (980 ± 1.7), broiled beef top loin excluding fat (980 ± 0.5), cooked whelk (976 ± 0.6), stewed chicken drumstick without skin (975 ± 1.1), broiled beef round excluding fat (974 ± 0.5), broiled beef tenderloin excluding fat (967 ± 0.5), well done baked extra lean ground beef (965 ± 0.5), broiled porterhouse steak excluding fat (964 ± 0.5), roasted chicken breast without skin (963 ± 0.6), broiled T-bone steak excluding fat (963 ± 0.5), broiled beef rib eye excluding fat (960 ± 0.5), fried chicken breast with skin (959 ± 0.5), roasted light meat chicken without skin (959 ± 0.5), braised beef flank excluding fat (959 ± 0.5), dry instant skim milk (952 ± 0.5), roasted soybean nuts (950 ± 0.6), roasted beef tenderloin excluding fat (943 ± 0.5), well done baked lean ground beef (942 ± 0.5), roasted beef rib roasted excluding fat (942 ± 0.5), braised beef flank including fat (942 ± 0.5), fried beef sirloin including fat (940 ± 0.5), raw pheasant without skin (939 ± 0.5), roasted soybean flour (938 ± 0.6), broiled beef sirloin including fat (938 ± 0.5), roasted turkey light meat without skin (933 ± 0.5), soybean flour (931 ± 0.6), roasted pumpkin and squash seeds (929 ± 1.8), braised beef chuck arm pot roast including fat (927 ± 0.5), well done baked ground beef (917 ± 0.5), roasted eye of beef round including fat (916 ± 0.5), american cheese (914 ± 1.8), pimento cheese spread (914 ± 1.8), well done broiled extra lean ground beef (910 ± 0.5), roasted beef round tip including fat (906 ± 0.5), fried light meat chicken with skin (904 ± 0.5), braised beef spleen (900 ± 0.5), stewed chicken breast without skin (900 ± 0.5), well done broiled lean ground beef (898 ± 0.5), roasted chicken breast with skin (897 ± 0.5), fried dark meat chicken without skin (897 ± 0.5), stewed light meat chicken without skin (896 ± 0.5), roasted cured pork arm excluding fat (894 ± 0.5), roasted turkey dark meat without skin (893 ± 0.5), well done fried extra lean ground beef (891 ± 0.5), broiled beef tenderloin including fat (889 ± 0.5), broiled beef top loin including fat (881 ± 0.5), well done fried lean ground beef (879 ± 0.5), cooked bacon (877 ± 0.4), cheddar cheese (874 ± 0.5), broiled beef round including fat (874 ± 0.5), fried pork loin blade including fat (873 ± 0.5), monterey cheese (871 ± 1.8), canned tuna in water (871 ± 0.6), braised beef chuck blade roast including fat (871 ± 0.5), broiled beef flank excluding fat (870 ± 0.5), grilled canadian bacon (870 ± 1.1), broiled beef rib eye including fat (869 ± 0.5), well done broiled ground beef (866 ± 0.5), canned corned beef (863 ± 0.5), well done fried ground beef (860 ± 0.5), roasted turkey light meat with skin (859 ± 0.5), broiled porterhouse steak including fat (859 ± 0.5), roasted light meat chicken with skin (858 ± 0.5), dry buttermilk (857 ± 7.1), broiled beef flank including fat (857 ± 0.5), chipped beef (854 ± 1.8), anchovy canned in olive oil (850 ± 2.5), roasted dark meat chicken without skin (849 ± 0.5), broiled beef rib excluding fat (844 ± 0.5), colby cheese (843 ± 1.8), muenster cheese (839 ± 1.8), roasted beef tenderloin including fat (838 ± 0.5), roasted canned lean ham (836 ± 0.5), yellow mustard seed (833 ± 16.7), cheshire cheese (832 ± 1.8), brick cheese (832 ± 1.8), creamy peanut butter (831 ± 3.1), roasted turkey dark meat with skin (828 ± 0.5), raw quail without skin (825 ± 0.5), stewed chicken breast with skin (824 ± 0.5), broiled T-bone steak including fat (821 ± 0.5), dried butternuts (818 ± 1.8), fried chicken back with skin (815 ± 0.7), roasted canned ham (811 ± 0.5), medium-rare broiled extra lean ground beef (809 ± 0.5), cooked coho salmon (806 ± 0.6), fried chicken drumstick with skin (806 ± 1), stewed dark meat chicken without skin (806 ± 0.5), roasted chicken thigh without skin (806 ± 1), cooked sockeye salmon (804 ± 0.6), fried dark meat chicken with skin (804 ± 0.5), roasted chicken drumstick with skin (802 ± 1), medium-rare fried extra lean ground beef (795 ± 0.5), oil roasted peanuts (793 ± 1.8), simmered beef heart (792 ± 0.5), fried chicken thigh with skin (790 ± 0.8), medium-rare broiled lean ground beef (787 ± 0.5), braised beef brisket including fat (787 ± 0.5), dried watermelon seeds (786 ± 1.8), milk chocolate (786 ± 1.8), cooked halibut (786 ± 0.6), canned tuna in oil (781 ± 0.6), medium-rare baked extra lean ground beef (779 ± 0.5), roasted beef rib including fat (778 ± 0.5), cooked rainbow trout (775 ± 0.6), cooked snapper (774 ± 0.6), stewed light meat chicken with skin (774 ± 0.5), medium-rare fried lean ground beef (772 ± 0.5), bleu cheese (768 ± 1.8), cooked sheepshead (766 ± 0.6), medium-rare broiled ground beef (766 ± 0.5), medium-rare baked lean ground beef (762 ± 0.5), medium-rare fried ground beef (762 ± 0.5), simmered chicken neck without skin (761 ± 2.8), fried chicken giblets (759 ± 0.5), roasted dark meat chicken with skin (759 ± 0.5), dried peanuts (757 ± 1.8), stewed chicken drumstick with skin (753 ± 0.9), roasted chicken wing with skin (750 ± 1.5), roasted lean ham (750 ± 0.5), cooked swordfish (747 ± 0.6), unheated canadian bacon (746 ± 0.9), smoked haddock (744 ± 0.6), mozzarella cheese (739 ± 1.8), roasted chicken thigh with skin (739 ± 0.8), braised beef shortribs including fat (738 ± 0.5), medium-rare baked ground beef (733 ± 0.5), fried beef liver (731 ± 0.5), cooked grouper (731 ± 0.6), cooked mullet (731 ± 0.6), canned lean ham (731 ± 0.5), fried chicken wing with skin (728 ± 1.6), cooked pike (727 ± 0.6), brie cheese (725 ± 1.8), sardines (725 ± 2.1), kippered herring (725 ± 1.3), cooked tilefish (721 ± 0.6), tilsit cheese (714 ± 1.8), cooked haddock (714 ± 0.6), dry whole milk (713 ± 1.6), cooked flatfish (711 ± 0.6), roasted ham (711 ± 0.5), braised pork liver (708 ± 0.5), cooked rockfish (708 ± 0.6), oil roasted spanish peanuts (707 ± 1.8), cooked perch (704 ± 0.6), simmered turkey heart (702 ± 0.5), cooked mackerel (702 ± 0.6), roasted goose with skin (700 ± 0.5), turkey breast meat summer sausage (700 ± 2.4), hard pork and beef salami (700 ± 5), pork link sausage (699 ± 0.7), cooked pompano (696 ± 0.6), cooked eel (696 ± 0.6), cooked sea bass (696 ± 0.6), chopped ham lunch meat (695 ± 2.4), port du salut cheese (693 ± 1.8),

HISTIDINE

simmered chicken heart (693 ± 0.5), camembert cheese (693 ± 1.8), unheated lean ham (693 ± 0.5), cooked whiting (691 ± 0.6), dried pumpkin and squash seeds (689 ± 1.8), dried black walnuts (689 ± 1.8), smoked whitefish (689 ± 0.6), broiled beef rib including fat (689 ± 0.5), stewed dark meat chicken with skin (687 ± 0.5), oil roasted valencia peanuts (686 ± 1.8), ham and cheese roll (686 ± 1.8), stewed chicken thigh with skin (685 ± 0.7), roasted cured pork arm including fat (683 ± 0.5), raw yellowtail (681 ± 0.6), cooked herring (678 ± 0.6), dried sesame kernels (675 ± 6.3), cooked carp (673 ± 0.6), braised pork spleen (672 ± 0.5), cooked cod (672 ± 0.6), canned cod (671 ± 0.6), canned ham (671 ± 0.5), braised beef liver (667 ± 0.5), simmered beef kidneys (665 ± 0.5), cooked rainbow smelt (665 ± 0.6), pork luxury loaf (664 ± 1.8), fried chicken neck with skin (658 ± 1.4), pork and beef peppered loaf (657 ± 1.8), oil roasted virginia peanuts (654 ± 1.8), caviar (650 ± 3.1), simmered chicken liver (647 ± 0.5), stewed chicken wing with skin (640 ± 1.3), dried sunflower seeds (639 ± 1.8), simmered turkey liver (637 ± 0.5), raw coho salmon (636 ± 0.6), unheated ham (629 ± 0.5), raw sockeye salmon (627 ± 0.6), simmered turkey giblets (625 ± 0.5), roasted duck without skin (620 ± 0.5), braised beef lungs (620 ± 0.5), roasted pork blade roll (619 ± 0.5), raw shark (618 ± 0.6), pork and beef honey loaf (618 ± 1.8), pork and beef pepperoni (617 ± 8.3), raw halibut (613 ± 0.6), raw trout (611 ± 0.6), braised pork kidneys (610 ± 0.5), hard pork salami (610 ± 5), cooked sturgeon (609 ± 0.6), crunchy peanut butter (609 ± 1.6), jellied corned beef loaf (607 ± 1.8), braised pork tongue (605 ± 0.5), raw rainbow trout (605 ± 0.6), minced ham lunch meat (605 ± 2.4), raw milkfish (604 ± 0.6), raw snapper (604 ± 0.6), canned sockeye salmon (604 ± 0.6), corn germ (604 ± 0.5), simmered chicken giblets (602 ± 0.5), braised pork heart (600 ± 0.4), oil roasted sunflower seeds (600 ± 1.8), dry roasted peanuts (600 ± 1.8), raw anchovy (599 ± 0.6), raw sheepshead (595 ± 0.6), simmered turkey gizzard (593 ± 0.5), raw chum salmon (593 ± 0.6), chopped smoked beef (593 ± 1.8), raw chinook salmon (591 ± 0.6), raw bluefish (591 ± 0.6), turkey ham (588 ± 0.9), raw pink salmon (587 ± 0.6), canned pink salmon (587 ± 0.6), berliner (587 ± 2.2), limburger cheese (586 ± 1.8), raw swordfish (584 ± 0.6), dried pignolia pine nuts (582 ± 1.8), oil roasted almonds (579 ± 1.8), toasted almonds (579 ± 1.8), italian pork sausage (578 ± 0.7), cooked corned beef brisket (578 ± 0.5), simmered beef tongue (573 ± 0.5), raw pollock (572 ± 0.6), raw sunfish (571 ± 0.6), raw grouper (571 ± 0.6), raw mullet (571 ± 0.6), raw burbot (569 ± 0.6), raw pike (567 ± 0.6), pork sausage (567 ± 1.9), dried almonds (564 ± 1.8), raw whitefish (562 ± 0.6), lebanon bologna (561 ± 2.2), raw cusk (559 ± 0.6), raw cisco (559 ± 0.6), raw ling (559 ± 0.6), beef pastrami (557 ± 1.8), raw haddock (556 ± 0.6), raw scup (556 ± 0.6), raw freshwater bass (555 ± 0.6), raw flatfish (555 ± 0.6), braised pork pancreas (552 ± 0.5), raw rockfish (552 ± 0.6), raw perch (548 ± 0.6), raw mackerel (548 ± 0.6), simmered chicken gizzard (547 ± 0.5), turkey pastrami (546 ± 0.9), raw spot (545 ± 0.6), raw dolphinfish (545 ± 0.6), raw pompano (544 ± 0.6), raw eel (544 ± 0.6), raw sea bass (544 ± 0.6), dried pistachio nuts (543 ± 1.8), dry roasted sunflower seeds (543 ± 1.8), raw whiting (539 ± 0.6), battered and fried shark (538 ± 0.6), smoked chinook salmon (538 ± 0.6), raw catfish (535 ± 0.6), braised beef pancreas (533 ± 0.5), breaded and fried catfish (531 ± 0.6), raw herring (529 ± 0.6), breaded and fried croaker (525 ± 0.6), raw carp (525 ± 0.6), raw cod (524 ± 0.6), raw croaker (524 ± 0.6), whole dried sesame seeds (522 ± 5.6), raw striped bass (522 ± 0.6), raw lingcod (520 ± 0.6), pork and beef mortadella (520 ± 3.3), raw rainbow smelt (519 ± 0.6), simmered chicken neck with skin (516 ± 1.3), raw freshwater drum (516 ± 0.6), raw tilefish (515 ± 0.6), raw wolffish (515 ± 0.6), natto (513 ± 0.6), raw butterfish (509 ± 0.6), toasted sesame kernels (507 ± 1.8), whole toasted sesame seeds (507 ± 1.8), beef lunch meat (504 ± 1.8), poppy seed (500 ± 16.7), sesame butter (500 ± 3.3), fried tofu (500 ± 3.8), summer sausage (500 ± 2.2), raw duck liver (498 ± 0.5), tempeh (498 ± 0.6), raw shad (498 ± 0.6), pork and beef luncheon sausage (496 ± 2.2), raw white sucker (493 ± 0.6), raw seatrout (493 ± 0.6), cooked clams (491 ± 0.6), raw whelk (488 ± 0.6), cooked crayfish (486 ± 0.6), pork bologna (483 ± 2.2), smoked cisco (482 ± 0.6), grilled ham patties (482 ± 0.8), beef salami (478 ± 2.2), raw sturgeon (475 ± 0.6), raw turbot (473 ± 0.6), pork and beef link sausage with american cheese (472 ± 1.2), unheated ham patties (468 ± 0.8), tuna salad (467 ± 0.2), cashew butter (464 ± 1.8), dry roasted almonds (464 ± 1.8), roasted duck with skin (462 ± 0.5), firm raw tofu (459 ± 0.4), ricotta cheese (459 ± 0.4), cooked mussels (456 ± 0.6), pork liverwurst (450 ± 2.8), polish pork sausage (450 ± 1.8), boiled soybeans (449 ± 0.3), smoked beef sausage (449 ± 1.2), boiled lupins (443 ± 0.3), pork and beef picnic loaf (443 ± 1.8), breaded and fried shrimp (436 ± 0.6), pork and beef brotwurst (436 ± 0.7), raw goose liver (435 ± 0.5), raw monkfish (426 ± 0.6), oil roasted cashews (425 ± 1.8), cooked shrimp (425 ± 0.6), almond butter (425 ± 3.1), braised pork lungs (420 ± 0.5), pickled herring (420 ± 3.3), pork and beef old fashioned loaf (418 ± 1.8), pork beerwurst salami (417 ± 2.2), canned blue crab (416 ± 0.6), cooked lobster (416 ± 0.6), cottage cheese (415 ± 0.2), pork and beef sausage (415 ± 3.8), crab cakes (413 ± 0.8), raw shrimp (413 ± 0.6), cooked blue crab (411 ± 0.6), pork and beef lunch meat (411 ± 1.8), dried brazil nuts (407 ± 1.8), pork bratwurst (406 ± 0.6), dry roasted cashews (404 ± 1.8), basil (400 ± 50), beef beerwurst salami (396 ± 2.2), pork and beef link sausage (396 ± 0.7), chicken egg yolk (394 ± 2.9), dry roasted pistachio nuts (393 ± 1.8), cooked alaska king crab (393 ± 0.6), dried hickory nuts (393 ± 1.8), pork livercheese (392 ± 1.3), beef bologna (391 ± 2.2), puffed wheat (386 ± 3.6), braised beef thymus (385 ± 0.5), raw lobster (382 ± 0.6), beef frankfurter (382 ± 0.9), raw crayfish (379 ± 0.6), fried abalone (378 ± 0.6), raw queen crab (376 ± 0.6), freeze-dried chives (375 ± 62.5), freeze-dried red sweet peppers (375 ± 31.3), freeze-dried green sweet peppers (375 ± 31.3), raw alaska king crab (372 ± 0.6), raw blue crab (367 ± 0.6), falafel (365 ± 1), oil roasted filberts (364 ± 1.8), dried english/persian walnuts (364 ± 1.8), raw beef tripe (363 ± 0.5), canned pork lunch meat (362 ± 2.4), beef and pork salami (361 ± 2.2), neufchatel cheese (361 ± 1.8), pork and beef knockwurst (360 ± 0.7), breaded and fried scallops (355 ± 1.6), raw dungeness crab (354 ± 0.6), ham salad (354 ± 1.8), stir-fried sprouted soybeans (352 ± 0.5), beef and pork frankfurter (351 ± 1.1), dill seed (350 ± 25), fennel seed (350 ± 25), breaded fried squid (348 ± 0.6), bran (346 ± 1.8), pork pickle and pimento loaf (346 ± 1.8), boiled peanuts (341 ± 1.6), almond paste (336 ± 1.8), dried filberts (332 ± 1.8), boiled green soybeans (332 ± 0.6), miso (329 ± 0.4), raw abalone (328 ± 0.6), braised pork brains (326 ± 0.5), raw scallops (322 ± 0.6), pork liver sausage (322 ± 2.8), duck egg (320 ± 0.7), fried beef brains (320 ± 0.5), beef and pork bologna (317 ± 2.2), pork and beef kielbasa/kolbassy (315 ± 1.9), raw cuttlefish (312 ± 0.6), simmered pork tail (306 ± 0.5), pork headcheese (300 ± 1.8), garlic powder (300 ± 16.7), raw squid (299 ± 0.6), gjetost cheese (296 ± 1.8), pork olive loaf (296 ± 1.8), chicken egg (294 ± 1), surimi scallops (294 ± 0.6), bran flakes (293 ± 1.8), raw octopus (286 ± 0.6), surimi shrimp (285 ± 0.6), dried pinyon pine nuts (282 ± 1.8), simmered beef brains (282 ± 0.5), breaded and fried clams (280 ± 0.6), vienna sausage (275 ± 3.1), cream cheese (275 ± 1.8), frozen and boiled black-eyed peas (274 ± 0.6), cooked oysters (271 ± 0.6), boiled white beans (271 ± 0.3), gefiltefish with broth (262 ± 1.2), boiled cranberry beans (260 ± 0.3), dried pilinuts (257 ± 1.8), shredded wheat (257 ± 1.8), boiled yardlong bean (257 ± 0.3), boiled yellow beans (255 ± 0.3), dry roasted filberts (254 ± 1.8), boiled lentils (254 ± 0.3), boiled pink beans (252 ± 0.3), stir-fried sprouted lentils (252 ± 0.5), boiled catjang black-eyed peas (252 ± 0.3), raw bacon (249 ± 0.7), boiled black beans (247 ± 0.3), raw pork stomach (246 ± 0.5), dried ginko nuts (246 ± 1.8), boiled baby lima beans (246 ± 0.3), raw clams (245 ± 0.6), boiled chickpeas (244 ± 0.3), chicken egg white (242 ± 1.5), boiled navy beans (242 ± 0.3), boiled kidney beans (242 ± 0.3), boiled winged beans (241 ± 0.3), boiled pigeon peas (241 ± 0.3), boiled black-eyed peas (240 ± 0.3), boiled lima beans (238 ± 0.3), dry roasted pecans (236 ± 1.8), raw tofu (235 ± 0.4), boiled hyacinth beans (233 ± 0.3), boiled great northern beans (232 ± 0.3), dried acorns (229 ± 1.8), boiled pinto beans (229 ± 0.3), dried pecans (229 ± 1.8), raw mussels (228 ± 0.6), steamed sprouted soybeans (228 ± 1.1), boiled black turtle beans (228 ± 0.3), boiled and frozen baby lima beans (226 ± 0.6), wheat flakes (221 ± 1.8), corn flakes (221 ± 1.8), boiled sprouted green peas (217 ± 0.5), sweetened condensed milk (216 ± 1.3), simmered pork chitterlings (215 ± 0.5), simmered pork feet (211 ± 0.5), acorn flour (211 ± 1.8), boiled mungo beans (211 ± 0.3), dry roasted macadamia nuts (211 ± 1.8), canned navy beans (209 ± 0.2), raisin bran (207 ± 1.8), oil roasted pecans (207 ± 1.8), frozen and boiled fordhook lima beans (206 ± 0.6), canned great northern beans (205 ± 0.2), boiled mung beans (205 ± 0.2), tomato powder (204 ± 0.5), evaporated skim milk (203 ± 1.6), boiled split peas (203 ± 0.3), canned white beans (202 ± 0.2), freeze-dried shallots (200 ± 12.5), dried chinese chestnuts (200 ± 1.8), boiled adzuki beans (198 ± 0.2), boiled french beans (196 ± 0.3), puffed rice (193 ± 3.6), boiled broad beans (193 ± 0.3), potato flour (192 ± 0.6), simmered pork ears (189 ± 0.5), evaporated whole milk (185 ± 0.4), dried european chestnuts (179 ± 1.8), breaded and fried oysters (175 ± 0.6), refried beans (173 ± 0.2), raw acorns (171 ± 1.8), dried macadamia nuts (171 ± 1.8), canned black turtle beans (168 ± 0.2), sheep milk (167 ± 0.2), dried coconut (161 ± 1.8), boiled succotash (161 ± 0.5), dried shitake mushrooms (160 ± 3.3), canned cranberry beans (154 ± 0.2), au gratin potatoes (151 ± 0.4), canned lima beans (151 ± 0.2), ginger (150 ± 25), oil roasted macadamia nuts (150 ± 1.8), onion powder (150 ± 25), pickled pork feet (149 ± 0.5), dried oriental radish (148 ± 0.9), baked beans with pork and sweet sauce (148 ± 0.2), canned black-eyed peas (147 ± 0.2), evaporated lowfat milk (146 ± 0.4), canned kidney beans (145 ± 0.2), baked beans with pork and tomato sauce (144 ± 0.2), dehydrated onion flakes (143 ± 3.6), skim yogurt (142 ± 0.2), raw laver/nori seaweed (140 ± 0.5), potato pancakes (139 ± 0.7), canned broad beans (139 ± 0.2), raw oysters (136 ± 0.6), canned oysters (136 ± 0.6), frozen and boiled succotash (136 ± 0.6), canned chickpeas (136 ± 0.2), vegetarian baked beans (133 ± 0.2), dried japanese chestnuts (132 ± 1.8), roasted chinese chestnuts (132 ± 1.8), raw dock (131 ± 0.7), lowfat yogurt (130 ± 0.2), canned pinto beans (127 ± 0.2), dried toasted coconut (125 ± 1.8), hummus (125 ± 0.2), raw chinese chestnuts (121 ± 1.8), yellow corn pudding (115 ± 0.2), raw spirulina seaweed (112 ± 0.5), raw garlic (111 ± 5.6), stir-fried sprouted mung beans (110 ± 0.8), protein fortified skim milk (107 ± 0.2), protein fortified lowfat 2% milk (107 ± 0.2), protein fortified lowfat 1% milk (107 ± 0.2), raw green peas (106 ± 0.6), restaurant vanilla milkshake (105 ± 0.2), boiled green peas (105 ± 0.6), ginko nuts (104 ± 1.8), malted milk (103 ± 0.2), frozen and boiled green peas (101 ± 0.6), hot chocolate (99 ± 0.2), noodles (98 ± 0.3), buttermilk (95 ± 0.2), vanilla milkshake (95 ± 0.2), eggnog (94 ± 0.2), skim milk (93 ± 0.2), retaurant chocolate milkshake (93 ± 0.2), okara tofu (93 ± 0.8), restaurant strawberry milkshake (92 ± 0.2), boiled yellow corn (91 ± 0.6), lowfat 2% milk (90 ± 0.2), filled milk (90 ± 0.2), goat milk (89 ± 0.2), lowfat 1% milk (89 ± 0.2), whole milk (89 ± 0.2), roasted european chestnuts (89 ± 1.8), lowfat 1% chocolate milk (88 ± 0.2), boiled lamb's-quarters (88 ± 0.6), lowfat 2% chocolate milk (87 ± 0.2), canned green peas (87 ± 0.6), whole yogurt (86 ± 0.2), whole chocolate milk (86 ± 0.2), boiled jute potherb (86 ± 1.2), boiled and steamed chinese chestnuts (86 ± 1.8), carob milk (85 ± 0.2), canned succotash with creamed corn (84 ± 0.4), coconut cream (83 ± 0.2), chocolate milkshake (83 ± 0.2), frozen and boiled yellow corn (83 ± 0.6), canned succotash (83 ± 0.4), frozen and boiled brussels sprouts (82 ± 0.6), bananas (81 ± 0.4), frozen and boiled turnip greens (80 ± 0.6), half and half cream (80 ± 3.3), macaroni (80 ± 0.4), spaghetti (80 ± 0.4), raw coconut (78 ± 1.1), indian buffalo milk (77 ± 0.2), dried sweetened flaked coconut (76 ± 0.7), lard (75 ± 1.8), roasted japanese chestnuts (75 ± 1.8), light cream (73 ± 3.3), raw pork jowl (72 ± 0.5), soy milk (71 ± 0.2), scalloped potatoes (70 ± 0.4), frozen and boiled spinach (69 ± 0.5), raw sprouted mung beans (69 ± 1), french fried potatoes (68 ± 1), raw european chestnuts (68 ± 1.8), medium cream (67 ± 3.3), dried sweetened shredded coconut (66 ± 0.5), boiled spinach (66 ± 0.6), raw spinach (64 ± 1.8), canned yellow corn with red and green peppers (63 ± 0.4), potato salad (62 ± 0.4), sweetened coconut cream (62 ± 0.2), canned spinach (62 ± 0.5), frozen and boiled kale (60 ± 0.8), whipping cream, light (60 ± 3.3), dried apricots (60 ± 1.4), boiled mushrooms (60 ± 0.6), tomato paste (60 ± 0.4), frozen and fried french fried potatoes (58 ± 1), oatmeal (58 ± 0.3), raw japanese chestnuts (57 ± 1.8), boiled and steamed european chestnuts (57 ± 1.8), raw mushrooms (57 ± 1.4), frozen and boiled collards (56 ± 0.6), canned yellow corn (56 ± 0.5), boiled brussels sprouts (56 ± 0.6), frozen and boiled broccoli (55 ± 0.5), microwaved potato with skin (54 ± 0.2), o'brien potatoes (54 ± 0.3), frozen and fried hash brown potatoes (53 ± 0.6), whipping cream, heavy (53 ± 3.3), coconut milk (53 ± 0.2), boiled broccoli (53 ± 0.6), cooked sprouted mung beans (52 ± 0.8), baked potato with skin (50 ± 0.2), raw broccoli (50 ± 1.1), boiled dock (49 ± 0.5), canned creamed yellow corn (48 ± 0.4), frozen and boiled asparagus (47 ± 0.8), raw swamp cabbage (46 ± 0.9), microwaved potato without skin (46 ± 0.3), dried longans (45 ± 0.5), frozen and boiled turnip greens and turnips (45 ± 0.5), raw potato without skin (45 ± 0.4), raw cassava (45 ± 0.5), boiled amaranth (44 ± 0.8), mashed potatoes (44 ± 0.5), dried figs (43 ± 0.3), frozen and boiled potato without skin (43 ± 0.5), baked potato without skin (43 ± 0.3), raw bamboo shoots (42 ± 0.7), sapote (42 ± 0.2), boiled burdock root (42 ± 0.4), raw watercress (41 ± 2.9), hash brown potatoes (41 ± 0.6), frozen and boiled mustard greens (41 ± 0.7), soybean oil and cottonseed oil margarine liquid (40 ± 10), raw shallots (40 ± 5), boiled kale (40 ± 0.8), boiled scotch kale (40 ± 0.8), raw cauliflower (40 ± 1), boiled asparagus (40 ± 0.6), boiled mustard greens (40 ± 0.7), raw savoy cabbage (40 ± 1.4), raw red chili peppers (40 ± 1.1), raw green chili peppers (40 ± 1.1), boiled cauliflower (39 ± 0.8), boiled beet greens (39 ± 0.7), corn grits (39 ± 0.2), raw vine spinach (39 ± 0.5), cooked plantain (39 ± 0.3), boiled swiss chard (38 ± 0.6), boiled potato without skin (38 ± 0.4), steamed hawaii mountain yam (38 ± 0.7), raw lotus root (38 ± 0.6), boiled swamp cabbage (37 ± 1), boiled savoy cabbage (37 ± 0.7), raw turnip greens (36 ± 1.8), boiled green beans (35 ± 0.8), boiled yellow snap beans (35 ± 0.8), cream of wheat (35 ± 0.3), canned sprouted mung beans (35 ± 0.8), raw chives (33 ± 16.7), frozen and boiled okra (33 ± 0.5), canned turnip greens (33 ± 0.4), raw welsh onions (33 ± 0.5), frozen and boiled cauliflower (33 ± 0.6), frozen and baked sweet potato (32 ± 0.6), baked sweet potato (32 ± 0.4), farina (32 ± 0.3), baked/boiled tropical yam (32 ± 0.7), canned sweet potato (31 ± 0.3), boiled sweet potato (31 ± 0.3), canned potato without skin (31 ± 0.6), boiled yambean (31 ± 0.5), raw california avocado (30 ± 0.3), raw spring onions with tops (30 ± 1), dried dates (30 ± 0.6), chocolate syrup (29 ± 1.3), raw romaine lettuce (29 ± 1.8), boiled okra (29 ± 0.6), raw chickory greens (29 ± 0.6), raw ginger root (29 ± 2.1), safflower oil and soybean oil mayonnaise (29 ± 3.6), soybean oil mayonnaise (29 ± 3.6), raw red cabbage (29 ± 1.4), boiled turnip greens (28 ± 0.7), whole chocolate malted milk (28 ± 0.2), canned bamboo shoots (28 ± 0.4), baked hubbard squash (28 ± 0.5), boiled hubbard squash (28 ± 0.4), coleslaw (28 ± 0.8), frozen and boiled crookneck squash (28 ± 0.5), boiled chinese cabbage (27 ± 0.6), raw leeks (27 ± 1.9), whipped butter (27 ± 4.5), apricots (27 ± 0.5), boiled rutabaga (27 ± 0.6), dried japanese persimmon (26 ± 1.5), frozen and boiled yellow snap beans (26 ± 0.7), frozen and boiled green beans (26 ± 0.7), raw chinese cabbage (26 ± 1.4), cream of rice (26 ± 0.3), raw scallop squash (26 ± 0.8), raw green cabbage (26 ± 1.4), tomato purée (26 ± 0.2), boiled bamboo shoots (25 ± 0.4), raw zucchini (25 ± 0.8), frozen and boiled zucchini (25 ± 0.4), raw kelp/kombu/tangle seaweed (24 ± 0.5), raw endive (24 ± 2), raw eggplant (24 ± 1.2), frozen and boiled turnips (24 ± 0.5), european grapes (24 ± 0.3), raw florida avocado (23 ± 0.2), human milk (23 ± 1.6), frozen and boiled butternut squash (23 ± 0.4), rice polish (23 ± 0.9), american grapes (23 ± 0.5), boiled purslane (22 ± 0.9), cooked shitake mushrooms (22 ± 0.7), canned green beans (22 ± 0.7), canned yellow snap beans (22 ± 0.7), boiled scallop squash (22 ± 0.6), canned zucchini in tomato juice (22 ± 0.4), boiled lotus root (22 ± 0.6), raw looseleaf lettuce (21 ± 1.8), beef suet (21 ± 1.8), boiled collards (21 ± 0.3), boiled and steamed japanese chestnuts (21 ± 1.8), baked acorn squash (21 ± 0.5), boiled red cabbage (21 ± 0.7), raw butterhead lettuce (20 ± 3.3), butter (20 ± 3.3), corn oil margarine stick (20 ± 10), safflower oil and soybean oil margarine stick (20 ± 10), soybean oil margarine stick (20 ± 10), soybean oil and cottonseed oil margarine stick (20 ± 10), raw iceberg lettuce (20 ± 2.5), boiled kohlrabi (20 ± 0.6), valencia orange (20 ± 0.4), raw crookneck squash (20 ± 0.8), boiled crookneck squash (20 ± 0.6), corn oil margarine tub (20 ± 10), safflower oil margarine tub (20 ± 10), soybean oil margarine tub (20 ± 10), soybean oil and cottonseed oil margarine tub (20 ± 10), corn oil diet margarine tub (20 ± 10), soybean oil and cottonseed oil diet margarine tub (20 ± 10), boiled eggplant (19 ± 1), navel orange (19 ± 0.4), boiled green cabbage (19 = 0.7), frozen and boiled red sweet peppers (19 ± 0.5), frozen and boiled green sweet peppers (19 ± 0.5), raw onions (19 ± 0.6), frozen and boiled carrots (18 ± 0.7), raw chickory (18 ± 1.1), canned red chili peppers (18 ± 0.7), canned green chili peppers (18 ± 0.7), raw red sweet peppers (18 ± 1), raw green sweet peppers (18 ± 1), raw green tomato (18 ± 0.4), candied sweet potato (17 ± 0.5), boiled carrots (17 ± 0.6), coconut water (17 ± 0.2), raw carrots (17 ± 0.7), boiled butternut squash (17 ± 0.5), canned pumpkin (17 ± 0.4), canned pumpkin pie mix (17 ± 0.4), boiled red tomato (17 ± 0.4), persimmon (16 ± 2), canned jalapeño peppers (16 ± 0.7), canned red sweet peppers (16 ± 0.7), canned green sweet peppers (16 ± 0.7), sapodilla (16 ± 0.3), raw wakame seaweed (15 ± 0.5), whey (15 ± 0.2), raw celtuce (15 ± 0.5), elderberries (15 ± 0.3), boiled leeks (15 ± 1.9), boiled onions (15 ± 0.5), boiled beets (14 ± 0.6), boiled oriental radish (14 ± 0.7), boiled zucchini (14 ± 0.6), canned peeled red tomatoes (14 ± 0.4), canned stewed red tomatoes (14 ± 0.4), plum (14 ± 0.8), thompson seedless grapes (14 ± 0.3), peach (13 ± 0.6), raw radish (13 ± 1.1), boiled acorn squash (13 ± 0.4), canned crookneck squash (13 ± 0.5), boiled red sweet peppers (13 ± 0.7), boiled green sweet peppers (13 ± 0.7), raw red tomato (13 ± 0.4), pear (13 ± 0.3), longans (12 ± 0.5), japanese persimmon (12 ± 0.3), figs (12 ± 1), cooked taro (12 ± 0.8), tangerine (12 ± 0.6), boiled/baked spaghetti squash (12 ± 0.6), mango (12 ± 0.2), boiled turnips (12 ± 0.6), strawberries (12 ± 0.3), frozen and boiled onions (12 ± 0.5), tomato juice (12 ± 0.3), boiled chayote (11 ± 0.6), pickled beets (11 ± 0.4), raw oriental radish (11 ± 1.1), mandarin oranges (11 ± 0.4), boiled pumpkin (11 ± 0.4), blueberries (10 ± 0.3), canned carrots (10 ± 0.7), raw celery (10 ± 1.3), pineapple (9 ± 0.3), boiled celery (8 ± 0.7), raw cucumber (8 ± 1), guava (7 ± 0.6), loquats (7 ± 0.5), grape juice (7 ± 0.2), crab apples (6 ± 0.5), watermelon (6 ± 0.3), papaya (5 ± 0.2), ale (5 ± 0.1), carambola (4 ± 0.4), boiled calabash gourd (4 ± 0.7), cooked apple without skin (4 ± 0.3), orange juice (3 ± 0.2), raw apple with skin (3 ± 0.4), applesauce (3 ± 0.4), screwdriver (2 ± 0.2), tangerine juice (2 ± 0.2), raw apple without skin (2 ± 0.4), coffee (2 ± 0.3), white grapefruit juice (2 ± 0.2), internal synthesis in adults

Applications: arthritis, tissue overloads of copper, iron, or other heavy metals, over-stimulated schizophrenics, allergies, cardiocirculatory conditions, anemia, auditory dysfunction, radiation

Daily Dosage: Nutritional — 0.3–1 g; therapeutic — 3; experimental — 6; toxic — none known up to 6 g

Warnings: Suicidally depressed schizophrenics can be worsened with histidine. Other depressions may also be worsened. Increased amounts are dangerous. See toxicity symptoms.

HYDROXYPROLINE

Names: hydroxyproline
Classification: nonessential amino acid
Deficiency: protein decreased
Side-effects: none known
Toxicity: none known
Inhibitors: increased amounts of other amino acids

Helpers: moderate amounts of other amino acids
Sources: internal synthesis
Applications: unknown
Daily Dosage: unknown
Warnings: none

ISOLEUCINE

Names: isoleucine, ISL, ISO, ILE, amino acid I, 2-amino-3-methyl-pentanoic acid
Forms: 2-amino-3-methyl pentanoic acid
Classification: essential amino acid
Deficiency: protein decreased, decreased hemoglobin formation
Side-effects: none known
Toxicity: none known
Inhibitors: increased amounts of other amino acids
Helpers: moderate amounts of other amino acids
Sources (mg per 100 grams of food): torula yeast (9500 ± 1.8), soybean protein isolate (4254 ± 1.8), dried spirulina seaweed (3209 ± 0.5), soybean protein concentrate (2943 ± 1.8), dried and salted cod (2895 ± 0.6), dried and frozen tofu (2376 ± 2.9), fenugreek seed (2300 ± 25), defatted soybean flour (2281 ± 0.5), low fat soybean flour (2257 ± 0.6), parmesan cheese (2200 ± 10), dry skim milk (2187 ± 1.7), defatted soy meal (2180 ± 0.4), dry instant skim milk (2123 ± 0.5), peanut flour (2025 ± 12.5), dry buttermilk (1929 ± 7.1), dry roasted soybean nuts (1920 ± 0.6), toasted soybean nuts (1911 ± 1.8), fried chicken breast without skin (1765 ± 0.6), almond meal (1739 ± 1.8), fried light meat chicken without skin (1732 ± 0.5), roasted pumpkin and squash seeds (1721 ± 1.8), roasted soybean nuts (1709 ± 0.6), roasted soybean flour (1688 ± 0.6), braised pork center loin excluding fat (1680 ± 0.5), soybean flour (1675 ± 0.6), braised pork top loin excluding fat (1664 ± 0.5), stewed chicken drumstick without skin (1659 ± 1.1), cooked whelk (1655 ± 0.6), roasted chicken breast without skin (1638 ± 0.6), fried chicken breast with skin (1632 ± 0.5), roasted light meat chicken without skin (1632 ± 0.5), gruyere cheese (1632 ± 1.8), fried chicken giblets (1630 ± 0.5), braised pork sirloin excluding fat (1619 ± 0.5), braised pork loin excluding fat (1594 ± 0.5), dry whole milk (1594 ± 1.6), fried beef top round excluding fat (1576 ± 0.5), swiss cheese (1557 ± 1.8), roasted turkey light meat without skin (1555 ± 0.5), broiled pork center loin excluding fat (1547 ± 0.5), cheddar cheese (1546 ± 0.5), monterey cheese (1539 ± 1.8), fried light meat chicken with skin (1537 ± 0.5), stewed chicken breast without skin (1531 ± 0.5), cooked flounder (1530 ± 0.5), fried dark meat chicken without skin (1528 ± 0.5), roasted chicken breast with skin (1526 ± 0.5), stewed light meat chicken without skin (1525 ± 0.5), simmered beef shank excluding fat (1514 ± 0.5), braised pork boston blade excluding fat (1505 ± 0.5), tilsit cheese (1504 ± 1.8), braised pork pancreas (1496 ± 0.5), colby cheese (1493 ± 1.8), braised beef chuck arm pot roast excluding fat (1485 ± 0.5), roasted turkey dark meat without skin (1485 ± 0.5), cheshire cheese (1468 ± 1.8), port du salut cheese (1464 ± 1.8), fried beef sirloin excluding fat (1460 ± 0.5), roasted light meat chicken with skin (1458 ± 0.5), roasted dark meat chicken without skin (1445 ± 0.5), braised pork loin blade excluding fat (1435 ± 0.5), simmered turkey heart (1435 ± 0.5), roasted turkey light meat with skin (1432 ± 0.5), milk chocolate (1429 ± 1.8), fried beef top round including fat (1425 ± 0.5), broiled beef top round excluding fat (1425 ± 0.5), simmered beef shank including fat (1421 ± 0.5), braised beef bottom round excluding fat (1420 ± 0.5), simmered chicken heart (1415 ± 0.5), roasted pork rump excluding fat (1408 ± 0.5), braised pork spareribs (1404 ± 0.5), stewed chicken breast with skin (1401 ± 0.5), braised beef chuck blade roast excluding fat

(1396 ± 0.5), braised pork center loin including fat (1393 ± 0.5), broiled pork center rib excluding fat (1393 ± 0.5), broiled pork top loin excluding fat (1393 ± 0.5), roasted pork tenderloin (1391 ± 0.5), fried pork center loin excluding fat (1390 ± 0.5), fried chicken back with skin (1390 ± 0.7), simmered turkey gizzard (1389 ± 0.5), broiled beef top round including fat (1386 ± 0.5), braised beef shortribs excluding fat (1383 ± 0.5), roasted turkey dark meat with skin (1379 ± 0.5), roasted pork center loin excluding fat (1377 ± 0.5), fried chicken drumstick with skin (1371 ± 1), stewed dark meat chicken without skin (1371 ± 0.5), braised beef pancreas (1370 ± 0.5), fried dark meat chicken with skin (1370 ± 0.5), roasted chicken thigh without skin (1369 ± 1), roasted pork leg excluding fat (1368 ± 0.5), broiled pork sirloin excluding fat (1365 ± 0.5), broiled beef sirloin excluding fat (1365 ± 0.5), canned tuna in water (1364 ± 0.6), roasted pork center rib excluding fat (1363 ± 0.5), roasted pork shank excluding fat (1363 ± 0.5), roasted pork top loin excluding fat (1363 ± 0.5), roasted chicken drumstick with skin (1362 ± 1), dried watermelon seeds (1361 ± 1.8), braised pork kidneys (1357 ± 0.5), fried pork center rib excluding fat (1351 ± 0.5), fried pork top loin excluding fat (1351 ± 0.5), braised pork center rib including fat (1350 ± 0.5), fried chicken thigh with skin (1347 ± 0.8), broiled pork loin excluding fat (1345 ± 0.5), braised beef bottom round including fat (1340 ± 0.5), simmered turkey giblets (1339 ± 0.5), anchovy canned in olive oil (1330 ± 2.5), roasted pork sirloin excluding fat (1328 ± 0.5), edam cheese (1325 ± 1.8), raw pheasant without skin (1324 ± 0.5), braised pork sirloin including fat (1322 ± 0.5), braised beef brisket excluding fat (1321 ± 0.5), gouda cheese (1321 ± 1.8), braised pork liver (1320 ± 0.5), stewed light meat chicken with skin (1316 ± 0.5), roasted eye of beef round excluding fat (1303 ± 0.5), braised pork top loin including fat (1301 ± 0.5), roasted pork loin excluding fat (1300 ± 0.5), broiled pork center loin including fat (1299 ± 0.5), well done baked extra lean ground beef (1299 ± 0.5), simmered chicken giblets (1297 ± 0.5), simmered chicken liver (1294 ± 0.5), simmered chicken neck without skin (1294 ± 2.8), roasted beef tip excluding fat (1291 ± 0.5), dried sesame kernels (1288 ± 6.3), roasted dark meat chicken with skin (1288 ± 0.5), broiled beef top loin excluding fat (1287 ± 0.5), dried pumpkin and squash seeds (1282 ± 1.8), braised pork loin including fat (1281 ± 0.5), simmered chicken gizzard (1281 ± 0.5), stewed chicken drumstick with skin (1281 ± 0.9), broiled beef round excluding fat (1279 ± 0.5), simmered turkey liver (1273 ± 0.5), roasted chicken wing with skin (1271 ± 1.5), roasted pork rump including fat (1270 ± 0.5), broiled beef tenderloin excluding fat (1270 ± 0.5), well done baked lean ground beef (1269 ± 0.5), broiled porterhouse steak excluding fat (1266 ± 0.5), broiled T-bone steak excluding fat (1265 ± 0.5), braised pork arm braised including fat (1264 ± 0.5), simmered beef heart (1262 ± 0.5), broiled beef rib eye excluding fat (1261 ± 0.5), cooked coho salmon (1261 ± 0.6), braised beef flank excluding fat (1260 ± 0.5), braised pork spleen (1259 ± 0.5), cooked sockeye salmon (1258 ± 0.6), roasted chicken thigh with skin (1256 ± 0.8), braised pork boston blade braised including fat (1247 ± 0.5), roasted beef tenderloin excluding fat (1238 ± 0.5), roasted beef rib roasted excluding fat (1238 ± 0.5), fried chicken wing with skin (1238 ± 1.6), braised beef flank including fat (1237 ± 0.5), cooked bacon (1237 ± 0.4), limburger cheese (1236 ± 1.8), well done baked ground beef (1235 ± 0.5), fried beef sirloin including fat (1234 ± 0.5), broiled beef sirloin including fat (1231 ± 0.5), cooked halibut (1231 ± 0.6), well done broiled extra lean ground beef (1226 ± 0.5), roasted pork shoulder excluding fat (1226 ± 0.5), canned tuna in oil

(1224 ± 0.6), fried beef liver (1222 ± 0.5), braised beef chuck arm pot roast including fat (1218 ± 0.5), broiled pork boston blade excluding fat (1216 ± 0.5), cooked rainbow trout (1214 ± 0.6), cooked snapper (1212 ± 0.6), well done broiled lean ground beef (1209 ± 0.5), roasted pork center loin including fat (1208 ± 0.5), chipped beef (1207 ± 1.8), roasted duck without skin (1206 ± 0.5), broiled pork loin blade excluding fat (1204 ± 0.5), roasted eye of beef round including fat (1203 ± 0.5), yellow mustard seed (1200 ± 16.7), well done fried extra lean ground beef (1200 ± 0.5), cooked sheepshead (1199 ± 0.6), dried butternuts (1196 ± 1.8), roasted pork sirloin including fat (1192 ± 0.5), roasted pork loin blade excluding fat (1192 ± 0.5), roasted beef round tip including fat (1190 ± 0.5), roasted pork leg including fat (1187 ± 0.5), raw quail without skin (1187 ± 0.5), well done fried lean ground beef (1183 ± 0.5), roasted goose with skin (1183 ± 0.5), roasted pork boston blade excluding fat (1177 ± 0.5), fried pork loin blade excluding fat (1175 ± 0.5), roasted pork center rib including fat (1172 ± 0.5), cooked swordfish (1171 ± 0.6), turkey breast meat summer sausage (1171 ± 2.4), canned corned beef (1170 ± 0.5), broiled beef tenderloin including fat (1167 ± 0.5), stewed dark meat chicken with skin (1167 ± 0.5), well done broiled ground beef (1166 ± 0.5), stewed chicken thigh with skin (1165 ± 0.7), smoked haddock (1162 ± 0.6), broiled pork center rib including fat (1161 ± 0.5), muenster cheese (1161 ± 1.8), cooked crayfish (1160 ± 0.6), well done fried ground beef (1158 ± 0.5), broiled beef top loin including fat (1157 ± 0.5), dried sunflower seeds (1154 ± 1.8), brick cheese (1150 ± 1.8), roasted pork shank including fat (1148 ± 0.5), broiled beef round including fat (1148 ± 0.5), cooked grouper (1145 ± 0.6), braised beef chuck blade roast including fat (1144 ± 0.5), cooked mullet (1144 ± 0.6), roasted pork top loin including fat (1142 ± 0.5), broiled beef flank excluding fat (1142 ± 0.5), broiled beef rib eye including fat (1141 ± 0.5), bleu cheese (1139 ± 1.8), broiled pork sirloin including fat (1138 ± 0.5), cooked pike (1138 ± 0.6), braised pork heart (1137 ± 0.4), sardines (1133 ± 2.1), kippered herring (1133 ± 1.3), broiled porterhouse steak including fat (1128 ± 0.5), cooked tilefish (1128 ± 0.6), broiled beef flank including fat (1126 ± 0.5), braised pork loin blade including fat (1121 ± 0.5), fried chicken neck with skin (1119 ± 1.4), braised beef liver (1116 ± 0.5), cooked haddock (1116 ± 0.6), broiled pork top loin including fat (1114 ± 0.5), cooked flatfish (1113 ± 0.6), cooked clams (1112 ± 0.6), broiled pork loin including fat (1110 ± 0.5), roasted pork loin including fat (1108 ± 0.5), broiled beef rib excluding fat (1108 ± 0.5), cooked rockfish (1108 ± 0.6), creamy peanut butter (1106 ± 3.1), provolone cheese (1104 ± 1.8), roasted beef tenderloin including fat (1100 ± 0.5), cooked perch (1100 ± 0.6), cooked mackerel (1099 ± 0.6), braised pork tongue (1099 ± 0.5), roasted cured pork arm excluding fat (1094 ± 0.5), cooked pompano (1092 ± 0.6), cooked eel (1091 ± 0.6), fried pork center loin including fat (1089 ± 0.5), medium-rare broiled extra lean ground beef (1089 ± 0.5), cooked sea bass (1089 ± 0.6), cooked black bass (1086 ± 0.4), stewed chicken wing with skin (1085 ± 1.3), cooked whiting (1082 ± 0.6), oil roasted sunflower seeds (1082 ± 1.8), hard pork salami (1080 ± 5), broiled T-bone steak including fat (1078 ± 0.5), smoked whitefish (1078 ± 0.6), medium-rare fried extra lean ground beef (1070 ± 0.5), pickled anchovy (1068 ± 1.8), raw yellowtail (1066 ± 0.6), cooked herring (1061 ± 0.6), medium-rare broiled lean ground beef (1060 ± 0.5), cooked carp (1054 ± 0.6), oil roasted peanuts (1054 ± 1.8), cooked cod (1052 ± 0.6), medium-rare baked extra lean ground beef (1049 ± 0.5), canned cod (1049 ± 0.6), roasted pork arm including fat (1047 ± 0.5), breaded and fried shrimp (1042 ± 0.6), cooked rainbow smelt (1041 ± 0.6), simmered beef kidneys (1040 ± 0.5), medium-rare fried lean ground beef (1039 ± 0.5), caviar (1038 ± 3.1), roasted pork shoulder including fat (1038 ± 0.5), cooked mussels (1036 ± 0.6), american cheese (1036 ± 1.8), pimento cheese spread (1036 ± 1.8), braised beef brisket including fat (1034 ± 0.5), medium-rare broiled ground beef (1032 ± 0.5), broiled pork boston blade including fat (1031 ± 0.5), roasted pork boston blade including fat (1031 ± 0.5), brie cheese (1029 ± 1.8), medium-rare baked lean ground beef (1026 ± 0.5), medium-rare fried ground beef (1026 ± 0.5), roasted beef rib including fat (1021 ± 0.5), anchovy paste (1014 ± 7.1), cooked shrimp (1014 ± 0.6), dried peanuts (1011 ± 1.8), fried pork center rib including fat (1002 ± 0.5), tempeh (1002 ± 0.6), fried pork top loin including fat (997 ± 0.5), crab cakes (997 ± 0.8), raw coho salmon (996 ± 0.6), raw duck liver (995 ± 0.5), canned blue crab (995 ± 0.6), cooked lobster (994 ± 0.6), dried black walnuts (993 ± 1.8), raw alewife (989 ± 0.5), dried pistachio nuts (989 ± 1.8), roasted pork loin blade including fat (988 ± 0.5), medium-rare baked ground beef (987 ± 0.5), oil roasted spanish peanuts (986 ± 1.8), raw shrimp (985 ± 0.6), raw sockeye salmon (982 ± 0.6), dry roasted sunflower seeds (982 ± 1.8), camembert cheese (982 ± 1.8), turkey ham (979 ± 0.9), raw black bass (979 ± 0.5), cooked blue crab (979 ± 0.6), braised beef lungs (973 ± 0.5), hard pork and beef salami (970 ± 5), braised beef shortribs including fat (969 ± 0.5), braised beef spleen (968 ± 0.5), raw shark (967 ± 0.6), broiled pork loin blade including fat (965 ± 0.5), raw halibut (959 ± 0.6), raw trout (956 ± 0.6), cooked sturgeon (954 ± 0.6), simmered beef tongue (952 ± 0.5), pork link sausage (951 ± 0.7), oil roasted valencia peanuts (950 ± 1.8), raw smelt (949 ± 0.5), raw rainbow trout (947 ± 0.6), raw milkfish (946 ± 0.6), dried pignolia pine nuts (946 ± 1.8), raw snapper (945 ± 0.6), canned sockeye salmon (944 ± 0.6), mozzarella cheese (943 ± 1.8), chicken egg yolk (941 ± 2.9), raw anchovy (938 ± 0.6), canned smelt (938 ± 0.5), cooked alaska king crab (938 ± 0.6), raw sheepshead (931 ± 0.6), natto (931 ± 0.6), raw chum salmon (928 ± 0.6), raw chinook salmon (924 ± 0.6), raw bluefish (924 ± 0.6), raw pink salmon (919 ± 0.6), canned pink salmon (919 ± 0.6), roasted lean ham (918 ± 0.5), turkey pastrami (914 ± 0.9), raw swordfish (912 ± 0.6), roasted canned lean ham (911 ± 0.5), oil roasted virginia peanuts (911 ± 1.8), raw lobster (911 ± 0.6), broiled beef rib including fat (905 ± 0.5), grilled canadian bacon (904 ± 1.1), raw crayfish (904 ± 0.6), oil roasted almonds (900 ± 1.8), raw pollock (896 ± 0.6), toasted almonds (896 ± 1.8), raw queen crab (896 ± 0.6), raw sunfish (894 ± 0.6), raw grouper (893 ± 0.6), raw mullet (892 ± 0.6), raw burbot (891 ± 0.6), raw pike (887 ± 0.6), raw alaska king crab (887 ± 0.6), roasted canned ham (883 ± 0.5), raw whitefish (880 ± 0.6), dried almonds (879 ± 1.8), freeze-dried chives (875 ± 62.5), raw cusk (875 ± 0.6), raw cisco (875 ± 0.6), raw ling (875 ± 0.6), raw blue crab (875 ± 0.6), roasted duck with skin (872 ± 0.5), jellied corned beef loaf (871 ± 1.8), raw haddock (871 ± 0.6), raw scup (871 ± 0.6), simmered chicken neck with skin (871 ± 1.3), roasted ham (870 ± 0.5), raw goose liver (870 ± 0.5), raw freshwater bass (869 ± 0.6), raw flatfish (868 ± 0.6), battered and fried shark (867 ± 0.6), fried pork loin blade including fat (864 ± 0.5), raw rockfish (864 ± 0.6), roasted cured pork arm including fat (861 ± 0.5), raw perch (858 ± 0.6), raw mackerel (856 ± 0.6), fried abalone (854 ± 0.6), fried tofu (854 ± 3.8), raw spot (853 ± 0.6), raw dolphinfish (852 ± 0.6), raw pompano (851 ± 0.6), raw eel (851 ± 0.6), cashew butter (850 ± 1.8), raw sea bass (849 ± 0.6), unheated lean ham (848 ± 0.5), breaded and fried croaker (847 ± 0.6),

ISOLEUCINE

crunchy peanut butter (844 ± 1.6), raw whiting (844 ± 0.6), raw dungeness crab (844 ± 0.6), smoked chinook salmon (842 ± 0.6), raw catfish (838 ± 0.6), chopped smoked beef (836 ± 1.8), breaded and fried catfish (836 ± 0.6), poppy seed (833 ± 16.7), pork and beef pepperoni (833 ± 8.3), dry roasted peanuts (832 ± 1.8), raw black bullhead (831 ± 0.5), raw whelk (828 ± 0.6), raw herring (828 ± 0.6), canned alewife (826 ± 0.5), pork luxury loaf (825 ± 1.8), raw carp (822 ± 0.6), raw cod (821 ± 0.6), raw croaker (819 ± 0.6), raw striped bass (816 ± 0.6), raw lingcod (814 ± 0.6), raw rainbow smelt (812 ± 0.6), miso (811 ± 0.4), raw freshwater drum (808 ± 0.6), boiled soybeans (807 ± 0.3), raw tilefish (806 ± 0.6), raw wolffish (806 ± 0.6), basil (800 ± 50), breaded and fried scallops (800 ± 1.6), dill seed (800 ± 25), canned lean ham (796 ± 0.5), raw butterfish (796 ± 0.6), pork and beef peppered loaf (793 ± 1.8), lebanon bologna (791 ± 2.2), pork and beef luncheon sausage (787 ± 2.2), cooked corned beef brisket (785 ± 0.5), firm raw tofu (782 ± 0.4), oil roasted cashews (782 ± 1.8), breaded fried squid (780 ± 0.6), raw shad (780 ± 0.6), unheated canadian bacon (775 ± 0.9), raw white sucker (772 ± 0.6), raw seatrout (772 ± 0.6), chopped ham lunch meat (771 ± 2.4), unheated ham (770 ± 0.5), whole dried sesame seeds (767 ± 5.6), ham and cheese roll (764 ± 1.8), chicken egg (760 ± 1), roasted pork blade roll (757 ± 0.5), raw flounder (756 ± 0.5), raw sole (756 ± 0.5), smoked cisco (754 ± 0.6), beef pastrami (754 ± 1.8), braised beef thymus (745 ± 0.5), raw abalone (744 ± 0.6), raw sturgeon (744 ± 0.6), dry roasted cashews (743 ± 1.8), raw turbot (740 ± 0.6), toasted sesame kernels (739 ± 1.8), whole toasted sesame seeds (739 ± 1.8), tuna salad (739 ± 0.2), pork and beef honey loaf (736 ± 1.8), cottage cheese (734 ± 0.2), sesame butter (733 ± 3.3), italian pork sausage (731 ± 0.7), canned ham (731 ± 0.5), raw scallops (731 ± 0.6), beef lunch meat (721 ± 1.8), pork sausage (719 ± 1.9), dry roasted almonds (718 ± 1.8), dry roasted pistachio nuts (718 ± 1.8), raw cuttlefish (707 ± 0.6), pork and beef mortadella (707 ± 3.3), minced ham lunch meat (700 ± 2.4), fennel seed (700 ± 25), boiled lupins (695 ± 0.3), berliner (683 ± 2.2), summer sausage (683 ± 2.2), raw squid (678 ± 0.6), beef and pork salami (674 ± 2.2), raw monkfish (667 ± 0.6), chicken egg white (667 ± 1.5), braised pork lungs (664 ± 0.5), pork liverwurst (661 ± 2.8), pork bologna (661 ± 2.2), almond butter (656 ± 3.1), stir-fried sprouted soybeans (654 ± 0.5), pickled herring (653 ± 3.3), beef salami (652 ± 2.2), pork and beef lunch meat (650 ± 1.8), raw octopus (649 ± 0.6), pork and beef kielbasa/kolbassy (638 ± 1.9), breaded and fried clams (636 ± 0.6), puffed wheat (636 ± 3.6), pork livercheese (632 ± 1.3), oil roasted filberts (629 ± 1.8), pork and beef link sausage with american cheese (619 ± 1.2), polish pork sausage (618 ± 1.8), cooked oysters (614 ± 0.6), dried brazil nuts (611 ± 1.8), smoked beef sausage (609 ± 1.2), pork and beef brotwurst (606 ± 0.7), garlic powder (600 ± 16.7), duck egg (599 ± 0.7), corn germ (597 ± 0.5), surimi scallops (596 ± 0.6), ricotta cheese (590 ± 0.4), pork and beef picnic loaf (589 ± 1.8), raw beef tripe (589 ± 0.5), dried hickory nuts (586 ± 1.8), surimi shrimp (579 ± 0.6), grilled ham patties (577 ± 0.8), canned pork lunch meat (576 ± 2.4), dried filberts (575 ± 1.8), dried english/persian walnuts (575 ± 1.8), falafel (567 ± 1), pork and beef old fashioned loaf (564 ± 1.8), freeze-dried red sweet peppers (563 ± 31.3), freeze-dried green sweet peppers (563 ± 31.3), braised pork brains (561 ± 0.5), unheated ham patties (560 ± 0.8), raw pork stomach (560 ± 0.5), raw clams (556 ± 0.6), vienna sausage (556 ± 3.1), pork headcheese (550 ± 1.8), boiled green soybeans (543 ± 0.6), beef beerwurst salami (535 ± 2.2), neufchatel cheese (532 ± 1.8), pork and beef sausage (531 ± 3.8), beef bologna (530 ± 2.2), gjetost cheese (525 ± 1.8), boiled mothbeans (525 ± 0.3), almond paste (521 ± 1.8), beef frankfurter (519 ± 0.9), raw mussels (518 ± 0.6), pork bratwurst (514 ± 0.6), beef and pork bologna (509 ± 2.2), dried ginko nuts (507 ± 1.8), pork pickle and pimento loaf (496 ± 1.8), dried pilinuts (489 ± 1.8), sweetened condensed milk (489 ± 1.3), fried beef brains (487 ± 0.5), pork beerwurst salami (487 ± 2.2), gefiltefish with broth (486 ± 1.2), pork and beef link sausage (485 ± 0.7), beef and pork frankfurter (484 ± 1.1), pork liver sausage (483 ± 2.8), boiled peanuts (475 ± 1.6), freeze-dried shallots (475 ± 12.5), pork and beef knockwurst (466 ± 0.7), dried pinyon pine nuts (457 ± 1.8), evaporated skim milk (456 ± 1.6), frozen and boiled black-eyed peas (455 ± 0.6), boiled winged beans (448 ± 0.3), dry roasted filberts (443 ± 1.8), shredded wheat (439 ± 1.8), simmered beef brains (429 ± 0.5), boiled white beans (429 ± 0.3), bran flakes (429 ± 1.8), boiled and frozen baby lima beans (428 ± 0.6), pork olive loaf (425 ± 1.8), steamed sprouted soybeans (423 ± 1.1), boiled baby lima beans (423 ± 0.3), simmered pork chitterlings (420 ± 0.5), boiled cranberry beans (412 ± 0.3), evaporated whole milk (412 ± 0.4), boiled lima beans (411 ± 0.3), dried shitake mushrooms (407 ± 3.3), ham salad (407 ± 1.8), boiled yellow beans (405 ± 0.3), bran (404 ± 1.8), cream cheese (404 ± 1.8), boiled pink beans (400 ± 0.3), raw tofu (400 ± 0.4), breaded and fried oysters (396 ± 0.6), simmered pork tail (391 ± 0.5), boiled black beans (391 ± 0.3), frozen and boiled fordhook lima beans (391 ± 0.6), boiled lentils (390 ± 0.3), boiled hyacinth beans (390 ± 0.3), boiled mungo beans (385 ± 0.3), boiled navy beans (384 ± 0.3), boiled kidney beans (383 ± 0.3), dried acorns (382 ± 1.8), boiled chickpeas (380 ± 0.3), wheat flakes (379 ± 1.8), boiled great northern beans (368 ± 0.3), boiled pinto beans (363 ± 0.3), simmered pork ears (362 ± 0.5), boiled black turtle beans (361 ± 0.3), acorn flour (354 ± 1.8), evaporated lowfat milk (352 ± 0.4), raw bacon (351 ± 0.7), potato flour (346 ± 0.6), dried oriental radish (345 ± 0.9), boiled split peas (344 ± 0.3), puffed rice (343 ± 3.6), sheep milk (338 ± 0.2), boiled yardlong bean (337 ± 0.3), dry roasted pecans (336 ± 1.8), canned navy beans (332 ± 0.2), raw spirulina seaweed (331 ± 0.5), boiled catjang black-eyed peas (330 ± 0.3), simmered pork feet (326 ± 0.5), canned great northern beans (325 ± 0.2), dried pecans (325 ± 1.8), stir-fried sprouted lentils (320 ± 0.5), canned white beans (320 ± 0.2), corn flakes (318 ± 1.8), boiled black-eyed peas (314 ± 0.3), dehydrated onion flakes (314 ± 3.6), skim yogurt (312 ± 0.2), boiled french beans (311 ± 0.3), potato pancakes (308 ± 0.7), raw oysters (307 ± 0.6), canned oysters (307 ± 0.6), raisin bran (307 ± 1.8), boiled broad beans (306 ± 0.3), boiled adzuki beans (300 ± 0.2), onion powder (300 ± 25), boiled mung beans (297 ± 0.2), dry roasted macadamia nuts (293 ± 1.8), oil roasted pecans (293 ± 1.8), raw acorns (289 ± 1.8), lowfat yogurt (286 ± 0.2), boiled succotash (286 ± 0.5), au gratin potatoes (284 ± 0.4), refried beans (275 ± 0.2), dried coconut (275 ± 1.8), canned black turtle beans (266 ± 0.2), dried japanese chestnuts (261 ± 1.8), raw laver/nori seaweed (259 ± 0.5), canned lima beans (259 ± 0.2), dried chinese chestnuts (257 ± 1.8), dried european chestnuts (257 ± 1.8), tomato powder (250 ± 0.5), ginger (250 ± 25), dried macadamia nuts (246 ± 1.8), boiled pigeon peas (245 ± 0.3), canned cranberry beans (245 ± 0.2), frozen and boiled succotash (244 ± 0.6), protein fortified skim milk (239 ± 0.2), protein fortified lowfat 2% milk (239 ± 0.2), protein fortified lowfat 1% milk (238 ± 0.2), yellow corn pudding (235 ± 0.2), baked beans with pork and sweet sauce (234 ± 0.2), restaurant vanilla milkshake (234 ± 0.2), pickled pork feet (230 ± 0.5), eggnog (230 ± 0.2), canned kidney beans (229 ± 0.2), baked beans with pork and tomato sauce (228 ± 0.2), raw garlic (222 ± 5.6), boiled sprouted green peas (221 ± 0.5), canned broad beans (221 ± 0.2), hot chocolate (220 ± 0.2), oil roasted macadamia nuts (218 ± 1.8), malted milk (214 ± 0.2), vegetarian baked beans (212 ± 0.2), canned chickpeas (212 ± 0.2), ginko nuts (211 ± 1.8), dried toasted coconut (211 ± 1.8), vanilla milkshake (211 ± 0.2), goat milk (207 ± 0.2), restaurant chocolate milkshake (206 ± 0.2), skim milk (206 ± 0.2), stir-fried sprouted mung beans (206 ± 0.8), buttermilk (204 ± 0.2), restaurant strawberry milkshake (204 ± 0.2), indian buffalo milk (203 ± 0.2), noodles (203 ± 0.3), lowfat 2% milk (202 ± 0.2), filled milk (202 ± 0.2), canned pinto beans (201 ± 0.2), lowfat 1% milk (199 ± 0.2), whole milk (199 ± 0.2), hummus (198 ± 0.2), lowfat 1% chocolate milk (196 ± 0.2), raw green peas (195 ± 0.6), lowfat 2% chocolate milk (194 ± 0.2), boiled green peas (193 ± 0.6), canned blackeyed peas (193 ± 0.2), boiled lamb's-quarters (193 ± 0.6), whole chocolate milk (192 ± 0.2), carob milk (190 ± 0.2), whole yogurt (189 ± 0.2), frozen and boiled green peas (185 ± 0.6), chocolate milkshake (185 ± 0.2), half and half cream (180 ± 3.3), lard (175 ± 1.8), french fried potatoes (174 ± 1), boiled jute potherb (174 ± 1.2), frozen and boiled turnip greens (173 ± 0.6), macaroni (170 ± 0.4), spaghetti (170 ± 0.4), frozen and boiled kale (169 ± 0.8), raw pork jowl (168 ± 0.5), roasted chinese chestnuts (168 ± 1.8), sour cream (167 ± 4.2), light cream (167 ± 3.3), raw dock (167 ± 0.7), raw chinese chestnuts (161 ± 1.8), frozen and boiled spinach (160 ± 0.5), canned green peas (159 ± 0.6), okara tofu (159 ± 0.8), boiled swiss chard (155 ± 0.6), boiled spinach (152 ± 0.6), frozen and fried french fried potatoes (150 ± 1), canned succotash with creamed corn (149 ± 0.4), medium cream (147 ± 3.3), canned succotash (147 ± 0.4), roasted japanese chestnuts (146 ± 1.8), raw spinach (146 ± 1.8), frozen and boiled brussels sprouts (144 ± 0.6), canned spinach (144 ± 0.5), scalloped potatoes (144 ± 0.4), soy milk (144 ± 0.2), raw sprouted alfalfa seeds (142 ± 1.5), coconut cream (142 ± 0.2), potato salad (141 ± 0.4), frozen and fried hash brown potatoes (136 ± 0.6), boiled yellow corn (133 ± 0.6), whipping cream, light (133 ± 3.3), raw chives (133 ± 16.7), raw sprouted mung beans (133 ± 1), raw coconut (131 ± 1.1), roasted european chestnuts (129 ± 1.8), dried sweetened flaked coconut (128 ± 0.7), whipping cream, heavy (127 ± 3.3), frozen and boiled yellow corn (121 ± 0.6), frozen and boiled collards (121 ± 0.6), frozen and boiled broccoli (121 ± 0.5), oatmeal (118 ± 0.3), boiled broccoli (115 ± 0.6), raw japanese chestnuts (114 ± 1.8), boiled kale (114 ± 0.8), dried sweetened shredded coconut (113 ± 0.5), boiled scotch kale (112 ± 0.8), dried apricots (111 ± 1.4), boiled and steamed chinese chestnuts (111 ± 1.8), raw shallots (110 ± 5), o'brien potatoes (110 ± 0.3), raw broccoli (109 ± 1.1), frozen and boiled asparagus (108 ± 0.8), sweetened coconut cream (106 ± 0.2), raw swamp cabbage (104 ± 0.9), hash brown potatoes (104 ± 0.6), raw romaine lettuce (104 ± 1.8), boiled amaranth (102 ± 0.8), frozen and boiled turnip greens and turnips (101 ± 0.5), raw chickory greens (101 ± 0.6), boiled brussels sprouts (100 ± 0.6), soybean oil and cottonseed oil margarine liquid (100 ± 10), raw savoy cabbage (100 ± 1.4), microwaved potato with skin (99 ± 0.2), cooked sprouted mung beans (98 ± 0.8), dried longans (97 ± 0.5), raw european chestnuts (96 ± 1.8), boiled asparagus (96 ± 0.6), raw watercress (94 ± 2.9), dried figs (93 ± 0.3), canned yellow corn with red and green peppers (93 ± 0.4), boiled dock (93 ± 0.5), baked potato with skin (93 ± 0.2), coconut milk (90 ± 0.2), boiled savoy cabbage (90 ± 0.7), boiled chinese cabbage (89 ± 0.6), boiled mushrooms (88 ± 0.6), raw bamboo shoots (88 ± 0.7), raw wakame seaweed (87 ± 0.5), mashed potatoes (87 ± 0.5), baked sweet potato (86 ± 0.4), frozen and baked sweet potato (86 ± 0.6), raw looseleaf lettuce (86 ± 1.8), raw chinese cabbage (86 ± 1.4), microwaved potato without skin (85 ± 0.3), boiled swamp cabbage (84 ± 1), raw potato without skin (84 ± 0.4), raw mushrooms (83 ± 1.4), canned sweet potato (83 ± 0.3), frozen and boiled mustard greens (83 ± 0.7), boiled kohlrabi (83 ± 0.6), boiled sweet potato (82 ± 0.3), canned yellow corn (82 ± 0.5), raw welsh onions (81 ± 0.5), boiled mustard greens (81 ± 0.7), frozen and boiled potato without skin (80 ± 0.5), baked potato without skin (80 ± 0.3), raw butterhead lettuce (80 ± 3.3), raw turnip greens (79 ± 1.8), boiled and steamed european chestnuts (79 ± 1.8), raw cauliflower (76 ± 1), raw kelp/kombu/tangle seaweed (76 ± 0.5), raw california avocado (75 ± 0.3), raw iceberg lettuce (75 ± 2.5), raw spring onions with tops (74 ± 1), tomato paste (73 ± 0.4), frozen and boiled okra (72 ± 0.5), raw endive (72 ± 2), boiled cauliflower (71 ± 0.8), raw red cabbage (71 ± 1.4), canned turnip greens (70 ± 0.4), canned creamed yellow corn (70 ± 0.4), boiled potato without skin (70 ± 0.4), boiled green beans (69 ± 0.8), boiled yellow snap beans (69 ± 0.8), canned sprouted mung beans (68 ± 0.8), cream of wheat (67 ± 0.3), boiled okra (65 ± 0.6), raw red chili peppers (64 ± 1.1), raw green chili peppers (64 ± 1.1), safflower oil and soybean oil mayonnaise (64 ± 3.6), soybean oil mayonnaise (64 ± 3.6), breadfruit (64 ± 0.5), chocolate syrup (63 ± 1.3), farina (63 ± 0.3), coleslaw (62 ± 0.8), frozen and boiled turnips (62 ± 0.5), raw cassava (61 ± 0.5), frozen and boiled cauliflower (61 ± 0.6), whole chocolate malted milk (61 ± 0.2), raw green cabbage (60 ± 1.4), butter (60 ± 3.3), raw chickory (60 ± 1.1), dried japanese persimmon (59 ± 1.5), boiled turnip greens (58 ± 0.7), steamed hawaii mountain yam (58 ± 0.7), canned bamboo shoots (58 ± 0.4), baked hubbard squash (58 ± 0.5), boiled hubbard squash (58 ± 0.4), raw florida avocado (57 ± 0.2), canned potato without skin (57 ± 0.6), cooked shitake mushrooms (56 ± 0.7), corn grits (56 ± 0.2), human milk (55 ± 1.6), raw celtuce (55 ± 0.5), whipped butter (55 ± 4.5), raw leeks (54 ± 1.9), raw lotus root (54 ± 0.6), boiled beet greens (53 ± 0.7), raw vine spinach (53 ± 0.5), boiled purslane (53 ± 0.9), boiled red cabbage (53 ± 0.7), boiled bamboo shoots (51 ± 0.4), frozen and boiled yellow snap beans (50 ± 0.7), frozen and boiled green beans (50 ± 0.7), baked/boiled tropical yam (50 ± 0.7), raw ginger root (50 ± 2.1), raw eggplant (49 ± 1.2), frozen and boiled butternut squash (48 ± 0.4), boiled green cabbage (48 ± 0.7), whey (47 ± 0.2), dried dates (47 ± 0.6), frozen and boiled carrots (47 ± 0.7), sapote (46 ± 0.2), boiled rutabaga (46 ± 0.6), frozen and boiled crookneck squash (46 ± 0.5), boiled collards (45 ± 0.3), candied sweet potato (44 ± 0.5), boiled carrots (44 ± 0.6), baked acorn squash (44 ± 0.5), canned green beans (43 ± 0.7), canned yellow snap beans (43 ± 0.7), raw scallop squash (43 ± 0.8), raw onions (43 ± 0.6), raw carrots (42 ± 0.7), raw zucchini (42 ± 0.8), apricots (41 ± 0.5), boiled burdock root (41 ± 0.4), frozen and boiled zucchini (41 ± 0.4), corn oil margarine stick (40 ± 10), safflower oil and soybean oil margarine stick (40 ± 10), soybean oil margarine stick (40 ± 10), soybean oil and cottonseed oil margarine stick (40 ± 10), corn oil margarine tub (40 ± 10), safflower oil margarine tub (40 ± 10), soybean oil margarine tub (40 ± 10), soybean oil and cottonseed oil margarine tub (40 ± 10), boiled and steamed japanese chestnuts (39 ± 1.8), boiled scallop squash (37 ± 0.6), canned zucchini in tomato juice (37 ± 0.4), persimmon (36 ± 2), boiled eggplant (35 ± 1), boiled butternut squash (35 ± 0.5), canned pumpkin (34 ± 0.4), canned pumpkin pie mix (34 ± 0.4), raw crookneck squash (34 ± 0.8), bananas (33 ± 0.4), boiled lotus root (33 ± 0.6), boiled chayote (33 ± 0.6), boiled crookneck squash (33 ± 0.6), tomato purée (32 ± 0.2), boiled beets (32 ± 0.6), boiled onions (32 ± 0.5), boiled calabash gourd (32 ± 0.7), frozen and boiled red sweet peppers (31 ± 0.5), frozen and boiled green sweet peppers (31 ± 0.5), raw radish (31 ± 1.1), guava (30 ± 0.6), beef suet (29 ± 1.8), canned red chili peppers (29 ± 0.7), canned green chili peppers (29 ± 0.7), raw green tomato (29 ± 0.4), boiled turnips (29 ± 0.6), raw red sweet peppers (28 ± 1), raw green sweet peppers (28 ± 1), boiled oriental radish (28 ± 0.7), coconut water (28 ± 0.2), navel orange (28 ± 0.4), valencia orange (28 ± 0.4), boiled leeks (27 ± 1.9), boiled red tomato (27 ± 0.4), elderberries (27 ± 0.3), longans (26 ± 0.5), boiled yambean (26 ± 0.5), canned jalapeño peppers (26 ± 0.7), canned red sweet peppers (26 ± 0.7), canned green sweet peppers (26 ± 0.7), boiled acorn squash (26 ± 0.4), japanese persimmon (25 ± 0.3), rice polish (25 ± 0.9), raw oriental radish (25 ± 1.1), canned carrots (25 ± 0.7), frozen and boiled onions (25 ± 0.5), pickled beets (24 ± 0.4), figs (24 ± 1), boiled/baked spaghetti squash (24 ± 0.6), carambola (23 ± 0.4), boiled pumpkin (23 ± 0.4), boiled zucchini (23 ± 0.6), canned peeled red tomatoes (22 ± 0.4), canned stewed red tomatoes (22 ± 0.4), cooked plantain (22 ± 0.3), canned crookneck squash (22 ± 0.5), boiled red sweet peppers (21 ± 0.7), boiled green sweet peppers (21 ± 0.7), raw red tomato (21 ± 0.4), blueberries (21 ± 0.3), peach (20 ± 0.6), cooked taro (20 ± 0.8), raw celery (20 ± 1.3), corn oil diet margarine tub (20 ± 10), soybean oil and cottonseed oil diet margarine tub (20 ± 10), pear (20 ± 0.3), watermelon (19 ± 0.3), mango (18 ± 0.2), plum (17 ± 0.8), raw cucumber (17 ± 1), mandarin oranges (17 ± 0.4), tangerine (17 ± 0.6), boiled celery (16 ± 0.7), crab apples (16 ± 0.5), cream of rice (15 ± 0.3), tomato juice (15 ± 0.3), loquats (15 ± 0.5), sapodilla (15 ± 0.3), strawberries (14 ± 0.3), pineapple (13 ± 0.3), cooked apple without skin (10 ± 0.3), papaya (8 ± 0.2), orange juice (8 ± 0.2), raw apple with skin (8 ± 0.4), grape juice (7 ± 0.2), applesauce (7 ± 0.4), screwdriver (6 ± 0.2), raw apple without skin (6 ± 0.4), european grapes (5 ± 0.3), american grapes (5 ± 0.5), tangerine juice (5 ± 0.2), ale (5 ± 0.1), thompson seedless grapes (3 ± 0.3), white grapefruit juice (2 ± 0.2), coffee (2 ± 0.3)

Daily Dosage: Nutritional — 1–3 g; therapeutic — ?; experimental — ?; toxic — none known up to 3 g
Warnings: none

LEUCINE

Names: leucine, LEU, amino acid L, 2-amino-4-methyl-pentanoic acid
Forms: 2-amino-4-methyl-pentanoic acid
Classification: essential amino acid
Deficiency: protein decreased
Side-effects: see toxicity symptoms
Toxicity: precipitates pellagra?, vitamin B_3 decreased
Inhibitors: increased amounts of other amino acids
Helpers: moderate amounts of other amino acids
Sources (mg per 100 grams of food): torula yeast (14761 ± 1.8), soybean protein isolate (6782 ± 1.8), dried and salted cod (5106 ± 0.6), dried spirulina seaweed (4947 ± 0.5), soybean protein con-

centrate (4918 ± 1.8), parmesan cheese (4020 ± 10), peanut flour (3925 ± 12.5), defatted soybean flour (3828 ± 0.5), cooked whelk (3807 ± 0.6), low fat soybean flour (3789 ± 0.6), defatted soy meal (3660 ± 0.4), dried and frozen tofu (3641 ± 2.9), dry skim milk (3543 ± 1.7), dry instant skim milk (3438 ± 0.5), fenugreek seed (3250 ± 25), dry roasted soybean nuts (3223 ± 0.6), gruyere cheese (3143 ± 1.8), almond meal (3118 ± 1.8), dry buttermilk (3114 ± 7.1), toasted soybean nuts (3057 ± 1.8), swiss cheese (2996 ± 1.8), roasted soybean nuts (2867 ± 0.6), roasted soybean flour (2834 ± 0.6), roasted pumpkin and squash seeds (2832 ± 1.8), braised pork center loin excluding fat (2826 ± 0.5), soybean flour (2812 ± 0.6), braised pork top loin excluding fat (2799 ± 0.5), fried beef top round excluding fat (2771 ± 0.5), braised pork sirloin excluding fat (2723 ± 0.5), braised pork loin excluding fat (2681 ± 0.5), simmered beef shank excluding fat (2662 ± 0.5), braised beef chuck arm pot roast excluding fat (2610 ± 0.5), fried chicken giblets (2601 ± 0.5), broiled pork center loin excluding fat (2601 ± 0.5), edam cheese (2600 ± 1.8), gouda cheese (2596 ± 1.8), tilsit cheese (2579 ± 1.8), dry whole milk (2578 ± 1.6), fried beef sirloin excluding fat (2567 ± 0.5), simmered beef heart (2547 ± 0.5), braised pork boston blade excluding fat (2532 ± 0.5), port du salut cheese (2514 ± 1.8), fried beef liver (2513 ± 0.5), fried chicken breast without skin (2509 ± 0.6), fried beef top round including fat (2505 ± 0.5), broiled beef top round excluding fat (2505 ± 0.5), simmered beef shank including fat (2498 ± 0.5), braised beef bottom round excluding fat (2497 ± 0.5), fried light meat chicken without skin (2463 ± 0.5), braised beef chuck blade roast excluding fat (2455 ± 0.5), broiled beef top round including fat (2436 ± 0.5), braised beef shortribs excluding fat (2431 ± 0.5), milk chocolate (2429 ± 1.8), well done baked extra lean ground beef (2429 ± 0.5), braised pork loin blade excluding fat (2414 ± 0.5), canned tuna in water (2404 ± 0.6), stewed chicken drumstick without skin (2402 ± 1.1), broiled beef sirloin excluding fat (2400 ± 0.5), cheddar cheese (2385 ± 0.5), roasted turkey light meat without skin (2383 ± 0.5), monterey cheese (2375 ± 1.8), braised pork center loin including fat (2375 ± 0.5), well done baked lean ground beef (2373 ± 0.5), roasted pork rump excluding fat (2368 ± 0.5), braised pork spareribs (2362 ± 0.5), braised beef bottom round including fat (2356 ± 0.5), fried chicken breast with skin (2353 ± 0.5), anchovy canned in olive oil (2350 ± 2.5), broiled pork center rib excluding fat (2342 ± 0.5), broiled pork top loin excluding fat (2342 ± 0.5), roasted pork tenderloin (2340 ± 0.5), fried pork center loin excluding fat (2338 ± 0.5), simmered turkey heart (2333 ± 0.5), roasted chicken breast without skin (2328 ± 0.6), provolone cheese (2325 ± 1.8), braised beef brisket excluding fat (2322 ± 0.5), roasted light meat chicken without skin (2319 ± 0.5), braised pork liver (2319 ± 0.5), roasted pork center loin excluding fat (2315 ± 0.5), well done baked ground beef (2309 ± 0.5), braised pork center rib including fat (2307 ± 0.5), braised pork spleen (2306 ± 0.5), colby cheese (2304 ± 1.8), simmered chicken heart (2303 ± 0.5), roasted pork leg excluding fat (2302 ± 0.5), broiled pork sirloin excluding fat (2296 ± 0.5), braised beef liver (2294 ± 0.5), roasted pork center rib excluding fat (2293 ± 0.5), roasted pork shank excluding fat (2293 ± 0.5), roasted pork top loin excluding fat (2293 ± 0.5), roasted eye of beef round excluding fat (2291 ± 0.5), well done broiled extra lean ground beef (2291 ± 0.5), muenster cheese (2289 ± 1.8), braised pork kidneys (2280 ± 0.5), roasted turkey dark meat without skin (2276 ± 0.5), fried pork center rib excluding fat (2273 ± 0.5), fried pork top loin excluding fat (2273 ± 0.5), brick cheese (2271 ± 1.8), roasted beef tip excluding fat (2269 ± 0.5), cheshire cheese (2268 ± 1.8), broiled pork loin excluding fat (2263 ± 0.5), broiled beef top loin excluding fat (2262 ± 0.5), well done broiled lean ground beef (2261 ± 0.5), braised pork sirloin including fat (2258 ± 0.5), cooked flounder (2250 ± 0.5), broiled beef round excluding fat (2249 ± 0.5), well done fried extra lean ground beef (2243 ± 0.5), roasted pork sirloin excluding fat (2235 ± 0.5), broiled beef tenderloin excluding fat (2233 ± 0.5), dried butternuts (2232 ± 1.8), fried light meat chicken with skin (2231 ± 0.5), braised pork top loin including fat (2229 ± 0.5), broiled porterhouse steak excluding fat (2226 ± 0.5), cooked coho salmon (2224 ± 0.6), broiled T-bone steak excluding fat (2223 ± 0.5), roasted turkey light meat with skin (2220 ± 0.5), cooked sockeye salmon (2219 ± 0.6), braised beef spleen (2217 ± 0.5), broiled beef rib eye excluding fat (2216 ± 0.5), broiled pork center loin including fat (2215 ± 0.5), braised beef flank excluding fat (2215 ± 0.5), well done fried lean ground beef (2212 ± 0.5), chipped beef (2200 ± 1.8), roasted chicken breast with skin (2198 ± 0.5), simmered chicken liver (2198 ± 0.5), braised pork loin including fat (2191 ± 0.5), roasted pork loin excluding fat (2186 ± 0.5), well done broiled ground beef (2180 ± 0.5), dried watermelon seeds (2179 ± 1.8), roasted beef tenderloin excluding fat (2177 ± 0.5), fried dark meat chicken without skin (2176 ± 0.5), roasted beef rib roasted excluding fat (2176 ± 0.5), stewed chicken breast without skin (2175 ± 0.5), braised beef flank including fat (2175 ± 0.5), fried beef sirloin including fat (2170 ± 0.5), cooked halibut (2169 ± 0.6), stewed light meat chicken without skin (2167 ± 0.5), broiled beef sirloin including fat (2165 ± 0.5), well done fried ground beef (2164 ± 0.5), simmered turkey liver (2163 ± 0.5), braised pork arm braised including fat (2163 ± 0.5), canned tuna in oil (2156 ± 0.6), roasted pork rump including fat (2155 ± 0.5), dried sesame kernels (2150 ± 6.3), braised beef chuck arm pot roast including fat (2141 ± 0.5), cooked rainbow trout (2141 ± 0.6), simmered turkey giblets (2140 ± 0.5), roasted turkey dark meat with skin (2138 ± 0.5), creamy peanut butter (2138 ± 3.1), cooked snapper (2136 ± 0.6), braised pork boston blade braised including fat (2131 ± 0.5), caviar (2131 ± 3.1), braised pork pancreas (2130 ± 0.5), braised pork heart (2130 ± 0.4), roasted light meat chicken with skin (2119 ± 0.5), cooked bacon (2119 ± 0.4), limburger cheese (2118 ± 1.8), braised beef pancreas (2116 ± 0.5), roasted eye of beef round including fat (2115 ± 0.5), cooked sheepshead (2115 ± 0.6), roasted goose with skin (2109 ± 0.5), dried pumpkin and squash seeds (2107 ± 1.8), roasted beef round tip including fat (2092 ± 0.5), simmered turkey gizzard (2067 ± 0.5), simmered chicken giblets (2066 ± 0.5), cooked swordfish (2064 ± 0.6), roasted pork shoulder excluding fat (2062 ± 0.5), roasted pork center loin including fat (2054 ± 0.5), roasted dark meat chicken without skin (2053 ± 0.5), broiled beef tenderloin including fat (2052 ± 0.5), smoked haddock (2051 ± 0.6), broiled pork boston blade excluding fat (2046 ± 0.5), simmered beef kidneys (2043 ± 0.5), oil roasted peanuts (2039 ± 1.8), medium-rare broiled extra lean ground beef (2036 ± 0.5), broiled beef top loin including fat (2033 ± 0.5), fried chicken back with skin (2032 ± 0.7), roasted pork sirloin including fat (2026 ± 0.5), broiled pork loin blade excluding fat (2025 ± 0.5), roasted pork leg including fat (2023 ± 0.5), stewed chicken breast with skin (2019 ± 0.5), cooked grouper (2019 ± 0.6), broiled beef round including fat (2018 ± 0.5), cooked mullet (2016 ± 0.6), braised beef chuck blade roast including fat (2011 ± 0.5), broiled beef flank excluding fat (2008 ± 0.5), cooked pike (2007 ± 0.6), roasted pork loin blade excluding fat (2006 ± 0.5), broiled beef rib eye including fat

(2006 ± 0.5), medium-rare fried extra lean ground beef (2001 ± 0.5), sardines (2000 ± 2.1), roasted pork center rib including fat (1999 ± 0.5), kippered herring (1998 ± 1.3), raw pheasant without skin (1995 ± 0.5), fried dark meat chicken with skin (1994 ± 0.5), cooked tilefish (1991 ± 0.6), canned corned beef (1990 ± 0.5), fried chicken drumstick with skin (1984 ± 1), roasted duck without skin (1983 ± 0.5), broiled pork center rib including fat (1983 ± 0.5), broiled porterhouse steak including fat (1983 ± 0.5), american cheese (1982 ± 1.8), pimento cheese spread (1982 ± 1.8), medium-rare broiled lean ground beef (1981 ± 0.5), roasted pork boston blade excluding fat (1980 ± 0.5), roasted cured pork arm excluding fat (1980 ± 0.5), broiled beef flank including fat (1979 ± 0.5), roasted chicken drumstick with skin (1977 ± 1), fried pork loin blade excluding fat (1977 ± 0.5), cooked haddock (1971 ± 0.6), yellow mustard seed (1967 ± 16.7), cooked flatfish (1964 ± 0.6), roasted pork shank including fat (1962 ± 0.5), medium-rare baked extra lean ground beef (1961 ± 0.5), fried chicken thigh with skin (1958 ± 0.8), dried peanuts (1957 ± 1.8), cooked rockfish (1954 ± 0.6), brie cheese (1954 ± 1.8), roasted pork top loin including fat (1953 ± 0.5), stewed dark meat chicken without skin (1949 ± 0.5), broiled beef rib excluding fat (1947 ± 0.5), roasted chicken thigh without skin (1946 ± 1), bleu cheese (1946 ± 1.8), broiled pork sirloin including fat (1946 ± 0.5), medium-rare fried lean ground beef (1942 ± 0.5), cooked perch (1941 ± 0.6), cooked mackerel (1938 ± 0.6), roasted beef tenderloin including fat (1934 ± 0.5), braised pork tongue (1932 ± 0.5), medium-rare broiled ground beef (1929 ± 0.5), braised pork loin blade including fat (1928 ± 0.5), cooked pompano (1925 ± 0.6), cooked eel (1922 ± 0.6), cooked sea bass (1921 ± 0.6), medium-rare baked lean ground beef (1918 ± 0.5), mozzarella cheese (1918 ± 1.8), medium-rare fried ground beef (1917 ± 0.5), broiled pork top loin including fat (1912 ± 0.5), stewed light meat chicken with skin (1910 ± 0.5), cooked whiting (1909 ± 0.6), simmered chicken gizzard (1907 ± 0.5), raw whelk (1904 ± 0.6), smoked whitefish (1902 ± 0.6), broiled pork loin including fat (1901 ± 0.5), cooked crayfish (1898 ± 0.6), roasted chicken wing with skin (1897 ± 1.5), broiled T-bone steak including fat (1896 ± 0.5), roasted pork loin including fat (1892 ± 0.5), roasted dark meat chicken with skin (1883 ± 0.5), raw yellowtail (1881 ± 0.6), fried pork center loin including fat (1873 ± 0.5), cooked herring (1872 ± 0.6), raw quail without skin (1866 ± 0.5), camembert cheese (1864 ± 1.8), cooked carp (1858 ± 0.6), cooked cod (1856 ± 0.6), stewed chicken drumstick with skin (1854 ± 0.9), canned cod (1851 ± 0.6), fried chicken wing with skin (1850 ± 1.6), corn germ (1849 ± 0.5), medium-rare baked ground beef (1845 ± 0.5), simmered chicken neck without skin (1844 ± 2.8), cooked rainbow smelt (1836 ± 0.6), roasted chicken thigh with skin (1827 ± 0.8), braised beef brisket including fat (1817 ± 0.5), oil roasted spanish peanuts (1814 ± 1.8), roasted pork arm including fat (1799 ± 0.5), cooked clams (1798 ± 0.6), turkey breast meat summer sausage (1795 ± 2.4), roasted beef rib including fat (1795 ± 0.5), roasted pork shoulder including fat (1776 ± 0.5), broiled pork boston blade including fat (1764 ± 0.5), roasted pork boston blade including fat (1759 ± 0.5), raw coho salmon (1756 ± 0.6), oil roasted valencia peanuts (1754 ± 1.8), dried pignolia pine nuts (1754 ± 1.8), fried pork center rib including fat (1735 ± 0.5), raw sockeye salmon (1731 ± 0.6), hard pork and beef salami (1730 ± 5), dried black walnuts (1729 ± 1.8), fried pork top loin including fat (1728 ± 0.5), stewed dark meat chicken with skin (1705 ± 0.5), raw shark (1705 ± 0.6), braised beef shortribs including fat (1704 ± 0.5), dried pistachio nuts (1700 ± 1.8), breaded and fried shrimp (1698 ± 0.6), stewed chicken thigh with skin (1696 ± 0.7), roasted pork loin blade including fat (1696 ± 0.5), pork link sausage (1693 ± 0.7), raw halibut (1692 ± 0.6), raw duck liver (1691 ± 0.5), fried chicken neck with skin (1689 ± 1.4), raw trout (1688 ± 0.6), grilled canadian bacon (1687 ± 1.1), dried sunflower seeds (1682 ± 1.8), cooked sturgeon (1682 ± 0.6), oil roasted virginia peanuts (1679 ± 1.8), cooked mussels (1676 ± 0.6), raw rainbow trout (1671 ± 0.6), raw milkfish (1669 ± 0.6), raw snapper (1667 ± 0.6), canned sockeye salmon (1664 ± 0.6), broiled pork loin blade including fat (1662 ± 0.5), roasted lean ham (1661 ± 0.5), cooked shrimp (1659 ± 0.6), raw anchovy (1654 ± 0.6), simmered beef tongue (1652 ± 0.5), cooked black bass (1650 ± 0.4), roasted canned lean ham (1645 ± 0.5), raw sheepshead (1644 ± 0.6), tempeh (1636 ± 0.6), raw chum salmon (1636 ± 0.6), raw chinook salmon (1631 ± 0.6), hard pork salami (1630 ± 5), raw bluefish (1629 ± 0.6), canned blue crab (1628 ± 0.6), cooked lobster (1627 ± 0.6), raw pink salmon (1621 ± 0.6), canned pink salmon (1621 ± 0.6), crab cakes (1617 ± 0.8), stewed chicken wing with skin (1615 ± 1.3), raw shrimp (1612 ± 0.6), raw swordfish (1609 ± 0.6), oil roasted almonds (1607 ± 1.8), toasted almonds (1607 ± 1.8), cooked blue crab (1604 ± 0.6), jellied corned beef loaf (1604 ± 1.8), roasted cured pork arm including fat (1602 ± 0.5), roasted canned ham (1596 ± 0.5), broiled beef rib including fat (1590 ± 0.5), pickled anchovy (1589 ± 1.8), raw pollock (1580 ± 0.6), oil roasted sunflower seeds (1579 ± 1.8), raw sunfish (1576 ± 0.6), raw grouper (1575 ± 0.6), dried almonds (1575 ± 1.8), roasted ham (1574 ± 0.5), raw mullet (1573 ± 0.6), raw burbot (1571 ± 0.6), raw pike (1565 ± 0.6), crunchy peanut butter (1559 ± 1.6), raw whitefish (1551 ± 0.6), raw cusk (1544 ± 0.6), raw cisco (1544 ± 0.6), raw ling (1544 ± 0.6), cooked alaska king crab (1536 ± 0.6), raw haddock (1536 ± 0.6), dry roasted peanuts (1536 ± 1.8), pork luxury loaf (1536 ± 1.8), unheated lean ham (1535 ± 0.5), raw scup (1534 ± 0.6), raw freshwater bass (1533 ± 0.6), raw flatfish (1532 ± 0.6), chopped smoked beef (1529 ± 1.8), raw rockfish (1524 ± 0.6), battered and fried shark (1515 ± 0.6), anchovy paste (1514 ± 7.1), raw perch (1514 ± 0.6), raw mackerel (1512 ± 0.6), natto (1509 ± 0.6), breaded and fried catfish (1508 ± 0.6), fried pork loin blade including fat (1505 ± 0.5), raw spot (1505 ± 0.6), raw dolphinfish (1504 ± 0.6), raw pompano (1502 ± 0.6), turkey ham (1500 ± 0.9), basil (1500 ± 50), raw eel (1499 ± 0.6), braised beef lungs (1498 ± 0.5), raw sea bass (1498 ± 0.6), cashew butter (1496 ± 1.8), raw lobster (1492 ± 0.6), raw whiting (1488 ± 0.6), smoked chinook salmon (1486 ± 0.6), raw crayfish (1481 ± 0.6), breaded and fried croaker (1480 ± 0.6), raw goose liver (1477 ± 0.5), raw catfish (1476 ± 0.6), raw alewife (1474 ± 0.5), raw queen crab (1468 ± 0.6), roasted duck with skin (1465 ± 0.5), raw herring (1460 ± 0.6), raw black bass (1459 ± 0.5), braised beef thymus (1458 ± 0.5), raw alaska king crab (1452 ± 0.6), pork and beef pepperoni (1450 ± 8.3), raw carp (1449 ± 0.6), raw cod (1447 ± 0.6), unheated canadian bacon (1446 ± 0.9), raw croaker (1445 ± 0.6), lebanon bologna (1443 ± 2.2), raw striped bass (1441 ± 0.6), canned lean ham (1438 ± 0.5), raw lingcod (1435 ± 0.6), raw blue crab (1433 ± 0.6), raw rainbow smelt (1433 ± 0.6), dry roasted sunflower seeds (1429 ± 1.8), raw freshwater drum (1425 ± 0.6), raw tilefish (1422 ± 0.6), raw wolffish (1422 ± 0.6), turkey pastrami (1414 ± 0.9), raw butterfish (1405 ± 0.6), pork and beef peppered loaf (1404 ± 1.8), poppy seed (1400 ± 16.7), raw smelt (1395 ± 0.5), chicken egg yolk (1394 ± 2.9), unheated ham (1394 ± 0.5), pork and beef honey loaf (1389 ± 1.8), fried abalone (1386 ± 0.6), raw dungeness crab (1381 ± 0.6), canned smelt (1380 ± 0.5), raw shad

(1376 ± 0.6), chopped ham lunch meat (1376 ± 2.4), oil roasted cashews (1375 ± 1.8), ham and cheese roll (1371 ± 1.8), roasted pork blade roll (1371 ± 0.5), raw white sucker (1362 ± 0.6), raw seatrout (1361 ± 0.6), whole dried sesame seeds (1356 ± 5.6), boiled soybeans (1355 ± 0.3), italian pork sausage (1343 ± 0.7), simmered chicken neck with skin (1342 ± 1.3), cooked corned beef brisket (1334 ± 0.5), pork livercheese (1332 ± 1.3), smoked cisco (1331 ± 0.6), beef lunch meat (1329 ± 1.8), canned ham (1320 ± 0.5), pork sausage (1319 ± 1.9), toasted sesame kernels (1318 ± 1.8), whole toasted sesame seeds (1318 ± 1.8), raw sturgeon (1312 ± 0.6), fried tofu (1308 ± 3.8), raw turbot (1305 ± 0.6), dry roasted cashews (1304 ± 1.8), sesame butter (1300 ± 3.3), breaded and fried scallops (1294 ± 1.6), tuna salad (1293 ± 0.2), dry roasted almonds (1289 ± 1.8), braised pork lungs (1288 ± 0.5), cottage cheese (1284 ± 0.2), beef pastrami (1282 ± 1.8), pork and beef luncheon sausage (1278 ± 2.2), breaded fried squid (1267 ± 0.6), minced ham lunch meat (1257 ± 2.4), freeze-dried chives (1250 ± 62.5), raw black bullhead (1239 ± 0.5), dry roasted pistachio nuts (1236 ± 1.8), canned alewife (1231 ± 0.5), corn flakes (1225 ± 1.8), ricotta cheese (1221 ± 0.4), oil roasted filberts (1218 ± 1.8), pork and beef mortadella (1213 ± 3.3), raw abalone (1204 ± 0.6), dried brazil nuts (1204 ± 1.8), berliner (1200 ± 2.2), firm raw tofu (1199 ± 0.4), raw scallops (1181 ± 0.6), boiled lupins (1181 ± 0.3), raw monkfish (1176 ± 0.6), almond butter (1175 ± 3.1), pork bologna (1170 ± 2.2), summer sausage (1161 ± 2.2), pork and beef picnic loaf (1161 ± 1.8), pickled herring (1153 ± 3.3), pork liverwurst (1150 ± 2.8), raw cuttlefish (1144 ± 0.6), miso (1129 ± 0.4), raw flounder (1125 ± 0.5), raw sole (1125 ± 0.5), dried filberts (1114 ± 1.8), beef salami (1104 ± 2.2), stir-fried sprouted soybeans (1100 ± 0.5), pork and beef link sausage with american cheese (1098 ± 1.2), duck egg (1097 ± 0.7), raw squid (1096 ± 0.6), polish pork sausage (1089 ± 1.8), pork and beef old fashioned loaf (1089 ± 1.8), pork and beef brotwurst (1080 ± 0.7), puffed wheat (1079 ± 3.6), chicken egg (1066 ± 1), pork and beef lunch meat (1061 ± 1.8), braised pork brains (1058 ± 0.5), raw octopus (1049 ± 0.6), dried hickory nuts (1043 ± 1.8), grilled ham patties (1042 ± 0.8), smoked beef sausage (1035 ± 1.2), pork liver sausage (1033 ± 2.8), breaded and fried clams (1024 ± 0.6), pork headcheese (1021 ± 1.8), unheated ham patties (1014 ± 0.8), surimi scallops (1011 ± 0.6), dried english/persian walnuts (1007 ± 1.8), gjetost cheese (1004 ± 1.8), fennel seed (1000 ± 25), cooked oysters (994 ± 0.6), raw pork stomach (990 ± 0.5), surimi shrimp (981 ± 0.6), neufchatel cheese (979 ± 1.8), pork and beef sausage (977 ± 3.8), garlic powder (967 ± 16.7), canned pork lunch meat (967 ± 2.4), pork pickle and pimento loaf (961 ± 1.8), simmered pork tail (952 ± 0.5), dill seed (950 ± 25), raw beef tripe (948 ± 0.5), pork bratwurst (944 ± 0.6), falafel (943 ± 1), fried beef brains (943 ± 0.5), pork beerwurst salami (943 ± 2.2), freeze-dried red sweet peppers (938 ± 31.3), freeze-dried green sweet peppers (938 ± 31.3), almond paste (936 ± 1.8), beef and pork salami (930 ± 2.2), beef beerwurst salami (913 ± 2.2), dried pilinuts (904 ± 1.8), beef bologna (900 ± 2.2), beef and pork bologna (900 ± 2.2), raw clams (899 ± 0.6), chicken egg white (897 ± 1.5), beef frankfurter (884 ± 0.9), boiled green soybeans (883 ± 0.6), pork olive loaf (879 ± 1.8), boiled peanuts (875 ± 1.6), pork and beef kielbasa/kolbassy (873 ± 1.9), simmered pork ears (868 ± 0.5), dry roasted filberts (857 ± 1.8), dried pinyon pine nuts (846 ± 1.8), simmered pork feet (845 ± 0.5), raw mussels (838 ± 0.6), simmered beef brains (831 ± 0.5), pork and beef knockwurst (821 ± 0.7), beef and pork frankfurter (820 ± 1.1), gefiltefish with broth (810 ± 1.2), pork and beef link sausage

(810 ± 0.7), simmered pork chitterlings (809 ± 0.5), vienna sausage (800 ± 3.1), sweetened condensed milk (779 ± 1.3), boiled white beans (776 ± 0.3), shredded wheat (768 ± 1.8), dried ginko nuts (764 ± 1.8), bran flakes (764 ± 1.8), boiled winged beans (762 ± 0.3), bran (750 ± 1.8), boiled cranberry beans (746 ± 0.3), cream cheese (739 ± 1.8), evaporated skim milk (738 ± 1.6), ham salad (736 ± 1.8), boiled yellow beans (732 ± 0.3), boiled pink beans (723 ± 0.3), steamed sprouted soybeans (711 ± 1.1), boiled black beans (708 ± 0.3), boiled navy beans (695 ± 0.3), boiled baby lima beans (694 ± 0.3), boiled kidney beans (693 ± 0.3), boiled hyacinth beans (691 ± 0.3), dried shitake mushrooms (680 ± 3.3), boiled lima beans (673 ± 0.3), evaporated whole milk (667 ± 0.4), boiled great northern beans (665 ± 0.3), wheat flakes (664 ± 1.8), boiled pinto beans (656 ± 0.3), boiled lentils (654 ± 0.3), dried acorns (654 ± 1.8), boiled black turtle beans (653 ± 0.3), freeze-dried shallots (650 ± 12.5), breaded and fried oysters (638 ± 0.6), boiled yardlong bean (635 ± 0.3), boiled adzuki beans (632 ± 0.2), boiled chickpeas (631 ± 0.3), boiled mungo beans (625 ± 0.3), boiled catjang black-eyed peas (623 ± 0.3), stir-fried sprouted lentils (617 ± 0.5), raw tofu (614 ± 0.4), frozen and boiled black-eyed peas (606 ± 0.6), acorn flour (604 ± 1.8), raw bacon (601 ± 0.7), canned navy beans (601 ± 0.2), boiled split peas (598 ± 0.3), pickled pork feet (595 ± 0.5), boiled black-eyed peas (592 ± 0.3), canned great northern beans (588 ± 0.2), sheep milk (587 ± 0.2), canned white beans (579 ± 0.2), skim yogurt (578 ± 0.2), dry roasted macadamia nuts (575 ± 1.8), boiled broad beans (572 ± 0.3), boiled french beans (563 ± 0.3), raisin bran (546 ± 1.8), boiled mung beans (544 ± 0.2), dry roasted pecans (543 ± 1.8), evaporated lowfat milk (541 ± 0.4), puffed rice (536 ± 3.6), dried pecans (529 ± 1.8), lowfat yogurt (529 ± 0.2), boiled and frozen baby lima beans (522 ± 0.6), dried coconut (518 ± 1.8), raw spirulina seaweed (509 ± 0.5), raw laver/nori seaweed (501 ± 0.5), refried beans (498 ± 0.2), raw oysters (496 ± 0.6), canned oysters (496 ± 0.6), raw acorns (496 ± 1.8), potato flour (492 ± 0.6), boiled pigeon peas (483 ± 0.3), canned black turtle beans (481 ± 0.2), frozen and boiled fordhook lima beans (476 ± 0.6), oil roasted pecans (475 ± 1.8), boiled sprouted green peas (473 ± 0.5), dried macadamia nuts (468 ± 1.8), lard (464 ± 1.8), boiled succotash (446 ± 0.5), raw pork jowl (446 ± 0.5), potato pancakes (443 ± 0.7), au gratin potatoes (443 ± 0.4), canned cranberry beans (443 ± 0.2), yellow corn pudding (430 ± 0.2), dried chinese chestnuts (429 ± 1.8), boiled mothbeans (425 ± 0.3), canned lima beans (425 ± 0.2), baked beans with pork and sweet sauce (424 ± 0.2), canned kidney beans (415 ± 0.2), dried oriental radish (414 ± 0.9), baked beans with pork and tomato sauce (412 ± 0.2), canned broad beans (411 ± 0.2), oil roasted macadamia nuts (411 ± 1.8), dried toasted coconut (400 ± 1.8), protein fortified skim milk (388 ± 0.2), protein fortified lowfat 2% milk (387 ± 0.2), protein fortified lowfat 1% milk (385 ± 0.2), vegetarian baked beans (383 ± 0.2), dried european chestnuts (382 ± 1.8), frozen and boiled succotash (379 ± 0.6), restaurant vanilla milkshake (378 ± 0.2), eggnog (369 ± 0.2), hummus (368 ± 0.2), indian buffalo milk (366 ± 0.2), canned pinto beans (364 ± 0.2), canned black-eyed peas (363 ± 0.2), tomato powder (359 ± 0.5), boiled yellow corn (359 ± 0.6), hot chocolate (356 ± 0.2), malted milk (356 ± 0.2), canned chickpeas (352 ± 0.2), onion powder (350 ± 25), ginger (350 ± 25), whole yogurt (350 ± 0.2), vanilla milkshake (341 ± 0.2), restaurant chocolate milkshake (334 ± 0.2), skim milk (334 ± 0.2), dried japanese chestnuts (329 ± 1.8), buttermilk (329 ± 0.2), restaurant strawberry milkshake (329 ± 0.2), lowfat 2% milk (326 ± 0.2), filled milk (326 ± 0.2), frozen and boiled yellow

corn (326 ± 0.6), raw green peas (323 ± 0.6), lowfat 1% milk (322 ± 0.2), whole milk (322 ± 0.2), ginko nuts (321 ± 1.8), boiled green peas (320 ± 0.6), lowfat 1% chocolate milk (317 ± 0.2), dehydrated onion flakes (314 ± 3.6), goat milk (314 ± 0.2), lowfat 2% chocolate milk (314 ± 0.2), raw garlic (311 ± 5.6), whole chocolate milk (310 ± 0.2), frozen and boiled green peas (308 ± 0.6), carob milk (307 ± 0.2), boiled jute potherb (307 ± 1.2), frozen and boiled turnip greens (307 ± 0.6), chocolate milkshake (299 ± 0.2), half and half cream (287 ± 3.3), roasted chinese chestnuts (279 ± 1.8), stir-fried sprouted mung beans (276 ± 0.8), noodles (273 ± 0.3), coconut cream (269 ± 0.2), boiled lamb's-quarters (267 ± 0.6), sour cream (267 ± 4.2), light cream (267 ± 3.3), raw sprouted alfalfa seeds (267 ± 1.5), raw chinese chestnuts (264 ± 1.8), canned green peas (264 ± 0.6), raw wakame seaweed (257 ± 0.5), raw coconut (247 ± 1.1), canned yellow corn with red and green peppers (246 ± 0.4), french fried potatoes (244 ± 1), frozen and boiled spinach (244 ± 0.5), okara tofu (244 ± 0.8), dried sweetened flaked coconut (243 ± 0.7), soy milk (241 ± 0.2), medium cream (240 ± 3.3), canned succotash with creamed corn (232 ± 0.4), boiled spinach (231 ± 0.6), canned succotash (229 ± 0.4), scalloped potatoes (225 ± 0.4), canned yellow corn (222 ± 0.5), raw spinach (221 ± 1.8), macaroni (220 ± 0.4), spaghetti (220 ± 0.4), canned spinach (219 ± 0.5), corn grits (215 ± 0.2), dried sweetened shredded coconut (214 ± 0.5), dried apricots (214 ± 1.4), whipping cream, light (213 ± 3.3), frozen and fried french fried potatoes (210 ± 1), potato salad (202 ± 0.4), dried longans (202 ± 0.5), whipping cream, heavy (200 ± 3.3), sweetened coconut cream (200 ± 0.2), frozen and boiled kale (198 ± 0.8), oatmeal (197 ± 0.3), frozen and fried hash brown potatoes (190 ± 0.6), roasted european chestnuts (189 ± 1.8), canned creamed yellow corn (188 ± 0.4), roasted japanese chestnuts (186 ± 1.8), frozen and boiled collards (184 ± 0.6), boiled and steamed chinese chestnuts (182 ± 1.8), soybean oil and cottonseed oil margarine liquid (180 ± 10), raw sprouted mung beans (175 ± 1), coconut milk (170 ± 0.2), o'brien potatoes (169 ± 0.3), raw chives (167 ± 16.7), boiled amaranth (167 ± 0.8), raw watercress (165 ± 2.9), frozen and boiled brussels sprouts (164 ± 0.6), frozen and boiled turnip greens and turnips (156 ± 0.5), boiled dock (152 ± 0.5), raw shallots (150 ± 5), microwaved potato with skin (147 ± 0.2), raw swamp cabbage (146 ± 0.9), hash brown potatoes (146 ± 0.6), raw european chestnuts (146 ± 1.8), frozen and boiled broccoli (145 ± 0.5), raw japanese chestnuts (139 ± 1.8), raw bamboo shoots (139 ± 0.7), boiled broccoli (138 ± 0.6), baked potato with skin (138 ± 0.2), boiled mushrooms (137 ± 0.6), raw turnip greens (136 ± 1.8), boiled swiss chard (135 ± 0.6), dried figs (133 ± 0.3), boiled kale (132 ± 0.8), boiled scotch kale (132 ± 0.8), raw broccoli (132 ± 1.1), mashed potatoes (132 ± 0.5), cooked sprouted mung beans (131 ± 0.8), raw california avocado (131 ± 0.3), raw mushrooms (129 ± 1.4), frozen and boiled asparagus (128 ± 0.8), baked sweet potato (126 ± 0.4), frozen and baked sweet potato (126 ± 0.6), microwaved potato without skin (126 ± 0.3), raw potato without skin (124 ± 0.4), canned turnip greens (124 ± 0.4), canned sweet potato (121 ± 0.3), boiled sweet potato (121 ± 0.3), boiled and steamed european chestnuts (121 ± 1.8), frozen and boiled potato without skin (119 ± 0.5), baked potato without skin (118 ± 0.3), boiled swamp cabbage (116 ± 1), raw cauliflower (116 ± 1), boiled green beans (116 ± 0.8), boiled yellow snap beans (116 ± 0.8), raw dock (115 ± 0.7), cream of wheat (115 ± 0.3), boiled beet greens (115 ± 0.7), boiled brussels sprouts (114 ± 0.6), raw welsh onions (113 ± 0.5), boiled asparagus (112 ± 0.6), boiled cauliflower (110 ± 0.8), frozen and boiled okra (109 ± 0.5), farina

(109 ± 0.3), steamed hawaii mountain yam (108 ± 0.7), boiled turnip greens (106 ± 0.7), tomato paste (105 ± 0.4), raw spring onions with tops (104 ± 1), raw red chili peppers (104 ± 1.1), raw green chili peppers (104 ± 1.1), raw savoy cabbage (103 ± 1.4), boiled potato without skin (103 ± 0.4), raw vine spinach (101 ± 0.5), raw endive (100 ± 2), whole chocolate malted milk (100 ± 0.2), dried japanese persimmon (100 ± 1.5), raw florida avocado (99 ± 0.2), boiled okra (98 ± 0.6), chocolate syrup (97 ± 1.3), raw romaine lettuce (96 ± 1.8), raw leeks (96 ± 1.9), frozen and boiled cauliflower (94 ± 0.6), human milk (94 ± 1.6), baked/boiled tropical yam (94 ± 0.7), boiled savoy cabbage (93 ± 0.7), safflower oil and soybean oil mayonnaise (93 ± 3.6), soybean oil mayonnaise (93 ± 3.6), canned bamboo shoots (93 ± 0.4), cooked shitake mushrooms (93 ± 0.7), boiled chinese cabbage (91 ± 0.6), boiled purslane (91 ± 0.9), raw cassava (90 ± 0.5), raw chinese cabbage (89 ± 1.4), canned sprouted mung beans (89 ± 0.8), dried dates (88 ± 0.6), canned potato without skin (86 ± 0.6), beef suet (86 ± 1.8), baked hubbard squash (84 ± 0.5), boiled hubbard squash (84 ± 0.4), frozen and boiled yellow snap beans (84 ± 0.7), frozen and boiled green beans (84 ± 0.7), sapote (84 ± 0.2), raw kelp/kombu/tangle seaweed (83 ± 0.5), coleslaw (82 ± 0.8), whipped butter (82 ± 4.5), boiled bamboo shoots (82 ± 0.4), raw butterhead lettuce (80 ± 3.3), butter (80 ± 3.3), corn oil margarine stick (80 ± 10), safflower oil and soybean oil margarine stick (80 ± 10), soybean oil margarine stick (80 ± 10), soybean oil and cottonseed oil margarine stick (80 ± 10), raw looseleaf lettuce (79 ± 1.8), whey (78 ± 0.2), apricots (77 ± 0.5), raw ginger root (75 ± 2.1), frozen and boiled crookneck squash (75 ± 0.5), raw chickory greens (74 ± 0.6), cream of rice (73 ± 0.3), frozen and boiled mustard greens (71 ± 0.7), boiled kohlrabi (71 ± 0.6), raw red cabbage (71 ± 1.4), canned green beans (71 ± 0.7), canned yellow snap beans (71 ± 0.7), raw scallop squash (71 ± 0.8), bananas (71 ± 0.4), raw iceberg lettuce (70 ± 2.5), frozen and boiled butternut squash (70 ± 0.4), boiled mustard greens (69 ± 0.7), raw lotus root (69 ± 0.6), raw eggplant (68 ± 1.2), boiled collards (68 ± 0.3), raw zucchini (68 ± 0.8), frozen and boiled zucchini (67 ± 0.4), breadfruit (65 ± 0.5), candied sweet potato (65 ± 0.5), baked acorn squash (64 ± 0.5), raw green cabbage (63 ± 1.4), boiled scallop squash (61 ± 0.6), corn oil margarine tub (60 ± 10), safflower oil margarine tub (60 ± 10), soybean oil margarine tub (60 ± 10), soybean oil and cottonseed oil margarine tub (60 ± 10), canned zucchini in tomato juice (60 ± 0.4), persimmon (60 ± 2), elderberries (60 ± 0.3), boiled chayote (58 ± 0.6), frozen and boiled turnips (57 ± 0.5), guava (56 ± 0.6), boiled red cabbage (55 ± 0.7), raw crookneck squash (55 ± 0.8), boiled leeks (54 ± 1.9), longans (54 ± 0.5), boiled crookneck squash (53 ± 0.6), coconut water (53 ± 0.2), raw celtuce (52 ± 0.5), boiled eggplant (52 ± 1), boiled butternut squash (51 ± 0.5), canned pumpkin (51 ± 0.4), boiled and steamed japanese chestnuts (50 ± 1.8), canned pumpkin pie mix (50 ± 0.4), boiled green cabbage (49 ± 0.7), frozen and boiled carrots (49 ± 0.7), frozen and boiled red sweet peppers (49 ± 0.5), frozen and boiled green sweet peppers (49 ± 0.5), rice polish (48 ± 0.9), canned red chili peppers (47 ± 0.7), canned green chili peppers (47 ± 0.7), boiled carrots (46 ± 0.6), tomato purée (46 ± 0.2), boiled beets (45 ± 0.6), raw chickory (44 ± 1.1), boiled burdock root (44 ± 0.4), raw green tomato (44 ± 0.4), raw red sweet peppers (44 ± 1), raw green sweet peppers (44 ± 1), raw carrots (43 ± 0.7), canned jalapeño peppers (43 ± 0.7), boiled lotus root (42 ± 0.6), japanese persimmon (42 ± 0.3), raw onions (41 ± 0.6), boiled red tomato (41 ± 0.4), canned red sweet peppers (41 ± 0.7), canned green sweet peppers (41 ± 0.7), boiled yambean

(40 ± 0.5), carambola (40 ± 0.4), blueberries (40 ± 0.3), peach (40 ± 0.6), corn oil diet margarine tub (40 ± 10), soybean oil and cottonseed oil diet margarine tub (40 ± 10), raw radish (38 ± 1.1), boiled acorn squash (38 ± 0.4), cooked taro (38 ± 0.8), boiled zucchini (37 ± 0.6), boiled calabash gourd (36 ± 0.7), cooked plantain (36 ± 0.3), canned crookneck squash (36 ± 0.5), boiled rutabaga (35 ± 0.6), boiled oriental radish (35 ± 0.7), boiled/baked spaghetti squash (35 ± 0.6), pickled beets (34 ± 0.4), figs (34 ± 1), boiled pumpkin (34 ± 0.4), canned peeled red tomatoes (34 ± 0.4), canned stewed red tomatoes (34 ± 0.4), raw red tomato (33 ± 0.4), boiled onions (32 ± 0.5), raw oriental radish (32 ± 1.1), boiled red sweet peppers (32 ± 0.7), boiled green sweet peppers (32 ± 0.7), mango (31 ± 0.2), strawberries (31 ± 0.3), raw celery (30 ± 1.3), canned carrots (27 ± 0.7), boiled turnips (26 ± 0.6), navel orange (26 ± 0.4), valencia orange (26 ± 0.4), loquats (26 ± 0.5), frozen and boiled onions (25 ± 0.5), crab apples (25 ± 0.5), boiled celery (24 ± 0.7), sapodilla (24 ± 0.3), raw cucumber (23 ± 1), plum (21 ± 0.8), tomato juice (21 ± 0.3), pineapple (19 ± 0.3), watermelon (18 ± 0.3), cooked apple without skin (16 ± 0.3), papaya (16 ± 0.2), mandarin oranges (15 ± 0.4), tangerine (15 ± 0.6), pear (14 ± 0.3), european grapes (14 ± 0.3), orange juice (13 ± 0.2), american grapes (13 ± 0.5), raw apple with skin (12 ± 0.4), grape juice (12 ± 0.2), applesauce (11 ± 0.4), screwdriver (10 ± 0.2), tangerine juice (10 ± 0.2), raw apple without skin (9 ± 0.4), thompson seedless grapes (8 ± 0.3), ale (6 ± 0.1), coffee (5 ± 0.3), white grapefruit juice (4 ± 0.2)
Applications: pain control?
Daily Dosage: Nutritional — 2.5–5 g; therapeutic — ?; experimental — ?; toxic — none known up to 5 g
Warnings: increased amounts are dangerous. See toxicity symptoms.

LYSINE

Names: lysine, LYS, amino acid K, 2-diamino-hexanoic acid
Forms: 2-diamino-hexanoic acid
Classification: essential amino acid
Deficiency: easily tired, fatigue, nausea, dizziness, deterioration of appetite, weight decreased, emotional agitation, mental health deterioration, decreased antibody formation, decreased immunity, slow growth, anemia, enzyme deterioration, reproductive systems deterioration, pneumonia, acidosis, blood-shot eyes, decreased vitamin B_T formation, protein decreased
Side-effects: none known
Toxicity: none known
Inhibitors: increased amounts of other amino acids
Helpers: ?moderate amounts of other amino acids
Sources (mg per 100 grams of food): torula yeast (14761 ± 1.8), dried and salted cod (5769 ± 0.6), soybean protein isolate (5329 ± 1.8), soybean protein concentrate (3929 ± 1.8), parmesan cheese (3840 ± 10), braised pork center loin excluding fat (3427 ± 0.5), braised pork top loin excluding fat (3394 ± 0.5), braised pork sirloin excluding fat (3302 ± 0.5), braised pork loin excluding fat (3251 ± 0.5), dried and frozen tofu (3159 ± 2.9), broiled pork center loin excluding fat (3154 ± 0.5), freeze-dried parsley (3143 ± 35.7), defatted soybean flour (3129 ± 0.5), fenugreek seed (3100 ± 25), low fat soy-

bean flour (3097 ± 0.6), braised pork boston blade excluding fat (3071 ± 0.5), dried spirulina seaweed (3025 ± 0.5), defatted soy meal (2991 ± 0.4), cooked whelk (2931 ± 0.6), braised pork loin blade excluding fat (2927 ± 0.5), fried beef top round excluding fat (2917 ± 0.5), braised pork center loin including fat (2878 ± 0.5), roasted pork rump excluding fat (2872 ± 0.5), dry skim milk (2867 ± 1.7), braised pork spareribs (2864 ± 0.5), broiled pork center rib excluding fat (2840 ± 0.5), broiled pork top loin excluding fat (2840 ± 0.5), roasted pork tenderloin (2837 ± 0.5), fried chicken breast without skin (2836 ± 0.6), fried pork center loin excluding fat (2835 ± 0.5), roasted turkey light meat without skin (2818 ± 0.5), roasted pork center loin excluding fat (2808 ± 0.5), simmered beef shank excluding fat (2802 ± 0.5), braised pork center rib including fat (2794 ± 0.5), roasted pork leg excluding fat (2791 ± 0.5), dry instant skim milk (2784 ± 0.5), fried light meat chicken without skin (2784 ± 0.5), broiled pork sirloin excluding fat (2784 ± 0.5), roasted pork center rib excluding fat (2781 ± 0.5), roasted pork top loin excluding fat (2781 ± 0.5), roasted pork shank excluding fat (2780 ± 0.5), fried pork center rib excluding fat (2756 ± 0.5), fried pork top loin excluding fat (2756 ± 0.5), braised beef chuck arm pot roast excluding fat (2747 ± 0.5), broiled pork loin excluding fat (2744 ± 0.5), gruyere cheese (2743 ± 1.8), braised pork sirloin including fat (2735 ± 0.5), canned tuna in water (2716 ± 0.6), roasted pork sirloin excluding fat (2710 ± 0.5), fried beef sirloin excluding fat (2702 ± 0.5), braised pork top loin including fat (2700 ± 0.5), stewed chicken drumstick without skin (2698 ± 1.1), edam cheese (2693 ± 1.8), roasted turkey dark meat without skin (2693 ± 0.5), gouda cheese (2686 ± 1.8), broiled pork center loin including fat (2683 ± 0.5), provolone cheese (2679 ± 1.8), anchovy canned in olive oil (2655 ± 2.5), braised pork loin including fat (2654 ± 0.5), roasted pork loin excluding fat (2651 ± 0.5), cooked flounder (2640 ± 0.5), broiled beef top round excluding fat (2637 ± 0.5), fried beef top round including fat (2636 ± 0.5), roasted chicken breast without skin (2635 ± 0.6), dry roasted soybean nuts (2634 ± 0.6), fried chicken breast with skin (2634 ± 0.5), simmered beef shank including fat (2630 ± 0.5), braised beef bottom round excluding fat (2628 ± 0.5), roasted light meat chicken without skin (2626 ± 0.5), toasted soybean nuts (2621 ± 1.8), braised pork arm braised including fat (2620 ± 0.5), swiss cheese (2618 ± 1.8), roasted pork rump including fat (2613 ± 0.5), roasted turkey light meat with skin (2599 ± 0.5), braised beef chuck blade roast excluding fat (2584 ± 0.5), braised pork boston blade braised including fat (2582 ± 0.5), broiled beef top round including fat (2565 ± 0.5), braised beef shortribs excluding fat (2559 ± 0.5), well done baked extra lean ground beef (2531 ± 0.5), dry buttermilk (2529 ± 7.1), broiled beef sirloin excluding fat (2527 ± 0.5), cooked coho salmon (2513 ± 0.6), cooked sockeye salmon (2508 ± 0.6), roasted turkey dark meat with skin (2503 ± 0.5), roasted pork shoulder excluding fat (2501 ± 0.5), roasted pumpkin and squash seeds (2496 ± 1.8), roasted pork center loin including fat (2490 ± 0.5), fried light meat chicken with skin (2487 ± 0.5), broiled pork boston blade excluding fat (2481 ± 0.5), braised beef bottom round including fat (2480 ± 0.5), roasted chicken breast with skin (2473 ± 0.5), well done baked lean ground beef (2472 ± 0.5), stewed chicken breast without skin (2462 ± 0.5), roasted pork sirloin including fat (2456 ± 0.5), broiled pork loin blade excluding fat (2455 ± 0.5), stewed light meat chicken without skin (2454 ± 0.5), cooked halibut (2451 ± 0.6), roasted pork leg including fat (2451 ± 0.5), braised beef brisket excluding fat (2445 ± 0.5), fried dark meat chicken without skin (2441 ± 0.5), canned tuna in oil (2436 ± 0.6), roasted pork loin blade excluding fat (2432 ± 0.5), roasted pork center rib including fat (2422 ± 0.5), cooked rainbow trout (2419 ± 0.6), cooked snapper (2415 ± 0.6), roasted eye of beef round excluding fat (2412 ± 0.5), well done baked ground beef (2405 ± 0.5), chipped beef (2404 ± 1.8), broiled pork center rib including fat (2403 ± 0.5), roasted pork boston blade excluding fat (2401 ± 0.5), fried pork loin blade excluding fat (2397 ± 0.5), cooked sheepshead (2391 ± 0.6), roasted beef tip excluding fat (2389 ± 0.5), well done broiled extra lean ground beef (2387 ± 0.5), broiled beef top loin excluding fat (2381 ± 0.5), roasted pork shank including fat (2376 ± 0.5), roasted light meat chicken with skin (2374 ± 0.5), simmered beef heart (2372 ± 0.5), broiled beef round excluding fat (2367 ± 0.5), roasted pork top loin including fat (2366 ± 0.5), broiled pork sirloin including fat (2357 ± 0.5), well done broiled lean ground beef (2355 ± 0.5), broiled beef tenderloin excluding fat (2350 ± 0.5), fried chicken giblets (2348 ± 0.5), roasted soybean nuts (2344 ± 0.6), broiled porterhouse steak excluding fat (2343 ± 0.5), broiled T-bone steak excluding fat (2341 ± 0.5), well done fried extra lean ground beef (2338 ± 0.5), braised pork loin blade including fat (2335 ± 0.5), broiled beef rib eye excluding fat (2333 ± 0.5), cooked swordfish (2332 ± 0.6), braised beef flank excluding fat (2331 ± 0.5), roasted dark meat chicken without skin (2325 ± 0.5), roasted soybean flour (2316 ± 0.6), smoked haddock (2316 ± 0.6), broiled pork top loin including fat (2315 ± 0.5), well done fried lean ground beef (2305 ± 0.5), broiled pork loin including fat (2303 ± 0.5), soybean flour (2298 ± 0.6), roasted pork loin including fat (2292 ± 0.5), roasted beef tenderloin excluding fat (2291 ± 0.5), roasted beef rib roasted excluding fat (2290 ± 0.5), braised beef flank including fat (2289 ± 0.5), fried beef sirloin including fat (2284 ± 0.5), cooked grouper (2282 ± 0.6), broiled beef sirloin including fat (2279 ± 0.5), cooked mullet (2278 ± 0.6), well done broiled ground beef (2272 ± 0.5), stewed chicken breast with skin (2272 ± 0.5), fried pork center loin including fat (2268 ± 0.5), cooked pike (2267 ± 0.6), cooked bacon (2261 ± 0.4), sardines (2258 ± 2.1), kippered herring (2258 ± 1.3), well done fried ground beef (2255 ± 0.5), braised beef chuck arm pot roast including fat (2254 ± 0.5), cooked tilefish (2249 ± 0.6), simmered turkey heart (2243 ± 0.5), roasted eye of beef round including fat (2227 ± 0.5), cooked haddock (2227 ± 0.6), american cheese (2225 ± 1.8), fried chicken back with skin (2221 ± 0.7), pimento cheese spread (2221 ± 1.8), cooked flatfish (2219 ± 0.6), roasted chicken drumstick with skin (2215 ± 1), simmered chicken heart (2214 ± 0.5), fried chicken drumstick with skin (2214 ± 1), cooked rockfish (2208 ± 0.6), stewed dark meat chicken without skin (2206 ± 0.5), roasted chicken thigh without skin (2204 ± 1), roasted beef round tip including fat (2202 ± 0.5), fried dark meat chicken with skin (2202 ± 0.5), cooked perch (2193 ± 0.6), roasted pork arm including fat (2178 ± 0.5), cooked pompano (2175 ± 0.6), cooked eel (2171 ± 0.6), cooked sea bass (2171 ± 0.6), fried chicken thigh with skin (2169 ± 0.8), muenster cheese (2164 ± 1.8), broiled beef tenderloin including fat (2160 ± 0.5), raw pheasant without skin (2157 ± 0.5), cooked whiting (2156 ± 0.6), roasted pork shoulder including fat (2152 ± 0.5), brick cheese (2150 ± 1.8), smoked whitefish (2149 ± 0.6), cooked mackerel (2144 ± 0.6), stewed light meat chicken with skin (2142 ± 0.5), broiled beef top loin including fat (2140 ± 0.5), broiled pork boston blade including fat (2137 ± 0.5), roasted pork boston blade including fat (2131 ± 0.5), raw yellowtail (2126 ± 0.6), broiled beef round including fat (2125 ± 0.5), medium-rare broiled extra lean ground beef (2121 ± 0.5), turkey breast meat summer sausage

(2119 ± 2.4), braised beef chuck blade roast including fat (2117 ± 0.5), roasted cured pork arm excluding fat (2115 ± 0.5), cooked herring (2115 ± 0.6), broiled beef flank excluding fat (2113 ± 0.5), broiled beef rib eye including fat (2111 ± 0.5), cooked black bass (2110 ± 0.4), braised pork spleen (2107 ± 0.5), roasted dark meat chicken with skin (2105 ± 0.5), roasted chicken wing with skin (2100 ± 1.5), cooked carp (2100 ± 0.6), fried pork center rib including fat (2100 ± 0.5), cooked cod (2096 ± 0.6), canned cod (2091 ± 0.6), fried pork top loin including fat (2091 ± 0.5), dry whole milk (2088 ± 1.6), broiled porterhouse steak including fat (2088 ± 0.5), medium-rare fried extra lean ground beef (2085 ± 0.5), broiled beef flank including fat (2084 ± 0.5), simmered chicken neck without skin (2083 ± 2.8), cooked crayfish (2082 ± 0.6), stewed chicken drumstick with skin (2082 ± 0.9), canned corned beef (2076 ± 0.5), cooked rainbow smelt (2076 ± 0.6), cheddar cheese (2072 ± 0.5), medium-rare broiled lean ground beef (2065 ± 0.5), tilsit cheese (2064 ± 1.8), monterey cheese (2064 ± 1.8), roasted pork loin blade including fat (2053 ± 0.5), broiled beef rib excluding fat (2050 ± 0.5), roasted chicken thigh with skin (2047 ± 0.8), medium-rare baked extra lean ground beef (2044 ± 0.5), roasted beef tenderloin including fat (2035 ± 0.5), simmered turkey gizzard (2034 ± 0.5), fried chicken wing with skin (2031 ± 1.6), peanut flour (2025 ± 12.5), medium-rare fried lean ground beef (2024 ± 0.5), broiled pork loin blade including fat (2012 ± 0.5), port du salut cheese (2011 ± 1.8), medium-rare broiled ground beef (2010 ± 0.5), roasted duck without skin (2009 ± 0.5), braised pork liver (2007 ± 0.5), colby cheese (2004 ± 1.8), braised beef pancreas (1999 ± 0.5), medium-rare baked lean ground beef (1999 ± 0.5), medium-rare fried ground beef (1998 ± 0.5), mozzarella cheese (1996 ± 1.8), broiled T-bone steak including fat (1996 ± 0.5), roasted goose with skin (1988 ± 0.5), raw coho salmon (1985 ± 0.6), braised pork tongue (1970 ± 0.5), cheshire cheese (1968 ± 1.8), braised pork pancreas (1965 ± 0.5), raw sockeye salmon (1956 ± 0.6), braised pork heart (1952 ± 0.4), simmered turkey giblets (1951 ± 0.5), raw shark (1926 ± 0.6), medium-rare baked ground beef (1923 ± 0.5), braised beef brisket including fat (1913 ± 0.5), raw halibut (1911 ± 0.6), cooked clams (1909 ± 0.6), raw trout (1907 ± 0.6), stewed dark meat chicken with skin (1906 ± 0.5), raw quail without skin (1905 ± 0.5), cooked sturgeon (1901 ± 0.6), stewed chicken thigh with skin (1899 ± 0.7), roasted beef rib including fat (1890 ± 0.5), grilled canadian bacon (1887 ± 1.1), raw rainbow trout (1887 ± 0.6), simmered chicken giblets (1886 ± 0.5), raw milkfish (1886 ± 0.6), raw snapper (1884 ± 0.6), canned sockeye salmon (1880 ± 0.6), hard pork salami (1880 ± 5), bleu cheese (1879 ± 1.8), simmered chicken gizzard (1877 ± 0.5), brie cheese (1875 ± 1.8), raw anchovy (1869 ± 0.6), dried pumpkin and squash seeds (1861 ± 1.8), raw sheepshead (1856 ± 0.6), fried beef liver (1855 ± 0.5), raw chum salmon (1849 ± 0.6), simmered chicken liver (1843 ± 0.5), raw chinook salmon (1842 ± 0.6), raw bluefish (1840 ± 0.6), pickled anchovy (1839 ± 1.8), caviar (1831 ± 3.1), raw pink salmon (1831 ± 0.6), canned pink salmon (1831 ± 0.6), braised pork kidneys (1829 ± 0.5), fried chicken neck with skin (1828 ± 1.4), fried pork loin blade including fat (1821 ± 0.5), hard pork and beef salami (1820 ± 5), cooked shrimp (1820 ± 0.6), roasted canned lean ham (1818 ± 0.5), raw swordfish (1818 ± 0.6), braised beef thymus (1818 ± 0.5), braised beef spleen (1815 ± 0.5), simmered turkey liver (1814 ± 0.5), braised beef shortribs including fat (1794 ± 0.5), stewed chicken wing with skin (1790 ± 1.3), camembert cheese (1789 ± 1.8), canned blue crab (1786 ± 0.6), raw pollock (1786 ± 0.6), cooked lobster (1784 ± 0.6), jellied corned beef loaf (1782 ±

1.8), raw sunfish (1782 ± 0.6), cooked mussels (1779 ± 0.6), raw grouper (1779 ± 0.6), raw mullet (1776 ± 0.6), roasted lean ham (1775 ± 0.5), turkey ham (1775 ± 0.9), raw burbot (1774 ± 0.6), raw shrimp (1768 ± 0.6), raw pike (1768 ± 0.6), breaded and fried shrimp (1764 ± 0.6), roasted canned ham (1764 ± 0.5), cooked blue crab (1758 ± 0.6), anchovy paste (1757 ± 7.1), raw whitefish (1753 ± 0.6), pork link sausage (1746 ± 0.7), raw cusk (1744 ± 0.6), raw cisco (1744 ± 0.6), raw ling (1744 ± 0.6), raw haddock (1736 ± 0.6), crab cakes (1733 ± 0.8), raw scup (1733 ± 0.6), raw freshwater bass (1732 ± 0.6), raw flatfish (1731 ± 0.6), roasted cured pork arm including fat (1728 ± 0.5), raw rockfish (1722 ± 0.6), milk chocolate (1714 ± 1.8), raw perch (1711 ± 0.6), raw mackerel (1708 ± 0.6), raw alewife (1707 ± 0.5), simmered beef tongue (1705 ± 0.5), raw spot (1700 ± 0.6), raw dolphinfish (1699 ± 0.6), limburger cheese (1696 ± 1.8), simmered beef kidneys (1696 ± 0.5), raw pompano (1696 ± 0.6), raw eel (1694 ± 0.6), braised beef liver (1693 ± 0.5), raw sea bass (1693 ± 0.6), raw black bass (1690 ± 0.5), cooked alaska king crab (1684 ± 0.6), roasted ham (1682 ± 0.5), raw whiting (1682 ± 0.6), smoked chinook salmon (1679 ± 0.6), broiled beef rib including fat (1674 ± 0.5), pork luxury loaf (1671 ± 1.8), raw catfish (1669 ± 0.6), chopped smoked beef (1668 ± 1.8), yellow mustard seed (1667 ± 16.7), turkey pastrami (1653 ± 0.9), raw herring (1651 ± 0.6), unheated lean ham (1640 ± 0.5), raw carp (1638 ± 0.6), raw smelt (1637 ± 0.5), raw lobster (1636 ± 0.6), raw cod (1635 ± 0.6), battered and fried shark (1634 ± 0.6), raw croaker (1633 ± 0.6), raw striped bass (1628 ± 0.6), raw crayfish (1624 ± 0.6), raw lingcod (1622 ± 0.6), raw rainbow smelt (1619 ± 0.6), canned smelt (1619 ± 0.5), unheated canadian bacon (1616 ± 0.9), raw queen crab (1611 ± 0.6), raw tilefish (1607 ± 0.6), raw wolffish (1607 ± 0.6), breaded and fried catfish (1596 ± 0.6), raw alaska king crab (1592 ± 0.6), canned lean ham (1589 ± 0.5), raw butterfish (1587 ± 0.6), lebanon bologna (1578 ± 2.2), raw blue crab (1572 ± 0.6), breaded and fried croaker (1571 ± 0.6), raw shad (1555 ± 0.6), raw freshwater drum (1540 ± 0.6), raw white sucker (1539 ± 0.6), raw seatrout (1538 ± 0.6), chopped ham lunch meat (1533 ± 2.4), pork and beef peppered loaf (1529 ± 1.8), ham and cheese roll (1529 ± 1.8), italian pork sausage (1522 ± 0.7), raw dungeness crab (1515 ± 0.6), smoked cisco (1504 ± 0.6), pork and beef pepperoni (1500 ± 8.3), pork and beef honey loaf (1500 ± 1.8), pork sausage (1493 ± 1.9), unheated ham (1489 ± 0.5), roasted duck with skin (1486 ± 0.5), raw sturgeon (1484 ± 0.6), beef lunch meat (1479 ± 1.8), raw turbot (1474 ± 0.6), simmered chicken neck with skin (1466 ± 1.3), raw whelk (1465 ± 0.6), roasted pork blade roll (1465 ± 0.5), canned ham (1458 ± 0.5), tuna salad (1457 ± 0.2), pork and beef luncheon sausage (1448 ± 2.2), braised beef lungs (1446 ± 0.5), raw black bullhead (1435 ± 0.5), fried abalone (1433 ± 0.6), canned alewife (1426 ± 0.5), raw duck liver (1418 ± 0.5), cooked corned beef brisket (1392 ± 0.5), minced ham lunch meat (1362 ± 2.4), almond meal (1339 ± 1.8), beef pastrami (1339 ± 1.8), ricotta cheese (1338 ± 0.4), raw monkfish (1331 ± 0.6), berliner (1317 ± 2.2), breaded fried squid (1311 ± 0.6), raw flounder (1306 ± 0.5), raw sole (1306 ± 0.5), pickled herring (1300 ± 3.3), dried pistachio nuts (1296 ± 1.8), breaded and fried scallops (1281 ± 1.6), raw abalone (1278 ± 0.6), pork and beef mortadella (1260 ± 3.3), raw scallops (1254 ± 0.6), raw goose liver (1239 ± 0.5), raw cuttlefish (1213 ± 0.6), braised pork lungs (1211 ± 0.5), summer sausage (1209 ± 2.2), pork and beef picnic loaf (1207 ± 1.8), pork bologna (1204 ± 2.2), pork and beef lunch meat (1200 ± 1.8), pork livercheese (1179 ± 1.3), pork liverwurst (1167 ± 2.8), surimi scallops

LYSINE

(1167 ± 0.6), raw squid (1164 ± 0.6), pork and beef link sausage with american cheese (1158 ± 1.2), beef salami (1152 ± 2.2), natto (1145 ± 0.6), pork and beef brotwurst (1139 ± 0.7), surimi shrimp (1132 ± 0.6), fried tofu (1131 ± 3.8), tempeh (1125 ± 0.6), polish pork sausage (1125 ± 1.8), grilled ham patties (1120 ± 0.8), raw octopus (1114 ± 0.6), chicken egg yolk (1112 ± 2.9), beef and pork salami (1109 ± 2.2), boiled soybeans (1108 ± 0.3), creamy peanut butter (1100 ± 3.1), dill seed (1100 ± 25), pork and beef old fashioned loaf (1089 ± 1.8), unheated ham patties (1089 ± 0.8), pork and beef sausage (1085 ± 3.8), smoked beef sausage (1081 ± 1.2), corn germ (1076 ± 0.5), pork bratwurst (1071 ± 0.6), cooked oysters (1055 ± 0.6), oil roasted peanuts (1050 ± 1.8), raw beef tripe (1044 ± 0.5), firm raw tofu (1039 ± 0.4), poppy seed (1033 ± 16.7), pork beerwurst salami (1030 ± 2.2), simmered pork tail (1020 ± 0.5), pork and beef kielbasa/kolbassy (1012 ± 1.9), cottage cheese (1010 ± 0.2), dried peanuts (1007 ± 1.8), oil roasted spanish peanuts (1004 ± 1.8), freeze-dried chives (1000 ± 62.5), breaded and fried clams (992 ± 0.6), pork headcheese (979 ± 1.8), oil roasted valencia peanuts (971 ± 1.8), braised pork brains (954 ± 0.5), raw clams (954 ± 0.6), beef beerwurst salami (952 ± 2.2), duck egg (951 ± 0.7), dried sunflower seeds (950 ± 1.8), cashew butter (950 ± 1.8), canned pork lunch meat (943 ± 2.4), dry roasted pistachio nuts (939 ± 1.8), beef bologna (939 ± 2.2), pork and beef knockwurst (932 ± 0.7), oil roasted virginia peanuts (929 ± 1.8), beef frankfurter (923 ± 0.9), stir-fried sprouted soybeans (916 ± 0.5), dried pignolia pine nuts (914 ± 1.8), pork liver sausage (911 ± 2.8), pork pickle and pimento loaf (907 ± 1.8), neufchatel cheese (904 ± 1.8), beef and pork frankfurter (904 ± 1.1), pork and beef link sausage (903 ± 0.7), dried watermelon seeds (900 ± 1.8), basil (900 ± 50), oil roasted sunflower seeds (893 ± 1.8), raw mussels (889 ± 0.6), beef and pork bologna (883 ± 2.2), oil roasted cashews (875 ± 1.8), raw pork stomach (874 ± 0.5), crunchy peanut butter (863 ± 1.6), falafel (857 ± 1), dry roasted peanuts (850 ± 1.8), gefiltefish with broth (843 ± 1.2), boiled lupins (832 ± 0.3), dry roasted cashews (829 ± 1.8), simmered pork feet (826 ± 0.5), dried sesame kernels (825 ± 6.3), gjetost cheese (825 ± 1.8), chicken egg (820 ± 1), pork olive loaf (818 ± 1.8), freeze-dried red sweet peppers (813 ± 31.3), freeze-dried green sweet peppers (813 ± 31.3), dry roasted sunflower seeds (807 ± 1.8), vienna sausage (794 ± 3.1), dried butternuts (782 ± 1.8), ham salad (782 ± 1.8), fried beef brains (752 ± 0.5), fennel seed (750 ± 25), boiled green soybeans (739 ± 0.6), dried black walnuts (732 ± 1.8), simmered pork ears (725 ± 0.5), stir-fried sprouted lentils (698 ± 0.5), oil roasted almonds (689 ± 1.8), toasted almonds (689 ± 1.8), cream cheese (686 ± 1.8), dried almonds (675 ± 1.8), chicken egg white (670 ± 1.5), boiled white beans (668 ± 0.3), simmered beef brains (662 ± 0.5), miso (660 ± 0.4), simmered pork chitterlings (656 ± 0.5), boiled winged beans (652 ± 0.3), raw bacon (643 ± 0.7), boiled cranberry beans (641 ± 0.3), sweetened condensed milk (632 ± 1.3), boiled lentils (630 ± 0.3), boiled yellow beans (629 ± 0.3), boiled pink beans (622 ± 0.3), boiled black beans (608 ± 0.3), boiled split peas (602 ± 0.3), evaporated skim milk (597 ± 1.6), boiled navy beans (597 ± 0.3), boiled kidney beans (595 ± 0.3), boiled chickpeas (593 ± 0.3), steamed sprouted soybeans (591 ± 1.1), breaded and fried oysters (582 ± 0.6), pickled pork feet (582 ± 0.5), boiled great northern beans (572 ± 0.3), whole dried sesame seeds (567 ± 5.6), boiled adzuki beans (567 ± 0.2), boiled pinto beans (564 ± 0.3), boiled black turtle beans (562 ± 0.3), boiled yardlong bean (561 ± 0.3), frozen and boiled black-eyed peas (558 ± 0.6), boiled hyacinth beans (556 ± 0.3), dry roasted almonds

(554 ± 1.8), toasted sesame kernels (550 ± 1.8), whole toasted sesame seeds (550 ± 1.8), dried brazil nuts (550 ± 1.8), freeze-dried shallots (550 ± 12.5), boiled catjang black-eyed peas (550 ± 0.3), sesame butter (547 ± 3.3), lard (546 ± 1.8), evaporated whole milk (540 ± 0.4), boiled baby lima beans (539 ± 0.3), garlic powder (533 ± 16.7), raw tofu (532 ± 0.4), raw oysters (529 ± 0.6), canned oysters (528 ± 0.6), raw pork jowl (528 ± 0.5), boiled lima beans (523 ± 0.3), boiled black-eyed peas (523 ± 0.3), canned navy beans (517 ± 0.2), skim yogurt (514 ± 0.2), sheep milk (513 ± 0.2), dried acorns (511 ± 1.8), evaporated lowfat milk (507 ± 0.4), almond butter (506 ± 3.1), canned great northern beans (506 ± 0.2), dried hickory nuts (504 ± 1.8), dried ginko nuts (500 ± 1.8), boiled mungo beans (500 ± 0.3), onion powder (500 ± 25), canned white beans (498 ± 0.2), boiled sprouted green peas (497 ± 0.5), boiled mung beans (490 ± 0.2), bran (489 ± 1.8), boiled broad beans (486 ± 0.3), boiled peanuts (484 ± 1.6), boiled french beans (484 ± 0.3), potato flour (479 ± 0.6), acorn flour (475 ± 1.8), boiled pigeon peas (474 ± 0.3), lowfat yogurt (470 ± 0.2), oil roasted filberts (443 ± 1.8), dried pinyon pine nuts (439 ± 1.8), boiled and frozen baby lima beans (439 ± 0.6), dehydrated onion flakes (429 ± 3.6), refried beans (428 ± 0.2), puffed wheat (414 ± 3.6), canned black turtle beans (414 ± 0.2), dried filberts (404 ± 1.8), frozen and boiled fordhook lima beans (401 ± 0.6), almond paste (400 ± 1.8), dried english/persian walnuts (393 ± 1.8), dried oriental radish (393 ± 0.9), raw acorns (389 ± 1.8), bran flakes (382 ± 1.8), dried european chestnuts (382 ± 1.8), au gratin potatoes (381 ± 0.4), canned cranberry beans (381 ± 0.2), dried pilinuts (375 ± 1.8), dried chinese chestnuts (375 ± 1.8), potato pancakes (371 ± 0.7), tomato powder (370 ± 0.5), baked beans with pork and sweet sauce (364 ± 0.2), canned kidney beans (356 ± 0.2), baked beans with pork and tomato sauce (354 ± 0.2), canned broad beans (350 ± 0.2), dried japanese chestnuts (346 ± 1.8), shredded wheat (343 ± 1.8), dried shitake mushrooms (340 ± 3.3), dry roasted macadamia nuts (332 ± 1.8), canned chickpeas (331 ± 0.2), canned lima beans (330 ± 0.2), dried macadamia nuts (329 ± 1.8), vegetarian baked beans (329 ± 0.2), canned black-eyed peas (321 ± 0.2), raw green peas (317 ± 0.6), protein fortified skim milk (314 ± 0.2), boiled green peas (314 ± 0.6), protein fortified lowfat 2% milk (313 ± 0.2), canned pinto beans (313 ± 0.2), raw spirulina seaweed (312 ± 0.5), protein fortified lowfat 1% milk (312 ± 0.2), hummus (312 ± 0.2), dry roasted filberts (311 ± 1.8), whole yogurt (311 ± 0.2), dried coconut (307 ± 1.8), restaurant vanilla milkshake (306 ± 0.2), dry roasted pecans (304 ± 1.8), frozen and boiled green peas (303 ± 0.6), eggnog (298 ± 0.2), boiled succotash (297 ± 0.5), wheat flakes (296 ± 1.8), dried pecans (296 ± 1.8), goat milk (290 ± 0.2), oil roasted macadamia nuts (289 ± 1.8), hot chocolate (289 ± 0.2), indian buffalo milk (280 ± 0.2), raw garlic (278 ± 5.6), buttermilk (277 ± 0.2), vanilla milkshake (276 ± 0.2), raisin bran (275 ± 1.8), puffed rice (271 ± 3.6), skim milk (271 ± 0.2), restaurant chocolate milkshake (270 ± 0.2), boiled lamb's-quarters (270 ± 0.6), malted milk (269 ± 0.2), restaurant strawberry milkshake (266 ± 0.2), oil roasted pecans (264 ± 1.8), lowfat 2% milk (264 ± 0.2), filled milk (264 ± 0.2), yellow corn pudding (263 ± 0.2), lowfat 1% milk (261 ± 0.2), whole milk (261 ± 0.2), stir-fried sprouted mung beans (261 ± 0.8), canned green peas (259 ± 0.6), lowfat 1% chocolate milk (257 ± 0.2), lowfat 2% chocolate milk (254 ± 0.2), dried apricots (254 ± 1.4), frozen and boiled succotash (252 ± 0.6), whole chocolate milk (252 ± 0.2), ginger (250 ± 25), carob milk (249 ± 0.2), roasted chinese chestnuts (246 ± 1.8), chocolate milkshake (242 ± 0.2), dried toasted coconut (236 ± 1.8), half

and half cream (233 ± 3.3), raw chinese chestnuts (232 ± 1.8), boiled mushrooms (227 ± 0.6), raw laver/nori seaweed (222 ± 0.5), raw parsley (220 ± 1.7), frozen and boiled turnip greens (218 ± 0.6), raw sprouted alfalfa seeds (215 ± 1.5), french fried potatoes (214 ± 1), light cream (213 ± 3.3), ginko nuts (211 ± 1.8), okara tofu (211 ± 0.8), raw mushrooms (211 ± 1.4), sour cream (208 ± 4.2), roasted japanese chestnuts (196 ± 1.8), medium cream (193 ± 3.3), frozen and boiled spinach (192 ± 0.5), scalloped potatoes (192 ± 0.4), roasted european chestnuts (189 ± 1.8), dried lychees (187 ± 0.5), frozen and fried french fried potatoes (184 ± 1), boiled spinach (182 ± 0.6), soy milk (179 ± 0.2), raw spinach (175 ± 1.8), whipping cream, light (173 ± 3.3), boiled jute potherb (172 ± 1.2), canned spinach (172 ± 0.5), dried longans (172 ± 0.5), potato salad (171 ± 0.4), frozen and boiled kale (169 ± 0.8), frozen and fried hash brown potatoes (168 ± 0.6), raw sprouted mung beans (165 ± 1), frozen and boiled brussels sprouts (165 ± 0.6), corn flakes (161 ± 1.8), coconut cream (160 ± 0.2), whipping cream, heavy (160 ± 3.3), boiled and steamed chinese chestnuts (157 ± 1.8), frozen and boiled broccoli (157 ± 0.5), canned succotash with creamed corn (154 ± 0.4), canned succotash (152 ± 0.4), raw japanese chestnuts (150 ± 1.8), boiled broccoli (150 ± 0.6), microwaved potato with skin (149 ± 0.2), raw coconut (147 ± 1.1), raw european chestnuts (146 ± 1.8), dried sweetened flaked coconut (145 ± 0.7), o'brien potatoes (144 ± 0.3), boiled yellow corn (141 ± 0.6), frozen and boiled collards (141 ± 0.6), raw broccoli (141 ± 1.1), soybean oil and cottonseed oil margarine liquid (140 ± 10), baked potato with skin (140 ± 0.2), frozen and boiled asparagus (140 ± 0.8), tamarind (139 ± 0.4), noodles (138 ± 0.3), raw watercress (135 ± 2.9), raw bamboo shoots (134 ± 0.7), raw chives (133 ± 16.7), raw shallots (130 ± 5), frozen and boiled yellow corn (128 ± 0.6), hash brown potatoes (128 ± 0.6), microwaved potato without skin (128 ± 0.3), dried sweetened shredded coconut (127 ± 0.5), raw potato without skin (126 = 0.4), mashed potatoes (124 ± 0.5), cooked sprouted mung beans (123 ± 0.8), boiled asparagus (123 ± 0.6), dried figs (122 ± 0.3), boiled and steamed european chestnuts (121 ± 1.8), frozen and boiled turnip greens and turnips (120 ± 0.5), frozen and boiled potato without skin (120 ± 0.5), sweetened coconut cream (119 ± 0.2), baked potato without skin (119 ± 0.3), boiled brussels sprouts (115 ± 0.6), boiled kale (114 ± 0.8), raw wakame seaweed (112 ± 0.5), boiled scotch kale (112 ± 0.8), macaroni (109 ± 0.4), spaghetti (109 ± 0.4), boiled amaranth (109 ± 0.8), raw swamp cabbage (109 ± 0.9), raw cauliflower (108 ± 1), tomato paste (108 ± 0.4), oatmeal (105 ± 0.3), boiled dock (105 ± 0.5), boiled potato without skin (104 ± 0.4), raw romaine lettuce (104 ± 1.8), boiled swiss chard (103 ± 0.6), frozen and boiled mustard greens (103 ± 0.7), boiled mustard greens (103 ± 0.7), coconut milk (101 ± 0.2), raw california avocado (100 ± 0.3), boiled cauliflower (100 ± 0.8), raw cassava (100 ± 0.5), canned yellow corn with red and green peppers (99 ± 0.4), apricots (97 ± 0.5), raw turnip greens (96 ± 1.8), sapote (96 ± 0.2), raw welsh onions (95 ± 0.5), raw savoy cabbage (94 ± 1.4), raw lotus root (94 ± 0.6), raw chinese cabbage (93 ± 0.6), boiled burdock root (92 ± 0.4), boiled green beans (90 ± 0.8), boiled yellow snap beans (90 ± 0.8), raw red chili peppers (89 ± 1.1), raw green chili peppers (89 ± 1.1), canned bamboo shoots (89 ± 0.4), raw chinese cabbage (89 ± 1.4), canned yellow corn (88 ± 0.5), canned turnip greens (88 ± 0.4), boiled swamp cabbage (88 ± 1), raw spring onions with tops (88 ± 1), raw butterhead lettuce (87 ± 3.3), raw vine spinach (86 ± 0.5), frozen and boiled cauliflower (86 ± 0.6), canned potato without skin (86 ± 0.6), raw looseleaf lettuce (86 ± 1.8), baked sweet potato (85 ± 0.4), boiled savoy cabbage (85 ± 0.7), canned sprouted mung beans (85 ± 0.8), frozen and baked sweet potato (84 ± 0.6), frozen and boiled okra (84 ± 0.5), chocolate syrup (82 ± 1.3), raw kelp/kombu/tangle seaweed (82 ± 0.5), whipped butter (82 ± 4.5), canned sweet potato (81 ± 0.3), boiled sweet potato (81 ± 0.3), whole chocolate malted milk (79 ± 0.2), dried japanese persimmon (79 ± 1.5), boiled bamboo shoots (79 ± 0.4), raw leeks (77 ± 1.9), boiled mothbeans (75 ± 0.3), boiled beet greens (75 ± 0.7), raw florida avocado (75 ± 0.2), boiled okra (75 ± 0.6), beef suet (75 ± 1.8), raw iceberg lettuce (75 ± 2.5), canned creamed yellow corn (74 ± 0.4), boiled turnip greens (74 ± 0.7), coleslaw (72 ± 0.8), safflower oil and soybean oil mayonnaise (71 ± 3.6), soybean oil mayonnaise (71 ± 3.6), frozen and boiled crookneck squash (71 ± 0.5), human milk (68 ± 1.6), raw scallop squash (68 ± 0.8), steamed hawaii mountain yam (67 ± 0.7), whey (67 ± 0.2), raw chickory greens (67 ± 0.6), boiled purslane (66 ± 0.9), frozen and boiled yellow snap beans (66 ± 0.7), frozen and boiled green beans (66 ± 0.7), raw red cabbage (66 ± 1.4), raw zucchini (65 ± 0.8), raw endive (64 ± 2), frozen and boiled zucchini (64 ± 0.4), watermelon (62 ± 0.3), frozen and boiled turnips (61 ± 0.5), dried dates (60 ± 0.6), butter (60 ± 3.3), corn oil margarine stick (60 ± 10), safflower oil and soybean oil margarine stick (60 ± 10), soybean oil margarine stick (60 ± 10), soybean oil and cottonseed oil margarine stick (60 ± 10), corn oil margarine tub (60 ± 10), safflower oil margarine tub (60 ± 10), soybean oil margarine tub (60 ± 10), soybean oil and cottonseed oil margarine tub (60 ± 10), canned pumpkin (60 ± 0.4), soursop (60 ± 0.2), boiled kohlrabi (59 ± 0.6), canned pumpkin pie mix (59 ± 0.4), raw ginger root (58 ± 2.1), baked/boiled tropical yam (57 ± 0.7), raw green cabbage (57 ± 1.4), boiled scallop squash (57 ± 0.6), canned zucchini in tomato juice (57 ± 0.4), boiled lotus root (57 ± 0.6), raw onions (56 ± 0.6), baked hubbard squash (55 ± 0.5), boiled hubbard squash (55 ± 0.4), raw celtuce (55 ± 0.5), sugar apple (55 ± 0.3), canned green beans (54 ± 0.7), canned yellow snap beans (54 ± 0.7), boiled and steamed japanese chestnuts (54 ± 1.8), boiled collards (53 ± 0.3), valencia orange (53 ± 0.4), raw crookneck squash (52 ± 0.8), navel orange (52 ± 0.4), raw eggplant (51 ± 1.2), boiled red cabbage (51 ± 0.7), boiled crookneck squash (50 ± 0.6), bananas (48 ± 0.4), tomato purée (48 ± 0.2), cooked shitake mushrooms (47 ± 0.7), frozen and boiled carrots (47 ± 0.7), longans (46 ± 0.5), frozen and boiled butternut squash (45 ± 0.4), boiled green cabbage (45 ± 0.7), candied sweet potato (44 ± 0.5), persimmon (44 ± 2), boiled carrots (44 ± 0.6), raw green tomato (44 ± 0.4), boiled onions (43 ± 0.5), boiled leeks (42 ± 1.9), frozen and boiled red sweet peppers (42 ± 0.5), frozen and boiled green sweet peppers (42 ± 0.5), boiled yambean (42 ± 0.5), baked acorn squash (41 ± 0.5), boiled red tomato (41 ± 0.4), mango (41 ± 0.2), lychees (41 ± 0.5), boiled eggplant (40 ± 1), canned red chili peppers (40 ± 0.7), canned green chili peppers (40 ± 0.7), raw chickory (40 ± 1.1), raw carrots (40 ± 0.7), carambola (40 ± 0.4), corn oil diet margarine tub (40 ± 10), soybean oil and cottonseed oil diet margarine tub (40 ± 10), cream of wheat (39 ± 0.3), boiled pumpkin (39 ± 0.4), sapodilla (39 ± 0.3), breadfruit (38 ± 0.5), boiled beets (38 ± 0.6), raw red sweet peppers (38 ± 1), raw green sweet peppers (38 ± 1), cream of rice (37 ± 0.3), cooked plantain (37 ± 0.3), mammy apple (37 ± 0.5), canned red sweet peppers (36 ± 0.7), canned green sweet peppers (36 ± 0.7), raw radish (36 ± 1.1), boiled zucchini (36 ± 0.6), boiled rutabaga (36 ± 0.6), canned jalapeño peppers (35 ± 0.7), raw dock (34 ± 0.7), canned crookneck squash (34 ± 0.5), canned peeled red tomatoes (34 ± 0.4), canned stewed red

tomatoes (34 ± 0.4), frozen and boiled onions (34 ± 0.5), farina (33 ± 0.3), boiled butternut squash (33 ± 0.5), japanese persimmon (33 ± 0.3), raw red tomato (33 ± 0.4), coconut water (32 ± 0.2), boiled oriental radish (32 ± 0.7), tangerine (32 ± 0.6), mandarin oranges (31 ± 0.4), boiled chayote (30 ± 0.6), figs (30 ± 1), raw oriental radish (30 ± 1.1), pickled beets (29 ± 0.4), corn grits (28 ± 0.2), boiled red sweet peppers (28 ± 0.7), boiled green sweet peppers (28 ± 0.7), boiled turnips (28 ± 0.6), elderberries (26 ± 0.3), boiled acorn squash (25 ± 0.4), strawberries (25 ± 0.3), raw celery (25 ± 1.3), canned carrots (25 ± 0.7), crab apples (25 ± 0.5), pineapple (25 ± 0.3), papaya (25 ± 0.2), guava (23 ± 0.6), peach (23 ± 0.6), cooked taro (23 ± 0.8), loquats (23 ± 0.5), boiled/baked spaghetti squash (22 ± 0.6), tomato juice (22 ± 0.3), boiled calabash gourd (21 ± 0.7), raw cucumber (21 ± 1), boiled celery (20 ± 0.7), white grapefruit (18 ± 0.4), plum (17 ± 0.8), cooked apple without skin (16 ± 0.3), european grapes (15 ± 0.3), american grapes (14 ± 0.5), pink grapefruit (14 ± 0.4), lime (13 ± 0.7), blueberries (12 ± 0.3), raw apple with skin (12 ± 0.4), applesauce (11 ± 0.4), grape juice (10 ± 0.2), orange juice (9 ± 0.2), raw apple without skin (9 ± 0.4), thompson seedless grapes (9 ± 0.3), screwdriver (7 ± 0.2), tangerine juice (7 ± 0.2), ale (7 ± 0.1), pear (5 ± 0.3), rice polish (4 ± 0.9), white grapefruit juice (3 ± 0.2), coffee (1 ± 0.3)
Applications: viral related diseases, herpes simplex virus, deterioration of growth rate in children, deterioration of gastric function, deterioration of appetite
Daily Dosage: Nutritional — 1–3 g; therapeutic — ?; experimental — ?; toxic — none known up to 3 g
Warnings: none

METHIONINE

Names: methionine, MET, amino acid M, 2-amino-4(methylthio)-buanoic acid
Forms: 2-amino-4(methylthio)-butanoic acid
Classification: essential amino acid, sulfur-containing amino acid
Deficiency: baldness, liver deterioration, rheumatic fever, toxemia of pregnancy, muscles paralyzation, protein decreased, decreased cysteine production, decreased cystine production, selenium decreased
Side-effects: none known
Toxicity: none known
Inhibitors: increased amounts of other amino acids
Helpers: vitamin B_6, vitamin B_{12}, vitamin B_C, magnesium
Sources (mg per 100 grams of food): dried and salted cod (1859 ± 0.6), cooked whelk (1205 ± 0.6), dried spirulina seaweed (1149 ± 0.5), soybean protein isolate (1129 ± 1.8), parmesan cheese (1120 ± 10), dried brazil nuts (1029 ± 1.8), torula yeast (1018 ± 1.8), fried chicken breast without skin (926 ± 0.6), fried light meat chicken without skin (908 ± 0.5), dry skim milk (907 ± 1.7), dried sesame kernels (900 ± 6.3), fried beef top round excluding fat (898 ± 0.5), stewed chicken drumstick without skin (882 ± 1.1), dry instant skim milk (880 ± 0.5), canned tuna in water (876 ± 0.6), cooked flounder (870 ± 0.5), roasted turkey light meat without skin (866 ± 0.5), simmered beef shank excluding fat (862 ± 0.5), fried chicken breast with skin (862 ± 0.5), roasted chicken breast without skin (859 ± 0.6), braised pork center loin excluding fat (857 ± 0.5), anchovy canned in olive oil (855 ± 2.5),

roasted light meat chicken without skin (855 ± 0.5), braised pork top loin excluding fat (849 ± 0.5), dried watermelon seeds (846 ± 1.8), braised beef chuck arm pot roast excluding fat (845 ± 0.5), gruyere cheese (832 ± 1.8), fried beef sirloin excluding fat (832 ± 0.5), roasted turkey dark meat without skin (828 ± 0.5), braised pork sirloin excluding fat (825 ± 0.5), fried light meat chicken with skin (815 ± 0.5), soybean protein concentrate (814 ± 1.8), braised pork loin excluding fat (813 ± 0.5), broiled beef top round excluding fat (811 ± 0.5), fried beef top round including fat (811 ± 0.5), cooked coho salmon (811 ± 0.6), fried chicken giblets (810 ± 0.5), simmered beef shank including fat (809 ± 0.5), braised beef bottom round excluding fat (809 ± 0.5), cooked sockeye salmon (808 ± 0.6), roasted chicken breast with skin (807 ± 0.5), roasted turkey light meat with skin (803 ± 0.5), stewed chicken breast without skin (802 ± 0.5), dry buttermilk (800 ± 7.1), stewed light meat chicken without skin (799 ± 0.5), fried dark meat chicken without skin (799 ± 0.5), braised beef chuck blade roast excluding fat (795 ± 0.5), swiss cheese (793 ± 1.8), cooked halibut (791 ± 0.6), broiled beef top round including fat (789 ± 0.5), broiled pork center loin excluding fat (788 ± 0.5), braised beef shortribs excluding fat (787 ± 0.5), canned tuna in oil (785 ± 0.6), cooked rainbow trout (780 ± 0.6), cooked snapper (778 ± 0.6), broiled beef sirloin excluding fat (777 ± 0.5), roasted light meat chicken with skin (776 ± 0.5), roasted turkey dark meat with skin (774 ± 0.5), simmered turkey gizzard (772 ± 0.5), cooked sheepshead (771 ± 0.6), braised pork boston blade excluding fat (768 ± 0.5), tilsit cheese (764 ± 1.8), braised beef bottom round including fat (763 ± 0.5), roasted dark meat chicken without skin (757 ± 0.5), braised beef brisket excluding fat (752 ± 0.5), cooked swordfish (751 ± 0.6), roasted pumpkin and squash seeds (750 ± 1.8), smoked haddock (747 ± 0.6), port du salut cheese (743 ± 1.8), roasted eye of beef round excluding fat (742 ± 0.5), stewed chicken breast with skin (741 ± 0.5), simmered beef heart (737 ± 0.5), roasted beef tip excluding fat (735 ± 0.5), cooked grouper (735 ± 0.6), fried chicken back with skin (735 ± 0.7), cooked mullet (734 ± 0.6), broiled beef top loin excluding fat (733 ± 0.5), braised pork loin blade excluding fat (732 ± 0.5), cooked pike (731 ± 0.6), edam cheese (729 ± 1.8), gouda cheese (729 ± 1.8), sardines (729 ± 2.1), broiled beef round excluding fat (728 ± 0.5), kippered herring (728 ± 1.3), cooked tilefish (725 ± 0.6), fried dark meat chicken with skin (725 ± 0.5), fried chicken drumstick with skin (724 ± 1), broiled beef tenderloin excluding fat (723 ± 0.5), roasted chicken drumstick with skin (723 ± 1), broiled porterhouse steak excluding fat (721 ± 0.5), broiled T-bone steak excluding fat (720 ± 0.5), stewed dark meat chicken without skin (719 ± 0.5), roasted pork rump excluding fat (718 ± 0.5), broiled beef rib eye excluding fat (718 ± 0.5), cooked haddock (718 ± 0.6), braised beef flank excluding fat (717 ± 0.5), roasted chicken thigh without skin (717 ± 1), braised pork spareribs (716 ± 0.5), cooked flatfish (715 ± 0.6), fried chicken thigh with skin (713 ± 0.8), braised pork center loin including fat (712 ± 0.5), cooked rockfish (712 ± 0.6), simmered chicken gizzard (712 ± 0.5), chipped beef (711 ± 1.8), broiled pork center rib excluding fat (710 ± 0.5), broiled pork top loin excluding fat (710 ± 0.5), roasted pork tenderloin (709 ± 0.5), fried pork center loin excluding fat (709 ± 0.5), well done baked extra lean ground beef (708 ± 0.5), cooked perch (707 ± 0.6), cooked mackerel (706 ± 0.6), roasted beef tenderloin excluding fat (705 ± 0.5), roasted beef rib roasted excluding fat (705 ± 0.5), braised beef flank including fat (704 ± 0.5), fried beef sirloin including fat (703 ± 0.5), roasted pork center loin excluding fat (702 ± 0.5), broiled beef sirloin including fat

(701 ± 0.5), cooked pompano (701 ± 0.6), cooked eel (700 ± 0.6), cooked sea bass (699 ± 0.6), stewed light meat chicken with skin (699 ± 0.5), roasted pork leg excluding fat (698 ± 0.5), broiled pork sirloin excluding fat (696 ± 0.5), roasted pork center rib excluding fat (695 ± 0.5), roasted pork top loin excluding fat (695 ± 0.5), roasted pork shank excluding fat (695 ± 0.5), cooked whiting (695 ± 0.6), provolone cheese (693 ± 1.8), braised beef chuck arm pot roast including fat (693 ± 0.5), smoked whitefish (693 ± 0.6), braised pork center rib including fat (691 ± 0.5), well done baked lean ground beef (691 ± 0.5), fried pork center rib excluding fat (689 ± 0.5), fried pork top loin excluding fat (689 ± 0.5), raw quail without skin (689 ± 0.5), roasted dark meat chicken with skin (688 ± 0.5), roasted chicken wing with skin (688 ± 1.5), broiled pork loin excluding fat (686 ± 0.5), raw pheasant without skin (686 ± 0.5), roasted eye of beef round including fat (685 ± 0.5), raw yellowtail (685 ± 0.6), cooked herring (682 ± 0.6), stewed chicken drumstick with skin (681 ± 0.9), roasted beef round tip including fat (678 ± 0.5), simmered chicken neck without skin (678 ± 2.8), roasted pork sirloin excluding fat (677 ± 0.5), cooked carp (676 ± 0.6), cooked cod (676 ± 0.6), braised pork sirloin including fat (675 ± 0.5), fried beef liver (675 ± 0.5), canned cod (674 ± 0.6), cooked crayfish (674 ± 0.6), well done baked ground beef (673 ± 0.5), cooked bacon (672 ± 0.4), cooked rainbow smelt (669 ± 0.6), roasted chicken thigh with skin (669 ± 0.8), fried chicken wing with skin (669 ± 1.6), well done broiled extra lean ground beef (668 ± 0.5), braised pork top loin including fat (666 ± 0.5), broiled beef tenderloin including fat (665 ± 0.5), broiled pork center loin including fat (663 ± 0.5), roasted pork loin excluding fat (663 ± 0.5), simmered turkey giblets (662 ± 0.5), well done broiled lean ground beef (659 ± 0.5), broiled beef top loin including fat (659 ± 0.5), roasted cured pork arm excluding fat (659 ± 0.5), dry whole milk (659 ± 1.6), braised pork loin including fat (655 ± 0.5), well done fried extra lean ground beef (654 = 0.5), broiled beef round including fat (654 ± 0.5), turkey breast meat summer sausage (652 ± 2.4), cheddar cheese (652 ± 0.5), braised beef chuck blade roast including fat (651 ± 0.5), grilled canadian bacon (651 ± 1.1), fenugreek seed (650 ± 25), broiled beef flank excluding fat (650 ± 0.5), broiled beef rib eye including fat (650 ± 0.5), monterey cheese (650 ± 1.8), roasted pork rump including fat (649 ± 0.5), braised pork arm braised including fat (647 ± 0.5), simmered turkey heart (646 ± 0.5), simmered chicken giblets (646 ± 0.5), well done fried lean ground beef (645 ± 0.5), braised pork liver (645 ± 0.5), caviar (644 ± 3.1), broiled porterhouse steak including fat (642 ± 0.5), broiled beef flank including fat (641 ± 0.5), raw coho salmon (640 ± 0.6), braised pork boston blade braised including fat (638 ± 0.5), simmered chicken heart (638 ± 0.5), well done broiled ground beef (635 ± 0.5), roasted duck without skin (635 ± 0.5), defatted soybean flour (634 ± 0.5), well done fried ground beef (631 ± 0.5), broiled beef rib excluding fat (631 ± 0.5), raw sockeye salmon (631 ± 0.6), canned corned beef (629 ± 0.5), colby cheese (629 ± 1.8), limburger cheese (629 ± 1.8), low fat soybean flour (627 ± 0.6), roasted beef tenderloin including fat (626 ± 0.5), roasted pork shoulder excluding fat (624 ± 0.5), stewed dark meat chicken with skin (623 ± 0.5), raw shark (621 ± 0.6), stewed chicken thigh with skin (621 ± 0.7), dried butternuts (621 ± 1.8), broiled pork boston blade excluding fat (620 ± 0.5), cheshire cheese (618 ± 1.8), roasted pork center loin including fat (617 ± 0.5), raw halibut (616 ± 0.6), braised beef liver (616 ± 0.5), raw trout (615 ± 0.6), broiled pork loin blade excluding fat (614 ± 0.5), broiled T-bone steak including fat

(614 ± 0.5), cooked sturgeon (613 ± 0.6), dried and frozen tofu (612 ± 2.9), roasted pork sirloin including fat (609 ± 0.5), roasted pork loin blade excluding fat (608 ± 0.5), roasted goose with skin (608 ± 0.5), raw rainbow trout (608 ± 0.6), raw milkfish (608 ± 0.6), roasted pork leg including fat (607 ± 0.5), raw snapper (607 ± 0.6), pickled anchovy (607 ± 1.8), defatted soy meal (606 ± 0.4), canned sockeye salmon (606 ± 0.6), fried chicken neck with skin (606 ± 1.4), braised pork heart (604 ± 0.4), raw whelk (604 ± 0.6), raw anchovy (602 ± 0.6), roasted pork boston blade excluding fat (600 ± 0.5), brie cheese (600 ± 1.8), roasted pork center rib including fat (599 ± 0.5), fried pork loin blade excluding fat (599 ± 0.5), raw sheepshead (598 ± 0.6), raw chum salmon (596 ± 0.6), pork link sausage (596 ± 0.7), raw chinook salmon (594 ± 0.6), broiled pork center rib including fat (593 ± 0.5), medium-rare broiled extra lean ground beef (593 ± 0.5), bleu cheese (593 ± 1.8), raw bluefish (593 ± 0.6), raw pink salmon (591 ± 0.6), canned pink salmon (591 ± 0.6), breaded and fried shrimp (591 ± 0.6), hard pork and beef salami (590 ± 5), braised beef brisket including fat (589 ± 0.5), cooked shrimp (589 ± 0.6), whole dried sesame seeds (589 ± 5.6), stewed chicken wing with skin (588 ± 1.3), roasted pork shank including fat (587 ± 0.5), raw swordfish (586 ± 0.6), anchovy paste (586 ± 7.1), cooked black bass (585 ± 0.4), roasted pork top loin including fat (584 ± 0.5), medium-rare fried extra lean ground beef (583 ± 0.5), broiled pork sirloin including fat (582 ± 0.5), roasted beef rib including fat (581 ± 0.5), american cheese (579 ± 1.8), pimento cheese spread (579 ± 1.8), canned blue crab (578 ± 0.6), medium-rare broiled lean ground beef (577 ± 0.5), simmered chicken liver (577 ± 0.5), cooked clams (576 ± 0.6), raw pollock (576 ± 0.6), cooked lobster (576 ± 0.6), duck egg (576 ± 0.7), muenster cheese (575 ± 1.8), braised pork loin blade including fat (574 ± 0.5), raw sunfish (574 ± 0.6), raw grouper (574 ± 0.6), raw mullet (573 ± 0.6), medium-rare baked extra lean ground beef (572 ± 0.5), raw burbot (572 ± 0.6), raw shrimp (572 ± 0.6), brick cheese (571 ± 1.8), camembert cheese (571 ± 1.8), raw pike (571 ± 0.6), milk chocolate (571 ± 1.8), broiled pork top loin including fat (570 ± 0.5), cooked blue crab (569 ± 0.6), broiled pork loin including fat (568 ± 0.5), simmered turkey liver (568 ± 0.5), crab cakes (568 ± 0.8), toasted sesame kernels (568 ± 1.8), whole toasted sesame seeds (568 ± 1.8), roasted pork loin including fat (566 ± 0.5), medium-rare fried lean ground beef (566 ± 0.5), raw whitefish (565 ± 0.6), raw alewife (563 ± 0.5), medium-rare broiled ground beef (562 ± 0.5), raw cusk (562 ± 0.6), raw cisco (562 ± 0.6), raw ling (562 ± 0.6), raw haddock (560 ± 0.6), sesame butter (560 ± 3.3), medium-rare baked lean ground beef (559 ± 0.5), medium-rare fried ground beef (559 ± 0.5), raw scup (559 ± 0.6), raw freshwater bass (558 ± 0.6), raw flatfish (558 ± 0.6), unheated canadian bacon (558 ± 0.9), fried pork center loin including fat (557 ± 0.5), dried pumpkin and squash seeds (557 ± 1.8), raw black bass (557 ± 0.5), raw rockfish (555 ± 0.6), roasted lean ham (553 ± 0.5), roasted canned lean ham (552 ± 0.5), braised beef shortribs including fat (552 ± 0.5), raw perch (551 ± 0.6), raw mackerel (551 ± 0.6), mozzarella cheese (550 ± 1.8), raw spot (548 ± 0.6), raw dolphinfish (548 ± 0.6), raw pompano (547 ± 0.6), turkey ham (546 ± 0.9), raw eel (546 ± 0.6), raw sea bass (546 ± 0.6), braised pork kidneys (545 ± 0.5), cooked alaska king crab (545 ± 0.6), raw whiting (542 ± 0.6), smoked chinook salmon (541 ± 0.6), battered and fried shark (541 ± 0.6), braised pork tongue (540 ± 0.5), raw smelt (539 ± 0.5), medium-rare baked ground beef (538 ± 0.5), raw catfish (538 ± 0.6), roasted pork arm including fat (536 ± 0.5), cooked mussels (536 ± 0.6), roasted canned ham

METHIONINE

(535 ± 0.5), dry roasted soybean nuts (534 ± 0.6), canned smelt (534 ± 0.5), yellow mustard seed (533 ± 16.7), raw herring (532 ± 0.6), roasted pork shoulder including fat (530 ± 0.5), simmered beef kidneys (530 ± 0.5), raw lobster (529 ± 0.6), raw carp (528 ± 0.6), broiled pork boston blade including fat (527 ± 0.5), roasted pork boston blade including fat (527 ± 0.5), raw cod (527 ± 0.6), breaded and fried catfish (527 ± 0.6), raw croaker (526 ± 0.6), peanut flour (525 ± 12.5), raw striped bass (525 ± 0.6), raw crayfish (525 ± 0.6), breaded and fried croaker (525 ± 0.6), roasted ham (524 ± 0.5), raw lingcod (524 ± 0.6), braised pork spleen (523 ± 0.5), raw rainbow smelt (522 ± 0.6), raw queen crab (521 ± 0.6), raw freshwater drum (519 ± 0.6), raw tilefish (518 ± 0.6), raw wolffish (518 ± 0.6), roasted cured pork arm including fat (516 ± 0.5), broiled beef rib including fat (515 ± 0.5), raw alaska king crab (515 ± 0.6), fried pork center rib including fat (513 ± 0.5), raw butterfish (512 ± 0.6), fried pork top loin including fat (511 ± 0.5), jellied corned beef loaf (511 ± 1.8), unheated lean ham (511 ± 0.5), turkey pastrami (509 ± 0.9), raw blue crab (508 ± 0.6), toasted soybean nuts (507 ± 1.8), roasted pork loin blade including fat (505 ± 0.5), raw shad (501 ± 0.6), dried sunflower seeds (500 ± 1.8), raw white sucker (496 ± 0.6), raw seatrout (496 ± 0.6), broiled pork loin blade including fat (494 ± 0.5), chopped smoked beef (493 ± 1.8), raw dungeness crab (491 ± 0.6), braised beef pancreas (490 ± 0.5), italian pork sausage (487 ± 0.7), smoked cisco (484 ± 0.6), pork and beef pepperoni (483 ± 8.3), canned lean ham (482 ± 0.5), pork and beef honey loaf (482 ± 1.8), simmered chicken neck with skin (482 ± 1.3), dried black walnuts (479 ± 1.8), pork sausage (478 ± 1.9), raw sturgeon (478 ± 0.6), roasted soybean nuts (476 ± 0.6), pork luxury loaf (475 ± 1.8), roasted duck with skin (475 ± 0.5), raw turbot (475 ± 0.6), raw black bullhead (473 ± 0.5), braised pork pancreas (470 ± 0.5), hard pork salami (470 ± 5), tuna salad (470 ± 0.2), canned alewife (470 ± 0.5), roasted soybean flour (469 ± 0.6), oil roasted sunflower seeds (468 ± 1.8), simmered beef tongue (467 ± 0.5), soybean flour (466 ± 0.6), lebanon bologna (465 ± 2.2), unheated ham (464 ± 0.5), braised beef spleen (462 ± 0.5), minced ham lunch meat (457 ± 2.4), almond meal (457 ± 1.8), roasted pork blade roll (456 ± 0.5), pork and beef peppered loaf (454 ± 1.8), chopped ham lunch meat (452 ± 2.4), raw duck liver (444 ± 0.5), fried pork loin blade including fat (443 ± 0.5), ham and cheese roll (443 ± 1.8), canned ham (443 ± 0.5), fried abalone (441 ± 0.6), dried pignolia pine nuts (436 ± 1.8), raw flounder (434 ± 0.5), raw sole (434 ± 0.5), surimi scallops (433 ± 0.6), poppy seed (433 ± 16.7), raw monkfish (429 ± 0.6), beef lunch meat (425 ± 1.8), dry roasted sunflower seeds (425 ± 1.8), cooked corned beef brisket (421 ± 0.5), pickled herring (420 ± 3.3), surimi shrimp (420 ± 0.6), chicken egg yolk (418 ± 2.9), pork bologna (413 ± 2.2), braised beef lungs (408 ± 0.5), breaded and fried scallops (406 ± 1.6), beef pastrami (404 ± 1.8), breaded fried squid (404 ± 0.6), dried pilinuts (400 ± 1.8), pork and beef mortadella (393 ± 3.3), chicken egg (392 ± 1), berliner (391 ± 2.2), raw goose liver (388 ± 0.5), dried pistachio nuts (386 ± 1.8), raw abalone (386 ± 0.6), pork and beef picnic loaf (382 ± 1.8), polish pork sausage (382 ± 1.8), raw scallops (379 ± 0.6), chicken egg white (379 ± 1.5), cottage cheese (376 ± 0.2), pork and beef brotwurst (369 ± 0.7), raw cuttlefish (366 ± 0.6), summer sausage (365 ± 2.2), pork and beef link sausage (365 ± 0.7), pork and beef link sausage with american cheese (358 ± 1.2), pork beerwurst salami (357 ± 2.2), pork and beef luncheon sausage (352 ± 2.2), raw squid (351 ± 0.6), beef salami (348 ± 2.2), grilled ham patties (348 ± 0.8), oil roasted spanish peanuts (343 ± 1.8), pork

III. Amino Acids — METHIONINE

livercheese (342 ± 1.3), pork bratwurst (342 ± 0.6), unheated ham patties (338 ± 0.8), canned pork lunch meat (338 ± 2.4), raw octopus (336 ± 0.6), oil roasted valencia peanuts (332 ± 1.8), pork and beef sausage (331 ± 3.8), smoked beef sausage (328 ± 1.2), pork and beef old fashioned loaf (325 ± 1.8), breaded and fried clams (324 ± 0.6), gjetost cheese (321 ± 1.8), cooked oysters (319 ± 0.6), cashew butter (318 ± 1.8), oil roasted virginia peanuts (318 ± 1.8), raw beef tripe (315 ± 0.5), pork liver sausage (311 ± 2.8), simmered pork tail (306 ± 0.5), braised beef thymus (304 ± 0.5), pork olive loaf (304 ± 1.8), dried hickory nuts (304 ± 1.8), beef and pork salami (300 ± 2.2), basil (300 ± 50), fennel seed (300 ± 25), garlic powder (300 ± 16.7), creamy peanut butter (294 ± 3.1), crunchy peanut butter (294 ± 1.6), pork and beef lunch meat (293 ± 1.8), oil roasted cashews (293 ± 1.8), dry roasted macadamia nuts (293 ± 1.8), pork liverwurst (289 ± 2.8), dry roasted peanuts (289 ± 1.8), raw clams (288 ± 0.6), raw pork stomach (288 ± 0.5), beef beerwurst salami (287 ± 2.2), pork and beef knockwurst (287 ± 0.7), dried english/persian walnuts (286 ± 1.8), beef bologna (283 ± 2.2), ricotta cheese (281 ± 0.4), oil roasted peanuts (279 ± 1.8), dry roasted pistachio nuts (279 ± 1.8), beef frankfurter (279 ± 0.9), dry roasted cashews (279 ± 1.8), beef and pork bologna (278 ± 2.2), pork and beef kielbasa/kolbassy (277 ± 1.9), corn germ (269 ± 0.5), braised pork lungs (268 ± 0.5), dried peanuts (268 ± 1.8), pork headcheese (268 ± 1.8), raw mussels (268 ± 0.6), tempeh (265 ± 0.6), vienna sausage (263 ± 3.1), fried beef brains (261 ± 0.5), pork pickle and pimento loaf (257 ± 1.8), puffed wheat (257 ± 3.6), gefiltefish with broth (255 ± 1.2), freeze-dried chives (250 ± 62.5), neufchatel cheese (243 ± 1.8), braised pork brains (241 ± 0.5), oil roasted almonds (236 ± 1.8), toasted almonds (236 ± 1.8), ham salad (232 ± 1.8), simmered beef brains (230 ± 0.5), beef and pork frankfurter (229 ± 1.1), dried almonds (229 ± 1.8), boiled soybeans (224 ± 0.3), fried tofu (223 ± 3.8), freeze-dried parsley (214 ± 35.7), simmered pork feet (211 ± 0.5), dried pinyon pine nuts (211 ± 1.8), natto (208 ± 0.6), firm raw tofu (202 ± 0.4), sweetened condensed milk (200 ± 1.3), breaded and fried oysters (199 ± 0.6), simmered pork chitterlings (194 ± 0.5), shredded wheat (193 ± 1.8), dry roasted pecans (193 ± 1.8), puffed rice (193 ± 3.6), corn flakes (193 ± 1.8), raw bacon (191 ± 0.7), dry roasted almonds (189 ± 1.8), dried pecans (189 ± 1.8), freeze-dried red sweet peppers (188 ± 31.3), freeze-dried green sweet peppers (188 ± 31.3), evaporated skim milk (188 ± 1.6), falafel (186 ± 1), cream cheese (182 ± 1.8), dried shitake mushrooms (180 ± 3.3), bran (179 ± 1.8), oil roasted filberts (179 ± 1.8), bran flakes (179 ± 1.8), almond butter (175 ± 3.1), evaporated whole milk (171 ± 0.4), skim yogurt (169 ± 0.2), dried chinese chestnuts (168 ± 1.8), wheat flakes (168 ± 1.8), oil roasted pecans (168 ± 1.8), boiled peanuts (166 ± 1.6), dried filberts (164 ± 1.8), raw oysters (160 ± 0.6), canned oysters (159 ± 0.6), evaporated lowfat milk (158 ± 0.4), sheep milk (155 ± 0.2), lowfat yogurt (155 ± 0.2), dried european chestnuts (154 ± 1.8), dill seed (150 ± 25), boiled green soybeans (150 ± 0.6), miso (149 ± 0.4), pickled pork feet (149 ± 0.5), stir-fried sprouted soybeans (147 ± 0.5), boiled white beans (146 ± 0.3), raw laver/nori seaweed (145 ± 0.5), potato pancakes (141 ± 0.7), boiled cranberry beans (140 ± 0.3), dried acorns (139 ± 1.8), boiled yellow beans (138 ± 0.3), boiled pink beans (136 ± 0.3), dried ginko nuts (136 ± 1.8), almond paste (136 ± 1.8), boiled black beans (133 ± 0.3), dried coconut (132 ± 1.8), raisin bran (132 ± 1.8), boiled navy beans (131 ± 0.3), boiled kidney beans (130 ± 0.3), acorn flour (129 ± 1.8), dried japanese chestnuts (129 ± 1.8), dry roasted filberts (129 ± 1.8), simmered pork ears (127 ± 0.5), boiled great northern beans (125 ± 0.3), freeze-dried shallots (125 ± 12.5), boiled pinto beans (124 ± 0.3), potato flour (124 ± 0.6), boiled black turtle beans (123 ± 0.3), frozen and boiled black-eyed peas (121 ± 0.6), boiled yardlong bean (118 ± 0.3), raw spirulina seaweed (118 ± 0.5), au gratin potatoes (117 ± 0.4), yellow corn pudding (117 ± 0.2), boiled chickpeas (116 ± 0.3), boiled catjang black-eyed peas (116 ± 0.3), canned navy beans (113 ± 0.2), canned great northern beans (111 ± 0.2), roasted chinese chestnuts (111 ± 1.8), boiled lupins (110 ± 0.3), boiled black-eyed peas (110 ± 0.3), boiled mungo beans (110 ± 0.3), boiled winged beans (109 ± 0.3), canned white beans (109 ± 0.2), boiled and frozen baby lima beans (108 ± 0.6), boiled french beans (106 ± 0.3), raw acorns (104 ± 1.8), raw chinese chestnuts (104 ± 1.8), stir-fried sprouted lentils (103 ± 0.5), raw tofu (103 ± 0.4), boiled baby lima beans (102 ± 0.3), whole yogurt (102 ± 0.2), onion powder (100 ± 25), dried toasted coconut (100 ± 1.8), boiled lima beans (99 ± 0.3), protein fortified skim milk (99 ± 0.2), protein fortified lowfat 2% milk (99 ± 0.2), protein fortified lowfat 1% milk (98 ± 0.2), restaurant vanilla milkshake (97 ± 0.2), indian buffalo milk (97 ± 0.2), steamed sprouted soybeans (96 ± 1.1), lard (96 ± 1.8), raw pork jowl (95 ± 0.5), refried beans (94 ± 0.2), dried macadamia nuts (93 ± 1.8), canned black turtle beans (91 ± 0.2), hot chocolate (91 ± 0.2), malted milk (90 ± 0.2), boiled sprouted green peas (89 ± 0.5), eggnog (87 ± 0.2), vanilla milkshake (87 ± 0.2), skim milk (86 ± 0.2), boiled split peas (85 ± 0.3), restaurant chocolate milkshake (85 ± 0.2), boiled mung beans (84 ± 0.2), restaurant strawberry milkshake (84 ± 0.2), lowfat 2% milk (84 ± 0.2), filled milk (84 ± 0.2), canned cranberry beans (83 ± 0.2), raw green peas (82 ± 0.6), oil roasted macadamia nuts (82 ± 1.8), lowfat 1% milk (82 ± 0.2), whole milk (82 ± 0.2), baked beans with pork and sweet sauce (81 ± 0.2), boiled green peas (81 ± 0.6), buttermilk (81 ± 0.2), lowfat 1% chocolate milk (81 ± 0.2), goat milk (80 ± 0.2), lowfat 2% chocolate milk (80 ± 0.2), whole chocolate milk (80 ± 0.2), boiled adzuki beans (79 ± 0.2), carob milk (79 ± 0.2), canned kidney beans (78 ± 0.2), baked beans with pork and tomato sauce (78 ± 0.2), frozen and boiled green peas (78 ± 0.6), raw garlic (78 ± 5.6), boiled lentils (77 ± 0.3), boiled pigeon peas (76 ± 0.3), dried oriental radish (76 ± 0.9), chocolate milkshake (76 ± 0.2), frozen and boiled turnip greens (76 ± 0.6), roasted european chestnuts (75 ± 1.8), half and half cream (73 ± 3.3), vegetarian baked beans (72 ± 0.2), dehydrated onion flakes (71 ± 3.6), roasted japanese chestnuts (71 ± 1.8), boiled and steamed chinese chestnuts (71 ± 1.8), boiled yellow corn (70 ± 0.6), noodles (70 ± 0.3), canned pinto beans (69 ± 0.2), boiled succotash (68 ± 0.5), coconut cream (68 ± 0.2), canned black-eyed peas (67 ± 0.2), canned green peas (67 ± 0.6), light cream (67 ± 3.3), sour cream (67 ± 4.2), tomato powder (66 ± 0.5), potato salad (66 ± 0.4), boiled hyacinth beans (65 ± 0.3), canned chickpeas (65 ± 0.2), frozen and boiled yellow corn (63 ± 0.6), raw wakame seaweed (63 ± 0.5), boiled broad beans (62 ± 0.3), canned lima beans (62 ± 0.2), raw coconut (62 ± 1.1), dried sweetened flaked coconut (61 ± 0.7), frozen and boiled fordhook lima beans (60 ± 0.6), medium cream (60 ± 3.3), frozen and boiled succotash (58 ± 0.6), frozen and boiled spinach (58 ± 0.5), scalloped potatoes (58 ± 0.4), ginko nuts (57 ± 1.8), raw european chestnuts (57 ± 1.8), boiled spinach (56 ± 0.6), raw spinach (54 ± 1.8), raw japanese chestnuts (54 ± 1.8), dried sweetened shredded coconut (54 ± 0.5), stir-fried sprouted mung beans (53 ± 0.8), whipping cream, light (53 ± 3.3), whipping cream, heavy (53 ± 3.3), canned spinach (52 ± 0.5), boiled jute potherb (51 ± 1.2), macaroni (51 ± 0.4), spaghetti (51 ± 0.4), ginger (50 ± 25), sweetened coconut cream (50 ±

0.2), dried longans (49 ± 0.5), canned yellow corn with red and green peppers (48 ± 0.4), french fried potatoes (46 ± 1), boiled and steamed european chestnuts (46 ± 1.8), canned broad beans (45 ± 0.2), raw swamp cabbage (45 ± 0.9), boiled mushrooms (44 ± 0.6), o'brien potatoes (43 ± 0.3), oatmeal (43 ± 0.3), coconut milk (43 ± 0.2), canned yellow corn (43 ± 0.5), dried lychees (42 ± 0.5), baked sweet potato (42 ± 0.4), frozen and baked sweet potato (42 ± 0.6), hummus (41 ± 0.2), okara tofu (41 ± 0.8), frozen and boiled turnip greens and turnips (41 ± 0.5), canned sweet potato (41 ± 0.3), boiled sweet potato (41 ± 0.3), raw mushrooms (40 ± 1.4), frozen and fried french fried potatoes (40 ± 1), soy milk (40 ± 0.2), frozen and boiled collards (40 ± 0.6), soybean oil and cottonseed oil margarine liquid (40 ± 10), boiled mothbeans (40 ± 0.3), microwaved potato with skin (39 ± 0.2), raw california avocado (39 ± 0.3), boiled lamb's-quarters (37 ± 0.6), frozen and boiled broccoli (37 ± 0.5), boiled broccoli (36 ± 0.6), baked potato with skin (36 ± 0.2), raw turnip greens (36 ± 1.8), canned creamed yellow corn (36 ± 0.4), safflower oil and soybean oil mayonnaise (36 ± 3.6), soybean oil mayonnaise (36 ± 3.6), frozen and fried hash brown potatoes (35 ± 0.6), raw sprouted mung beans (35 ± 1), frozen and boiled brussels sprouts (35 ± 0.6), canned succotash with creamed corn (35 ± 0.4), canned succotash (35 ± 0.4), mashed potatoes (35 ± 0.5), boiled swamp cabbage (35 ± 1), raw broccoli (34 ± 1.1), corn grits (34 ± 0.2), raw chives (33 ± 16.7), microwaved potato without skin (33 ± 0.3), raw potato without skin (33 ± 0.4), boiled dock (32 ± 0.5), frozen and boiled potato without skin (31 ± 0.5), baked potato without skin (31 ± 0.3), canned turnip greens (31 ± 0.4), raw bamboo shoots (30 ± 0.7), raw shallots (30 ± 5), boiled amaranth (30 ± 0.8), raw florida avocado (29 ± 0.2), frozen and boiled kale (28 ± 0.8), frozen and boiled asparagus (28 ± 0.8), raw cauliflower (28 ± 1), cream of wheat (28 ± 0.3), hash brown potatoes (27 ± 0.6), boiled potato without skin (27 ± 0.4), whipped butter (27 ± 4.5), cooked sprouted mung beans (26 ± 0.8), boiled asparagus (26 ± 0.6), boiled cauliflower (26 ± 0.8), raw cassava (26 ± 0.5), whole chocolate malted milk (26 ± 0.2), boiled turnip greens (26 ± 0.7), cream of rice (26 ± 0.3), farina (26 ± 0.3), dried figs (25 ± 0.3), raw kelp/kombu/tangle seaweed (25 ± 0.5), cooked shitake mushrooms (25 ± 0.7), boiled brussels sprouts (24 ± 0.6), raw red chili peppers (24 ± 1.1), raw green chili peppers (24 ± 1.1), steamed hawaii mountain yam (24 ± 0.7), boiled green beans (23 ± 0.8), boiled yellow snap beans (23 ± 0.8), frozen and boiled cauliflower (23 ± 0.6), raw lotus root (22 ± 0.6), canned potato without skin (22 ± 0.6), frozen and boiled okra (22 ± 0.5), dried dates (22 ± 0.6), valencia orange (22 ± 0.4), navel orange (22 ± 0.4), european grapes (22 ± 0.3), raw romaine lettuce (21 ± 1.8), frozen and boiled mustard greens (21 ± 0.7), boiled mustard greens (21 ± 0.7), raw welsh onions (21 ± 0.5), boiled beet greens (21 ± 0.7), baked/boiled tropical yam (21 ± 0.7), boiled and steamed japanese chestnuts (21 ± 1.8), candied sweet potato (21 ± 0.5), american grapes (21 ± 0.5), boiled swiss chard (20 ± 0.6), raw savoy cabbage (20 ± 1.4), canned bamboo shoots (20 ± 0.4), raw spring onions with tops (20 ± 1), boiled okra (20 ± 0.6), butter (20 ± 3.3), corn oil margarine stick (20 ± 10), safflower oil and soybean oil margarine stick (20 ± 10), soybean oil margarine stick (20 ± 10), soybean oil and cottonseed oil margarine stick (20 ± 10), corn oil margarine tub (20 ± 10), safflower oil margarine tub (20 ± 10), soybean oil margarine tub (20 ± 10), soybean oil and cottonseed oil margarine tub (20 ± 10), corn oil diet margarine tub (20 ± 10), soybean oil and cottonseed oil diet margarine tub (20 ± 10), tomato paste (19 ± 0.4), raw vine spinach (19 ± 0.5), raw leeks (19 ± 1.9), human milk

(19 ± 1.6), frozen and boiled turnips (19 ± 0.5), raw watercress (18 ± 2.9), boiled kale (18 ± 0.8), boiled scotch kale (18 ± 0.8), boiled savoy cabbage (18 ± 0.7), canned sprouted mung beans (18 ± 0.8), beef suet (18 ± 1.8), coleslaw (18 ± 0.8), frozen and boiled crookneck squash (18 ± 0.5), frozen and boiled yellow snap beans (18 ± 0.7), frozen and boiled green beans (18 ± 0.7), baked hubbard squash (18 ± 0.5), boiled hubbard squash (18 ± 0.4), dried apricots (17 ± 1.4), raw parsley (17 ± 1.7), boiled bamboo shoots (17 ± 0.4), raw scallop squash (17 ± 0.8), raw zucchini (17 ± 0.8), peach (17 ± 0.6), sapote (16 ± 0.2), chocolate syrup (16 ± 1.3), whey (16 ± 0.2), raw endive (16 ± 2), frozen and boiled zucchini (16 ± 0.4), boiled scallop squash (16 ± 0.6), raw iceberg lettuce (15 ± 2.5), canned zucchini in tomato juice (15 ± 0.4), canned green beans (15 ± 0.7), canned yellow snap beans (15 ± 0.7), boiled collards (15 ± 0.3), frozen and boiled butternut squash (15 ± 0.4), tamarind (14 ± 0.4), raw looseleaf lettuce (14 ± 1.8), boiled purslane (14 ± 0.9), raw red cabbage (14 ± 1.4), raw crookneck squash (14 ± 0.8), baked acorn squash (14 ± 0.5), elderberries (14 ± 0.3), rice polish (14 ± 0.9), raw butterhead lettuce (13 ± 3.3), boiled kohlrabi (13 ± 0.6), raw ginger root (13 ± 2.1), boiled lotus root (13 ± 0.6), boiled crookneck squash (13 ± 0.6), longans (13 ± 0.5), coconut water (13 ± 0.2), tangerine (13 ± 0.6), mandarin oranges (13 ± 0.4), thompson seedless grapes (13 ± 0.3), boiled burdock root (12 ± 0.4), dried japanese persimmon (12 ± 1.5), canned pumpkin (12 ± 0.4), canned pumpkin pie mix (12 ± 0.4), raw eggplant (12 ± 1.2), boiled leeks (12 ± 1.9), boiled beets (12 ± 0.6), raw green cabbage (11 ± 1.4), boiled red cabbage (11 ± 0.7), bananas (11 ± 0.4), boiled green cabbage (11 ± 0.7), frozen and boiled red sweet peppers (11 ± 0.5), frozen and boiled green sweet peppers (11 ± 0.5), boiled yambean (11 ± 0.5), carambola (11 ± 0.4), boiled butternut squash (11 ± 0.5), pineapple (11 ± 0.3), blueberries (11 ± 0.3), raw chickory greens (10 ± 0.6), raw onions (10 ± 0.6), raw celtuce (10 ± 0.5), raw green tomato (10 ± 0.4), boiled red tomato (10 ± 0.4), canned red chili peppers (10 ± 0.7), canned green chili peppers (10 ± 0.7), breadfruit (10 ± 0.5), raw red sweet peppers (10 ± 1), raw green sweet peppers (10 ± 1), cooked plantain (10 ± 0.3), canned red sweet peppers (10 ± 0.7), canned green sweet peppers (10 ± 0.7), canned jalapeño peppers (10 ± 0.7), boiled chinese cabbage (9 ± 0.6), raw chinese cabbage (9 ± 1.4), tomato purée (9 ± 0.2), lychees (9 ± 0.5), boiled zucchini (9 ± 0.6), boiled rutabaga (9 ± 0.6), canned crookneck squash (9 ± 0.5), pickled beets (9 ± 0.4), boiled turnips (9 ± 0.6), frozen and boiled carrots (8 ± 0.7), persimmon (8 ± 2), boiled eggplant (8 ± 1), boiled pumpkin (8 ± 0.4), canned peeled red tomatoes (8 ± 0.4), canned stewed red tomatoes (8 ± 0.4), raw red tomato (8 ± 0.4), boiled acorn squash (8 ± 0.4), cooked taro (8 ± 0.8), soursop (7 ± 0.2), sugar apple (7 ± 0.3), boiled onions (7 ± 0.5), raw chickory (7 ± 1.1), raw carrots (7 ± 0.7), raw radish (7 ± 1.1), raw oriental radish (7 ± 1.1), boiled red sweet peppers (7 ± 0.7), boiled green sweet peppers (7 ± 0.7), apricots (6 ± 0.5), watermelon (6 ± 0.3), boiled carrots (6 ± 0.6), mammy apple (6 ± 0.5), frozen and boiled onions (6 ± 0.5), figs (6 ± 1), guava (6 ± 0.6), boiled/baked spaghetti squash (6 ± 0.6), plum (6 ± 0.8), mango (5 ± 0.2), japanese persimmon (5 ± 0.3), boiled oriental radish (5 ± 0.7), raw celery (5 ± 1.3), canned carrots (4 ± 0.7), crab apples (4 ± 0.5), loquats (4 ± 0.5), tomato juice (4 ± 0.3), boiled calabash gourd (4 ± 0.7), raw cucumber (4 ± 1), boiled celery (4 ± 0.7), pear (4 ± 0.3), sapodilla (3 ± 0.3), cooked apple without skin (3 ± 0.3), orange juice (3 ± 0.2), screwdriver (3 ± 0.2), papaya (2 ± 0.2), white grapefruit (2 ± 0.4), pink grapefruit (2 ± 0.4), raw apple with skin

(2 ± 0.4), applesauce (2 ± 0.4), raw apple without skin (2 ± 0.4), tangerine juice (2 ± 0.2), boiled chayote (1 ± 0.6), strawberries (1 ± 0.3), lime (1 ± 0.7), grape juice (1 ± 0.2), ale (1 ± 0.1), internal conversion of vitamin B_C and vitamin B_{12}
Applications: deterioration of nucleic acid structure formation, deterioration of collagen formation, low protein synthesis, toxicity, heavy metal poisoning, radiation, free radical oxidation, alcohol poisoning, histadelic schizophrenics, histamine toxicity, vitamin B_P decreased, selenium therapy, cancer, sun sensitivity, free radical oxidation, life span extension, radiation, decreased radiation resistance, aging, antioxidant therapy
Daily Dosage: Nutritional—0.5–1 g; therapeutic—?; experimental—?; toxic—none known up to 1 g
Warnings: none

ORNITHINE

Names: ornithine
Classification: nonessential amino acid
Deficiency: (not required)
Side-effects: see toxicity symptoms
Toxicity: thickening of skin, enlarged joints, larynx growth
Daily Dosage: Nutritional—?; therapeutic—3 g; experimental—5 g; toxic—10 g
Inhibitors: increased amounts of other amino acids
Helpers: moderate amounts of other amino acids
Sources: internal conversion of arginine
Applications: growth hormone releaser
Daily Dosage: Nutritional—?; therapeutic—3 g; experimental—5 g; toxic—none known up to 5 g
Warnings: Stimulates growth hormone release, which is undesirable for those still growing. Increases schizophrenia symptoms. May increase severity of herpes simplex infections.

PHENYLALANINE

Names: phenylalanine, PHA, PHE, amino acid F, 2-amino-3-phenyl propanoic acid
Forms: 2-amino-3-phenyl propanoic acid
Classification: essential amino acid
Deficiency: nerves, clouded thought, emotional agitation, depression, decreased alertness, memory decreased, behavioral changes, decreased sexual interest, bloodshot eyes, cataracts, decreased insulin, decreased skin melanin, eating increased, protein decreased, decreased tyrosine formation
Side-effects: see toxicity symptoms
Toxicity: increased blood pressure, emotional agitation, lack of sleep, headaches, tyrosine toxicity
Inhibitors: increased amounts of other amino acids
Helpers: vitamin B_6, vitamin C
Sources (mg per 100 grams of food):

torula yeast (9332 ± 1.8), soybean protein isolate (4593 ± 1.8), soybean protein concentrate (3279 ± 1.8), peanut flour (2975 ± 12.5), dried spirulina seaweed (2777 ± 0.5), defatted soybean flour (2453 ± 0.5), dried and salted cod (2452 ± 0.6), low fat soybean flour (2428 ± 0.6), defatted soy meal (2346 ± 0.4), dried and frozen tofu (2335 ± 2.9), parmesan cheese (2240 ± 10), almond meal (2236 ± 1.8), dry roasted soybean nuts (2066 ± 0.6), dried watermelon seeds (2061 ± 1.8), fenugreek seed (2000 ± 25), toasted soybean nuts (1957 ± 1.8), roasted soybean nuts (1838 ± 0.6), roasted soybean flour (1816 ± 0.6), soybean flour (1802 ± 0.6), gruyere cheese (1764 ± 1.8), dry skim milk (1747 ± 1.7), dry instant skim milk (1695 ± 0.5), swiss cheese (1682 ± 1.8), roasted pumpkin and squash seeds (1664 ± 1.8), cooked whelk (1648 ± 0.6), creamy peanut butter (1625 ± 3.1), oil roasted peanuts (1554 ± 1.8), dry buttermilk (1543 ± 7.1), dried sesame kernels (1525 ± 6.3), dried peanuts (1489 ± 1.8), fried chicken giblets (1480 ± 0.5), dried butternuts (1464 ± 1.8), milk chocolate (1464 ± 1.8), oil roasted spanish peanuts (1454 ± 1.8), edam cheese (1450 ± 1.8), gouda cheese (1450 ± 1.8), fried beef liver (1423 ± 0.5), oil roasted valencia peanuts (1404 ± 1.8), braised pork center loin excluding fat (1391 ± 0.5), braised pork top loin excluding fat (1378 ± 0.5), tilsit cheese (1375 ± 1.8), fried beef top round excluding fat (1369 ± 0.5), braised pork sirloin excluding fat (1340 ± 0.5), oil roasted virginia peanuts (1339 ± 1.8), port du salut cheese (1339 ± 1.8), fried chicken breast without skin (1328 ± 0.6), braised pork loin excluding fat (1319 ± 0.5), simmered beef shank excluding fat (1315 ± 0.5), cheddar cheese (1311 ± 0.5), monterey cheese (1304 ± 1.8), provolone cheese (1304 ± 1.8), fried light meat chicken without skin (1303 ± 0.5), simmered beef heart (1303 ± 0.5), braised beef liver (1299 ± 0.5), braised beef chuck arm pot roast excluding fat (1289 ± 0.5), stewed chicken drumstick without skin (1280 ± 1.1), broiled pork center loin excluding fat (1280 ± 0.5), braised pork liver (1274 ± 0.5), dry whole milk (1272 ± 1.6), fried beef sirloin excluding fat (1268 ± 0.5), colby cheese (1268 ± 1.8), muenster cheese (1257 ± 1.8), fried chicken breast with skin (1253 ± 0.5), crunchy peanut butter (1247 ± 1.6), braised pork boston blade excluding fat (1246 ± 0.5), brick cheese (1246 ± 1.8), cheshire cheese (1246 ± 1.8), dried pumpkin and squash seeds (1239 ± 1.8), broiled beef top round excluding fat (1237 ± 0.5), fried beef top round including fat (1237 ± 0.5), simmered beef shank including fat (1234 ± 0.5), braised beef bottom round excluding fat (1233 ± 0.5), roasted chicken breast without skin (1231 ± 0.6), dry roasted peanuts (1229 ± 1.8), roasted light meat chicken without skin (1226 ± 0.5), simmered turkey gizzard (1224 ± 0.5), simmered beef kidneys (1223 ± 0.5), braised pork pancreas (1222 ± 0.5), braised beef chuck blade roast excluding fat (1212 ± 0.5), simmered chicken liver (1212 ± 0.5), simmered turkey giblets (1207 ± 0.5), braised pork spleen (1205 ± 0.5), broiled beef top round including fat (1203 ± 0.5), braised beef shortribs excluding fat (1201 ± 0.5), dried pistachio nuts (1200 ± 1.8), braised pork kidneys (1199 ± 0.5), simmered turkey heart (1199 ± 0.5), simmered turkey liver (1193 ± 0.5), fried light meat chicken with skin (1192 ± 0.5), braised pork loin blade excluding fat (1188 ± 0.5), roasted turkey light meat without skin (1187 ± 0.5), dried sunflower seeds (1186 ± 1.8), broiled beef sirloin excluding fat (1186 ± 0.5), simmered chicken heart (1183 ± 0.5), cooked bacon (1174 ± 0.4), braised pork center loin including fat (1173 ± 0.5), brie cheese (1171 ± 1.8), simmered chicken giblets (1170 ± 0.5), roasted chicken breast with skin (1169 ± 0.5), yellow mustard seed (1167 ± 16.7), roasted pork rump excluding fat (1166 ± 0.5), braised beef bottom round including fat (1164 ± 0.5), braised pork spareribs (1162 ± 0.5), fried dark meat chicken without skin (1156 ± 0.5), canned tuna in water (1155 ± 0.6), oil roasted almonds (1154 ± 1.8), toasted almonds (1154 ± 1.8), broiled pork center rib excluding fat (1153 ± 0.5), broiled pork top loin excluding fat (1153 ± 0.5), roasted pork tenderloin (1152 ± 0.5), fried pork center loin excluding fat (1151 ± 0.5), stewed chicken breast without skin (1151 ± 0.5), well done baked extra lean ground beef (1149 ± 0.5), braised beef brisket excluding fat (1147 ± 0.5), stewed light meat chicken without skin (1146 ± 0.5), roasted pork center loin excluding fat (1140 ± 0.5), braised pork center rib including fat (1140 ± 0.5), american cheese (1139 ± 1.8), pimento cheese spread (1139 ± 1.8), roasted turkey dark meat without skin (1134 ± 0.5), roasted pork leg excluding fat (1133 ± 0.5), roasted eye of beef round excluding fat (1132 ± 0.5), roasted light meat chicken with skin (1130 ± 0.5), broiled pork sirloin excluding fat (1130 ± 0.5), anchovy canned in olive oil (1130 ± 2.5), roasted pork center rib excluding fat (1129 ± 0.5), roasted pork top loin excluding fat (1129 ± 0.5), dried almonds (1129 ± 1.8), simmered chicken gizzard (1129 ± 0.5), limburger cheese (1129 ± 1.8), roasted pork shank excluding fat (1128 ± 0.5), braised beef pancreas (1127 ± 0.5), well done baked lean ground beef (1122 ± 0.5), dried black walnuts (1121 ± 1.8), roasted beef tip excluding fat (1121 ± 0.5), fried pork center rib excluding fat (1119 ± 0.5), fried pork top loin excluding fat (1119 ± 0.5), camembert cheese (1118 ± 1.8), broiled beef top loin excluding fat (1117 ± 0.5), roasted turkey light meat with skin (1117 ± 0.5), braised pork sirloin including fat (1115 ± 0.5), broiled pork loin excluding fat (1114 ± 0.5), oil roasted sunflower seeds (1111 ± 1.8), broiled beef round excluding fat (1111 ± 0.5), cooked flounder (1110 ± 0.5), chipped beef (1104 ± 1.8), bleu cheese (1104 ± 1.8), broiled beef tenderloin excluding fat (1103 ± 0.5), braised pork top loin including fat (1102 ± 0.5), roasted pork sirloin excluding fat (1100 ± 0.5), broiled porterhouse steak excluding fat (1099 ± 0.5), broiled T-bone steak excluding fat (1098 ± 0.5), fried chicken back with skin (1097 ± 0.7), broiled beef rib eye excluding fat (1095 ± 0.5), braised beef flank excluding fat (1094 ± 0.5), broiled pork center loin including fat (1093 ± 0.5), well done baked ground beef (1092 ± 0.5), roasted dark meat chicken without skin (1086 ± 0.5), well done broiled extra lean ground beef (1084 ± 0.5), braised pork loin including fat (1082 ± 0.5), roasted cured pork arm excluding fat (1078 ± 0.5), roasted pork loin excluding fat (1076 ± 0.5), roasted turkey dark meat with skin (1076 ± 0.5), roasted beef tenderloin excluding fat (1075 ± 0.5), roasted beef rib roasted excluding fat (1075 ± 0.5), stewed chicken breast with skin (1074 ± 0.5), braised beef flank including fat (1074 ± 0.5), fried beef sirloin including fat (1072 ± 0.5), fried dark meat chicken with skin (1071 ± 0.5), caviar (1069 ± 3.1), braised pork arm braised including fat (1069 ± 0.5), broiled beef sirloin including fat (1069 ± 0.5), well done broiled lean ground beef (1069 ± 0.5), cooked coho salmon (1068 ± 0.6), cooked sockeye salmon (1066 ± 0.6), roasted pork rump including fat (1063 ± 0.5), well done fried extra lean ground beef (1061 ± 0.5), fried chicken drumstick with skin (1057 ± 1), braised beef chuck arm pot roast including fat (1057 ± 0.5), roasted goose with skin (1055 ± 0.5), roasted chicken drumstick with skin (1054 ± 1), fried chicken thigh with skin (1052 ± 0.8), braised pork boston blade braised including fat (1052 ± 0.5), well done fried lean ground beef (1047 ± 0.5), roasted eye of beef round including fat (1045 ± 0.5), braised pork heart (1042 ± 0.4), cooked halibut (1042 ± 0.6), canned tuna in oil (1036 ± 0.6), roasted beef round tip including fat (1033 ± 0.5), well done broiled ground beef (1031 ± 0.5), stewed dark meat chicken without skin (1030 ± 0.5), roasted chicken thigh without skin (1029 ± 1), cooked rainbow trout (1028 ± 0.6), cooked snapper (1027 ± 0.6), mozzarella cheese (1025 ± 1.8), roasted chicken wing with skin (1024 ± 1.5), well done fried ground beef (1024 ± 0.5), stewed light meat chicken with skin (1019 ± 0.5), cooked sheepshead (1016 ± 0.6), roasted pork shoulder excluding fat (1015 ± 0.5), roasted pork center loin including fat (1014 ± 0.5), broiled beef tenderloin including fat (1013 ± 0.5), tempeh (1012 ± 0.6), cooked crayfish (1011 ± 0.6), braised beef spleen (1008 ± 0.5), dry roasted sunflower seeds (1007 ± 1.8), roasted dark meat chicken with skin (1007 ± 0.5), broiled pork boston blade excluding fat (1007 ± 0.5), broiled beef top loin including fat (1004 ± 0.5), fried chicken wing with skin (1003 ± 1.6), basil (1000 ± 50), roasted pork sirloin including fat (999 ± 0.5), braised pork tongue (999 ± 0.5), roasted pork leg including fat (998 ± 0.5), broiled pork loin blade excluding fat (997 ± 0.5), broiled beef round including fat (997 ± 0.5), braised beef chuck blade roast including fat (993 ± 0.5), broiled beef flank excluding fat (992 ± 0.5), broiled beef rib eye including fat (991 ± 0.5), cooked swordfish (991 ± 0.6), stewed chicken drumstick with skin (988 ± 0.9), roasted pork loin blade excluding fat (987 ± 0.5), roasted pork center rib including fat (987 ± 0.5), roasted duck without skin (984 ± 0.5), smoked haddock (984 ± 0.6), broiled pork center rib including fat (980 ± 0.5), broiled porterhouse steak including fat (980 ± 0.5), broiled beef flank including fat (978 ± 0.5), roasted chicken thigh with skin (976 ± 0.8), canned corned beef (975 ± 0.5), roasted pork boston blade excluding fat (975 ± 0.5), fried pork loin blade excluding fat (973 ± 0.5), simmered chicken neck without skin (972 ± 2.8), cooked grouper (971 ± 0.6), roasted pork shank including fat (969 ± 0.5), cooked mullet (968 ± 0.6), roasted pork top loin including fat (965 ± 0.5), cooked pike (964 ± 0.6), sardines (963 ± 2.1), medium-rare broiled extra lean ground beef (963 ± 0.5), broiled beef rib excluding fat (962 ± 0.5), broiled pork sirloin including fat (961 ± 0.5), kippered herring (960 ± 1.3), cooked tilefish (956 ± 0.6), roasted beef tenderloin including fat (955 ± 0.5), braised pork loin blade including fat (954 ± 0.5), medium-rare fried extra lean ground beef (947 ± 0.5), cooked haddock (946 ± 0.6), broiled pork top loin including fat (945 ± 0.5), raw quail without skin (944 ± 0.5), whole dried sesame seeds (944 ± 5.6), cooked flatfish (944 ± 0.6), natto (941 ± 0.6), broiled pork loin including fat (940 ± 0.5), hard pork salami (940 ± 5), cooked rockfish (938 ± 0.6), medium-rare broiled lean ground beef (937 ± 0.5), broiled T-bone steak including fat (936 ± 0.5), roasted pork loin including fat (934 ± 0.5), dried pignolia pine nuts (932 ± 1.8), cooked perch (932 ± 0.6), raw duck liver (932 ± 0.5), cooked mackerel (931 ± 0.6), breaded and fried shrimp (928 ± 0.6), medium-rare baked extra lean ground beef (928 ± 0.5), fried pork center loin including fat (927 ± 0.5), dry roasted almonds (925 ± 1.8), cooked pompano (925 ± 0.6), cooked eel (924 ± 0.6), cooked sea bass (924 ± 0.6), cashew butter (921 ± 1.8), raw pheasant without skin (920 ± 0.5), fried chicken neck with skin (919 ± 1.4), medium-rare fried lean ground beef (919 ± 0.5), cooked whiting (916 ± 0.6), cooked clams (915 ± 0.6), smoked whitefish (914 ± 0.6), simmered beef tongue (913 ± 0.5), medium-rare broiled ground beef (913 ± 0.5), toasted sesame kernels (911 ± 1.8), whole toasted sesame seeds (911 ± 1.8), stewed dark meat chicken with skin (911 ± 0.5), medium-rare baked lean ground beef (907 ± 0.5), medium-rare fried ground beef (907 ± 0.5), roasted lean ham (904 ± 0.5), stewed chicken thigh with skin (904 ± 0.7), raw yellowtail (904 ± 0.6), sesame butter (900 ± 3.3), cooked herring (899 ± 0.6), braised beef brisket including fat (898 ± 0.5), turkey breast meat summer sausage (895 ± 2.4), cooked carp (893 ± 0.6), cooked cod (891 ± 0.6), roasted pork arm including fat (889 ± 0.5), canned cod (889 ± 0.6), roasted beef rib including fat (887 ± 0.5), cooked shrimp (884 ± 0.6), cooked rainbow smelt (882 ± 0.6), roasted pork shoulder including fat (878 ± 0.5), medium-rare baked ground beef (873 ± 0.5), broiled pork boston blade including fat (872 ± 0.5), dry roasted pistachio nuts (871 ± 1.8), roasted cured pork arm including fat (871 ± 0.5), stewed chicken wing with skin (870 ± 1.3), hard pork and beef salami (870 ± 5), crab cakes (870 ± 0.8), roasted pork boston blade including fat (869 ± 0.5), boiled soybeans (869 ± 0.3), canned blue crab (867 ± 0.6), cooked lobster (866 ± 0.6), fried pork center rib including fat (860 ± 0.5), raw shrimp (858 ± 0.6), roasted ham (857 ± 0.5), fried pork top loin including fat (856 ± 0.5), cooked mussels (853 ± 0.6), cooked blue crab (853 ± 0.6), pork link sausage (849 ± 0.7), oil roasted cashews (846 ± 1.8), almond butter (844 ± 3.1), raw coho salmon (844 ± 0.6), braised beef shortribs including fat (842 ± 0.5), duck egg (840 ± 0.7), roasted pork loin blade including fat (838 ± 0.5), fried tofu (838 ± 3.8), unheated lean ham (836 ± 0.5), poppy seed (833 ± 16.7), raw sockeye salmon (832 ± 0.6), braised beef lungs (829 ± 0.5), raw whelk (824 ± 0.6), broiled pork loin blade including fat (823 ± 0.5), jellied corned beef loaf (821 ± 1.8), raw shark (819 ± 0.6), roasted canned lean ham (816 ± 0.5), cooked alaska king crab (816 ± 0.6), raw goose liver (815 ± 0.5), raw halibut (813 ± 0.6), raw trout (811 ± 0.6), cooked sturgeon (808 ± 0.6), dry roasted cashews (804 ± 1.8), raw rainbow trout (802 ± 0.6), raw milkfish (802 ± 0.6), raw snapper (801 ± 0.6), canned sockeye salmon (799 ± 0.6), raw anchovy (794 ± 0.6), raw lobster (794 ± 0.6), cooked black bass (794 ± 0.4), roasted canned ham (791 ± 0.5), raw sheepshead (789 ± 0.6), raw crayfish (788 ± 0.6), raw chum salmon (786 ± 0.6), broiled beef rib including fat (785 ± 0.5), raw chinook salmon (784 ± 0.6), raw bluefish (782 ± 0.6), raw queen crab (781 ± 0.6), grilled canadian bacon (779 ± 1.1), raw pink salmon (778 ± 0.6), canned pink salmon (778 ± 0.6), raw swordfish (773 ± 0.6), raw alaska king crab (773 ± 0.6), puffed wheat (771 ± 3.6), pickled anchovy (771 ± 1.8), firm raw tofu (768 ± 0.4), chopped smoked beef (764 ± 1.8), raw blue crab (764 ± 0.6), oil roasted filberts (761 ± 1.8), unheated ham (759 ± 0.5), raw pollock (759 ± 0.6), dried brazil nuts (757 ± 1.8), raw sunfish (756 ± 0.6), raw grouper (756 ± 0.6), raw mullet (755 ± 0.6), raw burbot (754 ± 0.6), roasted duck with skin (752 ± 0.5), raw pike (752 ± 0.6), battered and fried shark (751 ± 0.6), fried pork loin blade including fat (747 ± 0.5), turkey ham (747 ± 0.9), roasted pork blade roll (746 ± 0.5), raw whitefish (745 ± 0.6), anchovy paste (743 ± 7.1), breaded and fried croaker (742 ± 0.6), raw cusk (742 ± 0.6), raw cisco (741 ± 0.6), raw ling (741 ± 0.6), raw haddock (738 ± 0.6), raw scup (736 ± 0.6), raw freshwater bass (736 ± 0.6), raw flatfish (736 ± 0.6), raw dungeness crab (735 ± 0.6), simmered chicken neck with skin (732 ± 1.3), raw rockfish (732 ± 0.6), pork luxury loaf (729 ± 1.8), raw perch (727 ± 0.6), lebanon bologna (726 ± 2.2), breaded and fried catfish (726 ± 0.6), raw mackerel (726 ± 0.6), raw spot (724 ± 0.6), raw dolphinfish (722 ± 0.6), dried hickory nuts (721 ± 1.8), raw pompano (721 ± 0.6), raw eel (720 ± 0.6), raw sea bass (720 ± 0.6), raw alewife (718 ± 0.5), pork and beef pepperoni (717 ± 8.3), pork livercheese (716 ± 1.3), fried abalone (715 ± 0.6), raw whiting (715 ± 0.6), smoked chinook salmon (714 ± 0.6), canned lean ham (713 ± 0.5), chicken egg yolk (712 ± 2.9), turkey pastrami (711 ±

0.9), raw catfish (711 ± 0.6), raw black bass (710 ± 0.5), falafel (708 ± 1), pork and beef peppered loaf (704 ± 1.8), raw herring (701 ± 0.6), ham and cheese roll (700 ± 1.8), boiled peanuts (700 ± 1.6), dill seed (700 ± 25), dried filberts (696 ± 1.8), raw carp (696 ± 0.6), raw cod (695 ± 0.6), raw croaker (694 ± 0.6), raw striped bass (692 ± 0.6), braised pork lungs (691 ± 0.5), chopped ham lunch meat (690 ± 2.4), raw lingcod (689 ± 0.6), raw rainbow smelt (688 ± 0.6), raw smelt (688 ± 0.5), breaded and fried scallops (687 ± 1.6), chicken egg (686 ± 1), raw freshwater drum (685 ± 0.6), raw tilefish (684 ± 0.6), raw wolffish (684 ± 0.6), beef lunch meat (682 ± 1.8), canned smelt (681 ± 0.5), pork and beef honey loaf (679 ± 1.8), raw butterfish (675 ± 0.6), chicken egg white (673 ± 1.5), cottage cheese (673 ± 0.2), almond paste (671 ± 1.8), italian pork sausage (670 ± 0.7), unheated canadian bacon (667 ± 0.9), raw shad (661 ± 0.6), breaded fried squid (656 ± 0.6), pork sausage (656 ± 1.9), canned ham (655 ± 0.5), cooked corned beef brisket (654 ± 0.5), raw white sucker (654 ± 0.6), raw seatrout (654 ± 0.6), fennel seed (650 ± 25), minced ham lunch meat (643 ± 2.4), smoked cisco (639 ± 0.6), dried english/persian walnuts (636 ± 1.8), fried beef brains (635 ± 0.5), raw sturgeon (631 ± 0.6), beef pastrami (629 ± 1.8), raw turbot (627 ± 0.6), tuna salad (626 ± 0.2), braised beef thymus (626 ± 0.5), freeze-dried chives (625 ± 62.5), pork and beef luncheon sausage (622 ± 2.2), pork liverwurst (622 ± 2.8), boiled lupins (618 ± 0.3), braised pork brains (618 ± 0.5), pork headcheese (614 ± 1.8), raw abalone (613 ± 0.6), raw black bullhead (603 ± 0.5), raw scallops (601 ± 0.6), berliner (600 ± 2.2), pork and beef mortadella (600 ± 3.3), canned alewife (599 ± 0.5), miso (596 ± 0.4), pork bologna (587 ± 2.2), raw cuttlefish (582 ± 0.6), boiled pigeon peas (579 ± 0.3), summer sausage (570 ± 2.2), raw monkfish (565 ± 0.6), pork and beef link sausage with american cheese (565 ± 1.2), freeze-dried red sweet peppers (563 ± 31.3), freeze-dried green sweet peppers (563 ± 31.3), neufchatel cheese (561 ± 1.8), simmered beef brains (560 ± 0.5), grilled ham patties (560 ± 0.8), boiled green soybeans (559 ± 0.6), raw squid (558 ± 0.6), simmered pork feet (557 ± 0.5), pork liver sausage (556 ± 2.8), ricotta cheese (556 ± 0.4), pickled herring (553 ± 3.3), raw flounder (553 ± 0.5), raw sole (553 ± 0.5), breaded and fried clams (549 ± 0.6), polish pork sausage (546 ± 1.8), gjetost cheese (546 ± 1.8), unheated ham patties (543 ± 0.8), pork and beef picnic loaf (543 ± 1.8), pork and beef brotwurst (541 ± 0.7), beef salami (539 ± 2.2), shredded wheat (536 ± 1.8), dry roasted filberts (536 ± 1.8), raw octopus (534 ± 0.6), corn germ (527 ± 0.5), boiled white beans (526 ± 0.3), bran flakes (525 ± 1.8), raw pork stomach (525 ± 0.5), stir-fried sprouted soybeans (524 ± 0.5), pork and beef lunch meat (518 ± 1.8), pork and beef old fashioned loaf (518 ± 1.8), simmered pork tail (510 ± 0.5), smoked beef sausage (507 ± 1.2), cooked oysters (506 ± 0.6), boiled cranberry beans (505 ± 0.3), simmered pork ears (505 ± 0.5), dried pilinuts (504 ± 1.8), surimi scallops (501 ± 0.6), pork and beef kielbasa/kolbassy (500 ± 1.9), boiled yellow beans (496 ± 0.3), canned pork lunch meat (495 ± 2.4), gefiltefish with broth (493 ± 1.2), boiled pink beans (490 ± 0.3), dried shitake mushrooms (487 ± 3.3), bran (486 ± 1.8), surimi shrimp (486 ± 0.6), boiled yardlong bean (484 ± 0.3), beef and pork salami (483 ± 2.2), boiled black beans (479 ± 0.3), pork and beef sausage (477 ± 3.8), boiled chickpeas (475 ± 0.3), boiled catjang black-eyed peas (475 ± 0.3), raw beef tripe (471 ± 0.5), pork bratwurst (471 ± 0.6), boiled navy beans (470 ± 0.3), boiled kidney beans (469 ± 0.3), garlic powder (467 ± 16.7), frozen and boiled black-eyed peas (466 ± 0.6), wheat flakes (464 ± 1.8), boiled baby lima beans (463 ± 0.3), beef and pork bologna (461 ± 2.2), raw clams (458 ±

0.6), pork beerwurst salami (452 ± 2.2), boiled great northern beans (451 ± 0.3), boiled black-eyed peas (451 ± 0.3), dried pinyon pine nuts (450 ± 1.8), boiled lima beans (449 ± 0.3), beef beerwurst salami (448 ± 2.2), boiled lentils (445 ± 0.3), boiled pinto beans (444 ± 0.3), pork pickle and pimento loaf (443 ± 1.8), boiled black turtle beans (442 ± 0.3), boiled mungo beans (440 ± 0.3), beef bologna (439 ± 2.2), boiled winged beans (436 ± 0.3), stir-fried sprouted lentils (434 ± 0.5), beef frankfurter (433 ± 0.9), dry roasted pecans (429 ± 1.8), raw mussels (426 ± 0.6), vienna sausage (425 ± 3.1), pork olive loaf (425 ± 1.8), cream cheese (425 ± 1.8), boiled mung beans (425 ± 0.2), corn flakes (418 ± 1.8), dried pecans (414 ± 1.8), dried ginko nuts (414 ± 1.8), boiled hyacinth beans (410 ± 0.3), simmered pork chitterlings (410 ± 0.5), pork and beef knockwurst (407 ± 0.7), canned navy beans (407 ± 0.2), pork and beef link sausage (404 ± 0.7), canned great northern beans (399 ± 0.2), boiled adzuki beans (398 ± 0.2), raw tofu (393 ± 0.4), canned white beans (392 ± 0.2), pickled pork feet (392 ± 0.5), boiled split peas (384 ± 0.3), sweetened condensed milk (384 ± 1.3), boiled french beans (381 ± 0.3), raisin bran (375 ± 1.8), oil roasted pecans (371 ± 1.8), potato flour (366 ± 0.6), evaporated skim milk (363 ± 1.6), dried acorns (361 ± 1.8), beef and pork frankfurter (360 ± 1.1), ham salad (357 ± 1.8), dried coconut (354 ± 1.8), breaded and fried oysters (352 ± 0.6), freeze-dried shallots (350 ± 12.5), steamed sprouted soybeans (338 ± 1.1), refried beans (337 ± 0.2), raw bacon (334 ± 0.7), acorn flour (332 ± 1.8), evaporated whole milk (329 ± 0.4), boiled and frozen baby lima beans (328 ± 0.6), canned black turtle beans (326 ± 0.2), boiled sprouted green peas (325 ± 0.5), dry roasted macadamia nuts (321 ± 1.8), boiled broad beans (321 ± 0.3), evaporated lowfat milk (317 ± 0.4), dried chinese chestnuts (314 ± 1.8), skim yogurt (312 ± 0.2), potato pancakes (304 ± 0.7), canned cranberry beans (300 ± 0.2), frozen and boiled fordhook lima beans (299 ± 0.6), baked beans with pork and sweet sauce (287 ± 0.2), raw spirulina seaweed (286 ± 0.5), lowfat yogurt (286 ± 0.2), canned lima beans (284 ± 0.2), sheep milk (284 ± 0.2), canned kidney beans (281 ± 0.2), baked beans with pork and tomato sauce (279 ± 0.2), canned black-eyed peas (277 ± 0.2), dried european chestnuts (275 ± 1.8), raw laver/nori seaweed (273 ± 0.5), tomato powder (273 ± 0.5), raw acorns (271 ± 1.8), puffed rice (271 ± 3.6), dried toasted coconut (271 ± 1.8), canned chickpeas (265 ± 0.2), dried macadamia nuts (264 ± 1.8), dried oriental radish (262 ± 0.9), vegetarian baked beans (259 ± 0.2), raw oysters (254 ± 0.6), au gratin potatoes (254 ± 0.4), canned oysters (253 ± 0.6), onion powder (250 ± 25), canned pinto beans (247 ± 0.2), lard (246 ± 1.8), boiled succotash (245 ± 0.5), raw pork jowl (239 ± 0.5), canned broad beans (231 ± 0.2), dehydrated onion flakes (229 ± 3.6), oil roasted macadamia nuts (229 ± 1.8), yellow corn pudding (224 ± 0.2), frozen and boiled succotash (208 ± 0.6), dried japanese chestnuts (207 ± 1.8), roasted chinese chest-nuts (207 ± 1.8), hummus (207 ± 0.2), frozen and boiled turnip greens (206 ± 0.6), ginger (200 ± 25), raw green peas (200 ± 0.6), noodles (198 ± 0.3), boiled green peas (198 ± 0.6), raw chinese chestnuts (193 ± 1.8), protein fortified skim milk (191 ± 0.2), protein fortified lowfat 2% milk (191 ± 0.2), protein fortified lowfat 1% milk (190 ± 0.2), frozen and boiled green peas (190 ± 0.6), whole yogurt (189 ± 0.2), restaurant vanilla milkshake (186 ± 0.2), coconut cream (184 ± 0.2), eggnog (182 ± 0.2), stir-fried sprouted mung beans (182 ± 0.8), malted milk (181 ± 0.2), raw garlic (178 ± 5.6), macaroni (177 ± 0.4), spaghetti (177 ± 0.4), hot chocolate (176 ± 0.2), ginko nuts (175 ± 1.8), buttermilk (174 ± 0.2), french fried potatoes (172 ± 1), raw coconut (169 ± 1.1), vanilla milkshake (168 ± 0.2), boiled jute potherb (167 ± 1.2), dried sweetened flaked coconut (166 ± 0.7), skim milk (164 ± 0.2), restaurant chocolate milkshake (164 ± 0.2), canned green peas (164 ± 0.6), restaurant strawberry milkshake (162 ± 0.2), indian buffalo milk (161 ± 0.2), lowfat 2% milk (161 ± 0.2), filled milk (161 ± 0.2), lowfat 1% milk (159 ± 0.2), whole milk (159 ± 0.2), okara tofu (157 ± 0.8), lowfat 1% chocolate milk (156 ± 0.2), goat milk (155 ± 0.2), lowfat 2% chocolate milk (155 ± 0.2), boiled yellow corn (155 ± 0.6), whole chocolate milk (153 ± 0.2), carob milk (152 ± 0.2), soy milk (151 ± 0.2), dried apricots (151 ± 1.4), frozen and fried french fried potatoes (148 ± 1), chocolate milkshake (147 ± 0.2), dried sweetened shredded coconut (146 ± 0.5), frozen and boiled kale (146 ± 0.8), frozen and boiled spinach (142 ± 0.5), oatmeal (141 ± 0.3), frozen and boiled yellow corn (141 ± 0.6), half and half cream (140 ± 3.3), sweetened coconut cream (137 ± 0.2), roasted european chestnuts (136 ± 1.8), potato salad (135 ± 0.4), scalloped potatoes (135 ± 0.4), frozen and fried hash brown potatoes (135 ± 0.6), boiled spinach (134 ± 0.6), light cream (133 ± 3.3), boiled and steamed chinese chestnuts (132 ± 1.8), raw spinach (129 ± 1.8), canned spinach (127 ± 0.5), raw swamp cabbage (127 ± 0.9), canned succotash with creamed corn (127 ± 0.4), boiled lamb's-quarters (126 ± 0.6), canned succotash (126 ± 0.4), sour cream (125 ± 4.2), medium cream (120 ± 3.3), roasted japanese chestnuts (118 ± 1.8), raw sprouted mung beans (117 ± 1), coconut milk (116 ± 0.2), boiled amaranth (114 ± 0.8), boiled swiss chard (114 ± 0.6), raw wakame seaweed (112 ± 0.5), raw watercress (112 ± 2.9), dried longans (112 ± 0.5), microwaved potato with skin (109 ± 0.2), whipping cream, light (107 ± 3.3), canned yellow corn with red and green peppers (107 ± 0.4), o'brien potatoes (106 ± 0.3), frozen and boiled brussels sprouts (106 ± 0.6), frozen and boiled collards (105 ± 0.6), raw european chestnuts (104 ± 1.8), boiled dock (104 ± 0.5), hash brown potatoes (103 ± 0.6), baked sweet potato (103 ± 0.4), frozen and baked sweet potato (103 ± 0.6), baked potato with skin (102 ± 0.2), boiled swamp cabbage (102 ± 1), frozen and boiled turnip greens and turnips (102 ± 0.5), whipping cream, heavy (100 ± 3.3), raw chives (100 ± 16.7), canned sweet potato (99 ± 0.3), boiled kale (97 ± 0.8), boiled scotch kale (97 = 0.8), canned yellow corn (95 ± 0.5), microwaved potato without skin (93 ± 0.3), frozen and boiled broccoli (93 ± 0.5), raw turnip greens (93 ± 1.8), raw potato without skin (92 ± 0.4), boiled broccoli (90 ± 0.6), raw japanese chestnuts (89 ± 1.8), raw bamboo shoots (89 ± 0.7), frozen and boiled potato without skin (88 ± 0.5), boiled mushrooms (88 ± 0.6), baked potato without skin (87 ± 0.3), boiled and steamed european chestnuts (86 ± 1.8), mashed potatoes (86 ± 0.5), raw vine spinach (85 ± 0.5), cooked sprouted mung beans (85 ± 0.8), raw dock (84 ± 0.7), raw broccoli (84 ± 1.1), canned turnip greens (83 ± 0.4), cream of wheat (82 ± 0.3), steamed hawaii mountain yam (81 ± 0.7), canned creamed yellow corn (81 ± 0.4), tomato paste (80 ± 0.4), soybean oil and cottonseed oil margarine liquid (80 ± 10), raw mushrooms (80 ± 1.4), raw shallots (80 ± 5), chocolate syrup (79 ± 1.3), farina (78 ± 0.3), boiled potato without skin (76 ± 0.4), dried figs (74 ± 0.3), boiled brussels sprouts (74 ± 0.6), corn grits (73 ± 0.2), raw cauliflower (72 ± 1), raw california avocado (72 ± 0.3), frozen and boiled asparagus (70 ± 0.8), boiled green beans (69 ± 0.8), boiled yellow snap beans (69 ± 0.8), baked/boiled tropical yam (69 ± 0.7), boiled turnip greens (69 ± 0.7), boiled beet greens (68 ± 0.7), boiled cauliflower (68 ± 0.8), frozen and boiled okra (68 ± 0.5), raw romaine lettuce (68 ± 1.8), cooked shitake mushrooms (67 ± 0.7), canned potato without skin (63 ± 0.6), raw savoy cabbage (63 ±

1.4), raw red chili peppers (62 ± 1.1), raw green chili peppers (62 ± 1.1), dried japanese persimmon (62 ± 1.5), boiled asparagus (61 ± 0.6), boiled okra (61 ± 0.6), raw welsh onions (61 ± 0.5), raw cassava (60 ± 0.5), frozen and boiled mustard greens (60 ± 0.7), boiled mustard greens (60 ± 0.7), canned bamboo shoots (60 ± 0.4), canned sprouted mung beans (60 ± 0.8), boiled purslane (59 ± 0.9), frozen and boiled cauliflower (58 ± 0.6), boiled savoy cabbage (58 ± 0.7), baked hubbard squash (58 ± 0.5), boiled hubbard squash (58 ± 0.4), safflower oil and soybean oil mayonnaise (57 ± 3.6), soybean oil mayonnaise (57 ± 3.6), raw spring onions with tops (56 ± 1), dried dates (55 ± 0.6), whipped butter (55 ± 4.5), raw leeks (54 ± 1.9), raw florida avocado (54 ± 0.2), raw looseleaf lettuce (54 ± 1.8), boiled bamboo shoots (53 ± 0.4), raw butterhead lettuce (53 ± 3.3), sapote (53 ± 0.2), raw endive (52 ± 2), candied sweet potato (52 ± 0.5), apricots (52 ± 0.5), whole chocolate malted milk (51 ± 0.2), raw iceberg lettuce (50 ± 2.5), frozen and boiled yellow snap beans (50 ± 0.7), frozen and boiled green beans (50 ± 0.7), beef suet (50 ± 1.8), frozen and boiled butternut squash (48 ± 0.4), coleslaw (47 ± 0.8), boiled chinese cabbage (46 ± 0.6), raw ginger root (46 ± 2.1), raw eggplant (46 ± 1.2), human milk (45 ± 1.6), frozen and boiled crookneck squash (45 ± 0.5), boiled burdock root (45 ± 0.4), raw lotus root (44 ± 0.6), baked acorn squash (44 ± 0.5), raw kelp/kombu/tangle seaweed (43 ± 0.5), raw chinese cabbage (43 ± 1.4), raw red cabbage (43 ± 1.4), canned green beans (43 ± 0.7), canned yellow snap beans (43 ± 0.7), raw scallop squash (42 ± 0.8), raw zucchini (42 ± 0.8), boiled kohlrabi (41 ± 0.6), raw chickory greens (41 ± 0.6), elderberries (40 ± 0.3), frozen and boiled zucchini (40 ± 0.4), raw green cabbage (40 ± 1.4), butter (40 ± 3.3), corn oil margarine stick (40 ± 10), safflower oil and soybean oil margarine stick (40 ± 10), soybean oil margarine stick (40 ± 10), soybean oil and cottonseed oil margarine stick (40 ± 10), corn oil margarine tub (40 ± 10), safflower oil margarine tub (40 ± 10), soybean oil margarine tub (40 ± 10), soybean oil and cottonseed oil margarine tub (40 ± 10), boiled collards (39 ± 0.3), bananas (38 ± 0.4), boiled sweet potato (38 ± 0.3), cream of rice (37 ± 0.3), coconut water (37 ± 0.2), frozen and boiled carrots (37 ± 0.7), persimmon (36 ± 2), boiled scallop squash (36 ± 0.6), canned zucchini in tomato juice (36 ± 0.4), raw celtuce (36 ± 0.5), boiled chayote (36 ± 0.6), tomato purée (35 ± 0.2), boiled red cabbage (35 ± 0.7), boiled carrots (35 ± 0.6), boiled butternut squash (35 ± 0.5), boiled eggplant (35 ± 1), canned pumpkin (35 ± 0.4), canned pumpkin pie mix (35 ± 0.4), valencia orange (34 ± 0.4), navel orange (34 ± 0.4), boiled and steamed japanese chestnuts (32 ± 1.8), raw crookneck squash (32 ± 0.8), boiled crookneck squash (32 ± 0.6), raw carrots (32 ± 0.7), raw green tomato (31 ± 0.4), boiled leeks (31 ± 1.9), boiled beets (31 ± 0.6), boiled green cabbage (31 ± 0.7), raw onions (30 ± 0.6), frozen and boiled turnips (30 ± 0.5), rice polish (30 ± 0.9), longans (30 ± 0.5), frozen and boiled red sweet peppers (29 ± 0.5), frozen and boiled green sweet peppers (29 ± 0.5), boiled red tomato (29 ± 0.4), boiled rutabaga (29 ± 0.6), canned red chili peppers (28 ± 0.7), canned green chili peppers (28 ± 0.7), boiled lotus root (28 ± 0.6), whey (27 ± 0.2), cooked plantain (27 ± 0.3), cooked taro (27 ± 0.8), boiled yambean (27 ± 0.5), raw red sweet peppers (26 ± 1), raw green sweet peppers (26 ± 1), canned red sweet peppers (26 ± 0.7), canned green sweet peppers (26 ± 0.7), japanese persimmon (26 ± 0.3), breadfruit (26 ± 0.5), boiled acorn squash (26 ± 0.4), canned jalapeño peppers (25 ± 0.7), canned peeled red tomatoes (24 ± 0.4), canned stewed red tomatoes (24 ± 0.4), blueberries (24 ± 0.3), boiled/baked spaghetti squash (24 ± 0.6), boiled onions (23 ±

0.5), raw red tomato (23 ± 0.4), pickled beets (23 ± 0.4), boiled pumpkin (23 ± 0.4), boiled zucchini (22 ± 0.6), peach (22 ± 0.6), canned crookneck squash (22 ± 0.5), boiled oriental radish (22 ± 0.7), raw radish (22 ± 1.1), tangerine (21 ± 0.6), canned carrots (21 ± 0.7), mandarin oranges (20 ± 0.4), raw oriental radish (20 ± 1.1), raw celery (20 ± 1.3), corn oil diet margarine tub (20 ± 10), soybean oil and cottonseed oil diet margarine tub (20 ± 10), boiled red sweet peppers (19 ± 0.7), boiled green sweet peppers (19 ± 0.7), carambola (19 ± 0.4), frozen and boiled onions (18 ± 0.5), figs (18 ± 1), strawberries (18 ± 0.3), plum (17 ± 0.8), mango (17 ± 0.2), tomato juice (16 ± 0.3), raw cucumber (15 ± 1), boiled celery (15 ± 0.7), watermelon (15 ± 0.3), european grapes (14 ± 0.3), loquats (14 ± 0.5), boiled turnips (14 ± 0.6), boiled calabash gourd (14 ± 0.7), american grapes (13 ± 0.5), sapodilla (13 ± 0.3), pineapple (12 ± 0.3), grape juice (12 ± 0.2), crab apples (11 ± 0.5), orange juice (9 ± 0.2), papaya (9 ± 0.2), thompson seedless grapes (8 ± 0.3), cooked apple without skin (7 ± 0.3), screwdriver (7 ± 0.2), tangerine juice (6 ± 0.2), ale (6 ± 0.1), raw apple with skin (5 ± 0.4), applesauce (5 ± 0.4), raw apple without skin (4 ± 0.4), white grapefruit juice (4 ± 0.2), pear (3 ± 0.3), coffee (3 ± 0.3), raw chickory (2 ± 1.1), guava (2 ± 0.6)

Applications: pain control (dl-phenylalanine)
Daily Dosage: Nutritional—0.75–2 g; therapeutic—2–3 g; experimental—4–5 g; toxic—5 g
Warnings: Hypertensives should check with doctor. Don't use with MAO inhibitors. People with increased blood pressure should start with low doses and monitor blood pressure. Increased amounts are toxic. See toxicity symptoms.

PROLINE

Names: proline, PRO, amino acid P
Classification: nonessential amino acid
Deficiency: protein decreased
Side-effects: none known
Toxicity: none known
Helpers: vitamin C
Sources: internal synthesis
Applications: cosmetic deterioration of aged tissues, wound healing, deterioration of collagen status, soft tissue strains, hypermobile joints, sagging aged tissues
Daily Dosage: Nutritional—1–2 g; therapeutic—?; experimental—?; toxic—none known up to 2 g
Warnings: none

SERINE

Names: serine, SER, amino acid S, 2-amino-3-hydroxy-propanoic acid
Forms: 2-amino-3-hydroxy-propanoic acid
Classification: nonessential amino acid
Deficiency: protein decreased
Side-effects: none known
Toxicity: none known
Inhibitors: increased amounts of other amino acids

Helpers: moderate amounts of other amino acids
Sources: internal synthesis
Daily Dosage: Nutritional — 1–3 g; therapeutic — ?; experimental — ?; toxic — none known up to 3 g
Warnings: none

TAURINE

Classification: nonessential amino acid
Deficiency: bile salts formation deterioration, decreased cholesterol solubility, protein decreased, epilepsy, vision decreased, decreased potassium in heart, decreased osmotic control of calcium and potassium in heart
Side-effects: see toxicity symptoms
Toxicity: decreased cell membrane integrity, gastro-intestinal pain, gallbladder deterioration, cardiac arrhythmias
Inhibitors: female hormone estradiol
Helpers: vitamin A, vitamin B_6, vitamin E, full spectrum light
Sources: animal protein, internal synthesis of methionine and cysteine
Applications: cholestasis, convulsions, epilepsy, increased blood sugar, muscular dystrophy, Down's syndrome
Daily Dosage: Nutritional — ?; therapeutic — 1 g; experimental — ?; toxic — none known up to 1 g
Warnings: increased amounts are dangerous. See toxicity symptoms.

THREONINE

Names: threonine, THR, amino acid T, 2-amino-3-hydroxy-butanoic acid
Forms: 2-amino-3-hydroxy-butanoic acid
Classification: essential amino acid
Deficiency: emotional agitation, mental health deterioration, deterioration of digestion, intestinal malfunctions, increased liver fat, deterioration of nutrient absorption, protein decreased
Side-effects: none known
Toxicity: none known
Inhibitors: increased amounts of other amino acids
Helpers: moderate amounts of other amino acids
Sources (mg per 100 grams of food): torula yeast (9161 ± 1.8), soybean protein isolate (3136 ± 1.8), dried spirulina seaweed (2970 ± 0.5), dried and salted cod (2754 ± 0.6), soybean protein concentrate (2475 ± 1.8), cooked whelk (2136 ± 0.6), defatted soybean flour (2042 ± 0.5), low fat soybean flour (2020 ± 0.6), dried and frozen tofu (1959 ± 2.9), defatted soy meal (1952 ± 0.4), dry roasted soybean nuts (1719 ± 0.6), fenugreek seed (1650 ± 25), braised pork center loin excluding fat (1636 ± 0.5), dry skim milk (1633 ± 1.7), toasted soybean nuts (1629 ± 1.8), braised pork top loin excluding fat (1620 ± 0.5), dry instant skim milk (1585 ± 0.5), braised pork sirloin excluding fat (1576 ± 0.5), braised pork loin excluding fat (1551 ± 0.5), parmesan cheese (1540 ± 10), fried beef top round excluding fat (1532 ± 0.5),

roasted soybean nuts (1530 ± 0.6), roasted soybean flour (1511 ± 0.6), broiled pork center loin excluding fat (1505 ± 0.5), peanut flour (1500 ± 12.5), soybean flour (1500 ± 0.6), almond meal (1486 ± 1.8), simmered beef shank excluding fat (1471 ± 0.5), fried chicken giblets (1467 ± 0.5), braised pork boston blade excluding fat (1466 ± 0.5), dry buttermilk (1443 ± 7.1), braised beef chuck arm pot roast excluding fat (1442 ± 0.5), fried beef sirloin excluding fat (1419 ± 0.5), fried chicken breast without skin (1412 ± 0.6), braised pork loin blade excluding fat (1397 ± 0.5), fried light meat chicken without skin (1386 ± 0.5), fried beef top round including fat (1384 ± 0.5), broiled beef top round excluding fat (1384 ± 0.5), simmered beef shank including fat (1381 ± 0.5), braised beef bottom round excluding fat (1380 ± 0.5), roasted pork rump excluding fat (1371 ± 0.5), braised pork spareribs (1367 ± 0.5), braised pork center loin including fat (1366 ± 0.5), stewed chicken drumstick without skin (1361 ± 1.1), simmered beef heart (1359 ± 0.5), braised beef chuck blade roast excluding fat (1357 ± 0.5), broiled pork center rib excluding fat (1356 ± 0.5), broiled pork top loin excluding fat (1356 ± 0.5), simmered turkey gizzard (1356 ± 0.5), roasted pork tenderloin (1354 ± 0.5), fried pork center loin excluding fat (1353 ± 0.5), broiled beef top round including fat (1346 ± 0.5), braised beef shortribs excluding fat (1343 ± 0.5), roasted pork center loin excluding fat (1340 ± 0.5), roasted pork leg excluding fat (1332 ± 0.5), roasted turkey light meat without skin (1330 ± 0.5), broiled pork sirloin excluding fat (1329 ± 0.5), roasted pork center rib excluding fat (1327 ± 0.5), roasted pork shank excluding fat (1327 ± 0.5), roasted pork top loin excluding fat (1327 ± 0.5), fried chicken breast with skin (1327 ± 0.5), broiled beef sirloin excluding fat (1327 ± 0.5), braised pork center rib including fat (1324 ± 0.5), fried pork center rib excluding fat (1315 ± 0.5), fried pork top loin excluding fat (1315 ± 0.5), roasted chicken breast without skin (1310 ± 0.6), broiled pork loin excluding fat (1309 ± 0.5), roasted light meat chicken without skin (1305 ± 0.5), braised beef bottom round including fat (1302 ± 0.5), braised pork sirloin including fat (1296 ± 0.5), canned tuna in water (1296 ± 0.6), roasted pork sirloin excluding fat (1293 ± 0.5), cooked flounder (1290 ± 0.5), braised beef brisket excluding fat (1283 ± 0.5), braised pork pancreas (1281 ± 0.5), braised pork top loin including fat (1278 ± 0.5), broiled pork center loin including fat (1272 ± 0.5), roasted turkey dark meat without skin (1271 ± 0.5), well done baked extra lean ground beef (1270 ± 0.5), roasted eye of beef round excluding fat (1266 ± 0.5), roasted pork loin excluding fat (1265 ± 0.5), anchovy canned in olive oil (1265 ± 2.5), caviar (1263 ± 3.1), fried light meat chicken with skin (1262 ± 0.5), braised beef pancreas (1257 ± 0.5), braised pork loin including fat (1257 ± 0.5), roasted beef tip excluding fat (1254 ± 0.5), simmered chicken gizzard (1251 ± 0.5), broiled beef top loin excluding fat (1250 ± 0.5), roasted turkey light meat with skin (1247 ± 0.5), roasted chicken breast with skin (1244 ± 0.5), broiled beef round excluding fat (1243 ± 0.5), roasted pork rump including fat (1242 ± 0.5), well done baked lean ground beef (1241 ± 0.5), braised pork arm braised including fat (1241 ± 0.5), chipped beef (1236 ± 1.8), broiled beef tenderloin excluding fat (1234 ± 0.5), simmered beef kidneys (1231 ± 0.5), broiled porterhouse steak excluding fat (1230 ± 0.5), roasted pumpkin and squash seeds (1229 ± 1.8), broiled T-bone steak excluding fat (1229 ± 0.5), broiled beef rib eye excluding fat (1225 ± 0.5), braised pork boston blade braised including fat (1224 ± 0.5), stewed chicken breast without skin (1224 ± 0.5), braised beef flank excluding fat (1224 ± 0.5), fried beef liver (1222 ± 0.5), fried dark meat chicken without skin (1221 ± 0.5), stewed light meat chicken without skin (1220 ± 0.5), simmered turkey heart (1212 ± 0.5), well done baked ground beef (1207 ± 0.5), simmered turkey giblets (1204 ± 0.5), roasted beef tenderloin excluding fat (1203 ± 0.5), roasted light meat chicken with skin (1202 ± 0.5), braised beef flank including fat (1202 ± 0.5), roasted beef rib roasted excluding fat (1202 ± 0.5), yellow mustard seed (1200 ± 16.7), cooked coho salmon (1200 ± 0.6), roasted turkey dark meat with skin (1200 ± 0.5), fried beef sirloin including fat (1199 ± 0.5), well done broiled extra lean ground beef (1198 ± 0.5), simmered chicken heart (1196 ± 0.5), broiled beef sirloin including fat (1196 ± 0.5), cooked sockeye salmon (1196 ± 0.6), roasted pork shoulder excluding fat (1193 ± 0.5), dry whole milk (1188 ± 1.6), broiled pork boston blade excluding fat (1184 ± 0.5), braised beef chuck arm pot roast including fat (1183 ± 0.5), well done broiled lean ground beef (1182 ± 0.5), roasted pork center loin including fat (1182 ± 0.5), raw pheasant without skin (1180 ± 0.5), dried sesame kernels (1175 ± 6.3), well done fried extra lean ground beef (1173 ± 0.5), broiled pork loin blade excluding fat (1172 ± 0.5), simmered chicken giblets (1172 ± 0.5), cooked halibut (1171 ± 0.6), roasted eye of beef round including fat (1169 ± 0.5), cooked bacon (1169 ± 0.4), roasted pork sirloin including fat (1166 ± 0.5), canned tuna in oil (1164 ± 0.6), roasted pork leg including fat (1163 ± 0.5), roasted pork loin blade excluding fat (1161 ± 0.5), well done fried lean ground beef (1157 ± 0.5), roasted dark meat chicken without skin (1156 ± 0.5), roasted beef round tip including fat (1156 ± 0.5), cooked rainbow trout (1155 ± 0.6), cooked snapper (1153 ± 0.6), roasted pork center rib including fat (1149 ± 0.5), roasted pork boston blade excluding fat (1146 ± 0.5), fried pork loin blade excluding fat (1144 ± 0.5), stewed chicken breast with skin (1143 ± 0.5), fried chicken back with skin (1142 ± 0.7), cooked sheepshead (1141 ± 0.6), well done broiled ground beef (1140 ± 0.5), broiled pork center rib including fat (1139 ± 0.5), broiled beef tenderloin including fat (1134 ± 0.5), well done fried ground beef (1132 ± 0.5), dried watermelon seeds (1129 ± 1.8), braised pork spleen (1128 ± 0.5), roasted pork shank including fat (1126 ± 0.5), broiled beef top loin including fat (1124 ± 0.5), fried dark meat chicken with skin (1123 ± 0.5), roasted goose with skin (1123 ± 0.5), roasted pork top loin including fat (1121 ± 0.5), roasted chicken drumstick with skin (1121 ± 1), fried chicken drumstick with skin (1120 ± 1), broiled pork sirloin including fat (1117 ± 0.5), braised beef liver (1116 ± 0.5), broiled beef round including fat (1115 ± 0.5), cooked swordfish (1113 ± 0.6), braised beef chuck blade roast including fat (1111 ± 0.5), roasted cured pork arm excluding fat (1109 ± 0.5), broiled beef flank excluding fat (1109 ± 0.5), broiled beef rib eye including fat (1108 ± 0.5), braised pork liver (1107 ± 0.5), smoked haddock (1106 ± 0.6), fried chicken thigh with skin (1105 ± 0.8), gruyere cheese (1104 ± 1.8), braised pork loin blade including fat (1103 ± 0.5), cooked clams (1099 ± 0.6), stewed dark meat chicken without skin (1097 ± 0.5), roasted chicken thigh without skin (1096 ± 1), broiled porterhouse steak including fat (1096 ± 0.5), broiled pork top loin including fat (1095 ± 0.5), broiled beef flank including fat (1094 ± 0.5), raw quail without skin (1090 ± 0.5), broiled pork loin including fat (1090 ± 0.5), cooked grouper (1089 ± 0.6), roasted chicken wing with skin (1088 ± 1.5), cooked mullet (1088 ± 0.6), roasted pork loin including fat (1086 ± 0.5), stewed light meat chicken with skin (1084 ± 0.5), simmered chicken liver (1083 ± 0.5), cooked pike (1082 ± 0.6), sardines (1079 ± 2.1), kippered herring (1078 ± 1.3), broiled beef rib excluding fat (1076 ± 0.5), cooked tilefish (1074 ±

0.6), milk chocolate (1071 ± 1.8), fried pork center loin including fat (1071 ± 0.5), roasted dark meat chicken with skin (1071 ± 0.5), roasted beef tenderloin including fat (1069 ± 0.5), raw whelk (1068 ± 0.6), simmered turkey liver (1066 ± 0.5), medium-rare broiled extra lean ground beef (1065 ± 0.5), cooked haddock (1064 ± 0.6), cooked flatfish (1059 ± 0.6), fried chicken wing with skin (1056 ± 1.6), cooked rockfish (1054 ± 0.6), braised pork kidneys (1053 ± 0.5), stewed chicken drumstick with skin (1051 ± 0.9), swiss cheese (1050 ± 1.8), broiled T-bone steak including fat (1048 ± 0.5), cooked perch (1047 ± 0.6), medium-rare fried extra lean ground beef (1046 ± 0.5), cooked mackerel (1045 ± 0.6), simmered chicken neck without skin (1039 ± 2.8), cooked pompano (1038 ± 0.6), roasted chicken thigh with skin (1037 ± 0.8), medium-rare broiled lean ground beef (1036 ± 0.5), cooked sea bass (1036 ± 0.6), cooked eel (1036 ± 0.6), braised pork heart (1035 ± 0.4), roasted pork arm including fat (1030 ± 0.5), cooked whiting (1029 ± 0.6), medium-rare baked extra lean ground beef (1026 ± 0.5), smoked whitefish (1026 ± 0.6), cooked mussels (1025 ± 0.6), canned corned beef (1023 ± 0.5), roasted pork shoulder including fat (1019 ± 0.5), braised pork tongue (1018 ± 0.5), medium-rare fried lean ground beef (1016 ± 0.5), raw yellowtail (1015 ± 0.6), broiled pork boston blade including fat (1012 ± 0.5), roasted pork boston blade including fat (1011 ± 0.5), cooked herring (1011 ± 0.6), fried chicken neck with skin (1011 ± 1.4), hard pork salami (1010 ± 5), medium-rare broiled ground beef (1009 ± 0.5), braised beef brisket including fat (1004 ± 0.5), roasted duck without skin (1003 ± 0.5), medium-rare baked lean ground beef (1003 ± 0.5), medium-rare fried ground beef (1003 ± 0.5), cooked carp (1002 ± 0.6), cooked cod (1001 ± 0.6), turkey breast meat summer sausage (1000 ± 2.4), canned cod (998 ± 0.6), provolone cheese (993 ± 1.8), roasted beef rib including fat (992 ± 0.5), cooked rainbow smelt (991 ± 0.6), fried pork center rib including fat (989 ± 0.5), braised beef spleen (988 ± 0.5), fried pork top loin including fat (985 ± 0.5), roasted pork loin blade including fat (971 ± 0.5), stewed dark meat chicken with skin (969 ± 0.5), cooked crayfish (968 ± 0.6), medium-rare baked ground beef (965 ± 0.5), stewed chicken thigh with skin (962 ± 0.7), grilled canadian bacon (962 ± 1.1), simmered beef tongue (962 ± 0.5), oil roasted spanish peanuts (961 ± 1.8), hard pork and beef salami (960 ± 5), dried butternuts (954 ± 1.8), broiled pork loin blade including fat (950 ± 0.5), raw coho salmon (947 ± 0.6), roasted canned lean ham (944 ± 0.5), dried sunflower seeds (943 ± 1.8), edam cheese (943 ± 1.8), gouda cheese (943 ± 1.8), braised beef shortribs including fat (942 ± 0.5), cooked black bass (940 ± 0.4), raw sockeye salmon (934 ± 0.6), roasted lean ham (931 ± 0.5), pork link sausage (929 ± 0.7), oil roasted valencia peanuts (925 ± 1.8), stewed chicken wing with skin (925 ± 1.3), raw shark (920 ± 0.6), roasted canned ham (916 ± 0.5), dried pumpkin and squash seeds (914 ± 1.8), raw halibut (912 ± 0.6), tilsit cheese (911 ± 1.8), raw trout (911 ± 0.6), cooked sturgeon (907 ± 0.6), jellied corned beef loaf (904 ± 1.8), raw rainbow trout (901 ± 0.6), muenster cheese (900 ± 1.8), raw milkfish (900 ± 0.6), pickled anchovy (900 ± 1.8), raw snapper (899 ± 0.6), canned sockeye salmon (896 ± 0.6), brick cheese (893 ± 1.8), raw anchovy (892 ± 0.6), pork luxury loaf (889 ± 1.8), chicken egg yolk (888 ± 2.9), port du salut cheese (886 ± 1.8), cheddar cheese (886 ± 0.5), oil roasted virginia peanuts (886 ± 1.8), raw sheepshead (886 ± 0.6), roasted cured pork arm including fat (885 ± 0.5), raw chum salmon (884 ± 0.6), oil roasted sunflower seeds (882 ± 1.8), monterey cheese (882 ± 1.8), roasted ham (882 ± 0.5), raw chinook salmon (879 ± 0.6), broiled beef rib including fat (879 ± 0.5), raw bluefish (878 ± 0.6), freeze-dried chives (875 ± 62.5), raw pink salmon (874 ± 0.6), canned pink salmon (874 ± 0.6), raw swordfish (868 ± 0.6), breaded and fried shrimp (860 ± 0.6), unheated lean ham (860 ± 0.5), colby cheese (857 ± 1.8), anchovy paste (857 ± 7.1), chopped smoked beef (857 ± 1.8), fried pork loin blade including fat (856 ± 0.5), raw pollock (852 ± 0.6), raw sunfish (851 ± 0.6), raw grouper (849 ± 0.6), raw mullet (848 ± 0.6), raw burbot (847 ± 0.6), cooked shrimp (846 ± 0.6), raw pike (844 ± 0.6), battered and fried shark (844 ± 0.6), cheshire cheese (843 ± 1.8), fried abalone (838 ± 0.6), turkey ham (837 ± 0.9), raw whitefish (836 ± 0.6), raw alewife (834 ± 0.5), raw duck liver (833 ± 0.5), poppy seed (833 ± 16.7), raw cusk (833 ± 0.6), raw cisco (832 ± 0.6), raw ling (832 ± 0.6), canned blue crab (831 ± 0.6), cooked lobster (831 ± 0.6), raw haddock (829 ± 0.6), crab cakes (828 ± 0.8), raw freshwater bass (827 ± 0.6), raw scup (827 ± 0.6), raw flatfish (826 ± 0.6), canned lean ham (826 ± 0.5), raw black bass (826 ± 0.5), creamy peanut butter (825 ± 3.1), unheated canadian bacon (825 ± 0.9), raw shrimp (822 ± 0.6), crunchy peanut butter (822 ± 1.6), raw rockfish (822 ± 0.6), cooked blue crab (818 ± 0.6), raw perch (816 ± 0.6), raw mackerel (815 ± 0.6), natto (813 ± 0.6), raw spot (812 ± 0.6), dry roasted peanuts (811 ± 1.8), raw dolphinfish (811 ± 0.6), raw pompano (811 ± 0.6), raw eel (809 ± 0.6), lebanon bologna (809 ± 2.2), raw sea bass (808 ± 0.6), raw whiting (804 ± 0.6), smoked chinook salmon (801 ± 0.6), dry roasted sunflower seeds (800 ± 1.8), basil (800 ± 50), raw smelt (800 ± 0.5), bleu cheese (796 ± 1.8), raw catfish (796 ± 0.6), pork and beef honey loaf (796 ± 1.8), turkey pastrami (793 ± 0.9), italian pork sausage (793 ± 0.7), breaded and fried catfish (791 ± 0.6), canned smelt (791 ± 0.5), braised beef thymus (790 ± 0.5), breaded and fried croaker (788 ± 0.6), raw herring (787 ± 0.6), oil roasted peanuts (786 ± 1.8), cooked alaska king crab (784 ± 0.6), pork and beef pepperoni (783 ± 8.3), raw carp (782 ± 0.6), unheated ham (781 ± 0.5), raw cod (781 ± 0.6), raw croaker (780 ± 0.6), pork sausage (778 ± 1.9), simmered chicken neck with skin (776 ± 1.3), raw striped bass (776 ± 0.6), raw lingcod (774 ± 0.6), roasted duck with skin (773 ± 0.5), raw rainbow smelt (773 ± 0.6), dried pignolia pine nuts (771 ± 1.8), pork and beef peppered loaf (771 ± 1.8), tempeh (770 ± 0.6), raw freshwater drum (769 ± 0.6), oil roasted almonds (768 ± 1.8), toasted almonds (768 ± 1.8), breaded and fried scallops (768 ± 1.6), roasted pork blade roll (768 ± 0.5), raw tilefish (767 ± 0.6), raw wolffish (767 ± 0.6), breaded fried squid (764 ± 0.6), chopped ham lunch meat (762 ± 2.4), brie cheese (761 ± 1.8), raw lobster (761 ± 0.6), braised beef lungs (761 ± 0.5), stir-fried sprouted soybeans (759 ± 0.5), raw butterfish (758 ± 0.6), canned ham (758 ± 0.5), raw crayfish (755 ± 0.6), dried peanuts (754 ± 1.8), dried almonds (750 ± 1.8), beef lunch meat (750 ± 1.8), mozzarella cheese (750 ± 1.8), raw queen crab (749 ± 0.6), limburger cheese (746 ± 1.8), raw shad (742 ± 0.6), raw alaska king crab (741 ± 0.6), dried black walnuts (739 ± 1.8), duck egg (736 ± 0.7), raw abalone (736 ± 0.6), raw white sucker (735 ± 0.6), raw seatrout (734 ± 0.6), whole dried sesame seeds (733 ± 5.6), minced ham lunch meat (733 ± 2.4), dried pistachio nuts (732 ± 1.8), raw blue crab (731 ± 0.6), american cheese (729 ± 1.8), pimento cheese spread (729 ± 1.8), ham and cheese roll (729 ± 1.8), raw goose liver (728 ± 0.5), camembert cheese (725 ± 1.8), boiled soybeans (723 ± 0.3), raw scallops (722 ± 0.6), smoked cisco (716 ± 0.6), toasted sesame kernels (714 ± 1.8), whole toasted sesame seeds (714 ± 1.8), raw sturgeon (708 ± 0.6), sesame butter (707 ± 3.3), raw dungeness crab (705 ± 0.6), raw turbot (704 ± 0.6), tuna salad (701 ± 0.2), raw black bullhead (701 ± 0.5), fried tofu (700 ± 3.8), raw cuttlefish (699 ± 0.6), canned alewife (697 ± 0.5), cashew butter (689 ± 1.8), freeze-dried red sweet peppers (688 ± 31.3), freeze-dried green sweet peppers (688 ± 31.3), cooked corned beef brisket (686 ± 0.5), pork liverwurst (678 ± 2.8), corn germ (675 ± 0.5), raw squid (671 ± 0.6), pork and beef luncheon sausage (665 ± 2.2), pork and beef picnic loaf (664 ± 1.8), beef pastrami (661 ± 1.8), berliner (652 ± 2.2), pork livercheese (650 ± 1.3), dried ginko nuts (650 ± 1.8), raw flounder (646 ± 0.5), raw sole (646 ± 0.5), firm raw tofu (644 ± 0.4), raw octopus (642 ± 0.6), pork bologna (639 ± 2.2), miso (639 ± 0.4), oil roasted cashews (636 ± 1.8), raw monkfish (635 ± 0.6), pork and beef mortadella (633 ± 3.3), pork and beef old fashioned loaf (625 ± 1.8), pickled herring (620 ± 3.3), surimi scallops (616 ± 0.6), dry roasted almonds (614 ± 1.8), cooked oysters (608 ± 0.6), breaded and fried clams (602 ± 0.6), fennel seed (600 ± 25), dry roasted cashews (600 ± 1.8), polish pork sausage (600 ± 1.8), dill seed (600 ± 25), pork and beef brotwurst (599 ± 0.7), surimi shrimp (599 ± 0.6), fried beef brains (597 ± 0.5), chicken egg (596 ± 1), summer sausage (596 ± 2.2), simmered pork tail (595 ± 0.5), grilled ham patties (587 ± 0.8), braised pork lungs (584 ± 0.5), boiled lupins (573 ± 0.3), unheated ham patties (571 ± 0.8), beef salami (570 ± 2.2), braised pork brains (567 ± 0.5), pork beerwurst salami (565 ± 2.2), almond butter (556 ± 3.1), pork bratwurst (556 ± 0.6), cottage cheese (554 ± 0.2), pork and beef sausage (554 ± 3.8), raw clams (551 ± 0.6), pork and beef lunch meat (550 ± 1.8), pork and beef link sausage with american cheese (542 ± 1.2), pork liver sausage (533 ± 2.8), smoked beef sausage (533 ± 1.2), dry roasted pistachio nuts (532 ± 1.8), pork pickle and pimento loaf (529 ± 1.8), simmered beef brains (526 ± 0.5), beef and pork salami (522 ± 2.2), simmered pork feet (518 ± 0.5), ricotta cheese (517 ± 0.4), beef and pork bologna (513 ± 2.2), raw mussels (512 ± 0.6), raw pork stomach (510 ± 0.5), raw beef tripe (503 ± 0.5), dried shitake mushrooms (500 ± 3.3), oil roasted filberts (496 ± 1.8), canned pork lunch meat (495 ± 2.4), boiled green soybeans (492 ± 0.6), falafel (492 ± 1), steamed sprouted soybeans (491 ± 1.1), gefiltefish with broth (488 ± 1.2), pork and beef knockwurst (479 ± 0.7), pork olive loaf (479 ± 1.8), simmered pork ears (473 ± 0.5), beef beerwurst salami (470 ± 2.2), dried brazil nuts (468 ± 1.8), pork and beef link sausage (465 ± 0.7), boiled peanuts (463 ± 1.6), beef bologna (461 ± 2.2), puffed wheat (457 ± 3.6), dried filberts (454 ± 1.8), dried english/persian walnuts (454 ± 1.8), beef frankfurter (454 ± 0.9), simmered pork chitterlings (451 ± 0.5), pork headcheese (450 ± 1.8), almond paste (446 ± 1.8), chicken egg white (445 ± 1.5), garlic powder (433 ± 16.7), pork and beef kielbasa/kolbassy (431 ± 1.9), dried hickory nuts (429 ± 1.8), neufchatel cheese (429 ± 1.8), freeze-dried shallots (425 ± 12.5), ham salad (414 ± 1.8), dried pilinuts (414 ± 1.8), bran (411 ± 1.8), boiled white beans (409 ± 0.3), beef and pork frankfurter (407 ± 1.1), gjetost cheese (396 ± 1.8), boiled cranberry beans (393 ± 0.3), boiled mothbeans (388 ± 0.3), boiled yellow beans (386 ± 0.3), boiled pink beans (381 ± 0.3), boiled black beans (373 ± 0.3), shredded wheat (371 ± 1.8), dried pinyon pine nuts (371 ± 1.8), bran flakes (368 ± 1.8), boiled navy beans (366 ± 0.3), breaded and fried oysters (365 ± 0.6), boiled kidney beans (365 ± 0.3), pickled pork feet (365 ± 0.5). boiled winged beans (360 ± 0.3), sweetened condensed milk (358 ± 1.3), vienna sausage (356 ± 3.1), boiled great northern beans (351 ± 0.3), dry roasted filberts (350 ± 1.8), boiled baby lima beans (347 ± 0.3), boiled pinto beans (346 ± 0.3), boiled black turtle beans (344 ± 0.3), evaporated skim milk (341 ± 1.6), boiled lima beans (337 ± 0.3), raw bacon (332 ± 0.7), raw tofu (330 ± 0.4), boiled chickpeas (329 ± 0.3), dried oriental radish (326 ± 0.9), cream cheese (325 ± 1.8), potato flour (324 ± 0.6), boiled lentils (323 ± 0.3), stir-fried sprouted lentils (322 ± 0.5), wheat flakes (321 ± 1.8), puffed rice (321 ± 3.6), dried acorns (318 ± 1.8), canned navy beans (317 ± 0.2), boiled yardlong bean (316 ± 0.3), frozen and boiled black-eyed peas (316 ± 0.6), boiled hyacinth beans (315 ± 0.3), canned great northern beans (310 ± 0.2), boiled catjang black-eyed peas (309 ± 0.3), evaporated whole milk (307 ± 0.4), raw spirulina seaweed (306 ± 0.5), canned white beans (305 ± 0.2), raw oysters (304 ± 0.6), canned oysters (304 ± 0.6), dry roasted macadamia nuts (304 ± 1.8), boiled french beans (297 ± 0.3), boiled split peas (296 ± 0.3), tomato powder (295 ± 0.5), boiled black-eyed peas (294 ± 0.3), acorn flour (293 ± 1.8), corn flakes (286 ± 1.8), boiled and frozen baby lima beans (282 ± 0.6), dried chinese chestnuts (275 ± 1.8), ginko nuts (271 ± 1.8), boiled broad beans (270 ± 0.3), sheep milk (268 ± 0.2), dried macadamia nuts (268 ± 1.8), dry roasted pecans (264 ± 1.8), raisin bran (264 ± 1.8), boiled mungo beans (262 ± 0.3), refried beans (262 ± 0.2), dried pecans (257 ± 1.8), frozen and boiled fordhook lima beans (256 ± 0.6), boiled adzuki beans (255 ± 0.2), evaporated lowfat milk (254 ± 0.4), dried coconut (254 ± 1.8), canned black turtle beans (254 ± 0.2), potato pancakes (253 ± 0.7), boiled sprouted green peas (240 ± 0.5), raw acorns (239 ± 1.8), boiled pigeon peas (239 ± 0.3), skim yogurt (235 ± 0.2), canned cranberry beans (233 ± 0.2), dried european chestnuts (232 ± 1.8), raw laver/nori seaweed (232 ± 0.5), oil roasted macadamia nuts (232 ± 1.8), boiled mung beans (230 ± 0.2), oil roasted pecans (229 ± 1.8), baked beans with pork and sweet sauce (223 ± 0.2), canned kidney beans (219 ± 0.2), lard (218 ± 1.8), baked beans with pork and tomato sauce (217 ± 0.2), lowfat yogurt (215 ± 0.2), dehydrated onion flakes (214 ± 3.6), dried japanese chestnuts (214 ± 1.8), canned lima beans (213 ± 0.2), boiled succotash (211 ± 0.5), raw pork jowl (210 ± 0.5), raw green peas (203 ± 0.6), vegetarian baked beans (202 ± 0.2), boiled green peas (201 ± 0.6), onion powder (200 ± 25), yellow corn pudding (198 ± 0.2), dried toasted coconut (196 ± 1.8), canned broad beans (194 ± 0.2), frozen and boiled green peas (193 ± 0.6), au gratin potatoes (192 ± 0.4), canned pinto beans (192 ± 0.2), frozen and boiled turnip greens (184 ± 0.6), french fried potatoes (184 ± 1), canned chickpeas (184 ± 0.2), indian buffalo milk (182 ± 0.2), roasted chinese chestnuts (182 ± 1.8), canned black-eyed peas (180 ± 0.2), protein fortified skim milk (179 ± 0.2), frozen and boiled succotash (179 ± 0.6), protein fortified lowfat 2% milk (178 ± 0.2), protein fortified lowfat 1% milk (177 ± 0.2), eggnog (175 ± 0.2), restaurant vanilla milkshake (174 ± 0.2), hummus (173 ± 0.2), noodles (173 ± 0.3), raw chinese chestnuts (168 ± 1.8), raw wakame seaweed (165 ± 0.5), canned green peas (165 ± 0.6), hot chocolate (164 ± 0.2), malted milk (163 ± 0.2), goat milk (163 ± 0.2), vanilla milkshake (158 ± 0.2), frozen and fried french fried potatoes (158 ± 1), buttermilk (158 ± 0.2), raw garlic (156 ± 5.6), restaurant chocolate milkshake (154 ± 0.2), skim milk (154 ± 0.2), restaurant strawberry milkshake (152 ± 0.2), ginger (150 ± 25), lowfat 2% milk (150 ± 0.2), filled milk (150 ± 0.2), lowfat 1% milk (148 ± 0.2), whole milk (148 ± 0.2), lowfat 1% chocolate milk (146 ± 0.2), lowfat 2% chocolate milk (145 ± 0.2), frozen and fried hash brown potatoes (144 ± 0.6), whole chocolate milk (143 ± 0.2), carob milk (142 ± 0.2), whole yogurt (142 ± 0.2), raw swamp cabbage (139 ± 0.9), chocolate milkshake (138 ± 0.2), raw watercress (135 ± 2.9), frozen and boiled spinach (134 ± 0.5), macaroni (133 ± 0.4), spaghetti (133 ± 0.4), half and half cream (133 ± 3.3), boiled yellow corn (133 ± 0.6), raw sprouted alfalfa seeds (133 ± 1.5), coconut cream

(132 ± 0.2), dried apricots (131 ± 1.4), okara tofu (131 ± 0.8), boiled jute potherb (130 ± 1.2), frozen and boiled brussels sprouts (129 ± 0.6), frozen and boiled kale (128 ± 0.8), dried longans (128 ± 0.5), boiled spinach (127 ± 0.6), sour cream (125 ± 4.2), boiled lamb's-quarters (124 ± 0.6), stir-fried sprouted mung beans (123 ± 0.8), roasted japanese chestnuts (121 ± 1.8), raw spinach (121 ± 1.8), frozen and boiled yellow corn (121 ± 0.6), light cream (120 ± 3.3), raw coconut (120 ± 1.1), canned spinach (120 ± 0.5), dried sweetened flaked coconut (119 ± 0.7), boiled and steamed chinese chestnuts (118 ± 1.8), potato salad (116 ± 0.4), scalloped potatoes (115 ± 0.4), roasted european chestnuts (114 ± 1.8), soy milk (113 ± 0.2), medium cream (113 ± 3.3), boiled swamp cabbage (112 ± 1), hash brown potatoes (110 ± 0.6), canned succotash with creamed corn (110 ± 0.4), canned succotash (108 ± 0.4), frozen and boiled collards (105 ± 0.6), dried sweetened shredded coconut (105 ± 0.5), raw dock (101 ± 0.7), boiled mushrooms (101 ± 0.6), frozen and boiled broccoli (101 ± 0.5), whipping cream, light (100 ± 3.3), raw chives (100 ± 16.7), raw shallots (100 ± 5), dried figs (100 ± 0.3), sweetened coconut cream (98 ± 0.2), frozen and boiled turnip greens and turnips (97 ± 0.5), boiled broccoli (96 ± 0.6), raw mushrooms (94 ± 1.4), raw japanese chestnuts (93 ± 1.8), whipping cream, heavy (93 ± 3.3), canned yellow corn with red and green peppers (93 ± 0.4), raw broccoli (91 ± 1.1), boiled brussels sprouts (91 ± 0.6), oatmeal (90 ± 0.3), microwaved potato with skin (89 ± 0.2), o'brien potatoes (89 ± 0.3), raw european chestnuts (86 ± 1.8), raw bamboo shoots (86 ± 0.7), tomato paste (86 ± 0.4), baked sweet potato (86 ± 0.4), boiled swiss chard (86 ± 0.6), boiled dock (86 ± 0.5), boiled amaranth (85 ± 0.8), boiled kale (85 ± 0.8), boiled scotch kale (85 ± 0.8), frozen and baked sweet potato (85 ± 0.6), baked potato with skin (84 ± 0.2), coconut milk (83 ± 0.2), frozen and boiled asparagus (82 ± 0.8), raw turnip greens (82 ± 1.8), boiled sweet potato (82 ± 0.3), canned sweet potato (82 ± 0.3), boiled green beans (82 ± 0.8), boiled yellow snap beans (82 ± 0.8), canned yellow corn (82 ± 0.5), soybean oil and cottonseed oil margarine liquid (80 ± 10), raw sprouted mung beans (79 ± 1), boiled beet greens (76 ± 0.7), microwaved potato without skin (76 ± 0.3), raw potato without skin (75 ± 0.4), raw romaine lettuce (75 ± 1.8), canned turnip greens (74 ± 0.4), raw welsh onions (74 ± 0.5), mashed potatoes (73 ± 0.5), raw red chili peppers (73 ± 1.1), raw green chili peppers (73 ± 1.1), frozen and boiled potato without skin (72 ± 0.5), boiled asparagus (72 ± 0.6), raw cauliflower (72 ± 1), baked potato without skin (71 ± 0.3), dried japanese persimmon (71 ± 1.5), boiled and steamed european chestnuts (71 ± 1.8), raw california avocado (70 ± 0.3), canned creamed yellow corn (70 ± 0.4), raw savoy cabbage (69 ± 1.4), boiled cauliflower (68 ± 0.8), raw spring onions with tops (68 ± 1), frozen and boiled okra (68 ± 0.5), cooked shitake mushrooms (68 ± 0.7), raw cassava (65 ± 0.5), chocolate syrup (63 ± 1.3), boiled turnip greens (63 ± 0.7), boiled potato without skin (62 ± 0.4), boiled savoy cabbage (62 ± 0.7), raw leeks (62 ± 1.9), boiled okra (61 ± 0.6), steamed hawaii mountain yam (61 ± 0.7), raw looseleaf lettuce (61 ± 1.8), boiled mustard greens (60 ± 0.7), frozen and boiled mustard greens (60 ± 0.7), raw butterhead lettuce (60 ± 3.3), frozen and boiled cauliflower (59 ± 0.6), frozen and boiled yellow snap beans (59 ± 0.7), frozen and boiled green beans (59 ± 0.7), cooked sprouted mung beans (58 ± 0.8), sapote (58 ± 0.2), canned bamboo shoots (57 ± 0.4), safflower oil and soybean oil mayonnaise (57 ± 3.6), soybean oil mayonnaise (57 ± 3.6), raw kelp/kombu/tangle seaweed (55 ± 0.5), raw vine spinach (55 ± 0.5), raw iceberg lettuce (55 ± 2.5), whey (54 ± 0.2), raw florida avocado (53 ± 0.2), dried dates (52 ±

0.6), boiled kohlrabi (52 ± 0.6), raw endive (52 ± 2), breadfruit (52 ± 0.5), canned potato without skin (51 ± 0.6), raw lotus root (51 ± 0.6), boiled chinese cabbage (51 ± 0.6), baked/boiled tropical yam (51 ± 0.7), boiled bamboo shoots (50 ± 0.4), boiled purslane (50 ± 0.9), canned green beans (50 ± 0.7), canned yellow snap beans (50 ± 0.7), corn grits (50 ± 0.2), raw chinese cabbage (49 ± 1.4), raw red cabbage (49 ± 1.4), cream of wheat (48 ± 0.3), coleslaw (48 ± 0.8), raw chickory greens (47 ± 0.6), apricots (47 ± 0.5), whole chocolate malted milk (47 ± 0.2), human milk (45 ± 1.6), baked hubbard squash (44 ± 0.5), boiled hubbard squash (44 ± 0.4), farina (44 ± 0.3), cream of rice (44 ± 0.3), raw green cabbage (43 ± 1.4), candied sweet potato (43 ± 0.5), beef suet (43 ± 1.8), frozen and boiled turnips (42 ± 0.5), frozen and boiled carrots (42 ± 0.7), boiled rutabaga (42 ± 0.6), butter (40 ± 3.3), corn oil margarine stick (40 ± 10), safflower oil and soybean oil margarine stick (40 ± 10), soybean oil margarine stick (40 ± 10), soybean oil and cottonseed oil margarine stick (40 ± 10), corn oil margarine tub (40 ± 10), safflower oil margarine tub (40 ± 10), soybean oil margarine tub (40 ± 10), soybean oil and cottonseed oil margarine tub (40 ± 10), canned sprouted mung beans (40 ± 0.8), persimmon (40 ± 2), boiled carrots (40 ± 0.6), boiled collards (39 ± 0.3), raw eggplant (39 ± 1.2), raw celtuce (39 ± 0.5), raw ginger root (38 ± 2.1), raw carrots (38 ± 0.7), tomato purée (38 ± 0.2), frozen and boiled butternut squash (37 ± 0.4), boiled red cabbage (36 ± 0.7), frozen and boiled red sweet peppers (35 ± 0.5), frozen and boiled green sweet peppers (35 ± 0.5), boiled burdock root (35 ± 0.4), boiled leeks (35 ± 1.9), bananas (34 ± 0.4), longans (34 ± 0.5), baked acorn squash (33 ± 0.5), boiled green cabbage (33 ± 0.7), canned pumpkin (32 ± 0.4), canned red chili peppers (32 ± 0.7), canned green chili peppers (32 ± 0.7), raw red sweet peppers (32 ± 1), raw green sweet peppers (32 ± 1), boiled and steamed japanese chestnuts (32 ± 1.8), canned pumpkin pie mix (31 ± 0.4), boiled beets (31 ± 0.6), boiled lotus root (31 ± 0.6), frozen and boiled crookneck squash (31 ± 0.5), boiled chayote (31 ± 0.6), guava (31 ± 0.6), japanese persimmon (30 ± 0.3), raw green tomato (30 ± 0.4), boiled yambean (30 ± 0.5), raw chickory (29 ± 1.1), raw scallop squash (29 ± 0.8), canned jalapeño peppers (29 ± 0.7), canned red sweet peppers (29 ± 0.7), canned green sweet peppers (29 ± 0.7), boiled eggplant (29 ± 1), raw radish (29 ± 1.1), rice polish (29 ± 0.9), raw onions (28 ± 0.6), raw zucchini (28 ± 0.8), frozen and boiled zucchini (28 ± 0.4), boiled red tomato (28 ± 0.4), boiled oriental radish (28 ± 0.7), elderberries (27 ± 0.3), boiled butternut squash (27 ± 0.5), watermelon (27 ± 0.3), whipped butter (27 ± 4.5), boiled scallop squash (26 ± 0.6), coconut water (26 ± 0.2), peach (26 ± 0.6), canned zucchini in tomato juice (25 ± 0.4), raw oriental radish (25 ± 1.1), pickled beets (24 ± 0.4), cooked taro (24 ± 0.8), boiled red sweet peppers (24 ± 0.7), boiled green sweet peppers (24 ± 0.7), canned peeled red tomatoes (24 ± 0.4), figs (24 ± 1), raw crookneck squash (23 ± 0.8), canned carrots (23 ± 0.7), canned stewed red tomatoes (23 ± 0.4), carambola (23 ± 0.4), boiled onions (22 ± 0.5), boiled crookneck squash (22 ± 0.6), raw red tomato (22 ± 0.4), cooked plantain (21 ± 0.3), boiled pumpkin (21 ± 0.4), boiled turnips (21 ± 0.6), raw celery (20 ± 1.3), boiled acorn squash (20 ± 0.4), corn oil diet margarine tub (20 ± 10), soybean oil and cottonseed oil diet margarine tub (20 ± 10), mango (19 ± 0.2), strawberries (19 ± 0.3), boiled/baked spaghetti squash (18 ± 0.6), blueberries (18 ± 0.3), european grapes (18 ± 0.3), navel orange (17 ± 0.4), valencia orange (17 ± 0.4), frozen and boiled onions (17 ± 0.5), tomato juice (17 ± 0.3), american grapes (17 ± 0.5), plum (17 ± 0.8), boiled zucchini (16 ± 0.6), boiled calabash gourd (16 ± 0.7), grape juice (16 ± 0.2), boiled celery (15 ± 0.7), loquats (15 ± 0.5), canned crookneck squash (15 ± 0.5), raw cucumber (15 ± 1), crab apples (14 ± 0.5), pineapple (12 ± 0.3), sapodilla (12 ± 0.3), pear (11 ± 0.3), papaya (11 ± 0.2), thompson seedless grapes (11 ± 0.3), mandarin oranges (10 ± 0.4), tangerine (10 ± 0.6), cooked apple without skin (9 ± 0.3), orange juice (8 ± 0.2), raw apple with skin (7 ± 0.4), applesauce (7 ± 0.4), screwdriver (6 ± 0.2), tangerine juice (6 ± 0.2), ale (5 ± 0.1), raw apple without skin (5 ± 0.4), white grapefruit juice (5 ± 0.2), coffee (1 ± 0.3)

Applications: none known
Daily Dosage: Nutritional — 1–3 g; therapeutic — ?; experimental — ?; toxic — none known up to 3 g
Warnings: none known

TRYPTOPHAN

Names: tryptophan, TRP, TRY, amino acid W, 2-amino-3-(3 indoyl) propanoic acid
Forms: 2-amino-3-(3 indoyl) propanoic acid
Classification: essential amino acid
Deficiency: sterility, testicle deterioration, weight decreased, dry skin, bloodshot eyes, hair decreased, slow growth, deterioration of vitamin B_{3b} production, digestion upsets, decreased clearance of blood clots, nervousness, lack of sleep, memory decreased, aggression, emotional agitation, compulsion, hallucinations, depression, emotional deterioration, schizophrenia, protein decreased
Side-effects: vitamin B_6 decreased, induces sleep
Toxicity: none known
Inhibitors: milk, protein, taking with food
Helpers: vitamin B_6, vitamin C, taking with starches or carbohydrates
Sources (mg per 100 grams of food): torula yeast (1868 ± 1.8), soybean protein isolate (1114 ± 1.8), dried spirulina seaweed (929 ± 0.5), soybean protein concentrate (836 ± 1.8), dried and frozen tofu (747 ± 2.9), almond meal (718 ± 1.8), dried and salted cod (704 ± 0.6), fenugreek seed (700 ± 25), defatted soybean flour (683 ± 0.5), low fat soybean flour (676 ± 0.6), defatted soy meal (653 ± 0.4), braised pork pancreas (625 ± 0.5), peanut flour (625 ± 12.5), cooked whelk (618 ± 0.6), roasted pumpkin and squash seeds (586 ± 1.8), dry roasted soybean nuts (576 ± 0.6), toasted soybean nuts (571 ± 1.8), yellow mustard seed (567 ± 16.7), parmesan cheese (560 ± 10), roasted soybean nuts (512 ± 0.6), dry skim milk (510 ± 1.7), roasted soybean flour (506 ± 0.6), soybean flour (502 ± 0.6), freeze-dried parsley (500 ± 35.7), dry instant skim milk (496 ± 0.5), dried sesame kernels (475 ± 6.3), braised pork center loin excluding fat (467 ± 0.5), braised pork top loin excluding fat (463 ± 0.5), braised pork sirloin excluding fat (450 ± 0.5), braised pork loin excluding fat (443 ± 0.5), dry buttermilk (443 ± 7.1), dried pumpkin and squash seeds (436 ± 1.8), broiled pork center loin excluding fat (431 ± 0.5), milk chocolate (429 ± 1.8), gruyere cheese (425 ± 1.8), braised pork boston blade excluding fat (419 ± 0.5), swiss cheese (407 ± 1.8), braised pork loin blade excluding fat (399 ± 0.5), dried watermelon seeds (396 ± 1.8), fried beef top round excluding fat (393 ± 0.5), roasted pork rump excluding fat (392 ± 0.5), braised

pork spareribs (391 ± 0.5), fried chicken breast without skin (390 ± 0.6), whole dried sesame seeds (389 ± 5.6), fried pork center loin excluding fat (387 ± 0.5), broiled pork center rib excluding fat (387 ± 0.5), roasted pork tenderloin (387 ± 0.5), broiled pork top loin excluding fat (387 ± 0.5), fried beef liver (385 ± 0.5), roasted pork center loin excluding fat (383 ± 0.5), fried light meat chicken without skin (383 ± 0.5), braised pork center loin including fat (382 ± 0.5), roasted pork leg excluding fat (381 ± 0.5), broiled pork sirloin excluding fat (380 ± 0.5), roasted pork center rib excluding fat (379 ± 0.5), roasted pork shank excluding fat (379 ± 0.5), roasted pork top loin excluding fat (379 ± 0.5), simmered beef shank excluding fat (377 ± 0.5), fried pork center rib excluding fat (376 ± 0.5), fried pork top loin excluding fat (376 ± 0.5), toasted sesame kernels (375 ± 1.8), whole toasted sesame seeds (375 ± 1.8), broiled pork loin excluding fat (374 ± 0.5), well done baked extra lean ground beef (373 ± 0.5), sesame butter (373 ± 3.3), fried chicken giblets (373 ± 0.5), dry whole milk (372 ± 1.6), oil roasted almonds (371 ± 1.8), toasted almonds (371 ± 1.8), dried butternuts (371 ± 1.8), braised beef chuck arm pot roast excluding fat (370 ± 0.5), braised pork center rib including fat (370 ± 0.5), roasted pork sirloin excluding fat (370 ± 0.5), stewed chicken drumstick without skin (370 ± 1.1), fried chicken back with skin (369 ± 0.7), braised pork liver (366 ± 0.5), well done baked lean ground beef (365 ± 0.5), fried beef sirloin excluding fat (364 ± 0.5), dried almonds (364 ± 1.8), fried chicken breast with skin (364 ± 0.5), roasted pork loin excluding fat (362 ± 0.5), braised pork sirloin including fat (362 ± 0.5), roasted chicken breast without skin (362 ± 0.6), roasted light meat chicken without skin (361 ± 0.5), tilsit cheese (357 ± 1.8), broiled pork center loin including fat (357 ± 0.5), well done baked ground beef (355 ± 0.5), fried beef top round including fat (355 ± 0.5), broiled beef top round excluding fat (355 ± 0.5), braised pork top loin including fat (355 ± 0.5), braised beef bottom round excluding fat (354 ± 0.5), simmered beef shank including fat (354 ± 0.5), dried sunflower seeds (354 ± 1.8), well done broiled extra lean ground beef (352 ± 0.5), braised beef liver (351 ± 0.5), braised beef pancreas (351 ± 0.5), braised pork loin including fat (350 ± 0.5), roasted pork rump including fat (350 ± 0.5), braised beef chuck blade roast excluding fat (348 ± 0.5), well done broiled lean ground beef (347 ± 0.5), simmered beef kidneys (347 ± 0.5), port du salut cheese (346 ± 1.8), braised pork arm braised including fat (346 ± 0.5), well done fried extra lean ground beef (345 ± 0.5), broiled beef top round including fat (345 ± 0.5), braised beef shortribs excluding fat (344 ± 0.5), creamy peanut butter (344 ± 3.1), fried light meat chicken with skin (344 ± 0.5), simmered turkey heart (343 ± 0.5), simmered chicken liver (343 ± 0.5), braised pork boston blade braised including fat (342 ± 0.5), roasted pork shoulder excluding fat (341 ± 0.5), raw quail without skin (341 ± 0.5), well done fried lean ground beef (340 ± 0.5), broiled beef sirloin excluding fat (340 ± 0.5), fried dark meat chicken without skin (340 ± 0.5), roasted chicken breast with skin (340 ± 0.5), roasted turkey light meat without skin (340 ± 0.5), stewed chicken breast without skin (339 ± 0.5), broiled pork boston blade excluding fat (338 ± 0.5), simmered chicken heart (338 ± 0.5), simmered turkey liver (338 ± 0.5), stewed light meat chicken without skin (337 ± 0.5), well done broiled ground beef (335 ± 0.5), broiled pork loin blade excluding fat (335 ± 0.5), braised beef bottom round including fat (334 ± 0.5), cooked crayfish (333 ± 0.6), well done fried ground beef (333 ± 0.5), muenster cheese (332 ± 1.8), roasted pork center loin including fat (332 ± 0.5), roasted pork loin blade excluding fat (332 ± 0.5), oil roasted sunflower seeds (332 ± 1.8),

TRYPTOPHAN

canned tuna in water (331 ± 0.6), american cheese (329 ± 1.8), brick cheese (329 ± 1.8), braised beef brisket excluding fat (329 ± 0.5), roasted pork sirloin including fat (329 ± 0.5), braised pork kidneys (329 ± 0.5), oil roasted peanuts (329 ± 1.8), raw pheasant without skin (328 ± 0.5), roasted pork boston blade excluding fat (327 ± 0.5), fried pork loin blade excluding fat (327 ± 0.5), roasted duck without skin (327 ± 0.5), roasted pork leg including fat (326 ± 0.5), roasted light meat chicken with skin (326 ± 0.5), brie cheese (325 ± 1.8), pimento cheese spread (325 ± 1.8), anchovy canned in olive oil (325 ± 2.5), caviar (325 ± 3.1), roasted eye of beef round excluding fat (325 ± 0.5), dried black walnuts (325 ± 1.8), roasted turkey dark meat without skin (325 ± 0.5), roasted beef tip excluding fat (322 ± 0.5), simmered beef heart (322 ± 0.5), broiled beef top loin excluding fat (321 ± 0.5), roasted pork center rib including fat (321 ± 0.5), cheddar cheese (320 ± 0.5), roasted dark meat chicken without skin (320 ± 0.5), broiled beef round excluding fat (319 ± 0.5), bleu cheese (318 ± 1.8), monterey cheese (318 ± 1.8), broiled pork center rib including fat (317 ± 0.5), broiled beef tenderloin excluding fat (316 ± 0.5), broiled porterhouse steak excluding fat (315 ± 0.5), broiled T-bone steak excluding fat (315 ± 0.5), roasted turkey light meat with skin (315 ± 0.5), braised beef flank excluding fat (314 ± 0.5), broiled beef rib eye excluding fat (314 ± 0.5), roasted pork shank including fat (314 ± 0.5), dried peanuts (314 ± 1.8), medium-rare broiled extra lean ground beef (313 ± 0.5), roasted pork top loin including fat (312 ± 0.5), stewed chicken breast with skin (312 ± 0.5), camembert cheese (311 ± 1.8), colby cheese (311 ± 1.8), broiled pork sirloin including fat (311 ± 0.5), raw whelk (309 ± 0.6), braised beef flank including fat (308 ± 0.5), medium-rare fried extra lean ground beef (308 ± 0.5), roasted beef rib roasted excluding fat (308 ± 0.5), roasted beef tenderloin excluding fat (308 ± 0.5), fried beef sirloin including fat (308 ± 0.5), fried dark meat chicken with skin (308 ± 0.5), broiled beef sirloin including fat (307 ± 0.5), dried pignolia pine nuts (307 ± 1.8), simmered turkey giblets (307 ± 0.5), cooked coho salmon (306 ± 0.6), cooked sockeye salmon (306 ± 0.6), fried chicken drumstick with skin (306 ± 1), medium-rare broiled lean ground beef (305 ± 0.5), braised pork loin blade including fat (305 ± 0.5), cheshire cheese (304 ± 1.8), roasted pork loin including fat (304 ± 0.5), roasted chicken drumstick with skin (304 ± 1), roasted chicken thigh without skin (304 ± 1), roasted turkey dark meat with skin (304 ± 0.5), braised beef chuck arm pot roast including fat (303 ± 0.5), broiled pork loin including fat (303 ± 0.5), broiled pork top loin including fat (303 ± 0.5), stewed dark meat chicken without skin (303 ± 0.5), fried chicken thigh with skin (302 ± 0.8), medium-rare baked extra lean ground beef (301 ± 0.5), cooked flounder (300 ± 0.5), roasted eye of beef round including fat (300 ± 0.5), dry roasted sunflower seeds (300 ± 1.8), basil (300 ± 50), stir-fried sprouted soybeans (300 ± 0.5), cooked halibut (299 ± 0.6), breaded and fried shrimp (299 ± 0.6), medium-rare fried lean ground beef (299 ± 0.5), roasted cured pork arm excluding fat (299 ± 0.5), medium-rare broiled ground beef (297 ± 0.5), canned tuna in oil (296 ± 0.6), roasted beef round tip including fat (296 ± 0.5), fried pork center loin including fat (296 ± 0.5), dry roasted almonds (296 ± 1.8), cooked rainbow trout (295 ± 0.6), medium-rare baked lean ground beef (295 ± 0.5), medium-rare fried ground beef (295 ± 0.5), simmered chicken giblets (295 ± 0.5), cooked snapper (294 ± 0.6), stewed light meat chicken with skin (294 ± 0.5), limburger cheese (293 ± 1.8), cooked bacon (292 ± 0.4), cooked sheepshead (291 ± 0.6), cooked shrimp (291 ± 0.6), broiled beef tenderloin including fat (291 ± 0.5), braised pork spleen (289 ± 0.5), roasted dark meat chicken with skin (289 ± 0.5), simmered chicken neck without skin (289 ± 2.8), broiled beef top loin including fat (288 ± 0.5), roasted chicken wing with skin (288 ± 1.5), cooked clams (286 ± 0.6), canned blue crab (286 ± 0.6), broiled beef round including fat (286 ± 0.5), roasted pork arm including fat (286 ± 0.5), dried pistachio nuts (286 ± 1.8), stewed chicken drumstick with skin (286 ± 0.9), cooked lobster (285 ± 0.6), braised beef chuck blade roast including fat (285 ± 0.5), smoked haddock (284 ± 0.6), raw shrimp (284 ± 0.6), cooked swordfish (284 ± 0.6), broiled beef flank excluding fat (284 ± 0.5), medium-rare baked ground beef (284 ± 0.5), broiled beef rib eye including fat (284 ± 0.5), roasted pork shoulder including fat (284 ± 0.5), crab cakes (282 ± 0.8), broiled pork boston blade including fat (282 ± 0.5), roasted pork boston blade including fat (282 ± 0.5), tempeh (282 ± 0.6), cooked blue crab (281 ± 0.6), broiled porterhouse steak including fat (281 ± 0.5), roasted chicken thigh with skin (281 ± 0.8), fried chicken wing with skin (281 ± 1.6), broiled beef flank including fat (280 ± 0.5), cooked grouper (278 ± 0.6), cooked mullet (278 ± 0.6), braised pork tongue (278 ± 0.5), cooked pike (276 ± 0.6), broiled beef rib excluding fat (276 ± 0.5), kippered herring (275 ± 1.3), sardines (275 ± 2.1), cashew butter (275 ± 1.8), cooked tilefish (274 ± 0.6), roasted beef tenderloin including fat (274 ± 0.5), cooked haddock (272 ± 0.6), braised pork heart (272 ± 0.4), cooked flatfish (271 ± 0.6), fried pork center rib including fat (271 ± 0.5), oil roasted spanish peanuts (271 ± 1.8), roasted pork loin blade including fat (270 ± 0.5), cooked alaska king crab (269 ± 0.6), cooked rockfish (269 ± 0.6), broiled T-bone steak including fat (269 ± 0.5), fried pork top loin including fat (269 ± 0.5), almond butter (269 ± 3.1), fried tofu (269 ± 3.8), cooked mackerel (267 ± 0.6), cooked mussels (267 ± 0.6), cooked perch (267 ± 0.6), cooked pompano (265 ± 0.6), cooked sea bass (265 ± 0.6), cooked eel (265 ± 0.6), cooked whiting (264 ± 0.6), dried brazil nuts (264 ± 1.8), oil roasted valencia peanuts (264 ± 1.8), simmered turkey gizzard (264 ± 0.5), raw duck liver (264 ± 0.5), raw lobster (262 ± 0.6), smoked whitefish (262 ± 0.6), broiled pork loin blade including fat (262 ± 0.5), braised beef spleen (261 ± 0.5), stewed dark meat chicken with skin (261 ± 0.5), duck egg (260 ± 0.7), raw crayfish (260 ± 0.6), stewed chicken thigh with skin (260 ± 0.7), raw yellowtail (259 ± 0.6), raw queen crab (258 ± 0.6), cooked herring (258 ± 0.6), braised beef brisket including fat (258 ± 0.5), turkey breast meat summer sausage (257 ± 2.4), cooked carp (256 ± 0.6), cooked cod (256 ± 0.6), fried chicken neck with skin (256 ± 1.4), canned cod (255 ± 0.6), raw alaska king crab (255 ± 0.6), roasted beef rib including fat (254 ± 0.5), oil roasted cashews (254 ± 1.8), cooked rainbow smelt (253 ± 0.6), raw blue crab (251 ± 0.6), roasted lean ham (251 ± 0.5), hard pork salami (250 ± 5), oil roasted virginia peanuts (250 ± 1.8), fennel seed (250 ± 25), freeze-dried chives (250 ± 62.5), freeze-dried red sweet peppers (250 ± 31.3), freeze-dried green sweet peppers (250 ± 31.3), canned corned beef (247 ± 0.5), firm raw tofu (246 ± 0.4), stewed chicken wing with skin (245 ± 1.3), simmered chicken gizzard (243 ± 0.5), raw dungeness crab (242 ± 0.6), raw coho salmon (242 ± 0.6), braised beef shortribs including fat (242 ± 0.5), boiled soybeans (242 ± 0.3), chicken egg yolk (241 ± 2.9), roasted canned lean ham (240 ± 0.5), raw sockeye salmon (239 ± 0.6), chipped beef (239 ± 1.8), dry roasted cashews (239 ± 1.8), oil roasted filberts (239 ± 1.8), grilled canadian bacon (238 ± 1.1), roasted ham (238 ± 0.5), raw shark (235 ± 0.6), crunchy peanut butter (234 ± 1.6), raw halibut (233 ± 0.6), raw trout (233 ± 0.6), roasted canned ham (233 ± 0.5), poppy seed

(233 ± 16.7), boiled winged beans (233 ± 0.3), cooked sturgeon (232 ± 0.6), unheated lean ham (232 ± 0.5), fried pork loin blade including fat (232 ± 0.5), roasted duck with skin (232 ± 0.5), raw milkfish (231 ± 0.6), raw snapper (231 ± 0.6), raw rainbow trout (231 ± 0.6), raw goose liver (230 ± 0.5), puffed wheat (229 ± 3.6), canned sockeye salmon (229 ± 0.6), dry roasted peanuts (229 ± 1.8), raw anchovy (228 ± 0.6), roasted cured pork arm including fat (227 ± 0.5), raw chum salmon (226 ± 0.6), raw sheepshead (226 ± 0.6), raw chinook salmon (225 ± 0.6), broiled beef rib including fat (225 ± 0.5), fried abalone (224 ± 0.6), raw bluefish (224 ± 0.6), raw pink salmon (224 ± 0.6), canned pink salmon (224 ± 0.6), natto (223 ± 0.6), raw swordfish (222 ± 0.6), raw pollock (218 ± 0.6), pork luxury loaf (218 ± 1.8), dried filberts (218 ± 1.8), raw burbot (216 ± 0.6), raw grouper (216 ± 0.6), raw mullet (216 ± 0.6), raw pike (216 ± 0.6), raw sunfish (216 ± 0.6), pork link sausage (216 ± 0.7), shredded wheat (214 ± 1.8), raw whitefish (214 ± 0.6), turkey ham (214 ± 0.9), almond paste (214 ± 1.8), raw cisco (213 ± 0.6), raw cusk (213 ± 0.6), raw ling (213 ± 0.6), raw haddock (212 ± 0.6), battered and fried shark (212 ± 0.6), pickled anchovy (211 ± 1.8), raw freshwater bass (211 ± 0.6), raw flatfish (211 ± 0.6), raw rockfish (211 ± 0.6), raw scup (211 ± 0.6), unheated ham (211 ± 0.5), ham and cheese roll (211 ± 1.8), dry roasted pistachio nuts (211 ± 1.8), breaded and fried scallops (210 ± 1.6), canned lean ham (210 ± 0.5), chopped ham lunch meat (210 ± 2.4), hard pork and beef salami (210 ± 5), raw perch (209 ± 0.6), breaded and fried croaker (208 ± 0.6), raw mackerel (208 ± 0.6), raw dolphinfish (207 ± 0.6), raw pompano (207 ± 0.6), raw spot (207 ± 0.6), roasted pork blade roll (207 ± 0.5), raw eel (207 ± 0.6), dry roasted pecans (207 ± 1.8), raw sea bass (206 ± 0.6), smoked chinook salmon (205 ± 0.6), raw whiting (205 ± 0.6), pork livercheese (205 ± 1.3), raw catfish (204 ± 0.6), unheated canadian bacon (204 ± 0.9), dried pecans (204 ± 1.8), breaded fried squid (202 ± 0.6), turkey pastrami (202 ± 0.9), raw herring (201 ± 0.6), bran flakes (200 ± 1.8), anchovy paste (200 ± 7.1), raw carp (200 ± 0.6), breaded and fried catfish (200 ± 0.6), pork and beef peppered loaf (200 ± 1.8), simmered chicken neck with skin (200 ± 1.3), garlic powder (200 ± 16.7), raw striped bass (199 ± 0.6), raw cod (199 ± 0.6), raw croaker (199 ± 0.6), raw lingcod (198 ± 0.6), bran (196 ± 1.8), raw freshwater drum (196 ± 0.6), raw rainbow smelt (196 ± 0.6), raw tilefish (196 ± 0.6), raw wolffish (196 ± 0.6), chicken egg (194 ± 1), raw alewife (194 ± 0.5), raw butterfish (194 ± 0.6), steamed sprouted soybeans (194 ± 1.1), canned ham (193 ± 0.5), dried english/persian walnuts (193 ± 1.8), raw abalone (192 ± 0.6), raw black bass (192 ± 0.5), raw shad (191 ± 0.6), cooked black bass (188 ± 0.4), raw scallops (188 ± 0.6), raw seatrout (188 ± 0.6), raw white sucker (188 ± 0.6), raw smelt (186 ± 0.5), braised beef lungs (186 ± 0.5), pork and beef honey loaf (186 ± 1.8), smoked cisco (184 ± 0.6), canned smelt (184 ± 0.5), pork and beef pepperoni (183 ± 8.3), wheat flakes (182 ± 1.8), raw cuttlefish (182 ± 0.6), oil roasted pecans (182 ± 1.8), raw sturgeon (181 ± 0.6), tuna salad (180 ± 0.2), raw turbot (180 ± 0.6), raw squid (174 ± 0.6), berliner (174 ± 2.2), dried ginko nuts (171 ± 1.8), chicken egg white (170 ± 1.5), simmered beef tongue (170 ± 0.5), breaded and fried clams (168 ± 0.6), braised beef thymus (168 ± 0.5), chopped smoked beef (168 ± 1.8), jellied corned beef loaf (168 ± 1.8), dry roasted filberts (168 ± 1.8), raw octopus (167 ± 0.6), cooked corned beef brisket (166 ± 0.5), raw black bullhead (163 ± 0.5), canned alewife (162 ± 0.5), raw monkfish (162 ± 0.6), italian pork sausage (161 ± 0.7), beef pastrami (161 ± 1.8), pork and beef luncheon sausage (161 ± 2.2), pickled herring (160 ± 3.3), cooked oysters

(158 ± 0.6), minced ham lunch meat (157 ± 2.4), lebanon bologna (157 ± 2.2), pork sausage (156 ± 1.9), pork liverwurst (156 ± 2.8), braised pork brains (155 ± 0.5), grilled ham patties (153 ± 0.8), pork and beef mortadella (153 ± 3.3), pork and beef old fashioned loaf (150 ± 1.8), onion powder (150 ± 25), boiled green soybeans (150 ± 0.6), unheated ham patties (149 ± 0.8), pork and beef link sausage with american cheese (149 ± 1.2), raw flounder (148 ± 0.5), raw sole (148 ± 0.5), pork bologna (148 ± 2.2), braised pork lungs (146 ± 0.5), pork and beef picnic loaf (146 ± 1.8), dried pinyon pine nuts (146 ± 1.8), raw clams (144 ± 0.6), pork liver sausage (144 ± 2.8), summer sausage (143 ± 2.2), miso (143 ± 0.4), raisin bran (139 ± 1.8), cottage cheese (139 ± 0.2), beef lunch meat (139 ± 1.8), polish pork sausage (139 ± 1.8), beef salami (139 ± 2.2), dried hickory nuts (139 ± 1.8), pork and beef kielbasa/kolbassy (138 ± 1.9), gjetost cheese (136 ± 1.8), pork and beef lunch meat (136 ± 1.8), raw mussels (133 ± 0.6), potato flour (133 ± 0.6), falafel (133 ± 1), pork and beef sausage (131 ± 3.8), pork and beef brotwurst (131 ± 0.7), boiled peanuts (131 ± 1.6), dehydrated onion flakes (129 ± 3.6), smoked beef sausage (128 ± 1.2), raw tofu (126 ± 0.4), boiled lupins (125 ± 0.3), freeze-dried shallots (125 ± 12.5), canned pork lunch meat (124 ± 2.4), pork pickle and pimento loaf (118 ± 1.8), boiled white beans (115 ± 0.3), raw beef tripe (114 ± 0.5), beef bologna (113 ± 2.2), pork bratwurst (113 ± 0.6), beef beerwurst salami (113 ± 2.2), pork beerwurst salami (113 ± 2.2), beef and pork salami (113 ± 2.2), sweetened condensed milk (113 ± 1.3), beef frankfurter (111 ± 0.9), boiled cranberry beans (111 ± 0.3), boiled yellow beans (108 ± 0.3), pork and beef knockwurst (107 ± 0.7), pork and beef link sausage (107 ± 0.7), boiled pink beans (107 ± 0.3), vienna sausage (106 ± 3.1), evaporated skim milk (106 ± 1.6), breaded and fried oysters (105 ± 0.6), boiled black beans (105 ± 0.3), beef and pork bologna (104 ± 2.2), pork olive loaf (104 ± 1.8), fried beef brains (103 ± 0.5), boiled kidney beans (103 ± 0.3), boiled navy beans (103 ± 0.3), simmered pork tail (102 ± 0.5), boiled yardlong bean (102 ± 0.3), dried acorns (100 ± 1.8), boiled catjang black-eyed peas (100 ± 0.3), boiled great northern beans (99 ± 0.3), raw pork stomach (98 ± 0.5), frozen and boiled black-eyed peas (98 ± 0.6), boiled black turtle beans (97 ± 0.3), boiled pinto beans (97 ± 0.3), evaporated whole milk (96 ± 0.4), raw spirulina seaweed (96 ± 0.5), boiled black-eyed peas (95 ± 0.3), boiled baby lima beans (95 ± 0.3), potato pancakes (95 ± 0.7), raw dock (94 ± 0.7), puffed rice (93 ± 3.6), acorn flour (93 ± 1.8), boiled split peas (93 ± 0.3), evaporated lowfat milk (92 ± 0.4), boiled lima beans (92 ± 0.3), simmered beef brains (90 ± 0.5), neufchatel cheese (89 ± 1.8), ham salad (89 ± 1.8), canned navy beans (89 ± 0.2), tomato powder (89 ± 0.5), canned great northern beans (87 ± 0.2), boiled and frozen baby lima beans (87 ± 0.6), gefiltefish with broth (86 ± 1.2), pork headcheese (86 ± 1.8), canned white beans (86 ± 0.2), boiled chickpeas (85 ± 0.3), sheep milk (84 ± 0.2), boiled french beans (83 ± 0.3), raw bacon (82 ± 0.7), beef and pork frankfurter (82 ± 1.1), dried chinese chestnuts (82 ± 1.8), dried coconut (82 ± 1.8), boiled lentils (81 ± 0.3), frozen and boiled fordhook lima beans (80 ± 0.6), raw oysters (79 ± 0.6), canned oysters (79 ± 0.6), surimi scallops (78 ± 0.6), boiled mungo beans (78 ± 0.3), boiled mung beans (76 ± 0.2), surimi shrimp (75 ± 0.6), raw acorns (75 ± 1.8), dried japanese chestnuts (75 ± 1.8), refried beans (74 ± 0.2), boiled adzuki beans (72 ± 0.2), boiled broad beans (72 ± 0.3), dried european chestnuts (71 ± 1.8), ginko nuts (71 ± 1.8), canned black turtle beans (71 ± 0.2), au gratin potatoes (70 ± 0.4), cream cheese (68 ± 1.8), boiled hyacinth beans (68 ± 0.3), raw garlic (67 ± 5.6), dried apricots

(66 ± 1.4), canned cranberry beans (66 ± 0.2), boiled pigeon peas (66 ± 0.3), dried toasted coconut (64 ± 1.8), baked beans with pork and sweet sauce (63 ± 0.2), simmered pork chitterlings (61 ± 0.5), dry roasted macadamia nuts (61 ± 1.8), baked beans with pork and tomato sauce (61 ± 0.2), canned kidney beans (61 ± 0.2), frozen and boiled turnip greens (59 ± 0.6), canned black-eyed peas (58 ± 0.2), canned lima beans (58 ± 0.2), stir-fried sprouted mung beans (58 ± 0.8), vegetarian baked beans (57 ± 0.2), boiled succotash (57 ± 0.5), protein fortified lowfat 2% milk (56 ± 0.2), protein fortified skim milk (56 ± 0.2), protein fortified lowfat 1% milk (55 ± 0.2), indian buffalo milk (54 ± 0.2), eggnog (54 ± 0.2), malted milk (54 ± 0.2), restaurant vanilla milkshake (54 ± 0.2), roasted chinese chestnuts (54 ± 1.8), canned pinto beans (54 ± 0.2), french fried potatoes (54 ± 1), yellow corn pudding (53 ± 0.2), canned broad beans (52 ± 0.2), dried dates (51 ± 0.6), hot chocolate (51 ± 0.2), boiled mushrooms (51 ± 0.6), okara tofu (51 ± 0.8), corn flakes (50 ± 1.8), raw chinese chestnuts (50 ± 1.8), ginger (50 ± 25), boiled mothbeans (50 ± 0.3), vanilla milkshake (49 ± 0.2), restaurant chocolate milkshake (48 ± 0.2), skim milk (48 ± 0.2), canned chickpeas (48 ± 0.2), frozen and fried french fried potatoes (48 ± 1), raw kelp/kombu/tangle seaweed (48 ± 0.5), frozen and boiled succotash (48 ± 0.6), restaurant strawberry milkshake (47 ± 0.2), lowfat 2% milk (47 ± 0.2), filled milk (47 ± 0.2), hummus (47 ± 0.2), noodles (46 ± 0.3), lowfat 1% milk (46 ± 0.2), whole milk (46 ± 0.2), lowfat 1% chocolate milk (46 ± 0.2), raw mushrooms (46 ± 1.4), lowfat 2% chocolate milk (45 ± 0.2), whole chocolate milk (45 ± 0.2), carob milk (44 ± 0.2), goat milk (43 ± 0.2), soy milk (43 ± 0.2), chocolate milkshake (43 ± 0.2), roasted japanese chestnuts (43 ± 1.8), raw cassava (43 ± 0.5), dried oriental radish (43 ± 0.9), raw laver/nori seaweed (43 ± 0.5), coconut cream (42 ± 0.2), frozen and fried hash brown potatoes (42 ± 0.6), potato salad (42 ± 0.4), scalloped potatoes (42 ± 0.4), frozen and boiled spinach (42 ± 0.5), macaroni (41 ± 0.4), spaghetti (41 ± 0.4), half and half cream (40 ± 3.3), light cream (40 ± 3.3), raw coconut (40 ± 1.1), boiled beet greens (40 ± 0.7), frozen and boiled brussels sprouts (40 ± 0.6), boiled spinach (40 ± 0.6), raw spinach (39 ± 1.8), simmered pork feet (38 ± 0.5), dried sweetened flaked coconut (38 ± 0.7), frozen and boiled collards (38 ± 0.6), boiled green peas (38 ± 0.6), microwaved potato with skin (38 ± 0.2), canned spinach (38 ± 0.5), raw sprouted mung beans (37 ± 1), raw parsley (37 ± 1.7), raw green peas (37 ± 0.6), buttermilk (36 ± 0.2), boiled and steamed chinese chestnuts (36 ± 1.8), roasted european chestnuts (36 ± 1.8), baked potato with skin (36 ± 0.2), oatmeal (35 ± 0.3), frozen and boiled kale (35 ± 0.8), frozen and boiled green peas (35 ± 0.6), o'brien potatoes (35 ± 0.3), raw wakame seaweed (35 ± 0.5), dried sweetened shredded coconut (34 ± 0.5), medium cream (33 ± 3.3), sour cream (33 ± 4.2), whipping cream, light (33 ± 3.3), dried lychees (33 ± 0.5), raw chives (33 ± 16.7), dried shitake mushrooms (33 ± 3.3), hash brown potatoes (33 ± 0.6), microwaved potato without skin (33 ± 0.3), skim yogurt (32 ± 0.2), raw japanese chestnuts (32 ± 1.8), frozen and boiled broccoli (32 ± 0.5), raw potato without skin (32 ± 0.4), simmered pork ears (31 ± 0.5), sweetened coconut cream (31 ± 0.2), boiled broccoli (31 ± 0.6), raw chickory greens (31 ± 0.6), canned green peas (31 ± 0.6), frozen and boiled potato without skin (31 ± 0.5), frozen and boiled turnip greens and turnips (31 ± 0.5), lowfat yogurt (30 ± 0.2), raw broccoli (30 ± 1.1), baked potato without skin (30 ± 0.3), raw shallots (30 ± 5), canned succotash with creamed corn (30 ± 0.4), raw european chestnuts (29 ± 1.8), boiled lamb's-quarters (29 ± 0.6), mashed potatoes (29 ± 0.5), canned suc-

cotash (29 ± 0.4), raw watercress (29 ± 2.9), frozen and boiled asparagus (28 ± 0.8), raw bamboo shoots (28 ± 0.7), boiled brussels sprouts (28 ± 0.6), raw vine spinach (28 ± 0.5), whipping cream, heavy (27 ± 3.3), pickled pork feet (27 ± 0.5), coconut milk (27 ± 0.2), boiled amaranth (27 ± 0.8), cooked sprouted mung beans (27 ± 0.8), raw red chili peppers (27 ± 1.1), raw green chili peppers (27 ± 1.1), boiled potato without skin (27 ± 0.4), dried figs (26 ± 0.3), chocolate syrup (26 ± 1.3), boiled asparagus (26 ± 0.6), raw cauliflower (26 ± 1), boiled cauliflower (26 ± 0.8), boiled mustard greens (26 ± 0.7), tomato paste (26 ± 0.4), frozen and boiled mustard greens (25 ± 0.7), raw turnip greens (25 ± 1.8), dried japanese persimmon (24 ± 1.5), sapote (23 ± 0.2), boiled yellow corn (23 ± 0.6), boiled jute potherb (23 ± 1.2), boiled kale (23 ± 0.8), boiled scotch kale (23 ± 0.8), canned turnip greens (23 ± 0.4), raw california avocado (22 ± 0.3), canned potato without skin (22 ± 0.6), cream of wheat (21 ± 0.3), lard (21 ± 1.8), raw pork jowl (21 ± 0.5), boiled and steamed european chestnuts (21 ± 1.8), frozen and boiled cauliflower (21 ± 0.6), frozen and boiled yellow corn (21 ± 0.6), raw welsh onions (21 ± 0.5), baked hubbard squash (21 ± 0.5), boiled hubbard squash (21 ± 0.4), baked sweet potato (21 ± 0.4), farina (20 ± 0.3), butter (20 ± 3.3), soybean oil and cottonseed oil margarine liquid (20 ± 10), corn oil margarine stick (20 ± 10), safflower oil and soybean oil margarine stick (20 ± 10), soybean oil margarine stick (20 ± 10), soybean oil and cottonseed oil margarine stick (20 ± 10), corn oil margarine tub (20 ± 10), safflower oil margarine tub (20 ± 10), soybean oil margarine tub (20 ± 10), soybean oil and cottonseed oil margarine tub (20 ± 10), raw savoy cabbage (20 ± 1.4), raw lotus root (20 ± 0.6), raw spring onions with tops (20 ± 1), boiled sweet potato (20 ± 0.3), canned sweet potato (20 ± 0.3), frozen and baked sweet potato (20 ± 0.6), whole yogurt (19 ± 0.2), boiled green beans (19 ± 0.8), boiled yellow snap beans (19 ± 0.8), canned sprouted mung beans (19 ± 0.8), boiled turnip greens (19 ± 0.7), tamarind (18 ± 0.4), canned bamboo shoots (18 ± 0.4), boiled savoy cabbage (18 ± 0.7), boiled swiss chard (18 ± 0.6), raw chickory (18 ± 1.1), raw onions (18 ± 0.6), raw florida avocado (17 ± 0.2), coleslaw (17 ± 0.8), canned yellow corn with red and green peppers (17 ± 0.4), frozen and boiled okra (17 ± 0.5), frozen and boiled butternut squash (17 ± 0.4), persimmon (16 ± 2), human milk (16 ± 1.6), boiled bamboo shoots (16 ± 0.4), boiled okra (16 ± 0.6), boiled purslane (16 ± 0.9), baked acorn squash (16 ± 0.5), apricots (15 ± 0.5), boiled chinese cabbage (15 ± 0.6), frozen and boiled yellow snap beans (15 ± 0.7), frozen and boiled green beans (15 ± 0.7), frozen and boiled turnips (15 ± 0.5), whole chocolate malted milk (14 ± 0.2), safflower oil and soybean oil mayonnaise (14 ± 3.6), soybean oil mayonnaise (14 ± 3.6), raw chinese cabbage (14 ± 1.4), raw red cabbage (14 ± 1.4), boiled collards (14 ± 0.3), canned yellow corn (14 ± 0.5), steamed hawaii mountain yam (14 ± 0.7), cream of rice (13 ± 0.3), elderberries (13 ± 0.3), raw ginger root (13 ± 2.1), boiled onions (13 ± 0.5), canned pumpkin (13 ± 0.4), canned pumpkin pie mix (13 ± 0.4), boiled butternut squash (13 ± 0.5), whey (13 ± 0.2), bananas (12 ± 0.4), boiled beets (12 ± 0.6), boiled carrots (12 ± 0.6), frozen and boiled carrots (12 ± 0.7), canned creamed yellow corn (12 ± 0.4), canned green beans (12 ± 0.7), canned yellow snap beans (12 ± 0.7), raw leeks (12 ± 1.9), boiled lotus root (12 ± 0.6), canned red chili peppers (12 ± 0.7), canned green chili peppers (12 ± 0.7), raw red sweet peppers (12 ± 1), raw green sweet peppers (12 ± 1), frozen and boiled red sweet peppers (12 ± 0.5), frozen and boiled green sweet peppers (12 ± 0.5), boiled rutabaga (12 ± 0.6), baked/boiled tropical yam (12 ± 0.7), soursop (11 ± 0.2), boiled and steamed japanese chestnuts (11 ± 1.8), raw green cabbage (11 ± 1.4), boiled green cabbage (11 ± 0.7), boiled red cabbage (11 ± 0.7), raw carrots (11 ± 0.7), boiled kohlrabi (11 ± 0.6), raw romaine lettuce (11 ± 1.8), raw looseleaf lettuce (11 ± 1.8), frozen and boiled crookneck squash (11 ± 0.5), raw scallop squash (11 ± 0.8), raw zucchini (11 ± 0.8), candied sweet potato (11 ± 0.5), tomato purée (11 ± 0.2), navel orange (10 ± 0.4), valencia orange (10 ± 0.4), pear (10 ± 0.3), japanese persimmon (10 ± 0.3), sugar apple (10 ± 0.3), raw celery (10 ± 1.3), raw eggplant (10 ± 1.2), raw iceberg lettuce (10 ± 2.5), frozen and boiled onions (10 ± 0.5), canned jalapeño peppers (10 ± 0.7), canned red sweet peppers (10 ± 0.7), canned green sweet peppers (10 ± 0.7), frozen and boiled zucchini (10 ± 0.4), boiled acorn squash (10 ± 0.4), cooked plantain (9 ± 0.3), pickled beets (9 ± 0.4), boiled pumpkin (9 ± 0.4), boiled scallop squash (9 ± 0.6), canned zucchini in tomato juice (9 ± 0.4), boiled/baked spaghetti squash (9 ± 0.6), raw green tomato (9 ± 0.4), corn grits (8 ± 0.2), mango (8 ± 0.2), papaya (8 ± 0.2), coconut water (8 ± 0.2), boiled burdock root (8 ± 0.4), boiled chayote (8 ± 0.6), boiled eggplant (8 ± 1), boiled leeks (8 ± 1.9), raw crookneck squash (8 ± 0.8), boiled crookneck squash (8 ± 0.6), cooked taro (8 ± 0.8), boiled red tomato (8 ± 0.4), guava (7 ± 0.6), lychees (7 ± 0.5), strawberries (7 ± 0.3), watermelon (7 ± 0.3), canned carrots (7 ± 0.7), boiled celery (7 ± 0.7), raw butterhead lettuce (7 ± 3.3), boiled red sweet peppers (7 ± 0.7), boiled green sweet peppers (7 ± 0.7), raw red tomato (7 ± 0.4), canned stewed red tomatoes (7 ± 0.4), canned peeled red tomatoes (7 ± 0.4), figs (6 ± 1), mandarin oranges (6 ± 0.4), tangerine (6 ± 0.6), raw celtuce (6 ± 0.5), boiled zucchini (6 ± 0.6), boiled turnips (6 ± 0.6), tomato juice (5 ± 0.3), loquats (5 ± 0.5), mammy apple (5 ± 0.5), pineapple (5 ± 0.3), sapodilla (5 ± 0.3), canned crookneck squash (5 ± 0.5), carambola (4 ± 0.4), crab apples (4 ± 0.5), raw cucumber (4 ± 1), raw endive (4 ± 2), cooked shitake mushrooms (4 ± 0.7), raw radish (4 ± 1.1), boiled oriental radish (4 ± 0.7), ale (3 ± 0.1), blueberries (3 ± 0.3), american grapes (3 ± 0.5), european grapes (3 ± 0.3), lime (3 ± 0.7), boiled calabash gourd (3 ± 0.7), orange juice (2 ± 0.2), raw apple with skin (2 ± 0.4), cooked apple without skin (2 ± 0.3), applesauce (2 ± 0.4), pink grapefruit (2 ± 0.4), white grapefruit (2 ± 0.4), thompson seedless grapes (2 ± 0.3), peach (2 ± 0.6), raw oriental radish (2 ± 1.1), screwdriver (1 ± 0.2), tangerine juice (1 ± 0.2), raw apple without skin (1 ± 0.4)

Applications: lack of sleep, depression, weight increased, pain control, cancer, antioxidant therapy, sun sensitivity, free radical oxidation, life span extension, radiation, decreased radiation resistance, aging

Daily Dosage: (1 niacin equivalent = 1 mg niacin = 60 mg tryptophan) 1974 PCS — .600–1.800 g as vitamin B_3; 1980 RDA — 0.780–1.140 g as vitamin B_3; 1980 USRDA — 1.200 g as vitamin B_3; 1989 RDA — 0.780–1.140 g as vitamin B_3; 1989 USRDA — 1.200 g as vitamin B_3; nutritional — 1.200–1.800 g; therapeutic — 1–3 g; experimental — 3–11 g; toxic — 11 g +

Warnings: Don't use if excitation increases. High doses cause decreased fetal weights and higher fetal death rates in animals. Do not use with MAO inhibiting drugs. Increased amounts can cause side-effects. See side-effects symptoms.

TYROSINE

Names: tyrosine, TYR, amino acid Y, 2-amino-3-(4 hydroxyphenyl) propanoic acid
Forms: 2-amino-3-(4 hydroxyphenyl) propanoic acid
Classification: nonessential amino acid
Deficiency: paleness, decreased skin melanin, protein decreased
Side-effects: none known
Toxicity: none known
Inhibitors: increased amounts of other amino acids
Helpers: moderate amounts of other amino acids
Sources (mg per 100 grams of food): torula yeast (5939 ± 1.8), soybean protein isolate (3221 ± 1.8), dried spirulina seaweed (2584 ± 0.5), peanut flour (2500 ± 12.5), parmesan cheese (2320 ± 10), soybean protein concentrate (2300 ± 1.8), dried and salted cod (2121 ± 0.6), gruyere cheese (1796 ± 1.8), defatted soybean flour (1778 ± 0.5), low fat soybean flour (1760 ± 0.6), dry skim milk (1747 ± 1.7), swiss cheese (1714 ± 1.8), defatted soy meal (1700 ± 0.4), dry instant skim milk (1695 ± 0.5), dried and frozen tofu (1606 ± 2.9), dry buttermilk (1543 ± 7.1), provolone cheese (1539 ± 1.8), cooked whelk (1518 ± 0.6), dry roasted soybean nuts (1497 ± 0.6), toasted soybean nuts (1489 ± 1.8), edam cheese (1475 ± 1.8), tilsit cheese (1475 ± 1.8), gouda cheese (1471 ± 1.8), port du salut cheese (1439 ± 1.8), almond meal (1414 ± 1.8), fenugreek seed (1400 ± 25), roasted pumpkin and squash seeds (1389 ± 1.8), creamy peanut butter (1369 ± 3.1), roasted soybean nuts (1333 ± 0.6), roasted soybean flour (1316 ± 0.6), bleu cheese (1314 ± 1.8), soybean flour (1306 ± 0.6), oil roasted peanuts (1304 ± 1.8), dry whole milk (1272 ± 1.6), dried peanuts (1250 ± 1.8), braised pork center loin excluding fat (1241 ± 0.5), braised pork top loin excluding fat (1229 ± 0.5), american cheese (1229 ± 1.8), pimento cheese spread (1225 ± 1.8), brie cheese (1214 ± 1.8), limburger cheese (1211 ± 1.8), cheddar cheese (1202 ± 0.5), monterey cheese (1196 ± 1.8), braised pork sirloin excluding fat (1195 ± 0.5), braised pork pancreas (1195 ± 0.5), braised beef pancreas (1184 ± 0.5), roasted turkey light meat without skin (1182 ± 0.5), fried beef top round excluding fat (1178 ± 0.5), braised pork loin excluding fat (1177 ± 0.5), simmered pork tail (1173 ± 0.5), colby cheese (1161 ± 1.8), camembert cheese (1161 ± 1.8), cheshire cheese (1143 ± 1.8), broiled pork center loin excluding fat (1141 ± 0.5), oil roasted spanish peanuts (1139 ± 1.8), muenster cheese (1136 ± 1.8), mozzarella cheese (1136 ± 1.8), simmered beef shank excluding fat (1132 ± 0.5), brick cheese (1129 ± 1.8), roasted turkey dark meat without skin (1129 ± 0.5), fried chicken breast without skin (1128 ± 0.6), dried sesame kernels (1125 ± 6.3), braised pork boston blade excluding fat (1112 ± 0.5), braised beef chuck arm pot roast excluding fat (1109 ± 0.5), fried light meat chicken without skin (1108 ± 0.5), oil roasted valencia peanuts (1100 ± 1.8), fried beef sirloin excluding fat (1091 ± 0.5), roasted turkey light meat with skin (1084 ± 0.5), stewed chicken drumstick without skin (1068 ± 1.1), fried chicken giblets (1067 ± 0.5), broiled beef top round excluding fat (1065 ± 0.5), fried beef top round including fat (1065 ± 0.5), simmered beef shank including fat (1062 ± 0.5), braised beef bottom round excluding fat (1061 ± 0.5), fried beef liver (1060 ± 0.5), braised pork loin blade excluding fat (1060 ± 0.5), oil roasted virginia peanuts (1054 ± 1.8), fried chicken breast with skin (1049 ± 0.5), roasted chicken breast without skin

TYROSINE

(1047 ± 0.6), simmered beef heart (1046 ± 0.5), roasted turkey dark meat with skin (1044 ± 0.5), roasted light meat chicken without skin (1043 ± 0.5), braised beef chuck blade roast excluding fat (1043 ± 0.5), roasted pork rump excluding fat (1040 ± 0.5), braised pork spareribs (1037 ± 0.5), broiled beef top round including fat (1036 ± 0.5), braised beef shortribs excluding fat (1033 ± 0.5), dried watermelon seeds (1032 ± 1.8), dried pumpkin and squash seeds (1032 ± 1.8), broiled pork center rib excluding fat (1028 ± 0.5), broiled pork top loin excluding fat (1028 ± 0.5), roasted pork tenderloin (1027 ± 0.5), fried pork center loin excluding fat (1026 ± 0.5), braised pork center loin including fat (1025 ± 0.5), broiled beef sirloin excluding fat (1020 ± 0.5), roasted pork center loin excluding fat (1016 ± 0.5), roasted pork leg excluding fat (1010 ± 0.5), raw quail without skin (1010 ± 0.5), broiled pork sirloin excluding fat (1008 ± 0.5), roasted pork center rib excluding fat (1007 ± 0.5), roasted pork top loin excluding fat (1007 ± 0.5), roasted pork shank excluding fat (1006 ± 0.5), braised beef bottom round including fat (1001 ± 0.5), canned tuna in water (999 ± 0.6), fried pork center rib excluding fat (998 ± 0.5), fried pork top loin excluding fat (998 ± 0.5), braised pork center rib including fat (993 ± 0.5), broiled pork loin excluding fat (993 ± 0.5), fried light meat chicken with skin (991 ± 0.5), dried butternuts (989 ± 1.8), braised beef brisket excluding fat (987 ± 0.5), roasted pork sirloin excluding fat (981 ± 0.5), roasted chicken breast with skin (980 ± 0.5), crunchy peanut butter (978 ± 1.6), fried dark meat chicken without skin (978 ± 0.5), stewed chicken breast without skin (978 ± 0.5), stewed light meat chicken without skin (975 ± 0.5), anchovy canned in olive oil (975 ± 2.5), roasted eye of beef round excluding fat (974 ± 0.5), braised pork sirloin including fat (971 ± 0.5), caviar (969 ± 3.1), braised beef liver (967 ± 0.5), roasted beef tip excluding fat (965 ± 0.5), dry roasted peanuts (964 ± 1.8), broiled beef top loin excluding fat (962 ± 0.5), roasted pork loin excluding fat (960 ± 0.5), simmered turkey heart (959 ± 0.5), simmered beef kidneys (958 ± 0.5), broiled beef round excluding fat (956 ± 0.5), braised pork top loin including fat (956 ± 0.5), broiled pork center loin including fat (954 ± 0.5), broiled beef tenderloin excluding fat (949 ± 0.5), simmered chicken heart (946 ± 0.5), broiled porterhouse steak excluding fat (946 ± 0.5), well done baked extra lean ground beef (945 ± 0.5), broiled T-bone steak excluding fat (945 ± 0.5), broiled beef rib eye excluding fat (942 ± 0.5), braised beef flank excluding fat (941 ± 0.5), braised pork loin including fat (941 ± 0.5), roasted light meat chicken with skin (940 ± 0.5), roasted pork rump including fat (936 ± 0.5), braised pork arm braised including fat (929 ± 0.5), roasted beef tenderloin excluding fat (925 ± 0.5), roasted beef rib roasted excluding fat (925 ± 0.5), well done baked lean ground beef (924 ± 0.5), roasted dark meat chicken without skin (924 ± 0.5), braised beef flank including fat (924 ± 0.5), cooked coho salmon (924 ± 0.6), fried beef sirloin including fat (923 ± 0.5), cooked sockeye salmon (922 ± 0.6), broiled beef sirloin including fat (920 ± 0.5), braised pork boston blade braised including fat (918 ± 0.5), braised pork kidneys (914 ± 0.5), braised beef chuck arm pot roast including fat (910 ± 0.5), roasted pork shoulder excluding fat (905 ± 0.5), cooked halibut (901 ± 0.6), stewed chicken breast with skin (900 ± 0.5), fried chicken back with skin (899 ± 0.7), well done baked ground beef (899 ± 0.5), roasted eye of beef round including fat (899 ± 0.5), broiled pork boston blade excluding fat (898 ± 0.5), canned tuna in oil (896 ± 0.6), simmered turkey gizzard (895 ± 0.5), roasted duck without skin (894 ± 0.5), well done broiled extra lean ground beef (892 ± 0.5), turkey breast meat summer sausage (890 ± 2.4), chipped beef (889 ± 1.8), roasted beef round tip including fat (889 ± 0.5), cooked rainbow trout (889 ± 0.6), broiled pork loin blade excluding fat (889 ± 0.5), dried pignolia pine nuts (889 ± 1.8), cooked snapper (888 ± 0.6), roasted pork center loin including fat (888 ± 0.5), braised pork liver (887 ± 0.5), cooked bacon (887 ± 0.4), fried dark meat chicken with skin (884 ± 0.5), canned corned beef (884 ± 0.5), fried chicken drumstick with skin (882 ± 1), well done broiled lean ground beef (880 ± 0.5), roasted pork loin blade excluding fat (880 ± 0.5), cooked sheepshead (878 ± 0.6), roasted pork sirloin including fat (878 ± 0.5), simmered turkey giblets (877 ± 0.5), roasted chicken drumstick with skin (877 ± 1), stewed dark meat chicken without skin (877 ± 0.5), roasted chicken thigh without skin (877 ± 1), well done fried extra lean ground beef (873 ± 0.5), roasted pork leg including fat (873 ± 0.5), broiled beef tenderloin including fat (872 ± 0.5), fried chicken thigh with skin (869 ± 0.8), roasted pork boston blade excluding fat (869 ± 0.5), fried pork loin blade excluding fat (868 ± 0.5), broiled beef top loin including fat (864 ± 0.5), roasted pork center rib including fat (863 ± 0.5), well done fried lean ground beef (861 ± 0.5), broiled beef round including fat (858 ± 0.5), simmered chicken liver (857 ± 0.5), cooked swordfish (856 ± 0.6), braised beef chuck blade roast including fat (855 ± 0.5), broiled beef rib eye including fat (853 ± 0.5), broiled pork center rib including fat (853 ± 0.5), broiled beef flank excluding fat (852 ± 0.5), smoked haddock (852 ± 0.6), well done broiled ground beef (849 ± 0.5), simmered chicken giblets (848 ± 0.5), stewed light meat chicken with skin (848 ± 0.5), simmered turkey liver (844 ± 0.5), roasted pork shank including fat (844 ± 0.5), broiled porterhouse steak including fat (843 ± 0.5), well done fried ground beef (842 ± 0.5), broiled beef flank including fat (841 ± 0.5), roasted pork top loin including fat (840 ± 0.5), cooked grouper (839 ± 0.6), cooked mullet (836 ± 0.6), broiled pork sirloin including fat (836 ± 0.5), yellow mustard seed (833 ± 16.7), cooked pike (833 ± 0.6), roasted dark meat chicken with skin (832 ± 0.5), kippered herring (830 ± 1.3), sardines (829 ± 2.1), simmered chicken neck without skin (828 ± 2.8), broiled beef rib excluding fat (828 ± 0.5), cooked tilefish (827 ± 0.6), roasted chicken wing with skin (826 ± 1.5), simmered chicken gizzard (825 ± 0.5), stewed chicken drumstick with skin (825 ± 0.9), braised pork loin blade including fat (823 ± 0.5), roasted beef tenderloin including fat (822 ± 0.5), roasted cured pork arm excluding fat (818 ± 0.5), cooked haddock (818 ± 0.6), broiled pork top loin including fat (818 ± 0.5), cooked flatfish (816 ± 0.6), cooked clams (816 ± 0.6), broiled pork loin including fat (815 ± 0.5), roasted pork loin including fat (814 ± 0.5), cooked rockfish (812 ± 0.6), roasted chicken thigh with skin (810 ± 0.8), braised pork heart (808 ± 0.4), fried chicken wing with skin (806 ± 1.6), broiled T-bone steak including fat (806 ± 0.5), cooked perch (806 ± 0.6), roasted goose with skin (805 ± 0.5), cooked mackerel (805 ± 0.6), cooked pompano (800 ± 0.6), fried pork center loin including fat (799 ± 0.5), cooked eel (798 ± 0.6), cooked sea bass (798 ± 0.6), cooked crayfish (796 ± 0.6), cooked whiting (793 ± 0.6), medium-rare broiled extra lean ground beef (792 ± 0.5), smoked whitefish (791 ± 0.6), braised pork spleen (790 ± 0.5), raw yellowtail (781 ± 0.6), medium-rare fried extra lean ground beef (779 ± 0.5), cooked herring (778 ± 0.6), raw pheasant without skin (773 ± 0.5), braised beef brisket including fat (773 ± 0.5), cooked carp (772 ± 0.6), medium-rare broiled lean ground beef (771 ± 0.5), cooked cod (771 ± 0.6), roasted pork arm including fat (769 ± 0.5), canned cod (768 ± 0.6), cooked rainbow smelt (764 ± 0.6), medium-rare baked extra lean ground beef (763 ± 0.5), roasted beef rib including fat (763 ± 0.5), roasted pork shoulder including fat (763 ± 0.5), cooked mussels (762 ± 0.6), dried black walnuts (761 ± 1.8), raw whelk (759 ± 0.6), roasted pork boston blade including fat (758 ± 0.5), broiled pork boston blade including fat (757 ± 0.5), medium-rare fried lean ground beef (756 ± 0.5), stewed dark meat chicken with skin (754 ± 0.5), medium-rare broiled ground beef (751 ± 0.5), stewed chicken thigh with skin (751 ± 0.7), medium-rare baked lean ground beef (747 ± 0.5), medium-rare fried ground beef (746 ± 0.5), whole dried sesame seeds (744 ± 5.6), turkey ham (744 ± 0.9), fried pork center rib including fat (734 ± 0.5), tempeh (733 ± 0.6), fried chicken neck with skin (733 ± 1.4), raw coho salmon (731 ± 0.6), fried pork top loin including fat (730 ± 0.5), oil roasted almonds (729 ± 1.8), toasted almonds (729 ± 1.8), grilled canadian bacon (726 ± 1.1), dried pistachio nuts (725 ± 1.8), braised beef shortribs including fat (725 ± 0.5), roasted pork loin blade including fat (725 ± 0.5), toasted sesame kernels (721 ± 1.8), whole toasted sesame seeds (721 ± 1.8), breaded and fried shrimp (719 ± 0.6), raw sockeye salmon (719 ± 0.6), medium-rare baked ground beef (718 ± 0.5), braised beef spleen (715 ± 0.5), simmered beef tongue (715 ± 0.5), milk chocolate (714 ± 1.8), dried almonds (714 ± 1.8), sesame butter (713 ± 3.3), hard pork and beef salami (710 ± 5), broiled pork loin blade including fat (708 ± 0.5), raw shark (708 ± 0.6), chicken egg yolk (706 ± 2.9), stewed chicken wing with skin (705 ± 1.3), raw halibut (704 ± 0.6), raw trout (701 ± 0.6), pork link sausage (699 ± 0.7), cooked sturgeon (699 ± 0.6), cooked shrimp (696 ± 0.6), roasted canned lean ham (694 ± 0.5), raw rainbow trout (694 ± 0.6), raw milkfish (693 ± 0.6), raw snapper (692 ± 0.6), canned sockeye salmon (691 ± 0.6), hard pork salami (690 ± 5), turkey pastrami (689 ± 0.9), roasted lean ham (687 ± 0.5), raw anchovy (687 ± 0.6), canned blue crab (684 ± 0.6), crab cakes (682 ± 0.8), cooked lobster (682 ± 0.6), raw sheepshead (682 ± 0.6), raw chum salmon (680 ± 0.6), raw shrimp (676 ± 0.6), broiled beef rib including fat (676 ± 0.5), raw chinook salmon (676 ± 0.6), raw bluefish (676 ± 0.6), dried sunflower seeds (675 ± 1.8), roasted canned ham (673 ± 0.5), raw pink salmon (673 ± 0.6), canned pink salmon (673 ± 0.6), cooked blue crab (672 ± 0.6), raw swordfish (668 ± 0.6), cottage cheese (666 ± 0.2), raw duck liver (660 ± 0.5), raw pollock (656 ± 0.6), raw sunfish (655 ± 0.6), raw grouper (654 ± 0.6), pork luxury loaf (654 ± 1.8), raw mullet (653 ± 0.6), raw burbot (652 ± 0.6), roasted ham (651 ± 0.5), raw pike (651 ± 0.6), cooked alaska king crab (644 ± 0.6), raw whitefish (644 ± 0.6), raw cusk (641 ± 0.6), raw cisco (641 ± 0.6), raw ling (641 ± 0.6), roasted duck with skin (640 ± 0.5), raw haddock (638 ± 0.6), roasted cured pork arm including fat (637 ± 0.5), oil roasted sunflower seeds (636 ± 1.8), raw scup (636 ± 0.6), raw freshwater bass (636 ± 0.6), raw flatfish (636 ± 0.6), unheated lean ham (634 ± 0.5), poppy seed (633 ± 16.7), battered and fried shark (633 ± 0.6), raw rockfish (633 ± 0.6), boiled soybeans (630 ± 0.3), raw perch (629 ± 0.6), raw mackerel (628 ± 0.6), fried abalone (627 ± 0.6), raw lobster (626 ± 0.6), jellied corned beef loaf (625 ± 1.8), raw spot (625 ± 0.6), raw dolphinfish (625 ± 0.6), freeze-dried chives (625 ± 62.5), raw pompano (624 ± 0.6), raw eel (624 ± 0.6), fried pork loin blade including fat (623 ± 0.5), breaded and fried catfish (622 ± 0.6), raw sea bass (622 ± 0.6), raw crayfish (621 ± 0.6), breaded and fried croaker (621 ± 0.6), unheated canadian bacon (621 ± 0.9), chopped smoked beef (618 ± 1.8), raw whiting (618 ± 0.6), pork and beef pepperoni (617 ± 8.3), raw queen crab (616 ± 0.6), smoked chinook salmon (616 ± 0.6), raw catfish (614 ± 0.6), duck egg (613 ± 0.7), pork and beef luncheon sausage (613 ± 2.2), raw alaska king crab (609 ± 0.6), canned lean ham (607 ± 0.5), raw herring (606 ± 0.6), raw carp (602 ± 0.6), raw blue crab (601 ± 0.6), raw cod (601 ± 0.6), basil (600 ± 50), raw croaker (600 ± 0.6), raw striped bass (599 ± 0.6), raw lingcod (596 ± 0.6), raw rainbow smelt (595 ± 0.6), cooked corned beef brisket (593 ± 0.5), raw freshwater drum (592 ± 0.6), raw tilefish (591 ± 0.6), raw wolffish (591 ± 0.6), ricotta cheese (590 ± 0.4), lebanon bologna (587 ± 2.2), breaded and fried scallops (587 ± 1.6), dry roasted almonds (586 ± 1.8), pork and beef honey loaf (586 ± 1.8), boiled lupins (585 ± 0.3), raw butterfish (584 ± 0.6), raw dungeness crab (579 ± 0.6), italian pork sausage (578 ± 0.7), fried tofu (577 ± 3.8), raw goose liver (576 ± 0.5), unheated ham (576 ± 0.5), breaded fried squid (576 ± 0.6), simmered chicken neck with skin (574 ± 1.3), raw shad (572 ± 0.6), dry roasted sunflower seeds (571 ± 1.8), cashew butter (571 ± 1.8), beef pastrami (571 ± 1.8), ham and cheese roll (568 ± 1.8), roasted pork blade roll (567 ± 0.5), pork sausage (567 ± 1.9), raw white sucker (566 ± 0.6), raw seatrout (565 ± 0.6), pork and beef peppered loaf (557 ± 1.8), canned ham (557 ± 0.5), natto (556 ± 0.6), smoked cisco (552 ± 0.6), beef and pork salami (552 ± 2.2), boiled peanuts (550 ± 1.6), gjetost cheese (550 ± 1.8), raw abalone (547 ± 0.6), raw sturgeon (545 ± 0.6), raw turbot (542 ± 0.6), tuna salad (539 ± 0.2), minced ham lunch meat (538 ± 2.4), raw scallops (536 ± 0.6), chopped ham lunch meat (533 ± 2.4), pork and beef mortadella (533 ± 3.3), almond butter (531 ± 3.1), firm raw tofu (528 ± 0.4), dry roasted pistachio nuts (525 ± 1.8), oil roasted cashews (525 ± 1.8), raw cuttlefish (520 ± 0.6), beef lunch meat (518 ± 1.8), pork and beef link sausage with american cheese (516 ± 1.2), surimi scallops (515 ± 0.6), summer sausage (513 ± 2.2), braised pork brains (509 ± 0.5), pork and beef lunch meat (507 ± 1.8), chicken egg (506 ± 1), oil roasted filberts (504 ± 1.8), surimi shrimp (499 ± 0.6), raw squid (498 ± 0.6), dry roasted cashews (496 ± 1.8), dry roasted macadamia nuts (493 ± 1.8), beef salami (491 ± 2.2), raw monkfish (489 ± 0.6), pork and beef kielbasa/kolbassy (488 ± 1.9), pork bologna (483 ± 2.2), neufchatel cheese (482 ± 1.8), pickled herring (480 ± 3.3), berliner (478 ± 2.2), raw octopus (476 ± 0.6), pork headcheese (471 ± 1.8), pork livercheese (466 ± 1.3), dried brazil nuts (464 ± 1.8), breaded and fried clams (464 ± 0.6), pork and beef picnic loaf (464 ± 1.8), dried hickory nuts (461 ± 1.8), dried filberts (461 ± 1.8), braised beef lungs (460 ± 0.5), smoked beef sausage (460 ± 1.2), cooked oysters (452 ± 0.6), puffed wheat (450 ± 3.6), polish pork sausage (450 ± 1.8), pork and beef old fashioned loaf (450 ± 1.8), dried english/persian walnuts (446 ± 1.8), fried beef brains (446 ± 0.5), boiled winged beans (445 ± 0.3), boiled green soybeans (443 ± 0.6), pork and beef brotwurst (443 ± 0.7), grilled ham patties (432 ± 0.8), stir-fried sprouted soybeans (432 ± 0.5), dried pinyon pine nuts (429 ± 1.8), pork liver sausage (428 ± 2.8), almond paste (425 ± 1.8), unheated ham patties (420 ± 0.8), raw pork stomach (412 ± 0.5), raw clams (409 ± 0.6), pork and beef sausage (408 ± 3.8), pork bratwurst (406 ± 0.6), pork beerwurst salami (404 ± 2.2), beef beerwurst salami (404 ± 2.2), chicken egg white (403 ± 1.5), fennel seed (400 ± 25), beef bologna (400 ± 2.2), raw beef tripe (396 ± 0.5), pork pickle and pimento loaf (396 ± 1.8), simmered beef brains (393 ± 0.5), beef frankfurter (393 ± 0.9), pork olive loaf (389 ± 1.8), dried pilinuts (386 ± 1.8), canned pork lunch meat (386 ± 2.4), pork and beef link sausage (385 ± 0.7), sweetened condensed milk (384 ± 1.3), gefiltefish with broth (381 ± 1.2), raw mussels (381 ± 0.6), simmered pork chitterlings (379 ± 0.5), freeze-dried red sweet peppers (375 ± 31.3), freeze-dried green sweet peppers (375 ± 31.3), pork liverwurst (367 ± 2.8), cream cheese (364 ± 1.8), evaporated skim milk (363 ± 1.6), miso (362 ± 0.4), bran (361 ± 1.8), beef and pork bologna (361 ± 2.2), pork and beef knockwurst (360 ± 0.7), puffed rice (357 ± 3.6), dry roasted filberts (354 ± 1.8), corn germ (354 ± 0.5), frozen and boiled black-eyed peas (348 ± 0.6), corn flakes (346 ± 1.8), vienna sausage (344 ± 3.1), bran flakes (343 ± 1.8), dried macadamia nuts (343 ± 1.8), falafel (339 ± 1), evaporated lowfat milk (331 ± 0.4), evaporated whole milk (329 ± 0.4), shredded wheat (325 ± 1.8), freeze-dried shallots (325 ± 12.5), dried shitake mushrooms (320 ± 3.3), simmered pork ears (316 ± 0.5), beef and pork frankfurter (313 ± 1.1), simmered pork feet (307 ± 0.5), oil roasted macadamia nuts (300 ± 1.8), dry roasted pecans (296 ± 1.8), boiled hyacinth beans (291 ± 0.3), breaded and fried oysters (291 ± 0.6), dried pecans (289 ± 1.8), skim yogurt (289 ± 0.2), boiled baby lima beans (284 ± 0.3), wheat flakes (282 ± 1.8), ham salad (282 ± 1.8), sheep milk (281 ± 0.2), steamed sprouted soybeans (279 ± 1.1), boiled lima beans (276 ± 0.3), boiled white beans (274 ± 0.3), raw tofu (270 ± 0.4), boiled yardlong bean (268 ± 0.3), raw spirulina seaweed (266 ± 0.5), lowfat yogurt (265 ± 0.2), boiled cranberry beans (263 ± 0.3), boiled catjang black-eyed peas (263 ± 0.3), potato flour (260 ± 0.6), boiled yellow beans (258 ± 0.3), oil roasted pecans (257 ± 1.8), boiled pink beans (255 ± 0.3), raw laver/nori seaweed (254 ± 0.5), raw bacon (251 ± 0.7), boiled black beans (250 ± 0.3), boiled black-eyed peas (250 ± 0.3), dried acorns (250 ± 1.8), onion powder (250 ± 25), boiled mothbeans (250 ± 0.3), stir-fried sprouted lentils (248 ± 0.5), boiled navy beans (245 ± 0.3), boiled kidney beans (244 ± 0.3), raisin bran (243 ± 1.8), boiled split peas (242 ± 0.3), boiled lentils (241 ± 0.3), boiled broad beans (241 ± 0.3), boiled great northern beans (235 ± 0.3), boiled mungo beans (234 ± 0.3), potato pancakes (234 ± 0.7), boiled pinto beans (231 ± 0.3), boiled black turtle beans (230 ± 0.3), au gratin potatoes (230 ± 0.4), acorn flour (229 ± 1.8), raw oysters (226 ± 0.6), canned oysters (226 ± 0.6), boiled adzuki beans (224 ± 0.2), dehydrated onion flakes (221 ± 3.6), boiled chickpeas (220 ± 0.3), pickled pork feet (216 ± 0.5), dried coconut (214 ± 1.8), boiled and frozen baby lima beans (214 ± 0.6), canned navy beans (212 ± 0.2), boiled mung beans (210 ± 0.2), canned great northern beans (208 ± 0.2), dried chinese chestnuts (207 ± 1.8), canned white beans (204 ± 0.2), garlic powder (200 ± 16.7), boiled french beans (199 ± 0.3), frozen and boiled fordhook lima beans (195 ± 0.6), protein fortified skim milk (191 ± 0.2), protein fortified lowfat 2% milk (191 ± 0.2), protein fortified lowfat 1% milk (190 ± 0.2), raw acorns (189 ± 1.8), restaurant vanilla milkshake (186 ± 0.2), indian buffalo milk (183 ± 0.2), yellow corn pudding (182 ± 0.2), eggnog (182 ± 0.2), dried european chestnuts (179 ± 1.8), goat milk (179 ± 0.2), hot chocolate (176 ± 0.2), refried beans (175 ± 0.2), whole yogurt (175 ± 0.2), malted milk (175 ± 0.2), canned lima beans (174 ± 0.2), tomato powder (173 ± 0.5), boiled succotash (173 ± 0.5), canned broad beans (173 ± 0.2), canned black turtle beans (170 ± 0.2), boiled pigeon peas (168 ± 0.3), dried toasted coconut (168 ± 1.8), vanilla milkshake (168 ± 0.2), boiled sprouted green peas (164 ± 0.5), skim milk (164 ± 0.2), restaurant chocolate milkshake (164 ± 0.2), restaurant strawberry milkshake (162 ± 0.2), lowfat 2% milk (161 ± 0.2), filled milk (161 ± 0.2), lowfat 1% milk (159 ± 0.2), whole milk (159 ± 0.2), canned cranberry beans (156 ± 0.2), lowfat 1% chocolate milk (156 ± 0.2), hummus (155 ± 0.2), lowfat 2% chocolate milk (155 ± 0.2), dried japanese chestnuts (154 ± 1.8), canned black-eyed peas (153 ± 0.2), whole chocolate milk (153 ± 0.2), carob milk (152 ± 0.2), dried oriental radish (150 ± 0.9), baked beans with pork and sweet sauce (149 ± 0.2), frozen and boiled succotash (147 ± 0.6), chocolate milkshake (147 ± 0.2), dried ginko nuts (146 ± 1.8), canned kidney beans (146 ± 0.2), baked beans with pork and tomato sauce (145 ± 0.2), half and half cream (140 ± 3.3), buttermilk (138 ± 0.2), roasted chinese chestnuts (136 ± 1.8), vegetarian baked beans (135 ± 0.2), boiled lamb's-quarters (134 ± 0.6), light cream (133 ± 3.3), raw dock (133 ± 0.7), dried figs (132 ± 0.3), frozen and boiled turnip greens (130 ± 0.6), raw chinese chestnuts (129 ± 1.8), canned pinto beans (128 ± 0.2), boiled yellow corn (126 ± 0.6), canned chickpeas (123 ± 0.2), scalloped potatoes (121 ± 0.4), medium cream (120 ± 3.3), frozen and boiled mustard greens (120 ± 0.7), frozen and boiled spinach (119 ± 0.5), boiled mustard greens (119 ± 0.7), boiled jute potherb (116 ± 1.2), frozen and boiled yellow corn (115 ± 0.6), raw green peas (113 ± 0.6), boiled green peas (113 ± 0.6), macaroni (113 ± 0.4), spaghetti (113 ± 0.4), boiled spinach (113 ± 0.6), coconut cream (112 ± 0.2), soy milk (112 ± 0.2), frozen and boiled green peas (108 ± 0.6), okara tofu (108 ± 0.8), lard (107 ± 1.8), raw spinach (107 ± 1.8), whipping cream, light (107 ± 3.3), canned spinach (106 ± 0.5), raw pork jowl (104 ± 0.5), potato salad (104 ± 0.4), noodles (103 ± 0.3), french fried potatoes (102 ± 1), raw coconut (102 ± 1.1), frozen and boiled kale (102 ± 0.8), dried sweetened flaked coconut (101 ± 0.7), ginger (100 ± 25), whipping cream, heavy (100 ± 3.3), dried longans (94 ± 0.5), canned green peas (93 ± 0.6), oatmeal (92 ± 0.3), o'brien potatoes (92 ± 0.3), microwaved potato with skin (91 ± 0.2), frozen and boiled okra (91 ± 0.5), canned succotash with creamed corn (90 ± 0.4), dried sweetened shredded coconut (89 ± 0.5), roasted european chestnuts (89 ± 1.8), canned succotash (89 ± 0.4), frozen and fried french fried potatoes (88 ± 1), canned yellow corn with red and green peppers (87 ± 0.4), dried apricots (86 ± 1.4), boiled and steamed chinese chestnuts (86 ± 1.8), roasted japanese chestnuts (86 ± 1.8), baked potato with skin (85 ± 0.2), sweetened coconut cream (83 ± 0.2), stir-fried sprouted mung beans (81 ± 0.8), boiled okra (81 ± 0.6), raw swamp cabbage (80 ± 0.9), frozen and boiled collards (80 ± 0.6), soybean oil and cottonseed oil margarine liquid (80 ± 10), frozen and fried hash brown potatoes (79 ± 0.6), raw garlic (78 ± 5.6), canned yellow corn (78 ± 0.5), microwaved potato without skin (78 ± 0.3), raw potato without skin (77 ± 0.4), mashed potatoes (76 ± 0.5), boiled dock (75 ± 0.5), frozen and boiled potato without skin (73 ± 0.5), baked potato without skin (73 ± 0.3), coconut milk (71 ± 0.2), baked sweet potato (71 ± 0.4), frozen and baked sweet potato (70 ± 0.6), frozen and boiled broccoli (70 ± 0.5), raw shallots (70 ± 5), boiled amaranth (68 ± 0.8), raw european chestnuts (68 ± 1.8), canned sweet potato (68 ± 0.3), boiled kale (68 ± 0.8), boiled scotch kale (68 ± 0.8), boiled sweet potato (68 ± 0.3), raw chives (67 ± 16.7), boiled broccoli (67 ± 0.6), frozen and boiled turnip greens and turnips (66 ± 0.5), canned creamed yellow corn (66 ± 0.4), raw watercress (65 ± 2.9), raw japanese chestnuts (64 ± 1.8), raw broccoli (64 ± 1.1), boiled potato without skin (64 ± 0.4), boiled swamp cabbage (63 ± 1), chocolate syrup (63 ± 1.3), hash brown potatoes (62 ± 0.6), ginko nuts (61 ± 1.8), corn grits (61 ± 0.2), boiled beet greens (61 ± 0.7), raw turnip greens (57 ± 1.8), boiled and steamed european chestnuts (57 ± 1.8), raw welsh onions (55 ± 0.5), whipped butter (55 ± 4.5), sapote (55 ± 0.2), canned turnip greens (53 ± 0.4), raw sprouted mung beans (52 ± 1), raw california avocado (52 ± 0.3), canned potato without skin (52 ± 0.6), human milk (52 ± 1.6), tomato paste (51 ± 0.4), elderberries (51 ± 0.3), baked hubbard squash (50 ± 0.5), boiled hubbard squash

(50 ± 0.4), raw spring onions with tops (50 ± 1), raw wakame seaweed (49 ± 0.5), whole chocolate malted milk (49 ± 0.2), cream of rice (49 ± 0.3), raw vine spinach (48 ± 0.5), cream of wheat (48 ± 0.3), boiled mushrooms (47 ± 0.6), frozen and boiled asparagus (47 ± 0.8), steamed hawaii mountain yam (46 ± 0.7), raw mushrooms (46 ± 1.4), canned pumpkin (46 ± 0.4), farina (45 ± 0.3), canned pumpkin pie mix (45 ± 0.4), raw cauliflower (44 ± 1), boiled green beans (44 ± 0.8), boiled yellow snap beans (44 ± 0.8), boiled turnip greens (44 ± 0.7), cooked shitake mushrooms (44 ± 0.7), safflower oil and soybean oil mayonnaise (43 ± 3.6), soybean oil mayonnaise (43 ± 3.6), raw red chili peppers (42 ± 1.1), raw green chili peppers (42 ± 1.1), raw leeks (42 ± 1.9), boiled asparagus (41 ± 0.6), frozen and boiled butternut squash (41 ± 0.4), baked/boiled tropical yam (40 ± 0.7), boiled cauliflower (40 ± 0.8), raw cassava (40 ± 0.5), raw endive (40 ± 2), butter (40 ± 3.3), corn oil margarine stick (40 ± 10), safflower oil and soybean oil margarine stick (40 ± 10), soybean oil margarine stick (40 ± 10), soybean oil and cottonseed oil margarine stick (40 ± 10), corn oil margarine tub (40 ± 10), safflower oil margarine tub (40 ± 10), soybean oil margarine tub (40 ± 10), soybean oil and cottonseed oil margarine tub (40 ± 10), cooked sprouted mung beans (39 ± 0.8), raw romaine lettuce (39 ± 1.8), raw florida avocado (39 ± 0.2), dried japanese persimmon (38 ± 1.5), baked acorn squash (38 ± 0.5), frozen and boiled cauliflower (36 ± 0.6), candied sweet potato (36 ± 0.5), raw savoy cabbage (34 ± 1.4), frozen and boiled crookneck squash (34 ± 0.5), raw butterhead lettuce (33 ± 3.3), coleslaw (33 ± 0.8), boiled savoy cabbage (32 ± 0.7), raw looseleaf lettuce (32 ± 1.8), frozen and boiled yellow snap beans (32 ± 0.7), frozen and boiled green beans (32 ± 0.7), beef suet (32 ± 1.8), raw scallop squash (32 ± 0.8), figs (32 ± 1), boiled chinese cabbage (31 ± 0.6), raw zucchini (31 ± 0.8), frozen and boiled zucchini (31 ± 0.4), dried dates (30 ± 0.6), raw iceberg lettuce (30 ± 2.5), boiled collards (30 ± 0.3), boiled butternut squash (30 ± 0.5), boiled pumpkin (30 ± 0.4), apricots (29 ± 0.5), raw eggplant (29 ± 1.2), raw chinese cabbage (29 ± 1.4), raw onions (29 ± 0.6), rice polish (29 ± 0.9), raw lotus root (28 ± 0.6), boiled scallop squash (28 ± 0.6), canned zucchini in tomato juice (27 ± 0.4), canned sprouted mung beans (26 ± 0.8), raw kelp/kombu/tangle seaweed (26 ± 0.5), canned green beans (26 ± 0.7), canned yellow snap beans (26 ± 0.7), boiled and steamed japanese chestnuts (25 ± 1.8), raw crookneck squash (25 ± 0.8), boiled beets (25 ± 0.6), longans (25 ± 0.5), boiled purslane (24 ± 0.9), boiled burdock root (24 ± 0.4), bananas (24 ± 0.4), persimmon (24 ± 2), boiled chayote (24 ± 0.6), boiled crookneck squash (24 ± 0.6), whey (24 ± 0.2), raw red cabbage (23 ± 1.4), frozen and boiled carrots (23 ± 0.7), boiled eggplant (23 ± 1), boiled leeks (23 ± 1.9), frozen and boiled turnips (23 ± 0.5), boiled acorn squash (23 ± 0.4), carambola (23 ± 0.4), coconut water (22 ± 0.2), tomato purée (22 ± 0.2), boiled onions (22 ± 0.5), raw ginger root (21 ± 2.1), raw celtuce (21 ± 0.5), boiled carrots (21 ± 0.6), raw green tomato (21 ± 0.4), boiled rutabaga (21 ± 0.6), boiled/baked spaghetti squash (21 ± 0.6), strawberries (21 ± 0.3), raw green cabbage (20 ± 1.4), frozen and boiled red sweet peppers (20 ± 0.5), frozen and boiled green sweet peppers (20 ± 0.5), cooked plantain (20 ± 0.3), cooked taro (20 ± 0.8), corn oil diet margarine tub (20 ± 10), soybean oil and cottonseed oil diet margarine tub (20 ± 10), boiled red cabbage (19 ± 0.7), raw carrots (19 ± 0.7), boiled red tomato (19 ± 0.4), canned red chili peppers (19 ± 0.7), canned green chili peppers (19 ± 0.7), boiled yambean (19 ± 0.5), breadfruit (19 ± 0.5), pickled beets (19 ± 0.4), valencia orange (18 ± 0.4), raw red sweet peppers (18 ± 1), raw green sweet peppers (18 ± 1),

canned jalapeño peppers (18 ± 0.7), peach (18 ± 0.6), frozen and boiled onions (18 ± 0.5), navel orange (17 ± 0.4), boiled lotus root (17 ± 0.6), canned red sweet peppers (17 ± 0.7), canned green sweet peppers (17 ± 0.7), boiled zucchini (17 ± 0.6), boiled green cabbage (16 ± 0.7), japanese persimmon (16 ± 0.3), canned peeled red tomatoes (16 ± 0.4), canned stewed red tomatoes (16 ± 0.4), canned crookneck squash (16 ± 0.5), raw red tomato (15 ± 0.4), ale (15 ± 0.1), boiled oriental radish (14 ± 0.7), sapodilla (14 ± 0.3), pear (14 ± 0.3), raw radish (13 ± 1.1), boiled red sweet peppers (13 ± 0.7), boiled green sweet peppers (13 ± 0.7), loquats (13 ± 0.5), canned carrots (12 ± 0.7), watermelon (12 ± 0.3), european grapes (12 ± 0.3), boiled turnips (12 ± 0.6), pineapple (12 ± 0.3), tangerine (11 ± 0.6), raw oriental radish (11 ± 1.1), american grapes (11 ± 0.5), mandarin oranges (10 ± 0.4), raw celery (10 ± 1.3), mango (10 ± 0.2), tomato juice (10 ± 0.3), raw cucumber (10 ± 1), guava (10 ± 0.6), blueberries (8 ± 0.3), crab apples (8 ± 0.5), boiled celery (7 ± 0.7), plum (6 ± 0.8), thompson seedless grapes (6 ± 0.3), papaya (5 ± 0.2), cooked apple without skin (5 ± 0.3), orange juice (4 ± 0.2), raw apple with skin (4 ± 0.4), grape juice (3 ± 0.2), screwdriver (3 ± 0.2), tangerine juice (3 ± 0.2), applesauce (3 ± 0.4), raw apple without skin (3 ± 0.4), coffee (2 ± 0.3), white grapefruit juice (1 ± 0.2)

Applications: none known
Daily Dosage: Nutritional — 1–2 g; therapeutic — ?; experimental — ?; toxic — none known up to 2 g
Warnings: none

VALINE

Names: valine, VAL, amino acid V, 2-amino-3-methyl-butanoic acid
Forms: 2-amino-3-methyl-butanoic acid
Classification: essential amino acid
Deficiency: decreased coordination, deterioration of muscle function, mental health deterioration, lack of sleep, nervousness, skin hypersensitivity, protein decreased
Side-effects: see toxicity symptoms
Toxicity: headaches, emotional agitation, crawling skin feeling
Sources (mg per 100 grams of food): torula yeast (11029 ± 1.8), soybean protein isolate (4096 ± 1.8), dried spirulina seaweed (3512 ± 0.5), dried and salted cod (3236 ± 0.6), soybean protein concentrate (3064 ± 1.8), parmesan cheese (2860 ± 10), roasted pumpkin and squash seeds (2689 ± 1.8), dry skim milk (2420 ± 1.7), dried and frozen tofu (2418 ± 2.9), peanut flour (2350 ± 12.5), dry instant skim milk (2349 ± 0.5), defatted soybean flour (2346 ± 0.5), low fat soybean flour (2322 ± 0.6), gruyere cheese (2271 ± 1.8), defatted soy meal (2243 ± 0.4), swiss cheese (2164 ± 1.8), dry buttermilk (2129 ± 7.1), cooked whelk (2075 ± 0.6), almond meal (2064 ± 1.8), fenugreek seed (2050 ± 25), dried pumpkin and squash seeds (2000 ± 1.8), dry roasted soybean nuts (1976 ± 0.6), braised beef thymus (1947 ± 0.5), toasted soybean nuts (1875 ± 1.8), braised pork center loin excluding fat (1864 ± 0.5), braised pork top loin excluding fat (1846 ± 0.5), edam cheese (1832 ± 1.8), gouda cheese (1829 ± 1.8), braised pork sirloin excluding fat (1796 ± 0.5), tilsit cheese (1775 ± 1.8), braised pork loin excluding fat (1768 ± 0.5), dry whole milk (1763 ± 1.6), roasted soybean nuts

(1758 ± 0.6), fried chicken giblets (1737 ± 0.5), roasted soybean flour (1736 ± 0.6), port du salut cheese (1729 ± 1.8), soybean flour (1724 ± 0.6), broiled pork center loin excluding fat (1715 ± 0.5), fried beef top round excluding fat (1705 ± 0.5), braised pork boston blade excluding fat (1670 ± 0.5), cheddar cheese (1663 ± 0.5), provolone cheese (1661 ± 1.8), fried chicken breast without skin (1659 ± 0.6), monterey cheese (1654 ± 1.8), fried beef liver (1650 ± 0.5), simmered beef shank excluding fat (1638 ± 0.5), fried light meat chicken without skin (1628 ± 0.5), colby cheese (1607 ± 1.8), braised pork liver (1607 ± 0.5), braised beef chuck arm pot roast excluding fat (1606 ± 0.5), stewed chicken drumstick without skin (1600 ± 1.1), braised pork loin blade excluding fat (1592 ± 0.5), simmered beef kidneys (1590 ± 0.5), cooked flounder (1590 ± 0.5), roasted turkey light meat without skin (1588 ± 0.5), fried beef sirloin excluding fat (1580 ± 0.5), bleu cheese (1579 ± 1.8), cheshire cheese (1579 ± 1.8), dried watermelon seeds (1579 ± 1.8), milk chocolate (1571 ± 1.8), braised pork center loin including fat (1569 ± 0.5), dried butternuts (1564 ± 1.8), fried chicken breast with skin (1562 ± 0.5), roasted pork rump excluding fat (1562 ± 0.5), braised pork spareribs (1557 ± 0.5), broiled pork center rib excluding fat (1545 ± 0.5), broiled pork top loin excluding fat (1545 ± 0.5), roasted pork tenderloin (1543 ± 0.5), broiled beef top round excluding fat (1542 ± 0.5), fried pork center loin excluding fat (1542 ± 0.5), fried beef top round including fat (1541 ± 0.5), roasted chicken breast without skin (1540 ± 0.6), braised pork pancreas (1537 ± 0.5), simmered beef shank including fat (1537 ± 0.5), braised beef bottom round excluding fat (1536 ± 0.5), simmered chicken liver (1535 ± 0.5), braised pork spleen (1534 ± 0.5), roasted light meat chicken without skin (1533 ± 0.5), roasted pork center loin excluding fat (1527 ± 0.5), canned tuna in water (1524 ± 0.6), braised pork center rib including fat (1523 ± 0.5), roasted turkey dark meat without skin (1518 ± 0.5), roasted pork leg excluding fat (1518 ± 0.5), simmered turkey heart (1516 ± 0.5), broiled pork sirloin excluding fat (1514 ± 0.5), roasted pork center rib excluding fat (1512 ± 0.5), roasted pork top loin excluding fat (1512 ± 0.5), roasted pork shank excluding fat (1512 ± 0.5), braised beef chuck blade roast excluding fat (1511 ± 0.5), simmered turkey liver (1511 ± 0.5), braised beef spleen (1510 ± 0.5), braised beef liver (1506 ± 0.5), simmered beef heart (1502 ± 0.5), muenster cheese (1500 ± 1.8), broiled beef top round including fat (1499 ± 0.5), fried pork center rib excluding fat (1499 ± 0.5), fried pork top loin excluding fat (1499 ± 0.5), braised beef shortribs excluding fat (1496 ± 0.5), simmered chicken heart (1496 ± 0.5), broiled pork loin excluding fat (1492 ± 0.5), braised pork sirloin including fat (1491 ± 0.5), anchovy canned in olive oil (1490 ± 2.5), brick cheese (1489 ± 1.8), roasted turkey light meat with skin (1487 ± 0.5), fried light meat chicken with skin (1485 ± 0.5), broiled beef sirloin excluding fat (1477 ± 0.5), dried sesame kernels (1475 ± 6.3), roasted pork sirloin excluding fat (1474 ± 0.5), braised pork top loin including fat (1473 ± 0.5), well done baked extra lean ground beef (1469 ± 0.5), stewed chicken breast without skin (1468 ± 0.5), yellow mustard seed (1467 ± 16.7), cooked bacon (1466 ± 0.4), braised pork kidneys (1463 ± 0.5), broiled pork center loin including fat (1462 ± 0.5), roasted chicken breast with skin (1461 ± 0.5), limburger cheese (1457 ± 1.8), braised beef pancreas (1453 ± 0.5), braised beef bottom round including fat (1450 ± 0.5), braised pork loin including fat (1447 ± 0.5), roasted pork loin excluding fat (1442 ± 0.5), fried dark meat chicken without skin (1438 ± 0.5), well done baked lean ground beef (1435 ± 0.5), stewed light meat chicken without skin

III. Amino Acids

VALINE

(1433 ± 0.5), roasted turkey dark meat with skin (1432 ± 0.5), simmered turkey giblets (1430 ± 0.5), braised beef brisket excluding fat (1429 ± 0.5), dried pistachio nuts (1429 ± 1.8), braised pork arm braised including fat (1428 ± 0.5), roasted pork rump including fat (1423 ± 0.5), roasted light meat chicken with skin (1412 ± 0.5), cooked coho salmon (1411 ± 0.6), roasted eye of beef round excluding fat (1410 ± 0.5), cooked sockeye salmon (1407 ± 0.6), braised pork boston blade braised including fat (1407 ± 0.5), roasted beef tip excluding fat (1397 ± 0.5), well done baked ground beef (1396 ± 0.5), broiled beef top loin excluding fat (1392 ± 0.5), well done broiled extra lean ground beef (1386 ± 0.5), broiled beef round excluding fat (1384 ± 0.5), simmered chicken giblets (1379 ± 0.5), cooked halibut (1375 ± 0.6), broiled beef tenderloin excluding fat (1374 ± 0.5), broiled porterhouse steak excluding fat (1370 ± 0.5), broiled T-bone steak excluding fat (1368 ± 0.5), canned tuna in oil (1367 ± 0.6), well done broiled lean ground beef (1367 ± 0.5), broiled beef rib eye excluding fat (1364 ± 0.5), braised beef flank excluding fat (1363 ± 0.5), roasted pork shoulder excluding fat (1360 ± 0.5), brie cheese (1357 ± 1.8), roasted dark meat chicken without skin (1357 ± 0.5), well done fried extra lean ground beef (1357 ± 0.5), cooked rainbow trout (1356 ± 0.6), roasted pork center loin including fat (1356 ± 0.5), cooked snapper (1355 ± 0.6), chipped beef (1354 ± 1.8), fried chicken back with skin (1351 ± 0.7), broiled pork boston blade excluding fat (1349 ± 0.5), american cheese (1343 ± 1.8), pimento cheese spread (1343 ± 1.8), dried sunflower seeds (1343 ± 1.8), stewed chicken breast with skin (1342 ± 0.5), cooked sheepshead (1340 ± 0.6), roasted beef tenderloin excluding fat (1339 ± 0.5), roasted beef rib roasted excluding fat (1339 ± 0.5), braised beef flank including fat (1338 ± 0.5), roasted pork sirloin including fat (1338 ± 0.5), well done fried lean ground beef (1338 ± 0.5), fried beef sirloin including fat (1335 ± 0.5), broiled pork loin blade excluding fat (1335 ± 0.5), roasted pork leg including fat (1335 ± 0.5), broiled beef sirloin including fat (1332 ± 0.5), fried dark meat chicken with skin (1326 ± 0.5), roasted pork loin blade excluding fat (1323 ± 0.5), roasted pork center rib including fat (1320 ± 0.5), well done broiled ground beef (1319 ± 0.5), braised beef chuck arm pot roast including fat (1318 ± 0.5), simmered turkey gizzard (1318 ± 0.5), fried chicken drumstick with skin (1318 ± 1), roasted chicken drumstick with skin (1315 ± 1), broiled pork center rib including fat (1310 ± 0.5), well done fried ground beef (1309 ± 0.5), cooked swordfish (1308 ± 0.6), roasted pork boston blade excluding fat (1306 ± 0.5), raw pheasant without skin (1305 ± 0.5), dried black walnuts (1304 ± 1.8), fried chicken thigh with skin (1303 ± 0.8), fried pork loin blade excluding fat (1303 ± 0.5), roasted eye of beef round including fat (1302 ± 0.5), smoked haddock (1300 ± 0.6), roasted pork shank including fat (1296 ± 0.5), camembert cheese (1293 ± 1.8), roasted pork top loin including fat (1290 ± 0.5), creamy peanut butter (1288 ± 3.1), stewed dark meat chicken without skin (1288 ± 0.5), roasted beef round tip including fat (1287 ± 0.5), roasted chicken thigh without skin (1287 ± 1), broiled pork sirloin including fat (1285 ± 0.5), cooked grouper (1280 ± 0.6), roasted chicken wing with skin (1279 ± 1.5), cooked mullet (1278 ± 0.6), braised pork loin blade including fat (1274 ± 0.5), stewed light meat chicken with skin (1273 ± 0.5), cooked pike (1272 ± 0.6), sardines (1267 ± 2.1), kippered herring (1265 ± 1.3), caviar (1263 ± 3.1), broiled beef tenderloin including fat (1263 ± 0.5), broiled pork top loin including fat (1263 ± 0.5), cooked tilefish (1262 ± 0.6), roasted dark meat chicken with skin (1258 ± 0.5), dried pignolia pine nuts (1257 ± 1.8), broiled pork loin including fat (1256 ± 0.5), braised pork tongue (1253 ± 0.5), broiled beef top loin including fat (1251 ± 0.5), braised pork heart (1250 ± 0.4), oil roasted sunflower seeds (1250 ± 1.8), cooked haddock (1249 ± 0.6), roasted pork loin including fat (1249 ± 0.5), cooked flatfish (1245 ± 0.6), fried chicken wing with skin (1244 ± 1.6), broiled beef round including fat (1242 ± 0.5), braised beef chuck blade roast including fat (1238 ± 0.5), cooked rockfish (1238 ± 0.6), fried pork center loin including fat (1238 ± 0.5), broiled beef flank excluding fat (1235 ± 0.5), stewed chicken drumstick with skin (1235 ± 0.9), broiled beef rib eye including fat (1234 ± 0.5), roasted goose with skin (1232 ± 0.5), cooked perch (1231 ± 0.6), medium-rare broiled extra lean ground beef (1231 ± 0.5), oil roasted peanuts (1229 ± 1.8), mozzarella cheese (1229 ± 1.8), roasted duck without skin (1228 ± 0.5), cooked mackerel (1228 ± 0.6), broiled porterhouse steak including fat (1221 ± 0.5), cooked pompano (1220 ± 0.6), broiled beef flank including fat (1218 ± 0.5), roasted chicken thigh with skin (1218 ± 0.8), cooked eel (1218 ± 0.6), simmered chicken neck without skin (1217 ± 2.8), simmered chicken gizzard (1216 ± 0.5), cooked sea bass (1216 ± 0.6), cooked whiting (1211 ± 0.6), cashew butter (1211 ± 1.8), medium-rare fried extra lean ground beef (1210 ± 0.5), smoked whitefish (1206 ± 0.6), poppy seed (1200 ± 16.7), dill seed (1200 ± 25), broiled beef rib excluding fat (1198 ± 0.5), medium-rare broiled lean ground beef (1198 ± 0.5), turkey breast meat summer sausage (1195 ± 2.4), canned corned beef (1192 ± 0.5), raw yellowtail (1192 ± 0.6), roasted beef tenderloin including fat (1190 ± 0.5), roasted pork arm including fat (1188 ± 0.5), cooked herring (1187 ± 0.6), medium-rare baked extra lean ground beef (1186 ± 0.5), raw duck liver (1181 ± 0.5), raw quail without skin (1180 ± 0.5), dried peanuts (1179 ± 1.8), cooked carp (1178 ± 0.6), cooked cod (1176 ± 0.6), oil roasted spanish peanuts (1175 ± 1.8), medium-rare fried lean ground beef (1175 ± 0.5), canned cod (1173 ± 0.6), roasted pork shoulder including fat (1173 ± 0.5), broiled T-bone steak including fat (1167 ± 0.5), medium-rare broiled ground beef (1167 ± 0.5), broiled pork boston blade including fat (1165 ± 0.5), cooked rainbow smelt (1164 ± 0.6), roasted pork boston blade including fat (1161 ± 0.5), medium-rare baked lean ground beef (1160 ± 0.5), medium-rare fried ground beef (1160 ± 0.5), cooked black bass (1149 ± 0.4), fried pork center rib including fat (1147 ± 0.5), fried pork top loin including fat (1142 ± 0.5), stewed dark meat chicken with skin (1139 ± 0.5), fried chicken neck with skin (1139 ± 1.4), oil roasted valencia peanuts (1136 ± 1.8), dry roasted sunflower seeds (1132 ± 1.8), stewed chicken thigh with skin (1129 ± 0.7), cooked crayfish (1125 ± 0.6), roasted pork loin blade including fat (1120 ± 0.5), hard pork salami (1120 ± 5), braised beef brisket including fat (1118 ± 0.5), cooked clams (1116 ± 0.6), medium-rare baked ground beef (1116 ± 0.5), oil roasted cashews (1114 ± 1.8), raw coho salmon (1113 ± 0.6), pickled anchovy (1111 ± 1.8), roasted beef rib including fat (1105 ± 0.5), broiled pork loin blade including fat (1098 ± 0.5), raw sockeye salmon (1096 ± 0.6), stewed chicken wing with skin (1088 ± 1.3), oil roasted virginia peanuts (1086 ± 1.8), roasted cured pork arm excluding fat (1082 ± 0.5), raw shark (1081 ± 0.6), hard pork and beef salami (1080 ± 5), raw halibut (1072 ± 0.6), raw trout (1071 ± 0.6), pork link sausage (1068 ± 0.7), cooked sturgeon (1066 ± 0.6), oil roasted almonds (1064 ± 1.8), toasted almonds (1064 ± 1.8), simmered beef tongue (1058 ± 0.5), raw rainbow trout (1058 ± 0.6), raw milkfish (1058 ± 0.6), anchovy paste (1057 ± 7.1), raw snapper (1056 ± 0.6), canned sockeye salmon (1055 ± 0.6), dry roasted cashews (1054 ± 1.8), braised beef shortribs including fat (1049 ± 0.5), raw anchovy (1048 ± 0.6), raw sheepshead (1041 ± 0.6), cooked mussels (1040 ± 0.6), dry roasted pistachio nuts (1039 ± 1.8), raw whelk (1036 ± 0.6), raw chum salmon (1036 ± 0.6), raw chinook salmon (1033 ± 0.6), raw bluefish (1032 ± 0.6), raw goose liver (1032 ± 0.5), raw alewife (1028 ± 0.5), raw pink salmon (1027 ± 0.6), canned pink salmon (1027 ± 0.6), breaded and fried shrimp (1025 ± 0.6), raw swordfish (1020 ± 0.6), natto (1018 ± 0.6), raw black bass (1018 ± 0.5), crunchy peanut butter (1009 ± 1.6), dried almonds (1007 ± 1.8), jellied corned beef loaf (1007 ± 1.8), braised beef lungs (1005 ± 0.5), raw pollock (1002 ± 0.6), turkey ham (1000 ± 0.9), chicken egg yolk (1000 ± 2.9), basil (1000 ± 50), raw sunfish (999 ± 0.6), raw grouper (998 ± 0.6), raw mullet (996 ± 0.6), raw burbot (995 ± 0.6), fried pork loin blade including fat (995 ± 0.5), dry roasted peanuts (993 ± 1.8), raw pike (992 ± 0.6), whole dried sesame seeds (989 ± 5.6), braised pork lungs (988 ± 0.5), raw smelt (986 ± 0.5), cooked shrimp (984 ± 0.6), raw whitefish (984 ± 0.6), tempeh (980 ± 0.6), crab cakes (980 ± 0.8), broiled beef rib including fat (979 ± 0.5), raw cusk (979 ± 0.6), raw cisco (978 ± 0.6), raw ling (978 ± 0.6), canned smelt (975 ± 0.5), raw haddock (974 ± 0.6), corn germ (974 ± 0.5), raw scup (972 ± 0.6), raw freshwater bass (971 ± 0.6), raw flatfish (971 ± 0.6), raw rockfish (966 ± 0.6), canned blue crab (965 ± 0.6), battered and fried shark (965 ± 0.6), cooked lobster (964 ± 0.6), toasted sesame kernels (961 ± 1.8), whole toasted sesame seeds (961 ± 1.8), raw perch (960 ± 0.6), raw mackerel (958 ± 0.6), raw shrimp (956 ± 0.6), grilled canadian bacon (955 ± 1.1), raw spot (954 ± 0.6), sesame butter (953 ± 3.3), raw dolphinfish (953 ± 0.6), raw pompano (952 ± 0.6), cooked blue crab (951 ± 0.6), raw eel (951 ± 0.6), raw sea bass (951 ± 0.6), roasted canned lean ham (948 ± 0.5), turkey pastrami (947 ± 0.9), breaded and fried croaker (947 ± 0.6), breaded and fried catfish (944 ± 0.6), raw whiting (944 ± 0.6), smoked chinook salmon (942 ± 0.6), chopped smoked beef (939 ± 1.8), roasted duck with skin (938 ± 0.5), raw catfish (936 ± 0.6), raw herring (925 ± 0.6), dried brazil nuts (925 ± 1.8), roasted canned ham (920 ± 0.5), raw carp (919 ± 0.6), raw cod (916 ± 0.6), raw croaker (916 ± 0.6), raw striped bass (914 ± 0.6), simmered chicken neck with skin (913 ± 1.3), cooked alaska king crab (911 ± 0.6), raw lingcod (911 ± 0.6), roasted lean ham (908 ± 0.5), raw rainbow smelt (908 ± 0.6), pork luxury loaf (904 ± 1.8), raw freshwater drum (904 ± 0.6), raw tilefish (902 ± 0.6), raw wolffish (902 ± 0.6), pork and beef pepperoni (900 ± 8.3), fennel seed (900 ± 25), roasted cured pork arm including fat (894 ± 0.5), raw butterfish (891 ± 0.6), lebanon bologna (887 ± 2.2), duck egg (886 ± 0.7), raw lobster (884 ± 0.6), raw crayfish (878 ± 0.6), freeze-dried chives (875 ± 62.5), chicken egg (874 ± 1), raw shad (872 ± 0.6), raw queen crab (871 ± 0.6), pork and beef luncheon sausage (870 ± 2.2), fried tofu (869 ± 3.8), pork liverwurst (867 ± 2.8), raw white sucker (864 ± 0.6), raw seatrout (864 ± 0.6), raw black bullhead (864 ± 0.5), raw alaska king crab (861 ± 0.6), roasted ham (860 ± 0.5), fried abalone (860 ± 0.6), canned alewife (859 ± 0.5), dry roasted almonds (854 ± 1.8), raw blue crab (849 ± 0.6), smoked cisco (844 ± 0.6), unheated lean ham (839 ± 0.5), pork and beef peppered loaf (836 ± 1.8), beef lunch meat (836 ± 1.8), raw sturgeon (832 ± 0.6), boiled soybeans (831 ± 0.3), canned lean ham (829 ± 0.5), raw turbot (827 ± 0.6), tuna salad (824 ± 0.2), breaded and fried scallops (819 ± 1.6), raw dungeness crab (819 ± 0.6), unheated canadian bacon (818 ± 0.9), ham and cheese roll (811 ± 1.8), chicken egg white (809 ± 1.5), pork livercheese (805 ± 1.3), italian pork sausage (804 ± 0.7), chopped ham lunch meat (800 ± 2.4), cooked corned beef brisket (799 ± 0.5), firm raw tofu (796 ± 0.4), raw flounder

VALINE

(794 ± 0.5), raw sole (794 ± 0.5), pork sausage (789 ± 1.9), pork and beef honey loaf (786 ± 1.8), breaded fried squid (786 ± 0.6), gjetost cheese (775 ± 1.8), almond butter (775 ± 3.1), cottage cheese (773 ± 0.2), beef pastrami (768 ± 1.8), unheated ham (762 ± 0.5), canned ham (760 ± 0.5), freeze-dried red sweet peppers (750 ± 31.3), freeze-dried green sweet peppers (750 ± 31.3), roasted pork blade roll (749 ± 0.5), minced ham lunch meat (748 ± 2.4), raw abalone (747 ± 0.6), raw monkfish (746 ± 0.6), miso (742 ± 0.4), pork bologna (739 ± 2.2), dried hickory nuts (739 ± 1.8), stir-fried sprouted soybeans (734 ± 0.5), raw scallops (733 ± 0.6), pork and beef mortadella (733 ± 3.3), pickled herring (733 ± 3.3), oil roasted filberts (732 ± 1.8), dried english/persian walnuts (732 ± 1.8), pork and beef lunch meat (721 ± 1.8), puffed wheat (714 ± 3.6), dried pilinuts (711 ± 1.8), raw cuttlefish (709 ± 0.6), berliner (704 ± 2.2), summer sausage (696 ± 2.2), pork and beef link sausage with american cheese (693 ± 1.2), ricotta cheese (692 ± 0.4), raw pork stomach (692 ± 0.5), braised pork brains (691 ± 0.5), polish pork sausage (686 ± 1.8), dried ginko nuts (686 ± 1.8), raw squid (680 ± 0.6), pork and beef brotwurst (676 ± 0.7), dried filberts (671 ± 1.8), beef and pork salami (670 ± 2.2), pork headcheese (668 ± 1.8), garlic powder (667 ± 16.7), canned pork lunch meat (662 ± 2.4), beef salami (661 ± 2.2), raw octopus (651 ± 0.6), breaded and fried clams (651 ± 0.6), boiled lupins (650 ± 0.3), pork and beef picnic loaf (650 ± 1.8), surimi scallops (648 ± 0.6), pork and beef kielbasa/kolbassy (638 ± 1.9), simmered pork ears (632 ± 0.5), surimi shrimp (628 ± 0.6), pork and beef old fashioned loaf (625 ± 1.8), beef and pork bologna (622 ± 2.2), smoked beef sausage (621 ± 1.2), almond paste (621 ± 1.8), fried beef brains (617 ± 0.5), pork liver sausage (617 ± 2.8), cooked oysters (616 ± 0.6), raw beef tripe (613 ± 0.5), dried pinyon pine nuts (607 ± 1.8), neufchatel cheese (593 ± 1.8), bran (593 ± 1.8), pork and beef sausage (592 ± 3.8), grilled ham patties (575 ± 0.8), vienna sausage (575 ± 3.1), boiled peanuts (566 ± 1.6), pork bratwurst (566 ± 0.6), pork pickle and pimento loaf (564 ± 1.8), falafel (563 ± 1), unheated ham patties (558 ± 0.8), raw clams (558 ± 0.6), bran flakes (557 ± 1.8), shredded wheat (550 ± 1.8), boiled green soybeans (549 ± 0.6), beef beerwurst salami (548 ± 2.2), gefiltefish with broth (548 ± 1.2), simmered beef brains (544 ± 0.5), beef bologna (539 ± 2.2), sweetened condensed milk (532 ± 1.3), beef frankfurter (530 ± 0.9), raw mussels (520 ± 0.6), pork and beef knockwurst (515 ± 0.7), pork olive loaf (514 ± 1.8), dry roasted filberts (514 ± 1.8), simmered pork tail (510 ± 0.5), pork beerwurst salami (509 ± 2.2), boiled white beans (509 ± 0.3), evaporated skim milk (503 ± 1.6), simmered pork chitterlings (502 ± 0.5), freeze-dried shallots (500 ± 12.5), frozen and boiled black-eyed peas (492 ± 0.6), boiled cranberry beans (489 ± 0.3), dried shitake mushrooms (487 ± 3.3), boiled baby lima beans (484 ± 0.3), simmered pork feet (480 ± 0.5), boiled yellow beans (479 ± 0.3), wheat flakes (475 ± 1.8), skim yogurt (474 ± 0.2), steamed sprouted soybeans (474 ± 1.1), boiled pink beans (474 ± 0.3), beef and pork frankfurter (471 ± 1.1), boiled lima beans (469 ± 0.3), boiled winged beans (467 ± 0.3), pork and beef link sausage (465 ± 0.7), boiled black beans (464 ± 0.3), dried acorns (461 ± 1.8), evaporated whole milk (456 ± 0.4), boiled navy beans (455 ± 0.3), ham salad (454 ± 1.8), boiled kidney beans (454 ± 0.3), sheep milk (448 ± 0.2), boiled lentils (448 ± 0.3), cream cheese (446 ± 1.8), boiled great northern beans (436 ± 0.3), lowfat yogurt (434 ± 0.2), boiled pinto beans (430 ± 0.3), acorn flour (429 ± 1.8), boiled black turtle beans (428 ± 0.3), boiled mungo beans (423 ± 0.3), boiled hyacinth beans (422 ± 0.3), dried coconut (421 ± 1.8), corn flakes (418 ± 1.8),

raw bacon (418 ± 0.7), boiled and frozen baby lima beans (416 ± 0.6), puffed rice (414 ± 3.6), potato flour (413 ± 0.6), breaded and fried oysters (409 ± 0.6), raw tofu (408 ± 0.4), dry roasted pecans (404 ± 1.8), raw laver/nori seaweed (402 ± 0.5), raisin bran (400 ± 1.8), boiled yardlong bean (395 ± 0.3), boiled split peas (394 ± 0.3), canned navy beans (394 ± 0.2), dried pecans (393 ± 1.8), stir-fried sprouted lentils (391 ± 0.5), boiled catjang black-eyed peas (387 ± 0.3), boiled adzuki beans (387 ± 0.2), canned great northern beans (386 ± 0.2), canned white beans (380 ± 0.2), evaporated lowfat milk (379 ± 0.4), frozen and boiled fordhook lima beans (379 ± 0.6), potato pancakes (376 ± 0.7), boiled chickpeas (372 ± 0.3), boiled french beans (369 ± 0.3), boiled black-eyed peas (368 ± 0.3), dried oriental radish (366 ± 0.9), boiled mung beans (364 ± 0.2), dried chinese chestnuts (364 ± 1.8), raw spirulina seaweed (362 ± 0.5), dried european chestnuts (361 ± 1.8), oil roasted pecans (350 ± 1.8), raw acorns (350 ± 1.8), ginger (350 ± 25), dry roasted macadamia nuts (343 ± 1.8), boiled broad beans (338 ± 0.3), pickled pork feet (338 ± 0.5), refried beans (326 ± 0.2), dried macadamia nuts (325 ± 1.8), au gratin potatoes (325 ± 0.4), dried toasted coconut (325 ± 1.8), dried japanese chestnuts (318 ± 1.8), canned black turtle beans (315 ± 0.2), lard (314 ± 1.8), raw oysters (308 ± 0.6), canned oysters (308 ± 0.6), boiled succotash (308 ± 0.5), raw pork jowl (305 ± 0.5), canned lima beans (296 ± 0.2), boiled pigeon peas (292 ± 0.3), canned cranberry beans (290 ± 0.2), raw garlic (289 ± 5.6), yellow corn pudding (287 ± 0.2), whole yogurt (287 ± 0.2), oil roasted macadamia nuts (286 ± 1.8), ginko nuts (286 ± 1.8), boiled sprouted green peas (285 ± 0.5), baked beans with pork and sweet sauce (278 ± 0.2), canned kidney beans (272 ± 0.2), baked beans with pork and tomato sauce (270 ± 0.2), protein fortified skim milk (265 ± 0.2), protein fortified lowfat 2% milk (264 ± 0.2), tomato powder (264 ± 0.5), protein fortified lowfat 1% milk (263 ± 0.2), frozen and boiled succotash (262 ± 0.6), restaurant vanilla milkshake (258 ± 0.2), eggnog (253 ± 0.2), vegetarian baked beans (251 ± 0.2), onion powder (250 ± 25), hot chocolate (244 ± 0.2), canned broad beans (243 ± 0.2), buttermilk (243 ± 0.2), noodles (243 ± 0.3), goat milk (240 ± 0.2), roasted chinese chestnuts (239 ± 1.8), canned pinto beans (239 ± 0.2), malted milk (238 ± 0.2), raw green peas (235 ± 0.6), vanilla milkshake (233 ± 0.2), boiled green peas (233 ± 0.6), skim milk (228 ± 0.2), restaurant chocolate milkshake (228 ± 0.2), frozen and boiled turnip greens (228 ± 0.6), canned black-eyed peas (226 ± 0.2), restaurant strawberry milkshake (225 ± 0.2), lowfat 2% milk (223 ± 0.2), filled milk (223 ± 0.2), frozen and boiled green peas (223 ± 0.6), raw chinese chestnuts (221 ± 1.8), lowfat 1% milk (220 ± 0.2), whole milk (220 ± 0.2), coconut cream (220 ± 0.2), indian buffalo milk (219 ± 0.2), hummus (218 ± 0.2), lowfat 1% chocolate milk (217 ± 0.2), dried longans (217 ± 0.5), lowfat 2% chocolate milk (215 ± 0.2), whole chocolate milk (212 ± 0.2), carob milk (210 ± 0.2), raw wakame seaweed (209 ± 0.5), canned chickpeas (208 ± 0.2), dehydrated onion flakes (207 ± 3.6), french fried potatoes (206 ± 1), chocolate milkshake (204 ± 0.2), stir-fried sprouted mung beans (203 ± 0.8), raw coconut (202 ± 1.1), half and half cream (200 ± 3.3), dried sweetened flaked coconut (199 ± 0.7), boiled jute potherb (195 ± 1.2), canned green peas (192 ± 0.6), boiled yellow corn (191 ± 0.6), sour cream (183 ± 4.2), roasted european chestnuts (182 ± 1.8), light cream (180 ± 3.3), roasted japanese chestnuts (179 ± 1.8), frozen and boiled spinach (177 ± 0.5), frozen and fried french fried potatoes (176 ± 1), dried sweetened shredded coconut (175 ± 0.5), scalloped potatoes (174 ± 0.4), frozen and boiled yellow corn (173 ± 0.6), boiled lamb's-quarters

(172 ± 0.6), potato salad (172 ± 0.4), boiled spinach (168 ± 0.6), medium cream (167 ± 3.3), frozen and boiled brussels sprouts (167 ± 0.6), sweetened coconut cream (163 ± 0.2), okara tofu (162 ± 0.8), frozen and fried hash brown potatoes (162 ± 0.6), raw spinach (161 ± 1.8), canned succotash with creamed corn (160 ± 0.4), canned spinach (158 ± 0.5), canned succotash (158 ± 0.4), frozen and boiled kale (155 ± 0.8), boiled and steamed chinese chestnuts (154 ± 1.8), oatmeal (151 ± 0.3), whipping cream, light (147 ± 3.3), frozen and boiled collards (145 ± 0.6), raw sprouted alfalfa seeds (145 ± 1.5), frozen and boiled broccoli (142 ± 0.5), soy milk (141 ± 0.2), whipping cream, heavy (140 ± 3.3), coconut milk (139 ± 0.2), microwaved potato with skin (138 ± 0.2), raw swamp cabbage (136 ± 0.9), raw european chestnuts (136 ± 1.8), boiled broccoli (136 ± 0.6), raw japanese chestnuts (136 ± 1.8), raw watercress (135 ± 2.9), dried apricots (134 ± 1.4), o'brien potatoes (133 ± 0.3), raw chives (133 ± 16.7), canned yellow corn with red and green peppers (132 ± 0.4), raw sprouted mung beans (131 ± 1), baked potato with skin (130 ± 0.2), raw broccoli (127 ± 1.1), hash brown potatoes (123 ± 0.6), boiled dock (121 ± 0.5), soybean oil and cottonseed oil margarine liquid (120 ± 10), frozen and boiled turnip greens and turnips (120 ± 0.5), canned yellow corn (118 ± 0.5), microwaved potato without skin (118 ± 0.3), boiled amaranth (118 ± 0.8), raw potato without skin (117 ± 0.4), boiled brussels sprouts (117 ± 0.6), dried figs (115 ± 0.3), boiled and steamed european chestnuts (114 ± 1.8), boiled swiss chard (114 ± 0.6), frozen and baked sweet potato (113 ± 0.6), frozen and boiled asparagus (113 ± 0.8), mashed potatoes (112 ± 0.5), baked sweet potato (112 ± 0.4), frozen and boiled potato without skin (111 ± 0.5), baked potato without skin (110 ± 0.3), raw shallots (110 ± 5), canned sweet potato (108 ± 0.3), boiled sweet potato (108 ± 0.3), boiled swamp cabbage (108 ± 1), raw dock (107 ± 0.7), raw bamboo shoots (107 ± 0.7), boiled kale (105 ± 0.8), boiled scotch kale (105 ± 0.8), raw turnip greens (104 ± 1.8), raw california avocado (103 ± 0.3), boiled mushrooms (103 ± 0.6), canned creamed yellow corn (100 ± 0.4), raw cauliflower (100 ± 1), boiled asparagus (100 ± 0.6), raw mushrooms (97 ± 1.4), cooked sprouted mung beans (97 ± 0.8), boiled potato without skin (96 ± 0.4), frozen and boiled okra (95 ± 0.5), chocolate syrup (95 ± 1.3), boiled green beans (94 ± 0.8), boiled yellow snap beans (94 ± 0.8), boiled cauliflower (94 ± 0.8), canned turnip greens (92 ± 0.4), boiled mustard greens (89 ± 0.7), frozen and boiled mustard greens (88 ± 0.7), raw romaine lettuce (86 ± 1.8), raw savoy cabbage (86 ± 1.4), boiled okra (85 ± 0.6), raw welsh onions (84 ± 0.5), raw red chili peppers (84 ± 1.1), raw green chili peppers (84 ± 1.1), frozen and boiled cauliflower (81 ± 0.6), canned potato without skin (80 ± 0.6), raw cassava (79 ± 0.5), raw spring onions with tops (78 ± 1), boiled turnip greens (78 ± 0.7), raw florida avocado (78 ± 0.2), sapote (77 ± 0.2), tomato paste (77 ± 0.4), boiled savoy cabbage (77 ± 0.7), raw chickory greens (77 ± 0.6), boiled beet greens (76 ± 0.7), raw ginger root (75 ± 2.1), corn grits (73 ± 0.2), cream of wheat (73 ± 0.3), raw kelp/kombu/tangle seaweed (72 ± 0.5), boiled purslane (72 ± 0.9), whole chocolate malted milk (71 ± 0.2), safflower oil and soybean oil mayonnaise (71 ± 3.6), soybean oil mayonnaise (71 ± 3.6), dried japanese persimmon (71 ± 1.5), raw looseleaf lettuce (71 ± 1.8), canned bamboo shoots (71 ± 0.4), steamed hawaii mountain yam (69 ± 0.7), farina (69 ± 0.3), boiled chinese cabbage (69 ± 0.6), frozen and boiled yellow snap beans (68 ± 0.7), frozen and boiled green beans (68 ± 0.7), canned sprouted mung beans (68 ± 0.8), cooked shitake mushrooms (67 ± 0.7), raw butterhead lettuce (67 ± 3.3), dried dates (66 ± 0.6), raw chinese

cabbage (66 ± 1.4), raw vine spinach (65 ± 0.5), baked hubbard squash (64 ± 0.5), boiled hubbard squash (64 ± 0.4), raw endive (64 ± 2), coleslaw (62 ± 0.8), boiled bamboo shoots (62 ± 0.4), human milk (61 ± 1.6), baked/boiled tropical yam (60 ± 0.7), butter (60 ± 3.3), corn oil margarine stick (60 ± 10), safflower oil and soybean oil margarine stick (60 ± 10), soybean oil margarine stick (60 ± 10), soybean oil and cottonseed oil margarine stick (60 ± 10), raw iceberg lettuce (60 ± 2.5), raw red cabbage (60 ± 1.4), raw leeks (58 ± 1.9), frozen and boiled crookneck squash (58 ± 0.5), longans (58 ± 0.5), cream of rice (57 ± 0.3), candied sweet potato (57 ± 0.5), canned green beans (57 ± 0.7), canned yellow snap beans (57 ± 0.7), raw eggplant (56 ± 1.2), raw lotus root (56 ± 0.6), whipped butter (55 ± 4.5), beef suet (54 ± 1.8), raw scallop squash (54 ± 0.8), boiled collards (54 ± 0.3), frozen and boiled butternut squash (53 ± 0.4), raw zucchini (52 ± 0.8), frozen and boiled zucchini (52 ± 0.4), boiled kohlrabi (52 ± 0.6), frozen and boiled carrots (51 ± 0.7), frozen and boiled turnips (51 ± 0.5), raw green cabbage (51 ± 1.4), boiled and steamed japanese chestnuts (50 ± 1.8), baked acorn squash (48 ± 0.5), boiled chayote (48 ± 0.6), apricots (47 ± 0.5), boiled scallop squash (47 ± 0.6), bananas (47 ± 0.4), breadfruit (47 ± 0.5), canned zucchini in tomato juice (46 ± 0.4), boiled burdock root (46 ± 0.4), whey (46 ± 0.2), raw celtuce (46 ± 0.5), boiled carrots (46 ± 0.6), boiled red cabbage (45 ± 0.7), persimmon (44 ± 2), boiled eggplant (44 ± 1), coconut water (44 ± 0.2), boiled rutabaga (44 ± 0.6), raw carrots (44 ± 0.7), valencia orange (44 ± 0.4), navel orange (44 ± 0.4), raw chickory (44 ± 1.1), raw crookneck squash (42 ± 0.8), boiled crookneck squash (41 ± 0.6), boiled green cabbage (41 ± 0.7), corn oil margarine tub (40 ± 10), safflower oil margarine tub (40 ± 10), soybean oil margarine tub (40 ± 10), soybean oil and cottonseed oil margarine tub (40 ± 10), frozen and boiled red sweet peppers (40 ± 0.5), frozen and boiled green sweet peppers (40 ± 0.5), corn oil diet margarine tub (40 ± 10), soybean oil and cottonseed oil diet margarine tub (40 ± 10), boiled butternut squash (39 ± 0.5), canned pumpkin (38 ± 0.4), canned pumpkin pie mix (38 ± 0.4), canned red chili peppers (38 ± 0.7), canned green chili peppers (38 ± 0.7), peach (38 ± 0.6), rice polish (36 ± 0.9), boiled beets (36 ± 0.6), raw red sweet peppers (36 ± 1), raw green sweet peppers (36 ± 1), boiled yambean (35 ± 0.5), tomato purée (34 ± 0.2), canned jalapeño peppers (34 ± 0.7), boiled lotus root (34 ± 0.6), canned red sweet peppers (34 ± 0.7), canned green sweet peppers (34 ± 0.7), elderberries (33 ± 0.3), boiled leeks (31 ± 1.9), raw green tomato (31 ± 0.4), boiled oriental radish (31 ± 0.7), raw radish (31 ± 1.1), japanese persimmon (30 ± 0.3), boiled acorn squash (29 ± 0.4), boiled red tomato (29 ± 0.4), boiled zucchini (29 ± 0.6), figs (28 ± 1), raw onions (28 ± 0.6), cooked plantain (28 ± 0.3), pickled beets (28 ± 0.4), canned crookneck squash (28 ± 0.5), guava (28 ± 0.6), blueberries (28 ± 0.3), cooked taro (27 ± 0.8), canned carrots (27 ± 0.7), tangerine (27 ± 0.6), raw oriental radish (27 ± 1.1), carambola (26 ± 0.4), boiled/baked spaghetti squash (26 ± 0.6), boiled red sweet peppers (26 ± 0.7), boiled green sweet peppers (26 ± 0.7), mandarin oranges (26 ± 0.4), mango (26 ± 0.2), boiled calabash gourd (26 ± 0.7), boiled pumpkin (25 ± 0.4), raw celery (25 ± 1.3), canned peeled red tomatoes (24 ± 0.4), canned stewed red tomatoes (24 ± 0.4), raw red tomato (23 ± 0.4), boiled turnips (23 ± 0.6), boiled onions (21 ± 0.5), loquats (21 ± 0.5), boiled celery (20 ± 0.7), plum (20 ± 0.8), crab apples (19 ± 0.5), strawberries (18 ± 0.3), european grapes (18 ± 0.3), american grapes (17 ± 0.5), raw cucumber (17 ± 1), frozen and boiled onions (16 ± 0.5), sapodilla (16 ± 0.3), watermelon (16 ± 0.3), pineapple (16 ± 0.3), tomato juice

VALINE

(15 ± 0.3), cooked apple without skin (12 ± 0.3), thompson seedless grapes (11 ± 0.3), orange juice (11 ± 0.2), papaya (10 ± 0.2), grape juice (10 ± 0.2), ale (9 ± 0.1), raw apple with skin (9 ± 0.4), screwdriver (9 ± 0.2), tangerine juice (8 ± 0.2), applesauce (8 ± 0.4), pear (7 ± 0.3), raw apple without skin (7 ± 0.4), coffee (3 ± 0.3), white grapefruit juice (3 ± 0.2)

Applications: none known
Daily Dosage: Nutritional—1–3 g; therapeutic—?; experimental—?; toxic—none known up to 3 g
Warnings: Valine replaces glutamic acid in hemoglobin resulting in sickle cell anemia in those genetically predestined. Increased amounts are dangerous. See toxicity symptoms.

PART IV

Macronutrients

CARBOHYDRATE

Classification: macronutrient
Forms: cellulose, fructose, glucose, glycogen, lactose, maltose, starch, sucrose
Deficiency: fatigue
Side-effects: fat accumulation, tooth decay, diabetes, heart disease
Toxicity: none known
Sources (grams per 100 grams of food): powdered sugar (99.5 ± 0.04), white granulated sugar (99.5 ± 0.03), brown sugar (96.4 ± 0.03), savory (95 ± 5), jelly beans (94.3 ± 0.18), ry-krisp crackers (92.9 ± 0.36), puffed rice (91.4 ± 0.36), maple sugar (91.1 ± 0.18), basil (90 ± 5), cinnamon (90 ± 2.5), corn starch (90 ± 0.63), carob flour (88.9 ± 0.05), gum drops (88.6 ± 0.18), corn flakes (87.1 ± 0.18), white pepper (85 ± 2.5), onion powder (85 ± 2.5), dehydrated onion flakes (83.6 ± 0.36), dried japanese chestnuts (82.5 ± 0.18), honey (82.4 ± 0.24), peanut brittle (82.1 ± 0.18), dried chinese chestnuts (81.1 ± 0.18), dried agar seaweed (80.9 ± 0.05), dried pepeao (80.8 ± 0.42), puffed wheat (80.7 ± 0.36), shredded wheat (80.7 ± 0.18), wheat flakes (80.7 ± 0.18), rice flour (80.5 ± 0.04), pretzels (80 ± 0.18), animal crackers (80 ± 0.19), marshmallow (80 ± 0.83), potato flour (79.9 ± 0.06), light buckwheat flour (79.5 ± 0.05), golden seedless raisins (79.5 ± 0.05), wheat cake (79.4 ± 0.04), bran flakes (79.3 ± 0.18), seedless raisins (79.1 ± 0.05), light pearled barley (78.8 ± 0.03), dried european chestnuts (78.6 ± 0.18), seeded raisins (78.5 ± 0.05), enriched white degermed corn meal (78.4 ± 0.04), unenriched corn meal (78.4 ± 0.04), enriched yellow degermed corn meal (78.4 ± 0.04), unenriched yellow degermed corn meal (78.4 ± 0.04), light rye flour (77.9 ± 0.05), pot barley pearled (77.2 ± 0.03), graham crackers (77.1 ± 0.36), white corn flour (76.8 ± 0.04), yellow corn flour (76.8 ± 0.04), popcorn (76.7 ± 0.83), whole grain rye crackers (76.2 ± 0.38), all purpose enriched wheat flour

CARBOHYDRATE

(76.1 ± 0.04), bread sticks (75.5 ± 0.25), bran (75.4 ± 0.18), raisin bran (75.4 ± 0.14), raw wild rice (75.3 ± 0.03), dried shitake mushrooms (75.3 ± 0.33), tomato powder (74.7 ± 0.05), enriched wheat bread (74.7 ± 0.04), white bolted corn meal (74.5 ± 0.04), zwieback crackers (74.3 ± 0.71), yellow bolted corn meal (73.8 ± 0.04), white whole ground corn meal (73.7 ± 0.04), yellow whole ground corn meal (73.7 ± 0.04), dried ginko nuts (73.6 ± 0.18), dried jujube (73.6 ± 0.05), dried dates (73.5 ± 0.06), dried japanese persimmon (73.5 ± 0.15), dry breadcrumbs (73.4 ± 0.05), saltine crackers (73.3 ± 0.83), corn syrup (73.3 ± 0.24), sorghum grain (73 ± 0.05), millit (72.9 ± 0.05), freeze-dried shallots (72.5 ± 1.25), dark buckwheat flour (72 ± 0.05), soda crackers (71.8 ± 0.18), rusk crackers (71.1 ± 0.56), whole wheat flour (71 ± 0.04), dried lychees (70.7 ± 0.05), strawberries (70.5 ± 0.03), toaster pastry (70.4 ± 0.1), dried parsley (70 ± 5), turmeric (70 ± 2.5), cardamom (70 ± 2.5), black pepper (70 ± 2.5), allspice (70 ± 2.5), citrus marmalade (70 ± 0.25), barbados molasses (70 ± 0.25), freeze-dried red sweet peppers (68.8 ± 3.13), freeze-dried green sweet peppers (68.8 ± 3.13), oatmeal cookie (68.5 ± 0.38), dark rye flour (68.1 ± 0.04), sugar cookie (68.1 ± 0.31), gingersnaps cookie (67.1 ± 0.71), garlic powder (66.7 ± 1.67), granola bar (66.7 ± 0.21), sorghum (66.7 ± 0.24), oyster crackers (66.3 ± 0.63), taco shell (65.5 ± 0.45), tostada shell (65.5 ± 0.45), dried figs (65.3 ± 0.03), ginger (65 ± 2.5), cloves (65 ± 2.5), light molasses (65 ± 0.25), cane syrup (64 ± 0.25), maple syrup (64 ± 0.25), dried oriental radish (63.4 ± 0.09), brownies (63 ± 0.25), dried prunes (62.7 ± 0.06), freeze-dried chives (62.5 ± 6.25), tamarind (62.5 ± 0.04), dried apricots (61.7 ± 0.14), shortbread cookie (61.3 ± 0.63), german sweet bakers chocolate (60.4 ± 0.18), dried dill weed (60 ± 5), curry powder (60 ± 2.5), paprika (60 ± 2.5), dill seed (60 ± 2.5), toasted raisin bread (60 ± 0.24), pumpkin pie spice (60 ± 2.5), medium molasses (60 ± 0.25), fruitcake, dark (59.8 ± 0.12), angel food cake (59.5 ± 0.08), caramel cake (59.1 ± 0.05), rice polish (58.9 ± 0.09), sweet chocolate (58.6 ± 0.18), semi-sweet bakers chocolate (58.2 ± 0.18), chocolate syrup (58.2 ± 0.13), semi-sweet chocclate (57.9 ± 0.18), milk chocolate (57.5 ± 0.18), fruitcake, light (57.4 ± 0.12), toasted cracked wheat bread (57.1 ± 0.24), cheese crackers with peanut butter (56.7 ± 0.12), french rolls (56.6 ± 0.1), raisin bun (56.5 ± 0.08), bagels (56.2 ± 0.09), toasted english muffin (56.2 ± 0.1), dry cocoa powder (56 ± 1), toasted white bread (55.7 ± 0.24), acorn flour (55.4 ± 0.18), fenugreek seed (55 ± 1.25), anise seed (55 ± 2.5), caraway seed (55 ± 2.5), fennel seed (55 ± 2.5), toasted pumpernickel bread (55 ± 0.18), nutmeg (55 ± 2.5), blackstrap molasses (55 ± 0.25), sweetened condensed milk (54.7 ± 0.13), toasted rye bread (54.5 ± 0.23), dried acorns (54.3 ± 0.18), cottage pudding (54.3 ± 0.09), pita bread (54.2 ± 0.13), sponge cake (54.1 ± 0.08), roasted european chestnuts (53.6 ± 0.18), italian bread (53.2 ± 0.18), roasted chinese chestnuts (53.2 ± 0.18), french bread (52.9 ± 0.18), raisin bread (52.8 ± 0.2), potato chips (52.5 ± 0.18), white cake (52.4 ± 0.06), frozen and boiled kale (52.3 ± 0.08), dry instant skim milk (52.2 ± 0.05), whole wheat roll (52.1 ± 0.18), dry skim milk (52 ± 0.17), cheese crackers (52 ± 0.33), toasted whole wheat bread (51.9 ± 0.24), yellow cake (51.9 ± 0.07), peach pie (51.3 ± 0.05), pecan pie (51.3 ± 0.05), gingerbread (51.1 ± 0.08), sweet roll (51 ± 0.12), devil's food cake (50.7 ± 0.08), toasted wheat germ (50.4 ± 0.18), hamburger bun (50.3 ± 0.13), hot dog bun (50.3 ± 0.13), saffron (50 ± 5), cayenne pepper (50 ± 2.5), oregano (50 ± 2.5), coriander seed (50 ± 2.5), bay leaf (50 ± 5), cracked wheat bread (50 ± 0.2), dinner rolls (50 ± 0.18), poultry seasoning

(50 ± 2.5), boston cream pie (49.9 ± 0.05), pineapple upside-down cake (49.9 ± 0.07), raw chinese chestnuts (49.6 ± 0.18), white bread (48.8 ± 0.21), pumpernickel bread (48.1 ± 0.16), rye bread (48 ± 0.2), corn meal muffins (48 ± 0.13), dried sweetened shredded coconut (47.6 ± 0.05), bittersweet chocolate (47.5 ± 0.18), wheat gluten (47.2 ± 0.04), pound cake (47 ± 0.17), 53 proof coffee liqueur (46.9 ± 0.1), mixed grain bread (46.8 ± 0.2), chili powder (46.7 ± 1.67), oatmeal bread (46.4 ± 0.18), raw european chestnuts (46.1 ± 0.18), english muffin (46 ± 0.09), dry buttermilk (45.7 ± 0.71), enriched biscuits (45.7 ± 0.18), unenriched biscuits (45.7 ± 0.18), roasted japanese chestnuts (45.7 ± 0.18), rice bran (45.6 ± 0.05), whole wheat bread (45.6 ± 0.2), cumin seed (45 ± 2.5), dried toasted coconut (45 ± 0.18), mace (45 ± 2.5), danish pastry (44.8 ± 0.12), almond paste (44.3 ± 0.18), lemon chiffon pie (43.8 ± 0.06), pie crust (43.8 ± 0.03), chocolate chiffon pie (43.7 ± 0.06), gjetost cheese (43.2 ± 0.18), raisin pie (43 ± 0.04), freeze-dried parsley (42.9 ± 3.57), corn tortilla (42.7 ± 0.17), whole ground corn meal muffins (42.5 ± 0.13), corn germ (42 ± 0.05), blueberry muffins (42 ± 0.13), bran muffins (41.8 ± 0.13), creme de menthe (41.6 ± 0.1), raw acorns (41.4 ± 0.18), mince pie (41.2 ± 0.04), pancakes with butter and syrup (41.1 ± 0.03), yellow mustard seed (40 ± 1.67), tarragon (40 ± 2.5), celery seed (40 ± 2.5), sage (40 ± 5), dried rosemary (40 ± 2.5), french fried potatoes (40 ± 0.1), onion rings (39.4 ± 0.06), lemon meringue pie (39.4 ± 0.04), dry bakers yeast (39.3 ± 0.18), dried sweetened flaked coconut (39.2 ± 0.2), pineapple chiffon pie (39.1 ± 0.06), brewers yeast (38.9 ± 0.18), hush puppies (38.9 ± 0.11), french fries (38.9 ± 0.06), dry whole milk (38.4 ± 0.16), cherry pie (38.4 ± 0.04), butterscotch pie (38.3 ± 0.04), ginko nuts (38.2 ± 0.18), rhubarb pie (38.2 ± 0.04), pineapple pie (38.1 ± 0.04), torula yeast (37.5 ± 0.18), prune pudding (36.9 ± 0.04), sweet pickles (36.7 ± 0.33), apple pie (36.4 ± 0.04), hamburger pickle relish (36.4 ± 0.18), defatted soy meal (35.9 ± 0.04), raw japanese chestnuts (35.4 ± 0.18), peanut flour (35 ± 1.25), bulgur (35 ± 0.04), blueberry pie (34.9 ± 0.04), potato pancakes (34.7 ± 0.07), cooked taro (34.5 ± 0.08), blackberry pie (34.4 ± 0.04), waffles (34.3 ± 0.07), boiled and steamed chinese chestnuts (34.3 ± 0.18), pancakes (34.1 ± 0.19), frozen and french fried potatoes without skin (34 ± 0.1), defatted soybean flour (33.9 ± 0.05), sapote (33.8 ± 0.02), low fat soybean flour (33.6 ± 0.06), roasted soybean nuts (33.6 ± 0.06), persimmon (33.6 ± 0.2), raw garlic (33.3 ± 0.56), dry roasted cashews (33.2 ± 0.18), dry roasted soybean nuts (32.6 ± 0.06), cheese pizza (32.6 ± 0.04), caramel sundae (32.1 ± 0.03), pineapple custard pie (32.1 ± 0.04), 63 proof coffee liqueur (32.1 ± 0.1), soybean flour (31.9 ± 0.06), falafel (31.8 ± 0.1), cooked plantain (31.2 ± 0.03), frozen and cooked sweetened rhubarb (31.2 ± 0.04), toasted soybean nuts (31.1 ± 0.18), cornbread (30.9 ± 0.06), corn pone (30.9 ± 0.06), strawberry pie (30.9 ± 0.05), banana custard pie (30.7 ± 0.04), pepperoni pizza (30.6 ± 0.04), roasted soybean flour (30.4 ± 0.06), unsweetened baking chocolate (30 ± 0.18), dried coriander leaf (30 ± 5), dried chervil (30 ± 5), burrito (29.8 ± 0.03), apple brown betty (29.7 ± 0.02), chocolate cream pie (29.5 ± 0.04), apple tapioca pudding (29.4 ± 0.02), almond meal (29.3 ± 0.18), hamburger sandwich (29 ± 0.05), oil roasted cashews (28.9 ± 0.18), cheesecake (28.6 ± 0.06), bread pudding (28.4 ± 0.02), piña colada (28.3 ± 0.04), hot fudge sundae (28.2 ± 0.03), boiled and steamed european chestnuts (28.2 ± 0.18), frozen and fried hash brown potatoes (28.1 ± 0.06), cooked prunes (28.1 ± 0.05), low calorie russian salad dressing (28.1 ± 0.31), miso (28 ± 0.04), cashew butter (27.9 ± 0.18), dry

roasted pistachio nuts (27.9 ± 0.18), boiled pink beans (27.9 ± 0.03), strawberry sundae (27.9 ± 0.03), candied sweet potato (27.9 ± 0.05), baked/boiled tropical yam (27.6 ± 0.07), boiled chickpeas (27.4 ± 0.03), poi (27.3 ± 0.04), breadfruit (27.1 ± 0.05), raw cassava (26.9 ± 0.05), french toast (26.5 ± 0.08), toasted sesame kernels (26.4 ± 0.18), canned pumpkin pie mix (26.4 ± 0.04), boiled navy beans (26.3 ± 0.03), whole toasted sesame seeds (26.1 ± 0.18), popover roll (25.8 ± 0.13), boiled pinto beans (25.7 ± 0.03), chocolate pudding (25.7 ± 0.02), cooked brown rice (25.5 ± 0.03), soybean protein concentrate (25.4 ± 0.18), dried pistachio nuts (25.4 ± 0.18), cheeseburger (25.3 ± 0.04), boiled yellow beans (25.3 ± 0.03), tomato catsup (25.3 ± 0.33), baked potato with skin (25.2 ± 0.02), boiled white beans (25.1 ± 0.03), boiled yellow corn (25.1 ± 0.06), orange peel (25 ± 0.83), coconut custard pie (24.9 ± 0.04), boiled adzuki beans (24.8 ± 0.02), chili sauce (24.7 ± 0.33), dry roasted almonds (24.6 ± 0.18), boiled cranberry beans (24.5 ± 0.03), pumpkin pie (24.5 ± 0.04), rennin (24.5 ± 0.45), boiled black turtle beans (24.4 ± 0.03), boiled succotash (24.4 ± 0.05), dry roasted sunflower seeds (24.3 ± 0.18), baked sweet potato (24.3 ± 0.04), boiled sweet potato (24.3 ± 0.03), fish sandwich (24.2 ± 0.03), cooked instant white rice (24.2 ± 0.03), cooked white rice (24.2 ± 0.02), macaroni (24.1 ± 0.04), spaghetti (24.1 ± 0.04), microwaved potato with skin (24.1 ± 0.02), boiled french beans (24 ± 0.03), jackfruit (24 ± 0.05), cherimoya (24 ± 0.01), dried spirulina seaweed (23.9 ± 0.05), dried coconut (23.9 ± 0.18), raw chinese water chestnuts (23.9 ± 0.08), frozen and boiled black-eyed peas (23.8 ± 0.06), cooked extra long grain white rice (23.8 ± 0.06), boiled black beans (23.7 ± 0.03), sweet potato pie (23.7 ± 0.04), sugar apple (23.6 ± 0.03), custard pie (23.4 ± 0.04), frozen and baked sweet potato (23.4 ± 0.06), bananas (23.4 ± 0.04), whole dried sesame seeds (23.3 ± 0.56), poppy seed (23.3 ± 1.67), boiled baby lima beans (23.3 ± 0.03), boiled pigeon peas (23.3 ± 0.03), passion fruit (23.3 ± 0.28), cooked parboiled white rice (23.3 ± 0.03), microwaved potato without skin (23.3 ± 0.03), toasted almonds (23.2 ± 0.18), dry roasted red pistachio nuts (23.2 ± 0.18), frozen white corn (22.9 ± 0.07), hot dog pickle relish (22.9 ± 0.18), boiled kidney beans (22.8 ± 0.03), canned chickpeas (22.6 ± 0.02), roast beef sandwich (22.5 ± 0.03), dry roasted pecans (22.5 ± 0.18), cooked parboiled extra long grain white rice (22.5 ± 0.06), canned white beans (21.9 ± 0.02), boiled sprouted green peas (21.9 ± 0.05), low calorie french salad dressing (21.9 ± 0.31), crunchy peanut butter (21.6 ± 0.16), cooked wild long grain rice (21.5 ± 0.04), baked potato without skin (21.5 ± 0.03), dry roasted peanuts (21.4 ± 0.18), sesame butter (21.3 ± 0.33), almond butter (21.3 ± 0.31), boiled peanuts (21.3 ± 0.16), stir-fried sprouted lentils (21.3 ± 0.05), hash brown potatoes (21.3 ± 0.06), chocolate milkshake (21.2 ± 0.02), canned sweet potato (21.2 ± 0.03), boiled great northern beans (21.1 ± 0.03), boiled split peas (21.1 ± 0.03), boiled yardlong bean (21.1 ± 0.03), boiled burdock root (21.1 ± 0.04), boiled mothbeans (21 ± 0.03), canned great northern beans (21 ± 0.02), baked beans with pork and sweet sauce (21 ± 0.02), ham and cheese sandwich (20.9 ± 0.03), boiled lima beans (20.9 ± 0.03), boiled black-eyed peas (20.8 ± 0.03), dried almonds (20.7 ± 0.18), boiled hyacinth beans (20.7 ± 0.03), canned white corn (20.7 ± 0.05), canned navy beans (20.5 ± 0.02), cream puff (20.5 ± 0.04), vegetarian baked beans (20.5 ± 0.02), restaurant chocolate milkshake (20.5 ± 0.02), frozen and boiled yellow corn (20.5 ± 0.06), boiled catjang black-eyed peas (20.3 ± 0.03), boiled lentils (20.2 ± 0.03), hummus (20.2 ± 0.02), jujube (20.2 ±

0.05), frozen and boiled succotash (20 ± 0.06), frozen creamed yellow corn (20 ± 0.04), boiled potato without skin (20 ± 0.04), steamed hawaii mountain yam (20 ± 0.07), sapodilla (19.9 ± 0.03), crab apples (19.9 ± 0.05), applesauce (19.9 ± 0.04), bread stuffing (19.7 ± 0.03), dried pinyon pine nuts (19.6 ± 0.18), boiled broad beans (19.6 ± 0.03), boiled parsnips (19.5 ± 0.06), bacon cheeseburger (19.4 ± 0.03), boiled and frozen baby lima beans (19.4 ± 0.06), baked beans with pork and tomato sauce (19.4 ± 0.02), canned yellow corn (19.4 ± 0.05), oil roasted filberts (19.3 ± 0.18), boiled mung beans (19.2 ± 0.02), figs (19.2 ± 0.1), oil roasted peanuts (18.9 ± 0.18), dried sunflower seeds (18.9 ± 0.18), fried chicken (18.9 ± 0.06), tomato paste (18.9 ± 0.04), restaurant strawberry milkshake (18.9 ± 0.02), frozen and boiled fordhook lima beans (18.8 ± 0.06), dried english/persian walnuts (18.6 ± 0.18), dried hickory nuts (18.6 ± 0.18), dried pecans (18.6 ± 0.18), japanese persimmon (18.6 ± 0.03), refried beans (18.5 ± 0.02), elderberries (18.4 ± 0.03), boiled mungo beans (18.3 ± 0.03), dried pumpkin and squash seeds (18.2 ± 0.18), dry roasted filberts (18.2 ± 0.18), canned yellow corn with red and green peppers (18.2 ± 0.04), canned creamed yellow corn (18.1 ± 0.04), restaurant vanilla milkshake (18 ± 0.02), bread and butter pickles (18 ± 0.33), raw potato without skin (17.9 ± 0.04), vanilla milkshake (17.8 ± 0.02), european grapes (17.8 ± 0.03), thompson seedless grapes (17.8 ± 0.03), canned succotash with creamed corn (17.6 ± 0.04), oil roasted spanish peanuts (17.5 ± 0.18), raw jerusalem artichoke (17.5 ± 0.07), prune juice (17.5 ± 0.02), raw lotus root (17.3 ± 0.06), american grapes (17.2 ± 0.05), tapioca cream pudding (17.1 ± 0.03), pomegranate (17.1 ± 0.03), tempeh (17 ± 0.06), raw shallots (17 ± 0.5), mango (17 ± 0.02), soursop (16.8 ± 0.02), mashed potatoes (16.7 ± 0.05), lemon peel (16.7 ± 0.83), low calorie thousand island salad dressing (16.7 ± 0.33), hot dog (16.6 ± 0.06), cherries (16.6 ± 0.07), canned black turtle beans (16.5 ± 0.02), lychees (16.5 ± 0.05), oil roasted valencia peanuts (16.4 ± 0.18), dried peanuts (16.4 ± 0.18), oil roasted pecans (16.4 ± 0.18), kumquats (16.3 ± 0.26), pickled beets (16.3 ± 0.04), breaded and fried fish (16.2 ± 0.07), oil roasted virginia peanuts (16.1 ± 0.18), oil roasted almonds (16.1 ± 0.18), teriyaki sauce (16.1 ± 0.28), boiled lotus root (16.1 ± 0.06), red beans (16 ± 0.04), vanilla pudding (15.9 ± 0.02), maltex (15.9 ± 0.03), boiled arrowhead (15.8 ± 0.42), pear nectar (15.8 ± 0.02), dried watermelon seeds (15.7 ± 0.18), dried filberts (15.7 ± 0.18), taco (15.7 ± 0.06), light mayonnaise (15.7 ± 0.36), creamy peanut butter (15.6 ± 0.31), boiled green peas (15.6 ± 0.06), java plum (15.6 ± 0.04), cooked whelk (15.5 ± 0.06), o'brien potatoes (15.5 ± 0.03), boiled salsify (15.4 ± 0.07), black currants (15.4 ± 0.09), raw coconut (15.3 ± 0.11), quince (15.3 ± 0.05), raw apple with skin (15.3 ± 0.04), enchilada (15.1 ± 0.02), canned cranberry beans (15.1 ± 0.02), longans (15.1 ± 0.05), pear (15.1 ± 0.03), oil roasted sunflower seeds (15 ± 0.18), canned butter beans (15 ± 0.04), raw ginger root (15 ± 0.21), thousand island salad dressing (15 ± 0.31), grape juice (15 ± 0.02), boiled winged beans (14.9 ± 0.03), canned kidney beans (14.9 ± 0.02), canned lima beans (14.9 ± 0.02), kiwi fruit (14.9 ± 0.07), cranberry juice cocktail (14.9 ± 0.02), raw apple without skin (14.8 ± 0.04), dried and frozen tofu (14.7 ± 0.29), baked acorn squash (14.6 ± 0.05), raw green peas (14.5 ± 0.06), canned pinto beans (14.5 ± 0.02), frozen and boiled potatoes without skin (14.5 ± 0.05), yellow passion fruit juice (14.5 ± 0.02), papaya nectar (14.5 ± 0.02), boiled dishcloth gourd (14.4 ± 0.06), apricot nectar (14.4 ± 0.02), restaurant cranberry juice cocktail (14.4 ± 0.03), dried pignolia pine nuts (14.3 ± 0.18), natto (14.3 ± 0.06), frozen and boiled green peas (14.3 ± 0.06), cooked shitake mushrooms (14.3 ± 0.07), raw leeks (14.2 ± 0.19), blueberries (14.1 ± 0.03), canned succotash (14 ± 0.04), dried macadamia nuts (13.9 ± 0.18), peach nectar (13.9 ± 0.02), red currants (13.8 ± 0.09), white currants (13.8 ± 0.09), pineapple juice (13.8 ± 0.02), canned potato without skin (13.7 ± 0.06), roasted pumpkin and squash seeds (13.6 ± 0.18), canned black-eyed peas (13.6 ± 0.02), purple passion fruit juice (13.6 ± 0.02), cooked apple without skin (13.6 ± 0.03), eggnog (13.5 ± 0.02), carissa (13.5 ± 0.25), maypo (13.3 ± 0.03), oil roasted macadamia nuts (13.2 ± 0.18), loganberries (13 ± 0.03), instant corn grits (13 ± 0.04), corn grits (13 ± 0.02), plum (13 ± 0.08), dried brazil nuts (12.9 ± 0.18), boiled and steamed japanese chestnuts (12.9 ± 0.18), yellow corn pudding (12.8 ± 0.02), blackberries (12.8 ± 0.07), cranberry sauce (12.7 ± 0.05), canned green peas (12.6 ± 0.06), okara tofu (12.6 ± 0.08), barbecue sauce (12.5 ± 0.31), coleslaw (12.5 ± 0.08), mammy apple (12.5 ± 0.05), canned broad beans (12.4 ± 0.02), canned chinese water chestnuts (12.4 ± 0.07), pineapple (12.4 ± 0.03), raw irish moss seaweed (12.3 ± 0.05), orange soda (12.3 ± 0.01), boysenberries (12.2 ± 0.04), white hominy (12.2 ± 0.02), dried butternuts (12.1 ± 0.18), dried black walnuts (12.1 ± 0.18), pork and beef spread (12.1 ± 0.18), mayonnaise (12.1 ± 0.36), loquats (12.1 ± 0.05), valencia orange (11.9 ± 0.04), guava (11.9 ± 0.06), sweet dessert wine (11.9 ± 0.08), yellow hominy with red and green peppers (11.8 ± 0.02), nectarine (11.8 ± 0.04), cooked kingfish (11.7 ± 0.05), restaurant coleslaw (11.7 ± 0.05), apple juice (11.7 ± 0.02), breaded and fried oysters (11.6 ± 0.06), navel orange (11.6 ± 0.04), breaded and fried shrimp (11.5 ± 0.06), yellow hominy (11.5 ± 0.02), cream of rice (11.5 ± 0.03), raspberries (11.5 ± 0.04), roselle (11.4 ± 0.09), evaporated skim milk (11.3 ± 0.16), au gratin potatoes (11.2 ± 0.04), potato salad (11.2 ± 0.04), ground-cherries (11.2 ± 0.04), tangerine (11.2 ± 0.06), fried abalone (11.1 ± 0.06), boiled green soybeans (11.1 ± 0.06), compressed bakers yeast (11.1 ± 0.18), custard (11.1 ± 0.02), cream of wheat (11.1 ± 0.03), apricots (11.1 ± 0.05), peach (11.1 ± 0.06), whole chocolate malted milk (11 ± 0.02), cola (11 ± 0.01), fried tofu (10.8 ± 0.38), scalloped potatoes (10.8 ± 0.04), oatmeal (10.8 ± 0.03), baked hubbard squash (10.8 ± 0.05), restaurant orange juice (10.8 ± 0.04), ham salad (10.7 ± 0.18), russian salad dressing (10.7 ± 0.33), cooked enriched white degermed corn meal (10.7 ± 0.02), cooked unenriched corn meal (10.7 ± 0.02), cooked enriched yellow degermed corn meal (10.7 ± 0.02), cooked unenriched yellow degermed corn meal (10.7 ± 0.02), surimi scallops (10.6 ± 0.06), stir-fried sprouted mung beans (10.6 ± 0.08), farina (10.6 ± 0.03), lime (10.6 ± 0.07), boiled carrots (10.5 ± 0.06), boiled butternut squash (10.5 ± 0.05), lemon-lime soda (10.5 ± 0.01), breaded and fried clams (10.4 ± 0.06), whole chocolate milk (10.4 ± 0.02), lowfat 2% chocolate milk (10.4 ± 0.02), lowfat 1% chocolate milk (10.4 ± 0.02), boiled artichoke hearts (10.4 ± 0.06), boiled yambean (10.4 ± 0.05), orange juice (10.4 ± 0.02), hot chocolate (10.3 ± 0.02), instant oatmeal (10.2 ± 0.03), marinara sauce (10.2 ± 0.02), gooseberries (10.2 ± 0.03), evaporated whole milk (10.1 ± 0.04), frozen and boiled butternut squash (10.1 ± 0.04), raw carrots (10.1 ± 0.07), tangerine juice (10.1 ± 0.02), light cream (10 ± 0.17), dried sesame kernels (10 ± 0.63), breaded and fried scallops (10 ± 0.16), pickled herring (10 ± 0.33), malted milk (10 ± 0.02), tomato purée (10 ± 0.02), italian salad dressing (10 ± 0.33), boiled soybeans (9.9 ± 0.03), boiled lupins (9.9 ± 0.03), mulberries (9.8 ± 0.04), papaya (9.8 ± 0.02), raw red chili peppers (9.6 ± 0.11), raw green chili peppers (9.6 ± 0.11), raw kelp/kombu/

tangle seaweed (9.6 ± 0.05), prickly pear (9.6 ± 0.05), pummelo (9.6 ± 0.03), mandarin oranges (9.6 ± 0.04), tuna salad (9.4 ± 0.02), stir-fried sprouted soybeans (9.4 ± 0.05), evaporated lowfat milk (9.4 ± 0.04), fried chicken back with skin (9.3 ± 0.07), pork olive loaf (9.3 ± 0.18), raw dandelion greens (9.3 ± 0.18), horseradish sauce (9.3 ± 0.33), lemon (9.3 ± 0.09), carrot juice (9.3 ± 0.03), surimi shrimp (9.2 ± 0.06), chili (9.2 ± 0.02), white grapefruit juice (9.2 ± 0.02), red grapefruit juice (9.2 ± 0.02), boiled artichoke (9.1 ± 0.06), raw wakame seaweed (9.1 ± 0.05), lime juice (9 ± 0.02), dry roasted macadamia nuts (8.9 ± 0.18), raw florida avocado (8.9 ± 0.02), carob milk (8.8 ± 0.02), boiled acorn squash (8.8 ± 0.04), boiled brussels sprouts (8.7 ± 0.06), screwdriver (8.6 ± 0.02), lemon juice (8.6 ± 0.02), tequila sunrise (8.5 ± 0.03), shoyu sauce (8.4 ± 0.09), cantaloupe (8.4 ± 0.03), white grapefruit (8.4 ± 0.04), frozen and boiled brussels sprouts (8.3 ± 0.06), sweetened coconut cream (8.3 ± 0.02), frozen and boiled okra (8.2 ± 0.05), frozen and boiled carrots (8.2 ± 0.07), canned pumpkin (8.1 ± 0.04), breaded and fried catfish (8 ± 0.06), fried beef liver (7.9 ± 0.05), cooked oysters (7.9 ± 0.06), boiled green beans (7.9 ± 0.08), boiled yellow snap beans (7.9 ± 0.08), raw whelk (7.8 ± 0.06), breaded fried squid (7.8 ± 0.06), soy sauce (7.8 ± 0.09), boiled rutabaga (7.8 ± 0.06), carambola (7.8 ± 0.04), skim yogurt (7.7 ± 0.02), honeydew melon (7.7 ± 0.05), boiled leeks (7.7 ± 0.19), pink grapefruit (7.7 ± 0.04), raw acerola (7.7 ± 0.05), pork mothers loaf (7.6 ± 0.24), breaded and fried croaker (7.5 ± 0.06), pitanga (7.5 ± 0.03), cooked mussels (7.4 ± 0.06), gefiltefish with broth (7.4 ± 0.12), dried longans (7.4 ± 0.05), raw onions (7.4 ± 0.06), boiled okra (7.3 ± 0.06), boiled jute potherb (7.2 ± 0.12), frozen and boiled collards (7.2 ± 0.06), tomato sauce (7.2 ± 0.04), watermelon (7.2 ± 0.03), chicken and turkey salad (7.1 ± 0.18), lowfat yogurt (7 ± 0.02), boiled black-eyed-pea pods (7 ± 0.11), raw parsley (7 ± 0.17), gin & tonic (7 ± 0.02), chicken frankfurter (6.9 ± 0.11), cooked tahitian taro (6.9 ± 0.07), raw california avocado (6.9 ± 0.03), oheloberries (6.9 ± 0.04), canned chicken liver pâté (6.8 ± 0.18), canned zucchini in tomato juice (6.8 ± 0.04), human milk (6.8 ± 0.16), raw pepeao (6.8 ± 0.05), raw agar seaweed (6.8 ± 0.05), daiquiri (6.8 ± 0.08), coconut cream (6.7 ± 0.02), boiled kohlrabi (6.7 ± 0.06), boiled beets (6.7 ± 0.06), boiled eggplant (6.7 ± 0.1), frozen and boiled onions (6.7 ± 0.05), steamed sprouted soybeans (6.6 ± 0.11), barbecue loaf (6.5 ± 0.22), canned shitake mushrooms (6.5 ± 0.04), raw welsh onions (6.5 ± 0.05), canned stewed red tomatoes (6.5 ± 0.04), pickled rip manzanillo/mission olives (6.5 ± 1.09), battered and fried shark (6.4 ± 0.06), boiled hubbard squash (6.4 ± 0.04), boiled/baked spaghetti squash (6.4 ± 0.06), raw goose liver (6.3 ± 0.05), boiled dandelion greens (6.3 ± 0.1), raw eggplant (6.3 ± 0.12), boiled onions (6.3 ± 0.05), cheese soufflé (6.2 ± 0.05), canned shellie beans (6.2 ± 0.04), frozen and boiled yellow snap beans (6.2 ± 0.07), frozen and boiled green beans (6.2 ± 0.07), casaba melon (6.2 ± 0.03), canned red chili peppers (6.2 ± 0.07), canned green chili peppers (6.2 ± 0.07), pork pickle and pimento loaf (6.1 ± 0.18), raw abalone (6 ± 0.06), brown mustard (6 ± 1), yellow mustard (6 ± 1), raw sprouted mung beans (6 ± 0.1), raw savoy cabbage (6 ± 0.14), raw red cabbage (6 ± 0.14), cider vinegar (6 ± 0.33), raw celeriac (5.9 ± 0.05), pork and beef old fashioned loaf (5.7 ± 0.18), boiled kale (5.7 ± 0.08), boiled scotch kale (5.7 ± 0.08), raw turnip greens (5.7 ± 0.18), boiled red tomato (5.7 ± 0.04), rose apple (5.7 ± 0.05), protein fortified skim milk (5.6 ± 0.02), raw garden cress (5.6 ± 0.2), raw spring onions with tops (5.6 ± 0.1), pickled rip sevillano/ascolano olives (5.6 ± 0.31), whiskey sour

CARBOHYDRATE

(5.6 ± 0.06), tamari sauce (5.5 ± 0.09), protein fortified lowfat 2% milk (5.5 ± 0.02), protein fortified lowfat 1% milk (5.5 ± 0.02), boiled broccoli (5.5 ± 0.06), coconut milk (5.5 ± 0.02), boiled savoy cabbage (5.5 ± 0.07), frozen and boiled crookneck squash (5.5 ± 0.05), canned carrots (5.5 ± 0.07), pork and beef honey loaf (5.4 ± 0.18), chicken salad (5.4 ± 0.18), frozen and boiled spinach (5.4 ± 0.05), boiled beet greens (5.4 ± 0.07), raw green cabbage (5.4 ± 0.14), restaurant hot chocolate (5.4 ± 0.02), raw red sweet peppers (5.4 ± 0.1), raw green sweet peppers (5.4 ± 0.1), sheep milk (5.3 ± 0.02), frozen and boiled broccoli (5.3 ± 0.05), raw bamboo shoots (5.3 ± 0.07), boiled cardoon (5.3 ± 0.05), distilled vinigar (5.3 ± 0.33), indian buffalo milk (5.2 ± 0.02), raw broccoli (5.2 ± 0.11), raw laver/nori seaweed (5.1 ± 0.05), boiled mushrooms (5.1 ± 0.06), raw green tomato (5.1 ± 0.04), whey (5.1 ± 0.02), boiled chayote (5.1 ± 0.06), raw frozen rhubarb (5.1 ± 0.04), cheshire cheese (5 ± 0.18), pork luxury loaf (5 ± 0.18), pork and beef picnic loaf (5 ± 0.18), frozen and boiled turnip greens (5 ± 0.06), boiled lamb's-quarters (5 ± 0.06), raw cauliflower (5 ± 0.1), skim milk (4.9 ± 0.02), boiled turnips (4.9 ± 0.06), canned jalapeño peppers (4.9 ± 0.07), boiled pumpkin (4.9 ± 0.04), lowfat 2% milk (4.8 ± 0.02), lowfat 1% milk (4.8 ± 0.02), buttermilk (4.8 ± 0.02), filled milk (4.8 ± 0.02), frozen and boiled asparagus (4.8 ± 0.08), boiled green cabbage (4.8 ± 0.07), acerola juice (4.8 ± 0.02), whole yogurt (4.7 ± 0.02), whole milk (4.7 ± 0.02), boiled cauliflower (4.7 ± 0.08), raw chickory greens (4.7 ± 0.06), boiled red cabbage (4.7 ± 0.07), low calorie italian salad dressing (4.7 ± 0.33), pork and beef peppered loaf (4.6 ± 0.18), canned goose liver pâté (4.6 ± 0.38), canned smoked goose liver pâté (4.6 ± 0.18), raw mushrooms (4.6 ± 0.14), canned green beans (4.6 ± 0.07), canned yellow snap beans (4.6 ± 0.07), goat milk (4.5 ± 0.02), fried chicken giblets (4.4 ± 0.05), boiled asparagus (4.4 ± 0.06), frozen and boiled turnips (4.4 ± 0.05), anchovy paste (4.3 ± 0.71), firm raw tofu (4.3 ± 0.04), feta cheese (4.3 ± 0.18), boiled turnip greens (4.3 ± 0.07), boiled crookneck squash (4.3 ± 0.06), canned peeled red tomatoes (4.3 ± 0.04), raw red tomato (4.3 ± 0.04), canned sauerkraut (4.3 ± 0.04), fried chicken neck with skin (4.2 ± 0.14), sour cream (4.2 ± 0.42), cooked sprouted mung beans (4.2 ± 0.08), restaurant tomato juice (4.2 ± 0.04), tomato juice (4.2 ± 0.03), fried dark meat chicken with skin (4.1 ± 0.05), boiled amaranth (4.1 ± 0.08), boiled swiss chard (4.1 ± 0.06), raw oriental radish (4.1 ± 0.11), dry dessert wine (4.1 ± 0.08), parmesan cheese (4 ± 1), half and half cream (4 ± 0.33), raw crookneck squash (4 ± 0.08), boiled red sweet peppers (4 ± 0.07), boiled green sweet peppers (4 ± 0.07), dried pilinuts (3.9 ± 0.18), canned oysters (3.9 ± 0.06), raw oysters (3.9 ± 0.06), raw sprouted alfalfa seeds (3.9 ± 0.15), raw mustard spinach (3.9 ± 0.07), frozen and boiled red sweet peppers (3.9 ± 0.05), frozen and boiled green sweet peppers (3.9 ± 0.05), canned red sweet peppers (3.9 ± 0.07), canned green sweet peppers (3.9 ± 0.07), boiled zucchini (3.9 ± 0.06), braised pork liver (3.8 ± 0.05), caviar (3.8 ± 0.31), turkey salad (3.8 ± 0.09), tuna salad spread (3.8 ± 0.09), boiled spinach (3.8 ± 0.06), boiled garden cress (3.8 ± 0.07), frozen and boiled cauliflower (3.8 ± 0.06), raw scallop squash (3.8 ± 0.08), wax beans (3.8 ± 0.04), raw celery (3.8 ± 0.13), boiled swamp cabbage (3.7 ± 0.1), raw celtuce (3.7 ± 0.05), coconut water (3.7 ± 0.02), boiled calabash gourd (3.7 ± 0.07), beer (3.7 ± 0.01), romano cheese (3.6 ± 0.18), swiss cheese (3.6 ± 0.18), raw mussels (3.6 ± 0.06), raw spinach (3.6 ± 0.18), boiled purslane (3.6 ± 0.09), raw looseleaf lettuce (3.6 ± 0.18), frozen and boiled zucchini (3.6 ± 0.04), raw radish (3.6 ± 0.11), french salad dressing (3.6 ± 0.36), raw duck liver (3.5 ± 0.05), boiled celery (3.5 ± 0.07), braised beef liver (3.4 ± 0.05), simmered turkey liver (3.4 ± 0.05), canned spinach (3.4 ± 0.05), raw vine spinach (3.4 ± 0.05), boiled oriental radish (3.4 ± 0.07), pork and beef pepperoni (3.3 ± 0.83), pork and beef mortadella (3.3 ± 0.33), pork liver sausage (3.3 ± 0.28), raw chives (3.3 ± 1.67), medium cream (3.3 ± 0.33), boiled scallop squash (3.3 ± 0.06), fried chicken thigh with skin (3.2 ± 0.08), caraway cheese (3.2 ± 0.18), beef pastrami (3.2 ± 0.18), raw swamp cabbage (3.2 ± 0.09), canned bamboo shoots (3.2 ± 0.04), raw endive (3.2 ± 0.2), bloody mary (3.2 ± 0.03), malt liquor (3.2 ± 0.01), manhattan (3.2 ± 0.09), unheated ham (3.1 ± 0.05), raw squid (3.1 ± 0.06), pork and beef sausage (3.1 ± 0.38), ricotta cheese (3.1 ± 0.04), frozen and boiled mustard greens (3.1 ± 0.07), raw dock (3.1 ± 0.07), raw chickory (3.1 ± 0.11), ale (3.1 ± 0.01), hard pork and beef salami (3 ± 0.5), beef salami (3 ± 0.22), beef summer sausage (3 ± 0.22), pork and beef brotwurst (3 ± 0.07), boiled pokeberry shoots (3 ± 0.06), canned crookneck squash (3 ± 0.05), boiled waxgourd (3 ± 0.06), broiled lamb liver (2.9 ± 0.11), brick cheese (2.9 ± 0.18), beef lunch meat (2.9 ± 0.18), cottage cheese (2.9 ± 0.18), neufchatel cheese (2.9 ± 0.18), cream cheese (2.9 ± 0.18), frozen and boiled turnip greens and turnips (2.9 ± 0.05), boiled dock (2.9 ± 0.05), safflower oil and soybean oil mayonnaise (2.9 ± 0.36), soybean oil mayonnaise (2.9 ± 0.36), raw zucchini (2.9 ± 0.08), raw cucumber (2.9 ± 0.1), fruit pectin (2.9 ± 0.36), boiled mustard spinach (2.8 ± 0.06), cooked black bass (2.7 ± 0.04), beef and pork frankfurter (2.7 ± 0.11), whipping cream, heavy (2.7 ± 0.33), whipping cream, light (2.7 ± 0.33), raw butterhead lettuce (2.7 ± 0.33), fried dark meat chicken without skin (2.6 ± 0.05), lebanon bologna (2.6 ± 0.22), berliner (2.6 ± 0.22), raw clams (2.6 ± 0.06), beef and pork bologna (2.6 ± 0.22), boiled collards (2.6 ± 0.03), fried chicken wing with skin (2.5 ± 0.16), colby cheese (2.5 ± 0.18), bleu cheese (2.5 ± 0.18), light meat chicken roll (2.5 ± 0.05), chicken roll (2.5 ± 0.09), pork and beef lunch meat (2.5 ± 0.18), raw coriander (2.5 ± 1.25), raw romaine lettuce (2.5 ± 0.18), raw new zealand spinach (2.5 ± 0.18), raw scallops (2.4 ± 0.06), raw spirulina seaweed (2.4 ± 0.05), canned turnip greens (2.4 ± 0.04), smoked beef sausage (2.3 ± 0.12), pork and beef kielbasa/kolbassy (2.3 ± 0.19), raw chinese cabbage (2.3 ± 0.14), beef honey roll sausage (2.2 ± 0.22), beef and pork summer sausage (2.2 ± 0.22), raw octopus (2.2 ± 0.06), pork beerwurst salami (2.2 ± 0.22), pork liverwurst (2.2 ± 0.28), beef and pork salami (2.2 ± 0.22), boiled new zealand spinach (2.2 ± 0.06), dill pickles (2.2 ± 0.08), boiled butterbur (2.2 ± 0.05), simmered turkey heart (2.1 ± 0.05), simmered turkey giblets (2.1 ± 0.05), provolone cheese (2.1 ± 0.18), gouda cheese (2.1 ± 0.18), pork link sausage (2.1 ± 0.07), roquefort cheese (2.1 ± 0.18), mozzarella cheese (2.1 ± 0.18), raw smelt (2.1 ± 0.05), light and dark meat turkey roll (2.1 ± 0.09), ham and cheese spread (2.1 ± 0.18), pork bratwurst (2.1 ± 0.06), turkey bologna (2.1 ± 0.18), boiled mustard greens (2.1 ± 0.07), canned sprouted mung beans (2.1 ± 0.08), hard pork salami (2 ± 0.5), raw iceberg lettuce (2 ± 0.25), sour pickles (2 ± 0.08), minced ham lunch meat (1.9 ± 0.24), canned pork lunch meat (1.9 ± 0.24), vienna sausage (1.9 ± 0.31), raw tofu (1.9 ± 0.04), boiled bamboo shoots (1.9 ± 0.04), fried light meat chicken with skin (1.8 ± 0.05), tilsit cheese (1.8 ± 0.18), american cheese (1.8 ± 0.18), pimento cheese spread (1.8 ± 0.18), unheated canadian bacon (1.8 ± 0.09), chopped smoked beef (1.8 ± 0.18), polish pork sausage (1.8 ± 0.18), beef frankfurter (1.8 ± 0.09), pork and beef knockwurst (1.8 ± 0.07), soy milk (1.8 ± 0.02), boiled chinese cabbage (1.8 ± 0.06), pork and beef luncheon sausage (1.7 ±

0.22), grilled ham patties (1.7 ± 0.08), unheated ham patties (1.7 ± 0.08), beef beerwurst salami (1.7 ± 0.22), red table wine (1.7 ± 0.05), fried chicken breast with skin (1.6 ± 0.05), fried chicken drumstick with skin (1.6 ± 0.1), turkey pastrami (1.6 ± 0.09), roasted lean ham (1.5 ± 0.05), italian pork sausage (1.5 ± 0.07), pork and beef link sausage (1.5 ± 0.07), rose table wine (1.5 ± 0.05), chipped beef (1.4 ± 0.18), fontina cheese (1.4 ± 0.18), cheddar cheese (1.4 ± 0.18), edam cheese (1.4 ± 0.18), lobster paste (1.4 ± 0.71), turkey summer sausage (1.4 ± 0.18), ham and cheese roll (1.4 ± 0.18), pork and beef link sausage with american cheese (1.4 ± 0.12), duck egg (1.4 ± 0.07), tom collins (1.4 ± 0.02), grilled canadian bacon (1.3 ± 0.11), cooked lobster (1.3 ± 0.06), goose egg (1.3 ± 0.03), turkey frankfurter (1.3 ± 0.11), chicken egg white (1.2 ± 0.15), raw watercress (1.2 ± 0.29), simmered chicken gizzard (1.1 ± 0.05), muenster cheese (1.1 ± 0.18), havarti cheese (1.1 ± 0.18), pork sausage (1.1 ± 0.19), turkey egg (1.1 ± 0.06), pickled green olives (1.1 ± 0.11), braised lamb heart (1 ± 0.03), simmered chicken giblets (1 ± 0.05), simmered beef kidneys (1 ± 0.05), unheated lean ham (1 ± 0.05), simmered chicken liver (0.9 ± 0.05), raw shrimp (0.9 ± 0.06), pork bologna (0.9 ± 0.22), beef bologna (0.9 ± 0.22), raw cuttlefish (0.8 ± 0.06), chicken egg (0.8 ± 0.1), white table wine (0.8 ± 0.05), monterey cheese (0.7 ± 0.18), port du salut cheese (0.7 ± 0.18), raw dungeness crab (0.7 ± 0.06), cooked bacon (0.6 ± 0.04), simmered turkey gizzard (0.6 ± 0.05), fried chicken breast without skin (0.5 ± 0.06), roasted canned lean ham (0.5 ± 0.05), crab cakes (0.5 ± 0.08), raw lobster (0.5 ± 0.06), light meat turkey roll (0.5 ± 0.09), turkey salami (0.5 ± 0.09), coffee (0.5 ± 0.03), fried light meat chicken without skin (0.4 ± 0.05), gruyere cheese (0.4 ± 0.18), simmered beef heart (0.4 ± 0.05), braised pork heart (0.4 ± 0.04), brie cheese (0.4 ± 0.18), roasted canned ham (0.4 ± 0.05), limburger cheese (0.4 ± 0.18), camembert cheese (0.4 ± 0.18), pickled anchovy (0.4 ± 0.18), turkey ham (0.4 ± 0.09), roasted pork blade roll (0.4 ± 0.05), pork headcheese (0.4 ± 0.18), summer sausage (0.4 ± 0.22), simmered beef tongue (0.3 ± 0.05), martini (0.3 ± 0.07), pork livercheese (0.2 ± 0.13), chamomile tea (0.2 ± 0.03), black tea (0.2 ± 0.03), instant tea (0.2 ± 0.02), 86 proof whiskey (0.2 ± 0.12), roasted veal leg (0.1 ± 0.06), simmered chicken heart (0.1 ± 0.05), cooked corned beef brisket (0.1 ± 0.05), raw bacon (0.1 ± 0.07), clam liquid (0.1 ± 0.02)

Applications: constipation, diarrhea, diverticulitis, fatigue

Daily Dosage: (1 g of carbohydrate = 4 calories) Nutritional — 277–390 g; therapeutic — ?; experimental — ?; toxic — none known

Warnings: Keep levels low with epilepsy. Don't use cellulose during gastritis or hemorrhoids. Keep levels low with tuberculosis. Excessive amounts can cause side-effects. See side-effects symptoms.

ENERGY

Names: energy, heat, calories, Kilocalories

Classification: energy
Deficiency: fatigue

IV. Macronutrients

ENERGY

Side-effects: see *toxicity* symptoms
Toxicity: fat accumulation, overweight, heart disease
Sources (kilocalories per 100 grams of food): (Note that kilocalories are often abbreviated "calories" when talking about food.) beef tallow (892 ± 3.8), mutton tallow (892 ± 3.8), duck fat (885 ± 3.8), chicken fat (885 ± 3.8), goose fat (885 ± 3.8), lard (885 ± 3.8), turkey fat (885 ± 3.8), beef suet (864 ± 1.8), almond oil (857 ± 3.6), coconut oil (857 ± 3.6), corn oil (857 ± 3.6), cottonseed oil (857 ± 3.6), palm oil (857 ± 3.6), palm kernel oil (857 ± 3.6), safflower oil (857 ± 3.6), sesame oil (857 ± 3.6), soybean oil (857 ± 3.6), soybean lecithin (857 ± 3.6), sunflower oil (857 ± 3.6), wheat germ oil (857 ± 3.6), olive oil (850 ± 3.6), peanut oil (850 ± 3.6), whipped butter (736 ± 4.5), oil roasted macadamia nuts (729 ± 1.8), dried pilinuts (729 ± 1.8), butter (720 ± 3.3), sweet butter (720 ± 3.3), dried macadamia nuts (711 ± 1.8), safflower oil and soybean oil mayonnaise (707 ± 3.6), soybean oil mayonnaise (707 ± 3.6), oil roasted pecans (696 ± 1.8), dry roasted macadamia nuts (689 ± 1.8), soybean oil and cottonseed oil margarine liquid (680 ± 10), corn oil margarine stick (680 ± 10), safflower oil and soybean oil margarine stick (680 ± 10), soybean oil margarine stick (680 ± 10), soybean oil and cottonseed oil margarine stick (680 ± 10), corn oil margarine tub (680 ± 10), safflower oil margarine tub (680 ± 10), soybean oil margarine tub (680 ± 10), soybean oil and cottonseed oil margarine tub (680 ± 10), dried pecans (679 ± 1.8), dry roasted filberts (671 ± 1.8), dried coconut (668 ± 1.8), oil roasted filberts (668 ± 1.8), dried hickory nuts (668 ± 1.8), dry roasted pecans (668 ± 1.8), dried brazil nuts (664 ± 1.8), raw pork jowl (655 ± 0.5), potato pancakes (651 ± 0.7), dried english/persian walnuts (650 ± 1.8), dried filberts (639 ± 1.8), almond butter (631 ± 3.1), french salad dressing (629 ± 3.6), oil roasted almonds (629 ± 1.8), oil roasted sunflower seeds (625 ± 1.8), dried butternuts (621 ± 1.8), dry roasted pistachio nuts (614 ± 1.8), dried black walnuts (614 ± 1.8), dried toasted coconut (600 ± 1.8), nutmeg (600 ± 25), dried almonds (596 ± 1.8), dry roasted almonds (596 ± 1.8), toasted almonds (596 ± 1.8), cashew butter (596 ± 1.8), creamy peanut butter (594 ± 3.1), sesame butter (593 ± 3.3), oil roasted peanuts (589 ± 1.8), oil roasted valencia peanuts (589 ± 1.8), dry roasted sunflower seeds (589 ± 1.8), crunchy peanut butter (588 ± 1.6), dried sesame kernels (588 ± 6.3), dry roasted peanuts (586 ± 1.8), dried pistachio nuts (586 ± 1.8), dry roasted cashews (582 ± 1.8), oil roasted cashews (582 ± 1.8), oil roasted spanish peanuts (579 ± 1.8), dried sunflower seeds (579 ± 1.8), whole dried sesame seeds (578 ± 5.6), cooked bacon (576 ± 0.4), dried peanuts (575 ± 1.8), oil roasted virginia peanuts (575 ± 1.8), dried pinyon pine nuts (575 ± 1.8), toasted sesame kernels (575 ± 1.8), whole toasted sesame seeds (575 ± 1.8), dry roasted red pistachio nuts (564 ± 1.8), dried watermelon seeds (564 ± 1.8), raw bacon (556 ± 0.7), dried pumpkin and squash seeds (550 ± 1.8), cheese crackers (540 ± 3.3), sweet chocolate (536 ± 1.8), potato chips (529 ± 1.8), roasted pumpkin and squash seeds (529 ± 1.8), milk chocolate (525 ± 1.8), shortbread cookie (525 ± 6.3), dried pignolia pine nuts (521 ± 1.8), dried acorns (518 ± 1.8), semi-sweet chocolate (514 ± 1.8), russian salad dressing (507 ± 3.3), acorn flour (507 ± 1.8), german sweet bakers chocolate (504 ± 1.8), dried sweetened shredded coconut (501 ± 0.5), pie crust (500 ± 0.3), yellow mustard seed (500 ± 16.7), poppy seed (500 ± 16.7), cheese crackers with peanut butter (498 ± 1.2), dry whole milk (497 ± 1.6), unsweetened baking chocolate (496 ± 1.8), mayonnaise (493 ± 3.6), corn germ (490 ± 0.5), gingersnaps cookie (486 ± 7.1), bittersweet chocolate (482 ± 1.8), dried and frozen tofu (482 ± 2.9), semi-sweet bakers choco-

late (479 ± 1.8), oatmeal cookie (477 ± 3.8), pound cake (473 ± 1.7), gjetost cheese (471 ± 1.8), braised beef shortribs including fat (471 ± 0.5), roasted soybean nuts (471 ± 0.6), canned smoked goose liver pâté (468 ± 1.8), canned goose liver pâté (462 ± 3.8), toasted soybean nuts (461 ± 1.8), parmesan cheese (460 ± 10), italian salad dressing (460 ± 3.3), dried sweetened flaked coconut (456 ± 2), taco shell (455 ± 4.5), tostada shell (455 ± 4.5), granola bar (454 ± 2.1), almond paste (454 ± 1.8), pork and beef pepperoni (450 ± 8.3), dry roasted soybean nuts (450 ± 0.6), soda crackers (446 ± 1.8), sugar cookie (444 ± 3.1), roasted soybean flour (439 ± 0.6), rice bran (438 ± 0.5), soybean flour (433 ± 0.6), saltine crackers (433 ± 8.3), havarti cheese (432 ± 1.8), animal crackers (431 ± 1.9), brownies (430 ± 2.5), graham crackers (429 ± 3.6), zwieback crackers (429 ± 7.1), peanut brittle (425 ± 1.8), rusk crackers (422 ± 5.6), hard pork and beef salami (420 ± 5), gruyere cheese (418 ± 1.8), peach pie (418 ± 0.5), pecan pie (418 ± 0.5), fried pork loin blade including fat (414 ± 0.5), almond meal (414 ± 1.8), oyster crackers (413 ± 6.3), braised pork loin blade including fat (410 ± 0.5), hard pork salami (410 ± 5), puffed rice (407 ± 3.6), cheddar cheese (407 ± 1.8), broiled lamb rib chop including fat (407 ± 0.7), colby cheese (400 ± 1.8), pork and beef sausage (400 ± 3.8), basil (400 ± 50), celery seed (400 ± 25), cumin seed (400 ± 25), mace (400 ± 25), dried parsley (400 ± 50), savory (400 ± 50), turmeric (400 ± 25), braised pork spareribs (397 ± 0.5), pretzels (396 ± 1.8), simmered pork tail (396 ± 0.5), corn flakes (393 ± 1.8), cheshire cheese (393 ± 1.8), fontina cheese (393 ± 1.8), romano cheese (393 ± 1.8), rice polish (393 ± 0.9), broiled pork loin blade including fat (393 ± 0.5), dry breadcrumbs (392 ± 0.5), fried pork top loin including fat (392 ± 0.5), braised beef brisket including fat (391 ± 0.5), toaster pastry (390 ± 1), fried pork center rib including fat (390 ± 0.5), pork link sausage (390 ± 0.7), fruitcake, light (388 ± 1.2), toasted wheat germ (386 ± 1.8), bread sticks (385 ± 2.5), powdered sugar (385 ± 0.4), white granulated sugar (385 ± 0.3), popcorn (383 ± 8.3), danish pastry (383 ± 1.2), braised beef chuck blade roast including fat (383 ± 0.5), caraway cheese (382 ± 1.8), swiss cheese (382 ± 1.8), roasted beef rib including fat (381 ± 0.5), braised pork top loin including fat (381 ± 0.5), american cheese (379 ± 1.8), monterey cheese (379 ± 1.8), pimento cheese spread (379 ± 1.8), caramel cake (379 ± 0.5), fruitcake, dark (379 ± 1.2), dried european chestnuts (379 ± 1.8), devil's food cake (378 ± 0.8), wheat gluten (378 ± 0.4), yellow cake (377 ± 0.7), brick cheese (375 ± 1.8), roquefort cheese (375 ± 1.8), fried pork center loin including fat (375 ± 0.5), raw acorns (375 ± 1.8), corn starch (375 ± 6.3), brown sugar (373 ± 0.3), creme de menthe (372 ± 1), jelly beans (371 ± 1.8), puffed wheat (371 ± 3.6), muenster cheese (371 ± 1.8), chicken egg yolk (371 ± 2.9), braised pork boston blade braised including fat (371 ± 0.5), low fat soybean flour (370 ± 0.6), pork sausage (370 ± 1.9), thousand island salad dressing (369 ± 3.1), white corn flour (368 ± 0.4), yellow corn flour (368 ± 0.4), enriched biscuits (368 ± 1.8), braised pork loin including fat (368 ± 0.5), dried chinese chestnuts (368 ± 1.8), sweet roll (367 ± 1.2), braised pork center rib including fat (367 ± 0.5), white cake (365 ± 0.6), enriched wheat bread (365 ± 0.4), shredded wheat (364 ± 1.8), enriched white degermed corn meal (364 ± 0.4), unenriched corn meal (364 ± 0.4), enriched yellow degermed corn meal (364 ± 0.4), unenriched yellow degermed corn meal (364 ± 0.4), all purpose enriched wheat flour (364 ± 0.4), wheat cake (364 ± 0.4), unenriched biscuits (364 ± 1.8), roasted pork loin blade including fat (364 ± 0.5), dried japanese chestnuts (364 ± 1.8), dry skim milk (363 ± 1.7), white bolted corn meal (362 ± 0.4), yellow bolted corn meal (362 ± 0.4), broiled beef rib including fat (362 ± 0.5), edam cheese (361 ± 1.8), gouda cheese (361 ± 1.8), pork liver sausage (361 ± 2.8), broiled pork top loin including fat (360 ± 0.5), broiled lamb loin chop including fat (359 ± 0.7), dry instant skim milk (358 ± 0.5), bleu cheese (357 ± 1.8), port du salut cheese (357 ± 1.8), provolone cheese (357 ± 1.8), light rye flour (357 ± 0.5), pork and beef lunch meat (357 ± 1.8), dry buttermilk (357 ± 7.1), white whole ground corn meal (355 ± 0.4), yellow whole ground corn meal (355 ± 0.4), maple sugar (354 ± 1.8), wheat flakes (354 ± 1.8), cream cheese (354 ± 1.8), braised pork center loin including fat (354 ± 0.5), beef pastrami (354 ± 1.8), dried ginko nuts (354 ± 1.8), raw coconut (353 ± 1.1), raw wild rice (353 ± 0.3), rice flour (352 ± 0.4), braised pork sirloin including fat (352 ± 0.5), potato flour (351 ± 0.6), gum drops (350 ± 1.8), braised beef chuck arm pot roast including fat (350 ± 0.5), broiled pork boston blade including fat (350 ± 0.5), anise seed (350 ± 25), caraway seed (350 ± 25), cloves (350 ± 25), fennel seed (350 ± 25), onion powder (350 ± 25), white pepper (350 ± 25), light pearled barley (349 ± 0.3), pot barley pearled (348 ± 0.3), beef and pork summer sausage (348 ± 2.2), whipping cream, heavy (347 ± 3.3), light buckwheat flour (347 ± 0.5), whole grain rye crackers (346 ± 3.8), broiled pork loin including fat (346 ± 0.5), braised pork arm braised including fat (345 ± 0.5), cottage pudding (344 ± 0.9), tilsit cheese (343 ± 1.8), broiled pork center rib including fat (343 ± 0.5), hush puppies (340 ± 1.1), corn oil diet margarine tub (340 ± 10), soybean oil and cottonseed oil diet margarine tub (340 ± 10), brie cheese (339 ± 1.8), fried beef sirloin including fat (339 ± 0.5), roasted lamb shoulder including fat (338 ± 0.6), grilled ham patties (338 ± 0.8), defatted soy meal (337 ± 0.4), pork and beef link sausage (337 ± 0.7), roasted duck with skin (337 ± 0.5), soybean protein isolate (336 ± 1.8), 53 proof coffee liqueur (335 ± 1), onion rings (335 ± 0.6), summer sausage (335 ± 2.2), dark buckwheat flour (333 ± 0.5), whole wheat flour (333 ± 0.4), canned pork lunch meat (333 ± 2.4), falafel (333 ± 1), bran flakes (332 ± 1.8), limburger cheese (332 ± 1.8), restaurant hash brown potatoes (332 ± 0.6), sorghum grain (332 ± 0.5), roasted pork arm including fat (331 ± 0.5), broiled pork sirloin including fat (331 ± 0.5), fried chicken back with skin (331 ± 0.7), fried chicken neck with skin (331 ± 1.4), roasted pork top loin including fat (330 ± 0.5), beef beerwurst salami (330 ± 2.2), coconut cream (330 ± 0.2), polish pork sausage (329 ± 1.8), soybean protein concentrate (329 ± 1.8), chocolate chiffon pie (328 ± 0.6), pork liverwurst (328 ± 2.8), pork and beef link sausage with american cheese (328 ± 1.2), millit (327 ± 0.5), dark rye flour (327 ± 0.4), defatted soybean flour (327 ± 0.5), waffles (327 ± 0.7), roasted pork shoulder including fat (326 ± 0.5), peanut flour (325 ± 12.5), freeze-dried shallots (325 ± 12.5), broiled T-bone steak including fat (324 ± 0.5), italian pork sausage (324 ± 0.7), sweetened condensed milk (324 ± 1.3), pork and beef brotwurst (323 ± 0.7), french fries (322 ± 0.6), fried chicken wing with skin (322 ± 1.6), roasted pork boston blade including fat (321 ± 0.5), dehydrated onion flakes (321 ± 3.6), beef and pork frankfurter (320 ± 1.1), roasted pork loin including fat (319 ± 0.5), braised beef thymus (319 ± 0.5), boiled peanuts (319 ± 1.6), roasted pork center rib including fat (318 ± 0.5), marshmallow (317 ± 8.3), well done baked ground beef (317 ± 0.5), unheated ham patties (317 ± 0.8), beef and pork bologna (317 ± 2.2), broiled pork center loin including fat (316 ± 0.5), beef frankfurter (316 ± 0.9), french fried potatoes (316 ± 1), pineapple upside-down cake (315 ± 0.7), corn meal muffins (315 ± 1.3), lemon chiffon pie (314 ± 0.6), toasted raisin bread (314 ± 2.4), light mayonnaise (314 ± 3.6), braised pork loin blade excluding fat (313 ± 0.5), beef bologna (313 ± 2.2), pork and beef mortadella (313 ± 3.3), freeze-dried red sweet peppers (313 ± 31.3), freeze-dried green sweet peppers (313 ± 31.3), smoked beef sausage (312 ± 1.2), pork and beef kielbasa/kolbassy (312 ± 1.9), raisin bran (311 ± 1.4), beef lunch meat (311 ± 1.8), bacon cheeseburger (309 ± 0.3), 63 proof coffee liqueur (308 ± 1), pork and beef knockwurst (307 ± 0.7), medium-rare fried ground beef (306 ± 0.5), dried agar seaweed (306 ± 0.5), toasted white bread (305 ± 2.4), roasted pork center loin including fat (305 ± 0.5), roasted goose with skin (305 ± 0.5), honey (305 ± 2.4), camembert cheese (304 ± 1.8), dinner rolls (304 ± 1.8), braised/stewed veal breast plate (304 ± 0.6), beef summer sausage (304 ± 2.2), roasted beef tenderloin including fat (303 ± 0.5), roasted pork shank including fat (303 ± 0.5), simmered pork chitterlings (303 ± 0.5), pork livercheese (303 ± 1.3), boston cream pie (302 ± 0.5), cheesecake (302 ± 0.6), golden seedless raisins (302 ± 0.5), tomato powder (302 ± 0.5), pork bratwurst (301 ± 0.6), seedless raisins (300 ± 0.5), toasted cracked wheat bread (300 ± 2.4), toasted rye bread (300 ± 2.3), broiled pork loin blade excluding fat (300 ± 0.5), cardamom (300 ± 25), cinnamon (300 ± 25), curry powder (300 ± 25), dill seed (300 ± 25), dried dill weed (300 ± 50), fenugreek seed (300 ± 12.5), garlic powder (300 ± 16.7), ginger (300 ± 25), paprika (300 ± 25), cayenne pepper (300 ± 25), pumpkin pie spice (300 ± 25), dried pepeao (300 ± 4.2), broiled porterhouse steak including fat (299 ± 0.5), seeded raisins (296 ± 0.5), bagels (296 ± 0.9), 100 proof gin (295 ± 1.2), 100 proof rum (295 ± 1.2), 100 proof vodka (295 ± 1.2), 100 proof whiskey (295 ± 1.2), broiled beef rib eye including fat (295 ± 0.5), braised beef shortribs excluding fat (295 ± 0.5), braised pork boston blade excluding fat (294 ± 0.5), roasted pork leg including fat (294 ± 0.5), whipping cream, light (293 ± 3.3), toasted pumpernickel bread (293 ± 1.8), dried shitake mushrooms (293 ± 3.3), well done baked lean ground beef (292 ± 0.5), well done broiled ground beef (292 ± 0.5), roasted pork sirloin including fat (291 ± 0.5), roasted chicken wing with skin (291 ± 1.5), toasted english muffin (290 ± 1), fried beef top round including fat (290 ± 0.5), dried spirulina seaweed (290 ± 0.5), dried and salted cod (289 ± 0.6), french bread (289 ± 1.8), medium-rare broiled ground beef (289 ± 0.5), pineapple chiffon pie (288 ± 0.6), whole ground corn meal muffins (288 ± 1.3), dried jujube (287 ± 0.5), medium-rare baked ground beef (287 ± 0.5), roasted pork blade roll (287 ± 0.5), mozzarella cheese (286 ± 1.8), dried longans (286 ± 0.5), ry-krisp crackers (286 ± 3.6), well done fried ground beef (286 ± 0.5), freeze-dried parsley (286 ± 35.7), dry bakers yeast (286 ± 1.8), brewers yeast (286 ± 1.8), sponge cake (285 ± 0.8), hamburger bun (285 ± 1.3), hot dog bun (285 ± 1.3), fried dark meat chicken with skin (285 ± 0.5), fried pork loin blade excluding fat (283 ± 0.5), simmered beef tongue (283 ± 0.5), torula yeast (282 ± 1.8), toasted whole wheat bread (281 ± 2.4), pork mothers loaf (281 ± 2.4), vienna sausage (281 ± 3.1), corn syrup (281 ± 2.4), raisin bread (280 ± 2), blueberry muffins (280 ± 1.3), bran muffins (280 ± 1.3), well done broiled lean ground beef (280 ± 0.5), broiled beef top loin including fat (280 ± 0.5), broiled beef sirloin including fat (280 ± 0.5), roasted cured pork arm including fat (280 ± 0.5), dry cocoa powder (280 ± 10), italian bread (279 ± 1.8), pita bread (279 ± 1.3), roasted leg of lamb including fat (279 ± 0.6), roasted pork loin blade excluding fat (279 ± 0.5), dried lychees (277 ± 0.5), well done fried lean ground beef (277 ± 0.5), braised pork center rib excluding fat (277 ± 0.5), braised pork top loin excluding fat (277 ± 0.5), fried chicken giblets (277 ± 0.5), 94 proof gin (276 ± 1.2), 94 proof rum (276 ± 1.2), 94 proof vodka (276 ± 1.2), 94 proof whiskey (276 ± 1.2), gingerbread (276 ± 0.8), fish sandwich (276 ± 0.3), raisin bun (275 ± 0.8), dried dates (275 ± 0.6), medium-rare fried lean ground beef (275 ± 0.5), cooked whelk (274 ± 0.6), dried japanese persimmon (274 ± 1.5), french rolls (274 ± 1), well done baked extra lean ground beef (274 ± 0.5), broiled beef round including fat (274 ± 0.5), broiled pork boston blade excluding fat (274 ± 0.5), roasted pork rump including fat (274 ± 0.5), braised pork loin excluding fat (273 ± 0.5), medium-rare broiled lean ground beef (272 ± 0.5), braised pork center loin excluding fat (272 ± 0.5), mince pie (271 ± 0.4), braised beef pancreas (271 ± 0.5), braised pork tongue (271 ± 0.5), dried oriental radish (271 ± 0.9), raisin pie (270 ± 0.4), braised beef chuck blade roast excluding fat (270 ± 0.5), barbados molasses (270 ± 2.5), roasted veal rib roast (269 ± 0.6), fried tofu (269 ± 3.8), feta cheese (268 ± 1.8), angel food cake (268 ± 0.8), breaded and fried fish (268 ± 0.7), medium-rare baked lean ground beef (268 ± 0.5), butterscotch pie (267 ± 0.4), cheeseburger (267 ± 0.4), white bread (267 ± 2.1), chili powder (267 ± 16.7), broiled beef tenderloin including fat (266 ± 0.5), fried pork center loin excluding fat (266 ± 0.5), well done broiled extra lean ground beef (265 ± 0.5), neufchatel cheese (264 ± 1.8), chocolate cream pie (264 ± 0.4), fried chicken (264 ± 0.6), cracked wheat bread (264 ± 2), rye bread (264 ± 2), pork pickle and pimento loaf (264 ± 1.8), well done fried extra lean ground beef (263 ± 0.5), 90 proof gin (262 ± 1.2), cooked mackerel (262 ± 0.6), minced ham lunch meat (262 ± 2.4), cherry pie (261 ± 0.4), hot dog (261 ± 0.6), braised beef bottom round including fat (261 ± 0.5), braised pork sirloin excluding fat (261 ± 0.5), ham and cheese roll (261 ± 1.8), pork and beef luncheon sausage (261 ± 2.2), beef salami (261 ± 2.2), fried chicken thigh with skin (261 ± 0.8), pickled herring (260 ± 3.3), braised lamb heart (260 ± 0.3), broiled lamb liver (260 ± 1.1), broiled pork center rib excluding fat (258 ± 0.5), broiled pork top loin excluding fat (258 ± 0.5), chicken frankfurter (258 ± 1.1), canned bonito (257 ± 0.5), whole wheat roll (257 ± 1.8), braised beef flank including fat (257 ± 0.5), fried pork center rib excluding fat (257 ± 0.5), broiled pork loin excluding fat (257 ± 0.5), fried pork top loin excluding fat (257 ± 0.5), mixed grain bread (256 ± 2), pumpernickel bread (256 ± 1.6), medium-rare broiled extra lean ground beef (256 ± 0.5), roasted pork boston blade excluding fat (256 ± 0.5), pepperoni pizza (255 ± 0.4), cooked kingfish (255 ± 0.5), dried figs (255 ± 0.3), medium-rare fried extra lean ground beef (255 ± 0.5), roasted racoon (255 ± 0.5), citrus marmalade (255 ± 2.5), bran (254 ± 1.8), cooked black bass (254 ± 0.4), oatmeal bread (254 ± 1.8), broiled beef flank including fat (254 ± 0.5), pineapple pie (253 ± 0.4), rhubarb pie (253 ± 0.4), roasted dark meat chicken with skin (253 ± 0.5), sorghum (252 ± 2.4), cooked corned beef brisket (251 ± 0.5), roasted beef round tip including fat (251 ± 0.5), 86 proof whiskey (250 ± 1.2), lemon meringue pie (250 ± 0.4), hamburger sandwich (250 ± 0.5), caviar (250 ± 3.1), canned corned beef (250 ± 0.5), medium-rare baked extra lean ground beef (250 ± 0.5), roasted european chestnuts (250 ± 1.8), stewed chicken wing with skin (250 ± 1.3), allspice (250 ± 25), coriander seed (250 ± 25), oregano (250 ± 25), black pepper (250 ± 25), poultry seasoning (250 ± 25), tarragon (250 ± 25), light molasses (250 ± 2.5), cane syrup (250 ± 2.5), maple syrup (250 ± 2.5), freeze-dried chives (250 ± 62.5), braised pork arm excluding fat (248 ± 0.5), roasted beaver (248 ± 0.6), pork bologna (248 ± 2.2), beef and pork salami (248 ± 2.2), medium cream (247 ± 3.3), simmered chicken neck with skin (247 ± 1.3), roasted chicken thigh with skin (247 ± 0.8), ham and cheese spread (246 ± 1.8), fried light meat chicken with skin (246 ± 0.5), roasted pork center rib excluding fat (245 ± 0.5), roasted pork top loin excluding fat (245 ± 0.5), fried chicken drumstick with skin (245 ± 1), whole wheat bread (244 ± 2), braised beef flank excluding fat (244 ± 0.5), simmered beef shank including fat (244 ± 0.5), roasted pork shoulder excluding fat (244 ± 0.5), blackberry pie (243 ± 0.4), broiled beef flank excluding fat (243 ± 0.5), roasted eye of beef round including fat (243 ± 0.5), broiled pork sirloin excluding fat (243 ± 0.5), pork and beef old fashioned loaf (243 ± 1.8), roasted chinese chestnuts (243 ± 1.8), blueberry pie (242 ± 0.4), cheese pizza (242 ± 0.4), breaded and fried shrimp (242 ± 0.6), braised beef brisket excluding fat (241 ± 0.5), ham and cheese sandwich (240 ± 0.3), roasted beef rib roasted excluding fat (240 ± 0.5), roasted pork center loin excluding fat (240 ± 0.5), roasted pork loin excluding fat (240 ± 0.5), apple pie (239 ± 0.4), dried prunes (239 ± 0.6), tamarind (239 ± 0.4), pork and beef spread (239 ± 1.8), pork olive loaf (239 ± 1.8), pork beerwurst salami (239 ± 2.2), fried dark meat chicken without skin (239 ± 0.5), fried beef sirloin excluding fat (238 ± 0.5), roasted goose without skin (238 ± 0.5), dried apricots (237 ± 1.4), english muffin (237 ± 0.9), roasted pork sirloin excluding fat (236 ± 0.5), pork and beef picnic loaf (236 ± 1.8), coconut custard pie (235 ± 0.4), french toast (235 ± 0.8), braised/roasted/stewed veal chuck (235 ± 0.6), cooked eel (235 ± 0.6), pancakes with butter and syrup (234 ± 0.3), braised/broiled veal loin (234 ± 0.6), cream puff (233 ± 0.4), stewed dark meat chicken with skin (233 ± 0.5), stewed chicken thigh with skin (232 ± 0.7), 80 proof rum (231 ± 1.2), 80 proof vodka (231 ± 1.2), roast beef sandwich (231 ± 0.3), taco (231 ± 0.6), braised beef chuck arm pot roast excluding fat (231 ± 0.5), broiled pork center loin excluding fat (231 ± 0.5), pancakes (230 ± 1.9), berliner (230 ± 2.2), coconut milk (230 ± 0.2), medium molasses (230 ± 2.5), chopped ham lunch meat (229 ± 2.4), raw chinese chestnuts (229 ± 1.8), breaded and fried catfish (228 ± 0.6), broiled beef rib excluding fat (228 ± 0.5), roasted pork arm excluding fat (228 ± 0.5), battered and fried shark (228 ± 0.6), fried beef top round excluding fat (227 ± 0.5), roasted canned ham (226 ± 0.5), manhattan (225 ± 0.9), burrito (225 ± 0.3), popover roll (225 ± 1.3), broiled beef rib eye excluding fat (225 ± 0.5), martini (223 ± 0.7), corn tortilla (223 ± 1.7), braised beef bottom round excluding fat (222 ± 0.5), turkey frankfurter (222 ± 1.1), roasted light meat chicken with skin (222 ± 0.5), fried chicken breast with skin (222 ± 0.5), frozen and french fried potatoes without skin (222 ± 1), banana custard pie (221 ± 0.4), breaded and fried croaker (221 ± 0.6), cornbread (221 ± 0.6), corn pone (221 ± 0.6), roasted pork rump excluding fat (221 ± 0.5), roasted opossum (221 ± 0.5), roasted turkey dark meat with skin (221 ± 0.5), pineapple custard pie (220 ± 0.4), roasted pork leg excluding fat (220 ± 0.5), roasted beef tenderloin excluding fat (219 ± 0.5), braised pork pancreas (219 ± 0.5), custard pie (218 ± 0.4), cheese soufflé (218 ± 0.5), kippered herring (218 ± 1.3), broiled porterhouse steak excluding fat (218 ± 0.5), ham salad (218 ± 1.8), frozen and fried hash brown potatoes (218 ± 0.6), sour cream (217 ± 4.2), fried beef liver (217 ± 0.5), breaded and fried scallops (216 ± 1.6), braised/broiled veal round with rump (216 ± 0.6), stewed rabbit (216 ± 0.4), chocolate syrup (216 ± 1.3), roasted pork shank excluding fat (215 ± 0.5), roasted chicken drumstick with skin (215 ± 1), blackstrap molasses (215 ± 2.5), broiled T-bone steak excluding fat (214 ± 0.5), turkey bologna (214 ± 1.8), pork headcheese (214 ± 1.8), raw european chestnuts (214 ± 1.8), sweet potato pie (213 ± 0.4), lebanon bologna (213 ± 2.2), natto (213 ± 0.6), broiled lamb rib chop excluding fat (212 ± 1.2), pumpkin pie (211 ± 0.4), broiled beef top round including fat (211 ± 0.5), anchovy canned in olive oil (210 ± 2.5), roasted chicken thigh without skin (210 ± 1), hash brown potatoes (209 ± 0.6), sardines (208 ± 2.1), bread stuffing (208 ± 0.3), broiled beef sirloin excluding fat (208 ± 0.5), miso (206 ± 0.4), raw mackerel (205 ± 0.6), roasted lamb shoulder excluding fat (205 ± 0.6), roasted dark meat chicken without skin (205 ± 0.5), broiled beef tenderloin excluding fat (204 ± 0.5), canned chicken liver pâté (204 ± 1.8), chicken and turkey salad (204 ± 1.8), roasted japanese chestnuts (204 ± 1.8), stewed chicken drumstick with skin (204 ± 0.9), broiled beef top loin excluding fat (203 ± 0.5), pickled pork feet (203 ± 0.5), cooked flounder (202 ± 0.5), cooked herring (202 ± 0.6), breaded and fried clams (201 ± 0.6), simmered beef shank excluding fat (201 ± 0.5), stewed light meat chicken with skin (201 ± 0.5), roasted duck without skin (201 ± 0.5), anchovy paste (200 ± 7.1), canned smelt (200 ± 0.5), bay leaf (200 ± 50), dried coriander leaf (200 ± 50), marjoram (200 ± 50), dried rosemary (200 ± 25), saffron (200 ± 50), sage (200 ± 50), caramel sundae (199 ± 0.3), tempeh (199 ± 0.6), strawberry pie (198 ± 0.5), roasted chicken breast with skin (197 ± 0.5), roasted turkey light meat with skin (197 ± 0.5), breaded and fried oysters (196 ± 0.6), raw shad (196 ± 0.6), fried beef brains (196 ± 0.5), chicken salad (196 ± 1.8), turkey salami (195 ± 0.9), simmered pork feet (194 ± 0.5), goose egg (192 ± 0.3), sweetened coconut cream (192 ± 0.2), stewed dark meat chicken without skin (192 ± 0.5), fried light meat chicken without skin (192 ± 0.5), broiled beef top round excluding fat (191 ± 0.5), roasted beef tip excluding fat (190 ± 0.5), canned ham (190 ± 0.5), hot fudge sundae (189 ± 0.3), fried abalone (189 ± 0.6), turkey salad (189 ± 0.9), broiled lamb loin chop excluding fat (188 ± 1), bread pudding (187 ± 0.2), tuna salad (187 ± 0.2), fried chicken breast without skin (187 ± 0.6), roasted turkey dark meat without skin (187 ± 0.5), piña colada (186 ± 0.4), duck egg (186 ± 0.7), lobster paste (186 ± 7.1), canned tuna in oil (186 ± 0.6), roasted leg of lamb excluding fat (186 ± 0.6), turkey summer sausage (186 ± 1.8), ginko nuts (186 ± 1.8), daiquiri (185 ± 0.8), cooked coho salmon (185 ± 0.6), simmered chicken heart (185 ± 0.5), broiled beef round excluding fat (184 ± 0.5), raw eel (184 ± 0.6), stewed chicken breast with skin (184 ± 0.5), roasted eye of beef round excluding fat (183 ± 0.5), grilled canadian bacon (183 ± 1.1), beef honey roll sausage (183 ± 2.2), unheated ham (182 ± 0.5), raw chinook salmon (180 ± 0.6), carob flour (180 ± 0.5), smoked cisco (178 ± 0.6), roasted ham (178 ± 0.5), simmered chicken neck without skin (178 ± 2.8), simmered turkey heart (177 ± 0.5), raw california avocado (177 ± 0.3), pickled anchovy (175 ± 1.8), simmered beef heart (175 ± 0.5), breaded fried squid (175 ± 0.6), ricotta cheese (174 ± 0.4), barbecue loaf (174 ± 2.2), strawberry sundae (173 ± 0.3), cooked mussels (173 ± 0.6), roasted light meat chicken without skin (173 ± 0.5), stewed chicken drumstick without skin (173 ± 1.1), boiled soybeans (173 ± 0.3), enchilada (172 ± 0.2), turkey egg (171 ± 0.6), hummus (171 ± 0.2), roasted cured pork arm excluding fat (170 ± 0.5), tuna salad spread (170 ± 0.9), simmered turkey liver (169 ± 0.5), bulgur (168 ± 0.4), raw sockeye salmon (168 ± 0.6), chipped beef (168 ± 1.8), simmered turkey giblets (167 ± 0.5), roasted pork tenderloin (166 ± 0.5), raw pompano (165 ± 0.6), simmered pork ears (165 ± 0.5), braised pork liver (165 ± 0.5), roasted chicken breast without skin (165 ± 0.6), boiled chickpeas (164 ± 0.3), simmered turkey gizzard (163 ± 0.5), cooked carp (162 ± 0.6), braised beef liver (161 ± 0.5), low calorie thousand island salad dressing (160 ± 3.3), simmered beef brains (160 ± 0.5), roasted veal leg (159 ± 0.6), stewed light meat chicken without skin (159 ± 0.5), light meat chicken roll (159 ± 0.5), chicken egg (158 ± 1), raw herring (158 ± 0.6), chicken roll (158 ± 0.9), raw pork stomach (157 ± 0.5), boiled and steamed chinese chestnuts (157 ± 1.8), raw japanese chestnuts (157 ± 1.8), roasted turkey light meat without skin (157 ± 0.5), simmered chicken giblets (157 ± 0.5), simmered chicken liver (157 ± 0.5), prune pudding (156 ± 0.4), quail egg (156 ± 5.6), raw spiny dogfish (156 ± 0.5), unheated canadian bacon (156 ± 0.9), raw whale meat (156 ± 0.5), crab cakes (155 ± 0.8), cooked swordfish (155 ± 0.6), jellied corned beef loaf (154 ± 1.8), sweet dessert wine (153 ± 0.8), canned sockeye salmon (153 ± 0.6), roasted muskrat (153 ± 0.5), simmered chicken gizzard (153 ± 0.5), cooked rainbow trout (152 ± 0.6), stewed chicken breast without skin (152 ± 0.5), apple brown betty (151 ± 0.2), braised pork kidneys (151 ± 0.5), pork and beef peppered loaf (150 ± 1.8), cooked mullet (149 ± 0.6), braised pork spleen (149 ± 0.5), boiled pink beans (149 ± 0.3), chocolate pudding (148 ± 0.2), raw milkfish (148 ± 0.6), raw trout (148 ± 0.6), braised pork heart (148 ± 0.4), cooked tilefish (147 ± 0.6), light and dark meat turkey roll (147 ± 0.9), boiled winged beans (147 ± 0.3), sweet pickles (147 ± 3.3), raw butterfish (146 ± 0.6), raw coho salmon (146 ± 0.6), raw yellowtail (146 ± 0.6), light meat turkey roll (146 ± 0.9), roasted lean ham (145 ± 0.5), braised beef spleen (145 ± 0.5), firm raw tofu (145 ± 0.4), low calorie russian salad dressing (144 ± 3.1), simmered beef kidneys (144 ± 0.5), raw garlic (144 ± 5.6), boiled yellow beans (144 ± 0.3), pork luxury loaf (143 ± 1.8), potato salad (143 ± 0.4), boiled navy beans (142 ± 0.3), cooked taro (142 ± 0.8), canned alewife (141 ± 0.5), boiled green soybeans (141 ± 0.6), cooked halibut (140 ± 0.6), turkey pastrami (140 ± 0.9), canned pink salmon (139 ± 0.6), boiled white beans (139 ± 0.3), cooked oysters (138 ± 0.6), raw whelk (138 ± 0.6), low calorie french salad dressing (138 ± 3.1), braised pork brains (138 ± 0.5), whiskey sour (137 ± 0.6), boiled pinto beans (137 ± 0.3), candied sweet potato (137 ± 0.5), canned tuna in water (136 ± 0.6), roasted canned lean ham (136 ± 0.5), chopped smoked beef (136 ± 1.8), raw duck liver (136 ± 0.5), boiled cranberry beans (136 ± 0.3), hamburger pickle relish (136 ± 1.8), cooked sturgeon (135 ± 0.6), eggnog (135 ± 0.2), tapioca cream pudding (134 ± 0.3), raw whitefish (134 ± 0.6), sapote (134 ± 0.2), evaporated whole milk (134 ± 0.4), raw quail without skin (134 ± 0.5), half and half cream (133 ± 3.3), raw pheasant without skin (133 ± 0.5), raw goose liver (133 ± 0.5), boiled and steamed european chestnuts (132 ± 1.8), boiled black beans (132 ± 0.3), frozen and boiled black-eyed peas (132 ± 0.6), raw anchovy (131 ± 0.6), unheated lean ham (131 ± 0.5), raw shark (131 ± 0.6), au gratin potatoes (131 ± 0.4), boiled black turtle beans (130 = 0.3), pork and beef honey loaf (129 ± 1.8), boiled french beans (129 ± 0.3), cooked wild long grain rice (129 ± 0.4), cooked snapper (128 ± 0.6), persimmon (128 ± 2), turkey ham (128 ± 0.9), boiled adzuki beans (128 ± 0.2), restaurant chocolate milkshake (127 ± 0.2), raw alewife (127 ± 0.5), raw carp (127 ± 0.6), raw reindeer (127 ± 0.5), boiled kidney beans (127 ± 0.3), cooked sheepshead (126 ± 0.6), raw venison (126 ± 0.6), boiled baby lima beans (126 ± 0.3), dry dessert wine (125 ± 0.8), cooked rainbow smelt (125 ± 0.6), noodles (125 ± 0.3), stir-fried sprouted soybeans (125 ± 0.5), raw bluefish (124 ± 0.6), cooked sea bass (124 ± 0.6), raw spot (124 ± 0.6), restaurant coleslaw (122 ± 0.5), cooked perch (121 ± 0.6), cooked rockfish (121 ± 0.6), raw swordfish (121 ± 0.6), boiled pigeon peas (121 ± 0.3), raw chum salmon (120 ± 0.6), canned lean ham (120 ± 0.5), braised beef lungs (120 ± 0.5), raw cassava (120 ± 0.5), raw freshwater drum (119 ± 0.6), chocolate milkshake (119 ± 0.2), canned chickpeas (119 ± 0.2), boiled great northern beans (119 ± 0.3), boiled lupins (119 ± 0.3), cooked brown rice (119 ± 0.3), cooked grouper (118 ± 0.6), raw rainbow trout (118 ± 0.6), boiled hyacinth beans (118 ± 0.3), boiled sprouted green peas (118 ± 0.5), boiled split peas (118 ± 0.3), boiled yardlong bean (118 ± 0.3), apple tapioca pudding (117 ± 0.2), boiled catjang black-eyed peas (117 ± 0.3), boiled lentils (117 ± 0.3), boiled mothbeans (117 ± 0.3), canned white beans (117 ± 0.2), raw catfish (116 ± 0.6), cooked flatfish (116 ± 0.6), smoked haddock (116 ± 0.6), raw mullet (116 ± 0.6), smoked chinook salmon (116 ± 0.6), raw pink salmon (116 ± 0.6), cooked plantain (116 ± 0.3), boiled black-eyed peas (116 ± 0.3), frozen and cooked sweetened rhubarb (116 ± 0.4), boiled succotash (116 ± 0.5), baked/boiled tropical yam (116 ± 0.7), custard (115 ± 0.2), cooked whiting (115 ± 0.6), canned great northern beans (115 ± 0.2), boiled lima beans (115 ± 0.3), raw freshwater bass (114 ± 0.6), cooked crayfish (114 ± 0.6), macaroni (114 ± 0.4), spaghetti (114 ± 0.4), frozen white corn (114 ± 0.7), restaurant strawberry milkshake (113 ± 0.2), raw barracuda (113 ± 0.5), cooked pike (113 ± 0.6), canned navy beans (113 ± 0.2), cooked haddock (112 ± 0.6), vanilla milkshake (112 ± 0.2), raw florida avocado (112 ± 0.2), poi (112 ± 0.4), vanilla pudding (111 ± 0.2), restaurant vanilla milkshake (111 ± 0.2), baked beans with pork and sweet sauce (111 ± 0.2), tequila sunrise (110 ± 0.3), turkey breast meat summer sausage (110 ± 2.4), raw halibut (109 ± 0.6), boiled broad beans (109 ± 0.3), boiled yellow corn (109 ± 0.6), baked potato with skin (109 ± 0.2), cooked white rice (109 ± 0.2), cooked instant white rice (109 ± 0.3), pickled rip manzanillo/mission olives (109 ± 10.9), rennin (109 ± 4.5), raw sheepshead (108 ± 0.6), smoked whitefish (108 ± 0.6), sheep milk (108 ± 0.2), yellow corn pudding (108 ± 0.2), cooked prunes (107 ± 0.5), low calorie italian salad dressing (107 ± 3.3), tomato catsup (107 ± 3.3), chili sauce (107 ± 3.3), refried beans (107 ± 0.2), raw shrimp (106 ± 0.6), raw sturgeon (106 ± 0.6), boiled mungo beans (106 ± 0.3), mashed potatoes (106 ± 0.5), cooked parboiled white rice (106 ± 0.3), cooked extra long grain white rice (106 ± 0.6), raw chinese water chestnuts (106 ± 0.8), raw abalone (105 ± 0.6), cooked clams (105 ± 0.6), canned cod (105 ± 0.6), cooked cod (105 ± 0.6), raw croaker (105 ± 0.6), raw scup (105 ± 0.6), boiled mung beans (105 ± 0.2), microwaved potato with skin (105 ± 0.2), boiled sweet potato (105 ± 0.3), cottage cheese (104 ± 1.8), raw seatrout (104 ± 0.6), boiled and frozen baby lima beans (104 ± 0.6), canned pumpkin pie mix (104 ± 0.4), cooked parboiled extra long grain white rice (104 ± 0.6), baked sweet potato (104 ± 0.4), chili (103 ± 0.2), breadfruit (103 ± 0.5), cooked blue crab (102 ± 0.6), surimi shrimp (101 ± 0.6), stir-fried sprouted lentils (101 ± 0.5), light cream (100 ± 1.7), raw snapper (100 ± 0.6), passion fruit (100 ± 2.8), brown mustard (100 ± 10), dried chervil (100 ± 50), frozen and boiled fordhook lima beans (100 ± 0.6), microwaved potato without skin (100 ± 0.3), frozen and baked sweet potato (100 ± 0.6), raw cisco (99 ± 0.6), canned blue crab (99 ± 0.6), surimi scallops (99 ± 0.6), cooked shrimp (99 ± 0.6), braised pork lungs (99 ± 0.5), cooked lobster (98 ± 0.6), cooked sockeye salmon (98 ± 0.6), raw smelt (98 ± 0.5), raw rainbow smelt (98 ± 0.6), raw beef tripe (98 ± 0.5), baked beans with pork and tomato sauce (98 ± 0.2), pickled green olives (98 ± 1.1), indian buffalo milk (97 ± 0.2), raw striped bass (96 ± 0.6), cooked alaska king crab (96 ± 0.6), raw sea bass (96 ± 0.6), raw wolffish (96 ± 0.6), canned white corn (96 ± 0.5), raw tilefish (95 ± 0.6), raw turbot (95 ± 0.6),

ENERGY

raw perch (94 ± 0.6), raw rockfish (94 ± 0.6), cherimoya (94 ± 0.1), jackfruit (94 ± 0.5), sugar apple (94 ± 0.3), frozen creamed yellow corn (94 ± 0.4), raw black bass (93 ± 0.5), cooked pompano (93 ± 0.6), raw white sucker (93 ± 0.6), vegetarian baked beans (93 ± 0.2), baked potato without skin (93 ± 0.3), frozen and boiled succotash (93 ± 0.6), raw flatfish (92 ± 0.6), raw grouper (92 ± 0.6), raw pollock (92 ± 0.6), bananas (92 ± 0.4), raw squid (92 ± 0.6), canned sweet potato (92 ± 0.3), bourbon & soda (91 ± 0.4), raw lobster (91 ± 0.6), raw whiting (91 ± 0.6), canned black turtle beans (91 ± 0.2), raw burbot (89 ± 0.6), raw queen crab (89 ± 0.6), raw crayfish (89 ± 0.6), raw sunfish (89 ± 0.6), malted milk (89 ± 0.2), hot dog pickle relish (89 ± 1.8), raw pike (88 ± 0.6), raw scallops (88 ± 0.6), whole chocolate malted milk (88 ± 0.2), boiled burdock root (88 ± 0.4), raw blue crab (87 ± 0.6), raw cusk (87 ± 0.6), raw haddock (87 ± 0.6), raw ling (87 ± 0.6), evaporated lowfat milk (87 ± 0.4), hot chocolate (87 ± 0.2), raw dungeness crab (86 ± 0.6), raw dolphinfish (86 ± 0.6), raw mussels (86 ± 0.6), boiled potato without skin (86 ± 0.4), scalloped potatoes (86 ± 0.4), compressed bakers yeast (86 ± 1.8), raw lingcod (85 ± 0.6), red beans (85 ± 0.4), raw black bullhead (84 ± 0.5), raw alaska king crab (84 ± 0.6), boiled green peas (84 ± 0.6), tomato paste (84 ± 0.4), gefiltefish with broth (83 ± 1.2), teriyaki sauce (83 ± 2.8), whole chocolate milk (83 ± 0.2), canned cranberry beans (83 ± 0.2), screwdriver (82 ± 0.2), raw cod (82 ± 0.6), sapodilla (82 ± 0.3), raw octopus (82 ± 0.6), canned butter beans (82 ± 0.4), frozen and boiled yellow corn (82 ± 0.6), steamed hawaii mountain yam (82 ± 0.7), canned kidney beans (81 ± 0.2), boiled parsnips (81 ± 0.6), raw green peas (81 ± 0.6), o'brien potatoes (81 ± 0.3), steamed sprouted soybeans (81 ± 1.1), yellow mustard (80 ± 10), pickled rip sevillano/ascolano olives (80 ± 3.1), raw cuttlefish (79 ± 0.6), jujube (79 ± 0.5), canned yellow corn (79 ± 0.5), canned lima beans (79 ± 0.2), frozen and boiled green peas (79 ± 0.6), raw potato without skin (79 ± 0.4), bloody mary (78 ± 0.3), evaporated skim milk (78 ± 1.6), canned pinto beans (78 ± 0.2), canned black-eyed peas (77 ± 0.2), okara tofu (77 ± 0.8), canned succotash with creamed corn (77 ± 0.4), gin & tonic (76 ± 0.2), applesauce (76 ± 0.4), carob milk (76 ± 0.2), raw jerusalem artichoke (76 ± 0.7), raw tofu (76 ± 0.4), raw monkfish (75 ± 0.6), crab apples (75 ± 0.5), barbecue sauce (75 ± 3.1), boiled arrowhead (75 ± 4.2), canned yellow corn with red and green peppers (75 ± 0.4), raw clams (74 ± 0.6), figs (74 ± 1), raw frog legs (73 ± 0.5), canned creamed yellow corn (73 ± 0.4), bread and butter pickles (73 ± 3.3), red table wine (72 ± 0.5), maltex (72 ± 0.3), cherries (72 ± 0.7), elderberries (72 ± 0.3), lowfat 2% chocolate milk (72 ± 0.2), rose table wine (71 ± 0.5), maypo (71 ± 0.3), prune juice (71 ± 0.2), european grapes (71 ± 0.3), thompson seedless grapes (71 ± 0.3), canned broad beans (71 ± 0.2), raw ginger root (71 ± 2.1), japanese persimmon (70 ± 0.3), coleslaw (70 ± 0.8), raw shallots (70 ± 5), raw oysters (69 ± 0.6), goat milk (69 ± 0.2), canned green peas (69 ± 0.6), white table wine (68 ± 0.5), raw flounder (68 ± 0.5), canned oysters (68 ± 0.6), raw sole (68 ± 0.5), pomegranate (68 ± 0.3), marinara sauce (68 ± 0.2), human milk (68 ± 1.6), boiled salsify (68 ± 0.7), soursop (67 ± 0.2), lychees (66 ± 0.5), pickled beets (66 ± 0.4), boiled lotus root (66 ± 0.6), mango (65 ± 0.2), frozen and boiled potatoes without skin (65 ± 0.5), black currants (64 ± 0.9), american grapes (63 ± 0.5), kumquats (63 ± 2.6), filled milk (63 ± 0.2), lowfat yogurt (63 ± 0.2), lowfat 1% chocolate milk (63 ± 0.2), canned succotash (63 ± 0.4), oatmeal (62 ± 0.3), raw leeks (62 ± 1.9), grape juice (61 ± 0.2), java plum (61 ± 0.4), kiwi fruit (61 ± 0.7), whole milk (61 ± 0.2), whole yogurt (61 ± 0.2), instant corn grits (60 ± 0.4), corn grits (60 ± 0.2), yellow passion fruit juice (60 ± 0.2), pear nectar (60 ± 0.2), carissa (60 ± 2.5), longans (60 ± 0.5), tamari sauce (60 ± 0.9), canned potato without skin (60 ± 0.6), instant oatmeal (59 ± 0.3), raw apple with skin (59 ± 0.4), pear (59 ± 0.3), cranberry juice cocktail (58 ± 0.2), quince (58 ± 0.5), restaurant cranberry juice cocktail (57 ± 0.3), papaya nectar (57 ± 0.2), blueberries (57 ± 0.3), boiled and steamed japanese chestnuts (57 ± 1.8), apricot nectar (56 ± 0.2), pineapple juice (56 ± 0.2), raw apple without skin (56 ± 0.4), protein fortified lowfat 2% milk (56 ± 0.2), skim yogurt (56 ± 0.2), boiled dishcloth gourd (56 ± 0.6), white hominy (56 ± 0.2), raw lotus root (56 ± 0.6), cooked shitake mushrooms (56 ± 0.7), baked acorn squash (56 ± 0.5), tom collins (55 ± 0.2), red currants (55 ± 0.9), white currants (55 ± 0.9), plum (55 ± 0.8), peach nectar (54 ± 0.2), loganberries (54 ± 0.3), cream of wheat (53 ± 0.3), cooked apple without skin (53 ± 0.3), groundcherries (53 ± 0.4), yellow hominy (53 ± 0.2), yellow hominy with red and green peppers (53 ± 0.2), cream of rice (52 ± 0.3), shoyu sauce (52 ± 0.9), purple passion fruit juice (51 ± 0.2), blackberries (51 ± 0.7), mammy apple (51 ± 0.5), farina (50 ± 0.3), boysenberries (50 ± 0.4), guava (50 ± 0.6), pineapple (50 ± 0.3), raspberries (50 ± 0.4), cooked enriched white degermed corn meal (50 ± 0.2), cooked unenriched corn meal (50 ± 0.2), cooked enriched yellow degermed corn meal (50 ± 0.2), cooked unenriched yellow degermed corn meal (50 ± 0.2), lowfat 2% milk (50 ± 0.2), stir-fried sprouted mung beans (50 ± 0.8), baked hubbard squash (50 ± 0.5), canned chinese water chestnuts (50 ± 0.7), nectarine (49 ± 0.4), valencia orange (49 ± 0.4), roselle (49 ± 0.9), raw irish moss seaweed (49 ± 0.5), chicken egg white (48 ± 1.5), orange soda (48 ± 0.1), apricots (48 ± 0.5), cranberry sauce (48 ± 0.5), protein fortified lowfat 1% milk (48 ± 0.2), apple juice (47 ± 0.2), loquats (47 ± 0.5), navel orange (46 ± 0.4), raw dandelion greens (46 ± 1.8), boiled yambean (46 ± 0.5), restaurant orange juice (45 ± 0.4), orange juice (45 ± 0.2), gooseberries (45 ± 0.3), boiled artichoke (45 ± 0.6), boiled carrots (45 ± 0.6), raw wakame seaweed (45 ± 0.5), mulberries (44 ± 0.4), tangerine (44 ± 0.6), boiled artichoke hearts (44 ± 0.6), cooked tahitian taro (44 ± 0.7), ale (43 ± 0.1), cola (43 ± 0.1), tangerine juice (43 ± 0.2), peach (43 ± 0.6), raw carrots (43 ± 0.7), raw kelp/kombu/tangle seaweed (43 ± 0.5), lowfat 1% milk (42 ± 0.2), frozen and boiled brussels sprouts (42 ± 0.6), beer (41 ± 0.1), malt liquor (41 ± 0.1), lemon-lime soda (41 ± 0.1), prickly pear (41 ± 0.5), soy sauce (41 ± 0.9), protein fortified skim milk (41 ± 0.2), tomato purée (41 ± 0.2), carrot juice (40 ± 0.3), horseradish sauce (40 ± 3.3), buttermilk (40 ± 0.2), canned shitake mushrooms (40 ± 0.4), raw red chili peppers (40 ± 1.1), raw green chili peppers (40 ± 1.1), boiled butternut squash (40 ± 0.5), white grapefruit juice (39 ± 0.2), red grapefruit juice (39 ± 0.2), frozen and boiled butternut squash (39 ± 0.4), papaya (38 ± 0.2), boiled brussels sprouts (38 ± 0.6), mandarin oranges (37 ± 0.4), pummelo (37 ± 0.3), boiled jute potherb (37 ± 1.2), frozen and boiled okra (37 ± 0.5), cantaloupe (36 ± 0.3), frozen and boiled carrots (36 ± 0.7), frozen and boiled collards (36 ± 0.6), skim milk (35 ± 0.2), boiled green beans (35 ± 0.8), boiled yellow snap beans (35 ± 0.8), raw laver/nori seaweed (35 ± 0.5), boiled blackeyed-pea pods (34 ± 1.1), raw onions (34 ± 0.6), raw welsh onions (34 ± 0.5), canned pumpkin (34 ± 0.4), boiled rutabaga (34 ± 0.6), boiled acorn squash (34 ± 0.4), carambola (33 ± 0.4), white grapefruit (33 ± 0.4), honeydew melon (33 ± 0.5), pitanga (33 ± 0.3), soy milk (33 ± 0.2), raw chives (33 ± 16.7), boiled dandelion greens (33 ± 1), raw parsley (33 ± 1.7), raw acerola (32 ± 0.5), raw garden cress (32 ± 2), boiled kale (32 ± 0.8), boiled lamb's-quarters (32 ± 0.6), watermelon (31 ± 0.3), boiled beets (31 ± 0.6), frozen and boiled kale (31 ± 0.8), boiled leeks (31 ± 1.9), raw sprouted mung beans (31 ± 1), boiled okra (31 ± 0.6), pink grapefruit (30 ± 0.4), lime (30 ± 0.7), strawberries (30 ± 0.3), tomato sauce (30 ± 0.4), raw sprouted alfalfa seeds (30 ± 1.5), canned shellie beans (30 ± 0.4), boiled hubbard squash (30 ± 0.4), lemon (29 ± 0.9), boiled broccoli (29 ± 0.6), raw red cabbage (29 ± 1.4), raw savoy cabbage (29 ± 1.4), boiled kohlrabi (29 ± 0.6), canned zucchini in tomato juice (29 ± 0.4), boiled/baked spaghetti squash (29 ± 0.6), frozen and boiled turnip greens (29 ± 0.6), oheloberries (28 ± 0.4), frozen and boiled asparagus (28 ± 0.8), raw bamboo shoots (28 ± 0.7), boiled beet greens (28 ± 0.7), boiled scotch kale (28 ± 0.8), boiled onions (28 ± 0.5), frozen and boiled onions (28 ± 0.5), frozen and boiled spinach (28 ± 0.5), lime juice (27 ± 0.2), raw broccoli (27 ± 1.1), frozen and boiled broccoli (27 ± 0.5), raw eggplant (27 ± 1.2), boiled eggplant (27 ± 1), boiled mushrooms (27 ± 0.6), canned stewed red tomatoes (27 ± 0.4), whey (27 ± 0.2), casaba melon (26 ± 0.3), frozen and boiled yellow snap beans (26 ± 0.7), frozen and boiled green beans (26 ± 0.7), raw mushrooms (26 ± 1.4), raw spring onions with tops (26 ± 1), raw agar seaweed (26 ± 0.5), raw spirulina seaweed (26 ± 0.5), lemon juice (25 ± 0.2), rose apple (25 ± 0.5), boiled savoy cabbage (25 ± 0.7), raw celeriac (25 ± 0.5), raw coriander (25 ± 12.5), raw pepeao (25 ± 0.5), canned red chili peppers (25 ± 0.7), canned green chili peppers (25 ± 0.7), canned jalapeño peppers (25 ± 0.7), frozen and boiled crookneck squash (25 ± 0.5), boiled red tomato (25 ± 0.4), raw turnip greens (25 ± 1.8), boiled asparagus (24 ± 0.6), raw cauliflower (24 ± 1), boiled cauliflower (24 ± 0.8), boiled chayote (24 ± 0.6), boiled garden cress (24 ± 0.7), raw red sweet peppers (24 ± 1), raw green sweet peppers (24 ± 1), raw green tomato (24 ± 0.4), restaurant hot chocolate (23 ± 0.2), raw green cabbage (23 ± 1.4), canned carrots (23 ± 0.7), raw chickory greens (23 ± 0.6), raw mustard spinach (23 ± 0.7), boiled spinach (23 ± 0.6), canned spinach (23 ± 0.5), frozen and boiled turnips (23 ± 0.5), boiled cardoon (22 ± 0.5), raw celtuce (22 ± 0.5), raw dock (22 ± 0.7), acerola juice (21 ± 0.2), boiled amaranth (21 ± 0.8), boiled green cabbage (21 ± 0.7), boiled red cabbage (21 = 0.7), cooked sprouted mung beans (21 ± 0.8), raw frozen rhubarb (21 ± 0.4), raw spinach (21 ± 1.8), boiled turnip greens (21 ± 0.7), wax beans (21 ± 0.4), boiled swiss chard (20 ± 0.6), boiled dock (20 ± 0.5), boiled pokeberry shoots (20 ± 0.6), boiled pumpkin (20 ± 0.4), boiled crookneck squash (20 ± 0.6), raw swamp cabbage (20 ± 0.9), boiled swamp cabbage (20 ± 1), raw red tomato (20 ± 0.4), canned peeled red tomatoes (20 ± 0.4), coconut water (19 ± 0.2), canned bamboo shoots (19 ± 0.4), frozen and boiled cauliflower (19 ± 0.6), canned green beans (19 ± 0.7), canned yellow snap beans (19 ± 0.7), frozen and boiled mustard greens (19 ± 0.7), canned red sweet peppers (19 ± 0.7), canned green sweet peppers (19 ± 0.7), canned sauerkraut (19 ± 0.4), raw vine spinach (19 ± 0.5), tomato juice (18 ± 0.3), raw looseleaf lettuce (18 ± 1.8), boiled red sweet peppers (18 ± 0.7), boiled green sweet peppers (18 ± 0.7), frozen and boiled red sweet peppers (18 ± 0.5), frozen and boiled green sweet peppers (18 ± 0.5), raw oriental radish (18 ± 1.1), boiled oriental radish (18 ± 0.7), raw crookneck squash (18 ± 0.8), raw scallop squash (18 ± 0.8), boiled turnips (18 ± 0.6), restaurant tomato juice (17 ± 0.4), boiled purslane (17 ± 0.9), frozen and boiled zucchini (17 ± 0.4), frozen and boiled turnip greens and turnips (17 ± 0.5), raw chickory (16 ± 1.1), raw endive (16 ± 2), boiled mustard greens (16 ± 0.7), boiled mustard spinach (16 ± 0.6), raw radish (16 ± 1.1), boiled scallop squash (16 ± 0.6), boiled zucchini (16 ± 0.6), raw celery (15 ± 1.3), boiled celery (15 ± 0.7), boiled calabash gourd (15 ± 0.7), raw iceberg lettuce (15 ± 2.5), canned turnip greens (15 ± 0.4), raw chinese cabbage (14 ± 1.4), boiled collards (14 ± 0.3), raw romaine lettuce (14 ± 1.8), raw new zealand spinach (14 ± 1.8), raw zucchini (14 ± 0.8), fruit pectin (14 ± 3.6), boiled bamboo shoots (13 ± 0.4), raw cucumber (13 ± 1), raw butterhead lettuce (13 ± 3.3), canned sprouted mung beans (13 ± 0.8), canned crookneck squash (13 ± 0.5), boiled waxgourd (13 ± 0.6), cider vinegar (13 ± 3.3), distilled vinegar (13 ± 3.3), boiled chinese cabbage (12 ± 0.6), boiled new zealand spinach (12 ± 0.6), raw watercress (12 ± 2.9), dill pickles (11 ± 0.8), sour pickles (11 ± 0.8), boiled butterbur (8 ± 0.5), clam liquid (3 ± 0.2), coffee (2 ± 0.3), restaurant coffee (2 ± 0.3), black tea (1 ± 0.3), instant tea (1 ± 0.2), chamomile tea (1 ± 0.3), tea (1 ± 0.2)

Daily Dosage: Nutritional — 1600–2900 kilocalories; therapeutic — ?; experimental — ?; toxic — none known
Warnings: Increased amounts are dangerous. See toxicity symptoms.

FATS

Names: fats, lipids, oils
Classification: lipids
Deficiency: growth cessation, skin eczema, acne, dry skin, dandruff, dry hair, dull hair, brittle hair, hair decreased, nail deterioration, soft nails, brittle nails, flaking nails, dry eyes, dry mouth, diarrhea, allergies, varicose veins, weight decreased, weight increased, gallstones, decreased radiation resistance, heart disease, cancer, deterioration of skin, sterility, swollen joints, liver deterioration, fatigue, emotional agitation, decreased immunity, decreased T-cell blood count
Side-effects: see toxicity symptoms
Toxicity: metabolism decreased, abnormal weight gain
Inhibitors: saturated fats, saturated fatty acids, nonessential fatty acids, trans fatty acids, alcohol, diabetes, aging, cholesterol, heat, deodorizing oils, hydrogenating oils, oxygen, ferrous sulfate, viral infections, radiation, cancer, vitamin C decreased, zinc decreased
Helpers: vitamin A, vitamin B_{3a}, vitamin B_6, vitamin B_{12}, vitamin B_{15}, vitamin B_T, vitamin C, vitamin D, vitamin E, magnesium, phosphorus, selenium, zinc, polyunsaturated fats
Sources (grams per 100 grams of food): beef tallow (98.5 ± 0.38), mutton tallow (98.5 ± 0.38), duck fat (98.5 ± 0.38), chicken fat (98.5 ± 0.38), goose fat (98.5 ± 0.38), lard (98.5 ± 0.38), turkey fat (98.5 ± 0.38), almond oil (97.1 ± 0.36), coconut oil (97.1 ± 0.36), corn oil (97.1 ± 0.36), cottonseed oil (97.1 ± 0.36), palm oil (97.1 ± 0.36), palm kernel oil (97.1 ± 0.36), safflower oil (97.1 ± 0.36), sesame oil (97.1 ± 0.36), soybean oil (97.1 ± 0.36), soybean lecithin (97.1 ± 0.36), sunflower oil (97.1 ± 0.36), wheat germ oil (97.1 ± 0.36), olive oil (96.4 ± 0.36), peanut oil (96.4 ± 0.36), beef suet (95.4 ± 0.18), whipped butter (83.6 ± 0.45), butter (81.3 ± 0.33), sweet butter (81.3 ± 0.33), dried pilinuts (80.7 ± 0.18), safflower oil and soybean oil mayonnaise (78.6 ± 0.36), soybean oil mayonnaise (78.6 ± 0.36), oil roasted macadamia nuts (77.5 ± 0.18), corn oil margarine tub (76 ± 1), safflower oil margarine tub (76 ± 1), soybean oil

margarine tub (76 ± 1), soybean oil and cottonseed oil margarine tub (76 ± 1), soybean oil and cottonseed oil margarine liquid (76 ± 1), corn oil margarine stick (76 ± 1), safflower oil and soybean oil margarine stick (76 ± 1), soybean oil margarine stick (76 ± 1), soybean oil and cottonseed oil margarine stick (76 ± 1), dry roasted macadamia nuts (75.7 ± 0.18), dried macadamia nuts (74.6 ± 0.18), oil roasted pecans (72.1 ± 0.18), french salad dressing (70 ± 0.36), raw pork jowl (69.6 ± 0.05), dried pecans (68.6 ± 0.18), dried brazil nuts (67.1 ± 0.18), dry roasted filberts (67.1 ± 0.18), dry roasted pecans (65.7 ± 0.18), dried coconut (65.4 ± 0.18), dried hickory nuts (65.4 ± 0.18), oil roasted filberts (64.6 ± 0.18), dried filberts (63.6 ± 0.18), dried english/persian walnuts (62.9 ± 0.18), dried pinyon pine nuts (61.8 ± 0.18), almond butter (59.4 ± 0.31), oil roasted almonds (58.6 ± 0.18), oil roasted sunflower seeds (58.2 ± 0.18), dried butternuts (57.9 ± 0.18), raw bacon (57.5 ± 0.07), dried black walnuts (57.5 ± 0.18), dried sesame kernels (55 ± 0.63), sesame butter (54 ± 0.33), dry roasted pistachio nuts (53.6 ± 0.18), dried almonds (52.9 ± 0.18), dry roasted almonds (52.5 ± 0.18), unsweetened baking chocolate (52.1 ± 0.18), russian salad dressing (52 ± 0.33), dried pignolia pine nuts (51.4 ± 0.18), toasted almonds (51.4 ± 0.18), oil roasted valencia peanuts (51.4 ± 0.18), creamy peanut butter (51.3 ± 0.31), dried sunflower seeds (50.4 ± 0.18), dry roasted sunflower seeds (50.4 ± 0.18), dried peanuts (50 ± 0.18), whole dried sesame seeds (50 ± 0.56), cashew butter (50 ± 0.18), oil roasted peanuts (50 ± 0.18), crunchy peanut butter (50 ± 0.16), dry roasted peanuts (49.6 ± 0.18), mayonnaise (49.3 ± 0.36), cooked bacon (49.2 ± 0.04), dried pistachio nuts (48.9 ± 0.18), oil roasted cashews (48.9 ± 0.18), oil roasted spanish peanuts (48.9 ± 0.18), toasted sesame kernels (48.6 ± 0.18), whole toasted sesame seeds (48.6 ± 0.18), oil roasted virginia peanuts (48.6 ± 0.18), dried watermelon seeds (48.2 ± 0.18), dry roasted red pistachio nuts (47.9 ± 0.18), dried toasted coconut (47.9 ± 0.18), italian salad dressing (47.3 ± 0.33), dry roasted cashews (47.1 ± 0.18), dried pumpkin and squash seeds (46.4 ± 0.18), canned smoked goose liver pâté (44.3 ± 0.18), canned goose liver pâté (43.8 ± 0.38), poppy seed (43.3 ± 1.67), roasted pumpkin and squash seeds (42.9 ± 0.18), braised beef short-ribs including fat (42 ± 0.05), bittersweet chocolate (40.4 ± 0.18), pork and beef pepperoni (40 ± 0.83), nutmeg (40 ± 2.5), corn oil diet margarine tub (38 ± 1), soybean oil and cottonseed oil diet margarine tub (38 ± 1), havarti cheese (37.9 ± 0.18), whipping cream, heavy (37.3 ± 0.33), fried pork loin blade including fat (36.9 ± 0.05), pork and beef sausage (36.2 ± 0.38), potato chips (36.1 ± 0.18), semi-sweet chocolate (36.1 ± 0.18), simmered pork tail (35.8 ± 0.05), broiled lamb rib chop including fat (35.7 ± 0.07), sweet chocolate (35.7 ± 0.18), dried sweetened shredded coconut (35.5 ± 0.05), cream cheese (35.4 ± 0.18), thousand island salad dressing (35 ± 0.31), dried sweetened flaked coconut (34.8 ± 0.2), coconut cream (34.7 ± 0.02), braised pork loin blade including fat (34.1 ± 0.05), hard pork salami (34 ± 0.5), hard pork and beef salami (34 ± 0.5), broiled pork loin blade including fat (33.8 ± 0.05), raw coconut (33.6 ± 0.11), cheddar cheese (33.6 ± 0.18), german sweet bakers chocolate (33.6 ± 0.18), pie crust (33.4 ± 0.03), yellow mustard seed (33.3 ± 1.67), fried pork top loin including fat (33.2 ± 0.05), fried pork center rib including fat (33 ± 0.05), chicken egg yolk (32.9 ± 0.29), gruyere cheese (32.9 ± 0.18), milk chocolate (32.9 ± 0.18), cheese crackers (32.7 ± 0.33), pork and beef lunch meat (32.5 ± 0.18), colby cheese (32.5 ± 0.18), semi-sweet bakers chocolate (32.5 ± 0.18), braised beef brisket including fat (32.4 ± 0.05), pork liver sausage (32.2 ± 0.28), roasted beef rib

IV. Macronutrients 373 FATS

including fat (31.8 ± 0.05), american cheese (31.8 ± 0.18), pork link sausage (31.8 ± 0.07), dried acorns (31.8 ± 0.18), pimento cheese spread (31.4 ± 0.18), fontina cheese (31.4 ± 0.18), pork sausage (31.1 ± 0.19), roquefort cheese (31.1 ± 0.18), cheshire cheese (31.1 ± 0.18), whipping cream, light (30.7 ± 0.33), grilled ham patties (30.7 ± 0.08), monterey cheese (30.7 ± 0.18), acorn flour (30.7 ± 0.18), dried and frozen tofu (30.6 ± 0.29), canned pork lunch meat (30.5 ± 0.24), roasted pork loin blade including fat (30.5 ± 0.05), fried pork center loin including fat (30.5 ± 0.05), braised beef chuck blade roast including fat (30.4 ± 0.05), muenster cheese (30.4 ± 0.18), pork and beef link sausage (30.3 ± 0.07), braised pork spareribs (30.3 ± 0.05), beef beerwurst salami (30 ± 0.22), beef and pork summer sausage (30 ± 0.22), broiled beef rib including fat (30 ± 0.05), brick cheese (30 ± 0.18), parmesan cheese (30 ± 1), gjetost cheese (30 ± 0.18), mace (30 ± 2.5), pound cake (29.7 ± 0.17), summer sausage (29.6 ± 0.22), beef pastrami (29.6 ± 0.18), caraway cheese (29.6 ± 0.18), broiled lamb loin chop including fat (29.4 ± 0.07), bleu cheese (29.3 ± 0.18), braised pork top loin including fat (29.2 ± 0.05), beef and pork frankfurter (29.1 ± 0.11), pork and beef link sausage with american cheese (29.1 ± 0.12), polish pork sausage (28.9 ± 0.18), simmered pork chitterlings (28.8 ± 0.05), shortbread cookie (28.8 ± 0.63), beef bologna (28.7 ± 0.22), braised pork boston blade braised including fat (28.7 ± 0.05), beef frankfurter (28.6 ± 0.09), broiled pork top loin including fat (28.6 ± 0.05), port du salut cheese (28.6 ± 0.18), broiled pork boston blade including fat (28.5 ± 0.05), roasted duck with skin (28.4 ± 0.05), unheated ham patties (28.3 ± 0.08), beef and pork bologna (28.3 ± 0.22), pork liverwurst (28.3 ± 0.28), brie cheese (28.2 ± 0.18), edam cheese (28.2 ± 0.18), light mayonnaise (27.9 ± 0.36), pork and beef brotwurst (27.9 ± 0.07), braised pork loin including fat (27.9 ± 0.05), gouda cheese (27.9 ± 0.18), swiss cheese (27.9 ± 0.18), pork and beef knockwurst (27.8 ± 0.07), limburger cheese (27.5 ± 0.18), almond paste (27.5 ± 0.18), pork and beef kielbasa/kolbassy (27.3 ± 0.19), roasted lamb shoulder including fat (27.2 ± 0.06), broiled pork loin including fat (27.2 ± 0.05), braised pork center rib including fat (27.2 ± 0.05), provolone cheese (27.1 ± 0.18), romano cheese (27.1 ± 0.18), smoked beef sausage (27 ± 0.12), beef summer sausage (27 ± 0.22), dry whole milk (26.9 ± 0.16), beef lunch meat (26.4 ± 0.18), broiled pork center rib including fat (26.4 ± 0.05), tilsit cheese (26.4 ± 0.18), roasted pork arm including fat (26.1 ± 0.05), braised beef chuck arm pot roast including fat (26 ± 0.05), pork bratwurst (25.9 ± 0.06), roasted pork shoulder including fat (25.7 ± 0.05), italian pork sausage (25.7 ± 0.07), braised pork sirloin including fat (25.7 ± 0.05), pork livercheese (25.5 ± 0.13), braised pork arm braised including fat (25.5 ± 0.05), braised pork center loin including fat (25.4 ± 0.05), medium cream (25.3 ± 0.33), pork and beef mortadella (25.3 ± 0.33), roasted pork boston blade including fat (25.3 ± 0.05), broiled pork sirloin including fat (25.3 ± 0.05), roasted soybean nuts (25.3 ± 0.06), roasted pork top loin including fat (25.1 ± 0.05), corn germ (25 ± 0.05), vienna sausage (25 ± 0.31), braised beef thymus (25 ± 0.05), cumin seed (25 ± 2.5), celery seed (25 ± 2.5), camembert cheese (24.6 ± 0.18), broiled T-bone steak including fat (24.6 ± 0.05), fried beef sirloin including fat (24.6 ± 0.05), roasted pork loin including fat (24.3 ± 0.05), raw acorns (24.3 ± 0.18), toasted soybean nuts (24.3 ± 0.18), cheese crackers with peanut butter (24.3 ± 0.12), coconut milk (23.8 ± 0.02), neufchatel cheese (23.6 ± 0.18), roasted pork center rib including fat (23.6 ± 0.05), fried chicken neck with skin (23.6 ± 0.14), roasted pork blade roll (23.5 ± 0.05), peach pie (22.9 ± 0.05), pecan pie (22.9 ± 0.05), gingersnaps cookie (22.9 ± 0.71), medium-rare fried ground beef (22.6 ± 0.05), pork mothers loaf (22.4 ± 0.24), fried chicken wing with skin (22.2 ± 0.16), roasted pork shank including fat (22.1 ± 0.05), broiled pork center loin including fat (22.1 ± 0.05), roasted beef tenderloin including fat (22 ± 0.05), roasted goose with skin (21.9 ± 0.05), boiled peanuts (21.9 ± 0.16), roasted soybean flour (21.9 ± 0.06), mozzarella cheese (21.8 ± 0.18), roasted pork center loin including fat (21.8 ± 0.05), rice bran (21.7 ± 0.05), dry roasted soybean nuts (21.6 ± 0.06), broiled pork loin blade excluding fat (21.5 ± 0.05), well done baked ground beef (21.5 ± 0.05), pork pickle and pimento loaf (21.4 ± 0.18), feta cheese (21.4 ± 0.18), roasted cured pork arm including fat (21.4 ± 0.05), broiled porterhouse steak including fat (21.2 ± 0.05), braised/stewed veal breast plate (21.2 ± 0.06), danish pastry (21 ± 0.12), pork and beef luncheon sausage (20.9 ± 0.22), beef salami (20.9 ± 0.22), medium-rare baked ground beef (20.9 ± 0.05), sour cream (20.8 ± 0.42), simmered beef tongue (20.7 ± 0.05), medium-rare broiled ground beef (20.7 ± 0.05), roasted pork leg including fat (20.7 ± 0.05), fried chicken back with skin (20.7 ± 0.07), soybean flour (20.7 ± 0.06), broiled beef rib eye including fat (20.6 ± 0.05), braised pork loin blade excluding fat (20.6 ± 0.05), minced ham lunch meat (20.5 ± 0.24), ham and cheese roll (20.4 ± 0.18), roasted pork sirloin including fat (20.4 ± 0.05), pork bologna (20 ± 0.22), beef and pork salami (20 ± 0.22), fried tofu (20 ± 0.38), brownies (20 ± 0.25), cloves (20 ± 2.5), dry cocoa powder (20 ± 1), taco shell (20 ± 0.45), tostada shell (20 ± 0.45), oatmeal cookie (20 ± 0.38), fried pork loin blade excluding fat (19.8 ± 0.05), chicken frankfurter (19.6 ± 0.11), well done broiled ground beef (19.5 ± 0.05), roasted chicken wing with skin (19.4 ± 0.15), roasted pork loin blade excluding fat (19.3 ± 0.05), cheesecake (19.2 ± 0.06), canned bonito (19.1 ± 0.05), medium-rare fried lean ground beef (19.1 ± 0.05), cooked corned beef brisket (19 ± 0.05), ham and cheese spread (18.9 ± 0.18), roasted leg of lamb including fat (18.9 ± 0.06), well done fried ground beef (18.9 ± 0.05), broiled beef top loin including fat (18.8 ± 0.05), devil's food cake (18.8 ± 0.08), pork beerwurst salami (18.7 ± 0.22), braised pork tongue (18.6 ± 0.05), almond meal (18.6 ± 0.18), medium-rare broiled lean ground beef (18.5 ± 0.05), broiled pork boston blade excluding fat (18.4 ± 0.05), well done baked lean ground beef (18.4 ± 0.05), onion rings (18.4 ± 0.06), medium-rare baked lean ground beef (18.3 ± 0.05), simmered chicken neck with skin (18.2 ± 0.13), pork and beef old fashioned loaf (18.2 ± 0.18), broiled beef round including fat (18.2 ± 0.05), bacon cheeseburger (18.2 ± 0.03), braised beef shortribs excluding fat (18.1 ± 0.05), caviar (18.1 ± 0.31), turkey frankfurter (18 ± 0.11), pickled herring (18 ± 0.33), broiled beef sirloin including fat (18 ± 0.05), roasted pork rump including fat (17.8 ± 0.05), cooked mackerel (17.8 ± 0.06), falafel (17.8 ± 0.1), sweetened coconut cream (17.7 ± 0.02), well done fried lean ground beef (17.7 ± 0.05), well done broiled lean ground beef (17.6 ± 0.05), braised pork boston blade excluding fat (17.6 ± 0.05), pork and beef spread (17.5 ± 0.18), granola bar (17.5 ± 0.21), berliner (17.4 ± 0.22), raw california avocado (17.3 ± 0.03), cooked black bass (17.2 ± 0.04), braised beef pancreas (17.2 ± 0.05), broiled beef tenderloin including fat (17.2 ± 0.05), cheese soufflé (17.1 ± 0.05), chopped ham lunch meat (17.1 ± 0.24), fried beef top round including fat (17.1 ± 0.05), enriched biscuits (17.1 ± 0.18), unenriched biscuits (17.1 ± 0.18), roasted veal rib roast (16.9 ± 0.06), fried dark meat chicken with skin (16.9 ± 0.05), sugar cookie (16.9 ± 0.31), stewed chicken wing with skin (16.8 ± 0.13), pork and beef picnic loaf (16.8 ± 0.18), pork olive loaf (16.8 ± 0.18), roasted pork boston blade excluding fat (16.8 ± 0.05), waffles (16.8 ± 0.07), hot dog (16.7 ± 0.06), french fries (16.7 ± 0.06), potato pancakes (16.6 ± 0.07), french fried potatoes (16.6 ± 0.1), yellow cake (16.5 ± 0.07), fruitcake, light (16.5 ± 0.12), medium-rare fried extra lean ground beef (16.4 ± 0.05), medium-rare broiled extra lean ground beef (16.3 ± 0.05), broiled beef flank including fat (16.3 ± 0.05), sweet roll (16.2 ± 0.12), pickled pork feet (16.1 ± 0.05), turkey bologna (16.1 ± 0.18), pork headcheese (16.1 ± 0.18), medium-rare baked extra lean ground beef (16.1 ± 0.05), well done fried extra lean ground beef (16 ± 0.05), well done baked extra lean ground beef (16 ± 0.05), fried pork center loin excluding fat (15.9 ± 0.05), fried beef brains (15.8 ± 0.05), roasted dark meat chicken with skin (15.8 ± 0.05), well done broiled extra lean ground beef (15.8 ± 0.05), ham salad (15.7 ± 0.18), fish sandwich (15.7 ± 0.03), breaded and fried fish (15.6 ± 0.07), roasted chicken thigh with skin (15.5 ± 0.08), braised beef flank including fat (15.5 ± 0.05), roasted beef round tip including fat (15.3 ± 0.05), broiled pork loin excluding fat (15.3 ± 0.05), fried pork center rib excluding fat (15.3 ± 0.05), fried pork top loin excluding fat (15.3 ± 0.05), braised beef chuck blade roast excluding fat (15.3 ± 0.05), chocolate chiffon pie (15.3 ± 0.06), fruitcake, dark (15.3 ± 0.12), roasted canned ham (15.2 ± 0.05), chocolate cream pie (15.2 ± 0.04), roasted pork shoulder excluding fat (15 ± 0.05), broiled beef flank excluding fat (15 ± 0.05), fried chicken thigh with skin (15 ± 0.08), anise seed (15 ± 2.5), caraway seed (15 ± 2.5), curry powder (15 ± 2.5), dill seed (15 ± 2.5), paprika (15 ± 2.5), coriander seed (15 ± 2.5), cayenne pepper (15 ± 2.5), cooked eel (14.9 ± 0.06), canned corned beef (14.9 ± 0.05), broiled pork center rib excluding fat (14.9 ± 0.05), broiled pork top loin excluding fat (14.9 ± 0.05), white cake (14.9 ± 0.06), braised beef bottom round including fat (14.8 ± 0.05), caramel cake (14.8 ± 0.05), stewed chicken thigh with skin (14.7 ± 0.07), stewed dark meat chicken with skin (14.7 ± 0.05), braised pork loin excluding fat (14.6 ± 0.05), roasted racoon (14.5 ± 0.05), braised lamb heart (14.4 ± 0.03), braised pork center rib excluding fat (14.4 ± 0.05), braised pork top loin excluding fat (14.4 ± 0.05), roasted eye of beef round including fat (14.2 ± 0.05), fried chicken (14.2 ± 0.06), hash brown potatoes (14 ± 0.06), raw mackerel (13.9 ± 0.06), battered and fried shark (13.9 ± 0.06), cream puff (13.9 ± 0.04), roasted pork loin excluding fat (13.9 ± 0.05), raw shad (13.8 ± 0.06), roasted beef rib roasted excluding fat (13.8 ± 0.05), braised beef flank excluding fat (13.8 ± 0.05), roasted pork center rib excluding fat (13.8 ± 0.05), roasted pork top loin excluding fat (13.8 ± 0.05), duck egg (13.7 ± 0.07), turkey salami (13.7 ± 0.09), fried chicken drumstick with skin (13.7 ± 0.1), braised pork center loin excluding fat (13.7 ± 0.05), chicken and turkey salad (13.6 ± 0.18), broiled pork sirloin excluding fat (13.6 ± 0.05), roasted beaver (13.6 ± 0.06), canned smelt (13.5 ± 0.05), fried chicken giblets (13.5 ± 0.05), cheeseburger (13.5 ± 0.04), cooked kingfish (13.4 ± 0.05), braised/broiled veal loin (13.4 ± 0.06), goose egg (13.3 ± 0.03), breaded and fried catfish (13.3 ± 0.06), chili powder (13.3 ± 1.67), canned chicken liver pâté (13.2 ± 0.18), roasted pork sirloin excluding fat (13.2 ± 0.05), soda crackers (13.2 ± 0.18), roasted pork center loin excluding fat (13.1 ± 0.05), ricotta cheese (13 ± 0.04), canned ham (13 ± 0.05), lebanon bologna (13 ± 0.22), broiled beef rib excluding fat (13 ± 0.05), braised pork sirloin excluding fat (13 ± 0.05), bread stuffing (12.8 ± 0.03), braised/roasted/stewed veal chuck (12.8 ± 0.06), taco (12.8 ± 0.06), braised beef brisket excluding fat

(12.8 ± 0.05), bran muffins (12.8 ± 0.13), breaded and fried croaker (12.7 ± 0.06), roasted goose without skin (12.7 ± 0.05), breaded and fried oysters (12.6 ± 0.06), roasted pork arm excluding fat (12.6 ± 0.05), lemon chiffon pie (12.6 ± 0.06), simmered beef brains (12.5 ± 0.05), turkey summer sausage (12.5 ± 0.18), kippered herring (12.5 ± 0.13), coconut custard pie (12.5 ± 0.04), rice polish (12.5 ± 0.09), oyster crackers (12.5 ± 0.63), simmered pork feet (12.4 ± 0.05), broiled lamb liver (12.4 ± 0.11), braised pork arm excluding fat (12.2 ± 0.05), breaded and fried shrimp (12.2 ± 0.06), hush puppies (12.2 ± 0.11), cooked pompano (12.1 ± 0.06), fried light meat chicken with skin (12.1 ± 0.05), simmered beef shank including fat (12.1 ± 0.05), pineapple chiffon pie (12.1 ± 0.06), pineapple upside-down cake (12.1 ± 0.07), turkey egg (11.9 ± 0.06), smoked cisco (11.9 ± 0.06), chicken salad (11.8 ± 0.18), sardines (11.7 ± 0.21), raw eel (11.6 ± 0.06), cooked herring (11.6 ± 0.06), broiled beef rib eye excluding fat (11.6 ± 0.05), fried dark meat chicken without skin (11.6 ± 0.05), restaurant coleslaw (11.5 ± 0.05), roasted turkey dark meat with skin (11.5 ± 0.05), frozen and fried hash brown potatoes (11.5 ± 0.06), mince pie (11.5 ± 0.04), anchovy paste (11.4 ± 0.71), toaster pastry (11.4 ± 0.1), half and half cream (11.3 ± 0.33), turkey salad (11.3 ± 0.09), tuna salad spread (11.3 ± 0.09), roasted beef tenderloin excluding fat (11.3 ± 0.05), sweet potato pie (11.3 ± 0.04), cherry pie (11.3 ± 0.04), hamburger sandwich (11.3 ± 0.05), cottage pudding (11.3 ± 0.09), chicken egg (11.2 ± 0.1), roasted duck without skin (11.2 ± 0.05), roasted chicken drumstick with skin (11.2 ± 0.1), breaded and fried clams (11.2 ± 0.06), pumpkin pie (11.2 ± 0.04), quail egg (11.1 ± 0.56), braised/broiled veal round with rump (11.1 ± 0.06), custard pie (11.1 ± 0.04), roasted chicken thigh without skin (11 ± 0.1), roasted pork leg excluding fat (11 ± 0.05), breaded and fried scallops (11 ± 0.16), fried beef sirloin excluding fat (11 ± 0.05), natto (11 ± 0.06), blackberry pie (11 ± 0.04), butterscotch pie (11 ± 0.04), pickled rip manzanillo/mission olives (10.9 ± 1.09), cooked sockeye salmon (10.9 ± 0.06), roasted light meat chicken with skin (10.9 ± 0.05), braised pork pancreas (10.8 ± 0.05), broiled porterhouse steak excluding fat (10.8 ± 0.05), blueberry pie (10.8 ± 0.04), pickled green olives (10.7 ± 0.11), simmered pork ears (10.7 ± 0.05), low calorie thousand island salad dressing (10.7 ± 0.33), stewed chicken drumstick with skin (10.7 ± 0.09), roasted pork rump excluding fat (10.7 ± 0.05), pineapple pie (10.7 ± 0.04), rhubarb pie (10.7 ± 0.04), raisin pie (10.7 ± 0.04), toasted wheat germ (10.7 ± 0.18), graham crackers (10.7 ± 0.36), unheated ham (10.6 ± 0.05), raw chinook salmon (10.5 ± 0.06), roasted pork shank excluding fat (10.5 ± 0.05), broiled lamb rib chop excluding fat (10.5 ± 0.12), broiled pork center loin excluding fat (10.5 ± 0.05), beef honey roll sausage (10.4 ± 0.22), broiled T-bone steak excluding fat (10.4 ± 0.05), pickled anchovy (10.4 ± 0.18), peanut brittle (10.4 ± 0.18), french toast (10.3 ± 0.08), ham and cheese sandwich (10.3 ± 0.03), whole ground corn meal muffins (10.3 ± 0.13), roasted opossum (10.2 ± 0.05), stewed rabbit (10.1 ± 0.04), apple pie (10.1 ± 0.04), low calorie italian salad dressing (10 ± 0.33), stewed light meat chicken with skin (10 ± 0.05), roasted lamb shoulder excluding fat (10 ± 0.06), lobster paste (10 ± 0.71), braised beef chuck arm pot roast excluding fat (10 ± 0.05), light cream (10 ± 0.17), corn meal muffins (10 ± 0.13), turmeric (10 ± 2.5), basil (10 ± 5), dried parsley (10 ± 5), savory (10 ± 5), fennel seed (10 ± 2.5), allspice (10 ± 2.5), sage (10 ± 5), pumpkin pie spice (10 ± 2.5), oregano (10 ± 2.5), dried rosemary (10 ± 2.5), saltine crackers (10 ± 0.83), bay leaf (10 ± 5), roasted dark meat chicken without skin (9.7 ± 0.05), braised beef bottom round excluding fat (9.7 ± 0.05), raw pork stomach (9.6 ± 0.05), pepperoni pizza (9.6 ± 0.04), braised pork brains (9.5 ± 0.05), raw pompano (9.5 ± 0.06), anchovy canned in olive oil (9.5 ± 0.25), lemon meringue pie (9.4 ± 0.04), boston cream pie (9.4 ± 0.05), tuna salad (9.3 ± 0.02), broiled beef tenderloin excluding fat (9.3 ± 0.05), popover roll (9.3 ± 0.13), banana custard pie (9.3 ± 0.04), blueberry muffins (9.3 ± 0.13), animal crackers (9.2 ± 0.19), raw herring (9.1 ± 0.06), enchilada (9.1 ± 0.02), barbecue loaf (9.1 ± 0.22), raw spiny dogfish (9 ± 0.05), stewed dark meat chicken without skin (9 ± 0.05), roasted ham (9 ± 0.05), boiled soybeans (9 ± 0.03), raw florida avocado (8.9 ± 0.02), broiled beef top loin excluding fat (8.9 ± 0.05), fried chicken breast with skin (8.9 ± 0.05), roast beef sandwich (8.9 ± 0.03), rusk crackers (8.9 ± 0.56), broiled beef top round including fat (8.8 ± 0.05), cornbread (8.8 ± 0.06), corn pone (8.8 ± 0.06), frozen and french fried potatoes without skin (8.8 ± 0.1), firm raw tofu (8.7 ± 0.04), broiled beef sirloin excluding fat (8.7 ± 0.05), pineapple custard pie (8.7 ± 0.04), sweetened con- densed milk (8.7 ± 0.13), raw sockeye salmon (8.6 ± 0.06), fried beef top round excluding fat (8.6 ± 0.05), zwieback crackers (8.6 ± 0.71), hummus (8.5 ± 0.02), simmered chicken neck without skin (8.3 ± 0.28), roasted turkey light meat with skin (8.3 ± 0.05), grilled canadian bacon (8.3 ± 0.11), potato salad (8.2 ± 0.04), cooked flounder (8.2 ± 0.05), canned tuna in oil (8.1 ± 0.06), raw butterfish (8 ± 0.06), canned alewife (8 ± 0.05), fried beef liver (8 ± 0.05), simmered chicken heart (7.9 ± 0.05), roasted chicken breast with skin (7.8 ± 0.05), strawberry pie (7.8 ± 0.05), burrito (7.8 ± 0.03), tempeh (7.7 ± 0.06), dried spirulina seaweed (7.7 ± 0.05), au gratin potatoes (7.6 ± 0.04), broiled lamb loin chop excluding fat (7.6 ± 0.1), eggnog (7.5 ± 0.02), evaporated whole milk (7.5 ± 0.04), crab cakes (7.5 ± 0.08), raw whale meat (7.5 ± 0.05), stewed chicken breast with skin (7.5 ± 0.05), cooked coho salmon (7.5 ± 0.06), breaded fried squid (7.5 ± 0.06), roasted beef tip excluding fat (7.5 ± 0.05), dinner rolls (7.5 ± 0.18), light meat chicken roll (7.4 ± 0.05), chicken roll (7.4 ± 0.09), canned sockeye salmon (7.3 ± 0.06), light meat turkey roll (7.2 ± 0.09), cooked carp (7.2 ± 0.06), roasted turkey dark meat without skin (7.2 ± 0.05), cheese pizza (7.2 ± 0.04), stir-fried sprouted soybeans (7.1 ± 0.05), roasted leg of lamb excluding fat (7.1 ± 0.06), freeze-dried parsley (7.1 ± 3.57), sheep milk (7 ± 0.02), light and dark meat turkey roll (7 ± 0.09), unheated canadian bacon (7 ± 0.09), roasted cured pork arm excluding fat (7 ± 0.05), broiled beef round excluding fat (7 ± 0.05), pancakes (7 ± 0.19), indian buffalo milk (6.9 ± 0.02), pickled rip sevillano/ ascolano olives (6.8 ± 0.31), fried abalone (6.8 ± 0.06), gingerbread (6.8 ± 0.08), raw milkfish (6.7 ± 0.06), low fat soybean flour (6.7 ± 0.06), raw trout (6.6 ± 0.06), hot fudge sundae (6.6 ± 0.03), roasted eye of beef round excluding fat (6.5 ± 0.05), boiled green soybeans (6.4 ± 0.06), pork and beef peppered loaf (6.4 ± 0.18), simmered beef shank excluding fat (6.4 ± 0.05), freeze-dried red sweet peppers (6.3 ± 3.13), freeze-dried green sweet peppers (6.3 ± 3.13), broiled beef top round excluding fat (6.2 ± 0.05), turkey pastrami (6.1 ± 0.09), jellied corned beef loaf (6.1 ± 0.18), simmered turkey heart (6.1 ± 0.05), bread pudding (6.1 ± 0.02), miso (6.1 ± 0.04), brown mustard (6 ± 1), raw coho salmon (6 ± 0.06), canned pink salmon (6 ± 0.06), simmered turkey liver (6 ± 0.05), caramel sundae (6 ± 0.03), raw whitefish (5.9 ± 0.06), boiled winged beans (5.9 ± 0.03), stewed chicken drumstick without skin (5.7 ± 0.11), dry buttermilk (5.7 ± 0.71), raw carp (5.6 ± 0.06), low calorie french salad dressing (5.6 ± 0.31), simmered beef heart (5.6 ± 0.05), custard (5.5 ± 0.02), simmered chicken liver (5.5 ± 0.05), roasted lean

FATS

ham (5.5 ± 0.05), fried light meat chicken without skin (5.5 ± 0.05), strawberry sundae (5.4 ± 0.03), yellow corn pudding (5.3 ± 0.02), raw yellowtail (5.3 ± 0.06), hamburger bun (5.3 ± 0.13), hot dog bun (5.3 ± 0.13), cooked sturgeon (5.2 ± 0.06), cooked swordfish (5.2 ± 0.06), pancakes with butter and syrup (5.2 ± 0.03), toasted whole wheat bread (5.2 ± 0.24), tapioca cream pudding (5.1 ± 0.03), turkey ham (5.1 ± 0.09), simmered turkey giblets (5.1 ± 0.05), unheated lean ham (5 ± 0.05), pork luxury loaf (5 ± 0.18), braised pork heart (5 ± 0.04), white pepper (5 ± 2.5), cardamom (5 ± 2.5), cinnamon (5 ± 2.5), ginger (5 ± 2.5), black pepper (5 ± 2.5), fenugreek seed (5 ± 1.25), poultry seasoning (5 ± 2.5), tarragon (5 ± 2.5), popcorn (5 ± 0.83), raw alewife (4.9 ± 0.05), raw freshwater drum (4.9 ± 0.06), raw spot (4.9 ± 0.06), cooked oysters (4.9 ± 0.06), roasted canned lean ham (4.9 ± 0.05), braised beef liver (4.9 ± 0.05), raw tofu (4.8 ± 0.04), raw anchovy (4.8 ± 0.06), cooked mullet (4.8 ± 0.06), simmered chicken giblets (4.8 ± 0.05), roasted pork tenderloin (4.8 ± 0.05), fried chicken breast without skin (4.8 ± 0.06), cooked tilefish (4.7 ± 0.06), braised pork kidneys (4.7 ± 0.05), chocolate pudding (4.7 ± 0.02), sponge cake (4.7 ± 0.08), roasted veal leg (4.6 ± 0.06), cottage cheese (4.6 ± 0.18), canned lean ham (4.6 ± 0.05), raw duck liver (4.6 ± 0.05), pork and beef honey loaf (4.6 ± 0.18), chopped smoked beef (4.6 ± 0.18), dried european chestnuts (4.6 ± 0.18), dry breadcrumbs (4.6 ± 0.05), human milk (4.5 ± 0.16), steamed sprouted soybeans (4.5 ± 0.11), raw shark (4.5 ± 0.06), raw quail without skin (4.5 ± 0.05), roasted light meat chicken without skin (4.5 ± 0.05), cooked mussels (4.5 ± 0.06), smoked chinook salmon (4.4 ± 0.06), low calorie russian salad dressing (4.4 ± 0.31), braised pork liver (4.4 ± 0.05), cooked rainbow trout (4.4 ± 0.06), whole wheat bread (4.4 ± 0.2), raw goose liver (4.3 ± 0.05), oatmeal bread (4.3 ± 0.18), toasted white bread (4.3 ± 0.24), toasted raisin bread (4.3 ± 0.24), raw catfish (4.2 ± 0.06), mashed potatoes (4.2 ± 0.05), raw bluefish (4.2 ± 0.06), braised beef spleen (4.2 ± 0.05), goat milk (4.1 ± 0.02), roasted muskrat (4.1 ± 0.05), toasted rye bread (4.1 ± 0.23), raw beef tripe (4 ± 0.05), yellow mustard (4 ± 1), raw sturgeon (4 ± 0.06), raw swordfish (4 ± 0.06), raw venison (4 ± 0.06), stewed light meat chicken without skin (4 ± 0.05), raisin bread (4 ± 0.2), vanilla pudding (3.9 ± 0.02), simmered turkey gizzard (3.9 ± 0.05), chipped beef (3.9 ± 0.18), french bread (3.9 ± 0.18), toasted pumpernickel bread (3.9 ± 0.18), white whole ground corn meal (3.9 ± 0.04), yellow whole ground corn meal (3.9 ± 0.04), raw mullet (3.8 ± 0.06), raw chum salmon (3.8 ± 0.06), raw reindeer (3.8 ± 0.05), white bread (3.8 ± 0.21), toasted cracked wheat bread (3.8 ± 0.24), malted milk (3.7 ± 0.02), scalloped potatoes (3.7 ± 0.04), braised beef lungs (3.7 ± 0.05), restaurant chocolate milkshake (3.7 ± 0.02), simmered chicken gizzard (3.7 ± 0.05), corn tortilla (3.7 ± 0.17), hot chocolate (3.6 ± 0.02), raw seatrout (3.6 ± 0.06), raw freshwater bass (3.6 ± 0.06), raw pheasant without skin (3.6 ± 0.05), cooked wild long grain rice (3.6 ± 0.04), compressed bakers yeast (3.6 ± 0.18), roasted chicken breast without skin (3.6 ± 0.06), rye bread (3.6 ± 0.2), mixed grain bread (3.6 ± 0.2), cracked wheat bread (3.6 ± 0.2), soybean protein isolate (3.6 ± 0.18), pretzels (3.6 ± 0.18), chili (3.5 ± 0.02), apple brown betty (3.5 ± 0.02), whole milk (3.4 ± 0.02), filled milk (3.4 ± 0.02), marinara sauce (3.4 ± 0.02), whole chocolate milk (3.4 ± 0.02), whole chocolate malted milk (3.4 ± 0.02), raw pink salmon (3.4 ± 0.06), raw rainbow trout (3.4 ± 0.06), simmered beef kidneys (3.4 ± 0.05), white bolted corn meal (3.4 ± 0.04), yellow bolted corn meal (3.4 ± 0.04), whole yogurt (3.3 ± 0.02), sorghum grain (3.3 ± 0.05), carob milk (3.2 ± 0.02), raw croaker (3.2 ± 0.06), candied sweet potato (3.2 ± 0.05), braised pork spleen (3.2 ± 0.05), roasted turkey light meat without skin (3.2 ± 0.05), braised pork lungs (3.1 ± 0.05), cooked rainbow smelt (3.1 ± 0.06), stewed chicken breast without skin (3.1 ± 0.05), restaurant vanilla milkshake (3 ± 0.02), vanilla milkshake (3 ± 0.02), bread sticks (3 ± 0.25), raw turbot (2.9 ± 0.06), cooked halibut (2.9 ± 0.06), boiled lupins (2.9 ± 0.03), whole wheat roll (2.9 ± 0.18), raisin bun (2.9 ± 0.08), millit (2.9 ± 0.05), restaurant strawberry milkshake (2.8 ± 0.02), coleslaw (2.7 ± 0.08), raw scup (2.7 ± 0.06), chocolate milkshake (2.7 ± 0.02), raw barracuda (2.6 ± 0.05), cooked sea bass (2.6 ± 0.06), boiled chickpeas (2.6 ± 0.03), white corn flour (2.6 ± 0.04), yellow corn flour (2.6 ± 0.04), dark buckwheat flour (2.6 ± 0.05), dark rye flour (2.6 ± 0.04), canned oysters (2.5 ± 0.06), raw oysters (2.5 ± 0.06), raw rainbow smelt (2.5 ± 0.06), raw sheepshead (2.5 ± 0.06), canned tuna in water (2.5 ± 0.06), pumpernickel bread (2.5 ± 0.16), bagels (2.5 ± 0.09), raw wolffish (2.4 ± 0.06), raw white sucker (2.4 ± 0.06), raw striped bass (2.4 ± 0.06), raw tilefish (2.4 ± 0.06), raw halibut (2.4 ± 0.06), evaporated lowfat milk (2.4 ± 0.04), toasted english muffin (2.4 ± 0.1), dried and salted cod (2.4 ± 0.06), defatted soy meal (2.4 ± 0.04), raw mussels (2.2 ± 0.06), cooked perch (2.1 ± 0.06), raw european chestnuts (2.1 ± 0.18), roasted european chestnuts (2.1 ± 0.18), italian bread (2.1 ± 0.18), dried ginko nuts (2.1 ± 0.18), shredded wheat (2.1 ± 0.18), protein fortified lowfat 2% milk (2 ± 0.02), lowfat 2% chocolate milk (2 ± 0.02), raw sea bass (2 ± 0.06), cooked rockfish (2 ± 0.06), whole wheat flour (2 ± 0.04), soy milk (1.9 ± 0.02), lowfat 2% milk (1.9 ± 0.02), raw cisco (1.9 ± 0.06), barbecue sauce (1.9 ± 0.31), english muffin (1.9 ± 0.09), wheat gluten (1.9 ± 0.04), raisin bran (1.9 ± 0.14), okara tofu (1.8 ± 0.08), cooked blue crab (1.8 ± 0.06), raw shrimp (1.8 ± 0.06), cooked snapper (1.8 ± 0.06), piña colada (1.8 ± 0.04), ginko nuts (1.8 ± 0.18), dried chinese chestnuts (1.8 ± 0.18), dry bakers yeast (1.8 ± 0.18), wheat flakes (1.8 ± 0.18), bran flakes (1.8 ± 0.18), bran (1.8 ± 0.18), gefiltefish with broth (1.7 ± 0.12), raw black bullhead (1.6 ± 0.05), raw perch (1.6 ± 0.06), cooked whiting (1.6 ± 0.06), cooked sheepshead (1.6 ± 0.06), pita bread (1.6 ± 0.13), lowfat yogurt (1.5 ± 0.02), carissa (1.5 ± 0.25), raw monkfish (1.5 ± 0.06), raw rockfish (1.5 ± 0.06), cooked alaska king crab (1.5 ± 0.06), surimi shrimp (1.5 ± 0.06), cooked flatfish (1.5 ± 0.06), baked beans with pork and sweet sauce (1.5 ± 0.02), noodles (1.5 ± 0.03), whole grain rye crackers (1.5 ± 0.38), ry-krisp crackers (1.4 ± 0.36), raw squid (1.4 ± 0.06), cooked crayfish (1.4 ± 0.06), turkey breast meat summer sausage (1.4 ± 0.24), boiled and steamed european chestnuts (1.4 ± 0.18), dried japanese chestnuts (1.4 ± 0.18), puffed wheat (1.4 ± 0.36), raw whiting (1.3 ± 0.06), o'brien potatoes (1.3 ± 0.03), raw snapper (1.3 ± 0.06), cooked grouper (1.3 ± 0.06), boiled yellow corn (1.3 ± 0.06), dried shitake mushrooms (1.3 ± 0.33), protein fortified lowfat 1% milk (1.2 ± 0.02), raw queen crab (1.2 ± 0.06), raw black bass (1.2 ± 0.05), raw flatfish (1.2 ± 0.06), canned blue crab (1.2 ± 0.06), dried figs (1.2 ± 0.03), dried lychees (1.2 ± 0.05), enriched white degermed corn meal (1.2 ± 0.04), unenriched corn meal (1.2 ± 0.04), enriched yellow degermed corn meal (1.2 ± 0.04), unenriched yellow degermed corn meal (1.2 ± 0.04), light buckwheat flour (1.2 ± 0.05), defatted soybean flour (1.2 ± 0.05), lowfat 1% milk (1.1 ± 0.02), raw lingcod (1.1 ± 0.06), raw crayfish (1.1 ± 0.06), raw octopus (1.1 ± 0.06), raw grouper (1.1 ± 0.06), raw blue crab (1.1 ± 0.06), sapodilla (1.1 ± 0.03), cooked shrimp (1.1 ± 0.06), frozen white corn (1.1 ± 0.07), refried beans (1.1 ± 0.02), canned chickpeas (1.1 ± 0.02), boiled yellow beans (1.1 ± 0.03), raw garlic (1.1 ± 0.56), raw chinese chestnuts (1.1 ± 0.18), roasted chinese chestnuts (1.1 ± 0.18), dried jujube (1.1 ± 0.05), enriched wheat bread (1.1 ± 0.04), pot barley pearled (1.1 ± 0.03), torula yeast (1.1 ± 0.18), brewers yeast (1.1 ± 0.18), instant oatmeal (1 ± 0.03), oatmeal (1 ± 0.03), lowfat 1% chocolate milk (1 ± 0.02), maypo (1 ± 0.03), cherries (1 ± 0.07), baked beans with pork and tomato sauce (1 ± 0.02), all purpose enriched wheat flour (1 ± 0.04), light pearled barley (1 ± 0.03), light rye flour (1 ± 0.05), buttermilk (0.9 ± 0.02), raw clams (0.9 ± 0.06), raw dungeness crab (0.9 ± 0.06), raw pollock (0.9 ± 0.06), raw lobster (0.9 ± 0.06), cooked haddock (0.9 ± 0.06), frozen and boiled succotash (0.9 ± 0.06), tomato paste (0.9 ± 0.04), cooked pike (0.9 ± 0.06), smoked haddock (0.9 ± 0.06), smoked whitefish (0.9 ± 0.06), rennin (0.9 ± 0.45), raw garden cress (0.8 ± 0.2), raw ginger root (0.8 ± 0.21), raw burbot (0.8 ± 0.06), cooked cod (0.8 ± 0.06), cooked clams (0.8 ± 0.06), boiled succotash (0.8 ± 0.05), chocolate syrup (0.8 ± 0.13), cooked whelk (0.8 ± 0.06), french rolls (0.8 ± 0.1), wheat cake (0.8 ± 0.04), rice flour (0.8 ± 0.04), dried pepeao (0.8 ± 0.42), potato flour (0.8 ± 0.06), dry instant skim milk (0.8 ± 0.05), raw dock (0.7 ± 0.07), boiled lamb's-quarters (0.7 ± 0.06), roselle (0.7 ± 0.09), cooked tahitian taro (0.7 ± 0.07), raw dandelion greens (0.7 ± 0.18), ground-cherries (0.7 ± 0.04), raw cod (0.7 ± 0.06), raw cuttlefish (0.7 ± 0.06), raw haddock (0.7 ± 0.06), raw sunfish (0.7 ± 0.06), raw pike (0.7 ± 0.06), raw scallops (0.7 ± 0.06), raw dolphinfish (0.7 ± 0.06), raw cusk (0.7 ± 0.06), raw abalone (0.7 ± 0.06), boiled catjang black-eyed peas (0.7 ± 0.03), tomato catsup (0.7 ± 0.33), boiled french beans (0.7 ± 0.03), frozen and boiled black-eyed peas (0.7 ± 0.06), boiled and steamed chinese chestnuts (0.7 ± 0.18), raw japanese chestnuts (0.7 ± 0.18), sweet pickles (0.7 ± 0.33), hamburger pickle relish (0.7 ± 0.18), bulgur (0.7 ± 0.04), roasted japanese chestnuts (0.7 ± 0.18), dried oriental radish (0.7 ± 0.09), gum drops (0.7 ± 0.18), raw wild rice (0.7 ± 0.03), dehydrated onion flakes (0.7 ± 0.36), carob flour (0.7 ± 0.05), dry skim milk (0.7 ± 0.17), puffed rice (0.7 ± 0.36), boiled dock (0.6 ± 0.05), boiled garden cress (0.6 ± 0.07), raw mushrooms (0.6 ± 0.14), raw sprouted alfalfa seeds (0.6 ± 0.15), canned jalapeño peppers (0.6 ± 0.07), boiled dandelion greens (0.6 ± 0.1), gooseberries (0.6 ± 0.03), raspberries (0.6 ± 0.04), guava (0.6 ± 0.06), plum (0.6 ± 0.08), baked hubbard squash (0.6 ± 0.05), raw kelp/kombu/tangle seaweed (0.6 ± 0.05), european grapes (0.6 ± 0.03), thompson seedless grapes (0.6 ± 0.03), raw wakame seaweed (0.6 ± 0.05), raw ling (0.6 ± 0.06), raw alaska king crab (0.6 ± 0.06), cooked lobster (0.6 ± 0.06), passion fruit (0.6 ± 0.28), boiled mungo beans (0.6 ± 0.03), cooked brown rice (0.6 ± 0.03), boiled mothbeans (0.6 ± 0.03), boiled hyacinth beans (0.6 ± 0.03), sapote (0.6 ± 0.02), tamarind (0.6 ± 0.04), dried apricots (0.6 ± 0.14), dried japanese persimmon (0.6 ± 0.15), raw flounder (0.5 ± 0.05), raw sole (0.5 ± 0.05), boiled chayote (0.5 ± 0.06), canned spinach (0.5 ± 0.05), boiled kale (0.5 ± 0.08), frozen and boiled asparagus (0.5 ± 0.08), boiled scotch kale (0.5 ± 0.08), boiled mushrooms (0.5 ± 0.06), raw broccoli (0.5 ± 0.11), frozen and boiled kale (0.5 ± 0.08), frozen and boiled turnip greens (0.5 ± 0.06), frozen and boiled collards (0.5 ± 0.06), prickly pear (0.5 ± 0.05), boiled brussels sprouts (0.5 ± 0.06), pineapple (0.5 ± 0.03), boiled artichoke (0.5 ± 0.06), mammy apple (0.5 ± 0.05), canned succotash (0.5 ± 0.04), elderberries (0.5 ± 0.03), canned black-eyed peas (0.5 ± 0.02), canned succotash with creamed corn (0.5 ± 0.04), canned yellow corn with red and green peppers (0.5 ± 0.04), canned yellow corn (0.5 ± 0.05), frozen creamed yellow corn (0.5 ± 0.04), boiled sprouted green peas (0.5 ±

0.05), bananas (0.5 ± 0.04), surimi scallops (0.5 ± 0.06), macaroni (0.5 ± 0.04), spaghetti (0.5 ± 0.04), boiled black-eyed peas (0.5 ± 0.03), boiled great northern beans (0.5 ± 0.03), boiled yardlong bean (0.5 ± 0.03), stir-fried sprouted lentils (0.5 ± 0.05), boiled kidney beans (0.5 ± 0.03), boiled black beans (0.5 ± 0.03), boiled cranberry beans (0.5 ± 0.03), boiled pinto beans (0.5 ± 0.03), boiled navy beans (0.5 ± 0.03), boiled pink beans (0.5 ± 0.03), dried prunes (0.5 ± 0.06), dried dates (0.5 ± 0.06), seeded raisins (0.5 ± 0.05), seedless raisins (0.5 ± 0.05), golden seedless raisins (0.5 ± 0.05), raw romaine lettuce (0.4 ± 0.18), raw radish (0.4 ± 0.11), canned bamboo shoots (0.4 ± 0.04), raw endive (0.4 ± 0.2), raw looseleaf lettuce (0.4 ± 0.18), raw new zealand spinach (0.4 ± 0.18), whey (0.4 ± 0.02), boiled pokeberry shoots (0.4 ± 0.06), raw red sweet peppers (0.4 ± 0.1), raw green sweet peppers (0.4 ± 0.1), watermelon (0.4 ± 0.03), strawberries (0.4 ± 0.03), raw spinach (0.4 ± 0.18), raw turnip greens (0.4 ± 0.18), pitanga (0.4 ± 0.03), boiled leeks (0.4 ± 0.19), raw spirulina seaweed (0.4 ± 0.05), raw welsh onions (0.4 ± 0.05), mulberries (0.4 ± 0.04), boiled and steamed japanese chestnuts (0.4 ± 0.18), frozen and boiled brussels sprouts (0.4 ± 0.06), nectarine (0.4 ± 0.04), apricots (0.4 ± 0.05), blackberries (0.4 ± 0.07), cooked apple without skin (0.4 ± 0.03), blueberries (0.4 ± 0.03), raw apple with skin (0.4 ± 0.04), pear (0.4 ± 0.03), raw leeks (0.4 ± 0.19), kiwi fruit (0.4 ± 0.07), black currants (0.4 ± 0.09), lychees (0.4 ± 0.05), canned green peas (0.4 ± 0.06), maltex (0.4 ± 0.03), figs (0.4 ± 0.1), raw green peas (0.4 ± 0.06), canned creamed yellow corn (0.4 ± 0.04), canned white corn (0.4 ± 0.05), frozen and boiled fordhook lima beans (0.4 ± 0.06), cherimoya (0.4 ± 0.01), hot dog pickle relish (0.4 ± 0.18), boiled mung beans (0.4 ± 0.02), vegetarian baked beans (0.4 ± 0.02), boiled broad beans (0.4 ± 0.03), canned navy beans (0.4 ± 0.02), canned great northern beans (0.4 ± 0.02), boiled lima beans (0.4 ± 0.03), boiled lentils (0.4 ± 0.03), boiled split peas (0.4 ± 0.03), boiled pigeon peas (0.4 ± 0.03), raw cassava (0.4 ± 0.05), boiled baby lima beans (0.4 ± 0.03), raw whelk (0.4 ± 0.06), persimmon (0.4 ± 0.2), dried longans (0.4 ± 0.05), jelly beans (0.4 ± 0.18), soybean protein concentrate (0.4 ± 0.18), tomato powder (0.4 ± 0.05), corn flakes (0.4 ± 0.18), red beans (0.3 ± 0.04), canned butter beans (0.3 ± 0.04), canned shitake mushrooms (0.3 ± 0.04), honeydew melon (0.3 ± 0.05), boiled bamboo shoots (0.3 ± 0.04), raw chinese cabbage (0.3 ± 0.14), boiled oriental radish (0.3 ± 0.07), raw celery (0.3 ± 0.13), boiled red sweet peppers (0.3 ± 0.07), boiled green sweet peppers (0.3 ± 0.07), canned turnip greens (0.3 ± 0.04), raw celtuce (0.3 ± 0.05), boiled mustard greens (0.3 ± 0.07), acerola juice (0.3 ± 0.02), raw crookneck squash (0.3 ± 0.08), frozen and boiled mustard greens (0.3 ± 0.07), boiled crookneck squash (0.3 ± 0.06), canned peeled red tomatoes (0.3 ± 0.04), boiled red cabbage (0.3 ± 0.07), boiled turnip greens (0.3 ± 0.07), raw vine spinach (0.3 ± 0.05), rose apple (0.3 ± 0.05), raw green cabbage (0.3 ± 0.14), boiled red tomato (0.3 ± 0.04), boiled/baked spaghetti squash (0.3 ± 0.06), raw mustard spinach (0.3 ± 0.07), boiled asparagus (0.3 ± 0.06), raw chickory greens (0.3 ± 0.06), raw acerola (0.3 ± 0.05), raw red cabbage (0.3 ± 0.14), canned red sweet peppers (0.3 ± 0.07), canned green sweet peppers (0.3 ± 0.07), frozen and boiled okra (0.3 ± 0.05), boiled hubbard squash (0.3 ± 0.04), raw bamboo shoots (0.3 ± 0.07), raw onions (0.3 ± 0.06), carambola (0.3 ± 0.04), boiled broccoli (0.3 ± 0.06), cantaloupe (0.3 ± 0.03), boiled green beans (0.3 ± 0.08), boiled yellow snap beans (0.3 ± 0.08), lemon (0.3 ± 0.09), raw parsley (0.3 ± 0.17), white hominy (0.3 ± 0.02), valencia orange (0.3 ± 0.04), boysenberries (0.3 ± 0.04), raw laver/nori seaweed (0.3 ±

FATS

0.05), loganberries (0.3 ± 0.03), raw apple without skin (0.3 ± 0.04), boiled dishcloth gourd (0.3 ± 0.06), cooked shitake mushrooms (0.3 ± 0.07), raw frog legs (0.3 ± 0.05), mango (0.3 ± 0.02), american grapes (0.3 ± 0.05), soursop (0.3 ± 0.02), pomegranate (0.3 ± 0.03), frozen and boiled green peas (0.3 ± 0.06), evaporated skim milk (0.3 ± 0.16), crab apples (0.3 ± 0.05), canned pinto beans (0.3 ± 0.02), canned kidney beans (0.3 ± 0.02), boiled green peas (0.3 ± 0.06), boiled parsnips (0.3 ± 0.06), canned cranberry beans (0.3 ± 0.02), canned black turtle beans (0.3 ± 0.02), sugar apple (0.3 ± 0.03), boiled sweet potato (0.3 ± 0.03), boiled and frozen baby lima beans (0.3 ± 0.06), jackfruit (0.3 ± 0.05), canned white beans (0.3 ± 0.02), boiled black turtle beans (0.3 ± 0.03), boiled white beans (0.3 ± 0.03), dried agar seaweed (0.3 ± 0.05), wax beans (0.2 ± 0.04), boiled waxgourd (0.2 ± 0.06), raw cucumber (0.2 ± 0.1), boiled collards (0.2 ± 0.03), raw zucchini (0.2 ± 0.08), raw chickory (0.2 ± 0.11), coconut water (0.2 ± 0.02), boiled scallop squash (0.2 ± 0.06), boiled new zealand spinach (0.2 ± 0.06), sour pickles (0.2 ± 0.08), frozen and boiled red sweet peppers (0.2 ± 0.05), frozen and boiled green sweet peppers (0.2 ± 0.05), frozen and boiled zucchini (0.2 ± 0.04), boiled mustard spinach (0.2 ± 0.06), raw scallop squash (0.2 ± 0.08), frozen and boiled turnip greens and turnips (0.2 ± 0.05), raw red tomato (0.2 ± 0.04), frozen and boiled cauliflower (0.2 ± 0.06), frozen and boiled turnips (0.2 ± 0.05), cooked sprouted mung beans (0.2 ± 0.08), boiled purslane (0.2 ± 0.09), dill pickles (0.2 ± 0.08), raw green tomato (0.2 ± 0.04), boiled swamp cabbage (0.2 ± 0.1), boiled cauliflower (0.2 ± 0.08), raw swamp cabbage (0.2 ± 0.09), canned sauerkraut (0.2 ± 0.04), oheloberries (0.2 ± 0.04), boiled onions (0.2 ± 0.05), frozen and boiled crookneck squash (0.2 ± 0.05), raw cauliflower (0.2 ± 0.1), raw spring onions with tops (0.2 ± 0.1), raw celeriac (0.2 ± 0.05), boiled eggplant (0.2 ± 0.1), boiled amaranth (0.2 ± 0.08), canned stewed red tomatoes (0.2 ± 0.04), boiled spinach (0.2 ± 0.06), skim milk (0.2 ± 0.02), canned shellie beans (0.2 ± 0.04), raw sprouted mung beans (0.2 ± 0.1), boiled rutabaga (0.2 ± 0.06), canned pumpkin (0.2 ± 0.04), frozen and boiled spinach (0.2 ± 0.05), boiled blackeyed-pea pods (0.2 ± 0.11), protein fortified skim milk (0.2 ± 0.02), tomato sauce (0.2 ± 0.04), tangerine juice (0.2 ± 0.02), carrot juice (0.2 ± 0.03), orange juice (0.2 ± 0.02), raw red chili peppers (0.2 ± 0.11), raw green chili peppers (0.2 ± 0.11), cooked enriched white degermed corn meal (0.2 ± 0.02), cooked unenriched corn meal (0.2 ± 0.02), cooked enriched yellow degermed corn meal (0.2 ± 0.02), cooked unenriched yellow degermed corn meal (0.2 ± 0.02), tangerine (0.2 ± 0.06), yellow hominy (0.2 ± 0.02), boiled jute potherb (0.2 ± 0.12), cream of wheat (0.2 ± 0.03), yellow hominy with red and green peppers (0.2 ± 0.02), loquats (0.2 ± 0.05), cranberry sauce (0.2 ± 0.05), restaurant hot chocolate (0.2 ± 0.02), corn grits (0.2 ± 0.02), skim yogurt (0.2 = 0.02), papaya nectar (0.2 ± 0.02), stir-fried sprouted mung beans (0.2 ± 0.08), canned potato without skin (0.2 ± 0.06), yellow passion fruit juice (0.2 ± 0.02), red currants (0.2 ± 0.09), white currants (0.2 ± 0.09), java plum (0.2 ± 0.04), raw irish moss seaweed (0.2 ± 0.05), canned broad beans (0.2 ± 0.02), japanese persimmon (0.2 ± 0.03), applesauce (0.2 ± 0.04), jujube (0.2 ± 0.05), canned lima beans (0.2 ± 0.02), canned sweet potato (0.2 ± 0.03), boiled burdock root (0.2 ± 0.04), soy sauce (0.2 ± 0.09), raw chinese water chestnuts (0.2 ± 0.08), poi (0.2 ± 0.04), shoyu sauce (0.2 ± 0.09), breadfruit (0.2 ± 0.05), cooked prunes (0.2 ± 0.05), cooked plantain (0.2 ± 0.03), tamari sauce (0.2 ± 0.09), cooked taro (0.2 ± 0.08), prune pudding (0.2 ± 0.04), 63 proof coffee liqueur (0.2 = 0.1), angel food cake (0.2 ± 0.08), 53 proof coffee liqueur (0.2 ± 0.1), creme de menthe (0.2 ± 0.1), canned crookneck squash (0.1 ± 0.05), boiled chinese cabbage (0.1 ± 0.06), boiled celery (0.1 ± 0.07), boiled zucchini (0.1 ± 0.06), tomato juice (0.1 ± 0.03), restaurant tomato juice (0.1 ± 0.04), boiled pumpkin (0.1 ± 0.04), boiled green cabbage (0.1 ± 0.07), boiled turnips (0.1 ± 0.06), boiled cardoon (0.1 ± 0.05), raw frozen rhubarb (0.1 ± 0.04), canned green beans (0.1 ± 0.07), canned yellow snap beans (0.1 ± 0.07), canned carrots (0.1 ± 0.07), boiled swiss chard (0.1 ± 0.06), canned red chili peppers (0.1 ± 0.07), canned green chili peppers (0.1 ± 0.07), frozen and boiled onions (0.1 ± 0.05), boiled savoy cabbage (0.1 ± 0.07), casaba melon (0.1 ± 0.03), frozen and boiled yellow snap beans (0.1 ± 0.07), frozen and boiled green beans (0.1 ± 0.07), pink grapefruit (0.1 ± 0.04), frozen and boiled broccoli (0.1 ± 0.05), canned zucchini in tomato juice (0.1 ± 0.04), white grapefruit (0.1 ± 0.04), boiled kohlrabi (0.1 ± 0.06), lime juice (0.1 ± 0.02), white grapefruit juice (0.1 ± 0.02), red grapefruit juice (0.1 ± 0.02), frozen and boiled carrots (0.1 ± 0.07), boiled okra (0.1 ± 0.06), boiled acorn squash (0.1 ± 0.04), boiled beet greens (0.1 ± 0.07), pummelo (0.1 ± 0.03), papaya (0.1 ± 0.02), lime (0.1 ± 0.07), restaurant orange juice (0.1 ± 0.04), farina (0.1 ± 0.03), apple juice (0.1 ± 0.02), boiled yambean (0.1 ± 0.05), raw carrots (0.1 ± 0.07), boiled butternut squash (0.1 ± 0.05), frozen and boiled butternut squash (0.1 ± 0.04), peach (0.1 ± 0.06), cream of rice (0.1 ± 0.03), boiled carrots (0.1 ± 0.06), tomato purée (0.1 ± 0.02), navel orange (0.1 ± 0.04), boiled artichoke hearts (0.1 ± 0.06), bloody mary (0.1 ± 0.03), restaurant cranberry juice cocktail (0.1 ± 0.03), pineapple juice (0.1 ± 0.02), instant corn grits (0.1 ± 0.04), apricot nectar (0.1 ± 0.02), grape juice (0.1 ± 0.02), quince (0.1 ± 0.05), baked acorn squash (0.1 ± 0.05), frozen and boiled potatoes without skin (0.1 ± 0.05), longans (0.1 ± 0.05), pickled beets (0.1 ± 0.04), boiled lotus root (0.1 ± 0.06), boiled salsify (0.1 ± 0.07), tequila sunrise (0.1 ± 0.03), raw lotus root (0.1 ± 0.06), raw potato without skin (0.1 ± 0.04), boiled potato without skin (0.1 ± 0.04), steamed hawaii mountain yam (0.1 ± 0.07), whiskey sour (0.1 ± 0.06), frozen and boiled yellow corn (0.1 ± 0.06), canned cod (0.1 ± 0.06), baked potato without skin (0.1 ± 0.03), frozen and baked sweet potato (0.1 ± 0.06), microwaved potato without skin (0.1 ± 0.03), cooked parboiled white rice (0.1 ± 0.03), baked sweet potato (0.1 ± 0.04), cooked white rice (0.1 ± 0.02), microwaved potato with skin (0.1 ± 0.02), canned pumpkin pie mix (0.1 ± 0.04), cooked parboiled extra long grain white rice (0.1 ± 0.06), baked potato with skin (0.1 ± 0.02), baked/boiled tropical yam (0.1 ± 0.07), apple tapioca pudding (0.1 ± 0.02), frozen and cooked sweetened rhubarb (0.1 ± 0.04), boiled adzuki beans (0.1 ± 0.02)

Daily Dosage: (1 g fats = 9 calories) Nutritional—66–87 g; therapeutic—?; experimental—?; toxic—?

Warnings: Keep levels low with gallstones, digestion upset and skin psoriasis. Increased amounts are toxic. See toxicity symptoms.

FIBER

Classification: macronutrient, often not classified as a nutrient because it is not absorbed into the body
Deficiency: colon cancer, hemorrhoids, diverticulosis, colitis, ulcerative colitis, chronic constipation, appendicitis, vericose veins, hiatal hernia
Side-effects: lowered mineral absorption, bloating, cramps
Toxicity: none known
Sources (grams per 100 grams of food): bran (30 ± 1.8), ry-krisp crackers (18 ± 3.6), bran flakes (14 ± 1.8), roasted european chestnuts (12 ± 1.8), raisin bran (11 ± 1.4), defatted soy meal (11 ± 0.4), toasted wheat germ (11 ± 1.8), raw european chestnuts (10 ± 1.8), shredded wheat (9 ± 1.8), wheat flakes (7 ± 1.8), whole wheat bread (6 ± 2), toasted cracked wheat bread (5 ± 2.4), dried almonds (5 ± 1.8), boiled broad beans (5 ± 0.3), boiled pigeon peas (5 ± 0.3), puffed wheat (4 ± 3.6), cracked wheat bread (4 ± 2), mixed grain bread (4 ± 2), boiled black beans (4 ± 0.3), boiled kidney beans (4 ± 0.3), boiled lentils (4 ± 0.3), boiled baby lima beans (4 ± 0.3), boiled navy beans (4 ± 0.3), boiled green peas (4 ± 0.6), canned green peas (4 ± 0.6), frozen and boiled green peas (4 ± 0.6), boiled pink beans (4 ± 0.3), boiled pinto beans (4 ± 0.3), corn tortilla (3 ± 1.7), crunchy peanut butter (3 ± 1.6), raw california avocado (3 ± 0.3), raw bamboo shoots (3 ± 0.7), vegetarian baked beans (3 ± 0.2), boiled chickpeas (3 ± 0.3), boiled black-eyed peas (3 ± 0.3), boiled cranberry beans (3 ± 0.3), boiled great northern beans (3 ± 0.3), boiled lima beans (3 ± 0.3), boiled mung beans (3 ± 0.2), boiled parsnips (3 ± 0.6), raw green peas (3 ± 0.6), boiled sprouted green peas (3 ± 0.5), miso (3 ± 0.4), raw spinach (3 ± 1.8), canned spinach (3 ± 0.5), french bread (2 ± 1.8), oatmeal bread (2 ± 1.8), raisin bread (2 ± 2), raw sprouted alfalfa seeds (2 ± 1.5), baked beans with pork and tomato sauce (2 ± 0.2), frozen and boiled broccoli (2 ± 0.5), frozen and boiled brussels sprouts (2 ± 0.6), raw carrots (2 ± 0.7), boiled carrots (2 ± 0.6), frozen and boiled carrots (2 ± 0.7), boiled cauliflower (2 ± 0.8), frozen and boiled cauliflower (2 ± 0.6), frozen and boiled yellow corn (2 ± 0.6), boiled green beans (2 ± 0.8), boiled yellow snap beans (2 ± 0.8), frozen and boiled yellow snap beans (2 ± 0.7), frozen and boiled green beans (2 ± 0.7), boiled spinach (2 ± 0.6), frozen and boiled spinach (2 ± 0.5), baked sweet potato (2 ± 0.4), pickled rip manzanillo/mission olives (2 ± 10.9), pickled rip sevillano/ascolano olives (2 ± 3.1), oatmeal (1 ± 0.3), corn flakes (1 ± 1.8), puffed rice (1 ± 3.6), bagels (1 ± 0.9), italian bread (1 ± 1.8), pita bread (1 ± 1.3), soy milk (1 ± 0.2), raw broccoli (1 ± 1.1), boiled brussels sprouts (1 ± 0.6), raw green cabbage (1 ± 1.4), raw red cabbage (1 ± 1.4), canned carrots (1 ± 0.7), raw celery (1 ± 1.3), raw cucumber (1 ± 1), raw eggplant (1 ± 1.2), canned green beans (1 ± 0.7), canned yellow snap beans (1 ± 0.7), raw leeks (1 ± 1.9), raw butterhead lettuce (1 ± 3.3), raw iceberg lettuce (1 ± 2.5), raw sprouted mung beans (1 ± 1), raw onions (1 ± 0.6), raw red sweet peppers (1 ± 1), raw green sweet peppers (1 ± 1), raw crookneck squash (1 ± 0.8), boiled crookneck squash (1 ± 0.6), canned crookneck squash (1 ± 0.5), frozen and boiled crookneck squash (1 ± 0.5), frozen and baked sweet potato (1 ± 0.6), raw red tomato (1 ± 0.4), canned peeled red tomatoes (1 ± 0.4), light cream (0.34 ± 1.7)
Daily Dosage: Nutritional — 5 g; therapeutic — 10 g; experimental — 15 g; toxic — none known
Warnings: Excessive amounts can cause side-effects. See side-effects symptoms.

PROTEIN COMPLEX

(Different amino acids. See alanine, arginine, asparagine, aspartic acid, cysteine, cystine, glutamic acid, glutamine, glycine, histidine, hydroxyproline, isoleucine, leucine, lysine, methionine, phenylalanine, proline, serine, threonine, tryptophan, tyrosine and valine.)

Forms: combinations of various amino acids
essential amino acids: isoleucine, leucine, lysine, methionine, phenylalanine, threonine, tryptophan, valine
semi-essential amino acids (sometimes made internally): arginine, histidine
nonessential amino acids (made internally): alanine, asparagine, aspartic acid, cysteine, cystine, glutamic acid, glutamine, glycine, hydroxyproline, proline, serine, tyrosine.
Deficiency: appetite decreased, fatigue, retarded growth, alterations in skin/hair pigmentation, skin edema, fatty infiltration, cell deterioration, fibrosis of liver, nutritional skin dermatosis, digestion deterioration, emotional agitation, depression, amino acid deficiencies
Inhibitors: unbalanced amounts of various amino acids
Helpers: vitamin B_{12}, vitamin B_W, vitamin C, vitamin T, chromium, sleep
Sources (grams per 100 grams of food): soybean protein isolate (88.2 ± 0.18), soybean protein concentrate (63.6 ± 0.18), dried and salted cod (62.8 ± 0.06), dried spirulina seaweed (57.5 ± 0.05), peanut flour (52.5 ± 1.25), defatted soybean flour (51.5 ± 0.05), low fat soybean flour (50.9 ± 0.06), defatted soy meal (49.2 ± 0.04), dried and frozen tofu (48.2 ± 0.29), cooked whelk (47.6 ± 0.06), parmesan cheese (42 ± 1), wheat gluten (41.4 ± 0.04), marjoram (40 ± 5), almond meal (40 ± 0.18), dry roasted soybean nuts (39.5 ± 0.06), brewers yeast (39.3 ± 0.18), torula yeast (38.9 ± 0.18), light cream (38.3 ± 0.17), roasted soybean flour (38.1 ± 0.06), soybean flour (37.8 ± 0.06), dry bakers yeast (37.5 ± 0.18), toasted soybean nuts (37.5 ± 0.18), dry skim milk (36.3 ± 0.17), roasted soybean nuts (35.2 ± 0.06), fried beef top round excluding fat (35.1 ± 0.05), dry instant skim milk (35.1 ± 0.05), braised pork center loin excluding fat (34.8 ± 0.05), braised pork center rib excluding fat (34.4 ± 0.05), braised pork top loin excluding fat (34.4 ± 0.05), simmered beef shank excluding fat (33.7 ± 0.05), roasted pumpkin and squash seeds (33.6 ± 0.18), braised pork sirloin excluding fat (33.5 ± 0.05), fried chicken breast without skin (33.5 ± 0.06), braised pork loin excluding fat (33 ± 0.05), braised beef chuck arm pot roast excluding fat (33 ± 0.05), fried light meat chicken without skin (32.8 ± 0.05), fried chicken giblets (32.5 ± 0.05), fried beef sirloin excluding fat (32.5 ± 0.05), braised pork arm excluding fat (32.3 ± 0.05), broiled lamb liver (32.2 ± 0.11), romano cheese (32.1 ± 0.18), broiled pork center loin excluding fat (32 ± 0.05), fried chicken breast with skin (31.8 ± 0.05), fried beef top round including fat (31.7 ± 0.05), broiled beef top round excluding fat (31.7 ± 0.05), braised beef bottom round excluding fat (31.6 ± 0.05), simmered beef shank including fat (31.6 ± 0.05), dry buttermilk (31.4 ± 0.71), braised pork boston blade excluding fat

(31.2 ± 0.05), braised beef chuck blade roast excluding fat (31.1 ± 0.05), roasted chicken breast without skin (31 ± 0.06), roasted light meat chicken without skin (30.9 ± 0.05), braised beef shortribs excluding fat (30.8 ± 0.05), broiled beef top round including fat (30.8 ± 0.05), fried light meat chicken with skin (30.5 ± 0.05), cooked bacon (30.5 ± 0.04), gruyere cheese (30.4 ± 0.18), broiled beef sirloin excluding fat (30.4 ± 0.05), well done baked extra lean ground beef (30.3 ± 0.05), roasted opossum (30.2 ± 0.05), cooked flounder (30 ± 0.05), dried parsley (30 ± 5), roasted turkey light meat without skin (29.9 ± 0.05), braised beef bottom round including fat (29.8 ± 0.05), roasted chicken breast with skin (29.8 ± 0.05), braised pork loin blade excluding fat (29.7 ± 0.05), chipped beef (29.6 ± 0.18), well done baked lean ground beef (29.6 ± 0.05), oil roasted virginia peanuts (29.6 ± 0.18), toasted wheat germ (29.6 ± 0.18), braised lamb heart (29.5 ± 0.03), simmered turkey gizzard (29.4 ± 0.05), braised pork center loin including fat (29.4 ± 0.05), braised beef brisket excluding fat (29.4 ± 0.05), stewed rabbit (29.3 ± 0.04), roasted veal leg (29.3 ± 0.06), roasted racoon (29.2 ± 0.05), roasted beaver (29.2 ± 0.06), braised pork spareribs (29.1 ± 0.05), roasted pork rump excluding fat (29.1 ± 0.05), anchovy canned in olive oil (29 ± 0.25), fried dark meat chicken without skin (29 ± 0.05), roasted goose without skin (29 ± 0.05), roasted light meat chicken with skin (29 ± 0.05), roasted eye of beef round excluding fat (29 ± 0.05), swiss cheese (28.9 ± 0.18), stewed light meat chicken without skin (28.9 ± 0.05), stewed chicken breast without skin (28.9 ± 0.05), dried watermelon seeds (28.9 ± 0.18), simmered beef heart (28.8 ± 0.05), well done baked ground beef (28.8 ± 0.05), fried pork center loin excluding fat (28.8 ± 0.05), broiled pork center rib excluding fat (28.8 ± 0.05), broiled pork top loin excluding fat (28.8 ± 0.05), roasted pork tenderloin (28.8 ± 0.05), creamy peanut butter (28.8 ± 0.31), roasted leg of lamb excluding fat (28.7 ± 0.06), roasted beef tip excluding fat (28.7 ± 0.05), well done broiled extra lean ground beef (28.6 ± 0.05), braised pork center rib including fat (28.6 ± 0.05), roasted turkey dark meat without skin (28.6 ± 0.05), roasted turkey light meat with skin (28.6 ± 0.05), broiled beef top loin excluding fat (28.6 ± 0.05), freeze-dried parsley (28.6 ± 3.57), braised pork pancreas (28.5 ± 0.05), roasted pork center loin excluding fat (28.5 ± 0.05), broiled beef round excluding fat (28.5 ± 0.05), stewed chicken drumstick without skin (28.4 ± 0.11), broiled pork sirloin excluding fat (28.3 ± 0.05), roasted pork leg excluding fat (28.3 ± 0.05), broiled beef tenderloin excluding fat (28.3 ± 0.05), braised pork spleen (28.2 ± 0.05), well done broiled lean ground beef (28.2 ± 0.05), roasted pork shank excluding fat (28.2 ± 0.05), broiled lamb loin chop excluding fat (28.2 ± 0.1), broiled porterhouse steak excluding fat (28.2 ± 0.05), roasted pork center rib excluding fat (28.2 ± 0.05), roasted pork top loin excluding fat (28.2 ± 0.05), broiled T-bone steak excluding fat (28.1 ± 0.05), braised pork sirloin including fat (28 ± 0.05), well done fried extra lean ground beef (28 ± 0.05), fried pork center rib excluding fat (28 ± 0.05), fried pork top loin excluding fat (28 ± 0.05), broiled beef rib eye excluding fat (28 ± 0.05), braised beef flank excluding fat (28 ± 0.05), braised/roasted/stewed veal chuck (27.9 ± 0.06), oil roasted spanish peanuts (27.9 ± 0.18), broiled pork loin excluding fat (27.8 ± 0.05), fried chicken back with skin (27.8 ± 0.07), braised pork top loin including fat (27.7 ± 0.05), well done fried lean ground beef (27.6 ± 0.05), fried beef sirloin including fat (27.5 ± 0.05), roasted pork sirloin excluding fat (27.5 ± 0.05), roasted turkey dark meat with skin (27.5 ± 0.05), roasted beef tenderloin excluding fat (27.5 ± 0.05), braised beef flank including fat

(27.5 ± 0.05), broiled pork center loin including fat (27.4 ± 0.05), roasted dark meat chicken without skin (27.4 ± 0.05), broiled beef sirloin including fat (27.4 ± 0.05), stewed chicken breast with skin (27.4 ± 0.05), cooked coho salmon (27.4 ± 0.06), cooked sockeye salmon (27.3 ± 0.06), roasted veal rib roast (27.2 ± 0.06), broiled lamb rib chop excluding fat (27.2 ± 0.12), roasted muskrat (27.2 ± 0.05), simmered chicken gizzard (27.2 ± 0.05), braised pork loin including fat (27.2 ± 0.05), well done broiled ground beef (27.2 ± 0.05), fried dark meat chicken with skin (27.2 ± 0.05), roasted beef rib roasted excluding fat (27.2 ± 0.05), braised beef pancreas (27.1 ± 0.05), braised/broiled veal round with rump (27.1 ± 0.06), braised beef chuck arm pot roast including fat (27.1 ± 0.05), roasted chicken drumstick with skin (27.1 ± 0.1), canned corned beef (27.1 ± 0.05), oil roasted valencia peanuts (27.1 ± 0.18), oil roasted peanuts (27.1 ± 0.18), well done fried ground beef (27 ± 0.05), fried chicken drumstick with skin (26.9 ± 0.1), roasted pork loin excluding fat (26.9 ± 0.05), roasted lamb shoulder excluding fat (26.8 ± 0.06), simmered turkey heart (26.8 ± 0.05), braised pork arm braised including fat (26.8 ± 0.05), fried chicken thigh with skin (26.8 ± 0.08), roasted chicken wing with skin (26.8 ± 0.15), roasted eye of beef round including fat (26.8 ± 0.05), fried beef liver (26.7 ± 0.05), roasted pork arm excluding fat (26.7 ± 0.05), cooked halibut (26.7 ± 0.06), canned tuna in water (26.7 ± 0.06), yellow mustard seed (26.7 ± 1.67), simmered turkey giblets (26.6 ± 0.05), roasted pork rump including fat (26.6 ± 0.05), canned tuna in oil (26.6 ± 0.06), simmered chicken heart (26.5 ± 0.05), roasted beef round tip including fat (26.5 ± 0.05), braised/broiled veal loin (26.4 ± 0.06), braised pork boston blade braised including fat (26.4 ± 0.05), cooked rainbow trout (26.4 ± 0.06), cooked snapper (26.4 ± 0.06), dry whole milk (26.3 ± 0.16), fried chicken wing with skin (26.3 ± 0.16), dried sesame kernels (26.3 ± 0.63), braised/stewed veal breast plate (26.1 ± 0.06), fontina cheese (26.1 ± 0.18), stewed light meat chicken with skin (26.1 ± 0.05), provolone cheese (26.1 ± 0.18), dried peanuts (26.1 ± 0.18), cooked sheepshead (26 ± 0.06), braised pork liver (26 ± 0.05), roasted chicken thigh without skin (26 ± 0.1), roasted dark meat chicken with skin (26 ± 0.05), stewed dark meat chicken without skin (26 ± 0.05), broiled beef tenderloin including fat (26 ± 0.05), broiled beef rib excluding fat (26 ± 0.05), simmered chicken giblets (25.9 ± 0.05), broiled beef top loin including fat (25.7 ± 0.05), simmered beef kidneys (25.5 ± 0.05), broiled beef round including fat (25.5 ± 0.05), caraway cheese (25.4 ± 0.18), braised pork kidneys (25.4 ± 0.05), gouda cheese (25.4 ± 0.18), cheddar cheese (25.4 ± 0.18), braised beef chuck blade roast including fat (25.4 ± 0.05), roasted pork shoulder excluding fat (25.4 ± 0.05), roasted pork center loin including fat (25.4 ± 0.05), edam cheese (25.4 ± 0.18), medium-rare broiled extra lean ground beef (25.4 ± 0.05), broiled beef rib eye including fat (25.4 ± 0.05), broiled beef flank excluding fat (25.4 ± 0.05), cooked swordfish (25.4 ± 0.06), dried butternuts (25.4 ± 0.18), roasted leg of lamb including fat (25.3 ± 0.06), stewed chicken drumstick with skin (25.3 ± 0.09), broiled pork boston blade excluding fat (25.2 ± 0.05), roasted goose with skin (25.2 ± 0.05), smoked haddock (25.2 ± 0.06), braised beef spleen (25.1 ± 0.05), broiled porterhouse steak including fat (25.1 ± 0.05), broiled beef flank including fat (25.1 ± 0.05), roasted chicken thigh with skin (25 ± 0.08), roasted pork leg including fat (25 ± 0.05), roasted pork sirloin including fat (25 ± 0.05), medium-rare fried extra lean ground beef (25 ± 0.05), dried pumpkin and squash seeds (25 ± 0.18), freeze-dried chives (25 ± 6.25), broiled pork loin blade excluding fat (24.9 ± 0.05), roasted cured pork arm excluding fat (24.9 ± 0.05), cooked mullet (24.8 ± 0.06), cooked grouper (24.8 ± 0.06), roasted pork loin blade excluding fat (24.7 ± 0.05), medium-rare broiled lean ground beef (24.7 ± 0.05), roasted pork center rib including fat (24.7 ± 0.05), cooked pike (24.7 ± 0.06), monterey cheese (24.6 ± 0.18), sardines (24.6 ± 0.21), tilsit cheese (24.6 ± 0.18), broiled pork center rib including fat (24.6 ± 0.05), dried black walnuts (24.6 ± 0.18), cooked tilefish (24.5 ± 0.06), roasted beef tenderloin including fat (24.5 ± 0.05), kippered herring (24.5 ± 0.13), medium-rare baked extra lean ground beef (24.5 ± 0.05), simmered chicken liver (24.4 ± 0.05), caviar (24.4 ± 0.31), braised beef liver (24.4 ± 0.05), roasted pork boston blade excluding fat (24.4 ± 0.05), simmered chicken neck without skin (24.4 ± 0.28), fried pork loin blade excluding fat (24.3 ± 0.05), roasted pork shank including fat (24.3 ± 0.05), dried pignolia pine nuts (24.3 ± 0.18), broiled pork sirloin including fat (24.2 ± 0.05), medium-rare fried lean ground beef (24.2 ± 0.05), roasted pork top loin including fat (24.2 ± 0.05), cooked haddock (24.2 ± 0.06), braised pork tongue (24.1 ± 0.05), medium-rare broiled ground beef (24.1 ± 0.05), cooked flatfish (24.1 ± 0.06), crunchy peanut butter (24.1 ± 0.16), simmered turkey liver (24 ± 0.05), braised pork loin blade including fat (24 ± 0.05), broiled T-bone steak including fat (24 ± 0.05), grilled canadian bacon (24 ± 0.11), cooked rockfish (24 ± 0.06), cooked crayfish (23.9 ± 0.06), port du salut cheese (23.9 ± 0.18), colby cheese (23.9 ± 0.18), fried chicken neck with skin (23.9 ± 0.14), medium-rare fried ground beef (23.9 ± 0.05), medium-rare baked lean ground beef (23.9 ± 0.05), cooked mackerel (23.9 ± 0.06), raw whelk (23.9 ± 0.06), cooked perch (23.9 ± 0.06), cooked mussels (23.8 ± 0.06), broiled pork top loin including fat (23.7 ± 0.05), raw pheasant without skin (23.6 ± 0.05), braised pork heart (23.6 ± 0.04), cooked eel (23.6 ± 0.06), cheshire cheese (23.6 ± 0.18), muenster cheese (23.6 ± 0.18), brick cheese (23.6 ± 0.18), broiled pork loin including fat (23.6 ± 0.05), cooked pompano (23.6 ± 0.06), cooked sea bass (23.6 ± 0.06), dry roasted peanuts (23.6 ± 0.18), roasted duck without skin (23.5 ± 0.05), cooked whiting (23.5 ± 0.06), stewed dark meat chicken with skin (23.5 ± 0.05), roasted pork loin including fat (23.4 ± 0.05), smoked whitefish (23.4 ± 0.06), fried pork center loin including fat (23.3 ± 0.05), raw yellowtail (23.2 ± 0.06), stewed chicken thigh with skin (23.2 ± 0.07), jellied corned beef loaf (23.2 ± 0.18), dried sunflower seeds (23.2 ± 0.18), cooked herring (23.1 ± 0.06), hard pork salami (23 ± 0.5), braised beef brisket including fat (23 ± 0.05), medium-rare baked ground beef (23 ± 0.05), hard pork and beef salami (23 ± 0.5), cooked carp (22.8 ± 0.06), stewed chicken wing with skin (22.8 ± 0.13), cooked cod (22.8 ± 0.06), cooked clams (22.8 ± 0.06), canned cod (22.8 ± 0.06), cooked rainbow smelt (22.6 ± 0.06), roasted ham (22.6 ± 0.05), american cheese (22.5 ± 0.18), pimento cheese spread (22.5 ± 0.18), fenugreek seed (22.5 ± 1.25), turkey breast meat summer sausage (22.4 ± 0.24), cooked kingfish (22.3 ± 0.05), roasted pork arm including fat (22.3 ± 0.05), pork link sausage (22.2 ± 0.07), simmered beef tongue (22.1 ± 0.05), roasted pork shoulder including fat (22 ± 0.05), broiled lamb loin chop including fat (22 ± 0.07), braised beef thymus (21.9 ± 0.05), broiled pork boston blade including fat (21.9 ± 0.05), roasted beef rib including fat (21.9 ± 0.05), raw reindeer (21.8 ± 0.05), raw quail without skin (21.8 ± 0.05), havarti cheese (21.8 ± 0.18), roasted pork boston blade including fat (21.8 ± 0.05), roquefort cheese (21.8 ± 0.18), bleu cheese (21.8 ± 0.18), oil roasted sunflower seeds (21.8 ± 0.18), roasted lamb shoulder including fat (21.6 ± 0.06), braised beef shortribs including fat (21.6 ± 0.05), fried pork center rib including fat (21.6 ± 0.05), raw coho salmon (21.6 ± 0.06), broiled beef rib including fat (21.5 ± 0.05), fried pork top loin including fat (21.5 ± 0.05), lobster paste (21.4 ± 0.71), breaded and fried shrimp (21.4 ± 0.06), raw sockeye salmon (21.3 ± 0.06), roasted canned lean ham (21.2 ± 0.05), raw venison (21.1 ± 0.06), brie cheese (21.1 ± 0.18), roasted pork loin blade including fat (21.1 ± 0.05), raw barracuda (21 ± 0.05), cooked black bass (20.9 ± 0.04), cooked shrimp (20.9 ± 0.06), roasted lean ham (20.9 ± 0.05), raw shark (20.9 ± 0.06), raw trout (20.8 ± 0.06), raw halibut (20.8 ± 0.06), cooked sturgeon (20.7 ± 0.06), broiled pork loin blade including fat (20.7 ± 0.05), oil roasted almonds (20.7 ± 0.18), toasted almonds (20.7 ± 0.18), dried pistachio nuts (20.7 ± 0.18), raw whale meat (20.6 ± 0.05), raw rainbow trout (20.6 ± 0.06), raw milkfish (20.6 ± 0.06), canned blue crab (20.5 ± 0.06), cooked lobster (20.5 ± 0.06), roasted canned ham (20.5 ± 0.05), unheated canadian bacon (20.5 ± 0.09), canned sockeye salmon (20.5 ± 0.06), raw snapper (20.5 ± 0.06), raw anchovy (20.4 ± 0.06), braised beef lungs (20.4 ± 0.05), raw shrimp (20.4 ± 0.06), limburger cheese (20.4 ± 0.18), roasted cured pork arm including fat (20.4 ± 0.05), chopped smoked beef (20.4 ± 0.18), dry roasted red pistachio nuts (20.4 ± 0.18), dried almonds (20.4 ± 0.18), cooked blue crab (20.2 ± 0.06), raw sheepshead (20.2 ± 0.06), crab cakes (20.2 ± 0.08), broiled lamb rib chop including fat (20.1 ± 0.07), raw chum salmon (20.1 ± 0.06), raw chinook salmon (20.1 ± 0.06), anchovy paste (20 ± 0.71), pork and beef pepperoni (20 ± 0.83), italian pork sausage (20 ± 0.07), camembert cheese (20 ± 0.18), raw bluefish (20 ± 0.06), raw pink salmon (20 ± 0.06), cumin seed (20 ± 2.5), basil (20 ± 5), tarragon (20 ± 2.5), dried dill weed (20 ± 5), celery seed (20 ± 2.5), anise seed (20 ± 2.5), caraway seed (20 ± 2.5), canned bonito (19.8 ± 0.05), canned pink salmon (19.8 ± 0.06), raw swordfish (19.8 ± 0.06), simmered chicken neck with skin (19.7 ± 0.13), fried abalone (19.6 ± 0.06), pork sausage (19.6 ± 0.19), mozzarella cheese (19.6 ± 0.18), lebanon bologna (19.6 ± 0.22), dry roasted sunflower seeds (19.6 ± 0.18), light meat chicken roll (19.5 ± 0.05), chicken roll (19.5 ± 0.09), raw alewife (19.4 ± 0.05), raw pollock (19.4 ± 0.06), raw sunfish (19.4 ± 0.06), cooked alaska king crab (19.4 ± 0.06), raw mullet (19.4 ± 0.06), unheated lean ham (19.4 ± 0.05), raw grouper (19.4 ± 0.06), pickled anchovy (19.3 ± 0.18), raw burbot (19.3 ± 0.06), raw pike (19.3 ± 0.06), raw black bass (19.2 ± 0.05), simmered pork feet (19.2 ± 0.05), raw whitefish (19.1 ± 0.06), roasted duck with skin (19 ± 0.05), raw ling (18.9 ± 0.06), raw cisco (18.9 ± 0.06), raw haddock (18.9 ± 0.06), raw cusk (18.9 ± 0.06), tempeh (18.9 ± 0.06), raw scup (18.8 ± 0.06), raw lobster (18.8 ± 0.06), fried pork loin blade including fat (18.8 ± 0.05), raw freshwater bass (18.8 ± 0.06), turkey ham (18.8 ± 0.09), raw flatfish (18.8 ± 0.06), freeze-dried red sweet peppers (18.8 ± 3.13), freeze-dried green sweet peppers (18.8 ± 3.13), raw duck liver (18.7 ± 0.05), raw crayfish (18.7 ± 0.06), beef honey roll sausage (18.7 ± 0.22), raw rockfish (18.7 ± 0.06), raw smelt (18.6 ± 0.05), raw mackerel (18.6 ± 0.06), battered and fried shark (18.6 ± 0.06), light meat turkey roll (18.6 ± 0.09), raw perch (18.6 ± 0.06), pork luxury loaf (18.6 ± 0.18), raw spot (18.5 ± 0.06), raw eel (18.5 ± 0.06), raw dolphinfish (18.5 ± 0.06), raw queen crab (18.5 ± 0.06), raw pompano (18.5 ± 0.06), raw sea bass (18.5 ± 0.06), canned lean ham (18.5 ± 0.05), canned smelt (18.4 ± 0.05), raw whiting (18.4 ± 0.06), raw alaska king crab (18.4 ± 0.06), cooked corned beef brisket (18.2 ± 0.05), breaded and fried croaker (18.2 ± 0.06), raw catfish (18.2 ± 0.06), turkey pastrami (18.2 ± 0.09), smoked chinook salmon

(18.2 ± 0.06), breaded and fried catfish (18.1 ± 0.06), raw blue crab (18.1 ± 0.06), breaded and fried scallops (18.1 ± 0.16), light and dark meat turkey roll (18.1 ± 0.09), breaded fried squid (18 ± 0.06), raw herring (18 ± 0.06), dry cocoa powder (18 ± 1), raw carp (17.9 ± 0.06), cashew butter (17.9 ± 0.18), raw striped bass (17.8 ± 0.06), raw croaker (17.8 ± 0.06), raw cod (17.8 ± 0.06), whole dried sesame seeds (17.8 ± 0.56), natto (17.7 ± 0.06), raw spiny dogfish (17.6 ± 0.05), raw rainbow smelt (17.6 ± 0.06), unheated ham (17.6 ± 0.05), raw lingcod (17.6 ± 0.06), raw tilefish (17.5 ± 0.06), beef pastrami (17.5 ± 0.18), raw freshwater drum (17.5 ± 0.06), raw wolffish (17.5 ± 0.06), pork and beef peppered loaf (17.5 ± 0.18), raw dungeness crab (17.4 ± 0.06), raw butterfish (17.3 ± 0.06), roasted pork blade roll (17.3 ± 0.05), sesame butter (17.3 ± 0.33), raw abalone (17.1 ± 0.06), chopped ham lunch meat (17.1 ± 0.24), toasted sesame kernels (17.1 ± 0.18), whole toasted sesame seeds (17.1 ± 0.18), simmered pork tail (17 ± 0.05), canned ham (17 ± 0.05), corn germ (17 ± 0.05), raw shad (16.9 ± 0.06), fried tofu (16.9 ± 0.38), turkey summer sausage (16.8 ± 0.18), ham and cheese roll (16.8 ± 0.18), raw scallops (16.8 ± 0.06), raw seatrout (16.7 ± 0.06), pork and beef mortadella (16.7 ± 0.33), bacon cheeseburger (16.7 ± 0.03), raw white sucker (16.7 ± 0.06), garlic powder (16.7 ± 1.67), poppy seed (16.7 ± 1.67), braised pork lungs (16.6 ± 0.05), boiled soybeans (16.6 ± 0.03), chicken egg yolk (16.5 ± 0.29), raw pork stomach (16.5 ± 0.05), raw frog legs (16.4 ± 0.05), raw goose liver (16.4 ± 0.05), ham and cheese spread (16.4 ± 0.18), smoked cisco (16.4 ± 0.06), dry roasted almonds (16.4 ± 0.18), oil roasted cashews (16.4 ± 0.18), raw black bullhead (16.3 ± 0.05), turkey salami (16.3 ± 0.09), dark rye flour (16.3 ± 0.04), canned alewife (16.2 ± 0.05), raw cuttlefish (16.2 ± 0.06), minced ham lunch meat (16.2 ± 0.24), raw sturgeon (16.1 ± 0.06), pork headcheese (16.1 ± 0.18), beef and pork summer sausage (16.1 ± 0.22), pork and beef honey loaf (16.1 ± 0.18), raw turbot (16 ± 0.06), tuna salad (16 ± 0.02), simmered pork ears (15.9 ± 0.05), firm raw tofu (15.8 ± 0.04), chicken salad (15.7 ± 0.18), summer sausage (15.7 ± 0.22), barbecue loaf (15.7 ± 0.22), dry roasted cashews (15.7 ± 0.18), ham and cheese sandwich (15.6 ± 0.03), raw squid (15.5 ± 0.06), cheese crackers with peanut butter (15.5 ± 0.12), boiled lupins (15.5 ± 0.03), pork livercheese (15.3 ± 0.13), pork and beef luncheon sausage (15.2 ± 0.22), beef salami (15.2 ± 0.22), pork bologna (15.2 ± 0.22), berliner (15.2 ± 0.22), pork and beef picnic loaf (15 ± 0.18), curry powder (15 ± 2.5), white pepper (15 ± 2.5), puffed wheat (15 ± 0.36), almond butter (15 ± 0.31), paprika (15 ± 2.5), dry roasted pistachio nuts (15 ± 0.18), dill seed (15 ± 2.5), fennel seed (15 ± 2.5), raw flounder (14.9 ± 0.05), raw sole (14.9 ± 0.05), raw octopus (14.9 ± 0.06), roast beef sandwich (14.9 ± 0.03), fried chicken (14.8 ± 0.06), beef summer sausage (14.8 ± 0.22), raw beef tripe (14.6 ± 0.05), beef lunch meat (14.6 ± 0.18), dried english/persian walnuts (14.6 ± 0.18), dried brazil nuts (14.6 ± 0.18), oil roasted filberts (14.6 ± 0.18), raw monkfish (14.5 ± 0.06), breaded and fried fish (14.4 ± 0.07), feta cheese (14.3 ± 0.18), polish pork sausage (14.3 ± 0.18), pork and beef brotwurst (14.3 ± 0.07), pork beerwurst salami (14.3 ± 0.22), bran (14.3 ± 0.18), smoked beef sausage (14.2 ± 0.12), breaded and fried clams (14.2 ± 0.06), cooked oysters (14.1 ± 0.06), pork bratwurst (14.1 ± 0.06), raw wild rice (14.1 ± 0.03), pork and beef link sausage with american cheese (14 ± 0.12), pickled herring (14 ± 0.33), goose egg (13.9 ± 0.03), pork liverwurst (13.9 ± 0.28), turkey bologna (13.9 ± 0.18), beef and pork salami (13.9 ± 0.22), pork and beef sausage (13.8 ± 0.38), turkey egg (13.7 ± 0.06), canned chicken liver pâté (13.6 ± 0.18), pork and beef old fashioned loaf (13.6 ± 0.18), pickled pork feet (13.5 ± 0.05), pork and beef kielbasa/kolbassy (13.5 ± 0.19), pork and beef link sausage (13.4 ± 0.07), boiled peanuts (13.4 ± 0.16), quail egg (13.3 ± 0.56), pork liver sausage (13.3 ± 0.28), rusk crackers (13.3 ± 0.56), popcorn (13.3 ± 0.83), whole wheat flour (13.3 ± 0.04), falafel (13.3 ± 0.1), grilled ham patties (13.2 ± 0.08), cheeseburger (13.2 ± 0.04), dried filberts (13.2 ± 0.18), taco (13.1 ± 0.06), stir-fried sprouted soybeans (13.1 ± 0.05), whole grain rye crackers (13.1 ± 0.38), duck egg (12.9 ± 0.07), chicken frankfurter (12.9 ± 0.11), turkey frankfurter (12.9 ± 0.11), pork and beef lunch meat (12.9 ± 0.18), bran flakes (12.9 ± 0.18), dried hickory nuts (12.9 ± 0.18), tomato powder (12.9 ± 0.05), unheated ham patties (12.8 ± 0.08), raw clams (12.8 ± 0.06), surimi scallops (12.8 ± 0.06), fried beef brains (12.6 ± 0.05), beef beerwurst salami (12.6 ± 0.22), dry breadcrumbs (12.6 ± 0.05), cottage cheese (12.5 ± 0.18), rice polish (12.5 ± 0.09), chicken egg (12.4 ± 0.1), canned pork lunch meat (12.4 ± 0.24), surimi shrimp (12.4 ± 0.06), hamburger sandwich (12.3 ± 0.05), boiled green soybeans (12.3 ± 0.06), beef bologna (12.2 ± 0.22), cheese pizza (12.2 ± 0.04), compressed bakers yeast (12.1 ± 0.18), braised pork brains (12.1 ± 0.05), beef frankfurter (12.1 ± 0.09), pork olive loaf (12.1 ± 0.18), almond paste (12.1 ± 0.18), bread sticks (12 ± 0.25), pork and beef knockwurst (11.9 ± 0.07), pork mothers loaf (11.9 ± 0.24), raw mussels (11.9 ± 0.06), rice bran (11.9 ± 0.05), pork pickle and pimento loaf (11.8 ± 0.18), chicken and turkey salad (11.8 ± 0.18), enriched wheat bread (11.8 ± 0.04), dried pinyon pine nuts (11.8 ± 0.18), miso (11.8 ± 0.04), beef and pork bologna (11.7 ± 0.22), dark buckwheat flour (11.7 ± 0.05), canned goose liver pâté (11.5 ± 0.38), canned smoked goose liver pâté (11.4 ± 0.18), ricotta cheese (11.3 ± 0.04), beef and pork frankfurter (11.3 ± 0.11), turkey salad (11.3 ± 0.09), tuna salad spread (11.3 ± 0.09), simmered beef brains (11.1 ± 0.05), hot dog (11.1 ± 0.06), shredded wheat (11.1 ± 0.18), dried pilinuts (11.1 ± 0.18), unsweetened baking chocolate (11.1 ± 0.18), toasted whole wheat bread (11 ± 0.24), sorghum grain (11 ± 0.05), bagels (10.9 ± 0.09), pepperoni pizza (10.8 ± 0.04), raisin bran (10.8 ± 0.14), ry-krisp crackers (10.7 ± 0.36), fish sandwich (10.6 ± 0.03), vienna sausage (10.6 ± 0.31), boiled winged beans (10.6 ± 0.03), toasted cracked wheat bread (10.5 ± 0.24), pita bread (10.5 ± 0.13), all purpose enriched wheat flour (10.5 ± 0.04), tamari sauce (10.5 ± 0.09), toasted pumpernickel bread (10.4 ± 0.18), dried ginko nuts (10.4 ± 0.18), simmered pork chitterlings (10.3 ± 0.05), chicken egg white (10.3 ± 0.15), neufchatel cheese (10 ± 0.18), chili powder (10 ± 1.67), turmeric (10 ± 2.5), savory (10 ± 5), saltine crackers (10 ± 0.83), zwieback crackers (10 ± 0.71), mixed grain bread (10 ± 0.2), whole wheat roll (10 ± 0.18), italian bread (10 ± 0.18), saffron (10 ± 5), onion powder (10 ± 2.5), dried coriander leaf (10 ± 5), dried chervil (10 ± 5), cayenne pepper (10 ± 2.5), dry roasted filberts (10 ± 0.18), oregano (10 ± 2.5), coriander seed (10 ± 2.5), sage (10 ± 5), bay leaf (10 ± 5), cardamom (10 ± 2.5), ginger (10 ± 2.5), black pepper (10 ± 2.5), freeze-dried shallots (10 ± 1.25), cheese soufflé (9.9 ± 0.05), millit (9.9 ± 0.05), boiled white beans (9.7 ± 0.03), gjetost cheese (9.6 ± 0.18), granola bar (9.6 ± 0.21), whole wheat bread (9.6 ± 0.2), french bread (9.6 ± 0.18), toasted english muffin (9.6 ± 0.1), pot barley pearled (9.6 ± 0.03), wheat flakes (9.6 ± 0.18), toasted white bread (9.5 ± 0.24), toasted rye bread (9.5 ± 0.23), light rye flour (9.4 ± 0.05), cheese crackers (9.3 ± 0.33), soda crackers (9.3 ± 0.18), pretzels (9.3 ± 0.18), boiled cranberry beans (9.3 ± 0.03), dried shitake mushrooms (9.3 ± 0.33), dehydrated onion flakes

(9.3 ± 0.36), chili (9.2 ± 0.02), waffles (9.2 ± 0.07), white whole ground corn meal (9.2 ± 0.04), yellow whole ground corn meal (9.2 ± 0.04), cracked wheat bread (9.2 ± 0.2), boiled yellow beans (9.2 ± 0.03), taco shell (9.1 ± 0.45), tostada shell (9.1 ± 0.45), pumpernickel bread (9.1 ± 0.16), boiled pink beans (9.1 ± 0.03), gefiltefish with broth (9 ± 0.12), toasted raisin bread (9 ± 0.24), white bolted corn meal (9 ± 0.04), yellow bolted corn meal (9 ± 0.04), boiled lentils (9 ± 0.03), ham salad (8.9 ± 0.18), burrito (8.9 ± 0.03), breaded and fried oysters (8.8 ± 0.06), oyster crackers (8.8 ± 0.63), french toast (8.8 ± 0.08), popover roll (8.8 ± 0.13), boiled chickpeas (8.8 ± 0.03), stir-fried sprouted lentils (8.8 ± 0.05), boiled black beans (8.8 ± 0.03), raw bacon (8.7 ± 0.07), boiled kidney beans (8.7 ± 0.03), boiled navy beans (8.7 ± 0.03), dinner rolls (8.6 ± 0.18), oatmeal bread (8.6 ± 0.18), french rolls (8.6 ± 0.1), dry roasted macadamia nuts (8.6 ± 0.18), dried macadamia nuts (8.6 ± 0.18), hamburger bun (8.5 ± 0.13), hot dog bun (8.5 ± 0.13), steamed sprouted soybeans (8.5 ± 0.11), frozen and boiled black-eyed peas (8.5 ± 0.06), raisin bread (8.4 ± 0.2), rye bread (8.4 ± 0.2), boiled great northern beans (8.4 ± 0.03), boiled split peas (8.4 ± 0.03), white bread (8.3 ± 0.21), boiled yardlong bean (8.3 ± 0.03), light pearled barley (8.2 ± 0.03), corn flakes (8.2 ± 0.18), dry roasted pecans (8.2 ± 0.18), dried acorns (8.2 ± 0.18), boiled pinto beans (8.2 ± 0.03), boiled black turtle beans (8.2 ± 0.03), boiled hyacinth beans (8.1 ± 0.03), raw tofu (8.1 ± 0.04), boiled catjang black-eyed peas (8.1 ± 0.03), angel food cake (8 ± 0.08), potato flour (8 ± 0.06), boiled baby lima beans (8 ± 0.03), pork and beef spread (7.9 ± 0.18), sweetened condensed milk (7.9 ± 0.13), bittersweet chocolate (7.9 ± 0.18), milk chocolate (7.9 ± 0.18), english muffin (7.9 ± 0.09), enriched white degermed corn meal (7.9 ± 0.04), unenriched corn meal (7.9 ± 0.04), enriched yellow degermed corn meal (7.9 ± 0.04), unenriched yellow degermed corn meal (7.9 ± 0.04), dried pecans (7.9 ± 0.18), dried oriental radish (7.9 ± 0.09), white corn flour (7.8 ± 0.04), yellow corn flour (7.8 ± 0.04), boiled mothbeans (7.8 ± 0.03), boiled lima beans (7.8 ± 0.03), boiled black-eyed peas (7.7 ± 0.03), boiled mungo beans (7.6 ± 0.03), boiled broad beans (7.6 ± 0.03), enchilada (7.5 ± 0.02), cream cheese (7.5 ± 0.18), evaporated skim milk (7.5 ± 0.16), enriched biscuits (7.5 ± 0.18), unenriched biscuits (7.5 ± 0.18), bran muffins (7.5 ± 0.13), wheat cake (7.5 ± 0.04), boiled adzuki beans (7.5 ± 0.02), acorn flour (7.5 ± 0.18), oil roasted macadamia nuts (7.5 ± 0.18), canned navy beans (7.5 ± 0.02), canned great northern beans (7.4 ± 0.02), sponge cake (7.3 ± 0.08), whole ground corn meal muffins (7.3 ± 0.13), blueberry muffins (7.3 ± 0.13), canned white beans (7.3 ± 0.02), evaporated lowfat milk (7.1 ± 0.04), canned oysters (7.1 ± 0.06), hush puppies (7.1 ± 0.11), graham crackers (7.1 ± 0.36), oil roasted pecans (7.1 ± 0.18), dried coconut (7.1 ± 0.18), boiled french beans (7.1 ± 0.03), boiled sprouted green peas (7.1 ± 0.05), raw oysters (7 ± 0.06), lemon chiffon pie (7 ± 0.06), corn meal muffins (7 ± 0.13), pancakes (7 ± 0.19), corn tortilla (7 ± 0.17), boiled mung beans (7 ± 0.02), raisin bun (6.9 ± 0.08), evaporated whole milk (6.8 ± 0.04), chocolate chiffon pie (6.8 ± 0.06), dried chinese chestnuts (6.8 ± 0.18), boiled pigeon peas (6.8 ± 0.03), boiled and frozen baby lima beans (6.7 ± 0.06), raw garlic (6.7 ± 0.56), cottage pudding (6.5 ± 0.09), pineapple chiffon pie (6.5 ± 0.06), animal crackers (6.5 ± 0.19), cream puff (6.5 ± 0.04), raw pork jowl (6.4 ± 0.05), light buckwheat flour (6.4 ± 0.05), puffed rice (6.4 ± 0.36), potato chips (6.4 ± 0.18), raw acorns (6.4 ± 0.18), dried european chestnuts (6.4 ± 0.18), shortbread cookie (6.3 ± 0.63), sugar cookie

(6.3 ± 0.31), rice flour (6.3 ± 0.04), danish pastry (6.2 ± 0.12), oatmeal cookie (6.2 ± 0.38), sweet roll (6.2 ± 0.12), bulgur (6.2 ± 0.04), refried beans (6.2 ± 0.02), dried agar seaweed (6.2 ± 0.05), pie crust (6.1 ± 0.03), custard pie (6.1 ± 0.04), potato pancakes (6.1 ± 0.07), frozen and boiled fordhook lima beans (6.1 ± 0.06), teriyaki sauce (6.1 ± 0.28), fruitcake, light (6 ± 0.12), sheep milk (6 ± 0.02), coconut custard pie (6 ± 0.04), brown mustard (6 ± 1), canned black turtle beans (6 ± 0.02), white cake (5.9 ± 0.06), raw spirulina seaweed (5.9 ± 0.05), raw laver/nori seaweed (5.8 ± 0.05), pound cake (5.7 ± 0.17), devil's food cake (5.7 ± 0.08), yellow cake (5.7 ± 0.07), skim yogurt (5.7 ± 0.02), peanut brittle (5.7 ± 0.18), bread pudding (5.6 ± 0.02), canned cranberry beans (5.5 ± 0.02), canned broad beans (5.5 ± 0.02), cheesecake (5.4 ± 0.06), custard (5.4 ± 0.02), semi-sweet bakers chocolate (5.4 ± 0.18), dried toasted coconut (5.4 ± 0.18), dried japanese chestnuts (5.4 ± 0.18), raw green peas (5.4 ± 0.06), boiled green peas (5.4 ± 0.06), baked beans with pork and sweet sauce (5.3 ± 0.02), baked beans with pork and tomato sauce (5.2 ± 0.02), lowfat yogurt (5.2 ± 0.02), canned kidney beans (5.2 ± 0.02), shoyu sauce (5.2 ± 0.09), au gratin potatoes (5.1 ± 0.04), peach pie (5.1 ± 0.05), pecan pie (5.1 ± 0.05), boiled succotash (5.1 ± 0.05), frozen and boiled green peas (5.1 ± 0.06), brownies (5 ± 0.25), boston cream pie (5 ± 0.05), tapioca cream pudding (5 ± 0.03), cloves (5 ± 2.5), pumpkin pie spice (5 ± 2.5), dried rosemary (5 ± 2.5), poultry seasoning (5 ± 2.5), dried pepeao (5 ± 0.42), mace (5 ± 2.5), canned chickpeas (5 ± 0.02), nutmeg (5 ± 2.5), allspice (5 ± 2.5), cinnamon (5 ± 2.5), fruitcake, dark (4.9 ± 0.12), dried longans (4.9 ± 0.05), hummus (4.9 ± 0.02), canned lima beans (4.9 ± 0.02), canned black-eyed peas (4.8 ± 0.02), vegetarian baked beans (4.8 ± 0.02), cornbread (4.7 ± 0.06), corn pone (4.7 ± 0.06), canned butter beans (4.7 ± 0.04), carob flour (4.7 ± 0.05), chocolate cream pie (4.6 ± 0.04), roasted chinese chestnuts (4.6 ± 0.18), canned pinto beans (4.6 ± 0.02), onion rings (4.5 ± 0.06), sweet potato pie (4.5 ± 0.04), banana custard pie (4.5 ± 0.04), red beans (4.5 ± 0.04), canned green peas (4.5 ± 0.06), prune pudding (4.4 ± 0.04), yellow corn pudding (4.4 ± 0.02), caramel sundae (4.4 ± 0.03), hot fudge sundae (4.4 ± 0.03), bread stuffing (4.4 ± 0.03), butterscotch pie (4.4 ± 0.04), frozen and boiled succotash (4.4 ± 0.06), stir-fried sprouted mung beans (4.4 ± 0.08), semi-sweet chocolate (4.3 ± 0.18), sweet chocolate (4.3 ± 0.18), gingersnaps cookie (4.3 ± 0.71), ginko nuts (4.3 ± 0.18), raw chinese chestnuts (4.3 ± 0.18), strawberry sundae (4.2 ± 0.03), boiled arrowhead (4.2 ± 0.42), noodles (4.1 ± 0.03), malted milk (4.1 ± 0.02), cooked tahitian taro (4.1 ± 0.07), french fries (4 ± 0.06), pumpkin pie (4 ± 0.04), pineapple custard pie (4 ± 0.04), yellow mustard (4 ± 1), french fried potatoes (4 ± 0.1), vanilla milkshake (3.9 ± 0.02), protein fortified lowfat 2% milk (3.9 ± 0.02), protein fortified lowfat 1% milk (3.9 ± 0.02), protein fortified skim milk (3.9 ± 0.02), raw sprouted alfalfa seeds (3.9 ± 0.15), eggnog (3.8 ± 0.02), indian buffalo milk (3.8 ± 0.02), toaster pastry (3.8 ± 0.1), dried lychees (3.8 ± 0.05), tomato paste (3.8 ± 0.04), caramel cake (3.7 ± 0.05), dried jujube (3.7 ± 0.05), macaroni (3.7 ± 0.04), spaghetti (3.7 ± 0.04), dried apricots (3.7 ± 0.14), boiled jute potherb (3.7 ± 0.12), hot chocolate (3.6 ± 0.02), goat milk (3.6 ± 0.02), german sweet bakers chocolate (3.6 ± 0.18), coconut cream (3.6 ± 0.02), frozen and boiled brussels sprouts (3.6 ± 0.06), vanilla pudding (3.5 ± 0.02), pancakes with butter and syrup (3.5 ± 0.03), whole chocolate malted milk (3.5 ± 0.02), whole yogurt (3.5 ± 0.02), restaurant vanilla milkshake (3.5 ± 0.02), restaurant chocolate milkshake (3.4 ± 0.02), restaurant strawberry milkshake (3.4 ± 0.02), skim milk (3.4 ± 0.02), frozen and french fried potatoes without skin (3.4 ± 0.1), frozen and boiled turnip greens (3.4 ± 0.06), golden seedless raisins (3.4 ± 0.05), pineapple upside-down cake (3.3 ± 0.07), sour cream (3.3 ± 0.42), whole milk (3.3 ± 0.02), lowfat 2% milk (3.3 ± 0.02), lowfat 1% milk (3.3 ± 0.02), buttermilk (3.3 ± 0.02), filled milk (3.3 ± 0.02), okara tofu (3.3 ± 0.08), boiled yellow corn (3.3 ± 0.06), raw coconut (3.3 ± 0.11), raw chives (3.3 ± 1.67), gingerbread (3.2 ± 0.08), carob milk (3.2 ± 0.02), whole chocolate milk (3.2 ± 0.02), lowfat 2% chocolate milk (3.2 ± 0.02), lowfat 1% chocolate milk (3.2 ± 0.02), lemon meringue pie (3.2 ± 0.04), frozen and fried hash brown potatoes (3.2 ± 0.06), roasted european chestnuts (3.2 ± 0.18), dried sweetened flaked coconut (3.2 ± 0.2), boiled lamb's-quarters (3.2 ± 0.06), seedless raisins (3.2 ± 0.05), frozen and boiled spinach (3.2 ± 0.05), frozen and boiled broccoli (3.2 ± 0.05), chocolate pudding (3.1 ± 0.02), chocolate milkshake (3.1 ± 0.02), boiled artichoke (3.1 ± 0.06), raw cassava (3.1 ± 0.05), raw sprouted mung beans (3.1 ± 0.1), dried figs (3 ± 0.03), raw wakame seaweed (3 ± 0.05), frozen and boiled asparagus (3 ± 0.08), raw broccoli (3 ± 0.11), boiled spinach (3 ± 0.06), frozen and boiled yellow corn (3 ± 0.06), raw shallots (3 ± 0.5), scalloped potatoes (2.9 ± 0.04), frozen white corn (2.9 ± 0.07), raw dandelion greens (2.9 ± 0.18), frozen and boiled collards (2.9 ± 0.06), dried sweetened shredded coconut (2.9 ± 0.05), boiled and steamed chinese chestnuts (2.9 ± 0.18), roasted japanese chestnuts (2.9 ± 0.18), frozen and boiled kale (2.9 ± 0.08), boiled broccoli (2.9 ± 0.06), raw spinach (2.9 ± 0.18), boiled salsify (2.8 ± 0.07), soy milk (2.8 ± 0.02), raw garden cress (2.8 ± 0.2), canned spinach (2.8 ± 0.05), tamarind (2.8 ± 0.04), medium cream (2.7 ± 0.33), potato salad (2.7 ± 0.04), half and half cream (2.7 ± 0.33), canned shitake mushrooms (2.7 ± 0.04), raw swamp cabbage (2.7 ± 0.09), chili sauce (2.7 ± 0.33), sweetened coconut cream (2.7 ± 0.02), cherry pie (2.6 ± 0.04), blackberry pie (2.6 ± 0.04), raisin pie (2.6 ± 0.04), oatmeal (2.6 ± 0.03), boiled brussels sprouts (2.6 ± 0.06), canned succotash (2.6 ± 0.04), canned succotash with creamed corn (2.6 ± 0.04), dried prunes (2.6 ± 0.06), boiled asparagus (2.6 ± 0.06), raw bamboo shoots (2.6 ± 0.07), boiled beet greens (2.6 ± 0.07), boiled blackeyed-pea pods (2.6 ± 0.11), raw lotus root (2.6 ± 0.06), mince pie (2.5 ± 0.04), rhubarb pie (2.5 ± 0.04), instant oatmeal (2.5 ± 0.03), cooked brown rice (2.5 ± 0.03), raw coriander (2.5 ± 1.25), raw european chestnuts (2.5 ± 0.18), baked hubbard squash (2.5 ± 0.05), seeded raisins (2.5 ± 0.05), cooked parboiled extra long grain white rice (2.5 ± 0.06), cooked extra long grain white rice (2.5 ± 0.06), o'brien potatoes (2.4 ± 0.03), blueberry pie (2.4 ± 0.04), maypo (2.4 ± 0.03), hash brown potatoes (2.4 ± 0.06), canned yellow corn with red and green peppers (2.4 ± 0.04), cooked wild long grain rice (2.4 ± 0.04), canned yellow corn (2.4 ± 0.05), soy sauce (2.4 ± 0.09), microwaved potato with skin (2.4 ± 0.02), raw watercress (2.4 ± 0.29), frozen creamed yellow corn (2.3 ± 0.04), boiled pokeberry shoots (2.3 ± 0.06), maltex (2.3 ± 0.03), canned white corn (2.3 ± 0.05), raw mustard spinach (2.3 ± 0.07), raw parsley (2.3 ± 0.17), coconut milk (2.3 ± 0.02), boiled artichoke hearts (2.3 ± 0.06), boiled mustard greens (2.3 ± 0.07), frozen and boiled mustard greens (2.3 ± 0.07), baked potato with skin (2.3 ± 0.02), pineapple pie (2.2 ± 0.04), passion fruit (2.2 ± 0.28), cooked instant white rice (2.2 ± 0.03), boiled mushrooms (2.2 ± 0.06), sapote (2.1 ± 0.02), sugar apple (2.1 ± 0.03), boiled burdock root (2.1 ± 0.04), cooked parboiled white rice (2.1 ± 0.03), raw california avocado (2.1 ± 0.03), boiled and steamed european chestnuts (2.1 ± 0.18), boiled amaranth (2.1 ± 0.08), frozen and boiled okra (2.1 ± 0.05), frozen and boiled turnip greens and turnips (2.1 ± 0.05), microwaved potato without skin (2.1 ± 0.03), raw japanese chestnuts (2.1 ± 0.18), cooked sprouted mung beans (2.1 ± 0.08), raw potato without skin (2.1 ± 0.04), whipping cream, heavy (2 ± 0.33), whipping cream, light (2 ± 0.33), apple pie (2 ± 0.04), tomato catsup (2 ± 0.33), boiled swamp cabbage (2 ± 0.1), cooked white rice (2 ± 0.02), soybean oil and cottonseed oil margarine liquid (2 ± 1), raw mushrooms (2 ± 0.14), frozen and boiled potatoes without skin (2 ± 0.05), baked potato without skin (2 ± 0.03), raw cauliflower (2 ± 0.1), raw red chili peppers (2 ± 0.11), raw green chili peppers (2 ± 0.11), raw savoy cabbage (2 ± 0.14), raw jerusalem artichoke (2 ± 0.07), mashed potatoes (1.9 ± 0.05), strawberry pie (1.9 ± 0.05), raw dock (1.9 ± 0.07), ground-cherries (1.9 ± 0.04), boiled dandelion greens (1.9 ± 0.1), dried dates (1.9 ± 0.06), boiled swiss chard (1.9 ± 0.06), barbecue sauce (1.9 ± 0.31), raw welsh onions (1.9 ± 0.05), boiled green beans (1.9 ± 0.08), boiled yellow snap beans (1.9 ± 0.08), boiled cauliflower (1.9 ± 0.08), boiled garden cress (1.9 ± 0.07), boiled okra (1.9 ± 0.06), boiled dock (1.8 ± 0.05), hot dog pickle relish (1.8 ± 0.18), raw vine spinach (1.8 ± 0.05), boiled kale (1.8 ± 0.08), boiled scotch kale (1.8 ± 0.08), canned bamboo shoots (1.8 ± 0.04), baked sweet potato (1.8 ± 0.04), chocolate syrup (1.8 ± 0.13), raw romaine lettuce (1.8 ± 0.18), raw spring onions with tops (1.8 ± 0.1), boiled savoy cabbage (1.8 ± 0.07), boiled kohlrabi (1.8 ± 0.06), restaurant coleslaw (1.7 ± 0.05), boiled mustard spinach (1.7 ± 0.06), marshmallow (1.7 ± 0.83), canned creamed yellow corn (1.7 ± 0.04), raw chickory greens (1.7 ± 0.06), frozen and boiled cauliflower (1.7 ± 0.06), canned shellie beans (1.7 ± 0.04), canned sweet potato (1.7 ± 0.03), boiled potato without skin (1.7 ± 0.04), raw ginger root (1.7 ± 0.21), raw kelp/kombu/tangle seaweed (1.7 ± 0.05), tomato purée (1.7 ± 0.02), steamed hawaii mountain yam (1.7 ± 0.07), frozen and baked sweet potato (1.7 ± 0.06), lemon peel (1.7 ± 0.83), orange peel (1.7 ± 0.83), apple brown betty (1.6 ± 0.02), boiled purslane (1.6 ± 0.09), raw florida avocado (1.6 ± 0.02), marinara sauce (1.6 ± 0.02), boiled sweet potato (1.6 ± 0.03), boiled lotus root (1.6 ± 0.06), dried japanese persimmon (1.5 ± 0.15), loganberries (1.5 ± 0.03), jackfruit (1.5 ± 0.05), cream of wheat (1.5 ± 0.03), raw chinese water chestnuts (1.5 ± 0.08), instant corn grits (1.5 ± 0.04), boiled hubbard squash (1.5 ± 0.04), boiled bamboo shoots (1.5 ± 0.04), frozen and boiled turnips (1.5 ± 0.05), raw irish moss seaweed (1.5 ± 0.05), boiled chinese cabbage (1.5 ± 0.06), raw leeks (1.5 ± 0.19), cooked shitake mushrooms (1.5 ± 0.07), baked/boiled tropical yam (1.5 ± 0.07), canned sprouted mung beans (1.5 ± 0.08), beef suet (1.4 ± 0.18), mulberries (1.4 ± 0.04), corn grits (1.4 ± 0.02), farina (1.4 ± 0.03), safflower oil and soybean oil mayonnaise (1.4 ± 0.36), soybean oil mayonnaise (1.4 ± 0.36), black currants (1.4 ± 0.09), apricots (1.4 ± 0.05), canned turnip greens (1.4 ± 0.04), canned potato without skin (1.4 ± 0.06), raw looseleaf lettuce (1.4 ± 0.18), raw new zealand spinach (1.4 ± 0.18), raw turnip greens (1.4 ± 0.18), raw chinese cabbage (1.4 ± 0.14), raw red cabbage (1.4 ± 0.14), red currants (1.4 ± 0.09), white currants (1.4 ± 0.09), russian salad dressing (1.3 ± 0.33), coleslaw (1.3 ± 0.08), cherimoya (1.3 ± 0.01), longans (1.3 ± 0.05), boiled new zealand spinach (1.3 ± 0.06), frozen and boiled crookneck squash (1.3 ± 0.05), tomato sauce (1.3 ± 0.04), boiled parsnips (1.3 ± 0.06), frozen and boiled yellow snap beans (1.3 ± 0.07), frozen and boiled green beans (1.3 ± 0.07), frozen and boiled butternut squash (1.3 ± 0.04), raw butterhead lettuce (1.3 ± 0.33), horseradish sauce (1.3 ± 0.33), jujube (1.2 ± 0.05), boiled yambean

(1.2 ± 0.05), pickled rip sevillano/ ascolano olives (1.2 ± 0.31), cherries (1.2 ± 0.07), raw scallop squash (1.2 ± 0.08), frozen and boiled zucchini (1.2 ± 0.04), raw green tomato (1.2 ± 0.04), boiled carrots (1.2 ± 0.06), raw endive (1.2 ± 0.2), raw zucchini (1.2 ± 0.08), canned green beans (1.2 ± 0.07), canned yellow snap beans (1.2 ± 0.07), frozen and boiled carrots (1.2 ± 0.07), raw eggplant (1.2 ± 0.12), pickled green olives (1.1 ± 0.11), roselle (1.1 ± 0.09), kiwi fruit (1.1 ± 0.07), boysenberries (1.1 ± 0.04), wax beans (1.1 ± 0.04), boiled collards (1.1 ± 0.03), cooked enriched white degermed corn meal (1.1 ± 0.02), cooked unenriched corn meal (1.1 ± 0.02), cooked enriched yellow degermed corn meal (1.1 ± 0.02), cooked unenriched yellow degermed corn meal (1.1 ± 0.02), yellow hominy (1.1 ± 0.02), yellow hominy with red and green peppers (1.1 ± 0.02), kumquats (1.1 ± 0.26), bananas (1.1 ± 0.04), boiled red cabbage (1.1 ± 0.07), boiled turnip greens (1.1 ± 0.07), boiled red tomato (1.1 ± 0.04), raw onions (1.1 ± 0.06), valencia orange (1.1 ± 0.04), boiled rutabaga (1.1 ± 0.06), canned zucchini in tomato juice (1.1 ± 0.04), baked acorn squash (1.1 ± 0.05), raw green cabbage (1.1 ± 0.14), raw chickory (1.1 ± 0.11), canned pumpkin (1.1 ± 0.04), cooked prunes (1.1 ± 0.05), canned pumpkin pie mix (1.1 ± 0.04), boiled beets (1.1 ± 0.06), human milk (1 ± 0.16), nectarine (1 ± 0.04), soursop (1 ± 0.02), pomegranate (1 ± 0.03), raw celeriac (1 ± 0.05), breadfruit (1 ± 0.05), boiled scallop squash (1 ± 0.06), frozen and boiled red sweet peppers (1 ± 0.05), frozen and boiled green sweet peppers (1 ± 0.05), boiled onions (1 ± 0.05), raw carrots (1 ± 0.07), lemon (1 ± 0.09), navel orange (1 ± 0.04), raw iceberg lettuce (1 ± 0.25), whipped butter (0.9 ± 0.45), whey (0.9 ± 0.02), raw celtuce (0.9 ± 0.05), cantaloupe (0.9 ± 0.03), casaba melon (0.9 ± 0.03), cream of rice (0.9 ± 0.03), canned chinese water chestnuts (0.9 ± 0.07), raspberries (0.9 ± 0.04), candied sweet potato (0.9 ± 0.05), raw crookneck squash (0.9 ± 0.08), boiled crookneck squash (0.9 ± 0.06), canned peeled red tomatoes (0.9 ± 0.04), canned red sweet peppers (0.9 ± 0.07), canned green sweet peppers (0.9 ± 0.07), raw red tomato (0.9 ± 0.04), canned sauerkraut (0.9 ± 0.04), canned stewed red tomatoes (0.9 ± 0.04), carrot juice (0.9 ± 0.03), boiled green cabbage (0.9 ± 0.07), gooseberries (0.9 ± 0.03), canned red chili peppers (0.9 ± 0.07), canned green chili peppers (0.9 ± 0.07), boiled butternut squash (0.9 ± 0.05), prickly pear (0.8 ± 0.05), pitanga (0.8 ± 0.03), lychees (0.8 ± 0.05), persimmon (0.8 ± 0.2), honeydew melon (0.8 ± 0.05), white hominy (0.8 ± 0.02), dill pickles (0.8 ± 0.08), restaurant hot chocolate (0.8 ± 0.02), cooked plantain (0.8 ± 0.03), restaurant tomato juice (0.8 ± 0.04), guava (0.8 ± 0.06), plum (0.8 ± 0.08), raw red sweet peppers (0.8 ± 0.1), raw green sweet peppers (0.8 ± 0.1), figs (0.8 ± 0.1), boiled leeks (0.8 ± 0.19), raw celery (0.8 ± 0.13), boiled eggplant (0.8 ± 0.1), tomato juice (0.8 ± 0.03), boiled turnips (0.8 ± 0.06), boiled cardoon (0.8 ± 0.05), pickled beets (0.8 ± 0.04), italian salad dressing (0.7 ± 0.33), butter (0.7 ± 0.33), sweet butter (0.7 ± 0.33), mayonnaise (0.7 ± 0.36), light mayonnaise (0.7 ± 0.36), low calorie thousand island salad dressing (0.7 ± 0.33), sweet pickles (0.7 ± 0.33), hamburger pickle relish (0.7 ± 0.18), elderberries (0.7 ± 0.03), blackberries (0.7 ± 0.07), blueberries (0.7 ± 0.03), yellow passion fruit juice (0.7 ± 0.02), java plum (0.7 ± 0.04), pummelo (0.7 ± 0.03), restaurant orange juice (0.7 ± 0.04), bread and butter pickles (0.7 ± 0.33), canned jalapeño peppers (0.7 ± 0.07), european grapes (0.7 ± 0.03), thompson seedless grapes (0.7 ± 0.03), boiled oriental radish (0.7 ± 0.07), boiled dishcloth gourd (0.7 ± 0.06), american grapes (0.7 ± 0.05), canned carrots (0.7 ± 0.07), raw radish (0.7 ± 0.11), boiled and steamed japanese chestnuts

(0.7 ± 0.18), coconut water (0.7 ± 0.02), orange juice (0.7 ± 0.02), boiled zucchini (0.7 ± 0.06), boiled pumpkin (0.7 ± 0.04), frozen and boiled onions (0.7 ± 0.05), white grapefruit (0.7 ± 0.04), boiled acorn squash (0.7 ± 0.04), lime (0.7 ± 0.07), peach (0.7 ± 0.06), raw oriental radish (0.7 ± 0.11), thousand island salad dressing (0.6 ± 0.31), low calorie russian salad dressing (0.6 ± 0.31), boiled chayote (0.6 ± 0.06), watermelon (0.6 ± 0.03), rose apple (0.6 ± 0.05), carambola (0.6 ± 0.04), japanese persimmon (0.6 ± 0.03), raw frozen rhubarb (0.6 ± 0.04), strawberries (0.6 ± 0.03), boiled red sweet peppers (0.6 ± 0.07), boiled green sweet peppers (0.6 ± 0.07), boiled/baked spaghetti squash (0.6 ± 0.06), raw cucumber (0.6 ± 0.1), tangerine (0.6 ± 0.06), canned crookneck squash (0.6 ± 0.05), pink grapefruit (0.6 ± 0.04), papaya (0.6 ± 0.02), grape juice (0.6 ± 0.02), mandarin oranges (0.6 ± 0.04), screwdriver (0.6 ± 0.02), prune juice (0.6 ± 0.02), honey (0.5 ± 0.24), carissa (0.5 ± 0.25), mammy apple (0.5 ± 0.05), sour pickles (0.5 ± 0.08), raw pepeao (0.5 ± 0.05), boiled waxgourd (0.5 ± 0.06), mango (0.5 ± 0.02), tangerine juice (0.5 ± 0.02), cooked taro (0.5 ± 0.08), boiled celery (0.5 ± 0.07), white grapefruit juice (0.5 ± 0.02), red grapefruit juice (0.5 ± 0.02), bloody mary (0.5 ± 0.03), boiled calabash gourd (0.5 ± 0.07), raw agar seaweed (0.5 ± 0.05), citrus marmalade (0.5 ± 0.25), clam liquid (0.4 ± 0.02), sapodilla (0.4 ± 0.03), acerola juice (0.4 ± 0.02), raw acerola (0.4 ± 0.05), oheloberries (0.4 ± 0.04), cranberry sauce (0.4 ± 0.05), frozen and cooked sweetened rhubarb (0.4 ± 0.04), lemon juice (0.4 ± 0.02), piña colada (0.4 ± 0.04), pineapple (0.4 ± 0.03), pear (0.4 ± 0.03), crab apples (0.4 ± 0.05), loquats (0.4 ± 0.05), poi (0.4 ± 0.04), lime juice (0.4 ± 0.02), apricot nectar (0.4 ± 0.02), quince (0.4 ± 0.05), purple passion fruit juice (0.4 ± 0.02), cooked apple without skin (0.3 ± 0.03), pineapple juice (0.3 ± 0.02), tequila sunrise (0.3 ± 0.03), ale (0.3 ± 0.01), malt liquor (0.3 ± 0.01), beer (0.3 ± 0.01), peach nectar (0.3 ± 0.02), apple tapioca pudding (0.2 ± 0.02), boiled butterbur (0.2 ± 0.05), raw apple with skin (0.2 ± 0.04), raw apple without skin (0.2 ± 0.04), applesauce (0.2 ± 0.04), papaya nectar (0.2 ± 0.02), whiskey sour (0.2 ± 0.06), rose table wine (0.2 ± 0.05), red table wine (0.2 ± 0.05), sweet dessert wine (0.2 ± 0.08), dry dessert wine (0.2 ± 0.08), coffee (0.1 ± 0.03), apple juice (0.1 ± 0.02), chamomile tea (0.1 ± 0.03), coffee (0.1 ± 0.03), white table wine (0.1 ± 0.05), pear nectar (0.1 ± 0.02)

Applications: celiac disease
Daily Dosage: 1980 RDA—44–76 g; 1989 RDA—44–76 g; nutritional—44–56 g; therapeutic—56–76 g; experimental—80–100 g; toxic—?
Warnings: Keep levels low with epilepsy, decreased blood sugar, kidney inflammation and skin psoriasis. When combining incomplete proteins, all essential amino acids must be in diet. Contrary to former theories, all essential amino acids do not have to be consumed at the same time. The body can store them for a few days.

WATER

Classification: macronutrient
Deficiency: thirst, dehydration
Toxicity: none known
Side-effects: (1½ gallons within an hour could be dangerous)
Sources (grams per 100 grams of food):

Perrier water (99.9 ± 0.03), black tea (99.7 ± 0.03), instant tea (99.7 ± 0.02), chamomile tea (99.7 ± 0.03), restaurant coffee (99.6 ± 0.03), tea (99.6 ± 0.02), coffee (99.3 ± 0.03), clam liquid (97.7 ± 0.02), boiled butterbur (96.7 ± 0.05), canned sprouted mung beans (96.1 ± 0.08), boiled waxgourd (96.1 ± 0.06), raw iceberg lettuce (96 ± 0.25), raw cucumber (96 ± 0.1), canned crookneck squash (96 ± 0.05), boiled bamboo shoots (95.9 ± 0.04), boiled collards (95.7 ± 0.03), boiled chinese cabbage (95.5 ± 0.06), raw chinese cabbage (95.4 ± 0.14), boiled calabash gourd (95.3 ± 0.07), raw butterhead lettuce (95.3 ± 0.33), distilled vinegar (95.3 ± 0.33), raw watercress (95.3 ± 0.29), raw zucchini (95.2 ± 0.08), raw chickory (95.1 ± 0.11), boiled celery (95.1 ± 0.07), coconut water (95 ± 0.02), boiled oriental radish (95 ± 0.07), boiled scallop squash (95 ± 0.06), raw romaine lettuce (95 ± 0.18), raw radish (94.9 ± 0.11), boiled zucchini (94.8 ± 0.06), raw celery (94.8 ± 0.13), boiled new zealand spinach (94.8 ± 0.06), sour pickles (94.8 ± 0.08), boiled red sweet peppers (94.7 ± 0.07), boiled green sweet peppers (94.7 ± 0.07), frozen and boiled red sweet peppers (94.7 ± 0.05), frozen and boiled green sweet peppers (94.7 ± 0.05), frozen and boiled zucchini (94.7 ± 0.04), canned turnip greens (94.7 ± 0.04), boiled mustard spinach (94.6 ± 0.06), raw celtuce (94.5 ± 0.05), raw oriental radish (94.5 ± 0.11), canned bamboo shoots (94.4 ± 0.04), boiled mustard greens (94.4 ± 0.07), acerola juice (94.3 ± 0.02), raw crookneck squash (94.2 ± 0.08), raw scallop squash (94.2 ± 0.08), frozen and boiled turnip greens and turnips (94.2 ± 0.05), raw red tomato (94 ± 0.04), frozen and boiled cauliflower (94 ± 0.06), raw endive (94 ± 0.2), cider vinegar (94 ± 0.33), frozen and boiled mustard greens (93.9 ± 0.07), tomato juice (93.9 ± 0.03), raw looseleaf lettuce (93.9 ± 0.18), restaurant tomato juice (93.9 ± 0.04), raw new zealand spinach (93.9 ± 0.18), boiled pumpkin (93.7 ± 0.04), boiled crookneck squash (93.7 ± 0.06), canned peeled red tomatoes (93.7 ± 0.04), frozen and boiled turnips (93.6 ± 0.05), boiled green cabbage (93.6 ± 0.07), boiled red cabbage (93.6 ± 0.07), boiled dock (93.6 ± 0.05), boiled turnips (93.6 ± 0.06), fruit pectin (93.6 ± 0.36), boiled cardoon (93.5 ± 0.05), raw frozen rhubarb (93.5 ± 0.04), boiled chayote (93.4 ± 0.06), cooked sprouted mung beans (93.4 ± 0.08), boiled purslane (93.4 ± 0.09), soy milk (93.3 ± 0.02), raw chives (93.3 ± 1.67), boiled turnip greens (93.2 ± 0.07), canned green beans (93.2 ± 0.07), canned yellow snap beans (93.2 ± 0.07), dill pickles (93.2 ± 0.08), whey (93.1 ± 0.02), raw vine spinach (93.1 ± 0.05), rose apple (93 ± 0.05), raw green tomato (93 ± 0.04), canned carrots (93 ± 0.07), raw dock (93 ± 0.07), boiled pokeberry shoots (92.9 ± 0.06), boiled swamp cabbage (92.9 ± 0.1), raw red sweet peppers (92.8 ± 0.1), raw green sweet peppers (92.8 ± 0.1), raw pepeao (92.6 ± 0.05), boiled cauliflower (92.6 ± 0.08), raw green cabbage (92.6 ± 0.14), boiled swiss chard (92.6 ± 0.06), raw coriander (92.5 ± 1.25), canned red chili peppers (92.5 ± 0.07), canned green chili peppers (92.5 ± 0.07), boiled garden cress (92.5 ± 0.07), raw swamp cabbage (92.5 ± 0.09), canned sauerkraut (92.5 ± 0.04), beer (92.4 ± 0.01), boiled red tomato (92.4 ± 0.04), boiled/baked spaghetti squash (92.3 ± 0.06), oheloberries (92.3 ± 0.04), boiled onions (92.3 ± 0.05), frozen and boiled crookneck squash (92.3 ± 0.05), raw mustard spinach (92.3 ± 0.07), frozen and boiled onions (92.2 ± 0.05), raw cauliflower (92.2 ± 0.1), boiled savoy cabbage (92.1 ± 0.07), raw eggplant (92 ± 0.12), casaba melon (92 ± 0.03), raw spring onions with tops (92 ± 0.1), raw celeriac (92 ± 0.05), boiled asparagus (92 ± 0.06), raw chickory greens (92 ± 0.06), boiled eggplant (91.9 ± 0.1), frozen and boiled yellow snap beans (91.9 ± 0.07), frozen and boiled green beans (91.9 ± 0.07), canned spinach (91.8 ± 0.05), raw mushrooms (91.7 ± 0.14), watermelon (91.5 ± 0.03), strawberries (91.5 ± 0.03), boiled amaranth (91.5 ± 0.08), tom collins (91.4 ± 0.02), raw acerola (91.4 ± 0.05), pink grapefruit (91.4 ± 0.04), raw red cabbage (91.4 ± 0.14), raw spinach (91.4 ± 0.18), canned stewed red tomatoes (91.3 ± 0.04), raw agar seaweed (91.3 ± 0.05), canned red sweet peppers (91.3 ± 0.07), canned green sweet peppers (91.3 ± 0.07), boiled kale (91.2 ± 0.08), raw sprouted alfalfa seeds (91.2 ± 0.15), frozen and boiled asparagus (91.2 ± 0.08), boiled scotch kale (91.2 ± 0.08), boiled spinach (91.2 ± 0.06), frozen and boiled okra (91.1 ± 0.05), boiled hubbard squash (91.1 ± 0.04), raw savoy cabbage (91.1 ± 0.14), raw bamboo shoots (91.1 ± 0.07), raw turnip greens (91.1 ± 0.18), boiled mushrooms (91 ± 0.06), raw onions (90.9 ± 0.06), carambola (90.9 ± 0.04), boiled beets (90.9 ± 0.06), skim milk (90.8 ± 0.02), pitanga (90.8 ± 0.03), boiled leeks (90.8 ± 0.19), frozen and boiled broccoli (90.8 ± 0.05), canned shellie beans (90.7 ± 0.04), raw broccoli (90.7 ± 0.11), raw spirulina seaweed (90.7 ± 0.05), lemon juice (90.7 ± 0.02), canned zucchini in tomato juice (90.6 ± 0.04), raw welsh onions (90.5 ± 0.05), white grapefruit (90.5 ± 0.04), frozen and boiled kale (90.5 ± 0.08), raw sprouted mung beans (90.4 ± 0.1), boiled kohlrabi (90.4 ± 0.06), frozen and boiled turnip greens (90.4 ± 0.06), boiled broccoli (90.3 ± 0.06), boiled rutabaga (90.2 ± 0.06), lime juice (90.2 ± 0.02), lowfat 1% milk (90.1 ± 0.02), buttermilk (90.1 ± 0.02), white grapefruit juice (90 ± 0.02), red grapefruit juice (90 ± 0.02), canned pumpkin (90 ± 0.04), frozen and boiled spinach (90 ± 0.05), frozen and boiled carrots (89.9 ± 0.07), boiled okra (89.9 ± 0.06), canned jalapeño peppers (89.9 ± 0.07), cantaloupe (89.8 ± 0.03), boiled dandelion greens (89.8 ± 0.1), boiled acorn squash (89.7 ± 0.04), white table wine (89.6 ± 0.05), boiled blackeyed-pea pods (89.6 ± 0.11), raw garden cress (89.6 ± 0.2), mandarin oranges (89.5 ± 0.04), lemon-lime soda (89.4 ± 0.01), protein fortified skim milk (89.3 ± 0.02), lowfat 2% milk (89.2 ± 0.02), boiled green beans (89.2 ± 0.08), boiled yellow snap beans (89.2 ± 0.08), boiled beet greens (89.2 ± 0.07), pummelo (89.1 ± 0.03), tomato sauce (89.1 ± 0.04), lemon (89 ± 0.09), rose table wine (88.9 ± 0.05), cola (88.9 ± 0.01), tangerine juice (88.9 ± 0.02), carrot juice (88.9 ± 0.03), boiled lamb's-quarters (88.9 ± 0.06), papaya (88.8 ± 0.02), protein fortified lowfat 1% milk (88.7 ± 0.02), frozen and boiled collards (88.5 ± 0.06), red table wine (88.4 ± 0.05), orange juice (88.3 ± 0.02), raw parsley (88.3 ± 0.17), chicken egg white (88.2 ± 0.15), lime (88.2 ± 0.07), restaurant orange juice (88.1 ± 0.04), whole milk (88 ± 0.02), whole yogurt (87.9 ± 0.02), farina (87.9 ± 0.03), apple juice (87.9 ± 0.02), boiled yambean (87.9 ± 0.05), gooseberries (87.9 ± 0.03), raw carrots (87.8 ± 0.07), raw red chili peppers (87.8 ± 0.11), raw green chili peppers (87.8 ± 0.11), boiled butternut squash (87.8 ± 0.05), frozen and boiled butternut squash (87.8 ± 0.04), filled milk (87.7 ± 0.02), protein fortified lowfat 2% milk (87.7 ± 0.02), cooked enriched white degermed corn meal (87.7 ± 0.02), cooked unenriched corn meal (87.7 ± 0.02), cooked enriched yellow degermed corn meal (87.7 ± 0.02), cooked unenriched yellow degermed corn meal (87.7 ± 0.02), mulberries (87.7 ± 0.04), peach (87.7 ± 0.06), orange soda (87.6 ± 0.01), tangerine (87.6 ± 0.06), prickly pear (87.6 ± 0.05), cream of rice (87.5 ± 0.03), boiled carrots (87.4 ± 0.06), tomato purée (87.3 ± 0.02), horseradish sauce (87.3 ± 0.33), boiled brussels sprouts (87.3 ± 0.06), yellow hominy (87.2 ± 0.02), pineapple (87.2 ± 0.03), boiled jute potherb (87.2 ± 0.12), human milk (87.1 ± 0.16), boiled and steamed japanese chestnuts (87.1 ± 0.18), cream of wheat (87.1 ± 0.03), goat milk (87 ± 0.02), yellow hominy with red and green peppers (87 ± 0.02), bourbon & soda (86.9 ± 0.04), navel orange (86.8 ± 0.04), frozen and boiled brussels sprouts (86.8 ± 0.06), white hominy (86.7 ± 0.02), roselle (86.7 ± 0.09), loquats (86.7 ± 0.05), raspberries (86.6 ± 0.04), cranberry sauce (86.5 ± 0.05), boiled artichoke (86.5 ± 0.06), boiled artichoke hearts (86.5 ± 0.06), cooked tahitian taro (86.5 ± 0.07), canned chinese water chestnuts (86.4 ± 0.07), valencia orange (86.4 ± 0.04), nectarine (86.3 ± 0.04), apricots (86.3 ± 0.05), restaurant hot chocolate (86.3 ± 0.02), mammy apple (86.2 ± 0.05), guava (86.1 ± 0.06), bloody mary (86 ± 0.03), boysenberries (85.9 ± 0.04), gin & tonic (85.8 ± 0.02), instant oatmeal (85.8 ± 0.03), blackberries (85.7 ± 0.07), raw dandelion greens (85.7 ± 0.18), peach nectar (85.6 ± 0.02), purple passion fruit juice (85.6 ± 0.02), restaurant cranberry juice cocktail (85.5 ± 0.03), pineapple juice (85.5 ± 0.02), cooked apple without skin (85.5 ± 0.03), ground-cherries (85.4 ± 0.04), oatmeal (85.3 ± 0.03), corn grits (85.3 ± 0.02), canned oysters (85.2 ± 0.06), skim yogurt (85.2 ± 0.02), plum (85.2 ± 0.08), raw oysters (85.1 ± 0.06), lowfat yogurt (85.1 ± 0.02), baked hubbard squash (85.1 ± 0.05), instant corn grits (85 ± 0.04), cranberry juice cocktail (85 ± 0.02), papaya nectar (85 ± 0.02), raw laver/nori seaweed (85 ± 0.05), apricot nectar (84.9 ± 0.02), pickled rip sevillano/ascolano olives (84.6 ± 0.31), blueberries (84.6 ± 0.03), loganberries (84.6 ± 0.03), raw tofu (84.5 ± 0.04), lowfat 1% chocolate milk (84.5 ± 0.02), raw apple without skin (84.5 ± 0.04), stir-fried sprouted mung beans (84.4 ± 0.08), canned potato without skin (84.3 ± 0.06), boiled dishcloth gourd (84.3 ± 0.06), yellow passion fruit juice (84.2 ± 0.02), grape juice (84.1 ± 0.02), carob milk (84 ± 0.02), pear nectar (84 ± 0.02), carissa (84 ± 0.25), raw apple with skin (83.9 ± 0.04), red currants (83.9 ± 0.09), white currants (83.9 ± 0.09), screwdriver (83.8 ± 0.02), pear (83.8 ± 0.03), quince (83.8 ± 0.05), lowfat 2% chocolate milk (83.6 ± 0.02), cooked shitake mushrooms (83.5 ± 0.07), indian buffalo milk (83.4 ± 0.02), raw monkfish (83.3 ± 0.06), raw leeks (83.1 ± 0.19), java plum (83.1 ± 0.04), kiwi fruit (83 ± 0.07), baked acorn squash (82.9 ± 0.05), frozen and boiled potatoes without skin (82.8 ± 0.05), longans (82.8 ± 0.05), maypo (82.7 ± 0.03), marinara sauce (82.5 ± 0.02), whole chocolate milk (82.3 ± 0.02), low calorie italian salad dressing (82 ± 0.33), black currants (82 ± 0.09), canned succotash (82 ± 0.04), raw clams (81.9 ± 0.06), raw frog legs (81.9 ± 0.05), lychees (81.8 ± 0.05), pickled beets (81.8 ± 0.04), lemon peel (81.7 ± 0.83), raw ginger root (81.7 ± 0.21), mango (81.7 ± 0.02), hot chocolate (81.6 ± 0.02), okara tofu (81.6 ± 0.08), canned green peas (81.6 ± 0.06), kumquats (81.6 ± 0.26), raw kelp/kombu/tangle seaweed (81.6 ± 0.05), coleslaw (81.5 ± 0.08), boiled lotus root (81.5 ± 0.06), raw beef tripe (81.4 ± 0.05), raw black bullhead (81.3 ± 0.05), prune juice (81.3 ± 0.02), american grapes (81.3 ± 0.05), raw irish moss seaweed (81.3 ± 0.05), malted milk (81.2 ± 0.02), whole chocolate malted milk (81.2 ± 0.02), raw cod (81.2 ± 0.06), soursop (81.2 ± 0.02), raw lingcod (81.1 ± 0.06), scalloped potatoes (81 ± 0.04), maltex (81 ± 0.03), pomegranate (81 ± 0.03), boiled salsify (81 ± 0.07), raw crayfish (80.8 ± 0.06), half and half cream (80.7 ± 0.33), sheep milk (80.7 ± 0.02), cherries (80.7 ± 0.07), raw queen crab (80.6 ± 0.06), raw mussels (80.6 ± 0.06), raw cuttlefish (80.6 ± 0.06), european grapes (80.6 ± 0.03), thompson seedless grapes (80.6 ± 0.03), pickled rip manzanillo/mission olives (80.4 ± 1.09), canned broad beans (80.3 ± 0.02), japanese persimmon (80.3 ± 0.03), raw whiting (80.2 ± 0.06), gefiltefish with broth (80.2 ± 0.12), raw octopus (80.2 ± 0.06), braised pork lungs (80 ± 0.05), yellow mustard (80 ± 1), raw shallots (80 ± 0.5), raw wakame seaweed (80 ± 0.05), raw wolffish (79.9 ± 0.06), raw haddock (79.9 ± 0.06), tequila sunrise (79.8 ± 0.03), raw white sucker (79.8 ± 0.06), elderberries (79.8 ± 0.03), raw florida avocado (79.7 ± 0.02), cottage cheese (79.6 ± 0.18), raw ling (79.6 ± 0.06), o'brien potatoes (79.6 ± 0.03), canned black-eyed peas (79.6 ± 0.02), applesauce (79.6 ± 0.04), raw sunfish (79.5 ± 0.06), raw alaska king crab (79.5 ± 0.06), frozen and boiled green peas (79.5 ± 0.06), raw alewife (79.4 ± 0.05), steamed sprouted soybeans (79.4 ± 0.11), raw rockfish (79.3 ± 0.06), raw black bass (79.3 ± 0.05), raw burbot (79.3 ± 0.06), raw striped bass (79.2 ± 0.06), raw grouper (79.2 ± 0.06), raw dungeness crab (79.2 ± 0.06), figs (79.2 ± 0.1), raw flatfish (79.1 ± 0.06), raw blue crab (79.1 ± 0.06), evaporated skim milk (79.1 ± 0.16), raw lotus root (79.1 ± 0.06), raw smelt (79 ± 0.05), raw cisco (78.9 ± 0.06), raw tilefish (78.9 ± 0.06), raw pike (78.9 ± 0.06), raw potato without skin (78.9 ± 0.04), crab apples (78.9 ± 0.05), raw rainbow smelt (78.8 ± 0.06), raw green peas (78.8 ± 0.06), canned pinto beans (78.8 ± 0.02), barbecue sauce (78.8 ± 0.31), canned creamed yellow corn (78.8 ± 0.04), raw perch (78.7 ± 0.06), bread and butter pickles (78.7 ± 0.33), raw squid (78.6 ± 0.06), raw scallops (78.6 ± 0.06), pickled green olives (78.3 ± 0.11), raw sea bass (78.2 ± 0.06), raw pollock (78.2 ± 0.06), canned succotash with creamed corn (78.2 ± 0.04), raw seatrout (78.1 ± 0.06), raw sheepshead (78 ± 0.06), raw croaker (78 ± 0.06), brown mustard (78 ± 1), sapodilla (78 ± 0.03), canned kidney beans (78 ± 0.02), raw jerusalem artichoke (78 ± 0.07), raw halibut (77.9 ± 0.06), boiled green peas (77.9 ± 0.06), jujube (77.9 ± 0.05), boiled parsnips (77.7 ± 0.06), cooked alaska king crab (77.5 ± 0.06), raw dolphinfish (77.5 ± 0.06), boiled potato without skin (77.5 ± 0.04), canned cranberry beans (77.5 ± 0.02), boiled arrowhead (77.5 ± 0.42), canned yellow corn with red and green peppers (77.5 ± 0.04), cooked blue crab (77.4 ± 0.06), raw freshwater drum (77.3 ± 0.06), cooked shrimp (77.3 ± 0.06), custard (77.2 ± 0.02), raw mullet (77.1 ± 0.06), steamed hawaii mountain yam (77.1 ± 0.07), canned lima beans (77.1 ± 0.02), whiskey sour (77 ± 0.06), raw turbot (76.9 ± 0.06), raw snapper (76.8 ± 0.06), raw lobster (76.7 ± 0.06), raw sturgeon (76.6 ± 0.06), evaporated lowfat milk (76.6 ± 0.04), canned yellow corn (76.6 ± 0.05), raw carp (76.4 ± 0.06), braised beef lungs (76.4 ± 0.05), raw catfish (76.4 ± 0.06), raw pink salmon (76.4 ± 0.06), chili (76.4 ± 0.02), raw cusk (76.4 ± 0.06), yellow corn pudding (76.3 ± 0.02), mashed potatoes (76.3 ± 0.05), frozen creamed yellow corn (76.2 ± 0.04), canned blue crab (76.1 ± 0.06), canned sweet potato (76.1 ± 0.03), potato salad (76 ± 0.04), raw spot (76 ± 0.06), vanilla pudding (76 ± 0.02), cooked lobster (76 ± 0.06), braised pork brains (75.9 ± 0.05), raw shrimp (75.9 ± 0.06), cooked cod (75.9 ± 0.06), boiled burdock root (75.7 ± 0.04), frozen and boiled yellow corn (75.7 ± 0.06), soy sauce (75.7 ± 0.09), raw swordfish (75.6 ± 0.06), raw freshwater bass (75.6 ± 0.06), cooked clams (75.6 ± 0.06), canned cod (75.6 ± 0.06), canned black turtle beans (75.6 ± 0.02), raw chum salmon (75.4 ± 0.06), cooked crayfish (75.4 ± 0.06), raw barracuda (75.4 ± 0.05), raw scup (75.4 ± 0.06), baked potato without skin (75.4 ± 0.03), canned white corn (75 ± 0.05), surimi shrimp (74.9 ± 0.06), chicken egg (74.8 ± 0.1), cooked whiting (74.7 ± 0.06), restaurant vanilla milkshake (74.7 ± 0.02), raw abalone (74.6 ± 0.06), raw yellowtail (74.5 ± 0.06), quail egg (74.4 ± 0.56), eggnog (74.4 ± 0.02), boiled sprouted green peas (74.4 ± 0.05), vanilla milkshake (74.4 ± 0.02), bananas (74.3 ± 0.04), cooked haddock (74.2 ± 0.06), raw butterfish (74.1 ± 0.06), restaurant strawberry milkshake (74.1 ± 0.02), frozen and boiled succotash (74.1 ± 0.06), evaporated whole milk (74 ± 0.04), au gratin potatoes (74 ± 0.04), raw venison (74 ± 0.06), tomato paste (74 ± 0.04), surimi scallops (73.9 ± 0.06), frozen and baked sweet potato (73.8 ± 0.06), raw pork stomach (73.6 ± 0.05), raw shark (73.5 ± 0.06), canned lean ham (73.5 ± 0.05), frozen and boiled fordhook lima beans (73.5 ± 0.06), microwaved potato without skin (73.5 ± 0.03), cherimoya (73.5 ± 0.01), raw anchovy (73.4 ± 0.06), cooked rockfish (73.4 ± 0.06), cooked grouper (73.4 ± 0.06), cooked parboiled white rice (73.4 ± 0.03), raw chinese water chestnuts (73.4 ± 0.08), orange peel (73.3 ± 0.83), simmered beef brains (73.3 ± 0.05), raw reindeer (73.3 ± 0.05), raw chinook salmon (73.2 ± 0.06), cooked flatfish (73.2 ± 0.06), sugar apple (73.2 ± 0.03), hot dog pickle relish (73.2 ± 0.18), canned alewife (73 ± 0.05), cooked pike (72.9 ± 0.06), cooked instant white rice (72.9 ± 0.03), boiled sweet potato (72.9 ± 0.03), raw whitefish (72.8 ± 0.06), raw pheasant without skin (72.8 ± 0.05), cooked rainbow smelt (72.8 ± 0.06), restaurant coleslaw (72.8 ± 0.05), baked sweet potato (72.8 ± 0.04), passion fruit (72.8 ± 0.28), cooked perch (72.7 ± 0.06), boiled mung beans (72.7 ± 0.02), baked beans with pork and tomato sauce (72.7 ± 0.02), raw coho salmon (72.6 ± 0.06), frozen white corn (72.6 ± 0.07), cooked white rice (72.6 ± 0.02), vegetarian baked beans (72.6 ± 0.02), raw california avocado (72.5 ± 0.03), turkey egg (72.5 ± 0.06), boiled mungo beans (72.5 ± 0.03), sweet dessert wine (72.4 ± 0.08), dry dessert wine (72.4 ± 0.08), raw spiny dogfish (72.3 ± 0.05), refried beans (72.3 ± 0.02), boiled and frozen baby lima beans (72.3 ± 0.06), jackfruit (72.3 ± 0.05), chocolate milkshake (72.2 ± 0.02), cooked sea bass (72.1 ± 0.06), simmered pork ears (72 ± 0.05), raw herring (72 ± 0.06), smoked chinook salmon (72 ± 0.06), microwaved potato with skin (72 ± 0.02), turkey breast meat summer sausage (71.9 ± 0.24), raw duck liver (71.8 ± 0.05), tapioca cream pudding (71.8 ± 0.03), raw goose liver (71.8 ± 0.05), ricotta cheese (71.7 ± 0.04), poi (71.7 ± 0.04), cooked halibut (71.6 ± 0.06), restaurant chocolate milkshake (71.5 ± 0.02), raw rainbow trout (71.5 ± 0.06), smoked haddock (71.5 ± 0.06), macaroni (71.5 ± 0.04), spaghetti (71.5 ± 0.04), boiled broad beans (71.5 ± 0.03), canned pumpkin pie mix (71.5 ± 0.04), raw trout (71.4 ± 0.06), pork and beef honey loaf (71.4 ± 0.18), cooked extra long grain white rice (71.3 ± 0.06), cooked parboiled extra long grain white rice (71.3 ± 0.06), sweetened coconut cream (71.2 ± 0.02), raw pompano (71.2 ± 0.06), light meat turkey roll (71.2 ± 0.09), cooked wild long grain rice (71.2 ± 0.04), baked potato with skin (71.2 ± 0.02), turkey ham (71.1 ± 0.09), boiled lupins (71.1 ± 0.03), compressed bakers yeast (71.1 ± 0.18), crab cakes (71 ± 0.08), shoyu sauce (71 ± 0.09), duck egg (70.9 ± 0.07), raw whale meat (70.9 ± 0.05), sour cream (70.8 ± 0.42), fried beef brains (70.8 ± 0.05), raw milkfish (70.8 ± 0.06), raw bluefish (70.8 ± 0.06), smoked whitefish (70.8 ± 0.06), low calorie thousand island salad dressing (70.7 ± 0.33), baked beans with pork and sweet sauce (70.7 ± 0.02), low calorie french salad dressing (70.6 ± 0.31), breadfruit (70.6 ± 0.05), cooked mullet (70.5 ± 0.06), unheated lean ham (70.5 ± 0.05), canned navy beans (70.5 ± 0.02), goose egg (70.4 ± 0.03), turkey pastrami (70.4 ± 0.09), cooked snapper (70.4 ± 0.06), noodles (70.4 ± 0.03), cooked brown rice (70.3 ± 0.03), raw sockeye salmon (70.2 ± 0.06), cooked tilefish (70.2 ± 0.06), cooked oysters (70.2 ± 0.06), canned white beans (70.1 ± 0.02), boiled black-eyed peas (70.1 ± 0.03), baked/boiled tropical yam (70.1 ± 0.07), jellied corned beef loaf (70 ± 0.18), braised beef spleen

(70 ± 0.05), cooked sturgeon (70 ± 0.06), raw quail without skin (70 ± 0.05), apple tapioca pudding (70 ± 0.02), canned great northern beans (69.9 ± 0.02), daiquiri (69.8 ± 0.08), smoked cisco (69.8 ± 0.06), light and dark meat turkey roll (69.8 ± 0.09), firm raw tofu (69.8 ± 0.04), boiled lima beans (69.8 ± 0.03), canned chickpeas (69.7 ± 0.02), boiled catjang black-eyed peas (69.7 ± 0.03), cooked prunes (69.7 ± 0.05), cooked carp (69.6 ± 0.06), chopped smoked beef (69.6 ± 0.18), boiled lentils (69.6 ± 0.03), canned tuna in water (69.5 ± 0.06), roasted canned lean ham (69.5 ± 0.05), boiled split peas (69.5 ± 0.03), boiled yellow corn (69.5 ± 0.06), pork luxury loaf (69.3 ± 0.18), boiled and steamed european chestnuts (69.3 ± 0.18), boiled mothbeans (69.2 ± 0.03), cooked sheepshead (69.1 ± 0.06), boiled hyacinth beans (69.1 ± 0.03), boiled great northern beans (69 ± 0.03), simmered beef kidneys (68.8 ± 0.05), canned pink salmon (68.8 ± 0.06), boiled yardlong bean (68.8 ± 0.03), medium cream (68.7 ± 0.33), cooked swordfish (68.7 ± 0.06), canned sockeye salmon (68.7 ± 0.06), braised pork kidneys (68.7 ± 0.05), tomato catsup (68.7 ± 0.33), stir-fried sprouted lentils (68.7 ± 0.05), pickled pork feet (68.6 ± 0.05), light meat chicken roll (68.6 ± 0.05), boiled green soybeans (68.6 ± 0.06), boiled pigeon peas (68.6 ± 0.03), raw cassava (68.5 ± 0.05), enchilada (68.3 ± 0.02), simmered chicken liver (68.3 ± 0.05), stewed chicken breast without skin (68.3 ± 0.05), boiled succotash (68.3 ± 0.05), raw shad (68.2 ± 0.06), raw eel (68.2 ± 0.06), chicken roll (68.2 ± 0.09), pork and beef peppered loaf (68.2 ± 0.18), braised pork heart (68.1 ± 0.04), stewed light meat chicken without skin (68 ± 0.05), chili sauce (68 ± 0.33), frozen and cooked sweetened rhubarb (67.8 ± 0.04), teriyaki sauce (67.8 ± 0.28), roasted lean ham (67.7 ± 0.05), coconut milk (67.6 ± 0.02), martini (67.6 ± 0.07), simmered chicken giblets (67.6 ± 0.05), roasted muskrat (67.3 ± 0.05), simmered chicken gizzard (67.3 ± 0.05), cooked plantain (67.3 ± 0.03), simmered chicken neck without skin (67.2 ± 0.28), boiled winged beans (67.2 ± 0.03), stir-fried sprouted soybeans (67.2 ± 0.05), boiled baby lima beans (67.1 ± 0.03), candied sweet potato (67 ± 0.05), boiled kidney beans (66.9 ± 0.03), stewed chicken drumstick without skin (66.8 ± 0.11), 80 proof rum (66.7 ± 0.12), 80 proof vodka (66.7 ± 0.12), unheated canadian bacon (66.7 ± 0.09), braised pork spleen (66.7 ± 0.05), boiled french beans (66.6 ± 0.03), canned ham (66.5 ± 0.05), roasted turkey light meat without skin (66.3 ± 0.05), low calorie russian salad dressing (66.3 ± 0.31), boiled adzuki beans (66.3 ± 0.02), stewed chicken breast with skin (66.2 ± 0.05), manhattan (66.1 ± 0.09), turkey bologna (66.1 ± 0.18), frozen and boiled black-eyed peas (66.1 ± 0.06), simmered pork feet (66 ± 0.05), raw whelk (66 ± 0.06), tamari sauce (66 ± 0.09), braised beef liver (65.9 ± 0.05), stewed dark meat chicken without skin (65.8 ± 0.05), turkey salad spread (65.8 ± 0.09), tuna salad (65.8 ± 0.09), chocolate pudding (65.8 ± 0.02), boiled black beans (65.8 ± 0.03), pork headcheese (65.7 ± 0.18), boiled black turtle beans (65.7 ± 0.03), simmered turkey liver (65.6 ± 0.05), turkey salami (65.4 ± 0.09), turkey summer sausage (65.4 ± 0.18), cooked coho salmon (65.4 ± 0.06), simmered turkey giblets (65.4 ± 0.05), simmered turkey gizzard (65.4 ± 0.05), roasted pork tenderloin (65.2 ± 0.05), roasted chicken breast without skin (65.2 ± 0.06), cheese soufflé (65.1 ± 0.05), stewed chicken drumstick with skin (65.1 ± 0.09), stewed light meat chicken with skin (65.1 ± 0.05), piña colada (65 ± 0.04), simmered chicken heart (64.9 ± 0.05), hummus (64.9 ± 0.02), beef honey roll sausage (64.8 ± 0.22), barbecue loaf (64.8 ± 0.22), roasted light meat chicken without skin (64.8 ± 0.05), breaded and fried oysters (64.7 ± 0.06),

cooked herring (64.6 ± 0.06), unheated ham (64.6 ± 0.05), breaded fried squid (64.6 ± 0.06), boiled cranberry beans (64.6 ± 0.03), roasted ham (64.5 ± 0.05), apple brown betty (64.5 ± 0.02), persimmon (64.4 ± 0.2), braised pork liver (64.3 ± 0.05), boiled pinto beans (64.3 ± 0.03), roasted duck without skin (64.2 ± 0.05), simmered turkey heart (64.2 ± 0.05), simmered beef heart (64.1 ± 0.05), turkey frankfurter (64 ± 0.11), canned tuna in oil (64 ± 0.06), roasted cured pork arm excluding fat (63.9 ± 0.05), 86 proof whiskey (63.8 ± 0.12), chopped ham lunch meat (63.8 ± 0.24), cooked taro (63.8 ± 0.08), ham salad (63.6 ± 0.18), raw mackerel (63.5 ± 0.06), broiled beef round excluding fat (63.4 ± 0.05), cooked rainbow trout (63.4 ± 0.06), whipping cream, light (63.3 ± 0.33), tuna salad (63.2 ± 0.02), boiled navy beans (63.2 ± 0.03), stewed chicken thigh with skin (63.1 ± 0.07), roasted dark meat chicken without skin (63.1 ± 0.05), roasted turkey dark meat without skin (63.1 ± 0.05), boiled white beans (63.1 ± 0.03), stewed dark meat chicken with skin (63 ± 0.05), boiled yellow beans (63 ± 0.03), neufchatel cheese (62.9 ± 0.18), roasted chicken thigh without skin (62.9 ± 0.1), cooked pompano (62.9 ± 0.06), roasted turkey light meat with skin (62.8 ± 0.05), roasted beef tip excluding fat (62.8 ± 0.05), roasted eye of beef round excluding fat (62.8 ± 0.05), roasted chicken drumstick with skin (62.7 ± 0.1), canned smelt (62.7 ± 0.05), boiled soybeans (62.6 ± 0.03), boiled and steamed chinese chestnuts (62.5 ± 0.18), simmered pork chitterlings (62.4 ± 0.05), roasted chicken breast with skin (62.4 ± 0.05), sapote (62.4 ± 0.02), stewed chicken wing with skin (62.3 ± 0.13), restaurant hash brown potatoes (62.2 ± 0.06), roasted leg of lamb excluding fat (62.2 ± 0.06), 90 proof gin (62.1 ± 0.12), raw japanese chestnuts (62.1 ± 0.18), broiled lamb loin chop excluding fat (62 ± 0.1), cooked sockeye salmon (61.9 ± 0.06), simmered chicken neck with skin (61.8 ± 0.13), strawberry sundae (61.6 ± 0.03), hash brown potatoes (61.5 ± 0.06), breaded and fried clams (61.5 ± 0.06), bread stuffing (61.4 ± 0.03), roasted lamb shoulder excluding fat (61.4 ± 0.06), lobster paste (61.4 ± 0.71), pork beerwurst salami (61.3 ± 0.22), cooked mussels (61.2 ± 0.06), boiled pink beans (61.2 ± 0.03), pork and beef spread (61.1 ± 0.18), pork and beef picnic loaf (61.1 ± 0.18), grilled canadian bacon (61.1 ± 0.11), berliner (60.9 ± 0.22), roasted canned ham (60.9 ± 0.05), lebanon bologna (60.9 ± 0.22), broiled beef top round excluding fat (60.9 ± 0.05), sweet pickles (60.7 ± 0.33), hamburger pickle relish (60.7 ± 0.18), roasted light meat chicken with skin (60.5 ± 0.05), pork bologna (60.4 ± 0.22), beef and pork salami (60.4 ± 0.22), braised/ broiled veal round with rump (60.4 ± 0.06), roasted pork shank excluding fat (60.4 ± 0.05), broiled beef tenderloin excluding fat (60.4 ± 0.05), roasted pork arm excluding fat (60.3 ± 0.05), braised pork pancreas (60.3 ± 0.05), broiled T-bone steak excluding fat (60.3 ± 0.05), 94 proof gin (60.2 ± 0.12), 94 proof rum (60.2 ± 0.12), 94 proof vodka (60.2 ± 0.12), 94 proof whiskey (60.2 ± 0.12), roasted turkey dark meat with skin (60.2 ± 0.05), broiled lamb rib chop excluding fat (60.2 ± 0.12), broiled beef top loin excluding fat (60.2 ± 0.05), fried chicken breast without skin (60.2 ± 0.06), boiled chickpeas (60.2 ± 0.03), battered and fried shark (60.1 ± 0.06), fried light meat chicken without skin (60.1 ± 0.05), fried abalone (60.1 ± 0.06), vienna sausage (60 ± 0.31), ham and cheese spread (60 ± 0.18), pork and beef old fashioned loaf (60 ± 0.18), roasted beef tenderloin excluding fat (60 ± 0.05), cooked corned beef brisket (59.8 ± 0.05), breaded and fried croaker (59.8 ± 0.06), kippered herring (59.8 ± 0.13), stewed rabbit (59.8 ± 0.04), roasted pork leg excluding fat (59.7 ± 0.05), hot fudge sundae (59.7 ± 0.03), sardines (59.6 ± 0.21), broiled beef rib excluding fat (59.5 ± 0.05), broiled porterhouse steak excluding fat (59.5 ± 0.05), roasted chicken thigh with skin (59.4 ± 0.08), cooked eel (59.3 ± 0.06), roasted pork rump excluding fat (59.3 ± 0.05), sweet potato pie (59.3 ± 0.04), pumpkin pie (59.2 ± 0.04), broiled beef top round including fat (59.2 ± 0.05), pork olive loaf (58.9 ± 0.18), braised/broiled veal loin (58.9 ± 0.06), raw garlic (58.9 ± 0.56), roasted pork shoulder excluding fat (58.8 ± 0.05), breaded and fried catfish (58.8 ± 0.06), pork and beef luncheon sausage (58.7 ± 0.22), roasted pork sirloin excluding fat (58.7 ± 0.05), ham and cheese roll (58.6 ± 0.18), roasted dark meat chicken with skin (58.6 ± 0.05), medium-rare baked extra lean ground beef (58.6 ± 0.05), broiled beef sirloin excluding fat (58.6 ± 0.05), bread pudding (58.6 ± 0.02), pickled anchovy (58.6 ± 0.18), braised/roasted/stewed veal chuck (58.5 ± 0.06), broiled beef rib eye excluding fat (58.4 ± 0.05), breaded and fried scallops (58.4 ± 0.16), strawberry pie (58.4 ± 0.05), beef salami (58.3 ± 0.22), cream puff (58.3 ± 0.04), simmered beef shank excluding fat (58.2 ± 0.05), roasted pork loin excluding fat (58.1 ± 0.05), custard pie (58.1 ± 0.04), cooked flounder (58.1 ± 0.05), whipping cream, heavy (58 ± 0.33), pork pickle and pimento loaf (57.9 ± 0.18), broiled beef flank excluding fat (57.8 ± 0.05), roasted pork boston blade excluding fat (57.7 ± 0.05), canned corned beef (57.7 ± 0.05), roasted beef rib roasted excluding fat (57.7 ± 0.05), 100 proof gin (57.6 ± 0.12), 100 proof rum (57.6 ± 0.12), 100 proof vodka (57.6 ± 0.12), 100 proof whiskey (57.6 ± 0.12), chicken frankfurter (57.6 ± 0.11), medium-rare fried extra lean ground beef (57.6 ± 0.05), roasted eye of beef round including fat (57.5 ± 0.05), medium-rare broiled extra lean ground beef (57.3 ± 0.05), roasted beef round tip including fat (57.3 ± 0.05), braised beef flank excluding fat (57.3 ± 0.05), roasted pork center loin excluding fat (57.3 ± 0.05), roasted opossum (57.3 ± 0.05), roasted goose without skin (57.2 ± 0.05), braised beef bottom round excluding fat (57.2 ± 0.05), prune pudding (57.2 ± 0.04), minced ham lunch meat (57.1 ± 0.24), chipped beef (57.1 ± 0.18), roasted pork center rib excluding fat (57 ± 0.05), roasted pork top loin excluding fat (57 ± 0.05), braised pork tongue (56.9 ± 0.05), medium-rare baked lean ground beef (56.9 ± 0.05), broiled beef flank including fat (56.9 ± 0.05), broiled pork center loin excluding fat (56.8 ± 0.05), fried chicken drumstick with skin (56.7 ± 0.1), broiled pork sirloin excluding fat (56.7 ± 0.05), fried chicken breast with skin (56.6 ± 0.05), caramel sundae (56.5 ± 0.03), roasted pork blade roll (56.2 ± 0.05), braised beef flank including fat (56.2 ± 0.05), roasted beaver (56.2 ± 0.06), taco (56.2 ± 0.06), frozen and fried hash brown potatoes (56.2 ± 0.06), pork bratwurst (56.1 ± 0.06), simmered beef tongue (56.1 ± 0.05), feta cheese (56.1 ± 0.18), ginko nuts (56.1 ± 0.18), corn oil diet margarine tub (56 ± 1), soybean oil and cottonseed oil diet margarine tub (56 ± 1), bulgur (56 ± 0.04), fried beef sirloin excluding fat (55.9 ± 0.05), roasted pork loin blade excluding fat (55.8 ± 0.05), medium-rare broiled lean ground beef (55.7 ± 0.05), braised beef brisket excluding fat (55.7 ± 0.05), fried dark meat chicken without skin (55.7 ± 0.05), fried beef liver (55.7 ± 0.05), medium-rare fried lean ground beef (55.6 ± 0.05), braised beef pancreas (55.6 ± 0.05), broiled pork loin excluding fat (55.5 ± 0.05), fried beef top round excluding fat (55.5 ± 0.05), pork and beef knockwurst (55.4 ± 0.07), coconut custard pie (55.4 ± 0.04), pickled herring (55.3 ± 0.33), beef bologna (55.2 ± 0.22), braised beef chuck arm pot roast excluding fat (55.2 ± 0.05), medium-rare baked ground beef (55.1 ± 0.05), broiled beef round including fat (55.1 ± 0.05), broiled pork boston blade excluding fat (55.1 ± 0.05), fried pork center rib excluding fat (55.1 ± 0.05), fried pork top loin excluding fat (55.1 ± 0.05), roasted chicken wing with skin (55 ± 0.15), mozzarella cheese (55 ± 0.18), broiled beef tenderloin including fat (55 ± 0.05), popover roll (55 ± 0.13), natto (55 ± 0.06), tempeh (54.9 ± 0.06), pork mothers loaf (54.8 ± 0.24), roasted racoon (54.8 ± 0.05), beef frankfurter (54.7 ± 0.09), roasted cured pork arm including fat (54.7 ± 0.05), roasted pork rump including fat (54.7 ± 0.05), fried light meat chicken with skin (54.7 ± 0.05), unheated ham patties (54.6 ± 0.08), roasted veal rib roast (54.6 ± 0.06), simmered beef shank including fat (54.6 = 0.05), broiled pork center rib excluding fat (54.5 ± 0.05), broiled pork top loin excluding fat (54.5 ± 0.05), fried pork center loin excluding fat (54.4 ± 0.05). banana custard pie (54.4 ± 0.04), cream cheese (54.3 ± 0.18), beef and pork bologna (54.3 ± 0.22), light mayonnaise (54.3 ± 0.36), braised pork arm excluding fat (54.3 ± 0.05), pineapple custard pie (54.3 ± 0.04), medium-rare broiled ground beef (54.2 ± 0.05), fried chicken thigh with skin (54.2 ± 0.08), braised lamb heart (54.1 ± 0.03), roasted pork sirloin including fat (54 ± 0.05), roasted leg of lamb including fat (54 ± 0.06), coconut cream (53.9 ± 0.02), polish pork sausage (53.9 ± 0.18), well done broiled extra lean ground beef (53.9 ± 0.05), well done fried extra lean ground beef (53.9 ± 0.05), braised beef bottom round including fat (53.9 ± 0.05), beef and pork frankfurter (53.8 ± 0.11), pork and beef kielbasa/kolbassy (53.8 ± 0.19), cornbread (53.8 ± 0.06), corn pone (53.8 ± 0.06), pork livercheese (53.7 ± 0.13), well done fried lean ground beef (53.7 ± 0.05), smoked beef sausage (53.5 ± 0.12), fried pork loin blade excluding fat (53.5 ± 0.05), broiled beef top loin including fat (53.5 ± 0.05), cooked mackerel (53.3 ± 0.06), hot dog (53.3 ± 0.06), beef lunch meat (53.2 ± 0.18), beef beerwurst salami (53 ± 0.22), frozen and french fried potatoes without skin (53 ± 0.1), well done broiled lean ground beef (52.9 ± 0.05), braised beef thymus (52.8 ± 0.05), roasted pork shank including fat (52.8 ± 0.05), breaded and fried shrimp (52.8 ± 0.06), french toast (52.8 ± 0.08), pork and beef mortadella (52.7 ± 0.33), well done fried ground beef (52.7 ± 0.05), well done baked extra lean ground beef (52.7 ± 0.05), braised beef chuck blade roast excluding fat (52.7 ± 0.05), pork and beef link sausage with american cheese (52.6 ± 0.12), broiled beef sirloin including fat (52.6 ± 0.05), camembert cheese (52.5 ± 0.18), roasted beef tenderloin including fat (52.5 ± 0.05), broiled pork loin blade excluding fat (52.4 ± 0.05), broiled porterhouse steak including fat (52.4 ± 0.05), roasted pork leg including fat (52.4 ± 0.05), medium-rare fried ground beef (52.3 ± 0.05), broiled beef rib eye including fat (52.3 ± 0.05), pork and beef link sausage (52.2 ± 0.07), pork liverwurst (52.2 ± 0.28), braised/stewed veal breast plate (52.1 ± 0.06), breaded and fried fish (52.1 ± 0.07), braised pork sirloin excluding fat (52.1 ± 0.05), roasted pork boston blade including fat (52 ± 0.05), roasted goose with skin (52 ± 0.05), well done broiled ground beef (52 ± 0.05), roast beef sandwich (52 ± 0.03), roasted duck with skin (51.8 ± 0.05), roasted pork center loin including fat (51.8 ± 0.05), roasted pork shoulder including fat (51.6 ± 0.05), canned pork lunch meat (51.4 ± 0.24), pork and beef brotwurst (51.3 ± 0.07), roasted pork loin including fat (51.3 ± 0.05), well done baked lean ground beef (51.3 ± 0.05), burrito (51.3 ± 0.03), roasted pork arm including fat (51.1 ± 0.05), braised pork loin excluding fat (51.1 ± 0.05), grilled ham patties (51 ± 0.08), blackberry pie (51 ± 0.04), blueberry pie (51 ± 0.04), summer sausage (50.9 ± 0.22), beef summer sausage (50.9 ± 0.22), fried dark meat chicken with skin (50.8 ± 0.05), fried tofu (50.8 ± 0.38), apple pie (50.8 ± 0.04), roasted pork center rib including fat (50.7 ± 0.05), ham and cheese sandwich (50.7 ± 0.03), roasted japanese chestnuts (50.7 ± 0.18), anchovy canned in olive oil (50.5 ± 0.25), broiled T-bone steak including fat (50.4 ± 0.05), fried beef top round including fat (50.4 ± 0.05), fried chicken (50.4 ± 0.06), broiled lamb liver (50.4 ± 0.11), braised beef shortribs excluding fat (50.2 ± 0.05), braised pork center loin excluding fat (50.2 ± 0.05), pork and beef lunch meat (50 ± 0.18), italian pork sausage (50 ± 0.07), pancakes (50 ± 0.19), braised pork center rib excluding fat (49.9 ± 0.05), braised pork top loin excluding fat (49.9 ± 0.05), broiled pork center loin including fat (49.8 ± 0.05), roasted pork top loin including fat (49.7 ± 0.05), roasted lamb shoulder including fat (49.6 ± 0.06), chocolate cream pie (49.5 ± 0.04), braised pork boston blade excluding fat (49.4 ± 0.05), broiled pork sirloin including fat (49.3 ± 0.05), raw european chestnuts (49.3 ± 0.18), brie cheese (48.9 ± 0.18), limburger cheese (48.9 ± 0.18), well done baked ground beef (48.9 ± 0.05), pancakes with butter and syrup (48.9 ± 0.03), chicken egg yolk (48.8 ± 0.29), fried chicken wing with skin (48.8 ± 0.16), braised pork loin blade excluding fat (48.6 ± 0.05), broiled pork boston blade including fat (48.5 ± 0.05), roasted pork loin blade including fat (48.2 ± 0.05), fish sandwich (48.1 ± 0 .03), broiled pork loin including fat (48 ± 0.05), pineapple pie (48 ± 0.04), fried chicken giblets (47.9 ± 0.05), pork liver sausage (47.8 ± 0.28), beef and pork summer sausage (47.8 ± 0.22), beef pastrami (47.5 ± 0.18), broiled pork center rib including fat (47.5 ± 0.05), fried chicken neck with skin (47.5 ± 0.14), caviar (47.5 ± 0.31), broiled beef rib including fat (47.4 ± 0.05), rhubarb pie (47.4 ± 0.04), lemon meringue pie (47.4 ± 0.04), fried beef sirloin including fat (47.3 ± 0.05), raw coconut (47.1 ± 0.11), broiled lamb loin chop including fat (47 ± 0.07), pepperoni pizza (46.8 ± 0.04), simmered pork tail (46.7 ± 0.05), braised pork arm braised including fat (46.6 ± 0.05), cherry pie (46.6 ± 0.04), broiled pork top loin including fat (46.1 ± 0.05), port du salut cheese (46.1 ± 0.18), cheesecake (45.9 ± 0.06), cheeseburger (45.8 ± 0.04), hamburger sandwich (45.7 ± 0.05), braised beef chuck arm pot roast including fat (45.5 ± 0.05), cheese pizza (45.5 ± 0.04), fried pork center loin including fat (45.4 ± 0.05), corn tortilla (45.3 ± 0.17), roasted beef rib including fat (45.2 ± 0.05), braised pork sirloin including fat (45.1 ± 0.05), butterscotch pie (45.1 ± 0.04), thousand island salad dressing (45 ± 0.31), pork and beef sausage (44.6 ± 0.38), raw chinese chestnuts (44.6 ± 0.18), broiled pork loin blade including fat (44.4 ± 0.05), pork sausage (44.4 ± 0.19), fried pork center rib including fat (44.1 ± 0.05), braised pork center loin including fat (44.1 ± 0.05), fried pork top loin including fat (44 ± 0.05), fried chicken back with skin (44 ± 0.07), braised pork loin including fat (43.8 ± 0.05), tilsit cheese (43.6 ± 0.18), bacon cheeseburger (43.6 ± 0.03), braised beef chuck blade roast including fat (43.3 ± 0.05), braised pork boston blade braised including fat (43.3 ± 0.05), braised pork center rib including fat (43.2 ± 0.05), braised beef brisket including fat (43 ± 0.05), mince pie (43 ± 0.04), bleu cheese (42.9 ± 0.18), broiled lamb rib chop including fat (42.8 ± 0.07), fried pork loin blade including fat (42.5 ± 0.05), raisin pie (42.5 ± 0.04), braised pork top loin including fat (42.1 ± 0.05), muenster cheese (42.1 ± 0.18), edam cheese (42.1 ± 0.18), gouda cheese (42.1 ± 0.18), boiled peanuts (41.9 ± 0.16), english muffin (41.9 ± 0.09), brick cheese (41.8 ± 0.18), monterey cheese (41.4 ± 0.18), provolone cheese (41.4 ± 0.18), miso (41.4 ± 0.04), 63 proof coffee liqueur (41.3 ± 0.1), pineapple chiffon pie (41.1 ± 0.06), roasted european chestnuts (41.1 ± 0.18), braised pork loin blade including fat (41 ± 0.05), roasted chinese chestnuts (40.7 ± 0.18), braised pork spareribs (40.4 ± 0.05), roquefort cheese (40 ± 0.18), potato pancakes (39.9 ± 0.07), hush puppies (39.8 ± 0.11), caraway cheese (39.6 ± 0.18), american cheese (39.6 ± 0.18), pimento cheese spread (39.6 ± 0.18), pork link sausage (39.3 ± 0.07), blueberry muffins (39 ± 0.13), colby cheese (38.6 ± 0.18), fontina cheese (38.6 ± 0.18), french fries (38.6 ± 0.06), whole wheat bread (38.4 ± 0.2), light cream (38.3 ± 0.17), cheshire cheese (38.2 ± 0.18), french fried potatoes (38 ± 0.1), swiss cheese (37.9 ± 0.18), whole ground corn meal muffins (37.8 ± 0.13), rye bread (37.6 ± 0.2), mixed grain bread (37.6 ± 0.2), canned smoked goose liver pâté (37.5 ± 0.18), italian salad dressing (37.3 ± 0.33), pumpernickel bread (37.2 ± 0.16), havarti cheese (37.1 ± 0.18), cheddar cheese (37.1 ± 0.18), white bread (37.1 ± 0.21), gingerbread (37 ± 0.08), canned goose liver pâté (36.9 ± 0.38), waffles (36.8 ± 0.07), oatmeal bread (36.8 ± 0.18), chocolate syrup (36.6 ± 0.13), mayonnaise (36.4 ± 0.36), hard pork salami (36 ± 0.5), onion rings (35.8 ± 0.06), braised beef shortribs including fat (35.7 ± 0.05), lemon chiffon pie (35.6 ± 0.06), bran muffins (35.5 ± 0.13), russian salad dressing (35.3 ± 0.33), cracked wheat bread (35.2 ± 0.2), hard pork and beef salami (35 ± 0.5), falafel (34.7 ± 0.1), boston cream pie (34.5 ± 0.05), hamburger bun (34 ± 0.13), hot dog bun (34 ± 0.13), gruyere cheese (33.6 ± 0.18), pineapple upside-down cake (33.3 ± 0.07), raisin bread (33.2 ± 0.2), chocolate chiffon pie (33 ± 0.06), corn meal muffins (32.8 ± 0.13), sponge cake (32.7 ± 0.08), cane syrup (32.5 ± 0.25), maple syrup (32.5 ± 0.25), dried prunes (32.4 ± 0.06), dinner rolls (32.1 ± 0.18), whole wheat roll (32.1 ± 0.18), raisin bun (32 ± 0.08), cooked whelk (32 ± 0.06), french rolls (32 ± 0.1), italian bread (31.8 ± 0.18), raw bacon (31.6 ± 0.07), angel food cake (31.5 ± 0.08), romano cheese (31.4 ± 0.18), tamarind (31.4 ± 0.04), pita bread (31.3 ± 0.13), dried apricots (31.1 ± 0.14), 53 proof coffee liqueur (31 ± 0.1), french bread (29.6 ± 0.18), bagels (29.1 ± 0.09), toasted english muffin (29 ± 0.1), toasted whole wheat bread (29 ± 0.24), citrus marmalade (29 ± 0.25), toasted rye bread (28.6 ± 0.23), creme de menthe (28.4 ± 0.1), dried figs (28.4 ± 0.03), raw acorns (28.2 ± 0.18), toasted pumpernickel bread (28.2 ± 0.18), toasted white bread (28.1 ± 0.24), enriched biscuits (27.5 ± 0.18), unenriched biscuits (27.5 ± 0.18), sweetened condensed milk (27.4 ± 0.13), danish pastry (26.7 ± 0.12), cottage pudding (26.7 ± 0.09), toasted cracked wheat bread (26.2 ± 0.24), pork and beef pepperoni (25 ± 0.83), sweet roll (25 ± 0.12), white cake (24.9 ± 0.06), french salad dressing (24.3 ± 0.36), toasted raisin bread (24.3 ± 0.24), barbados molasses (24 ± 0.25), light molasses (24 ± 0.25), medium molasses (24 ± 0.25), blackstrap molasses (24 ± 0.25), yellow cake (23.9 ± 0.07), corn syrup (23.3 ± 0.24), devil's food cake (23 ± 0.08), dried japanese persimmon (22.9 ± 0.15), dried dates (22.5 ± 0.06), sorghum (22.4 ± 0.24), dried lychees (22.3 ± 0.05), raw pork jowl (22.2 ± 0.05), dried sweetened flaked coconut (22 ± 0.2), caramel cake (20.9 ± 0.05), dried jujube (19.7 ± 0.05), dried oriental radish (19.7 ± 0.09), peach pie (19.5 ± 0.05), pecan pie (19.5 ± 0.05), fruitcake, light (18.6 ± 0.12), fruitcake, dark (18.1 ± 0.12), parmesan cheese (18 ± 1), dried longans (17.6 ± 0.05), pound cake (17.3 ± 0.17), honey (17.1 ± 0.24), marshmallow (16.7 ± 0.83), seeded raisins (16.6 ± 0.05), whipped butter (16.4 ± 0.45), dried and salted cod (16.1 ± 0.06), butter (16 ± 0.33), sweet butter (16 ± 0.33), corn oil margarine tub (16 ± 1), safflower oil margarine tub (16 ± 1), soybean oil margarine tub (16 ± 1), soybean oil and cottonseed oil margarine tub (16 ± 1), almond paste (15.7 ± 0.18), seedless raisins (15.4 ± 0.05), safflower oil and soybean oil mayonnaise (15 ± 0.36), soybean oil mayonnaise (15 ± 0.36), turmeric (15 ± 2.5), white pepper (15 ± 2.5), pie crust (14.9 ± 0.03), golden seedless raisins (14.9 ± 0.05), soybean oil and cottonseed oil margarine liquid (14 ± 1), corn oil margarine stick (14 ± 1), safflower oil and soybean oil margarine stick (14 ± 1), soybean oil margarine stick (14 ± 1), soybean oil and cottonseed oil margarine stick (14 ± 1), gjetost cheese (13.6 ± 0.18), cooked bacon (12.9 ± 0.04), toaster pastry (12.8 ± 0.1), dried sweetened shredded coconut (12.6 ± 0.05), dried ginko nuts (12.5 ± 0.18), white corn flour (12 ± 0.04), yellow corn flour (12 ± 0.04), enriched wheat bread (12 ± 0.04), enriched white degermed corn meal (12 ± 0.04), unenriched corn meal (12 ± 0.04), enriched yellow degermed corn meal (12 ± 0.04), unenriched yellow degermed corn meal (12 ± 0.04), all purpose enriched wheat flour (12 ± 0.04), wheat cake (12 ± 0.04), white bolted corn meal (12 ± 0.04), yellow bolted corn meal (12 ± 0.04), white whole ground corn meal (12 ± 0.04), yellow whole ground corn meal (12 ± 0.04), rice flour (12 ± 0.04), light buckwheat flour (12 ± 0.05), dark buckwheat flour (12 ± 0.05), whole wheat flour (12 ± 0.04), gum drops (11.8 ± 0.18), millit (11.8 ± 0.05), light pearled barley (11.1 ± 0.03), light rye flour (11 ± 0.05), sorghum grain (11 ± 0.05), dark rye flour (11 ± 0.04), pot barley pearled (10.8 ± 0.03), dried pepeao (10.8 ± 0.42), brownies (10.5 ± 0.25), basil (10 ± 5), cumin seed (10 ± 2.5), dried parsley (10 ± 5), savory (10 ± 5), corn starch (10 ± 0.63), dried japanese chestnuts (10 ± 0.18), anise seed (10 ± 2.5), caraway seed (10 ± 2.5), fennel seed (10 ± 2.5), cardamom (10 ± 2.5), cinnamon (10 ± 2.5), curry powder (10 ± 2.5), dill seed (10 ± 2.5), dried dill weed (10 ± 5), ginger (10 ± 2.5), paprika (10 ± 2.5), allspice (10 ± 2.5), coriander seed (10 ± 2.5), black pepper (10 ± 2.5), saffron (10 ± 5), sage (10 ± 5), rice polish (9.8 ± 0.09), dried european chestnuts (9.6 ± 0.18), dried shitake mushrooms (9.3 ± 0.33), rennin (9.1 ± 0.45), dried chinese chestnuts (8.9 ± 0.18), dried agar seaweed (8.7 ± 0.05), wheat gluten (8.5 ± 0.04), raw wild rice (8.5 ± 0.03), raisin bran (8.4 ± 0.14), sugar cookie (8.1 ± 0.31), maple sugar (7.9 ± 0.18), potato flour (7.6 ± 0.06), peanut flour (7.5 ± 1.25), fenugreek seed (7.5 ± 1.25), defatted soybean flour (7.3 ± 0.05), dried pumpkin and squash seeds (7.1 ± 0.18), roasted pumpkin and squash seeds (7.1 ± 0.18), almond meal (7.1 ± 0.18), defatted soy meal (7 ± 0.04), dried peanuts (6.8 ± 0.18), dried pignolia pine nuts (6.8 ± 0.18), yellow mustard seed (6.7 ± 1.67), poppy seed (6.7 ± 1.67), garlic powder (6.7 ± 1.67), chili powder (6.7 ± 1.67), dry breadcrumbs (6.5 ± 0.05), jelly beans (6.4 ± 0.18), whole grain rye crackers (6.2 ± 0.38), dried pinyon pine nuts (6.1 ± 0.18), acorn flour (6.1 ± 0.18), torula yeast (6.1 ± 0.18), dried and frozen tofu (5.9 ± 0.29), zwieback crackers (5.7 ± 0.71), toasted wheat germ (5.7 ± 0.18), soybean protein concentrate (5.7 ± 0.18), dried filberts (5.4 ± 0.18), dried sunflower seeds (5.4 ± 0.18), shredded wheat (5.4 ± 0.18), rice bran (5.2 ± 0.05), soybean flour (5.2 ± 0.06), dried pecans (5 ± 0.18), nutmeg (5 ± 2.5), dried sesame kernels (5 ± 0.63), toasted sesame kernels (5 ± 0.18), dried watermelon seeds (5 ± 0.18), dried acorns (5 ± 0.18), celery seed (5 ± 2.5), mace (5 ± 2.5), bread sticks (5 ± 0.25), cloves (5 ± 2.5), onion powder (5 ± 2.5), soybean protein isolate (5 ± 0.18), cayenne pepper (5 ± 2.5), pumpkin pie spice (5 ± 2.5), dry bakers yeast (5 ± 0.18), brewers yeast (5 ± 0.18), oregano (5 ± 2.5), poultry seasoning (5 ± 2.5), tarragon (5 ± 2.5), dried rosemary (5 ± 2.5), dried spirulina seaweed (4.7 ± 0.05), dried almonds (4.6 ± 0.18), wheat flakes (4.6 ± 0.18), whole dried sesame seeds (4.4 ± 0.56), rusk crackers (4.4 ± 0.56), oil roasted pecans (4.3 ± 0.18), dried black walnuts (4.3 ± 0.18), toasted soybean nuts (4.3 ± 0.18), dehydrated onion flakes

(4.3 ± 0.36), granola bar (4.2 ± 0.21), cheese crackers (4 ± 0.33), dry instant skim milk (4 ± 0.05), dry cocoa powder (4 ± 1), beef suet (3.9 ± 0.18), dried pistachio nuts (3.9 ± 0.18), oil roasted cashews (3.9 ± 0.18), soda crackers (3.9 ± 0.18), roasted soybean flour (3.8 ± 0.06), oyster crackers (3.8 ± 0.63), dried brazil nuts (3.6 ± 0.18), dried english/persian walnuts (3.6 ± 0.18), dried butternuts (3.6 ± 0.18), taco shell (3.6 ± 0.45), tostada shell (3.6 ± 0.45), graham crackers (3.6 ± 0.36), carob flour (3.6 ± 0.05), sesame butter (3.3 ± 0.33), saltine crackers (3.3 ± 0.83), popcorn (3.3 ± 0.83), dry skim milk (3.3 ± 0.17), dried coconut (3.2 ± 0.18), oil roasted almonds (3.2 ± 0.18), dry roasted almonds (3.2 ± 0.18), whole toasted sesame seeds (3.2 ± 0.18), bran flakes (3.2 ± 0.18), bran (3.2 ± 0.18), oatmeal cookie (3.1 ± 0.38), animal crackers (3.1 ± 0.19), tomato powder (3.1 ± 0.05), dried pilinuts (2.9 ± 0.18), dried macadamia nuts (2.9 ± 0.18), dried hickory nuts (2.9 ± 0.18), cashew butter (2.9 ± 0.18), gingersnaps cookie (2.9 ± 0.71), puffed rice (2.9 ± 0.36), puffed wheat (2.9 ± 0.36), dry buttermilk (2.9 ± 0.71), low fat soybean flour (2.7 ± 0.06), oil roasted sunflower seeds (2.5 ± 0.18), toasted almonds (2.5 ± 0.18), dry roasted red pistachio nuts (2.5 ± 0.18), potato chips (2.5 ± 0.18), shortbread cookie (2.5 ± 0.63), dry whole milk (2.5 ± 0.16), pretzels (2.5 ± 0.18), corn flakes (2.5 ± 0.18), freeze-dried shallots (2.5 ± 1.25), cheese crackers with peanut butter (2.4 ± 0.12), dry roasted pistachio nuts (2.1 ± 0.18), oil roasted peanuts (2.1 ± 0.18), oil roasted valencia peanuts (2.1 ± 0.18), oil roasted virginia peanuts (2.1 ± 0.18), peanut brittle (2.1 ± 0.18), brown sugar (2.1 ± 0.03), roasted soybean nuts (2 ± 0.06), oil roasted macadamia nuts (1.8 ± 0.18), dry roasted filberts (1.8 ± 0.18), dry roasted cashews (1.8 ± 0.18), oil roasted spanish peanuts (1.8 ± 0.18), bittersweet chocolate (1.8 ± 0.18), dry roasted macadamia nuts (1.4 ± 0.18), dry roasted peanuts (1.4 ± 0.18), unsweetened baking chocolate (1.4 ± 0.18), almond butter (1.3 ± 0.31), creamy peanut butter (1.3 ± 0.31), crunchy peanut butter (1.3 ± 0.16), oil roasted filberts (1.1 ± 0.18), dry roasted pecans (1.1 ± 0.18), dried toasted coconut (1.1 ± 0.18), dry roasted sunflower seeds (1.1 ± 0.18), sweet chocolate (1.1 ± 0.18), milk chocolate (1.1 ± 0.18), semi-sweet chocolate (1.1 ± 0.18), semi-sweet bakers chocolate (1.1 ± 0.18), dry roasted soybean nuts (0.8 ± 0.06), german sweet bakers chocolate (0.7 ± 0.18), powdered sugar (0.5 ± 0.04), white granulated sugar (0.5 ± 0.03)

Daily Dosage: Nutritional — 3 litres; therapeutic — 5 litres; experimental — 7 + litres; toxic — none known

Warnings: Excessive amounts can cause side-effects. See side-effects symptoms.

Appendix A: INDICATIONS FOR USE

abdominal cramps
deficiency of vitamin B_5
toxicity of vitamin B_p, fluorine

abdominal pains
deficiency of vitamin B_1
toxicity of vitamin A, boron, copper, zinc

abscess
deficiency of calcium, sulfur
helped by vitamin A, vitamin C, vitamin E

acetylcholine receptors, decreased
deficiency of lithium

acidosis
deficiency of lysine

acne
deficiency of vitamin A, vitamin B_{3a}, vitamin E, vitamin F, potassium, zinc
helped by vitamin A, vitamin B_2, vitamin B_{3a}, vitamin B_5, vitamin B_6, vitamin C, vitamin D, vitamin E, vitamin F, calcium, potassium
toxicity of bromine, iodine

acrid taste
toxicity of vitamin B_{17}

ADP, decreased synthesis
deficiency of vitamin B_8

adrenal deterioration
deficiency of vitamin B_5, bromine, nickel
helped by vitamin B_2, vitamin B_5, vitamin B_{12}, vitamin B_C, vitamin C, sodium

Adriamycin treatments for cancer
helped by vitamin Q_{10}

aged tissues sagging
helped by proline

aggression
deficiency of tryptophan

aging
helped by provitamin A, vitamin

A, vitamin B_1, vitamin B_3, vitamin B_5, vitamin B_6, vitamin B_H, vitamin B_P, vitamin B_X, vitamin C, vitamin D, vitamin E, vitamin K, vitamin P, vitamin Q_{10}, calcium, iron, selenium, zinc, cysteine, methionine, tryptophan

aging, premature
deficiency of vitamin B_{13}, vitamin B_{15}, cysteine
helped by vitamin M_i™?

AIDS
helped by vitamin C, vitamin Q_{10}

albumin in urine, slight
toxicity of lead

alcohol cravings
deficiency of lithium, glutamine
helped by vitamin A, vitamin B_1, vitamin B_2, vitamin B_{3a}, vitamin B_{3b}, vitamin B_5, vitamin B_6, vitamin B_{12}, vitamin B_{15}, vitamin B_C, vitamin B_P, vitamin C, vitamin D, vitamin K, iron, lithium, magnesium, potassium, zinc

alcohol poisoning
helped by methionine

alertness, decreased
deficiency of glutamine, phenylalanine

allergies
deficiency of vitamin F
helped by vitamin A, vitamin B_5, vitamin B_{12}, vitamin C, vitamin D, vitamin E, vitamin F, vitamin Q_{10}, calcium, manganese, potassium, histidine
toxicity of vitamin B_1

Alzheimer's disease
deficiency of vitamin B_P?
helped by vitamin Q_{10}

Amanita poisoning
helped by vitamin N

amino acid deficiencies
deficiency of vitamin B_{3a}, vitamin B_{3b}, protein

amnesia
toxicity of bromine

anemia
deficiency of vitamin B_2, vitamin B_6, vitamin B_{13}, vitamin B_C, vitamin B_W, vitamin C, vitamin E, cobalt, copper, iron, zinc, lysine
helped by vitamin B_1, vitamin B_5, vitamin B_6, vitamin B_{12}, vitamin B_{13}, vitamin B_{14}, vitamin B_C, vitamin B_X, vitamin C, vitamin E, calcium, copper, iron, moybdenum, histidine
toxicity of vitamin K, arsenic, chlorine, lead, mercury, molybdenum, silver, vanadium
warning about iron

anemia, dialysis induced
helped by vitamin B_T

anemia, pernicious
deficiency of vitamin B_{12}, vitamin B_{14}, vitamin T
helped by vitamin B_6, vitamin B_{12}, vitamin B_C, vitamin C, vitamin E, calcium, cobalt, iron

anemia, sickle cell
deficiency of vitamin B_{17}
helped by vitamin B_{17}
warning about iron

anesthesia
toxicity of magnesium

angina
deficiency of vitamin B_{15}, vitamin B_T, vitamin E, vitamin H_3, vitamin Q_{10}, silicon
helped by vitamin A, vitamin B_{12}, vitamin B_{15}, vitamin B_P, vitamin B_T, vitamin C, vitamin E, vitamin F, vitamin Q_{10}, iodine, potassium

ankles, swelling of
toxicity of arsenic

A. Indications for Use

anorexia, infantile
 helped by vitamin T
antibody formation, decreased
 deficiency of vitamin B_2, vitamin B_5, lysine
antibody G levels, decreased
 deficiency of vitamin Q_{10}
antimony resistance
 deficiency of vitamin B_X
antioxidant therapy
 helped by provitamin A, vitamin A, vitamin B_1, vitamin B_3, vitamin B_5, vitamin B_6, vitamin B_H, vitamin B_P, vitamin B_X, vitamin C, vitamin E, vitamin P, vitamin Q_{10}, selenium, vanadium, zinc, cysteine, methionine, tryptophan
antiviral activity, decreased
 deficiency of vitamin C_3, vitamin P
anxiety
 deficiency of phosphorus, potassium, sodium
 toxicity of vitamin B_{17}, lead, mercury, potassium
aorta, increased incidence of plaques on
 deficiency of trivalent chromium
apoplexy
 helped by germanium
appetite, decreased
 deficiency of vitamin A, vitamin B_1, vitamin B_2, vitamin B_{3a}, vitamin B_{3b}, vitamin B_5, vitamin B_{17}, vitamin B_W, vitamin C, vitamin T, vitamin U, phosphorus, sodium, zinc, protein, lysine
 helped by lysine
 toxicity of vitamin A, vitamin D, aluminum, arsenic, lead, mercury
arms, coldness of
 toxicity of bromine
arms, paralysis of
 toxicity of barium

arms, spasms in
 toxicity of antimony
arsenic, decreased resistance
 deficiency of vitamin B_X
arterial fibrillation
 deficiency of vitamin H_3
arteries, hardening of
 deficiency of vitamin B_6, vitamin B_{15}, vitamin B_P, vitamin H_3, trivalent chromium
 helped by vitamin A, vitamin B_{3a}, vitamin B_{3b}, vitamin B_{12}, vitamin B_{15}, vitamin B_C, vitamin B_H, vitamin B_P, vitamin C, vitamin E, vitamin P, calcium, iodine, magnesium, phosphorus, silicon, vanadium, zinc, arginine
arteries, inflammation of
 deficiency of vitamin H_3
arteries in brain, hardening of
 deficiency of vitamin H_3
arthritis
 deficiency of vitamin B_6, vitamin H_3
 helped by vitamin A, vitamin B_2, vitamin B_{3a}, vitamin B_{3b}, vitamin B_5, vitamin B_6, vitamin B_{12}, vitamin B_C, vitamin C, vitamin D, vitamin E, vitamin F, vitamin H_3, vitamin P, calcium, germanium, iodine, magnesium, phosphorus, potassium, silicon, sulfur, histidine
 toxicity of fluorine
arthritis, rheumatoid
 helped by germanium
asthma
 deficiency of vitamin H_3
 helped by vitamin A, vitamin B_5, vitamin B_6, vitamin B_{12}, vitamin B_{15}, vitamin B_H, vitamin B_P, vitamin F, vitamin H_3, vitamin Q_{10}, manganese
 toxicity of nickel
atherosclerosis
 deficiency of vitamin B_6

athlete's foot
: helped by vitamin A, vitamin B_{3a}, vitamin B_{3b}, vitamin B_{12}, vitamin B_{15}, vitamin B_C, vitamin B_P, vitamin B_H, vitamin E, vitamin P, calcium, iodine, magnesium, phosphorus, zinc, arginine

athlete's foot
: helped by vitamin A, vitamin C, vitamin E

ATP, decreased synthesis
: deficiency of vitamin B_8

autism
: helped by vitamin B_{15}

backache
: deficiency of vitamin B_{3a}, vitamin B_{3b}, vitamin C
: helped by vitamin C, vitamin D, vitamin E, magnesium, phosphorus

bacterial resistance, decreased
: deficiency of sulfur

bad breath
: deficiency of vitamin B_{3a}, vitamin B_{3b}, vitamin B_{17}, zinc
: toxicity of mercury, selenium

balance, decreased
: deficiency of vitamin B_5
: toxicity of lead

baldness
: deficiency of vitamin B_2, vitamin H_3, vitamin B_X?, methionine
: helped by vitamin B_2, vitamin B_{3a}, vitamin B_5, vitamin B_6, vitamin B_C, vitamin B_H, vitamin B_X, vitamin B_T, vitamin C, vitamin E, copper

bedsores
: helped by vitamin A, vitamin B_2, vitamin B_{3a}, vitamin B_{3b}, vitamin C, vitamin D, vitamin E, copper

Bell's palsy
: helped by vitamin B_1, vitamin B_6

beriberi
: deficiency of vitamin B_1
: helped by vitamin C

beryllium in organs
: toxicity of beryllium

bile salts formation, deterioration
: deficiency of vitamin B_5, taurine
: helped by taurine

bilirubin in blood, increased
: toxicity of vitamin K

birth defects
: deficiency of vitamin A
: toxicity of lithium

Bitiot's spots
: helped by vitamin A, vitamin D

bladder inflammation
: helped by vitamin A, vitamin B_5, vitamin B_6, vitamin C, vitamin D, vitamin E

bladder stones
: toxicity of cysteine

bleeding, increased
: deficiency of vitamin A
: helped by vitamin A, vitamin B_1, vitamin B_2, vitamin B_{3a}, vitamin B_{3b}
: toxicity of vitamin E

blood, decreased hemoglobin levels in
: deficiency of vitamin B_T

blood, increased bilirubin in
: toxicity of vitamin K

blood, increased calcium in
: toxicity of calcium
: deficiency of vitamin D

blood, increased lead in
: deficiency of vitamin D

blood, increased phosphorus in
: toxicity of phosphorus

blood, increased pyruvic acid in
: deficiency of vitamin B_1

blood, keytones increased in
: deficiency of vitamin B_T

blood, thin
: toxicity of vitamin E

A. *Indications for Use*

blood cell count, decreased
deficiency of vitamin B₁₄, vitamin B₁₇, arsenic

blood cell rupture
deficiency of vitamin E, copper

blood cells, aggregation of
deficiency of vitamin C₃, vitamin P

blood cells, decreased survival time
deficiency of arsenic, vitamin E

blood cells, large abnormal
deficiency of vitamin B₁₂, vitamin B₁₃, vitamin B$_c$

blood clots
deficiency of vitamin C₃, vitamin P, vitamin Q, tryptophan

blood clotting, decreased
deficiency of vitamin B₁₂, vitamin K, vitamin Q, vitamin T, manganese
helped by vitamin A, vitamin B₃ₐ, vitamin B₃ᵦ, vitamin C, vitamin P, vitamin Q₁₀, vitamin T, calcium

blood fats, slightly increased
toxicity of vitamin E

blood in chicks, deterioration of
deficiency of vitamin B₁₀, vitamin B₁₁

blood iron level deceptively increased
deficiency of vitamin B₁₂

blood lipids, increased
deficiency of vitamin B$_T$

blood oxygenation, decreased
deficiency of vitamin B₁₅

blood platelets, decreased production of
deficiency of vitamin T

blood pressure, decreased
deficiency of vitamin B₁, vitamin B₅
helped by vitamin B₁₇
toxicity of mercury, potassium

blood pressure, increased
deficiency of vitamin B₁₇, vitamin B$_p$, vitamin H₃, vitamin K, vitamin Q₁₀, trivalent chromium, germanium, magnesium
helped by vitamin B₃ₐ, vitamin B₃ᵦ, vitamin B₁₅, vitamin B₁₇, vitamin B$_p$, vitamin C, vitamin E, vitamin F, vitamin Q₁₀, vitamin P, calcium, germanium, magnesium, potassium
side-effect of phenylalanine
toxicity of copper, sodium
warning about vitamin B₃ₐ, vitamin B₃ᵦ, vitamin E, sodium, phenylalanine

blood pressure, unregulated
deficiency of vitamin B₁₇

blood sugar, decreased
deficiency of vitamin B₅, vitamin B₆, potassium
helped by vitamin B₅, vitamin B₆, vitamin B₁₂, vitamin B$_p$, vitamin C
toxicity of vitamin B₃ₐ, vitamin B₃ᵦ, vitamin E
warning about vitamin B₃ᵦ, protein

blood sugar, increased
helped by taurine

blood triglycerides, increased
deficiency of vitamin B$_T$, vanadium
toxicity of vitamin E

blood vessels, calcium salts deposits in
toxicity of vitamin D

blood vessels, collapse of
deficiency of sodium

blood vessels, fragile
deficiency of vitamin C₃, vitamin P

body odor
deficiency of vitamin B₁₂
toxicity of arsenic

boils
helped by vitamin A, vitamin C, vitamin E

bone, decreased development of
deficiency of copper, zinc

bone deformities
deficiency of magnesium, manganese, silicon
toxicity of vitamin C, vitamin D

bone demineralization
deficiency of vitamin D, boron, phosphorus, silicon, strontium, vanadium
toxicity of vitamin A

bone deterioration
deficiency of boron, germanium

bone disease
deficiency of copper
toxicity of molybdenum

bone fracture
helped by vitamin A, vitamin B_5, vitamin C, vitamin D, calcium, magnesium, phosphorus, potassium

bone growth, projecting of spine
toxicity of fluorine

bone marrow red blood cell production, deterioration of
deficiency of vitamin B_{14}

bone thickening
toxicity of vitamin A

bones, broken
helped by vitamin D, calcium, silicon, vanadium

bones, numb
toxicity of vitamin D

bones, rickets
deficiency of vitamin D
helped by vitamin A, vitamin C, vitamin D, calcium, magnesium, phosphorus

bones, separation of ends of
deficiency of vitamin C

bones, softening
deficiency of calcium, vitamin D
helped by vitamin A, vitamin C, phosphorus

bones, thinning
deficiency of calcium, flourine
helped by vitamin B_{12}, vitamin C, vitamin D, vitamin E, copper, germanium, magnesium, phosphorus, strontium

bones, weak in children
deficiency of vitamin C

brain, decreased zinc to
toxicity of copper

brain arteries, hardening of
deficiency of vitamin H_3

brain deterioration
deficiency of bromine, vitamin B_1, vitamin B_{12}
helped by vitamin F

brain waves, abnormal
deficiency of vitamin B_6

breast cancer
helped by vitamin F

breasts, sore
deficiency of calcium, phosphorus

breath, bad
deficiency of vitamin B_{3a}, vitamin B_{3b}, vitamin B_{17}, zinc
toxicity of arsenic, bromine, mercury, selenium
helped by vitamin A, vitamin B_{3a}, vitamin B_{3b}, vitamin B_6, vitamin C

breathing, difficult
deficiency of vitamin B_1, vitamin B_{15}, vitamin C, copper
toxicity of beryllium, chlorine, iron, lead, nickel

bronchitis
helped by vitamin A, vitamin C, vitamin D, vitamin E, vitamin F

bruises
deficiency of vitamin C, vitamin C_3, vitamin P, vitamin Q_{10}
helped by vitamin B_c, vitamin C,

A. Indications for Use

vitamin E, vitamin K, vitamin P, iron
toxicity of lead

burns
helped by vitamin A, vitamin B_x, vitamin C, vitamin D, vitamin E, potassium, zinc

bursitis
helped by vitamin B_{12}, vitamin C, vitamin E

cadmium retention
deficiency of calcium

calcium, decreased
deficiency of vitamin D, boron, calcium, magnesium
toxicity of aluminum, phosphorus

calcium, decreased in urine
toxicity of phosphorus

calcium, decreased osmotic control of in heart
deficiency of taurine

calcium, increased levels in blood
deficiency of vitamin D
toxicity of calcium

calcium deposits in tissues
deficiency of magnesium

calcium salts deposits in blood vessels
toxicity of vitamin D

cancer
deficiency of vitamin B_{17}, vitamin F, vanadium
helped by provitamin A, vitamin A, vitamin B_2, vitamin B_{3a}, vitamin B_{3b}, vitamin B_5, vitamin B_6, vitamin B_{14}, vitamin B_{15}, vitamin B_{17}, vitamin B_H, vitamin B_P, vitamin B_X, vitamin C, vitamin D, vitamin E, vitamin K, vitamin P, vitamin Q_{10}, germanium, iron, phosphorus, potassium, rubidium, selenium, tin, vanadium, zinc, cysteine, methionine, tryptophan

toxicity of vitamin A
warning about vitamin B_H

cancer, Adriamycin treatments for
helped by vitamin Q_{10}

cancer, bone
toxicity of strontium 90

cancer, breast
helped by vitamin F

cancer, esophagus
toxicity of hexavalent chromium

cancer, lung
toxicity of hexavalent chromium

cancer, thyroid
toxicity of iodine 131

cancer cells, accelerated reproduction of
deficiency of vitamin B_{14}

cancer resistance, decreased
deficiency of provitamin A, vitamin A, vitamin B_1, vitamin C, vitamin E, vitamin N, molybdenum, rubidium, tin
toxicity of chlorine

canker sore
helped by vitamin A, vitamin B_{3a}, vitamin B_{3b}, vitamin C, vitamin D

carbohydrate metabolism, disordered
deficiency of vitamin B_8, vitamin B_{12}, arginine

carbuncle
deficiency of calcium, silicon
helped by vitamin A, vitamin C, vitamin E, vitamin D
toxicity of fluorine

cardiac arrest
toxicity of potassium

cardiac arrhythmias
toxicity of taurine

cardiac deterioration
deficiency of vitamin B_1, vitamin E, vitamin Q_{10}, magnesium
helped by silicon

cardiac enlargement
toxicity of cobalt

cardiac ischemia
helped by vitamin B_T

cardiac output decreased
helped by histidine
toxicity of cobalt

cardiospasm
deficiency of vitamin H_3

cardiovascular disease, severe
helped by silicon
warning about lithium

carnitine formation, decreased
deficiency of lysine

carotene deposition in tissues
side-effect of provitamin A

cataracts
deficiency of phenylalanine
helped by vitamin A, vitamin B_2, vitamin B_5, vitamin C, vitamin D, vitamin E, calcium

cavities
deficiency of flourine

celiac disease
helped by vitamin A, vitamin B_6, vitamin B_{12}, vitamin B_C, vitamin C, vitamin D, vitamin E, vitamin K, calcium, iron, magnesium, protein

cell deterioration
deficiency of vitamin B_{13}, protein

cell life shortened
deficiency of vitamin B_{15}, vitamin Q_{10}
helped by vitamin M_i™

cell membrane integrity, decreased
deficiency of nickel
helped by vitamin M_i™?
toxicity of taurine

cell oxygenation, decreased
deficiency of vitamin B_{15}, germanium

cell pH, decreased
deficiency of rubidium

cell regeneration, decreased
deficiency of vitamin H_3

cellular necrosis
deficiency of protein

central nervous system deterioration
deficiency of vitamin B_1

cerebral circulation deterioration
helped by cdp-vitamin B_P

ceroid deposits in muscle
deficiency of vitamin E

charley horse
deficiency of vitamin E

chemotherapy
helped by vitamin Q_{10}

chest pain
deficiency of vitamin B_1, vitamin B_{15}
toxicity of vitamin B_{17}, vitamin K, cadmium

chicken pox
helped by vitamin A, vitamin C

chicks, decreased development
deficiency of vitamin B_{10}, vitamin B_{11}

chicks, deterioration of blood in
deficiency of vitamin B_{10}, vitamin B_{11}

children, deterioration of growth in
helped by lysine

chills
toxicity of mercury

choking feeling
toxicity of vitamin B_{17}

cholesterol levels, decreased
deficiency of manganese
cholesterol solubility decreased
deficiency of taurine
cholesterol levels, increased
deficiency of vitamin B_{3a}, vitamin B_6, vitamin B_{15}, vitamin B_H, vitamin B_T, vitamin B_W, vitamin C, vitamin D, vitamin H_3, vitamin U, trivalent chromium, nickel, vanadium
helped by vitamin B_{3a}, vitamin B_6, vitamin B_{15}, vitamin B_H, vitamin B_P, vitamin C, vitamin D, vitamin P, vitamin U, germanium, magnesium, silicon, vanadium, arginine
warning about vitamin B_{3b}
cigarette smoke, protection from
helped by vitamin C_3
circulatory deterioration
deficiency of vitamin B_{15}
helped by vitamin H_3, vitamin Q, vitamin T
clots, blood
deficiency of vitamin C_3, vitamin P, vitamin Q, tryptophan
clotting of blood, slow
deficiency of vitamin B_{12}, vitamin K, vitamin Q, vitamin T, manganese
helped by vitamin Q_{10}
coenzyme A, decreased levels of
deficiency of vanadium
coenzyme pyridoxal phosphate levels, decreased
deficiency of vitamin B_6
coenzyme Q_{10}, lost source of
deficiency of vitamin Q_7, vitamin Q_8, vitamin Q_9
cold, common
helped by vitamin A, vitamin B_6, vitamin C, vitamin D, vitamin E, vitamin F, vitamin P, calcium
cold, sensitive to
deficiency of vitamin C, vitamin T

toxicity of iodine
colic
deficiency of magnesium, phosphorus
toxicity of vitamin C, aluminum
colitis
helped by vitamin A, vitamin B_6, vitamin C, vitamin E, vitamin F, calcium, iron, magnesium, phosphorus, potassium
toxicity of mercury
collagen deterioration
deficiency of silicon
helped by methionine, proline
collapse
toxicity of antimony, silver
color perception, decreased
deficiency of vitamin B_{12}
coma
deficiency of vitamin B_5
toxicity of lead, magnesium, sodium
compulsion
deficiency of tryptophan
confusion
deficiency of vitamin B_{3a}, vitamin B_{3b}, vitamin B_6, potassium, sodium
toxicity of vitamin B_{17}, bromine, vanadium
connective tissue, increased
deficiency of vitamin B_6
constipation
deficiency of vitamin B_1, vitamin B_5, vitamin D, iron, phosphorus, potassium
helped by vitamin A, vitamin B_1, vitamin B_{3a}, vitamin B_{3b}, vitamin B_H, vitamin B_P, vitamin B_X, vitamin C, vitamin D, vitamin E, vitamin F, calcium, potassium
toxicity of vitamin D, aluminum, arsenic, lead
convulsions
deficiency of vitamin B_5, vitamin B_6, calcium, glutamine
helped by taurine

coordination, decreased
    toxicity of vitamin B_{17}, vitamin H_3, fluorine, lead
    warning about vitamin B_H

coordination, decreased
    deficiency of vitamin B_5, manganese, glutamine, valine

copper, decreased
    deficiency of copper
    toxicity of molybdenum, zinc

copper, increased
    helped by histidine
    warning about copper

coronary thrombosis
    helped by vitamin E, vitamin F

cough
    toxicity of arsenic, mercury, nickel

cramps, abdominal
    deficiency of vitamin B_5, vitamin B_6, vitamin D, bromine, magnesium, phosphorus
    toxicity of vitamin B_P, barium, fluorine

cranium pressure, increased
    toxicity of vitamin A

creatine in urine, increased
    deficiency of vitamin E

cretinism
    helped by iodine

croup
    helped by vitamin A, vitamin C

cysteine production, decreased
    deficiency of methionine

cystic fibrosis
    helped by vitamin A, vitamin C, vitamin D, vitamin E, vitamin K, sodium

cystine production, decreased
    deficiency of methionine

dandruff
    deficiency of vitamin B_6, vitamin B_{12}, vitamin F, selenium
    helped by vitamin A, vitamin B_6, vitamin E

deafness
    deficiency of iodine, histidine

dehydration
    helped by sodium
    toxicity of zinc

depression
    deficiency of vitamin B_1, vitamin B_2, vitamin B_{3a}?, vitamin B_{3b}?, vitamin B_5, vitamin B_6, vitamin B_{12}, vitamin B_C, vitamin B_W, vitamin B_X, vitamin C, iron, lithium, potassium, phosphorus, rubidium, sodium, protein, phenylalanine, tryptophan
    helped by vitamin B_5, vitamin B_T, vitamin F, lithium, tryptophan
    toxicity of vitamin D, lead, mercury, histidine
    warning about histidine

depression, period of manic depressive psychosis
    warning about vitamin B_P
    helped by vanadium

depression, psychosis
    helped by germanium

depression taking tryptophan
    warning about vitamin B_6

depression/emotional agitation, alternating
    deficiency of vitamin B_{3a}, vitamin B_{3b}

diabetes
    helped by vitamin A, vitamin B_1, vitamin B_2, vitamin B_{3a}, vitamin B_{3b}, vitamin B_6, vitamin B_{12}, vitamin B_{15}, vitamin C, vitamin D, vitamin E, calcium, trivalent chromium, iron, magnesium, manganese, potassium, vanadium, zinc
    warning about provitamin A, vitamin B_{3a}, vitamin B_{3b}, vitamin C, vitamin E, cysteine

A. Indications for Use

diarrhea
    deficiency of vitamin B_{3a}, vitamin B_{3b}, vitamin B_{12}, vitamin B_C, vitamin F, vitamin K, iron, phosphorus
    helped by vitamin A, vitamin B_1, vitamin B_2, vitamin B_{3a}, vitamin B_{3b}, vitamin B_5, vitamin B_6, vitamin B_C, vitamin C, vitamin F, calcium, chlorine, iron, magnesium, potassium, sodium
    side-effect of vitamin C
    toxicity of vitamin A, vitamin B_P, vitamin B_T, vitamin D, arsenic, boron, magnesium, molybdenum, potassium

digestion, improper carbohydrate
    deficiency of vitamin B_1

digestion, improper fat
    deficiency of phosphorus, potassium, vanadium

digestion deterioration in pigeons
    deficiency of vitamin B_7

digestion upset
    deficiency of vitamin B_{3a}, vitamin B_{3b}, vitamin B_X, vitamin H_3, vitamin Q_{10}, iron, potassium, tryptophan, threonine
    helped by vitamin B_1, vitamin B_2, vitamin B_{3a}, vitamin B_{3b}, vitamin B_5, vitamin B_6, vitamin B_C, lysine
    toxicity of tin
    warning about vitamin F

diverticulitis
    helped by vitamin B_C

dizziness
    deficiency of vitamin B_2, iron, lysine, magnesium, manganese, sodium
    helped by vitamin B_2, vitamin B_{3a}, vitamin B_{3b}, vitamin B_5, vitamin B_6, vitamin B_{12}, vitamin B_H, vitamin B_P, vitamin C, calcium
    toxicity of vitamin B_{17}, vitamin E, iron, lead, nickel, zinc

DNA decreased synthesis
    deficiency of nickel

Down's syndrome
    helped by taurine

dream recall, increased
    side-effect of vitamin B_6, vitamin B_C

dream recall deterioration
    deficiency of vitamin B_6

dreaming color
    side-effect of vitamin B_{12}

ear infection
    helped by vitamin A, vitamin C

ear noises
    deficiency of vitamin H_3, manganese, potassium

eating, increased
    deficiency of phenylalanine

embryonic development, deterioration of
    deficiency of silicon

emotional agitation
    deficiency of vitamin B_1, vitamin B_2, vitamin B_{3a}, vitamin B_{3b}, vitamin B_5, vitamin B_6, vitamin B_{12}, vitamin C, vitamin D, vitamin F, calcium, protein, lysine, phenylalanine, threonine, tryptophan
    helped by lithium, vitamin F
    side-effect of phenylalanine
    toxicity of vitamin A, vitamin B_6, vitamin B_P, bromine, copper, lead, mercury, valine

emotional agitation/depression alternating
    deficiency of vitamin B_{3a}, vitamin B_{3b}

emotional deterioration
    deficiency of vitamin B_1, vitamin B_2, vitamin B_{3a}, vitamin B_{3b}, vitamin B_5, vitamin B_{12}, tryptophan, valine
    helped by vitamin H_3, germanium
    toxicity of bromine, calcium
    warning about vitamin B_P

emphysema

emphysema
   helped by vitamin A, vitamin B_{15}, vitamin B_C, vitamin C, vitamin D, vitamin E
   toxicity of cadmium

endocrine glands, degenerating
   deficiency of vitamin E

enzyme deterioration
   deficiency of rubidium, lysine
   toxicity of boron, beryllium, tungsten

epilepsy
   deficiency of taurine
   helped by vitamin A, vitamin B_{3a}, vitamin B_{3b}, vitamin B_5, vitamin B_6, vitamin B_{12}, vitamin C, vitamin D, vitamin E, calcium, magnesium, taurine
   warning about protein

epinephrine levels, decreased
   deficiency of vitamin C_3, vitamin P

estrogen synthesis, decreased
   deficiency of boron

excitation, increased
   warning about tryptophan

extremities, aching
   deficiency of vitamin C

extremeties, decreased feeling in
   toxicity of vitamin B_6

extremeties, pain in
   deficiency of potassium, sulfur

eyebrow hair, decreased
   toxicity of vitamin A

eyelid membranes, ashen-gray color of
   toxicity of silver

eyelids, granulation of
   deficiency of vitamin B_2, chlorine, potassium

eyelids, increased color in
   toxicity of arsenic

A. Indications for Use

eyelids, swelling of
   toxicity of arsenic

eyes, bloodshot
   deficiency of vitamin A, vitamin B_2, lysine, phenylalanine, tryptophan

eyes, burning
   deficiency of vitamin B_2

eyes, congestion of
   toxicity of arsenic

eyes, cornea deterioration
   deficiency of vitamin B_5

eyes, cornea opacities
   deficiency of trivalent chromium

eyes, cornea softening
   deficiency of vitamin A

eyes, cornea thickening
   deficiency of vitamin A

eyes, cracks around
   deficiency of vitamin B_6

eyes, dialated
   deficiency of vitamin B_2

eyes, dry
   deficiency of vitamin A, vitamin F

eyes, dull/glazed
   deficiency of silicon

eyes, fatigue
   deficiency of vitamin B_2

eyes, focus deterioration
   helped by vitamin C, vitamin D

eyes, glaucoma
   helped by vitamin A, vitamin B_2, vitamin B_H, vitamin B_P, vitamin C, vitamin D
   warning about vitamin B_{3a}, vitamin B_{3b}

eyes, hypersensitivity to light
   deficiency of vitamin B_{3a}, vitamin B_{3b}

eyes, inflammation of
   helped by vitamin A, vitamin B_2,

A. *Indications for Use* — feet, numb/tingling

vitamin B_{3a}, vitamin B_{3b}, vitamin B_6, vitamin C
toxicity of vanadium

eyes, inflammation of optic nerve
deficiency of vitamin B_1

eyes, itching
deficiency of vitamin A, vitamin B_2
toxicity of arsenic

eyes, lesions in
toxicity of fluorine

eyes, night blindness
deficiency of vitamin A
helped by vitamin A, vitamin B_1, vitamin B_2, vitamin B_{3a}, vitamin B_{3b}

eyes, protruding
toxicity of vitamin A

eyes, pupil dilation
toxicity of vitamin B_T

eyes, senile keratosism
deficiency of vitamin H_3

eyes, sties
deficiency of vitamin A, vitamin B_2

eyes, watering
deficiency of chlorine, sodium
toxicity of arsenic

eyestrain
helped by vitamin A, vitamin D, vitamin E

eyestrain, glaucoma
helped by vitamin C

face, increased color in
toxicity of arsenic

face muscles, very fine twitching of
toxicity of lead

facial expression, decreased
toxicity of manganese

facial skin scaling
deficiency of vitamin B_2

fatigue
deficiency of vitamin A, vitamin B_1, vitamin B_2, vitamin B_{3a}, vitamin B_{3b}, vitamin B_5, vitamin B_6, vitamin B_8, vitamin B_{12}, vitamin B_{15}, vitamin B_T, vitamin B_W, vitamin B_X, vitamin C, vitamin D, vitamin F, vitamin Q_{10}, vitamin T, iodine, iron, magnesium, molybdenum, nickel, phosphorus, potassium, silicon, sulfur, zinc, lysine, protein
helped by vitamin A, vitamin B_5, vitamin B_{17}, vitamin B_C, vitamin C, iodine, manganese
side-effect of vitamin B_{3a}, vitamin B_{3b}
toxicity of vitamin A, vitamin E, aluminum, bromine, iodine, iron, lead, magnesium, mercury, selenium, zinc

fats, deterioration or breakdown of
deficiency of vitamin B_8, vanadium

fatty acid conversion to prostaglandins, increased
deficiency of lithium

fatty acid metabolizing, slow
deficiency of vitamin B_T

fatty acid oxidation, increased
deficiency of molybdenum

feather growth in chicks, deterioration of
deficiency of vitamin B_{10}

feces, increased fat in
toxicity of copper

feet, burning
deficiency of vitamin B_1, vitamin B_2
toxicity of arsenic

feet, cold
deficiency of calcium, iodine, phosphorus

feet, numb/tingling
deficiency of vitamin B_5, vitamin B_{12}, vitamin B_C
toxicity of arsenic

feet, peeling of
  toxicity of arsenic
fetal abnormalities
  deficiency of vitamin B_5, selenium
  warning about tryptophan
fetor
  deficiency of vitamin B_{17}
fever
  deficiency of phosphorus, sodium, methionine
  helped by vitamin A, vitamin B_1, vitamin C, vitamin D, calcium, potassium, sodium
  toxicity of mercury, nickel
fingernail deterioration
  deficiency of cysteine, sulfur
  toxicity of arsenic
fingers, slight trembling in
  toxicity of lead
fingers, spasms in
  toxicity of antimony
fingertips, numb
  toxicity of vitamin D
flatulence
  deficiency of vitamin U, vitamin B_{17}, potassium
  side-effect of vitamin C
fluid accumulation in tissues
  deficiency of magnesium
  toxicity of vitamin B_1, lithium
frostbite, decreased resistance
  deficiency of vitamin C_3, vitamin P
gall bladder deterioration
  deficiency of phosphorus, sodium
  toxicity of taurine
gallstones
  deficiency of vitamin C, vitamin F
  helped by vitamin A, vitamin C, vitamin D, vitamin E, vitamin K
  warning about vitamin F
gastritis
  helped by vitamin A, vitamin B_5, vitamin B_6, vitamin B_{12}, vitamin B_C, vitamin B_H, vitamin C, vitamin E, iron, potassium
gastrointestinal deterioration
  deficiency of vitamin A, vitamin B_1, vitamin B_2, vitamin B_5, vitamin B_{17}, vitamin B_C, vitamin C, chlorine, protein
  helped by vitamin B_{17}, vitamin H_3, vitamin U
  toxicity of vitamin B_P, vitamin E, cadmium
gastrointestinal hemorrhage
  toxicity of hexavalent chromium
gastrointestinal pain
  toxicity of taurine
gastrointestinal tone, decreased
  toxicity of calcium
gastrointestinal tract, lesions in
  toxicity of fluorine
genital rashes
  deficiency of vitamin B_2
gingivitus
  deficiency of vitamin Q_{10}
  helped by vitamin Q_{10}
girdlemuscular dystrophy
  helped by vitamin B_T
gland deterioration
  deficiency of vitamin B_{15}, nickel
glands, swollen
  helped by vitamin A, vitamin C
glandular myoma
  helped by germanium
glucose tolerance, decreased
  deficiency of vitamin B_6, trivalent chromium, manganese, rubidium
glutamic acid in hemoglobin
  warning about valine
goiter
  deficiency of iodine
  helped by vitamin A, vitamin C, iodine

A. Indications for Use

gonad function, decreased
deficiency of zinc

gout
helped by vitamin A, vitamin B$_5$, vitamin C, vitamin E, iron, potassium
toxicity of vitamin B$_{3a}$, vitamin B$_{3b}$, molybdenum
warning about vitamin B$_{3a}$, vitamin B$_{3b}$

growth deterioration
deficiency of vitamin A, vitamin B$_2$, vitamin B$_5$, vitamin B$_{10}$, vitamin B$_{12}$, vitamin B$_{13}$, vitamin B$_C$, vitamin F, vitamin T, arsenic, trivalent chromium, cobalt, iodine, manganese, nickel, silicon, tin, vanadium, zinc, glycine, lysine, tryptophan, protein
helped by vitamin B$_2$, vitamin B$_5$, phosphorus, zinc
toxicity of vitamin A, molybdenum

growth hormone release
helped by arginine, ornithine
warning about arginine, ornithine

growth in chicks, deterioration of
deficiency of vitamin B$_{11}$

gum deterioration
deficiency of vitamin B$_{3a}$, vitamin B$_{3b}$, vitamin B$_6$, vitamin Q$_{10}$
helped by vitamin A, vitamin C, vitamin F, vitamin Q$_{10}$, calcium, fluorine, iron, phosphorus, potassium, sodium
toxicity of mercury, nickel

gums, bleeding
deficiency of vitamin C, vitamin Q$_{10}$
helped by vitamin Q$_{10}$

hair, brittle
deficiency of vitamin A, vitamin F, iodine
toxicity of vitamin A

hair, coarse
deficiency of arsenic, nickel
toxicity of vitamin A

hair, decreased
deficiency of vitamin B$_2$, vitamin B$_6$, vitamin B$_C$, vitamin B$_H$, vitamin B$_W$, vitamin C, vitamin F, silicon, chlorine, iron, zinc, tryptophan
toxicity of vitamin A, arsenic, selenium

hair, decreased in eyebrow
toxicity of vitamin A

hair, deterioration
helped by vitamin A, vitamin B$_P$, vitamin C, vitamin H$_3$, iodine, silicon

hair, dry
deficiency of vitamin A, vitamin B$_X$, iodine

hair, dull
deficiency of vitamin F, sulfur

hair, grey
deficiency of vitamin B$_5$

hair, oily
deficiency of vitamin B$_6$

hair, sparse
toxicity of vitamin A

hair color, changed
deficiency of manganese

hair color decreased
deficiency of vitamin H$_3$

hair follicles, enlarged
deficiency of vitamin A

hair follicles, hardening
deficiency of vitamin C

hair follicles, swelling
deficiency of vitamin C

hair growth, deterioration of
deficiency of manganese, cysteine

hair pigmentation, abnormal
deficiency of copper, nickel, protein

hallucinations
deficiency of vitamin B$_{3a}$, vitamin B$_{3b}$, vitamin B$_X$, tryptophan
toxicity of bromine, lead, mercury

hands, burning
: deficiency of vitamin B_2
: toxicity of arsenic

hands, cold
: deficiency of calcium, phosphorus

hands, numb/tingling
: deficiency of vitamin B_5, vitamin B_{12}, vitamin B_C
: toxicity of arsenic

hands, peeling of
: toxicity of arsenic

hands, slight trembling in
: toxicity of lead

hangover
: deficiency of vitamin B_{12}
: helped by vitamin F

hay fever
: helped by vitamin A, vitamin C, vitamin E
: deficiency of chlorine, sodium

headache
: deficiency of vitamin B_{3a}, vitamin B_5, vitamin B_{15}, calcium, iron, sodium
: helped by vitamin A, vitamin B_1, vitamin B_{3a}, vitamin B_{3b}, vitamin B_5, vitamin B_6, vitamin B_{15}, vitamin B_X, vitamin C, vitamin E, potassium
: toxicity of vitamin A, vitamin B_{3a}, vitamin B_{3b}, vitamin B_{17}, vitamin D, vitamin E, bromine, cadmium, iron, iodine, valine
: side-effect of phenylalanine
: warning about vitamin B_{3b}

healing, deterioration of
: deficiency of vitamin A, vitamin B_5, vitamin B_{15}, vitamin B_W, vitamin C, vitamin E, vitamin U, vitamin T, silicon, zinc
: helped by arginine, proline

hearing deterioration
: deficiency of vitamin H_3, manganese, histidine
: helped by vitamin H_3, histidine
: toxicity of lead

heart, cardiac ischemia
: helped by vitamin B_T

heart, decreased osmotic control of calcium in
: deficiency of taurine

heart, decreased osmotic control of potassium in
: deficiency of taurine

heart, decreased potassium in
: deficiency of taurine

heart, endocardium degenerating
: deficiency of magnesium

heart, enlarged
: deficiency of vitamin B_T

heart, fatty changes in
: toxicity of vitamin B_X

heart, myocardial infarction
: helped by vitamin A, vitamin E, potassium

heart, tin accumulation in
: toxicity of tin

heart arrhythmia
: deficiency of vitamin B_{13}, vitamin H_3

heart attack
: helped by vitamin Q_{10}

heart deterioration
: deficiency of vitamin B_6, vitamin B_{13}, vitamin B_{15}, vitamin B_T, vitamin B_W, vitamin F, vitamin H_3, vitamin Q_{10}, germanium, potassium, vanadium
: helped by vitamin B_T, vitamin Q_{10}, vitamin H_3, vitamin U, germanium
: toxicity of cobalt, fluorine
: warning about vitamin C, vitamin E

heart failure, congestive
: helped by vitamin A, vitamin B_1, vitamin E, potassium
: warning about sodium

heart palpitation
: deficiency of vitamin B_{12}
: toxicity of vitamin B_1

A. Indications for Use

heartbeat, rapid
   toxicity of vitamin B_1
heartbeat, slow
   toxicity of magnesium
heartburn
   toxicity of lead
heat tolerance, decreased
   deficiency of vitamin C
heaviness, feeling of
   deficiency of potassium, sulfur
heavy metal poisoning
   helped by methionine
hematocrit, decreased
   deficiency of nickel
hemoglobin, glutamic acid in
   warning about valine
hemoglobin formation, decreased
   deficiency of isoleucine
hemoglobin levels in blood, decreased
   deficiency of vitamin B_6, vitamin B_{17}, vitamin B_T
hemorrhage
   deficiency of vitamin C, vitamin K
   helped by vitamin K
hemorrhage, gastrointestinal
   toxicity of hexavalent chromium
hemorrhage, kidney
   deficiency of vitamin B_5, vitamin B_p
hemorrhoids
   deficiency of vitamin C_3, vitamin P
   helped by vitamin A, vitamin B_6, vitamin C, vitamin E, vitamin P, calcium
herpes simplex virus
   helped by lysine
   warning about arginine
histamine toxicity
   helped by methionine
Hodgkin's disease
   helped by zinc

insulin therapy

hormone function, breakdown of
   deficiency of boron
hormone imbalance
   deficiency of nickel
hyperkinetic
   helped by vitamin F, vitamin H_3
immunity, decreased
   deficiency of vitamin A, vitamin B_1, vitamin B_6, vitamin B_{12}, vitamin B_{13}, vitamin B_{15}, vitamin B_C, vitamin B_P, vitamin B_W, vitamin C, vitamin F, vitamin Q_{10}, vitamin T, germanium, iron, silicon, zinc, arginine, cysteine, lysine
   helped by vitamin Q_{10}
   toxicity of vitamin A
immunoglobulin levels, decreased
   deficiency of vitamin Q_{10}
incontinence
   toxicity of vitamin B_{17}
infection
   deficiency of vitamin A, vitamin B_5
   helped by vitamin A, vitamin B_5, vitamin C
inflammation, internal
   deficiency of zinc
   toxicity of iron
influenza
   helped by vitamin A, vitamin B_1, vitamin B_2, vitamin B_6, vitamin C
insulin, sensitivity to
   deficiency of vitamin B_5
insulin levels, decreased
   deficiency of phenylalanine
insulin levels in diabetics
   warning about cysteine, vitamin E
insulin production altered
   toxicity of vitamin B_1
insulin therapy
   helped by vitamin F

interferon production, decreased
deficiency of germanium

iodine, decreased
deficiency of iodine
side-effect of vitamin U?

iron, decreased
toxicity of zinc

iron, decreased utilization
toxicity of manganese

iron increase in spleen
deficiency of arsenic

iron level, deceptively increased in blood
deficiency of vitamin B_{12}

iron overloads in tissue
helped by histidine

jaundice
helped by vitamin A, vitamin B_6, vitamin C, vitamin D, vitamin K
toxicity of arsenic, copper

joint pain
deficiency of vitamin C, vitamin B_{17}
toxicity of vitamin A, lead, mercury

joints, hypermobile
helped by proline

joints, limited movement of
deficiency of vitamin B_{17}

joints, swollen
deficiency of vitamin C, vitamin F
toxicity of arginine, ornithine

keytones in blood, increased
deficiency of vitamin B_T
helped by vitamin B_T

kidney deterioration
deficiency of vitamin D, magnesium
helped by vitamin B_T
toxicity of vitamin A, vitamin B_X, vitamin D, aluminum, arsenic, cadmium, calcium, phosphorus
warning about lithium, magnesium

kidney hemorrhage
deficiency of vitamin B_5, vitamin B_F

kidney inflammation
helped by vitamin A, vitamin B_2, vitamin C, vitamin E, calcium, iron, magnesium
warning about protein

kidney stones
helped by vitamin A, vitamin B_6, vitamin C, vitamin E, magnesium
toxicity of vitamin C?, cysteine
warning about calcium

kwashiorkor
helped by vitamin A, vitamin B_C, vitamin C, vitamin D, vitamin K, trivalent chromium, magnesium, selenium

lactation deterioration
deficiency of vitamin A, vitamin L
helped by vitamin L
toxicity of vitamin B_6

laryngitis
toxicity of arsenic

larynx growth
toxicity of arginine, ornithine

L-dopa treatments
warning about vitamin B_6

lead retention
deficiency of vitamin D, calcium

leg cramp
deficiency of vitamin B_5
helped by vitamin B_1, vitamin B_2, vitamin B_5, vitamin C, vitamin D, vitamin E, vitamin F, calcium, magnesium, phosphorus, sodium

leg muscles tender
deficiency of vitamin B_1

leg muscles, twitching
toxicity of aluminum

legs, coldness of
toxicity of bromine

legs, paralysis of
toxicity of barium

legs, spasms in
toxicity of antimony

leukemia
helped by vitamin B_C, vitamin C, vitamin P, vitamin Q_{10}, copper, germanium, iron
toxicity of strontium 90

life span extension
helped by provitamin A, vitamin A, vitamin B_1, vitamin B_3, vitamin B_5, vitamin B_6, vitamin B_H, vitamin B_P, vitamin B_X, vitamin C, vitamin E, vitamin P, vitamin Q_{10}, selenium, vanadium, zinc, cysteine, methionine, tryptophan

life span shortened
deficiency of vitamin T
toxicity of tin

ligament difficulties
deficiency of vitamin B_6

light, hypersensitivity to
deficiency of vitamin B_2, vitamin B_{3a}, vitamin B_{3b}

light, need of brightness to see
deficiency of vitamin B_2

lips, chapped
toxicity of vitamin E

lips, scaling
deficiency of vitamin B_6, iron

lips, sore
toxicity of vitamin A

liver, cirrhosis of
deficiency of vitamin B_P
helped by vitamin A, vitamin B_{12}, vitamin B_{15}, vitamin B_H, vitamin B_P, vitamin C, vitamin D, vitamin K, vitamin U
toxicity of copper

liver, enlarged
toxicity of vitamin A, arsenic, copper

liver, fatty
deficiency of vitamin B_P, threonine
toxicity of vitamin B_X

liver, fibrosis of
deficiency of protein

liver, hepatitis
helped by vitamin A, vitamin B_{15}, vitamin B_P, vitamin C, vitamin K

liver, increased cholesterol formation in
deficiency of trivalent chromium

liver deterioration
deficiency of vitamin B_{13}, vitamin F, bromine, vanadium, zinc, glycine, methionine
helped by vitamin B_{13}, vitamin N, vitamin U
toxicity of vitamin B_{3a}, vitamin B_{3b}, vitamin B_5, vitamin D, aluminum
warning about vitamin B_{3a}, vitamin B_{3b}

lung, calcium salts deposits in
toxicity of vitamin D

lung, inflammation of
toxicity of vanadium

lung deterioration
deficiency of vitamin B_5

lymph cell count, decreased
deficiency of vitamin B_6, lithium

magnesium, decreased
deficiency of boron, magnesium
toxicity of vitamin D, beryllium

measles
helped by vitamin A, vitamin E

memory deterioration
deficiency of vitamin B_1, vitamin B_{3a}, vitamin B_{3b}, vitamin B_{12}, vitamin B_P, vitamin T, vitamin H_3, phosphorus, potassium, silicon, phenylalanine, tryptophan

Ménièr's syndrome
helped by vitamin B_1, vitamin B_2, vitamin B_{3a}, vitamin B_{3b}, vitamin E, vitamin F, calcium

meningitis
helped by vitamin A, vitamin C, vitamin D

menses, increased
deficiency of vitamin A, iron, phosphorus

menses, irregular
toxicity of vitamin A, iodine

mental deterioration
deficiency of vitamin B_1, vitamin B_6, vitamin B_{15}, vitamin T, bromine, magnesium, glutamine, lysine, threonine, valine
helped by vitamin B_1, vitamin B_{3a}, vitamin B_{3b}, vitamin B_5, vitamin B_6, vitamin B_C, vitamin C, vitamin E, vitamin F, vitamin H_3, calcium, lithium, magnesium, phosphorus
toxicity of lead, mercury

mental retardation
deficiency of vitamin B_6, vitamin B_{13}, iodine
toxicity of lead

metabolism, disordered carbohydrate
deficiency of arginine

metabolism deterioration
deficiency of vitamin B_{12}
toxicity of vitamin F

metabolism rate, decreased
toxicity of vitamin E

metabolizing of fatty acids, decreased
deficiency of vitamin B_T

mitochondria, decreased cell energy from
deficiency of vitamin Q_{10}

mitochondria deterioration
deficiency of strontium

mobility, decreased
toxicity of manganese

mononucleosis
helped by vitamin A, vitamin C, potassium

motion sickness
deficiency of vitamin B_6

mouth, cracks around
deficiency of vitamin B_2, vitamin B_6

mouth, dry
deficiency of vitamin F

mouth, frothing of
toxicity of vitamin B_{17}

mouth, inflamed
toxicity of vitamin E, arsenic

mouth, metallic taste in
toxicity of antimony, lead, mercury

mouth, pain in
toxicity of antimony, mercury

mouth sores
deficiency of vitamin B_2, vitamin B_6

mouth watering
toxicity of vitamin B_T

mucous membranes, grayish
deficiency of vitamin B_W

multiple sclerosis
helped by vitamin B_1, vitamin B_2, vitamin B_{3a}, vitamin B_{3b}, vitamin B_5, vitamin B_6, vitamin B_{12}, vitamin B_{13}, vitamin B_{15}, vitamin B_P, vitamin C, vitamin E, vitamin F, vitamin Q_{10}, magnesium, manganese

muscle, calcification of
deficiency of selenium

muscle, ceroid deposits in
deficiency of vitamin E

muscle, decreased development of
deficiency of zinc

muscle, fat accumulation in
deficiency of vitamin B_T

muscle activity, decreased
helped by potassium

muscle cramps
deficiency of vitamin B_T, chlorine, potassium, sodium
helped by calcium
toxicity of lead

A. Indications for Use

muscle deterioration
deficiency of vitamin E, selenium, valine

muscle disease
helped by vitamin B_T

muscle fibers, knotting of
deficiency of magnesium

muscle pain
deficiency of vitamin B_1, vitamin B_6, vitamin B_T, vitamin E, vitamin B_W, vitamin C
helped by vitamin B_T
toxicity of fluorine

muscle shrinkage
deficiency of sodium

muscle spasms
helped by vitamin D, calcium
toxicity of boron

muscle stiffness
deficiency of vitamin B_{12}
toxicity of vitamin D, lead, manganese

muscle tendon strain
deficiency of calcium
toxicity of fluorine

muscle tone, decreased
deficiency of vitamin D, vitamin H_3

muscle tremors
deficiency of vitamin B_2, calcium, sodium
toxicity of vitamin B_1, aluminum, boron, lead, magnesium, manganese, mercury

muscle weakness
deficiency of vitamin B_{3a}, vitamin B_{3b}, vitamin B_4, vitamin B_5, vitamin B_{12}, vitamin B_C, vitamin B_T, vitamin C, vitamin D, vitamin E, copper, iron, magnesium, potassium, sodium
toxicity of vitamin A, vitamin D, vitamin E, arsenic, cadmium, magnesium, manganese, potassium

muscular coordination, decreased
deficiency of manganese
toxicity of zinc

muscular dystrophy
helped by vitamin A, vitamin B_5, vitamin B_6, vitamin B_{12}, vitamin B_P, vitamin B_T, vitamin C, vitamin E, vitamin Q_{10}, potassium, taurine

muscular dystrophy, Duchenne-type
helped by vitamin B_T

muscular excitability
helped by magnesium

nail deterioration
deficiency of vitamin F
helped by vitamin A, vitamin B_C, calcium, iron

nails, brittle
deficiency of vitamin F
toxicity of vitamin A, iodine

nails, deformed
deficiency of zinc

nails, edges turned up
deficiency of iron

nails, flaking
deficiency of vitamin F

nails, ingrown
deficiency of silicon

nails, loss of
toxicity of selenium

nails, ribbed
deficiency of silicon

nails, soft
deficiency of vitamin F

nails, thin
deficiency of iron
toxicity of iodine

nails, white spots on
deficiency of zinc

nausea
  deficiency of vitamin B_{3a}, vitamin B_{3b}, vitamin B_5, vitamin B_6, vitamin B_W, manganese, potassium, sodium, lysine
  helped by vitamin B_5, chlorine, magnesium
  toxicity of vitamin A, vitamin B_{17}, vitamin B_P, vitamin B_T, vitamin B_X, vitamin D, vitamin E, vitamin H_3, aluminum, antimony, arsenic, barium, boron, cadmium, fluorine, lead, mercury, nickel, silver, zinc
  side-effect of vitamin B_{3a}, vitamin B_{3b}

nausea of pregnancy
  helped by vitamin B_6

neck, increased color on
  toxicity of arsenic

neck pains
  toxicity of lead

nerve pain
  deficiency of vitamin B_6, magnesium, phosphorus
  helped by vanadium

nerve regeneration, decreased
  deficiency of vitamin T

nerves, deterioration of myelin sheaths
  deficiency of histidine

nervous system, decreased development of
  deficiency of zinc

nervous system, lesions in
  toxicity of fluorine

nervous system deterioration
  deficiency of vitamin B_6, vitamin B_{12}, vitamin B_{15}, vitamin E, copper, phosphorus, potassium, phenylalanine
  helped by vitamin B_T, vitamin T
  toxicity of vitamin K, lead

nervous weakness
  deficiency of vitamin D

nervousness
  deficiency of vitamin B_{3a}, vitamin B_{3b}, vitamin B_6, vitamin B_{12}, vitamin B_X, vitamin Q_{10}, calcium, magnesium, tryptophan, valine
  helped by magnesium
  toxicity of vitamin B_1, vitamin B_P, mercury

neuritis
  helped by vitamin B_1, vitamin B_2, vitamin B_{3a}, vitamin B_{3b}, vitamin B_5, vitamin B_6, vitamin B_{12}, magnesium

niacin production, deterioration of
  deficiency of tryptophan

nipples, increased color in
  toxicity of arsenic

nose, inflammation of
  helped by vitamin A
  toxicity of arsenic

nose, runny
  deficiency of chlorine, sodium

nose, sinusitis
  helped by vitamin A, vitamin E

nosebleeds
  deficiency of iron, phosphorus, vitamin C, vitamin C_3, vitamin K, vitamin P

nucleic acid structure formation, deterioration of
  helped by methionine

numbness, general
  deficiency of vitamin B_5, vitamin B_6, vitamin B_{12}, vitamin B_C, calcium, phosphorus, potassium
  toxicity of vitamin B_2, vitamin B_{17}, aluminum, lead, mercury

nutrient absorption, deterioration of
  deficiency of zinc, threonine

organ failure
  toxicity of beryllium

A. *Indications for Use*

organs, ashen-grey color in
  toxicity of silver
organs, beryllium in
  toxicity of beryllium
organs, ulceration of
  deficiency of vitamin U
oxidation, free radical
  helped by provitamin A, vitamin A, vitamin B₁, vitamin B₃, vitamin B₅, vitamin B₆, vitamin B_H, vitamin B_p, vitamin B_x, vitamin C, vitamin E, vitamin P, vitamin Q₁₀, selenium, vanadium, zinc, cysteine, methionine, tryptophan
oxygen, decreased
  deficiency of vitamin B₁₅, vitamin H₃, germanium
  helped by vitamin B₁₅, vitamin B_T, vitamin C, vitamin P, vitamin Q₁₀
oxygen, need for increased
  deficiency of vitamin E
ozone, decreased resistance to
  deficiency of vitamin B_x, vitamin E, selenium
Paget's disease
  helped by silicon
pain control
  helped by leucine?, phenylalanine, tryptophan
paralysis
  deficiency of vitamin B₆, vitamin B_w, potassium, methionine
  toxicity of aluminum, arsenic, barium, lead, mercury, selenium
Parkinson's disease
  deficiency of vitamin H₃
  helped by vitamin B₂, vitamin B₃a, vitamin B₃b, vitamin B₆, vitamin C, vitamin E, vitamin H₃, calcium, magnesium
  warning about vitamin B₆
pellagra
  helped by vitamin B₁, vitamin B₂, vitamin B₁₂, vitamin B_c

  deficiency of vitamin B₃a, vitamin B₃b
  toxicity of leucine?
peroxidase elimination, decreased
  deficiency of rubidium
phlebitis
  helped by vitamin E
phosphorus, decreased levels of
  deficiency of vitamin D, boron
  toxicity of aluminum
phosphorus, increased in blood
  toxicity of phosphorus
pituitary gland deterioration
  deficiency of bromine
pneumonia
  helped by vitamin C, vitamin C₂, vitamin C₃, vitamin P
  deficiency of lysine
polio
  helped by potassium, sodium
potassium, decreased
  deficiency of magnesium, taurine
  toxicity of rubidium, sodium
potassium, decreased osmotic control of in heart
  deficiency of taurine
pregnancy
  helped by vitamin A, vitamin C, vitamin D, vitamin E, vitamin T, iron, phosphorus
  warning about lithium, tryptophan
pregnancy, miscarriage of
  helped by vitamin E
pregnancy, nausea of
  helped by vitamin B₆
pregnancy, toxemia of
  deficiency of methionine
prolactin regulation, deterioration of
  deficiency of nickel
prostaglandins, increased
  deficiency of lithium

prostate, inflammation of
helped by vitamin A, vitamin B6, vitamin C, vitamin E, vitamin F, zinc

protein, decreased
deficiency of vitamin B6, arginine. cysteine, glutamine, glycine, histidine, isoleucine, lysine, methionine, phenylalanine, taurine, threonine, tryptophan, tyrosine, valine
toxicity of cadmium, lead

protein, deterioration of
deficiency of vitamin B2, vitamin B8

protein synthesis depressed
deficiency of vitamin B1, vitamin B6, vitamin B15, vitamin T, zinc
helped by methionine

provitamin A conversion to vitamin A, decreased
toxicity of vitamin E

pulse, irregular
toxicity of potassium

pulse, rapid
toxicity of iodine

pulse, rapid/faint
deficiency of phosphorus, potassium

pulse, slow
toxicity of vitamin H3, potassium

pulse, weak
deficiency of vitamin B12
toxicity of chlorine, fluorine

P-waves depressed
toxicity of potassium

pyorrhea
helped by vitamin A, vitamin B3a, vitamin B3b, vitamin B6, vitamin C, vitamin D, vitamin P, calcium

pyridoxal phosphate supplies, inadequate levels
deficiency of vitamin B6

radiation
helped by provitamin A, vitamin A, vitamin B1, vitamin B3, vitamin B5, vitamin B6, vitamin B$_H$, vitamin B$_P$, vitamin B$_X$, vitamin C, vitamin E, vitamin P, vitamin Q10, selenium, vanadium, zinc, cysteine, histidine, methionine, tryptophan

radiation, decreased resistance
deficiency of vitamin F
helped by provitamin A, vitamin A, vitamin B1, vitamin B3, vitamin B5, vitamin B6, vitamin B$_H$, vitamin B$_P$, vitamin B$_X$, vitamin C, vitamin E, vitamin P, vitamin Q10, selenium, vanadium, zinc, cysteine, methionine, tryptophan

rashes, genital
deficiency of vitamin B2

reflexes, deterioration of
deficiency of vitamin B1, vitamin B5, vitamin B12
toxicity of magnesium

reproduction, idiopathic hypospermia
helped by arginine

reproduction, male sterility
deficiency of vitamin B$_T$, selenium, arginine
helped by vitamin B$_T$, vitamin C
toxicity of bromine

reproduction, sterility
deficiency of vitamin A, vitamin B12, vitamin E, vitamin F, magnesium, zinc, arginine, tryptophan

reproduction, testicle deterioration
deficiency of vitamin B5, zinc, tryptophan

reproduction, testicles hydrocele
deficiency of calcium, phosphorus

reproduction deterioration
deficiency of manganese, vanadium, lysine

A. Indications for Use

retardation
deficiency of vitamin B_6, vitamin B_{13}, glutamine

Reynaud's disease
helped by germanium

rheumatic fever
deficiency of methionine
helped by vitamin A, vitamin B_{15}, vitamin C, vitamin D, vitamin P
warning about vitamin E

rheumatism
deficiency of vitamin B_{17}
helped by vitamin B_6, vitamin B_{15}, vitamin B_{17}, vitamin C, vitamin E, vitamin H_3, vitamin P, potassium

RNA, decreased synthesis
deficiency of vitamin B_8, nickel

salivary glands, deterioration of
deficiency of vitamin A

salivation, decreased
deficiency of zinc

salivation, increased
toxicity of arsenic, iodine

salt cravings
deficiency of chlorine, sodium

sarcoma
helped by germanium

scars remaining
deficiency of vitamin E

schizophrenia
deficiency of vitamin H_3, tryptophan
helped by vitamin H_3, vitamin Q_{10}, histidine, methionine
warning about arginine, histidine

sciatica
helped by vitamin B_1, vitamin D, vitamin E, silicon

scurvy
deficiency of vitamin C
helped by vitamin A, vitamin B_c, vitamin C, vitamin P, vitamin T, iron

selenium, decreased
deficiency of selenium, methionine

selenium therapy
helped by methionine

sense of smell, hyperacute
deficiency of vitamin B_{3a}, vitamin B_{3b}

sense of taste, decreased
deficiency of copper

serotonin neurotransmitter levels, unstable
deficiency of lithium

sexual activity, decreased
deficiency of histidine, phenylalanine
helped by zinc
toxicity of vitamin E

sexual development, slowed
deficiency of zinc, arginine

shock
toxicity of boron

shoulder pains
toxicity of lead

sickle cell anemia
deficiency of vitamin B_{17}
helped by vitamin B_{17}
warning about iron, valine

skin, aging of
helped by vitamin A acid

skin, allergic reactions
toxicity of vitamin B_p?

skin, ashen-gray coloring of
toxicity of silver

skin, blue
deficiency of vitamin B_1

skin, bronze discolored
toxicity of iron

skin, burning sensations
deficiency of vitamin B_w
toxicity of vitamin B_2

skin, changes
deficiency of vitamin B_W

skin, cracked
deficiency of calcium
toxicity of fluorine

skin, crawling feeling
toxicity of valine

skin, dermatosclerosis
deficiency of vitamin H_3

skin, dermatosis nutritional
deficiency of protein

skin, dry
deficiency of vitamin A, vitamin C, vitamin F, potassium, tryptophan
helped by vitamin A
toxicity of vitamin A, iodine

skin, eczema
deficiency of vitamin A, vitamin B_{13}, vitamin B_H, vitamin B_W, vitamin F, chlorine, copper, iodine, potassium
helped by vitamin A, vitamin B_6, vitamin B_P, vitamin B_T, vitamin C, vitamin D, vitamin F, sulfur

skin, edema
deficiency of vitamin H_3, phosphorus, sodium, protein
helped by vitamin B_6, copper
warning about sodium

skin, greasy/cheezy oozing
deficiency of vitamin B_6

skin, hives
deficiency of vitamin H_3
toxicity of vitamin H_3

skin, hypersensitivity
deficiency of vitamin B_{3a}, valine

skin, ichthyosis
deficiency of vitamin H_3

skin, impetigo
helped by vitamin A, vitamin C, vitamin E

skin, iron increased in
warning about iron

skin, leukoderma
deficiency of vitamin H_3

skin, melanin decreased
deficiency of phenylalanine, tyrosine

skin, oily
deficiency of vitamin B_2

skin, pallor
deficiency of vitamin B_{12}, vitamin B_C, vitamin B_W, silicon, tyrosine

skin, peeling
toxicity of vitamin A

skin, premature aging of
helped by vitamin A acid, silicon

skin, prickling sensations
deficiency of vitamin B_W
toxicity of vitamin B_2

skin, psoriasis
deficiency of vitamin B_{13}, vitamin H_3
helped by vitamin A, vitamin B_5, vitamin B_6, vitamin B_{12}, vitamin B_C, vitamin C, vitamin D, vitamin F, magnesium, sulfur
warning about vitamin F, protein

skin, rough
deficiency of vitamin A

skin, scaling
toxicity of vitamin A

skin, shingles
helped by vitamin A, vitamin B_1, vitamin B_6, vitamin B_{12}, vitamin C, vitamin D

skin, small hemorrhages
deficiency of vitamin C

skin, small red lumps on
toxicity of nickel

skin, stretch marks remaining
deficiency of vitamin E

skin, sun sensitivity
deficiency of vitamin B_X

A. *Indications for Use*

helped by provitamin A, vitamin A, vitamin B₁, vitamin B₃, vitamin B₅, vitamin B₆, vitamin B_H, vitamin B_P, vitamin B_X, vitamin C, vitamin E, vitamin P, selenium, zinc, cysteine, methionine, tryptophan

skin, sunburn
helped by vitamin B_X, vitamin E, calcium

skin, wrinkled
deficiency of vitamin E, vitamin H₃

skin complexion deterioration
deficiency of sulfur

skin deterioration
deficiency of vitamin B₅, vitamin B₁₁, vitamin B₁₃, vitamin B₁₅, vitamin F, magnesium, silicon
helped by vitamin H₃, vitamin U
toxicity of vitamin B₃ₐ, vitamin B₃ᵦ, vitamin D, aluminum
warning about vitamin B₃ᵦ

skin eruptions
deficiency of calcium, potassium, sulfur

skin flabbiness
deficiency of silicon

skin flushing
side-effect of vitamin B₁₅?
toxicity of vitamin K
warning about vitamin B₃ₐ, vitamin B₃ᵦ

skin inflammation
deficiency of vitamin B₂, vitamin B₃ₐ, vitamin B₆, vitamin B_W, vitamin H₃, manganese
helped by vitamin A, vitamin B₂, vitamin B₃ₐ, vitamin B₆, vitamin B_T, vitamin F, potassium, sulfur
toxicity of vitamin B₁, boron, selenium

skin inflammation, infant
helped by vitamin B_T

skin itching
deficiency of iron

side-effect of vitamin B₃ₐ, vitamin B₃ᵦ
toxicity of vitamin A, vitamin B₂

skin lesions
deficiency of zinc
helped by vitamin Q₁₀

skin on face scaling
deficiency of vitamin B₂

skin on palms thickening
toxicity of arsenic

skin on soles thickening
toxicity of arsenic

skin pigmentation, abnormal
deficiency of copper, nickel, protein

skin rashes
deficiency of vitamin A, vitamin B_W, vitamin H₃
toxicity of vitamin A, vitamin C, arsenic

skin reactions, allergic
toxicity of vitamin B_P?

skin sensitivity, increased
deficiency of vitamin B_W

skin sores
deficiency of copper

skin thickening
toxicity of vitamin A, arginine, ornithine

skin ulcers
helped by vitamin A, vitamin B₂, vitamin B₁₂, vitamin B_C, vitamin E, vitamin K, vitamin P, iron
toxicity of hexavalent chromium

skin vitiligo
helped by vitamin B_X

skin yellowing
side-effect of provitamin A
toxicity of vitamin K

sleep, lack of
deficiency of vitamin B₃ₐ, vitamin B₃ᵦ, vitamin B₅, vitamin B₆, vita-

min B_{15}, vitamin B_H, vitamin B_W, calcium, phosphorus, potassium, tryptophan, valine
helped by vitamin B_{3a}, vitamin B_{3b}, vitamin B_5, vitamin B_6, vitamin B_{12}, vitamin C, vitamin D, calcium, potassium, tryptophan
side-effect of phenylalanine
toxicity of bromine, copper, lead, mercury

sleep induced
side-effect of tryptophan

sleep pattern altered
toxicity of vitamin B_p

smell, hyperacute sense of
deficiency of vitamin B_{3a}, vitamin B_{3b}

sneezing
toxicity of arsenic

sodium deposits in tissues
deficiency of magnesium
toxicity of lithium

sperm, decreased mobility and formation
deficiency of arginine

spinal cord deterioration
deficiency of vitamin B_5

spine bone growth, projecting of
toxicity of fluorine

spleen, enlarged
toxicity of vitamin A, copper

spleen, iron increased in
deficiency of arsenic

squalene synthetase, increased levels of
deficiency of vanadium

steroid, decreased synthesis
deficiency of boron

stomach, calcium deposits in
toxicity of vitamin D

stomach, pain in
toxicity of antimony, mercury, silver

stomach ailments
deficiency of vitamin U
toxicity of lead

strain, soft tissue
helped by proline

stress
deficiency of vitamin B_5, vitamin B_6, vitamin B_{15}, vitamin C, magnesium
helped by vitamin A, vitamin B_1, vitamin B_2, vitamin B_{3a}, vitamin B_{3b}, vitamin B_5, vitamin B_6, vitamin B_C, vitamin C, vitamin D, phosphorus, potassium

stroke
helped by vitamin A, vitamin B_H, vitamin B_p, vitamin C, vitamin E, vitamin P, vitamin Q_{10}, potassium
warning about sodium

sunstroke
toxicity of vitamin D?

suppressor cell activity, increased
deficiency of lithium

sweating
deficiency of calcium, phosphorus
toxicity of vitamin A, vitamin K, aluminum, arsenic

swelling, general
deficiency of vitamin B_6, vitamin C

swollen glands
helped by vitamin A, vitamin C

swollen joints
deficiency of vitamin C, vitamin F

taste, acrid
toxicity of vitamin B_{17}

taste in mouth, metallic
toxicity of antimony, lead, mercury

T-cell blood count, decreased
deficiency of vitamin F

tear ducts, deterioration of
deficiency of vitamin A

A. *Indications for Use* 439 tissue, increased connective

tear glands, drying
deficiency of vitamin C

teeth, demineralization of
deficiency of phosphorus, vanadium

teeth, grinding
deficiency of calcium

teeth, loose
deficiency of vitamin C
toxicity of mercury

teeth, loss of
deficiency of chlorine
toxicity of selenium

teeth, mottling of
toxicity of fluorine

teeth, pitting of
toxicity of fluorine

teeth, softening
deficiency of vitamin D, calcium

teeth, weak in children
deficiency of vitamin C

teeth decay
deficiency of vitamin B_{17}, silicon, strontium, vanadium
helped by fluorine, vitamin B_{17}

teeth dentition, late
deficiency of calcium
toxicity of fluorine

teeth deterioration
deficiency of magnesium, silicon
helped by vitamin A, vitamin C, vitamin F, calcium, fluorine, iron, phosphorus, potassium, sodium

teeth development, deterioration of
deficiency of silicon, tin

telangiectasia
helped by vitamin Q

thirst
toxicity of vitamin D

throat, dry
toxicity of cadmium

throat, hoarseness
toxicity of arsenic

throat, muteness
deficiency of iodine

throat, pain in
toxicity of antimony, silver

throat, sore
toxicity of arsenic, mercury

throat, swallowing difficult
deficiency of vitamin B_{12}, iron
toxicity of lead

throat, voice changes
toxicity of lead

thymus gland deterioration
deficiency of zinc
helped by arginine

thyroid, enlarged
toxicity of cobalt

thyroid, overactive
helped by vitamin A, vitamin B_p, vitamin E, iodine
warning about vitamin E

thyroid, underactive
helped by vitamin B_T, iodine

thyroid cancer
toxicity of iodine 131

thyroid deterioration
deficiency of vitamin B_{17}, bromine, nickel
helped by vitamin B_{17}

thyroid hormone levels, decreased
toxicity of vitamin E

thyroid hyperplasia
helped by vitamin C_2, vitamin C_3

thyroid production altered
toxicity of vitamin B_1

tin accumulation in heart
toxicity of tin

tissue, increased connective
deficiency of vitamin B_6

tissue, overloads of copper
helped by histidine

tissue, overloads of iron
helped by histidine

tissue elasticity, decreased
deficiency of selenium
helped by proline

tissue strains, soft
helped by proline

tissues, calcium deposits in
deficiency of magnesium

tissues, carotene deposition in
side-effect of provitamin A

tissues, fluid accumulation in
deficiency of magnesium
toxicity of vitamin B_1

tissues, sodium deposits in
deficiency of magnesium
toxicity of lithium

toes, tingling
deficiency of vitamin B_1

tongue, cracked
deficiency of vitamin B_W

tongue, glossy
deficiency of vitamin B_6

tongue, green
toxicity of vanadium

tongue, inflamed
deficiency of vitamin B_C, vitamin B_W
toxicity of arsenic

tongue, numb
toxicity of potassium

tongue, sore
deficiency of vitamin B_2, vitamin B_{12}, vitamin B_C, vitamin B_W
toxicity of nickel

tongue, yellow coating on back of
deficiency of potassium

tonsillitis
helped by vitamin B_C

touch, diminished sense of
deficiency of vitamin B_1

toxicity, general
deficiency of strontium
helped by methionine

tuberculosis
helped by vitamin A, vitamin B_{3a}, vitamin B_{3b}, vitamin B_5, vitamin B_6, vitamin B_{12}, vitamin C, vitamin D, calcium, iron

T-waves, increased
toxicity of potassium

tyrosine formation, decreased
deficiency of phenylalanine

tyrosine toxicity
side-effect of phenylalanine

ulceration of internal organs
deficiency of vitamin U

ulcers
deficiency of vitamin B_5, vitamin Q_{10}, vitamin U, vitamin B_P, sulfur
helped by vitamin A, vitamin B_2, vitamin B_{12}, vitamin C, vitamin E, vitamin Q_{10}, calcium, iron, silicon
warning about vitamin B_{3a}, vitamin B_{3b}
toxicity of vitamin B_{3a}, vitamin B_{3b}

ulcers, skin
deficiency of vitamin B_{3a}, calcium
helped by vitamin A, vitamin B_2, vitamin B_{12}, vitamin B_C, vitamin E, vitamin K, vitamin P, iron
toxicity of hexavalent chromium

ulcers, trophic
deficiency of vitamin H_3

urinary deterioration
toxicity of mercury

urinary oxalate excretion, increased
toxicity of vitamin C

urination, increased
deficiency of vitamin B_6

A. *Indications for Use*

side-effect of vitamin C
toxicity of vitamin D, iron

urination, urgency of
toxicity of vitamin D

urination changes
toxicity of arsenic

urination inability
deficiency of vitamin B_2

urine, calcium decreased in
toxicity of phosphorus

urine, crystals in
deficiency of vitamin B_{13}

urine, decreased
deficiency of molybdenum

urine, increased vitamin C in
deficiency of vitamin B_6

urine, increased creatine in
deficiency of vitamin E

urine, increased xanthurenic acid in
deficiency of vitamin B_6

urine, protein in
toxicity of cadmium, lead

urine, slight albumin in
toxicity of lead

urine turns yellow
side-effect of vitamin B_2

vaginal itching
deficiency of vitamin B_2
helped by vitamin A, vitamin B_2, vitamin B_6, vitamin D, vitamin E

vascular system, peripheral degenerating
deficiency of vitamin E

veins, coronary thrombosis
helped by vitamin E, vitamin F

veins, deterioration
helped by vitamin B_{3a}, vitamin B_{3b}, vitamin B_5, vitamin C, vitamin E, vitamin P

vitamin B_{12}, decreased

veins, varicose
deficiency of vitamin B_1, vitamin C_3, vitamin F, vitamin H_3, vitamin P, calcium
helped by vitamin A, vitamin C, vitamin E, vitamin P
toxicity of fluorine

veins, vascular tone deterioration of
deficiency of vitamin T

viral related diseases
helped by lysine

vision, blurred
deficiency of vitamin H_3
toxicity of vitamin A, vitamin E

vision deterioration
deficiency of vitamin B_1, vitamin B_{3a}, vitamin B_{3b}, vitamin B_6, vitamin H_3, taurine
helped by vitamin A, vitamin B_1, vitamin C, vitamin D, vitamin E, germanium
toxicity of lead, mercury

vitamin A absorption, decreased
deficiency of vitamin C_3, vitamin P

vitamin A conversion from provitamin A, decreased
deficiency of provitamin A
toxicity of vitamin E
warning about provitamin A

vitamin B_3, decreased
deficiency of vitamin B_3
toxicity of leucine

vitamin B_3 therapy
helped by vitamin C nicotinamide complex

vitamin B_6, decreased
deficiency of vitamin B_6, vitamin B_p
side-effect of tryptophan
toxicity of vitamin B_p

vitamin B_{12}, decreased
deficiency of vitamin B_{12}, vitamin B_{13}, cobalt
side-effect of vitamin C

vitamin B_C, decreased
deficiency of vitamin B_{12}, vitamin B_{13}, vitamin B_C
side-effect of vitamin C

vitamin B_P, decreased
deficiency of vitamin B_P
helped by methionine

vitamin C, decreased
deficiency of vitamin A, vitamin B_6, vitamin C, vitamin C_3, vitamin P
toxicity of antivitamin C
warning about vitamin C

vitamin C therapy
helped by vitamin C_2, vitamin C_3

vitamin P therapy
helped by vitamin C_2

vitamin Q_{10}, decreased
deficiency of vitamin Q_7, vitamin Q_8, vitamin Q_9, vitamin Q_{10}

walking difficulties
deficiency of vitamin B_{12}
toxicity of mercury

warts
deficiency of chlorine, potassium
helped by vitamin A, vitamin E

water retention
deficiency of vitamin B_6

weight, decreased
deficiency of vitamin A, vitamin B_2, vitamin C, vitamin F, manganese, sodium, lysine, tryptophan
helped by vitamin F
toxicity of iron, lead

weight, increased
deficiency of vitamin B_{13}, vitamin F
helped by vitamin B_6, vitamin B_{12}, vitamin B_H, vitamin B_T, vitamin C, vitamin E, vitamin F, vitamin Q_{10}, calcium, magnesium, tryptophan
toxicity of vitamin F

weight of fetus, decreased
warning about tryptophan

weight reduction
helped by arginine

white blood cell abnormalities
toxicity of fluorine

white blood cells, decrease in number
deficiency of vitamin B_C

Wilson's disease
warning about copper

worms
helped by vitamin A, vitamin B_1, vitamin B_2, vitamin B_5, vitamin B_6, vitamin B_{12}, vitamin C, vitamin D, vitamin K, calcium, iron, potassium, sulfur

wound healing deterioration
deficiency of vitamin B_{15}, vitamin B_W, vitamin U
helped by vitamin F, zinc, arginine, proline

xanthine oxidase enzyme disrupted
toxicity of tungsten

xanthurenic acid increased in urine
deficiency of vitamin B_6

zinc, decreased
deficiency of zinc
toxicity of vitamin B_P, copper

zinc absorption inhibited
side-effect of vitamin B_C

Appendix B: VITAMIN B COMPLEX DESIGNATIONS

This Book	Original Vitamin Letter	Original Factor Letter	Numeric Subscript	Alphabetic Subscript	Renumbered Subscript	Erroneous References	Chemical Name
B_1	F	F	B_1		B_1		thiamine
B_2	G	G	B_2		B_2		riboflavin
B_3	PP	PP	B_3		B_3	B_5, B_7	niacin
B_5			B_5		B_5	B_3, B_X	pantothenate
B_6	Y	Y	B_6		B_6	H_1	pyridoxine
B_7	I	I	B_7				(rice polish factor)
B_{12}		B,X	B_{12}		B_{12}		cobalamin
B_C	M	U,CF,LC		B_C	B_9		folacin
B_H				B_H	B_8		inositol
B_P	J	J		B_P	B_4		choline
B_T				B_T			carnitine
B_W	H	AN,X	H_1	B_W	B_7	B_4, B_T, I	biotin
B_X	V	V	H_2	B_X	B_{10}	H_1	para-aminobenzoic acid

Appendix C: NUTRIENT RATIOS

VITAMIN B_1

at least 100mg per every 1000mg vitamin B_{3a}

VITAMIN B_{3a}

at least 100mg vitamin B_1 per every 1000mg
at least 100mg vitamin B_6 per every 1000mg
at least 1000mg vitamin C per every 1000mg

VITAMIN B_5

at least 1mg vitamin B_C per every 300mg
at least 100mg vitamin B_W per every 300mg

VITAMIN B_6

at least 100mg per every 1000mg vitamin B_{3a}

VITAMIN B_C

at least 1mg per every 300mg vitamin B_5

VITAMIN B_W

at least 100mg per every 300mg vitamin B_5

VITAMIN C

at least 1000mg per every 1000mg vitamin B_{3a}
at least 3000mg per every 1000mg cysteine

CALCIUM

at least 50mg magnesium per every 100mg
at least 1300mg phosphorus per every 1000mg
not more than 2000mg phosphorus per every 1000mg

COPPER

at least 14mg zinc per every 1mg

MAGNESIUM

at least 50mg per 100mg calcium

PHOSPHORUS

at least 1300mg per every 1000mg calcium

not more than 2000mg per every 1000mg calcium

POTASSIUM

1700mg per every 1000mg sodium

SODIUM

1700mg potassium per every 1000mg

ZINC

at least 14mg per every 1mg copper

CYSTEINE

at least 3000mg vitamin C per 1000mg

Appendix D: NUTRIENT UNIT CONVERSIONS

PROVITAMIN A

1 retinol equivalent
 = 6 µg beta-carotene
 = 5 iu
 = 5 usp

VITAMIN A

1 retinol equivalent
 = 1 µg retinol
 = 5 iu
 = 5 usp

VITAMIN B$_3$

1 niacin equivalent
 = 1 mg niacin
 = 60 mg tryptophan

VITAMIN C

1 mg vitamin C palmitate
 = 0.42 mg vitamin C activity

VITAMIN D

10 µg cholecalciferol
 = 400 iu
 = 400 usp

VITAMIN E

1 alpha-tocopherol equivalent
 = 1 mg alpha-tocopherol
 = 3 iu
 = 3 usp

VITAMIN F

1 g fats
 = 9 calories

Bibliography

Adams, Ruth. *The Complete Home Guide to All the Vitamins.* New York: Larchmont Books, 1972.
Adams, Ruth and Frank Murray. *Body, Mind and the B Vitamins.* New York: Larchomont Books, 1975.
Adams, Ruth and Frank Murray. *Megavitamin Therapy.* New York: Larchmont Books, 1973.
Adams, Ruth and Frank Murray. *Minerals: Kill or Cure?* New York: Larchmont Books, 1976.
Adams, Ruth and Frank Murray. *Vitamin E: Wonder Worker of the 70's.* New York: Larchmont Books, 1972.
Airola, Paavo. *How to Get Well.* Phoenix, Arizona: Health Plus, 1974.
Alsleben, Rudolph H., M.D. and Wilfrid E. Shute, M.D. *How to Survive the New Health Catastrophes.* Anaheim, California: Survival Publications, Inc., 1973.
Altschul, A.M. *Proteins: Their Chemistry and Politics.* New York: Basic Books, 1965.
Anderson, Linnea and Marjorie Dibble and Helen S. Mitchell and Hendrika Rynbergen. *Nutrition in Nursing.* Philadelphia: J. B. Lippincott Co., 1972.
Arehart-Treichel, Joan. *Trace Elements.* New York: Holiday House, 1974.
Atkins, Robert C., M.D. and Shirley Linde. *Dr. Atkins' Super-Energy Diet.* New York: Bantam Books, 1977.
Atkins, Robert C., M.D. *Dr. Atkins' Nutritional Breakthrough: How to Treat Your Medical Condition Without Drugs.* New York: William Morrow and Co., 1981.
Ayers, J. C. and A. A. Kraft and H. E. Snyder and H. W. Walker eds. *Chemical and Biological Hazards in Food.* Amea, Iowa: Iowa State University Press, 1962.
Bailey, Herbert. *GH3: Will It Keep You Looking Younger Longer?* New York: Bantam Books, Inc., 1977.
Bailey, Herbert. *Vitamin E: Your Key to a Healthy Heart.* New York: Arc Books, 1968.
Baily, Herbert. *The Vitamin Pioneers.* Emmaus, PA: Rodale Press, 1968.

Bailey, Herbert. *The Vitamin Pioneers*. New York: Pyramid Books, 1969.
Barker, B. and D. Bender. *Vitamins in Medicine*. Vol. 1, 4th ed. London: William Heinemann Medical Books, Ltd., 1980.
Beers, R. and E. Basset eds. *Nutritional Factors: Modulating Effects on Metabolic Processes*. New York: Raven Press, 1981.
Bieler, Henry G. *Food Is Your Best Medicine*. New York: Random House, 1973.
Bircher-Benner, M. *Prevention of Incurable Disease*. New Canaan, Connecticut: Keats Publishing, Inc., 1978.
Blaine, Tom R. *Mental Health through Nutrition*. New York: Citidel Press, 1974.
Blaine, Tom R. *Nutrition and Your Heart*. New Canaan, Connecticut: Keats Publishing, Inc., 1979.
Bliznakov, Emile G., M.D. and Gerald L. Hunt. *The Miracle Nutrient Coenzyme Q_{10}*. New York: Bantam Books, Inc., 1987.
Borsaak, Henry. *Vitamins: What They Are and How They Can Benefit You*. New York: Pyramid Books, 1971.
Bosco, Dominick. *The Peoples Guide to Vitamins and Minerals, from A to Zinc*. Chicago, Illinois: Contemporary Books, Inc., 1980.
Bressler, Joel V., B.S., Ph.D. *Nutrition: The Case of Misplaced Emphasis*. Jacksonville, Florida: Bressler, Inc., 1986.
Brewster, Letitia and Michael F. Jacobsen. *The Changing American Diet*. Center for Science in the Public Interest, 1978.
Bricklin, Mark. *The Practical Encyclopedia of Natural Healing*. Emmaus, Pennsylvania: Rodale Press, Inc., 1976.
Briggs, M. ed. *Vitamins in Human Biology and Medicine*. Boca Raton, Florida: CRC Press, 1981.
Brody, Jane. *Jane Brody's Nutrition Book*. New York: W. W. Norton and Co., 1981.
Burland, W. and P. Samuel and J. Yudkin eds. *Obesity*. Edinburgh: Churchill Livingstone, 1974.
Burton, Benjamin T. *Human Nutrition*. New York: McGraw-Hill, 1976
Cameron, E. and Linus Pauling. *Cancer and Vitamin C*. Menlo Park California: The Linus Pauling Institute of Science and Medicine, 1979.
Carey, Ruth L. and Irma B. Vyhmeister and Jennie S. Hudson. *Commonsense Nutrition*. Omaha: Pacific Press, 1971.
Carson, Rachel. *Silent Spring*. Boston: Houghton Mifflin, 1962.
Chaitow, Leon. *Amino Acids in Therapy*. New York: Thorsons Publishers, Inc., 1985.
Cheraskin, E., M.D. and Kalita, B. K. *A Physician's Handbook on Orthomolecular Medicine*. New York: Pergamon Press, 1977.
Cheraskin, E., M.D., D.M.D. and W. M. Ringsdorf Jr., D.M.D., and J. W. Clark, D.D.S. *Diet and Disease*. New Canaan, Connecticut: Keats Publishing, Inc., 1968.
Cheraskin, E., M.D., D.M.D. and W. M. Ringsdorf, Jr. *New Hope for Incurable Disease*. New York: Arco Publishing, 1973.
Cheraskin, E., M.D., D.M.D. and W. M. Ringsdorf, Jr. *Predictive Medicine*. New Canaan, Connecticut: Keats Publishing, Inc., 1973.
Cheraskin, E., M.D., D.M.D. and W. M. Ringsdorf, Jr. and Arline Brecher. *Psychodietetics*. New York: Stein and Day, 1975.
Cheraskin, E., M.D., D.M.D. and W. M. Ringsdorf, Jr. and Arline Brecher. *Psychodietetics*. New York: Bantam Books, 1976.
Clark, Guy W. *A Vitamin Digest*. Springfield, Illinois: Charles C. Thomas, 1953.
Clark, Linda. *Get Well Naturally*. Old Greenwich, Connecticut: Devin-Adair, 1965.
Clark, Linda. *Get Well Naturally*. New York: Arco Publishing, 1972.

Bibliography

Clark, Linda. *Know Your Nutrition*. New Canaan, Connecticut: Keats Publishing, Inc., 1973.
Clark, Linda. *Light on Your Health Problems*. New Canaan, Connecticut: Keats Publishing, Inc., 1972.
Clark, Linda. *Stay Young Longer*. New York: Pyramid Communications, 1968.
Clarke, J. H. *The Prescriber*. 8th ed. Rustington, England: Health Science Press, 1968.
Clymer, R. Swinburne. *Nature's Healing Agents*. Philadelphia: Dorrance Co., 1963.
Cott, Allan, M.D. *The Orthomolecular Approach to Learning Disabilities*. New York: Academic Press, 1977.
Crain, Lloyd. *Magic Vitamins and Organic Foods*. Crandrich Studios, 1971.
Davidson, Stanley and R. Passmore and J. F. Brack. *Human Nutrition and Dietetics*. 5th ed. Baltimore: Williams and Wilkins Co., 1972.
Davis, Adelle. *Let's Cook It Right*. New York: New American Library, 1970.
Davis, Adelle. *Let's Eat Right to Keep Fit*. New York: Harcourt, Brace and Company, 1954.
Davis, Adelle. *Let's Eat Right to Keep Fit*. New York: New American Library, 1970.
Davis, Adelle. *Let's Eat Right to Keep Fit*. New York: New American Library, 1972.
Davis, Adelle. *Let's Get Well*. New York: Harcourt, Brace and World, 1968.
Davis, Adelle. *Let's Have Healthy Children*. 2nd ed. New York: Harcourt, Brace and World, 1959.
Di Cyan, Erwin. *Vitamin E and Aging*. New York: Pyramid Books, 1972.
Dorland's Illustrated Medical Dictionary. Twenty-fourth edition. Philadelphia: W.B. Saunders and Co., 1974.
Dorland's Illustrated Medical Dictionary. Twenty-fifth edition. Philadelphia: W.B. Saunders and Co., 1974.
Dubos, Rene and Maya Pines. *Health and Disease*. New York: Time-Life Books, 1965.
Ebon, Martin. *Which Vitamin Do You Need?* New York: Bantam Books, 1974.
Ebon, Martin. *The Truth About Vitamin E*. New York: Bantam Books, 1972.
Ellis, John M., M.D. and James Presley. *Vitamin B_6: The Doctor's Report*. New York: Harper and Row, 1973.
Elwood, Catharyn. *Feel Like a Million!* New York: Devin-Adair, 1965.
Epstein, Samuel S., M.D. *The Politics of Cancer*. San Francisco: Sierra Club, 1978.
Ershoff, Benjamin, M.D. *Proceedings of the Society of Experimental Biology and Medicine*. July 1951.
Faelton, Sharon. *The Complete Book of Minerals for Health*. Emmaus, PA: Rodale Press, 1981.
Fleck, Henrietta. *Introduction to Nutrition*. 2nd ed. New York: Macmillan Co., 1971.
Forman, Brenda. *B15: The "Miracle Vitamin."* New York: Grosset and Dunlap, 1979.
Fox, E. and D. Mathews. *The Physiological Basis of Physical Education and Athletics*. 3rd ed. Saunders College Publishing, 1981.
Frank, Benjamin S., M.D. *Dr. Frank's No Aging Diet*. New York: Dell, 1976.
Fredericks, Carlton, Ph.D. *Breast Cancer: A Nutritional Approach*. New York: Grosset and Dunlap, 1977.
Frederick, Carlton, Ph.D. *Carlton Fredericks' High-Fiber Way to Health*. New York: Pocket Books, 1976.
Fredericks, Carlton, Ph.D. *Carlton Fredericks' Program for Living Longer*. New York: Aimon and Schuster, 1983.
Fredericks, Carlton, Ph.D. *Eating Right for You*. New York: Grosset and Dunlap, 1975.

Fredericks, Carlton, Ph.D. *Eat Well, Get Well, Stay Well.* New York: Grosset and Dunlap, 1980.
Fredericks, Carlton, Ph.D. *Nutrition Guide for the Prevention & Cure of Common Ailments & Diseases.* New York: Simon and Schuster, 1982.
Fredericks, Carlton, Ph.D. *Nutrition Handbook: Your Key to Good Health.* Canoga Park, California: Manor Books, 1976.
Fredericks, Carlton, Ph.D. *Psycho-Nutrition.* New York: Grosset and Dunlap, 1976.
Fredericks, Carlton, Ph.D. and Herbert Bailey. *Food Facts and Fallacies.* New York: Arc Books, 1965.
Fredericks, Carlton, Ph.D. and Herman Goodman. *Low Blood Sugar and You.* New York: Constellation International, 1976.
Frenkel, R. and J. McGarry eds. *Carnitine Biosynthesis, Metabolism, and Functions.* New York: Academic Press, 1980.
Garrow, J. *Energy Balance and Obesity in Man.* 2nd ed. Amsterdam: North Holland Biomedical Press, 1978.
Georgakas, D. *The Methuselah Factors: Living Long and Living Well.* New York: Simon and Schuster, 1980.
Glasser, Ronald J. *The Body Is the Hero.* New York: Random House, 1976.
Gomez, Joan. *A Dictionary of Symptoms.* New York: Bantam Books, 1967.
Goodhart, Robert S. and Maurice E. Shils. *Modern Nutrition in Health and Disease.* 5th ed. Philadelphia: Lea and Febiger, 1973.
Goodhart, Robert S. and Maurice E. Shils. *Modern Nutrition in Health and Disease.* 5th ed. Philadelphia: Lea and Febiger, 1980.
Goodman, A. G. and L. S. Gilman. *Goodman and Gilman's The Pharmacological Basis of Therapeutics.* 6th ed. New York: Macmillan, 1980.
Goodwin, R. *Chemical Additives in Food.* London, 1967.
Graedon, Joe. *The People's Pharmacy.* New York: St. Martin's Press, 1976.
Graedon, Joe. *The People's Pharmacy.* New York: Avon Books, 1977.
Graedon, Joe and Teresa. *The New People's Pharmacy.* New York: Bantam Books, 1985.
Graham, F. *Since Silent Spring.* Boston: Houghton Mifflin, 1970.
Grant, Doris. *Recipe for Survival.* New Canaan, Connecticut: Keats Publishing, Inc., 1974.
Griffin, G. Edward. *World Without Cancer: The Story of Vitamin B_{17}.* Westlake Village, CA: American Media, 1974.
Griffin, LaDean. *Health in the Space Age.* Orem, Utah: BiWorld Publishers, 1982.
Guthrie, Helen Andrews. *Introductory Nutrition.* St. Louis: C. V. Mosby, 1975.
Harper, Harold W., M.D. and Michael L. Culbert. *Now You Can Beat the Killer Diseases.* New Rochelle, New York: Arlington House, 1977.
GLA and DGLA (Vitamin F). St. Albans, Vermont: Health Guides, 1981.
Hartmann, Ernest L., M.D. *The Functions of Sleep.* New Haven and London: Yale University Press, 1973.
Haas, Robert. *Eat to Win.* New York: Rawson Associates, 1983.
Hausman, Patricia. *The Right Dose.* Emmaus, Pennsylvania: Rodale Press, 1987.
Hayden, Naura. *Everything You've Always Wanted to Know About Energy, But Were Too Weak to Ask.* New York: Simon and Schuster, 1976.
Hayden, Naura. *Everything You've Always Wanted to Know About Energy, But Were Too Weak to Ask.* New York: Pocket Books, 1977.
Heinz, H. J. and Company. *Nutritional Data.* 3rd ed. Pittsburgh, PA: H. J. Heinz Company, 1956.
Henderson's Dictionary of Biological Terms. Ninth edition. New York: Van Nostrand Reinhold Co., 1979.

Hendler, Sheldon Saul, M.D., Ph.D. *The Complete Guide to Anti-Aging Nutrients.* New York: Simon and Schuster, 1985.
Heritage, Ford. *Composition and Facts about Food.* Mokelunine Hill, California: Health Research Center, 1968.
Hietanen, E. ed. *Regulation of Serum Lipids by Physical Exercise.* Boca Raton, Florida: CRC Press, 1982.
Hightower, Jim. *Eat Your Heart Out.* New York: Random House, 1975.
Hightower, Jim. *Eat Your Heart Out.* New York: Vintage Books, 1976.
Hill, Howard E. *Introduction to Lecithin.* Los Angeles: Nash Publishing, 1972.
Hoffer, Abram, M.D., Ph.D. and Humphrey Osmond. *How to Live with Schizophrenia.* New York: University Books, 1966.
Hoffer, Abram, M.D., Ph.D. and Morton Walker, D.P.M. *Nutrients to Age Without Senility.* New Canaan, Connecticut: Keats Publishing, Inc., 1980.
Hoffer, Abram, M.D., Ph.D. and Morton Walker, D.P.M. *Orthomolecular Nutrition.* New Canaan, Connecticut: Keats Publishing, Inc., 1978.
Holmes, Ann H. *Nutrition and Vitamins.* Vessals: Facts on File, Inc., 1983.
Holmes, Marjorie. *God and Vitamins.* New York: Doubleday and Company, Inc., 1980.
Hoover, John E. ed. *Remington's Pharmaceutical Sciences.* 14th ed. Easton, PA: Mack Publishing Co., 1970.
Howe, Phyllis S. *Basic Nutrition in Health and Disease.* 5th ed. Philadelphia: W.B. Saunders Co., 1971.
Hunter, Beatrice Trum. *The Great Nutrition Robbery.* New York: Charles Scribner's Sons, 1978.
Hunter, Beatrice Trum. *The Natural Foods Primer.* New York: Simon and Schuster, 1973.
Hunter, Beatrice Trum. *The Natural Foods Primer.* Revised Ed. New Canaan, Connecticut: Keats Publishing, Inc., 1980.
Illich, Ivan. *Medical Nemesis.* New York: Random House, 1976.
Illich, Ivan. *Medical Nemesis.* New York: Bantam Books, 1977.
Jacobson, Michael F. *Eater's Digest.* Garden City: New York: Doubleday Anchor, 1972.
Jacobson, Michael F. *Nutrition Scoreboard.* New York: Avon Books, 1975.
Jayson, Malcolm, I. V., M.D. and Allan St. J. Dixon, M.D. *Understanding Arthritis and Rheumatism.* New York: Dell Books, 1976.
Jennings, Isobel. *Vitamins in Endocrine Metabolism.* New York: Charles C. Thomas, 1970.
Jensen, Bernard, Ph.D. *The Chemistry of Man.* Escondido, California: Bernard Jensen Publisher, 1983.
Jolliffe, Norman ed. *Clinical Nutrition.* 2nd ed. New York: Harper and Brothers, 1962.
Kahn, Harold A., M.D., et al. *The Framingham Eye Study.* The Johns Hopkins University School of Hygiene and Public Health, 1977.
Katz, Deborah and Mary T. Goodwin. *Food: Where Nutrition, Politics and Culture Meet.* Center for Science in the Public Interest, 1976.
Kent, Saul. *The Life Extension Revolution.* New York: William Morrow and Company, 1979.
Kent, Saul. *Your Personal Life-Extension Program.* New York: William Morrow and Company, Inc., 1985.
Klenner, R. F., M.D. and Fred H. Bartz. *The Key to Good Health: Vitamin C.* Graphic Arts Research Foundation, 1959.
Kime, Zane R., M.D., M.S. *Sunlight Could Save Your Life.* Penryn, California: World Health Publications, 1980.

Kordel, Lelord. *Health Through Nutrition.* Cleveland, OH: World Books, 1950.
Kordel, Lelord. *Health Through Nutrition.* Canoga Park, California: Manor Books, 1971.
Kotschevar, Lendal H. and Margaret McWilliams. *Understanding Food.* New York: Hohn Wiley and Sons, 1969.
Kraus, Barbara. *Carbohydrate Guide to Brand Names and Basic Foods.* New York: The New American Library, Inc., 1979.
Krause, Marie V. and Martha A. Hunscher. *Food, Nutrition and Diet Therapy.* 5th ed. Philadelphia: W. B. Saunders Co., 1972.
Krebs, Ernest, M.D., Jr. *The Laetriles-Nitrilosides in the Prevention and Control of Cancer.* The McNaughton Foundation, 1964.
Kugler, Hans J. *Slowing Down the Aging Process.* New York: Pyramid, 1973.
Kuhne, Paul. *Home Medical Encyclopedia.* Greenwich, Connecticut: Fawcett, 1960.
Kulvinskas, Viktoras. *Survival into the 21st Century.* Wethersfield, Connecticut: Omangod Press, 1975.
Kurtzman, Joel and P. Gordon. *No More Dying.* New York: Dell, 1976.
Lappé, Frances M. *Diet for a Small Planet.* New York: Ballantin Books, 1971.
Lappé, Frances M. *Diet for a Small Planet.* New York: Ballantine Books, 1972.
Lappé, Frances M. *Diet for a Small Planet.* New York: Ballantine Books, 1975.
Larson, Gena. *Better Food for Better Babies.* New Canaan, Connecticut: Keats Publishing, Inc., 1972.
Legler, Henry. *How to Make the Rest of Your Life the Best of Your Life.* New York: Simon and Schuster, 1967.
Legler, Henry. *How to Make the Rest of Your Life the Best of Your Life.* New York: Pocket Books, 1970.
Leibovits, Brian, M.S. *Carnitine: The Vitamin B_T Phenomenon.* New York: Dell Publishing Co., Inc., 1984.
Leonard, Jon N. and J. L. Hofer and N. Pritikin. *Live Longer Now.* New York: Grosset and Dunlap, 1974.
Levy, R. and B. Rifkind and B. Dennis and N. Ernst eds. *Nutrition, Lipids, and Coronary Heart Disease.* New York: Raven Press, 1979.
Lindsay, Rae. *The Pursuit of Youth.* New York: Pinnacle Books, 1976.
Livingston, Virginia W., M.D. *Cancer: A New Breakthrough.* San Diego: Production House, 1972.
Locke, David M. *Enzymes: The Agents of Life.* New York: Crown Press, 1971.
Longwood, William. *The Poisons in Your Food.* New York: Pyramid House, 1969.
Lucas, Richard. *Nature's Medicines.* New York: Award Books, 1969.
McCollum, E. *A History of Nutrition — The Sequence of Ideas in Nutrition Investigations.* Boston: Houghton Mifflin Co., 1957.
McDermott, Irene E. and Mabel B. Trilling and Florence W. Nicolas. *Food for Better Living.* 3rd ed. Chicago: J. B. Lippincott Co., 1960.
McGrady, Patrick. *The Youth Doctors.* New York: Coward-McCann, 1968.
Macia, Rafael. *The Natural Foods and Nutrition Handbook.* New York: Harper and Row, 1972.
Mann, John A. *Secrets of Life Extension.* New York: Bantam Books, 1980
Mannersberg, D. and J. Roth. *Aerobic Nutrition.* New York: Hawthorn/Dutton, 1981
Manner, Harold W. *The Death of Cancer.* Chicago: Advanced Century Publishing Co., 1978.
Margolius, S. *The Great American Food Hoax.* New York: Walker and Co., 1971.
Martin, Ethel A. *Nutrition in Action.* 2nd ed. New York: Holt, Rinehart and Winston, 1967.

Martin, Marvin. *Great Vitamin Mystery*. Rosemont, Illinois: National Dairy Council, 1978.
Mayer, Jean. *Eater's Digest*. New York: Double Day and Company, Inc., 1976.
Mayo Clinic Committee on Dietetics. *Mayo Clinic Diet Manual*. 4th ed. Philadelphia: Saunders Company, 1971.
Melloni, Biagio John, Ph.D. and Gilbert M. Eisner, M.D. F.A.C.P. *Melloni's Illustrated Medical Dictionary*. Baltimore: Williams and Wilkins Co., 1979.
The Merck Index. Sixth edition. Rahway, New Jersey: Merck and Co., Inc., 1952.
The Merck Index. Eighth edition. Rahway, New Jersey: Merck and Co., Inc., 1968.
The Merck Index. Ninth edition. Rahway, New Jersey: Merck and Co., Inc., 1976.
The Merck Index. Tenth edition. Rahway, New Jersey: Merck and Co., Inc., 1983.
The Merck Manual of Diagnosis and Therapy. 13th ed. Merck Sharp and Dohme Research Laboratories, 1977.
Mervyn, Len. *Minerals and Your Health*. New Canaan, Connecticut: Keats Publishing, Inc., 1981.
Miller, Benjamin F., M.D. and Claire Brackman Keane, R.N., B.S., M.Ed. *Encyclopedia and Dictionary of Medicine, Nursing, and Allied Health*. Philadelphia: W. B. Saunders Co., 1983.
Mindell, Earl L. *Earl Mindell's Vitamin Bible*. New York: Rawson, Wade Publishers, Inc., 1979.
Mitchell, Helen S. and Hendrika J. Rynbergen and Linnea Anderson and Marjorie V. Dibble. *Cooper's Nutrition in Health and Diease*. Philadelphia: J. B. Lippincott Co., 1972.
Morehouse, L. and A. Miller. *Physiology of Exercise*. 7th ed. St. Louis: C. V. Mosby Co., 1976.
Moss, Ralph W. *The Cancer Syndrome*. New York: Grove Press, 1980.
Muramoto, Noboru B. *Natural Immunity*. Oroville, California: George Ohsawa Macrobiotic Foundation, 1988.
Newbold, H. L., M.D. *Meganutrients for Your Nerves*. New York: Peter H. Wyden, 1975.
Nittler, Alan H., M.D. *A New Breed of Doctors*. New York: Pyramid House, 1972.
Null, Gary and Steven. *The Complete Handbook of Nutrition*. New York: Robert Speller and Sons, 1972.
Null, Gary and Steven. *Poisons in Your Body*. New York: Arco Publishing, 1977.
Nutrition Reviews. *Present Knowledge in Nutrition*. 3rd ed. New York: The Nutrition Foundation, 1967.
Nutrition Reviews. *Present Knowledge in Nutrition*. 4th ed. Washington: The Nutrition Foundation, 1976.
Nutrition Search, Inc. *Nutrition Almanac*. New York: McGraw-Hill Book Company, 1975.
Nutrition Search, Inc. *Nutrition Almanac*. Revised Edition. New York: McGraw-Hill Book Company, 1979.
Nieper, H. A., M.D. and L. Blumberger, M.D. *Electrolytes and Cardiovascular Disease*. Baltimore, Md.: Williams and Wilkins, 1966.
Orten, J. M. and O. W. Neuhaus. *Human Biochemistry*. 10th ed. St. Louis: C. V. Mosby, 1982.
Ott, John. *Health and Light*. New York: Devin-Adair Co., 1973.
Ott, John. *Health and Light*. New York: Pocket Books, 1976.
Padus, Emrika. *The Complete Guide to Your Emotions and Your Health*. Emmaus, Pennsylvania: Rodale Press, 1986.
Page, Melvin E., D.D.S. *Degeneration-Regeneration*. St. Petersburg, Florida: Biochemical Research Foundation, 1949.

Page, Melvin E., D.D.S. and H. Leon Abrams Jr. *Your Body Is Your Best Doctor.* New Canaan, Connecticut: Keats Publishing, Inc., 1972.
Passwater, Richard. *Cancer and Its Nutritional Therapies.* New Canaan, Connecticut: Keats Publishing, Inc., 1978.
Passwater, Richard. *Supernutrition for Healthy Hearts.* New York: Jove, 1978.
Passwater, Richard. *Supernutrition—Megavitamin Revolution.* New York: Dial Press, 1975.
Passwater, Richard A., Ph.D. and Elmer M. Cranton, M.D. *Trace Elements, Hair Analysis and Nutrition.* New Canaan, CT Keats Publishing, Inc., 1983.
Pauling, Linus. *Vitamin C and the Common Cold.* New York: Bantam, 1971.
Pauling, Linus. *Vitamin C, the Common Cold and the Flu.* San Francisco: W. H. Freeman, 1970.
Pearsall, Paul, Ph.D. *Superimmunity.* New York: McGraw-Hill Book Company, 1987.
Pearson, Durk and Sandy Shaw. *Life Extension: A Practical Scientific Approach.* New York: Warner Books, Inc., 1980.
Pearson, Durk and Sandy Shaw. *The Life Extension Companion.* New York: Warner Books, Inc., 1984.
Pfeiffer, Carl C. *Mental and Elemental Nutrients.* New Canaan, Connecticut: Keats Publishing, Inc., 1975.
Philpott, William H. and Dwight K. Kalita. *Brain Allergies: The Psycho-Nutrient Connection.* New Canaan, Connecticut: Keats Publishing, Inc., 1980.
Philpott, William H., M.D. *The Physiology of Violence: The Role of the Central Nervous System.* The Huxley Institute for Biosocial Research, April 23, 1976.
Physician's Desk Reference. 33rd ed. Medical Economics Co., 1979.
Pickney, Edward and Cathy. *The Cholesterol Controversy.* Los Angeles, California: Sherbourne Press, 1973.
Pike, R. and M. Brown. *Nutrition: An Integrated Approach.* 2nd ed. New York: John Wiley and Sons, 1975.
Prehoda, Robert W. *Extended Youth.* New York: Putnam, 1968.
Prevention Magazine editors. *Future Youth.* Emmaus, Pennsylvania: Rodale Press, 1987.
Price, Weston A., D.D.S. *Nutrition and Physical Degeneration.* The Price-Pottenger Nutrition Foundation, 1945.
Rechcigl, M. ed. *CRC Handbook of Nutritional Supplements.* Vols 1 and 2. Boca Raton, Florida: CRC Press, 1983.
Reuben, David, M.D. *The Save Your Life Diet.* New York: Random House, 1975.
Reuben, David, M.D. *The Save Your Life Diet.* New York: Ballantine Books, 1976.
Richardson, John A., M.D. and Patricia Griffin, R.N. *Laetrile Case Histories.* New York: Bantam Books, 1977.
Ritchason, Jack. *Vitamin and Health Encyclopedia.* No publisher, no date.
Robert, S.E. *Ear, Nose and Throat Dysfunctions due to Deficiencies and Imbalances.* Springfield, IL: Charles C. Thomas, 1957.
Robinson, Corinne H. *Basic Nutrition and Diet Therapy.* 2nd ed. New York: Macmillan Co., 1970.
Robinson, Corinne H. *Normal and Therapeutic Nutrition.* 14th ed. New York: Macmillan Co., 1972.
Rockstein, Morris et al. eds. *Theoretical Aspects of Aging.* New York: Academic Press, 1974.
Rodale, J. I. *Cancer Facts and Fallacies.* Emmaus, Pennsylvania: Rodale Press, 1969.
Rodale, J. I. *The Compete Book of Food and Nutrition.* Emmaus, Pennsylvania: Rodale Press, 1961.

Rodale, J. I. *The Compete Book of Minerals for Health*. Emmaus, Pennsylvania: Rodale Press, 1976.
Rodale, J. I. *The Complete Book of Vitamins*. Emmaus, Pennsylvania: Rodale Press, 1975.
Rodale, J. I. *The Encyclopedia of Common Diseases*. Emmaus, Pennsylvania: Rodale Press, 1969.
Rodale, J. I. *The Enclopedia of Healthful Living*. Emmaus, Pennsylvania: Rodale Press, 1970.
Rodale, J. I. *The Health Builder*. Emmaus, Pennsylvania: Rodale Press, 1962.
Roe, D. A. *Handbook: Interactions of Selected Drugs and Nutrients in Patients*. 3rd ed. Chicago: American Dietetic Association, 1982.
Rosenberg, Harold and Feldzaman A. N. *The Doctor's Book of Vitamin Therapy: Megavitamins for Health*. New York: G. P. Putnam, 1974.
Rosenberg, Harold and Feldzaman A. N. *The Doctor's Book of Vitamin Therapy: Megavitamins for Health*. New York: Berkley, 1975.
Rosenfeld, Albert. *Prolongevity*. New York: Knopf, 1976.
Rothenberg, Robert E. *The New American Medical Dictionary and Health Manual*. New York: New American Library, 1968.
Samuels, Mike and Hal Bennett. *The Well Body Book*. New York: Random House, 1973.
Schoden, R. and W. Griffin. *Fundamentals of Clinical Nutrition*. New York: McGraw-Hill Book Co., 1980.
Schroeder, Henry A., M.D. *The Poisons Around Us*. New Canaan, Connecticut: Keats Publishing, Inc., 1978.
Schroeder, Henry A., M.D. *Pollution, Profits and Progress*. Brattleboro, Vermont: Stephen Greene Press, 1971.
Schroeder, Henry A., M.D. *The Trace Elements and Man*. Old Greenwich, Connecticut: Devin-Adair, 1973.
Sebrell, W. H., Jr. and Robert S. Harris. *The Vitamins*. Vol. 1. New York: Academic Press, 1973.
Sebrell, W. H., Jr. and Robert S. Harris. *The Vitamins*. Vol. 2. New York: Academic Press, 1973.
Sebrell, W. H., Jr. and Robert S. Harris. *The Vitamins*. Vol. 3. New York: Academic Press, 1973.
Sebrell, W. H., Jr. and Robert S. Harris. *The Vitamins*. Vol. 4. New York: Academic Press, 1973.
Sebrell, W. H., Jr. and Robert S. Harris. *The Vitamins*. Vol. 5. New York: Academic Press, 1973.
Segerberg, Osborn, Jr. *The Immortality Factor*. New York: Dutton, 1974.
Segerberg, Osborn, Jr. *The Immortality Factor*. New York: Bantam, 1975.
Segerberg, Osborn, Jr. *Living to Be 100*. New York: Charles Scribner's Sons, 1982.
Selye, Hans. *The Stress of Life*. New York: McGraw-Hill, 1956.
Shackelton, Alberta D. *Practical Nurse Nutrition Education*. 3rd ed. Philadelphia: W. B. Saunders Co., 1972.
Shute, Evan V. *The Heart and Vitamin E*. New Canaan, Connecticut: Keats Publishing, Inc., 1977.
Shute, Wilfrid E., M.D. *The Complete Vitamin E. Book*. New Canaan, Connecticut: Keats Publishing, Inc., 1978.
Shute, Wilfrid E., M.D. *Health Preserver*. Emmaus, PA: Rodale Press, 1977.
Sidhwa, Kekir. *Fit for Anything*. Lewes, Sussex, England: Health for All Publishing Co., 1969.
Simenton, O. Carl, M.D. and Stephanie Matthews-Simonton and James Creighton. *Getting Well Again*. New York: St. Martins Press, no date.

Soyka, Fred. *The Ion Effect.* New York: Bantam Books, Inc., 1977.
Steadman's Medical Dictionary. Twenty-second edition. Baltimore, MD: Williams & Wilkins Co., 1972.
Stone, Irwin. *The Healing Factor: Vitamin C Against Disease.* New York: Grosset and Dunlap, 1970.
Stone, Irwin. *The Healing Factor: Vitamin C Against Disease.* New York: Grosset and Dunlap, 1972.
Synder, Arthur W. *Vitamins and Minerals.* Los Angeles: Hansens, 1969.
Taber, Clarence Wilbur. *Taber's Cyclopedic Medical Dictionary.* Eleventh Edition. Philadelphia: F.A. Davis Company, 1969.
Thurston, Emory W. *Parents' Guide to Better Nutrition.* New Canaan, Connecticut: Keats Publishing, Inc., 1979.
Turner, J. *The Chemical Feast.* New York: Grossman, 1970.
U.S. Government. Consumer and Food Economics Research Division, Agricultural Research Service, U.S. Department of Agriculture. "Composition of Foods: Raw, Processed, Prepared." *Agriculture Handbook.* Washington, DC: U.S. Government Printing Office, December 1963.
U.S. Government. Department of Agriculture, *Natural, Organic, and Health Foods.* Extension Folder No. 280. St. Paul: University of Minnesota, 1973.
U.S. Government. Department of Health, Education and Welfare. *First Health and Nutritional Examination Survey.* Washington DC: U.S. Government Printing Office, 1972.
U.S. Government. Department of Health, Education and Welfare. *Framingham (Massachusetts) Heart Study.* Washington DC: U.S. Government Printing Office, 1978.
U.S. Government. Department of Health, Education and Welfare. *Report of the Secretary's Commission on Pesticides and Their Relationship to Environmental Health.* Washington DC: U.S. Government Printing Office, 1969.
U.S. Government. Department of Health, Education and Welfare. *The Swine Flu Affair.* Washington DC: U.S. Government Printing Office, 1978.
U.S. Government. *Dietary Goals for the United States.* Washington, DC: U.S. Government Printing Office, 1977.
U.S. Government. Food and Drug Administration. *Vitamin A Talk Paper.* Washington DC: U.S. Government Printing Office, August 23, 1979.
U.S. Government. National Academy of Sciences. *Recommended Dietary Allowances.* 7th ed. Washington, DC, 1974.
U.S. Government. National Academy of Sciences. *Recommended Dietary Allowances.* Washington, DC, 1980.
U.S. Government. National Academy of Sciences. *Recommended Dietary Allowances.* Washington, DC, 1989.
U.S. Government. National Academy of Sciences. *Evaluating the Safety of Food Chemicals.* Pub. No. 1859. Washington, DC, 1970.
U.S. Government. *Toxicants Occurring Naturally in Foods.* Washington DC: U.S. Government Printing Office, 1973.
U.S. Government. *Myths of Vitamins.* HEW Publication No. (FDA) 77-2045. Washington DC: U.S. Government Printing Office, 1977.
U.S. Goverment. Select Committee on Nutrition and Human Needs. U.S. Senate. *Diet Related to Killer Diseases.* Washington, DC: U.S. Government Printing Office, 1977.
U.S. Government and Adams, Catherine F. *Nutritive Value of American Foods in Common Units.* Handbook No. 456. Washington, DC: Superintendent of Documents, 1975.

Wade, Carlson. *Carlson Wade's Lecithin Book.* New Canaan, Connecticut: Keats Publishing, Inc., 1980.
Wade, Carlson. *Helping Your Health with Enzymes.* New York: Universal-Award House, 1971.
Wade, Carlson. *Hypertension and Your Diet.* New Canaan, Connecticut: Keats Publishing, Inc., 1975.
Wade, Carlson. *Magic Minerals: Key to Better Health.* West Nyack, New York: Parker Publishing Co., 1967.
Wade, Carlson. *Vitamin E: The Rejuvenation Vitamin.* New York: Award Books, 1970.
Wagner, Arthur F., Ph.D. and Karl Folkers, Ph.D., D.Sc. *Vitamins and Coenzymes.* Huntington, NY: R. E. Krieger Publishing Company, 1964.
Walford, R. L. *Maximum Lifespan.* New York: W. W. Norton and Company, 1983.
Walker, Morton. *Chelation Therapy: How to Prevent or Reverse Hardening of the Arteries.* Seal Beach, California: '76 Press, 1980.
Walker, Morton. *Now Not to Have a Heart Attack.* New York: Franklin Watts, Inc., 1980.
Walker, Morton. *Total Health.* New York: Everest Watt, Bernice K. and Annabel I. Merrill et al. "Composition of Foods." *Agriculture Handbook.* No. 8. U.S. Department of Agriculture, 1975.
Wayler, Thelma J. and Rose S. Klein. *Applied Nutrition.* New York: Macmillan Co., 1965.
Webster, James. *Vitamin C: The Protective Vitamin.* New York: Universal-Award House, 1971.
Wellford, H. *Sowing the Wind.* New York: Grossman, 1972.
Wheatly, Michael. *About Nutrition.* London: Thorsons, 1971.
Whelan, Elizabeth M., Dr. and Dr. Fredrick J. Stare. *The One-hundred-percent Natural, Purely Organic, Cholesterol-free, Megavitamin, Low-carbohydrate Nutrition Hoax.* New York: Atheneum, 1983.
White, Philip ed. *Let's Talk About Food.* 2nd ed. Chicago: American Medical Association, 1970.
White, P. and T. Mondeika ed. *Diet and Exercise: Synergism in Health Maintenance.* Chicago: American Medical Association, 1982.
Williams, M. H. *Nutritional Aspects of Human Physical and Athletic Performance.* Springfield, Illinois: Charles C. Thomas, 1976.
Williams, Roger J., Ph.D. *Nutrition Against Disease.* New York: Pitman, 1971.
Williams, Roger J., Ph.D. *Nutrition Against Disease.* New York: Bantam, 1973.
Williams, Roger J., Ph.D. *The Wonderful World Within You.* New York: Bantam Books, 1977.
Williams, Roger J., Ph.D. and Dwight K. Kalita. *A Physician's Handbook on Orthomolecular Medicine.* Elmsford, New York: Pergamon Press, 1977.
Williams, Sue R. *Review of Nutrition and Diet Therapy.* 2nd ed. St. Louis: C. V. Mosby Company, 1973.
Wilson, Eva D. and Katherine H. Fischer and Mary E. Fugue. *Principles of Nutrition.* 2nd ed. New York: Pitman Publishing, 1971.
Winick, M. ed. *Nutrition and Aging.* New York: John Wiley and Sons, 1976.
Winick, M. ed. *Nutrition and Cancer.* New York: John Wiley and Sons, 1977.
Winick, M. ed. *Nutrition and the Killer Diseases.* New York: John Wiley and Sons, 1981.
Winter, Ruth. *Ageless Aging.* New York: Crown, 1973.
Winter, Ruth. *Beware of the Food You Eat.* Revised ed. New York: Signet Books, 1971.

Winter, Ruth. *Beware of the Food You Eat.* Revised ed. New York: New American Library, 1972.

Wintrobe, M. M. et al. *Harrison's Principles of Internal Medicine.* 6th ed. New York: McGraw-Hill Book Co., 1970.

Wright, Jonathan V., M.D. *Dr. Wright's Book of Nutritional Therapy.* Emmaus, Pennsylvania: Rodale Press, 1979.

Yepsen, Roger B., Jr. *How to Boost Your Brain Power.* Emmaus, Pennsylvania: Rodale Press, 1987.

Index

abalone, fried 83, 131, 143, 155, 169, 215, 234, 243, 252, 263, 271, 280, 289, 298, 307, 317, 326, 334, 343, 354, 365, 377, 389, 404
abalone, raw 50, 83, 131, 143, 155, 170, 182, 216, 234, 243, 253, 263, 272, 281, 289, 298, 308, 317, 326, 335, 344, 355, 367, 380, 390, 401
abdominal and chest pains 11
abdominal cramps 50, 86, 151
abdominal pain 6, 41, 125, 228
ABEREL 10
abscess 9, 96, 101, 127, 226
AC 38_{555} 87
acerola juice 3, 22, 33, 42, 56, 91, 136, 164, 176, 199, 210, 224, 356, 370, 381, 397, 398
acerola, raw 3, 22, 33, 42, 55, 67, 91, 135, 165, 174, 198, 209, 223, 355, 369, 381, 397, 399
ACETATE REPLACING FACTOR 112
ACETIAMINE 23
acetoacetyl-coenzyme A levels, decreased 227
acetylcholine receptors, decreased 167
ACETYLCHOLINESTERASE 88
aches 98, 167
acidosis 285
ACIDS 69, 89, 140
acne 6, 9, 36, 37, 43, 58, 67, 96, 99, 101, 104, 105, 126, 138, 152, 199, 211, 228, 371
ACON 10
acorn flour 131, 142, 160, 169, 194, 200, 234, 244, 254, 264, 273, 282, 291, 299, 309, 319, 327, 336, 344, 350, 359, 373, 392, 409
acorns, dried 15, 31, 38, 130, 142, 161, 169, 194, 200, 234, 244, 254, 264, 273, 282, 291, 299, 309, 319, 327, 336, 344, 350, 359, 373, 392, 409
acorns, raw 16, 32, 39, 131, 142, 162, 169, 196, 201, 235, 244, 254, 264, 273, 282, 291, 300, 309, 319, 327, 336, 345, 351, 360, 373, 392, 408
acrid taste 77
ACTASAL 88
ACTIMIDE 76
ACTIVATED ERGOSTEROL 98
ACTIVATED 7-DEHYDRO-CHOLESTEROL 98, 99
additives 124
ADEMIDE 76
ADENEX 91, 97
ADENINE 49
ADENOSINE MONOPHOSPHATE 68
ADENYL 68
ADENYL THIOMETHYLPENTOSE 110
ADENYLIC ACID 67, 68
ADERMIN 58
ADP and ATP synthesis, decreased 68
adrenal deterioration 36, 50, 58, 74, 84, 96, 126, 187, 225
ADRENALINE 91
Adriamycin treatments for cancer 117
adzuki beans, boiled 5, 16, 33, 40, 132, 143, 158, 170, 179, 192, 201, 223, 232, 244, 255, 264, 273, 282, 290, 300, 309, 319, 327, 336, 345, 352, 366, 383, 392, 403
AFAXIN 10
AFLATOXIN 6
AG 213
agar seaweed, dried 22, 29, 42, 127, 154, 168, 178, 196, 200, 217, 349, 362, 382, 393, 409
agar seaweed, raw 22, 36, 43, 130, 158, 169, 179, 199, 206, 223, 355, 370, 397, 399
aged tissues, cosmetic deterioration of 312
aged tissues, sagging 312
aggression 322
aging 5, 9, 23, 43, 49, 58, 67, 85, 86, 90, 96, 97, 99, 101, 108, 111, 113, 117, 138, 166, 212, 238, 257, 303, 330
aging of skin, premature 10, 213
aging, premature 76, 111, 249
AGIOLAN 10
AIDS 96, 117
air pollution, nuclear power 153
AIROL 10
AKNOTEN 10

461

Index 462

AKOTIN 37
AL 123
AL+++ 123
AL+² 123
ALA 239
ALANINE 239
albumin in urine, slight 167
ALCOHOL 24, 37, 44, 50, 59, 69, 85, 86, 89, 90, 103, 104, 140, 168, 199, 371
alcohol cravings 9, 23, 36, 43, 49, 58, 67, 74, 77, 84, 86, 96, 99, 108, 166, 167, 167, 177, 211, 238, 258
alcohol poisoning 303
ALDOSTERONE DRUGS 199
ale 138, 166, 211, 223, 248, 257, 267, 276, 285, 294, 303, 312, 322, 330, 339, 348, 357, 369, 397
alertness, decreased 258, 303
alewife, canned 272, 281, 289, 298, 308, 318, 326, 343, 366, 377, 390, 402
alewife, raw 191, 271, 280, 289, 297, 307, 317, 326, 343, 366, 378, 389, 401
alfalfa 85, 86, 88, 89, 107, 114, 118, 213
alfalfa leaves 78
alfalfa seeds, raw sprouted 4, 17, 31, 40, 52, 66, 80, 94, 132, 144, 161, 171, 181, 195, 210, 223, 234, 274, 283, 292, 319, 346, 356, 370, 380, 384, 393, 399
alfalfa sprouts 78
algae 105
ALINAMIN 24
ALKALI 89
ALKALIES 24, 59, 69, 140
ALLERCORB 91, 97
allergic skin reactions 78
allergies 9, 11, 58, 74, 96, 99, 101, 104, 105, 117, 138, 185, 211, 267, 371
ALLOCAINE 106
ALLOXAZINE MONONUCLEOTIDE 37
allspice 3, 92, 127, 154, 169, 194, 200, 221, 234, 350, 363, 376, 393, 409
almond butter 16, 25, 38, 55, 65, 79, 127, 141, 155, 168, 178, 188, 200, 222, 230, 241, 252, 263, 272, 281, 291, 299, 307, 318, 325, 335, 344, 352, 359, 372, 390, 410
almond meal 14, 24, 38, 127, 154, 188, 200, 223, 240, 249, 260, 268, 277, 289, 298, 304, 313, 322, 331, 339, 351, 360, 374, 385, 409
almond oil 101, 104, 359, 371
almond paste 14, 25, 38, 53, 64, 80, 127, 141, 155, 168, 178, 188, 200, 222, 231, 242, 253, 263, 273, 281, 291, 299, 308, 318, 326, 335, 344, 351, 360, 373, 391, 408
almonds 89, 101, 103, 104, 116, 117, 125, 212
almonds, bitter 77, 78
almonds, dried 14, 25, 38, 53, 64, 79, 127, 141, 155, 168, 178, 188, 200, 222, 231, 240, 250, 262, 271, 280, 290, 299, 305, 317, 323, 334, 343, 352, 359, 372, 384, 389, 409
almonds, dry roasted 15, 25, 38, 55, 65, 79, 127, 141, 155, 168, 178, 188, 200, 222, 229, 240, 252, 263, 272, 281, 290, 299, 306, 318, 324, 335, 343, 352, 359, 372, 390, 410
almonds, mature 78
almonds, oil roasted 15, 25, 38, 55, 65, 79, 127, 141, 155, 168, 178, 188, 200, 222, 229, 240, 250, 262, 271, 280, 290, 299, 305, 317, 323, 334, 342, 353, 359, 372, 389, 410
almonds, toasted 15, 25, 38, 55, 65, 79, 127, 141, 155, 168, 178, 188, 200, 222, 229, 240, 250, 262, 271, 280, 290, 299, 305, 317, 323, 334, 342, 352, 359, 372, 389, 410
almonds, young 78
ALOCAINE 106
ALPHA COBIONE 75
ALPHA-CAROTENE 1
ALPHALIN 10
ALPHA-LIPOIC ACID 112
ALPHA-REDISOL 75
ALPHA-RUVITE 75
ALPHA-TOCOPHEROL 100, 102
ALPHA-TOCOPHEROL ACETATE 100, 102
ALPHA-TOCOPHEROL ACID SUCCINATE 100, 102
ALPHA-VITAMIN E 102
ALPHA-VITAMIN E ACETATE 102
ALPHA-VITAMIN E ACID SUCCINATE 102
altered thyroid and insulin production 11
ALUMINUM 123
ALUMINUM 123, 127, 187
aluminum cookware 123
ALUMINUM SALTS OF FLUORIDE 151
ALUMIUM 123
Alzheimer's disease 86, 117
Amanita poisoning 112
AMANITINE 86
amaranth, boiled 2, 22, 30, 41, 92, 128, 157, 169, 195, 201, 222, 246, 256, 265, 274, 283, 292, 301, 309, 320, 329, 337, 346, 356, 370, 382, 394, 399
AMBEN 90
AMIDE PP 43
AMINICOTIN 43
AMINO ACETIC ACID 258
amino acid 37, 43, 49, 88, 239, 248, 249, 257, 258, 259, 267, 268, 276, 285, 285, 294, 294, 303, 303, 303, 303, 303, 312, 312, 312, 312, 313, 313, 313, 322, 331, 331, 339, 385, 385, 397
AMINO ACID A 239
AMINO ACID B 248, 249
AMINO ACID C 249
AMINO ACID D 249
amino acid deficiencies 385
amino acid, essential 88, 239, 248, 249, 257, 258, 259, 267, 268, 276, 285, 294, 303, 312, 313, 322, 331, 339, 385, 397
AMINO ACID F 303
AMINO ACID G 258
AMINO ACID H 259
AMINO ACID I 268
AMINO ACID K 285
AMINO ACID L 276
AMINO ACID M 294
AMINO ACID N 248
AMINO ACID P 249, 312
AMINO ACID Q 258
AMINO ACID R 239
AMINO ACID S 312
amino acid, semi-essential 239, 259, 385
amino acid, sulfur-containing 249, 294
AMINO ACID T 313
amino acid utilization, deterioration of 37, 43
AMINO ACID V 339
AMINO ACID W 322
AMINO ACID Y 331
AMINO ACID Z 258
AMINO ACIDS 239, 248, 249, 257, 258, 259, 267, 268, 276, 285, 294, 303, 312, 313, 331, 385
AMINOCAINE 106
AMMONIUM CHLORIDE 91
amnesia 126
AMP 68
AMYGDALIN 77
AN FACTOR 90
ANABASI 76
ANATOLA 10
anchovy canned in olive oil 16, 26, 44, 63, 72, 127, 142, 155, 169, 190, 201, 214, 231, 241, 251, 261, 269, 277, 286, 294, 305, 314, 324, 332, 340, 365, 377, 386, 407

anchovy paste 271, 280, 289, 297, 307, 317, 326, 342, 356, 365, 376, 389
anchovy, pickled 128, 191, 270, 280, 288, 297, 307, 316, 326, 342, 358, 365, 376, 389, 405
anchovy, raw 19, 27, 44, 63, 72, 128, 143, 155, 170, 192, 202, 217, 232, 243, 253, 262, 271, 280, 288, 297, 307, 316, 326, 334, 343, 366, 378, 389, 402
ANECTINE CHLORIDE 87
anemia 23, 24, 58, 67, 74, 76, 78, 84, 89, 90, 91, 96, 100, 101, 107, 118, 124, 138, 139, 140, 141, 151, 153, 166, 167, 186, 186, 186, 213, 227, 228, 267, 285
anemia, dialysis-induced 88
anemia, pernicious 67, 69, 74, 76, 84, 96, 101, 138, 140, 166
anesthesia 168
ANESTHESOL 106
ANESTIL 106
ANEURIMEC 24
ANEURIN 11
ANEURIN-1, 5 SALT 24
ANEURINE DISULFIDE 23
ANEURINE HYDROCHLORIDE 23
ANEURINE MONONITRATE 23
angina 76, 88, 100, 105, 212
angina pectoris 9, 74, 77, 86, 88, 96, 101, 106, 116, 117, 152, 211
animal 68
animal fats, fresh 118
ANIMAL GALACTOSE FACTOR 76
animal glands 126
animal protein 313
ANIMAL PROTEIN FACTOR 68
animal symptoms possibly similar in humans 49, 67, 68, 119
anise seed 127, 153, 168, 188, 200, 229, 350, 361, 375, 389, 409
anorexia, infantile 118
antacids 91, 123
ANTHRANILIC ACID 110
ANTIBERIBERI FACTOR/VITAMIN 11
ANTIBIOTICS 24, 89, 107
antibody formation, decreased 285
antibody G, decreased 117
antibody production, decreased 24, 49
ANTICANITIC VITAMIN 90
ANTI-CHROMOTRICHIA FACTOR 90
ANTICOAGULANTS 91, 107
ANTICONVULSANTS 98
ANTIDEPRESSANTS 91

ANTI-DIABETIC AGENTS 69, 140
ANTI-EGG WHITE INJURY FACTOR 89
ANTI-GIZZARD EROSION FACTOR 118
ANTIHEMORRHAGIC FACTOR/VITAMIN 107
ANTIHISTAMINES 91
ANTI-INFECTIVE FACTOR/VITAMIN 6
ANTIMONY 123-124
antimony resistance, decreased 90
ANTINEURITIC FACTOR/VITAMIN 11
antioxidant 1, 5, 6, 9, 11, 23, 24, 37, 43, 49, 58, 67, 85, 86, 90, 91, 96, 97, 97, 98, 100, 101, 112, 113, 113, 114, 114, 115, 115, 116, 116, 117, 212, 227, 228, 238, 257, 303, 330
antioxidant therapy 5, 9, 23, 43, 49, 58, 67, 85, 86, 90, 96, 97, 101, 113, 117, 212, 228, 238, 257, 303, 330
ANTIOXIDANT THERAPY, FAT-SOLUBLE 97
ANTIOXIDANTS 90
ANTIPELLAGRA FACTOR/VITAMIN 37
ANTIPERNICIN 68
ANTIPERNICIOUS FACTOR 68
ANTIRACHITIC FACTOR 98
ANTISCROBUTIC VITAMIN 91, 97
ANTISTERILITY FACTOR/VITAMIN 100
antiviral activity, decreased 98, 112
ANTIVITAMIN C 90
ANTIXEROPHTHALMIC FACTOR/VITAMIN 6
anxiety 77, 167, 186, 187, 199, 214
AORAL 10
APEXOL 10
apoplexy 152
APOSTAVIT 10
appendicitis 384
appetite decreased 6, 11, 24, 37, 43, 49, 77, 89, 91, 98, 118, 123, 124, 167, 186, 187, 214, 228, 285, 294, 385
apple brown betty 4, 19, 35, 41, 96, 133, 163, 197, 210, 217, 351, 366, 378, 395, 404
apple juice 5, 22, 36, 43, 66, 96, 125, 137, 150, 164, 177, 181, 199, 209, 224, 238, 354, 369, 383, 397, 399
apple, mammy 3, 22, 34, 42, 57, 93, 135, 163, 198, 210, 222, 293, 302, 330, 354, 369, 380, 397, 400
apple pie 5, 16, 33, 40, 56, 66, 83, 136, 149, 161, 176,

197, 210, 217, 237, 351, 364, 376, 395, 406
apple seeds 78
apple, sugar 5, 16, 31, 40, 56, 63, 92, 132, 163, 173, 197, 206, 222, 293, 302, 330, 352, 368, 382, 394, 402
apple with skin, raw 4, 23, 36, 43, 57, 66, 84, 94, 137, 149, 166, 177, 182, 199, 209, 224, 238, 248, 257, 267, 276, 285, 294, 302, 312, 322, 330, 339, 348, 353, 369, 381, 397, 400
apple without skin, cooked 4, 22, 36, 43, 57, 66, 84, 138, 150, 165, 177, 181, 199, 210, 224, 238, 248, 257, 267, 276, 285, 294, 302, 312, 322, 330, 339, 348, 354, 369, 381, 397, 400
apple without skin, raw 4, 22, 36, 43, 57, 66, 84, 95, 138, 150, 166, 177, 183, 199, 209, 238, 248, 257, 267, 276, 285, 294, 303, 312, 322, 330, 339, 348, 353, 369, 382, 397, 400
apples 85, 86, 89, 101, 112, 139, 187, 227
applesauce 5, 22, 35, 42, 57, 67, 84, 95, 125, 138, 149, 165, 177, 181, 199, 210, 224, 238, 248, 257, 267, 276, 285, 294, 303, 312, 322, 330, 339, 348, 353, 368, 382, 397, 401
apricot kernels 77, 78
apricot nectar 2, 23, 36, 42, 84, 137, 148, 164, 177, 199, 209, 223, 237, 353, 369, 383, 397, 400
apricot oil 101
apricots 2, 22, 34, 41, 56, 65, 82, 94, 112, 134, 147, 163, 176, 181, 186, 187, 198, 205, 224, 236, 247, 257, 266, 275, 284, 292, 302, 311, 321, 329, 338, 347, 354, 369, 381, 395, 400
apricots, dried 2, 31, 38, 51, 63, 81, 95, 130, 142, 155, 170, 180, 194, 200, 223, 234, 246, 256, 265, 274, 283, 291, 302, 309, 320, 327, 337, 346, 350, 364, 380, 393, 408
APRIKERN 77
AQUOCOBALAMIN 69, 75
AQUOCOBAMIDE 75
ARACHIDONIC ACID 104, 105
ARDESYL 118, 119
ARG 239
ARGININE 49, 239-248
arginine, internal conversion of 303
ARLIFLAV 112

Index

arm excluding fat, braised pork 9, 12, 26, 44, 51, 60, 72, 83, 136, 144, 158, 173, 184, 190, 202, 217, 229, 363, 376, 385, 406
arm excluding fat, roasted cured pork 12, 28, 45, 52, 60, 71, 84, 135, 145, 161, 175, 182, 190, 205, 214, 231, 242, 250, 260, 270, 279, 288, 296, 305, 315, 324, 333, 342, 365, 377, 388, 404
arm exluding fat, roasted pork 9, 12, 26, 46, 52, 60, 72, 83, 136, 145, 159, 174, 182, 190, 203, 218, 230, 364, 376, 387, 404
arm including fat, braised pork 8, 13, 26, 45, 52, 62, 72, 84, 136, 144, 158, 174, 185, 192, 204, 218, 230, 241, 251, 259, 269, 278, 286, 296, 305, 314, 323, 332, 341, 361, 373, 387, 407
arm including fat, roasted cured pork 12, 29, 46, 52, 62, 72, 84, 135, 145, 161, 175, 183, 190, 206, 215, 231, 243, 252, 262, 271, 280, 289, 298, 307, 316, 326, 334, 343, 362, 374, 389, 406
arm including fat, roasted pork 9, 13, 27, 46, 53, 61, 72, 84, 136, 145, 160, 175, 182, 191, 205, 219, 230, 242, 252, 262, 270, 279, 287, 297, 307, 316, 325, 333, 342, 361, 373, 388, 406
armpits, increased color on 124
arms and legs, coldness of 126
arms, paralysis of 124
arms, spasms in 124
ARRET 88
arrowhead, boiled 15, 33, 40, 136, 160, 170, 191, 200, 222, 353, 368, 393, 401
ARSENIC 124
arsenic resistance, decreased 90
arterial fibrillation 106
arteries, hardening of 9, 43, 49, 74, 76, 85, 86, 96, 101, 112, 138, 139, 152, 177, 199, 238
arteries, inflammation of 106
arteries of the brain, hardening of 106
arthritis 9, 36, 43, 49, 58, 67, 74, 85, 96, 99, 101, 105, 106, 113, 138, 151, 153, 177, 199, 211, 213, 226, 267
ARTHROPAN 88
artichoke, boiled 4, 19, 30, 40, 56, 64, 79, 94, 132, 148, 163, 171, 180, 196, 205, 220, 236, 355, 369, 380, 394, 400
artichoke hearts, boiled 4, 19, 34, 40, 56, 65, 80, 94, 131, 148, 159, 170, 180, 196, 205, 219, 236, 354, 369, 383, 394, 400
artichoke, raw jerusalem 5, 14, 33, 39, 95, 134, 155, 175, 181, 195, 353, 368, 395, 401
ARTROBIONE 88
AS 124
ASCORBIC ACID 91, 97
ASCORBIC ACID CALCIUM SALT 91, 97
ASCORBIC ACID MAGNESIUM SALT 91, 97
ASCORBIC ACID NICOTINAMIDE COMPLEX 91, 97
ASCORBIC ACID POTASSIUM SALT 91, 97
ASCORBIC ACID SODIUM SALT 91, 97
ASCORBYL PALMITATE 91, 97
ASCORIN 91, 97
ASCORTEAL 91, 97
ASCORVIT 91, 97
ASN 248
ASP 249
ASPARAGINE 248
asparagus 86, 101, 107
asparagus, boiled 3, 17, 31, 40, 56, 63, 79, 93, 132, 146, 163, 174, 180, 196, 204, 223, 235, 246, 255, 265, 274, 283, 292, 301, 311, 320, 329, 338, 346, 356, 370, 381, 394, 398
asparagus, frozen and boiled 3, 18, 32, 40, 56, 66, 79, 92, 132, 144, 163, 175, 181, 196, 207, 224, 235, 246, 255, 265, 274, 283, 292, 301, 310, 320, 329, 338, 346, 356, 370, 380, 394, 399
ASPARTIC ACID 240, 249
Aspergillus niger cultures 90
ASPIRIN 69, 78, 91, 107, 127, 140, 199
aspirin, children's 123
ASPIRIN SUBSTITUTES 69, 91, 107, 140
asthma 9, 58, 67, 74, 77, 85, 86, 105, 106, 117, 185
asthma, chronic 187
ATAV 10
atherosclerosis 9, 43, 49, 58, 74, 77, 85, 86, 101, 112, 138, 153, 177, 199, 213, 228, 238, 248
athlete's foot 9, 96, 101
ATOXICOCAINE 106
ATP and ADP synthesis, decreased 68
ATP production, decreased 112
ATRIPINE 91
AU 152
AUDES 87
auditory dysfunction 267
AUSOVIT B₁ 24
autism 77
AVIBON 10
AVITA 10
AVITOL 10
avocado, raw california 3, 16, 31, 39, 51, 62, 79, 94, 135, 143, 160, 170, 180, 196, 201, 222, 235, 247, 256, 266, 275, 283, 292, 301, 310, 320, 329, 337, 346, 355, 365, 374, 384, 394, 402
avocado, raw florida 3, 16, 31, 39, 51, 62, 80, 94, 135, 143, 163, 170, 181, 196, 201, 223, 235, 247, 256, 266, 275, 284, 293, 301, 311, 320, 329, 338, 346, 355, 367, 377, 395, 401
avocados 89, 101, 103, 104
AXEROL 10
AXEROPHTHOL 6
AXLON 75
azuki beans 116, 117

B 125
BA 124
BA⁺⁺ 124
BA⁺² 124
back trouble 91
back with skin, fried chicken 7, 16, 28, 44, 50, 61, 73, 82, 132, 147, 158, 172, 181, 192, 206, 218, 231, 241, 250, 261, 269, 278, 287, 295, 305, 315, 323, 332, 341, 355, 361, 374, 386, 407
backache 37, 43, 96, 99, 101, 177, 199
bacon cheeseburger 7, 16, 30, 47, 56, 63, 71, 81, 96, 129, 147, 157, 172, 191, 206, 216, 230, 353, 362, 374, 390, 407
bacon, cooked 12, 27, 44, 50, 61, 71, 83, 134, 144, 158, 172, 182, 189, 201, 214, 230, 241, 251, 260, 269, 278, 287, 296, 304, 315, 324, 333, 340, 358, 359, 372, 386, 409
bacon, grilled canadian 12, 29, 44, 53, 59, 72, 84, 135, 149, 162, 173, 182, 189, 202, 214, 232, 243, 252, 261, 271, 280, 288, 296, 307, 316, 325, 334, 343, 358, 365, 377, 388, 404
bacon, raw 13, 32, 47, 54, 63, 72, 84, 136, 148, 163,

Index

176, 185, 193, 209, 215, 233, 244, 254, 264, 273, 282, 290, 299, 309, 318, 327, 336, 345, 358, 359, 372, 392, 408
bacon, unheated canadian 12, 30, 44, 53, 60, 72, 84, 136, 149, 163, 174, 183, 190, 203, 214, 233, 244, 252, 261, 272, 280, 289, 297, 308, 317, 326, 334, 343, 357, 366, 377, 389, 403
bacteria, internal synthesis from 74, 85, 108
bacterial resistance, decreased 226
BACTERIAL VITAMIN H₁ 89
bad breath 9, 37, 43, 49, 67, 77, 96, 186, 212, 228
bagels 13, 27, 38, 54, 66, 80, 131, 147, 156, 173, 195, 210, 216, 235, 350, 362, 379, 384, 391, 408
baker's yeast, compressed 12, 25, 44, 134, 155, 188, 201, 222, 354, 368, 378, 391, 402
baker's yeast, dry 11, 24, 44, 131, 154, 188, 200, 220, 351, 362, 379, 385, 409
baking powders 123
BAKING SODA 11
balance deterioration 49, 167
baldness 24, 36, 43, 58, 67, 85, 89, 90, 96, 101, 106, 151, 294
bamboo shoots, boiled 22, 34, 42, 135, 165, 177, 198, 201, 224, 247, 256, 266, 275, 284, 293, 302, 311, 321, 329, 347, 357, 371, 381, 395, 398
bamboo shoots, canned 5, 22, 36, 43, 96, 136, 165, 177, 197, 210, 223, 247, 256, 266, 275, 284, 292, 301, 311, 320, 329, 346, 357, 370, 381, 395, 398
bamboo shoots, raw 5, 15, 33, 40, 95, 134, 164, 177, 196, 201, 223, 246, 256, 265, 274, 283, 292, 301, 310, 320, 329, 346, 356, 370, 381, 384, 394, 399
bamboo sprouts 78
banana custard pie 3, 20, 31, 42, 96, 130, 163, 195, 207, 217, 351, 364, 377, 393, 406
bananas 4, 21, 32, 41, 55, 59, 80, 89, 94, 101, 108, 137, 139, 146, 165, 171, 181, 198, 202, 224, 237, 247, 256, 265, 276, 284, 293, 302, 311, 321, 329, 338, 347, 352, 368, 381, 396, 401
barbecue loaf 13, 27, 47, 50, 62, 71, 93, 130, 147, 160,

174, 182, 193, 204, 214, 231, 355, 365, 377, 390, 403
barbecue sauce 3, 19, 41, 65, 94, 133, 150, 162, 177, 198, 208, 215, 238, 354, 368, 379, 395, 401
BARBITURATES 91, 127
BARIUM 124
barley 77, 86, 89, 101, 103, 104, 116, 186, 187, 212
barley, cooked 85
barley, light pearled 16, 34, 38, 133, 157, 192, 208, 224, 349, 361, 380, 392, 409
barley pearled, pot 14, 33, 38, 131, 156, 189, 205, 349, 361, 380, 391, 409
barracuda, raw 367, 379, 389, 401
basal metabolism rate, decreased 100
basella 116
basil 2, 37, 91, 127, 153, 168, 188, 200, 228, 244, 253, 263, 272, 280, 290, 299, 306, 317, 324, 335, 343, 349, 360, 376, 389, 409
bass 99
bass, cooked black 8, 18, 31, 47, 129, 161, 190, 206, 220, 270, 280, 288, 297, 307, 316, 326, 342, 357, 363, 374, 389
bass, cooked sea 8, 134, 150, 164, 170, 190, 204, 218, 235, 243, 252, 261, 270, 279, 287, 296, 306, 316, 325, 333, 342, 366, 379, 388, 402
bass, raw black 206, 219, 271, 280, 289, 297, 308, 317, 326, 343, 368, 379, 389, 401
bass, raw freshwater 129, 146, 159, 171, 179, 191, 203, 219, 234, 244, 253, 262, 271, 280, 289, 297, 307, 317, 326, 334, 343, 367, 378, 389, 401
bass, raw sea 8, 135, 151, 165, 170, 184, 192, 206, 219, 236, 244, 253, 262, 271, 280, 289, 297, 307, 317, 326, 334, 343, 367, 379, 389, 401
bass, raw striped 70, 150, 162, 184, 219, 236, 244, 253, 262, 272, 280, 289, 298, 308, 317, 326, 335, 343, 367, 379, 390, 401
bay leaf 2, 127, 154, 169, 194, 204, 232, 350, 365, 376, 391
BC 16 87
BE 125
BE⁺⁺ 125
BE⁺² 125
bean nuts, toasted soy 4, 16,

31, 39, 53, 61, 79, 94, 128, 141, 155, 168, 188, 200, 223, 230, 240, 249, 260, 268, 277, 286, 298, 304, 313, 322, 331, 340, 351, 360, 373, 385, 409
bean sprouts, mung 78
beans 89, 101, 103, 104
beans, azuki 116, 117
beans, black 78
beans, boiled adzuki 5, 16, 33, 40, 132, 143, 158, 170, 179, 192, 201, 223, 232, 244, 255, 264, 273, 282, 290, 300, 309, 319, 327, 336, 345, 352, 366, 383, 392, 403
beans, boiled and frozen baby lima 4, 18, 33, 40, 56, 64, 79, 94, 132, 143, 158, 169, 179, 194, 202, 221, 235, 245, 254, 264, 273, 282, 291, 300, 309, 319, 327, 336, 345, 353, 367, 382, 392, 402
beans, boiled baby lima 15, 34, 40, 53, 65, 79, 132, 143, 157, 170, 179, 194, 202, 224, 234, 244, 254, 264, 273, 282, 291, 300, 308, 318, 327, 336, 344, 352, 366, 381, 384, 392, 403
beans, boiled black 5, 14, 34, 41, 56, 65, 79, 132, 143, 157, 169, 179, 193, 203, 225, 233, 244, 254, 264, 273, 282, 290, 299, 308, 318, 327, 336, 344, 352, 366, 381, 384, 392, 403
beans, boiled black turtle 5, 14, 34, 41, 55, 65, 79, 130, 143, 156, 170, 180, 193, 201, 224, 234, 244, 254, 264, 273, 282, 290, 300, 309, 318, 327, 336, 344, 352, 366, 382, 392, 403
beans, boiled cranberry 14, 33, 41, 56, 65, 79, 130, 143, 157, 170, 179, 193, 202, 225, 233, 244, 254, 264, 273, 282, 290, 299, 308, 318, 327, 336, 344, 352, 366, 381, 384, 391, 404
beans, boiled french 5, 16, 34, 40, 56, 64, 79, 96, 130, 145, 160, 169, 179, 194, 203, 223, 234, 245, 255, 264, 273, 282, 291, 300, 309, 319, 327, 336, 345, 352, 366, 380, 392, 403
beans, boiled great northern 5, 15, 34, 40, 55, 64, 79, 96, 129, 143, 157, 170, 179, 192, 202, 224, 234, 244, 254, 264, 273, 282, 290, 299, 309, 318, 327, 336, 344, 352, 367, 381, 384, 392, 403

Index

beans, boiled green 3, 18, 32, 40, 57, 65, 80, 94, 130, 146, 160, 172, 180, 196, 205, 224, 236, 247, 256, 266, 275, 283, 292, 301, 310, 320, 329, 338, 346, 355, 369, 381, 384, 395, 399
beans, boiled hyacinth 14, 35, 41, 131, 142, 155, 169, 194, 204, 223, 231, 244, 254, 264, 273, 282, 290, 300, 309, 319, 327, 336, 344, 352, 367, 380, 392, 403
beans, boiled kidney 15, 34, 40, 56, 64, 79, 96, 132, 143, 156, 170, 179, 193, 202, 224, 233, 244, 254, 264, 273, 282, 290, 299, 308, 318, 327, 336, 344, 352, 366, 381, 384, 392, 403
beans, boiled lima 15, 34, 41, 53, 63, 79, 133, 143, 157, 170, 179, 194, 201, 224, 234, 244, 254, 264, 273, 282, 291, 300, 309, 318, 327, 336, 344, 352, 367, 381, 384, 392, 403
beans, boiled moth 5, 16, 36, 40, 96, 138, 144, 156, 169, 179, 193, 204, 235, 245, 251, 273, 282, 293, 301, 318, 328, 336, 352, 367, 380, 392, 403
beans, boiled mung 5, 15, 34, 40, 54, 65, 79, 96, 132, 144, 159, 170, 180, 195, 205, 224, 234, 244, 255, 264, 273, 282, 291, 300, 309, 319, 327, 336, 345, 353, 367, 381, 384, 392, 402
beans, boiled mungo 5, 15, 33, 39, 53, 65, 79, 96, 130, 144, 158, 169, 179, 193, 206, 223, 234, 244, 255, 264, 273, 282, 291, 300, 309, 319, 327, 336, 344, 353, 367, 380, 392, 402
beans, boiled navy 5, 15, 34, 41, 55, 63, 79, 96, 129, 143, 157, 169, 179, 193, 203, 225, 233, 244, 254, 264, 273, 282, 290, 299, 308, 318, 327, 336, 344, 352, 366, 381, 384, 392, 404
beans, boiled pink 14, 33, 40, 55, 63, 79, 130, 143, 157, 169, 179, 192, 201, 224, 234, 244, 254, 264, 273, 282, 290, 299, 308, 318, 327, 336, 344, 352, 366, 381, 384, 392, 404
beans, boiled pinto 5, 15, 32, 41, 55, 63, 79, 95, 130, 143, 156, 169, 179, 193, 201, 224, 233, 244, 254, 264, 273, 282, 290, 300, 309, 318, 327, 336, 344, 352, 366, 381, 384, 392, 404

beans, boiled soy 5, 15, 27, 41, 56, 62, 80, 95, 117, 129, 142, 154, 169, 179, 190, 201, 225, 233, 243, 252, 263, 272, 281, 290, 299, 307, 317, 325, 334, 343, 354, 365, 377, 390, 404
beans, boiled white 16, 35, 42, 56, 65, 79, 129, 143, 155, 169, 179, 194, 201, 223, 233, 244, 254, 264, 273, 282, 290, 299, 308, 318, 327, 336, 344, 352, 366, 382, 391, 404
beans, boiled winged 14, 31, 40, 66, 81, 128, 142, 155, 169, 178, 193, 205, 222, 233, 244, 254, 264, 273, 282, 290, 300, 309, 318, 326, 335, 344, 353, 366, 377, 391, 403
beans, boiled yam 21, 35, 42, 93, 133, 165, 174, 197, 208, 223, 247, 257, 266, 276, 284, 293, 302, 311, 321, 338, 347, 354, 369, 383, 395, 399
beans, boiled yardlong 5, 14, 34, 41, 54, 65, 79, 96, 131, 143, 156, 169, 179, 192, 204, 223, 233, 244, 254, 264, 273, 282, 290, 300, 308, 319, 327, 336, 345, 352, 367, 381, 392, 403
beans, boiled yellow 5, 15, 32, 40, 56, 64, 79, 95, 130, 143, 157, 169, 179, 192, 204, 223, 233, 244, 254, 264, 273, 282, 290, 299, 308, 318, 327, 336, 344, 352, 366, 379, 392, 404
beans, boiled yellow snap 4, 18, 32, 40, 57, 65, 80, 94, 130, 146, 160, 172, 180, 197, 205, 224, 236, 247, 256, 266, 275, 283, 292, 301, 310, 320, 329, 338, 346, 355, 369, 381, 384, 395, 399
beans, canned black turtle 5, 15, 31, 40, 56, 66, 79, 95, 131, 143, 158, 170, 180, 194, 204, 216, 235, 245, 255, 264, 273, 282, 291, 300, 309, 319, 327, 336, 345, 353, 368, 382, 393, 401
beans, canned broad 5, 22, 34, 40, 57, 66, 80, 95, 132, 145, 161, 171, 180, 195, 206, 216, 234, 244, 255, 264, 274, 282, 291, 301, 309, 319, 328, 336, 345, 354, 368, 382, 393, 400
beans, canned butter 4, 20, 35, 41, 131, 159, 195, 207, 216, 353, 368, 381, 393
beans, canned cranberry 21,

35, 41, 56, 66, 79, 96, 131, 144, 158, 171, 180, 195, 205, 216, 234, 245, 255, 264, 273, 282, 291, 300, 309, 319, 328, 336, 345, 353, 368, 382, 393, 401
beans, canned great northern 5, 15, 34, 41, 55, 64, 79, 96, 130, 144, 158, 170, 170, 179, 193, 203, 224, 234, 245, 254, 264, 273, 282, 291, 300, 309, 319, 327, 336, 345, 352, 367, 381, 392, 403
beans, canned green 3, 23, 33, 43, 80, 95, 132, 150, 161, 175, 180, 198, 209, 217, 236, 247, 257, 266, 275, 284, 293, 302, 311, 321, 329, 338, 347, 356, 370, 383, 384, 396, 398
beans, canned kidney 16, 33, 41, 56, 65, 80, 96, 132, 144, 160, 171, 180, 194, 206, 216, 235, 245, 255, 264, 273, 282, 291, 300, 309, 319, 328, 337, 345, 353, 368, 382, 393, 401
beans, canned lima 20, 35, 42, 55, 65, 80, 132, 143, 158, 170, 179, 195, 207, 216, 234, 245, 255, 264, 273, 282, 291, 300, 309, 319, 328, 336, 345, 353, 368, 382, 393, 401
beans, canned navy 5, 15, 34, 41, 56, 64, 79, 96, 130, 143, 158, 170, 179, 193, 205, 216, 234, 245, 254, 264, 273, 282, 291, 300, 309, 319, 327, 336, 345, 352, 367, 381, 392, 402
beans, canned pinto 5, 17, 34, 42, 56, 65, 80, 96, 131, 144, 158, 171, 180, 195, 204, 216, 234, 245, 255, 264, 274, 282, 291, 300, 309, 319, 328, 337, 345, 353, 368, 382, 393, 401
beans, canned shellie 4, 21, 33, 42, 95, 132, 161, 209, 216, 355, 370, 382, 395, 399
beans, canned sprouted mung 5, 22, 34, 42, 81, 133, 144, 164, 176, 181, 197, 211, 246, 257, 266, 275, 284, 293, 302, 311, 321, 329, 338, 346, 357, 371, 395, 398
beans, canned white 17, 35, 42, 56, 65, 79, 129, 143, 156, 170, 179, 195, 201, 223, 233, 245, 255, 264, 273, 282, 291, 300, 309, 319, 327, 336, 345, 352, 367, 382, 392, 402
beans, canned yellow snap 4, 23, 33, 43, 80, 95, 132, 150,

467   *Index*

161, 175, 180, 198, 209, 217, 236, 247, 257, 266, 275, 284, 293, 302, 311, 321, 329, 338, 347, 356, 370, 383, 384, 396, 398
beans, cooked sprouted mung 5, 20, 32, 40, 56, 93, 135, 145, 163, 175, 181, 197, 210, 223, 235, 246, 256, 265, 274, 283, 292, 301, 310, 320, 329, 338, 346, 356, 370, 382, 395, 398
beans, dried 226
beans, dry 187
beans, dry navy 139
beans, fava 78
beans, frozen and boiled fordhook lima 4, 18, 33, 39, 56, 64, 79, 93, 132, 149, 159, 170, 180, 196, 202, 220, 235, 245, 255, 264, 273, 282, 291, 300, 309, 319, 327, 336, 345, 353, 367, 381, 393, 402
beans, frozen and boiled green 3, 21, 33, 41, 57, 65, 94, 130, 148, 162, 173, 179, 197, 209, 222, 235, 247, 256, 266, 275, 284, 293, 302, 311, 320, 329, 338, 346, 355, 370, 383, 384, 395, 399
beans, frozen and boiled yellow snap 4, 21, 33, 41, 57, 65, 94, 130, 148, 162, 173, 179, 197, 209, 222, 235, 247, 256, 266, 275, 284, 293, 302, 311, 320, 329, 338, 346, 355, 370, 383, 384, 395, 398
beans, garbanzo 78, 85, 86, 89
beans, green 86, 108, 125, 139, 186, 187, 212, 227
beans, green lima 77
beans, kidney 78, 212
beans, lima 85, 103, 104
beans, navy 78, 85
beans, raw sprouted mung 5, 18, 31, 40, 54, 64, 79, 93, 134, 144, 161, 173, 181, 196, 208, 223, 236, 246, 256, 265, 274, 283, 292, 301, 309, 320, 328, 337, 346, 355, 370, 382, 384, 394, 399
beans, red 5, 18, 32, 40, 131, 159, 195, 209, 216, 353, 368, 381, 393
beans, refried 20, 33, 41, 94, 130, 142, 158, 170, 195, 202, 216, 233, 245, 255, 264, 273, 282, 291, 300, 309, 319, 327, 336, 345, 353, 367, 379, 393, 402
beans, soy 85, 86, 89, 103, 104, 107, 114, 227
beans, steamed sprouted soy 5, 14, 33, 39, 51, 94, 130, 142, 159, 169, 179, 193, 203, 222, 234, 245, 255, 264, 273, 282, 290, 300, 309, 318, 326, 336, 344, 355, 368, 378, 392, 401
beans, stir-fried sprouted mung 15, 30, 39, 93, 134, 158, 234, 245, 255, 264, 274, 283, 291, 300, 309, 320, 328, 337, 345, 354, 369, 382, 393, 400
beans, stir-fried sprouted soy 5, 13, 29, 39, 50, 93, 129, 142, 164, 169, 178, 191, 201, 232, 244, 255, 263, 272, 281, 290, 299, 308, 317, 324, 335, 344, 355, 366, 377, 391, 403
beans, United States white lima 78
beans, vegetarian baked 4, 15, 33, 41, 57, 64, 80, 130, 143, 165, 170, 179, 194, 205, 216, 233, 245, 255, 264, 274, 282, 291, 300, 309, 319, 328, 337, 345, 352, 368, 381, 384, 393, 402
beans, wax 4, 21, 35, 42, 94, 129, 156, 198, 207, 217, 356, 370, 382, 396
beans with pork and sweet sauce, baked 7, 20, 34, 48, 57, 64, 74, 80, 95, 130, 146, 158, 170, 170, 179, 194, 205, 216, 233, 245, 255, 264, 273, 282, 291, 300, 309, 319, 328, 337, 345, 352, 367, 379, 393, 402
beans with pork and tomato sauce, baked 7, 20, 34, 48, 53, 65, 74, 80, 95, 130, 143, 155, 170, 170, 179, 194, 204, 216, 229, 245, 255, 264, 274, 282, 291, 300, 309, 319, 328, 337, 345, 353, 367, 380, 384, 393, 402
bear liver, polar 6
beaver, roasted 17, 26, 363, 375, 386, 405
BEDOCE 68
BEDODEKA 68
BEDOME 23
BEDOZ 68
bedsores 9, 36, 43, 49, 96, 99, 101, 151
bee pollen 89, 99, 101
beef 85, 86, 88, 89, 101, 103, 104, 105, 107, 108, 117, 152, 186, 187
beef and pork bologna 15, 31, 47, 55, 63, 71, 84, 93, 134, 147, 159, 176, 182, 195, 208, 215, 232, 244, 254, 263, 273, 281, 290, 299, 308, 318, 327, 336, 344, 357, 361, 373, 391, 406
beef and pork bratwurst 14, 28, 47, 57, 64, 71, 92, 130, 148, 161, 175, 182, 193, 205, 215, 232, 244, 254, 263, 272, 281, 290, 298, 308, 318, 327, 335, 344, 357, 361, 373, 390, 406
beef and pork honey loaf 13, 28, 47, 51, 61, 71, 93, 133, 148, 159, 174, 182, 193, 203, 214, 231, 244, 254, 262, 272, 280, 289, 298, 308, 317, 326, 335, 344, 356, 366, 378, 390, 402
beef and pork kielbasa/ kolbassy 14, 28, 47, 51, 63, 71, 93, 131, 145, 159, 175, 182, 193, 205, 215, 232, 244, 253, 263, 272, 281, 290, 299, 308, 318, 327, 335, 344, 357, 362, 373, 391, 406
beef and pork knockwurst 13, 30, 47, 55, 63, 71, 92, 135, 148, 161, 176, 195, 207, 215, 233, 244, 254, 263, 273, 281, 290, 299, 309, 318, 327, 336, 344, 357, 362, 373, 391, 405
beef and pork link sausage 14, 30, 47, 53, 63, 71, 93, 135, 148, 159, 176, 182, 194, 207, 215, 232, 244, 254, 263, 273, 281, 290, 298, 309, 318, 327, 335, 344, 358, 361, 373, 391, 406
beef and pork link sausage with american cheese 14, 30, 47, 51, 64, 71, 93, 130, 148, 161, 176, 183, 192, 207, 215, 232, 244, 254, 263, 272, 281, 290, 298, 308, 318, 327, 335, 344, 358, 361, 373, 390, 406
beef and pork lunch meat 14, 30, 47, 52, 63, 71, 83, 93, 135, 150, 162, 175, 182, 195, 207, 214, 232, 244, 253, 263, 272, 281, 289, 299, 308, 318, 327, 335, 344, 357, 361, 372, 391, 407
beef and pork luncheon sausage 14, 28, 47, 54, 62, 71, 93, 134, 147, 159, 176, 182, 194, 206, 214, 231, 243, 252, 263, 272, 281, 289, 298, 308, 318, 326, 334, 343, 357, 363, 374, 390, 405
beef and pork mortadella 16, 31, 47, 64, 71, 92, 133, 148, 159, 175, 182, 195, 208, 214, 232, 244, 253, 262, 272, 281, 289, 298, 308, 318, 327, 335, 344, 357, 362, 373, 390, 406
beef and pork old fashioned loaf 14, 27, 47, 52, 62, 71,

Index

84, 93, 129, 148, 160, 173, 182, 192, 202, 214, 232, 244, 254, 263, 272, 281, 290, 299, 308, 318, 327, 335, 344, 355, 364, 374, 391, 404
beef and pork peppered loaf 13, 26, 47, 52, 61, 71, 84, 92, 130, 146, 161, 173, 181, 192, 202, 214, 230, 243, 252, 262, 272, 280, 289, 298, 308, 317, 326, 335, 343, 356, 366, 377, 390, 403
beef and pork pepperoni 13, 30, 45, 50, 63, 70, 133, 159, 174, 194, 204, 214, 231, 243, 253, 262, 272, 280, 289, 298, 307, 317, 326, 334, 343, 357, 360, 372, 389, 408
beef and pork picnic loaf 13, 28, 47, 51, 61, 71, 84, 93, 130, 148, 161, 175, 182, 194, 205, 214, 232, 244, 254, 263, 272, 281, 289, 298, 308, 318, 327, 335, 344, 356, 364, 375, 390, 404
beef and pork salami 14, 25, 47, 51, 62, 70, 93, 134, 143, 156, 176, 181, 194, 207, 215, 232, 244, 253, 263, 272, 281, 290, 299, 308, 318, 327, 335, 344, 357, 363, 374, 390, 404
beef and pork salami (hard) 12, 27, 45, 50, 59, 71, 92, 135, 146, 159, 174, 193, 202, 214, 230, 242, 252, 261, 271, 279, 288, 297, 307, 316, 326, 334, 342, 357, 360, 372, 388, 408
beef and pork sausage 13, 30, 47, 53, 65, 73, 160, 176, 194, 215, 232, 244, 254, 263, 272, 281, 290, 299, 308, 318, 327, 335, 344, 357, 360, 372, 390, 407
beef and pork spread 15, 31, 48, 53, 64, 71, 135, 144, 162, 177, 182, 196, 209, 215, 234, 354, 364, 374, 392, 404
beef and pork summer sausage 15, 27, 46, 52, 61, 69, 84, 93, 136, 147, 157, 176, 182, 195, 206, 214, 232, 357, 361, 373, 390, 407
beef beerwurst salami 17, 31, 47, 54, 63, 71, 84, 93, 136, 150, 159, 176, 195, 208, 215, 231, 244, 254, 263, 272, 281, 290, 299, 309, 318, 327, 335, 344, 358, 361, 373, 391, 406
beef bologna 20, 31, 47, 55, 64, 71, 84, 93, 134, 149,

158, 176, 182, 195, 208, 215, 232, 244, 254, 263, 273, 281, 290, 299, 309, 318, 327, 335, 344, 358, 362, 373, 391, 405
beef bottom round excluding fat, braised 17, 27, 46, 53, 60, 70, 81, 137, 145, 155, 172, 183, 189, 204, 221, 229, 240, 250, 260, 268, 277, 286, 295, 304, 313, 323, 331, 340, 364, 376, 385, 405
beef bottom round including fat, braised 18, 28, 46, 54, 61, 70, 81, 137, 145, 155, 173, 184, 189, 205, 220, 229, 241, 251, 260, 269, 277, 286, 295, 305, 314, 323, 332, 340, 363, 375, 386, 406
beef brains, fried 15, 27, 46, 52, 60, 69, 83, 95, 136, 143, 157, 175, 182, 188, 203, 217, 233, 244, 253, 263, 273, 281, 290, 299, 308, 318, 327, 335, 344, 365, 375, 391, 402
beef brains, simmered 17, 30, 47, 52, 62, 69, 83, 95, 136, 143, 157, 175, 182, 188, 206, 217, 233, 244, 253, 264, 273, 281, 290, 299, 308, 318, 327, 335, 344, 366, 376, 391, 402
beef brisket, cooked corned 21, 30, 47, 53, 62, 71, 136, 144, 158, 176, 183, 194, 209, 214, 229, 244, 253, 262, 272, 281, 289, 298, 308, 318, 326, 335, 343, 358, 363, 374, 389, 404
beef brisket excluding fat, braised 18, 28, 46, 54, 61, 70, 82, 137, 145, 156, 173, 173, 184, 190, 205, 219, 228, 241, 251, 260, 269, 277, 286, 295, 305, 314, 324, 332, 341, 364, 375, 386, 405
beef brisket including fat, braised 19, 30, 47, 55, 62, 70, 83, 136, 146, 157, 174, 184, 192, 206, 220, 229, 243, 252, 261, 271, 279, 288, 297, 307, 316, 325, 333, 342, 360, 372, 388, 407
beef, canned corned 22, 30, 47, 64, 71, 95, 148, 157, 175, 185, 194, 209, 215, 230, 242, 251, 261, 270, 279, 288, 296, 306, 316, 325, 333, 342, 363, 375, 387, 405
beef, chipped 136, 144, 155, 171, 192, 201, 214, 229, 240, 251, 261, 270, 278, 287,

295, 305, 314, 325, 333, 341, 358, 365, 378, 386, 405
beef, chopped smoked 18, 30, 46, 52, 60, 71, 93, 156, 173, 192, 202, 214, 230, 243, 253, 262, 272, 280, 289, 298, 307, 317, 326, 334, 343, 357, 366, 378, 389, 403
beef chuck arm pot roast excluding fat, braised 17, 27, 46, 54, 61, 70, 81, 136, 144, 155, 172, 172, 183, 189, 205, 219, 228, 240, 250, 260, 268, 277, 286, 295, 304, 313, 323, 331, 340, 364, 376, 385, 405
beef chuck arm pot roast including fat, braised 18, 28, 47, 54, 62, 70, 82, 135, 145, 156, 174, 174, 184, 191, 206, 220, 228, 242, 251, 260, 270, 278, 287, 296, 305, 315, 324, 332, 341, 361, 373, 387, 407
beef chuck blade roast excluding fat, braised 17, 27, 47, 54, 61, 70, 83, 134, 144, 155, 172, 172, 183, 190, 205, 219, 228, 240, 251, 260, 268, 277, 286, 295, 304, 313, 323, 332, 340, 363, 375, 386, 406
beef chuck blade roast including fat, braised 18, 28, 47, 55, 62, 70, 83, 134, 145, 156, 174, 174, 184, 192, 207, 219, 228, 242, 252, 261, 270, 278, 288, 296, 306, 315, 325, 333, 342, 360, 373, 387, 407
beef flank excluding fat, braised 15, 29, 46, 54, 60, 70, 82, 137, 145, 155, 172, 172, 183, 189, 203, 219, 229, 241, 251, 260, 269, 278, 287, 295, 305, 314, 324, 332, 341, 364, 375, 386, 405
beef flank excluding fat, broiled 16, 29, 45, 53, 60, 70, 82, 137, 146, 156, 172, 183, 191, 202, 218, 229, 242, 252, 261, 270, 278, 288, 296, 306, 315, 325, 333, 342, 364, 375, 387, 405
beef flank including fat, braised 15, 30, 46, 54, 61, 70, 82, 136, 145, 155, 173, 183, 189, 203, 219, 229, 241, 251, 260, 269, 278, 287, 295, 305, 315, 324, 332, 341, 363, 375, 386, 405
beef flank including fat, broiled 16, 29, 45, 53, 60,

70, 82, 137, 146, 157, 172, 183, 191, 202, 218, 229, 242, 252, 261, 270, 279, 288, 296, 306, 315, 325, 333, 342, 363, 375, 387, 405
beef frankfurter 19, 32, 47, 55, 64, 71, 84, 92, 133, 149, 159, 177, 182, 195, 208, 215, 232, 244, 254, 263, 273, 281, 290, 299, 309, 318, 327, 335, 344, 357, 361, 373, 391, 406
beef heart, simmered 15, 25, 46, 62, 69, 84, 95, 137, 142, 154, 172, 181, 190, 206, 220, 230, 241, 250, 261, 269, 277, 287, 295, 304, 313, 324, 332, 340, 358, 365, 377, 386, 404
beef honey roll sausage 17, 30, 46, 53, 62, 70, 93, 136, 147, 157, 175, 182, 193, 205, 214, 230, 357, 365, 376, 389, 403
beef kidney 186
beef kidneys, simmered 6, 15, 24, 44, 50, 59, 69, 79, 95, 133, 142, 154, 174, 181, 189, 208, 217, 230, 242, 253, 262, 271, 278, 289, 298, 304, 314, 323, 332, 340, 358, 366, 378, 387, 403
beef, lean 226
beef liver 85, 89, 107, 152, 186, 212
beef liver, braised 6, 14, 24, 44, 50, 59, 69, 79, 93, 137, 141, 154, 174, 179, 188, 206, 219, 229, 242, 250, 262, 270, 277, 289, 296, 304, 315, 323, 332, 340, 357, 365, 378, 388, 403
beef liver extract 110
beef liver, fried 6, 14, 24, 44, 50, 59, 69, 79, 93, 135, 141, 154, 173, 179, 188, 203, 217, 229, 242, 250, 261, 270, 277, 288, 296, 304, 314, 323, 331, 340, 355, 364, 377, 387, 405
beef lunch meat 16, 29, 47, 52, 63, 70, 93, 135, 146, 157, 175, 182, 194, 207, 214, 231, 243, 253, 263, 272, 281, 289, 298, 308, 317, 327, 335, 343, 357, 362, 373, 390, 406
beef lungs, braised 8, 20, 30, 47, 52, 66, 70, 82, 92, 135, 143, 154, 176, 184, 192, 208, 218, 233, 243, 251, 262, 271, 280, 289, 298, 307, 317, 326, 335, 343, 366, 378, 389, 401
beef, medium-rare baked extra lean ground 20, 28,
46, 55, 62, 71, 82, 137, 148, 157, 175, 184, 194, 207, 221, 229, 242, 253, 261, 270, 279, 288, 297, 306, 316, 324, 333, 342, 363, 375, 388, 405
beef, medium-rare baked ground 21, 30, 45, 56, 62, 70, 82, 135, 148, 157, 175, 184, 193, 207, 220, 229, 242, 253, 261, 271, 279, 288, 297, 307, 316, 325, 334, 342, 362, 374, 388, 405
beef, medium-rare baked lean ground 19, 29, 46, 55, 63, 71, 82, 136, 148, 157, 175, 184, 194, 207, 220, 229, 242, 253, 261, 271, 279, 288, 297, 306, 316, 324, 334, 342, 363, 374, 388, 405
beef, medium-rare broiled extra lean ground 19, 27, 45, 54, 62, 71, 82, 137, 148, 157, 173, 184, 193, 204, 219, 229, 241, 252, 261, 270, 278, 287, 297, 306, 316, 324, 333, 342, 363, 375, 387, 405
beef, medium-rare broiled ground 21, 29, 45, 54, 62, 70, 82, 135, 147, 157, 174, 174, 184, 192, 205, 218, 229, 242, 253, 261, 271, 279, 288, 297, 306, 316, 324, 334, 342, 362, 374, 388, 406
beef, medium-rare broiled lean ground 19, 29, 45, 54, 62, 70, 82, 135, 148, 157, 173, 184, 193, 204, 218, 229, 242, 253, 261, 270, 279, 288, 297, 306, 316, 324, 333, 342, 363, 374, 388, 405
beef, medium-rare fried extra lean ground 19, 27, 46, 55, 62, 71, 82, 137, 147, 157, 173, 184, 193, 204, 219, 229, 242, 253, 261, 270, 279, 288, 297, 306, 316, 324, 333, 342, 363, 375, 387, 405
beef, medium-rare fried ground 21, 29, 45, 54, 62, 70, 82, 135, 147, 157, 174, 184, 192, 204, 218, 229, 242, 253, 261, 271, 279, 288, 297, 306, 316, 324, 334, 342, 362, 374, 388, 406
beef, medium-rare fried lean ground 19, 28, 45, 54, 62, 70, 82, 135, 147, 157, 174, 184, 193, 204, 218, 229, 242, 253, 261, 271, 279, 288, 297, 306, 316, 324,
334, 342, 363, 374, 388, 405
beef pancreas, braised 15, 25, 46, 50, 63, 69, 93, 133, 147, 156, 173, 180, 188, 206, 220, 229, 242, 262, 269, 278, 288, 298, 305, 314, 323, 331, 340, 363, 374, 387, 405
beef pastrami 16, 30, 45, 63, 71, 94, 136, 158, 174, 193, 206, 214, 230, 244, 253, 262, 272, 281, 289, 298, 308, 318, 326, 335, 344, 357, 361, 373, 390, 407
beef rib excluding fat, broiled 17, 29, 46, 54, 60, 70, 82, 135, 146, 156, 172, 172, 184, 191, 202, 219, 228, 242, 252, 261, 270, 279, 288, 296, 306, 315, 325, 333, 342, 364, 375, 387, 405
beef rib excluding fat, roasted 17, 29, 46, 53, 61, 70, 82, 135, 146, 156, 172, 184, 191, 202, 218, 228, 241, 251, 260, 269, 278, 287, 295, 305, 315, 324, 332, 341, 364, 375, 387, 405
beef rib eye excluding fat, broiled 16, 29, 45, 54, 60, 70, 82, 134, 146, 156, 171, 184, 191, 202, 219, 228, 241, 251, 260, 269, 278, 287, 295, 305, 314, 324, 332, 341, 364, 376, 386, 405
beef rib eye including fat, broiled 17, 29, 46, 55, 60, 70, 82, 134, 146, 157, 172, 172, 184, 192, 203, 219, 229, 242, 252, 261, 270, 278, 288, 296, 306, 315, 325, 333, 342, 362, 374, 387, 406
beef rib including fat, broiled 17, 30, 47, 55, 61, 70, 83, 135, 147, 157, 174, 174, 185, 192, 204, 220, 229, 243, 253, 262, 271, 280, 289, 298, 307, 316, 326, 334, 343, 361, 373, 389, 407
beef rib including fat, roasted 18, 30, 47, 54, 62, 70, 82, 135, 146, 157, 174, 185, 192, 205, 220, 229, 243, 252, 261, 271, 279, 288, 297, 307, 316, 325, 334, 342, 360, 372, 388, 407
beef round 139
beef round excluding fat, broiled 16, 28, 46, 54, 59, 70, 81, 137, 146, 156, 171, 184, 190, 202, 219, 229, 241, 251, 260, 269, 278, 287, 295, 305, 314, 324,

Index

332, 341, 365, 377, 386, 404
beef round excluding fat, roasted eye of 17, 30, 46, 53, 60, 71, 82, 137, 146, 158, 171, 184, 190, 202, 220, 229, 241, 251, 260, 269, 277, 287, 295, 305, 314, 324, 332, 341, 365, 377, 386, 404
beef round including fat, broiled 17, 29, 46, 54, 59, 70, 82, 137, 146, 157, 172, 184, 191, 203, 220, 230, 242, 252, 260, 270, 278, 287, 296, 306, 315, 325, 333, 342, 363, 374, 387, 405
beef round including fat, roasted eye of 17, 30, 47, 53, 61, 71, 82, 137, 146, 158, 172, 184, 191, 203, 220, 230, 242, 251, 260, 270, 278, 287, 296, 305, 315, 324, 332, 341, 364, 375, 387, 405
beef round tip including fat, roasted 17, 28, 47, 53, 60, 70, 82, 137, 145, 156, 172, 184, 190, 203, 220, 229, 242, 252, 260, 270, 278, 287, 296, 305, 315, 324, 333, 341, 363, 375, 387, 405
beef salami 17, 30, 47, 51, 63, 70, 93, 136, 145, 157, 176, 181, 194, 207, 214, 232, 244, 253, 263, 272, 281, 290, 298, 308, 318, 327, 335, 344, 357, 363, 374, 390, 405
beef sausage, smoked 19, 31, 47, 64, 71, 136, 158, 175, 194, 208, 214, 231, 244, 254, 263, 272, 281, 290, 299, 308, 318, 327, 335, 344, 357, 362, 373, 390, 406
beef shank excluding fat, simmered 15, 29, 44, 53, 60, 70, 81, 131, 144, 155, 171, 183, 189, 201, 219, 228, 240, 250, 259, 268, 277, 286, 294, 304, 313, 323, 331, 340, 365, 377, 385, 405
beef shank including fat, simmered 16, 29, 45, 54, 61, 70, 82, 132, 144, 155, 171, 183, 190, 202, 220, 228, 240, 250, 260, 268, 277, 286, 295, 304, 313, 323, 331, 340, 364, 376, 385, 406
beef shortribs excluding fat, braised 18, 29, 47, 54, 62, 70, 82, 135, 146, 155, 173, 173, 183, 190, 204, 220,

228, 241, 251, 260, 269, 277, 286, 295, 304, 313, 323, 332, 340, 362, 374, 386, 407
beef shortribs including fat, braised 19, 30, 47, 55, 62, 70, 83, 135, 146, 157, 175, 185, 193, 207, 221, 229, 243, 253, 261, 271, 279, 288, 297, 307, 316, 325, 334, 342, 360, 372, 388, 408
beef sirloin excluding fat, broiled 16, 27, 46, 54, 59, 70, 81, 135, 144, 155, 171, 171, 184, 190, 202, 219, 228, 241, 251, 260, 269, 277, 286, 295, 304, 314, 323, 332, 340, 365, 377, 386, 405
beef sirloin excluding fat, fried 15, 26, 46, 53, 59, 70, 81, 135, 144, 155, 170, 183, 189, 201, 218, 229, 240, 250, 260, 268, 277, 286, 295, 304, 313, 323, 331, 340, 364, 376, 385, 405
beef sirloin including fat, broiled 16, 27, 46, 54, 60, 70, 82, 135, 144, 156, 171, 171, 184, 191, 203, 220, 229, 241, 251, 260, 269, 278, 287, 295, 305, 315, 324, 332, 341, 362, 374, 387, 406
beef sirloin including fat, fried 16, 27, 46, 54, 60, 70, 82, 135, 144, 155, 171, 184, 190, 202, 219, 229, 241, 251, 260, 269, 278, 287, 295, 305, 315, 324, 332, 341, 361, 373, 386, 407
beef spleen, braised 19, 27, 45, 66, 69, 92, 135, 141, 153, 174, 181, 189, 205, 220, 231, 243, 249, 260, 271, 278, 288, 298, 306, 316, 325, 334, 340, 366, 378, 387, 402
beef suet 211, 246, 266, 276, 284, 293, 302, 311, 321, 338, 347, 359, 371, 395, 410
beef summer sausage 15, 27, 46, 62, 69, 84, 93, 136, 147, 157, 176, 194, 206, 214, 232, 357, 362, 373, 390, 406
beef tallow 371
beef tenderloin excluding fat, broiled 16, 27, 46, 54, 59, 70, 82, 137, 143, 155, 171, 171, 184, 190, 202, 220, 229, 241, 251, 260, 269, 278, 287, 295, 305, 314, 324, 332, 341, 365, 377, 386, 404
beef tenderloin excluding fat, roasted 16, 26, 47, 53, 60,

70, 82, 137, 144, 155, 171, 184, 190, 202, 220, 229, 241, 251, 260, 269, 278, 287, 295, 305, 315, 324, 332, 341, 364, 376, 386, 404
beef tenderloin including fat, broiled 16, 27, 47, 54, 60, 70, 82, 136, 144, 155, 171, 171, 184, 191, 202, 220, 229, 242, 252, 260, 270, 278, 287, 296, 306, 315, 324, 333, 341, 363, 374, 387, 406
beef tenderloin including fat, roasted 17, 27, 47, 54, 61, 70, 82, 136, 144, 155, 173, 185, 191, 203, 220, 230, 242, 252, 261, 270, 279, 288, 296, 306, 316, 325, 333, 342, 362, 374, 388, 406
beef thymus, braised 71, 92, 159, 188, 202, 217, 243, 263, 272, 280, 288, 299, 308, 317, 326, 340, 361, 373, 388, 406
beef tip excluding fat, roasted 16, 27, 46, 53, 60, 70, 82, 137, 145, 156, 171, 183, 190, 202, 219, 228, 241, 251, 260, 269, 277, 287, 295, 305, 314, 324, 332, 341, 365, 377, 386, 404
beef tongue, simmered 21, 26, 47, 53, 63, 69, 83, 95, 137, 143, 155, 175, 182, 193, 208, 220, 229, 243, 252, 262, 271, 280, 289, 298, 306, 316, 326, 334, 342, 358, 362, 374, 388, 405
beef top loin excluding fat, broiled 17, 29, 45, 54, 60, 71, 82, 136, 146, 157, 171, 184, 191, 202, 219, 229, 241, 251, 260, 269, 278, 287, 295, 305, 314, 324, 332, 341, 365, 377, 386, 404
beef top loin including fat, broiled 17, 30, 45, 54, 60, 71, 82, 136, 146, 157, 172, 184, 192, 203, 220, 229, 242, 252, 260, 270, 278, 287, 296, 306, 315, 325, 333, 342, 362, 374, 387, 406
beef top round excluding fat, broiled 16, 27, 44, 53, 59, 70, 81, 137, 145, 156, 171, 171, 184, 190, 201, 220, 229, 240, 250, 260, 268, 277, 286, 295, 304, 313, 323, 331, 340, 365, 377, 385, 404
beef top round excluding fat, fried 16, 27, 45, 53, 59, 70, 81, 137, 145, 155, 170, 183,

188, 201, 219, 229, 240, 250, 259, 268, 277, 286, 294, 304, 313, 322, 331, 340, 364, 377, 385, 405
beef top round including fat, broiled 16, 27, 44, 53, 59, 70, 81, 137, 145, 156, 171, 184, 190, 202, 220, 229, 241, 251, 260, 269, 277, 286, 295, 304, 313, 323, 332, 340, 365, 377, 386, 405
beef top round including fat, fried 16, 28, 45, 53, 59, 70, 81, 137, 145, 156, 171, 183, 189, 201, 219, 230, 240, 250, 260, 268, 277, 286, 295, 304, 313, 323, 331, 340, 362, 374, 385, 407
beef tripe, raw 22, 30, 48, 71, 84, 95, 147, 158, 176, 195, 205, 221, 231, 244, 254, 263, 272, 281, 290, 299, 308, 318, 327, 335, 344, 367, 378, 390, 400
beef, well done baked extra lean ground 19, 26, 45, 54, 61, 71, 81, 136, 146, 156, 173, 183, 193, 205, 219, 228, 240, 252, 260, 269, 277, 286, 295, 305, 314, 323, 332, 340, 363, 375, 386, 406
beef, well done baked ground 20, 29, 45, 55, 61, 70, 81, 134, 147, 156, 174, 183, 192, 205, 218, 229, 241, 252, 260, 269, 277, 287, 296, 305, 315, 323, 332, 341, 361, 374, 386, 407
beef, well done baked lean ground 18, 28, 45, 54, 62, 70, 81, 134, 146, 156, 173, 183, 192, 205, 219, 228, 240, 252, 260, 269, 277, 286, 296, 305, 314, 323, 332, 340, 362, 374, 386, 406
beef, well done broiled extra lean ground 18, 26, 44, 53, 61, 70, 81, 136, 147, 156, 172, 183, 192, 203, 218, 228, 241, 252, 260, 269, 277, 287, 296, 305, 315, 323, 332, 341, 363, 375, 386, 406
beef, well done broiled ground 20, 29, 44, 54, 61, 70, 81, 134, 146, 156, 173, 183, 192, 204, 218, 229, 241, 252, 261, 270, 278, 287, 296, 306, 315, 323, 333, 341, 362, 374, 387, 406
beef, well done broiled lean ground 19, 28, 44, 53, 61, 70, 81, 134, 147, 157, 172, 184, 192, 203, 218, 229, 241, 252, 260, 270, 278, 287, 296, 305, 315, 323, 333, 341, 362, 374, 386, 406
beef, well done fried extra lean ground 18, 27, 45, 55, 61, 70, 81, 136, 146, 156, 172, 183, 192, 203, 218, 229, 241, 252, 260, 270, 278, 287, 296, 305, 315, 323, 333, 341, 363, 375, 386, 406
beef, well done fried ground 20, 29, 44, 54, 62, 70, 81, 134, 147, 156, 173, 183, 192, 204, 218, 229, 241, 252, 261, 270, 278, 287, 296, 306, 315, 323, 333, 341, 362, 374, 387, 406
beef, well done fried lean ground 19, 28, 45, 54, 61, 70, 81, 135, 147, 157, 173, 184, 192, 204, 218, 229, 241, 252, 260, 270, 278, 287, 296, 305, 315, 323, 333, 341, 362, 374, 386, 406
beer 22, 35, 42, 57, 66, 74, 137, 140, 151, 166, 177, 185, 198, 211, 212, 213, 223, 238, 356, 369, 397, 398
beerwurst saiami, beef 17, 31, 47, 54, 63, 71, 84, 93, 136, 150, 159, 176, 195, 208, 215, 231, 244, 254, 263, 272, 281, 290, 299, 309, 318, 327, 335, 344, 358, 361, 373, 391, 406
beerwurst salami, pork 12, 30, 47, 53, 64, 72, 84, 92, 136, 149, 162, 176, 182, 194, 206, 214, 232, 244, 254, 263, 273, 281, 290, 298, 309, 318, 327, 335, 344, 357, 364, 374, 390, 404
beet greens, boiled 2, 16, 27, 41, 54, 63, 92, 128, 143, 158, 169, 196, 200, 217, 235, 247, 256, 265, 275, 283, 293, 301, 310, 320, 328, 337, 346, 356, 370, 383, 394, 399
beet pulp, sugar 213
beet tops 78
beets 187, 227
beets, boiled 5, 21, 36, 42, 57, 66, 80, 94, 135, 149, 163, 170, 180, 197, 204, 221, 236, 247, 256, 267, 276, 284, 293, 302, 311, 321, 329, 338, 347, 355, 370, 396, 399
beets, pickled 5, 21, 34, 42, 95, 135, 145, 164, 175, 198, 208, 217, 236, 248, 257, 267, 276, 285, 294, 302, 312, 321, 330, 338, 347, 353, 368, 383, 396, 400

BEFLAVINE 24
BEGIOLAN 23
behavioral changes 303
BEHEPAN 68
Bell's palsy 23, 67
BENERVA 23
BENICOT 43
BENZOIC ACID 106
BEQUIN 23
beriberi 11, 96
BERIN 23
berliner 13, 28, 47, 62, 70, 94, 134, 147, 160, 175, 182, 193, 205, 214, 231, 244, 253, 262, 272, 281, 289, 298, 308, 318, 326, 335, 344, 357, 364, 374, 390, 404
BERNACAINE 106
BERUBI 68
BERUBIGEN 68
BERYLLIUM 125
beryllium in organs 125
BETABION HYDROCHLORIDE 23
BETABION MONONITRATE 23
BETA-CAROTENE 1
BETALIN S 23
BETALIN-12 68
BETARIN 76
BETATRON 24
BETAXIN 23
BETHIAZINE 23
BETA-TOCOPHEROL 100, 102
BEVATINE-12 68
BEVIDOX 68
BETA-VITAMIN E 102
BEVITEX 23
BEWON 23
BEXII 68
BEXIL 68
bile salts formation deterioration 49, 313
BILETAN 112
BILINEURINE 86
bilirubin in blood, increased 107
BINOVA 24
BIOCOBALAMINE 68
BIOCOLINA 87
BIOCOLORIN 113
BIOCRES 68
BIOEPIDERM 89
BIOFLAVINOIDS 112
BIOS I 85
BIOS II 89
BIOSTEROL 6
BIOTIN 89
BIOTIN SULFOXIDE 89, 90
BIRTH CONTROL PILLS 228
birth defects 6, 167
BIRUTAN 113
BISCOLAN 87
biscuits, enriched 14, 29, 39, 128, 159, 192, 210, 215, 351, 360, 374, 392, 408
biscuits, unenriched 21, 32, 41, 128, 164, 192, 209, 215, 351, 360, 374, 392, 408

Index 472

BITEVAN 68
Bitiot's spots 9, 99
bitter almonds 77, 78
BIUNO 23
BIVATIN 23
BIVITA 23
black beans 78
black beans, boiled 5, 14, 34, 41, 56, 65, 79, 132, 143, 157, 169, 179, 193, 203, 225, 233, 244, 254, 264, 273, 282, 290, 299, 308, 318, 327, 336, 344, 352, 366, 381, 384, 392, 403
blackberries 4, 22, 34, 42, 56, 65, 93, 112, 131, 144, 163, 174, 178, 197, 207, 236, 354, 369, 381, 396, 400
blackberry, commercial 78
blackberry pie 4, 22, 36, 42, 95, 133, 164, 197, 210, 217, 351, 364, 376, 394, 406
blackberry, wild 78
black-eyed peas 78, 85, 86, 89
black-eyed peas, boiled 5, 14, 34, 41, 54, 64, 79, 96, 132, 143, 156, 170, 179, 193, 205, 224, 233, 244, 254, 264, 273, 282, 291, 300, 309, 319, 327, 336, 345, 352, 367, 381, 384, 392, 402
black-eyed peas, boiled catjang 5, 15, 34, 40, 54, 65, 79, 96, 132, 143, 156, 169, 169, 179, 193, 202, 222, 232, 244, 254, 264, 273, 282, 291, 300, 308, 319, 327, 336, 345, 352, 367, 380, 392, 403
black-eyed peas, canned 5, 18, 32, 41, 56, 66, 80, 95, 133, 145, 161, 171, 180, 195, 208, 216, 234, 245, 255, 264, 274, 282, 291, 300, 309, 319, 328, 336, 345, 354, 368, 380, 393, 401
black-eyed peas, frozen and boiled 4, 14, 34, 40, 56, 65, 79, 95, 132, 143, 157, 170, 179, 194, 202, 223, 233, 244, 254, 264, 273, 282, 290, 300, 308, 319, 327, 336, 344, 352, 366, 380, 392, 403
black-eyed pea pods, boiled 2, 17, 32, 40, 93, 130, 163, 196, 207, 224, 355, 369, 382, 394, 399
bladder inflammation 9, 58, 67, 96, 99, 101
bladder/kidney stones 249
blade roll, roasted pork 13, 27, 47, 51, 62, 71, 84, 136, 147, 162, 176, 183, 193, 207, 215, 231, 244, 252, 262, 272, 281, 289, 298, 307, 317, 326, 335, 344, 358, 362, 373, 390, 405
bleeding, increased 100
blindness, night 6, 9, 23, 36, 43, 49
bloating 384
blood abnormalities 187
blood calcium, diminished 127
blood calcium, increased 98, 127
blood cell breakdown 107
blood cell rupture 100, 141
blood cells, aggregation of 98, 112
blood cells, large abnormal 69, 76, 78
blood cholesterol, decreased 178
blood cholesterol, increased 76, 88, 139, 152, 187, 227
blood clots 114
blood clots, decreased clearance of 322
blood clots, predisposition for 98, 112
blood clotting 117
blood clotting, decreased 69, 107, 114, 118, 178
blood, deterioration of 68
blood fats, increased slightly 100
blood in chicks, deterioration of 68
blood iron level, deceptively increased 69
blood keytones, increased 88
blood lead, increased 98
blood levels of calcium, decreased 168
blood levels of hemoglobin, decreased 88
blood lipids, increased 88
blood oxygen, decreased 113
blood oxygenation, decreased 76
blood phosphorus, increased 187
blood platelets production, decreased 118
blood pressure, decreased 11, 49, 78, 186, 199
blood pressure, increased 77, 86, 98, 105, 106, 107, 117, 126, 139, 141, 152, 168, 186, 214, 303
blood pressure, unregulated 77
blood pyruvic acid, increased 11
blood sugar, decreased 37, 43, 49, 58, 67, 74, 86, 96, 100, 199
blood sugar, increased 313
blood T-cell count, decreased 104, 371
blood, thinner 100
blood triglycerides in women, increased 100
blood triglycerides, increased 88
blood vessels, fragile 98, 112
bloody mary 3, 22, 36, 42, 56, 65, 81, 93, 136, 148, 164, 177, 181, 198, 209, 217, 237, 357, 368, 383, 397, 400
blueberries 4, 20, 34, 42, 57, 66, 83, 93, 137, 139, 148, 166, 177, 180, 198, 210, 223, 237, 247, 257, 267, 276, 285, 294, 302, 311, 321, 330, 339, 347, 354, 369, 381, 396, 400
blueberry muffins 4, 15, 29, 39, 129, 159, 193, 209, 215, 351, 362, 377, 392, 408
blueberry pie 5, 22, 36, 42, 95, 135, 163, 197, 210, 217, 351, 364, 376, 394, 406
bluefish, raw 7, 19, 32, 44, 51, 60, 69, 84, 137, 149, 164, 170, 183, 190, 203, 220, 234, 243, 253, 262, 271, 280, 288, 297, 307, 317, 326, 334, 343, 366, 378, 389, 402
bodily weakness 153
body building 248
body fluid imbalances 167
body odor 69
body odor, garlic 124, 212
boils 9, 96, 101
bologna, beef 20, 31, 47, 55, 64, 71, 84, 93, 134, 149, 158, 176, 182, 195, 208, 215, 232, 244, 254, 263, 273, 281, 290, 299, 309, 318, 327, 335, 344, 358, 362, 373, 391, 405
bologna, lebanon 20, 30, 46, 53, 62, 70, 84, 93, 134, 147, 157, 175, 181, 193, 204, 214, 230, 243, 253, 262, 272, 280, 289, 298, 307, 317, 327, 335, 343, 357, 365, 375, 389, 404
bologna, pork 13, 30, 46, 51, 62, 72, 84, 92, 134, 148, 162, 175, 182, 193, 205, 214, 232, 244, 254, 263, 272, 281, 289, 298, 308, 318, 327, 335, 344, 358, 363, 374, 390, 404
bologna, pork and beef 15, 31, 47, 55, 63, 71, 84, 93, 134, 147, 159, 176, 182, 195, 208, 215, 232, 244, 254, 263, 273, 281, 290, 299, 308, 318, 327, 336, 344, 357, 361, 373, 391, 406
bologna, turkey 20, 29, 46, 63, 71, 129, 150, 160, 175, 193, 207, 215, 232, 357, 364, 375, 390, 403

bone 213
bone and teeth mineralization, deterioration of 227
bone cancers 225
bone deformities 168, 178
bone demineralization 6, 98, 125, 187, 213, 227
bone deterioration 125, 152
bone development, decreased 228
bone development, faulty 141
bone disease 141, 186
bone growth, abnormal 227
bone growth, deterioration of 213
bone growth in children, abnormal 98
bone growth of spine, projecting 151
bone marrow, deterioration of red blood cell production in 76
bone meal 151
bone muscle 168
bone thickening 6
bones, aching 6
bones and fingertips, numb/tingling 98
bones, broken 213, 228
bones, deformed 213
bones, depletion of 225
bones in children, weak 91
bones, softening of 9, 96, 98, 99, 127, 138, 199
bones, thinning of 74, 96, 99, 101, 127, 138, 151, 152, 177, 199, 225
bonito, canned 22, 32, 44, 136, 161, 171, 192, 204, 216, 363, 374, 389
BORON 125, 127, 168, 187
boston blade excluding fat, braised pork 8, 13, 25, 46, 52, 62, 72, 83, 136, 144, 157, 173, 184, 191, 202, 218, 229, 240, 250, 259, 268, 277, 286, 295, 304, 313, 322, 331, 340, 362, 374, 385, 407
boston blade excluding fat, broiled pork 9, 12, 25, 46, 51, 61, 71, 83, 137, 145, 159, 171, 185, 190, 202, 218, 230, 241, 251, 259, 270, 278, 286, 296, 306, 315, 323, 332, 341, 363, 374, 387, 405
boston blade excluding fat, roasted pork 9, 12, 26, 46, 52, 61, 72, 83, 136, 145, 158, 174, 185, 190, 203, 219, 230, 242, 251, 259, 270, 279, 287, 297, 306, 315, 324, 333, 341, 363, 375, 388, 405
boston blade including fat, braised pork 8, 13, 26, 46, 57, 62, 72, 84, 137, 144,

158, 174, 185, 192, 203, 219, 229, 241, 251, 259, 269, 278, 286, 296, 305, 314, 323, 332, 341, 360, 373, 387, 407
boston blade including fat, broiled pork 8, 12, 26, 46, 51, 62, 72, 84, 137, 145, 160, 172, 185, 191, 203, 218, 230, 242, 252, 260, 271, 279, 287, 298, 307, 316, 325, 334, 342, 361, 373, 388, 407
boston blade including fat, roasted pork 9, 13, 26, 46, 53, 62, 72, 84, 137, 145, 159, 174, 185, 192, 204, 219, 230, 242, 252, 260, 271, 279, 287, 298, 307, 316, 325, 334, 342, 361, 373, 388, 406
boston cream pie 7, 21, 32, 48, 130, 164, 194, 210, 217, 351, 362, 377, 393, 408
bottom round excluding fat, braised beef 17, 27, 46, 53, 60, 70, 81, 137, 145, 155, 172, 183, 189, 204, 221, 229, 240, 250, 260, 268, 277, 286, 295, 304, 313, 323, 331, 340, 364, 376, 385, 405
bottom round including fat, braised beef 18, 28, 46, 54, 61, 70, 81, 137, 145, 155, 173, 184, 189, 205, 220, 229, 241, 251, 260, 269, 277, 286, 295, 305, 314, 323, 332, 340, 363, 375, 386, 406
bourbon & soda 138, 177, 199, 211, 222, 238, 368, 400
boysenberries 4, 20, 35, 40, 55, 66, 78, 79, 95, 132, 147, 162, 175, 179, 197, 209, 224, 237, 354, 369, 381, 396, 400
BR 126
BR⁻ 126
brain 112
brain arteries, hardening of the 106
brain deterioration 11, 69, 126
brain dysfunction, minimal 105
brain protein supply, decreased 58
brain waves, abnormal 58
brain zinc supply, decreased 141
brains, braised pork 17, 28, 47, 50, 63, 71, 93, 136, 143, 158, 176, 181, 190, 207, 218, 233, 244, 263, 272, 281, 290, 299, 308, 318, 327, 335, 344, 366, 377, 391, 401
brains, fried beef 15, 27, 46,

52, 60, 69, 83, 95, 136, 143, 157, 175, 182, 188, 203, 217, 233, 244, 253, 263, 273, 281, 290, 299, 308, 318, 327, 335, 344, 365, 375, 391, 402
brains, simmered beef 17, 30, 47, 52, 62, 69, 83, 95, 136, 143, 157, 175, 182, 188, 206, 217, 233, 244, 253, 264, 273, 281, 290, 299, 308, 318, 327, 335, 344, 366, 376, 391, 402
bran 2, 11, 25, 37, 50, 59, 78, 86, 89, 90, 92, 101, 107, 129, 141, 154, 168, 188, 200, 212, 214, 228, 244, 252, 263, 273, 282, 291, 299, 308, 318, 326, 336, 344, 350, 363, 379, 384, 390, 410
bran flakes 2, 11, 25, 37, 59, 69, 78, 130, 142, 154, 168, 188, 201, 215, 228, 244, 253, 263, 273, 282, 291, 299, 308, 318, 326, 336, 344, 349, 361, 379, 384, 391, 410
bran muffins 3, 14, 27, 38, 52, 61, 74, 80, 94, 128, 143, 156, 169, 189, 206, 216, 231, 351, 362, 376, 392, 408
bratwurst, pork 13, 29, 47, 55, 62, 72, 95, 131, 146, 160, 175, 182, 193, 207, 216, 231, 244, 254, 263, 273, 281, 290, 299, 308, 318, 327, 335, 344, 357, 362, 373, 390, 405
bratwurst, pork and beef 14, 28, 47, 57, 64, 71, 92, 130, 148, 161, 175, 182, 193, 205, 215, 232, 244, 254, 263, 272, 281, 290, 298, 308, 318, 327, 335, 344, 357, 361, 373, 390, 406
brazilnuts 101, 103, 104, 212
brazilnuts, dried 11, 31, 39, 55, 62, 84, 128, 141, 155, 168, 179, 188, 201, 229, 240, 250, 263, 272, 281, 291, 294, 307, 318, 325, 335, 343, 354, 359, 372, 390, 410
bread, cracked wheat 13, 25, 38, 52, 65, 130, 156, 170, 194, 209, 216, 350, 363, 378, 384, 392, 408
bread, dark rye 101
bread, enriched wheat 13, 27, 38, 133, 156, 195, 210, 224, 350, 360, 380, 391, 409
bread, french 13, 28, 38, 53, 66, 80, 129, 143, 155, 173, 179, 194, 209, 216, 234, 350, 362, 378, 384, 391, 408
bread, italian 13, 29, 38, 54,

Index

66, 80, 130, 143, 156, 172, 179, 195, 209, 216, 234, 350, 362, 379, 384, 391, 408
bread, mixed grain 13, 25, 38, 52, 64, 79, 129, 143, 155, 170, 178, 191, 207, 216, 233, 351, 363, 378, 384, 391, 408
bread, oatmeal 13, 29, 38, 54, 66, 80, 130, 143, 156, 170, 179, 194, 208, 216, 234, 351, 363, 378, 384, 392, 408
bread, pita 13, 27, 38, 53, 66, 80, 129, 143, 157, 172, 179, 195, 209, 216, 234, 350, 362, 379, 384, 391, 408
bread pudding 7, 18, 29, 48, 96, 129, 160, 194, 207, 217, 351, 365, 377, 393, 405
bread, pumpernickel 13, 25, 38, 53, 63, 129, 156, 169, 190, 201, 216, 233, 351, 363, 379, 392, 408
bread, raisin 14, 25, 38, 53, 66, 80, 129, 144, 156, 175, 179, 195, 206, 216, 234, 350, 362, 378, 384, 392, 408
bread, rye 13, 26, 38, 53, 65, 80, 129, 139, 146, 156, 172, 186, 187, 193, 207, 215, 233, 351, 363, 378, 392, 408
bread sticks 19, 34, 40, 132, 161, 195, 210, 215, 350, 360, 379, 391, 409
bread stuffing 3, 19, 32, 40, 131, 161, 196, 210, 216, 353, 365, 375, 393, 404
bread, toasted cracked wheat 13, 25, 38, 51, 64, 129, 142, 156, 170, 178, 193, 208, 216, 232, 350, 362, 378, 384, 391, 408
bread, toasted pumpernickel 14, 25, 38, 53, 63, 129, 156, 169, 190, 201, 215, 233, 350, 362, 378, 391, 408
bread, toasted raisin 14, 25, 38, 53, 65, 80, 128, 144, 155, 171, 194, 205, 216, 234, 350, 362, 378, 392, 408
bread, toasted rye 13, 26, 38, 53, 64, 80, 129, 145, 156, 171, 192, 206, 215, 233, 350, 362, 378, 391, 408
bread, toasted white 15, 26, 38, 53, 66, 80, 128, 144, 155, 172, 194, 209, 215, 234, 350, 362, 378, 391, 408
bread, toasted whole wheat 13, 28, 38, 51, 63, 79, 129, 142, 155, 169, 189,

207, 215, 232, 350, 362, 378, 391, 408
bread, white 13, 27, 38, 53, 66, 80, 125, 128, 144, 156, 173, 194, 209, 216, 234, 351, 363, 378, 392, 408
bread, whole wheat 13, 29, 38, 51, 63, 80, 85, 101. 129, 139, 142, 152, 155, 169, 178, 186, 187, 189, 208, 212, 215, 232, 351, 364, 378, 384, 391, 408
breadcrumbs, dry 14, 27, 38, 128, 155, 193, 208, 215, 350, 360, 378, 391, 409
breadfruit 5, 16, 35, 40, 53, 92, 133, 147, 163, 172, 181, 197, 201, 224, 237, 257, 275, 284, 293, 302, 311, 321, 338, 347, 352, 367, 382, 396, 402
breast cancer 105
breast, chicken 125
breast plate, braised/stewed veal 19, 28, 46, 135, 155, 193, 207, 221, 362, 374, 387, 406
breast with skin, fried chicken 8, 17, 31, 44, 51, 59, 73, 83, 133, 149, 160, 171, 184, 190, 206, 218, 233, 241, 250, 260, 268, 277, 286, 294, 304, 314, 323, 331, 340, 358, 364, 377, 385, 405
breast with skin, roasted chicken 8, 18, 31, 44, 51, 59, 73, 84, 134, 149, 161, 171, 183, 191, 206, 219, 234, 241, 250, 260, 268, 278, 286, 295, 304, 314, 323, 332, 340, 365, 377, 386, 404
breast with skin, stewed chicken 8, 19, 31, 44, 52, 61, 74, 84, 134, 149, 161, 173, 183, 193, 208, 220, 234, 242, 250, 261, 268, 278, 287, 295, 305, 315, 324, 332, 341, 365, 377, 387, 403
breast without skin, fried chicken 8, 17, 31, 44, 50, 59, 73, 83, 133, 149, 160, 171, 183, 190, 205, 218, 233, 240, 250, 260, 268, 277, 286, 294, 304, 313, 323, 331, 340, 358, 365, 378, 385, 404
breast without skin, roasted chicken 8, 18, 31, 44, 51, 59, 73, 84, 133, 149, 161, 171, 184, 190, 206, 219, 234, 241, 250, 260, 268, 277, 286, 294, 304, 314, 323, 331, 340, 365, 378, 386, 403
breast without skin, stewed chicken 8, 20, 31, 44, 52,

61, 74, 84, 134, 149, 162, 172, 183, 192, 207, 220, 234, 241, 250, 260, 268, 278, 286, 295, 305, 314, 323, 332, 340, 366, 379, 386, 403
breasts, sore 127, 187
breath, fetid 126
breath, garlicky 124
breath, shortness of 76, 91, 153, 187
breathing deterioration 167
breathing, difficult 125, 139, 141
breathing, labored 11
brewer's yeast 11, 24, 44, 85, 86, 89, 90, 127, 139, 154, 186, 188, 200, 217, 351, 362, 380, 385, 409
brisket, cooked corned beef 21, 30, 47, 53, 62, 71, 136, 144, 158, 176, 183, 194, 209, 214, 229, 244, 253, 262, 272, 281, 289, 298, 308, 318, 326, 335, 343, 358, 363, 374, 389, 404
brisket excluding fat, braised beef 18, 28, 46, 54, 61, 70, 82, 137, 145, 156, 173, 173, 184, 190, 205, 219, 228, 241, 251, 260, 269, 277, 286, 295, 305, 314, 324, 332, 341, 364, 375, 386, 405
brisket including fat, braised beef 19, 30, 47, 55, 62, 70, 83, 136, 146, 157, 174, 184, 192, 206, 220, 229, 243, 252, 261, 271, 279, 288, 297, 307, 316, 325, 333, 342, 360, 372, 388, 407
broadbeans, boiled 5, 16, 32, 40, 56, 65, 79, 96, 131, 143, 159, 170, 179, 194, 205, 223, 234, 244, 254, 264, 273, 282, 291, 300, 309, 319, 327, 336, 345, 353, 367, 381, 384, 392, 402
broadbeans, canned 5, 22, 34, 40, 57, 66, 80, 95, 132, 145, 161, 171, 180, 195, 206, 216, 234, 244, 255, 264, 274, 282, 291, 301, 309, 319, 328, 336, 345, 354, 368, 382, 393, 400
broccoli 101, 107, 112, 116, 117, 125
broccoli, boiled 2, 17, 29, 40, 55, 63, 79, 92, 128, 148, 160, 169, 180, 196, 208, 222, 237, 246, 256, 265, 274, 283, 292, 301, 310, 320, 328, 337, 346, 356, 370, 381, 394, 399
broccoli, frozen and boiled 2, 20, 32, 41, 55, 64, 80, 92, 130, 149, 163, 173, 180, 196, 208, 222, 236, 246, 256,

Index

265, 274, 283, 292, 301, 310, 320, 328, 337, 346, 356, 370, 383, 384, 394, 399
broccoli, raw 2, 18, 32, 40, 52, 63, 79, 91, 130, 149, 162, 172, 180, 196, 204, 222, 236, 246, 256, 265, 274, 283, 292, 301, 310, 320, 328, 337, 346, 356, 370, 380, 384, 394, 399
BROMIN 126
BROMINE 126
bronchitis 9, 96, 99, 101, 105
brown sugar 22, 35, 42, 129, 155, 198, 203, 221, 349, 360, 410
brownies 7, 15, 32, 48, 131, 158, 193, 208, 217, 350, 360, 374, 393, 409
bruising 85, 91, 96, 98, 101, 108, 112, 117, 166, 167
brussels sprouts 107, 226
brussels sprouts, boiled 3, 17, 32, 40, 55, 63, 79, 92, 131, 147, 160, 173, 180, 196, 204, 222, 236, 246, 256, 265, 274, 283, 292, 301, 310, 320, 329, 346, 355, 369, 380, 384, 394, 399
brussels sprouts, frozen and boiled 3, 16, 31, 41, 54, 61, 79, 92, 132, 147, 162, 172, 180, 196, 204, 222, 236, 245, 256, 265, 274, 283, 292, 301, 310, 320, 328, 346, 355, 369, 381, 384, 393, 400
B-TWELV 68
buckwheat 78, 116, 117, 187, 227
buckwheat flour, dark 12, 30, 38, 131, 156, 189, 350, 361, 379, 391, 409
buckwheat flour, light 17, 35, 41, 135, 161, 195, 204, 349, 361, 379, 392, 409
buckwheat products 213
buffalo milk, indian 7, 19, 31, 48, 56, 66, 73, 83, 95, 128, 166, 171, 194, 208, 220, 237, 246, 255, 265, 274, 282, 291, 300, 309, 319, 328, 336, 345, 356, 367, 377, 393, 400
bulgur 20, 35, 38, 133, 159, 191, 210, 215, 351, 365, 380, 393, 405
bullhead, raw black 272, 281, 289, 298, 308, 317, 326, 343, 368, 379, 390, 400
bun, hamburger 13, 26, 38, 53, 66, 80, 128, 144, 156, 173, 195, 210, 215, 234, 350, 362, 378, 392, 408
bun, hot dog 13, 26, 38, 53, 66, 80, 128, 144, 156, 173, 195, 210, 215, 234, 350, 362, 378, 392, 408

bun, raisin 19, 32, 40, 129, 159, 195, 206, 216, 350, 363, 379, 392, 408
burbot, raw 130, 143, 161, 170, 179, 191, 202, 218, 234, 243, 253, 262, 271, 280, 289, 297, 307, 317, 326, 334, 343, 368, 380, 389, 401
burdock, edible 116
burdock root, boiled 21, 33, 42, 130, 162, 170, 195, 203, 224, 246, 257, 265, 275, 284, 292, 302, 311, 321, 330, 338, 347, 352, 368, 382, 394, 401
burning and tingling sensations 58
burning sensation 37, 43
burns 9, 90, 96, 99, 101, 211, 238
burrito 7, 14, 30, 47, 56, 64, 73, 81, 96, 129, 143, 158, 170, 193, 205, 215, 233, 351, 364, 377, 392, 406
BURSINE 86
bursitis 74, 96, 101
butter 6, 33, 86, 89, 99, 101, 103, 104, 107, 117, 133, 139, 152, 165, 186, 198, 211, 212, 215, 247, 266, 275, 284, 293, 301, 311, 321, 329, 338, 347, 359, 371, 396, 408
butter beans, canned 4, 20, 35, 41, 131, 159, 195, 207, 216, 353, 368, 381, 393
butter, sweet 6, 33, 133, 165, 198, 211, 222, 359, 371, 396, 408
butter, whipped 6, 132, 165, 197, 211, 215, 247, 266, 275, 284, 293, 301, 311, 321, 337, 347, 359, 371, 396, 408
butterbur, boiled 5, 23, 36, 43, 93, 130, 166, 176, 199, 203, 224, 357, 371, 397, 398
butterfish, raw 149, 164, 184, 202, 218, 234, 244, 254, 263, 272, 280, 289, 298, 308, 317, 326, 335, 343, 366, 377, 390, 402
buttermilk 8, 21, 30, 55, 66, 74, 96, 128, 166, 176, 195, 208, 217, 235, 246, 255, 265, 274, 282, 291, 300, 309, 319, 328, 337, 345, 356, 369, 380, 394, 399
buttermilk, dry 7, 13, 25, 48, 50, 61, 70, 80, 127, 165, 169, 188, 200, 216, 230, 243, 251, 261, 268, 277, 286, 295, 304, 313, 322, 331, 339, 351, 361, 377, 385, 410
butternuts, dried 130, 142,

155, 168, 178, 188, 202, 230, 240, 250, 261, 270, 278, 290, 296, 304, 316, 323, 332, 340, 354, 359, 372, 387, 410
butterscotch pie 3, 21, 32, 42, 129, 162, 195, 210, 217, 351, 363, 376, 393, 407
BUTYRYLCHOLINESTERASE 88
BYLADOCE 68

C-VIMIN 91, 97
C.V.P. 112 CA 126
CA^{++} 126
CA^{+2} 126
CABAGIN 118
CABAGIN-U 118
cabbage 85, 86, 90, 97, 101, 107, 112, 117, 139, 186, 187, 212, 226, 227
cabbage, boiled chinese 2, 21, 33, 41, 92, 129, 161, 176, 197, 203, 221, 246, 256, 266, 274, 284, 292, 302, 311, 321, 329, 338, 346, 357, 371, 383, 395, 398
cabbage, boiled green 4, 20, 34, 42, 57, 65, 80, 93, 131, 150, 164, 175, 181, 197, 207, 222, 237, 247, 257, 266, 275, 284, 293, 302, 311, 321, 330, 339, 347, 356, 370, 383, 396, 398
cabbage, boiled red 5, 21, 35, 42, 56, 63, 81, 92, 131, 148, 164, 176, 181, 197, 209, 223, 237, 247, 257, 266, 275, 284, 293, 302, 311, 321, 330, 338, 347, 356, 370, 381, 396, 398
cabbage, boiled savoy 3, 20, 35, 63, 93, 132, 164, 172, 197, 207, 222, 246, 256, 266, 274, 284, 293, 302, 311, 320, 329, 338, 346, 356, 370, 383, 395, 398
cabbage, boiled swamp 2, 22, 32, 42, 93, 130, 159, 171, 196, 205, 217, 246, 256, 266, 275, 283, 292, 301, 310, 320, 337, 346, 356, 370, 382, 395, 398
cabbage, chinese 117
cabbage juice 118
cabbage, raw 118
cabbage, raw chinese 2, 21, 32, 41, 92, 129, 162, 173, 197, 206, 219, 247, 256, 266, 275, 284, 292, 302, 311, 321, 329, 338, 346, 357, 371, 381, 395, 398
cabbage, raw green 4, 19, 35, 42, 57, 65, 80, 92, 130, 150, 163, 175, 181, 197, 206, 222, 237, 247, 257, 266, 275, 284, 293, 302, 311, 321, 330, 338, 347, 356, 370, 381, 384, 396, 398

cabbage, raw red 5, 19, 35, 42, 55, 63, 80, 92, 130, 146, 164, 175, 181, 196, 207, 222, 237, 247, 257, 266, 275, 284, 293, 302, 311, 321, 329, 338, 347, 355, 370, 381, 384, 395, 399
cabbage, raw savoy 3, 17, 35, 42, 63, 92, 131, 164, 171, 196, 206, 221, 246, 256, 265, 274, 284, 292, 301, 310, 320, 329, 338, 346, 355, 370, 395, 399
cabbage, raw swamp 2, 21, 32, 40, 92, 129, 158, 169, 196, 204, 217, 246, 255, 265, 274, 283, 292, 301, 309, 319, 337, 346, 357, 370, 382, 394, 398
cabbage, savoy 213
CADMIUM 126
cadmium-plated food containers 126
cadmium retention 126
CAESIUM 138
CAFFEINE 50, 85, 106, 153
cake, angel food 16, 28, 48, 56, 66, 74, 82, 134, 150, 162, 176, 197, 210, 217, 237, 350, 363, 382, 392, 408
cake, caramel 7, 7, 22, 22, 33, 33, 48, 48, 129, 129, 159, 159, 195, 195, 210, 210, 217, 217, 350, 350, 360, 360, 375, 375, 393, 393, 408, 408
cake, devil's food 8, 16, 30, 48, 54, 66, 74, 81, 128, 143, 158, 170, 194, 208, 217, 234, 350, 360, 374, 393, 408
cake, pineapple upside-down 7, 15, 32, 48, 54, 65, 74, 81, 94, 129, 145, 158, 174, 196, 208, 217, 235, 351, 361, 376, 394, 408
cake, pound 7, 21, 32, 48, 133, 163, 195, 210, 217, 351, 360, 373, 393, 408
cake, sponge 7, 15, 29, 48, 51, 65, 73, 80, 131, 149, 158, 176, 195, 210, 217, 233, 350, 362, 378, 392, 408
cake, white 8, 15, 29, 48, 55, 66, 74, 83, 129, 150, 161, 175, 195, 210, 216, 236, 350, 360, 375, 393, 408
cake, yellow 7, 15, 29, 48, 53, 66, 74, 81, 128, 149, 159, 175, 194, 210, 216, 235, 350, 360, 375, 393, 408
CALCIFEROL 99
CALCIUM 1, 6, 76, 91, 99, 126-138, 153, 167, 168, 178, 187, 213, 225, 228
CALCIUM 2-AEP 111
calcium absorption, decreased 98

CALCIUM ASCORBATE 97
calcium decreased 123, 125, 187
calcium decreased in urine 187
calcium deposits in tissues 168
calcium in blood, decreased 127, 168
calcium in blood, increased 98, 127
calcium osmotic control in heart, decreased 313
CALCIUM PANGAMATE 76
CALCIUM PANTOTHENATE 49, 58
calcium salts deposits in blood vessels 98
calcium salts deposits in kidney 98
calcium salts deposits in liver 98
calcium salts deposits in lung 98
calcium salts deposits in skin 98
calcium salts deposits in stomach 98
calf liver 139
CALOMIDE 76
CALORIES 358
CALPANATE 58
canadian bacon, grilled 12, 29, 44, 53, 59, 72, 84, 135, 149, 162, 173, 182, 189, 202, 214, 232, 243, 252, 261, 271, 280, 288, 296, 307, 316, 325, 334, 343, 358, 365, 377, 388, 404
canadian bacon, unheated 12, 30, 44, 53, 60, 72, 84, 136, 149, 163, 174, 183, 190, 203, 214, 233, 244, 252, 261, 272, 280, 289, 297, 308, 317, 326, 334, 343, 357, 366, 377, 389, 403
cancer 5, 6, 9, 23, 36, 43, 49, 58, 67, 76, 77, 78, 85, 86, 90, 96, 97, 99, 101, 104, 108, 113, 117, 152, 166, 199, 211, 212, 228, 238, 257, 303, 330, 371
cancer, bone 225
cancer, breast 105
cancer cells, accelerated reproduction of 76
cancer, colon 384
cancer, esophagus 140
cancer, lung 140
cancer resistance, decreased 1, 6, 11, 91, 100, 112, 139, 186, 211, 226, 227
cancer, thyroid 153
cane pulp, sugar 213
cane syrup 134, 211, 350, 363, 408
canker sore 9, 43, 49, 96, 99
cantaloupe 2, 21, 35, 41, 57,

64, 81, 85, 92, 112, 135, 149, 165, 176, 181, 186, 198, 204, 223, 237, 355, 369, 381, 396, 399
CANTAN 91, 97
CANTAXIN 91, 97
car exhaust 167, 187
carambola 3, 21, 36, 42, 93, 138, 145, 165, 176, 181, 198, 208, 224, 237, 248, 267, 276, 285, 293, 302, 312, 321, 330, 338, 347, 355, 369, 381, 397, 399
caraway seed 3, 13, 25, 38, 127, 154, 168, 188, 200, 229, 350, 361, 375, 389, 409
CARBACHOL 87
CARBAMYLCHOLINE CHLORIDE 87
CARBOCHOLINE 87
CARBOHYDRATE 322, 349-358
carbohydrate metabolism, deterioration of 68, 69, 239
carbuncles 9, 96, 99, 101, 127, 151, 213
CARCHOLIN 87
cardamom 127, 154, 168, 191, 200, 228, 350, 362, 378, 391, 409
cardiac arrest 199
cardiac arrhythmias 313
cardiac deterioration 11, 116
cardiac enlargement 140
cardiac ischemia 88
cardiac muscle 168
cardiac output, decreased 140
cardiocirculatory conditions 267
CARDIOMONE 68
CARDIOMONE SODIUM SALT 68
cardiospasm 106
cardiovascular disease 213
cardoon, boiled 4, 22, 35, 42, 95, 129, 162, 170, 197, 202, 217, 356, 370, 383, 396, 398
carissa 5, 20, 34, 92, 136, 143, 160, 175, 199, 205, 223, 354, 369, 379, 397, 400
CARNITINE 88
carob flour 5, 19, 25, 39, 57, 60, 80, 127, 142, 156, 169, 179, 195, 200, 221, 234, 349, 365, 380, 393, 410
carob milk 7, 20, 30, 48, 55, 66, 73, 83, 96, 128, 151, 165, 175, 185, 195, 209, 220, 236, 246, 255, 265, 274, 283, 291, 300, 309, 319, 328, 336, 345, 355, 368, 378, 394, 400
CAROTENE 1
carotene deposition in tissues 1
carp, cooked 8, 62, 84, 95,

130, 148, 158, 170, 188, 202, 219, 232, 243, 252, 262, 270, 279, 288, 296, 307, 316, 325, 333, 342, 365, 377, 388, 403
carp, raw 9, 63, 71, 95, 131, 149, 160, 171, 188, 204, 221, 233, 244, 253, 262, 272, 280, 289, 298, 308, 317, 326, 335, 343, 366, 377, 390, 401
carrot juice 1, 17, 34, 41, 56, 62, 84, 94, 132, 149, 164, 175, 181, 196, 205, 221, 237, 355, 369, 382, 396, 399
carrots 85, 86, 101, 108, 112, 117, 125, 139, 186, 187, 212, 213, 227
carrots, boiled 1, 21, 34, 41, 55, 62, 81, 95, 131, 145, 163, 175, 179, 197, 206, 219, 236, 247, 257, 266, 275, 284, 293, 302, 311, 321, 329, 338, 347, 354, 369, 383, 384, 396, 399
carrots, canned 2, 23, 35, 41, 56, 64, 81, 95, 132, 146, 163, 176, 179, 197, 208, 217, 236, 248, 257, 267, 276, 285, 294, 302, 312, 321, 330, 339, 347, 356, 370, 383, 384, 396, 398
carrots, frozen and boiled 2, 22, 35, 42, 56, 64, 81, 95, 132, 148, 164, 176, 179, 197, 208, 220, 236, 247, 257, 266, 275, 284, 293, 302, 311, 321, 329, 338, 347, 355, 369, 383, 384, 396, 399
carrots, raw 1, 17, 33, 40, 56, 63, 81, 94, 132, 149, 164, 175, 181, 196, 204, 221, 237, 247, 257, 266, 275, 284, 293, 302, 311, 321, 330, 338, 347, 354, 369, 383, 384, 396, 399
cashew butter 14, 30, 39, 50, 62, 79, 131, 141, 154, 168, 188, 201, 222, 229, 240, 251, 263, 271, 280, 290, 299, 306, 318, 325, 335, 342, 351, 359, 372, 390, 410
cashews 78, 89, 101, 103, 104
cashews, dry roasted 14, 29, 39, 50, 62, 79, 130, 141, 154, 168, 188, 201, 222, 229, 241, 252, 263, 272, 281, 290, 299, 307, 318, 325, 335, 342, 351, 359, 372, 390, 410
cashews, oil roasted 13, 30, 39, 50, 62, 79, 131, 141, 155, 168, 179, 188, 201, 222, 229, 241, 251, 263, 272, 281, 290, 299, 307, 318, 325, 335, 342, 351, 359, 372, 390, 410

cassava, bitter 77
cassava, raw 5, 14, 32, 39, 92, 129, 155, 169, 195, 200, 223, 245, 255, 265, 275, 284, 292, 301, 311, 320, 328, 338, 346, 352, 367, 381, 394, 403
cataracts 9, 36, 58, 96, 99, 101, 138, 227, 303
CATAVIN C 91, 97
CATECHOL 97
catfish, breaded and fried 9, 18, 31, 47, 131, 146, 159, 171, 191, 203, 216, 234, 244, 253, 262, 272, 280, 289, 298, 307, 317, 326, 334, 343, 355, 364, 375, 390, 405
catfish, raw 19, 31, 47, 131, 146, 161, 172, 184, 191, 203, 219, 234, 244, 253, 262, 272, 280, 289, 297, 308, 317, 326, 334, 343, 367, 378, 389, 401
catsup, tomato 2, 18, 33, 39, 93, 133, 163, 196, 203, 215, 352, 367, 380, 395, 403
cattlefish 116, 117
cauliflower 85, 88, 89, 108, 117, 186
cauliflower, boiled 5, 19, 34, 41, 57, 63, 80, 92, 132, 147, 164, 176, 181, 197, 204, 223, 237, 246, 256, 265, 275, 283, 292, 301, 310, 320, 329, 338, 346, 356, 370, 382, 384, 395, 398
cauliflower, frozen and boiled 5, 22, 33, 42, 57, 64, 80, 92, 133, 150, 164, 176, 181, 197, 209, 222, 237, 247, 256, 266, 275, 284, 292, 301, 311, 320, 329, 338, 346, 356, 370, 382, 384, 395, 398
cauliflower, raw 5, 17, 33, 40, 56, 62, 79, 92, 132, 150, 163, 175, 180, 196, 203, 222, 237, 246, 256, 265, 275, 283, 292, 301, 310, 320, 329, 338, 346, 356, 370, 382, 395, 398
caviar 86, 214, 242, 250, 262, 271, 278, 288, 296, 305, 314, 324, 332, 341, 356, 363, 374, 388, 407
cavities, predisposition to 151
CC 99
CCC 87
CD 126
CDP-CHOLINE 87
CDP-VITAMIN B$_P$ 87
CE-VI-SOL 91, 97
CEBICURE 91, 97
CEBID 91, 97
CEBION 91, 97
CECON 91, 97
CEGIOLAN 91, 97

celeriac, raw 22, 35, 42, 67, 95, 132, 164, 176, 196, 208, 220, 355, 370, 382, 396, 398
celery 118
celery, boiled 4, 22, 35, 42, 56, 66, 83, 94, 131, 150, 166, 176, 181, 197, 203, 219, 237, 248, 257, 267, 276, 285, 294, 302, 312, 322, 330, 339, 347, 357, 371, 383, 397, 398
celery, raw 4, 22, 35, 42, 56, 66, 82, 94, 131, 150, 164, 175, 181, 197, 205, 218, 237, 248, 257, 267, 276, 285, 294, 302, 312, 321, 330, 339, 347, 356, 371, 381, 384, 396, 398
celery seed 4, 127, 153, 168, 188, 200, 217, 228, 351, 360, 373, 389, 409
celiac disease 9, 67, 74, 85, 96, 99, 101, 108, 138, 166, 177, 397
CELIN 91, 97
cell dehydration 211
cell deterioration 76, 385
cell energy from mitochondria, decreased 117
cell life-span, decreased 76, 111
cell membrane integrity, deterioration of 111, 187, 313
cell oxygenation, decreased 76, 151
cell pH, decreased 211
cell regeneration, decreased 106
CELLULOSE 349
cellulose powder 213
celtuce, raw 2, 19, 33, 40, 93, 131, 163, 171, 196, 204, 225, 247, 257, 267, 275, 284, 293, 302, 311, 321, 330, 338, 347, 356, 370, 381, 396, 398
CENETONE 91, 97
center loin excluding fat, braised pork 8, 12, 27, 44, 51, 59, 73, 83, 137, 146, 161, 173, 173, 182, 190, 203, 220, 231, 240, 250, 259, 268, 277, 285, 294, 304, 313, 322, 331, 340, 363, 375, 385, 407
center loin excluding fat, broiled pork 9, 11, 26, 45, 51, 59, 72, 83, 137, 147, 161, 171, 185, 190, 202, 218, 232, 240, 250, 259, 268, 277, 285, 295, 304, 313, 322, 331, 340, 364, 376, 385, 405
center loin excluding fat, fried pork 9, 11, 26, 44, 51, 59, 72, 83, 137, 147, 161,

Index

171, 185, 189, 201, 218, 231, 240, 250, 259, 269, 277, 286, 295, 305, 313, 323, 332, 340, 363, 375, 386, 406
center loin excluding fat, roasted pork 9, 12, 27, 45, 52, 59, 73, 84, 137, 147, 160, 173, 184, 190, 203, 219, 232, 240, 250, 259, 269, 277, 286, 295, 305, 313, 323, 332, 340, 364, 375, 386, 405
center loin including fat, braised pork 8, 12, 28, 44, 52, 60, 73, 84, 137, 146, 162, 174, 182, 191, 204, 221, 231, 240, 250, 259, 269, 277, 286, 295, 304, 313, 323, 332, 340, 361, 373, 386, 407
center loin including fat, broiled pork 8, 11, 27, 45, 52, 60, 72, 83, 138, 147, 162, 172, 172, 185, 191, 203, 219, 232, 241, 251, 259, 269, 278, 286, 296, 305, 314, 323, 332, 340, 361, 374, 387, 407
center loin including fat, fried pork 8, 11, 27, 45, 51, 60, 72, 83, 137, 147, 162, 172, 185, 191, 203, 219, 232, 242, 252, 260, 270, 279, 287, 297, 306, 316, 324, 333, 342, 360, 373, 388, 407
center loin including fat, roasted pork 9, 12, 28, 45, 51, 60, 73, 84, 137, 147, 161, 174, 185, 191, 204, 219, 232, 241, 251, 259, 270, 278, 286, 296, 306, 315, 323, 333, 341, 362, 374, 387, 406
center rib excluding fat, braised pork 9, 12, 26, 44, 52, 60, 73, 82, 362, 375, 385, 407
center rib excluding fat, broiled pork 9, 12, 26, 45, 51, 60, 72, 82, 134, 147, 162, 171, 171, 183, 189, 201, 219, 231, 240, 250, 259, 269, 277, 286, 295, 305, 313, 323, 332, 340, 363, 375, 386, 406
center rib excluding fat, fried pork 9, 12, 26, 45, 51, 59, 72, 82, 136, 148, 162, 171, 185, 189, 201, 221, 232, 241, 250, 259, 269, 277, 286, 296, 305, 314, 323, 332, 340, 363, 375, 386, 406
center rib excluding fat, roasted pork 9, 12, 26, 45, 52, 60, 73, 82, 135, 147, 161, 173, 185, 189, 202, 221, 232, 241, 250, 259, 269, 277, 286, 296, 305, 313, 323, 332, 340, 364, 375, 386, 405
center rib excluding fat, braised pork 135, 146, 160, 172, 185, 190, 201, 220, 231
center rib including fat, braised pork 8, 13, 27, 45, 52, 61, 73, 83, 136, 147, 161, 174, 185, 191, 202, 221, 231, 240, 250, 259, 269, 277, 286, 296, 305, 314, 323, 332, 340, 360, 373, 386, 407
center rib including fat, broiled pork 9, 12, 27, 45, 52, 61, 72, 82, 134, 148, 162, 172, 172, 184, 190, 203, 220, 232, 241, 251, 259, 270, 279, 287, 297, 306, 315, 324, 333, 341, 361, 373, 388, 407
center rib including fat, fried pork 8, 12, 27, 46, 52, 61, 72, 83, 137, 148, 163, 173, 173, 185, 191, 203, 221, 233, 242, 252, 260, 271, 279, 288, 298, 307, 316, 325, 334, 342, 360, 372, 389, 407
center rib including fat, roasted pork 9, 12, 27, 45, 51, 60, 73, 82, 136, 148, 162, 174, 185, 190, 203, 221, 232, 241, 251, 259, 270, 279, 287, 297, 306, 315, 324, 333, 341, 361, 373, 388, 406
central nervous system deterioration 11
cereal grasses 118
cereal, whole wheat 86, 89
cereals 125
cereals, whole grain 77
CEREB 87
cerebral circulation deterioration 87
CEREON 91, 97
CERGONA 91, 97
ceroid deposits in muscle 100
CESCORBAT 91, 97
CESIUM 138-139, 212
cesium poisoning 138
CETAIN 106
CETAMID 91, 97
CETEMICAN 91, 97
CEVALIN 91, 97
CEVATINE 91, 97
CEVEX 91, 97
CEVIMIN 91, 97
CEVITAMIC ACID 91, 97
CEVITAMIN 91, 97
CEVITAN 91, 97
CEVITEX 91, 97
CEWIN 91, 97
CF 78
chamomile tea 5, 23, 36, 58, 84, 138, 151, 166, 177, 182, 211, 224, 238, 358, 371, 397, 398
chapped lips 100
chard, boiled swiss 2, 21, 32, 42, 56, 93, 130, 157, 169, 197, 201, 217, 246, 265, 274, 283, 292, 301, 309, 320, 329, 346, 356, 370, 383, 395, 398 chard, red swiss 212
charley horse 100
chayote, boiled 4, 21, 34, 42, 54, 94, 134, 165, 176, 197, 208, 225, 248, 267, 276, 284, 294, 303, 311, 321, 330, 338, 347, 356, 370, 380, 397, 398
cheese 85, 89, 99, 101, 103, 104, 107, 117, 187
cheese, american 6, 20, 26, 53, 65, 72, 83, 127, 164, 173, 188, 208, 214, 230, 244, 254, 260, 271, 279, 287, 297, 305, 317, 324, 331, 341, 357, 360, 373, 388, 408
cheese, bleu 7, 20, 25, 48, 50, 63, 71, 80, 127, 165, 172, 188, 205, 214, 231, 244, 254, 261, 270, 279, 288, 297, 305, 317, 324, 331, 340, 357, 361, 373, 388, 407
cheese, brick 6, 26, 55, 65, 71, 80, 127, 165, 172, 188, 209, 216, 231, 244, 254, 261, 270, 277, 287, 297, 304, 316, 324, 331, 340, 357, 360, 373, 388, 407
cheese, brie 7, 18, 25, 48, 51, 62, 71, 79, 128, 164, 192, 208, 215, 244, 254, 261, 271, 279, 288, 297, 304, 317, 324, 331, 341, 358, 361, 373, 389, 407
cheese, camembert 7, 20, 25, 48, 50, 62, 71, 79, 107, 127, 165, 173, 189, 207, 215, 231, 244, 254, 262, 271, 279, 288, 297, 305, 317, 324, 331, 341, 358, 362, 373, 389, 406
cheese, caraway 7, 20, 25, 48, 56, 73, 127, 173, 188, 215, 357, 360, 373, 387, 408
cheese, cheddar 7, 20, 25, 53, 65, 72, 81, 86, 107, 127, 152, 163, 171, 188, 210, 215, 230, 244, 254, 260, 268, 277, 288, 296, 304, 316, 324, 331, 340, 358, 360, 372, 387, 408
cheese, cheshire 7, 20, 27, 127, 165, 173, 188, 210, 215, 244, 254, 261, 268, 278, 288, 296, 304, 317, 324,

331, 340, 356, 360, 373, 388, 408
cheese, colby 7, 25, 56, 65, 72, 127, 162, 172, 188, 209, 215, 230, 244, 254, 261, 268, 277, 288, 296, 304, 317, 324, 331, 340, 357, 360, 372, 388, 408
cheese, cottage 7, 56, 65, 72, 81, 99, 101, 130, 152, 165, 177, 186, 193, 210, 212, 216, 234, 244, 254, 263, 272, 281, 290, 298, 308, 318, 327, 334, 344, 357, 367, 378, 391, 401
cheese, cream 6, 20, 29, 55, 66, 73, 129, 160, 177, 194, 209, 216, 235, 245, 255, 264, 273, 282, 290, 299, 309, 319, 327, 336, 344, 357, 361, 372, 392, 406
cheese, edam 7, 20, 25, 55, 65, 71, 81, 127, 164, 171, 188, 207, 215, 230, 244, 260, 269, 277, 286, 295, 304, 316, 331, 340, 358, 361, 373, 387, 407
cheese, feta 127, 163, 174, 189, 210, 214, 231, 356, 363, 374, 390, 405
cheese, fontina 6, 20, 29, 127, 165, 175, 230, 358, 360, 373, 387, 408
cheese, gjetost 48, 127, 188, 215, 245, 255, 263, 273, 281, 290, 299, 308, 318, 327, 335, 344, 351, 360, 373, 391, 409
cheese, gouda 7, 20, 26, 54, 65, 80, 127, 165, 171, 188, 209, 215, 230, 244, 260, 269, 277, 286, 295, 304, 316, 331, 340, 357, 361, 373, 387, 407
cheese, gruyere 6, 18, 27, 52, 65, 71, 81, 127, 188, 210, 216, 244, 251, 260, 268, 277, 286, 295, 304, 315, 322, 331, 339, 358, 360, 372, 386, 408
cheese, havarti 6, 20, 26, 55, 66, 73, 80, 127, 149, 166, 174, 183, 188, 210, 216, 230, 358, 360, 372, 388, 408
cheese, limburger 6, 18, 25, 48, 50, 65, 71, 80, 127, 166, 173, 188, 209, 215, 232, 244, 262, 269, 278, 289, 296, 305, 317, 324, 331, 340, 358, 361, 373, 389, 407
cheese, monterey 7, 25, 127, 162, 171, 188, 210, 216, 230, 244, 254, 260, 268, 277, 288, 296, 304, 316, 324, 331, 340, 358, 360, 373, 388, 407
cheese, mozzarella 7, 28, 57, 65, 72, 83, 127, 165, 174, 188, 210, 216, 232, 244, 254, 261, 271, 279, 288, 297, 306, 317, 331, 342, 357, 362, 374, 389, 406
cheese, muenster 6, 26, 56, 65, 71, 81, 127, 164, 171, 188, 209, 215, 231, 244, 254, 261, 270, 277, 287, 297, 304, 316, 323, 331, 340, 358, 360, 373, 388, 407
cheese, neufchatel 6, 29, 52, 66, 73, 81, 129, 165, 177, 193, 209, 216, 235, 245, 254, 263, 272, 281, 290, 299, 308, 318, 327, 335, 344, 357, 363, 373, 391, 404
cheese, parmesan 7, 25, 52, 63, 127, 161, 169, 188, 210, 214, 230, 242, 252, 259, 268, 277, 285, 294, 304, 313, 322, 331, 339, 356, 360, 373, 385, 408
cheese, port du salut 6, 28, 56, 65, 71, 81, 127, 188, 216, 244, 261, 268, 277, 288, 295, 304, 316, 323, 331, 340, 358, 361, 373, 388, 407
cheese, provolone 7, 20, 26, 53, 65, 71, 81, 127, 163, 171, 188, 209, 215, 230, 244, 254, 260, 270, 277, 286, 296, 304, 316, 331, 340, 357, 361, 373, 387, 407
cheese, ricotta 7, 22, 29, 48, 66, 73, 128, 164, 176, 193, 209, 218, 233, 244, 254, 263, 272, 281, 289, 299, 308, 318, 335, 344, 357, 365, 375, 391, 402
cheese, romano 7, 25, 82, 127, 188, 214, 356, 360, 373, 385, 408
cheese, roquefort 7, 20, 25, 48, 50, 63, 72, 80, 127, 163, 171, 188, 210, 214, 232, 357, 360, 373, 388, 408
cheese spread, pimento 6, 20, 26, 53, 65, 72, 83, 127, 164, 173, 188, 208, 214, 231, 244, 254, 260, 271, 279, 287, 297, 305, 317, 324, 331, 341, 357, 360, 373, 388, 408
cheese, swiss 7, 20, 26, 53, 65, 71, 83, 127, 139, 165, 170, 188, 209, 212, 217, 230, 244, 252, 260, 268, 277, 286, 295, 304, 316, 322, 331, 339, 356, 360, 373, 386, 408
cheese, tilsit 7, 18, 26, 48, 54, 71, 127, 165, 175, 188, 210, 215, 230, 244, 261, 268, 277, 288, 295, 304, 316, 323, 331, 340, 357, 361, 373, 388, 407
cheeseburger 7, 14, 29, 47, 56, 64, 72, 81, 96, 128, 146, 157, 173, 193, 207, 215, 232, 352, 363, 375, 391, 407
cheeseburger, bacon 7, 16, 30, 47, 56, 63, 71, 81, 96, 129, 147, 157, 172, 191, 206, 216, 230, 353, 362, 374, 390, 407
cheesecake 7, 20, 31, 48, 52, 65, 73, 81, 94, 130, 148, 164, 176, 195, 210, 217, 235, 351, 362, 374, 393, 407
chemotherapy 117
cherimoya 5, 16, 32, 39, 94, 132, 164, 196, 352, 368, 381, 395, 402
cherries 4, 21, 33, 41, 57, 66, 84, 94, 112, 125, 134, 146, 164, 176, 181, 198, 207, 238, 353, 368, 380, 396, 400
cherry bark, wild 77
cherry pie 3, 22, 36, 41, 134, 165, 197, 209, 216, 351, 363, 376, 394, 407
cherry seeds 78
chervil, dried 59, 127, 154, 169, 189, 200, 229, 351, 367, 391
chest and abdominal pains 11
chest pains 76
chest tightness 77, 107, 126
chestnuts 116, 117
chestnuts, boiled and steamed chinese 135, 143, 161, 169, 178, 195, 204, 223, 235, 245, 255, 265, 274, 283, 292, 300, 309, 320, 328, 337, 346, 351, 366, 380, 394, 404
chestnuts, boiled and steamed european 130, 142, 158, 169, 179, 194, 200, 221, 236, 246, 255, 265, 275, 283, 292, 301, 310, 320, 329, 337, 346, 351, 366, 379, 394, 403
chestnuts, boiled and steamed japanese 5, 15, 33, 40, 57, 93, 135, 143, 163, 174, 179, 197, 209, 223, 236, 247, 256, 266, 275, 284, 293, 301, 311, 321, 329, 338, 347, 354, 369, 381, 396, 399
chestnuts, canned chinese water 5, 22, 35, 42, 95, 138, 146, 162, 177, 198, 209, 223, 236, 354, 369, 396, 400
chestnuts, dried chinese 132, 142, 157, 169, 178, 193, 200, 223, 233, 244, 254, 264, 273, 282, 291, 299, 309, 319, 327, 336, 345, 349, 360, 379, 392, 409

chestnuts, dried european 14, 26, 40, 51, 59, 79, 93, 129, 142, 157, 169, 178, 192, 200, 221, 236, 245, 253, 264, 273, 282, 291, 299, 309, 319, 327, 336, 345, 349, 360, 378, 392, 409
chestnuts, dried japanese 4, 12, 25, 38, 53, 92, 129, 141, 155, 169, 178, 192, 200, 221, 231, 245, 254, 264, 273, 282, 291, 299, 309, 319, 327, 336, 345, 349, 360, 379, 393, 409
chestnuts, raw chinese 4, 15, 30, 40, 92, 133, 142, 159, 169, 178, 195, 201, 223, 234, 245, 254, 264, 274, 283, 292, 300, 309, 319, 328, 337, 345, 351, 364, 380, 393, 407
chestnuts, raw chinese water 15, 29, 39, 94, 135, 163, 173, 196, 201, 222, 352, 367, 382, 395, 402
chestnuts, raw european 5, 14, 30, 39, 53, 60, 79, 92, 132, 142, 161, 171, 178, 195, 201, 223, 235, 246, 255, 265, 274, 283, 292, 300, 310, 320, 328, 337, 346, 351, 364, 379, 384, 394, 407
chestnuts, raw japanese 5, 13, 30, 39, 56, 92, 131, 142, 159, 170, 178, 195, 204, 222, 233, 246, 255, 265, 274, 283, 292, 300, 310, 320, 328, 337, 346, 351, 366, 380, 395, 404
chestnuts, roasted chinese 5, 15, 31, 39, 133, 142, 159, 169, 194, 201, 223, 234, 245, 254, 264, 274, 283, 291, 300, 309, 319, 328, 337, 345, 350, 364, 380, 393, 407
chestnuts, roasted european 5, 14, 30, 39, 52, 59, 79, 92, 132, 142, 161, 171, 178, 194, 201, 223, 235, 245, 254, 265, 274, 283, 292, 300, 309, 320, 328, 337, 345, 350, 363, 379, 384, 394, 407
chestnuts, roasted japanese 4, 13, 28, 40, 92, 131, 142, 157, 169, 178, 195, 233, 246, 254, 265, 274, 283, 292, 300, 309, 320, 328, 337, 345, 351, 365, 380, 394, 407
CHICK ANTIDERMATITIS FACTOR 49
chick development, deterioration of 119
CHICK FACTOR 119
chicken 85, 88, 89, 101, 103, 104, 117, 139, 186, 212

chicken and turkey salad 7, 20, 33, 48, 55, 64, 73, 84, 135, 163, 197, 207, 216, 355, 365, 375, 391
chicken back with skin, fried 7, 16, 28, 44, 50, 61, 73, 82, 132, 147, 158, 172, 181, 192, 206, 218, 231, 241, 250, 261, 269, 278, 287, 295, 305, 315, 323, 332, 341, 355, 361, 374, 386, 407
chicken breast 125
chicken breast with skin, fried 8, 17, 31, 44, 51, 59, 73, 83, 133, 149, 160, 171, 184, 190, 206, 218, 233, 241, 250, 260, 268, 277, 286, 294, 304, 314, 323, 331, 340, 358, 364, 377, 385, 405
chicken breast with skin, roasted 8, 18, 31, 44, 51, 59, 73, 84, 134, 149, 161, 171, 183, 191, 206, 219, 234, 241, 250, 260, 268, 278, 286, 295, 304, 314, 323, 332, 340, 365, 377, 386, 404
chicken breast with skin, stewed 8, 19, 31, 44, 52, 61, 74, 84, 134, 149, 161, 173, 183, 193, 208, 220, 234, 242, 250, 261, 268, 278, 287, 295, 305, 315, 324, 332, 341, 365, 377, 387, 403
chicken breast without skin, fried 8, 17, 31, 44, 50, 59, 73, 83, 133, 149, 160, 171, 183, 190, 205, 218, 233, 240, 250, 260, 268, 277, 286, 294, 304, 313, 323, 331, 340, 358, 365, 378, 385, 404
chicken breast without skin, roasted 8, 18, 31, 44, 51, 59, 73, 84, 133, 149, 161, 171, 184, 190, 206, 219, 234, 241, 250, 260, 268, 277, 286, 294, 304, 314, 323, 331, 340, 365, 378, 386, 403
chicken breast without skin, stewed 8, 20, 31, 44, 52, 61, 74, 84, 134, 149, 162, 172, 183, 192, 207, 220, 234, 241, 250, 260, 268, 278, 286, 295, 305, 314, 323, 332, 340, 366, 379, 386, 403
chicken drumstick with skin, fried 8, 17, 28, 44, 50, 60, 73, 82, 134, 147, 159, 173, 182, 192, 206, 218, 231, 242, 250, 261, 269, 279, 287, 295, 305, 315, 324, 333, 341, 358, 364, 375, 387, 405

chicken drumstick with skin, roasted 7, 17, 29, 44, 50, 60, 73, 82, 134, 147, 159, 172, 183, 192, 206, 218, 231, 242, 250, 261, 269, 279, 287, 295, 305, 315, 324, 333, 341, 364, 376, 387, 404
chicken drumstick with skin, stewed 8, 19, 29, 46, 51, 63, 74, 83, 134, 148, 159, 174, 183, 193, 207, 218, 231, 242, 251, 261, 269, 279, 288, 296, 306, 316, 325, 333, 342, 365, 376, 387, 403
chicken drumstick without skin, stewed 8, 18, 28, 44, 50, 60, 73, 82, 135, 147, 160, 172, 183, 192, 206, 218, 230, 240, 250, 260, 268, 277, 286, 294, 304, 313, 323, 331, 340, 365, 377, 386, 403
chicken fat 238, 359, 371
chicken frankfurter 18, 31, 47, 129, 157, 214, 355, 363, 374, 391, 405
chicken, fried 8, 17, 32, 44, 61, 74, 81, 134, 148, 163, 172, 192, 206, 216, 234, 353, 363, 375, 390, 407
chicken giblets, fried 6, 16, 25, 44, 50, 59, 69, 78, 94, 133, 142, 154, 172, 180, 189, 204, 217, 229, 240, 250, 261, 268, 277, 287, 295, 304, 313, 323, 331, 340, 356, 363, 375, 385, 407
chicken giblets, simmered 6, 17, 25, 46, 50, 61, 69, 78, 94, 135, 143, 154, 174, 181, 190, 208, 220, 229, 241, 251, 262, 269, 278, 288, 296, 304, 315, 324, 333, 341, 358, 366, 378, 387, 403
chicken gizzard, simmered 7, 21, 28, 46, 51, 64, 71, 80, 95, 135, 145, 155, 174, 181, 193, 208, 219, 230, 241, 250, 262, 269, 279, 288, 295, 305, 314, 325, 333, 342, 358, 366, 378, 387, 403
chicken heart, simmered 8, 18, 25, 47, 50, 61, 69, 79, 95, 133, 142, 154, 173, 181, 191, 209, 221, 228, 242, 250, 262, 268, 277, 287, 296, 304, 315, 323, 332, 340, 358, 365, 377, 387, 403
chicken liver 88, 108
chicken liver pâté, canned 7, 18, 25, 44, 93, 135, 154, 355, 365, 375, 390
chicken liver, simmered 6,

15, 24, 46, 50, 59, 69, 78, 93, 134, 142, 154, 173, 180, 189, 209, 221, 230, 242, 251, 262, 269, 278, 288, 297, 304, 315, 323, 333, 340, 358, 366, 377, 388, 403
chicken neck with skin, fried 7, 17, 28, 45, 51, 62, 73, 83, 131, 145, 157, 174, 181, 193, 208, 218, 230, 242, 251, 262, 270, 280, 288, 297, 306, 316, 325, 334, 342, 356, 361, 373, 388, 407
chicken neck with skin, simmered 7, 20, 28, 47, 53, 64, 74, 84, 132, 146, 157, 175, 182, 194, 209, 220, 231, 243, 252, 263, 271, 281, 289, 298, 307, 317, 326, 335, 343, 363, 374, 389, 404
chicken neck without skin, simmered 7, 19, 27, 46, 51, 63, 74, 83, 131, 145, 156, 174, 181, 194, 209, 219, 230, 243, 251, 261, 269, 279, 288, 296, 306, 316, 325, 333, 342, 365, 377, 388, 403
chicken pox 9, 96
chicken roll 18, 31, 45, 131, 150, 161, 174, 193, 206, 216, 234, 357, 366, 377, 389, 403
chicken roll, light meat 18, 31, 45, 131, 149, 161, 174, 193, 206, 215, 234, 357, 366, 377, 389, 403
chicken salad 31, 47, 128, 157, 356, 365, 376, 390
chicken thigh with skin, fried 8, 16, 28, 44, 50, 61, 73, 82, 134, 147, 159, 176, 182, 192, 206, 218, 231, 242, 250, 261, 269, 279, 287, 295, 305, 315, 324, 333, 341, 357, 363, 375, 387, 406
chicken thigh with skin, roasted 7, 19, 29, 44, 50, 61, 73, 83, 134, 147, 159, 172, 183, 192, 207, 218, 231, 242, 251, 261, 269, 279, 288, 296, 306, 316, 325, 333, 342, 364, 375, 387, 405
chicken thigh with skin, stewed 7, 19, 29, 45, 51, 63, 74, 83, 134, 148, 159, 174, 183, 193, 208, 219, 232, 243, 251, 262, 270, 280, 288, 296, 306, 316, 325, 334, 342, 364, 375, 388, 404
chicken thigh without skin, roasted 8, 17, 28, 44, 50, 60, 73, 82, 134, 147, 159, 172, 183, 192, 206, 218, 231, 242, 251, 261, 269, 279, 287, 295, 306, 315, 324, 333, 341, 365, 376, 387, 404
chicken wing with skin, fried 7, 19, 31, 44, 60, 73, 84, 133, 148, 160, 174, 182, 193, 208, 218, 232, 242, 250, 261, 269, 279, 288, 296, 306, 316, 325, 333, 342, 357, 361, 374, 387, 407
chicken wing with skin, roasted 7, 21, 31, 44, 60, 73, 84, 133, 149, 160, 173, 183, 193, 207, 218, 232, 241, 250, 261, 269, 279, 288, 296, 306, 315, 325, 333, 341, 362, 374, 387, 406
chicken wing with skin, stewed 7, 20, 32, 46, 53, 62, 74, 84, 134, 149, 160, 175, 183, 194, 209, 219, 233, 243, 251, 262, 270, 280, 288, 297, 307, 316, 325, 334, 342, 363, 374, 388, 404
chicken with skin, fried dark meat 7, 16, 28, 44, 50, 61, 73, 82, 133, 147, 159, 172, 182, 192, 206, 218, 231, 242, 250, 261, 269, 279, 287, 295, 305, 315, 324, 333, 341, 356, 362, 374, 387, 406
chicken with skin, fried light meat 8, 17, 31, 44, 51, 59, 73, 83, 133, 148, 160, 171, 171, 182, 191, 206, 218, 233, 241, 250, 260, 268, 278, 286, 295, 304, 314, 323, 332, 340, 357, 364, 376, 386, 406
chicken with skin, roasted dark meat 7, 18, 29, 44, 50, 61, 73, 82, 133, 147, 159, 173, 183, 192, 207, 218, 231, 242, 251, 261, 269, 279, 288, 296, 306, 316, 325, 333, 341, 363, 375, 387, 405
chicken with skin, roasted light meat 7, 19, 31, 44, 51, 59, 73, 84, 133, 149, 160, 172, 183, 191, 206, 218, 233, 241, 250, 261, 268, 278, 287, 295, 305, 315, 324, 332, 341, 364, 376, 386, 404
chicken with skin, stewed dark meat 7, 19, 30, 46, 51, 63, 74, 83, 134, 148, 160, 174, 183, 193, 208, 219, 232, 243, 251, 262, 270, 279, 288, 296, 306, 316, 325, 334, 342, 364, 375, 388, 404
chicken with skin, stewed light meat 8, 20, 32, 44, 52, 62, 74, 84, 134, 149, 161, 174, 183, 193, 208, 220, 233, 242, 251, 261, 269, 279, 287, 296, 306, 315, 324, 333, 341, 365, 376, 387, 403
chicken without skin, fried dark meat 8, 17, 28, 44, 50, 60, 73, 82, 133, 147, 159, 172, 182, 192, 206, 218, 231, 241, 250, 260, 268, 278, 286, 295, 305, 314, 323, 332, 340, 357, 364, 376, 386, 405
chicken without skin, fried light meat 8, 18, 31, 44, 50, 59, 73, 83, 133, 149, 160, 171, 183, 190, 205, 218, 233, 240, 250, 260, 268, 277, 286, 294, 304, 313, 323, 331, 340, 358, 365, 378, 385, 404
chicken without skin, roasted dark meat 8, 18, 28, 44, 50, 60, 73, 82, 133, 147, 159, 172, 184, 192, 206, 218, 231, 242, 250, 261, 268, 278, 287, 295, 305, 315, 324, 332, 341, 365, 376, 387, 404
chicken without skin, roasted light meat 8, 18, 31, 44, 51, 59, 69, 83, 133, 149, 161, 171, 184, 191, 206, 218, 233, 241, 250, 260, 268, 277, 286, 295, 304, 314, 323, 332, 340, 365, 378, 386, 403
chicken without skin, stewed dark meat 8, 19, 29, 46, 62, 74, 83, 134, 148, 159, 173, 183, 193, 208, 219, 231, 242, 251, 261, 269, 279, 287, 295, 306, 315, 324, 333, 341, 365, 377, 387, 403
chicken without skin, stewed light meat 8, 20, 31, 44, 52, 61, 74, 84, 134, 149, 161, 173, 183, 193, 208, 219, 233, 241, 250, 260, 268, 278, 286, 295, 305, 315, 323, 332, 340, 366, 378, 386, 403
chickory greens, raw 2, 19, 32, 40, 93, 129, 161, 171, 196, 202, 221, 246, 266, 274, 284, 293, 302, 311, 321, 328, 346, 356, 370, 381, 395, 398
chickory, raw 18, 31, 41, 66, 93, 164, 175, 198, 207, 223, 247, 266, 275, 284, 293, 302, 312, 321, 329, 347, 357, 370, 382, 396, 398
chickpeas, boiled 5, 16, 34,

Index 482

41, 55, 64, 79, 96, 130, 142, 156, 170, 178, 192, 205, 223, 233, 244, 254, 264, 273, 282, 290, 300, 308, 318, 327, 336, 345, 352, 365, 379, 384, 392, 404
chickpeas, canned 5, 22, 35, 43, 55, 59, 79, 95, 131, 144, 159, 171, 179, 195, 208, 216, 233, 245, 255, 264, 274, 282, 291, 300, 309, 319, 328, 337, 345, 352, 367, 379, 393, 403
children, deterioration of growth in 294
children's aspirin 123
chili 7, 18, 19, 25, 33, 39, 48, 64, 73, 81, 93, 96, 132, 132, 144, 158, 163, 173, 195, 196, 202, 205, 214, 216, 233, 352, 355, 367, 367, 378, 392, 394, 401, 403
chili, fresh 139
chili powder 1, 13, 25, 37, 92, 127, 154, 169, 189, 200, 215, 231, 351, 363, 375, 391, 409
chili sauce 2
chills 186
chipped beef 136, 144, 155, 171, 192, 201, 214, 229, 240, 251, 261, 270, 278, 287, 295, 305, 314, 325, 333, 341, 358, 365, 378, 386, 405
chitterlings, simmered pork 33, 48, 132, 143, 155, 176, 196, 211, 221, 229, 244, 264, 273, 282, 290, 299, 309, 318, 328, 335, 344, 362, 373, 391, 404
chives, freeze-dried 1, 11, 25, 91, 127, 154, 188, 200, 242, 263, 271, 281, 290, 299, 308, 317, 325, 334, 343, 350, 363, 387
chives, raw 2, 26, 54, 61, 92, 130, 158, 169, 195, 205, 246, 266, 274, 283, 292, 301, 310, 320, 328, 337, 346, 357, 369, 394, 398
CHLORAMINE T 89
CHLORAMPHENICOL 69, 140
CHLORIDES 126
CHLORIN 139
CHLORINE 101, 106, 139
CHLOROGENIC ACID 11
chocolate, bittersweet 5, 21, 30, 39, 130, 155, 189, 201, 223, 351, 359, 372, 392, 410
chocolate chiffon pie 3, 22, 32, 42, 132, 160, 195, 209, 217, 351, 361, 375, 392, 408
chocolate cream pie 3, 16, 30, 40, 53, 66, 73, 82, 129, 145, 160, 172, 194, 209, 217, 234, 351, 363, 375, 393, 407

chocolate, german sweet baker's 5, 33, 42, 57, 84, 132, 142, 157, 169, 193, 205, 223, 233, 350, 359, 372, 393, 410
chocolate, milk 3, 18, 26, 41, 127, 161, 190, 202, 218, 244, 252, 261, 268, 277, 289, 297, 304, 316, 322, 334, 340, 350, 359, 372, 392, 410
chocolate milkshake, 3 8, 19, 29, 48, 54, 66, 73, 83, 128, 165, 175, 194, 206, 217, 235, 246, 255, 265, 274, 283, 291, 300, 309, 319, 328, 337, 345, 352, 366, 367, 378, 379, 393, 394, 402, 402
chocolate, semi-sweet 5, 33, 42, 131, 157, 193, 204, 223, 350, 359, 372, 393, 410
chocolate, semi-sweet baker's 5, 21, 32, 40, 57, 84, 131, 141, 155, 169, 191, 202, 223, 232, 350, 359, 372, 393, 410
chocolate, sweet 21, 31, 41, 129, 159, 193, 205, 221, 350, 359, 372, 393, 410
chocolate syrup 5, 34, 42, 57, 83, 134, 142, 157, 169, 179, 194, 207, 218, 234, 246, 256, 266, 275, 284, 293, 302, 310, 320, 329, 337, 346, 350, 364, 380, 395, 408
chocolate, unsweetened baking 4, 21, 28, 39, 56, 66, 81, 129, 141, 154, 168, 188, 200, 224, 351, 359, 372, 391, 410
choking feeling 77
CHOLECALCIFEROL 98
CHOLECALCIFEROL 99
CHOLEDYL 88
CHOLERGOL 88
cholestasis 313
CHOLESTEROL 103, 104, 371
cholesterol formation in liver, increased 139
cholesterol, decreased 178
cholesterol, increased 37, 43, 49, 58, 67, 76, 77, 85, 86, 88, 89, 91, 96, 98, 99, 106, 112, 118, 119, 139, 152, 177, 187, 213, 227, 238
cholesterol solubility, decreased 313
CHOLESTYRAMINE 69, 140
CHOLINE 86
CHOLINE BROMIDE HEX-AMETHYLENEDICAR-BAMATE 87
CHOLINE CHLORIDE 86, 87
CHOLINE CHLORIDE CAR-BAMATE 87
CHOLINE CHLORIDE DIHYDROGEN PHOSPHATE 87

CHOLINE CHLORIDE PHOSPHATE 87
CHOLINE CHLORIDE SUC-CINATE 87
CHOLINE DEHYDRO-CHOLATE 86, 87
CHOLINE DICHLORIDE 87
CHOLINE DIHYDROGEN CITRATE 86, 88
CHOLINE ESTERASE 86
CHOLINE OROTATE 88
CHOLINE PHOSPHATE CHLORIDE 87
CHOLINE PHOSPHORIC ACID ESTER 87
CHOLINE PHOSPHORIC ACID ESTER CHLORIDE 87
CHOLINE SALICYLATE 86
CHOLINE SALICYLATE ACID SALT 88
CHOLINE SUCCINATE DICHLORIDE 87
CHOLINE THEOPHYLLINATE 86, 88
CHOLINESTERASE 106
CHOLINOPHYLLINE 88
chondroitin-4-sulphate 213
chop, lamb 139, 212
chop, pork 139
CHOTHYN 88
CHROMIUM DECREASED 214
CHROMIUM, HEXAVALENT 140
CHROMIUM, TRIVALENT 139-140, 212, 214, 385
CHROMOTRICHIA FACTOR 90
chuck arm pot roast excluding fat, braised beef 17, 27, 46, 54, 61, 70, 81, 136, 144, 155, 172, 172, 183, 189, 205, 219, 228, 240, 250, 260, 268, 277, 286, 295, 304, 313, 323, 331, 340, 364, 376, 385, 405
chuck arm pot roast including fat, braised beef 18, 28, 47, 54, 62, 70, 82, 135, 145, 156, 174, 174, 184, 191, 206, 220, 228, 242, 251, 260, 270, 278, 287, 296, 305, 315, 324, 332, 341, 361, 373, 387, 407
chuck blade roast excluding fat, braised beef 17, 27, 47, 54, 61, 70, 83, 134, 144, 155, 172, 172, 183, 190, 205, 219, 228, 240, 251, 260, 268, 277, 286, 295, 304, 313, 323, 332, 340, 363, 375, 386, 406
chuck blade roast including fat, braised beef 18, 28, 47, 55, 62, 70, 83, 134, 145, 156, 174, 174, 184, 192, 207, 219, 228, 242, 252, 261, 270, 278, 288, 296, 306, 315, 325, 333, 342, 360, 373, 387, 407

chuck, braised/roasted/stewed veal 17, 27, 44, 135, 155, 193, 206, 221, 364, 375, 386, 405
CIAMIN 91, 97
CICHORIGENIN 113
cigarette smoke 140
cigarette smoke, protection from 98
cigarette smoking 126
cinnamon 3, 92, 127, 153, 170, 196, 201, 221, 231, 349, 362, 378, 393, 409
CIPCA 91, 97
CIPLAMIN H 75
CIRANTIN 113
circulatory deterioration 76, 106, 114, 118
cirrhosis of liver 9, 74, 77, 85, 86, 88, 96, 99, 108, 119
CIRROCOLINA 88
CIS-LINOLEIC ACID 103
cisco, raw 164, 203, 220, 243, 253, 262, 271, 280, 289, 297, 307, 317, 326, 334, 343, 367, 379, 389, 401
cisco, smoked 7, 19, 30, 47, 55, 62, 70, 84, 132, 143, 164, 175, 183, 193, 205, 216, 236, 244, 254, 263, 272, 281, 289, 298, 308, 317, 326, 335, 343, 365, 376, 390, 403
CITICOLINE 87
CITRACHOLINE 88
CITRIN 112
CITROVORUM FACTOR 78
CITRULLINE 240
citrulline, internal conversion of 248
citrus marmalade 94, 131, 164, 196, 210, 222, 350, 363, 397, 408
citrus pectin 213
CL 139
CL⁻ 139
clam liquid 134, 142, 176, 181, 217, 237, 358, 371, 397, 398
clams 29, 89, 104, 140, 152, 212
clams, breaded and fried 8, 28, 47, 69, 130, 142, 154, 175, 192, 204, 216, 233, 244, 253, 264, 272, 281, 290, 299, 308, 318, 326, 335, 344, 354, 365, 376, 390, 404
clams, cooked 8, 17, 33, 47, 62, 71, 95, 132, 150, 164, 170, 189, 201, 217, 235, 241, 251, 263, 270, 279, 288, 297, 306, 315, 325, 333, 342, 367, 380, 388, 401
clams, raw 8, 48, 54, 69, 130, 142, 154, 176, 179, 192, 204, 220, 233, 244, 254, 264, 272, 281, 290, 299,

308, 318, 327, 335, 344, 357, 368, 380, 391, 400
CLARETIN-12 68
clearance of blood clots, decreased 322
CLINOLAMIDE 103
CLOROCHOLINE CHLORIDE 87
CLOTIAMINA 23
clover 114
cloves 3, 25, 91, 127, 154, 168, 194, 200, 217, 234, 350, 361, 374, 393, 409
CO 116, 140
CO^{++} 140
CO^{+++} 140
CO^{+2} 140
COB(I)ALAMIN 69, 75
COB(II)ALAMIN 69, 75
COBALAMIN 68
COBALEX 75
COBALIN 68
COBALION 76
COBALT 140, 141, 153
COBALTAMIN S 76
COBAMAMIDE 76
COBAMAMIDUM 76
COBAMIN 68
COBAMINE 68
COBANZYME 76
COBAZYMASE 76
COBINAMIDE DICYANIDE 75
COBIONE 68
cocoa butter 103, 104
cocoa powder, dry 14, 25, 39, 128, 154, 168, 188, 200, 350, 362, 374, 390, 410
coconut 89, 101, 104, 186
coconut cream 21, 40, 95, 135, 142, 157, 178, 194, 204, 223, 234, 244, 255, 265, 274, 283, 292, 300, 309, 319, 328, 337, 345, 355, 361, 372, 393, 406
coconut custard pie 4, 18, 29, 42, 129, 163, 194, 208, 217, 352, 364, 376, 393, 405
coconut, dried 18, 32, 40, 51, 61, 81, 132, 142, 155, 169, 178, 191, 201, 221, 232, 243, 254, 264, 273, 282, 291, 299, 309, 319, 327, 336, 344, 352, 359, 372, 392, 410
coconut, dried sweetened flaked 21, 35, 41, 54, 63, 84, 135, 142, 158, 170, 194, 205, 222, 234, 244, 255, 265, 274, 283, 292, 300, 309, 320, 328, 337, 345, 351, 360, 372, 394, 408
coconut, dried sweetened shredded 21, 36, 42, 95, 133, 142, 158, 170, 178, 194, 204, 217, 232, 245, 255, 265, 274, 283, 292, 300, 309, 320, 328, 337, 345, 351, 359, 372, 394, 409

coconut, dried toasted 132, 142, 155, 169, 178, 191, 201, 221, 232, 244, 254, 264, 274, 282, 291, 300, 309, 319, 328, 336, 345, 351, 359, 372, 393, 410
coconut milk 21, 40, 95, 133, 143, 158, 170, 179, 195, 205, 222, 234, 245, 255, 265, 274, 283, 292, 301, 309, 320, 329, 337, 346, 356, 364, 373, 394, 403
coconut oil 21, 40, 95, 101, 103, 104, 133, 143, 158, 170, 179, 195, 205, 222, 234, 245, 255, 265, 274, 283, 292, 301, 309, 320, 329, 337, 346, 356, 359, 364, 371, 373, 394, 403
coconut, raw 18, 36, 41, 55, 66, 80, 95, 134, 142, 157, 171, 178, 194, 203, 222, 233, 244, 255, 265, 274, 283, 292, 300, 309, 320, 328, 337, 345, 353, 361, 372, 394, 407
coconut water 22, 34, 43, 57, 66, 95, 132, 150, 165, 172, 198, 206, 217, 237, 246, 256, 266, 276, 284, 294, 302, 311, 321, 330, 338, 347, 356, 370, 382, 397, 398
cod 104, 105, 212
cod, canned 8, 17, 33, 47, 62, 71, 95, 132, 150, 164, 170, 189, 201, 217, 235, 243, 252, 262, 270, 279, 288, 296, 307, 316, 325, 333, 342, 367, 383, 388, 401
cod, cooked 8, 17, 33, 47, 62, 71, 95, 134, 150, 164, 170, 193, 206, 218, 235, 243, 252, 262, 270, 279, 288, 296, 307, 316, 325, 333, 342, 367, 380, 388, 401
cod, dried and salted 8, 14, 28, 44, 50, 59, 69, 95, 128, 144, 156, 169, 188, 200, 214, 233, 240, 249, 259, 268, 276, 285, 294, 304, 313, 322, 331, 339, 362, 379, 385, 408
cod liver oil 6, 99, 101
cod, raw 8, 17, 33, 47, 56, 62, 72, 95, 134, 150, 164, 171, 184, 191, 202, 220, 235, 244, 253, 262, 272, 280, 289, 298, 308, 317, 326, 335, 343, 368, 380, 390, 400
CODEINE 69, 140
CODROXOMIN 75
COENZYME B₁₂ 76
coenzyme pyridoxal phosphate levels decreased 58

Index

COENZYME Q 115, 116
COENZYME Q₀ 114
COENZYME Q₁ 114, 116
COENZYME Q₂ 114
COENZYME Q₃ 114
COENZYME Q₄ 115
COENZYME Q₅ 115
COENZYME Q₆ 115
COENZYME Q₇ 115
COENZYME Q₈ 116
COENZYME Q₉ 116
COENZYME Q₁₀ 116
coenzyme Q₁₀, decreased 117
COENZYME R 89
COFFEE 199
coffee 42, 107, 126, 138, 151, 151, 164, 177, 182, 199, 210, 224, 238, 248, 257, 267, 276, 285, 294, 312, 322, 339, 348, 358, 371, 397, 398
coffee, freeze dried 125
coffee, restaurant 36, 42, 138, 151, 166, 177, 199, 210, 224, 238, 397, 398
COFLAVINASE 37
cola 138, 150, 166, 177, 198, 211, 223, 238, 354, 369, 399
COLASCOR 91, 97
COLCHICINE 69, 140
cold, common 9, 67, 96, 99, 101, 105, 113, 138
cold, decreased 118
cold sensitivity 91, 152
coleslaw 3, 18, 33, 42, 57, 64, 74, 80, 92, 131, 150, 163, 176, 181, 197, 207, 222, 237, 247, 256, 266, 275, 284, 293, 302, 311, 321, 329, 338, 347, 354, 368, 379, 395, 400
coleslaw, restaurant 21, 36, 56, 64, 74, 80, 95, 131, 150, 163, 175, 207, 217, 237, 354, 366, 376, 395, 402
COLETYL 87
colic 91, 123, 168, 187
COLITE 87
colitis 9, 67, 96, 101, 105, 138, 166, 177, 186, 199, 211, 384
collagen formation, deterioration of 213, 303
collagen status, deterioration of 312
collapse of blood vessels 214
collards, boiled 2, 22, 34, 42, 57, 66, 83, 94, 129, 144, 164, 176, 180, 198, 210, 222, 234, 247, 256, 266, 275, 284, 293, 302, 311, 321, 329, 338, 347, 357, 371, 382, 396, 398
collards, frozen and boiled 2, 20, 31, 41, 57, 64, 79, 92, 127, 149, 160, 171, 179, 197, 206, 221, 236, 246, 255, 265, 274, 283, 292, 301,

310, 320, 328, 337, 346, 355, 369, 380, 394, 399
colon cancer 384
color dreaming 69
color on neck, increased 124
color perception, decreased 69
coma 50, 167, 168, 214
comfrey 213
common cold 9, 67, 96, 99, 101, 105, 123, 138
compulsion 322
CONCEMIN 91, 97
CONDOL 99
confusion 37, 43, 58, 77, 126, 199, 214, 227
conjunctivitis 9, 36, 43, 49, 67, 96
connective tissue 213
connective tissue, increased 58
constipation 9, 11, 23, 43, 49, 85, 86, 90, 96, 98, 99, 101, 105, 123, 124, 138, 153, 167, 187, 199, 211, 358
constipation, chronic 384
contaminants 123, 124, 125, 126, 138, 139, 140, 152, 153, 167, 186, 187, 211, 213, 225, 226, 227
convulsions 50, 58, 77, 106, 127, 151, 167, 258, 313
cookie, oatmeal 5, 14, 30, 39, 53, 65, 74, 81, 129, 145, 156, 170, 193, 206, 216, 234, 350, 360, 374, 393, 410
cookie, shortbread 3, 14, 28, 38, 55, 81, 132, 148, 158, 175, 194, 210, 216, 235, 350, 359, 373, 392, 410
cookie, sugar 4, 16, 31, 39, 129, 159, 195, 210, 216, 350, 360, 374, 392, 409
coordination, decreased 49, 228, 258, 339
COPPER 140, 141-151, 153, 228
copper decreased 186, 228
copper, tissue overloads of 267
COPPERNICKEL 187
CO-Q 115, 116
CO-Q₀ 114
CO-Q₁ 114, 116
CO-Q₂ 114
CO-Q₃ 114
CO-Q₄ 115
CO-Q₅ 115
CO-Q₆ 115
CO-Q₇ 115
CO-Q₈ 116
CO-Q₉ 116
CO-Q₁₀ 116
CORDES VAS 10
CORENALIN 87
coriander leaf, dried 11, 25, 37, 91, 127, 154, 168, 189, 200, 218, 351, 365, 391

coriander, raw 2, 28, 129, 158, 172, 197, 201, 222, 357, 370, 394, 398
coriander seed 25, 127, 154, 168, 188, 200, 221, 230, 350, 363, 375, 391, 409
corn 77, 85, 86, 89, 101, 103, 104, 108, 186, 187, 227
corn, boiled yellow 4, 14, 33, 39, 65, 80, 94, 138, 149, 163, 171, 180, 194, 206, 222, 235, 246, 255, 265, 274, 282, 292, 300, 309, 319, 329, 337, 345, 352, 367, 379, 394, 403
corn, canned creamed yellow 4, 22, 34, 40, 56, 65, 80, 94, 138, 149, 164, 175, 182, 196, 209, 216, 235, 247, 256, 265, 275, 283, 293, 301, 310, 320, 329, 337, 346, 353, 368, 381, 395, 401
corn, canned white 3, 19, 33, 39, 94, 137, 165, 196, 207, 217, 352, 367, 381, 394, 401
corn, canned yellow 3, 21, 32, 39, 51, 65, 80, 94, 137, 149, 164, 173, 181, 196, 207, 217, 235, 246, 256, 265, 275, 283, 292, 301, 310, 320, 329, 337, 346, 353, 368, 380, 394, 401
corn flakes 2, 11, 25, 37, 56, 59, 78, 92, 125, 138, 148, 154, 176, 181, 196, 210, 214, 236, 245, 254, 264, 273, 281, 292, 299, 309, 319, 328, 336, 344, 349, 360, 381, 384, 392, 410
corn flour, yellow 3, 14, 34, 39, 137, 158, 195, 224, 349, 360, 379, 392, 409
corn flour, white 14, 34, 39, 137, 158, 195, 224, 349, 360, 379, 392, 409
corn, frozen and boiled yellow 3, 18, 33, 39, 56, 64, 80, 95, 138, 150, 165, 174, 181, 196, 209, 223, 236, 246, 255, 265, 274, 282, 292, 300, 309, 320, 329, 337, 345, 352, 368, 383, 384, 394, 401
corn, frozen creamed yellow 4, 18, 35, 41, 95, 136, 165, 196, 209, 217, 353, 368, 380, 394, 401
corn, frozen white 5, 21, 33, 39, 94, 136, 164, 195, 206, 224, 352, 367, 379, 394, 402
corn germ 4, 11, 25, 38, 59, 79, 95, 142, 154, 168, 188, 200, 221, 228, 242, 262, 272, 279, 290, 299, 308,

318, 336, 343, 351, 359, 373, 390
corn grits 22, 36, 42, 67, 77, 151, 165, 177, 184, 198, 211, 238, 247, 255, 265, 275, 283, 294, 301, 310, 321, 330, 337, 346, 354, 369, 382, 395, 400
corn grits, instant 16, 33, 40, 55, 66, 84, 137, 151, 162, 177, 198, 211, 217, 238, 354, 368, 383, 395, 400
corn meal 139
corn meal, cooked enriched white degermed 19, 35, 41, 138, 164, 198, 211, 217, 354, 369, 382, 396, 399
corn meal, cooked enriched yellow degermed 4, 19, 35, 41, 138, 164, 198, 211, 217, 354, 369, 382, 396, 399
corn meal, cooked unenriched 22, 36, 43, 138, 165, 198, 211, 217, 354, 369, 382, 396, 399
corn meal, cooked unenriched yellow degermed 4, 22, 36, 43, 138, 165, 198, 211, 217, 354, 369, 382, 396, 399
corn meal, enriched white degermed 13, 27, 38, 137, 156, 195, 209, 224, 349, 360, 379, 392, 409
corn meal, enriched yellow degermed 3, 13, 27, 38, 137, 156, 195, 209, 224, 349, 360, 379, 392, 409
corn meal muffins 3, 14, 28, 39, 129, 158, 192, 209, 216, 351, 361, 376, 392, 408
corn meal muffins, whole ground 3, 15, 30, 39, 128, 159, 191, 209, 216, 351, 362, 376, 392, 408
corn meal, unenriched 15, 34, 39, 137, 160, 195, 209, 224, 349, 360, 379, 392, 409
corn meal, unenriched yellow degermed 3, 15, 34, 39, 137, 160, 195, 209, 224, 349, 360, 379, 392, 409
corn meal, white bolted 14, 33, 39, 133, 158, 190, 206, 224, 350, 360, 378, 392, 409
corn meal, white whole ground 13, 32, 38, 133, 157, 189, 205, 224, 350, 361, 378, 392, 409
corn meal, yellow bolted 3, 14, 33, 39, 133, 158, 190, 206, 224, 350, 361, 378, 392, 409
corn meal, yellow whole ground 3, 13, 32, 38, 133, 157, 189, 205, 224, 350, 361, 378, 392, 409

corn oil 99, 101, 103, 104, 108, 116, 117, 227, 359, 371
corn pone 4, 15, 31, 39, 54, 64, 74, 82, 129, 148, 161, 172, 194, 209, 217, 235, 351, 364, 377, 393, 406
corn, raw sweet 116
corn starch 349, 360, 409
corn syrup 131, 155, 198, 211, 219, 350, 362, 408
corn tortilla 12, 25, 38, 56, 61, 80, 128, 143, 155, 169, 179, 192, 208, 217, 233, 351, 364, 378, 384, 392, 407
corn, whole grain 116
corn with red and green peppers, canned yellow 3, 21, 32, 40, 94, 138, 148, 162, 172, 196, 208, 216, 236, 246, 256, 265, 274, 283, 292, 301, 309, 320, 329, 337, 346, 353, 368, 380, 394, 401
cornbread 4, 15, 31, 39, 54, 64, 74, 82, 129, 148, 161, 172, 194, 209, 217, 235, 351, 364, 377, 393, 406
cornea deterioration 50
cornea, softening of 6
cornea, thickening of 6
corneal opacities 139
corned beef brisket, cooked 21, 30, 47, 53, 62, 71, 136, 144, 158, 176, 183, 194, 209, 214, 229, 244, 253, 262, 272, 281, 289, 298, 308, 318, 326, 335, 343, 358, 363, 374, 389, 404
corned beef, canned 22, 30, 47, 64, 71, 95, 148, 157, 175, 185, 194, 209, 215, 230, 242, 251, 261, 270, 279, 288, 296, 306, 316, 325, 333, 342, 363, 375, 387, 405
corned beef loaf, jellied 32, 48, 56, 64, 71, 84, 94, 135, 148, 157, 176, 182, 195, 210, 215, 230, 242, 253, 262, 271, 280, 288, 298, 307, 316, 326, 334, 343, 366, 377, 388, 402
coronary thrombosis 101, 105
CORTICOSTEROIDS 98, 228
CORTISONE 59, 91, 199
cottage cheese 7, 56, 65, 72, 81, 99, 101, 130, 152, 165, 177, 186, 193, 210, 212, 216, 234, 244, 254, 263, 272, 281, 290, 298, 308, 318, 327, 334, 344, 357, 367, 378, 391, 401
cotton 213
cottonseed oil 101, 103, 104, 116, 117, 359, 371
cough 124, 186, 187
COVIT 68

COVITOL 100
cow milk 88, 89, 99, 101, 103, 104, 108, 116
CR 139, 140
crab 186
crab cakes 69, 129, 142, 160, 170, 191, 204, 216, 230, 241, 253, 263, 271, 280, 289, 297, 307, 317, 325, 334, 343, 358, 366, 377, 389, 402
crab, canned blue 32, 48, 73, 129, 142, 162, 170, 189, 202, 216, 230, 241, 253, 263, 271, 280, 288, 297, 307, 317, 325, 334, 343, 367, 379, 389, 401
crab, cooked alaska king 9, 19, 33, 48, 130, 141, 162, 189, 205, 215, 228, 242, 253, 263, 271, 280, 289, 297, 307, 317, 325, 334, 343, 367, 379, 389, 401
crab, cooked blue 69, 91, 129, 142, 161, 170, 191, 204, 216, 230, 241, 253, 263, 271, 280, 289, 297, 307, 317, 325, 334, 343, 367, 379, 389, 401
crab, king 212
crab, raw alaska king 9, 19, 34, 48, 130, 141, 163, 182, 191, 207, 215, 229, 242, 253, 263, 271, 280, 289, 298, 307, 317, 325, 335, 343, 368, 380, 389, 401
crab, raw blue 129, 142, 162, 170, 178, 190, 204, 216, 230, 242, 253, 263, 271, 280, 289, 298, 307, 317, 325, 335, 343, 368, 379, 390, 401
crab, raw dungeness 19, 30, 47, 130, 142, 164, 170, 181, 192, 203, 216, 230, 242, 253, 263, 272, 280, 289, 298, 307, 317, 325, 335, 343, 358, 368, 380, 390, 401
crab, raw queen 132, 193, 208, 216, 242, 253, 263, 271, 280, 289, 298, 307, 317, 325, 334, 343, 368, 379, 389, 400
crabapples 5, 21, 36, 43, 94, 133, 148, 165, 177, 181, 198, 207, 224, 248, 257, 267, 276, 285, 294, 302, 312, 322, 330, 339, 347, 353, 368, 382, 397, 410
crackers, animal 4, 21, 31, 41, 130, 164, 194, 210, 216, 349, 360, 377, 392, 410
crackers, cheese 13, 25, 37, 129, 142, 155, 173, 191, 207, 214, 234, 350, 359, 372, 391, 410
crackers, graham 13, 27, 38,

Index

53, 65, 81, 131, 145, 156, 170, 194, 208, 216, 234, 349, 360, 376, 392, 410
crackers, oyster 39, 132, 160, 195, 209, 215, 350, 360, 376, 392, 410
crackers, rusk 4, 16, 28, 39, 132, 160, 194, 208, 217, 350, 360, 377, 391, 409
crackers, ry-krisp 14, 25, 38, 50, 59, 80, 129, 141, 156, 168, 188, 201, 215, 229, 349, 362, 379, 384, 391
crackers, rye & wheat 101
crackers, saltine 13, 25, 38, 54, 81, 129, 144, 156, 170, 195, 209, 214, 234, 350, 360, 376, 391, 410
crackers, soda 35, 39, 132, 159, 195, 209, 215, 350, 360, 375, 391, 410
crackers, whole grain rye 14, 28, 39, 130, 155, 188, 201, 215, 349, 361, 379, 391, 409
crackers with peanut butter, cheese 4, 22, 33, 38, 130, 162, 192, 206, 215, 350, 359, 373, 390, 410
crackers, zwieback 39, 134, 195, 208, 217, 350, 360, 377, 391, 409
cramps 58, 98, 124, 126, 168, 187, 384
cramps, abdominal 50, 86, 151
cramps, muscle 88, 139, 199, 214
cramps, occasional 167
cranberry 78
cranberry beans, boiled 14, 33, 41, 56, 65, 79, 130, 143, 157, 170, 179, 193, 202, 225, 233, 244, 254, 264, 273, 282, 290, 299, 308, 318, 327, 336, 344, 352, 366, 381, 384, 391, 404
cranberry beans, canned 21, 35, 41, 56, 66, 79, 96, 131, 144, 158, 171, 180, 195, 205, 216, 234, 245, 255, 264, 273, 282, 291, 300, 309, 319, 328, 336, 345, 353, 368, 382, 393, 401
cranberry juice cocktail 35, 57, 92, 138, 151, 166, 177, 181, 199, 211, 224, 238, 353, 369, 400
cranberry juice cocktail, restaurant 5, 23, 36, 43, 57, 67, 84, 92, 138, 151, 166, 177, 180, 199, 211, 223, 238, 353, 369, 383, 400
cranberry sauce 4, 22, 36, 43, 56, 65, 84, 93, 137, 148, 165, 177, 181, 199, 210, 224, 237, 354, 369, 382, 397, 400
cranium pressure, increased 6

crayfish, cooked 32, 47, 70, 95, 131, 142, 156, 171, 189, 203, 219, 232, 240, 252, 263, 270, 279, 288, 296, 306, 316, 323, 333, 342, 367, 379, 388, 401
crayfish, raw 33, 47, 54, 70, 95, 132, 142, 157, 172, 189, 205, 220, 233, 242, 253, 263, 271, 280, 289, 298, 307, 317, 325, 334, 343, 368, 379, 389, 400
cream 99, 101, 103, 104, 152
cream, half and half 7, 18, 31, 55, 65, 73, 129, 166, 175, 195, 209, 221, 235, 246, 255, 265, 274, 283, 292, 300, 309, 319, 328, 337, 345, 356, 366, 376, 394, 400
cream, heavy whipping 6, 31, 55, 74, 83, 130, 177, 196, 210, 221, 237, 247, 256, 265, 274, 283, 292, 300, 310, 320, 329, 337, 346, 357, 361, 372, 395, 405
cream, light 8, 33, 48, 66, 74, 130, 144, 155, 177, 196, 211, 222, 237, 246, 255, 265, 274, 283, 292, 300, 309, 320, 328, 337, 345, 354, 367, 376, 384, 385, 408
cream, light whipping 6, 31, 55, 74, 83, 130, 177, 196, 210, 221, 236, 247, 256, 265, 274, 283, 292, 300, 309, 320, 328, 337, 346, 357, 362, 373, 395, 404
cream, medium 7, 31, 55, 65, 74, 129, 166, 177, 195, 209, 221, 236, 246, 256, 265, 274, 283, 292, 300, 309, 320, 328, 337, 346, 357, 363, 373, 394, 403
cream puff 3, 21, 30, 43, 129, 143, 154, 199, 210, 218, 352, 364, 375, 392, 405
cream, sour 7, 30, 54, 73, 82, 128, 166, 176, 195, 209, 221, 236, 274, 283, 292, 300, 309, 320, 328, 345, 356, 364, 374, 394, 402
cream, sweetened coconut 22, 34, 138, 143, 164, 179, 197, 210, 221, 235, 245, 255, 265, 274, 283, 292, 300, 309, 320, 328, 337, 346, 355, 365, 374, 394, 402
creatine in urine, increased 100
creme de menthe 147, 166, 182, 223, 351, 360, 383, 408
cress, boiled garden 2, 19, 30, 40, 93, 130, 162, 196, 203, 223, 356, 370, 380, 395, 398
cress, raw garden 2, 17, 27,

39, 62, 92, 129, 159, 355, 369, 380, 394, 399
cretinism 153
croaker, breaded and fried 131, 148, 162, 170, 191, 203, 216, 235, 244, 253, 262, 271, 280, 289, 298, 307, 317, 326, 334, 343, 355, 364, 376, 389, 404
croaker, raw 134, 149, 164, 170, 183, 191, 203, 220, 235, 244, 253, 262, 272, 280, 289, 298, 308, 317, 326, 335, 343, 367, 378, 390, 401
croup 9, 96
crystals in urine 76
CRYSTAMIN 68
CS 138
CS$^+$ 138
CU 141
CU$^+$ 141
CU^{++} 141
CU^{+2} 141
cucumber 116, 187
cucumber, raw 4, 21, 36, 41, 55, 65, 81, 95, 134, 150, 165, 176, 181, 198, 208, 224, 237, 247, 257, 267, 276, 285, 294, 302, 312, 322, 330, 339, 347, 357, 371, 382, 384, 397, 398
cumin seed 2, 13, 25, 38, 127, 153, 168, 188, 200, 217, 229, 351, 360, 373, 389, 409
CUMOTOCOPHEROL 102
currant 78
currants, black 4, 20, 34, 41, 54, 65, 91, 112, 130, 147, 159, 172, 180, 196, 204, 224, 236, 353, 368, 381, 395, 400
currants, red 4, 21, 34, 42, 57, 65, 92, 131, 146, 161, 175, 181, 196, 205, 224, 237, 354, 369, 382, 395, 400
currants, white 4, 21, 34, 42, 57, 65, 92, 131, 146, 161, 175, 181, 196, 205, 224, 237, 354, 369, 382, 395, 400
curry powder 3, 13, 25, 38, 127, 154, 168, 188, 200, 213, 221, 230, 350, 362, 375, 390, 409
cusk, raw 31, 47, 55, 60, 71, 135, 151, 162, 171, 184, 191, 202, 221, 236, 243, 253, 262, 271, 280, 289, 297, 307, 317, 326, 334, 343, 368, 380, 389, 401
custard 9, 21, 29, 48, 129, 164, 194, 209, 218, 354, 367, 377, 393, 401
custard pie 4, 19, 30, 42, 129, 163, 194, 209, 216, 352, 364, 376, 393, 405

487   Index

cuttlefish, raw 22, 25, 48, 70,
  94, 129, 142, 154, 188, 203,
  216, 232, 243, 253, 263,
  272, 281, 289, 298, 308,
  317, 326, 335, 344, 358,
  368, 380, 390, 400
cyanide poisoning 77
CYANOCOBALAMIN 69, 74
CYANOCOBALAMIN ZINC TAN-
  NATE COMPLEX 75
CYCLOHEXANEHEXOL 85
CYCLOHEXITOL 85
CYCLOHEXYLAMIDE/
  LINOLEXAMIDE 103
CYCOBEMIN 68
CYCOCEL 87
CYCOGAN 87
CYCOLAMIN 68
CYKOBEMINET 68
CYS 249
CYSCHOLIN 87
CYSTEINE 77, 228, 249-257
cysteine, internal conversion
  of 257
*cysteine production,
  decreased* 294
cystic fibrosis 9, 96, 99, 101,
  108, 225
CYSTINE 49
cystine production, decreased
  294
CYTACON 68
CYTAMEN 68
CYTIDINE DIPHOSPHATE
  CHOLINE ESTER 87
CYTOBION 68
CYTOFLAV 37
CYTOFOL 78

D-CARNITINE 88
D-TRACETTEN 99
D3-VICOTRAT 99
daiquiri 5, 22, 58, 84, 95,
  138, 149, 166, 177, 199, 211,
  223, 238, 355, 365, 403
dairy 125
DAMBOSE 85
dandelion greens, boiled 2,
  16, 30, 93, 128, 158, 196,
  206, 221, 355, 369, 380,
  395, 399
dandelion greens, raw 2, 15,
  28, 92, 128, 156, 170, 196,
  202, 218, 355, 369, 380,
  394, 400
dandruff 9, 58, 67, 69, 101,
  104, 212, 371
danish pastry 4, 14, 28, 38,
  129, 158, 175, 194, 210, 216,
  234, 351, 360, 374, 393, 408
DASKIL 37
dates 125
dates, dried 4, 17, 32, 38, 51,
  63, 81, 131, 143, 160, 170,
  180, 196, 201, 224, 236,
  247, 255, 266, 275, 284,
  293, 301, 311, 320, 328,

338, 346, 350, 363, 381,
  395, 408
DAVITAMON C 91, 97
DAVITIN 99
DBC 76
DDT 6
deafness and muteness 152
DECAPS 99
*decreased radiation
  resistance* 5, 9, 23, 43, 49,
  50, 58, 67, 85, 86, 90, 96,
  97, 100, 101, 104, 113, 117,
  212, 228, 238, 257, 303,
  330, 371
*decreased red blood cells,
  number of* 88
DEE-RON 99
deficiency symptoms 1, 6, 11,
  24, 37, 43, 49, 58, 67, 68,
  69, 76, 77, 78, 85, 86, 88,
  89, 90, 91, 98, 100, 103,
  104, 106, 107, 110, 111, 112,
  114, 115, 116, 118, 119, 120,
  121, 123, 124, 125, 126, 138,
  139, 140, 141, 151, 152, 153,
  167, 168, 177, 186, 187, 199,
  211, 212, 213, 214, 225, 226,
  227, 228, 239, 248, 249,
  257, 258, 259, 267, 268,
  276, 285, 294, 303, 312,
  313, 322, 331, 339, 349,
  358, 371, 384, 385, 397
dehydration 225, 228, 397
DEHYDROASCORBIC ACID 90
DEHYDROCHOLIC ACID SALT
  OF CHOLINE 87
DEHYDRORETINOL 6, 10
DELSTEROL 99
DELTA-CAROTENE 1
DELTALIN 99
DELTA-TOCOPHEROL 100, 102
DELTA-VITAMIN E 102
demineralization of teeth 187
dentition, late 127, 151
DEPARAL 99
DEPOGAMMA 75
depression 11, 24, 37, 43, 49,
  58, 69, 78, 89, 90, 91, 98,
  105, 153, 167, 186, 187, 199,
  211, 214, 303, 322, 330, 385
*depression/emotional agita-
  tion, alternating* 37, 43
depression, suicidal 259
depressive states, manic 228
DE-RAT CONCENTRATE 99
DERATOL 99
DERMAIROL 10
dermatosclerosis 106
DETALUP 99
*development arrested,
  physical/mental* 152
*development in chicks,
  deterioration of* 119
diabetes 9, 23, 36, 43, 49,
  67, 74, 77, 96, 99, 101, 138,
  140, 166, 177, 185, 211, 228,
  238, 349
DIACTOL 99

*diagnostic aid with tracking
  of radioactive isotope* 74, 75
dialysis-induced anemia 88
diarrhea 6, 9, 23, 36, 37, 43,
  49, 58, 67, 69, 78, 84, 86,
  88, 91, 96, 98, 104, 105,
  107, 124, 125, 138, 139, 153,
  166, 168, 177, 186, 187, 199,
  211, 225, 358, 371
DIBASIC PHOSPHATE 106
DICUMARO 6
DICUMAROL 107
DIELRIN 6
difficulties, swallowing 167
DIFOSFOCIN 87
*digesting carbohydrates,
  difficult* 11
digestion deterioration 11, 23,
  24, 36, 37, 43, 49, 58, 67,
  77, 78, 85, 90, 91, 100, 106,
  117, 119, 139, 199, 313, 322,
  385
*digestion deterioration in
  pigeons* 67
DIHYDROTACHYSTEROL 98,
  100
DILCIT 86
DILEXPAL 86
dill pickles 4, 36, 94, 132,
  161, 197, 207, 214, 357, 371,
  382, 396, 398
dill seed 4, 13, 25, 38, 127,
  154, 168, 189, 200, 229,
  243, 263, 272, 281, 290,
  299, 308, 318, 342, 350,
  362, 375, 390, 409
dill weed, dried 59, 127, 153,
  168, 188, 200, 217, 231,
  350, 362, 389, 409
DIMETHYLGLYCINE 76
dinner rolls 13, 26, 38, 52,
  65, 80, 128, 147, 156, 173,
  193, 209, 216, 234, 350,
  362, 377, 392, 408
DIPEGYL 43
disease, muscle 88
disinfectant in community
  water supplies 213
DISTIVIT 68
DITHIOLAN-3, PEN-
  TANAMIDE 112
DITIOVIT 24
DIURETICS 91, 168, 199
DIURETICS, MERCURIAL 126
diverticulitis 85, 358
diverticulosis 384
DIVIT URTO 99
dizziness 24, 36, 43, 49, 58,
  67, 74, 77, 85, 86, 96, 100,
  138, 153, 167, 168, 178, 187,
  214, 228, 285
DMG 76
DNA 49, 68
*DNA/RNA production,
  deterioration of* 187
DOBETIN 68
DOCELAN 75
DOCEMINE 68

Index

DOCEVITA 75
DOCIGRAM 68
DOCIVIT 68
dock, boiled 2, 21, 32, 41, 92, 131, 157, 169, 196, 204, 224, 246, 265, 274, 283, 292, 301, 310, 320, 337, 346, 357, 370, 380, 395, 398
dock, raw 2, 21, 32, 41, 92, 131, 157, 169, 196, 202, 224, 247, 254, 264, 274, 283, 293, 310, 320, 327, 337, 346, 357, 370, 380, 395, 398
DODECABEE 68
DODECAVITE 68
DODEX 68
dogfish, raw spiny 19, 366, 377, 390, 402
DOHYFRAL A 10
DOLONEVRAN 76
dolphinfish, raw 149, 160, 184, 202, 218, 235, 244, 253, 262, 271, 280, 289, 297, 307, 317, 326, 334, 343, 368, 380, 389, 401
DORAL 99
DORYL 87
dosage, pre-1958 MDR 5, 9, 23, 36, 96, 97, 99
dosage, 1958 RDA 5, 9, 23, 36, 96, 97, 99
dosage, 1974 PCS 5, 9, 23, 36, 38, 43, 49, 52, 58, 67, 74, 76, 77, 85, 89, 91, 96, 97, 99, 101, 110, 111, 118, 119, 120, 123, 124, 126, 138, 139, 140, 151, 153, 166, 167, 177, 186, 199, 212, 213, 225, 226, 227, 238, 330
dosage, 1974 RDA 5, 23, 36, 43, 49, 67, 74, 85, 96, 97, 99, 101
dosage, 1974 USRDA 5, 23, 36, 43, 49, 67, 74, 85, 96, 97, 99, 101
dosage, 1980 RDA 5, 9, 23, 36, 43, 49, 58, 67, 74, 85, 89, 99, 101, 108, 138, 139, 140, 151, 153, 166, 167, 177, 185, 186, 187, 199, 211, 212, 213, 225, 226, 238, 330, 397
dosage, 1980 SADDI 58, 89, 108, 139, 140, 151, 185, 186, 187, 211, 212, 213, 225, 226
dosage, 1980 USRDA 6, 9, 23, 36, 43, 49, 58, 67, 74, 85, 89, 96, 97, 99, 101, 138, 153, 166, 167, 177, 199, 238, 330
dosage, 1989 RDA 6, 9, 23, 36, 43, 49, 58, 67, 68, 74, 76, 77, 78, 85, 86, 89, 91, 96, 97, 99, 101, 104, 105, 106, 108, 110, 111, 112, 113, 117, 118, 119, 120, 123, 124, 125, 126, 138, 139, 140, 151, 152, 153, 166, 167, 177, 185, 186, 187, 199, 211, 212, 213, 225, 226, 227, 238, 330, 397
dosage, 1989 SADDI 49, 58, 67, 68, 76, 77, 78, 85, 86, 88, 89, 90, 91, 104, 105, 106, 110, 111, 112, 113, 117, 118, 119, 120, 123, 124, 125, 126, 139, 140, 151, 152, 153, 167, 185, 186, 187, 211, 212, 213, 225, 226, 227, 228
dosage, 1989 USRDA 6, 9, 23, 36, 43, 49, 58, 67, 68, 74, 76, 77, 78, 85, 86, 89, 90, 91, 96, 97, 99, 101, 104, 105, 106, 108, 110, 111, 112, 113, 117, 118, 119, 120, 123, 124, 124, 125, 126, 138, 139, 140, 152, 153, 166, 167, 177, 186, 199, 212, 213, 225, 226, 227, 228, 238, 330
Down's syndrome 313
dream recall, deterioration of 58
dream recall, increased 58
dreaming, color 69
dreaming, vivid 78
DRISDOL 99
DROXOMIN 75
drum, raw freshwater 130, 143, 161, 171, 179, 192, 205, 218, 234, 244, 254, 263, 272, 280, 289, 298, 308, 317, 326, 335, 343, 367, 378, 390, 401
drumstick with skin, fried chicken 8, 17, 28, 44, 50, 60, 73, 82, 134, 147, 159, 173, 182, 192, 206, 218, 231, 242, 250, 261, 269, 279, 287, 295, 305, 315, 324, 333, 341, 358, 364, 375, 387, 405
drumstick with skin, roasted chicken 7, 17, 29, 44, 50, 60, 73, 82, 134, 147, 159, 172, 183, 192, 206, 218, 231, 242, 250, 261, 269, 279, 287, 295, 305, 315, 324, 333, 341, 364, 376, 387, 404
drumstick with skin, stewed chicken 8, 19, 29, 46, 51, 63, 74, 83, 134, 148, 159, 174, 183, 193, 207, 218, 231, 242, 251, 261, 269, 279, 288, 296, 306, 316, 325, 333, 342, 365, 376, 387, 403
drumstick without skin, stewed chicken 8, 18, 28, 44, 50, 60, 73, 82, 135, 147, 160, 172, 183, 192, 206, 218, 230, 240, 250, 260, 268, 277, 286, 294, 304, 313, 323, 331, 340, 365, 377, 386, 403

DTPT 24
Duchenne-type muscular dystrophy 88
duck fat 359, 371
duck liver, raw 6, 69, 135, 141, 153, 189, 243, 252, 263, 271, 280, 289, 298, 306, 317, 325, 334, 342, 356, 366, 378, 389, 402
duck with skin, roasted 7, 15, 27, 45, 50, 63, 73, 83, 135, 143, 156, 175, 193, 207, 220, 232, 243, 252, 263, 271, 280, 289, 298, 307, 317, 326, 334, 343, 361, 373, 389, 406
duck without skin, roasted 8, 14, 25, 45, 50, 62, 73, 81, 135, 143, 156, 174, 191, 206, 219, 231, 242, 250, 262, 270, 279, 288, 296, 306, 316, 324, 332, 342, 365, 376, 388, 404
DUCOBEE 68
DUCOBEE-HY 75
dulse 139
DUODECIBIN 68
duodenal ulcers 50
DUOSCORB 91, 97
DUPHAFRAL D$_3$ 1000 99
DURADOCE 75
DURALTA-12 75
dystrophy, myotonic 88

ear infection 9, 96
ear noises 106, 178, 199
ears, simmered pork 22, 33, 48, 133, 159, 177, 197, 210, 217, 243, 264, 273, 281, 290, 299, 308, 318, 328, 336, 344, 365, 376, 390, 402
eating, increased 303
EBIVIT 99
eel 117
eel, cooked 6, 15, 34, 46, 65, 70, 132, 150, 163, 189, 203, 219, 232, 243, 252, 261, 270, 279, 287, 296, 306, 316, 325, 333, 342, 364, 375, 388, 405
eel, raw 7, 15, 34, 47, 56, 65, 70, 133, 150, 164, 182, 191, 205, 221, 233, 244, 253, 262, 271, 280, 289, 297, 307, 317, 326, 334, 343, 365, 376, 389, 403
EFA 103, 104
egg, chicken 17, 26, 50, 64, 71, 76, 79, 85, 90, 99, 101, 103, 104, 108, 117, 130, 139, 148, 152, 158, 176, 182, 186, 187, 192, 209, 217, 225, 226, 227, 233, 244, 252, 263, 272, 281, 290, 298, 308, 318, 326, 335, 343, 358, 366, 376, 391, 401

egg, duck 6, 15, 25, 48, 62, 69, 79, 130, 155, 174, 190, 207, 217, 233, 244, 252, 263, 272, 281, 290, 297, 307, 317, 325, 334, 343, 358, 365, 375, 391, 402
egg, goose 358, 365, 375, 390, 402
egg noodles 125
egg, quail 7, 16, 25, 64, 129, 155, 190, 366, 376, 391, 401
egg, turkey 16, 25, 129, 155, 192, 358, 365, 376, 390, 402
egg white 86, 89
egg white, chicken 27, 56, 74, 81, 134, 150, 166, 176, 185, 198, 209, 217, 238, 244, 252, 264, 272, 281, 290, 298, 308, 318, 326, 335, 343, 358, 369, 391, 399
egg, whole 86, 89
egg yolk, chicken 6, 14, 25, 50, 61, 70, 79, 128, 144, 154, 174, 181, 188, 210, 221, 230, 243, 252, 263, 271, 280, 290, 298, 307, 316, 325, 334, 343, 360, 372, 390, 407
egg yolk, raw 118
egg yolks 86, 89, 99, 103, 104, 118, 212
eggnog 7, 20, 29, 48, 53, 66, 73, 84, 95, 128, 165, 174, 194, 208, 220, 235, 246, 255, 265, 273, 282, 291, 300, 309, 319, 328, 336, 345, 354, 366, 377, 393, 401
eggplant 117
eggplant, boiled 4, 18, 36, 40, 57, 65, 81, 95, 137, 145, 165, 176, 181, 197, 206, 224, 237, 247, 257, 266, 275, 284, 293, 302, 311, 321, 330, 338, 347, 355, 370, 382, 396, 398
eggplant, raw 4, 17, 36, 41, 57, 64, 81, 95, 131, 145, 163, 176, 181, 197, 207, 224, 237, 247, 257, 266, 275, 284, 293, 302, 311, 321, 330, 338, 347, 355, 370, 384, 396, 398
EKA-SILICON 151
elderberries 3, 18, 33, 41, 56, 62, 78, 92, 131, 158, 197, 205, 247, 256, 267, 276, 284, 294, 302, 311, 321, 329, 337, 347, 353, 368, 380, 396, 401
ELDRIN 113
electronic devices 125
ELEMENT 3 167
ELEMENT 4 125
ELEMENT 5 125
ELEMENT 9 151

ELEMENT 11 214
ELEMENT 12 168
ELEMENT 13 123
ELEMENT 14 212
ELEMENT 15 187
ELEMENT 16 226
ELEMENT 17 139
ELEMENT 19 199
ELEMENT 20 126
ELEMENT 22 226
ELEMENT 23 227
ELEMENT 24 139, 140
ELEMENT 25 177
ELEMENT 26 153
ELEMENT 27 140
ELEMENT 28 187
ELEMENT 29 141
ELEMENT 30 228
ELEMENT 32 151
ELEMENT 33 124
ELEMENT 34 212
ELEMENT 35 126
ELEMENT 37 211
ELEMENT 38 225
ELEMENT 42 186
ELEMENT 47 213
ELEMENT 48 126
ELEMENT 50 226
ELEMENT 53 152
ELEMENT 55 138
ELEMENT 56 124
ELEMENT 74 227
ELEMENT 79 152
ELEMENT 82 167
EMBIOL 68
embryonic development, deterioration of 213
EMICHOLIN 87
EMOCICLINA 68
emotional agitation 6, 11, 24, 37, 43, 50, 58, 69, 78, 91, 98, 104, 105, 126, 141, 167, 168, 186, 285, 303, 313, 322, 339, 371, 385
emotional agitation/depression, alternating 37, 43
emotional deterioration 11, 24, 49, 69, 106, 127, 152, 322
emotional instability 37, 43, 126
emphysema 9, 77, 85, 96, 99, 101, 126
ENALLACHROME 113
enamels 124
enchilada 17, 32, 55, 64, 72, 131, 145, 159, 170, 195, 205, 216, 235, 353, 365, 377, 392, 403
endive, raw 2, 17, 32, 41, 51, 66, 79, 94, 130, 146, 162, 175, 179, 197, 204, 222, 234, 247, 256, 266, 275, 284, 293, 302, 311, 321, 330, 338, 347, 357, 370, 381, 396, 398
endocrine glands, degenerating 100

ENERGY 358-371
english muffin 13, 26, 38, 53, 66, 80, 128, 142, 156, 174, 194, 201, 215, 234, 351, 364, 379, 392, 407
english muffin, toasted 13, 26, 38, 52, 66, 80, 128, 142, 155, 172, 193, 200, 215, 234, 350, 362, 379, 391, 408
ENSIGN 87
ENZICOBA 76
enzyme deterioration 285
enzyme inhibition 125
enzyme system interference 125
enzyme toxicity, increased 211
enzymes 227
EPADYN-U 118
EPHYNAL 100
EPI-ABEREL 10
epilepsy 9, 43, 49, 58, 67, 74, 96, 99, 101, 138, 177, 313
epinephrine, decreased 98, 112
EPITELIOL 10
EPROLIN-S 100
EPSILAN 100
EPSILON-CAROTENE 1
EPSILON-TOCOPHEROL 100, 102
EPSILON-VITAMIN E 102
ERGADENYLIC ACID 67
ERGOCALCIFEROL 98, 99
ERGORONE 99
ERGOSTEROL 98, 99
ERITRONE 68
ERTRON 99
ERYCYTOL 68
ERYTHROTIN 68
ESANTENE 86
ESCORB 100
ESCOSYL 113
ESCULETIN 112, 113
ESCULIN 113
ESCULOSIDE 113
esophagus cancer 140
essential amino acid 88, 239, 248, 249, 257, 258, 259, 267, 268, 268, 276, 285, 294, 303, 312, 313, 313, 322, 331, 339, 385, 397
ESSENTIAL FATTY ACIDS 103, 104
essential fatty acids to prostaglandins, unregulated conversion of 167
ESTRADIOL 313
ESTROGEN 11, 24, 37, 44, 50, 59, 69, 78, 85, 86, 89, 90, 91, 101, 140
estrogen synthesis, decreased 125
ETA-TOCOPHEROL 100, 102
ETAVIT 100
ETA-VITAMIN E 102
ETHOCAINE 105

Index

ETHYL LINOLEATE 103
ETIOCOBALAMIN 69, 75
EUDYNA 10
EUGERASE 106
EUHAEMON 68
evening primrose oil 103
EVION 100
E-VIMIN 100
EVIPHERO 100
extremities, decreased feeling in 58
extremities, pains in 199, 226
extremities, aching 91
EXTRINSIC FACTOR 68
eye fatigue 24
eye of beef round excluding fat, roasted 17, 30, 46, 53, 60, 71, 82, 137, 146, 158, 171, 184, 190, 202, 220, 229, 241, 251, 260, 269, 277, 287, 295, 305, 314, 324, 332, 341, 365, 377, 386, 404
eye of beef round including fat, roasted 17, 30, 47, 53, 61, 71, 82, 137, 146, 158, 172, 184, 191, 203, 220, 230, 242, 251, 260, 270, 278, 287, 296, 305, 315, 324, 332, 341, 364, 375, 387, 405
eye socket development, distorted 213
eyebrow hair, decreased 6
eyelid membranes problems 213
eyelids, granulation of 24, 139, 199
eyelids problems 124
eyelids, swelling 124
eyes, bloodshot 6, 24, 285, 303, 322
eyes, burning 24
eyes, congestion 124
eyes, cracks around 58
eyes, dilated 24
eyes, dry 6, 104, 371
eyes, dull/glazed 212
eyes, inflammation of 227
eyes, itching 6, 24, 124
eyes, lesions in 151
eyes, protruding 6
eyes, watering 124, 139, 214
eyestrain 9, 99, 101

F 151
F⁻ 151
face muscles, very fine twitching of 167
facial expression, decreased 178
FACTOR AN 90
FACTOR B 69, 75
FACTOR CF 78
FACTOR GT 139
FACTOR I 106
FACTOR LC 78
FACTOR PP 37
FACTOR P-ZYMA 113
FACTOR R 68
FACTOR S 68
FACTOR T 118
FACTOR U 78
FACTOR X 68, 89, 100
FACTOR Y 58
FAGINE 86
falafel 5, 15, 30, 40, 55, 64, 79, 95, 130, 143, 155, 169, 179, 192, 201, 216, 233, 243, 254, 263, 272, 281, 290, 299, 308, 318, 327, 336, 344, 351, 361, 374, 391, 408
FALSAURE 78
farina 23, 36, 43, 57, 67, 84, 138, 151, 166, 177, 198, 211, 224, 238, 247, 255, 266, 275, 283, 294, 301, 310, 321, 329, 338, 346, 354, 369, 383, 395, 399
fat accumulation 349, 359
fat accumulation in muscles 88
fat, chicken 238, 359, 371
fat digestion, improper 187, 199
fat, duck 359, 371
fat, goose 359, 371
fat in feces, increased 141
fat metabolism, deterioration of 227
fat, turkey 359, 371
fat-soluble antioxidant therapy 97
FAT-SOLUBLE VITAMIN C 97
fatigue 6, 9, 11, 24, 37, 43, 49, 50, 58, 69, 77, 78, 85, 88, 89, 90, 91, 96, 98, 99, 100, 104, 112, 117, 118, 123, 126, 152, 153, 167, 168, 185, 186, 187, 199, 212, 226, 228, 285, 349, 358, 371, 385
FATS 24, 127, 371-383
fats breakdown, deterioration 68
fats, fresh animal 118
fats, fungi 118
fatty acid oxidation, increased 186
FATTY ACIDS 103, 104, 371
fatty acids metabolizing, decreased 88
fatty infiltration 385
fava beans 78
fava seed 78
FAVONALS 112
FE 153
FE⁺⁺ 153
FE⁺⁺⁺ 153
FE⁺² 153
feather growth in chicks, deterioration of 68
feces, increased fat 141
feet and hands, cold 187
feet, burning 11, 24
feet, cold 127, 152
feet, numb/tingling 49, 69, 78, 199
feet, peeling of 124
feet, simmered pork 22, 34, 48, 131, 177, 196, 243, 264, 273, 281, 290, 299, 308, 318, 328, 336, 344, 365, 376, 389, 403
feet/hands, burning 124
feet/hands, numb 124
female sterility 168
fennel seed 4, 13, 25, 38, 127, 154, 168, 188, 200, 218, 230, 244, 253, 263, 272, 281, 290, 299, 308, 318, 325, 335, 343, 350, 361, 376, 390, 409
fenugreek seed 14, 28, 38, 80, 128, 153, 168, 189, 200, 221, 232, 240, 249, 259, 268, 277, 285, 296, 304, 313, 322, 331, 339, 350, 362, 378, 388, 409
FERROUS SULFATE 1, 6, 50, 91, 101, 103, 104, 371
fetal abnormalities 50
fetal death/reabsorption 212
fetor 77
fever 9, 23, 96, 99, 138, 186, 187, 211, 214, 225
FIBER 24, 384
fibrillation 88
fibrillation, arterial 106
fibrosis of liver 385
figs 4, 19, 34, 41, 55, 64, 95, 131, 148, 165, 175, 181, 198, 206, 224, 237, 248, 256, 267, 276, 285, 294, 302, 312, 321, 330, 338, 347, 353, 368, 381, 396, 401
figs, dried 4, 18, 32, 40, 53, 62, 83, 96, 128, 142, 157, 169, 179, 195, 200, 222, 235, 247, 255, 265, 274, 283, 292, 301, 310, 320, 329, 337, 346, 350, 363, 379, 394, 408
figs, seedy 213
filberts, dried 4, 13, 32, 39, 50, 59, 79, 128, 141, 155, 168, 178, 189, 201, 223, 231, 240, 253, 263, 272, 281, 291, 299, 308, 318, 326, 335, 344, 353, 359, 372, 391, 409
filberts, dry roasted 128, 141, 155, 168, 178, 189, 201, 223, 231, 242, 254, 264, 273, 281, 291, 299, 308, 318, 326, 336, 344, 353, 359, 372, 391, 410
filberts, oil roasted 128, 141, 155, 168, 178, 189, 201, 223, 231, 240, 252, 263, 272, 281, 291, 299, 307, 318,

325, 335, 344, 353, 359, 372, 390, 410
FILORAL 88
filter paper 213
FILTRATE FACTOR 49
fingernail growth, deterioration of 226, 249
fingernails, decreased 124
fingers, spasms in 124
fingers, slight trembling of 167
fish 88, 125, 186, 187, 226, 227
fish, breaded and fried 8, 18, 34, 48, 56, 64, 72, 81, 134, 165, 173, 192, 204, 216, 235, 353, 363, 375, 390, 406
fish, cattlefish 116, 117
fish sandwich 8, 14, 31, 47, 65, 72, 80, 131, 147, 160, 173, 193, 207, 216, 235, 352, 363, 375, 391, 407
5-CIS-VITAMIN A 10
FL 151
flank excluding fat, braised beef 15, 29, 46, 54, 60, 70, 82, 137, 145, 155, 172, 172, 183, 189, 203, 219, 229, 241, 251, 260, 269, 278, 287, 295, 305, 314, 324, 332, 341, 364, 375, 386, 405
flank excluding fat, broiled beef 16, 29, 45, 53, 60, 70, 82, 137, 146, 156, 172, 183, 191, 202, 218, 229, 242, 252, 261, 270, 278, 288, 296, 306, 315, 325, 333, 342, 364, 375, 387, 405
flank including fat, braised beef 15, 30, 46, 54, 61, 70, 82, 136, 145, 155, 173, 183, 189, 203, 219, 229, 241, 251, 260, 269, 278, 287, 295, 305, 315, 324, 332, 341, 363, 375, 386, 405
flank including fat, broiled beef 16, 29, 45, 53, 60, 70, 82, 137, 146, 157, 172, 183, 191, 202, 218, 229, 242, 252, 261, 270, 279, 288, 296, 306, 315, 325, 333, 342, 363, 375, 387, 405
flatfish 117
flatfish, cooked 8, 17, 31, 47, 62, 70, 133, 150, 165, 169, 189, 203, 217, 234, 243, 252, 261, 270, 279, 287, 295, 306, 316, 325, 333, 342, 367, 379, 388, 402
flatfish, raw 8, 17, 32, 47, 53, 63, 71, 133, 150, 165, 171, 184, 192, 203, 218, 235, 244, 253, 262, 271, 280, 289, 297, 307, 317, 326, 334, 343, 368, 379, 389, 401

flatulence 77
FLAVAXIN 24
FLAVINE MONONUCLEOTIDE 36
FLAVONES 112
flax seeds 77, 78, 86, 104
flounder, cooked 18, 33, 47, 95, 132, 159, 171, 189, 201, 217, 268, 278, 286, 294, 305, 314, 324, 340, 365, 377, 386, 405
flounder, raw 19, 34, 48, 130, 162, 171, 191, 203, 220, 272, 281, 289, 298, 308, 318, 327, 343, 368, 380, 390
flour, acorn 131, 142, 160, 169, 194, 200, 234, 244, 254, 264, 273, 282, 291, 299, 309, 319, 327, 336, 344, 350, 359, 373, 392, 409
flour, all purpose enriched wheat 13, 27, 38, 133, 156, 195, 210, 224, 349, 360, 380, 391, 409
flour, carob 5, 19, 25, 39, 57, 60, 80, 127, 142, 156, 169, 179, 195, 200, 221, 234, 349, 365, 380, 393, 410
flour, dark buckwheat 12, 30, 38, 131, 156, 189, 350, 361, 379, 391, 409
flour, dark rye 12, 29, 38, 130, 155, 188, 200, 224, 350, 361, 379, 390, 409
flour, defatted soybean 5, 12, 28, 38, 50, 59, 127, 141, 154, 168, 178, 188, 200, 222, 231, 240, 249, 259, 268, 277, 285, 296, 304, 313, 322, 331, 339, 351, 361, 379, 385, 409
flour, defatted soybean 79
flour, light buckwheat 17, 35, 41, 135, 161, 195, 204, 349, 361, 379, 392, 409
flour, light rye 15, 33, 40, 132, 160, 192, 208, 225, 349, 361, 380, 391, 409
flour, low fat soybean 5, 13, 27, 38, 50, 59, 78, 128, 141, 154, 168, 178, 188, 200, 222, 233, 240, 249, 259, 268, 277, 285, 296, 304, 313, 322, 331, 339, 351, 360, 377, 385, 410
flour, nutrisoy 213
flour, peanut 12, 25, 37, 128, 141, 157, 168, 178, 188, 200, 222, 229, 240, 249, 259, 268, 277, 288, 298, 304, 313, 322, 331, 339, 351, 361, 385, 409
flour, potato 13, 30, 38, 93, 131, 154, 192, 200, 221, 245, 254, 264, 273, 282, 291, 300, 309, 319, 327, 336, 345, 349, 361, 380, 392, 409

flour, rice 13, 25, 38, 135, 156, 170, 190, 208, 221, 349, 361, 380, 393, 409
flour, roasted soybean 4, 13, 25, 38, 50, 60, 79, 128, 141, 154, 168, 178, 188, 200, 222, 230, 240, 249, 260, 268, 277, 287, 298, 304, 313, 322, 331, 340, 351, 360, 374, 385, 410
flour, rye 139
flour, soybean 4, 12, 25, 38, 50, 59, 79, 85, 89, 128, 141, 154, 168, 178, 188, 200, 222, 230, 240, 249, 260, 268, 277, 287, 298, 304, 313, 322, 331, 340, 351, 360, 374, 385, 409
flour, wheat 213
flour, white corn 14, 34, 39, 137, 158, 195, 224, 349, 360, 379, 392, 409
flour, whole wheat 13, 31, 38, 85, 131, 155, 188, 203, 224, 350, 361, 379, 391, 409
flour, yellow corn 3, 14, 34, 39, 137, 158, 195, 224, 349, 360, 379, 392, 409
flours, white 123
fluid accumulation in tissues 11, 168
fluoridated water 151
FLUORIN 151
FLUORINE 151
flushing 77, 107
FOLACIN 78
FOLAEMIN 78
folate utilization, deterioration of 69
FOLDINE 78
FOLETTES 78
FOLIAMIN 78
FOLIC ACID 78
FOLICET 78
FOLIPAC 78
follicle, enlarged 6
FOLSAN 78
FOLVITE 78
food contamination 167
FORMALDEHYDE 89
FORTODYL 99
fracture 9, 58, 96, 99, 138, 177, 199, 211
frankfurter, beef 19, 32, 47, 55, 64, 71, 84, 92, 133, 149, 159, 177, 182, 195, 208, 215, 232, 244, 254, 263, 273, 281, 290, 299, 309, 318, 327, 335, 344, 357, 361, 373, 391, 406
frankfurter, beef and pork 14, 32, 47, 54, 64, 71, 84, 92, 135, 147, 160, 176, 183, 195, 208, 214, 232, 244, 254, 263, 273, 281, 290, 299, 309, 318, 327, 336, 344, 357, 361, 373, 391, 406
frankfurter, chicken 18, 31,

Index 492

47, 129, 157, 214, 355, 363, 374, 391, 405
frankfurter, pork and beef 14, 32, 47, 54, 64, 71, 84, 92, 135, 147, 160, 176, 183, 195, 208, 214, 232, 244, 254, 263, 273, 281, 290, 299, 309, 318, 327, 336, 344, 357, 361, 373, 391, 406
frankfurter, turkey 18, 29, 46, 63, 71, 128, 149, 158, 174, 192, 208, 215, 232, 358, 364, 374, 391, 404
free radical oxidation 5, 9, 23, 43, 49, 58, 67, 85, 86, 90, 96, 97, 101, 113, 117, 212, 228, 238, 257, 303, 330
french beans, boiled 5, 16, 34, 40, 56, 64, 79, 96, 130, 145, 160, 169, 179, 194, 203, 223, 234, 245, 255, 264, 273, 282, 291, 300, 309, 319, 327, 336, 345, 352, 366, 380, 392, 403
french fries 5, 16, 34, 38, 53, 62, 80, 94, 133, 146, 162, 170, 194, 200, 221, 236, 351, 361, 375, 393, 408
french toast 4, 15, 28, 39, 51, 65, 73, 80, 129, 147, 157, 174, 193, 209, 216, 234, 352, 364, 376, 392, 406
FRESMIN 68
fried chicken 8, 17, 32, 44, 61, 74, 81, 134, 148, 163, 172, 192, 206, 216, 234, 353, 363, 375, 390, 407
frog legs, raw 15, 28, 48, 133, 159, 193, 368, 382, 390, 400
frostbite resistance, decreased 98, 112
FRUCTOSE 349
fruit pectin 206, 223, 357, 371, 398
fruitcake, dark 7, 15, 31, 48, 129, 156, 194, 201, 217, 350, 360, 375, 393, 408
fruitcake, light 8, 17, 31, 48, 129, 158, 194, 206, 217, 350, 360, 375, 393, 408
fruits 85, 89, 101, 103, 104, 140
fruits, contaminated 153
fuels 126
fumes 125
fungi fats 118
fungicides 126, 186

GALAMILA 58
gall bladder deterioration 187, 214, 313
gallstones 9, 91, 96, 99, 101, 104, 108, 371
galvanoplating 126
GAMMA-CAROTENE 1

GAMMA-LINOLENIC ACID 103
GAMMA-TOCOPHEROL 100, 102
GAMMA-VITAMIN E 102
garbanzo beans 78, 85, 86, 89
garlic 116, 186, 212, 227
garlic powder 13, 130, 156, 169, 188, 200, 221, 231, 242, 254, 263, 272, 281, 291, 299, 308, 318, 326, 336, 344, 350, 362, 390, 409
garlic, raw 14, 32, 39, 92, 128, 158, 173, 193, 202, 222, 244, 255, 264, 274, 283, 291, 300, 309, 319, 327, 337, 345, 351, 366, 380, 392, 405
gas 91, 199
gas pains 118
gastric function, deterioration of 126, 294
gastritis 9, 58, 67, 74, 85, 96, 101, 166, 211
gastrointestinal hemorrhage 140
gastrointestinal pain 313
gastrointestinal tone, decreased 127
gastrointestinal tract, deterioration of 6
gastrointestinal tract, lesions in 151
gastrointestinal upsets 78
GE 151
gefiltefish with broth 8, 18, 33, 63, 72, 84, 132, 143, 157, 176, 181, 195, 210, 216, 234, 244, 254, 264, 273, 281, 290, 299, 308, 318, 327, 335, 344, 355, 368, 379, 392, 401
gelatin 212
genital rashes 24
GERMANIUM 151-152
GEROVITAL 105
GEROVITAL H₃ 105
GH₃ 105
giblets, fried chicken 6, 16, 25, 44, 50, 59, 69, 78, 94, 133, 142, 154, 172, 180, 189, 204, 217, 229, 240, 250, 261, 268, 277, 287, 295, 304, 313, 323, 331, 340, 356, 363, 375, 385, 407
giblets, simmered chicken 6, 17, 25, 46, 50, 61, 69, 78, 94, 135, 143, 154, 174, 181, 190, 208, 220, 229, 241, 251, 262, 269, 278, 288, 296, 304, 315, 324, 333, 341, 358, 366, 378, 387, 403
giblets, simmered turkey 6, 19, 25, 46, 50, 61, 69, 79, 95, 134, 142, 154, 175, 181, 191, 207, 220, 230, 241, 250, 262, 269, 278, 288,

296, 304, 315, 324, 333, 341, 357, 365, 378, 387, 403
gin & tonic 5, 138, 177, 199, 211, 223, 355, 368, 400
gin, 100 proof 150, 166, 183, 199, 211, 238, 362, 405
gin, 90 proof 151, 224, 363, 404
gin, 94 proof 150, 166, 183, 199, 211, 238, 363, 404
ginger 4, 38, 129, 154, 169, 193, 200, 221, 230, 246, 255, 264, 273, 282, 291, 300, 309, 319, 328, 337, 345, 350, 362, 378, 391, 409
ginger root, raw 21, 35, 40, 56, 63, 94, 133, 164, 170, 197, 202, 222, 247, 257, 266, 275, 284, 293, 302, 311, 321, 329, 338, 346, 353, 368, 380, 395, 400
gingerbread 21, 32, 48, 129, 158, 194, 205, 216, 350, 363, 377, 394, 408
gingersnap cookie 5, 15, 31, 39, 55, 74, 81, 131, 145, 157, 175, 196, 207, 216, 235, 350, 359, 374, 393, 410
gingivitus 117
ginko nuts 3, 14, 32, 38, 56, 93, 138, 143, 161, 171, 181, 194, 201, 223, 236, 245, 256, 265, 274, 283, 292, 300, 309, 319, 327, 337, 345, 351, 365, 379, 393, 405
ginko nuts, dried 3, 13, 30, 37, 50, 92, 132, 142, 158, 170, 180, 189, 200, 222, 234, 244, 255, 264, 273, 282, 291, 299, 309, 318, 326, 337, 344, 350, 361, 379, 391, 409
GINSENG 91
gizzard, simmered chicken 7, 21, 28, 46, 51, 64, 71, 80, 95, 135, 145, 155, 174, 181, 193, 208, 219, 230, 241, 250, 262, 269, 279, 288, 295, 305, 314, 325, 333, 342, 358, 366, 378, 387, 403
gizzard, simmered turkey 7, 21, 26, 47, 51, 64, 71, 80, 95, 134, 144, 154, 174, 181, 194, 207, 220, 230, 240, 250, 262, 269, 278, 288, 295, 304, 313, 325, 332, 341, 358, 365, 378, 386, 403
glands, animal 126
glands, swollen 9, 96
glandular deterioration 77, 187
glandular myoma 152
glaucoma 9, 36, 85, 86, 96, 99

GLN 258
GLOBULARICITRIN 113
glossy tongue 58
GLUCINUM 125
GLUCOSE 349
glucose tolerance, decreased 58, 139, 177, 211
GLUCOSE TOLERANCE FACTOR 139
GLUTAMATE 258
glutamate, internal conversion of 258
GLUTAMIC ACID 240
GLUTAMINE 258
GLUTETHIMIDE 98
GLY 258
GLYCINE 49, 258
GLYCOGEN 349
GOETSCH'S VITAMIN 118
goiter 9, 96, 152, 153
GOLD 152
gold eating utensils 152
gonad function, decreased 228
goose fat 359, 371
goose liver pâté, canned 17, 26, 47, 356, 360, 372, 391, 408
goose liver pâté, canned smoked 16, 26, 47, 69, 356, 360, 372, 391, 408
goose liver, raw 6, 12, 25, 44, 59, 131, 141, 172, 189, 206, 217, 244, 253, 263, 271, 280, 289, 298, 307, 317, 326, 335, 343, 355, 366, 378, 390, 402
goose with skin, roasted 8, 17, 26, 46, 60, 84, 134, 143, 156, 173, 189, 204, 219, 242, 261, 270, 278, 288, 297, 305, 315, 333, 342, 362, 374, 387, 406
goose without skin, roasted 17, 25, 46, 59, 134, 143, 156, 172, 189, 202, 218, 364, 376, 386, 405
gooseberries 3, 21, 35, 42, 55, 65, 78, 92, 132, 148, 165, 176, 181, 197, 207, 224, 237, 354, 369, 380, 396, 399
GOSSYPINE 86
gourd, boiled calabash 21, 35, 42, 94, 132, 154, 176, 198, 208, 225, 248, 267, 276, 285, 294, 302, 312, 321, 330, 347, 356, 371, 397, 398
gourd, boiled dishcloth 3, 21, 34, 42, 94, 136, 164, 174, 197, 201, 222, 353, 369, 382, 396, 400
gourd, boiled wax 21, 42, 94, 133, 164, 198, 211, 217, 357, 371, 382, 397, 398
gout 9, 37, 43, 58, 96, 101, 166, 186, 211
grains 225

grains, refined 126
grains, whole 76, 213, 258
granola bar 11, 31, 53, 130, 155, 189, 204, 216, 350, 360, 374, 391, 410
grape juice 5, 22, 35, 42, 57, 65, 84, 125, 136, 150, 165, 176, 179, 198, 209, 212, 224, 238, 247, 267, 276, 285, 294, 303, 312, 322, 339, 348, 353, 368, 383, 397, 400
grapefruit 85, 112
grapefruit juice, red 3, 21, 36, 42, 92, 136, 150, 165, 176, 183, 198, 208, 224, 238, 355, 369, 383, 397, 399
grapefruit juice, white 5, 21, 36, 42, 92, 136, 150, 165, 176, 183, 198, 208, 224, 238, 248, 267, 276, 285, 294, 312, 322, 339, 348, 355, 369, 383, 397, 399
grapefruit, white 5, 22, 36, 42, 55, 66, 81, 92, 135, 149, 166, 176, 185, 199, 208, 238, 294, 302, 330, 355, 369, 383, 397, 399
grapes 112, 187
grapes, american 4, 17, 34, 42, 58, 64, 84, 95, 134, 150, 165, 177, 179, 198, 207, 224, 238, 247, 257, 266, 276, 285, 294, 301, 312, 321, 330, 339, 347, 353, 368, 382, 396, 400
grapes, european 4, 17, 33, 42, 57, 64, 84, 94, 135, 147, 165, 177, 181, 198, 207, 224, 238, 247, 256, 266, 276, 285, 294, 301, 312, 321, 330, 339, 347, 353, 368, 380, 396, 400
grapes, thompson seedless 4, 17, 33, 42, 57, 64, 84, 94, 135, 147, 165, 177, 181, 198, 207, 224, 238, 247, 257, 267, 276, 285, 294, 302, 312, 322, 330, 339, 348, 353, 368, 380, 396, 400
grasses, cereal 118
great northern beans, canned 5, 15, 34, 41, 55, 64, 79, 96, 130, 144, 158, 170, 170, 179, 193, 203, 224, 234, 245, 254, 264, 273, 282, 291, 300, 309, 319, 327, 336, 345, 352, 367, 381, 392, 403
green beans 86, 108, 125, 139, 186, 187, 212, 227
green beans, boiled 3, 18, 32, 40, 57, 65, 80, 94, 130, 146, 160, 172, 180, 196, 205, 224, 236, 247, 256, 266, 275, 283, 292, 301, 310, 320, 329, 338, 346, 355, 369, 381, 384, 395, 399

green beans, canned 3, 23, 33, 43, 80, 95, 132, 150, 161, 175, 180, 198, 209, 217, 236, 247, 257, 266, 275, 284, 293, 302, 311, 321, 329, 338, 347, 356, 370, 383, 384, 396, 398
green beans, frozen and boiled 3, 21, 33, 41, 57, 65, 94, 130, 148, 162, 173, 179, 197, 209, 222, 235, 247, 256, 266, 275, 284, 293, 302, 311, 320, 329, 338, 346, 355, 370, 383, 384, 395, 399
greens, dark 213
greens, fresh 118
greens, leafy 139
greens, raw 118
greens, sea 139
grits, corn 22, 36, 42, 67, 77, 151, 165, 177, 184, 198, 211, 238, 247, 255, 265, 275, 283, 294, 301, 310, 321, 330, 337, 346, 354, 369, 382, 395, 400
grits, instant corn 16, 33, 40, 55, 66, 84, 137, 151, 162, 177, 198, 211, 217, 238, 354, 368, 383, 395, 400
grits, oat 77
ground beef, medium-rare baked 21, 30, 45, 56, 62, 70, 82, 135, 148, 157, 175, 184, 193, 207, 220, 229, 242, 253, 261, 271, 279, 288, 297, 307, 316, 325, 334, 342, 362, 374, 388, 405
ground beef, medium-rare baked extra lean 20, 28, 46, 55, 62, 71, 82, 137, 148, 157, 175, 184, 194, 207, 221, 229, 242, 253, 261, 270, 279, 288, 297, 306, 316, 324, 333, 342, 363, 375, 388, 405
ground beef, medium-rare baked lean 19, 29, 46, 55, 63, 71, 82, 136, 148, 157, 175, 184, 194, 207, 220, 229, 242, 253, 261, 271, 279, 288, 297, 306, 316, 324, 334, 342, 363, 374, 388, 405
ground beef, medium-rare broiled 21, 29, 45, 54, 62, 70, 82, 135, 147, 157, 174, 174, 184, 192, 205, 218, 229, 242, 253, 261, 271, 279, 288, 297, 306, 316, 324, 334, 342, 362, 374, 388, 406
ground beef, medium-rare broiled extra lean 19, 27, 45, 54, 62, 71, 82, 137, 148, 157, 173, 184, 193, 204, 219, 229, 241, 252, 261, 270,

278, 287, 297, 306, 316, 324, 333, 342, 363, 375, 387, 405
ground beef, medium-rare broiled lean 19, 29, 45, 54, 62, 70, 82, 135, 148, 157, 173, 184, 193, 204, 218, 229, 242, 253, 261, 270, 279, 288, 297, 306, 316, 324, 333, 342, 363, 374, 388, 405
ground beef, medium-rare fried 21, 29, 45, 54, 62, 70, 82, 135, 147, 157, 174, 184, 192, 204, 218, 229, 242, 253, 261, 271, 279, 288, 297, 306, 316, 324, 334, 342, 362, 374, 388, 406
ground beef, medium-rare fried extra lean 19, 27, 46, 55, 62, 71, 82, 137, 147, 157, 173, 184, 193, 204, 219, 229, 242, 253, 261, 270, 279, 288, 297, 306, 316, 324, 333, 342, 363, 375, 387, 405
ground beef, medium-rare fried lean 19, 28, 45, 54, 62, 70, 82, 135, 147, 157, 174, 184, 193, 204, 218, 229, 242, 253, 261, 271, 279, 288, 297, 306, 316, 324, 334, 342, 363, 374, 388, 405
ground beef, well done baked 20, 29, 45, 55, 61, 70, 81, 134, 147, 156, 174, 183, 192, 205, 218, 229, 241, 252, 260, 269, 277, 287, 296, 305, 315, 323, 332, 341, 361, 374, 386, 407
ground beef, well done baked extra lean 19, 26, 45, 54, 61, 71, 81, 136, 146, 156, 173, 183, 193, 205, 219, 228, 240, 252, 260, 269, 277, 286, 295, 305, 314, 323, 332, 340, 363, 375, 386, 406
ground beef, well done baked lean 18, 28, 45, 54, 62, 70, 81, 134, 146, 156, 173, 183, 192, 205, 219, 228, 240, 252, 260, 269, 277, 286, 296, 305, 314, 323, 332, 340, 362, 374, 386, 406
ground beef, well done broiled 20, 29, 44, 54, 61, 70, 81, 134, 146, 156, 173, 183, 192, 204, 218, 229, 241, 252, 261, 270, 278, 287, 296, 306, 315, 323, 333, 341, 362, 374, 387, 406
ground beef, well done broiled extra lean 18, 26, 44, 53, 61, 70, 81, 136, 147, 156, 172, 183, 192, 203, 218, 228, 241, 252, 260, 269, 277, 287, 296, 305, 315, 323, 332, 341, 363, 375, 386, 406
ground beef, well done broiled lean 19, 28, 44, 53, 61, 70, 81, 134, 147, 157, 172, 184, 192, 203, 218, 229, 241, 252, 260, 270, 278, 287, 296, 305, 315, 323, 333, 341, 362, 374, 386, 406
ground beef, well done fried 20, 29, 44, 54, 62, 70, 81, 134, 147, 156, 173, 183, 192, 204, 218, 229, 241, 252, 261, 270, 278, 287, 296, 306, 315, 323, 333, 341, 362, 374, 387, 406
ground beef, well done fried extra lean 18, 27, 45, 55, 61, 70, 81, 136, 146, 156, 172, 183, 192, 203, 218, 229, 241, 252, 260, 270, 278, 287, 296, 305, 315, 323, 333, 341, 363, 375, 386, 406
ground beef, well done fried lean 19, 28, 45, 54, 61, 70, 81, 135, 147, 157, 173, 184, 192, 204, 218, 229, 241, 252, 260, 270, 278, 287, 296, 305, 315, 323, 333, 341, 362, 374, 386, 406
groundcherries 3, 16, 34, 38, 94, 136, 161, 196, 354, 369, 380, 395, 400
grouper, cooked 17, 36, 48, 72, 132, 149, 160, 170, 185, 193, 201, 220, 235, 243, 252, 261, 270, 278, 287, 295, 306, 315, 325, 333, 341, 367, 379, 388, 402
grouper, raw 18, 48, 69, 132, 150, 162, 171, 185, 193, 201, 220, 235, 243, 253, 262, 271, 280, 289, 297, 307, 317, 326, 334, 343, 368, 379, 389, 401
growing bone, deterioration of 91
growth, deterioration of 6, 50, 68, 69, 76, 104, 118, 124, 139, 186, 187, 199, 212, 226, 227, 228, 258, 285, 322, 371
growth hormone release 248, 303
growth rate in children, deterioration of 294
GTF 139
guar gum 213
guava 3, 19, 33, 39, 56, 63, 91, 133, 146, 165, 176, 181, 197, 205, 224, 237, 248, 267, 276, 284, 294, 302, 312, 321, 330, 339, 347, 354, 369, 380, 396, 400
guava seeds 78

gum deterioration 9, 96, 105, 138, 151, 166, 199, 211, 225
gum disease 117
gum drops 137, 165, 211, 221, 349, 361, 380, 409
gums and throat problems 186
gums, bleeding 91, 117
gums, inflamed 58
gums, sore 187
gums, tender 37, 43

haddock 103, 104, 105, 152
haddock, cooked 8, 20, 34, 46, 61, 71, 131, 150, 159, 170, 190, 202, 218, 235, 243, 252, 261, 270, 279, 287, 295, 306, 316, 325, 333, 342, 367, 380, 388, 402
haddock, raw 8, 20, 34, 46, 57, 61, 71, 131, 150, 161, 170, 183, 192, 204, 219, 236, 243, 253, 262, 271, 280, 289, 297, 307, 317, 326, 334, 343, 368, 380, 389, 401
haddock, smoked 8, 19, 34, 45, 60, 71, 130, 149, 159, 169, 190, 202, 215, 235, 242, 252, 261, 270, 278, 287, 295, 306, 315, 325, 333, 341, 367, 380, 387, 402
hair, brittle 6, 104, 152, 371
hair, coarse 6, 124, 187
hair color, changed 178
hair color, decreased 106
hair, decreased 6, 24, 58, 78, 85, 89, 91, 104, 124, 139, 153, 212, 213, 228, 322, 371
hair deterioration 9, 86, 96, 106, 152, 213
hair, dry 6, 90, 91, 104, 152, 371
hair, dull 104, 226, 371
hair follicles, hardening 91
hair follicles, swelling 91
hair, graying 50, 90
hair growth, deterioration of 178, 249
hair, oily 58
hair pigmentation, abnormal 141
hair/skin pigmentation, alterations in 385
half and half cream 7, 18, 31, 55, 65, 73, 129, 166, 175, 195, 209, 221, 235, 246, 255, 265, 274, 283, 292, 300, 309, 319, 328, 337, 345, 356, 366, 376, 394, 400
halibut 85, 89, 101, 152
halibut, cooked 8, 18, 32, 44, 60, 71, 130, 150, 161, 169, 189, 201, 219, 235,

242, 252, 261, 269, 278, 286, 295, 305, 315, 324, 332, 341, 366, 379, 387, 402
halibut, raw 8, 19, 33, 45, 54, 61, 71, 130, 150, 162, 169, 184, 190, 201, 220, 235, 243, 253, 262, 271, 280, 288, 296, 307, 316, 325, 334, 342, 367, 379, 389, 401
hallucinations 37, 43, 90, 126, 167, 186, 322
ham 86
ham and cheese roll 12, 30, 46, 52, 62, 72, 92, 130, 148, 161, 174, 182, 189, 205, 214, 232, 243, 253, 262, 272, 281, 289, 298, 308, 317, 326, 335, 343, 358, 363, 374, 390, 405
ham and cheese sandwich 7, 16, 25, 48, 96, 128, 156, 207, 215, 352, 364, 376, 390, 407
ham and cheese spread 14, 29, 47, 52, 63, 72, 94, 127, 146, 162, 174, 182, 188, 208, 214, 231, 357, 364, 374, 390, 404
ham, canned 12, 28, 47, 54, 59, 72, 83, 137, 148, 162, 175, 183, 192, 204, 214, 232, 244, 253, 262, 272, 281, 289, 298, 308, 317, 326, 335, 344, 365, 375, 390, 403
ham, canned lean 12, 28, 45, 53, 59, 72, 83, 137, 147, 161, 175, 182, 190, 203, 214, 232, 243, 253, 261, 272, 280, 289, 298, 307, 317, 326, 335, 343, 366, 378, 389, 402
ham lunch meat, chopped 12, 29, 46, 55, 61, 72, 93, 137, 148, 162, 175, 182, 193, 204, 214, 232, 243, 252, 261, 272, 281, 289, 298, 308, 317, 326, 335, 343, 364, 374, 390, 404
ham lunch meat, minced 12, 29, 46, 56, 61, 72, 92, 135, 148, 162, 175, 182, 193, 204, 214, 232, 244, 254, 262, 272, 281, 289, 298, 308, 317, 327, 335, 344, 357, 363, 374, 390, 405
ham patties, grilled 13, 30, 47, 55, 63, 72, 136, 146, 158, 176, 195, 206, 215, 232, 244, 254, 263, 272, 281, 290, 298, 308, 318, 327, 335, 344, 358, 361, 373, 391, 406
ham patties, unheated 13, 30, 47, 55, 63, 71, 136, 148, 161, 176, 193, 206, 215, 233, 244, 254, 263, 272, 281,

290, 299, 308, 318, 327, 335, 344, 358, 361, 373, 391, 406
ham, roasted 12, 26, 44, 51, 61, 72, 136, 144, 159, 173, 182, 189, 202, 214, 231, 243, 252, 261, 271, 280, 289, 298, 307, 316, 325, 334, 343, 365, 377, 388, 404
ham, roasted canned 12, 27, 45, 51, 61, 71, 83, 136, 145, 159, 175, 182, 190, 203, 215, 231, 243, 252, 261, 271, 280, 289, 297, 307, 316, 325, 334, 343, 358, 364, 375, 389, 404
ham, roasted canned lean 11, 28, 45, 52, 59, 72, 83, 137, 149, 161, 173, 183, 191, 203, 214, 232, 243, 252, 261, 271, 280, 288, 297, 307, 316, 325, 334, 343, 358, 366, 378, 389, 403
ham, roasted lean 12, 29, 46, 54, 60, 72, 84, 136, 147, 159, 175, 181, 191, 205, 214, 231, 243, 251, 261, 271, 280, 289, 297, 306, 316, 325, 334, 343, 358, 366, 377, 389, 403
ham salad 13, 32, 47, 55, 63, 72, 94, 136, 148, 163, 176, 194, 208, 215, 233, 244, 255, 263, 273, 282, 290, 299, 309, 318, 327, 336, 344, 354, 364, 375, 392, 404
ham, turkey 19, 28, 47, 62, 71, 136, 146, 156, 174, 192, 204, 215, 231, 243, 253, 262, 271, 280, 289, 297, 307, 317, 326, 334, 343, 358, 366, 378, 389, 402
ham, unheated 12, 28, 45, 53, 61, 72, 84, 136, 146, 161, 174, 182, 190, 204, 214, 232, 243, 252, 262, 272, 280, 289, 298, 307, 317, 326, 335, 344, 357, 365, 376, 390, 404
ham, unheated lean 12, 29, 45, 53, 59, 72, 84, 136, 148, 162, 175, 182, 191, 203, 214, 232, 243, 252, 262, 271, 280, 289, 298, 307, 317, 326, 334, 343, 358, 366, 378, 389, 402
hamburger sandwich 8, 14, 27, 46, 55, 64, 72, 81, 95, 130, 146, 157, 173, 194, 207, 216, 232, 351, 363, 376, 391, 407
HAMOVANNID 86
hands and feet, cold 187
hands, burning 24
hands, cold 127
hands, numb/tingling 49, 69, 78, 199

hands, peeling of 124
hands, slight trembling of 167
hands/feet, burning 124
hands/feet, numb 124
hands/feet, tingling 124
hangover 69, 105
HAOCOLIN 87
hardening of the arteries 9, 43, 49, 74, 76, 85, 86, 96, 101, 112, 138, 139, 152, 177, 199, 238
hardening of the brain arteries 106
hay fever 9, 96, 101, 139, 214
hazelnuts 101, 103, 104, 117, 125, 187, 212
headache 6, 9, 23, 37, 43, 49, 50, 58, 67, 77, 90, 96, 98, 100, 101, 126, 153, 167, 211, 214, 303, 339
headcheese, pork 20, 30, 48, 56, 63, 71, 84, 93, 134, 145, 160, 176, 183, 196, 211, 214, 233, 243, 253, 263, 272, 281, 290, 299, 308, 318, 327, 335, 344, 358, 364, 375, 390, 403
healing 104, 238, 248, 312
healing, deterioration of 6, 50, 91, 100, 118, 212, 228
hearing deterioration 106, 167, 178, 259
heart 51
heart arrhythmia 76, 106
heart attack 117 heart, braised lamb 7, 14, 25, 44, 134, 190, 358, 363, 375, 386, 406
heart, braised pork 8, 13, 25, 44, 50, 60, 70, 84, 95, 137, 142, 154, 172, 181, 192, 207, 221, 230, 242, 250, 262, 270, 278, 288, 297, 305, 316, 325, 333, 342, 358, 366, 378, 388, 403
heart deterioration 76, 88, 88, 89, 106, 116, 152, 199
heart disease 58, 76, 88, 104, 116, 117, 119, 140, 152, 227, 349, 359, 371
heart endocardium, deterioration of 168
heart, enlarged 88
heart failure, congestive 9, 23, 101, 211
heart, fatty changes in 90
heart muscle, deterioration of 151
heart palpitation 11, 69
heart, simmered beef 15, 25, 46, 62, 69, 84, 95, 137, 142, 154, 172, 181, 190, 206, 220, 230, 241, 250, 261, 269, 277, 287, 295, 304, 313, 324, 332, 340, 358, 365, 377, 386, 404
heart, simmered chicken 8,

Index

18, 25, 47, 50, 61, 69, 79, 95, 133, 142, 154, 173, 181, 191, 209, 221, 228, 242, 250, 262, 268, 277, 287, 296, 304, 315, 323, 332, 340, 358, 365, 377, 387, 403
heart, simmered turkey 8, 18, 25, 47, 50, 61, 69, 79, 95, 134, 142, 154, 173, 181, 191, 207, 220, 229, 241, 250, 261, 268, 277, 287, 296, 304, 315, 323, 332, 340, 357, 365, 377, 387, 404
heart, tin accumulation in 226
heartbeat, slow 168
heartburn 167
HEAT 358
heat tolerance, decreased 91
heavy metal poisoning 152, 303
hematocrit, decreased 187
HEMO-B-DOZE 68
hemoglobin, decreased 77, 88
hemoglobin production, decreased 58, 268
HEMOMIN 68
hemophilia 9, 43, 49, 96, 112, 118, 138
hemorrhage 91, 107, 108
hemorrhoids 9, 67, 96, 98, 101, 112, 113, 138, 384
HEPACHOLINE 87
HEPAGON 68
hepatitis 9, 77, 86, 96, 108
HEPATOFLAVIN 24
HEPAVIS 68
HEPAXANTHIN 10
HEPCOVITE 68
HERACLENE 76
hernia, hiatal 384
herpes simplex virus 294
herring 103, 104
herring, cooked 8, 16, 27, 46, 60, 69, 95, 129, 145, 159, 170, 189, 202, 217, 233, 243, 252, 262, 270, 279, 288, 296, 307, 316, 325, 333, 342, 365, 376, 388, 404
herring, kippered 8, 15, 26, 46, 60, 69, 129, 145, 159, 170, 189, 201, 215, 233, 243, 252, 261, 270, 279, 287, 295, 306, 315, 325, 333, 341, 364, 376, 388, 404
herring, pickled 7, 18, 31, 57, 70, 129, 146, 160, 177, 182, 195, 210, 215, 235, 244, 254, 263, 272, 281, 289, 298, 308, 318, 326, 335, 344, 354, 363, 374, 390, 405
herring, raw 8, 17, 28, 47, 52, 61, 69, 95, 130, 147,

160, 170, 182, 190, 204, 218, 234, 244, 253, 262, 272, 280, 289, 298, 308, 317, 326, 335, 343, 366, 377, 390, 402
herring, smoked 212
HESPERETIN-7-RUTINOSIDE 113
HESPERIDIN 112, 113
HEXACARBACHOLINE BROMIDE 87
HEXAHYDROXYCYCLOHEXANE 85
HEXAMETHYLENEDICARBAMIC ACID CHOLINE BROMIDE DIESTER 87
HEXANICIT 86
HEXANICOTINOYL INOSITOL 86
HEXANICOTOL 86
HEXAVALENT CHROMIUM 140
HEXOPAL 86
HEXURONIC ACID 91, 97
HG 186
hiatal hernia 384
hickory nuts, dried 130, 142, 157, 168, 189, 201, 230, 240, 252, 263, 272, 281, 291, 299, 307, 318, 327, 335, 344, 353, 359, 372, 391, 410
HI-DERATOL 99
HI-FRESMIN 76
high blood pressure 43, 49, 77, 78, 86, 96, 101, 112, 117, 138, 177, 211
HIS 259
histadelic schizophrenics 303
histamine toxicity 303
HISTIDINE 259-267
hoarseness 124
Hodgkin's disease 238
hominy, white 22, 35, 138, 163, 211, 216, 354, 369, 381, 396, 400
hominy with red and green peppers, yellow 4, 35, 43, 138, 163, 211, 216, 354, 369, 382, 396, 400
hominy, yellow 4, 35, 138, 161, 211, 216, 354, 369, 382, 396, 399
honey 34, 48, 107, 125, 137, 164, 199, 210, 223, 349, 362, 397, 408
honey loaf, beef and pork 13, 28, 47, 51, 61, 71, 93, 133, 148, 159, 174, 182, 193, 203, 214, 231, 244, 254, 262, 272, 280, 289, 298, 308, 317, 326, 335, 344, 356, 366, 378, 390, 402
hormone function, deterioration of 68, 125
hormone imbalance 187
hormone precursor 98
HORNBEST 87
horseradish sauce 130, 163, 197, 205, 218, 355, 369, 395, 399

horsetail 213
hot chocolate 7, 20, 30, 48, 55, 66, 73, 83, 96, 128, 165, 173, 194, 207, 221, 235, 246, 255, 265, 274, 282, 291, 300, 309, 319, 328, 336, 345, 354, 368, 378, 393, 400
hot chocolate, restaurant 5, 35, 57, 59, 74, 84, 132, 149, 166, 176, 197, 210, 221, 237, 356, 370, 382, 396, 400
hot dog 14, 27, 46, 131, 145, 158, 175, 194, 208, 215, 231, 353, 363, 375, 391, 406
household products, common 125
HR 113
huckleberry 78
human milk 7, 35, 48, 56, 74, 83, 85, 89, 95, 99, 101, 103, 104, 105, 131, 165, 177, 198, 210, 222, 237, 247, 256, 266, 275, 284, 293, 301, 311, 321, 329, 337, 347, 355, 368, 378, 396, 399
hummus 5, 17, 34, 41, 55, 60, 80, 94, 130, 143, 158, 171, 179, 194, 208, 217, 233, 245, 255, 264, 274, 282, 291, 301, 309, 319, 328, 336, 345, 352, 365, 377, 393, 403
hush puppies 36, 38, 55, 64, 74, 80, 351, 361, 376, 392, 408
hyacinth beans, boiled 14, 35, 41, 131, 142, 155, 169, 194, 204, 223, 231, 244, 254, 264, 273, 282, 290, 300, 309, 319, 327, 336, 344, 352, 367, 380, 392, 403
HYBRIN 91, 97
HYCOBAL 76
HYDOXAMIN 68
HYDRALAZINE 69, 140
HYDRIDOCOBALAMIN 75
hydrocele 127, 187
HYDROCHLORIC ACID 127, 153
HYDROCHLORIC ACID DECREASED 127
HYDROGRISEVIT 75
HYDROVIT 75
HYDROXOBASE 68
HYDROXOCOBALAMIN 69, 75
HYDROXYPROLINE 267
hypercholesterolaemia 248
hyperkinesis 105
hyperkinetic children 106
hyperthyroidism 86, 101, 153
hypospermia, idiopathic 248
HYXOBAMINE 75

I 152
I- 152
I-131 153
I.Q., decreased 167, 258
ice cream, vanilla 125
idiopathic hypospermia 248
IDROGRISEOVIT 75
I-INOSITOL 85
ILE 268
ILXATHIN 113
IMBRETIL 87
immunity decreased 6, 11, 58, 69, 76, 78, 86, 89, 91, 104, 117, 118, 151, 153, 213, 228, 239, 249, 285, 371
immunoglobulin/antibody G, decreased 117
impotency 239
INCAFOLIC 78
incontinence 77
indian buffalo milk 7, 19, 31, 48, 56, 66, 73, 83, 95, 128, 166, 171, 194, 208, 220, 237, 246, 255, 265, 274, 282, 291, 300, 309, 319, 328, 336, 345, 356, 367, 377, 393, 400
INDOMETHACIN 91
INDUSIL 76
industry 124, 139, 226, 227
infection 6, 9, 49, 58, 96
inflammation 105
inflammation, internal 153, 228
influenza 9, 23, 36, 67, 96
INFRON 99
INOSITE 85 INOSITOL 85
INOSITOL HEXANICOTI-
  NATE 86
INOSITOL
  MONOPHOSPHATE 85
INOSITOL NIACINATE 85, 86
insecticides 151, 167
instant tea 58, 138, 151, 166, 177, 180, 199, 211, 224, 238, 358, 371, 398
insulin, decreased 303
insulin sensitivity 50
insulin therapy 105
interferon production, decreased 152
internal conversion of arginine 303
internal conversion of citrulline 248
internal conversion of cysteine 257
internal conversion of glutamate 258
internal conversion of lecithin 85
internal conversion of linoleic acid 105
internal conversion of lysine 88
internal conversion of methionine 86, 257
internal conversion of oleic acid 103

internal conversion of serine 258
internal conversion of vitamin Bc 90, 303
internal conversion of vitamin H₃ 90
internal conversion of vitamins/coenzymes Q₇, Q₈, Q₉ and possibly others 117
internal organs, ulceration of 118
internal synthesis 90, 112, 239, 248, 249, 267, 312, 313
internal synthesis from intestinal bacteria 74, 85, 108
internal synthesis from lecithin 86
internal synthesis from vitamin E 117
internal synthesis from vitamin H₃ 86
internal synthesis in adults 267
internal synthesis of methionine and cysteine 313
internal synthesis using vitamin B₆ 258
INTESTINAL BACTERIA 107
intestinal bacteria, internal synthesis from 74, 85, 108
intestinal diseases 153
intestinal flora 74, 85, 108
intestinal malfunctions 313
intestine, tin accumulation in 226
intractable sciatica 213
IODIN 152
IODINE 152-153
iodine, decreased 118
IODINE-131 152, 153
irishmoss seaweed, raw 13, 25, 42, 129, 144, 154, 179, 193, 210, 219, 232, 354, 369, 382, 395, 400
IROCAINE 106
IRON 88, 126, 127, 140, 141, 153-166, 167, 187, 267
iron, decreased 228
iron increased in spleen 124
iron level in blood deceptively increased 69
iron utilization, decreased 178
IRRADIATED 7-DEHYDRO-
  SITOSTEROL 98, 100
ISL 268
ISO 268
ISOCAINE-ASID 106
ISOCAINE-HEISLER 106
ISOLEUCINE 268-276
ISONIAZID/INH 59
itching 6, 24, 37, 43
itching, overall 153

jackfruit 3, 22, 42, 64, 94, 131, 143, 163, 170, 180, 197, 204, 224, 235, 352, 368, 382, 395, 402

jaundice 9, 67, 96, 99, 108, 124, 141
java plum 5, 23, 36, 42, 66, 93, 133, 165, 175, 198, 210, 222, 353, 368, 382, 396, 400
jelly beans 135, 161, 199, 222, 349, 360, 381, 409
JENACAIN 106
Job's tears 116, 117
joint pain 6, 77, 91, 167, 186
joints, enlarged 91, 104, 239, 303, 371
joints, hypermobile 312
joints, limited movement of 77
jowl, raw pork 9, 13, 28, 46, 55, 64, 72, 84, 138, 150, 164, 177, 185, 195, 208, 222, 244, 255, 265, 274, 282, 291, 300, 309, 319, 329, 337, 345, 359, 372, 392, 408
jujube 5, 22, 34, 40, 65, 92, 132, 148, 164, 176, 181, 197, 206, 224, 238, 352, 368, 382, 395, 401
jujube, dried 14, 26, 41, 93, 129, 143, 158, 170, 180, 194, 201, 223, 237, 350, 362, 380, 393, 408
jute potherb, boiled 2, 17, 29, 40, 92, 127, 156, 169, 195, 201, 222, 246, 255, 265, 274, 283, 292, 300, 309, 320, 329, 337, 345, 355, 369, 382, 393, 399
JUVOCAINE 106

K 199
K⁺ 199
kale, boiled 2, 20, 32, 41, 57, 63, 81, 92, 129, 144, 161, 174, 179, 197, 206, 222, 237, 246, 256, 265, 274, 283, 292, 302, 310, 320, 329, 337, 346, 355, 369, 380, 395, 399
kale, boiled scotch 2, 20, 34, 40, 57, 64, 81, 92, 128, 144, 158, 169, 179, 197, 205, 221, 237, 246, 256, 265, 274, 283, 292, 302, 310, 320, 329, 337, 346, 355, 370, 380, 395, 399
kale, frozen and boiled 2, 20, 32, 41, 57, 64, 81, 92, 128, 149, 161, 174, 179, 197, 204, 222, 237, 246, 255, 265, 274, 283, 292, 301, 309, 320, 328, 337, 346, 350, 370, 380, 394, 399
kelp 126, 139, 213
kelp/kombu/tangle seaweed, raw 4, 20, 30, 41, 79, 128, 145, 156, 169, 180, 196, 210, 217, 233, 247, 254, 266,

275, 284, 293, 301, 311,
320, 328, 338, 346, 354,
369, 380, 395, 400
keratosism, senile 106
KEROCAINE 106
keytones in blood, increased 88
KH₃ 105
kidney beans 78, 212
kidney beans, boiled 15, 34,
40, 56, 64, 79, 96, 132, 143,
156, 170, 179, 193, 202,
224, 233, 244, 254, 264,
273, 282, 290, 299, 308,
318, 327, 336, 344, 352,
366, 381, 384, 392, 403
kidney beans, canned 16, 33,
41, 56, 65, 80, 96, 132, 144,
160, 171, 180, 194, 206, 216,
235, 245, 255, 264, 273,
282, 291, 300, 309, 319,
328, 337, 345, 353, 368,
382, 393, 401
kidney, beef 186
kidney deterioration 98, 124,
126, 168
kidney disease 88
kidney failure 98, 127, 187
kidney hemorrhage 49, 86
kidney inflammation 9, 36,
96, 101, 138, 166, 177
kidney stones 9, 67, 91, 96,
101, 177
kidneys, braised pork 7, 13,
25, 45, 50, 59, 69, 80, 94,
134, 142, 154, 174, 181, 190,
209, 218, 230, 242, 249,
262, 269, 277, 288, 297,
304, 316, 324, 332, 340,
366, 378, 387, 403
kidneys, enlarged 6
*kidneys, fatty deterioration
of* 90, 123
kidneys, simmered beef 6, 15,
24, 44, 50, 59, 69, 79, 95,
133, 142, 154, 174, 181, 189,
208, 217, 230, 242, 253,
262, 271, 278, 289, 298,
304, 314, 323, 332, 340,
358, 366, 378, 387, 403
kielbasa/kolbassy, beef and
pork 14, 28, 47, 51, 63, 71,
93, 131, 145, 159, 175, 182,
193, 205, 215, 232, 244,
253, 263, 272, 281, 290,
299, 308, 318, 327, 335,
344, 357, 362, 373, 391,
406
kielbasa/kolbassy, pork and
beef 14, 28, 47, 51, 63, 71,
93, 131, 145, 159, 175, 182,
193, 205, 215, 232, 244,
253, 263, 272, 281, 290,
299, 308, 318, 327, 335,
344, 357, 362, 373, 391,
406
KILOCALORIES 358
kinako 116, 117

kingfish, cooked 8, 16, 31,
47, 129, 158, 169, 189, 205,
218, 354, 363, 375, 388
kiwifruit 4, 22, 34, 41, 91,
132, 164, 171, 196, 204, 223,
353, 368, 381, 396, 400
K-JECT 108
knockwurst, beef and
pork 13, 30, 47, 55, 63, 71,
92, 135, 148, 161, 176, 195,
207, 215, 233, 244, 254,
263, 273, 281, 290, 299,
309, 318, 327, 336, 344,
357, 362, 373, 391, 405
kohlrabi, boiled 5, 21, 35,
41, 92, 132, 164, 173, 196,
203, 222, 246, 257, 266,
275, 284, 293, 302, 311,
321, 330, 347, 355, 370,
383, 395, 399
KONAKION 108
kumquats 3, 16, 32, 92, 131,
146, 164, 176, 181, 197, 207,
223, 237, 353, 368, 396,
400
kwashiorkor 9, 85, 96, 99,
108, 140, 177, 212

lactation, deterioration of 6,
58, 110
LACTOBACILLUS CASEI
FACTOR 78
LACTOFLAVIN 24
LACTOSE 349
LAETRILE 77
lamb 85, 86, 88, 89, 101,
103, 104, 105, 152, 186, 187,
212
lamb chop 139, 212
lamb excluding fat, roasted
leg 15, 26, 44, 134, 157,
190, 204, 219, 365, 377,
386, 404
lamb heart, braised 7, 14, 25,
44, 134, 190, 358, 363, 375,
386, 406
lamb including fat, roasted
leg 15, 27, 45, 135, 158,
191, 205, 220, 362, 374,
387, 406
lamb liver 88, 89
lamb liver, broiled 6, 13, 24,
44, 92, 133, 154, 188, 204,
218, 357, 363, 376, 385, 407
lamb loin chop excluding fat,
broiled 15, 27, 44, 135, 157,
191, 204, 219, 365, 377,
386, 404
lamb loin chop including fat,
broiled 15, 28, 45, 136, 160,
192, 206, 220, 361, 373,
388, 407
lamb rib chop excluding fat,
broiled 15, 27, 45, 135, 158,
191, 204, 219, 365, 376,
387, 404
lamb rib chop including fat,

broiled 16, 29, 46, 136, 161,
193, 206, 221, 360, 372,
389, 407
lamb shoulder excluding fat,
roasted 15, 27, 45, 135, 158,
191, 204, 219, 365, 376,
387, 404
lamb shoulder including fat,
roasted 15, 28, 45, 135, 160,
192, 206, 220, 361, 373,
388, 407
lamb's-quarters, boiled 2, 16,
27, 40, 92, 127, 163, 196,
246, 255, 265, 274, 283,
291, 301, 309, 320, 328,
337, 345, 356, 369, 380,
394, 399
lard 101, 103, 104, 117, 117,
238, 244, 255, 265, 274,
282, 291, 300, 309, 319,
329, 337, 345, 359, 371
LARD FACTOR 6
LAROSCORBINE 91, 97
laryngitis 124
larynx growth 239, 303
laver/nori seaweed, raw 2,
16, 25, 39, 63, 92, 129, 143,
158, 177, 178, 196, 203, 221,
233, 245, 254, 264, 273,
282, 292, 299, 309, 319,
328, 336, 345, 356, 369,
381, 393, 400
LAXATIVES 1, 6, 97, 98, 101,
103, 107, 110, 112, 114, 115,
116, 117, 118
L-CARNITINE 88
LEAD 167
lead in blood, increased 98
lead retention 126
lecithin 85, 86
lecithin, internal conversion
of 85
lecithin, internal synthesis
from 86
lecithin, soybean 101, 107,
359, 371
leeks 101
leeks, boiled 4, 21, 35, 41,
80, 94, 132, 160, 175, 198,
210, 222, 247, 256, 267,
276, 284, 293, 302, 311,
321, 330, 338, 347, 355,
370, 381, 396, 399
leeks, raw 4, 17, 34, 41, 79,
93, 130, 157, 171, 197, 207,
222, 247, 255, 266, 275,
284, 293, 301, 311, 320,
329, 338, 347, 354, 368,
381, 384, 395, 400
leg cramp 23, 36, 58, 96, 99,
101, 105, 138, 177, 199, 225
leg excluding fat, roasted
pork 9, 12, 26, 45, 51, 59,
72, 81, 137, 145, 160, 172,
182, 189, 202, 219, 230,
240, 250, 259, 269, 277,
286, 296, 305, 313, 323,
332, 340, 364, 376, 386, 404

leg including fat, roasted
  pork 9, 12, 26, 46, 52, 60,
  72, 81, 137, 146, 161, 173,
  182, 190, 204, 220, 231,
  241, 251, 259, 270, 278,
  286, 297, 306, 315, 324,
  333, 341, 362, 374, 387,
  406
leg muscle cramps 50
leg muscles, tender 11
leg muscles, twitching 123
leg of lamb excluding fat,
  roasted 15, 26, 44, 134, 157,
  190, 204, 219, 365, 377,
  386, 404
leg of lamb including fat,
  roasted 15, 27, 45, 135, 158,
  191, 205, 220, 362, 374,
  387, 406
leg, roasted veal 15, 28, 44,
  137, 156, 204, 221, 358,
  366, 378, 386
*legs and arms, coldness
  of* 126
legs, paralysis of 124
legs, spasms in 124
legumes 76
LEMASCORB 91, 97
lemon 5, 22, 35, 43, 56, 65,
  81, 92, 132, 150, 163, 198,
  209, 224, 238, 355, 370,
  381, 396, 399
lemon chiffon pie 4, 22, 33,
  42, 95, 132, 162, 195, 210,
  217, 351, 361, 376, 392, 408
lemon juice 5, 22, 36, 43,
  57, 66, 81, 92, 97, 112, 137,
  150, 166, 177, 185, 199, 209,
  224, 238, 355, 370, 397,
  399
lemon meringue pie 4, 17,
  32, 42, 55, 66, 74, 82, 95,
  134, 150, 162, 177, 196, 210,
  217, 236, 351, 363, 377,
  394, 407
lemon pectin 213
lemon peel 4, 30, 54, 63, 91,
  128, 162, 174, 198, 208,
  353, 395, 400
lentils 78, 85, 86, 89, 186,
  187
lentils, boiled 5, 15, 33, 39,
  52, 63, 79, 95, 133, 143,
  155, 170, 179, 192, 203,
  224, 233, 244, 254, 264,
  273, 282, 290, 300, 309,
  319, 327, 336, 344, 352,
  367, 381, 384, 392, 403
lentils, stir-fried sprouted 5,
  14, 32, 39, 52, 93, 134, 142,
  156, 170, 179, 193, 205,
  233, 244, 251, 264, 273,
  282, 290, 300, 309, 319,
  336, 345, 352, 367, 381,
  392, 403
LENTIN 87
lettuce 85, 107, 112, 116, 139,
  152, 187, 213, 227

lettuce, iceberg 125
lettuce, raw butterhead 3, 18,
  33, 79, 94, 150, 165, 181,
  205, 223, 237, 247, 256,
  266, 275, 284, 292, 302,
  311, 320, 330, 338, 346,
  357, 371, 384, 395, 398
lettuce, raw iceberg 3, 20,
  34, 57, 66, 80, 94, 133, 150,
  164, 176, 181, 198, 208, 222,
  237, 247, 256, 266, 275,
  284, 293, 302, 311, 320,
  330, 338, 347, 357, 371,
  384, 396, 398
lettuce, raw looseleaf 2, 21,
  33, 41, 56, 65, 93, 129, 159,
  176, 197, 205, 222, 247,
  256, 266, 275, 284, 292,
  302, 311, 320, 330, 338,
  346, 356, 370, 381, 395,
  398
lettuce, raw romaine 2, 16,
  32, 41, 79, 92, 131, 160,
  177, 196, 205, 223, 246,
  256, 266, 274, 284, 292,
  301, 310, 320, 330, 338,
  346, 357, 371, 381, 395, 398
LEU 276
LEUCINE 276-285
LEUCO-4 49
leukemia 84, 96, 112, 117,
  151, 152, 166, 225
LI 167
LI⁺ 167
life span extension 5, 9, 23,
  43, 49, 58, 67, 85, 86, 90,
  96, 97, 101, 113, 117, 212,
  228, 238, 257, 303, 330
life span, shortened 118, 226
light, hypersensitivity to 37,
  43
light sensitivity 24
lima beans 85, 103, 104
lima beans, boiled 15, 34, 41,
  53, 63, 79, 133, 143, 157,
  170, 179, 194, 201, 224,
  234, 244, 254, 264, 273,
  282, 291, 300, 309, 318,
  327, 336, 344, 352, 367,
  381, 384, 392, 403
lima beans, boiled and frozen
  baby 4, 18, 33, 40, 56, 64,
  79, 94, 132, 143, 158, 169,
  179, 194, 202, 221, 235,
  245, 254, 264, 273, 282,
  291, 300, 309, 319, 327,
  336, 345, 353, 367, 382,
  392, 402
lima beans, boiled baby 15,
  34, 40, 53, 65, 79, 132, 143,
  157, 170, 179, 194, 202,
  224, 234, 244, 254, 264,
  273, 282, 291, 300, 308,
  318, 327, 336, 344, 352,
  366, 381, 384, 392, 403
lima beans, canned 20, 35,
  42, 55, 65, 80, 132, 143,
  158, 170, 179, 195, 207, 216,

234, 245, 255, 264, 273,
  282, 291, 300, 309, 319,
  328, 336, 345, 353, 368,
  382, 393, 401
lima beans, frozen and boiled
  fordhook 4, 18, 33, 39, 56,
  64, 79, 93, 132, 149, 159,
  170, 180, 196, 202, 220,
  235, 245, 255, 264, 273,
  282, 291, 300, 309, 319,
  327, 336, 345, 353, 367,
  381, 393, 402
lima beans, green 77
limas, black 77
limas, Burma white 77
limas, United States white 78
limb paralysis, temporary 58
*limb-girdlemuscular
  dystrophy* 88
lime 5, 22, 36, 43, 56, 82,
  92, 131, 148, 163, 198, 210,
  224, 237, 294, 303, 330,
  354, 370, 383, 397, 399
lime juice 5, 22, 36, 43, 57,
  66, 92, 136, 150, 166, 177,
  185, 199, 209, 224, 238,
  355, 370, 383, 397, 399
ling, raw 8, 16, 29, 47, 55,
  61, 73, 131, 163, 169, 191,
  202, 217, 243, 253, 262,
  271, 280, 289, 297, 307,
  317, 326, 334, 343, 368,
  380, 389, 401
lingcod, raw 20, 31, 47, 134,
  150, 165, 172, 183, 191, 201,
  220, 235, 244, 254, 262,
  272, 280, 289, 298, 308,
  317, 326, 335, 343, 368,
  379, 390, 400
link sausage, pork 12, 27, 46,
  51, 60, 71, 95, 132, 148,
  160, 174, 193, 204, 214, 231,
  243, 252, 261, 271, 280,
  289, 297, 307, 316, 326,
  334, 342, 357, 360, 373,
  388, 408
LINODIL 86
LINOLEIC ACID 103
LINOLEIC ACID CONCEN-
  TRATED MIXTURE 105
linoleic acid, internal conversion of 105
LINOLENIC ACID 103, 105
linseed oil 103
LIPIDS 371
lipids in blood, increased 88
lipid-soluble vitamin 1, 6,
  23, 97, 98, 100, 103, 104,
  105, 107, 108, 109, 110, 112,
  114, 115, 116, 118
LIPID-SOLUBLE VITAMIN C 97
LIPOIC ACID 112
LIPOICIN 112
LIPONEURINA 24
LIPOTRIL 87
lips, scaling 58, 153
lips, sore 6
liqueur, 53 proof coffee 36,

42, 138, 150, 166, 177, 199, 211, 223, 238, 351, 361, 383, 408
liqueur, 63 proof coffee 36, 42, 138, 150, 166, 177, 199, 211, 223, 238, 351, 362, 382, 407
liquor, malt 138, 166, 211, 223, 357, 369, 397
LITHIUM 86, 167
liver 77, 86, 99, 101, 103, 110, 187, 213, 227
liver, beef 85, 89, 107, 152, 186, 212
liver, braised beef 6, 14, 24, 44, 50, 59, 69, 79, 93, 137, 141, 154, 174, 179, 188, 206, 219, 229, 242, 250, 262, 270, 277, 289, 296, 304, 315, 323, 332, 340, 357, 365, 378, 388, 403
liver, braised pork 6, 14, 24, 44, 50, 59, 69, 79, 93, 136, 142, 154, 175, 180, 190, 208, 221, 228, 242, 250, 261, 269, 277, 288, 296, 304, 315, 323, 333, 340, 356, 365, 378, 387, 404
liver, broiled lamb 6, 13, 24, 44, 92, 133, 154, 188, 204, 218, 357, 363, 376, 385, 407
liver, calf 139
liver, chicken 88, 108
liver, cirrhosis of 9, 74, 77, 85, 86, 88, 96, 99, 108, 119, 141
liver deterioration 37, 43, 50, 76, 104, 112, 126, 227, 228, 258, 294, 371
liver disease 119
liver, enlarged 6, 124, 141
liver extract, beef 110
liver, fatty 86, 90, 313
liver, increased cholesterol formation in 139
liver, fried beef 6, 14, 24, 44, 50, 59, 69, 79, 93, 135, 141, 154, 173, 179, 188, 203, 217, 229, 242, 250, 261, 270, 277, 288, 296, 304, 314, 323, 331, 340, 355, 364, 377, 387, 405
LIVER LACTOBACILLUS CASEI FACTOR 78
liver, lamb 88, 89
liver oil, shark 6
liver oil, tuna 99
liver pâté, canned chicken 7, 18, 25, 44, 93, 135, 154, 355, 365, 375, 390
liver pâté, canned goose 17, 26, 47, 356, 360, 372, 391, 408
liver, polar bear 6
liver, pork 89, 107
liver, raw duck 6, 69, 135, 141, 153, 189, 243, 252, 263, 271, 280, 289, 298, 306, 317, 325, 334, 342, 356, 366, 378, 389, 402
liver, raw goose 6, 25, 44, 59, 131, 141, 172, 189, 206, 217, 244, 253, 263, 271, 280, 289, 298, 307, 317, 326, 335, 343, 355, 366, 378, 390, 402
liver sausage, pork 6, 14, 25, 44, 50, 61, 69, 93, 135, 143, 154, 176, 181, 192, 207, 214, 231, 244, 252, 263, 273, 281, 290, 299, 308, 318, 327, 335, 344, 357, 361, 372, 391, 407
liver, simmered chicken 6, 15, 24, 46, 50, 59, 69, 78, 93, 134, 142, 154, 173, 180, 189, 209, 221, 230, 242, 251, 262, 269, 278, 288, 297, 304, 315, 323, 333, 340, 358, 366, 377, 388, 403
liver, simmered turkey 6, 19, 25, 45, 50, 59, 69, 78, 95, 135, 142, 154, 175, 180, 189, 207, 219, 230, 243, 251, 262, 269, 278, 288, 297, 304, 316, 323, 333, 340, 357, 365, 377, 388, 403
livercheese, pork 6, 14, 24, 44, 50, 59, 69, 95, 136, 142, 154, 176, 180, 191, 206, 214, 230, 244, 251, 263, 272, 281, 289, 298, 307, 318, 326, 335, 343, 358, 362, 373, 390, 406
liverwurst, pork 14, 25, 50, 63, 69, 80, 132, 154, 190, 244, 254, 263, 272, 281, 289, 299, 308, 318, 327, 336, 343, 357, 361, 373, 390, 406
L-METHIONINE METHYL-SULFONIUM SALT 118
lobster 212, 227
lobster, cooked 8, 22, 33, 48, 55, 65, 70, 81, 130, 141, 164, 170, 181, 192, 203, 216, 231, 241, 253, 263, 271, 280, 288, 297, 307, 317, 325, 334, 343, 358, 367, 380, 389, 401
lobster paste 15, 27, 129, 159, 192, 358, 365, 376, 389, 404
lobster, raw 34, 48, 50, 72, 141, 181, 231, 242, 253, 263, 271, 280, 289, 298, 307, 317, 325, 334, 343, 358, 368, 380, 389, 401
lobster tail 139
loganberries 5, 20, 35, 40, 56, 65, 78, 80, 93, 132, 145, 163, 173, 178, 197, 209, 224, 236, 354, 369, 382, 395, 400
Lögic™ brand Nutritional Supplements 111
loin blade excluding fat, braised pork 9, 13, 26, 45, 52, 60, 72, 83, 133, 145, 158, 173, 173, 184, 191, 202, 218, 229, 240, 250, 259, 268, 277, 286, 295, 304, 313, 322, 331, 340, 362, 374, 386, 407
loin blade excluding fat, broiled pork 9, 12, 26, 45, 51, 60, 72, 83, 134, 146, 160, 172, 185, 190, 202, 218, 230, 241, 251, 259, 270, 278, 286, 296, 306, 315, 323, 333, 341, 362, 374, 387, 406
loin blade excluding fat, fried pork 9, 12, 26, 46, 51, 60, 72, 134, 146, 161, 172, 185, 190, 202, 218, 230, 242, 251, 259, 270, 279, 287, 297, 306, 315, 324, 333, 341, 362, 374, 388, 406
loin blade excluding fat, roasted pork 9, 12, 26, 45, 53, 59, 72, 83, 134, 146, 160, 175, 185, 191, 203, 219, 230, 241, 251, 259, 270, 278, 286, 297, 306, 315, 323, 333, 341, 362, 374, 388, 405
loin blade including fat, braised pork 9, 13, 27, 45, 53, 61, 72, 84, 134, 145, 160, 175, 185, 192, 204, 219, 230, 241, 251, 259, 270, 279, 287, 297, 306, 315, 324, 333, 341, 360, 372, 388, 407
loin blade including fat, broiled pork 9, 12, 26, 46, 52, 61, 72, 84, 135, 147, 161, 173, 173, 185, 191, 204, 219, 230, 243, 252, 260, 271, 280, 288, 298, 307, 316, 325, 334, 342, 360, 372, 389, 407
loin blade including fat, fried pork 8, 12, 27, 46, 52, 61, 72, 84, 135, 147, 162, 174, 185, 192, 205, 220, 231, 243, 253, 260, 271, 280, 288, 298, 307, 317, 326, 334, 343, 360, 372, 389, 407
loin blade including fat, roasted pork 13, 27, 46, 52, 60, 72, 84, 135, 147, 161, 175, 185, 192, 205, 220, 231, 242, 252, 260, 271, 280, 288, 298, 307, 316, 325, 334, 342, 360, 373, 389, 407
loin, braised/broiled veal 18, 28, 45, 135, 155, 190, 205, 219, 364, 375, 387, 405

Index

loin chop excluding fat, broiled lamb 15, 27, 44, 135, 157, 191, 204, 219, 365, 377, 386, 404
loin chop including fat, broiled lamb 15, 28, 45, 136, 160, 192, 206, 220, 361, 373, 388, 407
loin excluding fat, braised pork 9, 12, 26, 44, 51, 59, 72, 83, 135, 145, 159, 172, 183, 190, 202, 218, 230, 240, 250, 259, 268, 277, 285, 295, 304, 313, 322, 331, 340, 363, 375, 385, 406
loin excluding fat, broiled pork 9, 11, 25, 44, 59, 71, 83, 136, 146, 161, 171, 171, 185, 189, 202, 218, 231, 241, 250, 259, 269, 278, 286, 296, 305, 314, 323, 332, 340, 363, 375, 386, 405
loin excluding fat, roasted pork 9, 12, 26, 44, 52, 59, 72, 83, 136, 146, 160, 173, 184, 189, 203, 219, 230, 241, 251, 259, 269, 278, 286, 296, 305, 314, 323, 332, 340, 364, 375, 387, 405
loin including fat, braised pork 8, 12, 27, 44, 52, 60, 72, 84, 136, 145, 160, 174, 184, 191, 203, 219, 231, 241, 251, 259, 269, 278, 286, 296, 305, 314, 323, 332, 340, 360, 373, 387, 407
loin including fat, broiled pork 8, 12, 26, 45, 51, 60, 72, 83, 137, 147, 162, 172, 185, 190, 203, 219, 231, 242, 252, 260, 270, 279, 287, 297, 306, 315, 324, 333, 341, 361, 373, 388, 407
loin including fat, roasted pork 9, 12, 26, 45, 52, 60, 72, 83, 136, 146, 161, 174, 185, 190, 204, 220, 231, 242, 252, 260, 270, 279, 287, 297, 306, 315, 324, 333, 342, 361, 373, 388, 406
long bones, separation of ends of 91
longans 21, 30, 42, 92, 138, 144, 166, 176, 181, 197, 205, 238, 247, 267, 276, 284, 293, 302, 311, 321, 338, 347, 353, 369, 383, 395, 400
longans, dried 20, 92, 131, 142, 154, 170, 180, 191, 200, 221, 237, 246, 265, 274, 283, 292, 301, 309, 320, 337, 345, 355, 362, 381, 393, 408
loose teeth 91, 186

loquats 2, 22, 36, 42, 95, 133, 150, 165, 175, 181, 197, 205, 224, 238, 248, 257, 267, 276, 285, 294, 302, 312, 322, 330, 339, 347, 354, 369, 382, 397, 400
lotus root, boiled 16, 36, 42, 92, 132, 161, 173, 195, 203, 221, 247, 256, 266, 276, 284, 293, 302, 311, 321, 329, 339, 347, 353, 368, 383, 395, 400
lotus root, raw 15, 28, 41, 92, 131, 160, 173, 194, 201, 221, 246, 256, 266, 275, 284, 292, 301, 311, 321, 329, 338, 347, 353, 369, 383, 394, 401
LUMISTEROL 98, 99
lunch meat, beef 16, 29, 47, 52, 63, 70, 93, 135, 146, 157, 175, 182, 194, 207, 214, 231, 243, 253, 263, 272, 281, 289, 298, 308, 317, 327, 335, 343, 357, 362, 373, 390, 406
lunch meat, beef and pork 14, 30, 47, 52, 63, 71, 83, 93, 135, 150, 162, 175, 182, 195, 207, 214, 232, 244, 253, 263, 272, 281, 289, 299, 308, 318, 327, 335, 344, 357, 361, 372, 391, 407
lunch meat, canned pork 13, 29, 47, 53, 63, 72, 83, 137, 150, 163, 176, 183, 195, 207, 214, 233, 244, 253, 263, 272, 281, 290, 299, 308, 318, 327, 335, 344, 357, 361, 373, 391, 406
lunch meat, chopped ham 12, 29, 46, 55, 61, 72, 93, 137, 148, 162, 175, 182, 193, 204, 214, 232, 243, 252, 261, 272, 281, 289, 298, 308, 317, 326, 335, 343, 364, 374, 390, 404
lunch meat, minced ham 12, 29, 46, 56, 61, 72, 92, 135, 148, 162, 175, 182, 193, 204, 214, 232, 244, 254, 262, 272, 281, 289, 298, 308, 317, 327, 335, 344, 357, 363, 374, 390, 405
lunch meat, pork and beef 14, 30, 47, 52, 63, 71, 83, 93, 135, 150, 162, 175, 182, 195, 207, 214, 232, 244, 253, 263, 272, 281, 289, 299, 308, 318, 327, 335, 344, 357, 361, 372, 391, 407
lung cancer 140
lung infections 50
lung, inflammation of 227
lungs, braised beef 8, 20, 30, 47, 52, 66, 70, 82, 92, 135, 143, 154, 176, 184, 192, 208, 218, 233, 243, 251, 262, 271, 280, 289, 298, 307, 317, 326, 335, 343, 366, 378, 389, 401
lungs, braised pork 17, 26, 47, 52, 65, 71, 94, 136, 154, 176, 192, 208, 218, 231, 244, 263, 272, 281, 289, 299, 308, 318, 327, 343, 367, 379, 390, 401
lupins, boiled 16, 34, 41, 130, 143, 160, 169, 194, 206, 224, 233, 242, 253, 263, 272, 281, 290, 300, 308, 318, 327, 335, 344, 354, 367, 379, 390, 402
LURIDINE 86
luxury loaf, pork 12, 27, 46, 52, 61, 71, 84, 93, 131, 146, 161, 173, 182, 192, 202, 214, 230, 243, 254, 262, 272, 280, 289, 298, 307, 316, 326, 334, 343, 356, 366, 378, 389, 403
LYCEDAN 68
lychees 22, 33, 41, 92, 138, 144, 165, 176, 181, 197, 208, 224, 238, 293, 302, 330, 353, 368, 381, 396, 400
lychees, dried 22, 25, 38, 91, 131, 142, 158, 170, 180, 192, 200, 224, 236, 292, 301, 328, 350, 362, 379, 393, 408
lymph cell count, decreased 58
lymphocyte levels, decreased 167
LYOCHROME 24
LYOVIT-H 75
LYS 285
LYSINE 88, 285-294
lysine, internal conversion of 88
LYSTHENON 87

macadamia nuts 78
macadamia nuts, dried 13, 31, 38, 129, 143, 157, 169, 193, 203, 223, 232, 244, 254, 264, 273, 282, 291, 300, 309, 319, 336, 345, 354, 359, 372, 392, 410
macadamia nuts, dry roasted 3, 14, 32, 38, 55, 63, 79, 129, 141, 157, 169, 188, 205, 216, 232, 243, 253, 264, 273, 282, 291, 299, 309, 319, 328, 335, 345, 355, 359, 372, 392, 410
macadamia nuts, oil roasted 5, 14, 32, 38, 131, 142, 158, 169, 191, 204, 223, 233, 244, 254, 264, 274, 282, 291, 300, 309, 319,

Index

336, 345, 354, 359, 371, 392, 410
macaroni 15, 33, 39, 57, 67, 84, 136, 150, 158, 174, 180, 196, 210, 225, 235, 246, 255, 265, 274, 283, 292, 300, 309, 319, 328, 337, 352, 367, 381, 393, 402
mace 3, 13, 25, 128, 154, 169, 194, 202, 221, 232, 351, 360, 373, 393, 409
mackerel 89, 104, 117
mackerel, cooked 8, 15, 25, 44, 59, 69, 133, 146, 158, 169, 189, 202, 218, 234, 243, 252, 261, 270, 279, 287, 295, 306, 316, 325, 333, 342, 363, 374, 388, 406
mackerel, horse 117
mackerel, raw 8, 15, 26, 44, 51, 60, 69, 134, 148, 158, 169, 184, 191, 204, 218, 234, 244, 253, 262, 271, 280, 289, 297, 307, 317, 326, 334, 343, 365, 375, 389, 404
MACRABIN 68
MAGNESIUM 59, 76, 91, 103, 104, 127, 168-177, 200, 294, 371
MAGNESIUM ASCORBATE 97
magnesium, decreased 98, 125, 127
male sterility 88, 96, 126
malt liquor 138, 166, 211, 223, 357, 369, 397
malted milk 7, 17, 29, 48, 55, 65, 73, 82, 96, 128, 150, 166, 173, 182, 194, 207, 218, 235, 255, 265, 274, 282, 291, 300, 309, 319, 328, 336, 345, 354, 368, 378, 393, 400
malted milk, whole chocolate 7, 19, 30, 48, 55, 66, 73, 83, 96, 128, 150, 165, 174, 182, 195, 207, 220, 235, 256, 266, 275, 284, 293, 301, 311, 321, 329, 338, 346, 354, 368, 378, 393, 400
maltex 16, 34, 40, 57, 66, 82, 137, 145, 162, 173, 195, 209, 223, 234, 353, 368, 381, 394, 400
MALTOSE 349
mammy apple 3, 22, 34, 42, 57, 93, 135, 163, 198, 210, 222, 293, 302, 330, 354, 369, 380, 397, 400
MANGANESE 11, 86, 101, 127, 177-185, 187, 228, 240
mango 2, 19, 33, 40, 56, 63, 92, 136, 145, 166, 176, 182, 198, 208, 224, 238, 248, 267, 276, 285, 293, 302, 312, 321, 330, 339, 347, 353, 368, 382, 397, 400

manhattan 22, 42, 58, 138, 150, 166, 177, 182, 199, 211, 223, 238, 357, 364, 403
manic depressive states 228
maple sugar 128, 159, 198, 206, 222, 349, 361, 409
maple syrup 129, 161, 198, 208, 222, 350, 363, 408
margarine 101, 103
margarine liquid, soybean oil and cottonseed oil 2, 56, 74, 130, 196, 210, 215, 247, 256, 265, 274, 283, 292, 301, 310, 320, 329, 337, 346, 359, 372, 395, 409
margarine stick, corn oil 2, 133, 198, 210, 222, 248, 266, 275, 284, 293, 301, 311, 321, 329, 338, 347, 359, 372, 409
margarine stick, safflower oil and soybean oil 2, 133, 198, 210, 222, 248, 266, 275, 284, 293, 301, 311, 321, 329, 338, 347, 359, 372, 409
margarine stick, soybean oil 2, 133, 198, 210, 222, 248, 266, 275, 284, 293, 301, 311, 321, 329, 338, 347, 359, 372, 409
margarine stick, soybean oil and cottonseed oil 2, 133, 198, 210, 222, 248, 266, 275, 284, 293, 301, 311, 321, 329, 338, 347, 359, 372, 409
margarine tub, corn oil 2, 133, 198, 210, 221, 248, 266, 275, 284, 293, 301, 311, 321, 329, 338, 347, 359, 371, 408
margarine tub, corn oil diet 2, 133, 198, 211, 215, 248, 266, 276, 285, 293, 301, 312, 321, 338, 347, 361, 372, 405
margarine tub, safflower oil 2, 133, 198, 210, 221, 248, 266, 275, 284, 293, 301, 311, 321, 329, 338, 347, 359, 371, 408
margarine tub, soybean oil 2, 133, 198, 210, 221, 248, 266, 275, 284, 293, 301, 311, 321, 329, 338, 347, 359, 371, 408
margarine tub, soybean oil and cottonseed oil 2, 133, 198, 210, 221, 248, 266, 275, 284, 293, 301, 311, 321, 329, 338, 347, 359, 371, 408
margarine tub, soybean oil and cottonseed oil diet 2, 133, 198, 211, 215, 248, 266, 276, 285, 293, 301, 312, 321, 338, 347, 361, 372, 405

marinara sauce 3, 21, 33, 39, 93, 133, 144, 162, 172, 197, 202, 215, 236, 354, 368, 378, 395, 400
MARINEURINA 24
marjoram 2, 127, 153, 168, 191, 200, 232, 365, 385
marmalade, citrus 94, 131, 164, 196, 210, 222, 350, 363, 397, 408
marshmallow 133, 158, 221, 349, 361, 395, 408
martini 138, 151, 166, 177, 199, 211, 224, 238, 358, 364, 403
mayonnaise 7, 56, 74, 80, 101, 136, 150, 165, 198, 211, 215, 236, 354, 359, 372, 396, 408
mayonnaise, light 7, 57, 73, 80, 136, 150, 165, 197, 211, 215, 236, 353, 362, 373, 396, 406
mayonnaise, safflower oil and soybean oil 134, 163, 197, 210, 216, 237, 247, 256, 266, 275, 284, 293, 301, 311, 320, 329, 338, 346, 357, 359, 371, 395, 408
mayonnaise, soybean oil 94, 134, 163, 197, 210, 216, 237, 247, 256, 266, 275, 284, 293, 301, 311, 320, 329, 338, 346, 357, 359, 371, 395, 408
maypo 3, 14, 26, 38, 56, 60, 71, 84, 93, 130, 148, 155, 173, 194, 210, 224, 235, 354, 368, 380, 394, 400
MB 186
MDR, pre-1958 5, 9, 23, 36, 96, 97, 99
measles 9, 101
meat 125, 139, 225, 227
MEAT SUGAR 85
meats, organ 76, 140
medical testing 124
MEGABION 69
MEGALOVEL 69
melanin in skin, decreased 303, 331
MELIN 113
melon, casaba 5, 19, 36, 41, 93, 137, 164, 177, 199, 207, 222, 355, 370, 383, 396, 398
melon, honeydew 5, 20, 35, 41, 93, 134, 164, 198, 206, 222, 355, 369, 381, 396
MEMBRANE INTEGRITY FACTOR 111
membrane integrity of cell, decreased 313
memory, deterioration of 11, 37, 43, 69, 86, 106, 118, 187, 199, 213, 303, 322
MENADIOL 107
MENADIOL DIACETATE 109

Index

MENADIOL DIBUTYRATE 109
MENADIOL DIPHOSPHATE (TETRASODIUM SALT) 109
MENADIOL DISULFATE 109
MENAPHTHONE 107, 109
MENAQUINONE 107
Ménièr's syndrome 23, 36, 43, 49, 101, 105, 138
meningitis 9, 96, 99
MENODOINE 107, 109
menses, increased 6, 153, 187
menses, irregular 6, 152
menstrual deterioration 69
menstruation 166
mental and physical vigor, decreased 153
mental health deterioration 11, 58, 77, 106, 118, 126, 167, 168, 186, 285, 313, 339
mental illness 23, 43, 49, 58, 67, 84, 96, 101, 105, 138, 177, 199
mental imbalances 167
mental retardation 76
mental retardation in infants 58
mental/physical development, arrested 152
MEPHYTON 108
MERCURIAL DIURETICS 126
MERCURY 186
MERPRESS 97
MESOINOSITE 85
MESOINOSITOL 85
MESOINOSITOL HEXANICOTINATE 86
MESONEX 86
MESOTAL 86
MET 294
METABOLIN 23
metabolism rate, decreased 100, 103, 104, 371
metabolizing of fatty acids, decreased 88
METADEE 99
METAFORMIN 69, 140
metallic taste in mouth 123, 167, 186
metallurgy 167
METHENAMINE 91
METHIONINE 77, 88, 294-303
methionine and cysteine, internal synthesis of 313
methionine internal conversion 86, 257
METHIONINE METHYLSULFONIUM SALT 118
METHOTREXATE 69, 78, 140
METHYL BROMIDE 50
METHYL LINOLEATE 103
METHYLESCULIN 113
METHYLMETHIONINESULFONIUM BROMIDE 118, 119
METHYLMETHIONINESULFONIUM CHLORIDE 118, 119
MG 168
MG^{++} 168

MG^{+2} 168
MICRO-DEE 99
MIDARINE 87
MILBEDOCE 69
milk 85, 86, 104, 108, 117, 140, 152, 187, 212, 227
milk, carob 7, 20, 30, 48, 55, 66, 73, 83, 96, 128, 151, 165, 175, 185, 195, 209, 220, 236, 246, 255, 265, 274, 283, 291, 300, 309, 319, 328, 336, 345, 355, 368, 378, 394, 400
milk, coconut 21, 40, 95, 133, 143, 158, 170, 179, 195, 205, 222, 234, 245, 255, 265, 274, 283, 292, 301, 309, 320, 329, 337, 346, 356, 364, 373, 394, 403
milk, cow 88, 89, 99, 101, 103, 104, 108, 116, 140, 213, 225, 258
milk, dry buttermilk 7, 13, 25, 48, 50, 61, 70, 80, 127, 165, 169, 188, 200, 216, 230, 243, 251, 261, 268, 277, 286, 295, 304, 313, 322, 331, 339, 351, 361, 377, 385, 410
milk, dry instant skim 6, 13, 24, 48, 50, 61, 70, 80, 94, 127, 165, 169, 188, 200, 216, 230, 243, 251, 260, 268, 277, 286, 294, 304, 313, 322, 331, 339, 350, 361, 380, 385, 410
milk, dry skim 8, 13, 25, 48, 50, 60, 70, 80, 94, 127, 165, 169, 188, 200, 216, 230, 243, 251, 260, 268, 277, 286, 294, 304, 313, 322, 331, 339, 350, 360, 380, 385, 410
milk, dry whole 7, 14, 25, 48, 50, 61, 70, 80, 94, 127, 164, 169, 188, 200, 216, 230, 244, 252, 261, 268, 277, 288, 296, 304, 315, 323, 331, 340, 351, 359, 373, 387, 410
milk, evaporated lowfat 7, 21, 27, 48, 66, 74, 82, 96, 127, 165, 172, 191, 204, 217, 255, 264, 273, 282, 291, 299, 309, 319, 327, 336, 345, 355, 368, 379, 392, 401
milk, evaporated skim 7, 21, 26, 48, 51, 65, 74, 82, 127, 165, 171, 192, 204, 217, 234, 245, 255, 264, 273, 282, 290, 299, 309, 318, 327, 336, 344, 354, 368, 382, 392, 401
milk, evaporated whole 7, 20, 26, 48, 52, 66, 74, 82, 95, 127, 165, 172, 191, 204, 217, 234, 245, 255, 264,
273, 282, 291, 299, 309, 319, 327, 336, 344, 354, 366, 377, 392, 402
milk, filled 9, 21, 31, 48, 55, 66, 73, 83, 96, 128, 166, 175, 195, 209, 220, 236, 246, 255, 265, 274, 282, 291, 300, 309, 319, 328, 336, 345, 356, 368, 378, 394, 399
milk, goat 7, 20, 31, 48, 55, 66, 74, 89, 96, 103, 128, 166, 175, 186, 194, 207, 221, 236, 246, 255, 265, 274, 283, 291, 300, 309, 319, 328, 336, 345, 356, 368, 378, 393, 400
milk, human 7, 35, 48, 56, 74, 83, 85, 89, 95, 99, 101, 103, 104, 105, 131, 165, 177, 198, 210, 222, 237, 247, 256, 266, 275, 284, 293, 301, 311, 321, 329, 337, 347, 355, 368, 378, 396, 399
milk, indian buffalo 7, 19, 31, 48, 56, 66, 73, 83, 95, 128, 166, 171, 194, 208, 220, 237, 246, 255, 265, 274, 282, 291, 300, 309, 319, 328, 336, 345, 356, 367, 377, 393, 400
milk, lowfat 1% 7, 20, 30, 48, 55, 66, 73, 83, 96, 128, 166, 175, 195, 208, 221, 236, 246, 255, 265, 274, 283, 291, 300, 309, 319, 328, 336, 345, 356, 369, 379, 394, 399
milk, lowfat 1% chocolate 7, 20, 30, 48, 55, 66, 73, 83, 96, 128, 165, 175, 194, 208, 220, 236, 246, 255, 265, 274, 283, 291, 300, 309, 319, 328, 336, 345, 354, 368, 380, 394, 400
milk, lowfat 2% 7, 20, 30, 48, 55, 66, 73, 83, 96, 125, 128, 166, 175, 195, 208, 221, 236, 246, 255, 265, 274, 282, 291, 300, 309, 319, 328, 336, 345, 356, 369, 379, 394, 399
milk, lowfat 2% chocolate 7, 20, 30, 48, 55, 66, 73, 83, 96, 128, 165, 175, 194, 208, 220, 236, 246, 255, 265, 274, 283, 291, 300, 309, 319, 328, 336, 345, 354, 368, 379, 394, 400
milk, malted 7, 17, 29, 48, 55, 65, 73, 82, 96, 128, 150, 166, 173, 182, 194, 207, 218, 235, 255, 265, 274, 282, 291, 300, 309, 319, 328, 336, 345, 354, 368, 378, 393, 400
milk, protein fortified low-

Index 504

fat 1% 7, 20, 29, 48, 54, 66, 73, 83, 96, 128, 166, 175, 194, 208, 220, 235, 246, 255, 264, 273, 282, 291, 300, 309, 319, 328, 336, 345, 356, 369, 379, 393, 399
milk, protein fortified lowfat 2% 7, 20, 29, 48, 54, 66, 73, 83, 96, 128, 166, 175, 194, 207, 220, 235, 246, 255, 264, 273, 282, 291, 300, 309, 319, 328, 336, 345, 356, 369, 379, 393, 399
milk, protein fortified skim 7, 20, 29, 48, 54, 66, 73, 83, 96, 128, 166, 175, 194, 208, 220, 235, 246, 255, 264, 273, 282, 291, 300, 309, 319, 328, 336, 345, 355, 369, 382, 393, 399
milk, raw 118
milk residue concentrates 90
milk, sheep 7, 18, 26, 48, 54, 72, 95, 128, 166, 174, 193, 209, 221, 246, 255, 264, 273, 282, 291, 299, 309, 319, 327, 336, 344, 356, 367, 377, 393, 400
milk, skim 7, 20, 31, 48, 54, 66, 73, 83, 96, 128, 166, 176, 194, 208, 221, 236, 246, 255, 265, 274, 282, 291, 300, 309, 319, 328, 336, 345, 356, 369, 382, 394, 399
milk, soy 8, 15, 33, 48, 57, 66, 84, 138, 145, 163, 174, 181, 196, 209, 221, 237, 245, 255, 265, 274, 283, 292, 301, 309, 320, 328, 337, 346, 357, 369, 379, 384, 394, 398
milk, sweetened condensed 7, 17, 25, 48, 51, 65, 73, 81, 95, 127, 165, 172, 189, 202, 217, 233, 245, 255, 264, 273, 282, 290, 299, 309, 318, 327, 335, 344, 350, 361, 377, 392, 408
milk, whole 7, 20, 30, 48, 55, 66, 73, 83, 90, 96, 128, 166, 175, 186, 195, 208, 221, 236, 246, 255, 265, 274, 283, 291, 300, 309, 319, 328, 336, 345, 356, 368, 378, 394, 399
milk, whole chocolate 7, 20, 30, 48, 55, 66, 73, 83, 96, 129, 148, 165, 175, 181, 195, 208, 220, 236, 246, 255, 265, 274, 283, 291, 300, 309, 319, 328, 336, 345, 354, 368, 378, 394, 400
milk, whole chocolate malted 7, 19, 30, 48, 55,
66, 73, 83, 96, 128, 150, 165, 174, 182, 195, 207, 220, 235, 256, 266, 275, 284, 293, 301, 311, 321, 329, 338, 346, 354, 368, 378, 393, 400
milkfish, raw 22, 33, 44, 60, 70, 130, 150, 165, 193, 234, 243, 253, 262, 271, 280, 288, 297, 307, 316, 326, 334, 342, 366, 377, 389, 402
milkshake, restaurant chocolate 8, 19, 28, 48, 54, 66, 73, 84, 128, 148, 165, 174, 182, 194, 207, 218, 236, 246, 255, 265, 274, 282, 291, 300, 309, 319, 328, 336, 345
milkshake, restaurant vanilla 7, 19, 30, 48, 53, 66, 73, 84, 96, 128, 149, 166, 176, 185, 194, 208, 218, 236, 246, 255, 265, 273, 282, 291, 300, 309, 319, 328, 336, 345, 353, 367, 379, 393, 401
milkshake, vanilla 7, 21, 29, 48, 66, 73, 83, 128, 166, 176, 194, 207, 218, 236, 246, 255, 265, 274, 282, 291, 300, 309, 319, 328, 336, 345, 353, 367, 379, 393, 401
MILLAFOL 78
millet 12, 26, 38, 104, 116, 117, 132, 154, 168, 189, 201, 227, 350, 361, 379, 391, 409
MILLEVIT 69
MINA D2 99
mince pie 18, 35, 41, 96, 132, 161, 197, 208, 216, 351, 363, 376, 394, 407
mineral absorption, lowered 384
MINERAL OIL 1, 6, 97, 98, 101, 103, 107, 110, 112, 114, 115, 116, 117, 118
mineral, trace 123, 124, 125, 126, 138, 139, 140, 141, 151, 152, 153, 167, 177, 186, 187, 211, 212, 213, 225, 226, 227, 228
miso 4, 17, 28, 40, 55, 62, 74, 80, 129, 142, 156, 170, 179, 193, 208, 214, 230, 244, 254, 263, 272, 281, 290, 299, 308, 318, 327, 336, 344, 351, 365, 377, 384, 391, 407
mitochondria cell energy, decreased 117
mitochondria deterioration 225
MMSC 118
MN 177
mobility decreased 178
molasses 86, 186, 212
molasses, barbados 19, 29, 127, 196, 350, 363, 408
molasses, blackstrap 16, 29, 39, 85, 86, 89, 127, 154, 195, 200, 218, 350, 364, 408
molasses, light 19, 34, 128, 155, 196, 200, 222, 350, 363, 408
molasses, medium 32, 39, 89, 101, 127, 154, 195, 200, 221, 350, 364, 408
MOLYBDENUM 141, 186
molybdenum replaced by tungsten in enzymes 227
monkfish, raw 22, 33, 136, 150, 165, 173, 183, 222, 236, 244, 254, 263, 272, 281, 289, 298, 308, 318, 326, 335, 344, 368, 379, 390, 400
MONOEPOXYVITAMIN A 10
MONO-KAY 108
mononucleosis 9, 96, 211
mortadella, pork and beef 16, 31, 47, 64, 71, 92, 133, 148, 159, 175, 182, 195, 208, 214, 232, 244, 253, 262, 272, 281, 289, 298, 308, 318, 327, 335, 344, 357, 362, 373, 390, 406
MORYL 87
mothbeans, boiled 5, 16, 36, 40, 96, 138, 144, 156, 169, 179, 193, 204, 223, 235, 245, 251, 273, 282, 293, 301, 318, 328, 336, 352, 367, 380, 392, 403
mother's loaf, pork 12, 29, 47, 53, 63, 71, 131, 146, 159, 175, 181, 194, 207, 214, 233, 355, 362, 374, 391, 406
motion sickness 58
motor paralysis 123
motor vehicle exhaust 167, 187
MOUSE ANTIALOPECIA FACTOR 85
mouth, cracks around 24, 58
mouth, dry 104, 371
mouth, frothing 77
mouth, inflamed 100, 124
mouth, metallic taste in 123, 167, 186
mouth, pain in 124
mouth sores 24, 58, 186
mouth watering 88
mucous membranes, grayish 89
muffin, english 13, 26, 38, 53, 66, 80, 128, 142, 156, 174, 194, 201, 215, 234, 351, 364, 379, 392, 407
muffin, toasted english 13, 26, 38, 52, 66, 80, 128, 142, 155, 172, 193, 200, 215, 234, 350, 362, 379, 391, 408

muffins, blueberry 4, 15, 29, 39, 129, 159, 193, 209, 215, 351, 362, 377, 392, 408
muffins, bran 3, 14, 27, 38, 52, 61, 74, 80, 94, 128, 143, 156, 169, 189, 206, 216, 231, 351, 362, 376, 392, 408
muffins, corn meal 3, 14, 28, 39, 129, 158, 192, 209, 216, 351, 361, 376, 392, 408
muffins, whole ground corn meal 3, 15, 30, 39, 128, 159, 191, 209, 216, 351, 362, 376, 392, 408
mulberries 5, 21, 32, 41, 78, 92, 131, 158, 174, 197, 207, 222, 354, 369, 381, 395, 399
mullet, cooked 8, 59, 82, 95, 131, 144, 159, 170, 183, 190, 201, 219, 234, 243, 252, 261, 270, 278, 287, 295, 306, 315, 325, 333, 341, 366, 378, 388, 402
mullet, raw 8, 51, 60, 82, 95, 131, 149, 161, 171, 184, 190, 203, 219, 235, 243, 253, 262, 271, 280, 289, 297, 307, 317, 326, 334, 343, 367, 378, 389, 401
MULSIFEROL 99
multiple sclerosis 23, 36, 43, 49, 58, 67, 74, 76, 77, 86, 96, 101, 105, 117, 177, 185
MUNDISAL 88
mung bean sprouts 78
mung beans, boiled 5, 15, 34, 40, 54, 65, 79, 96, 132, 144, 159, 170, 180, 195, 205, 224, 234, 244, 255, 264, 273, 282, 291, 300, 309, 319, 327, 336, 345, 353, 367, 381, 384, 392, 402
mung beans, canned sprouted 5, 22, 34, 42, 81, 133, 144, 164, 176, 181, 197, 211, 246, 257, 266, 275, 284, 293, 302, 311, 321, 329, 338, 346, 357, 371, 395, 398
mung beans, cooked sprouted 5, 20, 32, 40, 56, 93, 135, 145, 163, 175, 181, 197, 210, 223, 235, 246, 256, 265, 274, 283, 292, 301, 310, 320, 329, 338, 346, 356, 370, 382, 395, 398
mung beans, raw sprouted 5, 18, 31, 40, 54, 64, 79, 93, 134, 144, 161, 173, 181, 196, 208, 223, 236, 246, 256, 265, 274, 283, 292, 301, 309, 320, 328, 337, 346, 355, 370, 382, 384, 394, 399
mung beans, stir-fried sprouted 15, 30, 39, 93,

134, 158, 234, 245, 255, 264, 274, 283, 291, 300, 309, 320, 328, 337, 345, 354, 369, 382, 393, 400
mungo beans, boiled 5, 15, 33, 39, 53, 65, 79, 96, 130, 144, 158, 169, 179, 193, 206, 223, 234, 244, 255, 264, 273, 282, 291, 300, 309, 319, 327, 336, 344, 353, 367, 380, 392, 402
muscle activity, decreased 211
MUSCLE ADENYLIC ACID 67
muscle, calcification 212
muscle, ceroid deposits in 100
muscle cramps 88, 138, 139, 199, 214
muscle cramps, leg 50
muscle, deterioration of 100, 212, 339
muscle disease 88
muscle fibers, knotting of 168
muscle pain 58, 88, 89, 91, 100, 151
muscle shrinkage 214
muscle spasm/twitching 11, 99, 125, 127, 138, 168, 178, 214
muscle stiffness 178
muscle strength, decreased 112
MUSCLE SUGAR 85
muscle tendon strain 127, 151
muscle tone, decreased 98, 106
muscle weakness 6, 37, 43, 88, 91, 98, 100, 126, 153, 199, 214
muscle weakness in rats and chicks 49
muscles in face, very fine twitching of 167
muscles in leg, tender 11
muscles in leg, twitching 123
muscles paralyzation 294
muscular coordination, decreased 178
muscular dystrophy 9, 58, 67, 74, 86, 96, 101, 117, 211, 313
muscular dystrophy, Duchenne-type 88
muscular dystrophy, limb-girdle 88
muscular response, exaggerated 177
mushrooms 89, 90, 99, 108, 140, 212, 213
mushrooms, boiled 18, 27, 38, 50, 64, 81, 95, 138, 142, 158, 176, 181, 195, 203, 224, 234, 246, 257, 265, 274, 283, 292, 301, 310, 320, 328, 338, 346, 356, 370, 380, 394, 399
mushrooms, canned shitake 22, 32, 38, 136, 161, 209, 219, 355, 369, 381, 394

mushrooms, cooked shitake 21, 30, 39, 138, 164, 175, 197, 209, 224, 246, 256, 266, 275, 284, 293, 301, 310, 320, 330, 338, 346, 354, 369, 382, 395, 400
mushrooms, dried shitake 13, 25, 37, 94, 134, 158, 169, 189, 200, 222, 244, 253, 264, 273, 282, 291, 299, 308, 318, 328, 336, 344, 350, 362, 379, 391, 409
mushrooms, raw 16, 25, 38, 50, 64, 80, 95, 137, 145, 160, 176, 194, 203, 224, 235, 246, 257, 265, 275, 283, 292, 301, 310, 320, 328, 338, 346, 356, 370, 380, 395, 399
muskrat, roasted 15, 29, 366, 378, 387, 403
mussels 105, 126
mussels, cooked 131, 141, 154, 170, 189, 205, 216, 231, 241, 251, 263, 271, 280, 289, 297, 307, 316, 325, 334, 343, 355, 365, 378, 388, 404
mussels, raw 132, 146, 155, 170, 191, 204, 216, 233, 244, 254, 264, 273, 281, 290, 299, 309, 318, 327, 335, 344, 356, 368, 379, 391, 400
mustard, brown 128, 158, 193, 209, 214, 355, 367, 377, 393, 401
mustard greens, boiled 2, 21, 33, 41, 57, 92, 129, 163, 175, 196, 207, 222, 246, 255, 265, 275, 284, 292, 301, 311, 320, 329, 337, 346, 357, 370, 381, 394, 398
mustard greens, frozen and boiled 2, 21, 34, 42, 58, 64, 93, 129, 148, 160, 175, 180, 197, 209, 222, 237, 246, 255, 265, 275, 284, 292, 301, 311, 320, 329, 337, 346, 357, 370, 381, 394, 398
mustard seed, yellow 4, 12, 26, 37, 127, 154, 168, 188, 200, 229, 241, 249, 261, 270, 279, 289, 298, 304, 315, 322, 333, 340, 351, 359, 372, 387, 409
mustard spinach, boiled 2, 92, 128, 162, 198, 357, 370, 382, 395, 398
mustard, yellow 129, 158, 195, 209, 214, 355, 368, 378, 393, 401
muteness and deafness 152
mutton tallow 359, 371
MY-B-DEN 68
MYCOINE 118

myelin sheaths around nerves, deterioration of 259
MYKOSTIN 99
myocardial infarction 9, 101, 211
MYOINOSITOL 85
myoma, glandular 152
MYOSTON 68
myotonic dystrophy 88
MYRTICOLORIN 113

NA 214
NA+ 214
NAD 119
NAGRAVON 69
nail deterioration 9, 85, 104, 138, 166, 371
nails, brittle 6, 104, 152, 371
nails, decreased growth in 212
nails, deformed 228
nails, edges of turned up 153
nails, flaking 104, 371
nails, ingrown 213
nails, ribbed 213
nails, soft 104, 371
nails, thin 152, 153
nails, white spots on 228
NAPHTHALENDIONE 108
NAPHTHALENEDIOL 109
NAPHTHALENEDIONE 108
NATRIUM 214
natto 15, 29, 93, 117, 127, 142, 154, 169, 178, 192, 200, 223, 230, 244, 253, 263, 271, 280, 290, 299, 306, 317, 326, 335, 343, 353, 365, 376, 390, 406
NAUCAINE 106
nausea 6, 37, 43, 49, 58, 77, 86, 88, 89, 90, 98, 100, 106, 123, 124, 125, 126, 139, 151, 167, 177, 178, 186, 187, 199, 213, 214, 228, 285
nausea of pregnancy 67
navel orange 4, 17, 34, 42, 55, 65, 80, 92, 131, 149, 166, 176, 182, 198, 208, 224, 238, 247, 256, 266, 276, 285, 293, 301, 311, 321, 330, 339, 347, 354, 369, 383, 396, 400
navy beans 78, 85
navy beans, boiled 5, 15, 34, 41, 55, 63, 79, 96, 129, 143, 157, 169, 179, 193, 203, 225, 233, 244, 254, 264, 273, 282, 290, 299, 308, 318, 327, 336, 344, 352, 366, 381, 384, 392, 404
navy beans, canned 5, 15, 34, 41, 56, 64, 79, 96, 130, 143, 158, 170, 179, 193, 205, 216, 234, 245, 254, 264, 273, 282, 291, 300, 309, 319, 327, 336, 345, 352, 367, 381, 392, 402

navy beans, dry 139
neck, color increased on 124
neck/shoulder pains 167
neck with skin, fried chicken 7, 17, 28, 45, 51, 62, 73, 83, 131, 145, 157, 174, 181, 193, 208, 218, 230, 242, 251, 262, 270, 280, 288, 297, 306, 316, 325, 334, 342, 356, 361, 373, 388, 407
neck with skin, simmered chicken 7, 20, 28, 47, 53, 64, 74, 84, 132, 146, 157, 175, 182, 194, 209, 220, 231, 243, 252, 263, 271, 281, 289, 298, 307, 317, 326, 335, 343, 363, 374, 389, 404
neck without skin, simmered chicken 7, 19, 27, 46, 51, 63, 74, 83, 131, 145, 156, 174, 181, 194, 209, 219, 230, 243, 251, 261, 269, 279, 288, 296, 306, 316, 325, 333, 342, 365, 377, 388, 403
nectarine 3, 23, 35, 40, 56, 66, 84, 94, 138, 148, 166, 177, 182, 198, 207, 237, 354, 369, 381, 396, 400
nectarine seeds 78
NEO-BETALIN 12 75
NEOCAINE 105
NEO-CYTAMEN 75
NEO DOHYFRAL D3 99
NEOLAMIN 23
NEO-MACRABIN 75
NEOMYCIN 69, 107, 127, 140, 153
neon signs 125
NEO-ROJAMIN 75
NEOTOCOPHEROL 102
NEOVITAMIN A 6, 10
nerve development, faulty 141
nerve pain 58, 168, 187
nerve regeneration, decreased 118
nerve tissue, destruction of 167
nerves, deterioration of myelin sheaths around 259
nervous system deterioration 58, 69, 77, 88, 100, 118, 187, 199, 228
nervous system, lesions in 151
nervousness 11, 37, 43, 58, 69, 78, 90, 117, 126, 168, 177, 186, 303, 322, 339
nettles 213
NEUCOLIS 87
neural symptoms, severe 107
neurasthenia 228
neuritis 23, 36, 43, 49, 58, 67, 74, 177
neurotransmitter levels of serotonin, unstable 167

NEUROTROPAN 88
NI 187
NIACIN 37
NIACINAMIDE 37, 43
NIACINIAMIDE ASCORBATE 97
NICACID 37
NICAGIN 37
NICAMINDON 43
NICASTUBINE 97
NICHOLIN 87
NICKEL 187
NICOBID 37
NICOFORT 43
NICOLAR 37
NICOLIN 87
NICONACID 37
NICOSCORBINE 97
NICO-SPAN 37
NICOTAMIDE 43
NICOTENE 37
NICOTILAMIDE 43
NICOTINAMIDE 37, 43
NICOTINAMIDE ADENINE DINUCLEOTIDE 119
NICOTINAMIDE ASCORBATE 97
NICOTINIC ACID 37
NICOTINIC ACID AMIDE 43
NICOTINIPCA 37
NICYL 37
1958 MDR, pre-1958 5, 9, 23, 36, 96, 97, 99
1958 RDA 5, 9, 23, 36, 96, 97, 99
1974 PCS 5, 9, 23, 36, 38, 43, 49, 52, 58, 67, 74, 76, 77, 85, 89, 91, 96, 97, 99, 101, 110, 111, 118, 119, 120, 123, 124, 126, 138, 139, 140, 151, 153, 166, 167, 177, 186, 199, 212, 213, 225, 226, 227, 238, 330
1974 RDA 5, 23, 36, 43, 49, 67, 74, 85, 96, 97, 99, 101
1974 USRDA 5, 23, 36, 43, 49, 67, 74, 85, 96, 97, 99, 101
1980 RDA 5, 9, 23, 36, 43, 49, 58, 67, 74, 85, 89, 99, 101, 108, 138, 139, 140, 151, 153, 166, 167, 177, 185, 186, 187, 199, 211, 212, 213, 225, 226, 238, 330, 397
1980 SADDI 58, 89, 108, 139, 140, 151, 185, 186, 187, 211, 212, 213, 225, 226
1980 USRDA 6, 9, 23, 36, 43, 49, 58, 67, 74, 85, 89, 96, 97, 99, 101, 138, 153, 166, 167, 177, 199, 238, 330
1989 RDA 6, 9, 23, 36, 43, 49, 58, 67, 68, 74, 76, 77, 78, 85, 86, 89, 91, 96, 97, 99, 101, 104, 105, 106, 108, 110, 111, 112, 113, 117, 118, 119, 120, 123, 124, 125, 126, 138, 139, 140, 151, 152, 153, 166, 167, 177, 185, 186, 187,

199, 211, 212, 213, 225, 226, 227, 238, 330, 397
1989 SADDI 49, 58, 67, 68, 76, 77, 78, 85, 86, 88, 89, 90, 91, 104, 105, 106, 110, 111, 112, 113, 117, 118, 119, 120, 123, 124, 125, 126, 139, 140, 151, 152, 153, 167, 185, 186, 187, 211, 212, 213, 225, 226, 227, 228
1989 USRDA 6, 9, 23, 36, 43, 49, 58, 67, 68, 74, 76, 77, 78, 85, 86, 89, 90, 91, 96, 97, 99, 101, 104, 105, 106, 108, 110, 111, 112, 113, 117, 118, 119, 120, 123, 124, 124, 125, 126, 138, 139, 140, 152, 153, 166, 167, 177, 186, 199, 212, 213, 225, 226, 227, 228, 238, 330
NIO-A-LET 10
NIOZYMIN 43
nipples 124
NITICOLIN 87
NITRILOSIDES 77
NITRITES 6
NITRITOCOBALAMIN 69, 75
NITROCOBALAMIN 75
NITROPRUSSIDE 69, 140
NITROSOCOBALAMIN 75
NITROUS ACID 89
NONESSENTIAL FATTY ACIDS 103, 104, 371
noodles 8, 15, 31, 48, 56, 67, 73, 84, 136, 147, 159, 174, 180, 196, 210, 224, 237, 254, 265, 274, 283, 292, 300, 309, 319, 328, 337, 345, 366, 379, 393, 402
noodles, egg 125
NORMOCYTIN 69
northern beans, boiled great 5, 15, 34, 40, 55, 64, 79, 96, 129, 143, 157, 170, 179, 192, 202, 224, 234, 244, 254, 264, 273, 282, 290, 299, 309, 318, 327, 336, 344, 352, 367, 381, 384, 392, 403
nose, inflammation of 9, 124
nose, runny 139, 214
nosebleeds 91, 98, 107, 112, 153, 187
NOVOCAIN 105
NSC-15780 77
NSC-20264 68
NUCITE 85, 89
nuclear power air pollution 153
nuclear technology 167
nucleic acid structure formation, deterioration of 303
numb hands/feet 124
numb/tingling bones and fingertips 98
numb/tingling feet 49, 69, 78, 199
numb/tingling hands 49, 69, 78, 199

numb/tingling tongue 199
numbness 24, 58, 77, 127, 167, 186, 187, 199
numbness, local 123
nutmeg 4, 13, 128, 155, 168, 190, 202, 231, 350, 359, 372, 393, 409
nutrient absorption, deterioration of 228, 313
nutrisoy flour 213
nuts, Brazil 101, 103, 104, 212

oat grits 77
oat hulls 213
oat straw 213
oatmeal 5, 16, 35, 42, 56, 67, 84, 85, 86, 89, 101, 136, 149, 163, 172, 179, 195, 210, 224, 235, 246, 255, 265, 274, 283, 292, 301, 309, 320, 328, 337, 346, 354, 368, 380, 384, 394, 400
oatmeal bread 13, 29, 38, 54, 66, 80, 130, 143, 156, 170, 179, 194, 208, 216, 234, 351, 363, 378, 384, 392, 408
oatmeal cookie 5, 14, 30, 39, 53, 65, 74, 81, 129, 145, 156, 170, 193, 206, 216, 234, 350, 360, 374, 393, 410
oatmeal, instant 3, 14, 30, 38, 56, 60, 79, 129, 155, 173, 195, 210, 217, 354, 369, 380, 394, 400
oats 77, 90, 103, 104, 107, 116, 186, 187, 212, 227
ocean 167
octopus, raw 20, 34, 47, 130, 142, 154, 182, 192, 232, 244, 253, 263, 272, 281, 290, 299, 308, 318, 326, 335, 344, 357, 368, 379, 390, 401
OH-DUPHAR 75
oheloberries 3, 22, 35, 42, 94, 137, 166, 177, 199, 210, 224, 355, 370, 382, 397, 398
oil roasted pecans 131, 141, 157, 169, 178, 189, 203, 229, 244, 254, 264, 273, 282, 291, 299, 309, 319, 326, 336, 345, 353, 359, 372, 392, 409
OILS 371
okra, boiled 3, 15, 34, 40, 56, 63, 80, 93, 130, 147, 164, 169, 179, 196, 204, 223, 235, 247, 256, 266, 275, 284, 293, 301, 311, 320, 329, 337, 346, 355, 370, 383, 395, 399
okra, frozen and boiled 3, 16, 31, 40, 56, 66, 79, 93,

129, 146, 163, 170, 178, 196, 206, 224, 234, 246, 256, 266, 275, 283, 293, 301, 310, 320, 329, 337, 346, 355, 369, 381, 394, 399
old fashioned loaf, beef and pork 14, 27, 47, 52, 62, 71, 84, 93, 129, 148, 160, 173, 182, 192, 202, 214, 232, 244, 254, 263, 272, 281, 290, 299, 308, 318, 327, 335, 344, 355, 364, 374, 391, 404
OLEIC ACID 103
oleic acid, internal conversion of 103
OLEOVITAMIN A 6
OLEOVITAMIN D_2 99
OLEOVITAMIN D_3 99
OLEOVITAMIN D_4 100
olive loaf, pork 14, 28, 48, 51, 62, 71, 84, 93, 129, 149, 163, 174, 182, 193, 204, 214, 233, 244, 254, 263, 273, 281, 290, 299, 309, 318, 327, 335, 344, 355, 364, 375, 391, 405
olive oil 101, 103, 104, 116, 117, 165, 227, 238, 359, 371
olives 103, 104
olives, pickled green 3, 130, 160, 198, 210, 214, 358, 367, 376, 396, 401
olives, pickled ripe manzanillo/mission 3, 129, 143, 155, 215, 355, 367, 376, 384, 400
olives, pickled ripe sevillano/ascolano 3, 129, 143, 155, 211, 215, 355, 368, 377, 384, 396, 400
olives, ripe 139
onion flakes, dehydrated 13, 33, 40, 50, 59, 79, 92, 127, 142, 158, 169, 178, 189, 200, 222, 232, 243, 254, 264, 273, 283, 291, 300, 309, 319, 327, 336, 345, 349, 361, 380, 391, 409
onion powder 13, 127, 157, 169, 188, 200, 221, 231, 243, 253, 264, 273, 282, 291, 300, 309, 319, 327, 336, 345, 349, 361, 391, 409
onion rings 5, 16, 31, 39, 56, 65, 73, 81, 95, 132, 147, 161, 174, 195, 208, 216, 235, 351, 361, 374, 393, 408
onions 85, 186, 187, 212, 213, 227
onions, boiled 21, 36, 43, 57, 63, 81, 94, 132, 150, 165, 176, 181, 197, 208, 223, 237, 246, 256, 267, 276, 285, 293, 302, 311, 321, 329, 338, 347, 355, 370, 382, 396, 398
onions, frozen and boiled 5,

Index 508

22, 36, 43, 57, 65, 81, 94, 132, 150, 165, 176, 182, 199, 210, 223, 237, 246, 256, 267, 276, 285, 294, 302, 312, 321, 330, 339, 347, 355, 370, 383, 397, 398
onions, raw 19, 36, 43, 57, 63, 80, 94, 132, 150, 164, 176, 181, 197, 208, 224, 237, 246, 256, 266, 275, 284, 293, 302, 311, 321, 329, 338, 347, 355, 369, 381, 384, 396, 399
onions, raw welsh 20, 32, 41, 92, 133, 196, 246, 266, 275, 283, 292, 301, 311, 320, 329, 337, 346, 355, 369, 381, 395, 399
onions with tops, raw spring 2, 18, 30, 42, 57, 81, 92, 130, 148, 158, 173, 197, 206, 224, 235, 246, 266, 275, 284, 292, 301, 311, 320, 329, 338, 346, 355, 370, 382, 395, 398
opossum, roasted 16, 26, 364, 376, 386, 405
optic nerve, inflammation of 11
orange juice 4, 17, 35, 56, 85, 86, 92, 97, 125, 135, 149, 165, 176, 185, 198, 212, 224, 238, 247, 257, 267, 276, 285, 294, 302, 312, 322, 330, 339, 348, 354, 369, 382, 397, 399
orange juice, restaurant 4, 18, 35, 42, 56, 66, 80, 92, 136, 149, 166, 176, 198, 207, 224, 238, 354, 369, 383, 396, 399
orange peel 3, 15, 30, 39, 53, 63, 91, 128, 162, 174, 198, 207, 352, 395, 402
orange, valencia 3, 17, 34, 42, 56, 65, 80, 92, 131, 150, 166, 176, 183, 198, 208, 238, 247, 256, 266, 276, 285, 293, 301, 311, 321, 330, 338, 347, 354, 369, 381, 396, 400
oranges 85, 139, 187, 212, 223
oranges, mandarin 3, 17, 35, 92, 135, 150, 165, 176, 198, 209, 223, 235, 247, 257, 267, 276, 285, 294, 302, 312, 322, 330, 339, 347, 355, 369, 397, 399
oregano 2, 13, 38, 127, 153, 168, 193, 200, 230, 350, 363, 376, 391, 409
organ failure 125
organ meats 76, 140
organs 213
ORNITHINE 303
OROBETINA 24
OROPUR 76

OROTIC ACID 76
OROTYL 76
ORTHO-AMINOBENZOIC ACID 110
ORYZAMIN 11
osmotic control of calcium and potassium in heart, decreased 313
OSTELIN 99
OSYRITIN 113
OSYRITRIN 113
overweight 359
OVOFLAVIN 24
OXALIC ACID 127
oxidation, free radical 5, 9, 23, 43, 49, 58, 67, 85, 86, 90, 96, 97, 101, 113, 117, 212, 228, 238, 257, 303, 330
OXOBEMIN 75
OXOLAMINE 75
OXTRIPHYLLINE 88
oxygen, decreased 76, 77, 88, 96, 106, 113, 117
oxygen, increased need for 100
oxygenation of cell, decreased 76, 151
O-XYLOTOCOPHEROL 102
OXYTRIMETHYLLINE 88
oysters 85, 89, 99, 104, 139, 140, 152, 212, 227
oysters, breaded and fried 29, 48, 65, 69, 81, 130, 141, 154, 169, 193, 206, 216, 228, 244, 254, 264, 273, 282, 290, 299, 309, 318, 327, 336, 345, 354, 365, 376, 392, 403
oysters, canned 30, 48, 64, 69, 82, 131, 141, 154, 170, 193, 206, 217, 228, 244, 254, 264, 273, 282, 291, 299, 309, 319, 327, 336, 345, 356, 368, 379, 392, 400
oysters, cooked 26, 47, 64, 69, 80, 129, 141, 154, 169, 189, 201, 217, 228, 244, 254, 264, 272, 281, 290, 299, 308, 318, 326, 335, 344, 355, 366, 378, 390, 402
oysters, raw 30, 48, 56, 66, 69, 81, 131, 141, 154, 169, 179, 193, 206, 217, 228, 244, 254, 264, 273, 282, 291, 299, 309, 319, 327, 336, 345, 356, 368, 379, 392, 400
ozone resistance, decreased 90, 100, 212

P 187
P⁻⁻ 187
P⁻² 187
PABA 90

Paget's disease 213
pain, abdominal 6, 125, 141, 228
pain control 285, 312, 330
pain in mouth 124
pain in stomach 124, 213
pain in throat 124, 213
pain, muscle 58, 88, 89, 151
pain, nerve 58, 168, 187
pain sensitivity 58
pains in extremities 199, 226
pains, shooting 168, 187
paints, white 226
paleness 69, 331
PALIUROSIDE 113
palm kernel oil 359, 371
palm oil 359, 371
palm oil, red 1
palms/soles, thickening of skin 124
PALOHEX 86
pancakes 4, 15, 28, 39, 129, 159, 193, 209, 216, 351, 364, 377, 392, 407
pancakes, potato 7, 16, 31, 47, 51, 60, 73, 80, 132, 142, 158, 171, 179, 194, 200, 216, 234, 245, 254, 264, 273, 282, 291, 299, 309, 319, 327, 336, 345, 351, 359, 375, 393, 408
pancakes with butter and syrup 7, 15, 29, 48, 57, 66, 74, 84, 95, 130, 149, 160, 175, 191, 209, 216, 236, 351, 364, 378, 393, 407
pancreas, braised beef 15, 25, 46, 50, 63, 69, 93, 133, 147, 156, 173, 180, 188, 206, 220, 229, 242, 262, 269, 278, 288, 298, 305, 314, 323, 331, 340, 363, 374, 387, 405
pancreas, braised pork 17, 25, 47, 50, 69, 94, 133, 145, 156, 172, 180, 189, 208, 221, 230, 242, 262, 268, 278, 288, 298, 304, 314, 322, 331, 340, 364, 376, 386, 404
PANGAMIC ACID 76
PANTHOJECT 58
PANTHOLIN 58
PANTOTHENATE 49
PANTOTHENIC ACID 49
PANTOTHENIC ACID CALCIUM SALT 58
papaya 2, 22, 35, 42, 56, 66, 92, 112, 132, 151, 166, 176, 185, 199, 206, 224, 238, 248, 267, 276, 285, 294, 302, 312, 322, 330, 339, 348, 354, 369, 383, 397, 399
papaya nectar 4, 23, 42, 57, 67, 84, 95, 136, 151, 165, 177, 185, 211, 223, 237, 353, 369, 382, 397, 400

paper, filter 213
paprika 1, 13, 24, 37, 92, 128, 154, 168, 188, 200, 221, 230, 350, 362, 375, 390, 409
PARA-AMINOBENZOATE HYDROCHLORIDE 105
PARA-AMINOBENZOIC ACID 90
PARA-AMINOBENZOIC DIETHYLAMINOETHANOL HYDROCHLORIDE 106
PARA-AMINOBEN-ZOYLDIETHYLAMINOETHANOL HYDROCHLORIDE 105
PARACAIN 106
paralysis 89, 123, 124, 167, 186, 199
paralysis of arms 124
paralysis of legs 124
paralysis of limb, temporary 58
paralysis, progressive 212
paralyzation, muscles 294
PARAMINOL 90
Parkinson's disease 36, 43, 49, 67, 96, 101, 106, 138, 177
PAROVEN 113
parsley 101, 112, 187, 227
parsley, dried 1, 24, 37, 59, 91, 127, 153, 168, 188, 200, 215, 229, 350, 360, 376, 386, 409
parsley, freeze-dried 1, 11, 24, 37, 50, 59, 78, 91, 128, 142, 153, 168, 178, 188, 200, 216, 228, 285, 299, 322, 351, 362, 377, 386
parsley, raw 2, 18, 32, 40, 55, 63, 79, 91, 128, 149, 154, 170, 181, 196, 201, 221, 234, 292, 302, 328, 355, 369, 381, 394, 399
parsnips 112, 139
parsnips, boiled 17, 34, 40, 52, 65, 80, 93, 131, 144, 163, 171, 180, 195, 203, 223, 236, 353, 368, 382, 384, 395, 401
passionfruit 3, 32, 39, 92, 135, 158, 171, 195, 203, 221, 352, 367, 380, 394, 402
passionfruit juice, yellow 2, 32, 38, 93, 138, 164, 175, 197, 205, 223, 353, 369, 382, 396, 400
pastrami, beef 16, 30, 45, 63, 71, 94, 136, 158, 174, 193, 206, 214, 230, 244, 253, 262, 272, 281, 289, 298, 308, 318, 326, 335, 344, 357, 361, 373, 390, 407
pastrami, turkey 19, 28, 47, 61, 71, 136, 149, 158, 175, 191, 206, 215, 232, 243, 253, 262, 271, 280, 289, 298, 307, 317, 326, 334, 343, 358, 366, 377, 389, 402

pastry, toaster 3, 14, 26, 38, 56, 60, 79, 128, 144, 155, 174, 192, 208, 216, 235, 350, 360, 376, 393, 409
PB 167
PBC 6
PCS, 1974 5, 9, 23, 36, 38, 43, 49, 52, 58, 67, 74, 76, 77, 85, 89, 91, 96, 97, 99, 101, 110, 111, 118, 119, 120, 123, 124, 126, 138, 139, 140, 151, 153, 166, 167, 177, 186, 199, 212, 213, 225, 226, 227, 238, 330
pea pods, boiled black-eyed 2, 17, 32, 40, 93, 130, 163, 196, 207, 224, 355, 369, 382, 394, 399
peach 3, 22, 34, 40, 56, 66, 84, 85, 94, 108, 125, 137, 148, 166, 177, 182, 198, 207, 237, 248, 257, 267, 276, 285, 294, 302, 312, 321, 330, 339, 347, 354, 369, 383, 397, 399
peach nectar 3, 36, 42, 94, 137, 148, 165, 177, 183, 199, 210, 223, 238, 354, 369, 397, 400
peach pie 4, 15, 33, 42, 130, 156, 194, 209, 217, 350, 360, 374, 393, 408
peach seed 78
peanut brittle 15, 35, 38, 131, 157, 195, 208, 221, 349, 360, 376, 393, 410
peanut butter 85, 86, 89, 101
peanut butter, creamy 16, 31, 37, 51, 60, 79, 131, 142, 158, 168, 178, 188, 200, 222, 231, 240, 250, 261, 270, 278, 290, 299, 304, 317, 323, 331, 341, 353, 359, 372, 386, 410
peanut butter, crunchy 16, 31, 37, 51, 59, 79, 131, 142, 158, 169, 178, 189, 200, 222, 231, 240, 251, 262, 272, 280, 290, 299, 304, 317, 325, 332, 343, 352, 359, 372, 384, 388, 410
peanut flour 12, 25, 37, 128, 141, 157, 168, 178, 188, 200, 222, 229, 240, 249, 259, 268, 277, 288, 298, 304, 313, 322, 331, 339, 351, 361, 385, 409
peanut oil 101, 103, 104, 227, 359, 371
peanuts 88, 101, 103, 104, 117, 125, 152, 186
peanuts, boiled 14, 34, 38, 51, 63, 79, 130, 142, 161, 169, 178, 191, 208, 215, 232, 242, 254, 263, 273, 281, 291, 299, 308, 318, 327, 335, 344, 352, 361, 374, 391, 407

peanuts, dried 12, 31, 37, 50, 61, 79, 130, 141, 155, 168, 178, 188, 200, 222, 230, 240, 251, 261, 271, 279, 290, 299, 304, 317, 324, 331, 342, 353, 359, 372, 387, 409
peanuts, dry roasted 13, 32, 37, 50, 62, 79, 130, 142, 157, 168, 178, 188, 200, 223, 230, 240, 251, 262, 272, 280, 290, 299, 304, 317, 326, 332, 343, 352, 359, 372, 388, 410
peanuts, oil roasted 14, 32, 37, 50, 60, 79, 129, 141, 158, 168, 178, 188, 200, 222, 228, 240, 251, 261, 270, 278, 290, 299, 304, 317, 324, 331, 342, 353, 359, 372, 387, 410
peanuts, oil roasted spanish 14, 33, 37, 50, 62, 79, 129, 142, 157, 168, 178, 188, 200, 223, 232, 240, 250, 261, 271, 279, 290, 298, 304, 316, 325, 331, 342, 353, 359, 372, 386, 410
peanuts, oil roasted valencia 16, 31, 37, 50, 62, 79, 130, 141, 158, 168, 178, 189, 201, 223, 230, 240, 251, 262, 271, 279, 290, 299, 304, 316, 325, 331, 342, 353, 359, 372, 387, 410
peanuts, oil roasted virginia 14, 32, 37, 50, 62, 79, 129, 141, 158, 168, 178, 188, 200, 223, 228, 240, 251, 262, 271, 280, 290, 299, 304, 316, 325, 331, 342, 353, 359, 372, 386, 410
peanuts, roasted 85, 86, 89, 101
pear nectar 36, 43, 96, 138, 148, 165, 177, 182, 199, 211, 223, 238, 353, 369, 397, 400
pear seed 78
pears 5, 22, 35, 43, 57, 66, 83, 95, 125, 135, 145, 165, 177, 181, 187, 198, 209, 224, 227, 237, 248, 257, 267, 276, 285, 294, 302, 312, 322, 330, 339, 348, 353, 369, 381, 397, 400
peas 78, 85, 86, 89, 101, 107, 112, 186, 187, 227
peas, black-eyed 78, 85, 86, 89
peas, boiled black-eyed 5, 14, 34, 41, 54, 64, 79, 96, 132, 143, 156, 170, 179, 193, 205, 224, 233, 244, 254, 264, 273, 282, 291, 300, 309, 319, 327, 336, 345, 352, 367, 381, 384, 392, 402

Index

peas, boiled catjang black-eyed 5, 15, 34, 40, 54, 65, 79, 96, 132, 143, 156, 169, 169, 179, 193, 202, 222, 232, 244, 254, 264, 273, 282, 291, 300, 308, 319, 327, 336, 345, 352, 367, 380, 392, 403
peas, boiled green 3, 14, 30, 39, 56, 63, 79, 93, 132, 144, 158, 170, 179, 194, 205, 224, 233, 245, 255, 265, 274, 283, 291, 300, 309, 319, 328, 337, 345, 353, 368, 382, 384, 393, 401
peas, boiled pigeon 5, 15, 34, 40, 55, 66, 79, 131, 143, 160, 170, 179, 194, 202, 223, 234, 245, 255, 264, 273, 282, 291, 300, 308, 319, 328, 336, 345, 352, 366, 381, 384, 392, 403
peas, boiled split 5, 15, 34, 40, 52, 66, 79, 96, 134, 143, 160, 170, 179, 195, 203, 224, 234, 244, 254, 264, 273, 282, 290, 300, 309, 319, 327, 336, 345, 352, 367, 381, 392, 403
peas, boiled sprouted green 4, 14, 27, 39, 51, 64, 80, 94, 132, 150, 158, 170, 170, 180, 197, 205, 224, 234, 244, 253, 264, 274, 282, 291, 300, 309, 319, 336, 345, 352, 367, 380, 384, 392, 401
peas, canned black-eyed 5, 18, 32, 41, 56, 66, 80, 95, 133, 145, 161, 171, 180, 195, 208, 216, 234, 245, 255, 264, 274, 282, 291, 300, 309, 319, 328, 336, 345, 354, 368, 380, 393, 401
peas, canned chick 5, 22, 35, 43, 55, 59, 79, 95, 131, 144, 159, 171, 179, 195, 208, 216, 233, 245, 255, 264, 274, 282, 291, 300, 309, 319, 328, 337, 345, 352, 367, 379, 393, 403
peas, frozen and boiled black-eyed 4, 14, 34, 40, 56, 65, 79, 95, 132, 143, 157, 170, 179, 194, 202, 223, 233, 244, 254, 264, 273, 282, 290, 300, 308, 319, 327, 336, 344, 352, 366, 380, 392, 403
peas, frozen and boiled green 3, 14, 32, 39, 56, 64, 80, 94, 132, 144, 158, 171, 179, 195, 208, 218, 234, 245, 255, 265, 274, 283, 291, 300, 309, 319, 328, 337, 345, 354, 368, 382, 384, 393, 401
peas, raw green 3, 14, 31, 38, 57, 63, 79, 92, 132, 144, 159, 170, 179, 194, 206, 223, 233, 245, 255, 265, 274, 283, 291, 300, 309, 319, 328, 337, 345, 353, 368, 381, 384, 393, 401
peas, split 85, 86, 89, 186, 187
pecan pie 4, 15, 33, 42, 130, 156, 194, 209, 217, 350, 360, 374, 393, 408
pecans 86, 89, 101, 103, 104, 212
pecans, dried 4, 12, 30, 39, 50, 63, 80, 94, 131, 141, 157, 169, 178, 189, 202, 229, 244, 253, 264, 273, 282, 291, 299, 309, 319, 326, 336, 345, 353, 359, 372, 392, 409
pecans, dry roasted 14, 31, 80, 131, 141, 157, 169, 178, 189, 202, 229, 243, 253, 264, 273, 282, 291, 299, 309, 319, 326, 336, 345, 352, 359, 372, 392, 410
PECITROL VEINOGENE 112
pectin, citrus 213
pectin, fruit 206, 223, 357, 371, 398
pectin, lemon 213
peeling of feet 124
peeling of hands 124
peeling skin 6
pellagra 23, 36, 37, 43, 74, 85, 276
PELLAGRA PREVENTIVE FACTOR/VITAMIN 37
PELMINE 43
PELONIN AMIDE 43
PENICILLAMINE 59
PENICIN 118
PENTANOIC ACID 112
pepeao, dried 26, 38, 128, 154, 169, 194, 200, 219, 349, 362, 380, 393, 409
pepeao, raw 17, 29, 42, 96, 133, 163, 172, 198, 210, 223, 355, 370, 397, 398
pepper, black 4, 25, 127, 153, 168, 191, 200, 221, 233, 350, 363, 378, 391, 409
pepper, cayenne 1, 13, 25, 37, 92, 128, 154, 169, 190, 200, 221, 231, 350, 362, 375, 391, 409
pepper, sweet 116, 117
pepper, white 127, 154, 169, 191, 210, 233, 349, 361, 378, 390, 409
peppered loaf, beef and pork 13, 26, 47, 52, 61, 71, 84, 92, 130, 146, 161, 173, 181, 192, 202, 214, 230, 243, 252, 262, 272, 280, 289, 298, 308, 317, 326, 335, 343, 356, 366, 377, 390, 403
pepperoni, beef and pork 13, 30, 45, 50, 63, 70, 133, 159, 174, 194, 204, 214, 231, 243, 253, 262, 272, 280, 289, 298, 307, 317, 326, 334, 343, 357, 360, 372, 389, 408
peppers 112
peppers, boiled green sweet 3, 19, 35, 42, 57, 64, 81, 91, 138, 148, 162, 176, 181, 198, 209, 225, 237, 247, 256, 267, 276, 285, 294, 302, 312, 321, 330, 339, 347, 356, 370, 381, 397, 398
peppers, boiled red sweet 2, 19, 35, 42, 57, 64, 81, 91, 138, 148, 162, 176, 181, 198, 209, 225, 237, 247, 256, 267, 276, 285, 294, 302, 312, 321, 330, 339, 347, 356, 370, 381, 397, 398
peppers, canned green chili 3, 22, 34, 40, 92, 136, 164, 198, 247, 256, 266, 276, 284, 293, 302, 311, 321, 329, 338, 347, 355, 370, 383, 396, 398
peppers, canned green sweet 4, 21, 35, 41, 92, 138, 145, 162, 176, 198, 209, 214, 237, 247, 256, 267, 276, 284, 293, 302, 311, 321, 330, 339, 347, 356, 370, 381, 396, 399
peppers, canned jalapeño 2, 21, 34, 42, 50, 93, 132, 144, 156, 176, 198, 209, 214, 237, 247, 256, 267, 276, 284, 293, 302, 311, 321, 330, 339, 347, 356, 370, 380, 396, 399
peppers, canned red chili 2, 22, 34, 40, 92, 136, 164, 198, 247, 256, 266, 276, 284, 293, 302, 311, 321, 329, 338, 347, 355, 370, 383, 396, 398
peppers, canned red sweet 3, 21, 35, 41, 92, 138, 145, 162, 176, 198, 209, 214, 237, 247, 256, 267, 276, 284, 293, 302, 311, 321, 330, 339, 347, 356, 370, 381, 396, 399
peppers, freeze-dried green sweet 2, 11, 25, 38, 52, 59, 79, 91, 128, 141, 154, 168, 178, 189, 200, 217, 231, 244, 250, 263, 272, 281, 290, 299, 308, 318, 325, 335, 344, 350, 362, 377, 389
peppers, freeze-dried red sweet 1, 11, 25, 38, 52, 59, 79, 91, 128, 141, 154, 168, 178, 189, 200, 217, 231,

244, 250, 263, 272, 281, 290, 299, 308, 318, 325, 335, 344, 350, 362, 377, 389
peppers, frozen and boiled green sweet 3, 20, 35, 39, 58, 64, 81, 92, 136, 149, 164, 177, 181, 198, 210, 224, 238, 247, 256, 266, 276, 284, 293, 302, 311, 321, 329, 338, 347, 356, 370, 382, 396, 398
peppers, frozen and boiled red sweet 2, 20, 35, 39, 58, 64, 81, 92, 136, 149, 164, 177, 181, 198, 210, 224, 238, 247, 256, 266, 276, 284, 293, 302, 311, 321, 329, 338, 347, 356, 370, 382, 396, 398
peppers, green 139, 152
peppers, raw green chili 3, 17, 32, 40, 57, 61, 80, 91, 133, 144, 160, 172, 180, 196, 203, 223, 236, 246, 255, 265, 275, 284, 292, 301, 311, 320, 329, 338, 346, 354, 369, 382, 395, 399
peppers, raw green sweet 3, 18, 33, 40, 57, 63, 81, 91, 137, 146, 160, 175, 181, 197, 207, 224, 237, 247, 256, 266, 276, 284, 293, 302, 311, 321, 329, 338, 347, 356, 370, 381, 384, 396, 398
peppers, raw red chili 3, 17, 32, 40, 57, 61, 80, 91, 133, 144, 160, 172, 180, 196, 203, 223, 236, 246, 255, 265, 275, 284, 292, 301, 311, 320, 329, 338, 346, 354, 369, 382, 395, 399
peppers, raw red sweet 2, 18, 33, 40, 57, 63, 81, 91, 137, 146, 160, 175, 181, 197, 207, 224, 237, 247, 256, 266, 276, 284, 293, 302, 311, 321, 329, 338, 347, 356, 370, 381, 384, 396, 398
peptic ulcer 9, 36, 74, 96, 101, 138, 166
peptide streptogenin activity in chicks 68
PERAEMON 69
perch, cooked 8, 31, 47, 71, 128, 150, 160, 170, 189, 203, 218, 235, 243, 252, 261, 270, 279, 287, 295, 306, 316, 325, 333, 342, 366, 379, 388, 402
perch, raw 8, 32, 47, 54, 72, 129, 150, 161, 171, 184, 191, 205, 218, 235, 244, 253, 262, 271, 280, 289, 297, 307, 317, 326, 334, 343, 368, 379, 389, 401
perilla leaf 117

peripheral vascular system, degenerating 100
peripheral vasodilator 86
PERNAEVIT 69
pernicious anemia 67, 69, 74, 76, 84, 96, 101, 138, 140, 166
PERNIPUR 69
peroxidase elimination, decreased 211
Perrier water 134, 177, 224, 398
persimmon 92, 132, 156, 197, 204, 247, 256, 266, 275, 284, 293, 302, 311, 321, 329, 338, 347, 351, 366, 381, 396, 404
persimmon, dried japanese 3, 35, 42, 132, 142, 162, 171, 178, 195, 200, 224, 235, 247, 255, 266, 275, 284, 293, 302, 311, 320, 329, 338, 346, 350, 363, 380, 395, 408
persimmon, japanese 2, 22, 36, 43, 82, 94, 136, 145, 166, 176, 179, 198, 208, 224, 237, 248, 256, 267, 276, 284, 294, 302, 311, 321, 330, 339, 347, 353, 368, 382, 397, 400
perspiration, increased 123
PGA 78
pH of cell, decreased 211
PHA 303
PHASEOMANNITE 85
PHE 303
pheasant without skin, raw 7, 17, 30, 44, 51, 59, 72, 94, 134, 148, 160, 174, 184, 190, 205, 221, 234, 243, 251, 260, 269, 279, 287, 296, 306, 315, 324, 333, 341, 366, 378, 388, 402
PHENOBARBITONE 98
PHENOFORMIN 69, 140
PHENOTHIAZINES 24
PHENTURIDE 98
PHENYLALANINE 303-312
PHENYLCHROMAN 112, 113
PHEYTOIN 98
phlebitis deterioration 101
PHOSADEN 68
PHOSPHOPROTEINS 153
PHOSPHORUS 1, 6, 24, 37, 44, 99, 101, 103, 104, 127, 153, 168, 178, 187-199, 228, 371
phosphorus 123
phosphorus absorption, decreased 98
phosphorus decreased 125, 228
phosphorus in blood, increased 187
PHOSPHORYLCHOLINE CHLORIDE 87
photographic agents 126
physical and mental vigor, decreased 153

physical/mental development, arrested 152
PHYTATES 153
PHYTIC ACID 127
PHYTOGERMINE 100
PHYTOMELIN 113
PHYTONADIONE 108
PHYTONADIONE OXIDE 108
pickle and pimento loaf, pork 7, 14, 28, 47, 51, 63, 71, 84, 93, 129, 145, 161, 174, 182, 193, 203, 214, 233, 244, 254, 263, 273, 281, 290, 299, 309, 318, 327, 335, 344, 355, 363, 374, 391, 405
pickle relish, hamburger 3, 20, 35, 40, 95, 136, 147, 160, 177, 198, 210, 215, 237, 351, 366, 380, 396, 404
pickle relish, hot dog 4, 20, 35, 41, 133, 147, 159, 174, 197, 210, 214, 237, 352, 368, 381, 395, 402
pickles, bread and butter 4, 94, 131, 158, 197, 215, 353, 368, 396, 401
pickles, sour 4, 36, 94, 133, 155, 198, 214, 357, 371, 382, 397, 398
pickles, sweet 4, 94, 134, 159, 198, 351, 366, 380, 396, 404
picnic loaf, beef and pork 13, 28, 47, 51, 61, 71, 84, 93, 130, 148, 161, 175, 182, 194, 205, 214, 232, 244, 254, 263, 272, 281, 289, 298, 308, 318, 327, 335, 344, 356, 364, 375, 390, 404
pie crust 14, 39, 78, 158, 196, 210, 222, 351, 359, 372, 393, 409
pigeon peas, boiled 5, 15, 34, 40, 55, 66, 79, 131, 143, 160, 170, 179, 194, 202, 223, 234, 245, 255, 264, 273, 282, 291, 300, 308, 319, 328, 336, 345, 352, 366, 381, 384, 392, 403
pigmentation, deterioration of 187
pignolia pine nuts, dried 12, 29, 38, 132, 141, 154, 188, 201, 223, 230, 240, 250, 262, 271, 279, 290, 298, 306, 317, 324, 333, 341, 353, 359, 372, 388, 409
pike, cooked 8, 18, 32, 63, 95, 129, 148, 162, 189, 204, 221, 234, 243, 252, 261, 270, 278, 287, 295, 306, 315, 325, 333, 341, 367, 380, 388, 402
pike, raw 8, 19, 33, 64, 95, 130, 149, 163, 190, 206, 221, 234, 243, 253, 262, 271, 280, 289, 297, 307, 317,

326, 334, 343, 368, 380, 389, 401
pilinuts, dried 4, 12, 31, 41, 128, 155, 188, 201, 223, 242, 264, 273, 281, 291, 298, 308, 318, 335, 344, 356, 359, 371, 391, 410
pimento cheese spread 6, 20, 26, 53, 65, 72, 83, 127, 164, 173, 188, 208, 214, 231, 244, 254, 260, 271, 279, 287, 297, 305, 317, 324, 331, 341, 357, 360, 373, 388, 408
piña colada 5, 22, 36, 43, 81, 94, 136, 147, 165, 199, 210, 223, 237, 351, 365, 379, 397, 403
pine nuts (pignolia), dried 12, 29, 38, 132, 141, 154, 188, 201, 223, 230, 240, 250, 262, 271, 279, 290, 298, 306, 317, 324, 333, 341, 353, 359, 372, 388, 409
pine nuts (pinyon), dried 5, 11, 29, 38, 94, 137, 141, 156, 168, 197, 201, 219, 230, 240, 253, 263, 273, 281, 291, 299, 309, 318, 327, 335, 344, 353, 359, 372, 391, 409
pine nuts 187
pineapple 5, 17, 34, 41, 56, 65, 81, 93, 137, 145, 152, 164, 175, 178, 199, 209, 224, 237, 248, 257, 267, 276, 285, 294, 302, 312, 322, 330, 339, 347, 354, 369, 380, 397, 399
pineapple chiffon pie 3, 20, 32, 41, 95, 132, 162, 195, 210, 217, 351, 362, 376, 392, 407
pineapple custard pie 4, 20, 32, 41, 96, 130, 164, 196, 210, 217, 351, 364, 377, 393, 406
pineapple juice 5, 19, 36, 42, 57, 64, 80, 94, 133, 147, 165, 175, 178, 199, 209, 224, 237, 354, 369, 383, 397, 400
pineapple pie 5, 20, 36, 41, 96, 134, 164, 197, 210, 217, 351, 363, 376, 394, 407
pink beans, boiled 14, 33, 40, 55, 63, 79, 130, 143, 157, 169, 179, 192, 201, 224, 234, 244, 254, 264, 273, 282, 290, 299, 308, 318, 327, 336, 344, 352, 366, 381, 384, 392, 404
pink grapefruit 3, 22, 36, 42, 55, 66, 81, 92, 135, 149, 166, 176, 185, 199, 209, 238, 294, 302, 330, 355, 370, 383, 397, 399

pinto beans, boiled 5, 15, 32, 41, 55, 63, 79, 95, 130, 143, 156, 169, 179, 193, 201, 224, 233, 244, 254, 264, 273, 282, 290, 300, 309, 318, 327, 336, 344, 352, 366, 381, 384, 392, 404
pinto beans, canned 5, 17, 34, 42, 56, 65, 80, 96, 131, 144, 158, 171, 180, 195, 204, 216, 234, 245, 255, 264, 274, 282, 291, 300, 309, 319, 328, 337, 345, 353, 368, 382, 393, 401
pinyon pine nuts, dried 5, 11, 29, 38, 94, 137, 141, 156, 168, 197, 201, 219, 230, 240, 253, 263, 273, 281, 291, 299, 309, 318, 327, 335, 344, 353, 359, 372, 391, 409
pipes, water 126
pistachio nuts, dried 3, 12, 29, 39, 79, 128, 141, 154, 168, 180, 188, 200, 223, 233, 240, 250, 262, 271, 279, 289, 298, 304, 317, 325, 334, 341, 352, 359, 372, 389, 410
pistachio nuts, dry roasted 13, 28, 39, 129, 141, 155, 169, 179, 188, 200, 223, 233, 242, 250, 263, 272, 281, 290, 299, 307, 318, 326, 335, 343, 351, 359, 372, 390, 410
pistachio nuts, dry roasted red 12, 26, 39, 129, 155, 200, 215, 352, 359, 372, 389, 410
pistachios 117
pita bread 13, 27, 38, 53, 66, 80, 129, 143, 157, 172, 179, 195, 209, 216, 234, 350, 362, 379, 384, 391, 408
pitanga 2, 21, 34, 42, 92, 136, 165, 176, 198, 210, 224, 355, 369, 381, 396, 399
pituitary gland deterioration 126
pizza, cheese 7, 14, 28, 47, 54, 64, 73, 80, 95, 128, 145, 159, 172, 192, 207, 215, 233, 351, 364, 377, 391, 407
pizza, pepperoni 7, 14, 28, 46, 54, 65, 73, 79, 95, 128, 157, 208, 215, 351, 363, 377, 391, 407
PLANAVIT C 91, 97
PLANOCAINE 106
plantain, cooked 3, 20, 34, 40, 56, 62, 80, 94, 138, 148, 163, 171, 197, 201, 223, 237, 247, 256, 265, 276, 285, 293, 302, 311, 321, 330, 338, 347, 351, 367, 382, 396, 403
plants and animals, all 49

plants, sea 126
plaques on aorta, increased incidence of 139
PLECYAMIN 69
plum 3, 20, 32, 41, 56, 65, 84, 94, 112, 138, 149, 166, 177, 181, 198, 208, 227, 237, 248, 257, 267, 276, 285, 294, 302, 312, 321, 339, 347, 354, 369, 380, 396, 400
plum, java 5, 23, 36, 42, 66, 93, 133, 165, 175, 198, 210, 222, 353, 368, 382, 396, 400
plum seed 78
pneumonia 96, 97, 98, 113, 285
POF 112
poi 5, 16, 35, 39, 95, 133, 162, 172, 197, 207, 222, 352, 367, 382, 397, 402
poisons, rodent 151
pokeberry shoots, boiled 2, 18, 27, 39, 92, 130, 160, 197, 357, 370, 381, 394, 398
polar bear liver 6
polio 211, 225
pollock, raw 8, 19, 29, 47, 54, 62, 70, 130, 149, 164, 169, 184, 190, 203, 218, 235, 243, 253, 262, 271, 280, 288, 297, 307, 317, 326, 334, 343, 368, 380, 389, 401
pollution 225
pollution, air 124, 126, 140, 151, 167, 186, 187, 226
pollution, nuclear power 153
pollution, water 123, 126, 140, 167, 186
POLYCHLORINATED BIPHENY 6
POLYCHROME 113
POLYNEURAMIN 11
POLYUNSATURATED FATS 1, 103, 104, 371
pomegranate 21, 35, 42, 52, 64, 94, 138, 165, 199, 206, 224, 353, 368, 382, 396, 400
pompano, cooked 131, 147, 163, 171, 182, 189, 201, 218, 234, 243, 252, 261, 270, 279, 287, 296, 306, 316, 325, 333, 342, 368, 376, 388, 404
pompano, raw 132, 150, 163, 171, 185, 191, 202, 219, 234, 244, 253, 262, 271, 280, 289, 297, 307, 317, 326, 334, 343, 365, 377, 389, 402
popcorn 30, 39, 133, 155, 189, 349, 360, 378, 391, 410
popover roll 3, 15, 28, 39, 129, 159, 193, 208, 217, 352, 364, 377, 392, 406
poppy seed 12, 26, 61, 127,

154, 168, 188, 200, 221, 228, 241, 250, 263, 272, 280, 290, 298, 307, 317, 325, 334, 342, 352, 359, 372, 390, 409
pork 85, 86, 89, 101, 103, 104, 107, 117, 152, 186, 187
pork and beef bologna 15, 31, 47, 55, 63, 71, 84, 93, 134, 147, 159, 176, 182, 195, 208, 215, 232, 244, 254, 263, 273, 281, 290, 299, 308, 318, 327, 336, 344, 357, 361, 373, 391, 406
pork and beef bratwurst 14, 28, 47, 57, 64, 71, 92, 130, 148, 161, 175, 182, 193, 205, 215, 232, 244, 254, 263, 272, 281, 290, 298, 308, 318, 327, 335, 344, 357, 361, 373, 390, 406
pork and beef frankfurter 14, 32, 47, 54, 64, 71, 84, 92, 135, 147, 160, 176, 183, 195, 208, 214, 232, 244, 254, 263, 273, 281, 290, 299, 309, 318, 327, 336, 344, 357, 361, 373, 391, 406
pork and beef honey loaf 13, 28, 47, 51, 61, 71, 93, 133, 148, 159, 174, 182, 193, 203, 214, 231, 244, 254, 262, 272, 280, 289, 298, 308, 317, 326, 335, 344, 356, 366, 378, 390, 402
pork and beef kielbasa/kolbassy 14, 28, 47, 51, 63, 71, 93, 131, 145, 159, 175, 182, 193, 205, 215, 232, 244, 253, 263, 272, 281, 290, 299, 308, 318, 327, 335, 344, 357, 362, 373, 391, 406
pork and beef knockwurst 13, 30, 47, 55, 63, 71, 92, 135, 148, 161, 176, 195, 207, 215, 233, 244, 254, 263, 273, 281, 290, 299, 309, 318, 327, 336, 344, 357, 362, 373, 391, 405
pork and beef link sausage 14, 30, 47, 53, 63, 71, 93, 135, 148, 159, 176, 182, 194, 207, 215, 232, 244, 254, 263, 273, 281, 290, 298, 309, 318, 327, 335, 344, 358, 361, 373, 391, 406
pork and beef link sausage with american cheese 14, 30, 47, 51, 64, 71, 93, 130, 148, 161, 176, 183, 192, 207, 215, 232, 244, 254, 263, 272, 281, 290, 298, 308, 318, 327, 335, 344, 358, 361, 373, 390, 406
pork and beef lunch meat 14, 30, 47, 52, 63, 71, 83, 93, 135, 150, 162, 175, 182, 195, 207, 214, 232, 244, 253, 263, 272, 281, 289, 299, 308, 318, 327, 335, 344, 357, 361, 372, 391, 407
pork and beef luncheon sausage 14, 28, 47, 54, 62, 71, 93, 134, 147, 159, 176, 182, 194, 206, 214, 231, 243, 252, 263, 272, 281, 289, 298, 308, 318, 326, 334, 343, 357, 363, 374, 390, 405
pork and beef mortadella 16, 31, 47, 64, 71, 92, 133, 148, 159, 175, 182, 195, 208, 214, 232, 244, 253, 262, 272, 281, 289, 298, 308, 318, 327, 335, 344, 357, 362, 373, 390, 406
pork and beef old fashioned loaf 14, 27, 47, 52, 62, 71, 84, 93, 129, 148, 160, 173, 182, 192, 202, 214, 232, 244, 254, 263, 272, 281, 290, 299, 308, 318, 327, 335, 344, 355, 364, 374, 391, 404
pork and beef peppered loaf 13, 26, 47, 52, 61, 71, 84, 92, 130, 146, 161, 173, 181, 192, 202, 214, 230, 243, 252, 262, 272, 280, 289, 298, 308, 317, 326, 335, 343, 356, 366, 377, 390, 403
pork and beef pepperoni 13, 30, 45, 50, 63, 70, 133, 159, 174, 194, 204, 214, 231, 243, 253, 262, 272, 280, 289, 298, 307, 317, 326, 334, 343, 357, 360, 372, 389, 408
pork and beef picnic loaf 13, 28, 47, 51, 61, 71, 84, 93, 130, 148, 161, 175, 182, 194, 205, 214, 232, 244, 254, 263, 272, 281, 289, 298, 308, 318, 327, 335, 344, 356, 364, 375, 390, 404
pork and beef salami 14, 25, 47, 51, 62, 70, 93, 134, 143, 156, 176, 181, 194, 207, 215, 232, 244, 253, 263, 272, 281, 290, 299, 308, 318, 327, 335, 344, 357, 363, 374, 390, 404
pork and beef salami, hard 12, 27, 45, 50, 59, 71, 92, 135, 146, 159, 174, 193, 202, 214, 230, 242, 252, 261, 271, 279, 288, 297, 307, 316, 326, 334, 342, 357, 362, 373, 388, 408
pork and beef sausage 13, 30, 47, 53, 65, 73, 160, 176, 194, 215, 232, 244, 254, 263, 272, 281, 290, 299, 308, 318, 327, 335, 344, 357, 360, 372, 390, 407
pork and beef spread 15, 31, 48, 53, 64, 71, 135, 144, 162, 177, 182, 196, 209, 215, 234, 354, 364, 374, 392, 404
pork and beef summer sausage 15, 27, 46, 52, 61, 69, 84, 93, 136, 147, 157, 176, 182, 195, 206, 214, 232, 357, 361, 373, 390, 407
pork arm excluding fat, braised 9, 12, 26, 44, 51, 60, 72, 83, 136, 144, 158, 173, 173, 184, 190, 202, 217, 229, 363, 376, 385, 406
pork arm excluding fat, roasted cured 12, 28, 45, 52, 60, 71, 84, 135, 145, 161, 175, 182, 190, 205, 214, 231, 242, 250, 260, 270, 279, 288, 296, 305, 315, 324, 333, 342, 365, 377, 388, 404
pork arm exluding fat, roasted 9, 12, 26, 46, 52, 60, 72, 83, 136, 145, 159, 174, 182, 190, 203, 218, 230, 364, 376, 387, 404
pork arm including fat, braised 8, 13, 26, 45, 52, 62, 72, 84, 136, 144, 158, 174, 185, 192, 204, 218, 230, 241, 251, 259, 269, 278, 286, 296, 305, 314, 323, 332, 341, 361, 373, 387, 407
pork arm including fat, roasted 9, 13, 27, 46, 53, 61, 72, 84, 136, 145, 160, 175, 182, 191, 205, 219, 230, 242, 252, 260, 270, 279, 287, 297, 307, 316, 325, 333, 342, 361, 373, 388, 406
pork arm including fat, roasted cured 12, 29, 46, 52, 62, 72, 84, 135, 145, 161, 175, 183, 190, 206, 215, 231, 243, 252, 262, 271, 280, 289, 298, 307, 316, 326, 334, 343, 362, 374, 389, 406
pork beerwurst salami 12, 30, 47, 53, 61, 72, 84, 92, 136, 149, 162, 176, 182, 194, 206, 214, 232, 244, 254, 263, 273, 281, 290, 298, 309, 318, 327, 335, 344, 357, 364, 374, 390, 404
pork blade roll, roasted 13, 27, 47, 51, 62, 71, 84, 136, 147, 162, 176, 183, 193, 207, 215, 231, 244, 252, 262, 272, 281, 289, 298, 307, 317, 326, 335, 344, 358, 362, 373, 390, 405

Index 514

pork bologna 13, 30, 46, 51, 62, 72, 84, 92, 134, 148, 162, 175, 182, 193, 205, 214, 232, 244, 254, 263, 272, 281, 289, 298, 308, 318, 327, 335, 344, 358, 363, 374, 390, 404
pork boston blade excluding fat, braised 8, 13, 25, 46, 52, 62, 72, 83, 136, 144, 157, 173, 184, 191, 202, 218, 229, 240, 250, 259, 268, 277, 286, 295, 304, 313, 322, 331, 340, 362, 374, 385, 407
pork boston blade excluding fat, broiled 9, 12, 25, 46, 51, 61, 71, 83, 137, 145, 159, 171, 185, 190, 202, 218, 230, 241, 251, 259, 270, 278, 286, 296, 306, 315, 323, 332, 341, 363, 374, 387, 405
pork boston blade excluding fat, roasted 9, 12, 26, 46, 52, 61, 72, 83, 136, 145, 158, 174, 185, 190, 203, 219, 230, 242, 251, 259, 270, 279, 287, 297, 306, 315, 324, 333, 341, 363, 375, 388, 405
pork boston blade including fat, braised 8, 13, 26, 46, 57, 62, 72, 84, 137, 144, 158, 174, 185, 192, 203, 219, 229, 241, 251, 259, 269, 278, 286, 296, 305, 314, 323, 332, 341, 360, 373, 387, 407
pork boston blade including fat, broiled 8, 12, 26, 46, 51, 62, 72, 84, 137, 145, 160, 172, 185, 191, 203, 218, 230, 242, 252, 260, 271, 279, 287, 298, 307, 316, 325, 334, 342, 361, 373, 388, 407
pork boston blade including fat, roasted 9, 13, 26, 46, 53, 62, 72, 84, 137, 145, 159, 174, 185, 192, 204, 219, 230, 242, 252, 260, 271, 279, 287, 298, 307, 316, 325, 334, 342, 361, 373, 388, 406
pork brains, braised 17, 28, 47, 50, 63, 71, 93, 136, 143, 158, 176, 181, 190, 207, 218, 233, 244, 263, 272, 281, 290, 299, 308, 318, 327, 335, 344, 366, 377, 391, 401
pork bratwurst 13, 29, 47, 55, 62, 72, 95, 131, 146, 160, 175, 182, 193, 207, 216, 231, 244, 254, 263, 273, 281, 290, 299, 308, 318, 327, 335, 344, 357, 362, 373, 390, 405

pork center loin excluding fat, braised 8, 12, 27, 44, 51, 59, 73, 83, 137, 146, 161, 173, 173, 182, 190, 203, 220, 231, 240, 250, 259, 268, 277, 285, 294, 304, 313, 322, 331, 340, 363, 375, 385, 407
pork center loin excluding fat, broiled 9, 11, 26, 45, 51, 59, 72, 83, 137, 147, 161, 171, 185, 190, 202, 218, 232, 240, 250, 259, 268, 277, 285, 295, 304, 313, 322, 331, 340, 364, 376, 385, 405
pork center loin excluding fat, fried 9, 11, 26, 44, 51, 59, 72, 83, 137, 147, 161, 171, 185, 189, 201, 218, 231, 240, 250, 259, 269, 277, 286, 295, 305, 313, 323, 332, 340, 363, 375, 386, 406
pork center loin excluding fat, roasted 9, 12, 27, 45, 52, 59, 73, 84, 137, 147, 160, 173, 184, 190, 203, 219, 232, 240, 250, 259, 269, 277, 286, 295, 305, 313, 323, 332, 340, 364, 375, 386, 405
pork center loin including fat, braised 8, 12, 28, 44, 52, 60, 73, 84, 137, 146, 162, 174, 182, 191, 204, 221, 231, 240, 250, 259, 269, 277, 286, 295, 304, 313, 323, 332, 340, 361, 373, 386, 407
pork center loin including fat, broiled 8, 11, 27, 45, 52, 60, 72, 83, 138, 147, 162, 172, 172, 185, 191, 203, 219, 232, 241, 251, 259, 269, 278, 286, 296, 305, 314, 323, 332, 340, 361, 374, 387, 407
pork center loin including fat, fried 8, 11, 27, 45, 51, 60, 72, 83, 137, 147, 162, 172, 185, 191, 203, 219, 232, 242, 252, 260, 270, 279, 287, 297, 306, 316, 324, 333, 342, 360, 373, 388, 407
pork center loin including fat, roasted 9, 12, 28, 45, 51, 60, 73, 84, 137, 147, 161, 174, 185, 191, 204, 219, 232, 241, 251, 259, 270, 278, 286, 296, 306, 315, 323, 333, 341, 362, 374, 387, 406
pork center rib excluding fat, braised 9, 12, 26, 44, 52, 60, 73, 82, 362, 375, 385, 407

pork center rib excluding fat, broiled 9, 12, 26, 45, 51, 60, 72, 82, 134, 147, 162, 171, 171, 183, 189, 201, 219, 231, 240, 250, 259, 269, 277, 286, 295, 305, 313, 323, 332, 340, 363, 375, 386, 406
pork center rib excluding fat, fried 9, 12, 26, 45, 51, 59, 72, 82, 136, 148, 162, 171, 185, 189, 201, 221, 232, 241, 250, 259, 269, 277, 286, 296, 305, 314, 323, 332, 340, 363, 375, 386, 406
pork center rib excluding fat, roasted 9, 12, 26, 45, 52, 60, 73, 82, 135, 147, 161, 173, 185, 189, 202, 221, 232, 241, 250, 259, 269, 277, 286, 296, 305, 313, 323, 332, 340, 364, 375, 386, 405
pork center rib excluding fat, braised 135, 146, 160, 172, 185, 190, 201, 220, 231
pork center rib including fat, braised 8, 13, 27, 45, 52, 61, 73, 83, 136, 147, 161, 174, 185, 191, 202, 221, 231, 240, 250, 259, 269, 277, 286, 296, 305, 314, 323, 332, 340, 360, 373, 386, 407
pork center rib including fat, broiled 9, 12, 27, 45, 52, 61, 72, 82, 134, 148, 162, 172, 172, 184, 190, 203, 220, 232, 241, 251, 259, 270, 279, 287, 297, 306, 315, 324, 333, 341, 361, 373, 388, 407
pork center rib including fat, fried 8, 12, 27, 46, 52, 61, 72, 83, 137, 148, 163, 173, 173, 185, 191, 203, 221, 233, 242, 252, 260, 271, 279, 288, 298, 307, 316, 325, 334, 342, 360, 372, 389, 407
pork center rib including fat, roasted 9, 12, 27, 45, 51, 60, 73, 82, 136, 148, 162, 174, 185, 190, 203, 221, 232, 241, 251, 259, 270, 279, 287, 297, 306, 315, 324, 333, 341, 361, 373, 388, 406
pork chitterlings, simmered 33, 48, 132, 143, 155, 176, 196, 211, 221, 229, 244, 264, 273, 282, 290, 299, 309, 318, 328, 335, 344, 362, 373, 391, 404
pork chop 139
pork ears, simmered 22, 33, 48, 133, 159, 177, 197, 210, 217, 243, 264, 273, 281,

Index

290, 299, 308, 318, 328, 336, 344, 365, 376, 390, 402
pork feet, pickled 22, 35, 48, 131, 177, 197, 244, 264, 273, 282, 290, 299, 309, 318, 329, 336, 345, 365, 375, 391, 403
pork feet, simmered 22, 34, 48, 131, 177, 196, 243, 264, 273, 281, 290, 299, 308, 318, 328, 336, 344, 365, 376, 389, 403
pork headcheese 20, 30, 48, 56, 63, 71, 84, 93, 134, 145, 160, 176, 183, 196, 211, 214, 233, 243, 253, 263, 272, 281, 290, 299, 308, 318, 327, 335, 344, 358, 364, 375, 390, 403
pork heart, braised 8, 13, 25, 44, 50, 60, 70, 84, 95, 137, 142, 154, 172, 181, 192, 207, 221, 230, 242, 250, 262, 270, 278, 288, 297, 305, 316, 325, 333, 342, 358, 366, 378, 388, 403
pork jowl, raw 9, 13, 28, 46, 55, 64, 72, 84, 138, 150, 164, 177, 185, 195, 208, 222, 244, 255, 265, 274, 282, 291, 300, 309, 319, 329, 337, 345, 359, 372, 392, 408
pork kidneys, braised 7, 13, 25, 45, 50, 59, 69, 80, 94, 134, 142, 154, 174, 181, 190, 209, 218, 230, 242, 249, 262, 269, 277, 288, 297, 304, 316, 324, 332, 340, 366, 378, 387, 403
pork leg excluding fat, roasted 9, 12, 26, 45, 51, 59, 72, 81, 137, 145, 160, 172, 182, 189, 202, 219, 230, 240, 250, 259, 269, 277, 286, 296, 305, 313, 323, 332, 340, 364, 376, 386, 404
pork leg including fat, roasted 9, 12, 26, 46, 52, 60, 72, 81, 137, 146, 161, 173, 182, 190, 204, 220, 231, 241, 251, 259, 270, 278, 286, 297, 306, 315, 324, 333, 341, 362, 374, 387, 406
pork link sausage 12, 27, 46, 51, 60, 71, 95, 132, 148, 160, 174, 193, 204, 214, 231, 243, 252, 261, 271, 280, 289, 297, 307, 316, 326, 334, 342, 357, 360, 373, 388, 408
pork liver 89, 107
pork liver, braised 6, 14, 24, 44, 50, 59, 69, 79, 93, 136, 142, 154, 175, 180, 190, 208,
221, 228, 242, 250, 261, 269, 277, 288, 296, 304, 315, 323, 333, 340, 356, 365, 378, 387, 404
pork liver sausage 6, 14, 25, 44, 50, 61, 69, 93, 135, 143, 154, 176, 181, 192, 207, 214, 231, 244, 252, 263, 273, 281, 290, 299, 308, 318, 327, 335, 344, 357, 361, 372, 391, 407
pork livercheese 6, 14, 24, 44, 50, 59, 69, 95, 136, 142, 154, 176, 180, 191, 206, 214, 230, 244, 251, 263, 272, 281, 289, 298, 307, 318, 326, 335, 343, 358, 362, 373, 390, 406
pork liverwurst 14, 25, 50, 63, 69, 80, 132, 154, 190, 244, 254, 263, 272, 281, 289, 299, 308, 318, 327, 336, 343, 357, 361, 373, 390, 406
pork loin blade excluding fat, braised 9, 13, 26, 45, 52, 60, 72, 83, 133, 145, 158, 173, 173, 184, 191, 202, 218, 229, 240, 250, 259, 268, 277, 286, 295, 304, 313, 322, 331, 340, 362, 374, 386, 407
pork loin blade excluding fat, broiled 9, 12, 26, 45, 51, 60, 72, 83, 134, 146, 160, 172, 185, 190, 202, 218, 230, 241, 251, 259, 270, 278, 286, 296, 306, 315, 323, 333, 341, 362, 374, 387, 406
pork loin blade excluding fat, fried 9, 12, 26, 46, 51, 60, 72, 134, 146, 161, 172, 185, 190, 202, 218, 230, 242, 251, 259, 270, 279, 287, 297, 306, 315, 324, 333, 341, 362, 374, 388, 406
pork loin blade excluding fat, roasted 9, 12, 26, 45, 53, 59, 72, 83, 134, 146, 160, 175, 185, 191, 203, 219, 230, 241, 251, 259, 270, 278, 286, 297, 306, 315, 323, 333, 341, 362, 374, 388, 405
pork loin blade including fat, braised 9, 13, 27, 45, 53, 61, 72, 84, 134, 145, 160, 175, 185, 192, 204, 219, 230, 241, 251, 259, 270, 279, 287, 297, 306, 315, 324, 333, 341, 360, 372, 388, 407
pork loin blade including fat, broiled 9, 12, 26, 46, 52, 61, 72, 84, 135, 147, 161, 173, 173, 185, 191, 204, 219, 230, 243, 252, 260, 271,
280, 288, 298, 307, 316, 325, 334, 342, 360, 372, 389, 407
pork loin blade including fat, fried 8, 12, 27, 46, 52, 61, 72, 84, 135, 147, 162, 174, 185, 192, 205, 220, 231, 243, 253, 260, 271, 280, 288, 298, 307, 317, 326, 334, 343, 360, 372, 389, 407
pork loin blade including fat, roasted 13, 27, 46, 52, 60, 72, 84, 135, 147, 161, 175, 185, 192, 205, 220, 231, 242, 252, 260, 271, 280, 288, 298, 307, 316, 325, 334, 342, 360, 373, 389, 407
pork loin excluding fat, braised 9, 12, 26, 44, 51, 59, 72, 83, 135, 145, 159, 172, 183, 190, 202, 218, 230, 240, 250, 259, 268, 277, 285, 295, 304, 313, 322, 331, 340, 363, 375, 385, 406
pork loin excluding fat, broiled 9, 11, 25, 44, 59, 71, 83, 136, 146, 161, 171, 171, 185, 189, 202, 218, 231, 241, 250, 259, 269, 278, 286, 296, 305, 314, 323, 332, 340, 363, 375, 386, 405
pork loin excluding fat, roasted 9, 12, 26, 44, 52, 59, 72, 83, 136, 146, 160, 173, 184, 189, 203, 219, 230, 241, 251, 259, 269, 278, 286, 296, 305, 314, 323, 332, 340, 364, 375, 387, 405
pork loin including fat, braised 8, 12, 27, 44, 52, 60, 72, 84, 136, 145, 160, 174, 184, 191, 203, 219, 231, 241, 251, 259, 269, 278, 286, 296, 305, 314, 323, 332, 340, 360, 373, 387, 407
pork loin including fat, broiled 8, 12, 26, 45, 51, 60, 72, 83, 137, 147, 162, 172, 185, 190, 203, 219, 231, 242, 252, 260, 270, 279, 287, 297, 306, 315, 324, 333, 341, 361, 373, 388, 407
pork loin including fat, roasted 9, 12, 26, 45, 52, 60, 72, 83, 136, 146, 161, 174, 185, 190, 204, 220, 231, 242, 252, 260, 270, 279, 287, 297, 306, 315, 324, 333, 342, 361, 373, 388, 406
pork lunch meat, canned 13, 29, 47, 53, 63, 72, 83, 137, 150, 163, 176, 183, 195, 207, 214, 233, 244, 253, 263, 272, 281, 290, 299, 308,

Index

pork lungs, braised 17, 26, 47, 52, 65, 71, 94, 136, 154, 176, 192, 208, 218, 231, 244, 263, 272, 281, 289, 299, 308, 318, 327, 343, 367, 379, 390, 401
pork luxury loaf 12, 27, 46, 52, 61, 71, 84, 93, 131, 146, 161, 173, 182, 192, 202, 214, 230, 243, 254, 262, 272, 280, 289, 298, 307, 316, 326, 334, 343, 356, 366, 378, 389, 403
pork mother's loaf 12, 29, 47, 53, 63, 71, 131, 146, 159, 175, 181, 194, 207, 214, 233, 355, 362, 374, 391, 406
pork olive loaf 14, 28, 48, 51, 62, 71, 84, 93, 129, 149, 163, 174, 182, 193, 204, 214, 233, 244, 254, 263, 273, 281, 290, 299, 309, 318, 327, 335, 344, 355, 364, 375, 391, 405
pork pancreas, braised 17, 25, 47, 50, 69, 94, 133, 145, 156, 172, 180, 189, 208, 221, 230, 242, 262, 268, 278, 288, 298, 304, 314, 322, 331, 340, 364, 376, 386, 404
pork pickle and pimento loaf 7, 14, 28, 47, 51, 63, 71, 84, 93, 129, 145, 161, 174, 182, 193, 203, 214, 233, 244, 254, 263, 273, 281, 290, 299, 309, 318, 327, 335, 344, 355, 363, 374, 391, 405
pork rump excluding fat, roasted 9, 12, 26, 45, 51, 61, 72, 83, 137, 145, 160, 171, 182, 189, 202, 219, 231, 240, 250, 259, 268, 277, 286, 295, 304, 313, 322, 332, 340, 364, 376, 386, 405
pork rump including fat, roasted 8, 12, 26, 45, 51, 62, 72, 83, 137, 146, 161, 172, 183, 189, 203, 220, 231, 241, 251, 259, 269, 278, 286, 296, 305, 314, 323, 332, 341, 363, 374, 387, 406
pork salami, hard 12, 27, 44, 59, 70, 135, 144, 160, 174, 181, 190, 214, 230, 243, 252, 262, 270, 280, 288, 298, 306, 316, 325, 334, 342, 357, 360, 372, 388, 408
pork sausage 12, 27, 46, 51, 61, 71, 131, 144, 160, 174, 181, 192, 203, 214, 231, 243, 253, 262, 272, 281, 289, 298, 308, 317, 327, 335, 344, 358, 360, 373, 389, 407
pork sausage, italian 12, 28, 46, 53, 61, 71, 95, 132, 147, 159, 174, 181, 192, 204, 215, 231, 243, 253, 262, 272, 281, 289, 298, 308, 317, 326, 335, 343, 358, 361, 373, 389, 407
pork sausage, polish 13, 31, 46, 53, 63, 72, 135, 146, 159, 175, 181, 193, 206, 215, 232, 244, 254, 263, 272, 281, 290, 298, 308, 318, 327, 335, 344, 357, 361, 373, 390, 406
pork shank excluding fat, roasted 9, 12, 26, 45, 51, 59, 72, 83, 137, 145, 160, 172, 182, 189, 203, 219, 230, 241, 250, 259, 269, 277, 286, 296, 305, 313, 323, 332, 340, 364, 376, 386, 404
pork shank including fat, roasted 9, 12, 27, 46, 52, 60, 72, 83, 137, 146, 161, 173, 182, 190, 204, 220, 231, 241, 251, 259, 270, 279, 287, 297, 306, 315, 324, 333, 341, 362, 374, 388, 406
pork shoulder excluding fat, roasted 9, 12, 26, 46, 52, 60, 72, 83, 136, 145, 159, 174, 182, 190, 203, 218, 230, 241, 251, 259, 269, 278, 286, 296, 306, 315, 323, 332, 341, 364, 375, 387, 405
pork shoulder including fat, roasted 9, 13, 26, 46, 53, 61, 72, 84, 137, 145, 159, 174, 183, 191, 204, 219, 230, 242, 252, 260, 271, 279, 287, 298, 307, 316, 325, 334, 342, 361, 373, 388, 406
pork sirloin excluding fat, braised 9, 12, 26, 45, 51, 59, 72, 83, 137, 145, 159, 171, 171, 184, 190, 201, 220, 230, 240, 250, 259, 268, 277, 285, 295, 304, 313, 322, 331, 340, 363, 375, 385, 404
pork sirloin excluding fat, broiled 9, 11, 25, 45, 59, 72, 83, 137, 146, 162, 170, 170, 185, 189, 201, 220, 231, 240, 250, 259, 269, 277, 286, 296, 305, 313, 323, 332, 340, 364, 375, 386, 405
pork sirloin excluding fat, roasted 9, 12, 26, 45, 52, 60, 72, 83, 135, 146, 160, 172, 182, 190, 203, 220, 231, 241, 250, 259, 269, 278, 286, 296, 305, 314, 323, 332, 340, 364, 375, 386, 405
pork sirloin including fat, braised 12, 27, 45, 52, 60, 72, 84, 137, 145, 160, 173, 185, 191, 203, 220, 231, 240, 250, 259, 269, 278, 286, 296, 305, 314, 323, 332, 340, 361, 373, 386, 407
pork sirloin including fat, broiled 9, 12, 26, 46, 51, 59, 72, 83, 137, 146, 162, 171, 185, 190, 203, 220, 232, 241, 251, 259, 270, 279, 287, 297, 306, 315, 324, 333, 341, 361, 373, 388, 407
pork sirloin including fat, roasted 9, 12, 26, 45, 52, 60, 72, 83, 136, 146, 161, 173, 182, 190, 204, 220, 232, 241, 251, 259, 270, 278, 286, 297, 306, 315, 324, 333, 341, 362, 374, 387, 406
pork spareribs, braised 8, 13, 26, 45, 51, 60, 71, 84, 130, 144, 158, 172, 184, 189, 204, 218, 229, 240, 250, 259, 268, 277, 286, 295, 305, 313, 322, 332, 340, 360, 373, 386, 408
pork spleen, braised 15, 27, 44, 65, 70, 93, 134, 145, 154, 182, 189, 206, 230, 242, 262, 269, 277, 288, 298, 304, 315, 324, 333, 340, 366, 379, 386, 403
pork stomach, raw 17, 31, 46, 66, 72, 136, 142, 157, 193, 207, 220, 232, 244, 264, 272, 281, 290, 299, 308, 318, 327, 335, 344, 366, 377, 390, 402
pork tail, simmered 18, 33, 48, 134, 177, 196, 243, 263, 273, 281, 290, 299, 308, 318, 327, 331, 344, 360, 372, 390, 407
pork tenderloin, roasted 9, 12, 25, 45, 51, 60, 73, 83, 136, 144, 159, 172, 182, 189, 201, 219, 231, 240, 250, 259, 269, 277, 286, 295, 305, 313, 323, 332, 340, 365, 378, 386, 403
pork tongue, braised 13, 25, 45, 62, 70, 95, 133, 155, 173, 192, 206, 217, 229, 243, 262, 270, 279, 288, 297, 306, 316, 325, 342, 363, 374, 388, 405
pork top loin excluding fat, braised 9, 12, 26, 44, 52,

60, 73, 82, 135, 146, 160, 173, 185, 190, 201, 220, 231, 240, 250, 259, 268, 277, 285, 295, 304, 313, 322, 331, 340, 363, 375, 385, 407
pork top loin excluding fat, broiled 9, 12, 26, 45, 51, 60, 72, 82, 134, 147, 162, 171, 183, 189, 201, 219, 231, 240, 250, 259, 269, 277, 286, 295, 305, 313, 323, 332, 340, 363, 375, 386, 406
pork top loin excluding fat, fried 9, 12, 26, 45, 51, 59, 72, 82, 136, 148, 162, 171, 185, 189, 201, 221, 232, 241, 250, 259, 269, 277, 286, 296, 305, 314, 323, 332, 340, 363, 375, 386, 406
pork top loin excluding fat, roasted 9, 12, 27, 45, 52, 60, 73, 82, 135, 147, 161, 173, 185, 189, 202, 221, 232, 241, 250, 259, 269, 277, 286, 296, 305, 314, 323, 332, 340, 364, 375, 386, 405
pork top loin including fat, braised 8, 13, 27, 45, 53, 61, 73, 83, 136, 147, 161, 174, 185, 191, 202, 221, 232, 240, 250, 259, 269, 278, 286, 296, 305, 314, 323, 332, 340, 360, 373, 386, 407
pork top loin including fat, broiled 9, 12, 27, 46, 52, 61, 72, 82, 135, 148, 163, 172, 184, 190, 203, 220, 232, 242, 251, 259, 270, 279, 287, 297, 306, 315, 324, 333, 341, 361, 373, 388, 407
pork top loin including fat, fried 9, 12, 27, 46, 52, 61, 73, 83, 137, 148, 163, 173, 173, 185, 191, 203, 221, 233, 242, 252, 260, 271, 279, 288, 298, 307, 316, 325, 334, 342, 360, 372, 389, 407
pork top loin including fat, roasted 9, 12, 27, 45, 53, 61, 73, 82, 136, 148, 162, 174, 185, 190, 203, 221, 232, 241, 251, 259, 270, 279, 287, 297, 306, 315, 324, 333, 341, 361, 373, 388, 407
port du salut cheese 6, 28, 56, 65, 71, 81, 127, 188, 216, 244, 261, 268, 277, 288, 295, 304, 316, 323, 331, 340, 358, 361, 373, 388, 407
porterhouse steak excluding fat, broiled 16, 28, 46, 54, 60, 70, 82, 137, 144, 156, 171, 184, 191, 202, 219, 229, 241, 251, 260, 269, 278, 287, 295, 305, 314, 324, 332, 341, 364, 376, 386, 405
porterhouse steak including fat, broiled 16, 29, 46, 55, 60, 71, 82, 136, 145, 156, 172, 184, 192, 203, 220, 229, 242, 252, 261, 270, 279, 288, 296, 306, 315, 325, 333, 342, 362, 374, 387, 406
POSORUTIN 113
POTASSIUM 59, 69, 76, 106, 199-211, 214
POTASSIUM ASCORBATE 97
POTASSIUM CHLORIDE, SLOW-RELEASE 69, 140
potassium decreased 214
potassium in heart, decreased 313
POTASSIUM METABISULFATE 106
potassium osmotic control in heart, decreased 313
potassium utilization, decreased 212
potato, boiled sweet 2, 20, 30, 40, 53, 62, 81, 93, 132, 144, 163, 176, 180, 197, 207, 222, 236, 247, 256, 266, 275, 283, 293, 301, 311, 320, 329, 337, 346, 352, 367, 382, 395, 402
potato, candied sweet 2, 22, 34, 42, 66, 74, 81, 94, 132, 146, 160, 176, 197, 207, 219, 237, 247, 257, 266, 275, 284, 293, 301, 311, 321, 330, 338, 347, 352, 366, 379, 396, 403
potato, canned sweet 2, 21, 33, 40, 53, 63, 81, 92, 132, 144, 162, 172, 179, 196, 204, 220, 237, 247, 256, 266, 275, 283, 293, 301, 310, 320, 329, 337, 346, 352, 368, 382, 395, 401
potato chips 15, 34, 38, 54, 59, 80, 92, 132, 143, 160, 169, 179, 193, 200, 216, 233, 350, 359, 372, 392, 410
potato flour 13, 30, 38, 93, 131, 154, 192, 200, 221, 245, 254, 264, 273, 282, 291, 300, 309, 319, 327, 336, 345, 349, 361, 380, 392, 409
potato, frozen and baked sweet 2, 18, 33, 40, 52, 63, 80, 94, 131, 143, 163, 173, 179, 196, 202, 223, 236, 247, 256, 266, 275, 283, 293, 301, 310, 320, 329, 337, 346, 352, 367, 383, 384, 395, 402
potato pancakes 7, 16, 31, 47, 51, 60, 73, 80, 132, 142, 158, 171, 179, 194, 200, 216, 234, 245, 254, 264, 273, 282, 291, 299, 309, 319, 327, 336, 345, 351, 359, 375, 393, 408
potato pie, sweet 2, 19, 31, 42, 95, 129, 163, 195, 208, 217, 352, 364, 376, 393, 405
potato salad 7, 17, 33, 48, 52, 64, 74, 83, 94, 133, 145, 163, 175, 181, 196, 206, 216, 236, 246, 255, 265, 274, 283, 292, 300, 309, 320, 328, 337, 346, 354, 366, 377, 394, 401
potatoes, sweet 78, 85, 86, 117
potato with skin, baked 16, 35, 39, 52, 61, 81, 93, 136, 142, 159, 171, 180, 196, 202, 223, 236, 246, 255, 265, 274, 283, 292, 301, 310, 320, 328, 337, 346, 352, 367, 383, 394, 402
potato with skin, microwaved 16, 35, 39, 61, 81, 93, 135, 142, 160, 171, 180, 194, 201, 223, 236, 246, 255, 265, 274, 283, 292, 301, 309, 320, 328, 337, 346, 352, 367, 383, 394, 402
potato without skin, baked 17, 36, 39, 52, 61, 82, 93, 138, 143, 165, 172, 181, 196, 202, 223, 236, 246, 256, 265, 275, 283, 292, 301, 310, 320, 328, 337, 346, 352, 368, 383, 395, 401
potato without skin, boiled 17, 36, 39, 53, 62, 82, 94, 137, 144, 165, 174, 181, 196, 204, 223, 236, 247, 256, 266, 275, 284, 292, 301, 310, 320, 329, 337, 346, 353, 368, 383, 395, 401
potato without skin, canned 18, 25, 40, 54, 63, 83, 94, 137, 149, 160, 176, 181, 197, 206, 236, 247, 256, 266, 275, 284, 292, 301, 310, 321, 329, 337, 346, 354, 369, 382, 395, 400
potato without skin, frozen and boiled 246, 256, 265, 275, 283, 292, 301, 310, 320, 328, 337, 346
potato without skin, microwaved 16, 35, 39, 52, 61, 81, 93, 137, 143, 164, 172, 181, 194, 202, 223,

236, 246, 255, 265, 275, 283, 292, 301, 310, 320, 328, 337, 346, 352, 367, 383, 395, 402
potato without skin, raw 17, 34, 39, 54, 62, 81, 93, 137, 143, 162, 173, 180, 196, 201, 223, 236, 246, 256, 265, 275, 283, 292, 301, 310, 320, 328, 337, 346, 353, 368, 383, 395, 401
potatoes 85, 86, 89, 107, 108, 112, 117, 139, 186, 187
potatoes, canned 125
potatoes, french fried 5, 15, 16, 34, 36, 38, 52, 53, 62, 80, 94, 133, 144, 146, 162, 170, 181, 194, 195, 200, 217, 221, 236, 246, 255, 265, 274, 283, 292, 301, 309, 319, 328, 337, 345, 351, 361, 375, 393, 408
potatoes, frozen and french fried 246, 256, 265, 274, 283, 292, 301, 309, 319, 328, 337, 345
potatoes, frozen and fried hash brown 16, 35, 38, 53, 64, 94, 134, 144, 159, 174, 180, 195, 201, 221, 236, 246, 256, 265, 274, 283, 292, 301, 309, 319, 328, 337, 346, 351, 364, 376, 394, 405
potatoes, hash brown 18, 35, 38, 53, 62, 82, 94, 135, 136, 143, 145, 162, 163, 173, 181, 195, 196, 204, 218, 222, 236, 246, 256, 265, 274, 283, 292, 301, 310, 320, 328, 337, 346, 352, 361, 365, 375, 394, 404
potatoes, mashed 7, 17, 35, 48, 52, 62, 74, 82, 94, 132, 144, 165, 174, 181, 196, 205, 216, 236, 247, 256, 265, 275, 283, 292, 301, 310, 320, 328, 337, 346, 353, 367, 378, 394, 401
potatoes, o'brien 7, 17, 33, 48, 53, 63, 74, 82, 93, 131, 145, 164, 174, 181, 196, 205, 217, 236, 246, 255, 265, 274, 283, 292, 301, 310, 320, 328, 337, 346, 353, 368, 379, 394, 401
potatoes, scalloped 7, 18, 32, 48, 53, 63, 74, 82, 94, 130, 144, 163, 174, 181, 196, 202, 216, 236, 246, 255, 265, 274, 283, 292, 300, 309, 320, 328, 337, 345, 354, 368, 378, 394, 400
potatoes without skin, frozen and boiled 17, 35, 39, 55, 63, 82, 94, 137, 147, 162, 176, 181, 197, 205, 222, 236, 353, 368, 383, 395, 400

potatoes without skin, frozen and french fried 16, 34, 38, 52, 62, 81, 93, 136, 144, 159, 173, 180, 195, 201, 221, 235, 351, 364, 377, 394, 406
poultry 140
poultry seasoning 2, 127, 154, 169, 193, 201, 231, 350, 363, 378, 393, 409
pound cake 7, 21, 32, 48, 133, 163, 195, 210, 217, 351, 360, 373, 393, 408
powdered sugar 166, 211, 225, 349, 360, 410
POYAMIN 69
PP FACTOR 37
precursor 1, 76, 98, 114, 115, 116
PREDNISONE 59, 91
pregnancy 9, 96, 99, 101, 118, 166, 199
pregnancy miscarriage 101
pregnancy, nausea of 67
premature aging 76, 111, 249
premature aging of skin 10, 213
premenstrual syndrome 105
PREPALIN 10
pretzels 14, 28, 38, 55, 66, 81, 132, 144, 158, 172, 195, 210, 214, 233, 349, 360, 378, 391, 410
pricklypear 4, 22, 33, 41, 93, 130, 165, 169, 197, 207, 223, 355, 369, 380, 396, 399
PRIMABALT RP 75
primrose oil, evening 103
PRO 312
PROCAINE HCL 106
procaine preparations, commercial 106
production of blood platelets, decreased 118
PROFECUNDIN 100
prolactin regulation, deterioration of 187
PROLINE 312
PROMIDONE 98
PRONEURIN 24
PROSCORBIN 91, 97
prostaglandins conversion from essential fatty acids, unregulated 167
prostate, inflammation of 9, 67, 96, 101, 105, 238
PROSULTIAMINE 24
PROTEIN 168, 187, 226, 322, 385-397
protein, animal 313
protein concentrate, soybean 14, 31, 40, 57, 63, 79, 127, 141, 154, 168, 178, 188, 200, 224, 230, 240, 249, 259, 268, 276, 285, 295, 304, 313, 322, 331, 339, 352, 361, 381, 385, 409

protein decreased 239, 248, 249, 257, 258, 258, 259, 267, 268, 276, 285, 294, 303, 312, 313, 322, 331, 339
protein formation, deterioration of 228
protein in urine, 126 167
protein isolate, soybean 15, 32, 39, 57, 79, 128, 141, 154, 170, 178, 188, 210, 215, 230, 240, 249, 259, 268, 276, 285, 294, 304, 313, 322, 331, 339, 361, 378, 385, 409
protein supply to brain, decreased 58
protein synthesis, decreased 11, 58, 77, 303
protein, textured vegetable 86, 89
protein utilization, deterioration of 24, 68, 118
PROTOGEN A 112
PROVITAMIN A 1–10, 90, 100
provitamin A to vitamin A conversion, decreased ability for 100
PROVITAMIN B$_3$ 322
PROVITAMIN B$_{3a}$ 322
PROVITAMIN B$_6$ 76
PROVITAMIN B$_P$ 88, 294
PROVITAMIN B$_T$ 285
PROVITAMIN F 103
PROVITAMIN Q 115, 116
PROVITINA 99
prune juice 5, 22, 33, 40, 95, 135, 148, 160, 175, 181, 197, 205, 223, 237, 353, 368, 397, 400
prune pudding 7, 22, 31, 48, 95, 132, 160, 197, 205, 217, 351, 366, 382, 393, 405
prune seed 78
prunes 112, 125
prunes, cooked 3, 22, 32, 40, 57, 62, 95, 132, 143, 160, 173, 181, 197, 204, 224, 236, 351, 367, 382, 396, 403
prunes, dried 2, 18, 30, 39, 53, 62, 84, 95, 130, 142, 157, 170, 180, 195, 200, 223, 235, 350, 364, 381, 394, 408
PSI-CAROTENE 1
PSI-VITAMIN B$_{12}$ 69
PSI-VITAMIN B$_{12d}$ 69
PTEROYLMONOGLUTAMIC ACID 68
pudding 7
pudding, apple tapioca 8, 138, 165, 199, 211, 220, 351, 367, 383, 397, 403
pudding, chocolate 7, 22, 31, 48, 129, 164, 195, 208, 220, 352, 366, 378, 394, 403
pudding, cottage 7, 15, 30, 48, 129, 159, 194, 210, 216, 350, 361, 376, 392, 408

Index

pudding, prune 7, 22, 31, 48, 95, 132, 160, 197, 205, 217, 351, 366, 382, 393, 405
pudding, tapioca cream 7, 20, 30, 48, 96, 129, 164, 194, 209, 217, 353, 366, 378, 393, 402
pudding, vanilla 7, 21, 30, 48, 96, 128, 195, 209, 219, 353, 367, 378, 393, 401
pudding, yellow corn 7, 13, 31, 48, 55, 64, 74, 80, 95, 131, 149, 163, 175, 196, 208, 220, 235, 245, 255, 264, 273, 282, 291, 300, 309, 319, 328, 336, 345, 354, 367, 378, 393, 401
pulse, faint/rapid 187, 199
pulse, rapid 152
pulse, rapid/irregular/feeble 77
pulse rate, decreased 106
pulse, slow/irregular 199
pulse, slow/weak 78
pulse, weak 69, 139, 151
pummelo 21, 35, 42, 66, 92, 138, 149, 166, 177, 184, 198, 207, 224, 238, 355, 369, 383, 396, 399
pumpkin 116
pumpkin, boiled 3, 21, 32, 41, 94, 134, 163, 176, 197, 206, 224, 247, 257, 267, 276, 285, 293, 302, 312, 321, 330, 338, 347, 356, 370, 383, 397, 398
pumpkin, canned 1, 22, 33, 42, 54, 65, 81, 95, 132, 146, 159, 172, 197, 207, 223, 237, 247, 257, 266, 275, 284, 293, 302, 311, 321, 329, 338, 347, 355, 369, 382, 396, 399
pumpkin pie 2, 21, 32, 41, 130, 163, 195, 208, 217, 352, 365, 376, 393, 405
pumpkin pie mix, canned 2, 22, 31, 42, 95, 131, 148, 161, 175, 196, 209, 217, 236, 247, 257, 266, 275, 284, 293, 302, 311, 321, 329, 338, 347, 352, 367, 383, 396, 402
pumpkin pie spice 4, 127, 154, 169, 194, 201, 221, 232, 350, 362, 376, 393, 409
pumpkin seeds 77
pumpkin seeds, dried 3, 14, 26, 39, 131, 141, 154, 168, 188, 200, 222, 228, 240, 251, 262, 269, 278, 288, 297, 304, 316, 322, 332, 339, 353, 359, 372, 387, 409
pumpkin seeds, roasted 64, 131, 141, 154, 168, 188, 200, 222, 228, 240, 250, 260, 268, 277, 286, 295, 304, 314, 322, 331, 339, 354, 359, 372, 385, 409
pupil dilation 88
purple passion fruit juice 3, 31, 39, 92, 138, 165, 198, 354, 369, 397, 400
purslane, boiled 2, 21, 32, 41, 94, 129, 162, 169, 197, 201, 221, 247, 257, 266, 275, 284, 293, 302, 311, 321, 329, 338, 346, 356, 370, 382, 395, 398
P-waves, decreased 199
P-XYLOTOCOPHEROL 102
pyorrhea 9, 43, 49, 67, 96, 99, 113, 138
PYRIDOXAL 58
pyridoxal phosphate levels decreased 58
PYRIDOXAMINE 58
PYRIDOXINE 58
pyrotechnics 126
PYRUVATE OXIDATION FACTOR 112
pyruvic acid in blood, increased 11

quail egg 7, 16, 25, 64, 129, 155, 190, 366, 376, 391, 401
quail without skin, raw 8, 14, 27, 44, 94, 134, 142, 155, 189, 206, 221, 243, 250, 261, 270, 279, 288, 296, 306, 315, 323, 332, 342, 366, 378, 388, 403
QUELICIN CHLORIDE 87
QUICKSILVER 186
QUICK'S VITAMIN 114
quince 5, 22, 35, 42, 57, 66, 78, 93, 135, 145, 163, 176, 198, 207, 223, 353, 369, 383, 397, 400
quince seeds 77

rabbit, stewed 19, 33, 44, 132, 159, 189, 203, 221, 364, 376, 386, 404
raccoon, roasted 12, 25, 363, 375, 386, 406
RACHITAMIN 98
RACHITASTEROL 98
RACOBALAMIN 74, 75
radiation 5, 9, 23, 43, 49, 58, 67, 85, 86, 90, 96, 97, 101, 113, 117, 212, 228, 238, 257, 267, 303, 330
RADIOACTIVE CYANOCOBALAMIN 74
RADIOACTIVE IODINE 153
RADIOACTIVE STRONTIUM 225
RADIOACTIVE VITAMIN B12 74
RADIOCYANOCOBALAMIN 74, 75
RADIOSTOL 99
radish, boiled oriental 35, 43, 93, 133, 166, 176, 197, 205, 222, 247, 257, 267, 276, 285, 294, 302, 312, 321, 330, 339, 347, 357, 370, 381, 396, 398
radish, dried oriental 14, 25, 38, 127, 154, 168, 191, 200, 216, 245, 255, 264, 273, 282, 291, 300, 309, 318, 328, 337, 345, 350, 363, 380, 392, 408
radish, raw 5, 34, 42, 57, 65, 80, 93, 133, 150, 165, 176, 181, 198, 206, 222, 236, 247, 257, 267, 276, 285, 293, 302, 312, 321, 330, 339, 347, 356, 370, 381, 396, 398
radish, raw oriental 22, 36, 42, 93, 132, 164, 175, 197, 206, 222, 247, 257, 267, 276, 285, 294, 302, 312, 321, 330, 339, 347, 356, 370, 397, 398
radishes 187, 212, 227
RADSTERIN 99
raisin bran 2, 11, 25, 37, 59, 70, 79, 131, 142, 154, 169, 188, 201, 215, 228, 245, 254, 264, 273, 282, 291, 299, 309, 319, 327, 336, 345, 350, 362, 379, 384, 391, 409
raisin pie 5, 21, 35, 42, 96, 133, 161, 196, 207, 216, 351, 363, 376, 394, 407
raisins 85, 108, 125, 152, 186
raisins, golden seedless 4, 23, 29, 39, 56, 61, 84, 95, 130, 142, 158, 170, 180, 194, 200, 222, 236, 349, 362, 381, 394, 409
raisins, seeded 16, 30, 39, 63, 84, 94, 132, 143, 156, 171, 180, 195, 200, 222, 237, 349, 362, 381, 394, 408
raisins, seedless 5, 15, 32, 40, 57, 62, 84, 95, 130, 142, 157, 170, 180, 195, 200, 222, 236, 349, 362, 381, 394, 408
rapeflower 117
rapeseed oil 116, 117
rashes 6, 106
rashes, genital 24
raspberries 4, 21, 32, 40, 56, 65, 78, 92, 132, 148, 163, 174, 178, 198, 208, 235, 354, 369, 380, 396, 400
RAT ANTISPECTACLED EYE FACTOR 85
RB 211
RB$^+$ 211
RDA, 1958 5, 9, 23, 36, 96, 97, 99
RDA, 1974 5, 23, 36, 43, 49, 67, 74, 85, 96, 97, 99, 101
RDA, 1980 5, 9, 23, 36, 43, 49, 58, 67, 74, 85, 89, 99,

Index

101, 108, 138, 139, 140, 151, 153, 166, 167, 177, 185, 186, 187, 199, 211, 212, 213, 225, 226, 238, 330, 397
RDA, 1989 6, 9, 23, 36, 43, 49, 58, 67, 68, 74, 76, 77, 78, 85, 86, 89, 91, 96, 97, 99, 101, 104, 105, 106, 108, 110, 111, 112, 113, 117, 118, 119, 120, 123, 124, 125, 126, 138, 139, 140, 151, 152, 153, 166, 167, 177, 185, 186, 187, 199, 211, 212, 213, 225, 226, 227, 238, 330, 397
rebound deficiency 58, 69, 89, 91, 100
RECOGNAN 87
red beans 5, 18, 32, 40, 131, 159, 195, 209, 216, 353, 368, 381, 393
red blood cell count, decreased 77
red blood cell life, decreased 100, 124
red blood cell production by bone marrow, deterioration of 76
red blood cells, large abnormal 69, 76, 78
REDAMINA 69
REDISOL 69
REDISOL H 75
REDOXON 91, 97
reflexes, decreased 49, 69, 168
regeneration of cell, decreased 106
reindeer, raw 13, 25, 45, 154, 366, 378, 388, 402
RELVENE 113
rennin 127, 191, 214, 352, 367, 380, 409
reproductive systems deterioration 227, 285
research on cancer 49
research on heredity 49
research on virus diseases 49
resistance to cold, decreased 118
retardation 258
retarded growth 6, 24, 36, 58, 78, 140, 177, 238, 385
RETIN-A 10
RETINOIC ACID 10
RETINOL 6, 10
REXORT 87
Reynaud's disease 152
rheumatic fever 9, 77, 96, 99, 113, 294
rheumatism 67, 77, 78, 96, 101, 106, 113, 211
rheumatoid arthritis 152
RHODACRYST 69
rhubarb, frozen and cooked sweetened 4, 22, 35, 42, 57, 67, 83, 95, 128, 150, 165, 175, 181, 199, 210, 224, 238, 351, 367, 383, 397, 403

rhubarb pie 4, 22, 35, 42, 95, 130, 163, 197, 208, 217, 351, 363, 376, 394, 407
rhubarb, raw frozen 4, 22, 35, 42, 57, 67, 82, 94, 128, 150, 165, 174, 181, 198, 209, 225, 237, 356, 370, 383, 397, 398
rib chop excluding fat, broiled lamb 15, 27, 45, 135, 158, 191, 204, 219, 365, 376, 387, 404
rib chop including fat, broiled lamb 16, 29, 46, 136, 161, 193, 206, 221, 360, 372, 389, 407
rib excluding fat, broiled beef 17, 29, 46, 54, 60, 70, 82, 135, 146, 156, 172, 172, 184, 191, 202, 219, 228, 242, 252, 261, 270, 279, 288, 296, 306, 315, 325, 333, 342, 364, 375, 387, 405
rib eye excluding fat, broiled beef 16, 29, 45, 54, 60, 70, 82, 134, 146, 156, 171, 184, 191, 202, 219, 228, 241, 251, 260, 269, 278, 287, 295, 305, 314, 324, 332, 341, 364, 376, 386, 405
rib eye including fat, broiled beef 17, 29, 46, 55, 60, 70, 82, 134, 146, 157, 172, 172, 184, 192, 203, 219, 229, 242, 252, 261, 270, 278, 288, 296, 306, 315, 325, 333, 342, 362, 374, 387, 406
rib including fat, broiled beef 17, 30, 47, 55, 61, 70, 83, 135, 147, 157, 174, 174, 185, 192, 204, 220, 229, 243, 253, 262, 271, 280, 289, 298, 307, 316, 326, 334, 343, 361, 373, 389, 407
rib including fat, roasted beef 18, 30, 47, 54, 62, 70, 82, 135, 146, 157, 174, 185, 192, 205, 220, 229, 243, 252, 261, 271, 279, 288, 297, 307, 316, 325, 334, 342, 360, 372, 388, 407
rib roast, veal 15, 26, 44, 135, 155, 190, 204, 219, 363, 374, 387, 406
rib roast excluding fat, beef 17, 29, 46, 53, 61, 70, 82, 135, 146, 156, 172, 184, 191, 202, 218, 228, 241, 251, 260, 269, 278, 287, 295, 305, 315, 324, 332, 341, 364, 375, 387, 405
RIBENA 91, 97
RIBIPCA 24
RIBOFLAVIN 24
RIBOFLAVIN PHOSPHATE 24, 36

rice 85, 89, 103, 104
rice bran 11, 27, 37, 77, 85, 86, 89, 116, 117, 127, 142, 154, 200, 223, 229, 351, 360, 374, 391, 409
rice bran oil 116
rice, brown 85, 86, 89, 101, 116, 186, 212
rice, cooked brown 17, 36, 39, 135, 164, 195, 210, 352, 367, 380, 394, 402
rice, cooked extra long grain white 39, 132, 162, 199, 211, 352, 367, 394, 402
rice, cooked instant white 16, 40, 138, 162, 198, 352, 367, 394, 402
rice, cooked parboiled extra long grain white 39, 133, 162, 196, 210, 352, 367, 383, 394, 402
rice, cooked parboiled white 16, 36, 39, 133, 162, 196, 210, 352, 367, 383, 394, 402
rice, cooked white 16, 36, 40, 136, 162, 197, 211, 352, 367, 383, 395, 402
rice, cooked wild long grain 4, 15, 36, 38, 57, 67, 74, 84, 96, 135, 149, 160, 176, 197, 210, 236, 352, 366, 378, 394, 402
rice, cream of 19, 41, 66, 84, 138, 150, 166, 177, 181, 198, 211, 224, 237, 247, 256, 266, 276, 284, 293, 301, 311, 321, 329, 338, 347, 354, 369, 383, 396, 399
rice flour 13, 25, 38, 135, 156, 170, 190, 208, 221, 349, 361, 380, 393, 409
rice germ 85, 86, 89
rice hulls 213
rice, minute 125
RICE POLISH FACTOR 67
rice polishings 11, 25, 41, 85, 86, 89, 130, 166, 200, 221, 247, 266, 276, 284, 294, 302, 311, 321, 338, 347, 350, 360, 376, 391, 409
rice, puffed 15, 33, 38, 54, 65, 80, 137, 144, 161, 173, 178, 195, 209, 233, 244, 254, 264, 273, 282, 291, 299, 309, 319, 327, 336, 345, 349, 360, 380, 384, 392, 410
rice, raw wild 13, 25, 38, 133, 155, 189, 207, 223, 350, 361, 380, 390, 409
rice straw 213
rice, well-milled 116
RICKETON 99
rickets 9, 96, 98, 99, 138, 177, 199
RIPRESIL 76
RNA 49, 68

RNA synthesis, decreased 68
RNA/DNA production, deterioration of 187
roaches 118
roast beef sandwich 7, 14, 28, 46, 53, 63, 72, 80, 95, 131, 145, 156, 172, 193, 206, 216, 231, 352, 364, 377, 390, 406
rock formations 167
rockfish, cooked 8, 19, 32, 46, 134, 150, 163, 170, 190, 201, 218, 235, 243, 252, 261, 270, 279, 287, 295, 306, 316, 325, 333, 342, 366, 379, 388, 402
rockfish, raw 8, 20, 33, 47, 136, 150, 164, 172, 184, 192, 202, 220, 236, 244, 253, 262, 271, 280, 289, 297, 307, 317, 326, 334, 343, 368, 379, 389, 401
rodent poisons 151
roll, chicken 18, 31, 45, 131, 150, 161, 174, 193, 206, 216, 234, 357, 366, 377, 389, 403
roll, sweet 13, 26, 38, 54, 64, 80, 132, 149, 158, 174, 195, 209, 216, 235, 350, 360, 375, 393, 408
roll, whole wheat 13, 31, 38, 129, 157, 189, 205, 216, 350, 363, 379, 391, 408
rolls, french 13, 26, 38, 53, 65, 80, 133, 146, 156, 172, 195, 210, 216, 235, 350, 363, 380, 392, 408
root vegetables 76
rose apple 3, 22, 35, 40, 93, 132, 151, 166, 177, 182, 199, 209, 238, 355, 370, 381, 397, 398
rose hips 91, 112, 125
roselle 3, 22, 34, 42, 93, 127, 159, 170, 197, 207, 223, 354, 369, 380, 396, 400
rosemary, dried 2, 13, 92, 127, 154, 169, 196, 201, 221, 232, 351, 365, 376, 393, 409
round, beef 139
round excluding fat, broiled beef 16, 28, 46, 54, 59, 70, 81, 137, 146, 156, 171, 184, 190, 202, 219, 229, 241, 251, 260, 269, 278, 287, 295, 305, 314, 324, 332, 341, 365, 377, 386, 404
round excluding fat, roasted eye of beef 17, 30, 46, 53, 60, 71, 82, 137, 146, 158, 171, 184, 190, 202, 220, 229, 241, 251, 260, 269, 277, 287, 295, 305, 314, 324, 332, 341, 365, 377, 386, 404
round including fat, broiled beef 17, 29, 46, 54, 59, 70, 82, 137, 146, 157, 172, 184, 191, 203, 220, 230, 242, 252, 260, 270, 278, 287, 296, 306, 315, 325, 333, 342, 363, 374, 387, 405
round including fat, roasted eye of beef 17, 30, 47, 53, 61, 71, 82, 137, 146, 158, 172, 184, 191, 203, 220, 230, 242, 251, 260, 270, 278, 287, 296, 305, 315, 324, 332, 341, 364, 375, 387, 405
round tip including fat, roasted beef 17, 28, 47, 53, 60, 70, 82, 137, 145, 156, 172, 184, 190, 203, 220, 229, 242, 252, 260, 270, 278, 287, 296, 305, 315, 324, 333, 341, 363, 375, 387, 405
round with rump, braised/broiled veal 18, 28, 45, 135, 155, 190, 204, 219, 364, 376, 387, 404
royal jelly 89
RUBESOL 69
RUBIDIUM 139, 199, 211-212
RUBIVITAN 69
RUBRAMIN 69
RUBRATOPE 74, 75
RUBRIPCA 69
RUBROCITOL 69
rum, 100 proof 150, 166, 183, 199, 211, 238, 362, 405
rum, 80 proof 149, 166, 199, 211, 238, 364, 403
rum, 94 proof 150, 166, 183, 199, 211, 238, 363, 404
rump excluding fat, roasted pork 9, 12, 26, 45, 51, 61, 72, 83, 137, 145, 160, 171, 182, 189, 202, 219, 231, 240, 250, 259, 268, 277, 286, 295, 304, 313, 322, 332, 340, 364, 376, 386, 405
rump including fat, roasted pork 8, 12, 26, 45, 51, 62, 72, 83, 137, 146, 161, 172, 183, 189, 203, 220, 231, 241, 251, 259, 269, 278, 286, 296, 305, 314, 323, 332, 341, 363, 374, 387, 406
rutabaga, boiled 18, 34, 40, 56, 65, 81, 93, 131, 150, 164, 173, 181, 196, 205, 222, 236, 246, 257, 266, 275, 285, 293, 302, 311, 321, 329, 338, 347, 355, 369, 382, 396, 399
RUTABION 113
RUTIN 112, 113
RUTOSIDE 113
RUTOZYD 113
RUVEN 113

rye 89, 103, 104
rye flour 139
rye flour, dark 12, 29, 38, 130, 155, 188, 200, 224, 350, 361, 379, 390, 409
rye flour, light 15, 33, 40, 132, 160, 192, 208, 225, 349, 361, 380, 391, 409
rye, whole 101

S 226
S^{--} 226
S^{-2} 226
SABALAMIN 76
sablefish 104
SADDI, 1980 58, 89, 108, 139, 140, 151, 185, 186, 187, 211, 212, 213, 225, 226
SADDI, 1989 49, 58, 67, 68, 76, 77, 78, 85, 86, 88, 89, 90, 91, 104, 105, 106, 110, 111, 112, 113, 117, 118, 119, 120, 123, 124, 125, 126, 139, 140, 151, 152, 153, 167, 185, 186, 187, 211, 212, 213, 225, 226, 227, 228
safflower nuts 101
safflower oil 101, 103, 104, 116, 117, 227, 359, 371
saffron 129, 154, 191, 200, 218, 350, 365, 391, 409
sage 2, 11, 127, 154, 168, 194, 200, 231, 351, 365, 376, 391, 409
salad dressing, french 136, 211, 215, 356, 359, 372, 408
salad dressing, italian 136, 199, 211, 215, 237, 354, 360, 372, 396, 408
salad dressing, low calorie french 134, 163, 198, 210, 215, 237, 352, 366, 377, 402
salad dressing, low calorie italian 199, 211, 215, 356, 367, 376, 400
salad dressing, low calorie russian 133, 163, 197, 208, 215, 351, 366, 378, 397, 403
salad dressing, low calorie thousand island 134, 163, 198, 209, 215, 353, 365, 376, 396, 402
salad dressing, russian 18, 33, 48, 94, 133, 163, 196, 208, 215, 235, 354, 359, 372, 395, 408
salad dressing, thousand island 134, 163, 198, 209, 215, 237, 353, 360, 372, 397, 407
salami, beef 17, 30, 47, 51, 63, 70, 93, 136, 145, 157, 176, 181, 194, 207, 214, 232, 244, 253, 263, 272, 281, 290, 298, 308, 318, 327, 335, 344, 357, 363, 374, 390, 405

Index 522

salami, beef and pork 14, 25, 47, 51, 62, 70, 93, 134, 143, 156, 176, 181, 194, 207, 215, 232, 244, 253, 263, 272, 281, 290, 299, 308, 318, 327, 335, 344, 357, 363, 374, 390, 404
salami, beef beerwurst 17, 31, 47, 54, 63, 71, 84, 93, 136, 150, 159, 176, 195, 208, 215, 231, 244, 254, 263, 272, 281, 290, 299, 309, 318, 327, 335, 344, 358, 361, 373, 391, 406
salami, hard pork 12, 27, 44, 59, 70, 135, 144, 160, 174, 181, 190, 214, 230, 243, 252, 262, 270, 280, 288, 298, 306, 316, 325, 334, 342, 357, 360, 372, 388, 408
salami, hard pork and beef 12, 27, 45, 50, 59, 71, 92, 135, 146, 159, 174, 193, 202, 214, 230, 242, 252, 261, 271, 279, 288, 297, 307, 316, 326, 334, 342, 357, 360, 372, 388, 408
salami, pork and beef 14, 25, 47, 51, 62, 70, 93, 134, 143, 156, 176, 181, 194, 207, 215, 232, 244, 253, 263, 272, 281, 290, 299, 308, 318, 327, 335, 344, 357, 363, 374, 390, 404
salami, pork beerwurst 12, 30, 47, 53, 61, 72, 84, 92, 136, 149, 162, 176, 182, 194, 206, 214, 232, 244, 254, 263, 273, 281, 290, 298, 309, 318, 327, 335, 344, 357, 364, 374, 390, 404
salami, turkey 18, 30, 47, 133, 149, 158, 175, 194, 206, 215, 232, 358, 365, 375, 390, 403
SALICYLIC ACID CHOLINE SALT 88
salivary glands, deterioration of 6
salivation, decreased 228
salivation, increased 124, 152
salmon 85, 99, 101, 103, 104, 152
salmon, canned 89
salmon, canned pink 8, 22, 29, 44, 81, 146, 162, 170, 189, 204, 218, 234, 243, 253, 262, 271, 280, 288, 297, 307, 317, 326, 334, 343, 366, 377, 389, 403
salmon, canned sockeye 8, 22, 29, 45, 82, 147, 161, 171, 189, 202, 218, 234, 243, 253, 262, 271, 280, 288, 297, 307, 316, 326, 334, 342, 366, 377, 389, 403

salmon, cooked coho 95, 148, 162, 201, 220, 235, 242, 252, 261, 269, 278, 286, 295, 305, 315, 324, 332, 341, 365, 377, 387, 403
salmon, cooked sockeye 8, 14, 29, 44, 62, 69, 137, 148, 163, 171, 189, 202, 219, 235, 242, 252, 261, 269, 278, 286, 295, 305, 315, 324, 332, 341, 367, 376, 387, 404
salmon, raw chinook 20, 31, 44, 95, 132, 149, 162, 184, 202, 221, 235, 243, 253, 262, 271, 280, 288, 297, 307, 316, 326, 334, 343, 365, 376, 389, 402
salmon, raw chum 8, 17, 29, 135, 149, 163, 184, 189, 202, 221, 235, 243, 253, 262, 271, 280, 288, 297, 307, 316, 326, 334, 343, 366, 378, 389, 401
salmon, raw coho 95, 149, 162, 184, 202, 221, 235, 243, 253, 262, 271, 279, 288, 296, 307, 316, 325, 334, 342, 366, 377, 389, 402
salmon, raw pink 8, 147, 162, 184, 204, 219, 235, 243, 253, 262, 271, 280, 288, 297, 307, 317, 326, 334, 343, 367, 378, 389, 401
salmon, raw sockeye 8, 14, 30, 45, 52, 63, 73, 137, 149, 164, 172, 185, 191, 202, 221, 235, 243, 253, 262, 271, 279, 288, 296, 307, 316, 325, 334, 342, 365, 377, 389, 402
salmon, smoked chinook 8, 22, 32, 46, 51, 62, 70, 84, 135, 143, 162, 174, 184, 192, 208, 215, 236, 244, 253, 262, 272, 280, 289, 297, 307, 317, 326, 334, 343, 367, 378, 389, 402
salsify, boiled 19, 30, 41, 95, 130, 163, 174, 196, 205, 222, 353, 368, 383, 394, 400
salt 126, 139
salt cravings 139, 214
sapodilla 4, 36, 42, 56, 66, 93, 132, 162, 198, 207, 222, 248, 267, 276, 285, 293, 302, 312, 322, 330, 339, 347, 353, 368, 379, 397, 401
sapote 3, 22, 36, 39, 93, 131, 161, 171, 197, 203, 223, 247, 265, 275, 284, 292, 302, 311, 320, 329, 337, 346, 351, 366, 380, 394, 404
sarcoma 152
sardines 8, 17, 29, 45, 52, 63, 69, 81, 104, 117, 143, 156, 170, 181, 188, 202, 216, 233, 243, 252, 261, 270, 279, 287, 295, 306, 315, 325, 333, 341, 365, 376, 388, 404
sardines, canned 89, 99, 152
SATURATED FATS 103, 104, 371
SATURATED FATTY ACIDS 103, 104, 371
sauerkraut, canned 5, 21, 35, 42, 57, 64, 93, 131, 146, 159, 175, 198, 208, 215, 237, 356, 370, 382, 396, 398
sausage, beef and pork 13, 30, 47, 53, 65, 73, 160, 176, 194, 215, 232, 244, 254, 263, 272, 281, 290, 299, 308, 318, 327, 335, 344, 357, 360, 372, 390, 407
sausage, beef and pork link 14, 30, 47, 53, 63, 71, 93, 135, 148, 159, 176, 182, 194, 207, 215, 232, 244, 254, 263, 273, 281, 290, 298, 309, 318, 327, 335, 344, 358, 361, 373, 391, 406
sausage, beef and pork luncheon 14, 28, 47, 54, 62, 71, 93, 134, 147, 159, 176, 182, 194, 206, 214, 231, 243, 252, 263, 272, 281, 289, 298, 308, 318, 326, 334, 343, 357, 363, 374, 390, 405
sausage, beef and pork summer 15, 27, 46, 52, 61, 69, 84, 93, 136, 147, 157, 176, 182, 195, 206, 214, 232, 357, 361, 373, 390, 407
sausage, beef honey roll 17, 30, 46, 53, 62, 70, 93, 136, 147, 157, 175, 182, 193, 205, 214, 230, 357, 365, 376, 389, 403
sausage, beef summer 15, 27, 46, 62, 69, 84, 93, 136, 147, 157, 176, 194, 206, 214, 232, 357, 362, 373, 390, 406
sausage, italian pork 12, 28, 46, 53, 61, 71, 95, 132, 147, 159, 174, 181, 192, 204, 215, 231, 243, 253, 262, 272, 281, 289, 298, 308, 317, 326, 335, 343, 358, 361, 373, 389, 407
sausage, polish pork 13, 31, 46, 53, 63, 72, 135, 146, 159, 175, 181, 193, 206, 215, 232, 244, 254, 263, 272, 281, 290, 298, 308, 318, 327, 335, 344, 357, 361, 373, 390, 406
sausage, pork 12, 27, 46, 51, 61, 71, 131, 144, 160, 174, 181, 192, 203, 214, 231, 243,

253, 262, 272, 281, 289, 298, 308, 317, 327, 335, 344, 358, 360, 373, 389, 407
sausage, pork and beef 13, 30, 47, 53, 65, 73, 160, 176, 194, 215, 232, 244, 254, 263, 272, 281, 290, 299, 308, 318, 327, 335, 344, 357, 360, 372, 390, 407
sausage, pork and beef link 14, 30, 47, 53, 63, 71, 93, 135, 148, 159, 176, 182, 194, 207, 215, 232, 244, 254, 263, 273, 281, 290, 298, 309, 318, 327, 335, 344, 358, 361, 373, 391, 406
sausage, pork and beef luncheon 14, 28, 47, 54, 62, 71, 93, 134, 147, 159, 176, 182, 194, 206, 214, 231, 243, 252, 263, 272, 281, 289, 298, 308, 318, 326, 334, 343, 357, 363, 374, 390, 405
sausage, pork link 12, 27, 46, 51, 60, 71, 95, 132, 148, 160, 174, 193, 204, 214, 231, 243, 252, 261, 271, 280, 289, 297, 307, 316, 326, 334, 342, 357, 360, 373, 388, 408
sausage, pork liver 6, 14, 25, 44, 50, 61, 69, 93, 135, 143, 154, 176, 181, 192, 207, 214, 231, 244, 252, 263, 273, 281, 290, 299, 308, 318, 327, 335, 344, 357, 361, 372, 391, 407
sausage, smoked beef 19, 31, 47, 64, 71, 136, 158, 175, 194, 208, 214, 231, 244, 254, 263, 272, 281, 290, 299, 308, 318, 327, 335, 344, 357, 362, 373, 390, 406
sausage, summer 15, 26, 46, 62, 69, 93, 134, 144, 156, 175, 194, 205, 214, 231, 244, 253, 263, 272, 281, 289, 298, 308, 318, 327, 335, 344, 358, 361, 373, 390, 406
sausage, turkey breast meat summer 19, 32, 44, 52, 61, 71, 137, 149, 164, 174, 190, 205, 214, 233, 242, 253, 261, 270, 279, 287, 296, 307, 316, 325, 332, 342, 367, 379, 388, 402
sausage, turkey summer 16, 25, 45, 62, 70, 134, 144, 157, 174, 181, 190, 206, 215, 231, 358, 365, 376, 390, 403
sausage, vienna 19, 31, 47, 64, 72, 134, 162, 177, 182, 196, 210, 215, 233, 244, 254, 264, 272, 282, 290,

299, 309, 318, 327, 336, 344, 357, 362, 373, 391, 404
sausage with american cheese, beef and pork link 14, 30, 47, 51, 64, 71, 93, 130, 148, 161, 176, 183, 192, 207, 215, 232, 244, 254, 263, 272, 281, 290, 298, 308, 318, 327, 335, 344, 358, 361, 373, 390, 406
savory 2, 11, 37, 127, 153, 168, 191, 200, 229, 349, 360, 376, 391, 409
SB 123
scallops 105, 139, 212
scallops, breaded and fried 21, 32, 48, 71, 131, 147, 162, 169, 190, 204, 216, 233, 243, 252, 263, 272, 281, 289, 298, 308, 317, 326, 335, 343, 354, 364, 376, 390, 405
scallops, raw 22, 33, 48, 56, 71, 132, 149, 165, 169, 181, 191, 204, 217, 234, 243, 253, 263, 272, 281, 289, 298, 308, 317, 326, 335, 344, 357, 368, 380, 390, 401
scallops, surimi 22, 36, 48, 136, 165, 210, 215, 244, 254, 263, 272, 281, 289, 298, 308, 318, 327, 335, 344, 354, 367, 381, 391, 402
scars, remaining 100
schizophrenia 106, 117, 322
schizophrenics, histadelic 303
schizophrenics, overstimulated 267
sciatica 23, 99, 101, 213
SCOLINE CHLORIDE 87
SCORBACID 91, 97
SCORBU-C 91, 97
screwdriver 4, 18, 36, 43, 57, 66, 80, 92, 136, 150, 166, 177, 185, 198, 208, 224, 238, 247, 257, 267, 276, 285, 294, 302, 312, 322, 330, 339, 348, 355, 368, 397, 400
scup, raw 131, 149, 163, 173, 182, 205, 221, 235, 244, 253, 262, 271, 280, 289, 297, 307, 317, 326, 334, 343, 367, 379, 389, 401
SCUROCAINE 106
scurvy 9, 85, 91, 96, 113, 118, 166
SE 212
SE$^-$ 212
SE^{-2} 212
sea plants 126, 139
seafood 139
seatrout, raw 133, 150, 165, 171, 184, 190, 203, 220, 235, 244, 254, 263, 272, 281, 289, 298, 308, 317,

326, 335, 343, 367, 378, 390, 401
seawater 126
SEDATIVES 127
SELENIUM 101, 103, 104, 114, 115, 116, 117, 126, 167, 186, 212, 228, 371
selenium decreased 294
selenium therapy 303
semi-essential amino acid 239, 259, 385
senile keratosism 106
senile Parkinsonism 106
sense of smell, decreased 6
sense of smell, hyperacute 37, 43
sense of taste, decreased 37, 43, 141
sense of touch, decreased 11
sensitivity, pain 58
SER 312
SERINE 312-313
serine, internal conversion of 258
serotonin neurotransmitter levels, unstable 167
sesame butter 11, 25, 38, 127, 141, 154, 169, 188, 202, 217, 229, 240, 251, 263, 272, 281, 291, 297, 307, 317, 323, 334, 343, 352, 359, 372, 390, 410
sesame kernels, dried 4, 12, 31, 38, 52, 128, 154, 168, 188, 202, 221, 228, 240, 250, 262, 269, 278, 290, 294, 304, 315, 322, 331, 340, 354, 359, 372, 387, 409
sesame kernels, toasted 11, 25, 38, 128, 154, 168, 188, 202, 221, 228, 240, 251, 263, 272, 281, 291, 297, 306, 317, 323, 334, 343, 352, 359, 372, 390, 409
sesame oil 101, 103, 104, 117, 359, 371
sesame seeds 103, 104, 117, 118
sesame seeds, whole dried 5, 12, 28, 38, 57, 59, 79, 127, 141, 154, 168, 178, 188, 201, 222, 228, 240, 250, 262, 272, 281, 290, 297, 306, 317, 323, 334, 343, 352, 359, 372, 390, 409
sesame seeds, whole toasted 127, 141, 154, 168, 178, 188, 201, 222, 228, 240, 251, 263, 272, 281, 291, 297, 306, 317, 323, 334, 343, 352, 359, 372, 390, 410
7-DEHYDROSITOSTEROL 99
SEVICAINE 106
sexual activity, decreased 238
sexual arousal, diminished 259, 303

Index 524

sexual development, slowed 228
sexual maturation, delayed 239
sexual organ function, decreased 100
shad, raw 130, 148, 161, 171, 182, 189, 202, 220, 236, 244, 254, 263, 272, 281, 289, 298, 308, 317, 326, 335, 343, 365, 375, 390, 403
shallots, freeze-dried 14, 50, 59, 79, 92, 128, 142, 154, 169, 178, 189, 200, 221, 232, 244, 264, 273, 282, 291, 300, 309, 318, 327, 336, 344, 350, 361, 391, 410
shallots, raw 16, 94, 131, 160, 196, 204, 223, 246, 265, 274, 283, 292, 301, 310, 320, 328, 337, 346, 353, 368, 394, 401
shank excluding fat, roasted pork 9, 12, 26, 45, 51, 59, 72, 83, 137, 145, 160, 172, 182, 189, 203, 219, 230, 241, 250, 259, 269, 277, 286, 296, 305, 313, 323, 332, 340, 364, 376, 386, 404
shank excluding fat, simmered beef 15, 29, 44, 53, 60, 70, 81, 131, 144, 155, 171, 183, 189, 201, 219, 228, 240, 250, 259, 268, 277, 286, 294, 304, 313, 323, 331, 340, 365, 377, 385, 405
shank including fat, roasted pork 9, 12, 27, 46, 52, 60, 72, 83, 137, 146, 161, 173, 182, 190, 204, 220, 231, 241, 251, 259, 270, 279, 287, 297, 306, 315, 324, 333, 341, 362, 374, 388, 406
shank including fat, simmered beef 16, 29, 45, 54, 61, 70, 82, 132, 144, 155, 171, 183, 190, 202, 220, 228, 240, 250, 260, 268, 277, 286, 295, 304, 313, 323, 331, 340, 364, 376, 385, 406
shark, battered and fried 8, 18, 32, 47, 71, 130, 149, 160, 170, 192, 208, 217, 235, 244, 253, 262, 271, 280, 289, 297, 307, 317, 326, 334, 343, 355, 364, 375, 389, 404
shark liver oil 6
shark, raw 8, 19, 33, 47, 51, 71, 131, 150, 162, 170, 184, 191, 208, 218, 235, 243, 253, 262, 271, 279, 288, 296, 307, 316, 325, 334, 342, 366, 378, 389, 402

sheep 88
sheep milk 7, 18, 26, 48, 54, 72, 95, 128, 166, 174, 193, 209, 221, 246, 255, 264, 273, 282, 291, 299, 309, 319, 327, 336, 344, 356, 367, 377, 393, 400
sheepshead, cooked 131, 145, 163, 170, 183, 189, 201, 219, 234, 242, 252, 261, 270, 278, 287, 295, 306, 315, 324, 333, 341, 366, 379, 387, 403
sheepshead, raw 132, 150, 164, 171, 185, 189, 202, 219, 236, 243, 253, 262, 271, 280, 288, 297, 307, 316, 326, 334, 343, 367, 379, 389, 401
shellie beans, canned 4, 21, 33, 42, 95, 132, 161, 209, 216, 355, 370, 382, 395, 399
shingles 9, 23, 67, 74, 96, 99
shock 125
SHOCK-FEROL 99
shortbread cookie 3, 14, 28, 38, 55, 81, 132, 148, 158, 175, 194, 210, 216, 235, 350, 359, 373, 392, 410
shortribs excluding fat, braised beef 18, 29, 47, 54, 62, 70, 82, 135, 146, 155, 173, 173, 183, 190, 204, 220, 228, 241, 251, 260, 269, 277, 286, 295, 304, 313, 323, 332, 340, 362, 374, 386, 407
shortribs including fat, braised beef 19, 30, 47, 55, 62, 70, 83, 135, 146, 157, 175, 185, 193, 207, 221, 229, 243, 253, 261, 271, 279, 288, 297, 307, 316, 325, 334, 342, 360, 372, 388, 408
shoulder excluding fat, roasted lamb 15, 27, 45, 135, 158, 191, 204, 219, 365, 376, 387, 404
shoulder excluding fat, roasted pork 9, 12, 26, 46, 52, 60, 72, 83, 136, 145, 159, 174, 182, 190, 203, 218, 230, 241, 251, 259, 269, 278, 286, 296, 306, 315, 323, 332, 341, 364, 375, 387, 405
shoulder including fat, roasted lamb 15, 28, 45, 135, 160, 192, 206, 220, 361, 373, 388, 407
shoulder including fat, roasted pork 9, 13, 26, 46, 53, 61, 72, 84, 137, 145, 159, 174, 183, 191, 204, 219, 230, 242, 252, 260, 271, 279, 287, 298, 307, 316,

325, 334, 342, 361, 373, 388, 406
shoulder/neck pains 167
shoyu sauce 20, 31, 38, 54, 63, 81, 133, 145, 157, 170, 194, 208, 214, 236, 355, 369, 382, 393, 402
shredded wheat 14, 27, 38, 51, 62, 80, 131, 142, 155, 169, 188, 203, 222, 230, 244, 253, 264, 273, 282, 291, 299, 308, 318, 326, 336, 344, 349, 360, 379, 384, 391, 409
shrimp 89, 99, 101, 139, 152, 212
shrimp, breaded and fried 16, 30, 47, 64, 71, 82, 129, 143, 160, 170, 191, 206, 216, 233, 241, 252, 263, 271, 280, 289, 297, 306, 317, 324, 334, 343, 354, 364, 376, 389, 406
shrimp, cooked 20, 34, 47, 64, 71, 84, 131, 143, 156, 170, 182, 193, 208, 217, 233, 241, 253, 263, 271, 280, 288, 297, 307, 317, 324, 334, 343, 367, 379, 389, 401
shrimp, raw 22, 34, 47, 55, 64, 71, 84, 130, 143, 157, 170, 181, 191, 207, 217, 233, 241, 253, 263, 271, 280, 289, 297, 307, 317, 325, 334, 343, 358, 367, 379, 389, 401
shrimp, surimi 22, 34, 48, 133, 163, 210, 215, 244, 254, 263, 272, 281, 290, 298, 308, 318, 327, 335, 344, 355, 367, 379, 391, 401
SI 212
sickle cell anemia 77, 78
signs, neon 125
SILICON 212-213
SILVER 213
silverware 213
SINCALINE 86
SINKRON 87
SINTOTIAMINA 24
sinusitis 9, 101
sirloin excluding fat, braised pork 9, 12, 26, 45, 51, 59, 72, 83, 137, 145, 159, 171, 171, 184, 190, 201, 220, 230, 240, 250, 259, 268, 277, 285, 295, 304, 313, 322, 331, 340, 363, 375, 385, 406
sirloin excluding fat, broiled beef 16, 27, 46, 54, 59, 70, 81, 135, 144, 155, 171, 171, 184, 190, 202, 219, 228, 241, 251, 260, 269, 277, 286, 295, 304, 314, 323, 332, 340, 365, 377, 386, 405

sirloin excluding fat, broiled
   pork 9, 11, 25, 45, 59, 72,
   83, 137, 146, 162, 170, 170,
   185, 189, 201, 220, 231,
   240, 250, 259, 269, 277,
   286, 296, 305, 313, 323,
   332, 340, 364, 375, 386,
   405
sirloin excluding fat, fried
   beef 15, 26, 46, 53, 59, 70,
   81, 135, 144, 155, 170, 183,
   189, 201, 218, 229, 240,
   250, 260, 268, 277, 286,
   295, 304, 313, 323, 331,
   340, 364, 376, 385, 405
sirloin excluding fat, roasted
   pork 9, 12, 26, 45, 52, 60,
   72, 83, 135, 146, 160, 172,
   182, 190, 203, 220, 231,
   241, 250, 259, 269, 278,
   286, 296, 305, 314, 323,
   332, 340, 364, 375, 386,
   405
sirloin including fat, braised
   pork 12, 27, 45, 52, 60, 72,
   84, 137, 145, 160, 173, 185,
   191, 203, 220, 231, 240,
   250, 259, 269, 278, 286,
   296, 305, 314, 323, 332,
   340, 361, 373, 386, 407
sirloin including fat, broiled
   beef 16, 27, 46, 54, 60, 70,
   82, 135, 144, 156, 171, 171,
   184, 191, 203, 220, 229,
   241, 251, 260, 269, 278,
   287, 295, 305, 315, 324,
   332, 341, 362, 374, 387,
   406
sirloin including fat, broiled
   pork 9, 12, 26, 46, 51, 59,
   72, 83, 137, 146, 162, 171,
   185, 190, 203, 220, 232,
   241, 251, 259, 270, 279,
   287, 297, 306, 315, 324,
   333, 341, 361, 373, 388, 407
sirloin including fat, fried
   beef 16, 27, 46, 54, 60, 70,
   82, 135, 144, 155, 171, 184,
   190, 202, 219, 229, 241,
   251, 260, 269, 278, 287,
   295, 305, 315, 324, 332,
   341, 361, 373, 386, 407
sirloin including fat, roasted
   pork 9, 12, 26, 45, 52, 60,
   72, 83, 136, 146, 161, 173,
   182, 190, 204, 220, 232,
   241, 251, 259, 270, 278,
   286, 297, 306, 315, 324,
   333, 341, 362, 374, 387,
   406
skin 213
skin ailments 123
*skin, ashen-gray coloring
   of* 213
skin, bronze discolored 153
skin burning sensations 24, 89
skin changes 89
skin color, bluish 11

*skin complexion
   deterioration* 226
skin, cracked 127, 151
skin, crawling feeling 339
skin dermatosis, nutritional 385
skin deterioration 37, 43, 50,
   68, 76, 104, 106, 119, 168,
   371
skin, dry 6, 9, 91, 104, 152,
   199, 322, 371
skin eczema 6, 9, 67, 76, 85,
   86, 89, 96, 99, 104, 105,
   141, 152, 226, 371
skin eczema, blistering 139,
   199
skin edema 67, 151, 187, 214,
   385
skin elasticity, decreased 213
skin eruptions 127, 199, 226
skin flabbiness 213
*skin, greasy/cheesy oozing
   of* 58
skin hemorrhages, small 91
skin hives 106
skin ichthyosis 106
skin impetigo 9, 96, 101
skin inflammation 9, 11, 24,
   36, 37, 43, 58, 67, 89, 105,
   106, 125, 178, 211, 212, 226
skin inflammation, infant 89
skin lesions 117, 228
skin leukoderma 106
skin melanin, decreased 303,
   331
skin, oily 24
skin on palms/soles, thickening of 124
skin pallor 78, 89, 213
skin, peeling 6
skin pigmentation, abnormal 141
skin prickling sensations 24, 89
skin psoriasis 9, 58, 67, 74,
   76, 85, 96, 99, 105, 106,
   177, 226
skin rashes 6, 89, 91, 124
skin, rough 6
skin, scaling 6
skin scaling on face 24
skin sensitivity, increased 37,
   89, 339
skin, small red lumps on 187
skin sores 141
skin, thickening of 6, 239, 303
skin ulcers 9, 36, 74, 85, 101,
   108, 113, 140, 166
skin vitiligo 90
skin, wrinkled 100, 106
skin, yellowing of 1, 107
*skin/hair pigmentation,
   alterations in* 385
sleep, induced 322
sleep, lack of 37, 43, 49, 50,
   58, 67, 74, 76, 85, 89, 96,
   99, 126, 138, 141, 167, 186,
   187, 199, 211, 303, 322,
   330, 339

sleep pattern, altered 78
SLEEPING PILLS 37, 44, 50,
   69, 86, 140
smell sense, decreased 6, 228
smell sense, hyperacute 37,
   43
smelt 104, 212
smelt, canned 127, 158, 188,
   271, 280, 289, 298, 308,
   317, 326, 343, 365, 375,
   389, 404
smelt, cooked rainbow 30,
   48, 70, 129, 144, 160, 170,
   189, 202, 218, 232, 243,
   253, 262, 271, 279, 288,
   296, 307, 316, 325, 333,
   342, 366, 379, 388, 402
smelt, raw 22, 31, 48, 164,
   189, 271, 280, 289, 297,
   308, 317, 326, 343, 357,
   367, 389, 401
smelt, raw rainbow 31, 48,
   52, 70, 130, 144, 161, 171,
   179, 190, 205, 220, 233,
   244, 254, 263, 272, 280,
   289, 298, 308, 317, 326,
   335, 343, 367, 379, 390,
   401
smoke, cigarette 126, 140
smoked beef, chopped 18,
   30, 46, 52, 60, 71, 93, 156,
   173, 192, 202, 214, 230,
   243, 253, 262, 272, 280,
   289, 298, 307, 317, 326,
   334, 343, 357, 366, 378,
   389, 403
smoked goose liver pâté,
   canned 16, 26, 47, 69, 356,
   360, 372, 391, 408
smokes 125
SN 226
snap beans, boiled yellow 4,
   18, 32, 40, 57, 65, 80, 94,
   130, 146, 160, 172, 180, 197,
   205, 224, 236, 247, 256,
   266, 275, 283, 292, 301,
   310, 320, 329, 338, 346,
   355, 369, 381, 384, 395,
   399
snap beans, canned yellow 4,
   23, 33, 43, 80, 95, 132, 150,
   161, 175, 180, 198, 209, 217,
   236, 247, 257, 266, 275,
   284, 293, 302, 311, 321,
   329, 338, 347, 356, 370,
   383, 384, 396, 398
snap beans, frozen and boiled
   yellow 4, 21, 33, 41, 57, 65,
   94, 130, 148, 162, 173, 179,
   197, 209, 222, 235, 247,
   256, 266, 275, 284, 293,
   302, 311, 320, 329, 338,
   346, 355, 370, 383, 384,
   395, 398
snapper, cooked 19, 48, 131,
   149, 165, 170, 184, 191, 201,
   220, 235, 242, 252, 261,
   270, 278, 287, 295, 306,

Index

315, 324, 333, 341, 366, 379, 387, 402
snapper, raw 19, 48, 131, 150, 165, 171, 185, 191, 202, 219, 236, 243, 253, 262, 271, 280, 288, 297, 307, 316, 326, 334, 342, 367, 379, 389, 401
sneezing 124
soda, lemon-lime 138, 151, 166, 177, 211, 223, 238, 354, 369, 399
soda, orange 138, 150, 166, 177, 199, 211, 222, 238, 354, 369, 399
SODIUM 59, 69, 106, 140, 167, 200, 214-225
SODIUM AMINOSALICYLATE 69, 140
SODIUM ASCORBATE 97
SODIUM BENZOAT 6
sodium deposits in tissues 168
SODIUM NITRATE 37, 91
SODIUM PANGAMATE 76
SODIUM PANTOTHENATE 49, 58
sodium pectate 213
sodium polypectate 213
sodium replacement in tissues 167
SODIUM SALICYLATE 91
sole 103, 104
sole, raw 19, 34, 48, 130, 162, 171, 192, 203, 220, 272, 281, 289, 298, 308, 318, 327, 344, 368, 380, 390
soles/palms, thickening of skin 124
SOLIPHYLLINE 88
SOPHORIN 113
SORBITOL 69
sorghum 13, 30, 32, 38, 128, 132, 154, 155, 189, 197, 203, 350, 350, 361, 363, 378, 391, 408, 409
souffle, cheese 7, 19, 28, 48, 128, 161, 192, 209, 216, 355, 364, 374, 391, 403
sour cream 7, 30, 54, 73, 82, 128, 166, 176, 195, 209, 221, 236, 274, 283, 292, 300, 309, 320, 328, 345, 356, 364, 374, 394, 402
source of vitamin A, lost 1
source of vitamin/coenzyme Q_{10} *lost* 114, 115, 116
source of vitamin F, lost 103
soursop 5, 18, 34, 40, 55, 65, 93, 134, 163, 174, 197, 205, 222, 293, 302, 329, 353, 368, 382, 396, 400
soy meal 125
soy meal, defatted 5, 12, 28, 38, 50, 59, 79, 127, 141, 154, 168, 178, 188, 200,

224, 229, 240, 249, 259, 268, 277, 286, 297, 304, 313, 322, 331, 339, 351, 361, 379, 384, 385, 409
soy milk 8, 15, 33, 48, 57, 66, 84, 138, 145, 163, 174, 181, 196, 209, 222, 237, 245, 255, 265, 274, 283, 292, 301, 309, 320, 328, 337, 346, 357, 369, 379, 384, 394, 398
soy oil 103, 104
soy sauce 22, 32, 38, 55, 63, 81, 137, 146, 159, 177, 195, 208, 214, 236, 355, 369, 382, 394, 401
soya fluff 213
soybean flour 4, 12, 25, 38, 50, 59, 79, 85, 89, 128, 141, 154, 168, 178, 187, 188, 200, 222, 230, 240, 249, 260, 268, 277, 287, 298, 304, 313, 322, 331, 340, 351, 360, 374, 385, 409
soybean flour, defatted 5, 12, 28, 38, 50, 59, 79, 127, 141, 154, 168, 178, 188, 200, 222, 231, 240, 249, 259, 268, 277, 285, 296, 304, 313, 322, 331, 339, 351, 361, 379, 385, 409
soybean flour, low fat 5, 13, 27, 38, 50, 59, 78, 128, 141, 154, 168, 178, 188, 200, 222, 233, 240, 249, 259, 268, 277, 285, 296, 304, 313, 322, 331, 339, 351, 360, 377, 385, 410
soybean flour, roasted 4, 13, 25, 38, 50, 60, 79, 128, 141, 154, 168, 178, 188, 200, 222, 230, 240, 249, 260, 268, 277, 287, 298, 304, 313, 322, 331, 340, 351, 360, 374, 385, 410
soybean lecithin 101, 107, 359, 371
soybean meal 213
soybean nuts, dry roasted 5, 13, 25, 39, 53, 62, 79, 94, 127, 141, 155, 168, 178, 188, 200, 224, 229, 240, 249, 260, 268, 277, 286, 298, 304, 313, 322, 331, 339, 351, 360, 374, 385, 410
soybean nuts, roasted 4, 16, 30, 39, 53, 63, 79, 95, 128, 142, 155, 169, 178, 188, 200, 217, 230, 240, 249, 260, 268, 277, 287, 298, 304, 313, 322, 331, 340, 351, 360, 373, 385, 410
soybean nuts, toasted 4, 16, 31, 39, 53, 61, 79, 94, 128, 141, 155, 168, 188, 200, 223, 230, 240, 249, 260, 268, 277, 286, 298, 304, 313, 322, 331, 340, 351, 360, 373, 385, 409

soybean oil 101, 116, 117, 359, 371
soybean protein concentrate 14, 31, 40, 57, 63, 79, 127, 141, 154, 168, 178, 188, 200, 224, 230, 240, 249, 259, 268, 276, 285, 295, 304, 313, 322, 331, 339, 352, 361, 381, 385, 409
soybean protein isolate 15, 32, 39, 57, 79, 128, 141, 154, 170, 178, 188, 210, 215, 230, 240, 249, 259, 268, 276, 285, 294, 304, 313, 322, 331, 339, 361, 378, 385, 409
soybeans 85, 86, 89, 103, 104, 107, 114, 227
soybeans, boiled 5, 15, 27, 41, 56, 62, 80, 95, 117, 129, 142, 154, 169, 179, 190, 201, 225, 233, 243, 252, 263, 272, 281, 290, 299, 307, 317, 325, 334, 343, 354, 365, 377, 390, 404
soybeans, boiled green 4, 14, 30, 39, 93, 128, 157, 193, 244, 254, 263, 272, 281, 290, 299, 308, 318, 327, 335, 344, 354, 366, 377, 391, 403
soybeans, dry 117, 187
soybeans, green raw 117
soybeans, steamed sprouted 5, 14, 33, 39, 51, 94, 130, 142, 159, 169, 179, 193, 203, 222, 234, 245, 255, 264, 273, 282, 290, 300, 309, 318, 326, 336, 344, 355, 368, 378, 392, 401
soybeans, stir-fried sprouted 5, 13, 29, 39, 50, 93, 129, 142, 164, 169, 178, 191, 201, 232, 244, 255, 263, 272, 281, 290, 299, 308, 317, 324, 335, 344, 355, 366, 377, 391, 403
spaghetti 15, 33, 39, 57, 67, 84, 125, 136, 150, 158, 174, 180, 196, 210, 225, 235, 246, 255, 265, 274, 283, 292, 300, 309, 319, 328, 337, 352, 367, 381, 393, 402
spareribs, braised pork 8, 13, 26, 45, 51, 60, 71, 84, 130, 144, 158, 172, 184, 189, 204, 218, 229, 240, 250, 259, 268, 277, 286, 295, 305, 313, 322, 332, 340, 360, 373, 386, 408
sperm mobility and formation, decreased 239
spices 103, 104
spinach 78, 86, 90, 101, 103, 104, 107, 112, 117, 139, 152, 186, 187
spinach, boiled 2, 16, 28, 41, 56, 62, 79, 94, 128, 144, 155, 169, 179, 196, 201, 219, 234, 246, 255, 265, 274, 283, 292, 300, 309, 320, 328, 337, 346, 356, 370, 382, 384, 394, 399
spinach, boiled new zealand 2, 21, 32, 41, 55, 93, 130, 163,

170, 197, 210, 217, 357, 371, 382, 395, 398
spinach, canned 2, 22, 30, 42, 57, 64, 79, 93, 128, 143, 157, 169, 179, 196, 203, 222, 235, 246, 255, 265, 274, 283, 292, 300, 309, 320, 328, 337, 346, 357, 370, 380, 384, 394, 399
spinach, frozen and boiled 2, 19, 30, 41, 57, 63, 79, 93, 128, 144, 159, 169, 178, 196, 205, 218, 234, 246, 255, 265, 274, 283, 292, 300, 309, 319, 328, 337, 345, 356, 370, 382, 384, 394, 399
spinach, raw 2, 18, 30, 40, 57, 63, 79, 92, 129, 145, 156, 169, 179, 196, 201, 218, 235, 246, 255, 265, 274, 283, 292, 300, 309, 320, 328, 337, 346, 356, 370, 381, 384, 394, 399
spinach, raw mustard 2, 91, 127, 159, 197, 356, 370, 381, 394, 398
spinach, raw new zealand 2, 21, 30, 41, 55, 92, 130, 162, 170, 197, 209, 217, 357, 371, 381, 395, 398
spinal cord deterioration 50
spine bone growth, projecting 151
spirulina seaweed, dried 11, 24, 37, 50, 60, 94, 154, 168, 194, 200, 215, 240, 249, 260, 268, 276, 286, 294, 304, 313, 322, 331, 339, 352, 362, 377, 385, 409
spirulina seaweed, raw 14, 26, 39, 54, 66, 96, 198, 209, 218, 245, 255, 264, 273, 282, 291, 300, 309, 319, 327, 336, 345, 357, 370, 381, 393, 399
spleen, braised beef 19, 27, 45, 66, 69, 92, 135, 141, 153, 174, 181, 189, 205, 220, 231, 243, 249, 260, 271, 278, 288, 298, 306, 316, 325, 334, 340, 366, 378, 387, 402
spleen, braised pork 15, 27, 44, 65, 70, 93, 134, 145, 154, 182, 189, 206, 230, 242, 262, 269, 277, 288, 298, 304, 315, 324, 333, 340, 366, 379, 386, 403
spleen, enlarged 6, 141
spot, raw 134, 149, 165, 170, 182, 192, 201, 221, 235, 244, 253, 262, 271, 280, 289, 297, 307, 317, 326, 334, 343, 366, 378, 389, 401
spread, beef and pork 15, 31, 48, 53, 64, 71, 135, 144,
162, 177, 182, 196, 209, 215, 234, 354, 364, 374, 392, 404
sprouts 85, 89
sprouts, bean 77
squalene synthetase levels, increased 227
squash, baked acorn 3, 15, 36, 40, 53, 63, 80, 94, 131, 147, 161, 170, 196, 201, 224, 237, 247, 257, 266, 275, 284, 293, 302, 311, 321, 329, 338, 347, 353, 369, 383, 396, 400
squash, baked hubbard 2, 18, 34, 40, 53, 63, 81, 94, 133, 149, 164, 173, 197, 203, 223, 237, 247, 256, 266, 275, 284, 293, 302, 311, 321, 329, 337, 347, 354, 369, 380, 394, 400
squash, boiled acorn 3, 17, 36, 41, 55, 64, 81, 94, 132, 149, 163, 172, 197, 205, 224, 237, 247, 257, 267, 276, 285, 294, 302, 311, 321, 330, 338, 347, 355, 369, 383, 397, 399
squash, boiled/baked spaghetti 4, 21, 35, 40, 54, 64, 82, 95, 132, 150, 165, 176, 198, 209, 222, 237, 247, 257, 267, 276, 285, 294, 302, 311, 321, 330, 338, 347, 355, 370, 381, 397, 398
squash, boiled butternut 2, 18, 36, 40, 54, 64, 80, 93, 131, 148, 163, 171, 197, 205, 224, 237, 247, 257, 266, 275, 284, 294, 302, 311, 321, 329, 338, 347, 354, 369, 383, 396, 399
squash, boiled crookneck 3, 21, 34, 41, 57, 64, 80, 94, 132, 146, 164, 172, 180, 197, 207, 225, 236, 247, 257, 266, 276, 284, 293, 302, 311, 321, 330, 338, 347, 356, 370, 381, 384, 396, 398
squash, boiled hubbard 2, 21, 35, 42, 55, 64, 81, 94, 136, 149, 165, 175, 198, 207, 223, 237, 247, 256, 266, 275, 284, 293, 302, 311, 321, 329, 337, 347, 355, 370, 381, 395, 399
squash, boiled scallop 4, 19, 36, 41, 57, 65, 80, 93, 133, 147, 165, 174, 181, 197, 209, 225, 237, 247, 256, 266, 275, 284, 293, 302, 311, 321, 330, 338, 347, 357, 370, 382, 396, 398
squash, canned crookneck 4, 22, 35, 41, 57, 66, 81, 95, 135, 147, 163, 176, 181, 198, 210, 223, 236, 248, 257,
267, 276, 285, 293, 302, 312, 322, 330, 339, 347, 357, 371, 383, 384, 397, 398
squash, frozen and boiled butternut 2, 20, 35, 41, 56, 65, 95, 133, 150, 163, 176, 198, 209, 224, 237, 247, 256, 266, 275, 284, 293, 302, 311, 321, 329, 338, 347, 354, 369, 383, 395, 399
squash, frozen and boiled crookneck 4, 21, 34, 41, 57, 64, 81, 94, 133, 148, 164, 171, 180, 196, 206, 223, 236, 247, 256, 266, 275, 284, 293, 302, 311, 321, 330, 338, 347, 356, 370, 382, 384, 395, 398
squash, raw crookneck 3, 20, 34, 41, 57, 64, 80, 94, 132, 146, 164, 173, 181, 197, 207, 224, 236, 247, 257, 266, 275, 284, 293, 302, 311, 321, 330, 338, 347, 356, 370, 381, 384, 396, 398
squash, raw scallop 4, 18, 35, 40, 57, 64, 80, 93, 133, 146, 164, 172, 181, 197, 207, 224, 236, 247, 256, 266, 275, 284, 293, 302, 311, 321, 330, 338, 347, 356, 370, 382, 396, 398
squash seed 78
squash seeds, dried 3, 14, 26, 39, 131, 141, 154, 168, 188, 200, 222, 228, 240, 251, 262, 269, 278, 288, 297, 304, 316, 322, 332, 339, 353, 359, 372, 387, 409
squash seeds, roasted 64, 131, 141, 154, 168, 188, 200, 222, 228, 240, 250, 260, 268, 277, 286, 295, 304, 314, 322, 331, 339, 354, 359, 372, 385, 409
squid, breaded fried 19, 25, 47, 65, 71, 94, 131, 141, 161, 170, 190, 205, 216, 232, 243, 253, 263, 272, 281, 289, 298, 308, 317, 326, 335, 344, 355, 365, 377, 390, 404
squid, raw 22, 25, 47, 65, 71, 94, 131, 141, 163, 170, 182, 190, 206, 221, 233, 243, 253, 263, 272, 281, 290, 298, 308, 318, 326, 335, 344, 357, 368, 379, 390, 401
SR 225
SR^{++} 225
SR^{+2} 225
STARCH 322, 349
steak excluding fat, broiled porterhouse 16, 28, 46, 54, 60, 70, 82, 137, 144, 156, 171, 184, 191, 202, 219, 229,

Index

241, 251, 260, 269, 278, 287, 295, 305, 314, 324, 332, 341, 364, 376, 386, 405
steak excluding fat, broiled T-bone 16, 28, 46, 54, 60, 70, 82, 137, 144, 156, 171, 184, 191, 202, 219, 229, 241, 251, 260, 269, 278, 287, 295, 305, 314, 324, 332, 341, 364, 376, 386, 404
steak including fat, broiled porterhouse 16, 29, 46, 55, 60, 71, 82, 136, 145, 156, 172, 184, 192, 203, 220, 229, 242, 252, 261, 270, 279, 288, 296, 306, 315, 325, 333, 342, 362, 374, 387, 406
steak including fat, broiled T-bone 17, 29, 46, 55, 61, 71, 82, 136, 145, 156, 172, 185, 192, 204, 220, 230, 242, 252, 261, 270, 279, 288, 296, 306, 316, 325, 333, 342, 361, 373, 388, 407
steel 125
sterility 6, 69, 100, 104, 228, 239, 322, 371
sterility, female 168
sterility, male 88, 96, 126, 212
STEROGYL 99
STEROID 6
steroid synthesis, decreased 125
sties 6, 24
stiffness 69, 98, 167, 178
STILBESTROL 91
stillbirths 168
stimulation, over-stimulation 259
stimulation, schizophrenics over-stimulated 267
stomach deterioration 118, 167
stomach extracts 112
stomach pain 124, 186, 213
stomach, raw pork 17, 31, 46, 66, 72, 136, 142, 157, 193, 207, 220, 232, 244, 264, 272, 281, 290, 299, 308, 318, 327, 335, 344, 366, 377, 390, 402
stomach ulcers 117, 118, 213
stomach upset 167
strains, soft tissue 312
straw, oat 213
straw, rice 213
straw, wheat 213
strawberries 5, 22, 33, 42, 54, 65, 81, 85, 92, 108, 134, 140, 149, 164, 176, 180, 186, 198, 208, 224, 237, 248, 257, 267, 276, 285, 294, 303, 312, 321, 330, 338, 347, 350, 370, 381, 397, 399

strawberry milkshake, restaurant 7, 19, 29, 48, 53, 66, 73, 84, 96, 128, 150, 166, 175, 184, 195, 207, 218, 236, 246, 255, 265, 274, 282, 291, 300, 309, 319, 328, 336, 345, 353, 367, 379, 393, 402
strawberry pie 5, 22, 34, 42, 92, 133, 162, 197, 209, 217, 351, 365, 377, 395, 405
strawberry, wild 213
streptogenin peptide activity in chicks 68
Streptomyces aureofaciens cultures 75
Streptomyces griseus cultures 75
STREPTOMYCIN 168
stress 9, 23, 36, 43, 49, 50, 58, 67, 76, 85, 91, 96, 99, 168, 199, 211
stretch marks, remaining 100
stroke 9, 85, 86, 96, 101, 112, 117, 211
STRONTIUM 127, 225
STRONTIUM-90 225
sturgeon, cooked 7, 149, 203, 235, 243, 253, 262, 271, 280, 288, 297, 307, 316, 326, 334, 342, 366, 378, 389, 403
sturgeon, raw 7, 149, 182, 205, 235, 244, 254, 263, 272, 281, 289, 298, 308, 317, 326, 335, 343, 367, 378, 390, 401
SUBVITAMIN B COMPLEX 78
SUCCICURAN 87
succotash, boiled 3, 15, 32, 39, 52, 64, 94, 133, 143, 159, 170, 194, 202, 222, 234, 245, 255, 264, 273, 282, 291, 300, 309, 319, 328, 336, 345, 352, 367, 380, 393, 403
succotash, canned 4, 22, 34, 41, 55, 66, 80, 94, 135, 145, 163, 174, 179, 196, 208, 217, 235, 246, 255, 265, 274, 283, 292, 301, 309, 320, 329, 337, 346, 354, 368, 380, 394, 400
succotash, frozen and boiled 3, 18, 33, 39, 56, 64, 80, 94, 134, 148, 162, 173, 180, 195, 205, 221, 235, 245, 255, 264, 273, 282, 291, 300, 309, 319, 328, 337, 345, 353, 368, 380, 393, 402
succotash with creamed corn, canned 4, 22, 33, 41, 56, 64, 80, 94, 135, 144, 163, 177, 179, 196, 207, 217, 246, 255, 265, 274, 283, 292, 301, 309, 320, 328, 337, 346, 353, 368, 380, 394, 401

sucker, raw white 129, 143, 159, 171, 179, 191, 202, 221, 234, 244, 254, 263, 272, 281, 289, 298, 308, 317, 326, 335, 343, 368, 379, 390, 401
SUCOSTRIN CHLORIDE 87
SUCROSE 349
suet, beef 211, 246, 266, 276, 284, 293, 302, 311, 321, 338, 347, 359, 371, 395, 410
SUGAR 199
sugar beet pulp 213
sugar, brown 22, 35, 42, 129, 155, 198, 203, 221, 349, 360, 410
sugar cane pulp 213
sugar, maple 128, 159, 198, 206, 222, 349, 361, 409
sugar, powdered 166, 211, 225, 349, 360, 410
sugar, white granulated 166, 211, 225, 349, 360, 410
SULFA DRUGS 24, 37, 44, 50, 78, 85, 86, 89, 90
SULFONAMIDES 91
sulfur-containing amino acid 249, 294
SULPHUR 11, 50, 89, 226
summer sausage 15, 26, 46, 62, 69, 93, 134, 144, 156, 175, 194, 205, 214, 231, 244, 253, 263, 272, 281, 289, 298, 308, 318, 327, 335, 344, 358, 361, 373, 390, 406
summer sausage, turkey 16, 25, 45, 62, 70, 134, 144, 157, 174, 181, 190, 206, 215, 231, 358, 365, 376, 390, 403
summer sausage, turkey breast meat 19, 32, 44, 52, 61, 71, 137, 149, 164, 174, 190, 205, 214, 233, 242, 253, 261, 270, 279, 287, 296, 307, 316, 325, 332, 342, 367, 379, 388, 402
sun sensitivity 5, 9, 23, 43, 49, 58, 67, 85, 86, 90, 96, 97, 101, 113, 212, 238, 257, 303, 330
SUNBRELLA 90
sunburn 90, 101, 138
SUNCHOLIN 87
sundae, caramel 7, 20, 29, 48, 56, 66, 73, 82, 95, 128, 149, 166, 174, 193, 207, 217, 235, 351, 365, 377, 393, 405
sundae, hot fudge 7, 20, 29, 48, 56, 65, 73, 83, 95, 130, 147, 166, 173, 193, 206, 217, 235, 351, 365, 377, 393, 404
sundae, strawberry 7, 20, 30, 48, 55, 66, 73, 81, 95, 129, 148, 165, 174, 194, 208, 220, 237, 352, 365, 378, 393, 404

sunfish, raw 129, 143, 160, 171, 179, 192, 203, 218, 233, 243, 253, 262, 271, 280, 289, 297, 307, 317, 326, 334, 343, 368, 380, 389, 401
sunflower oil 103, 104, 116, 117, 359, 371
sunflower seed oil 101, 227
sunflower seeds 77, 85, 86, 89, 90, 99, 101, 103, 104, 227
sunflower seeds, dried 4, 11, 28, 38, 128, 141, 154, 168, 178, 188, 200, 223, 229, 240, 250, 262, 270, 280, 290, 298, 304, 316, 323, 334, 341, 353, 359, 372, 388, 409
sunflower seeds, dry roasted 16, 28, 37, 129, 141, 155, 169, 178, 188, 200, 223, 229, 240, 250, 262, 271, 280, 290, 298, 306, 317, 324, 335, 342, 352, 359, 372, 389, 410
sunflower seeds, oil roasted 14, 27, 38, 79, 130, 141, 154, 169, 178, 188, 201, 223, 229, 240, 250, 262, 270, 280, 290, 298, 305, 316, 323, 334, 342, 353, 359, 372, 388, 410
sunshine 99
sunstroke 98
suppressor cell activity, increased 167
surimi scallops 22, 36, 48, 136, 165, 210, 215, 244, 254, 263, 272, 281, 289, 298, 308, 318, 327, 335, 344, 354, 367, 381, 391, 402
surimi shrimp 22, 34, 48, 133, 163, 210, 215, 244, 254, 263, 272, 281, 290, 298, 308, 318, 327, 335, 344, 355, 367, 379, 391, 401
SUXAMETHONIUM CHLORIDE 87
swallowing, difficult 153
sweating 107
sweating, increased 6, 124
sweats, night 127, 187
sweet potato, baked 1, 18, 31, 40, 52, 62, 80, 92, 132, 143, 164, 173, 179, 196, 203, 222, 236, 247, 256, 266, 275, 283, 293, 301, 310, 320, 329, 337, 346, 352, 367, 383, 384, 395, 402
sweet potato, boiled 2, 20, 30, 40, 53, 62, 81, 93, 132, 144, 163, 176, 180, 197, 207, 222, 236, 247, 256, 266, 275, 283, 293, 301, 311, 320, 329, 337, 346, 352, 367, 382, 395, 402
sweet potato, candied 2, 22,
34, 42, 66, 74, 81, 94, 132, 146, 160, 176, 197, 207, 219, 237, 247, 257, 266, 275, 284, 293, 301, 311, 321, 330, 338, 347, 352, 366, 379, 396, 403
sweet potato, canned 2, 21, 33, 40, 53, 63, 81, 92, 132, 144, 162, 172, 179, 196, 204, 220, 237, 247, 256, 266, 275, 283, 293, 301, 310, 320, 329, 337, 346, 352, 368, 382, 395, 401
sweet potato, frozen and baked 2, 18, 33, 40, 52, 63, 80, 94, 131, 143, 163, 173, 179, 196, 202, 223, 236, 247, 256, 266, 275, 283, 293, 301, 310, 320, 329, 337, 346, 352, 367, 383, 384, 395, 402
sweet potato pie 2, 19, 31, 42, 95, 129, 163, 195, 208, 217, 352, 364, 376, 393, 405
sweet potatoes/yams 78, 85, 86, 117
swelling, 58, 91
swordfish 105
swordfish, cooked 8, 19, 31, 44, 60, 71, 95, 137, 144, 161, 170, 189, 203, 217, 233, 242, 252, 261, 270, 278, 287, 295, 306, 315, 325, 333, 341, 366, 378, 387, 403
swordfish, raw 8, 20, 32, 44, 54, 61, 71, 95, 137, 145, 162, 171, 183, 189, 205, 218, 233, 243, 253, 262, 271, 280, 288, 297, 307, 317, 326, 334, 343, 366, 378, 389, 401
symptoms being researched 77, 123, 125, 138, 152, 213, 226, 227
SYNCAINE 105
synthetic 108
SYNTOPHEROL 100
syrup, cane 134, 211, 350, 363, 408
syrup, chocolate 5, 34, 42, 57, 83, 134, 142, 157, 169, 179, 194, 207, 218, 234, 246, 256, 266, 275, 284, 293, 302, 310, 320, 329, 337, 346, 350, 364, 380, 395, 408
syrup, corn 131, 155, 198, 211, 219, 350, 362, 408
syrup, maple 129, 161, 198, 208, 222, 350, 363, 408
SYTOBEX 69
SYTOBEX-H 75

taco 7, 16, 33, 48, 56, 63, 73, 81, 95, 128, 145, 159,
170, 192, 204, 216, 232, 353, 364, 375, 391, 405
taco shell 14, 29, 39, 128, 142, 156, 169, 190, 206, 215, 233, 350, 360, 374, 392, 410
T-ADENYLIC ACID 67
tail, simmered pork 18, 33, 48, 134, 177, 196, 243, 263, 273, 281, 290, 299, 308, 318, 327, 331, 344, 360, 372, 390, 407
tallow, beef 371
tallow, mutton 359, 371
tamari sauce 20, 30, 38, 54, 63, 80, 132, 145, 157, 170, 194, 207, 214, 235, 356, 369, 382, 391, 403
tamarind 5, 13, 30, 39, 56, 65, 95, 129, 156, 169, 194, 201, 222, 292, 302, 329, 350, 364, 380, 394, 408
tangerine 3, 16, 36, 42, 56, 65, 80, 92, 134, 150, 166, 176, 182, 198, 208, 224, 247, 257, 267, 276, 285, 294, 302, 312, 322, 330, 339, 347, 354, 369, 382, 397, 399
tangerine juice 3, 19, 36, 43, 92, 133, 150, 165, 176, 182, 198, 208, 224, 238, 247, 257, 267, 276, 285, 294, 303, 312, 322, 330, 339, 348, 354, 369, 382, 397, 399
TANNIC ACID 11
TANNIN 11
TANRUTIN 113
tapioca pudding, apple 8, 138, 165, 199, 211, 220, 351, 367, 383, 397, 403
taro, cooked 16, 35, 41, 94, 133, 162, 171, 195, 201, 222, 247, 257, 267, 276, 285, 294, 302, 311, 321, 330, 338, 347, 351, 366, 382, 397, 404
taro, cooked tahitian 2, 21, 29, 41, 92, 128, 158, 170, 196, 201, 220, 355, 369, 380, 393, 400
tarragon 2, 38, 127, 154, 168, 190, 200, 221, 231, 351, 363, 378, 389, 409
taste sense, decreased 37, 43, 141, 228
T-bone steak excluding fat, broiled 16, 28, 46, 54, 60, 70, 82, 137, 144, 156, 171, 184, 191, 202, 219, 229, 241, 251, 260, 269, 278, 287, 295, 305, 314, 324, 332, 341, 364, 376, 386, 404
T-bone steak including fat, broiled 17, 29, 46, 55, 61, 71, 82, 136, 145, 156, 172, 185, 192, 204, 220, 230,

242, 252, 261, 270, 279, 288, 296, 306, 316, 325, 333, 342, 361, 373, 388, 407
T-cell blood count, decreased 104, 371
tea 23, 126, 151, 166, 177, 199, 211, 224, 238, 371, 398
tea, black 36, 83, 151, 166, 177, 199, 210, 224, 238, 358, 371, 398
tea, chamomile 5, 23, 36, 58, 84, 138, 151, 166, 177, 182, 211, 224, 238, 358, 371, 397, 398
tea leaves 85
tear duct, deterioration of 6
tear glands, drying 91
teeth and bone mineralization, deterioration of 227
teeth, decreased 139
teeth, deterioration of 168
teeth grinding 127
teeth in children, weak 91
teeth, mottling of 151
teeth, pitting of 151
teeth, softening 98, 127
TEGOTIN 118
telangiectasia 114
TEMINA 118
tempeh 3, 16, 32, 38, 54, 61, 72, 80, 129, 142, 157, 169, 178, 191, 203, 223, 232, 243, 251, 263, 271, 280, 290, 299, 306, 317, 325, 334, 343, 353, 365, 377, 389, 406
tenderloin excluding fat, broiled beef 16, 27, 46, 54, 59, 70, 82, 137, 143, 155, 171, 171, 184, 190, 202, 220, 229, 241, 251, 260, 269, 278, 287, 295, 305, 314, 324, 332, 341, 365, 377, 386, 404
tenderloin excluding fat, roasted beef 16, 26, 47, 53, 60, 70, 82, 137, 144, 155, 171, 184, 190, 202, 220, 229, 241, 251, 260, 269, 278, 287, 295, 305, 315, 324, 332, 341, 364, 376, 386, 404
tenderloin including fat, broiled beef 16, 27, 47, 54, 60, 70, 82, 136, 144, 155, 171, 171, 184, 191, 202, 220, 229, 242, 252, 260, 270, 278, 287, 296, 306, 315, 324, 333, 341, 363, 374, 387, 406
tenderloin including fat, roasted beef 17, 27, 47, 54, 61, 70, 82, 136, 144, 155, 173, 185, 191, 203, 220, 230, 242, 252, 261, 270, 279, 288, 296, 306, 316, 325, 333, 342, 362, 374, 388, 406

tenderloin, roasted pork 9, 12, 25, 45, 51, 60, 73, 83, 136, 144, 159, 172, 182, 189, 201, 219, 231, 240, 250, 259, 269, 277, 286, 295, 305, 313, 323, 332, 340, 365, 378, 386, 403
tendon/ligament difficulties 58
tendon strain, muscle 127, 151
TEOFILCOLINA 88
tequila sunrise 4, 21, 36, 42, 66, 93, 137, 149, 165, 177, 198, 210, 223, 238, 355, 367, 383, 397, 401
teriyaki sauce 19, 34, 39, 56, 64, 80, 132, 146, 158, 169, 193, 206, 214, 237, 353, 368, 393, 403
termites 118
TERMITIN 118
TESTASCORBIC 91, 97
TESTAVOL 10
testicle deterioration 50, 228, 322
textured vegetable protein 86, 89
THEOBROMINE SODIUM SALICYLATE 91
THEOKOLIN 88
THEOPHYLLINE CHOLINATE 88
THEOPHYLLINE SALT OF CHOLINE 88
THEOXYLLINE 88
THIADOXINE 23
THIAMIN TRIPHOSPHORIC ACID SALT 24
THIAMINE 11
THIAMINE 1,5-SALT 11, 24
THIAMINE CHLORIDE HYDROCHLORIDE 23
THIAMINE CHLORIDE NAPHTHALENE-1,5-DISULFONIC ACID SALT 24
THIAMINE DISULFIDE 11
THIAMINE HYDROCHLORIDE 11, 23
THIAMINE MONONITRATE 11, 23
THIAMINE MONOPHOSPHATE ESTER PHOSPHORIC ACID SALT 23
THIAMINE ORTHOPHOSPHATE ESTER CHLORIDE 23
THIAMINE O,S-DIACETATE 11
THIAMINE PHOSPHORIC ACID ESTER CHLORIDE 11, 23
THIAMINE PHOSPHORIC ACID ESTER PHOSPHATE SALT 11
THIAMINE PROPYL DISULFIDE 24
THIAMINE TRIPHOSPHATE ESTER 24
THIAMINE TRIPHOSPHATE SALT 24
THIAMINE TRIPHOSPHORIC ACID ESTER 11, 24

THIAMINE TRIPHOSPHORIC ACID SALT 11
THIAMINIUM CHLORIDE HYDROCHLORIDE 23
THIANEURON 23
THIAVIT 23
thigh with skin, fried chicken 8, 16, 28, 44, 50, 61, 73, 82, 134, 147, 159, 176, 182, 192, 206, 218, 231, 242, 250, 261, 269, 279, 287, 295, 305, 315, 324, 333, 341, 357, 363, 375, 387, 406
thigh with skin, roasted chicken 7, 19, 29, 44, 50, 61, 73, 83, 134, 147, 159, 172, 183, 192, 207, 218, 231, 242, 251, 261, 269, 279, 288, 296, 306, 316, 325, 333, 342, 364, 375, 387, 405
thigh with skin, stewed chicken 7, 19, 29, 45, 51, 63, 74, 83, 134, 148, 159, 174, 183, 193, 208, 219, 232, 243, 251, 262, 270, 280, 288, 296, 306, 316, 325, 334, 342, 364, 375, 388, 404
thigh without skin, roasted chicken 8, 17, 28, 44, 50, 60, 73, 82, 134, 147, 159, 172, 183, 192, 206, 218, 231, 242, 251, 261, 269, 279, 287, 295, 306, 315, 324, 333, 341, 365, 376, 387, 404
THIMAINE DISULFIDE 23
THIMAINE PHOSPHORIC ACID ESTER SALT 23
THIOCTACID 112
THIOCTAN 112
THIOCTIC ACID 112
thirst 98, 397
thought, clouded 303
THR 313
THREONINE 313-322
throat, dry 126
throat, pain 124, 213
throat problems 186
thymus, braised beef 71, 92, 159, 188, 202, 217, 243, 263, 272, 280, 288, 299, 308, 317, 326, 340, 361, 373, 388, 406
thymus gland, deterioration 228, 248
THYROID 91
thyroid cancer 153
thyroid deterioration 77, 78, 88, 126, 153, 187
thyroid, enlarged 140
thyroid hormone levels, decreased 100
THYROID HORMONES 101
thyroid hyperplasia 97, 98
thyroid, overactive 9

Index

TI 226
TIAMIDON 23
TIAMINAL 23
tilefish, cooked 132, 149, 165, 170, 184, 190, 201, 220, 235, 243, 252, 261, 270, 279, 287, 295, 306, 315, 325, 333, 341, 366, 378, 388, 402
tilefish, raw 132, 149, 165, 171, 185, 192, 201, 220, 236, 244, 254, 263, 272, 280, 289, 298, 308, 317, 326, 335, 343, 367, 379, 390, 401
TIN 226
tin accumulation in heart and intestine 226
tin can flavoring in jarred foods 226
tin cans 226
tingling 167, 199
tingling and burning sensations 58
tingling hands/feet 124
tingling toes 11
TINIC 37
TIOCTAN 112
TIOCTIDASI 112
tip excluding fat, roasted beef 16, 27, 46, 53, 60, 70, 82, 137, 145, 156, 171, 183, 190, 202, 219, 228, 241, 251, 260, 269, 277, 287, 295, 305, 314, 324, 332, 341, 365, 377, 386, 404
TIPIDI 24
tired, easily 285
tissue, connective 213
tissue elasticity, decreased 212
tissues, sodium deposits in 168
tissues, sodium replacement in 167
TITANIUM 226-227
titanium poisoning 226
TMG 76
TOCOPHERYLQUINONE 100, 102
TOCOQUINONE 102
TOCOQUINONE-10 100, 102
toes, tingling 11
tofu, dried and frozen 3, 13, 27, 39, 54, 61, 79, 127, 141, 154, 169, 178, 188, 211, 223, 229, 240, 249, 259, 268, 277, 285, 297, 304, 313, 322, 331, 339, 353, 359, 373, 385, 409
tofu, firm raw 4, 15, 32, 41, 57, 64, 80, 128, 142, 154, 169, 178, 192, 206, 222, 233, 244, 253, 263, 272, 281, 290, 299, 307, 318, 325, 335, 343, 356, 366, 377, 390, 403
tofu, fried 15, 33, 56, 65, 80, 127, 142, 155, 169, 178, 189, 209, 222, 232, 243, 253, 263, 271, 281, 290, 299, 307, 317, 325, 335, 343, 354, 363, 374, 390, 406
tofu, okara 22, 36, 42, 129, 160, 172, 196, 207, 223, 245, 255, 265, 274, 283, 292, 301, 309, 320, 328, 337, 346, 354, 368, 379, 394, 400
tofu, raw 4, 18, 34, 42, 57, 66, 81, 129, 143, 154, 169, 179, 195, 209, 223, 234, 244, 254, 264, 273, 282, 291, 300, 309, 318, 327, 336, 345, 357, 368, 378, 392, 400
TOKOPHARM 100
tom collins 5, 84, 95, 137, 177, 211, 222, 238, 358, 369, 399
tomato 112
tomato, boiled red 2, 18, 33, 40, 55, 66, 82, 93, 136, 146, 163, 175, 181, 197, 205, 222, 237, 248, 256, 266, 276, 284, 293, 302, 311, 321, 330, 338, 347, 355, 370, 381, 396, 398
tomato catsup 2, 18, 33, 39, 93, 133, 163, 196, 203, 215, 352, 367, 380, 395, 403
tomato juice 3, 20, 35, 40, 55, 64, 80, 93, 136, 146, 163, 176, 181, 198, 207, 222, 237, 248, 257, 267, 276, 285, 294, 302, 312, 321, 330, 339, 347, 356, 370, 383, 396, 398
tomato juice, restaurant 3, 20, 35, 40, 55, 64, 80, 93, 136, 146, 163, 176, 198, 207, 216, 237, 356, 370, 383, 396, 398
tomato paste 2, 15, 29, 38, 51, 60, 92, 131, 142, 156, 170, 195, 200, 219, 234, 247, 256, 265, 275, 284, 292, 301, 310, 320, 329, 337, 346, 353, 368, 380, 393, 402
tomato powder 2, 12, 25, 37, 50, 59, 79, 91, 128, 141, 155, 168, 178, 189, 200, 217, 232, 245, 255, 264, 273, 282, 291, 300, 309, 319, 327, 336, 345, 350, 362, 381, 391, 410
tomato puree 2, 18, 33, 39, 53, 63, 92, 134, 144, 161, 172, 196, 202, 222, 237, 247, 257, 266, 276, 284, 293, 302, 311, 321, 330, 338, 347, 354, 369, 383, 395, 399
tomato, raw green 3, 19, 34, 41, 53, 93, 134, 147, 164, 176, 181, 197, 207, 222, 238, 247, 256, 266, 276, 284, 293, 302, 311, 321, 330, 338, 347, 356, 370, 382, 396, 398
tomato, raw red 2, 19, 34, 40, 56, 66, 81, 93, 137, 147, 164, 176, 181, 197, 207, 223, 237, 248, 256, 267, 276, 285, 294, 302, 312, 321, 330, 339, 347, 356, 370, 382, 384, 396, 398
tomato sauce 3, 18, 33, 39, 55, 93, 134, 143, 162, 174, 197, 203, 215, 236, 355, 370, 382, 395, 399
tomatoes 85, 101, 104, 107, 187, 213, 227
tomatoes, canned peeled red 3, 21, 35, 40, 56, 65, 93, 132, 145, 163, 176, 198, 207, 222, 237, 248, 256, 267, 276, 285, 293, 302, 311, 321, 330, 339, 347, 356, 370, 381, 384, 396, 398
tomatoes, canned stewed red 3, 20, 34, 40, 93, 131, 145, 162, 176, 198, 206, 217, 237, 248, 256, 267, 276, 285, 293, 302, 311, 321, 330, 339, 347, 355, 370, 382, 396, 399
tongue, braised pork 13, 25, 45, 62, 70, 95, 133, 155, 173, 192, 206, 217, 229, 243, 262, 270, 279, 288, 297, 306, 316, 325, 342, 363, 374, 388, 405
tongue, cracked 89
tongue, glossy 58
tongue, green 227
tongue, inflamed 78, 89, 124
tongue, numb/tingling 199
tongue, purplish 24
tongue, red 69
tongue, simmered beef 21, 26, 47, 53, 63, 69, 83, 95, 137, 143, 155, 175, 182, 193, 208, 220, 229, 243, 252, 262, 271, 280, 289, 298, 306, 316, 326, 334, 342, 358, 362, 374, 388, 405
tongue, sore 24, 69, 78, 89, 187
tongue, yellow coating on back of 199, 226
tonsillitis 85
tooth and gum deterioration 9, 96, 105, 138, 151, 166, 199, 211, 225
tooth dentition, late 127, 151
tooth deterioration 77, 78, 151, 212, 213, 225, 227, 349
tooth development, deterioration of 226
tooth enamel, abnormal 213
tooth mineralization, deterioration of 213

Index

top loin excluding fat, braised pork 9, 12, 26, 44, 52, 60, 73, 82, 135, 146, 160, 173, 185, 190, 201, 220, 231, 240, 250, 259, 268, 277, 285, 295, 304, 313, 322, 331, 340, 363, 375, 385, 407
top loin excluding fat, broiled beef 17, 29, 45, 54, 60, 71, 82, 136, 146, 157, 171, 184, 191, 202, 219, 229, 241, 251, 260, 269, 278, 287, 295, 305, 314, 324, 332, 341, 365, 377, 386, 404
top loin excluding fat, broiled pork 9, 12, 26, 45, 51, 60, 72, 82, 134, 147, 162, 171, 183, 189, 201, 219, 231, 240, 250, 259, 269, 277, 286, 295, 305, 313, 323, 332, 340, 363, 375, 386, 406
top loin excluding fat, fried pork 9, 12, 26, 45, 51, 59, 72, 82, 136, 148, 162, 171, 185, 189, 201, 221, 232, 241, 250, 259, 269, 277, 286, 296, 305, 314, 323, 332, 340, 363, 375, 386, 406
top loin excluding fat, roasted pork 9, 12, 27, 45, 52, 60, 73, 82, 135, 147, 161, 173, 185, 189, 202, 221, 232, 241, 250, 259, 269, 277, 286, 296, 305, 314, 323, 332, 340, 364, 375, 386, 405
top loin including fat, braised pork 8, 13, 27, 45, 53, 61, 73, 83, 136, 147, 161, 174, 185, 191, 202, 221, 232, 240, 250, 259, 269, 278, 286, 296, 305, 314, 323, 332, 340, 360, 373, 386, 407
top loin including fat, broiled beef 17, 30, 45, 54, 60, 71, 82, 136, 146, 157, 172, 184, 192, 203, 220, 229, 242, 252, 260, 270, 278, 287, 296, 306, 315, 325, 333, 342, 362, 374, 387, 406
top loin including fat, broiled pork 9, 12, 27, 46, 52, 61, 72, 82, 135, 148, 163, 172, 184, 190, 203, 220, 232, 242, 251, 259, 270, 279, 287, 297, 306, 315, 324, 333, 341, 361, 373, 388, 407
top loin including fat, fried pork 9, 12, 27, 46, 52, 61, 73, 83, 137, 148, 163, 173, 173, 185, 191, 203, 221, 233, 242, 252, 260, 271, 279, 288, 298, 307, 316, 325, 334, 342, 360, 372, 389, 407
top loin including fat, roasted pork 9, 12, 27, 45, 53, 61, 73, 82, 136, 148, 162, 174, 185, 190, 203, 221, 232, 241, 251, 259, 270, 279, 287, 297, 306, 315, 324, 333, 341, 361, 373, 388, 407
top round excluding fat, broiled beef 16, 27, 44, 53, 59, 70, 81, 137, 145, 156, 171, 171, 184, 190, 201, 220, 229, 240, 250, 260, 268, 277, 286, 295, 304, 313, 323, 331, 340, 365, 377, 385, 404
top round excluding fat, fried beef 16, 27, 45, 53, 59, 70, 81, 137, 145, 155, 170, 183, 188, 201, 219, 229, 240, 250, 259, 268, 277, 286, 294, 304, 313, 322, 331, 340, 364, 377, 385, 405
top round including fat, broiled beef 16, 27, 44, 53, 59, 70, 81, 137, 145, 156, 171, 184, 190, 202, 220, 229, 241, 251, 260, 269, 277, 286, 295, 304, 313, 323, 332, 340, 365, 377, 386, 405
top round including fat, fried beef 16, 28, 45, 53, 59, 70, 81, 137, 145, 156, 171, 183, 189, 201, 219, 230, 240, 250, 260, 268, 277, 286, 295, 304, 313, 323, 331, 340, 362, 374, 385, 407
TOPOKAIN 106
tortilla, corn 12, 25, 38, 56, 61, 80, 128, 143, 155, 169, 179, 192, 208, 217, 233, 351, 364, 378, 384, 392, 407
TORUTILIN 118
tostada shell 14, 29, 39, 128, 142, 156, 169, 190, 206, 215, 233, 350, 360, 374, 392, 410
touch, decreased sense of 11
toxemia in pregnancy 294
toxin resistance, decreased 225
TPD 24
trace mineral 123, 124, 125, 126, 138, 139, 140, 141, 151, 152, 153, 167, 177, 186, 187, 211, 212, 213, 225, 226, 227, 228
TRANS FATTY ACIDS 103, 104, 371
trembling 24, 186
tremors, muscle 11, 168, 178
TRETINOIN 10
TRICHOCHROMOGENIC FACTOR 90
triglyceride levels, increased 88, 227
triglycerides levels in women, increased 100

TRI(HYDROX-YETHYL)RUTOSIDE 113
TRIMETHYLGLYCINE 76
TRIOXYETHYLRUTIN 113
tripe, raw beef 22, 30, 48, 71, 84, 95, 147, 158, 176, 195, 205, 221, 231, 244, 254, 263, 272, 281, 290, 299, 308, 318, 327, 335, 344, 367, 378, 390, 400
TRIVALENT CHROMIUM 139-140, 212, 214, 385
TRIVITAN 99
trophic ulcers 106
trout 86, 104
trout, cooked rainbow 8, 17, 28, 95, 129, 144, 157, 170, 189, 201, 221, 233, 242, 252, 261, 270, 278, 287, 295, 306, 315, 324, 333, 341, 366, 378, 387, 404
trout, raw 8, 13, 26, 50, 69, 81, 131, 143, 159, 173, 179, 190, 203, 220, 234, 243, 253, 262, 271, 280, 288, 296, 307, 316, 325, 334, 342, 366, 377, 389, 402
trout, raw rainbow 8, 18, 29, 95, 130, 145, 158, 171, 179, 190, 201, 222, 233, 243, 253, 262, 271, 280, 288, 297, 307, 316, 326, 334, 342, 367, 378, 389, 402
TROXERUTIN 112, 113
TRP 322
TRY 322
TRYPTOPHAN 322-330
tuberculosis 9, 43, 49, 58, 67, 74, 96, 99, 138, 166
tuna 89, 99, 103, 104, 105
tuna, canned 152
tuna in oil, canned 22, 32, 44, 137, 145, 163, 170, 189, 204, 220, 235, 242, 252, 261, 269, 278, 286, 295, 305, 315, 324, 332, 341, 365, 377, 387, 404
tuna in water, canned 34, 45, 83, 143, 163, 205, 220, 241, 251, 260, 269, 277, 286, 294, 305, 314, 324, 332, 340, 366, 379, 387, 403
tuna liver oil 99
tuna salad 8, 21, 65, 83, 95, 133, 144, 161, 173, 192, 208, 216, 235, 244, 254, 263, 272, 281, 289, 298, 308, 317, 326, 335, 343, 355, 365, 377, 390, 404
tuna salad spread 8, 22, 35, 45, 58, 83, 135, 151, 162, 177, 195, 208, 215, 237, 356, 365, 376, 391, 403
TUNGSTEN 186, 227
tungsten replacing molybdenum in enzymes 227

turbot, raw 8, 18, 32, 47, 52, 71, 133, 150, 170, 199, 206, 217, 237, 244, 254, 263, 272, 281, 289, 298, 308, 317, 326, 335, 343, 367, 379, 390, 401
turkey 89, 101, 103, 104
turkey and chicken salad 7, 20, 33, 48, 55, 64, 73, 84, 135, 163, 197, 207, 216, 355, 365, 375, 391
turkey bologna 20, 29, 46, 63, 71, 129, 150, 160, 175, 193, 207, 215, 232, 357, 364, 375, 390, 403
turkey breast meat summer sausage 19, 32, 44, 52, 61, 71, 137, 149, 164, 174, 190, 205, 214, 233, 242, 253, 261, 270, 279, 287, 296, 307, 316, 325, 332, 342, 367, 379, 388, 402
turkey dark meat with skin, roasted 19, 28, 47, 50, 61, 73, 82, 131, 144, 157, 172, 183, 191, 205, 218, 230, 241, 251, 261, 269, 278, 286, 295, 305, 315, 324, 332, 341, 364, 376, 386, 404
turkey dark meat without skin, roasted 19, 28, 47, 50, 60, 73, 82, 131, 144, 157, 172, 183, 191, 205, 218, 230, 240, 252, 260, 268, 277, 286, 295, 305, 314, 324, 331, 340, 365, 377, 386, 404
turkey egg 16, 25, 129, 155, 192, 358, 365, 376, 390, 402
turkey fat 359, 371
turkey frankfurter 18, 29, 46, 63, 71, 128, 149, 158, 174, 192, 208, 215, 232, 358, 364, 374, 391, 404
turkey giblets, simmered 6, 19, 25, 46, 50, 61, 69, 79, 95, 134, 142, 154, 175, 181, 191, 207, 220, 230, 241, 250, 262, 269, 278, 288, 296, 304, 315, 324, 333, 341, 357, 365, 378, 387, 403
turkey gizzard, simmered 7, 21, 26, 47, 51, 64, 71, 80, 95, 134, 144, 154, 174, 181, 194, 207, 220, 230, 240, 250, 262, 269, 278, 288, 295, 304, 313, 325, 332, 341, 358, 365, 378, 386, 403
turkey ham 19, 28, 47, 62, 71, 136, 146, 156, 174, 192, 204, 215, 231, 243, 253, 262, 271, 280, 289, 297, 307, 317, 326, 334, 343, 358, 366, 378, 389, 402
turkey heart, simmered 8, 18, 25, 47, 50, 61, 69, 79, 95,

134, 142, 154, 173, 181, 191, 207, 220, 229, 241, 250, 261, 268, 277, 287, 296, 304, 315, 323, 332, 340, 357, 365, 377, 387, 404
turkey light meat with skin, roasted 19, 31, 44, 52, 59, 73, 83, 132, 149, 159, 172, 183, 191, 205, 220, 232, 240, 251, 261, 268, 278, 286, 295, 305, 314, 324, 331, 340, 365, 377, 386, 404
turkey light meat without skin, roasted 19, 31, 44, 51, 59, 73, 83, 133, 149, 159, 171, 183, 191, 204, 219, 232, 240, 251, 260, 268, 277, 286, 294, 304, 313, 323, 331, 340, 366, 379, 386, 403
turkey liver, simmered 6, 19, 25, 45, 50, 59, 69, 78, 95, 135, 142, 154, 175, 180, 189, 207, 219, 230, 243, 251, 262, 269, 278, 288, 297, 304, 316, 323, 333, 340, 357, 365, 377, 388, 403
turkey pastrami 19, 28, 47, 61, 71, 136, 149, 158, 175, 191, 206, 215, 232, 243, 253, 262, 271, 280, 289, 298, 307, 317, 326, 334, 343, 358, 366, 377, 389, 402
turkey roll, light and dark meat 17, 27, 45, 131, 148, 159, 174, 192, 205, 216, 232, 357, 366, 377, 390, 403
turkey roll, light meat 17, 28, 44, 131, 150, 160, 175, 192, 206, 216, 233, 358, 366, 377, 389, 402
turkey salad 9, 20, 33, 47, 54, 63, 74, 82, 134, 148, 161, 175, 195, 209, 216, 233, 356, 365, 376, 391, 403
turkey salami 18, 30, 47, 133, 149, 158, 175, 194, 206, 215, 232, 358, 365, 375, 390, 403
turkey summer sausage 16, 25, 45, 62, 70, 134, 144, 157, 174, 181, 190, 206, 215, 231, 358, 365, 376, 390, 403
turmeric 25, 38, 92, 128, 153, 168, 189, 200, 221, 229, 350, 360, 376, 391, 409
turnip greens 85, 107
turnip greens and turnips, frozen and boiled 2, 21, 33, 42, 94, 129, 159, 176, 198, 210, 222, 246, 256, 265, 274, 283, 292, 301, 310, 320, 328, 337, 346, 357, 370, 382, 395, 398

turnip greens, boiled 2, 21, 33, 41, 55, 63, 79, 92, 128, 143, 162, 173, 180, 197, 207, 221, 237, 247, 256, 266, 275, 284, 293, 301, 310, 320, 329, 338, 346, 356, 370, 381, 396, 398
turnip greens, canned 2, 23, 33, 42, 57, 66, 80, 93, 128, 147, 159, 173, 180, 197, 209, 217, 237, 246, 256, 266, 275, 283, 292, 301, 310, 320, 329, 337, 346, 357, 371, 381, 395, 398
turnip greens, frozen and boiled 2, 20, 33, 41, 57, 65, 80, 93, 128, 144, 158, 172, 179, 197, 207, 222, 235, 245, 255, 265, 274, 283, 292, 300, 309, 319, 328, 337, 345, 356, 370, 380, 394, 399
turnip greens, raw 2, 18, 32, 40, 54, 62, 79, 92, 128, 142, 160, 170, 179, 196, 205, 221, 237, 246, 256, 266, 275, 283, 292, 301, 310, 320, 329, 337, 346, 355, 370, 381, 395, 399
turnips 212
turnips, frozen and boiled 5, 21, 35, 41, 95, 131, 161, 175, 197, 207, 221, 247, 257, 266, 275, 284, 293, 302, 311, 321, 329, 338, 347, 356, 370, 382, 395, 398
turtle beans, boiled black 5, 14, 34, 41, 55, 65, 79, 130, 143, 156, 170, 180, 193, 201, 224, 234, 244, 254, 264, 273, 282, 290, 300, 309, 318, 327, 336, 344, 352, 366, 382, 392, 403
turtle beans, canned black 5, 15, 31, 40, 56, 66, 79, 95, 131, 143, 158, 170, 180, 194, 204, 216, 235, 245, 255, 264, 273, 282, 291, 300, 309, 319, 327, 336, 345, 353, 368, 382, 393, 401
T-waves, increased 199
2-AEP SALT 111
2-AMINO PROPANOIC ACID 239
2-AMINO-3(IMIDAZOLE)-PROPANOIC ACID 259
2-AMINO-3-(3 INDOYL) PROPANOIC ACID 322
2-AMINO-3-(4 HYDROXYPHENYL)-PROPANOIC ACID 331
2-AMINO-3-HYDROXY-BUTANOIC ACID 313
2-AMINO-3-HYDROXY-PROPANOIC ACID 312
2-AMINO-3-MERCATOPROPANOIC ACID 249

Index

2-AMINO-3-METHYL PENTANOIC ACID 268
2-AMINO-3-METHYL-BUTANOIC ACID 339
2-AMINO-3-METHYL-PENTANOIC ACID 268
2-AMINO-3-PHENYL PROPANOIC ACID 303
2-AMINO-4(METHYLTHIO)-BUANOIC ACID 294
2-AMINO-4(METHYLTHIO)-BUTANOIC ACID 294
2-AMINO-4-METHYL-PENTANOIC ACID 276
2-AMINO-5-GUANIDO-PENTANOIC ACID 239
2-AMINO-BUTANE-1,4-DIOIC ACID 249
2-AMINO-BUTANE-1,4-DIOIC ACID-4-AMIDE 248
2-AMINO-PENTANE-1,5-DIOIC ACID-5-AMIDE 258
2-DIAMINO-HEXANOIC ACID 285
twitching leg muscles 123
twitching, muscle 214
22,23-DIHYDROERGOCALCIFEROL 100
22:23-DIHYDROERGOSTEROL 99
22:23-DIHYDROVITAMIN D_2 100
22:23-DIHYDROVITAMIN D_2 98
TYR 331
TYROSINE 331-339
tyrosine formation, decreased 303
tyrosine toxicity 303

UBICHROMENOL(0) 114
UBICHROMENOL(5) 114
UBICHROMENOL(10) 114
UBICHROMENOL(15) 114, 115
UBICHROMENOL(20) 115
UBICHROMENOL(25) 115
UBICHROMENOL(30) 115
UBICHROMENOL(35) 115
UBICHROMENOL(40) 116
UBICHROMENOL(45) 116
UBICHROMENOL(50) 116
UBIQUINONE(0) 114
UBIQUINONE(5) 114
UBIQUINONE(10) 114
UBIQUINONE(15) 114, 115
UBIQUINONE(20) 115
UBIQUINONE(25) 115
UBIQUINONE(30) 115
UBIQUINONE(35) 115
UBIQUINONE(40) 116
UBIQUINONE(45) 116
UBIQUINONE(50) 116
ulcer, peptic 9, 36, 74, 96, 101, 138, 166
ulceration of internal organs 118
ulcerative colitis 384

ULCER-PREVENTIVE FACTOR 118
ulcers 37, 43, 50, 86
ulcers, chronic oozing 127, 226
ulcers, duodenal 50
ulcers, stomach 117, 118, 213
ulcers, trophic 106
ULTRAPAL CHLORIDE 87
UMBEON 23
UNKNOWN SUBSTANCE 67, 76, 77, 114
urinary deterioration 124, 186
urinary oxalate excretion, increased 91
urination inability 24
urination, increased 58, 91, 98, 153
urination, urgency of 98
urine albumin, slight 167
urine, crystals in 76
urine, lessened 186
urine, protein in 126, 167
urine vitamin C, increased 58
urine xanthurenic acid, increased 58
urine, yellow 24
UROFLAVIN 24
USRDA, 1974 5, 23, 36, 43, 49, 67, 74, 85, 96, 97, 99, 101
USRDA, 1980 6, 9, 23, 36, 43, 49, 58, 67, 74, 85, 89, 96, 97, 99, 101, 138, 153, 166, 167, 177, 199, 238, 330
USRDA, 1989 6, 9, 23, 36, 43, 49, 58, 67, 68, 74, 76, 77, 78, 85, 86, 89, 90, 91, 96, 97, 99, 101, 104, 105, 106, 108, 110, 111, 112, 113, 117, 118, 119, 120, 123, 124, 124, 125, 126, 138, 139, 140, 152, 153, 166, 167, 177, 186, 199, 212, 213, 225, 226, 227, 228, 238, 330

V 227
VACUALS 100
VAFLOL 10
vaginal itching 9, 24, 36, 67, 99, 101
VAL 339
VALERIC ACID 112
VALINE 339-348
VANADIUM 227-228
VAREMOID 113
varicose veins 9, 11, 96, 98, 101, 104, 106, 112, 127, 151, 371, 384
vascular tone of veins, deterioration of 118
vasodilator, peripheral 86
veal 85, 86, 103, 104
veal breast plate, braised/stewed 19, 28, 46, 135, 155, 193, 207, 221, 362, 374, 387, 406

veal chuck, braised/roasted/stewed 17, 27, 44, 135, 155, 193, 206, 221, 364, 375, 386, 405
veal leg, roasted 15, 28, 44, 137, 156, 204, 221, 358, 366, 378, 386
veal loin, braised/broiled 18, 28, 45, 135, 155, 190, 205, 219, 364, 375, 387, 405
veal rib roast, roasted 15, 26, 44, 135, 155, 190, 204, 219, 363, 374, 387, 406
veal round with rump, braised/broiled 18, 28, 45, 135, 155, 190, 204, 219, 364, 376, 387, 404
vegetable oils 86, 101
vegetable protein, textured 86, 89
vegetables 85, 89, 101, 125, 225
vegetables, contaminated 153
vegetables, green leafy 140
vegetables, root 76
vehicle exhaust 167, 187
VEINAMITOL 113
veins deterioration 43, 49, 58, 96, 101, 113, 118
veins, varicose 9, 11, 96, 98, 101, 104, 106, 112, 127, 151, 371, 384
venison 103
venison, raw 14, 25, 44, 135, 190, 366, 378, 389, 402
VENORUTON 113
VENORUTON P4 113
vetch 78
VI-ALPHA 10
VIBALT 69
VIBISONE 69
VICELAT 91, 97
VI-DE-3-HYDROSOL 99
VIDINE 86
VIGANTOL 99
VIGORSAN 99
vinegar, apple cider 212
vinegar, cider 137, 163, 199, 210, 355, 371, 398
vinegar, distilled 211, 356, 371, 398
vinespinach, raw 2, 20, 41, 91, 129, 160, 196, 247, 255, 265, 275, 284, 292, 301, 310, 320, 329, 338, 347, 357, 370, 381, 395, 398
VI-NICOTYL 43
VIO-D 99
VIOLAQUERCITRIN 113
VIOSTEROL 98, 99
viral related diseases 294
VIRUBRA 69
vision, blurred 6, 100, 106
vision deterioration 9, 11, 23, 24, 37, 43, 58, 96, 99, 101, 106, 152, 167, 186, 313
VITACEE 91, 97
VITACIMIN 91, 97

VITACIN 91, 97
VITADURIN 75
VITAMIN A 77, 86, 90, 98, 101, 103, 104, 114, 115, 116, 117, 127, 187, 228, 313, 371, 6-10
VITAMIN A ACETATE 6
VITAMIN A ACID 6, 10
VITAMIN A ALDEHYDE 6
VITAMIN A COMPLEX 6-10
VITAMIN A EPOXIDE 6, 10
VITAMIN A PALMITATE 6
vitamin A absorption, decreased 98, 112
vitamin A, lost source of 1
VITAMIN A_0 1
VITAMIN A_1 6, 10
VITAMIN A_2 6
VITAMIN B 11
VITAMIN B COMPLEX 1, 6, 10, 11, 37, 44, 50, 68, 77, 78, 85, 90, 101, 178, 226
VITAMIN B_1 1, 5-salt 24
VITAMIN B_1 37, 44, 59, 101, 178, 226, 249, 11-24
VITAMIN B_1 disulfide 23
VITAMIN B_1 hydrochloride 23
VITAMIN B_1 mononitrate 23
VITAMIN B_1 o, s-diacetate 23
VITAMIN B_1 phosphoric acid ester chloride 23
VITAMIN B_1 phosphoric acid ester salt 23
VITAMIN B_1 propyl disulfide 11, 24
VITAMIN B_1 triphosphoric acid ester 24
VITAMIN B_1 triphosphoric acid salt 24
VITAMIN B_2 11, 24-37, 44, 49, 59, 91
VITAMIN B_2 complex 24
VITAMIN B_2 phosphate 36
VITAMIN B_3 ascorbate 97
VITAMIN B_3 complex 11, 24, 37-49, 88, 119, 139
vitamin B_3 decreased 276
VITAMIN B_{3a} 37, 49, 103, 104, 371
VITAMIN B_{3a} 37-43
VITAMIN B_{3b} 43, 49
VITAMIN B_{3b} 43-49
VITAMIN B_{3b} ASCORBATE 97
vitamin B_{3b} production, deterioration of 322
vitamin B_{3b} therapy 97
VITAMIN B_4 49
VITAMIN B_5 37, 44, 49-58, 59, 78, 88, 89, 226
VITAMIN B_5 calcium salt 58
VITAMIN B_5 sodium salt 58
VITAMIN B_6 24, 49, 50, 58-67, 69, 88, 103, 104, 121, 153, 168, 200, 294, 303, 313, 322, 371
vitamin B_6 decreased 11, 86, 322
vitamin B_6 deficiency, symp-

toms similar to 121
VITAMIN B_7 67
VITAMIN B_8 67-68
VITAMIN B_9 68
VITAMIN B_{10} 68
VITAMIN B_{11} 68
VITAMIN B_{12} 50, 68-76, 78, 85, 86, 89, 103, 104, 106, 153, 294, 371, 385
VITAMIN B_{12} 68
VITAMIN B_{12} 74
VITAMIN B_{12} coenzyme 76
vitamin B_{12} decreased 91, 140
VITAMIN B_{12} deficiency 11
vitamin B_{12} usage, decreased 76
VITAMIN B_{12}-57co 74
VITAMIN B_{12}-60co 74
VITAMIN B_{12}-zinc tannate complex 75
VITAMIN B_{12a} 75
VITAMIN B_{12b} 75
VITAMIN B_{12c} 75
VITAMIN B_{12f} 69
VITAMIN B_{12III} 69
VITAMIN B_{12m} 69
VITAMIN B_{12p} 75
VITAMIN B_{12r} 75
VITAMIN B_{12s} 75
VITAMIN B_{13} 69, 76, 78
VITAMIN B_{14} 76
VITAMIN B_{15} 76-77, 103, 104, 371
VITAMIN B_{15H8} 76
VITAMIN B_{16} 77
VITAMIN B_{17} 69, 77-78
VITAMIN B_C 11, 50, 69, 78-85, 86, 89, 90, 106, 119, 153, 294
VITAMIN B_C COMPOUNDS 68
vitamin B_C decreased 91
vitamin B_C, internal conversion of 90, 303
vitamin B_C usage, decreased 76
VITAMIN B_H 69, 85-86, 101
VITAMIN B_H NIACINATE 86
VITAMIN B_P 1, 6, 69, 85, 86-88, 98, 101, 167
VITAMIN B_P BROMIDE HEXA-METHYLENEDICARBAMATE 87
VITAMIN B_P CHLORIDE 87
VITAMIN B_P CHLORIDE CARBAMATE 87
VITAMIN B_P CHLORIDE DIHYDROGEN PHOSPHATE 87
VITAMIN B_P CHLORIDE SUCCINATE 87
VITAMIN B_P DEHYDROCHOLATE 87
VITAMIN B_P DICHLORIDE 87
VITAMIN B_P DIHYDROGEN CITRATE 88
VITAMIN B_P ESTERASE 88
VITAMIN B_P OROTATE 88
VITAMIN B_P SALICYLATE 88
VITAMIN B_P THEOPHYLLINATE 88

vitamin B_P decreased 303
VITAMIN B_T 88-89, 103, 104, 371
vitamin B_T formation, decreased 285
VITAMIN B_W 120
VITAMIN B_W 120
VITAMIN B_W 50, 69, 78, 89-90, 120, 226, 385
VITAMIN B_W SULFOXIDE 90
vitamin B_W deficiency, symptoms similar to 120
VITAMIN B_X 90, 106, 124
VITAMIN C 1, 6, 11, 24, 37, 44, 50, 59, 69, 77, 78, 88, 89, 90-98, 101, 103, 104, 112, 126, 127, 139, 140, 153, 167, 168, 213, 228, 249, 303, 312, 322, 371, 385
VITAMIN C CALCIUM SALT 97
VITAMIN C COMPLEX 112
VITAMIN C DECREASED 103, 104, 371
VITAMIN C, FAT-SOLUBLE 97
VITAMIN C, LIPID-SOLUBLE 97
VITAMIN C MAGNESIUM SALT 97
VITAMIN C NICOTINAMIDE COMPLEX 75
VITAMIN C PALMITATE 97
VITAMIN C POTASSIUM SALT 97
VITAMIN C SODIUM SALT 97
vitamin C availability, decreased 98, 112
vitamin C, decreased 6, 90
vitamin C in urine, increased 58
vitamin C mixed with water 90
vitamin C therapy 97, 98
VITAMIN C_1 91, 97
VITAMIN C_2 91, 97
VITAMIN C_3 91
vitamin/coenzyme Q_{10} lost source 114, 115, 116
VITAMIN D 1, 6, 98-100, 103, 104, 127, 168, 187, 214, 371
VITAMIN D DECREASED 6, 127
vitamin D synthesis, decreased 125
VITAMIN D_1 99
VITAMIN D_2 99
VITAMIN D_3 99
VITAMIN D_4 100
VITAMIN D_5 100
VITAMIN D_C 98
VITAMIN D_M 98
VITAMIN E 1, 6, 11, 77, 90, 100-103, 104, 114, 115, 116, 117, 153, 167, 178, 186, 212, 228, 313, 371
VITAMIN E ACETATE 102
VITAMIN E ACID SUCCINATE 102
vitamin E, internal synthesis from 117

VITAMIN E$_1$ 100
VITAMIN E$_2$ 102
VITAMIN E$_2$(50) 102
VITAMIN F 59, 85, 86, 99, 101, 103–105, 106, 127, 187
vitamin F, lost source of 103
vitamin F therapy 104
VITAMIN F$_0$ 103
VITAMIN F$_1$ 104
VITAMIN F$_{99}$ 105
vitamin, fat-soluble 1, 6, 23, 97, 98, 100, 103, 104, 105, 107, 108, 109, 110, 112, 114, 115, 116, 118
VITAMIN G 24
VITAMIN H 89
VITAMIN H' 90
VITAMIN H$_1$ 89
VITAMIN H$_2$ 90
VITAMIN H$_3$ 86, 90, 105-106
vitamin H$_3$, internal conversion of 90
vitamin H$_3$, internal synthesis from 86
VITAMIN I 67
VITAMIN J 86, 97
VITAMIN K 107-110, 114, 115, 116, 117
VITAMIN K-S(II) 107
VITAMIN K$_1$ 108
VITAMIN K$_1$ oxide 108
VITAMIN K$_1$(20) 107, 108
VITAMIN K$_2$ 107
VITAMIN K$_2$(0) 107, 108
VITAMIN K$_2$(5) 107, 108
VITAMIN K$_2$(10) 107, 108
VITAMIN K$_2$(15) 107, 108
VITAMIN K$_2$(20) 107, 108
VITAMIN K$_2$(25) 107, 108
VITAMIN K$_2$(30) 107, 108
VITAMIN K$_2$(35) 107, 108
VITAMIN K$_2$(40) 107, 108
VITAMIN K$_2$(45) 107, 108
VITAMIN K$_2$(50) 107, 108
VITAMIN K$_3$ 107, 109
VITAMIN K$_4$ 107
VITAMIN K$_5$ 107
VITAMIN K$_6$ 107
VITAMIN K$_7$ 107
VITAMIN K$_8$ 107
VITAMIN K$_9$ 107
VITAMIN K$_9$(H) 107
VITAMIN L 110
VITAMIN L COMPLEX 110
VITAMIN L$_1$ 110
VITAMIN L$_2$ 110
vitamin, lipid-soluble 1, 6, 23, 97, 98, 100, 103, 104, 105, 107, 108, 109, 110, 112, 114, 115, 116, 118
VITAMIN M 78
VITAMIN M$_i$™ 111
VITAMIN MK$_1$ 107
VITAMIN MK$_2$ 107
VITAMIN MK$_3$ 107
VITAMIN MK$_4$ 107
VITAMIN MK$_5$ 107
VITAMIN MK$_6$ 107

VITAMIN MK$_7$ 107
VITAMIN MK$_8$ 107
VITAMIN MK$_9$ 107
VITAMIN MK$_{10}$ 107
VITAMIN N 112, 124, 167, 186
VITAMIN OHB$_{12}$ 69, 75, 76
VITAMIN P 90, 91, 106, 112-113
vitamin P therapy 97, 98
VITAMIN P$_1$ 113
VITAMIN P$_2$ 113
VITAMIN P$_3$ 113
VITAMIN P$_4$ 113
VITAMIN PP 37, 43
vitamin precursor 1, 76, 114, 115, 116
VITAMIN Q 113, 116
VITAMIN Q$_0$ 114
VITAMIN Q$_1$ 114, 116
VITAMIN Q$_2$ 114
VITAMIN Q$_3$ 114
VITAMIN Q$_4$ 115
VITAMIN Q$_5$ 115
VITAMIN Q$_6$ 115
VITAMIN Q$_7$ 115
VITAMIN Q$_8$ 116
VITAMIN Q$_9$ 116
VITAMIN Q$_{10}$ 116
VITAMIN R 68
VITAMIN S 68
VITAMIN T 91, 99, 118, 153, 385
VITAMIN T COMPLEX 118
VITAMIN T GOETSCH 118
VITAMIN U 118, 119
VITAMIN U BROMIDE 119
VITAMIN U CHICK FACTOR 119
VITAMIN U CHLORIDE 119
VITAMIN V 119
VITAMIN W 120
vitamin, water-soluble 10, 11, 24, 37, 43, 49, 58, 67, 68, 69, 76, 77, 78, 85, 86, 88, 89, 90, 91, 97, 98, 109, 111, 112, 113, 118, 119, 120, 121
VITAMIN X 120
VITAMIN Y 58, 121
vitamins/coenzymes Q$_7$, Q$_8$, Q$_9$ and possibly others, internal conversion of 117
VITANEURON 23
VITARUBIN 69
VITA-RUBRA 69
VITASCORBOL 91, 97
VITAS-U 118
VITEOLIN 100
VITPEX 10
VITRAL 69
vodka, 100 proof 150, 166, 183, 199, 211, 238, 362, 405
vodka, 80 proof 151, 199, 364, 403
vodka, 94 proof 150, 166, 183, 199, 211, 238, 363, 404
VOGAN 10
VOGAN-NEU 10
voice changes 167

W 227
waffles 4, 14, 26, 39, 51, 65, 73, 80, 128, 148, 158, 172, 192, 208, 215, 234, 351, 361, 375, 392, 408
wakame seaweed, raw 3, 19, 28, 39, 95, 128, 143, 157, 169, 178, 195, 210, 215, 236, 246, 255, 267, 275, 283, 292, 300, 309, 319, 328, 338, 345, 355, 369, 380, 394, 401
walking deterioration 96, 186
walnut oil 103, 104
walnuts 89, 101, 103, 104, 112, 117, 187
walnuts, dried black 3, 14, 31, 40, 130, 141, 156, 168, 178, 188, 201, 230, 240, 250, 262, 271, 279, 290, 298, 305, 317, 324, 334, 341, 354, 359, 372, 388, 409
walnuts, dried english/persian 4, 13, 31, 39, 52, 59, 79, 94, 129, 141, 157, 168, 168, 178, 189, 201, 222, 231, 240, 251, 263, 272, 281, 291, 299, 308, 318, 326, 335, 344, 353, 359, 372, 390, 410
WAMPOCAP 37
warts 9, 101, 139, 199
WATER 397-410
water, fluoridated 151
water, hard 213
water, Perrier 134, 177, 224, 398
water pipes 126
water pollution 123, 126, 140, 167, 186
water retention 58
water supplies, disinfectant 213
water-soluble vitamin 10, 11, 24, 37, 43, 49, 58, 67, 68, 69, 76, 77, 78, 85, 86, 88, 89, 90, 91, 97, 98, 109, 111, 112, 113, 118, 119, 120, 121
watercress 78, 107, 112
watercress, raw 2, 16, 31, 55, 64, 92, 128, 165, 172, 196, 204, 221, 246, 257, 265, 274, 283, 292, 302, 309, 319, 329, 337, 346, 358, 371, 394, 398
watermelon 3, 18, 36, 42, 56, 64, 84, 85, 94, 136, 150, 166, 176, 182, 187, 199, 209, 224, 238, 247, 257, 267, 276, 285, 293, 302, 312, 321, 330, 339, 347, 355, 370, 381, 397, 399
watermelon seeds, dried 15, 31, 38, 80, 130, 154, 168, 188, 200, 218, 240, 250, 261, 269, 278, 290, 295, 304, 315, 322, 332, 340, 353, 359, 372, 386, 409

Index

wax beans 4, 21, 35, 42, 94, 129, 156, 198, 207, 217, 356, 370, 382, 396
waxgourd, boiled 21, 42, 94, 133, 164, 198, 211, 217, 357, 371, 382, 397, 398
weakness 49, 69, 78, 91, 98, 124, 141, 168, 178
weakness, muscle 6, 37, 43, 88, 91, 98, 100, 126, 153, 199, 214
weed killers 124
weight decreased 6, 24, 91, 98, 104, 105, 153, 167, 178, 214, 285, 322, 371
weight increased 67, 74, 76, 85, 88, 96, 101, 103, 104, 105, 117, 138, 152, 177, 330, 371
weight reduction 248
WESTOCAINE 106
whale meat, raw 6, 17, 32, 94, 134, 193, 211, 218, 366, 377, 389, 402
wheat 88, 101, 103, 104
wheat bran 77, 139, 213
wheat cake 21, 35, 40, 133, 164, 195, 210, 224, 349, 360, 380, 392, 409
wheat, cream of 19, 34, 41, 57, 83, 133, 150, 155, 177, 198, 211, 224, 237, 247, 255, 266, 275, 283, 293, 301, 310, 321, 329, 338, 346, 354, 369, 382, 395, 400
wheat flakes 2, 11, 25, 37, 51, 59, 69, 80, 92, 128, 142, 154, 169, 178, 188, 202, 215, 232, 244, 254, 264, 273, 282, 291, 299, 308, 319, 326, 336, 344, 349, 361, 379, 384, 391, 409
wheat flour 213
wheat flour, all purpose enriched 13, 27, 38, 133, 156, 195, 210, 224, 349, 360, 380, 391, 409
wheat flour, whole 13, 31, 38, 85, 131, 155, 188, 203, 224, 350, 361, 379, 391, 409
wheat germ 77, 85, 86, 88, 89, 90, 101, 103, 104, 107, 116, 117, 139, 186, 212
wheat germ oil 101, 103, 104, 230, 359, 371
wheat germ, toasted 11, 25, 38, 50, 59, 78, 94, 131, 142, 154, 168, 178, 188, 200, 223, 228, 350, 360, 376, 384, 386, 409
wheat gluten 131, 193, 210, 224, 351, 360, 379, 385, 409
wheat grain 116
wheat, puffed 14, 29, 37, 53, 63, 80, 132, 142, 155, 169, 188, 203, 223, 231, 244, 252, 263, 272, 281, 291, 299, 307, 318, 326, 335, 344, 349, 360, 379, 384, 390, 410
wheat, shredded 14, 27, 38, 51, 62, 80, 131, 142, 155, 169, 188, 203, 222, 230, 244, 253, 264, 273, 282, 291, 299, 308, 318, 326, 336, 344, 349, 360, 379, 384, 391, 409
wheat straw 213
wheat, whole 85, 107, 186, 187, 227
whelk, cooked 8, 19, 29, 47, 59, 69, 81, 128, 141, 154, 168, 189, 200, 216, 230, 240, 250, 260, 268, 277, 286, 294, 304, 313, 322, 331, 339, 353, 363, 380, 385, 408
whelk, raw 8, 22, 32, 48, 56, 61, 69, 83, 130, 141, 154, 169, 179, 193, 203, 217, 233, 240, 254, 263, 272, 279, 289, 297, 307, 316, 324, 334, 343, 355, 366, 381, 388, 403
whey 8, 20, 30, 48, 54, 66, 73, 76, 84, 130, 166, 176, 196, 208, 220, 237, 248, 256, 267, 275, 284, 293, 302, 311, 320, 329, 338, 347, 356, 370, 381, 396, 398
WHEY FACTOR 76
whiskey, 100 proof 150, 166, 183, 199, 211, 238, 362, 405
whiskey, 86 proof 150, 166, 185, 199, 211, 238, 358, 363, 404
whiskey, 94 proof 150, 166, 183, 199, 211, 238, 363, 404
whiskey sour 5, 14, 36, 42, 57, 67, 83, 93, 137, 150, 166, 177, 184, 199, 210, 222, 238, 355, 366, 383, 397, 401
white beans, boiled 16, 35, 42, 56, 65, 79, 129, 143, 155, 169, 179, 194, 201, 223, 233, 244, 254, 264, 273, 282, 290, 299, 308, 318, 327, 336, 344, 352, 366, 382, 391, 404
white beans, canned 17, 35, 42, 56, 65, 79, 129, 143, 156, 170, 179, 195, 201, 223, 233, 245, 255, 264, 273, 282, 291, 300, 309, 319, 327, 336, 345, 352, 367, 382, 392, 402
white blood cell abnormalities 151
white blood cells, decreased number 78
whitefish 104
whitefish, raw 148, 164, 170, 204, 220, 234, 243, 253, 262, 271, 280, 289, 297, 307, 317, 326, 334, 343, 366, 377, 389, 402
whitefish, smoked 8, 20, 31, 47, 57, 60, 70, 82, 133, 142, 164, 173, 182, 193, 202, 215, 235, 243, 252, 262, 270, 279, 287, 296, 306, 316, 325, 333, 342, 367, 380, 388, 402
whiting, cooked 8, 18, 33, 48, 63, 70, 81, 130, 150, 164, 171, 189, 201, 217, 235, 243, 252, 262, 270, 279, 287, 296, 306, 316, 325, 333, 342, 367, 379, 388, 401
whiting, raw 8, 19, 34, 48, 56, 63, 70, 81, 130, 150, 165, 173, 181, 190, 206, 219, 234, 244, 253, 262, 272, 280, 289, 297, 307, 317, 326, 334, 343, 368, 379, 389, 400
wine 125, 212
wine, dry dessert 22, 36, 42, 57, 136, 149, 165, 176, 181, 198, 210, 223, 238, 356, 366, 397, 402
wine, red table 22, 35, 43, 57, 66, 74, 84, 136, 150, 164, 176, 179, 198, 209, 223, 237, 358, 368, 397, 399
wine, rose table 36, 43, 57, 66, 74, 84, 136, 149, 164, 176, 181, 198, 210, 223, 238, 358, 368, 397, 399
wine, sweet dessert 22, 36, 42, 57, 136, 149, 165, 176, 181, 198, 210, 223, 238, 354, 366, 397, 402
wine, white table 36, 43, 58, 67, 136, 150, 165, 176, 179, 198, 210, 223, 238, 358, 368, 397, 399
wing with skin, fried chicken 7, 19, 31, 44, 60, 73, 84, 133, 148, 160, 174, 182, 193, 208, 218, 232, 242, 250, 261, 269, 279, 288, 296, 306, 316, 325, 333, 342, 357, 361, 374, 387, 407
wing with skin, roasted chicken 7, 21, 31, 44, 60, 73, 84, 133, 149, 160, 173, 183, 193, 207, 218, 232, 241, 250, 261, 269, 279, 288, 296, 306, 315, 325, 333, 341, 362, 374, 387, 406
wing with skin, stewed chicken 7, 20, 32, 46, 53, 62, 74, 84, 134, 149, 160, 175, 183, 194, 209, 219, 233, 243, 251, 262, 270,

Index

280, 288, 297, 307, 316, 325, 334, 342, 363, 374, 388, 404
winged beans, boiled 14, 31, 40, 66, 81, 128, 142, 155, 169, 178, 193, 205, 222, 233, 244, 254, 264, 273, 282, 290, 300, 309, 318, 326, 335, 344, 353, 366, 377, 391, 403
wolffish, raw 7, 15, 32, 47, 52, 71, 150, 166, 184, 218, 234, 244, 254, 263, 272, 280, 289, 298, 308, 317, 326, 335, 343, 367, 379, 390, 401
WOLFRAM 227
worms 9, 23, 36, 58, 67, 74, 96, 99, 108, 138, 166, 211, 226
wound healing 104, 238, 248, 312
wound healing, slow 76, 89, 118

xanthine 227
xanthurenic acid in urine, increased 58
XITRIX 91, 97
XOBALINE 76

yam, baked/boiled tropical 17, 35, 40, 55, 62, 81, 93, 134, 144, 164, 174, 196, 200, 223, 237, 246, 256, 266, 275, 284, 293, 301, 310, 321, 329, 338, 347, 352, 367, 383, 395, 402
yam, steamed hawaii mountain 17, 36, 43, 137, 164, 176, 196, 201, 222, 246, 256, 266, 275, 284, 293, 301, 310, 320, 329, 338, 346, 353, 368, 383, 395, 401
yambean, boiled 21, 35, 42, 93, 133, 165, 174, 197, 208, 223, 247, 257, 266, 276, 284, 293, 302, 311, 321, 338, 347, 354, 369, 383, 395, 399
yams 89
yardlong bean, boiled 5, 14, 34, 41, 54, 65, 79, 96, 131, 143, 156, 169, 179, 192, 204, 223, 233, 244, 254, 264, 273, 282, 290, 300, 308, 319, 327, 336, 345, 352, 367, 381, 392, 403
YEAST ELUATE FACTOR 58
yeast, torula 11, 24, 44, 85, 86, 89, 118, 127, 154, 188, 200, 222, 240, 249, 259, 268, 276, 285, 294, 304, 313, 322, 331, 339, 351, 362, 380, 385, 409
yeasts 76, 77, 88, 88, 101, 110
yellow beans, boiled 5, 15, 32, 40, 56, 64, 79, 95, 130, 143, 157, 169, 179, 192, 204, 223, 233, 244, 254, 264, 273, 282, 290, 299, 308, 318, 327, 336, 344, 352, 366, 379, 392, 404
yellowtail 116, 117
yellowtail, raw 8, 15, 35, 44, 52, 63, 71, 84, 95, 149, 164, 184, 193, 221, 235, 243, 252, 262, 270, 279, 287, 296, 307, 316, 325, 333, 342, 366, 378, 388, 401
yogurt 103
yogurt, lowfat 8, 20, 28, 48, 52, 66, 73, 81, 96, 128, 166, 174, 193, 206, 219, 234, 246, 264, 273, 282, 291, 299, 309, 319, 328, 336, 344, 355, 368, 379, 393, 400
yogurt, skim 9, 19, 28, 48, 52, 66, 73, 81, 96, 128, 166, 174, 193, 206, 218, 234, 246, 264, 273, 282, 291, 299, 309, 319, 328, 336, 344, 355, 369, 382, 393, 400
yogurt, whole 7, 21, 31, 48, 54, 66, 73, 83, 128, 166, 176, 195, 208, 221, 235, 246, 265, 274, 282, 291, 300, 309, 319, 329, 336, 345, 356, 368, 378, 393, 399

ZETA$_1$-TOCOPHEROL 100, 102
ZETA$_2$-TOCOPHEROL 100, 102
ZETA$_1$-VITAMIN E 102
ZETA$_2$-VITAMIN E 102
ZINC 1, 6, 76, 103, 104, 126, 140, 141, 167, 228-238, 371
zinc absorption, inhibited 78
zinc decreased 103, 104, 371
zinc to brain, decreased 141
ZINC-VITAMIN B$_{12}$-TANNATE 75
ZN 228
zucchini, boiled 21, 35, 41, 57, 65, 81, 95, 134, 147, 164, 173, 181, 196, 206, 224, 237, 248, 257, 267, 276, 285, 293, 302, 312, 321, 330, 339, 347, 356, 370, 383, 397, 398
zucchini, frozen and boiled 3, 21, 35, 41, 55, 66, 82, 95, 133, 149, 164, 175, 180, 197, 207, 224, 237, 247, 256, 266, 275, 284, 293, 302, 311, 321, 330, 338, 347, 356, 370, 382, 396, 398
zucchini in tomato juice, canned 3, 21, 35, 41, 95, 133, 146, 163, 175, 197, 205, 216, 236, 247, 256, 266, 275, 284, 293, 302, 311, 321, 330, 338, 347, 355, 370, 383, 396, 399
zucchini, raw 3, 18, 35, 41, 57, 65, 80, 94, 134, 149, 164, 173, 181, 197, 206, 224, 237, 247, 256, 266, 275, 284, 293, 302, 311, 321, 330, 338, 347, 357, 371, 382, 396, 398